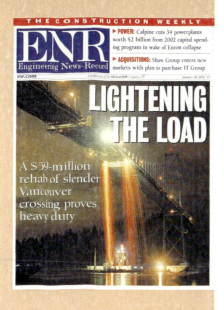

Student

Please enter my subscription for **Engineering News-Record.**

6 months ☐ $29.50 (Domestic)

Name
_____

Address
_____

City _____ State _____ Zip _____

☐ Payment enclosed ☐ Bill me later

McGraw_Hill CONSTRUCTION ENR

5EN2DMHE

‖‖‖‖‖‖‖‖‖‖‖‖‖‖‖‖‖
D0206783

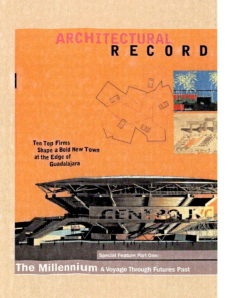

Friendly

Please enter my subscription for **Architectural Record.**

6 months ☐ $19.50 (Domestic)

Name
_____

Address
_____

City _____ State _____ Zip _____

☐ Payment enclosed ☐ Bill me later

McGraw_Hill CONSTRUCTION Architectural Record 5AR2DMHE

Savings

Please enter my subscription for **Aviation Week & Space Technology.**

6 months ☐ $29.95 (Domestic)

Name
_____

Address
_____

City _____ State _____ Zip _____

☐ Payment enclosed ☐ Bill me later

CAW34EDU

The McGraw-Hill Companies

NO POSTAGE
NECESSARY
IF MAILED
IN THE
UNITED STATES

BUSINESS REPLY MAIL
FIRST-CLASS MAIL   PERMIT NO. 42   HIGHTSTOWN NJ

POSTAGE WILL BE PAID BY ADDRESSEE

SUBSCRIPTION DEPT

McGraw_Hill
CONSTRUCTION ENR
PO BOX 516
HIGHTSTOWN NJ 08520-9467

The McGraw-Hill Companies

NO POSTAGE
NECESSARY
IF MAILED
IN THE
UNITED STATES

BUSINESS REPLY MAIL
FIRST-CLASS MAIL   PERMIT NO. 42   HIGHTSTOWN NJ

POSTAGE WILL BE PAID BY ADDRESSEE

McGraw_Hill
CONSTRUCTION Architectural
Record

PO BOX 564
HIGHTSTOWN NJ 08520-9890

The McGraw-Hill Companies

NO POSTAGE
NECESSARY
IF MAILED
IN THE
UNITED STATES

BUSINESS REPLY MAIL
FIRST-CLASS MAIL   PERMIT NO. 42   HIGHTSTOWN, NJ

POSTAGE WILL BE PAID BY ADDRESSEE

AVIATION WEEK'S
AVIATION
WEEK & SPACE TECHNOLOGY

PO BOX 503
HIGHTSTOWN NJ 08520-9475

Stay on top
of industry
developments
with the
leading
publications
in your
engineering field.

Today's economy
requires a
higher level
of knowledge
and adaptabilty—
increase yours
with

**McGraw-Hill
Construction
and Aviation
Week & Space
Technology:**
*Engineering
News-Record,
Architectural
Record,
and
Aviation Week.*

www.construction.com

www.AviationNow.com/awst

AutoCAD® 2004

2004

Instructor

A Student Guide to
Complete Coverage of AutoCAD's
Commands and Features

The McGraw-Hill Graphics Series

Providing you with the highest quality textbooks that meet your changing needs requires feedback, improvement, and revision. The team of authors and McGraw-Hill are committed to this effort. We invite you to become part of our team by offering your wishes, suggestions, and comments for future editions and new products and texts.

Please mail or fax your comments to: James A. Leach
c/o McGraw-Hill Higher Education
Engineering Department
420 Boylston St., 2nd Fl.
Boston, MA 02116
FAX: 617-867-9849

Titles in the McGraw-Hill Graphics Series

Consulting Editor
Gary R. Bertoline, Purdue University

Bertoline, *Graphic Drawing Workbook*

Bertoline, *Introduction to Graphics Communications for Engineers*

Bertoline/Wiebe, *Fundamentals of Graphics Communication*

Bertoline/Wiebe, *Technical Graphics Communication*

Condoor, *Mechanical Design Modeling Using Pro/Engineer*

Duff/Maxson, *The Complete Technical Illustrator*

Jensen, *Engineering Drawing and Design*

Kelley, *Pro/Engineer Assistant*

Kelley, *Pro/Engineer Instructor*

Leach, *AutoCAD Assistant*

Leach, *AutoCAD Companion*

Leach, *AutoCAD Instructor*

Leake, *Autodesk Inventor*

Lieu, *Graphics Interactive CD-ROM*

SDRC, *I-DEAS Student Guide*

Tickoo, *Mechanical Desktop Instructor*

Wohlers, *Applying AutoCAD*

AutoCAD® 2004
Instructor

A Student Guide to
Complete Coverage of AutoCAD's
Commands and Features

James A. LEACH
University of Louisville

Mc Graw Hill **Higher Education**

Boston   Burr Ridge, IL   Dubuque, IA   Madison, WI   New York   San Francisco   St. Louis
Bangkok   Bogotá   Caracas   Kuala Lumpur   Lisbon   London   Madrid   Mexico City
Milan   Montreal   New Delhi   Santiago   Seoul   Singapore   Sydney   Taipei   Toronto

McGraw Hill Higher Education

AUTOCAD 2004 INSTRUCTOR: A STUDENT GUIDE TO COMPLETE COVERAGE OF
AUTOCAD'S COMMANDS AND FEATURES

Published by McGraw-Hill, a business unit of The McGraw-Hill Companies, Inc., 1221 Avenue
of the Americas, New York, NY 10020. Copyright © 2004 by The McGraw-Hill Companies, Inc.
All rights reserved. No part of this publication may be reproduced or distributed in any form
or by any means, or stored in a database or retrieval system, without the prior written consent
of The McGraw-Hill Companies, Inc., including, but not limited to, in any network or other
electronic storage or transmission, or broadcast for distance learning.

Some ancillaries, including electronic and print components, may not be available to
customers outside the United States.

 This book is printed on recycled, acid-free paper containing 10% postconsumer waste.

4 5 6 7 8 9 0 QPD/QPD 0 9 8 7 6 5

ISBN 0–07–286854–6

Publisher: *Elizabeth A. Jones*
Senior sponsoring editor: *Suzanne Jeans*
Developmental editor: *Lisa Kalner Williams*
Marketing manager: *Sarah Martin*
Senior project manager: *Kay J. Brimeyer*
Lead production supervisor: *Sandy Ludovissy*
Media project manager: *Sandra M. Schnee*
Senior designer: *David W. Hash*
Page layout specialist: *Karen Collins*
Typeface: *10.5/12 Times Roman*
Printer: *Quebecor World Dubuque, IA*

Library of Congress Control Number: 2003108526

This book is dedicated to Stephanie Ann White, September 29, 1952–May 13, 2001. Stephanie was a great teacher, a brilliant musician, a wonderful person, and an inspiration to everyone.

DEDICATION

TABLE OF CONTENTS

TABLE OF CONTENTS

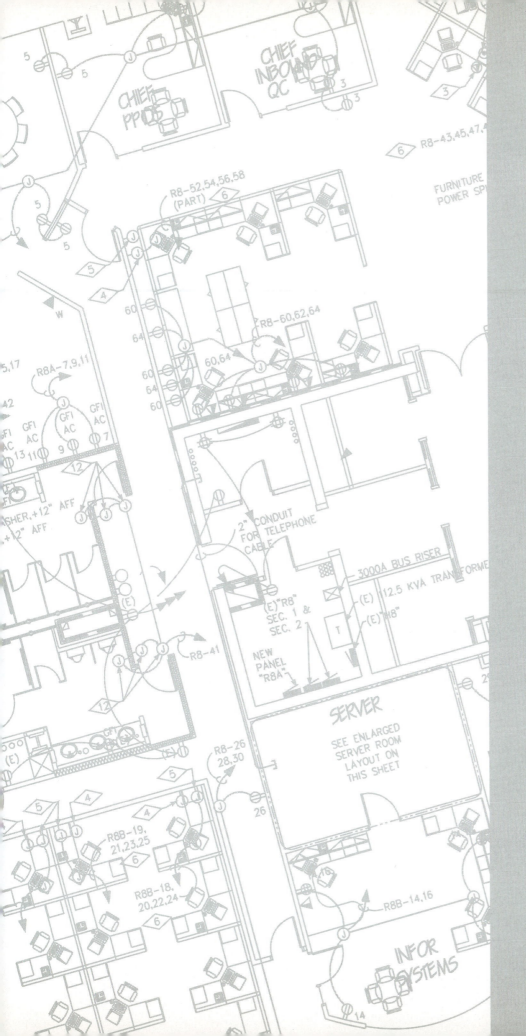

GUIDED
TOUR

GUIDED TOUR

Welcome to AutoCAD 2004 Instructor. Here are some features you will find in this book to help you learn AutoCAD 2004.

Chapter Objectives

Each chapter opens with a list of new concepts you can expect to learn. Having objectives in mind helps you focus on the important ideas as you move through each chapter.

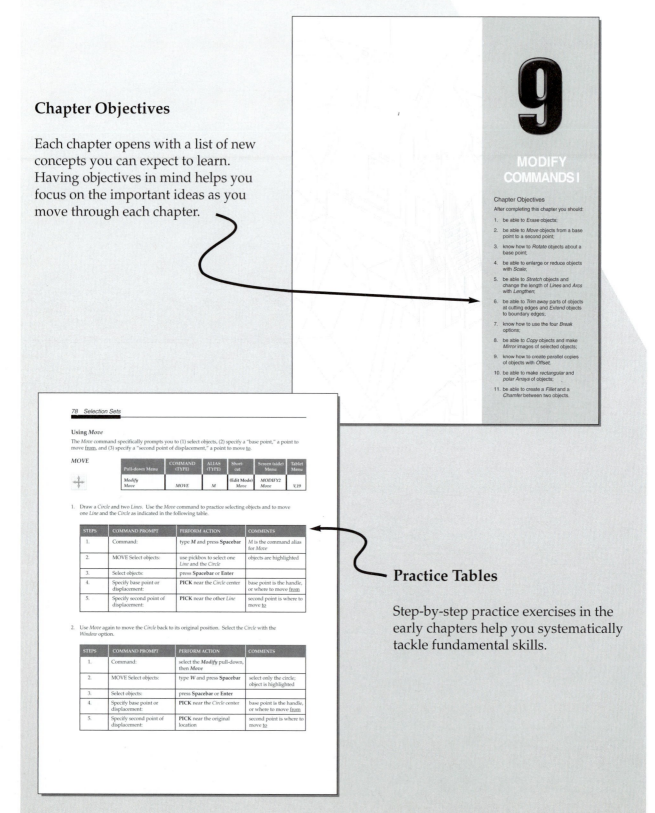

9

MODIFY COMMANDS I

Chapter Objectives

After completing this chapter you should:

1. be able to *Erase* objects;
2. be able to *Move* objects from a base point to a second point;
3. know how to *Rotate* objects about a base point;
4. be able to enlarge or reduce objects with *Scale*;
5. be able to *Stretch* objects and change the length of *Lines* and *Arcs* with *Lengthen*;
6. be able to *Trim* away parts of objects at cutting edges and *Extend* objects to boundary edges;
7. know how to use the four *Break* options;
8. be able to *Copy* objects and make *Mirror* images of selected objects;
9. know how to create parallel copies of objects with *Offset*;
10. be able to make *rectangular* and *polar Arrays* of objects;
11. be able to create a *Fillet* and a *Chamfer* between two objects.

---

78   Selection Sets

Using *Move*

The *Move* command specifically prompts you to (1) select objects, (2) specify a "base point," a point to move <u>from</u>, and (3) specify a "second point of displacement," a point to move <u>to</u>.

MOVE

Pull-down Menu	COMMAND (TYPE)	ALIAS (TYPE)	Short-cut	Screen (side) Menu	Tablet Menu
Modify *Move*	MOVE	M	(Edit Mode) *Move*	*MODIFY2* *Move*	V,19

1. Draw a *Circle* and two *Lines*. Use the *Move* command to practice selecting objects and to move one *Line* and the *Circle* as indicated in the following table.

STEPS	COMMAND PROMPT	PERFORM ACTION	COMMENTS
1.	Command:	type *M* and press **Spacebar**	*M* is the command alias for *Move*
2.	MOVE Select objects:	use pickbox to select one *Line* and the *Circle*	objects are highlighted
3.	Select objects:	press **Spacebar** or **Enter**	
4.	Specify base point or displacement:	**PICK** near the *Circle* center	base point is the handle, or where to move <u>from</u>
5.	Specify second point of displacement:	**PICK** near the other *Line*	second point is where to move <u>to</u>

2. Use *Move* again to move the *Circle* back to its original position. Select the *Circle* with the *Window* option.

STEPS	COMMAND PROMPT	PERFORM ACTION	COMMENTS
1.	Command:	select the *Modify* pull-down, then *Move*	
2.	MOVE Select objects:	type *W* and press **Spacebar**	select only the circle; object is highlighted
3.	Select objects:	press **Spacebar** or **Enter**	
4.	Specify base point or displacement:	**PICK** near the *Circle* center	base point is the handle, or where to move <u>from</u>
5.	Specify second point of displacement:	**PICK** near the original location	second point is where to move <u>to</u>

Practice Tables

Step-by-step practice exercises in the early chapters help you systematically tackle fundamental skills.

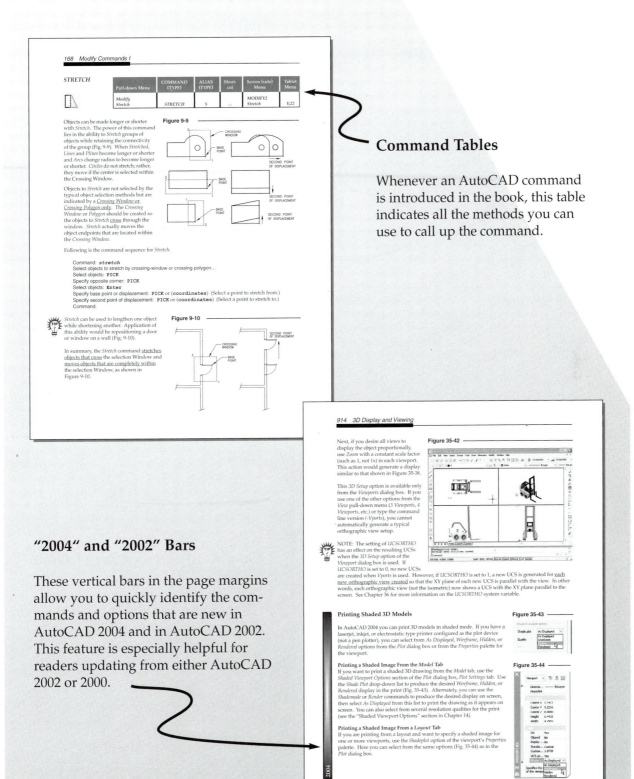

Command Tables

Whenever an AutoCAD command is introduced in the book, this table indicates all the methods you can use to call up the command.

"2004" and "2002" Bars

These vertical bars in the page margins allow you to quickly identify the commands and options that are new in AutoCAD 2004 and in AutoCAD 2002. This feature is especially helpful for readers updating from either AutoCAD 2002 or 2000.

"TIP!" Icons

"TIP!" icons point out time-saving tricks and tips that otherwise might be discovered only after much experience with AutoCAD.

Editing Multilines

Because a multiline is treated as one object and its complex configuration is defined by the multiline style, traditional editing commands such as *Trim* and *Extend*, cannot be used for typical editing functions. Instead, a special set of editing functions are designed for *Mlines*. The *Mledit* (multiline edit) command provides the editing functions for existing multilines and is discussed in Chapter 16. Your use for *Mlines* is limited until you can edit them with *Mledit*.

SKETCH

Pull-down Menu	COMMAND (TYPE)	ALIAS (TYPE)	Short-cut	Screen (side) Menu	Tablet Menu
...	SKETCH	...	...	...	...

The *Sketch* command is unlike other draw commands. *Sketch* quickly creates many short line segments (individual objects) by following the motions of the cursor. *Sketch* is used to give the appearance of a freehand "sketched" line, as used in Figure 15-32 for the tree and bushes. You do not specify individual endpoints, but rather draw a freehand line by placing the pen down, moving the cursor, and then picking up the pen. This action creates a large number of short *Line* segments. *Sketch* line segments are temporary (displayed in a different color) until *Recorded* or until *eXiting Sketch*.

Figure 15-32

 CAUTION: *Sketch* can increase the drawing file size greatly due to the relatively large number of line segment endpoints required to define the *Sketch* line.

```
Command: sketch
Record increment <0.1000>: (value) or Enter (Enter a value to specify the segment increment
length or press Enter to accept the default increment.)
Sketch. Pen eXit Quit Record Erase Connect. (letter) (Enter "p" or press button #1 to put the pen
down and begin drawing or enter another letter for another option. After drawing a sketch line,
enter "p" or press button #1 again to pick up the pen.)
```

It is important to specify an *increment* length for the short line segments that are created. This *increment* controls the "resolution" of the *Sketch* line (Fig. 15-33).

Figure 15-33

 Too large of an *increment* makes the straight line segments apparent, while too small of an *increment* unnecessarily increases file size. The default *increment* is 0.1 (appropriate for default *Limits* of 12 x 9) and should be changed proportionally with a change in *Limits*. Generally, multiply the default *increment* of 0.1 times the drawing scale factor.

Overlay

An *Xref Overlay* is similar to an *Attached Xref* with one main difference—an *Overlay* cannot be nested. In other words, if you *Attach* a drawing that (itself) has an *Overlay*, the *Overlay* does not appear in your drawing. On the other hand, if the first drawing is *Attached*, it appears when the parent drawing is *Attached* to another drawing (Fig. 30-18).

Figure 30-18

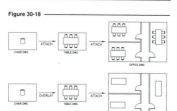

To produce an *Overlay* using the *Xref Manager*, you must first select the *Attach* button to produce the *Select File* dialog box. After selecting the desired file to overlay, press the *Overlay* button in the *External Reference* dialog box that appears (Fig. 30-19). If you type the *−Xref* command, simply use the *Overlay* option and the standard *Select File* dialog box appears.

Figure 30-19

The *Overlay* option prevents "circular" *Xrefs* from appearing by preventing unwanted nested *Xrefs*. This is helpful in a networking environment where many drawings *Xref* other drawings. For example, assume drawing B has drawing C as an *Overlay*. As you work on drawing C, you *Xref* and view only drawing B without drawing B's overlays—namely drawing A. This occurs because drawing A is an *Overlay* to B, not *Attached*. If all drawings are *Overlays*, no nesting occurs (Fig. 30-20).

Figure 30-20

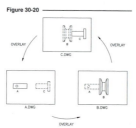

Graphically Driven Page Layout

Over 1500 illustrations are given in the book. Explanatory text is located directly next to the related figure.

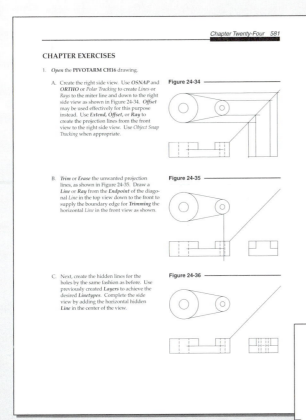

Chapter Twenty-Four   581

CHAPTER EXERCISES

1. *Open* the **PIVOTARM CH16** drawing.

   A. Create the right side view. Use *OSNAP* and *ORTHO* or *Polar Tracking* to create *Lines* or *Rays* to the miter line and down to the right side view as shown in Figure 24-34. *Offset* may be used effectively for this purpose instead. Use *Extend*, *Offset*, or *Ray* to create the projection lines from the front view to the right side view. Use *Object Snap Tracking* when appropriate.

   Figure 24-34

   B. *Trim* or *Erase* the unwanted projection lines, as shown in Figure 24-35. Draw a *Line* or *Ray* from the *Endpoint* of the diagonal *Line* in the top view down to the front to supply the boundary edge for *Trimming* the horizontal *Line* in the front view as shown.

   Figure 24-35

   C. Next, create the hidden lines for the holes by the same fashion as before. Use previously created *Layers* to achieve the desired *Linetypes*. Complete the side view by adding the horizontal hidden *Line* in the center of the view.

   Figure 24-36

Chapter Exercises

Exercises help you try out the commands discussed in each chapter. Exercises begin with simple, step-by-step use of commands and progress to more advanced drawings based on synthesis of earlier ideas.

Reference Material

Useful for self study or classroom use, tabbed pages help you quickly locate the Command Table Index, Shortcut Keys, Dimension Variables, System Variables, Tables of Limits Settings, Template Drawings and more.

1260   *Command Table Index*

Command Name (type)	Button	Pull-down Menu	ALIAS (type)	Short Cut	Screen (side) Menu	Tablet Menu	Chapter in this Text
HELP		Help / Help	?	F1	HELP / Help	Y,7	5
HIDE		View / Hide	HI	...	VIEW 2 / Hide	M,2	35
HLSETTINGS		...	...	...	...	...	35
HYPERLINK		Insert / Hyperlink...	...	Ctrl+K	...	...	19
HYPERLINKBASE		...	...	...	...	...	19
ID		Tools / Inquiry > / ID Point	...	...	TOOLS 1 / ID	U,9	17
IMAGE -IMAGE		Insert / Image Manager...	IM, -IM	...	INSERT / Image	T,3	32
IMAGEADJUST		Modify / Object> / Image> / Adjust...	IAD	...	MODIFY1 / Imageadj	X,20	32
IMAGEATTACH		Insert / Image Manager... / Attach...	IAT	...	INSERT / Image / Attach...	...	32
IMAGECLIP		Modify / Clip> / Image	ICL	...	MODIFY1 / Imageclp	X,22	32
IMAGEFRAME		Modify / Object> / Image> / Frame	...	...	MODIFY1 / Imagefrm	...	32
IMAGEQUALITY		Modify / Object> / Image> / Quality	...	...	MODIFY1 / Imagequa	...	32
IMPORT		...	IMP	...	...	T,2	32
INDEXCTL		File / Saveas... / Tools / Options... / Index type	...	...	FILE / Saveas... / Options... / Index type	...	30
INETLOCATION		Tools / Options / Files / Menu, Help, and / Misc. File Names / Default internet / Location	...	...	TOOLS2 / Options / Files / Menu, Help, and / Misc. File Names / Default internet / Location	...	19
INSERT -INSERT		Insert / Block...	I, -I	...	INSERT / Ddinsert	T,5	21
INSERTOBJ		Insert / OLE Object...	IO	...	INSERT / Insertob	T,1	31
INTERFERE		Draw / Solids > / Interference	INF	...	DRAW 2 / SOLIDS / Interfer	...	39

Supplements at www.mhhe.com/leach

An online learning center provides additional material for students and instructors. Students can download 25 test or review questions for each chapter, discipline-specific exercises for each chapter (architectural, mechanical and civil/electrical), and even 3 additional advanced chapters. Solutions for the chapter exercises (.DWG files) and review questions (Word format) are available for instructors.

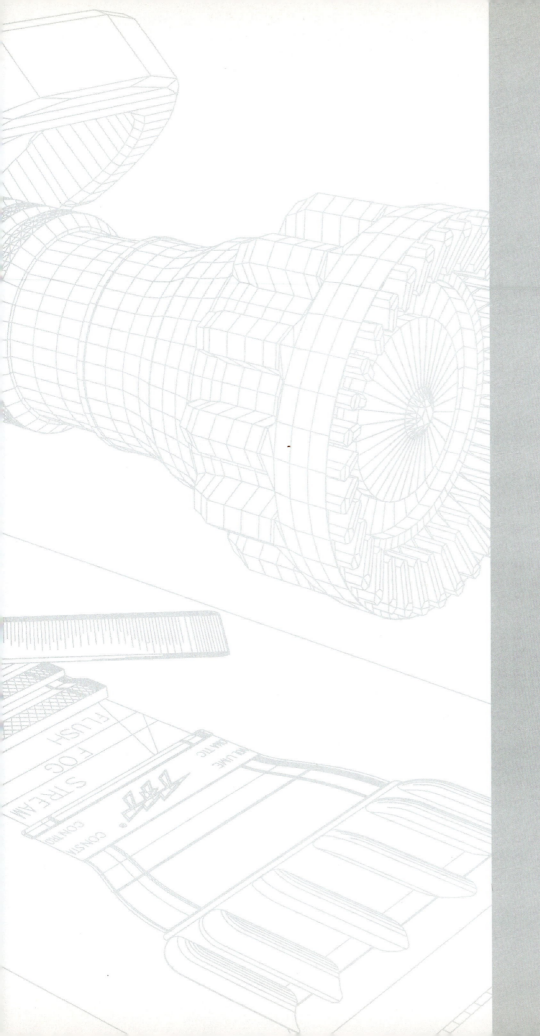

PREFACE

ABOUT THIS BOOK

This Book Is Your AutoCAD 2004 Instructor

The objective of *AutoCAD 2004 Instructor* is to provide you the best possible printed medium for learning AutoCAD, whether you are a professional or student learning AutoCAD on your own or whether you are attending an instructor-led course.

Complete Coverage

AutoCAD 2004 Instructor is written to instruct you in the full range of AutoCAD 2004 features. All commands, system variables, and features within AutoCAD are covered. This text can be used for a two-, three-, or four-course sequence.

Graphically Oriented

Because *AutoCAD 2004 Instructor* discusses concepts that are graphical by nature, many illustrations (approximately **1550**) are used to communicate the concepts, commands, and applications.

Pedagogical Progression

AutoCAD 2004 Instructor is presented in a pedagogical format by delivering the fundamental concepts first, then moving toward the more advanced and specialized features of AutoCAD. The book begins with small pieces of information explained in a simple form and then builds on that experience to deliver more complex ideas, requiring a synthesis of earlier concepts. The chapter exercises follow the same progression, beginning with a simple tutorial approach and ending with more challenging problems requiring a synthesis of earlier exercises.

Easy Update from AutoCAD 2000 and AutoCAD 2002

AutoCAD 2004 Instructor is helpful if you are already an AutoCAD user but are updating from AutoCAD 2000 or 2002. All new commands, concepts, features, and variables are denoted on the edges of the pages by a vertical "2002" bar (denoting an update since 2000) or a "2004" bar (denoting an update since 2002).

Important "Tips"

Tips, reminders, notes, and cautions are given in the book and denoted by a "TIP!" (light bulb) icon. This feature helps you identify and remember important concepts, commands, procedures, and tricks used by professionals that would otherwise be discovered only after much experience.

Valuable Reference Guide

AutoCAD 2004 Instructor is structured to be used as a reference guide to AutoCAD. Several important tables, lists, and variable settings are "tabbed" on the edge of the page for easy access. Every command throughout the book is given with a "command table" listing the possible methods of invoking the command. A complete index gives an alphabetical listing of all AutoCAD commands, command options, system variables, and concepts discussed.

For Professionals and Students in Diverse Areas

AutoCAD 2004 Instructor is written for professionals and students in the fields of engineering, architecture, design, construction, manufacturing, and any other field that has a use for AutoCAD. Applications and examples from many fields are given throughout the text. The applications and examples are not intended to have an inclination toward one particular field. Instead, applications to a particular field are used when they best explain an idea or use of a command.

www.mhhe.com/leach

Please visit our web page at the above address. Ancillary materials are available for reading or download. Questions for each chapter (25 true-false, multiple choice, and written answer) are available for review and testing. Over 400 drawing problems specifically for architectural, mechanical engineering, civil, and electrical applications are available. Solutions for drawing problems and questions can be downloaded by requesting a password from your McGraw-Hill representative.

Additional Chapters

Chapter 45, Menu Customization, Chapter 46, CAD Management, and Chapter 47, Express Tools and Batch Plotting, are available free at **www.mhhe.com/leach**.

Have Fun

I predict you will have a positive experience learning AutoCAD. Although learning AutoCAD is not a trivial endeavor, you will have fun learning this exciting technology. In fact, I predict that more than once in your learning experience you will say to yourself, "Cool!" (or something to that effect).

James A. Leach

ABOUT THE AUTHOR

James A. Leach (B.I.D., M.Ed.) is Professor of Engineering Graphics at the University of Louisville. He began teaching AutoCAD at Auburn University early in 1984 using Version 1.4, the first version of AutoCAD to operate on IBM personal computers. Jim is currently Director of the Autodesk Training Center (ATC) established at the University of Louisville in 1985, one of the first fifteen centers to be authorized by Autodesk.

In his 28 years of teaching Engineering Graphics and AutoCAD courses, Jim has published numerous journal and magazine articles, drawing workbooks, and textbooks about Autodesk and engineering graphics instruction. He has designed CAD facilities and written AutoCAD-related course materials for Auburn University, University of Louisville, the ATC at the University of Louisville, and several two-year and community colleges. Jim is the author of eleven AutoCAD textbooks published by Richard D. Irwin and McGraw-Hill.

CONTRIBUTING AUTHORS

Steven H. Baldock is an engineer at a consulting firm in Louisville and operates a CAD consulting firm, Infinity Computer Enterprises (ICE). Steve is an Autodesk Certified Instructor and teaches several courses at the University of Louisville AutoCAD Training Center. He has thirteen years experience using AutoCAD in architectural, civil, and structural design applications. Steve has degrees in engineering, computer science, and mathematics. Steve Baldock prepared material for several sections of *AutoCAD 2004 Instructor,* such as xrefs, groups, text, multiline drawing and editing, and dimensioning. Steve also created several hundred figures used in *AutoCAD 2004 Instructor.*

Michael E. Beall is the owner of Computer Aided Management and Planning in Shelbyville, Kentucky. Michael offers contract services and professional training on AutoCAD as well as CAP products from Sweets Group, a division of McGraw-Hill. He has co-authored *AutoCAD 14 Fundamentals,* and was a contributing author to *Inside AutoCAD 14* from New Riders Publishing. Other efforts include co-authoring *AutoCAD Release 13 for Beginners* and *Inside AutoCAD LT for Windows 95.* Mr. Beall has been presenting CAD training seminars to architects and engineers since 1982 and is currently an Autodesk Certified Instructor (ACI) at the University of Louisville ATC. He was also a presenter for the *Mastering Today's*

AutoCAD seminar series from Awareness Learning, Inc., an organization founded by Hugh Bathurst (www.awarenesslearning.com). Mr. Beall received a Bachelor of Architecture degree from the University of Cincinnati. Michael Beall assisted with several topics in *AutoCAD 2004 Instructor,* including the geometric calculator, slides and scripts, layer filters, toolbar customization, point filters, and express tools. You can contact Michael at 502.633.3994 or michael.beall@autocadtrainerguy.com.

Bruce Duffy has been an AutoCAD trainer and consultant for 15 years. He has aided many Fortune 500 companies in the United States with training, curriculum development, and productivity assessment/training. Bruce works primarily as an instructor at the University of Louisville Speed Scientific School and Authorized Autodesk Training Center. He consults and provides training for national and international clients. Bruce Duffy contributed much of the material for Chapters 31 and 32 of *AutoCAD 2004 Instructor.*

Patrick McCuistion, Ph.D., is an assistant professor of Industrial Technology at Ohio University. Dr. McCuistion taught three years at Texas A&M University and previously worked in various engineering design, drafting, and checking positions at several manufacturing industries. He has provided instruction in geometric dimensioning and tolerancing to many industry, military, and educational institutions, and has prepared several articles and presentations on the topic. Dr. McCuistion is an active member in several ANSI subcommittees, including Y14.5 Dimensioning and Tolerancing, Y14.3 Multiview and Section View Drawings, Y14.35 Revisions, Y14.36 Surface Texture, and B89.3.6 Functional Gages. Dr. McCuistion contributed the material on geometric dimensioning and tolerancing for *AutoCAD 2004 Instructor.*

ACKNOWLEDGMENTS

I want to thank all of the contributing authors for their assistance in writing *AutoCAD 2004 Instructor.* Without their help, this text could not have been as application-specific nor could it have been completed in the short time frame. I especially want to thank Steven H. Baldock for his hard work on early editions of this book and for his continual contributions and valuable input on new AutoCAD concepts and related industrial applications.

I am very grateful to Gary Bertoline for his foresight in conceiving the McGraw-Hill Graphics Series and for including my efforts in it.

I would like to give thanks to the excellent editorial and production group at McGraw-Hill who gave their talents and support during this project, especially Betsy Jones, Suzanne Jeans and Lisa Kalner-Williams.

A special thanks goes to Karen Collins for the layout and design of *AutoCAD 2004 Instructor.* She was instrumental in fulfilling my objective of providing the most direct and readable format for conveying concepts.

David Morgan and Brianne Duffy deserve credit for preparation of the contents of the "command tables" used throughout this text. David Hagg prepared the solution drawings for the Chapter Exercises that are available on the www.mhhe.com/leach Web site.

Again I want to thank Cecilia Crosby-Lampkin for the her precise, timely, and tireless proofing.

My thanks go to the following colleagues who provided detailed reviews of this text: Clark Cory (Purdue University), Gary Matthew (University of Florida), Fahmida Massom (University of Wisconsin, Platteville), and Moustafa R. Moustafa (Old Dominion University).

I also want to thank all of the readers that have contacted me with comments and suggestions on specific sections of this text. Your comments help me improve this book, assist me in developing new ideas, and keep me abreast of ways AutoCAD and this text are used in industrial and educational settings.

I also acknowledge: my colleague and friend Robert A. Matthews for his support of this project and for doing his job well; Charles Grantham of Contemporary Publishing Company of Raleigh, Inc., for generosity and consultation; and Speed Scientific School Dean's Office for support and encouragement.

Special thanks, once again, to my wife, Donna, for the many hours of copy editing required to produce this and the other texts.

TRADEMARK AND COPYRIGHT ACKNOWLEDGMENTS

The object used in the Wireframe Modeling Tutorial, Chapter 36 appears courtesy of James H. Earle, *Graphics for Engineers, Third Edition,* (pg. 120), ©1992 by Addison-Wesley Publishing Company, Inc. Reprinted by permission of the publisher.

The following drawings used for Chapter Exercises appear courtesy of James A. Leach, *Problems in Engineering Graphics Fundamentals, Series A* and *Series B,* ©1984 and 1985 by Contemporary Publishing Company of Raleigh, Inc.: Gasket A, Gasket B, Pulley, Holder, Angle Brace, Saddle, V-Block, Bar Guide, Cam Shaft, Bearing, Cylinder, Support Bracket, Corner Brace, and Adjustable Mount. Reprinted or redrawn by permission of the publisher.

Autodesk and the following names are either registered trademarks or trademarks of Autodesk, Inc., in the USA and/or other countries: 3D Studio®, 3D Studio MAX®, 3D Studio VIZ®, 3ds max™, ATC®, AutoCAD®, AutoCAD LT®, Autodesk®, Autodesk (logo)®, Autodesk Inventor®, Autodesk View®, AutoLISP®, AutoSnap™, AutoTrack™, Buzzsaw™, Buzzsaw.com™, Content Explorer™, DesignCenter™, Design Web Format™, DWF™, DXF™, Heidi®, HOOPS®, i-drop®, Kinetix®, Kinetix (logo)™, Mechanical Desktop®, ObjectARX®, ObjectDBX™, PolarSnap™, Volo®, WHIP!®, discreet®. Windows, Notepad, WordPad, Excel, and MS-DOS are registered trademarks of Microsoft Corporation. Corel WordPerfect is a registered trademark of Corel, Inc. Norton Editor is a registered trademark of S. Reifel & Company. City Blueprint, Country Blueprint, EuroRoman, EuroRoman-oblique, PanRoman, SuperFrench, Romantic, Romantic-bold, Sans Serif, Sans Serif-bold, Sans Serif-oblique, Sans Serif-BoldOblique, Technic Technic-light, and Technic-bold are Type 1 fonts, copyright 1992 P. B. Payne.

All other brand names, product names, or trademarks belong to their respective holders.

LEGEND

The following special treatment of characters and fonts in the textual content is intended to assist you in translating the meaning of words or sentences in *AutoCAD 2004 Instructor.*

<u>Underline</u>	Emphasis of a word or an idea.
Helvetica font	An AutoCAD prompt appearing on the <u>screen</u> at the command line or in a text window.
Italic (Upper and Lower)	An AutoCAD command, option, menu, toolbar, or dialog box name.
UPPER CASE	A file name.
UPPER CASE ITALIC	An AutoCAD system variable or a drawing aid (*OSNAP, SNAP, GRID, ORTHO*).

Anything in **Bold** represents user input:

Bold	What you should <u>type</u> or press on the keyboard.
Bold Italic	An AutoCAD <u>command</u> that you should type or <u>menu item</u> that you should select.
BOLD UPPER CASE	A <u>file name</u> that you should type.
BOLD UPPER CASE ITALIC	A <u>system variable</u> that you should type.
PICK	Move the cursor to the indicated position on the screen and press the <u>select</u> button (button #1 or left mouse button).

INTRODUCTION

WHAT IS CAD?

CAD is an acronym for Computer-Aided Design. CAD allows you to accomplish design and drafting activities using a computer. A CAD software package, such as AutoCAD, enables you to create designs and generate drawings to document those designs.

Design is a broad field involving the process of making an idea into a real product or system. The design process requires repeated refinement of an idea or ideas until a solution results—a manufactured product or constructed system. Traditionally, design involves the use of sketches, drawings, renderings, two-dimensional (2D) and three-dimensional (3D) models, prototypes, testing, analysis, and documentation. Drafting is generally known as the production of drawings that are used to document a design for manufacturing or construction or to archive the design.

CAD is a <u>tool</u> that can be used for design and drafting activities. CAD can be used to make "rough" idea drawings, although it is more suited to creating accurate finished drawings and renderings. CAD can be used to create a 2D or 3D computer model of the product or system for further analysis and testing by other computer programs. In addition, CAD can be used to supply manufacturing equipment such as lathes, mills, laser cutters, or rapid prototyping equipment with numerical data to manufacture a product. CAD is also used to create the 2D documentation drawings for communicating and archiving the design.

The tangible result of CAD activity is usually a drawing generated by a plotter or printer but can be a rendering of a model or numerical data for use with another software package or manufacturing device. Regardless of the purpose for using CAD, the resulting drawing or model is stored in a CAD file. The file consists of numeric data in binary form usually saved to a magnetic or optical device such as a diskette, hard disk, tape, or CD.

WHY SHOULD YOU USE CAD?

Although there are other methods used for design and drafting activities, CAD offers the following advantages over other methods in many cases:

1. Accuracy
2. Productivity for repetitive operations
3. Sharing the CAD file with other software programs

Accuracy

Since CAD technology is based on computer technology, it offers great accuracy. When you draw with a CAD system, the graphical elements, such as lines, arcs, and circles, are stored in the CAD file as numeric data. CAD systems store that numeric data with great precision. For example, AutoCAD stores values with fourteen significant digits. The value 1, for example, is stored in scientific notation as the equivalent of 1.0000000000000. This precision provides you with the ability to create designs and drawings that are 100% accurate for almost every case.

Productivity for Repetitive Operations

It may be faster to create a simple "rough" drawing, such as a sketch by hand (pencil and paper), than it would by using a CAD system. However, for larger and more complex drawings, particularly those involving similar shapes or repetitive operations, CAD methods are very efficient. Any kind of shape or operation accomplished with the CAD system can be easily duplicated since it is stored in a CAD file.

In short, it may take some time to set up the first drawing and create some of the initial geometry, but any of the existing geometry or drawing setups can be easily duplicated in the current drawing or for new drawings.

Likewise, making changes to a CAD file (known as editing) is generally much faster than creating the original geometry. Since all the graphical elements in a CAD drawing are stored, only the affected components of the design or drawing need to be altered, and the drawing can be plotted or printed again or converted to other formats.

As CAD and the associated technology advance and software becomes more interconnected, more productive developments are available. For example, it is possible to make a change to a 3D model that automatically causes a related change in the linked 2D engineering drawing. One of the main advantages of these technological advances is productivity.

Sharing the CAD File with Other Software Programs

Of course, CAD is not the only form of industrial activity that is making technological advances. Most industries use computer software to increase capability and productivity. Since software is written using digital information and may be written for the same or similar computer operating systems, it is possible and desirable to make software programs with the ability to share data or even interconnect, possibly appearing simultaneously on one screen.

For example, word processing programs can generate text that can be imported into a drawing file, or a drawing can be created and imported into a text file as an illustration. (This book is a result of that capability.) A drawing created with a CAD system such as AutoCAD can be exported to a finite element analysis program that can read the computer model and compute and analyze stresses. CAD files can be dynamically "linked" to spreadsheets or databases in such a way that changing a value in a spreadsheet or text in a database can automatically make the related change in the drawing, or vice versa.

Another advance in CAD technology is the automatic creation and interconnectivity of a 2D drawing and a 3D model in one CAD file. With this tool, you can design a 3D model and have the 2D drawings automatically generated. The resulting set has bi-directional associativity; that is, a change in either the 2D drawings or the 3D model is automatically updated in the other.

With the introduction of the new Web technologies, designers and related professionals can more easily collaborate by viewing and transferring drawings over the Internet. CAD drawings can contain Internet links to other drawings, text information, or other related Web sites. Multiple CAD users can even share a single CAD session from remote locations over the Internet.

CAD, however, may not be the best tool for every design related activity. For example, CAD may help develop ideas but probably won't replace the idea sketch, at least not with present technology. A 3D CAD model can save much time and expense for some analysis and testing but cannot replace the "feel" of an actual model, at least not until virtual reality technology is developed and refined.

With everything considered, CAD offers many opportunities for increased accuracy, productivity, and interconnectivity. Considering the speed at which this technology is advancing, many more opportunities are rapidly obtainable. However, we need to start with the basics. Beginning by learning to create an AutoCAD drawing is a good start.

WHY USE AutoCAD?

CAD systems are available for a number of computer platforms: laptops, personal computers (PCs), workstations, and mainframes. AutoCAD, offered to the public in late 1982, was one of the first PC-based CAD software products. Since that time, it has grown to be the world leader in market share for <u>all</u> CAD products. Autodesk, the manufacturer of AutoCAD, is the world's leading supplier of PC design software and multimedia tools. At the time of this writing, Autodesk is one of the largest software producers in the world and has over three million customers in more than 150 countries.

Learning AutoCAD offers a number of advantages to you. Since AutoCAD is the most widely used CAD software, using it gives you the highest probability of being able to share CAD files and related data and information with others.

As a student, learning AutoCAD, as opposed to learning another CAD software product, gives you a higher probability of using your skills in industry. Likewise, there are more employers who use AutoCAD than any other single CAD system. In addition, learning AutoCAD as a first CAD system gives you a good foundation for learning other CAD packages because many concepts and commands introduced by AutoCAD are utilized by other systems. In some cases, AutoCAD features become industry standards. The .DXF file format, for example, was introduced by Autodesk and has become an industry standard for CAD file conversion between systems.

As a professional, using AutoCAD gives you the highest possibility that you can share CAD files and related data with your colleagues, vendors, and clients. Compatibility of hardware and software is an important issue in industry. Maintaining compatible hardware and software allows you the highest probability for sharing data and information with others as well as offering you flexibility in experimenting with and utilizing the latest technological advancements. AutoCAD provides you with great compatibility in the CAD domain.

This introduction is not intended as a selling point but to remind you of the importance and potential of the task you are about to undertake. If you are a professional or a student, you have most likely already made up your mind that you want to learn to use AutoCAD as a design or drafting tool. If you have made up your mind, then you can accomplish anything. Let's begin.

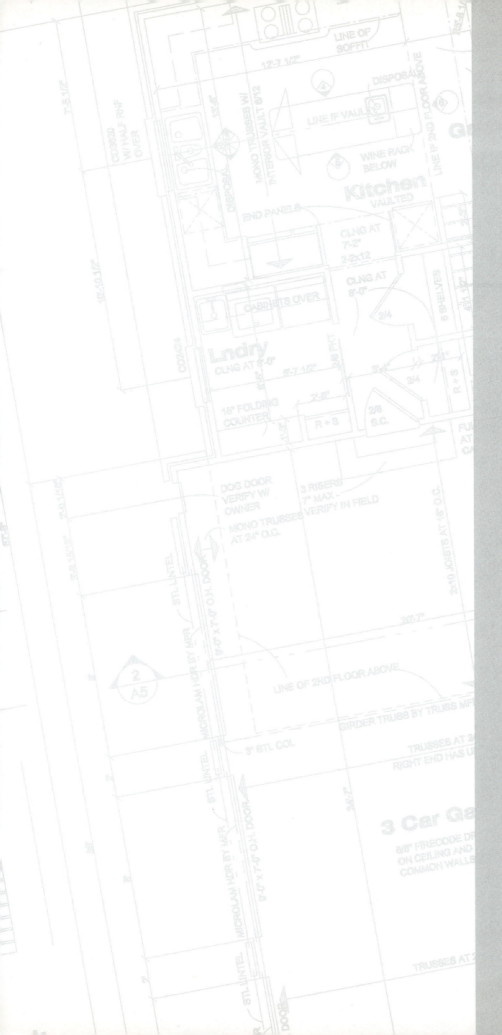

GETTING STARTED

Chapter Objectives

After completing this chapter you should:

1. understand how the X, Y, Z coordinate system is used to define the location of drawing elements in digital format in a CAD drawing file;

2. understand why you should create drawings full size in the actual units with CAD;

3. be able to start AutoCAD to begin drawing;

4. recognize the areas of the AutoCAD Drawing Editor and know the function of each;

5. be able to use the five methods of entering commands;

6. be able to turn on and off the *SNAP*, *GRID*, *ORTHO*, and *POLAR* drawing aids;

7. know how to customize the AutoCAD for Windows screen to your preferences.

CONCEPTS

Coordinate Systems

Any location in a drawing, such as the endpoint of a line, can be described in X, Y, and Z coordinate values (Cartesian coordinates). If a line is drawn on a sheet of paper, for example, its endpoints can be charted by giving the distance over and up from the lower-left corner of the sheet (Fig. 1-1).

These distances, or values, can be expressed as X and Y coordinates; X is the horizontal distance from the lower-left corner (origin) and Y is the vertical distance from that origin. In a three-dimensional coordinate system, the third dimension, Z, is measured from the origin in a direction perpendicular to the plane defined by X and Y.

Two-dimensional (2D) and three-dimensional (3D) CAD systems use coordinate values to define the location of drawing elements such as lines and circles (called objects in AutoCAD).

In a 2D drawing, a line is defined by the X and Y coordinate values for its two endpoints (Fig. 1-2).

In a 3D drawing, a line can be created and defined by specifying X, Y, and Z coordinate values (Fig. 1-3). Coordinate values are always expressed by the X value first separated by a comma, then Y, then Z.

Figure 1-1

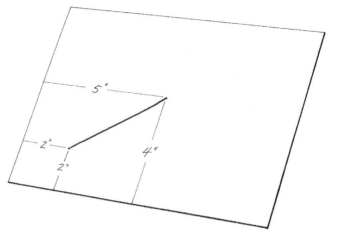

Figure 1-2

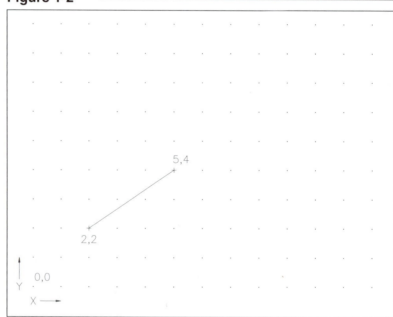

Figure 1-3

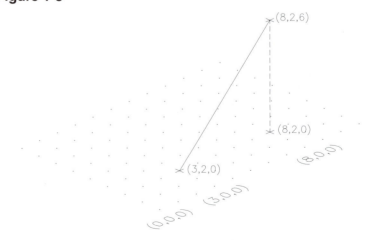

The CAD Database

A CAD (Computer-Aided Design) file, which is the electronically stored version of the drawing, keeps data in binary digital form. These digits describe coordinate values for all of the endpoints, center points, radii, vertices, etc. for all the objects composing the drawing, along with another code that describes the kinds of objects (line, circle, arc, ellipse, etc.). Figure 1-4 shows part of an AutoCAD DXF (Drawing Interchange Format) file giving numeric data defining lines and other objects. Knowing that a CAD system stores drawings by keeping coordinate data helps you understand the input that is required to create objects and how to translate the meaning of prompts on the screen.

Figure 1-4

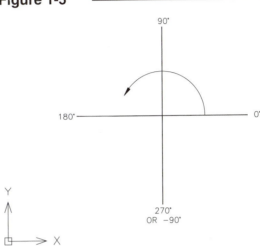

```
LINE
   8
0
  62
          8
  10
15.0
  20
8.105789
  30
0.0
  11
15.644291
  21
8.75
  31
0.0
   0
LINE
   8
0
  62
          8
  18
```

Angles in AutoCAD

Angles in AutoCAD are measured in a <u>counterclockwise direction</u>. Angle 0 is positioned in a positive X direction, that is, horizontally from left to right. Therefore, 90 degrees is in a positive Y direction, or straight up; 180 degrees is in a negative X direction, or to the left; and 270 degrees is in a negative Y direction, or straight down (Fig. 1-5).

The position and direction of measuring angles in AutoCAD can be changed; however, the defaults listed here are used in most cases.

Figure 1-5

Draw True Size

When creating a drawing with pencil and paper tools, you must first determine a scale to use so the drawing will be proportional to the actual object and will fit on the sheet (Fig. 1-6). However, when creating a drawing on a CAD system, there is no fixed size drawing area. The number of drawing units that appear on the screen is variable and is assigned to fit the application.

The CAD drawing is not scaled until it is physically transferred to a fixed size sheet of paper by plotter or printer.

Figure 1-6

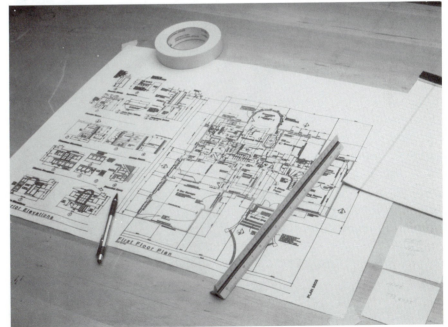

The rule for creating CAD drawings is that the drawing should be created <u>true size</u> using real-world units. The user specifies what units are to be used (architectural, engineering, etc.) and then specifies what size drawing area is needed (in X and Y values) to draw the necessary geometry.

Figure 1-7

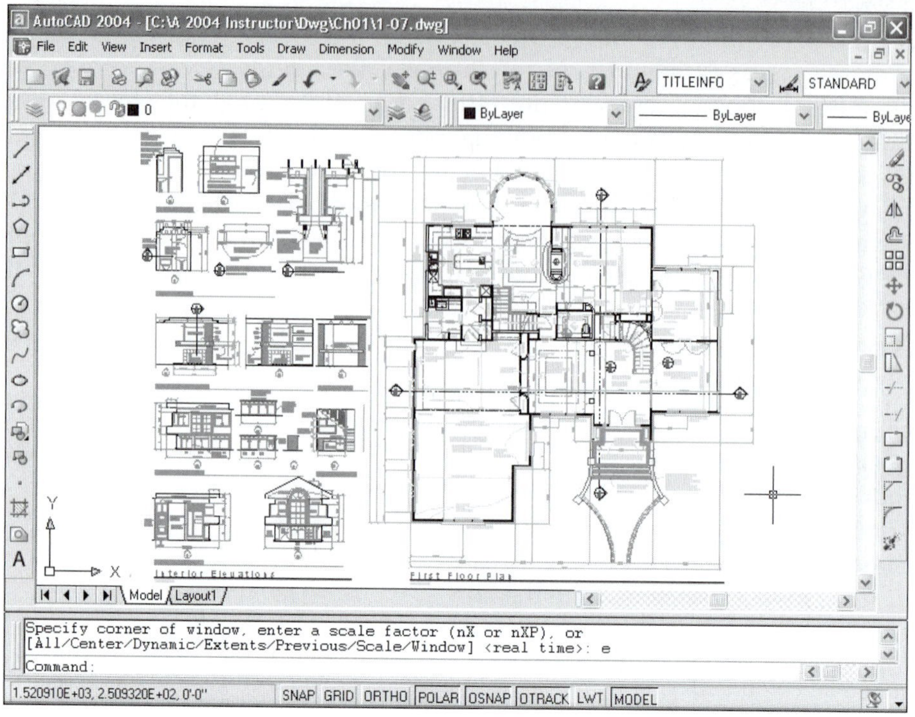

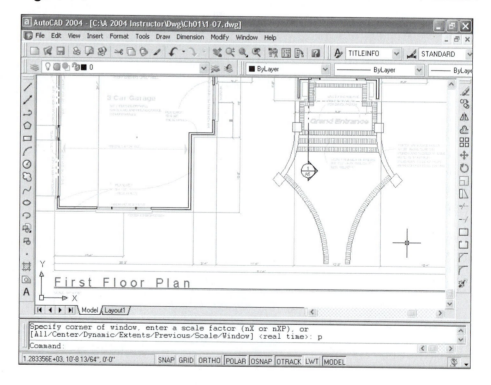

Whatever the specified size of the drawing area, it can be displayed on the screen in its entirety (Fig. 1-7) or as only a portion of the drawing area (Fig. 1-8).

Figure 1-8

Plot to Scale

As long as a drawing exists as a CAD file or is visible on the screen, it is considered a virtual, full-sized object. Only when the CAD drawing is transferred to paper by a plotter or printer is it converted (usually reduced) to a size that will fit on a sheet. A CAD drawing can be automatically scaled to fit on the sheet regardless of sheet size; however, this action results in a plotted drawing that is not to an accepted scale (not to a regular proportion of the real object). Usually it is desirable to plot a drawing so that the resulting drawing is a proportion of the actual object size. The scale to enter as the plot scale (Fig. 1-9) is simply the proportion of the plotted drawing size to the actual object.

Figure 1-9

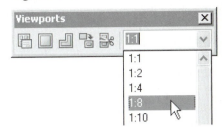

STARTING AutoCAD

Figure 1-10

Assuming that AutoCAD has been installed and configured properly for your system, you are ready to begin using AutoCAD.

To start AutoCAD 2004, locate the "AutoCAD 2004" shortcut icon on the desktop (Fig. 1-10). Double-clicking on the icon launches AutoCAD 2004.

If you cannot locate the AutoCAD 2004 shortcut icon on the desktop, select the "Start" button, highlight "Programs" and search for "Autodesk." From the list that appears select "AutoCAD 2004" (Fig. 1-11).

Figure 1-11

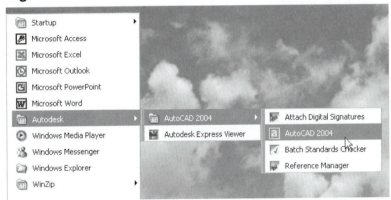

THE AutoCAD DRAWING EDITOR

Figure 1-12

The *Startup* Dialog Box

When you start AutoCAD 2004, you are presented with the *Startup* dialog box (Fig. 1-12) or a "blank" drawing screen, depending on how your system has been configured. By default, a blank drawing screen appears. You can turn on the *Startup* dialog box in the *Options* dialog box, *System* tab (or by typing in the *STARTUP* system variable and changing the setting to 1).

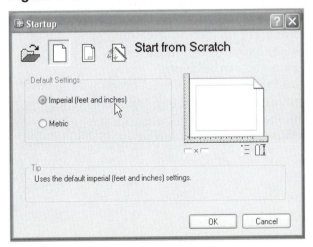

This *Startup* dialog box also allows you to begin a new drawing or open an existing drawing, including providing access to drawing templates and setup *Wizards*.

If you want to <u>start AutoCAD with the default English or metric settings</u> using the *Startup* dialog box, you can use one of the following options:

1. Select the *Start from Scratch* button in the *Startup* dialog box. Text appears displaying a choice of *Imperial (feet and inches)* or *Metric*. Make your selection, then press *OK*.

2. Select the *Template* option in the *Startup* dialog box, and then select the *Acad.dwt* (inches) or *Acadiso.dwt* (metric) template drawing.

3. Select the *Cancel* button in the *Startup* dialog box.

All three of these methods begin a new drawing using AutoCAD's default settings for use with inch or metric units. The inch settings (*Acad.dwt*) use a 12" x 9" drawing area (called *Limits*), and the metric settings (*Acadiso.dwt*) use a 420mm x 297mm drawing area. The *MEASUREINIT* system variable setting (0=inch, 1=metric) controls which of the two formats is used if you cancel the dialog boxes or if you use the startup *Wizards*.

NOTE: If your session starts with no dialog box and a template other than *Acad.dwt* or *Acadiso.dwt* appears, a specific template was specified by the *Qnew* setting in the *Options* dialog box (see "QNEW" in Chapter 2). The *Startup* dialog box and the *Create New Drawing* dialog boxes are explained in more detail in Chapters 2, 6, and 12. For the examples and exercises in Chapters 1-5, use the first method above to begin a drawing.

After specifying your choice in the *Startup* dialog box, the Drawing Editor appears on the screen and allows you to immediately begin drawing. Figure 1-13 displays the Drawing Editor in AutoCAD 2004.

Graphics Area

The large central area of the screen is the <u>Graphics area</u>. It displays the lines, circles, and other objects you draw that will make up the drawing. The <u>cursor</u> is the intersection of the crosshairs (vertical and horizontal lines that follow the mouse or puck movements). The default size of the graphics area for English settings is 12 units (X or horizontal) by 9 units (Y or vertical). This usable drawing area (12 x 9) is called the drawing *Limits* and can be changed to any size to fit the application. As you move the cursor, you will notice the numbers in the <u>Coordinate display</u> change (Fig. 1-13, bottom left).

Figure 1-13

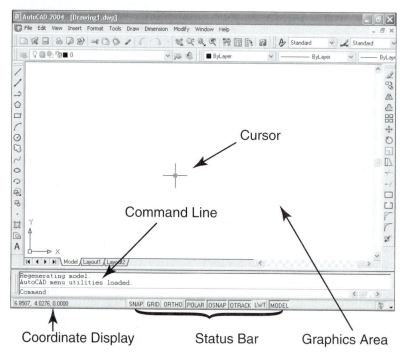

Cursor

Command Line

Coordinate Display   Status Bar   Graphics Area

Command Line

The <u>Command line</u> consists of the three text lines at the bottom of the screen (by default) and is the most important area other than the drawing itself (see Fig. 1-13). Any command that is entered or any prompt that AutoCAD issues appears here. The Command line is always visible and gives the current state of drawing activity. You should develop the habit of glancing at the Command line while you work in AutoCAD. The Command line can be set to display any number of lines and/or moved to another location (see "Customizing the AutoCAD Screen" later in this chapter).

Pressing the F2 key opens a text window displaying the command history. This window is like an expanded Command line because it displays many more than just the three text lines that normally appear at the bottom of the screen. See "AutoCAD Text Window" later in this chapter.

Toolbars

<u>AutoCAD provides a variety of tool-bars</u> (Fig. 1-14). Each toolbar contains a number of icon buttons (tools) that can be PICKed to invoke commands for drawing or editing objects (lines, arcs, circles, etc.) or for managing files and other functions.

Figure 1-14

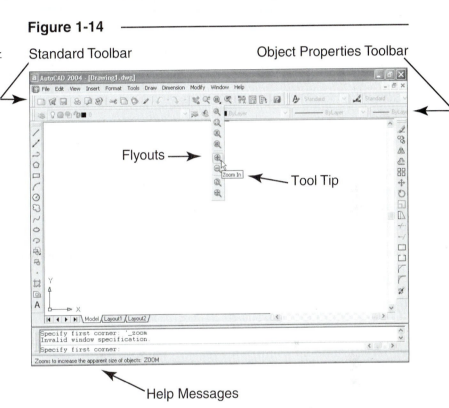

Standard Toolbar   Object Properties Toolbar

Flyouts →

Tool Tip ←

Help Messages

The <u>Standard toolbar</u> is the row of icons nearest the top of the screen. The Standard toolbar contains many standard icons used in other Windows applications (like *New, Open, Save, Print, Cut, Paste,* etc.) and other icons for AutoCAD-specific functions (like *Zoom* and *Pan*). The <u>Object Properties toolbar</u>, located beneath the standard toolbar, is used for managing properties of objects, such as *Layers* and *Linetypes.* The <u>Draw and Modify toolbars</u> also appear (by default) when you first use AutoCAD. As shown in Figure 1-14, the Draw and Modify toolbars (left and right side) and the Standard and Object Properties toolbars (top) are <u>docked</u>. Many other toolbars are available and can be made to float, resize, or dock (see "Customizing the AutoCAD Screen" later in this chapter).

If you place the pointer on an any icon and wait momentarily, a <u>Tool Tip</u> and a <u>Help message</u> appear. Tool Tips pop out by the pointer and give the command name (see Figure 1-14). The Help message appears at the bottom of the screen, giving a short description of the function. <u>Flyouts</u> are groups of related icons that pop out in a row or column when one of the group is selected. PICKing any icon that has a small black triangle in its lower-right corner causes the related icons to fly out.

Pull-down Menus

The <u>pull-down menu bar</u> is at the top of the screen just under the title bar (Fig. 1-15). Selecting any of the words in the menu bar activates, or pulls down, the respective menu. Selecting a word appearing with an arrow activates a cascading menu with other options. Selecting a word with ellipsis (. . .) activates a dialog box (see "Dialog Boxes"). Words in the pull-down menus are not necessarily the same as the formal command names used when typing commands. Menus can be canceled by pressing Escape or PICKing in the graphics area. The pull-down menus <u>do not contain</u> all of the AutoCAD commands and variables but contain the most commonly used ones.

Figure 1-15

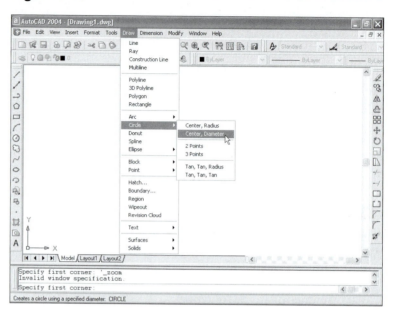

NOTE: Because the words that appear in the pull-down menus are <u>not always the same</u> as the formal commands you would type to invoke a command, AutoCAD can be confusing to learn. The Help message that appears just below the Command line (when a menu is pulled down or the pointer rests on an icon button) can be instrumental in avoiding this confusion. The Help message gives a description of the command followed by colon (:), then the <u>formal command name</u> (see Figures 1-14 and 1-15).

Screen (Side) Menu

The <u>screen menu</u> does not appear by default in AutoCAD, but can be made to appear by selecting *Options…* from the *Tools* pull-down menu. This selection causes the *Options* dialog box to appear (Fig. 1-16). Select the *Display* tab and check the *Display screen menu* option. Keep in mind that this menu <u>is not needed</u> if you prefer to use the icon buttons, pull-down menus, or keyboard to enter commands.

Not all of the AutoCAD commands can be accessed through the menu system located at the right side of the screen (see Figure 1-16). The screen menu has a tree structure.

Figure 1-16

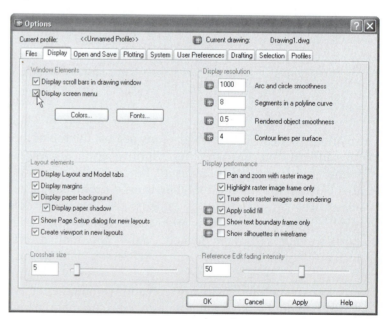

Menus and commands are accessed by branching out from the root or top-level menu (Fig. 1-17). Commands (in uppercase and lowercase) or menus (capital letters only) are selected by moving the cursor to the desired position until the word is highlighted and then pressing the PICK button. The root menu (top level) is accessed from any level of the structure by selecting the word "AutoCAD" at the top of any menu.

The commands in these menus are generally the formal command names that can also be typed at the keyboard. (Some commands names are truncated in the screen menu because only eight characters can be displayed.)

If commands are invoked by typing, pull-downs, or toolbars, the screen menu automatically changes to the current command.

Dialog Boxes

Dialog boxes provide an interface for controlling complex commands or a group of related commands. Depending on the command, the dialog boxes allow you to select among multiple options and sometimes give a preview of the effect of selections. The *Layer Properties Manager* dialog box (Fig. 1-18) gives complete control of layer colors, linetypes, and visibility.

Dialog boxes can be invoked by typing a command, selecting an icon button, or PICKing from the menus. For example, typing *Layer*, PICKing the *Layers* button, or selecting *Layer* from the *Format* pull-down menu causes the *Layer Properties Manager* dialog box to appear. In the pull-down menu, all commands that invoke a dialog box end with ellipsis points (…).

Figure 1-17

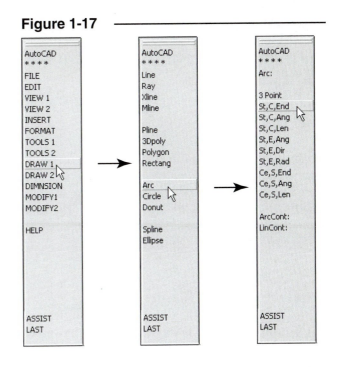

Figure 1-18

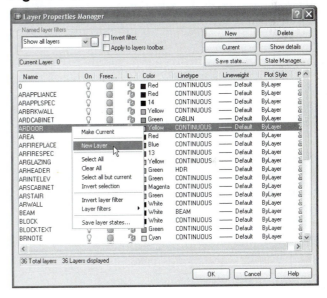

The basic element or smallest component of a dialog box is called a tile. Several types of tiles and the resulting actions of tile selection are listed below.

Button	Resembles a push button and triggers some type of action
Edit box	Allows typing or editing of a single line of text
Image tile	A button that displays a graphical image
List box	A list of text strings from which one or more can be selected
Drop-down list	A text string that drops down to display a list of selections
Radio button	A group of buttons, only one of which can be turned on at a time
Checkbox	A checkbox for turning a feature on or off (displays a check mark when on)

You can change dialog box border colors (for all applications) with the Windows Control Panel. The AutoCAD screen colors can also be customized (see "Customizing the AutoCAD Screen," this chapter). See "File Navigation Dialog Box Functions" in Chapter 2 for more information on dialog boxes that are used to access files.

Status Bar

The Status bar is a set of informative words or symbols that gives the status of the drawing aids. The Status bar appears at the very bottom of the screen (see Figure 1-13). The following drawing aids can be toggled on or off by single-clicking (pressing the left mouse button once) on the desired word or by using Function keys or Ctrl key sequences. The following drawing aids are explained here or in following chapters:
SNAP, GRID, ORTHO, POLAR, OSNAP, OTRACK, LWT, MODEL

Coordinate Display (*Coords*)

The Coordinate display is located in the lower-left corner of the AutoCAD screen (Fig. 1-19). The Coordinate display (*Coords*) displays the current position of the cursor in one of two possible formats explained below. This display can be very helpful when you draw because it can give the X, Y, and Z coordinate position of the cursor or give the cursor's distance and angle from the last point established. The format of *Coords* is controlled by toggling the F6 key, pressing Ctrl+D, or single-clicking on the *Coords* display (numbers) at the bottom left of the screen. *Coords* can also be toggled off.

Cursor tracking
When *Coords* is in this position, the values display the current location of the cursor in absolute (X, Y, and Z) coordinates (see Figure 1-19).

Relative polar display
This display is possible only if a draw or edit command is in use. The values give the distance and angle of the "rubberband" line from the last point established (not shown).

Figure 1-19

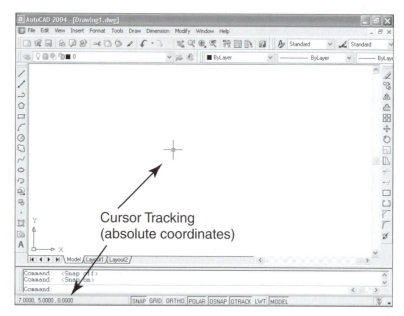

Cursor Tracking
(absolute coordinates)

Digitizing Tablet Menu

If you have a digitizing tablet, the AutoCAD commands are available by making the desired selection from the AutoCAD digitizing tablet menu (Fig. 1-20, on the next page). The open area located slightly to the right of center is called the Screen Pointing area. Locating the digitizing puck there makes the cursor appear on the screen. Locating the puck at any other location allows you to select a command.

The icons on the tablet menu are identical to the toolbar icons. Similar to the format of the toolbars, commands are located in groups such as Draw, Edit, View, and Dimension. The tablet menu has the columns numbered along the top and the rows lettered along the left side. The location of each command by column and row is given in the "command tables" in this book. For example, the *Line* command can be found at **10,J**.

Figure 1-20

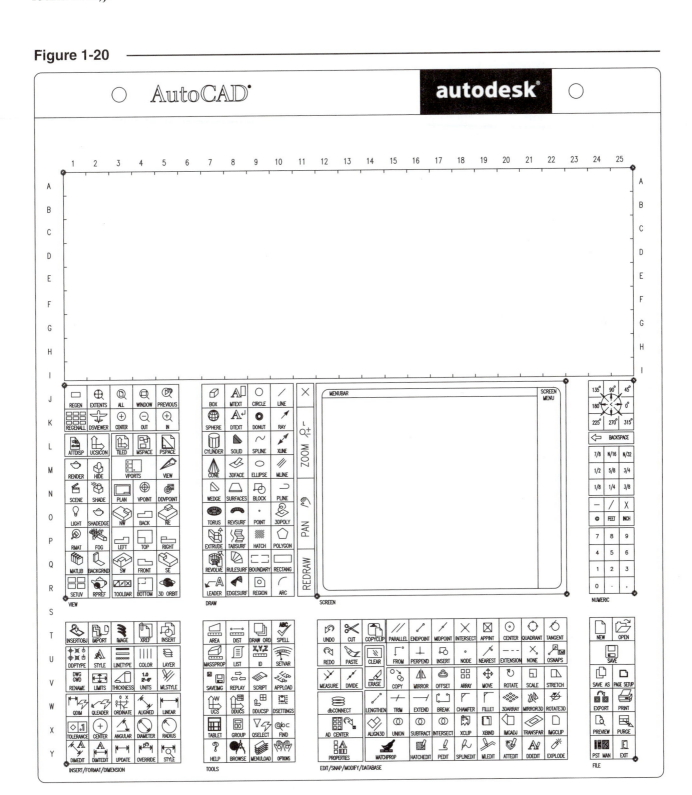

COMMAND ENTRY

Methods for Entering Commands

There are five possible methods for entering commands in AutoCAD depending on your *Options* setting (for the screen menu) and availability of a digitizing tablet. Generally, <u>any one</u> of the five methods can be used to invoke a particular command.

1. **Toolbars**	Select the command or dialog box by PICKing an icon (tool) from a toolbar.
2. **Pull-down menu**	Select the command or dialog box from a pull-down menu.
3. **Screen (side) menu**	Select the command or dialog box from the screen (side) menu.
4. **Keyboard**	Type the command name, command alias, or accelerator (Ctrl) keys at the keyboard.*
5. **Tablet menu**	Select the command from the digitizing tablet menu (if available).

*A command alias is a one- or two-letter shortcut. Accelerator keys use the Ctrl key plus another key and are used for utility operations. The command aliases and accelerator keys are given in the command tables (see below).

All five methods of entering commands accomplish the same goals; however, one method may offer a slightly different option or advantage to another depending on the command used.

All menus, including the digitizing tablet, can be customized by editing the ACAD.MNU file, and command aliases can be changed or added in the ACAD.PGP file so that command entry can be designed to your preference (see Chapter 44, Basic Customization, and Chapter 45, Menu Customization).

Using the "Command Tables" in This Book to Locate a Particular Command

Command tables, like the one below, are used throughout this book to show the possible methods for entering a particular command. The table shows the icon used in the toolbars and digitizing tablet, gives the selections to make for the pull-down and screen (side) menus, gives the correct spelling for entering commands and command aliases at the keyboard, and gives the command location (column, row) on the digitizing tablet menu. This example uses the *Copy* command.

COPY

Pull-down Menu	COMMAND (TYPE)	ALIAS (TYPE)	Short-cut	Screen (side) Menu	Tablet Menu
Modify *Copy*	*COPY*	*CO* or *CP*	*(Edit Mode)* *Copy Selection*	*MODIFY1* *Copy*	*V,15*

Shortcut Menus

AutoCAD makes use of shortcut menus that are activated by pressing the right mouse button (sometimes called right-click menus). Shortcut menus give quick access to command options. There are many shortcut menus to list since they are based on the active command or dialog box. The menus fall into five basic categories listed here.

Default Menu

The default menu appears when you right-click in the drawing area and no command is in progress.

Figure 1-21 ⸻

Edit-Mode Menu

This menu appears when you right-click when <u>objects have been selected</u> but no command is in progress.

NOTE: Edit mode shortcut menus do not appear if the *PICKFIRST* system variable is set to 0 (see "*PICKFIRST*," Chapter 20).

Figure 1-22 ⸻

Command-Mode Menu

These menus appear when you right-click <u>when a command is in progress</u>. This menu changes since the options are specific to the command.

Figure 1-23 ⸻

Dialog-Mode Menu

When the pointer is in a dialog box or tab and you right-click, this menu appears. The options on this menu can change based on the current dialog box.

Figure 1-24

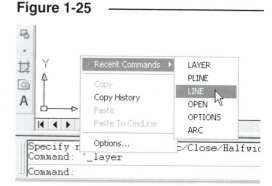

Other Menus

There are other menus that can be invoked. For example, this menu appears if you right-click in the Command line area.

Because there are so many shortcut menus, don't be too concerned about learning these until you have had some experience. The best advice at this time is just remember to experiment by right-clicking often to display the possible options.

Figure 1-25

Mouse and Digitizing Puck Buttons

Depending on the type of mouse or digitizing puck used for cursor control, a different number of buttons is available. In any case, the buttons perform the following tasks:

#1 (left button)	PICK	Used to select commands or pick locations on the screen.
#2 (right button)	Enter or shortcut menu	Depending on the status of the drawing or command, this button either performs the same function as the enter key or produces a shortcut menu.
#3 (middle button or wheel)	*Pan*	If you press and drag, you can pan the drawing about on the screen.
	Zoom	If you turn the wheel, you can zoom in and out centered on the location of the cursor.

Function Keys

Several function keys are usable with AutoCAD. They offer a quick method of turning on or off (toggling) drawing aids.

F1	*Help*	Opens a help window providing written explanations on commands and variables.
F2	*Flipscreen*	Activates a text window showing the previous command line activity (command history).
F3	*Osnap Toggle*	If Running Osnaps are set, toggling this tile temporarily turns the Running Osnaps off so that a point can be picked without using Osnaps. If no Running Osnaps are set, F3 produces the *Osnap Settings* dialog box (discussed in Chapter 7).
F4	*Tablet*	Turns the *TABMODE* variable on or off. If *TABMODE* is on, the digitizing tablet can be used to digitize an existing paper drawing into AutoCAD.

F5	*Isoplane*	When using an *Isometric* style *SNAP* and *GRID* setting, toggles the cursor (with *ORTHO* on) to draw on one of three isometric planes.
F6	*Coords*	Toggles the Coordinate Display between cursor tracking mode and off. If used transparently (during a command in operation), displays a polar coordinate format.
F7	*GRID*	Turns the *GRID* on or off (see "Drawing Aids").
F8	*ORTHO*	Turns *ORTHO* on or off (see "Drawing Aids").
F9	*SNAP*	Turns *SNAP* on or off (see "Drawing Aids").
F10	*POLAR*	Turns *POLAR* on or off (see "Drawing Aids").
F11	*Osnap Tracking*	Turns Object Snap Tracking on or off.

Control Key Sequences (Accelerator Keys)

Several control key sequences (holding down the Ctrl key or Alt key and pressing another key simultaneously) invoke regular AutoCAD commands or produce special functions. See Appendix D.

Special Key Functions

Esc	The Escape key cancels a command, menu, or dialog box or interrupts processing of plotting or hatching.
Space bar	In AutoCAD, the space bar performs the same action as the Enter key. Only when you are entering text into a drawing does the space bar create a space.
Enter	If Enter or Spacebar is pressed when no command is in use (the open Command: prompt is visible), the last command used is invoked again.

Drawing Aids

This section gives a brief introduction to AutoCAD's Drawing Aids. For a full explanation of the related commands and options, see Chapter 6.

SNAP **(F9 or Status Bar)**

SNAP has two modes in AutoCAD: Grid Snap and Polar Snap. Only one of the two modes can be active at one time. Grid Snap is a function that forces the cursor to "snap" to regular intervals (.5 units is the default English setting), which aids in creating geometry accurate to interval lengths. You can use the *Snap* command or the *Drafting Settings* dialog box to specify any value for the Grid Snap increment. Figure 1-26 displays a Snap setting of .25 (note the values in the coordinate display).

Figure 1-26

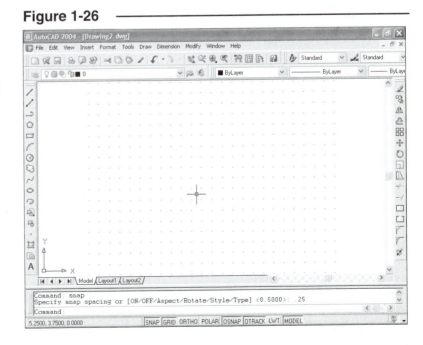

The other mode of Snap is Polar Snap. Polar Snap forces the cursor to snap to regular intervals along angular lines (Fig. 1-27). Polar Snap is functional only when *POLAR* is also toggled on since it works in conjunction with *POLAR*. The Polar Snap interval uses the Grid Snap setting by default but can be changed to any value using the *Snap* command or the *Drafting Settings* dialog box. Polar Snap is discussed in detail in Chapter 3.

Figure 1-27

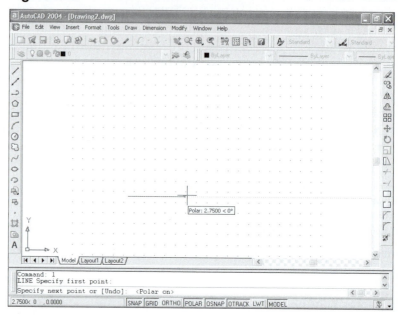

Since you can have only one *SNAP* mode on at a time (Grid Snap or Polar Snap), you can select which of the two is on by right-clicking on the word "SNAP" on the Status bar (Fig. 1-28). You can also access the *Drafting Settings* dialog box by selecting *Settings...* from the menu.

Figure 1-28

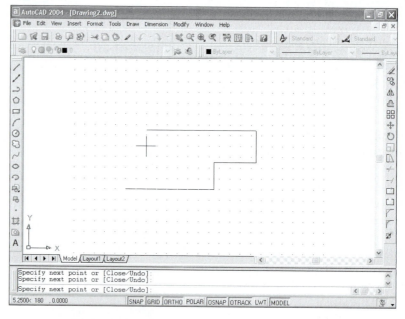

ORTHO (F8 or Status Bar)
If *ORTHO* is on, lines are forced to an orthogonal alignment (horizontal or vertical) when drawing (Fig. 1-29). *ORTHO* is often helpful since so many drawings are composed mainly of horizontal and vertical lines. *ORTHO* can be turned only on or off.

GRID (F7 or Status Bar)
A drawing aid called *GRID* can be used to give a visual reference of units of length. The *GRID* default value for English settings is .5 units. The *Grid* command or *Drawing Aids* dialog box allows you to change the interval to any value. The *GRID* is not part of the geometry and is not plotted.

Figure 1-29

Figure 1-29 displays a *GRID* of .5. *SNAP* and *GRID* are independent functions—they can be turned on or off independently. However, you can force the *GRID* to have the same interval as *SNAP* by entering a *GRID* value of 0 or you can use a proportion of *SNAP* by entering a *GRID* value followed by an "X".

Status Bar Control

Right-clicking in the Status bar produces the shortcut menu shown in Figure 1-30. This menu gives you control of which of these drawing aids appear in the Status bar. Clearing a check mark removes the associated drawing aid from the Status bar. It is recommended for most new AutoCAD users to keep all the drawing aids visible on the Status bar.

Figure 1-30

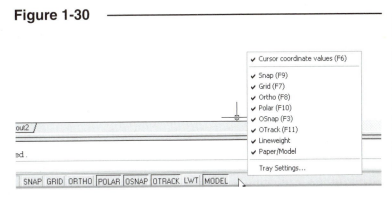

Other Drawing Aids

Other drawing aids that appear on the Status bar not discussed here are: *POLAR, OSNAP, OTRACK, LWT,* and *MODEL.* These features are explained fully in subsequent chapters. (See Chapter 3 for information on *POLAR, SNAP, GRID,* and *ORTHO.* See Chapter 7 for an explanation of *OSNAP* and *OTRACK. LWT* is discussed in Chapter 11, and the *MODEL/PAPER* toggle is explained in Chapter 13.)

AutoCAD Text Window (F2)

Pressing the F2 key activates the *AutoCAD Text Window,* sometimes called the Command History. Here you can see the text activity that occurred at the Command line—kind of an "expanded" Command line. Press F2 again to close the text window. The *Edit* pull-down menu in this text window provides several options. If you highlight text in the window (Fig. 1-31), you can then *Paste to Cmdline* (Command line), *Copy* it to another program such as a word processor, *Copy History* (entire command history) to another program, or *Paste* text into the window. The *Options* choice invokes the *Options* dialog box (discussed later).

Figure 1-31

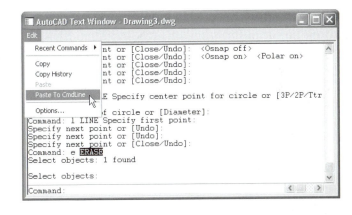

Commands When Multiple Drawings Are Open

In AutoCAD you can have several drawings open at the same time (Fig. 1-32). This feature offers several advantages. Most notably, you can *Copy* and *Paste* from one drawing to another.

You can use the *Syswindows* command to control how you want the drawings displayed in the graphics area. You can *Cascade* or *Tile* them or drag and resize them as you please. If you want each drawing to display as a full screen, use Ctrl+Tab to toggle between drawings. See *Syswindows* in Chapter 10 for more information.

Figure 1-32

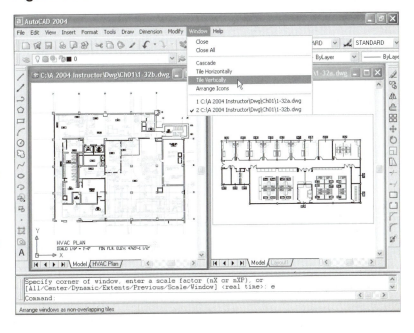

When multiple drawings are open, only one drawing can be active at a time. Any command that is issued (except *Syswindows*) affects <u>only the active drawing</u>. Each drawing has its own command history, so you can go from one drawing to the next and AutoCAD remembers the current command status for each drawing.

Model Tab and *Layout* Tabs

Model Tab

When you create a new drawing, the *Model* tab is the current tab (see Figure 1-33, bottom left). This area is also known as <u>model space</u>. In this area you <u>should create the geometry representing the subject of your drawing</u>, such as a floor plan, a mechanical part, or an electrical schematic. Dimensions are usually created and attached to your objects in model space.

Figure 1-33

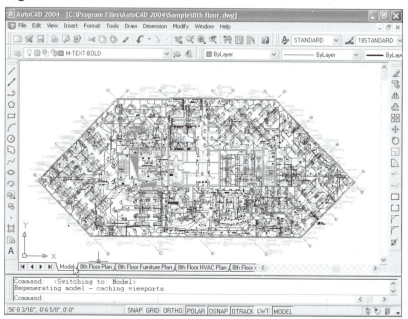

Layout Tabs

When you are finished with your drawing, you can plot it directly from the *Model* tab or switch to a *Layout* tab (Fig. 1-34). Layout tabs, sometimes known as paper space, represent sheets of paper that you plot on. You must use several commands to set up the layout to display the geometry and set all the plotting options such as scale, paper size, plot device, and so on. Plotting and layouts are discussed in detail in Chapters 13, 14, and 33.

Figure 1-34

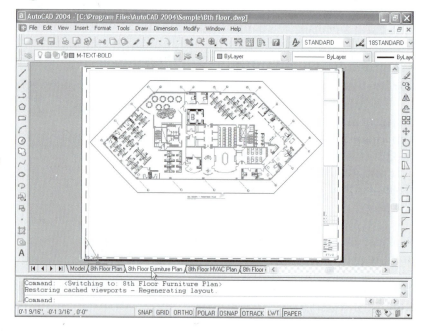

COMMAND ENTRY METHODS PRACTICE

Start AutoCAD. If the *Startup* dialog box appears, select *Start from Scratch,* choose *Imperial* as the default setting, and click the *OK* button. Invoke the *Line* command using each of the command entry methods as follows.

1. Type the command

STEPS	COMMAND PROMPT	PERFORM ACTION	COMMENTS
1.		press **Escape** if another command is in use	
2.	Command:	type *Line* and press **Enter**	
3.	LINE Specify first point:	**PICK** any point	a "rubberband" line appears
4.	Specify next point or [Undo]:	**PICK** any point	another "rubberband" line appears
5.	Specify next point or [Undo]:	press **Enter**	to complete command

2. Type the command alias

STEPS	COMMAND PROMPT	PERFORM ACTION	COMMENTS
1.		press **Escape** if another command is in use	
2.	Command:	type *L* and press **Enter**	
3.	LINE Specify first point:	**PICK** any point	a "rubberband" line appears
4.	Specify next point or [Undo]:	**PICK** any point	another "rubberband" line appears
5.	Specify next point or [Undo]:	press **Enter**	to complete command

3. Pull-down menu

STEPS	COMMAND PROMPT	PERFORM ACTION	COMMENTS
1.	Command:	select the *Draw* menu from the menu bar on top	
2.	Command:	select *Line*	menu disappears
3.	LINE Specify first point:	**PICK** any point	a "rubberband" line appears
4.	Specify next point or [Undo]:	**PICK** any point	another "rubberband" line appears
5.	Specify next point or [Undo]:	press **Enter**	to complete command

4.   Screen menu (if activated on your setup)

STEPS	COMMAND PROMPT	PERFORM ACTION	COMMENTS
1.	Command:	select **AutoCAD** from screen menu on the side	only if menu is not at root level
2.	Command:	select *DRAW1* from root screen menu	menu changes to *DRAW 1*
3.	Command:	select *Line*	
4.	LINE Specify first point:	**PICK** any point	a "rubberband" line appears
5.	Specify next point or [Undo]:	**PICK** any point	another "rubberband" line appears
6.	Specify next point or [Undo]:	press **Enter**	to complete command

5.   Toolbars

STEPS	COMMAND PROMPT	PERFORM ACTION	COMMENTS
1.	Command:	select the *Line* icon from the Draw toolbar (on the left side of the screen)	the *Line* tool should be located at the top of the toolbar
2.	LINE Specify first point:	**PICK** any point	a "rubberband" line appears
3.	Specify next point or [Undo]:	**PICK** any point	another "rubberband" line appears
4.	Specify next point or [Undo]:	press **Enter**	to complete command

6.   Digitizing tablet menu (if available)

STEPS	COMMAND PROMPT	PERFORM ACTION	COMMENTS
1.	Command:	select *LINE* on the digitizing tablet menu	located at *10,J*
2.	LINE Specify first point:	**PICK** any point	a "rubberband" line appears
3.	Specify next point or [Undo]:	**PICK** any point	another "rubberband" line appears
4.	Specify next point or [Undo]:	press **Enter**	to complete command

When you are finished practicing, use the *Files* pull-down menu and select *Exit* to exit AutoCAD. You do not have to "Save Changes."

CUSTOMIZING THE AutoCAD SCREEN

Toolbars

Many toolbars are available, each with a group of related commands for specialized functions. For example, when you are ready to dimension a drawing, you can activate the Dimension toolbar for efficiency. Right-clicking on any toolbar displays a list of all toolbars (Fig. 1-35). Selecting any toolbar name makes that toolbar appear on the screen. You can also select *Toolbars...* from the *View* pull-down menu or type the command *Toolbar* to display the *Customize* dialog box. Toolbars can be removed from the screen by clicking once on the "X" symbol in the upper right of a floating toolbar.

Figure 1-35

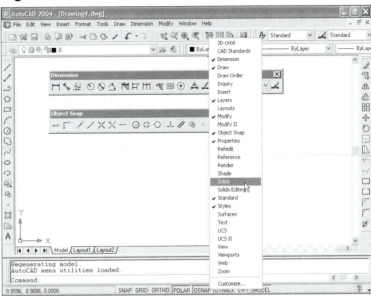

By default, the Object Properties and Standard toolbars (top) and the Draw and Modify toolbars (side) are docked, whereas toolbars that are newly activated are floating (see Fig. 1-35). A floating toolbar can be easily moved to any location on the screen if it obstructs an important area of a drawing. Placing the pointer in the title background allows you to move the toolbar by holding down the left button and dragging it to a new location (Fig. 1-36). Floating toolbars can also be resized by placing the pointer on the narrow border until a two-way arrow appears, then dragging left, right, up, or down (Fig 1-37).

Figure 1-36

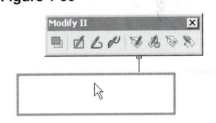

Figure 1-37

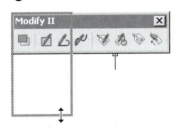

A floating toolbar can be docked against any border (right, left, top, bottom) by dragging it to the desired location (Fig. 1-38). Several toolbars can be stacked in a docked position. By the same method, docked toolbars can be dragged back onto the graphics area. The Object Properties and Standard toolbars can be moved onto the graphics area or docked on another border, although it is wise to keep these toolbars in their standard position. The Object Properties toolbar only displays its full options when located in a horizontal position.

Figure 1-38

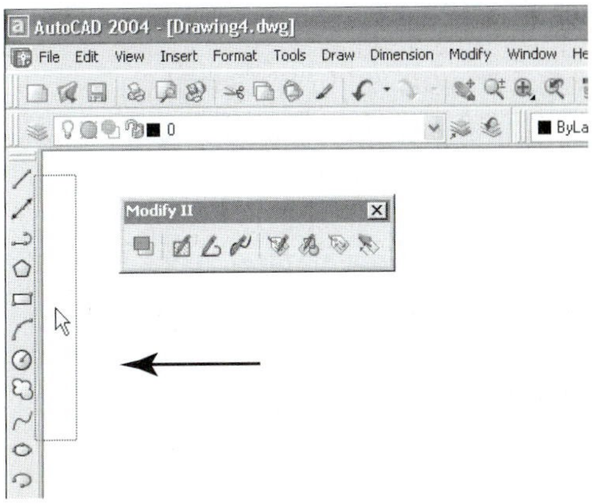

Holding down the **Ctrl** key while dragging a toolbar near a border of the drawing window prevents the toolbar from docking. This feature enables you to float a toolbar anywhere on the screen. New toolbars can be created and existing toolbars can be customized. Typically, you would create toolbars to include groups of related commands that you use most frequently or need for special activities. See "Customizing Toolbars" in Chapter 44.

Clean Screen

New for AutoCAD 2004, the *Cleanscreenon* and *Cleanscreenoff* commands can be used to change the AutoCAD window from a normal window showing all toolbars (for example, Figure 1-13 or Figure 1-26) to a window with <u>no</u> toolbars (Fig. 1-39). Notice that the Command line and pull-down menus are still visible. This is useful if you want to maximize the drawing area to view a drawing or to make a presentation. Instead of typing these commands or using the *View* pull-down menu, a shortcut keystroke of Ctrl+0 (zero) toggles between the two windows.

Figure 1-39

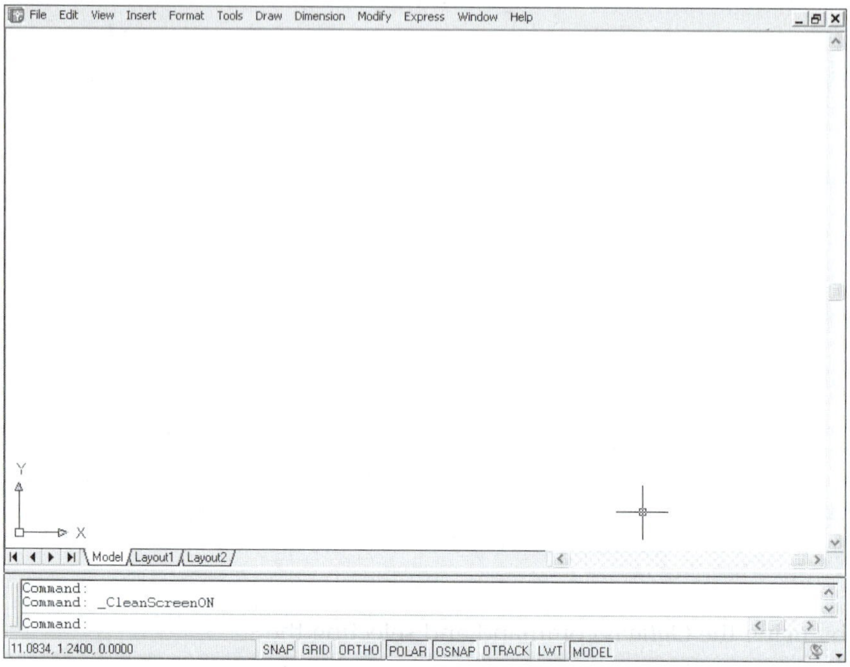

Communication Center

A new feature in AutoCAD 2004 called Communication Center provides up-to-date product information, software updates, and marketing announcements. When new information or software updates are available, an information bubble appears in the lower-right corner of the Status Bar. If the Communication Center icon is not visible on your system, the icon may be disabled (see "Tray Settings"), or this feature may not have been configured during the installation process. (For more information on Communication Center, see Chapter 19, Internet Tools.)

Tray Settings

Communication Center and other services are available through icons in the system tray (lower-left corner of the AutoCAD window). Use the *Traysettings* command or right-click in the Status bar area to produce the *Tray Settings* dialog box (Fig. 1-40). If you chose to *Display Icons from Services*, icons appear (for enabled services) such as Communication Center or Digital Signature Validation. If *Display Notifications from Services* is checked, you will receive alert messages, such as when Communication Center updates are available, digital signatures are invalid, or externally referenced drawings need reloading.

Figure 1-40

2004

Command Line

The Command line, normally located near the bottom of the screen, can be resized to display more or fewer lines of text. Moving the pointer to the border between the graphics screen and the Command line until two-way arrows appear allows you to slide the border up or down. A <u>minimum of two lines</u> of text is recommended. The Command window can also be moved to any location and can be floating or be docked. Point to the two vertical bars on the left of the Command window, then drag the window to the desired location. If the Command window is floating, you can specify transparency by right-clicking on the title bar, selecting *Transparency* from the menu, and setting the *Transparency Level* in the *Transparency* dialog box (Fig. 1-41).

Figure 1-41

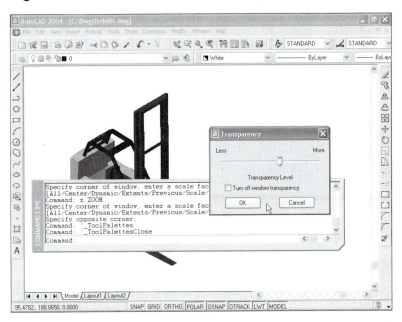

Options

Fonts, colors, and other features of the AutoCAD drawing editor can be customized to your liking by using the *Options* command and selecting the *Display* tab. Typing *Options* or selecting *Options...* from the *Tools* pull-down menu or default shortcut menu activates the dialog box shown in Figure 1-42. The screen menu can also be activated through this dialog box (second checkbox in the dialog box).

Figure 1-42

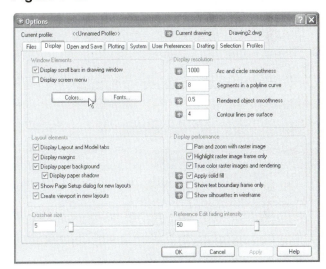

Selecting the *Color...* tile provides a dialog box for customizing the screen colors (Fig. 1-43).

All changes made to the Windows screen by any of the options discussed in this section are automatically saved for the next drawing session. The changes are saved as the current profile. However, if you are working in a school laboratory, it is likely that the computer systems are set up to present the same screen defaults each time you start AutoCAD.

Figure 1-43

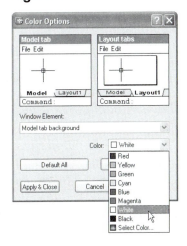

CHAPTER EXERCISES

1.   Starting and Exiting AutoCAD

Start AutoCAD by double-clicking the "AutoCAD 2004" shortcut icon or selecting "AutoCAD 2004" from the Programs menu. If the *Startup* dialog box appears, select *Start from Scratch,* choose *Imperial* as the default setting, and click the *OK* button. Draw a *Line.* Exit AutoCAD by selecting the *Exit* option from the *Files* pull-down menu. Answer *No* to the "Save changes to Drawing1.dwg?" prompt. Repeat these steps until you are confident with the procedure.

2.   Using Drawing Aids

Start AutoCAD. Turn on and off each of the following modes:

>   *SNAP, GRID, ORTHO, POLAR*

3.   Understanding Coordinates

Begin drawing a *Line* by PICKing a "Specify first point:". Toggle *Coords* to display each of the <u>three</u> formats. PICK several other points at the "Specify next point or [Undo]:" prompt. Pay particular attention to the coordinate values displayed for each point and visualize the relationship between that point and coordinate 0,0 (absolute value) or the last point established (relative polar value). Finish the command by pressing Enter.

4.   Using *Flipscreen*

Use *Flipscreen* (**F2**) to toggle between the text window and the graphics screen.

5.   Drawing with Drawing Aids

Draw four *Lines* using <u>each</u> Drawing Aid: ***GRID, SNAP, ORTHO, POLAR.*** Toggle on and off each of the drawing aids one at a time for each set of four *Lines.* Next, draw *Lines* using combinations of the Drawing Aids, particularly ***GRID + SNAP*** and ***GRID + SNAP + POLAR.***

WORKING WITH FILES

Chapter Objectives

After completing this chapter you should be able to:

1. name drawing files;

2. use file-related dialog boxes;

3. use the Windows right-click shortcut menus in file dialog boxes;

4. create *New* drawings;

5. *Open* and *Close* existing drawings;

6. *Save* drawings to disk;

7. use *SaveAs* to save a drawing under a different name, path, and/or format;

8. use *Partialopen* and *Partialload* to open part of a drawing;

9. practice good file management techniques.

AutoCAD DRAWING FILES

Naming Drawing Files

What is a drawing file? A CAD drawing file is the electronically stored data form of a drawing. The computer's hard disk is the principal magnetic storage device used for saving and restoring CAD drawing files. Diskettes, compact disks (CDs), networks, and the Internet are used to transport files from one computer to another, as in the case of transferring CAD files among clients, consultants, or vendors in industry. The AutoCAD commands used for saving drawings to and restoring drawings from files are explained in this chapter.

An AutoCAD drawing file has a name that you assign and a file extension of ".DWG." An example of an AutoCAD drawing file is:

PART-024.DWG
file name   extension

The file name you assign must be compliant with the Windows file name conventions; that is, it can have a <u>maximum</u> of 256 alphanumeric characters. File names and directory names (folders) can be in UPPER-CASE, Title Case, or lowercase letters. Characters such as _ - $ # ( ) ^ and spaces can be used in names, but other characters such as \ / : * ? < > | are not allowed. AutoCAD automatically appends the extension of .DWG to all AutoCAD-created drawing files.

NOTE: The chapter exercises and other examples in this book generally list file names in UPPERCASE letters for easy recognition. The file names and directory (folder) names on your system may appear as UPPERCASE, Title Case, or lowercase.

Beginning and Saving an AutoCAD Drawing

When you start AutoCAD, the drawing editor appears and allows you to begin drawing even before using any file commands.   As you draw, you should develop the habit of saving the drawing periodically (about every 15 or 20 minutes) using *Save*. *Save* stores the drawing in its most current state to disk.

The typical drawing session would involve using *New* to begin a new drawing or using *Open* to open an existing drawing. Alternately, the *Startup* dialog box that appears when starting AutoCAD would be used to begin a new drawing or open an existing one. *Save* would be used periodically, and *Exit* would be used for the final save and to end the session.

Figure 2-1

Accessing File Commands

Proper use of the file-related commands covered in this chapter allows you to manage your AutoCAD drawing files in a safe and efficient manner. Although the file-related commands can be invoked by several methods, they are easily accessible via the first pull-down menu option, *File* (Fig. 2-1). Most of the selections from this pull-down menu invoke dialog boxes for selection or specification of file names.

The Standard toolbar at the top of the AutoCAD screen has tools (icon buttons) for *New, Open,* and *Save*. File commands can also be entered at the keyboard by typing the formal command name, the command alias, or using the Ctrl keys sequences. File commands and related dialog boxes are also available from the *File* screen menu or by selection from the digitizing tablet menu.

File Navigation Dialog Box Functions

There are many dialog boxes appearing in AutoCAD that help you manage files. All of these dialog boxes operate in a similar manner. A few guidelines will help you use them. The *Save Drawing As* dialog box (Fig. 2-2) is used as an example.

- The top of the box gives the title describing the action to be performed. It is <u>very important</u> to glance at the title before acting, especially when saving or deleting files.
- The desired file can be selected by PICKing it, then PICKing *OK*. Double-clicking on the selection accomplishes the same action. File names can also be typed in the *File name:* edit box.
- Every file name has an extension (three letters following the period) called the <u>type.</u> File types can be selected from the *Files of Type:* section of the dialog boxes, or the desired file extension can be entered in the *File name:* edit box.
- The current folder (directory) is listed in the drop-down box near the top of the dialog box. You can select another folder (directory) or drive by using the drop-down list displaying the current path, by selecting the *Back* arrow, or by selecting the *Up One Level* icon to the right of the list. (Rest your pointer on an icon momentarily to make the tool tip appear.)
- Selecting one of the folders displayed on the left side of the file dialog boxes allows you to navigate to the following locations. The files or locations found appear in the list in the central area.

  History: Lists a history of files you have opened from most to least recent.

  My Documents: Lists all files and folders saved in the My Documents folder.

  Favorites: Lists all files and folders saved in the Favorites folder.

  Desktop: Shows files and folders and shortcuts on your desktop.

  FTP: Allows you to browse through the FTP sites you have saved.

  Buzzsaw: Connects you to Buzzsaw—an Autodesk Web site for collaboration of files and information.
- You can use the *Search the Web* icon to display the *Browse the Web* dialog box (not shown). The default site is http://www.autodesk.com/.
- Any highlighted file(s) can be deleted using the "X" (*Delete*) button.
- A new folder (subdirectory) can be created within the current folder (directory) by selecting the *New Folder* icon.
- The *View* drop-down list allows you to toggle the listing of files to a *List* or to show *Details*. The *List* option displays only file names (see Figure 2-2), whereas the *Details* option gives file-related information such as file size, file type, and time and date last modified (see Figure 2-3). The detailed list can also be sorted alphabetically by name, by file size, alphabetically by file type, or chronologically by time and date. Do this by clicking the *Name*, *Type*, *Size*, or *Date Modified* tiles immediately above the list. Double-clicking on one of the tiles reverses the order of the list. You

Figure 2-2

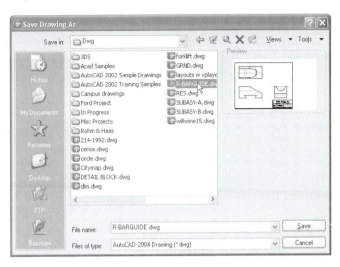

Figure 2-3

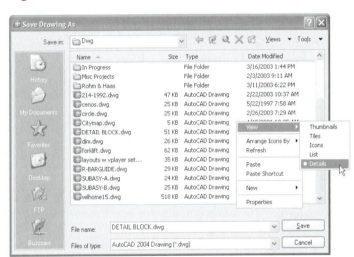

can resize the width of a column (*Name, Size, Type*, etc.) by moving the pointer to the "crack" between two columns, then sliding the double arrows that appear in either direction. You can also use the *Preview* option to show a preview of the highlighted file (see Figure 2-2) or to disable the preview (see Figure 2-3).

- Use the *Tools* drop-down list to add files or folders to the *Favorites* or *FTP* folder. *Places* adds files or folders to the folder listing on the left side of the dialog box. *Find* allows you to search for files or folders based on criteria that you enter.

Windows Right-Click Shortcut Menus

AutoCAD utilizes the right-click shortcut menus that operate with the Windows operating systems. Activate the shortcut menus by pressing the right mouse button (right-clicking) <u>inside the file list area</u> of a dialog box. There are two menus that give you additional file management capabilities.

Right-Click, No Files Highlighted

When no files are highlighted, right-clicking produces a menu for file list display and file management options (Fig. 2-3). The menu choices are as follows:

View	Shows *Thumbnails* (see Figure 2-4), *Tiles* (large icons), *Icons* (small icons), *List* (see Figure 2-3) or *Details* (see Figure 2-3).
Arrange Icons By	Sorts by file *Name*, file *Size*, extension *Type*, or date *Modified*.
Paste	If the *Copy* or *Cut* function was previously used (see "Right-Click, File Highlighted," Fig. 2-4), *Paste* can be used to place the copied file into the displayed folder.
Paste Shortcut	Use this option to place a *Shortcut* (to open a file) into the current folder.
New	Creates a new *Folder, Shortcut*, or document.
Properties	Displays a dialog box listing properties of the current folder.

Right-Click, File Highlighted

When a file is highlighted, right-clicking produces a menu with options for the selected file (Fig. 2-4). The menu choices are as follows:

Figure 2-4

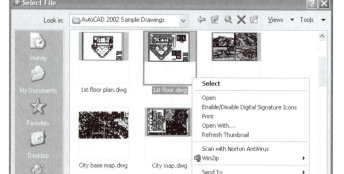

Select	Processes the file (like selecting the *OK* button) according to the dialog box function. For example, if the *SaveAs* dialog box is open, *Select* saves the highlighted file; or if the *Select File* dialog box is active, *Select* opens the drawing.
Open	Opens the application (AutoCAD or other program) and loads the selected file. Since AutoCAD can have multiple drawings open, you can select several drawings to open with this feature.
Print	Sends the selected file to the configured system printer.
Send To	Copies the selected file to the selected device. <u>This is an easy way to copy a file from your hard drive to a diskette in A: drive.</u>

Cut	In conjunction with *Paste,* allows you to move a file from one location to another.
Copy	In conjunction with *Paste,* allows you to copy the selected file to another location. You can copy the file to the same folder, but Windows renames the file to "Copy of . . .".
Create Shortcut	Use this option to create a *Shortcut* (to open the highlighted file) in the current folder. The shortcut can be moved to another location using drag and drop.
Delete	Sends the selected file to the Recycle Bin.
Rename	Allows you to rename the selected file. Move your cursor to the highlighted file name, click near the letters you want to change, then type or use the backspace, delete, space, or arrow keys.
Properties	Displays a dialog box listing properties of the selected file.

Specific features of other dialog boxes are explained in the following sections.

AutoCAD FILE COMMANDS

When you start AutoCAD, one of three situations occurs based on how the *Options* are set on your system:

1. The *Startup* dialog box appears.
2. No dialog box appears and the session starts with the ACAD.DWT or the ACADISO.DWT template (determined by the *MEASUREINIT* system variable setting, 0 or 1, respectively).
3. No dialog box appears and the session starts with a template specified by the *Qnew* setting in the *Options* dialog box (see"QNEW").

To control the appearance of the *Startup* dialog box, use the *Options* command to produce the *Options* dialog box, select the *System* tab, and make the desired selection from the *Startup* drop-down list (Fig. 2-5).

Figure 2-5

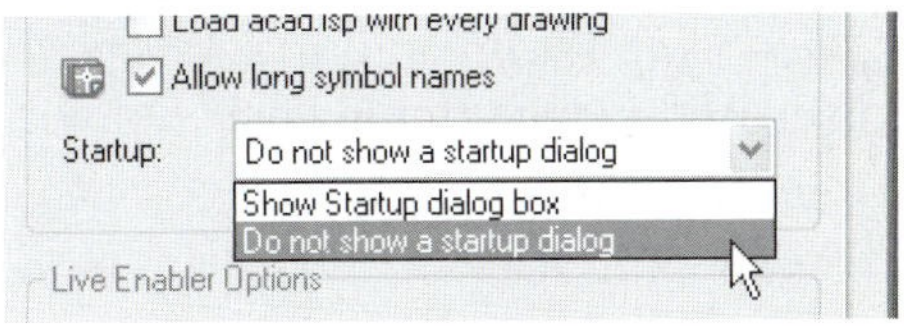

The *Startup* dialog box allows you to create new drawings and open existing drawings. These options are the same as using the *New* and *Open* commands. Therefore, a description of the *New* and *Open* commands explains these functions in the *Startup* dialog box.

NEW	**Pull-down Menu**	**COMMAND (TYPE)**	**ALIAS (TYPE)**	**Short-cut**	**Screen (side) Menu**	**Tablet Menu**
	File *New...*	NEW	...	*Ctrl+N*	*FILE* *New*	*T,24*

The *New* command begins a new drawing. The new drawing can be a completely "blank" drawing or it can be based on a template that may already have a title block and some additional desired settings. Based on the settings for your system, the *New* command produces one of two methods to start a new drawing:

1. The *Create New Drawing* dialog box appears (identical to the *Startup* dialog box).
2. The *Select Template* dialog box appears.

Control these two actions by the setting in the *System* tab of the *Options* dialog box (see Figure 2-5). If you select *Show Startup dialog box*, the *Create New Drawing* dialog box appears when you use *New*. If you select *Do not show a startup dialog*, the *Select Template* dialog box appears. These two dialog boxes are explained next.

The *Create New Drawing* Dialog Box

In the *Create New Drawing* dialog box (Fig. 2-6), the options for creating a new drawing are the same as in the *Startup* dialog box (shown previously in Figure 1-12): *Start from Scratch*, *Use a Template*, and *Use a Wizard*.

Figure 2-6

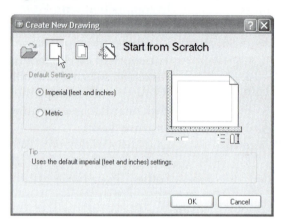

Start from Scratch
In the *Create New Drawing* dialog box select the *Start from Scratch* option (Fig. 2-6). Next, choose from either the *Imperial (feet and inches)* or *Metric* default settings. Selecting *Imperial* uses the default ACAD.DWT template drawing with a drawing area (called *Limits*) of 12 x 9 units. Choosing *Metric* uses the ACADISO.DWT template drawing which has *Limits* settings of 420 x 297. For more details on these templates, see "Table of AutoCAD-Supplied Template Drawing Settings" in Chapter 12.

Use a Wizard
In the *Create New Drawing* dialog box select the *Use a Wizard* option (Fig. 2-7). Next, choose between a *Quick Setup Wizard* and an *Advanced Setup Wizard*. The *Quick Setup Wizard* prompts you to select the type of drawing *Units* you want to use and to specify the drawing *Area* (*Limits*). The *Advanced Setup* offers options for units, angular direction and measurement, and area. The *Quick Setup Wizard* and the *Advanced Setup Wizard* are discussed in Chapter 6, Basic Drawing Setup.

Figure 2-7

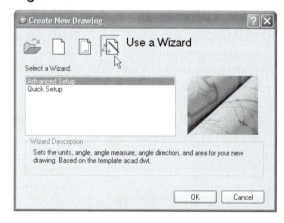

Use a Template
Use this option if you want to create a new drawing based on an existing template drawing (Fig. 2-8). A template drawing is one that may have some of the setup steps performed but contains no geometry (graphical objects).

Selecting the ACAD.DWT template begins a new drawing using the Imperial default settings with a drawing area of 12 x 9 units (identical to using the *Start from Scratch*, *Imperial* option). Selecting the ACADISO.DWT template begins a new drawing using the metric settings having a drawing area of 420 x 279 units (identical to using the *Start from Scratch*, *Metric* option). Details of these and other template drawings are discussed in Chapter 12, Advanced Drawing Setup.

Figure 2-8

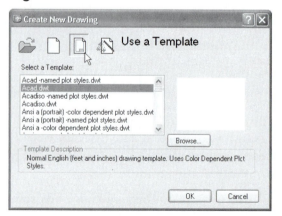

The *Select Template* Dialog Box

The *Select Template* dialog box (Fig. 2-9) appears when you use the *New* command and your system is set to *Do not show a startup dialog box* (see previous Figure 2-5). The *Select Template* dialog box has the identical outcome as using the *Use a Template* option in the *Create New Drawing* dialog box (see Figure 2-8); that is, you can select from all AutoCAD-supplied templates (see discussion above). See Chapter 12, Advanced Drawing Setup, for complete details of all template drawings.

Figure 2-9

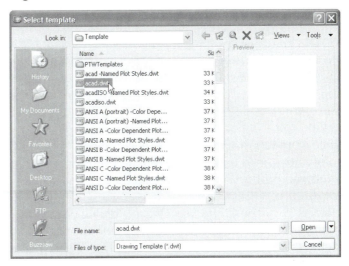

NOTE: For the purposes of learning AutoCAD starting with the basic principles and commands discussed in this text, it is helpful to use the Imperial (inch) settings when beginning a new drawing. Until you read Chapters 6 and 12, you can begin new drawings for completing the exercises and practicing the examples in Chapters 1 through 5 by any of the following methods.

Select *Imperial (feet and inches)* in the *Start from Scratch* option.
Select the ACAD.DWT template drawing using any method.
Cancel the *Startup* dialog when AutoCAD starts (the default setting of 0 for the *MEASURINIT* system variable uses the ACAD.DWT).

QNEW

Pull-down Menu	COMMAND (TYPE)	ALIAS (TYPE)	Short-cut	Screen (side) Menu	Tablet Menu
...	QNEW	...	...	...	...

The *Qnew* command is new for AutoCAD 2004. *Qnew* (Quick New) is intended to immediately begin a new drawing using the template file specified in the *Files* tab of the *Options* dialog box. If the proper settings are made, *Qnew* is faster to use than the *New* command since it does not force the *Create New Drawing* or the *Select Template* dialog box to open, as the *New* command does.

Figure 2-10

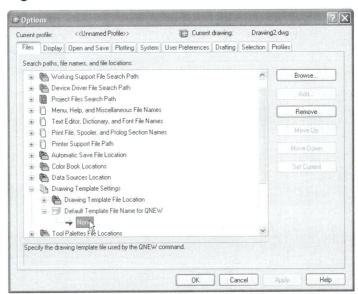

To set *Qnew* to operate quickly when opening a drawing, you must first make two settings (if you are using AutoCAD in a school or office setting, these settings may have already been made):

1. Specify a default template to use when *Qnew* is invoked. Open the *Options* dialog box by selecting *Options* from the *Tools* pull-down menu. Select the *Files* tab and expand the *Default Template File Name for QNEW* line (Fig. 2-10). The default setting is *None*. Highlight *None*, then pick the *Browse* button to locate the template file you want to use.

2. Set your system to *Do not show a startup dialog*. This is accomplished in the *System* tab of the *Options* dialog box (see previous Figure 2-5).

NOTE: The default setting for the *Default Template File Name for QNEW* line (in the *Options* dialog box) is "None." That means if you use *Qnew* without specifying a template drawing, as will be the case for all users upon first installing AutoCAD, it operates similarly to the *New* command by opening the *Create New Drawing* or *Select Template* dialog box, depending on your setting for *Show Startup dialog box* or *Do not show a startup dialog*.

Since Autodesk has made this new command relatively complex to understand and control, the three possibilities and controls are explained simply here. Based on the settings for your system, the *Qnew* command produces one of three methods to start a new drawing:

Method 1.   The *Create New Drawing* dialog box appears.
Method 2.   The *Select Template* dialog box appears.
Method 3.   A new drawing starts with the *Qnew* template specified in the *Options* dialog box.

Control the previous actions by the following settings:

Method 1.   Select *Show Startup dialog box* in the *System* tab of the *Options* dialog box.
Method 2.   Select *Do not show a startup dialog* in the *System* tab of the *Options* dialog box and ensure the *Drawing Template File Name* is set to "None" in the *Files* tab of the *Options* dialog box.
Method 3.   Select *Do not show a startup dialog* in the *System* tab of the *Options* dialog box and set the *Drawing Template File Name* in the *Files* tab of the *Options* dialog box to the desired template file.

OPEN

Pull-down Menu	COMMAND (TYPE)	ALIAS (TYPE)	Short-cut	Screen (side) Menu	Tablet Menu
File *Open...*	*OPEN*	...	Ctrl+O	FILE *Open*	T,25

Use *Open* to select an existing drawing to be loaded into AutoCAD. Normally you would open an existing drawing (one that is completed or partially completed) so you can continue drawing or to make a print or plot. You can *Open* multiple drawings at one time in AutoCAD 2000 or later versions.

The *Open* command produces the *Select File* dialog box (Fig. 2-11). In this dialog box you can select any drawing from the current directory list. PICKing a drawing name from the list displays a small bitmap image of the drawing in the *Preview* tile. Select the *Open* button or double-click on the file name to open the highlighted drawing. You could instead type the file name (and path) of the desired drawing in the *File name:* edit box and press Enter, but a preview of the typed entry will not appear.

Figure 2-11 ———

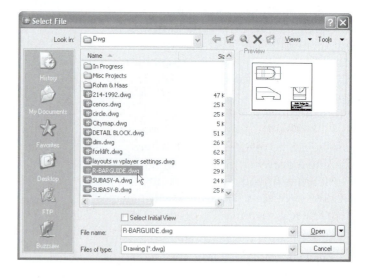

You can open multiple drawings at one time by holding down the Shift key to select a range (all files between and including the two selected) or holding down the Ctrl key to select multiple drawings not in a range.

Using the *Open* drop-down list (lower right) you can select from these options:

Open	Opens a drawing file and allows you to edit the file.
Open Read-Only	Opens a drawing file for viewing, but you cannot edit the file.
Partial Open	Allows you to open only a part of a drawing for editing. See *Partialopen*.
Partial Open Read-Only	Allows you to open only a part of a drawing for viewing, but you cannot edit the drawing. See *Partialopen*.

Other options in this dialog box are similar to the *Save Drawing As* dialog box described earlier in "File Navigation Dialog Box Functions." For example, to locate a drawing in another folder (directory) or on another drive on your computer or network, select the drop-down list on top of the dialog box next to *Look in:*, or use the *Up one level* button. Remember that you can also display the file details (size, type, modified) by toggling the *Details* option from the *Views* drop-down list. You can open a .DWG (drawing), .DWS (drawing standards), .DXF (drawing interchange format), or .DWT (drawing template) file by selecting from the *Files of type:* drop-down list.

Selecting the *Find...* option from the *Tools* drop-down list in the *Select File* dialog box (upper right) invokes the *Find* dialog box (Fig. 2-12). The dialog box offers two criteria to locate files: *Name & Location* and *Date Modified*.

Using the *Name & Location* tab, you can search for .DWG, .DWS, .DWT, or .DXF files in any directory, including subdirectories if desired. In the *Named:* edit box, enter a drawing name to find. Wildcards can be used (see Chapter 43, Miscellaneous Commands and Features, for valid wildcards in AutoCAD).

Figure 2-12

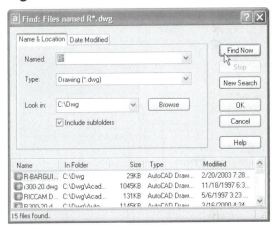

The *Date Modified* tab of the *Find* dialog box enables you to search for files meeting specific date criteria that you specify (Fig. 2-13). This feature helps you search for files *between* two dates, *during the previous months,* or *during the previous days*, based on when the files were last saved.

After specifying the search criteria in either tab, select the *Find Now* button. All file names that are found matching the criteria are displayed in the list at the bottom of the dialog box. Double-clicking on the name or using the *OK* button passes the name to the *File Name* section of the *Select File* dialog box.

Figure 2-13

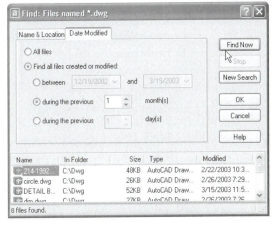

NOTE: It is considered <u>poor practice to *Open* a drawing</u> <u>from a diskette in A: drive</u>. Normally, drawings should be copied to a directory on a fixed (hard) drive, then opened from that directory. Opening a drawing from a diskette

and performing saves to a diskette are much slower and less reliable than operating from a hard drive. In addition, some temporary files may be written to the drive or directory from which the file is opened. See other "NOTE"s at the end of the discussion on *Save* and *Saveas*.

SAVE

Pull-down Menu	COMMAND (TYPE)	ALIAS (TYPE)	Short-cut	Screen (side) Menu	Tablet Menu
File *Save*	SAVE	...	Ctrl+S	...	U,24-U,25

The *Save* command is intended to be used periodically during a drawing session (every 15 to 20 minutes is recommended). When *Save* is selected from a menu, the current version of the drawing is saved to disk without interruption of the drawing session. The first time an unnamed drawing is saved, the *Save Drawing As* dialog box (see Figures 2-2 and 2-3) appears, which prompts you for a drawing name. Typically, however, the drawing already has an assigned name, in which case *Save* actually performs a *Qsave* (quick save). A *Qsave* gives no prompts or options, nor does it display a dialog box.

When the file has an assigned name, using *Save* by selecting the command from a menu or icon button automatically performs a quick save (*Qsave*). Therefore, *Qsave* automatically saves the drawing in the same drive and directory from which it was opened or where it was first saved. In contrast, typing *Save* always <u>produces the *Save Drawing As* dialog box,</u> where you can enter a new name and/or path to save the drawing or press Enter to keep the same name and path (see *SaveAs*).

NOTE: If you want to save a drawing directly to a diskette in A:, first *Save* the drawing to the hard drive, then use the *Send To* option in the right-click menu in the *Save, Save Drawing As*, or *Select File* dialog box (see "Windows Right-Click Shortcut Menus" and Figure 2-4).

QSAVE

Pull-down Menu	COMMAND (TYPE)	ALIAS (TYPE)	Short-cut	Screen (side) Menu	Tablet Menu
...	QSAVE	...	Ctrl+S	*FILE* *Qsave*	...

Qsave (quick save) is normally invoked automatically when *Save* is used (see *Save*) but can also be typed. *Qsave* saves the drawing under the previously assigned file name. No dialog boxes appear nor are any other inputs required. This is the same as using *Save* (from a menu), assuming the drawing name has been assigned. However, if the drawing has not been named when *Qsave* is invoked, the *Save Drawing As* dialog box appears.

SAVEAS

Pull-down Menu	COMMAND (TYPE)	ALIAS (TYPE)	Short-cut	Screen (side) Menu	Tablet Menu
File *Save As...*	SAVEAS	...	...	*FILE* *Saveas*	V,24

The *SaveAs* command can fulfill four functions:
1. save the drawing file under a new name if desired;
2. save the drawing file to a new path (drive and directory location) if desired;
3. in the case of either 1 or 2, assign the new file name and/or path to the current drawing (change the name and path of the current drawing);
4. save the drawing in a format other than the default AutoCAD 2004 format or as a different file type (.DWT, .DWS, or .DXF).

Therefore, assuming a name has previously been assigned, *SaveAs* allows you to save the current drawing under a different name and/or path; but, beware, <u>*SaveAs* sets the current drawing name and/or path to the last one entered.</u> This dialog box is shown in Figures 2-2 and 2-3.

SaveAs can be a benefit when creating two similar drawings. A typical scenario follows. A design engineer wants to make two similar but slightly different design drawings. During construction of the first drawing, the engineer periodically saves under the name DESIGN1 using *Save*. The first drawing is then completed and *Saved*. Instead of starting a *New* drawing, *SaveAs* is used to save the current drawing under the name DESIGN2. *SaveAs* also resets the current drawing name to DESIGN2. The designer then has two separate but identical drawing files on disk which can be further edited to complete the specialized differences. The engineer continues to work on the current drawing DESIGN2.

NOTE: If you want to save the drawing to a diskette in A: drive, do not use *SaveAs*. Invoking *SaveAs* by any method resets the drawing name and path to whatever is entered in the *Save Drawing As* dialog box, so entering A:NAME would set A: as the current drive. This could cause problems because of the speed and reliability of a diskette as opposed to a hard drive. Instead, you should save the drawing to the hard drive (usually C:), then close the drawing (by using *Close*). Next, use a right-click shortcut menu to copy the drawing file to A:. (See "Windows Right-Click Shortcut Menus" and Figure 2-4.)

You can save an AutoCAD 2004 drawing in several formats other than the default format (*AutoCAD 2004 Drawing *.dwg*). Use the drop-down list at the bottom of the *Save* (or *Save Drawing As*) dialog box (Fig. 2-14) to save the current drawing as an earlier version drawing (.DWG) of AutoCAD or LT, a template file (.DWT), or a .DWS or a .DXF file. In AutoCAD 2004, a drawing can be saved to an earlier release and later *Opened* in AutoCAD 2004 and all new features are retained during the "round trip."

Figure 2-14

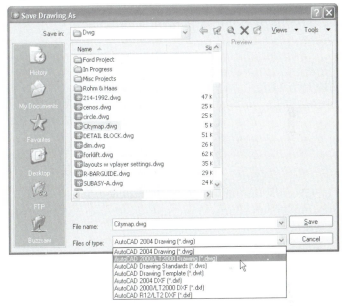

Passwords

An added security feature in AutoCAD 2004 allows you to save your drawings with password protection. If a password is assigned, the drawing can be opened again in AutoCAD <u>only</u> if you enter the correct password when using the *Open* command. Beware: If you lose or forget the password, the drawing cannot be opened or recovered in any way!

To assign a password during the *Save* or *SaveAs* command, select the *Tools* drop-down menu in the *Save Drawing As* dialog box, then select *Security Options…* (Fig. 2-15) to produce the *Security Options* dialog box. Passwords are not case sensitive.

Figure 2-15

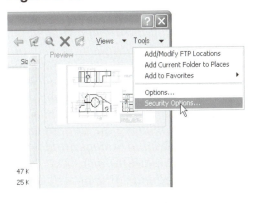

Digital Signatures

In AutoCAD 2004, you can also attach digital signatures to drawings using the *Security Options* dialog box. If a drawing has a digital signature assigned, it is designated as an original signed and dated drawing. Users are assured that the drawing is original and has not been changed in any way. If such a drawing is later modified, the digital signature is then invalidated, and anyone opening the drawing is notified that the drawing has been changed and the digital signature is invalid.

See Chapter 46, CAD Management, for more information on passwords, digital signatures, and other security options.

CLOSE

Pull-down Menu	COMMAND (TYPE)	ALIAS (TYPE)	Short-cut	Screen (side) Menu	Tablet Menu
File Close	CLOSE	...	...	...	...

Use the *Close* command to <u>close the current drawing</u>. Because AutoCAD allows you to have several drawings *Open, Close* gives you control to close one drawing while leaving others open. If the drawing has been changed but not saved, AutoCAD prompts you to save or discard the changes. In this case, a warning box appears (Fig. 2-16). *Yes* causes AutoCAD to *Save* then close the drawing; *No* closes the drawing without saving; and *Cancel* aborts the close operation so the drawing stays open.

Figure 2-16

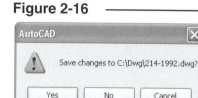

CLOSEALL

Pull-down Menu	COMMAND (TYPE)	ALIAS (TYPE)	Short-cut	Screen (side) Menu	Tablet Menu
Window Closeall	CLOSEALL	...	...	...	...

The *Closeall* command closes all drawings currently open in your AutoCAD session. If any of the drawings have been changed but not saved, you are prompted to save or discard the changes for each drawing.

EXIT

Pull-down Menu	COMMAND (TYPE)	ALIAS (TYPE)	Short-cut	Screen (side) Menu	Tablet Menu
File Exit	EXIT	...	...	...	Y,25

This is the simplest method to use when you want to exit AutoCAD. If any changes have been made to the drawings since the last *Save, Exit* invokes a warning box asking if you want to *Save changes to . . .?* before ending AutoCAD (see Figure 2-16).

An alternative to using the *Exit* option is to use the standard Windows methods for exiting an application. The two options are (1) select the **"X"** in the extreme upper-right corner of the AutoCAD window, or (2) select the AutoCAD logo in the extreme upper-left corner of the AutoCAD window. Selecting the logo in the upper-left corner produces a pull-down menu allowing you to *Minimize, Maximize,* etc., or to *Close* the window. Using this option is the same as using *Exit*.

QUIT

Pull-down Menu	COMMAND (TYPE)	ALIAS (TYPE)	Short-cut	Screen (side) Menu	Tablet Menu
...	*QUIT*	*EXIT*	...	*FILE* *Quit*	...

The *Quit* command accomplishes the same action as *Exit*. *Quit* discontinues (exits) the AutoCAD session and usually produces the dialog box shown in Figure 2-16 if changes have been made since the last *Save*.

PARTIALOPEN

Pull-down Menu	COMMAND (TYPE)	ALIAS (TYPE)	Short-cut	Screen (side) Menu	Tablet Menu
File *Open* *Partial Open*	*PARTIALOPEN*	...	...	...	...

This feature in AutoCAD speeds up and simplifies loading, viewing, and working with large drawings. With versions previous to AutoCAD 2000, you could only *Open* the entire contents of a drawing (all layers and all geometry) in order to work with the drawing. (The earlier *Xref* feature allowed you to view partial drawings, but you could not edit the geometry.) Since AutoCAD 2000, you can select which views and layers, and therefore which geometry, you want to load. This feature frees up system memory, speeds up regenerations, and prevents having to view and manipulate unneeded layers and geometry.

There are two ways you can use *Partialopen*: entering the command at the Command line or using the *Open* command to access the *Partial Open* dialog box interface. Generally you should use the *Open* command to access the *Select File* dialog box, then the *Partial Open* dialog box. Use the *Open* drop-down list in the lower-right corner of the *Select File* dialog box to access *Partial Open* (Fig. 2-17). The Command line version of *Partialopen* is explained at the end of this section.

Figure 2-17 ——————

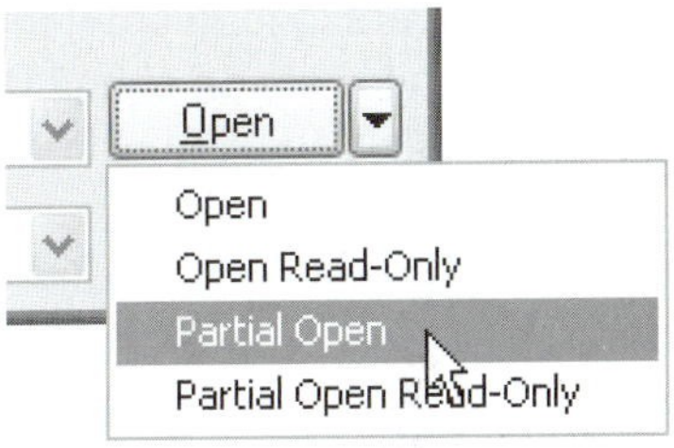

Figure 2-18 ——————

Choosing the *Partial Open* option from the *Select File* dialog box produces the *Partial Open* dialog box (Fig. 2-18). The dialog box allows you to select a *View* (if a named *View* has previously been saved in the drawing) and select which layers you want to open. (See Chapter 10 and Chapter 11 for information on views and layers.) You can select a range of layers by holding down Shift when selecting a layer at both ends of the range.

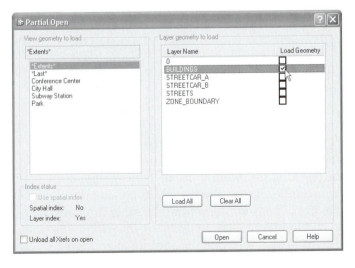

When a drawing is partially open, only the geometry on the selected layers is visible and can be edited. Although all of the layer names will appear in any list of layers in the drawing, <u>only the geometry on the selected layers is actually loaded</u>. In addition, all other named objects in the drawing (blocks, dimension styles, layouts, linetypes, text styles, UCSs, views, and viewport configurations) are loaded into AutoCAD.

For example, assume a drawing of the downtown area was partially opened. Selecting only the BUILDINGS layer (see Figure 2-18) may yield a display as shown in Figure 2-19 revealing only the objects on the BUILDINGS layer.

Figure 2-19

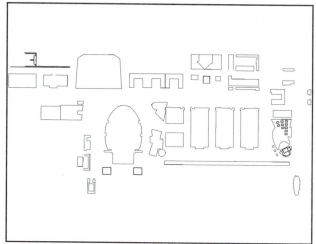

You can also use Spatial and Layer indexing (see Figure 2-18, lower-left corner) if the drawing's *INDEXCTL* variable is set to allow "demand loading." This feature speeds loading and saves system memory for large drawings. Also, any attached *Xref* drawings can be unloaded if desired (see Chapter 30, Xreferences).

If you choose to enter the *Partialopen* command at the Command line, a text interface appears instead of the dialog box.

```
Command: partialopen
Enter name of drawing to open: (Enter drawing name or enter the tilde symbol [~] to display the Select
File dialog box, then select the Partial Open button)
Enter view to load or [?] <*Extents*>: (Enter view name or press Enter to load the extents)
Enter layers to load or [?]<none>: (Enter desired layer names or an asterisk [*] to load all layers)
Unload all Xrefs on open? [Yes/No] <N>: (Entering "Y" loads all attached Xrefs)
Opening an AutoCAD 2004 format file.
Regenerating model.
AutoCAD menu utilities loaded.
```

PARTIALOAD

Pull-down Menu	COMMAND (TYPE)	ALIAS (TYPE)	Short-cut	Screen (side) Menu	Tablet Menu
File *Partial Load*	*PARTIALOAD*	...	...	...	...

You must have a partially opened drawing for *Partiaload* to operate (see *Partialopen* described previously). *Partiaload* loads additional geometry into a partially opened drawing.

Accessing this command by a menu or by typing *Partiaload* produces the *Partial Load* dialog. Here you select the additional layers or another area of the drawing you want to open. For example, Figure 2-20 displays the *Partial Load* dialog box displaying *Views* and *Layers* from the same drawing shown in Figure 2-19. In this case, the STREETS layer is loaded to reveal the additional geometry

Figure 2-20

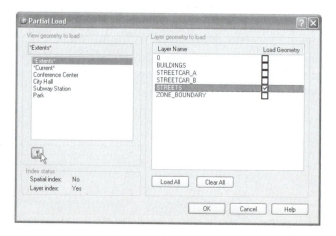

(see Figure 2-21). Any information that is loaded into the file using *Partialload* cannot be unloaded, not even with *Undo*. The *Partial Load* dialog box is essentially the same as the *Partial Open* dialog box (see Figure 2-18) with the exclusion of the *Index* and *Xref* options in the lower-left corner.

Figure 2-21

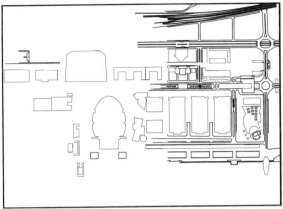

Notice the *Pick a Window* button in the lower-left corner of the dialog box. This feature allows you to return to the drawing and define an area (window) of the drawing to view. All geometry within and crossing the windowed area for the defined layers appears in the drawing. For example, the *Pick a Window* feature was used to select the right side of the drawing shown in Figure 2-21. Notice how the STREETS layer is displayed only on the right side of the drawing. This feature allows you to view two or more areas of the same drawing, each displaying a different set of layers.

If you enter *-Partialload* at the Command line (with the hyphen prefix), the text interface appears at the Command line.

RECOVER

Pull-down Menu	COMMAND (TYPE)	ALIAS (TYPE)	Short-cut	Screen (side) Menu	Tablet Menu
File *Drawing Utilities >* *Recover...*	*RECOVER*	...	...	*FILE* *Recover*	...

This option is used only in the case that *Open* does not operate on a particular drawing because of damage to the file. Damage to a drawing file can occur from improper exiting of AutoCAD (such as power failure) or from damage to a diskette. *Recover* can usually reassemble the file to a usable state and reload the drawing.

The *Recover* command produces the *Select File* dialog box. Select the drawing to recover. A recovery log is displayed in a text window reporting results of the recover operation.

DWGPROPS

Pull-down Menu	COMMAND (TYPE)	ALIAS (TYPE)	Short-cut	Screen (side) Menu	Tablet Menu
File *Drawing Properties*	*DWGPROPS*	...	...	...	...

Invoking *Dwgprops* by any method produces the *Drawing Properties* dialog box (Fig. 2-22, on the next page). This dialog box allows you to input details for better management and tracking of drawings such as title, author, subject, and other keywords. The information stored in the *Drawing Properties* dialog box should be specified by you while other information, such as time and date of last save, is read from the system. You must *Open* the drawing to input data into the user fields and *Save* the drawing to save the properties you enter.

The drawing properties can be read from Windows Explorer or My Computer by right-clicking a .DWG file and selecting *Properties*. Using an AutoCAD feature called DesignCenter, unopened drawings can be located by finding keywords stored in the drawing properties (see Chapter 21). The four tabs in the *Drawing Properties* dialog box are explained here.

Figure 2-22

General **Tab**
This tab (Fig. 2-22) displays the drawing type, location, size, and other information. All the information here is supplied by the operating system; therefore, all fields are read-only.

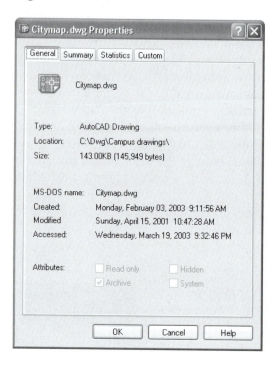

Summary **Tab**
This tab is used to input information you want to use to track the drawing (Fig. 2-23). Data you enter is viewable through AutoCAD's DesignCenter and Windows Explorer *Properties*. You can enter the drawing title, subject, author, keywords, comments, and a hyperlink base. The *Hyperlink Base* is a base address for all relative links you insert within the drawing using the *Hyperlink* command (see Chapter 19, Internet Tools). You can enter a URL (an Internet address) or a path to a folder on a network drive in the *Hyperlink Base* field.

Figure 2-23

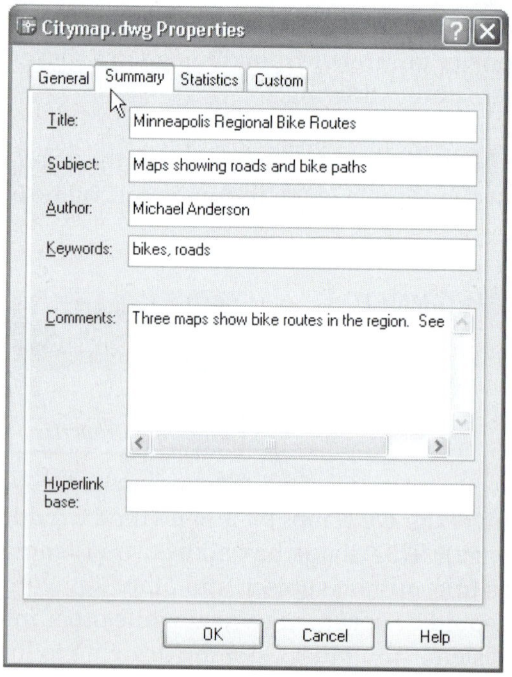

Statistics **Tab**

This tab shows read-only data such as file size and the dates that files were created and last modified (Fig. 2-24). You can search for all files created at a certain time or modified on a certain date. The *Created*, *Modified*, and *Total Editing Time* values are stored in the *TDCREATE*, *TDUPDATE*, and *TDINDWG* read-only system variables.

Custom **Tab**

The custom fields can also be used in searches to help locate the drawing using the *Find* dialog box of AutoCAD DesignCenter (not shown). AutoCAD also provides access to these properties using its native programming language, AutoLISP. Enter any data relevant to the drawing that may be useful in a search.

Figure 2-24

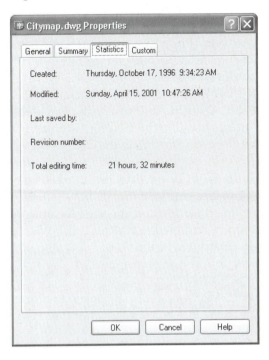

Created:	Thursday, October 17, 1996 9:34:23 AM		
Modified:	Sunday, April 15, 2001 10:47:26 AM		
Last saved by:			
Revision number:			
Total editing time:	21 hours, 32 minutes		

SAVETIME

Pull-down Menu	COMMAND (TYPE)	ALIAS (TYPE)	Short-cut	Screen (side) Menu	Tablet Menu
Tools *Options...* *Open and Save* *File Safety* *Precautions*	*SAVETIME*	...	...	*TOOLS 2* *Options* *Open and Save* *File Safety* *Precautions*	...

SAVETIME is actually a system variable that controls AutoCAD's temporary save feature. AutoCAD automatically saves a temporary drawing file for you at time intervals that you specify. The default time interval is 10 (minutes). A value of 0 disables this feature. The value for *SAVETIME* can also be set in the *Open and Save* tab of the *Options* dialog box (see Figure 2-25, on the next page).

When automatic saving occurs, the drawing is saved under the current name with a randomly generated suffix and a .SV$ extension is assigned. For example, if the ELECTRICAL.DWG drawing were automatically saved, the assigned file might be ELECTRICAL_1_1_9912.SV$. In this way the current drawing name is not overwritten. The path (location) of the automatically saved file can be specified using the *SAVEFILEPATH* system variable or the *Files* tab of the *Options* dialog box.

NOTE: The automatically saved file is only a temporary file. That is, once you save or close the drawing, the temporary save file is automatically deleted. Therefore, this feature is useful when you experience an improper shutdown of AutoCAD (such as a power outage). In this case, the temporary file can be retrieved and renamed so all of your work is not lost.

AutoCAD Backup Files

When a drawing is saved, AutoCAD creates a file with a .DWG extension. For example, if you name the drawing PART1, using *Save* creates a file named PART1.DWG. The next time you save, AutoCAD makes a new PART1.DWG and renames the old version to PART1.BAK. One .BAK (backup) file is kept automatically by AutoCAD by default. You can disable the automatic backup function by changing the *ISAVEBAK* system variable to 0 or by accessing the *Open and Save* tab in the *Options* dialog box (Fig. 2-25).

You cannot *Open* a .BAK file. It must be renamed to a .DWG file. Remember that you already have a .DWG file by the same name, so rename the extension <u>and</u> the filename. For example, PART1.BAK could be renamed to PART1OLD.DWG. Use the Windows Explorer, My Computer, or the *Select File* dialog box (use *Open*) with the right-click options to rename the file.

The .BAK files can also be deleted without affecting the .DWG files. The .BAK files accumulate after time, so you may want to periodically delete the unneeded ones to conserve disk space.

Figure 2-25

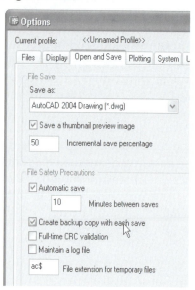

AutoCAD Drawing File Management

AutoCAD drawing files should be stored in folders (directories) used exclusively for that purpose. For example, you may use Windows Explorer or My Computer to create a new folder called "Drawings" or "DWG." This could be a folder in the main C: or D: drive, or could be a subdirectory such as "C:\My Documents\ AutoCAD Drawings." You can create folders for different types of drawings or different projects. For example, you may have a folder named "DWG" with several subdirectories for each project, drawing type, or client, such as "COMPONENTS" and "ASSEMBLIES," or "In Progress" and "Completed," or "TWA," "Ford," and "Dupont." If you are working in an office or school laboratory, a directory structure most likely already has been created for you to use.

Safety and organization are important considerations for storage of AutoCAD drawing files. AutoCAD drawings should be saved to designated folders to prevent accidental deletion of system or other important files. If you work with AutoCAD long enough or if you work in an office or laboratory, most likely many drawings are saved on the computer hard drive or network drives. It is imperative that a logical directory structure be maintained so important drawings can be located easily and saved safely.

Using Other File Formats

See Chapter 32 for file commands related to importing and exporting images and exchanging files in other than *.DWG format.

CHAPTER EXERCISES

Start AutoCAD. If the *Startup* dialog box appears, select *Start from Scratch*, then use the *Imperial* default settings. NOTE: The chapter exercises in this book list file names in UPPERCASE letters for easy recognition. The file names and directory (folder) names on your system may appear as UPPERCASE, Title Case, or lower case.

1. **Create or determine the name of the folder ("working" directory) for opening and saving AutoCAD files on your computer system**

   If you are working in an office or laboratory, a folder (directory) has most likely been created for saving your files. If you have installed AutoCAD yourself and are learning on your home or office system, you should create a folder for saving AutoCAD drawings. It should have a name like "C:\ACAD\DWG," "C:\My Documents\Dwgs," or "C:\Acad\Drawing Files." (HINT: Use the *SaveAs* command. The name of the folder last used for saving files appears at the top of the *Save Drawing As* dialog box.)

2. ***Save* and name a drawing file**

   Draw 2 vertical *Lines*. Select *Save* from the *File* pull-down, from the Standard toolbar, or by another method. (The *Save Drawing As* dialog box appears since a name has not yet been assigned.) Name the drawing **"CH2 VERTICAL."**

3. **Using *Qsave***

   Draw 2 more vertical *Lines*. Select *Save* again from the menu or by any other method except typing. (Notice that the *Qsave* command appears at the Command line since the drawing has already been named.) Draw 2 more vertical *Lines* (a total of six lines now). Select *Save* again. Do not *Close*.

4. **Start a *New* drawing**

   Invoke *New* from the *File* pull-down, the Standard toolbar, or by any other method. If the *Startup* dialog box appears, select **Start from Scratch,** then use the **Imperial** default settings. Draw 2 horizontal *Lines*. Use *Save*. Enter **"CH2 HORIZONTAL"** as the name for the drawing. Draw 2 more horizontal *Lines*, but <u>do not</u> *Save*. Continue to exercise 5.

5. ***Close* the current drawing**

   Use *Close* to close CH2 HORIZONTAL. Notice that AutoCAD first forces you to answer *Yes* or *No* to *Save changes to CH2 HORIZONTAL.DWG?* PICK *Yes* to save the changes.

6. **Using *SaveAs***

   The CH2 VERTICAL drawing should now be the current drawing. Draw 2 inclined (angled) *Lines* in the CH2 VERTICAL DRAWING. Invoke *SaveAs* to save the drawing under a new name. Enter **"CH2 INCLINED"** as the new name. Notice the current drawing name displayed in the AutoCAD title bar (at the top of the screen) is reset to the new name. Draw 2 more inclined *Lines* and *Save*.

7. *Open* **an AutoCAD sample drawing**

Open a drawing named **"COLORWH.DWG"** usually located in the "C:\Program Files\AutoCAD 2004\ Sample" directory. Use the *Look in:* section of the dialog box to change drive and directory if necessary. Do <u>not</u> *Save* the sample drawing after viewing it. Practice the *Open* command by looking at other sample drawings in the Sample directory. Finally, *Close* all the sample drawings. Be careful <u>not</u> to *Save* the drawings.

8. **Check the** *SAVETIME* **setting**

At the command prompt, enter *SAVETIME*. The reported value is the interval (in minutes) between automatic saves. Your system may be set to 10 (default setting). If you are using your own system, change the setting to **5.**

9. **Find and rename a backup file**

Close all open drawings. Use the *Open* command. When the *Select File* dialog box appears, locate the folder where your drawing files are saved, then enter **"*.BAK"** in the *File name:* edit box (* is a wildcard that means "all files"). All of the .BAK files should appear. Search for a backup file that was created the last time you saved **CH2 VERTICAL.DWG,** named **CH2 VERTICAL.BAK. Right-click** on the file name. Select *Rename* from the shortcut menu and rename the file **"CH2 VERT 2.DWG."** Next enter *.DWG in the *File name:* edit box to make all the .DWG files reappear. Highlight **CH2 VERT 2.DWG** and the bitmap image should appear in the preview display. You can now *Open* the file if you wish.

10. **Use** *Partialopen* **and** *Partiaload*

Use the *Open* command so the *Select File* dialog box appears. Locate **DB_SAMP.DWG** in the AutoCAD 2004\Sample folder. Highlight the file and pick the *Partial Open* button. When the *Partial Open* dialog box appears, select all six layers beginning with **E-B-** from the list, then select *Open*. The drawing should appear displaying only the outer shell and structural walls of the building. Next, use the *Partiaload* command. Load the layers **CHAIRS, CPU, FILE_CABINETS,** and **FURNITURE** by the same method you used previously. Now all the furniture should appear in the drawing.

Next, use *Partiaload* again and select the **PANELS_201** layer. Then use the *Pick a Window* button (in the lower-left corner of the dialog box) to make a window around the upper-left wing of the building. When the dialog box reappears, pick *OK*. The panel walls should appear only for the portion of the drawing you indicated with the window. Use *Partiaload* again and load the **PANELS_201** layer, but this time *Pick A Window* around the lower wing of the building. Finally, use *Partiaload* and select the *Load All* button to make the entire drawing load and appear. *Close* the drawing and *Exit* AutoCAD. <u>Do not save the changes.</u>

DRAW COMMAND CONCEPTS

Chapter Objectives

After completing this chapter you should be able to:

1. use *SNAP, GRID, ORTHO,* and *POLAR* while drawing *Lines* and *Circles* interactively;

2. create *Lines* by specifying <u>absolute</u> coordinates;

3. create *Lines* by specifying <u>relative rectangular</u> coordinates;

4. create *Lines* by specifying <u>relative polar</u> coordinates;

5. create *Lines* by specifying <u>direct distance entry</u> coordinates;

6. use Polar Tracking and Polar Snap to draw *Lines* interactively.

AutoCAD OBJECTS

Figure 3-1

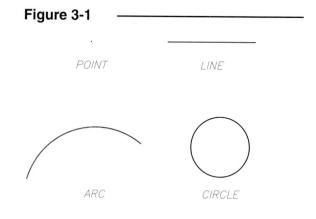

POINT   LINE

ARC   CIRCLE

The smallest component of a drawing in AutoCAD is called an <u>object</u> (sometimes referred to as an entity). An example of an object is a *Line,* an *Arc,* or a *Circle* (Fig. 3-1). A rectangle created with the *Line* command would contain four objects.

Draw commands <u>create</u> objects. The draw command names are the same as the object names.

Simple objects are *Point, Line, Arc,* and *Circle.*

Complex objects are shapes such as *Polygon, Rectangle, Ellipse, Polyline,* and *Spline* which are created with one command (Fig. 3-2). Even though they appear to have several segments, they are <u>treated</u> by AutoCAD as one object.

Figure 3-2

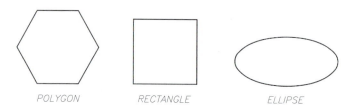

POLYGON   RECTANGLE   ELLIPSE

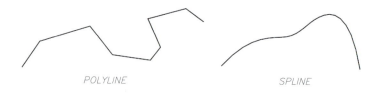

POLYLINE   SPLINE

It is not always apparent whether a shape is composed of one or more objects. However, if you pick an object with the "pickbox," an object is "highlighted," or shown in a broken line pattern (Fig. 3-3). This high-lighting reveals whether the shape is com-posed of one or several objects.

Figure 3-3

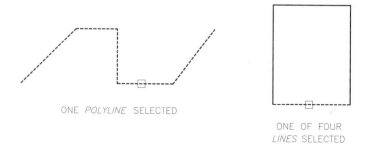

ONE *POLYLINE* SELECTED

ONE OF FOUR
LINES SELECTED

LOCATING THE DRAW COMMANDS

To invoke *Draw* commands, any of the five command entry methods can be used depending on your computer setup.

1. **Toolbars**   Select the icon from the *Draw* toolbar.
2. **Pull-down menu**   Select the command from the *Draw* pull-down menu.
3. **Screen menu**   Select the command from the *Draw* screen menu.
4. **Keyboard**   Type the command name, command alias, or accelerator keys at the keyboard (a command alias is a one- or two-letter shortcut).
5. **Tablet menu**   Select the icon from the digitizing tablet menu (if available).

For example, a draw command can be activated by PICKing its icon button from the *Draw* toolbar (Fig. 3-4).

Figure 3-4

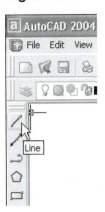

The *Draw* pull-down menu can also be used to select draw commands (Fig. 3-5). Options for a command are found on cascading menus.

Figure 3-5

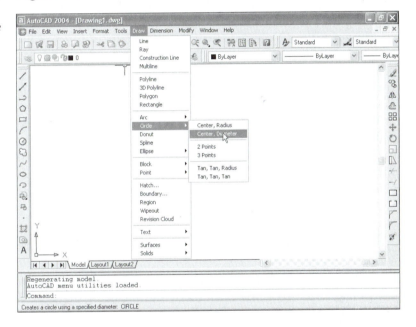

THE FIVE COORDINATE ENTRY METHODS

All drawing commands prompt you to specify points, or locations, in the drawing. For example, the *Line* command prompts you to give the "Specify first point:" and "Specify next point:," expecting you to specify locations for the first and second endpoints of the line. After you specify those points, AutoCAD stores the coordinate values to define the line. A two-dimensional line in AutoCAD is defined and stored in the database as two sets of X, Y, and Z values (with Z values of 0), one for each endpoint.

There are five ways to specify coordinates; that is, there are five ways to tell AutoCAD the location of points when you draw objects.

1.	**Interactive method**	**PICK**	Use the cursor to select points on the screen.
2.	**Absolute coordinates**	**X,Y**	Type explicit X and Y values relative to the origin at 0,0.
3.	**Relative rectangular coordinates**	**@X,Y**	Type explicit X and Y values relative to the last point (@ means "last point").
4.	**Relative polar coordinates**	**@dist<angle**	Type a distance value and angle value relative to the last point (< means "angle of").
5.	**Direct distance entry**	**dist,direction**	Type a distance value relative to the last point, indicate direction with the cursor, then press Enter.

The <u>interactive method</u> of coordinate entry is simple because you PICK the desired locations with your cursor when AutoCAD prompts for a point. AutoCAD 2000 introduced Polar Tracking and Polar Snap. Later in this chapter you will see how these new features greatly enhance your capabilities for using the interactive method.

<u>Absolute coordinates</u> are used when AutoCAD prompts for a point and you know the exact coordinates of the desired location. You simply key in the coordinate values at the keyboard (X and Y values are separated by a comma).

<u>Relative rectangular coordinates</u> are similar to absolute values except the X and Y distances are given in relation to the last point, instead of being relative to the origin. AutoCAD interprets the @ symbol as the "last point." Use this type of coordinate entry when you do not know the exact absolute values, but you know the X and Y distances from the last specified point.

<u>Relative polar coordinate entry</u> is often used when you want to draw a line or specify a point that is at an exact angle with respect to the last point. Interactive coordinate entry is not accurate enough to draw diagonal lines to a specific angle (unless Polar Tracking is used). AutoCAD interprets the @ symbol as last point and the < symbol as an angular designator for the following value. So, "@ 2<45" means "from the last point, 2 units at an angle of 45 degrees."

<u>Direct distance coordinate entry</u> is a combination of relative polar coordinates and interactive specification because a distance value is entered at the keyboard and the angle is specified by the direction of the cursor movement (from the last point). Direct distance entry is useful primarily for orthogonal (horizontal or vertical) operations by toggling *ORTHO* or *POLAR* on. For example, assume you wanted to draw a *Line* 7.5 units in a horizontal (positive X) direction from the last point. Using direct distance entry to establish the "Specify next point:" (second endpoint), you would turn on *ORTHO* or *POLAR* and move the cursor to the right any distance, then type "7.5" and press Enter.

DRAWING *LINES* USING THE FIVE COORDINATE ENTRY METHODS

To practice using the five coordinate entry methods in this section, you must <u>turn Grid Snap on</u> rather than Polar Snap. Turn Grid Snap on by right-clicking the word *SNAP* on the Status Bar and selecting *Grid Snap On* from the shortcut menu (Fig. 3-6). Also, Polar Tracking, Object Snap, and Object Snap Tracking should be turned off (these are on by default when you install AutoCAD). Turn off Polar Tracking, Object Snap, and Object Snap Tracking by toggling the *POLAR*, *OSNAP*, and *OTRACK* buttons on the Status Bar or using F10, F3, and F11, respectively. The Status Bar indicates the on/off status of these drawing aids such that the button in a <u>recessed</u> (or depressed) position means <u>on</u> and a <u>protruding</u> button means <u>off</u> (see Figure 3-6).

Figure 3-6

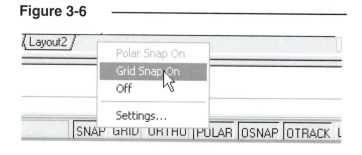

Begin a *New* drawing to complete the *Line* exercises. If the *Startup* dialog box appears, select *Start from Scratch*, choose *Imperial* as the default setting, and click the *OK* button. The *Line* command can be activated by any one of the methods shown in the command table below.

LINE

Pull-down Menu	COMMAND (TYPE)	ALIAS (TYPE)	Short-cut	Screen (side) Menu	Tablet Menu
Draw *Line*	LINE	L	...	DRAW 1 *Line*	J,10

Drawing Horizontal Lines

1. Draw a horizontal *Line* of 2 units length starting at point 2,2. Use the <u>interactive</u> method. See Figure 3-7.

Steps	Command Prompt	Perform Action	Comments
1.		turn on *GRID* (**F7**)	grid appears
2.		turn on *SNAP* (**F9**)	"SNAP" recessed on Status Bar
3.	Command:	select or type **Line**	use any method
4.	LINE Specify first point:	**PICK** location **2,2**	watch *Coords*
5.	Specify next point:	**PICK** location **4,2**	watch *Coords*
6.	Specify next point:	press **Enter**	completes command

The preceding steps produce a *Line* as shown in Figure 3-7.

Figure 3-7

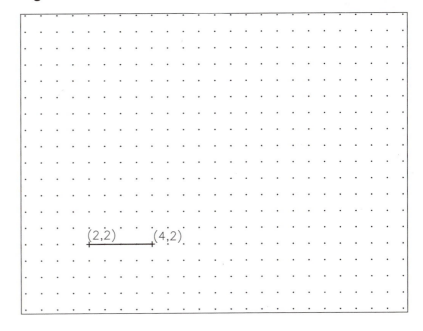

2. Draw a horizontal *Line* of 2 units length starting at point 2,3 using <u>absolute coordinates</u>.

Steps	Command Prompt	Perform Action	Comments
1.	Command:	select *Line*	use any method
2.	LINE Specify first point:	type **2,3** and press **Enter**	establishes the first endpoint
3.	Specify next point:	type **4,3** and press **Enter**	a *Line* should appear
4.	Specify next point:	press **Enter**	completes command

The above procedure produces the new *Line* above the first *Line* as shown (Fig. 3-8).

Figure 3-8

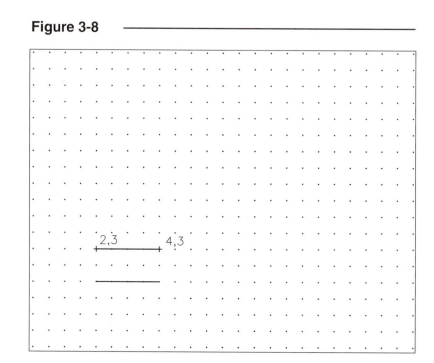

3. Draw a horizontal *Line* of 2 units length starting at point 2,4 using <u>relative rectangular coordinates</u>.

Steps	Command Prompt	Perform Action	Comments
1.	Command:	select *Line*	use any method
2.	LINE Specify first point:	type **2,4** and press **Enter**	establishes the first endpoint
3.	Specify next point:	type **@2,0** and press **Enter**	@ means "last point"
4.	Specify next point:	press **Enter**	completes command

The new *Line* appears above the previous two as shown in Figure 3-9.

Figure 3-9

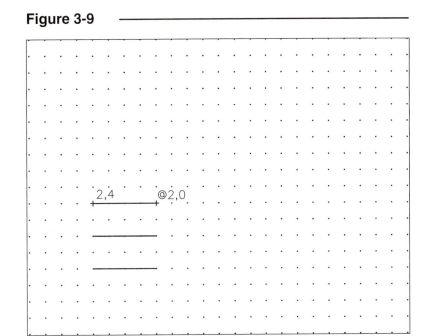

4. Draw a horizontal *Line* of 2 units length starting at point 2,5 using <u>relative polar coordinates</u>.

Steps	Command Prompt	Perform Action	Comments
1.	Command:	select *Line*	use any method
2.	LINE Specify first point:	type **2,5** and press **Enter**	establishes the first endpoint
3.	Specify next point:	type **@2<0** and press **Enter**	@ means "last point" < means "angle of"
4.	Specify next point:	press **Enter**	completes command

The new horizontal *Line* appears above the other three. See Figure 3-10.

Figure 3-10

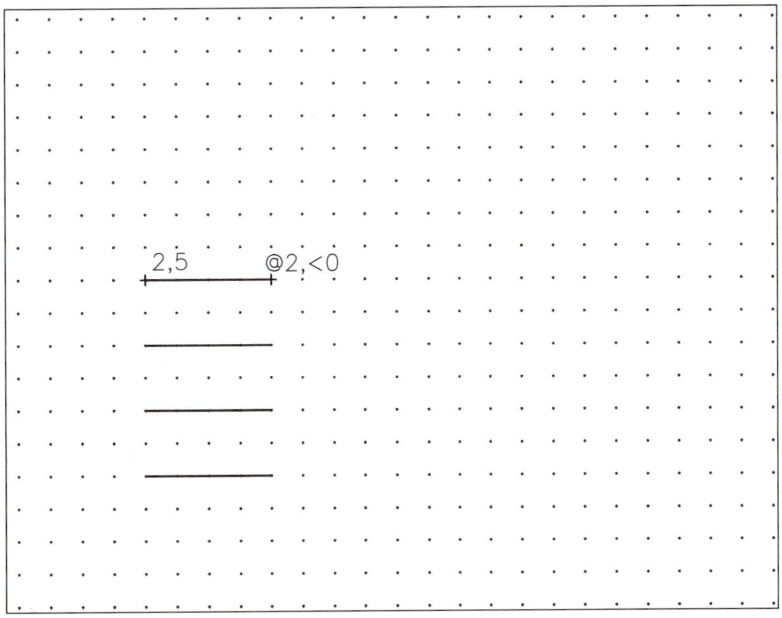

5. Draw a horizontal *Line* of 2 units length starting at point 2,6 using <u>direct distance</u> coordinate entry.

Steps	Command Prompt	Perform Action	Comments
1.	Command:	select *Line*	use any method
2.	LINE Specify first point:	type **2,6** and press **Enter**	establishes the first endpoint
3.	Specify next point:	turn on **ORTHO,** move the cursor to the right, type **2,** then press **Enter**	the cursor movement indicates direction, 2 is the distance
4.	Specify next point:	press **Enter**	completes command

The new line appears above the other four as shown in Figure 3-11.

Figure 3-11 ——————————————————

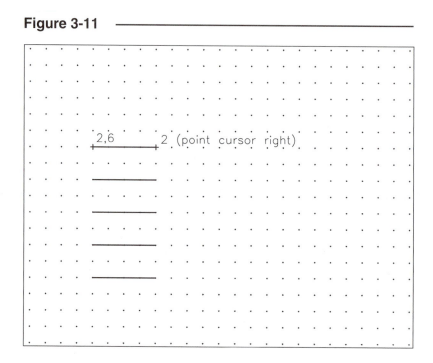

One of these methods may be more favorable than another in a particular situation. The interactive method is fast and easy, assuming that *OSNAP* or *POLAR* is used or that *SNAP* and *GRID* are used and set to appropriate values. *SNAP* and *GRID* are used successfully for small drawings where objects have regular interval lengths such as simple mechanical parts; however, *SNAP* and *GRID* are not used much for most drawings—those that use a variety of increments or occupy a large area, such as complex mechanical drawings, architectural drawings, or civil engineering drawings. Direct distance coordinate entry is fast, easy, and accurate when drawing horizontal or vertical lines. *ORTHO* or *POLAR* should be on for most applications of direct distance entry. Direct distance coordinate entry is preferred for horizontal or vertical drawing and editing in cases when *OSNAP* cannot be used (no geometry to *OSNAP* to) and when *SNAP* and *GRID* are not appropriate, such as an architectural or civil engineering application. In these applications, objects are relatively large (lines are long) and are not necessarily drawn to regular intervals; therefore, it is convenient and accurate to enter exact lengths as opposed to interactively PICKing points.

Drawing Vertical Lines

Below are listed the steps for drawing vertical lines using each of the five methods of coordinate entry. The following completed problems should look like those in Figure 3-12.

Figure 3-12

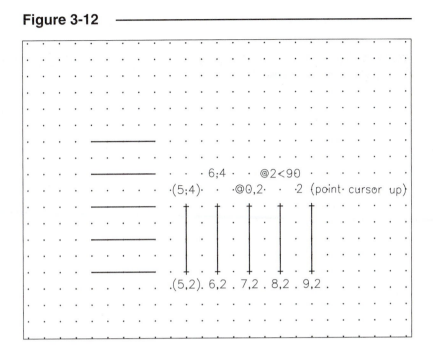

1. Draw a vertical *Line* of 2 units length starting at point 5,2 using the <u>interactive</u> method.

Steps	Command Prompt	Perform Action	Comments
1.		turn on *GRID* (**F7**)	grid appears
2.		turn on *SNAP* (**F9**)	"SNAP" recessed on Status Bar
3.	Command:	select or type *Line*	use any method
4.	LINE Specify first point:	**PICK** location **5,2**	watch *Coords*
5.	Specify next point:	**PICK** location **5,4**	watch *Coords*
6.	Specify next point:	press **Enter**	completes command

2. Draw a vertical *Line* of 2 units length starting at point 6,2 using <u>absolute coordinates</u>.

Steps	Command Prompt	Perform Action	Comments
1.	Command:	select *Line*	use any method
2.	LINE Specify first point:	type **6,2** and press **Enter**	establishes the first endpoint
3.	Specify next point:	type **6,4** and press **Enter**	a *Line* should appear
4.	Specify next point:	press **Enter**	completes command

3. Draw a vertical *Line* of 2 units length starting at point 7,2 using <u>relative rectangular coordinates</u>.

Steps	Command Prompt	Perform Action	Comments
1.	Command:	select *Line*	use any method
2.	LINE Specify first point:	type **7,2** and press **Enter**	establishes the first endpoint
3.	Specify next point:	type **@0,2** and press **Enter**	@ means "last point"
4.	Specify next point:	press **Enter**	completes command

4. Draw a vertical *Line* of 2 units length starting at point 8,2 using <u>relative polar coordinates</u>.

Steps	Command Prompt	Perform Action	Comments
1.	Command:	select *Line*	use any method
2.	LINE Specify first point:	type **8,2** and press **Enter**	establishes the first endpoint
3.	Specify next point:	type **@2<90** and press **Enter**	@ means "last point," < means "angle of"
4.	Specify next point:	press **Enter**	completes command

5. Draw a vertical *Line* of 2 units length starting at point 9,2 using <u>direct distance</u> coordinate entry.

Steps	Command Prompt	Perform Action	Comments
1.	Command:	select *Line*	use any method
2.	LINE Specify first point:	type **9,2** and press **Enter**	establishes the first endpoint
3.	Specify next point:	turn on *ORTHO*, move the cursor upward, type **2**, then press **Enter**	the cursor movement indicates direction, 2 is the distance
4.	Specify next point:	press **Enter**	completes command

The method used for drawing vertical lines depends on the application and the individual, much the same as it would for drawing horizontal lines. Refer to the discussion after the horizontal lines drawing exercises.

Drawing Inclined Lines

Following are listed the steps in drawing <u>inclined lines</u> using each of the five methods of coordinate entry. The following completed problems should look like those in Figure 3-13.

Figure 3-13

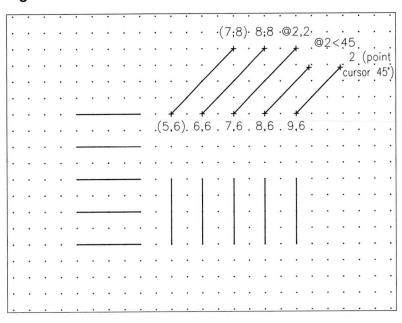

1. Draw an inclined *Line* of from 5,6 to 7,8. Use the <u>interactive</u> method.

Steps	Command Prompt	Perform Action	Comments
1.		turn on *GRID* (**F7**)	grid appears
2.		turn on *SNAP* (**F9**)	"SNAP" recessed on Status Bar
3.		turn off *ORTHO* (**F8**)	in order to draw inclined *Lines*
4.	Command:	select or type **Line**	use any method
5.	LINE Specify first point:	**PICK** location **5,6**	watch *Coords*
6.	Specify next point:	**PICK** location **7,8**	watch *Coords*
7.	Specify next point:	press **Enter**	completes command

2. Draw an inclined *Line* starting at 6,6 and ending at 8,8. Use <u>absolute coordinates</u>.

Steps	Command Prompt	Perform Action	Comments
1.	Command:	select **Line**	use any method
2.	LINE Specify first point:	type **6,6** and press **Enter**	establishes the first endpoint
3.	Specify next point:	type **8,8** and press **Enter**	a *Line* should appear
4.	Specify next point:	press **Enter**	completes command

3.   Draw an inclined *Line* starting at 7,6 and ending 2 units over (in a positive X direction) and 2 units up (in a positive Y direction). Use <u>relative rectangular coordinates</u>.

Steps	Command Prompt	Perform Action	Comments
1.	Command:	select *Line*	use any method
2.	LINE Specify first point:	type **7,6** and press **Enter**	establishes the first endpoint
3.	Specify next point:	type **@2,2** and press **Enter**	@ means "last point"
4.	Specify next point:	press **Enter**	completes command

4.   Draw an inclined *Line* of 2 units length at a 45 degree angle and starting at 8,6. Use <u>relative polar coordinates</u>.

Steps	Command Prompt	Perform Action	Comments
1.	Command:	select *Line*	use any method
2.	LINE Specify first point:	type **8,6** and press **Enter**	establishes the first endpoint
3.	Specify next point:	type **@2<45** and press **Enter**	@ means "last point" < means "angle of"
4.	To point	press **Enter**	completes command

5.   Draw an inclined *Line* of 2 units length at a 45 degree angle starting at point 9,6. Use <u>direct distance</u> coordinate entry.

Steps	Command Prompt	Perform Action	Comments
1.		Turn on *GRID* (**F7**)	grid appears
2.		Turn on *SNAP* (**F9**)	"SNAP" recessed on Status Bar
3.	Command:	select *Line*	use any method
4.	LINE Specify first point:	type **9,6** and press **Enter**	establishes the first endpoint
5.	Specify next point:	turn off **ORTHO,** move the cursor up and to the right (the coordinate display must indicate an angle of 45), type **2,** then press **Enter**	the cursor movement indicates direction, 2 is the distance
6.	Specify next point:	press **Enter**	completes command

Method 4 is suitable for drawing inclined lines when you know the exact length and angle. The other methods are not suitable for drawing angled lines for most cases. Instead, a feature called Polar Tracking should be used for most cases when many lines are to be drawn at regular angles such as 15, 30, or 45. Polar Tracking is described later in this chapter.

DRAWING *CIRCLES* USING THE FIVE COORDINATE ENTRY METHODS

Begin a *New* drawing to complete the *Circle* exercises. If the *Startup* dialog box appears, select *Start from Scratch,* and choose *Imperial* as the default setting. The *Circle* command can be invoked by any of the methods shown in the command table. Make sure *Grid Snap* is on and *POLAR, OSNAP,* and *OTRACK* are off for these exercises.

CIRCLE

Pull-down Menu	COMMAND (TYPE)	ALIAS (TYPE)	Short-cut	Screen (side) Menu	Tablet Menu
Draw *Circle >*	CIRCLE	C	...	DRAW 1 *Circle*	J,9

Below are listed the steps for drawing *Circles* using the *Center, Radius* method. The circles are to be drawn using each of the four coordinate entry methods and should look like those in Figure 3-14.

Figure 3-14

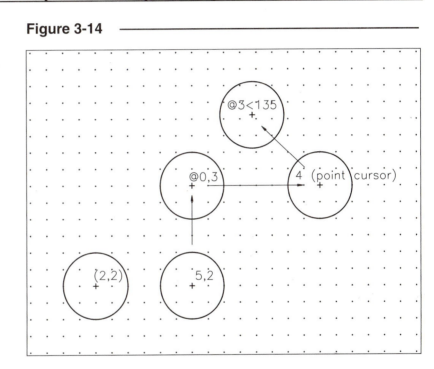

1. Draw a *Circle* of 1 unit radius with the center at point 2,2. Use the <u>interactive</u> method.

Steps	Command Prompt	Perform Action	Comments
1.		turn on *GRID* (**F7**)	grid appears
2.		turn on *SNAP* (**F9**)	"SNAP" recessed on Status Bar
3.		turn on *ORTHO* (**F8**)	"ORTHO" recessed on Status Bar
4.	Command:	select or type **Circle**	use *Center, Radius* method
5.	CIRCLE Specify center point for circle:	**PICK** location **2,2**	watch *Coords*
6.	Specify radius of circle:	move 1 unit and **PICK**	watch *Coords*

2.   Draw a circle of 1 unit radius with the center at point 5,2. Use <u>absolute coordinates</u>.

Steps	Command Prompt	Perform Action	Comments
1.	Command:	select *Circle*	use *Center, Radius* method
2.	CIRCLE Specify center point for circle:	type **5,2** and press **Enter**	a *Circle* should appear
3.	Specify radius of circle:	type **1** and press **Enter**	the correct *Circle* appears

3.   Draw a circle of 1 unit radius with the center 3 units above the last point (previous *Circle* center). Use <u>relative rectangular coordinates</u>.

Steps	Command Prompt	Perform Action	Comments
1.	Command:	select *Circle*	use *Center, Radius* method
2.	CIRCLE Specify center point for circle:	type **@0,3** and press **Enter**	a *Circle* should appear
3.	Specify radius of circle:	type **1** and press **Enter**	the correct *Circle* appears

(If the new *Circle* is not above the last, type "ID" and enter "5,2." The entered point becomes the last point. Then try again.)

4.   Draw a circle of 1 unit radius with the center 4 units to the right of the previous *Circle* center. Use <u>direct distance</u> coordinate entry.

Steps	Command Prompt	Perform Action	Comments
1.		turn on *ORTHO* (**F8**)	"ORTHO" recessed on Status Bar
2.	Command:	select *Circle*	use *Center, Radius* method
3.	CIRCLE Specify center point for circle:	type **4,** move the cursor to the right, and press **Enter**	a *Circle* should appear
4.	Specify radius of circle:	type **1,** then press **Enter**	the correct *Circle* appears

(If the new *Circle* is not directly to the right of the last, type "ID" and enter "5,5." The entered point becomes the last point. Then try again.)

5. Draw a circle of 1 unit radius with the center 3 units at a 135-degree angle above and to the left of the previous *Circle*. Use <u>relative polar coordinates</u>.

Steps	Command Prompt	Perform Action	Comments
1.	Command:	select **Circle**	use *Center, Radius* method
2.	CIRCLE Specify center point for circle:	type **@3<135** and press **Enter**	a *Circle* should appear
3.	Specify radius of circle:	type **1** and press **Enter**	the *Circle* appears

(If the new *Circle* is not correctly positioned with respect to the last, type "ID" and enter "9,5." The entered point becomes the last point. Then try again.)

POLAR TRACKING AND POLAR SNAP

Remember that all draw commands prompt you to specify locations, or coordinates, in the drawing. You can indicate these coordinates interactively (using the cursor), entering values (at the keyboard), or using a combination of both. Features in AutoCAD that provide an easy method for specifying coordinate locations interactively are Polar Tracking and Polar Snap. Another feature, Object Snap Tracking, is discussed in Chapter 7. These features help you draw objects at specific angles and in specific relationships to other objects.

Polar Tracking (*POLAR*) helps the rubberband line snap to angular increments such as 45, 30, or 15 degrees. Polar Snap can be used in conjunction to make the rubberband line snap to incremental lengths such as .5 or 1.0. You can toggle these features on and off with the *SNAP* and *POLAR* buttons on the Status Bar or by toggling F9 and F10, respectively. First, you should specify the settings in the *Drafting Settings* dialog box.

In the previous sections of this chapter you learned and practiced with the five coordinate entry methods listed below.

1. Interactive method
2. Absolute coordinates
3. Relative rectangular coordinates
4. Relative polar coordinates
5. Direct distance entry

Technically, <u>Polar Tracking and Polar Snap would be considered a variation of the Interactive method</u> since the settings are determined and specified beforehand, then points on the screen are PICKed using the cursor.

To practice with Polar Tracking and Polar Snap, you should turn both *OSNAP* and *OTRACK* off (these are on by default when you install AutoCAD). Since all of these new features generate alignment vectors, it can be difficult to determine which of the new features is operating when they are all on.

Polar Tracking

Figure 3-15

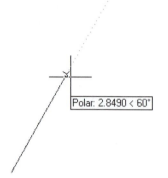

Polar Tracking (*POLAR*) simplifies drawing *Lines* or performing other operations such as *Move* or *Copy* at specific angle increments. For example, you can specify an increment angle of 30 degrees. Then when Polar Tracking is on, the rubberband line "snaps" to 30-degree increments when the cursor is in close proximity (within the *Aperture* box) to a specified angle. A dotted "tracking" line is displayed and a "tracking tip" appears at the cursor giving the current distance and angle (Fig. 3-15). In this case, it is simple to draw lines at 0, 30, 60, 90, or 120-degree angles and so on.

Figure 3-16 displays possible positions for Polar Tracking when a 30-degree angle is specified. Available angle options are 90, 45, 30, 22.5, 18, 15, 10, and 5 degrees, or you can specify any user-defined angle. This is a tremendous new aid for drawing angled lines since the only other method for precisely specifying angles is polar coordinate entry.

Figure 3-16

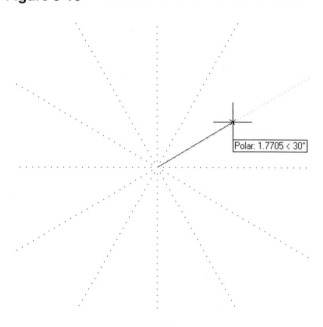

You can access the settings using the *Polar Tracking* tab of the *Drafting Settings* dialog box (Fig. 3-17). Invoke the dialog box by the following methods:

Figure 3-17

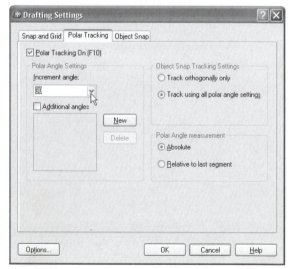

1. Enter *Dsettings* or *DS* at the Command line, then select the *Polar Tracking* tab.

2. Right-click on the word *POLAR* on the Status Bar, then choose *Settings…* from the menu.

3. Select *Drafting Settings* from the *Tools* pull-down menu, then select the *Polar Tracking* tab.

In this dialog box, you can select the *Increment Angle* from the drop-down list (as shown) or specify *Additional Angles*. Use the *New* button to create user-defined angles. Highlighting a user-defined angle and selecting the *Delete* button removes the angle from the list.

The *Object Snap Tracking Settings* are used only with Object Snap (discussed in Chapter 7). When the *Polar Angle Measurement* is set to *Absolute*, the angle reported on the "tracking tip" is an absolute angle (relative to angle 0 of the current coordinate system).

Polar Tracking with Direct Distance Entry

With normal Polar Tracking (Polar Snap is off), the line "snaps" to the set angles but can be drawn to any length (distance). This option is particularly useful in conjunction with Direct Distance Entry since you specify the angular increment in the *Polar Tracking* tab of the *Drafting Settings* dialog box, but indicate the distance by <u>entering values at the Command line</u>. In other words, when drawing a *Line*, move the cursor so it "snaps" to the desired angle, then enter the desired distance at the keyboard (Fig. 3-18).

Figure 3-18

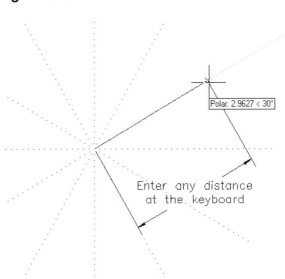

Polar Tracking Override

When Polar Tracking is on, you can override the previously specified angle increment and draw to another specific angle. The new angle is valid only for one point specification. To enter a polar override angle, enter the left angle bracket (<) and an angle value whenever a command asks you to specify a point. The following command prompt sequence shows a 12-degree override entered during a *Line* command.

```
Command: line
Specify first point: PICK
Specify next point or [Undo]: <12
Angle Override:  12
Specify next point or [Undo]: PICK
```

This action forces the line to a 12-degree angle for that segment only.

Figure 3-19

Polar Snap

<u>Polar Tracking</u> makes the rubberband line "snap" to <u>angular increments</u>, whereas <u>Polar Snap</u> makes the rubberband line "snap" to <u>distance increments</u> that you specify. For example, setting the distance increment to 2 allows you to draw lines at intervals of 2, 4, 6, 8, and so on. Therefore, Polar Tracking with Polar Snap allows you to draw at specific angular <u>and</u> distance increments (Fig. 3-19). Using Polar Tracking with Polar Snap off, as described in the previous section, allows you to draw at specified angles but not at any specific distance intervals.

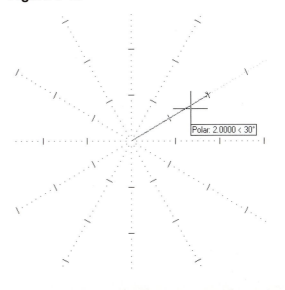

Only one type of Snap, Polar Snap or Grid Snap, can be used at any one time. Toggle either *Polar Snap* or *Grid Snap* on by setting the radio button in the *Snap and Grid* tab of the *Drafting Settings* dialog box (Fig. 3-20, lower right). Alternately, right-click on the word *SNAP* at the Status Bar and select either *Polar Snap On* or *Grid Snap On* from the shortcut menu (see Figure 3-21). Set the Polar Snap distance increment in the *Polar Distance* edit box (Fig. 3-20, lower left).

Figure 3-20

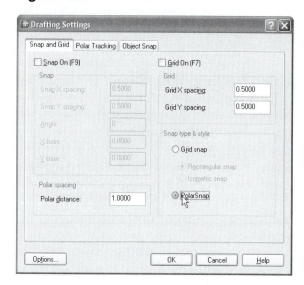

To utilize Polar Snap, both *SNAP* (F9) and *POLAR* (F10) must be turned on, Polar Snap must be on, and a Polar Distance value must be specified in the *Drafting Settings* dialog box. As a check, settings should be as shown in Figure 3-21 (*SNAP* and *POLAR* are recessed), with the addition of some Polar Distance setting in the dialog box. *SNAP* turns on or off whichever of the two snap types is active, Grid Snap or Polar Snap. For example, if Polar Snap is the active snap type, toggling F9 turns only Polar Snap off and on.

Figure 3-21

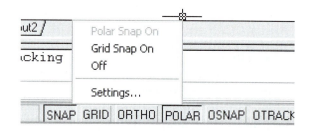

DRAWING LINES USING POLAR TRACKING AND POLAR SNAP

For these exercises you will use combinations of Polar Tracking with Polar Snap, Grid Snap and no snap type. Also, Object Snap and Object Snap Tracking should be turned off (these are on by default when you install AutoCAD). Turn off Object Snap and Object Snap Tracking by toggling the *OSNAP* and *OTRACK* buttons on the Status Bar to the protruding position or using F3 and F11 to do so.

Figure 3-22

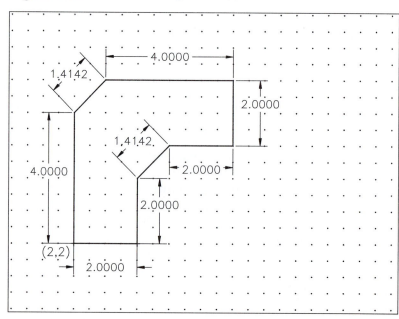

Using Polar Tracking and Grid Snap

Draw the shape in Figure 3-22 using Polar Tracking in conjunction with Grid Snap. Follow the steps below.

1. Begin a *New* drawing. Select **Start from Scratch** and the **Imperial** default settings.

2. Toggle on Grid Snap (rather than Polar Snap) by right-clicking on the word **SNAP** on the Status Bar, then selecting **Grid Snap On** from the menu (see Figure 3-6).

3. Turn on Polar Tracking and make the appropriate settings. Do this by right-clicking on the word *POLAR* on the Status Bar and selecting *Settings* from the menu (Figure 3-23). In the *Polar Tracking* tab of the *Drafting Settings* dialog box that appears, set the *Increment Angle* to **45.0** (see Figure 3-17). Select the *OK* button.

Figure 3-23

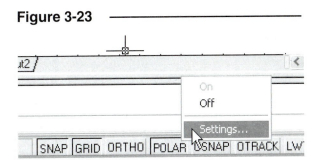

4. Make these other settings by single-clicking the words on the Status Bar or using the Function keys. The Status Bar indicates the on/off status of these drawing aids such that the button in a recessed position means <u>on</u> and a protruding button means <u>off</u> (Fig. 3-24). The command prompt also indicates the on or off status when they are changed.

Figure 3-24

SNAP (F9) is on
GRID (F7) is on
ORTHO (F8) is off
POLAR (F10) is on
OSNAP (F3) is off
OTRACK (F11) is off

The resulting Status Bar should look like that in Figure 3-24.

5. Use the *Line* command. Starting at location 2.000, 2.000, 0.000 (watch the *COORDS* display), begin drawing the shape shown in Figure 3-22. The tracking tip should indicate the current length and angle of the lines as you draw (Fig. 3-25).

6. Complete the shape. Compare your drawing to that in Figure 3-22. Use *Save* and name the drawing **PTRACK-GRIDSNAP**. *Close* the drawing.

Figure 3-25

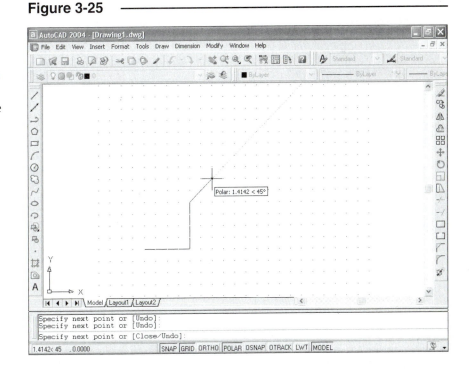

Using Polar Tracking and Polar Snap

Draw the shape in Figure 3-26. Since the lines are regular interval lengths, use Polar Tracking in conjunction with Polar Snap. Follow the steps below.

1. Begin a *New* drawing. Select *Start From Scratch* and the *Imperial* default settings.

2. Toggle on Polar Snap (rather than Grid Snap) by right-clicking on the word *SNAP* on the Status Bar, then selecting *Polar Snap On* from the menu.

Figure 3-26 —————————

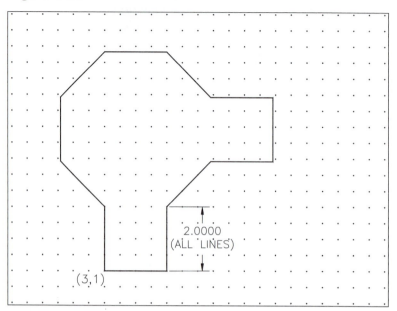

3. Make the appropriate Polar Snap settings. Do this by right-clicking on the word *SNAP* on the Status Bar and selecting *Settings* from the menu. In the *Snap and Grid* tab of the *Drafting Settings* dialog box that appears, set the *Polar Distance* to **1.0000** (see Figure 3-27). Next, access the *Polar Tracking* tab and ensure the *Increment Angle* is set to **45.0**. Select the *OK* button.

4. Ensure these other settings are correct by clicking the words on the Status Bar or using the Function keys.

   > *SNAP* (F9) is on
   > *GRID* (F7) is on
   > *ORTHO* (F8) is off
   > *POLAR* (F10) is on
   > *OSNAP* (F3) is off
   > *OTRACK* (F11) is off

5. Use the *Line* command. Starting at location 3.000, 1.000, 0.000 (enter absolute coordinates of **3,1**), begin drawing the shape shown in Figure 3-26. The tracking tip should indicate the current length and angle of the lines as you draw (Fig. 3-28).

6. Complete the shape. Compare your drawing to that in Figure 3-26. Use *Save* and name the drawing **PTRACK-POLARSNAP**. *Close* the drawing.

Figure 3-27 —————————

Figure 3-28 —————————

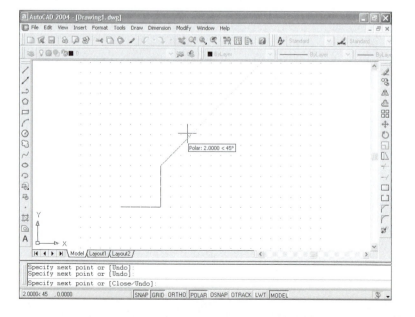

Using Polar Tracking and Direct Distance Entry

Draw the shape in Figure 3-29. Since the lines are at regular angles but irregular lengths, use Polar Tracking in conjunction with Direct Distance Entry. Follow the steps below.

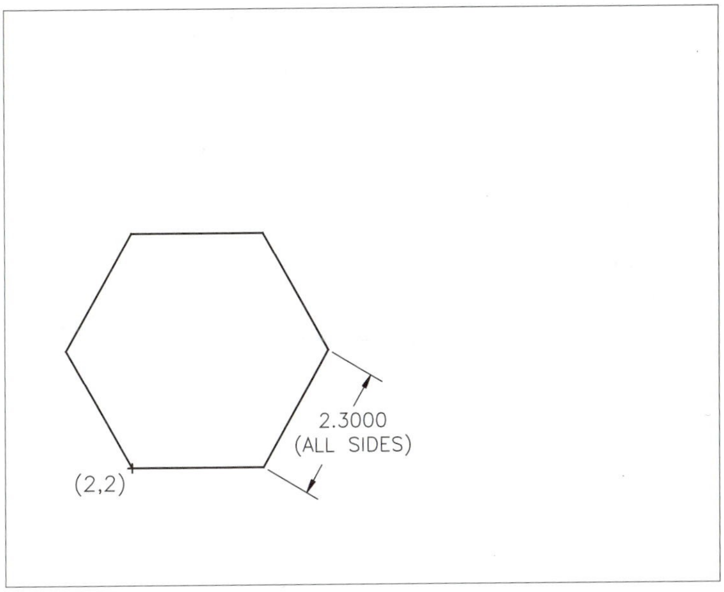

1. Begin a *New* drawing. Select *Start From Scratch* and the *Imperial* default settings.

2. Toggle off *SNAP* on the Status Bar. No Snap type is used since distances will be entered at the keyboard.

3. Turn on Polar Tracking and make the appropriate settings. Do this by right-clicking on the word *POLAR* on the Status Bar and selecting *Settings* from the menu. In the *Polar Tracking* tab of the *Drafting Settings* dialog box, ensure the *Increment Angle* is set to **30.0**, and *Polar Tracking* is *On*. Select the *OK* button.

Figure 3-29

4. Ensure these other settings are correct by clicking the words on the Status Bar or using the Function keys.

> *SNAP* (F9) is off
> *GRID* (F7) is off
> *ORTHO* (F8) is off
> *POLAR* (F10) is on
> *OSNAP* (F3) is off
> *OTRACK* (F11) is off

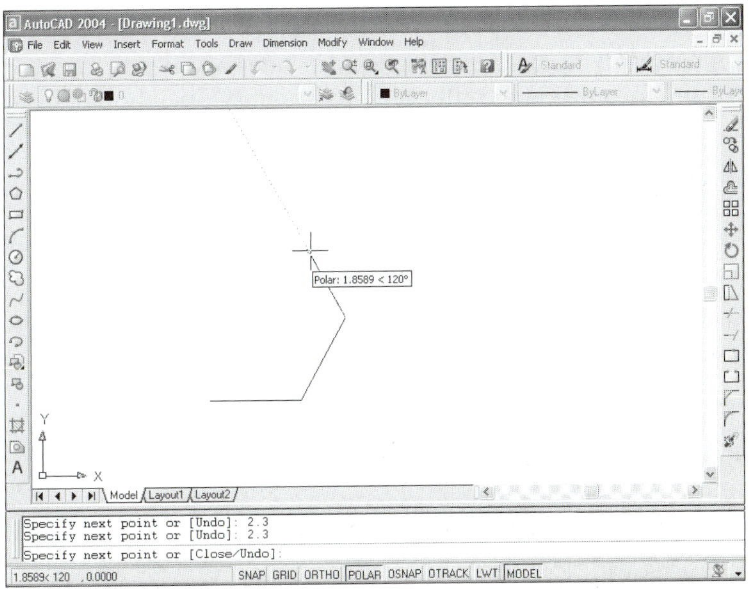

5. Use the *Line* command. Starting at location 2.000, 2.000, 0.000 (enter absolute coordinates of **2,2**), begin drawing the shape shown in Figure 3-29. Use Polar Tracking by moving the mouse in the desired direction (Fig. 3-30). When the tracking tip indicates the correct angle for each line, enter the distance for each line (**2.3**) at the keyboard.

6. Complete the shape. Compare your drawing to that in Figure 3-29. Use *Save* and name the drawing **PTRACK-DDE**. *Close* the drawing and *Exit* AutoCAD.

Figure 3-30

CHAPTER EXERCISES

1. **Start a *New* Drawing**

   Start a *New* drawing. Select the ***Start from Scratch*** option and select *Imperial* settings. Next, use the *Save* command and save the drawing as "**CH3EX1**." Remember to *Save* often as you complete the following exercises. The completed exercise should look like Figure 3-31.

Figure 3-31

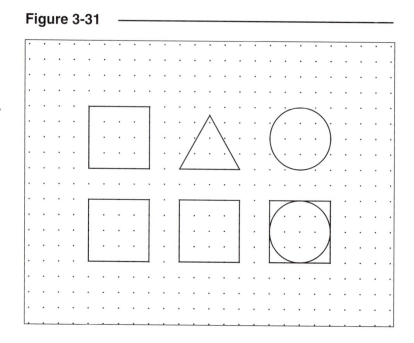

2. **Use interactive coordinate entry**

   Draw a square with sides of 2 units length. Locate the lower-left corner of the square at **2,2**. Use the *Line* command with interactive coordinate entry. (HINT: Turn on *SNAP, GRID,* and *ORTHO*.)

3. **Use absolute coordinates**

   Draw another square with 2 unit sides using the *Line* command. Enter absolute coordinates. Begin with the lower-left corner of the square at **5,2**.

4. **Use relative rectangular coordinates**

   Draw a third square (with 2 unit sides) using the *Line* command. Enter relative rectangular coordinates. Locate the lower-left corner at **8,2**.

5. **Use direct distance coordinate entry**

   Draw a fourth square (with 2 unit sides) beginning at the lower-left corner of **2,5**. Complete the square drawing *Lines* using direct distance coordinate entry. Don't forget to turn on *ORTHO* or *POLAR*.

6. **Use relative polar coordinates**

   Draw an equilateral triangle with sides of 2 units. Locate the lower-left corner at **5,5**. Use relative polar coordinates (<u>after</u> establishing the "Specify first point:"). HINT: An equilateral triangle has interior angles of 60 degrees.

7. **Use interactive coordinate entry**

   Draw a *Circle* with a 1 unit <u>radius</u>. Locate the center at **9,6**. Use the interactive method. (Turn on *SNAP* and *GRID*.)

8. **Use relative rectangular, relative polar, or direct distance entry coordinates**

   Draw another *Circle* with a 2 unit <u>diameter.</u> Using any method listed above, locate the center 3 units below the previous *Circle*.

9. ***Save* your drawing**

   Use *Save*. Compare your results with Figure 3-31. When you are finished, *Close* the drawing.

10. In the next series of steps, you will create the Stamped Plate shown in Figure 3-32. Begin a *New* drawing. Select the *Start from Scratch* option and select *Imperial* settings. Next, use the *Save* command and assign the name **CH3EX2**. Remember to *Save* often as you work.

Figure 3-32

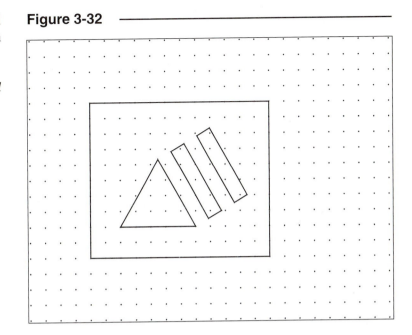

11. First, you will create the equilateral triangle as shown in Figure 3-33. All sides are equal and all angles are equal. To create the shape easily, you should use Polar Tracking with Direct Distance Entry. Access the *Polar Tracking* tab of the *Drafting Settings* dialog box and set the *Increment Angle* to **30**. (HINT: Right-click on the word *POLAR* and select *Settings* from the shortcut menu.) On the Status Bar, make sure *POLAR* is on (appears recessed), but not *SNAP, ORTHO, OSNAP,* or *OTRACK. GRID* is optional.

Figure 3-33

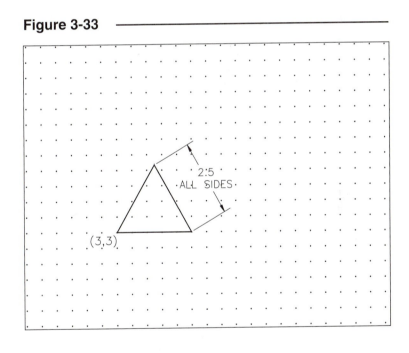

12. Use the *Line* command to create the equilateral triangle. Start at position **3,3** by entering absolute coordinates. All sides should be **2.5** units, so enter the distance values at the keyboard as you position the mouse in the desired direction for each line. The drawing at this point should look like that in Figure 3-33 (not including the notation). *Save* the drawing but do not *Close* it.

13. Next, you should create the outside shape (Fig. 3-34). Since the dimensions are all at even unit intervals, use Grid Snap to create the rectangle. (HINT: Right-click on the word *SNAP* and select *Grid Snap On*.) At the Status Bar, make sure that only *SNAP* and *POLAR* are on. Use the *Line* command to create the rectangular shape starting at point **2,2** (enter absolute coordinates).

Figure 3-34

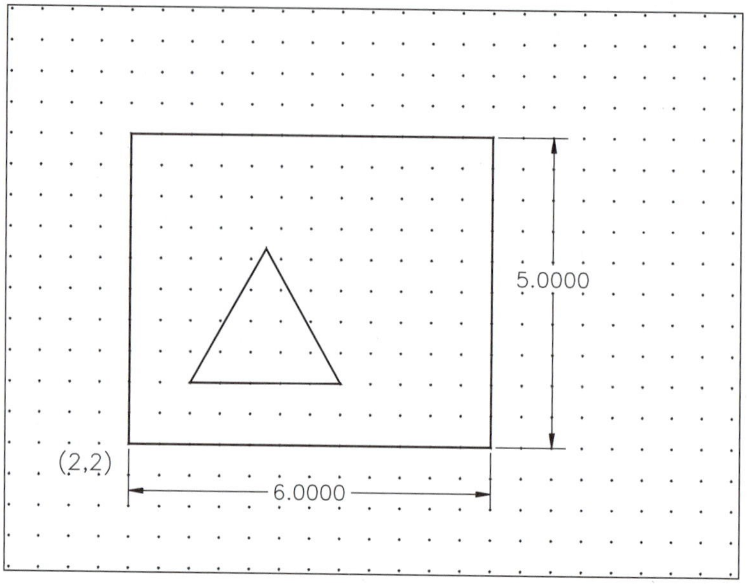

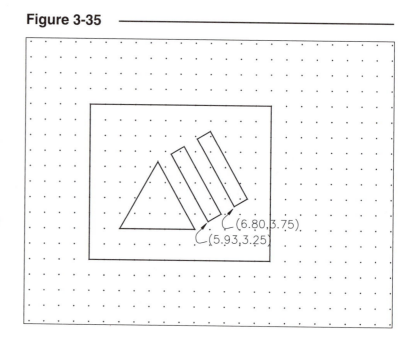

14. The two inside shapes are most easily created using Polar Tracking in combination with Polar Snap. Access the *Grid and Snap* tab of the *Drafting Settings* dialog box. (HINT: Right-click on the word *SNAP* and select *Settings*.) Set the *Snap Type* to *Polar Snap* and set the *Polar Distance* to **.5**. At the Status Bar, make sure that only *SNAP* and *POLAR* are on. Next, use the *Line* command and create the two shapes as shown in Figure 3-35. Specify the starting positions for each shape (**5.93,3.25** and **6.80,3.75**) by entering absolute coordinates.

Figure 3-35

15. When you are finished, *Save* the drawing and *Exit* AutoCAD.

SELECTION SETS

Chapter Objectives

After completing this chapter you should:

1. know that *Modify* commands and many other commands require you to select objects;

2. be able to create a selection set using each of the specification methods;

3. be able to *Erase* objects from the drawing;

4. be able to *Move* and *Copy* objects from one location to another;

5. understand Noun/Verb and Verb/Noun order of command syntax.

MODIFY COMMAND CONCEPTS

Draw commands create objects. Modify commands <u>change existing objects</u> or <u>use existing objects to create new ones</u>. Examples are *Copy* an existing *Circle, Move* a *Line,* or *Erase* a *Circle.*

Since all of the Modify commands use or modify <u>existing</u> objects, you must first select the objects that you want to act on. The process of selecting the objects you want to use is called building a <u>selection set</u>. For example, if you want to *Copy, Erase,* or *Move* several objects in the drawing, you must first select the set of objects that you want to act on.

Remember that any of the five command entry methods (depending on your setup) can be used to invoke Modify commands.

1. **Toolbars**   *Modify* or *ModifyII* toolbar
2. **Pull-down menu**   *Modify* pull-down menu
3. **Screen menu**   *MODIFY1* or *MODIFY2* screen menus
4. **Keyboard**   Type the command name <u>or</u> command alias
5. **Tablet menu**   Select the command icon

All of the Modify commands will be discussed in detail later, but for now we will focus on how to build selection sets.

SELECTION SETS

No matter which of the five methods you use to invoke a Modify command, you must specify a selection set during the command operation. There are two ways you can select objects: you can select the set of objects either (1) immediately before you invoke the command or (2) when the command prompts you to select objects. For example, examine the command syntax that would appear at the Command line when the *Erase* command is used (method 2).

    Command: **erase**
    Select objects:

Figure 4-1 ───────────────────────

The "Select objects:" prompt is your cue to use any of several methods to PICK the objects to erase. As a matter of fact, every Modify command begins with the same "Select objects:" prompt (unless you selected immediately before invoking the command).

When the "Select objects:" prompt appears, the "crosshairs" cursor disappears and only a small, square pickbox appears at the cursor (Fig. 4-1).

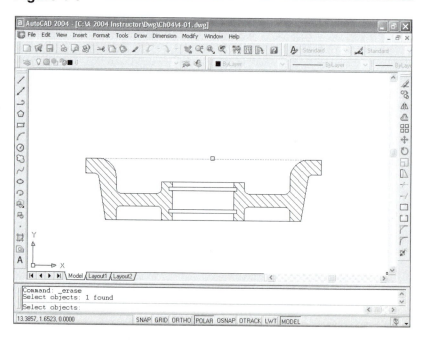

You can PICK objects using only the pickbox or any of several other methods illustrated in this chapter.

Only when you PICK objects at the "Select objects:" prompt can you use all of the selection methods shown here. If you PICK immediately before the command, called Noun/Verb order, only the *AUto* method (pickbox, window, or crossing window) can be used. (See "Noun/Verb Syntax" in this chapter.)

When the objects have been selected, they become highlighted (displayed as a broken line), which serves as a visual indication of the current selection set. Press Enter to indicate that you are finished selecting and are ready to proceed with the command.

Accessing Selection Set Options

When the "Select objects:" prompt appears, you should select objects using the pickbox, window, crossing window, or one of a variety of other methods. Any method can be used independently or in combination to achieve the desired set of objects because object selection is a cumulative process.

The pickbox is the default option, which can automatically be changed to a window or crossing window by PICKing in an open area (PICKing no objects). Since the pickbox, window, and crossing window (sometimes known as the *AUto* option) are the default selection methods, no action is taken on your part to activate these methods. The other methods can be selected from the screen (side) menu or by typing the capitalized letters shown in the option names following.

All the possible selection set options are listed if you enter a question mark symbol (?) at the "Select objects:" prompt.

```
Command: move
Select objects: ?
*Invalid selection*
Expects a point or
Window/Last/Crossing/BOX/ALL/Fence/WPolygon/CPolygon/Group/Add/Remove/Multiple/
Previous/Undo/AUto/SIngle
Select objects:
```

Use any option by typing the indicated uppercase letters and pressing Enter.

If the screen (side) menu is displayed, you can access many of the object selection options from the *ASSIST* menu. *ASSIST* is displayed at the bottom of all other menus.

In AutoCAD 2004, there are no icon buttons or pull-down menu selections that are easily available for the object selection options. However, a toolbar can be "customized" to display the object selection options (see Chapter 44, Basic Customization).

Figure 4-2

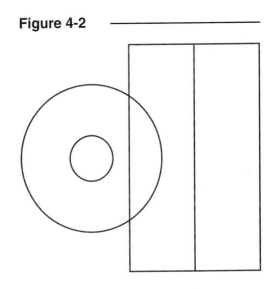

Selection Set Options

The options for creating selection sets (PICKing objects) are shown on this and the following pages. Two *Circles* and five *Lines* (as shown in Figure 4-2) are used for every example. In each case, the circles only are selected for editing. If you want to follow along and practice as you read, draw *Circles* and *Lines* in a configuration similar to this. Then use a *Modify* command. (Press Escape to cancel the command after selecting objects.)

AUto (pickbox)

This default option is used for selecting <u>one object</u> at a time. Locate the pickbox so that an object crosses through it and **PICK** (Fig. 4-3). You do not have to type or select anything to use this option.

Figure 4-3

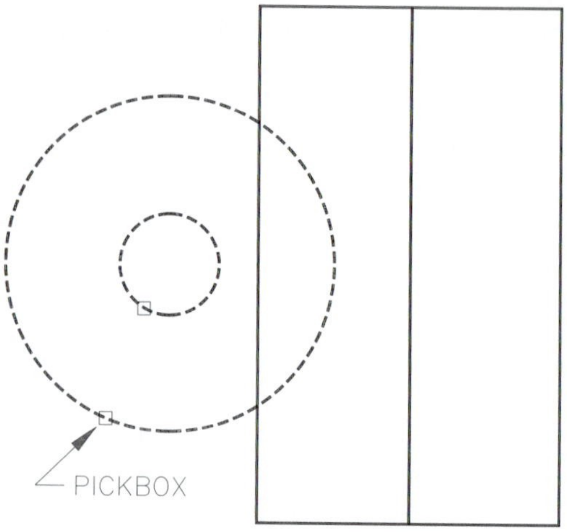

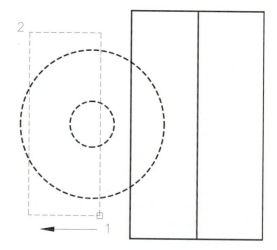

AUto (window)

To use this option, you do not have to type or select anything from the *Assist* screen menu. The pickbox must be positioned in an open area so that no objects cross through it; then **PICK** to start a window. If you drag to the <u>right</u>, a *Window* is created (Fig. 4-4). **PICK** the other corner.

Figure 4-4

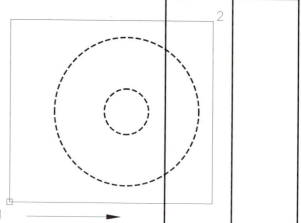

AUto (crossing window)

If you drag to the <u>left</u> instead, a *Crossing Window* forms (Fig. 4-5). (See "Window" and "Crossing Window" on the next page.)

Figure 4-5

Window

Only objects <u>completely within</u> the *Window* are selected. The *Window* is a solid linetype rectangular box. Select the first and second points (diagonal corners) in any direction as shown in Figure 4-6.

Figure 4-6

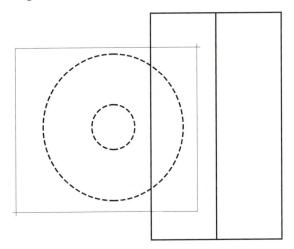

Crossing Window

All objects <u>within and crossing through</u> the window are selected. The *Crossing Window* is displayed as a broken linetype rectangular box (Fig. 4-7). Select two diagonal <u>corners</u>.

Figure 4-7

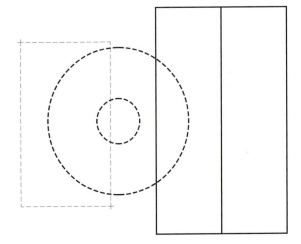

Window Polygon

The *Window Polygon* operates like a *Window,* but the box can be <u>any</u> irregular polygonal shape (Fig. 4-8). You can pick any number of corners to form any shape rather than picking just two corners to form a rectangle as with the *Window* option.

Figure 4-8

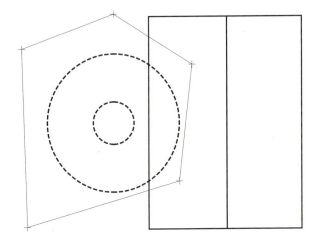

Crossing Polygon

The *Crossing Polygon* operates like a *Crossing Window,* but can have any number of corners like a *Window Polygon* (Fig. 4-9).

Figure 4-9

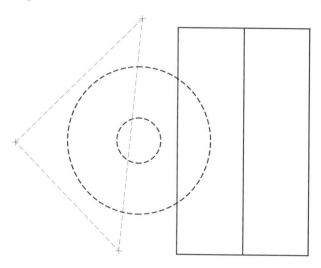

Fence

This option operates like a <u>crossing line</u>. Any objects crossing the *Fence* are selected. The *Fence* can have any number of segments (Fig. 4-10).

Figure 4-10

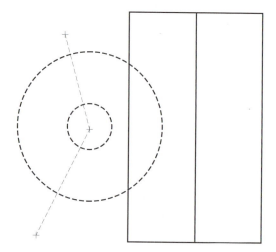

Last

This option automatically finds and selects <u>only</u> the last object <u>created</u>. *Last* does not find the last object modified (with *Move, Stretch,* etc.).

Previous

Previous finds and selects the <u>previous selection set</u>, that is, whatever was selected during the previous command (except after *Erase*). This option allows you to use several editing commands on the same set of objects without having to respecify the set.

ALL

This option selects <u>all objects</u> in the drawing except those on *Frozen* or *Locked* layers (*Layers* are covered in Chapter 11).

BOX

This option is equivalent to the *AUto* window/crossing window option <u>without</u> the pickbox. PICKing diagonal corners from left to right produces a window and PICKing diagonal corners from right to left produces a crossing window (see *AUto*).

Multiple
The *Multiple* option allows selection of objects with the pickbox only; however, the selected objects are not highlighted. Use this method to select very complex objects to save computing time required to change the objects' display to highlighted.

Single
This option allows only a <u>single selection</u> using one of the *AUto* methods (pickbox, window, or crossing window), then automatically continues with the command. Therefore, you can select multiple objects (if the window or crossing window is used), but only one selection is allowed. You do not have to press Enter after the selection is made.

Undo
Use *Undo* to cancel the selection of the object(s) most recently added to the selection set.

Remove
Selecting this option causes AutoCAD to <u>switch</u> to the "Remove objects:" mode. Any selection options used from this time on remove objects from the highlighted set (see Figure 4-11).

Figure 4-11

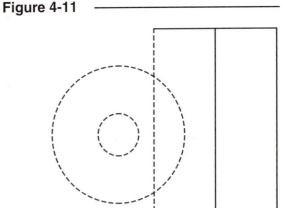

Add
The *Add* option switches back to the default "Select objects:" mode so additional objects can be added to the selection set.

Shift + Left Button
Holding down the **Shift** key and pressing the left button simultaneously <u>removes</u> objects selected from the highlighted set as shown in Figure 4-11. This method is generally quicker than, but performs the same action as, *Remove*. The advantage here is that the *Add* mode is in effect unless Shift is held down.

REMOVE OBJECT
FROM SELECTION SET

Group
The *Group* option selects groups of objects that were previously specified using the *Group* command. Groups are selection sets to which you can assign a name (see Chapter 20).

Ctrl + Left Button (Object Cycling)

Holding down the **Ctrl** key and then pressing the **PICK** (left) button several times causes AutoCAD to cycle through (highlight one at a time) two or more objects that may cross through the pickbox.

For example, if you attempted to PICK an object but other objects were in the pickbox, AutoCAD may not highlight the one you want (Fig. 4-12). In this case, hold down the **Ctrl** key and press the **PICK** button several times (the cursor can be located anywhere on the screen during cycling). All of the objects that passed through the pickbox will be highlighted one at a time. The "<Cycle on>" prompt appears at the Command line. When the object you want is highlighted, press Enter and AutoCAD adds the object to the selection set and returns to the "Select objects:" prompt.

Figure 4-12

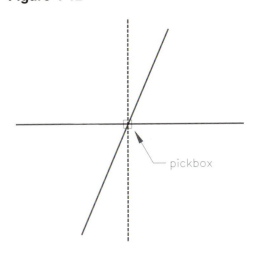

pickbox

SELECT

Pull-down Menu	COMMAND (TYPE)	ALIAS (TYPE)	Short-cut	Screen (side) Menu	Tablet Menu
...	*SELECT*	...	...	...	...

The *Select* command can be used to PICK objects to be saved in the selection set buffer for subsequent use with the *Previous* option. Any of the selection methods can be used to PICK the objects.

```
Command: select
Select objects: PICK (use any selection option)
Select objects: Enter (completes the selection process)
Command:
```

The selected objects become unhighlighted when you complete the command by pressing Enter. The objects become highlighted again and are used as the selection set if you use the *Previous* selection option in the next editing command.

NOUN/VERB SYNTAX

An object is the noun and a command is the verb. Noun/Verb syntax order means to pick objects (nouns) first, then use an editing command (verb) second. If you select objects first (at the open Command: prompt) and then immediately choose a Modify command, AutoCAD recognizes the selection set and passes through the "Select objects:" prompt to the next step in the command.

Verb/Noun means to invoke a command and then select objects within the command. For example, if the *Erase* command (verb) is invoked first, AutoCAD then issues the prompt to "Select objects:"; there-fore, objects (nouns) are PICKed second. In the previous examples, and with much older versions of AutoCAD, only Verb/Noun syntax order was used.

You can use either order you want (Noun/Verb or Verb/Noun) and AutoCAD automatically under-stands. If objects are selected first, the selection set is passed to the next editing command used, but if no objects are selected first, the editing command automatically prompts you to "Select objects:".

If you use Noun/Verb order, you are limited to using only the *AUto* options for object selection (pickbox, *Window*, and *Crossing Window*). You can only use the other options (e.g., *Crossing Polygon, Fence, Previous*) if you invoke the desired Modify command first, then select objects when the "Select objects:" prompt appears.

The *PICKFIRST* variable (a very descriptive name) enables Noun/Verb syntax. The default setting is 1 (*On*). If *PICKFIRST* is set to 0 (*Off*), Noun/Verb syntax is disabled and the selection set can be specified only within the editing commands (Verb/Noun).

Setting *PICKFIRST* to 1 provides two options: Noun/Verb and Verb/Noun. You can use either order you want. If objects are selected first, the selection set is passed to the next editing command, but if no objects are selected first, the editing command prompts you to select objects. See Chapter 20 for a com-plete explanation of *PICKFIRST* and advanced selection set features.

SELECTION SETS PRACTICE

NOTE: While learning and practicing with the editing commands, it is suggested that *GRIPS* be turned off. This can be accomplished by typing in **GRIPS** and setting the *GRIPS* variable to a value of **0**. The AutoCAD default has *GRIPS* on (set to 1). *GRIPS* are covered in Chapter 23.

Begin a *New* drawing to complete the selection set practice exercises. If the *Startup* dialog box appears, select *Start from Scratch*, choose *Imperial* as the default setting, and click the *OK* button. The *Erase, Move,* and *Copy* commands can be activated by any one of the methods shown in the command tables that follow.

Using *Erase*

Erase is the simplest editing command. *Erase* removes objects from the drawing. The only action required is the selection of objects to be erased.

ERASE

Pull-down Menu	COMMAND (TYPE)	ALIAS (TYPE)	Short-cut	Screen (side) Menu	Tablet Menu
Modify *Erase*	*ERASE*	*E*	(Edit Mode) *Erase*	*MODIFY1* *Erase*	*V,14*

1. Draw several *Lines* and *Circles*. Practice using the object selection options with the *Erase* command. The following sequence uses the pickbox, *Window*, and *Crossing Window*.

STEPS	COMMAND PROMPT	PERFORM ACTION	COMMENTS
1.	Command:	type *E* and press **Spacebar**	*E* is the alias for *Erase*, Spacebar can be used like Enter
2.	ERASE Select objects:	use pickbox to select one or two objects	objects are highlighted
3.	Select objects:	type *W* or use a **Window**, then select more objects	objects are highlighted
4.	Select objects:	type *C* or use a **Crossing Window**, then select objects	objects are highlighted
5.	Select objects:	press **Enter**	objects are erased

2. Draw several more *Lines* and *Circles*. Practice using the *Erase* command with the *AUto Window* and *AUto Crossing Window* options as indicated below.

STEPS	COMMAND PROMPT	PERFORM ACTION	COMMENTS
1.	Command:	select the *Modify* pull-down, then *Erase*	
2.	ERASE Select objects:	use pickbox to **PICK** an open area, drag *Window* to the <u>right</u> to select objects	objects inside *Window* are highlighted
3.	Select objects:	**PICK** an open area, drag *Crossing Window* to the <u>left</u> to Select objects	objects inside and crossing through *Window* are highlighted
4.	Select objects:	press **Enter**	objects are erased

Using *Move*

The *Move* command specifically prompts you to (1) select objects, (2) specify a "base point," a point to move <u>from</u>, and (3) specify a "second point of displacement," a point to move <u>to</u>.

MOVE

Pull-down Menu	COMMAND (TYPE)	ALIAS (TYPE)	Short-cut	Screen (side) Menu	Tablet Menu
Modify *Move*	*MOVE*	*M*	(Edit Mode) *Move*	*MODIFY2* *Move*	*V,19*

1. Draw a *Circle* and two *Lines*. Use the *Move* command to practice selecting objects and to move one *Line* and the *Circle* as indicated in the following table.

STEPS	COMMAND PROMPT	PERFORM ACTION	COMMENTS
1.	Command:	type *M* and press **Spacebar**	*M* is the command alias for *Move*
2.	MOVE Select objects:	use pickbox to select one *Line* and the *Circle*	objects are highlighted
3.	Select objects:	press **Spacebar** or **Enter**	
4.	Specify base point or displacement:	**PICK** near the *Circle* center	base point is the handle, or where to move <u>from</u>
5.	Specify second point of displacement:	**PICK** near the other *Line*	second point is where to move <u>to</u>

2. Use *Move* again to move the *Circle* back to its original position. Select the *Circle* with the *Window* option.

STEPS	COMMAND PROMPT	PERFORM ACTION	COMMENTS
1.	Command:	select the *Modify* pull-down, then *Move*	
2.	MOVE Select objects:	type *W* and press **Spacebar**	select only the circle; object is highlighted
3.	Select objects:	press **Spacebar** or **Enter**	
4.	Specify base point or displacement:	**PICK** near the *Circle* center	base point is the handle, or where to move <u>from</u>
5.	Specify second point of displacement:	**PICK** near the original location	second point is where to move <u>to</u>

Using *Copy*

The *Copy* command is similar to *Move* because you are prompted to (1) select objects, (2) specify a "base point," a point to copy <u>from,</u> and (3) specify a "second point of displacement," a point to copy <u>to.</u>

COPY

Pull-down Menu	COMMAND (TYPE)	ALIAS (TYPE)	Short-cut	Screen (side) Menu	Tablet Menu
Modify *Copy*	*COPY*	*CO* or *CP*	**(Edit Mode)** *Copy Selection*	*MODIFY1* *Copy*	*V,15*

1. Using the *Circle* and 2 *Lines* from the previous exercise, use the *Copy* command to practice selecting objects and to make copies of the objects as indicated in the following table.

STEPS	COMMAND PROMPT	PERFORM ACTION	COMMENTS
1.	Command:	type *CO* and press **Enter** or **Spacebar**	*CO* is the command alias for *Copy*
2.	COPY Select objects:	type *F* and press **Enter**	*F* invokes the *Fence* selection option
3.	First fence point:	**PICK** a point near two *Lines*	starts the "fence"
4.	Specify endpoint of line or [Undo]:	**PICK** a second point across the two *Lines*	lines are highlighted
5.	Specify endpoint of line or [Undo]:	Press **Enter**	completes *Fence* option
6.	Select objects:	Press **Enter**	completes object selection
7.	Specify base point or displacement:	**PICK** between the *Lines*	base point is the handle, or where to copy <u>from</u>
8.	Specify second point of displacement:	enter **@2<45** and press **Enter**	copies of the lines are created 2 units in distance at 45 degrees from the original location

2.   Practice removing objects from the selection set by following the steps given in the table below.

STEPS	COMMAND PROMPT	PERFORM ACTION	COMMENTS
1.	Command:	type *CO* and press **Enter** or **Spacebar**	*CO* is the command alias for *Copy*
2.	COPY Select objects:	**PICK** in an open area near the right of your drawing, drag *Window* to the left to select all objects	all objects within and crossing the *Window* are highlighted
3.	Select objects:	hold down **Shift** and **PICK** all highlighted objects except one *Circle*	holding down Shift while PICKing removes objects from the selection set
4.	Select objects:	press **Enter**	only one circle is highlighted
5.	Specify base point or displacement or [Multiple]:	**PICK** near the circle's center	base point is the handle, or where to copy <u>from</u>
6.	Specify second point of displacement:	turn on *ORTHO*, move the cursor to the right, type **3** and press **Enter**	a copy of the circle is created 3 units to the right of the original circle

Noun/Verb Command Syntax Practice

1.   Practice using the *Move* command using Noun/Verb syntax order by following the steps in the table below.

STEPS	COMMAND PROMPT	PERFORM ACTION	COMMENTS
1.	Command:	**PICK** one *Circle*	circle becomes highlighted (grips may appear if not disabled)
2.	Command:	type *M* and press **Enter** or **Spacebar**	*M* is the command alias for *Move*
3.	MOVE 1 found Specify base point or displacement:	**PICK** near the *Circle's* center	command skips the "Select objects:" prompt and proceeds
4.	Specify second point of displacement:	enter **@-3,-3** and press **Enter**	circle is moved -3 X units and -3 Y units from the original location

2. Practice using Noun/Verb syntax order by selecting objects for *Erase*.

STEPS	COMMAND PROMPT	PERFORM ACTION	COMMENTS
1.	Command:	**PICK** one *Line*	line becomes highlighted (grips may appear if not disabled)
2.	Command:	type *E* and press *Enter* or **Spacebar**	*E* is the command alias for *Erase*
3.	ERASE 1 found		command skips the "Select objects:" prompt and erases highlighted line

CHAPTER EXERCISES

Open drawing **CH3EX1** that you created in Chapter 3 Exercises. Turn off *SNAP* (**F9**) to make object selection easier.

1. **Use the pickbox to select objects**

   Invoke the *Erase* command by any method. Select the lower-left square with the pickbox (Fig. 4-13, highlighted). Each *Line* must be selected individually. Press **Enter** to complete *Erase*. Then use the *Oops* command to unerase the square. (Type *Oops*.)

Figure 4-13 —————

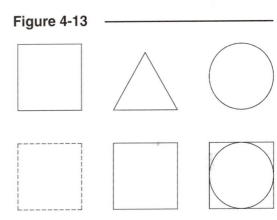

2. **Use the *AUto Window* and *AUto Crossing Window***

   Invoke *Erase*. Select the center square on the bottom row with the *AUto Window* and select the equilateral triangle with the *AUto Crossing Window*. Press **Enter** to complete the *Erase* as shown in Figure 4-14. Use *Oops* to bring back the objects.

Figure 4-14 —————

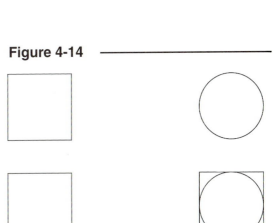

3. **Use the *Fence* selection option**

   Invoke *Erase* again. Use the **Fence** option to select all the vertical *Lines* and the *Circle* from the squares on the bottom row. Complete the *Erase* (see Fig. 4-15). Use *Oops* to unerase.

Figure 4-15

4. **Use the *ALL* option and deselect**

   Use *Erase*. Select all the objects with **ALL**. Remove the four *Lines* (shown highlighted in Fig. 4-16) from the selection set by pressing **Shift** while PICKing. Complete the *Erase* to leave only the four *Lines*. Finally, use *Oops*.

Figure 4-16

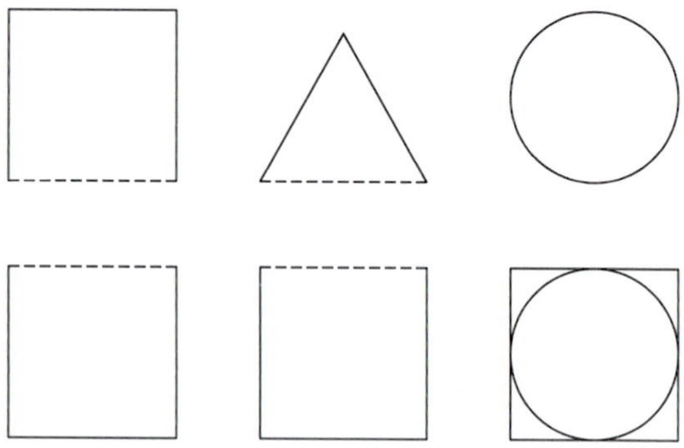

5. **Use Noun/Verb selection**

   Before invoking *Erase,* use the pickbox or *AUto Window* to select the triangle. (Make sure no other commands are in use.) Then invoke *Erase*. The triangle should disappear. Retrieve the triangle with *Oops*.

6. **Use *Move* with *Wpolygon***

   Invoke the *Move* command by any method. Use the **WP** option (*Window Polygon*) to select only the *Lines* comprising the triangle. Turn on *SNAP* and PICK the lower-left corner as the "Base point:". *Move* the triangle up 1 unit. (See Fig. 4-17.)

Figure 4-17

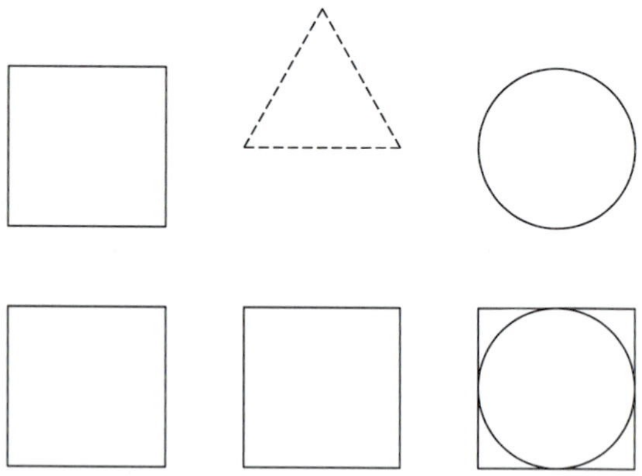

7. **Use *Previous* with *Move***

   Invoke *Move* again. At the "Select objects:" prompt, type **P** for *Previous*. The triangle should highlight. Using the same base point, move the triangle back to its original position.

8. *Exit* AutoCAD and do <u>not</u> save changes.

HELPFUL COMMANDS

Chapter Objectives

After completing this chapter you should be able to:

1. find *Help* for any command or system variable;

2. use the *Active Assistance* window to provide an active help window based on the *Settings* you prefer;

3. use *Oops* to unerase objects;

4. use *U* to undo one command or use *Undo* to undo multiple commands;

5. *Redo* commands that were undone;

6. regenerate the drawing with *Regen*.

CONCEPTS

There are several commands that do not draw or edit objects in AutoCAD, but are intended to assist you in using AutoCAD. These commands are used by experienced AutoCAD users and are particularly helpful to the beginner. The commands, as a group, are not located in any one menu, but are scattered throughout several menus.

HELP

Pull-down Menu	COMMAND (TYPE)	ALIAS (TYPE)	Short-cut	Screen (side) Menu	Tablet Menu
Help Help	HELP	?	F1	HELP Help	Y,7

Help gives you an explanation for any AutoCAD command or system variable as well as help for using the menus and toolbars. *Help* displays a window that gives a variety of methods for finding the information that you need. There is even help for using *Help*!

Help can be used two ways: (1) entered as a command at the open Command: prompt or (2) used transparently while a command is currently in use.

1. If the *Help* command is entered at an open Command: prompt (when no other commands are in use), the *Help* window appears (Fig. 5-1).

2. When *Help* is used transparently (when a command is in use), it is context sensitive; that is, help on the current command is given automatically. For example, Figure 5-2 displays the window that appears if *Help* is invoked during the *Line* command. (If typing a transparent command, an apostrophe (') symbol is typed as a prefix to the command, such as, '*HELP* or '?. If you PICK *Help* from the menus or press **F1**, it is automatically transparent.)

Figure 5-1

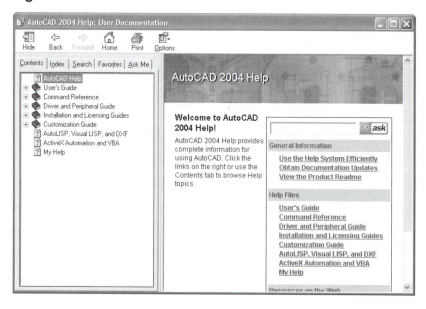

Figure 5-2

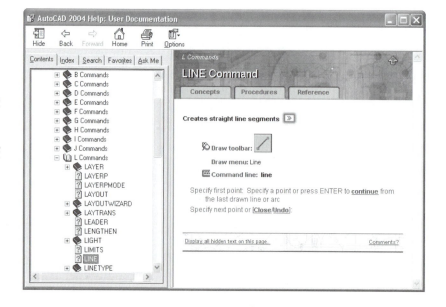

2002

Much of the text that appears in the window can be PICKed to reveal another level of help on that item. This feature, called hypertext, is activated by moving the pointer to a word (usually underlined) or an icon. When the pointer changes to a small hand, click on the item to activate the new information.

The five tabs in the AutoCAD 2004 Help window, *Contents*, *Index*, *Search*, *Favorites*, and *Ask Me*, are described next.

Contents

Several levels of the *Contents* tab are available, offering an overwhelming amount of information (Fig. 5-3). Each main level can be opened to reveal a second (chapter level) or third level of information. The main levels are:

AutoCAD Help	This section gives instructions and an overview of the other features listed below.

Figure 5-3 ————

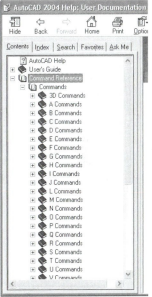

User's Guide	Ten "chapters" and a glossary are given, each with several sections explaining concepts and commands for AutoCAD 2004.

Command Reference	The following informative sections are available.
	    Commands
	    Command Aliases
	    System Variables
	    Command Modifiers
	    Dimension Variables
	    Utilities
	    Standard Libraries
	    AutoCAD Graphical Objects
	    Using *Unicode* fonts

Driver and Peripheral Guide	Get information on installing, configuring, and optimizing peripheral devices and drivers.
Installation and Licensing Guides	All information related to installation for stand-alone systems and for network deployment and network licensing is contained in this section.
Customization Guide	This section contains information on how to customize linetypes, shapes, fonts, menus, templates, and dialog boxes.
Visual Lisp, AutoLISP, and DXF	This area includes documentation for customizing AutoCAD using the AutoLISP and Visual LISP programming languages. The full reference guides and tutorials are provided. Full documentation on the AutoCAD DXF (Drawing Exchange) file format is included here.
ActiveX and VBA	Other support for these programming tools is given here with the *Developer's Guide*, *Reference Guide* and *Connectivity Automation Reference*.
My Help	You can customize help topics in this section. When you add the *My Help* topics to *AutoCAD Help*, any topics you add are available from the *Contents* tab, the *Index* tab (if you add index entries), and the *Search* tab.

Index

In this section, you can type two or three letters of a word that you want information about (Fig. 5-4). As you type, the list below displays the available contents beginning with the letters that you have entered. The word you type can be a command, system variable, term, or concept used in AutoCAD. Once the word you want information on is found, press Enter or click the *Display* button to display the related information on the right side of the window.

Figure 5-4

Search

Search operates like the *Index* function, except you can enter several words describing the topic you want help with (Fig. 5-5). There are three steps to using the *Search* function: (1) type in word or words in the top edit box, (2) select a topic from the list in the center to narrow your search, (3) select an option from below and click *Display*. The related information appears on the right.

Figure 5-5

Favorites

The *Favorites* section of the AutoCAD Help window (Fig. 5-6) is used to save sections of Help that you may want to refer to in the future. For example, tables or lists such as the list of command aliases or system variables make good candidates for the *Favorites* section. To save sections to the *Favorites* list: (1) use the *Contents*, *Index*, or *Search* tabs to locate the information you need, then (2) access the *Favorites* tab and press the *Add* button. To access the information in the *Topics* list, highlight the desired topic and press *Display*.

Figure 5-6

Ask Me

This tool (Fig. 5-7) allows you to access the same information that is available by the other methods, but you can ask for the information in sentence (query) format. In other words, type a question in the top edit box. You can specify which document you want AutoCAD to search in by making a selection from the *List of components to search*. Once the list of matching topics is found, select the desired topic from the list to display the related information.

Figure 5-7

ASSIST

Pull-down Menu	COMMAND (TYPE)	ALIAS (TYPE)	Short-cut	Screen (side) Menu	Tablet Menu
Help *Active Assistance*	*ASSIST*	...	...	...	...

An additional form of help is offered in AutoCAD called *Active Assistance*. Invoke *Active Assistance* by typing *Assist* or *'Assist* (for transparent use) or select it by using any of the methods shown in the command table above. *Active Assistance* is set to appear by default when you first install and start AutoCAD.

Active Assistance is an active, rather than passive, help screen on the current command (Fig. 5-8). In other words, *Active Assistance* can be set to appear automatically when you use a command instead of waiting for you to ask for help. *Active Assistance* displays a smaller, more comprehensive version of help than when the full *Help* screen is displayed (see Figures 5-1 and 5-2). When turned on, the *Active Assistance* window remains on the screen while you use the command and disappears when you finish the command.

Figure 5-8

Ask

The top section of the *Active Assistance* window is new for AutoCAD 2004. Entering a topic in the edit box and pressing the *Ask* button produces the full *Help* window (see Figures 5-1 and 5-2) with the specific topic displayed.

Figure 5-9

To set your preferences for *Active Assistance*, right-click inside the *Active Assistance* window or right-click on the *Active Assistance* icon in the system tray (in the lower-right corner of your screen). From the menu that appears, select *Settings...*. This action causes the *Active Assistance Settings* dialog box to appear (Fig. 5-9). *Active Assistance* can be made to appear when you start AutoCAD and remain on the screen until you

close it (*On Start*), appear with and disappear after each command (*All commands*), appear with only some commands (*New and enhanced commands* or *Dialogs only*), or made to appear only when you use the *Assist* command (*On demand*). For example, if you are an experienced AutoCAD user, you might want to set *Active Assistance* to appear only for *New and enhanced commands*, in which case the dialog box appears only for those features new with AutoCAD 2004. The *Hover Help* option in AutoCAD 2002 is no longer available in AutoCAD 2004.

OOPS

Pull-down Menu	COMMAND (TYPE)	ALIAS (TYPE)	Short-cut	Screen (side) Menu	Tablet Menu
...	OOPS	...	...	MODIFY1 *Erase* Oops:	...

The *Oops* command unerases whatever was erased with the <u>last</u> *Erase* command. *Oops* does not have to be used immediately after the *Erase*, but can be used at <u>any time after</u> the *Erase*. *Oops* is typically used after an accidental erase. However, *Erase* could be used intentionally to remove something from the screen temporarily to simplify some other action. For example, you can *Erase* a *Line* to simplify PICKing a group of other objects to *Move* or *Copy*, and then use *Oops* to restore the erased *Line*.

Oops can be used to restore the original set of objects after the *Block* or *Wblock* command is used to combine many objects into one object (explained in Chapter 21).

The *Oops* command is available only from the side menu; there is no icon button or option available in the pull-down menu. *Oops* appears on the side menu only <u>if</u> the *Erase* command is typed or selected. Otherwise, *Oops* must be typed.

U

Pull-down Menu	COMMAND (TYPE)	ALIAS (TYPE)	Short-cut	Screen (side) Menu	Tablet Menu
Edit *Undo*	U	...	*Crtl+Z* or (Default Menu) *Undo*	EDIT *Undo*	T,12

The *U* command undoes only the <u>last</u> command. *U* means "undo one command." If used after *Erase*, it unerases whatever was just erased. If used after *Line*, it undoes the group of lines drawn with the last *Line* command. Both *U* and *Undo* do not undo inquiry commands (like *Help*), the *Plot* command, or commands that cause a write-to-disk, such as *Save*.

If you type the letter **U**, select the icon button, or select **Undo** from the pull-down menu or screen (side) menu, only the last command is undone. Typing **Undo** invokes the full *Undo* command. The *Undo* command is explained next.

UNDO

Pull-down Menu	COMMAND (TYPE)	ALIAS (TYPE)	Short-cut	Screen (side) Menu	Tablet Menu
...	UNDO	...	...	ASSIST *Undo*	...

The full *Undo* command (as opposed to the *U* command) can be typed or can be selected using the *Undo* button drop-down list. This command has the same effect as the *U* command in that it undoes the previous command(s). However, the *Undo* command allows you to <u>undo multiple commands</u> in reverse chronological order.

New for AutoCAD 2004 is the drop-down list appearing with the *Undo* icon button. This feature allows you to select a range of commands to undo. Bring your pointer down to the last command you want to undo and select (Fig. 5-10).

Figure 5-10

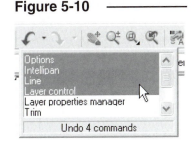

If you <u>type</u> the *Undo* command, the command prompt below appears. The default option of *Undo* is "<1>," which means to undo only the last command (the same action as using the *U* command). Do this by pressing Enter. All of the *Undo* options are listed next.

    Command: **undo**
    Enter the number of operations to undo or [Auto/Control/BEgin/End/Mark/Back]:

<number>
Enter a value for the number of commands to *Undo*. This is the default option.

Mark
This option sets a marker at that stage of the drawing. The marker is intended to be used by the *Back* option for future *Undo* commands.

Back
This option causes *Undo* to go back to the last marker encountered. Markers are created by the *Mark* option. If a marker is encountered, it is removed. If no marker is encountered, beware, because *Undo* goes back to the <u>beginning of the session</u>. A warning message appears in this case.

BEgin
This option sets the first designator for a group of commands to be treated as one *Undo*.

End
End sets the second designator for the end of a group.

Auto
If *On*, *Auto* treats each command as one group; for example, several lines drawn with one *Line* command would all be undone with *U*.

Control
This option allows you to disable the *Undo* command or limit it to one undo each time it is used.

REDO

	Pull-down Menu	COMMAND (TYPE)	ALIAS (TYPE)	Short-cut	Screen (side) Menu	Tablet Menu
	Edit *Redo*	*REDO*	...	*Crtl+Y* or (Default Menu) *Redo*	*EDIT* *Redo*	*U,12*

The *Redo* command undoes an *Undo*. *Redo* must be used as the <u>next</u> command after the *Undo*. The result of *Redo* is as if *Undo* was never used. *Redo* automatically reverses the action of the previous *Undo*, even when multiple commands were undone. Remember, *U* or *Undo* can be used at any time, but *Redo* has no effect unless used immediately after *U* or *Undo*.

MREDO

Pull-down Menu	COMMAND (TYPE)	ALIAS (TYPE)	Short-cut	Screen (side) Menu	Tablet Menu
...	MREDO	...	...	...	...

The *Redo* command reverses the total action of the last *Undo* command. In contrast, *Mredo* (multiple redo) allows you to select how many of the commands, in reverse chronological order, you want to redo. *Mredo* is useful only when the *Undo* command or drop-down list was just used (as the last command) and multiple commands were undone. Then you can immediately use *Mredo* to reverse the action of any number of the previous *Undos*. This new feature in AutoCAD 2004 is available (the icon is enabled) only if *Undo* was just used.

You can achieve a *Mredo* using the *Redo* drop-down list to select from the list of previously undone commands to reverse the action of the *Undo* (Fig. 5-11). Remember, the list includes only commands that were undone during the previous *Undo* command.

Figure 5-11

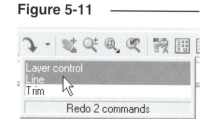

Typing *Mredo* yields the following prompt:

    Command: **mredo**
    Enter number of actions or [All/Last]:

Enter a value for the number of commands you want to redo. *All* will redo all of the undos (all the undos shown in the *Redo* drop-down list), the same as using the *Redo* command. The *Last* option generally has the same result as the *All* option; however, when the *U* command was used multiple times in succession, the *Last* option undoes only the last *U*.

REGEN

Pull-down Menu	COMMAND (TYPE)	ALIAS (TYPE)	Short-cut	Screen (side) Menu	Tablet Menu
View Regen	REGEN	RE	...	VIEW 1 Regen	J,1

The *Regen* command reads the database and redisplays the drawing accordingly. A *Regen* is caused by some commands automatically. Occasionally, the *Regen* command is required to update the drawing to display the latest changes made to some system variables. *Regenall* is used to regenerate all viewports when several viewports are being used.

2004

CHAPTER EXERCISES

Start AutoCAD. If the *Startup* dialog box appears, select **Start from Scratch**, choose **Imperial** as the default setting, and click the **OK** button. Complete the following exercises.

1. *Help*

    Use **Help** by any method to find information on the following commands. Use the *Contents* tab and select the *Command Reference* to locate information on each command. (Use *Help* at the open Command: prompt, not during a command in use.) Read the text screen for each command.

    > *New, Open, Save, SaveAs*
    > *Oops, Undo, U*

2. **Context-sensitive** *Help*

    Invoke each of the commands listed below. When you see the first prompt in each command, enter **'Help** or **'?** (transparently) or select **Help** from the menus or Standard toolbar. Read the explanation for each prompt. Select a hypertext item in each screen (underlined).

    > *Line, Arc, Circle, Point*

3. *Active Assistance*

    If the *Active Assistance* window is not currently visible on your screen, invoke it by using the *Active Assistance* icon button or by selecting it from the *Help* pull-down menu. Right-click inside the *Active Assistance* window to produce the shortcut menu and select the *All commands* option. Next, use the following commands and notice that the *Active Assistance* window appears and disappears with each command.

    > *Line, Circle, Move,* and *Copy*

    Next, access the *Active Assistance* window *Settings* dialog box again and select the *New and enhanced commands* option, then use the commands below (cancel after invoking each command by pressing Esc). The *Active Assistance* window should appear for only two of the following commands.

    > *Line, Mredo,* and *Color*

    Finally, select the *On demand* option in the *Settings* dialog box. Use a few more commands and notice that the *Active Assistance* window does not appear unless you invoke the *Active Assistance* command.

4. *Oops*

    Draw 3 vertical **Lines**. **Erase** one line; then use **Oops** to restore it. Next **Erase** two *Lines*, each with a separate use of the *Erase* command. Use **Oops**. Only the last *Line* is restored. **Erase** the remaining two *Lines*, but select both with a *Window*. Now use **Oops** to restore both *Lines* (since they were *Erased* at the same time).

5. **Delayed** *Oops*

    Oops can be used at any time, not only immediately after the *Erase*. Draw several horizontal **Lines** near the bottom of the screen. Draw a **Circle** on the **Lines**. Then **Erase** the *Circle*. Use **Move**, select the *Lines* with a *Window*, and displace the *Lines* to another location above. Now use **Oops** to make the *Circle* reappear.

6. *U*

    Press the letter *U* (make sure no other commands are in use). The *Circle* should disappear (*U* undoes the last command—*Oops*). Do this repeatedly to *Undo* one command at a time until the *Circle* and *Lines* are in their original position (when you first created them).

7. *Undo*

    Use the **Undo** command and select the **Back** option. Answer *Yes* to the warning message. This action should *Undo* everything.

    Draw a vertical **Line**. Next, draw a square with four **Line** segments (all drawn in the same *Line* command). Finally, draw a second vertical **Line**. **Erase** the first *Line*.

    Now type **Undo** and enter a value of **3**. You should have only one (the first) *Line* remaining. *Undo* reversed the following three commands:

Erase	The first vertical *Line* was unerased.
Line	The second vertical *Line* was removed.
Line	The four *Lines* comprising the square were removed.

8. *Multiple Undo*

    Draw two *Circles* and two more *Lines* so your drawing shows a total of two *Circles* and three *Lines*. Now use the *Multiple Undo* drop-down list. The list should contain *Line, Line, Circle, Circle, Line*. Highlight the first two lines and the two circles in the list, then left-click. Only one *Line* should remain in your drawing.

9. *Multiple Redo*

    Select the *Multiple Redo* drop-down list. The list should contain *Circle, Circle, Line, Line*. Highlight the first two *Circles* in the list, then left-click. The two *Circles* should reappear in your drawing, now showing a total of two *Circles* and one *Line*.

10. *Exit* AutoCAD and answer *No* to "Save Changes to…?"

BASIC DRAWING SETUP

Chapter Objectives

After completing this chapter you should:

1. know the basic steps for setting up a drawing;

2. know how to use the *Start from Scratch* option and the *Quick Setup* and *Advanced Setup* wizards that appear in the *Startup* and *Create New Drawing* dialog boxes;

3. be able to specify the desired *Units*, *Angles* format, and *Precision* for the drawing;

4. be able to specify the drawing *Limits*;

5. know how to specify the *Snap* increment;

6. know how to specify the *Grid* increment;

7. understand the basic function and use of a *Layout*.

STEPS FOR BASIC DRAWING SETUP

Assuming the general configuration (dimensions and proportions) of the geometry to be created is known, the following steps are suggested for setting up a drawing:

1. Determine and set the *Units* that are to be used.
2. Determine and set the drawing *Limits;* then *Zoom All.*
3. Set an appropriate *Snap* type and increment.
4. Set an appropriate *Grid* value to be used.

These additional steps for drawing setup are discussed also in Chapter 12, Advanced Drawing Setup.

5. Change the *LTSCALE* value based on the new *Limits.*
6. Create the desired *Layers* and assign appropriate *linetype* and *color* settings.
7. Create desired *Text Styles* (optional).
8. Create desired *Dimension Styles* (optional).
9. Activate a *Layout* tab, set it up for the plot or print device and paper size, and create a viewport (if not already existing).
10. Create or insert a title block and border in the layout.

When you start AutoCAD for the first time or use the *New* command to begin a new drawing, the *Startup, Create New Drawing,* or *Select Template* dialog box appears. These tools make available three options for setting up a new drawing. The three options represent three levels of automation/preparation for drawing setup. The *Start from Scratch* and *Use a Wizard* options are described in this chapter. *Use a Template* and creating template drawings are discussed in Chapter 12, Advanced Drawing Setup.

The *Start from Scratch* option requires you to step through each of the individual commands listed above to set up a drawing to your specifications. The *Use a Wizard* option provides two wizards, the *Quick Setup* and the *Advanced Setup* wizard. These two wizards lead you through the first two steps listed above.

STARTUP OPTIONS

When you start AutoCAD 2004 or use the *New* command, you are presented with one of three options, depending on how your system has been configured:

Figure 6-1

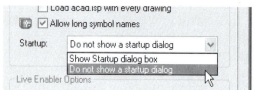

1. the *Startup* or *Create New Drawing* dialog box appears;
2. the *Select Template* dialog box appears; or
3. a new drawing starts with the *Qnew* template specified in the *Options* dialog box.

To control which of these actions occurs, make the following settings:

For method 1.   Select *Show Startup dialog box* in the *System* tab of the *Options* dialog box (Fig. 6-1).
For method 2.   Select *Do not show a startup dialog* in the *System* tab of the *Options* dialog box and ensure the *Drawing Template File Name* is set to "None" in the *Files* tab of the *Options* dialog box.
For method 3.   Select *Do not show a startup dialog* in the *System* tab of the *Options* dialog box and set the *Drawing Template File Name for Qnew* in the *Files* tab of the *Options* dialog box to the desired template file.

See "AutoCAD File Commands," "*New,*" and "*Qnew*" in Chapter 2 for more information on these settings.

DRAWING SETUP OPTIONS

There are three general options to set up a new drawing based on which of the three startup options you have configured. They are:

1.   *Start from Scratch*;
2.   *Use a Wizard*; and
3.   *Use a Template*.

All three methods, *Start From Scratch, Use a Wizard*, and *Use a Template*, are available in the *Startup* or *Create New Drawing* dialog boxes (Fig. 6-2). The *Select Template* dialog (Fig. 6-3) is essentially the same as the *Use a Template* option in the *Startup* or *Create New Drawing* dialog box.

Figure 6-2

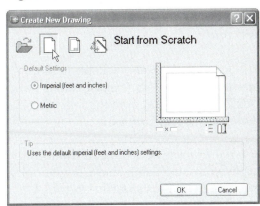

However, no matter which of the three methods you select, you can accomplish essentially the same setup to begin a drawing. Despite all the "bells and whistles" AutoCAD supplies, all of these startup options and drawing setup options generally start the drawing session with one of two setups, either the ACAD.DWT (inch) or the ACADISO.DWT (metric) drawing template. In fact, using *Start from Scratch* or selecting the defaults in either setup *Wizard* from any source, or selecting either of these templates by any method produces the same result—beginning with the ACAD.DWT (inch) or the ACADISO.DWT (metric) drawing template! The template used by the *Wizards* is determined by the *MEASUREINIT* system variable setting (0 for ACAD. DWT, 1 for ACADISO.DWT).

Start from Scratch

The *Start from Scratch* option is available in the *Startup* or *Create New Drawing* dialog box (see Figure 6-2). Use this option to begin a drawing with basic drawing settings, then determine your own system variable settings using the individual setup commands such as *Units, Limits, Snap* and *Grid*.

Imperial (feet and inches)
Use the *Imperial (feet and inches)* option if you want to begin with the traditional inch-based AutoCAD default drawing settings, such as *Limits* settings of 12 x 9. This option normally causes AutoCAD to use the ACAD.DWT template drawing. See the "Table of *Start from Scratch, Imperial* Settings (ACAD.DWT)." The same setup can be accomplished by any of the following methods.

Select the *Start from Scratch, Imperial* option in the *Startup* or *Create New Drawing* dialog box.
Select the ACAD.DWT by any *Template* method.
Configure your system for *Do not show a startup dialog*, then set the *Default Template File Name for Qnew* in the *Files* tab of the *Options* dialog box to ACAD.DWT.
Set the *MEASUREINIT* system variable to 0, then cancel the *Startup* dialog box that appears when AutoCAD starts.

Metric
Use the *Metric* option for setting up a drawing for use with metric units. This option causes AutoCAD to use the ACADISO.DWT template drawing. The drawing has *Limits* settings of 420 x 279, equal to a metric A3 sheet measured in mm. See the "Table of *Start from Scratch, Metric* Settings (ACADISO.DWT)." The same setup can be accomplished by any of the following methods.

Select the *Start from Scratch, Metric* option in the *Startup* or *Create New Drawing* dialog box.
Select the ACADISO.DWT by any *Template* method.

Configure your system for *Do not show a startup dialog*, then set the *Default Template File Name for Qnew* in the *Files* tab of the *Options* dialog box to ACADISO.DWT.

Set the *MEASUREINIT* system variable to 1, then cancel the *Startup* dialog box that appears when AutoCAD starts.

Template

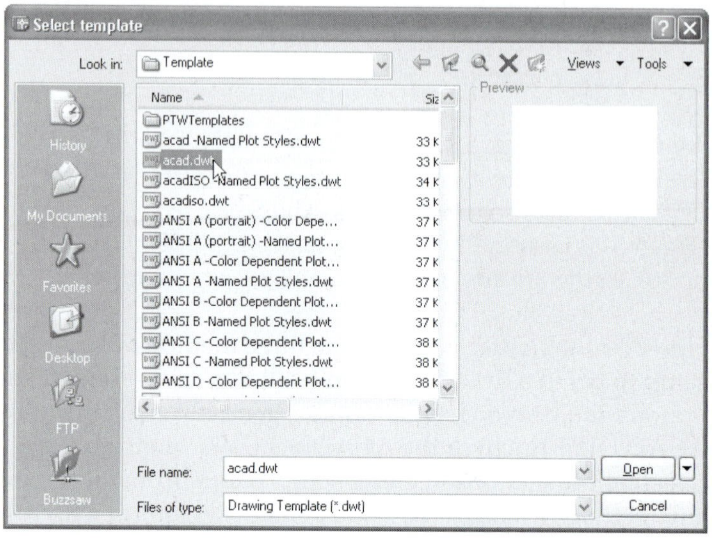

The *Template* option is available in the *Startup* or *Create New Drawing* dialog box (see Figure 6-2) and from the *Select Template* dialog box (see Figure 6-3).

Use the *Template* option if you want to begin a drawing using an existing template drawing (.DWT) as a starting point. A template drawing can have many of the drawing setup steps performed but contains no geometry. Several templates are provided by AutoCAD, including the default inch (*Imperial*) template ACAD.DWT and the default metric template ACADISO.DWT. See the "Table of *Start from Scratch*, *Imperial* Settings (ACAD.DWT)" and the "Table of *Start from Scratch*, *Metric* Settings (ACADISO.DWT)." More information on this option, including creating template drawings and using templates provided by AutoCAD, is given in Chapter 12, Advanced Drawing Setup.

Figure 6-3

The following two tables list the AutoCAD settings for the two template drawings, ACAD.DWT and ACADISO.DWT. Many of these settings, such as *Units*, *Limits*, *Snap*, and *Grid*, are explained in the following sections.

Table of *Start from Scratch Imperial* Settings (ACAD.DWT)

Related Command	Description	System Variable	Default Setting
Units	linear units	*LUNITS*	2 (decimal)
Limits	drawing area	*LIMMAX*	12.0000,9.0000
Snap	snap increment	*SNAPUNIT*	.5000, .5000
Grid	grid increment	*GRIDUNIT*	.5000, .5000
LTSCALE	linetype scale	*LTSCALE*	1.0000
DIMSCALE	dimension scale	*DIMSCALE*	1.0000
Dtext, Mtext	text height	*TEXTSIZE*	.2000
Hatch	hatch pattern scale	*HPSCALE*	1.0000

Table of *Start from Scratch* <u>*Metric*</u> **Settings (ACADISO.DWT)**

Related Command	Description	System Variable	Default Setting
Units	linear units	*LUNITS*	2 (decimal)
Limits	drawing area	*LIMMAX*	420.0000, 297.0000
Snap	snap increment	*SNAPUNIT*	10.0000, 10.0000
Grid	grid increment	*GRIDUNIT*	10.0000, 10.0000
LTSCALE	linetype scale	*LTSCALE*	1.0000
DIMSCALE	dimension scale	*DIMSCALE*	1.0000
Dtext, Mtext	text height	*TEXTSIZE*	2.5000
Hatch	hatch pattern scale	*HPSCALE*	1.0000

The metric drawing setup is intended to be used with ISO linetypes and ISO hatch patterns, which are pre-scaled for these *Limits,* hence the *LTSCALE* and hatch pattern scale of 1. The individual dimensioning variables for arrow size, dimension text size, gaps, and extensions, etc. are changed so the dimensions are drawn correctly with a *DIMSCALE* of 1 with the ISO-25 dimension style.

Wizard

Selecting the *Use a Wizard* option in the *Startup* or *Create New Drawing* dialog box (see Figure 6-2) gives a choice of using the *Quick Setup* or *Advanced Setup* wizard. The *Advanced Setup* wizard is an expanded version of the *Quick Setup* wizard.

The wizards use default settings based on either the ACAD.DWT or the ACADISO.DWT template. The template used by the wizards is determined by the *MEASUREINIT* system variable setting for your system (0 = ACAD.DWT, English template, and 1 = ACADISO.DWT, metric template). To change the setting, type *MEASUREINIT* at the Command prompt.

Quick Setup **Wizard**

The *Quick Setup* wizard automates only the first two steps listed under "Steps for Basic Drawing Setup" on the chapter's first page. Those functions, simply stated, are:

1. *Units*
2. *Limits*

Choosing the *Quick Setup* wizard invokes the *QuickSetup* dialog box (Fig. 6-4). There are two steps, *Units* and *Area*.

Figure 6-4

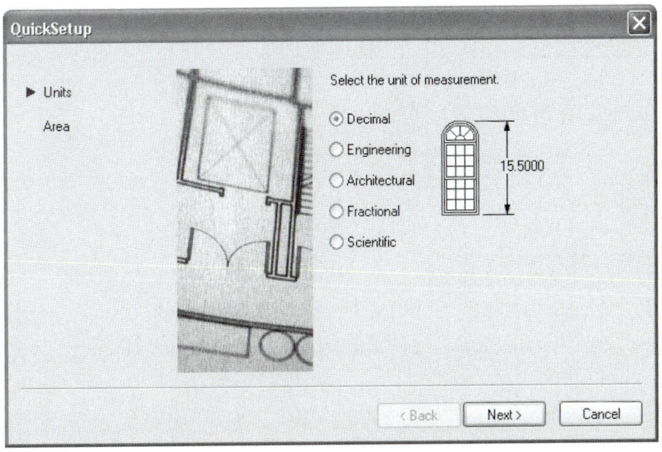

Units

Press the desired radio button to display the units you want to use for the drawing. The options are:

Decimal	Use generic decimal units with a precision 0.0000.
Engineering	Use feet and decimal inches with a precision of 0.0000.
Architectural	Use feet and fractional inches with a precision of 1/16 inch.
Fractional	Use generic fractional units with a precision of 1/16 units.
Scientific	Use generic decimal units showing a precision of 0.0000.

Use *Architectural* or *Engineering* units if you want to specify coordinate input using feet values with the apostrophe (') symbol. If you want to set additional parameters for units such as precision or system of angular measurement, use the *Units* or *-Units* command. Keep in mind the setting you select in this step changes only the display of units in the coordinate display area of the Status Bar (*Coords*) and in some dialog boxes, but not necessarily for the dimension text format. Select the *Next* button after specifying *Units*.

Area

Enter two values that constitute the X and Y measurements of the area you want to work in (Fig. 6-5). These values set the *Limits* of the drawing. The first edit box labeled *Width* specifies the X value for *Limits*. The X value is usually the longer of the two measurements for your drawing area and represents the distance across the screen in the X direction or along the long axis of a sheet of paper. The second edit box labeled *Length* specifies the Y value for *Limits* (the Y distance of the drawing area). (Beware: The terms "Width" and "Length" are misleading since "length" is defined as the measurement of something along its greatest dimension. When setting AutoCAD *Limits*, the Y measurement is generally the shorter of the two measurements.) The two values together specify the upper right corner of the drawing *Limits*. When you finish the two steps, you should use *Zoom All* to display the entire *Limits* area in the screen.

Figure 6-5 ———————————

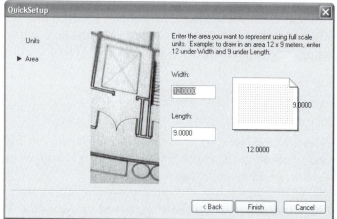

Advanced Setup Wizard

The *Advanced Setup* wizard performs the same tasks as the *Quick Setup* wizard with the addition of allowing you to select units precision and other units options (normally available in the *Units* dialog box). Selecting *Advanced Setup* produces a series of dialog boxes (not all shown). There are five steps involved in the series.

The following list indicates the "steps" in the *Advanced Setup* wizard and the related "Steps for Basic Drawing Setup" on the chapter's first page.

1.	*Units* and *Precision*	*Units* command
2.	*Angle*	*Units* command
3.	*Angle Measure*	*Units* command
4.	*Angle Direction*	*Units* command
5.	*Area*	*Limits* command

Units

You can select the units of measurement for the drawing as well as the unit's *Precision* in this first step (Fig. 6-6). These are the same options available in the *Drawing Units* dialog box (see "*Units*," this chapter). This is similar to the first step in the *Quick Setup* wizard but with the addition of *Precision*.

Similar to using the *Units* or *-Units* command, your choices in this and the next three dialog boxes determine the display of units for the coordinate display area of the Status Bar (*Coords*) and in dialog boxes. If you want to use feet units for coordinate input, select *Architectural* or *Engineering*.

Figure 6-6

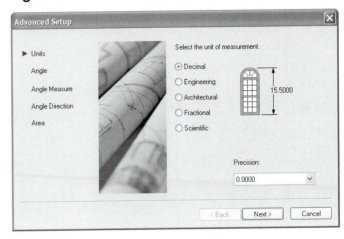

Angle

This step (not shown) provides for your input of the desired system of angular measurement. Select the drop-down list to select the angular *Precision*.

Angle Measure

This step sets the direction for angle 0. East (X positive) is the AutoCAD default. This has the same function as selecting the *Angle 0 Direction* in the *Drawing Units* dialog box.

Angle Direction

Select *Clockwise* if you want to change the AutoCAD default setting for <u>measuring</u> angles. This setting (identical to the *Drawing Units* dialog box option) affects the direction of positive and negative angles in commands such as *Rotate*, *Array Polar* and dimension commands that measure angles, but does not affect the direction *Arcs* are drawn (always counterclockwise).

Area

Enter values to define the upper right corner for the *Limits* of the drawing. The *Width* refers to the X *Limits* component and the *Length* refers to the Y component. Generally, the *Width* edit box contains the larger of the two values unless you want to set up a vertically oriented drawing area. If you plan to print or plot to a standard scale, your input for *Area* should be based on the intended plot scale and sheet size. See the "Tables of *Limits* Settings" in Chapter 14 for appropriate values to use.

SETUP COMMANDS

If you want to set up a drawing using individual commands instead of the *Quick Setup* or *Advanced Setup* wizard, use the commands given in this section. The *Quick Setup* and *Advanced Setup* wizards use only the first two commands discussed in this section, *Units* and *Limits*.

UNITS

Pull-down Menu	COMMAND (TYPE)	ALIAS (TYPE)	Short-cut	Screen (side) Menu	Tablet Menu
Format Units...	*UNITS* or *-UNITS*	*UN* or *-UN*	...	*FORMAT Units*	*V,4*

The *Units* command allows you to specify the type and precision of linear and angular units as well as the direction and orientation of angles to be used in the drawing. The current setting of *Units* determines the display of values by the coordinates display (*Coords*) and in some dialog boxes.

You can select *Units ...* from the *Format* pull-down or type *Units* (or command alias *UN*) to invoke the *Drawing Units* dialog box (Fig. 6-7). Type *-Units* (or *-UN*) to produce a text screen (Fig. 6-8).

Figure 6-7

The linear and angular units options are displayed in the dialog box format (Fig. 6-7) and in Command line format (Fig. 6-8). The choices for both linear and angular *Units* are shown in the figures.

Figure 6-8

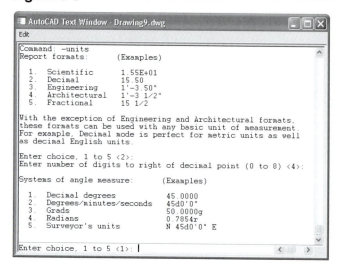

Units	Format	
1. Scientific	1.55E + 01	Generic decimal units with an exponent
2. Decimal	15.50	Generic decimal usually used for applications in metric or decimal inches
3. Engineering	1'-3.50"	Explicit feet and decimal inches with notation, one unit equals one inch
4. Architectural	1'-3 1/2"	Explicit feet and fractional inches with notation, one unit equals one inch
5. Fractional	15 1/2	Generic fractional units

Precision

When setting *Units,* you should also set the precision. *Precision* is the number of places to the right of the decimal or the denominator of the smallest fraction to display. The precision is set by making the desired selection from the *Precision* pop-up list in the *Drawing Units* dialog box (Fig. 6-7) or by keying in the desired selection in Command line format.

Precision controls only the <u>display</u> of *COORDS*. The <u>actual</u> <u>precision</u> of the drawing database is always the same in AutoCAD, that is, 14 significant digits.

Angles

You can specify a format other than the default (decimal degrees) for expression of angles. Format options for angular display and examples of each are shown in Figure 6-7 (dialog box format).

The orientation of angle 0 can be changed from the default position (east or X positive) to other options by selecting the *Direction* tile in the *Drawing Units* dialog box. This produces the *Direction Control* dialog box (Fig. 6-9). Alternately, the *Units* command can be typed to select these options in Command line format.

Figure 6-9

The direction of angular measurement can be changed from its default of counterclockwise to clockwise. The direction of angular measurement affects the direction of positive and negative angles in commands such as *Array Polar, Rotate* and dimension commands that <u>measure</u> angular values but does not change the direction *Arcs* are <u>drawn</u>, which is always counterclockwise.

Drawing Units for DesignCenter Blocks

This section in the *Drawing Units* dialog box specifies the units to use when you drag and drop *Blocks* from AutoCAD DesignCenter into the current drawing. This choice does not affect insertion of *Blocks* using the *Insert* command. There are many choices including *Unitless, Inches, Feet, Millimeters*, and so on. If you intend to insert *Blocks* into the drawing (you are currently setting units for) using drag-and-drop, select a unit from the list that matches the units of the *Blocks* to insert. If you are not sure, select *Unitless*. See Chapter 21, Blocks and DesignCenter, for more information on this subject.

Keyboard Input of *Units* Values

When AutoCAD prompts for a point or a distance, you can respond by entering values at the keyboard. The values can be in <u>any format</u>—integer, decimal, fractional, or scientific—regardless of the format of *Units* selected.

You can <u>type in explicit feet and inch values only if *Architectural* or *Engineering*</u> units have been specified as the drawing units. For this reason, specifying *Units* is the first step in setting up a drawing.

Type in explicit feet or inch values by using the apostrophe (') symbol after values representing feet and the quote (") symbol after values representing inches. If no symbol is used, the values are understood by AutoCAD to be <u>inches</u>.

Feet and inches input <u>cannot</u> contain a blank, so a hyphen (-) must be typed between inches and fractions. For example, with *Architectural* units, key in **6'2-1/2"**, which reads "six feet two and one-half inches." The standard engineering and architectural format for dimensioning, however, places the hyphen between feet and inches (as displayed by the default setting for the *Coords* display).

The *UNITMODE* variable set to **1** changes the display of *Coords* to remind you of the correct format for <u>input</u> of feet and inches (with the hyphen between inches and fractions) rather than displaying the standard format for feet and inch notation (standard format, *UNITMODE* of **0**, is the default setting). If options other than *Architectural* or *Engineering* are used, values are read as generic units.

LIMITS

Pull-down Menu	COMMAND (TYPE)	ALIAS (TYPE)	Short-cut	Screen (side) Menu	Tablet Menu
Format Drawing Limits	*LIMITS*	...	...	*FORMAT Limits*	V,2

The *Limits* command allows you to set the size of the drawing area by specifying the lower-left and upper-right corners in X,Y coordinate values.

Command: `limits`
Reset Model space limits
Specify lower left corner or [ON/OFF] <0.0000,0.0000>: `X,Y` or `Enter` (Enter an X,Y value or accept the 0,0 default—normally use 0,0 as lower-left corner.)
Specify upper right corner <12.0000,9.0000>: `X,Y` (Enter new values to change upper-right corner to allow adequate drawing area.)

The default *Limits* values in the ACAD.DWT are 12 and 9; that is, 12 units in the X direction and 9 units in the Y direction (Fig. 6-10). Starting a drawing by the following methods results in *Limits* of 12 x 9:

> Selecting the ACAD.DWT template drawing
> Selecting the *Imperial* defaults in the *Start from Scratch* option

If the *GRID* is turned on, the dots are displayed only over the *Limits*. The AutoCAD screen (default configuration) displays additional area on the right past the *Limits*. The units are generic decimal units that can be used to represent inches, feet, millimeters, miles, or whatever is appropriate for the intended drawing. Typically, however, decimal units are used to represent inches or millimeters. If the default units are used to represent inches, the default drawing size would be 12 x 9 inches.

Figure 6-10

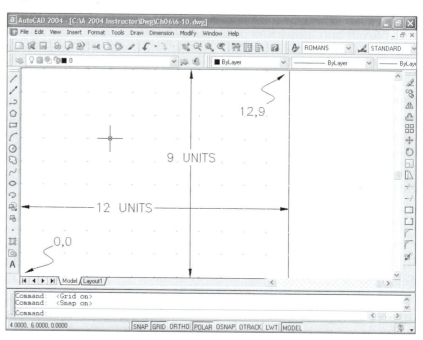

Remember that when a CAD system is used to create a drawing, the geometry should be drawn <u>full size</u> by specifying dimensions of objects in <u>real-world units</u>. A completed CAD drawing or model is virtually an exact dimensional replica of the actual object. Scaling of the drawing occurs only when plotting or printing the file to an actual fixed-size sheet of paper.

 Before beginning to create an AutoCAD drawing, determine the size of the drawing area needed for the intended geometry. After setting *Units,* appropriate *Limits* should be set in order to draw the object or geometry to the <u>real-world size in the actual units</u>. There are no practical maximum or minimum settings for *Limits.*

The X,Y values you enter as *Limits* are understood by AutoCAD as values in the units specified by the *Units* command. For example, if you previously specified *Architectural units,* then the values entered are understood as inches unless the notation for feet (') is given (**240,180** or **20',15'** would define the same coordinate). Remember, you can type in explicit feet and inch values only if *Architectural* or *Engineering* units have been specified as the drawing units.

 If you are planning to plot the drawing to scale, *Limits* should be set to a proportion of the <u>sheet size</u> you plan to plot on. For example, setting limits to 22 x 17 (2 times 11 by 8.5) would allow enough room for drawing an object about 20" x 15" and allow plotting at 1/2 size on the 11" x 8.5" sheet. Simply stated, <u>set *Limits* to a proportion of the paper</u>.

Limits also defines the display area for *GRID* as well as the minimum area displayed when a *Zoom All* is used. *Zoom All* forces the full display of the *Limits*. *Zoom All* can be invoked by typing **Z** (command alias) then *A* for the *All* option.

Changing *Limits* does <u>not</u> automatically change the display. As a general rule, you should make a habit of invoking a *Zoom All* <u>immediately following</u> a change in *Limits* to display the area defined by the new limits (Fig. 6-11).

When you <u>reduce</u> *Limits* while *Grid* is *ON*, it is apparent that a change in *Limits* does not automatically change the display. In this case, the area covered by the grid is reduced in size as *Limits* are reduced, yet the display remains unchanged.

Figure 6-11

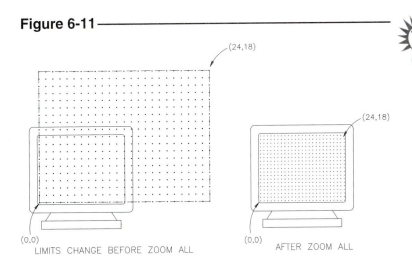

If you are already experimenting with drawing in different *Linetypes,* a change in *Limits* and *Zoom All* affects the display of the hidden and dashed lines. The *LTSCALE* variable controls the spacing of non-continuous lines. The *LTSCALE* is often changed proportionally with changes in *Limits*.

ON/OFF

If the *ON* option of *Limits* is used, limits checking is activated. Limits checking prevents you from drawing objects outside of the limits by issuing an outside-limits error. This is similar to drawing "off the paper." Limits checking is *OFF* by default.

SNAP

Pull-down Menu	COMMAND (TYPE)	ALIAS (TYPE)	Short-cut	Screen (side) Menu	Tablet Menu
Tools *Drafting Settings...* *Snap and Grid*	*SNAP*	SN	F9 or Ctrl+B	*TOOLS 2* *Grid* *Snap and Grid*	W,10

Snap in AutoCAD has two possible types, <u>*Grid Snap*</u> and <u>*Polar Snap*</u>. When you are setting up the drawing, you should set the desired *Snap type* and increment. You may decide to use both types of Snap, so set both increments initially. You can have only <u>one of the two *Snap types*</u> active at one time. (See Chapter 3, Draw Command Concepts, for further explanation and practice using both Snap types.)

<u>*Grid Snap*</u> forces the cursor to preset positions on the screen, similar to the *Grid*. *Grid Snap* is like an invisible grid that the cursor "snaps" to. This function can be of assistance if you are drawing *Lines* and other objects to set positions on the drawing, such as to every .5 unit. The default value for *Grid Snap* is .5, but it can be changed to any value.

<u>*Polar Snap*</u> forces the cursor to move in set intervals from the previously designated point (such as the "first point" selected during the *Line* command) to the next point. *Polar Snap* operates for cursor movement at any previously set *Polar Tracking* angle, whereas *Grid Snap* (since it is rectangular) forces regular intervals only in horizontal or vertical movements. *Polar Tracking* must also be on (*POLAR* or F10) for *Polar Snap* to operate.

Set the desired snap type and increments using the *Snap* command or the *Drafting Settings* dialog box. In the *Drafting Settings* dialog box, select the *Snap and Grid* tab (Fig. 6-12).

To set the *Polar Snap* increment, first select *Polar Snap* in the *Snap Type & Style* section (lower right). This action causes the *Polar Spacing* section to be enabled. Next enter the desired *Polar Distance* in the edit box.

To set the *Grid Snap* increment, first select *Grid Snap* in the *Snap Type & Style* section. This action causes the *Snap* section to be enabled. Next enter the desired *Snap X Spacing* value in the edit box. If you want a non-square snap grid, enter a different value in the *Snap Y Spacing* edit box. You can also rotate the snap grid by entering a value other than 0 in the *Angle* edit box. See the *Rotate* option of *Snap* (Command line version below).

Figure 6-12

The Command line format of *Snap* is as follows:

    Command: **snap**
    Specify snap spacing or [ON/OFF/Aspect/Rotate/Style/Type] <0.5000>:

Value
Entering a value at the Command line prompt sets the *Grid Snap* increment only. You must use the *Drafting Settings* dialog box to set the *Polar Snap* distance.

ON/OFF
Selecting *ON* or *OFF* accomplishes the same action as toggling the F9 key or selecting the word *SNAP* on the Status Bar.

Aspect
The *Aspect* option allows specification of unequal X and Y spacing for the snap grid. This is identical to entering unequal *Snap X Spacing* and *Snap Y Spacing* in the dialog box.

Rotate
The *Grid Snap* can be rotated about any point and set to any angle. When Snap has been rotated, the GRID, ORTHO, and the "crosshairs" automatically follow this alignment. This action facilitates creating objects oriented at the specified angle, for example, creating an auxiliary view of drawing part or a floor plan at an angle. To do this, use the *Rotate* option in Command line format or set the *Angle*, *X Base*, and *Y Base* (point to rotate about) in the *Drafting Settings* dialog box (see Chapter 27, Auxiliary Views).

Style
The *Style* option allows switching between a *Standard* snap pattern (square or rectangular) and an *Isometric* snap pattern. If using the dialog box, toggle *Isometric* snap. When *Snap Style* or *Rotate* (*Angle*) is changed, the GRID automatically aligns with it (see Chapter 25, Pictorial Drawings).

Type
This option switches between *Grid Snap* and *Polar Snap*. Remember, you can have only one of the two snap types active at one time.

Once your snap type and increment(s) are set, you can begin drawing using either snap type or no snap at all. While drawing, you can right-click on the word *SNAP* on the Status Bar to invoke the shortcut menu shown in Figure 6-13. Here you can toggle between *Grid Snap* and *Polar Snap* or turn both off (*POLAR* must also be on to use *Polar Snap*). Left-clicking on the word *SNAP* or pressing F9 toggles on or off whichever snap type is current.

Figure 6-13

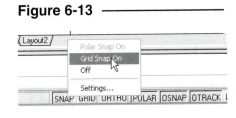

The process of setting the *Snap Type, Grid Snap Spacing, Polar Snap Spacing,* and *Polar Tracking Angle* is usually done during the initial stages of drawing setup, although these settings can be changed at any time. The *Grid Snap Spacing* value is stored in the *SNAPUNIT* system variable and saved in the <u>drawing file</u>. The other settings (*Snap Type, Polar Snap Spacing,* and *Polar Tracking Angle*) are saved in the <u>system registry</u> (as the *SNAPTYPE, POLARDIST,* and *POLARANG* system variables) so that the settings remain in affect for any drawing until changed.

GRID

Pull-down Menu	COMMAND (TYPE)	ALIAS (TYPE)	Short-cut	Screen (side) Menu	Tablet Menu
Tools *Drafting Settings...* *Snap and Grid*	GRID	...	*F7 or* *Ctrl+G*	*TOOLS 2* *Grid* *Snap and Grid*	*W,10*

GRID is visible on the screen, whereas *Grid Snap* is invisible. *GRID* is only a <u>visible</u> display of some regular interval. *GRID* and *Grid Snap* can be <u>independent</u> of each other. In other words, each can have separate spacing settings and the active state of each (*ON, OFF*) can be controlled independently. The *GRID* <u>follows</u> the *SNAP* if *SNAP* is rotated or changed to *Isometric Style*. Although the *GRID* spacing can be different than that of *SNAP,* it can also be forced to follow *SNAP* by using the *Snap* option. The default *GRID* setting is **0.5**.

The *GRID* <u>cannot</u> be plotted. It is <u>not</u> comprised of *Point* objects and therefore is not part of the current drawing. *GRID* is only a visual aid.

Grid can be accessed by Command line format (shown below) or set via the *Drafting Settings* dialog box (Fig. 6-12). The dialog box is invoked by menu selection or by typing *Dsettings* or *DS*. The dialog box allows only *X Spacing* and *Y Spacing* input for *Grid*.

    Command: **grid**
    Specify grid spacing(X) or [ON/OFF/Snap/Aspect] <0.5000>:

Grid Spacing (X)
If you supply a value for the *Grid spacing, GRID* is displayed at that spacing regardless of *SNAP* spacing. If you key in an *X* as a suffix to the value (for example, **2X**), the *GRID* is displayed as that value <u>times</u> the *SNAP* spacing (for example, "2 times" *SNAP*).

ON/OFF
The *ON* and *OFF* options simply make the *GRID* visible or not (like toggling the **F7** key, pressing **Ctrl+G**, or clicking **GRID** on the Status Bar).

Snap
The *Snap* option of the *Grid* command forces the *GRID* spacing to equal that of *SNAP,* even if *SNAP* is subsequently changed.

Aspect

The *Aspect* option of *GRID* allows different X and Y spacing (causing a rectangular rather than a square *GRID*).

DSETTINGS

Pull-down Menu	COMMAND (TYPE)	ALIAS (TYPE)	Short-cut	Screen (side) Menu	Tablet Menu
Tools *Drafting Settings...*	DSETTINGS	DS	*Status Bar* (right-click) *Settings...*	TOOLS 2 *Osnap...* *(Grid or Polar)*	*W,10*

You can access controls to *SNAP* and *GRID* features using the *Dsettings* command. *Dsettings* produces the *Drafting Settings* dialog box described earlier (see Figure 6-12). The three tabs in the dialog box are *Snap and Grid*, *Polar Tracking* and *Object Snap*. Use the *Snap and Grid* tab to control settings for *SNAP* and *GRID* as previously described. The *Polar Tracking* and *Object Snap* tabs are explained in Chapter 7.

Using *Snap* and *Grid*

Using *Snap* and *Grid* for drawing is a personal preference and should be used whenever appropriate for the drawing. Using *Snap* and *Grid* can be beneficial for some drawings, but may not be useful for others.

Generally, *Snap* and *Grid* can be useful in cases where many of the lines to be drawn or other measurements used are at some regular interval, such as 1 mm or 1/2". Typically, small mechanical drawings and some simple architectural drawings may fall into this category. On the other hand, if you anticipate that few of the measurements in the drawing will be at regular interval lengths, *Snap* and *Grid* may be of little value. Drawings such as civil engineering drawings involving site plans would fall into this category.

Also, *Grid* and *Snap* are useful only in cases where the interval is set to a large enough value relative to the screen size (or *Limits*) to facilitate seeing and PICKing points easily. For example, with *Limits* of 12 x 9 it would be relatively easy to see a *Grid* set to .5 and to PICK points at *Snap* intervals of .125. However, in cases where the *Limits* cover a large area and the desired increments are relatively small, *Snap* may not be of much usefulness. For example, an architectural drawing may have *Limits* that represent hundreds of feet, and although all measurements are to be drawn at 1/4" intervals, it would be almost impossible to interactively PICK points at such a small increment relative to the overall drawing size.

INTRODUCTION TO LAYOUTS AND PRINTING

The last several steps listed in the "Steps for Basic Drawing Setup" on this chapter's first page are also listed below.

5. Change the *LTSCALE* value based on the drawing *Limits*.
6. Create the desired *Layers* and assign appropriate *linetype* and *color* settings.
7. Create desired *Text Styles*.
8. Create desired *Dimension Styles*.
9. Activate a *Layout* tab, set it up for the plot or print device and paper size, and create a viewport (if not already created).
10. Create or insert a title block and border in the layout.

Part of the process of setting the *LTSCALE*, creating *Text*, and creating *Dimension Styles*, (steps 5, 7, and 8) is related to the size of the drawing. Steps 9 and 10 prepare the drawing for making a print or plot.

Although steps 9 and 10 are often performed after the drawing is complete and just before making a print or plot, it is wise to consider these steps in the drawing setup process. Because you want hidden line dashes, text, dimensions, hatch patterns, etc. to have the correct size in the finished print or plot, it would be sensible to consider the paper size of the print or plot and the drawing scale before you create text, dimensions, and hatch patterns in the drawing. In this way, you can more accurately set the necessary sizes and system variables before you draw, or as you draw, instead of changing multiple settings upon completion of the drawing.

To put it simply, to determine the size for linetypes (*LTSCALE*), text, and dimensions, use the proportion of the drawing to the paper size. In other words, if the size of the drawing area (*Limits*) is 22 x 17 and the paper size you will print on is 11 x 8.5, then the drawing is 2 times the size of the paper. Therefore, set the *LTSCALE* to 2 and create the text and dimensions twice as large as you want them to appear on the print.

This chapter gives an introduction to layouts and creating paper space viewports. Advanced features of drawing setup, layouts, viewports, and plotting are discussed in Chapter 12, Advanced Drawing Setup, Chapter 13, Layouts and Viewports, and Chapter 14, Printing and Plotting.

Model Tab and *Layout* Tabs

At the bottom of the drawing area you should see one *Model* tab and two *Layout* tabs. If the tabs have been turned off on your system, see "Setting Layout Options and Plotting Options" later in this section.

When you start AutoCAD and begin a drawing, the *Model* tab is active by default. Objects that represent the subject of the drawing (model geometry) are normally drawn in the *Model* tab, also known as "model space." Traditionally, dimensions and notes (text) are also created in model space.

A layout is activated by selecting the *Layout1*, *Layout2*, or other layout tab (Fig. 6-14). A layout represents the sheet of paper that you intend to print or plot on (sometimes called paper space). The dashed line around the "paper sheet" in Figure 6-14 represents the maximum printable area for the configured printer or plotter.

Figure 6-14

paper space   viewport

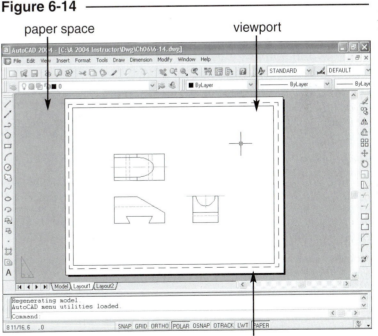

printable area

You can have multiple layouts (*Layout1*, *Layout2*, etc.), each representing a different sheet size and/or print or plot device. For example, you may have one layout to print with a laser printer on a 8.5 x 11 inch sheet and another layout set up to plot on an 24 x 18 ("C" size) sheet of paper.

With the default options set when AutoCAD is installed, a viewport is automatically created when you activate a layout. However, you can create layouts using the *Vports* command. A viewport is a "window" that looks into model space. Therefore, you first create the drawing objects in model space (the *Model* tab), then activate a layout and create a viewport to "look into" model space. Typically, only drawing objects such as a title block, border, and some text are created in a layout. Creating layouts and viewports is discussed in detail in Chapter 13, Layouts and Viewports.

Since a layout represents the actual printed sheet, you normally <u>print the layout full size (1:1)</u>. However, the view of the drawing objects appearing in the viewport is scaled to achieve the desired print scale. In other words, you can <u>control the scale of the drawing by setting the "viewport scale"</u>—the proportion of the drawing objects that appear in the viewport relative to the paper size, or simply stated, the proportion of model space to paper space. (This is the same idea as the proportion of the drawing area to the paper size, discussed earlier.)

One easy way to set the viewport scale is with the *Viewports* drop-down list (Fig. 6-15). To set the scale of the display of objects in the viewport relative to the paper size, select or enter the desired scale in the drop-down list in the *Viewports* toolbar. Setting the scale of the drawing for printing and plotting using this method and other methods is discussed in detail in Chapter 13, Layouts and Viewports, and Chapter 14, Printing and Plotting.

Using a *Layout* tab to set up a drawing for printing or plotting is recommended, although you can also make a print or plot directly from the *Model* tab (print from model space). This method is also discussed in Chapter 14, Printing and Plotting.

Step 10 in the "Steps for Basic Drawing Setup" is to create a titleblock and border for the layout. The titleblock and text that you want to appear only in the print but not in the drawing (in model space) are typically created in the layout (in paper space). Figure 6-16 shows the drawing after a titleblock has been added.

The advantage of creating a titleblock early in the drawing process is knowing how much space the titleblock occupies so you can plan around it. Note that in Figure 6-16 no drawing objects can be created in the lower-right corner because the titleblock occupies that area.

Often creating a titleblock is accomplished by using the *Insert* command to bring a previously

Figure 6-15 ————————————

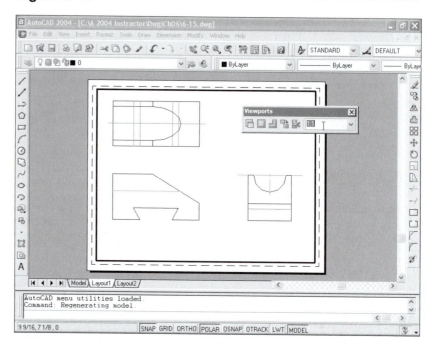

Figure 6-16 ————————————

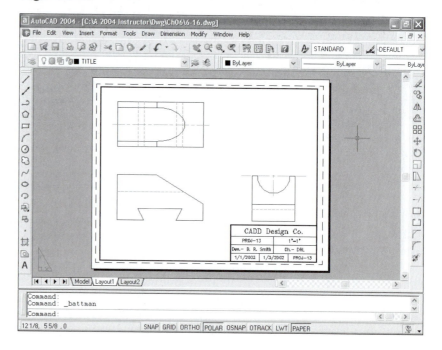

created *Block* into the layout. A *Block* in AutoCAD is a group of objects (*Lines*, *Arcs*, *Circles*, *Text*, etc.) that is saved as one object. In this way, you do not have to create the set of objects repeatedly—you need create them only once and save them as a *Block*, then use the *Insert* command to insert the *Block* into your drawing. For example, your school or office may have a standard titleblock saved as a *Block* that you can *Insert*. *Blocks* are covered in Chapter 21, Blocks and DesignCenter.

Why Set Up Layouts Before Drawing?

Knowing the intended drawing scale before completing the drawing helps you set the correct size for linetypes, text, dimensions, hatch patterns, and other size-related drawing objects. Since the text and dimensions must be readable in the final printed drawing (usually 1/8" to 1/4" or 3mm-6mm), you should know the drawing scale to determine how large to create the text and dimensions in the drawing. Although you can change the sizes when you are ready to print or plot, knowing the drawing scale early in the drawing process should save you time.

You may not have to go through the steps to create a layout and viewport to determine the drawing scale, although doing so is a very "visible" method to achieve this. <u>The important element is knowing ahead of time the intended drawing scale.</u> If you know the intended drawing scale, you can calculate the correct sizes for creating text and dimensions and setting the scale for linetypes. This topic is a major theme discussed in Chapters 12, 13, and 14.

Printing and Plotting

The *Plot* command allows you to print a drawing (using a printer) or plot a drawing (using a plotter). The *Plot* command invokes the *Plot* dialog box (Fig. 6-17). You can type the *Plot* command, select the *Plot* icon button from the Standard toolbar, or select *Plot* from the *File* pull-down menu, as well as use other methods (see "*Plot*," Chapter 14).

Figure 6-17

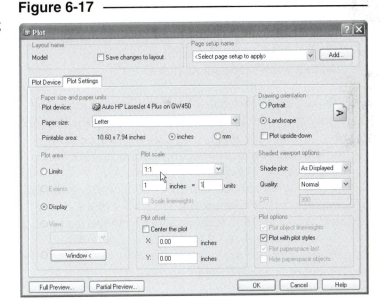

If you want to print the drawing as it appears in model space, invoke *Plot* while the *Model* tab is active. Likewise, if you want to print a layout, invoke *Plot* while the desired *Layout* tab is active.

The *Plot* dialog box allows you to specify several options with respect to printing and plotting, such as selecting the plot device, paper size, and scale for the plot. If you are printing the drawing from the *Model* tab, you would select an appropriate scale (such as *1:1*, *1:2*, or *Scaled to Fit*) in the *Scale* drop-down list so the drawing geometry would fit appropriately on the printed sheet (see Figure 6-17). If you are printing the drawing from a layout, you would select *1:1* in the *Scale* drop-down list. In this case, you want to print the layout (already set to the sheet size for the plot device) at full size. The drawing geometry that appears in the viewport can be scaled by setting the viewport scale (using the *Viewports* drop-down list), as described earlier. Any plot specifications you make with the *Plot* dialog box (or the *Page Setup* dialog box) are saved with each *Layout* tab and the *Model* tab; therefore, you can have several layouts, each saved with a particular print or plot setup (scale, device, etc.).

Details of printing and plotting, including all the options of the *Plot* dialog box, printing to scale, and configuring printers and plotters, are discussed in Chapter 14.

Setting Layout Options and Plot Options

In Chapters 12, 13, and 14 you will learn advanced steps in setting up a drawing, how to create layouts and viewports, and how to print and plot drawings. Until you study those chapters, and for printing drawings before that time, you may need to go through a simple process of configuring your system for a plot device and to automatically create a viewport in layouts. However, if you are at a school or office, some settings may have already been prepared for you, so the following steps may not be needed. Activating a *Layout* tab automatically creates a viewport and displays the *Page Setup* dialog box if you use AutoCAD's default options that appear when it is first installed.

As a check, start AutoCAD and draw a *Circle* in model space. When you activate a *Layout* tab for the first time in a drawing, you see either a "blank" sheet with no viewport, or a viewport that already exists, or one that is automatically created by AutoCAD. The viewport allows you to view the circle you created in model space. (If the *Page Setup* dialog box appears, select *Cancel* this time.) If a viewport exists, but no circle appears, double-click <u>inside</u> the viewport and type *Z* (for *Zoom*), press Enter, then type *A* (for *All*) and press Enter to make the circle appear.

If your system has not previously been configured for you at your school or office, follow the procedure given in "Configuring a Default Output Device" and "Creating Automatic Viewports." After doing so, you should be able to activate a *Layout* tab and AutoCAD will automatically match the layout size to the sheet size of your output device and automatically create a viewport.

Configuring a Default Output Device

If no default output device has been specified for your system or to check to see if one has already been specified for you, follow the steps below (Fig. 6-18).

1. Invoke the *Options* dialog box by right-clicking in the drawing area and selecting *Options...* from the bottom of
    the shortcut menu, or selecting *Options...* from the bottom of the *Tools* pull-down menu.
2. In the *Plotting* tab, locate the *Default plot settings for new drawings* cluster near the top-left corner of the dialog box. Select the desired plot device from the list (such as a laser printer), then select the *Use as default output device* button.
3. Select *OK*.

If output devices that are connected to your system do not appear in the list, see "Configuring Plotters and Printers" in Chapter 14.

Figure 6-18 ──────────────

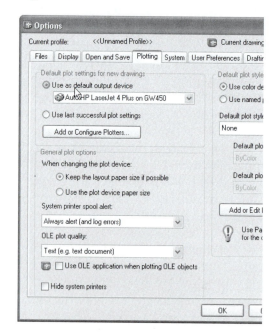

Creating Automatic Viewports

If no viewport exists on your screen, configure your system to automatically create a viewport by following these steps (Fig. 6-19).

1. Invoke the *Options* dialog box again.
2. In the *Display* tab, find the *Layout elements* cluster near the lower-left corner of the dialog box. Select the *Create viewport in new layouts* checkbox from the bottom of the list.
3. Select *OK*.

Printing the Drawing

Once an output device has been configured for your system, you are ready to print. For simplicity, you need to perform only three steps to make a print of your drawing, either from the *Model* tab or from a *Layout* tab.

1. Invoke the *Plot* dialog box. Set the desired scale in the *Plot Settings* tab, *Scale* drop-down list. If you are printing from a *Layout* tab, select *1:1*. If you are printing from the *Model* tab, select *Scaled to Fit* or other appropriate scale.
2. Select the *Full Preview* button to ensure the drawing will be printed as you expect.
3. Assuming the preview is as you expect, select the *OK* button to produce the print or plot.

Figure 6-19

This section, "Introduction to Layouts and Printing," will give you enough information to create prints or plots of drawings for practice and for the next several Chapter Exercises. You also have learned some important concepts that will help you understand all aspects of setting up drawings, creating layouts and viewports, and printing and plotting drawings, as discussed in Chapters 12, 13, and 14.

CHAPTER EXERCISES

1. A drawing is to be made to detail a mechanical part. The part is to be manufactured from sheet metal stock; therefore, only one view is needed. The overall dimensions are 16 x 10 inches, accurate to the nearest .125 inch. Complete the steps for drawing setup:

   A. The drawing will be automatically "scaled to fit" the paper (no standard scale).

      1. Begin a *New* drawing. When the *Create New Drawing* dialog box appears, select **Start from Scratch**. Select the **Imperial** default settings.
      2. **Units** should be **Decimal**. Set the **Precision** to **0.000**.
      3. Set **Limits** in order to draw full size. Make the lower-left corner **0,0** and the upper right at **24,18**. This is a 4 x 3 proportion and should allow space for the part and dimensions or notes.
      4. *Zoom All*. (Type **Z** for *Zoom*; then type **A** for *All*.)
      5. Set the *GRID* to **1**.
      6. Set the **Grid Snap** increment to **.125**. Since the *Polar Snap* increment and *Snap Type* are not saved with the drawing (but are saved in the system registry), it is of no use to set these options at this time.
      7. Save this drawing as **CH6EX1A** (to be used again later). (When plotting at a later time, "Scale to Fit" can be specified.)

B. The drawing will be printed from a layout to scale on engineering A size paper (11" x 8.5").

1. Begin a *New* drawing. When the *Create New Drawing* dialog box appears, select *Start from Scratch*. Select the *Imperial* default settings.
2. *Units* should be *Decimal*. Set the *Precision* to **0.000**.
3. Set *Limits* to a proportion of the paper size, making the lower-left corner **0,0** and the upper-right at **22,17**. This allows space for drawing full size and for dimensions or notes.
4. *Zoom All*. (Type **Z** for *Zoom*; then type **A** for *All*.)
5. Set the *GRID* to **1**.
6. Set the *Grid Snap* increment to **.125**. Since the *Polar Snap* increment and *Snap Type* are not saved with the drawing (but are saved in the system registry), it is of no use to set these options at this time.
7. Activate a *Layout* tab. Assuming your system is configured for a printer using an 11 x 8.5 sheet and is configured to automatically create a viewport (see "Setting Layout Options and Plot Options" in this chapter), a viewport should appear. If the *Page Setup* dialog box appears, select *OK*.
8. Double-click inside the viewport, then type **Z** for *Zoom* and press **Enter**, and type **A** for *All* and press **Enter**. Activate the *Viewports* drop-down list by right-clicking on any icon button and selecting *Viewports* from the list of available toolbars that appears. Select **1:2** from the list.
9. Activate the *Model* tab. *Save* the drawing as **CH6EX1B**.

2. A drawing is to be prepared for a house plan. Set up the drawing for a floor plan that is approximately 50' x 30'. Assume the drawing is to be automatically "Scaled to Fit" the sheet (no standard scale).

A. Begin a *New* drawing. When the *Create New Drawing* dialog box appears, select *Start from Scratch*. Select the *Imperial* default settings.
B. Set *Units* to *Architectural*. Set the *Precision* to **0'-0 1/4"**. Each unit equals 1 inch.
C. Set *Limits* to **0,0** and **80',60'**. Use the apostrophe (') symbol to designate feet. Otherwise, enter **0,0** and **960,720** (size in inch units is: 80x12=960 and 60x12=720).
D. *Zoom All*. (Type **Z** for *Zoom*; then type **A** for *All*.)
E. Set *GRID* to **24** (2 feet).
F. Set the *Grid Snap* increment to **6"**. Since the *Polar Snap* increment and *Snap Type* are not saved with the drawing (but are saved in the system registry), it is of no use to set these options at this time.
G. *Save* this drawing as **CH6EX2**.

3. A multiview drawing of a mechanical part is to be made. The part is 125mm in width, 30mm in height, and 60mm in depth. The plot is to be made on an A3 metric sheet size (420mm x 297mm). The drawing will use ISO linetypes and ISO hatch patterns, so AutoCAD's *Metric* default settings can be used.

A. Begin a *New* drawing. When the *Create New Drawing* dialog box appears, select *Start from Scratch*. Select the *Metric* default settings.
B. *Units* should be *Decimal*. Set the *Precision* to **0.00**.
C. Calculate the space needed for three views. If *Limits* are set to the sheet size, there should be adequate space for the views. Make sure the lower-left corner is at **0,0** and the upper right is at **420,297**. (Since the *Limits* are set to the sheet size, a plot can be made later at 1:1 scale.)
D. Set the *Grid Snap* increment to **10**. Make *Grid Snap* current.
E. *Save* this drawing as **CH6EX3** (to be used again later).

4. Assume you are working in an office that designs many mechanical parts in metric units. However, the office uses a standard laser jet printer for 11" x 8.5" sheets. Since AutoCAD does not have a setup for metric drawings on non-metric sheets, it would help to carry out the steps for drawing setup and save the drawing as a template to be used later.

   A. Begin a *New* drawing. When the *Create New Drawing* dialog box appears, select **Start from Scratch**. Select **Imperial** default settings.
   B. Set the **Units Precision** to **0.00**.
   C. Change the **Limits** to match an 11" x 8.5" sheet. Make the lower-left corner **0,0** and the upper right **279,216** (11 x 8.5 times 25.4, approximately). (Since the *Limits* are set to the sheet size, plots can easily be made at 1:1 scale.)
   D. **Zoom All.** (Type **Z** for *Zoom*, then **A** for *All*).
   E. Change the **Grid Snap** increment to **2**. When you begin the drawing in another exercise, you may want to set the *Polar Snap* increment to 2 and make *Polar Snap* current.
   F. Change **Grid** to **10**.
   G. At the Command prompt, type **LTSCALE**. Change the value to **25**.
   H. Activate a **Layout** tab. Assuming your system is configured for a printer using an 11 x 8.5 sheet and is configured to automatically create a viewport (see "Setting Layout Options and Plot Options" in this chapter), a viewport should appear. (If the *Page Setup* dialog box automatically appears, select **OK** and the viewport should appear.)
   I. Produce the *Page Setup* dialog box by selecting **Page Setup** from the **File** pull-down menu or type **Pagesetup** at the Command prompt. In the *Page Setup* dialog box, activate the **Layout Settings** tab. Select the **mm** button in the *Printable area* section (under the *Paper size* box). Next, select **1:1** in the *Scale* drop-down list near the center of the dialog box. Select the **OK** button. This action (using the *Page Setup* dialog box) saves the print settings for the layout for future use without making a print at this time.
   J. Examining the layout, it appears that the viewport has been reduced in size and exists in the lower-left corner of the sheet. Technically, the sheet size has been increased from 11 x 8.5 units to 279 x 216 units (approximately) and the viewport remained the original size. **Erase** the viewport.
   K. To make a new viewport, type **–Vports** at the Command prompt (don't forget the hyphen) and press **Enter**. Accept the default (**Fit**) option by pressing **Enter**. A viewport appears to fit the printable area.
   L. Double-click inside the viewport, then type **Z** for *Zoom* and press **Enter**, and type **A** for *All* and press **Enter**. Activate the *Viewports* drop-down list by right-clicking on any icon button and selecting *Viewports* from the list of available toolbars that appears. Select **1:1** from the list.
   M. Activate the *Model* tab. **Save** the drawing as **A-METRIC**.

5. Assume you are commissioned by the local parks and recreation department to provide a layout drawing for a major league sized baseball field. Follow these steps to set up the drawing:

   A. Begin a *New* drawing. When the *Create New Drawing* dialog box appears, select **Start from Scratch**, then **Imperial** default units.
   B. Set the **Units** to **Architectural** and the **Precision** to **1/2"**.
   C. Set **Limits** to an area of **512' x 384'** (make sure you key in the apostrophe to designate feet). *Zoom All.*
   D. Type **DS** to invoke the *Drafting Settings* dialog box. Change the **Grid Snap** to **10'** (don't forget the apostrophe). When you are ready to draw (at a later time), you may want to set the *Polar Snap* increment to 10' and make *Polar Snap* the current *Snap Type*.
   E. Use the **Grid** command and change the value to **20'**. Ensure *SNAP* and *GRID* are on.
   F. **Save** the drawing and assign the name **BALL FIELD CH6**.

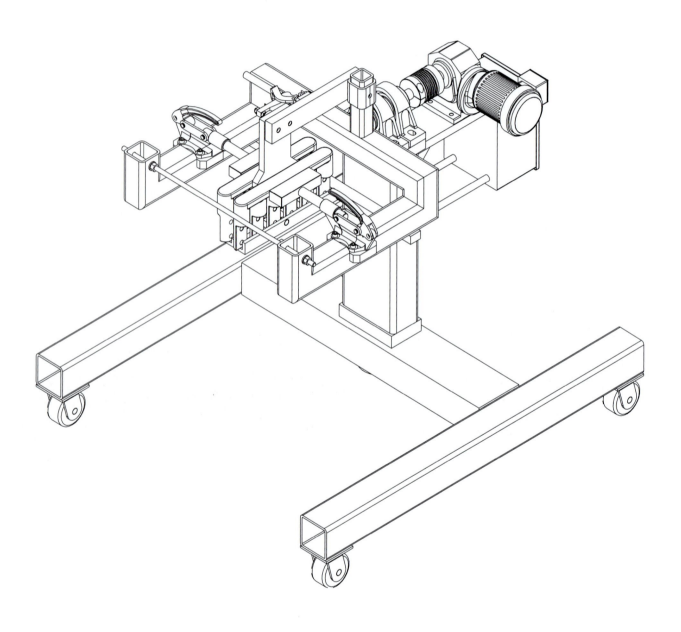

WELDING FIXTURE MODEL.DWG, Courtesy of Autodesk, Inc.

OBJECT SNAP AND OBJECT SNAP TRACKING

Chapter Objectives

After completing this chapter you should:

1. understand the importance of accuracy in CAD drawings;

2. know the function of each of the Object Snap (*OSNAP*) modes;

3. be able to recognize the AutoSnap Marker symbols;

4. be able to invoke *OSNAP*s for single point selection;

5. be able to operate Running Object Snap modes;

6. know that you can toggle Running Object Snap off to specify points not at object features;

7. be able to use Object Snap Tracking to create and edit objects that align with existing *OSNAP* points.

CAD ACCURACY

Because CAD databases store drawings as digital information with great precision (fourteen numeric places in AutoCAD), it is possible, practical, and desirable to create drawings that are 100% accurate; that is, a CAD drawing should be created as an exact dimensional replica of the actual object. For example, lines that appear to connect should actually connect by having the exact coordinate values for the matching line endpoints. Only by employing this precision can dimensions placed in a drawing automatically display the exact intended length, or can a CAD database be used to drive CNC (Computer Numerical Control) machine devices such as milling machines or lathes, or can the CAD database be used for rapid prototyping devices such as Stereo Lithography Apparatus. With CAD/CAM technology (Computer-Aided Design/Computer-Aided Manufacturing), the CAD database defines the configuration and accuracy of the finished part. <u>Accuracy is critical</u>. Therefore, in no case should you create CAD drawings with only visual accuracy such as one might do when sketching using the "eyeball method."

OBJECT SNAP

AutoCAD provides a capability called "Object Snap," or *OSNAP* for short, that enables you to "snap" to existing object endpoints, midpoints, centers, intersections, etc. When an *OSNAP* mode (*Endpoint, Midpoint, Center, Intersection,* etc.) is invoked, you can move the cursor <u>near</u> the desired object feature (endpoint, midpoint, etc.) and AutoCAD locates and calculates the coordinate location of the desired object feature. Available Object Snap modes are:

> *Center*
> *Endpoint*
> *Insert*
> *Intersection*
> *Midpoint*
> *Nearest*
> *Node* (Point)
> *Perpendicular*
> *Parallel*
> *Extension*
> *Temporary Tracking*
> *Quadrant*
> *Tangent*
> *From*
> *Apparent Intersection*
>   (for 3D use)

Figure 7-1

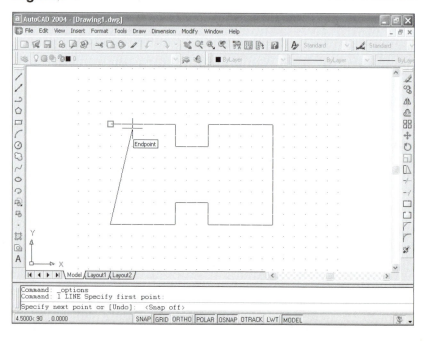

For example, when you want to draw a *Line* and connect its endpoint to an existing *Line,* you can invoke the *Endpoint OSNAP* mode at the "Specify next point or [Undo]:" prompt, then snap <u>exactly</u> to the desired line end by moving the cursor near it and PICKing (Fig. 7-1). *OSNAPs* can be used for any draw or modify operation— whenever AutoCAD prompts for a point (location).

A feature called "AutoSnap" displays a "Snap Marker" indicating the particular object feature (endpoint, midpoint, etc.) when you move the cursor <u>near</u> an object feature. Each *OSNAP* mode (*Endpoint, Midpoint, Center, Intersection,* etc.) has a distinct symbol (AutoSnap Marker) representing the object feature. This innovation allows you to preview and confirm the snap points before you PICK them (Fig. 7-2). The table in Figure 7-28 lists each Object Snap mode, the related AutoSnap Marker, and the icon button used to activate a single Object Snap selection.

Figure 7-2 ————————

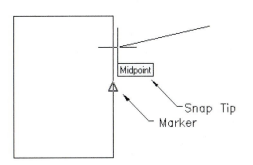

AutoSnap provides two other aids for previewing and confirming *OSNAP* points before you PICK. A "Snap Tip" appears shortly after the Snap Marker appears (hold the cursor still and wait one second). The Snap Tip gives the name of the found *OSNAP* point, such as an *Endpoint, Midpoint, Center,* or *Intersection,* (Fig. 7-2). In addition, a "Magnet" draws the cursor to the snap point if the cursor is within the confines of the Snap Marker. This Magnet feature helps confirm that you have the desired snap point before making the PICK.

A visible target box, or "Aperture," <u>can be displayed</u> at the cursor (invisible by default) whenever an *OSNAP* mode is in effect (Fig. 7-3). The Aperture is a square box larger than the pickbox (default size of 10 pixels square). Technically, this target box (visible or invisible) must be located <u>on an object</u> before a Snap Marker and related Snap Tip appear. The settings for the Aperture, Snap Markers, Snap Tips, and Magnet are controlled in the *Drafting* tab of the *Options* dialog box (discussed later).

Figure 7-3 —

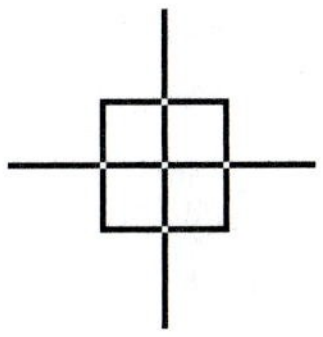

The Object Snap modes are explained in the next section. Each mode, and its relation to the AutoCAD objects, is illustrated.

Object Snaps must be activated in order for you to "snap" to the desired object features. Two methods for activating Object Snaps, Single Point Selection and Running Object Snaps, are discussed in the sections following Object Snap Modes. The *OSNAP* modes (*Endpoint, Midpoint, Center, Intersection,* etc.) operate identically for either method.

OBJECT SNAP MODES

AutoCAD provides the following Object Snap Modes.

Center

This *OSNAP* option finds the center of a *Circle, Arc,* or *Donut*. You can PICK the *Circle* <u>object</u> or where you think the center is.

Figure 7-4 ————————

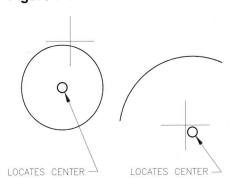

LOCATES CENTER   LOCATES CENTER

Endpoint

The *Endpoint* option snaps to the endpoint of a *Line, Pline, Spline,* or *Arc.* PICK the object <u>near</u> the desired end.

Figure 7-5

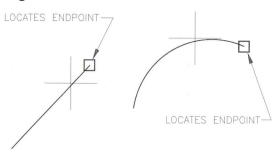

LOCATES ENDPOINT

LOCATES ENDPOINT

Insert

This option locates the insertion point of *Text* or a *Block.* PICK anywhere on the *Block* or line of *Text.*

Figure 7-6

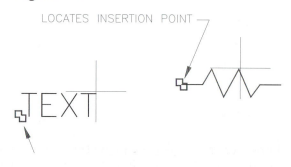

LOCATES INSERTION POINT

Intersection

Using this option causes AutoCAD to calculate and snap to the intersection of any two objects. You can locate the cursor (Aperture) so that <u>both</u> objects pass near (through) it, or you can PICK each object <u>individually</u>.

Figure 7-7

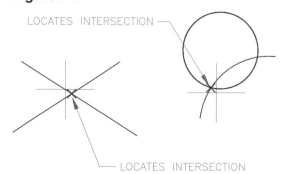

LOCATES INTERSECTION

LOCATES INTERSECTION

Even if the two objects that you PICK do not physically intersect, you can PICK each one individually with the *Intersection* mode and AutoCAD will find the <u>extended</u> intersection.

Figure 7-8

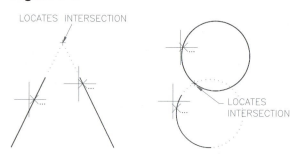

LOCATES INTERSECTION

LOCATES INTERSECTION

Midpoint

The *Midpoint* option snaps to the point of a *Line* or *Arc* that is <u>halfway</u> between the end-points. PICK anywhere on the object.

Figure 7-9

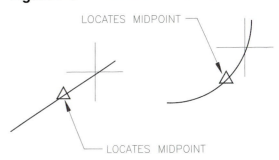

LOCATES MIDPOINT

LOCATES MIDPOINT

Nearest

The *Nearest* option locates the point on an object nearest to the <u>cursor position</u>. Place the cursor center nearest to the desired location, then PICK.

Nearest <u>cannot</u> be used effectively with *ORTHO* or *POLAR* to draw orthogonal lines because the *Nearest* point takes precedence over *ORTHO* and *POLAR*. However, <u>the *Intersection* mode (in Running Osnap mode only) in combination with *POLAR* does allow you to construct orthogonal lines (or lines at any Polar Tracking angle) that intersect with other objects.</u>

Figure 7-10

LOCATES NEAREST
POINT ON OBJECT

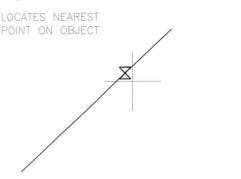

TIP

Node

This option snaps to a *Point* object. The *Point* must be within the Aperture (visible or invisible Aperture).

Figure 7-11

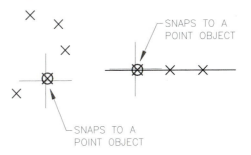

SNAPS TO A
POINT OBJECT

SNAPS TO A
POINT OBJECT

Perpendicular

Use this option to snap perpendicular to the selected object. PICK anywhere on a *Line* or straight *Pline* segment. The *Perpendicular* option is typically used for the second point ("Specify next point:" prompt) of the *Line* command.

Figure 7-12

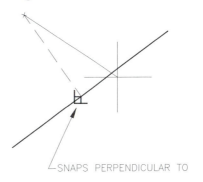

SNAPS PERPENDICULAR TO

Quadrant

The *Quadrant* option snaps to the 0, 90, 180, or 270 degree quadrant of a *Circle*. PICK <u>nearest</u> to the desired *Quadrant*.

Figure 7-13

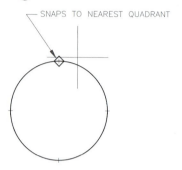

SNAPS TO NEAREST QUADRANT

Tangent

This option calculates and snaps to a tangent point of an *Arc* or *Circle*. PICK the *Arc* or *Circle* as near as possible to the expected *Tangent* point.

Figure 7-14

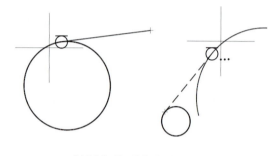

SNAPS TANGENT TO

From

The *From* option is designed to let you snap to a point <u>relative</u> to another point using relative rectangular, relative polar, or direct distance entry coordinates. There are two steps: first select a "Base-point:" (coordinates or another *OSNAP* may be used); then select an "Offset:" (enter relative rectangular, relative polar, or direct distance entry coordinates). The *From* option has effectively been replaced by the newer Object Snap Tracking feature (see "Object Snap Tracking").

Figure 7-15

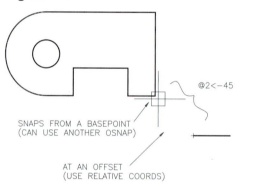

@2<-45

SNAPS FROM A BASEPOINT
(CAN USE ANOTHER OSNAP)

AT AN OFFSET
(USE RELATIVE COORDS)

Apparent intersection

Use this option when you are working with a 3D drawing and want to snap to a point in space where two objects appear to intersect (from your viewpoint) but do not actually physically intersect.

Figure 7-16

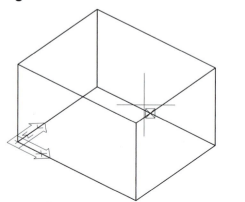

SNAPS TO APPARENT INTERSECTION IN 3D SPACE

Acquisition Object Snap Modes

Three new Object Snap modes were introduced with AutoCAD 2000: *Parallel, Extension,* and *Temporary Tracking*. When you use these Object Snap modes a dotted line appears, called an <u>alignment vector</u>, that indicates a vector along which the selected point will lie. In addition, these modes require an additional step—you must "acquire" (select) a point or object. For example, using the *Parallel* mode, you must "acquire" an object to be parallel to. The process used to "acquire" objects is explained here and is similar to that used for Object Snap Tracking (see "Object Snap Tracking" later in this chapter).

To Acquire an Object

Figure 7-17 ———

To acquire an object to use for an *Extension* or *Parallel* Object Snap mode, move the cursor over the desired object and pause briefly, <u>but do not pick the object</u>. A small plus sign (+) is displayed when AutoCAD acquires the object (Fig. 7-17). A dotted-line "alignment vector" appears as you move the cursor into a parallel or extension position (Fig. 7-18). <u>You can acquire multiple objects to generate multiple vectors</u>.

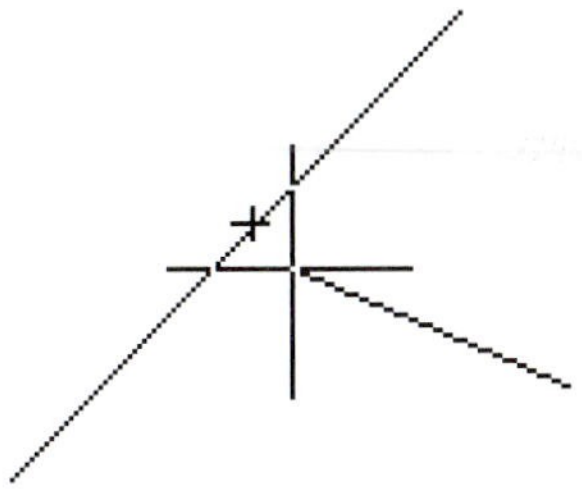

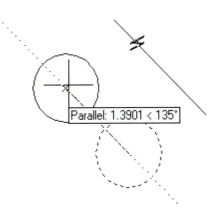

To clear an acquired object (in case you decide not to use that object), move the cursor back over the acquisition marker until the plus sign (+) disappears. Acquired points also clear automatically when another command is issued.

Parallel

Figure 7-18 ———

The *Parallel* Osnap option snaps the rubberband line into a parallel relationship with any acquired object (see "To Acquire an Object," earlier this section).

Start a *Line* by picking a start point. When AutoCAD prompts to "Specify next point or [Undo]:," acquire an object to use as the parallel source (a line to draw parallel to). Next, move the cursor to within a reasonably parallel position with the acquired line. The current rubberband line snaps into an exact parallel position (Fig. 7-18). A dotted-line parallel alignment vector appears as well as a tool tip indicating the current *Line* length and angle. The parallel Osnap symbol appears on the source parallel line. Pick to specify the current *Line* length.

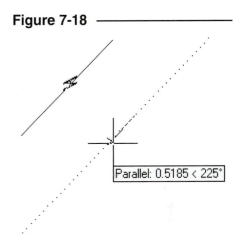

Keep in mind multiple lines can be acquired, giving you several parallel options. The acquired source objects lose their acquisition markers when each *Line* segment is completed.

Consider the use of *Parallel* Osnap with other commands such as *Move* or *Copy*. Figure 7-19 illustrates using *Move* with a *Circle* in a direction *Parallel* to the acquired *Line*.

Figure 7-19 ———

Extension

The *Extension* Osnap option snaps the rubberband line so that it intersects with an extension of any acquired object (see "To Acquire an Object," earlier this section).

For example, assume you use the *Line* command, then the *Extension* Osnap mode. When another existing object is acquired, the current *Line* segment intersects with an extension of the acquired object. Figure 7-20 depicts drawing a *Line* segment (upper left) to an *Extension* of an acquired *Line* (lower right). Notice the acquisition marker (plus symbol) on the acquired object.

Figure 7-20 ─────────────

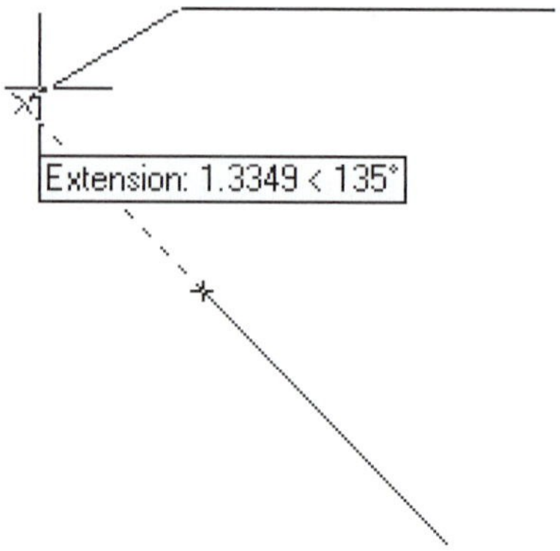

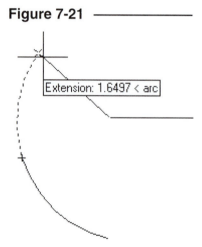

Consider drawing a *Line* to an *Extension* of other objects, such as shown in Figure 7-21. Here a *Line* is drawn to the *Extension* of an *Arc*.

Figure 7-21 ─────────────

If you set *Extension* as a <u>Running Osnap mode</u>, each newly created *Line* segment automatically becomes acquired; therefore, you can draw an extension of the previous segment (at the same angle) easily (Fig. 7-22). (See "Osnap Running Mode" later in this chapter.)

Figure 7-22 ─────────────

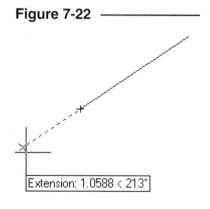

Temporary Tracking (TT)

Temporary Tracking sets up a temporary Polar Tracking point. This option allows you to "track" in a polar direction from any point you select. <u>Tracking</u> is the process of moving from a selected point in a preset angular direction. The preset angles are those specified in the *Increment Angle* section in the *Polar Tracking* tab of the *Drafting Settings* dialog box (see "Polar Tracking" in Chapter 3, Figure 3-17).

For example, if you use the *Line* command, then use the *Temporary Tracking* button or type **TT** at the "Specify first point:" prompt, you can select a point anywhere and that point becomes the temporary tracking point. The cursor moves in a preset angle from that point (Fig. 7-23). The next point selected along the alignment path becomes the *Line's* first point. *POLAR does not have to be on* to use *TT*. Note that the temporary tracking point is indicated by an acquisition marker, so it appears similar to any other acquired point.

Figure 7-23

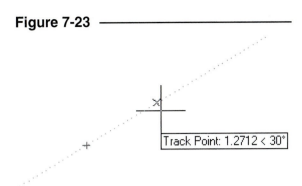

You can use *Temporary Tracking* in conjunction with Polar Snap. To do this, toggle *SNAP* on (assuming Polar Snap is the current Snap type) when you are in *Temporary Tracking* mode. Figure 7-24 illustrates this feature (note the tracking tip indicates an exact 1.5000).

Figure 7-24

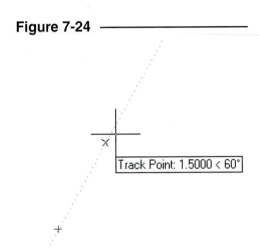

Temporary Tracking can be used in conjunction with another Osnap mode (use *TT*, then another *Osnap* mode to specify the point). As an example, assume you wanted to begin drawing a *Line* directly above the corner of an existing rectangular shape. To do this, use the *Line* command and at the "Specify first point:" prompt invoke *TT*. Then use *Endpoint* to acquire the desired corner of the rectangle. A temporary tracking alignment vector appears from the corner as indicated in Figure 7-25. To specify an exact distance, enter a value (Direct Distance Entry) or turn on Polar Snap. The resulting point is the point that satisfies the "Specify first point:" prompt for the *Line*.

Figure 7-25

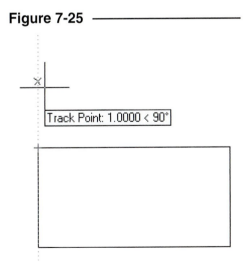

Object Snaps must be activated in order for you to snap to the desired object features. Two methods for activating Object Snaps, Single Point Selection and Running Object Snaps, are discussed next. The *OSNAP* modes (*Endpoint, Midpoint, Center, Intersection,* etc.) operate identically for either method.

OSNAP SINGLE POINT SELECTION

OBJECT SNAPS (SINGLE POINT)	Pull-down Menu	COMMAND (TYPE)	ALIAS (TYPE)	Short-cut	Screen (side) Menu	Tablet Menu	Cursor Menu (Shift+button 2)
	...	END, MID, etc. (first three letters)	...	...	****(asterisks)	T,15 - U,22	*Endpoint Midpoint*, etc.

There are many methods for invoking *OSNAP* modes for single point selection, as shown in the command table. If you prefer to type, enter only the <u>first three letters</u> of the *OSNAP* mode at the "Specify first point: " prompt, "Specify next point or [Undo]:" prompt, or <u>any time AutoCAD prompts for a point</u>.

If desired, an *Object Snap* toolbar (Fig. 7-26) can be activated to float or dock on the screen by right-clicking on any icon button and selecting *Object Snap* from the list of toolbars. The advantage of invoking the *Object Snap* toolbar is that only one PICK is required for an *OSNAP* mode, whereas the Standard toolbar requires two PICKs because the *OSNAP* icons are on a flyout.

In addition to these options, a special menu called the <u>cursor menu</u> can be used. The cursor menu pops up at the <u>current location</u> of the cursor and replaces the cursor when invoked (Fig. 7-27). This menu is activated by pressing **Shift + #2** (hold down the Shift key while clicking the right mouse button).

(If you have a wheel mouse, you can also change the *MBUTTON-PAN* system variable to 0. This action allows you to press the wheel to invoke the *Osnap* cursor menu, but disables your ability to *Pan* by holding down the wheel.)

Figure 7-26

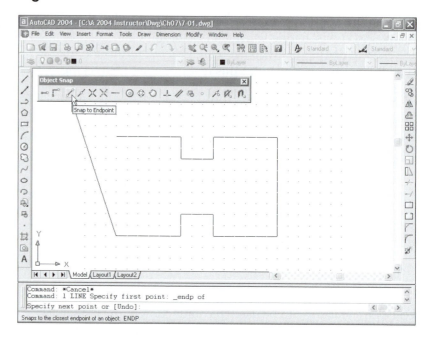

Figure 7-27

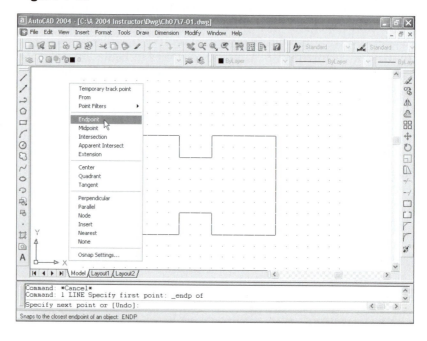

With any of these methods, *OSNAP* modes are active only for selection of a single point. If you want to *OSNAP* to another point, you must select an *OSNAP* mode again for the second point. With this method, the desired *OSNAP* mode is selected transparently (invoked during another command operation) immediately before selecting a point when prompted. In other words, whenever you are prompted for a point (for example, the "Specify first point:" prompt of the *Line* command), select or type an *OSNAP* option. Then PICK near the desired object feature (endpoint, center, etc.) with the cursor. AutoCAD snaps to the feature of the object and uses it for the point specification. Using *OSNAP* in this way allows the *OSNAP* mode to operate only for that single point selection.

For example, when using *OSNAP* during the *Line* command, the Command line reads as shown:

    Command: _line Specify first point: **endp of** (PICK)
    Specify next point or [Undo]: **endp of** (PICK)
    Specify next point or [Undo]: **Enter**
    Command:

If you prefer using *OSNAP* for single point specification, it may be helpful to associate the tools (buttons) with the Marker that appears on the drawing when you use the *OSNAP* mode. The table in Figure 7-28 displays the *OSNAP* icons and the respective Markers.

Figure 7-28

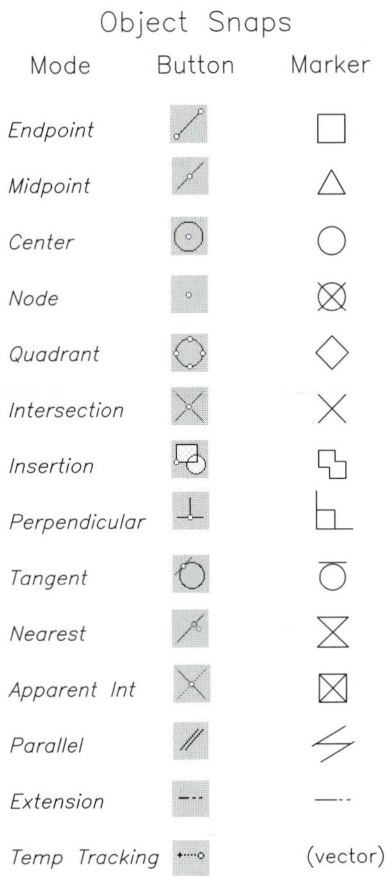

Object Snaps

Mode	Button	Marker
Endpoint		□
Midpoint		△
Center		○
Node		⊗
Quadrant		◇
Intersection		✕
Insertion		⌐⌐
Perpendicular		⌐
Tangent		○
Nearest		⊠
Apparent Int		⊠
Parallel		⫽
Extension		---
Temp Tracking		(vector)

OSNAP RUNNING MODE

OBJECT SNAPS (RUNNING)

Pull-down Menu	COMMAND (TYPE)	ALIAS (TYPE)	Short-cut	Screen (side) Menu	Tablet Menu	Cursor Menu (Shift+button 2)
Tools *Drafting Settings...* *Object Snap*	OSNAP or -OSNAP	OS or -OS	...	TOOLS 2 *Osnap...*	U,22	*Osnap* *Settings...*

Running Object Snap

A more effective method for using Object Snap is called "Running Object Snap" because one or more *OSNAP* modes (*Endpoint, Center, Midpoint,* etc.) can be turned on and kept running indefinitely. This method can obviously be more productive because you do not have to continually invoke an *OSNAP* mode each time you need to use one. For example, suppose you have several Endpoints to connect. It would be most efficient to turn on the running *Endpoint* OSNAP mode and leave it running during the multiple selections. This is faster than continually selecting the *OSNAP* mode each time before you PICK.

You can even have several *OSNAP* modes running at the same time. A common practice is to turn on the *Endpoint, Center, Midpoint* modes simultaneously. In that way, if you move your cursor near a *Circle,* the *Center* Marker appears; if you move the cursor near the end or the middle of a *Line,* the *Endpoint* or the *Midpoint* mode Markers appear.

Accessing Running Object Snap

All features of Running Object Snaps are controlled by the *Drafting Settings* dialog box (Fig. 7-29). This dialog box can be invoked by the following methods (see the previous Command Table):

1. type the *OSNAP* command;
2. type *OS*, the command alias;
3. select *Drafting Settings...* from the *Tools* pull-down menu;
4. select *Osnap Settings...* from the bottom of the cursor menu (Shift + #2 button);
5. select the *Object Snap Settings* icon button from the *Object Snap* toolbar;
6. right-click on the words *OSNAP* or *OTRACK* on the Status Bar, then select *Settings...* from the shortcut menu that appears (see Figure 7-30).

Figure 7-29 ————————

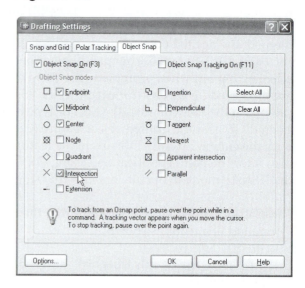

Use the *Object Snap* tab to select the desired Object Snap settings. Try using three or four commonly used modes together, such as *Endpoint, Midpoint, Center,* and *Intersection*. The AutoSnap Markers indicate which one of the modes would be used as you move the cursor near different object features. Using similar modes simultaneously, such as *Center, Quadrant,* and *Tangent,* can sometimes lead to difficulties since it requires the cursor to be placed almost in an exact snap spot, or in some cases it may not find one of the modes (*Quadrant* overrides *Center*). In these cases, the Tab key can be used to cycle through the options (see "Object Snap Cycling").

Figure 7-30 ——

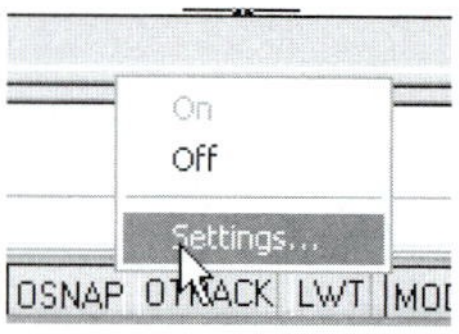

Object Snap Options

You can set other options for using Object Snap in the *Drafting* tab of the *Options* dialog box (Fig. 7-31). Here, more advanced preferences are set such as size and color for the markers and Aperture, and the display of tracking vectors and tool tips. Access the *Options* dialog box by typing *Options*, selecting the *Options* button from the *Drafting Settings* dialog box, or right-clicking in the drawing area while no commands are in progress, then selecting *Options...* from the shortcut menu. The possible settings are explained next.

Figure 7-31 ————————

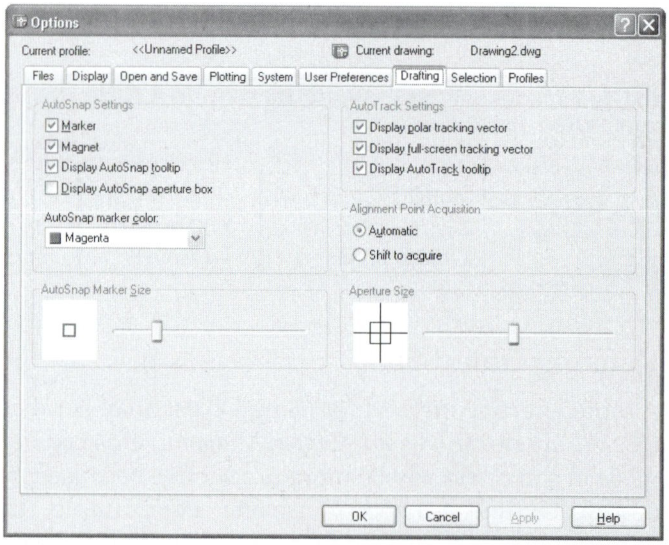

Marker

When this box is checked, an AutoSnap Marker appears when the cursor is moved near an object feature. Each *OSNAP* mode (*Endpoint, Center, Midpoint,* etc.) has a unique Marker (see Fig. 7-28). Normally this box should be checked, especially when using Running Object Snap, because the Markers help confirm when and which *OSNAP* mode is in effect.

Magnet

The Magnet feature causes the cursor to "lock" onto the object feature (*Endpoint, Center, Midpoint,* etc.) when the cursor is within the confines of the Marker. The Magnet helps confirm the exact location that will be snapped to for the subsequent PICK.

Display AutoSnap tooltip
The Snap Tip (similar to a Tool Tip) is helpful for beginners because it gives a verbal indication of the *OSNAP* mode in effect (*Endpoint, Center, Midpoint,* etc.) when the cursor is near the object feature. Experienced users may want to turn off the Snap Tips once the Marker symbols are learned.

Display AutoSnap Aperture box
This checkbox controls the visibility of the Aperture. Whether the Aperture is visible or not, its size determines how close the cursor must be to an object before a Marker appears and the displayed *OSNAP* mode takes effect. (See *Aperture size* in the *Running Osnap* dialog box.)

AutoSnap Marker size
This control determines the size of the Markers. Remember that if the Magnet is on, the cursor locks to the object feature when the cursor is within the confines of the Magnet. Therefore, the larger the Marker, the greater the Magnet effective area, and the smaller the Marker, the smaller the Magnet effective area.

AutoSnap Marker color
Use this drop-down box to select a color for the Markers. You may want to change Marker colors if you change the color of the Drawing Editor background. For example, if the background is changed to white, you may want to change the Marker color to dark blue instead of yellow.

Display polar tracking vector
Removing the check from this box disables the display of the alignment vector that appears only when you use Polar Tracking. This option does not affect the display of Object Snap alignment vectors. You can also change the setting of the *TRACKPATH* system variable to 2.

Display full-screen tracking vector
When the check is removed from this box, tracking vectors that appear are displayed from the acquired point to the cursor only, not across the entire screen. You can also change the setting of the *TRACKPATH* system variable to 0 (full-screen vectors) or 1 (short vectors).

Display AutoTracking tooltip
Removing this check disables the tool tips that appear for Polar Tracking, single point selection Object Snaps such as *Temporary Tracking* or *Extension*, and Object Snap Tracking.

Automatic (Alignment Point Acquisition)
Resting the cursor on a existing object for one second automatically acquires that point.

Shift to acquire (Alignment Point Acquisition)
With this option you must hold down the Shift key when resting the cursor on an object to acquire. This is sometimes helpful when you use Object Snap Tracking (discussed later in this chapter) since many unintended points can be automatically acquired.

Aperture size
Notice that the *Aperture Size* can be adjusted through this dialog box (or through the use of the *Aperture* command). Whether the Aperture is visible or not (see Figure 7-31, *Display AutoSnap aperture box*), its size determines how close the cursor must be to an object before the Marker appears and the displayed *OSNAP* mode takes effect. In other words, the smaller the Aperture, the closer the cursor must be to the object before a Marker appears and the *OSNAP* mode has an effect, and the larger the Aperture, the farther the cursor can be from the object while still having an effect. The Aperture default size is 10 pixels square. The Aperture size influences *Single Point Osnap* selection as well as *Running Osnaps*.

Running Object Snap Toggle

Another feature that makes Running Object Snap effective is Osnap Toggle. If you need to PICK a point without using *OSNAP*, use the Osnap Toggle to temporarily override (turn off) the modes. With Running Osnaps temporarily off, you can PICK any point without AutoCAD forcing your selection to an *Endpoint, Center,* or *Midpoint,* etc. When you toggle Running Object Snap on again, AutoCAD remembers which modes were previously set.

The following methods can be used to toggle Running Osnaps on and off:

1. click the word *OSNAP* that appears on the Status Bar (at the bottom of the screen)
2. press **F3**
3. press **Ctrl+F**

None

 The *None OSNAP* option is a Running Osnap <u>override effective for only one PICK</u>. *None* is similar to the Running Object Snap Toggle, except it is effective for one PICK, then Running Osnaps automatically come back on without having to use a toggle. If you have *OSNAP* modes running but want to deactivate them for a <u>single</u> PICK, use *None* in response to "Specify first point:" or other point selection prompt. In other words, using *None* <u>during a draw or edit command</u> overrides any Running Osnaps for that single point selection. *None* can be typed at the command prompt (when prompted for a point) or can be selected from the bottom of the Object Snap toolbar.

Object Snap Cycling

In cases when you have multiple Running Osnaps set, and it is difficult to get the desired AutoSnap Marker to appear, you can use the <u>Tab</u> key to cycle through the possible *OSNAP* modes for the highlighted object. In other words, pressing the Tab key makes AutoCAD highlight the object nearest the cursor, then cycles through the possible snap Markers (for running modes that are set) that affect the object.

For example, when the *Center* and *Quadrant* modes are both set as Running Osnaps, moving the cursor near a *Circle* makes the *Quadrant* Marker appear but not the *Center* Marker. In this case, pressing the Tab key highlights the *Circle,* then cycles through the four *Quadrant* and one *Center* snap candidates.

OBJECT SNAP TRACKING

A feature called Object Snap Tracking helps you draw objects at specific angles and in specific relationships to other objects. Object Snap Tracking works in conjunction with object snaps and displays temporary alignment paths called "tracking vectors" that help you create objects aligned at precise angular positions relative to other objects. You can toggle Object Snap Tracking on and off with the *OTRACK* button on the Status Bar or by toggling F11.

To practice with Object Snap Tracking, try turning the *Extension* and *Parallel* Osnap options off. Since these two new options require point acquisition and generate alignment vectors, it can be difficult to determine which of these features is operating when they are all on.

Object Snap Tracking is similar to Polar Tracking in that it displays and snaps to alignment vectors, but the alignment vectors are generated from <u>other existing objects</u>, not from the current object. These other objects are acquired by Osnapping to them. Once a point (*Endpoint, Midpoint,* etc.) is acquired, alignment vectors generate from them in proximity to the cursor location. This process allows you to construct geometry that has orthogonal or angular relationships to other existing objects.

Using Object Snap Tracking is essentially the same as using the *Temporary Tracking* Osnap option, then using another Osnap mode to acquire the tracking point. (See previous Figure 7-25 and related explanation to refresh your memory.) Object Snap Tracking can be used with either Single Point or Running Osnap mode.

For example, Figure 7-32 displays an alignment vector generated from an acquired *Endpoint*. The current *Line* can then be drawn to a point that is horizontally aligned with the acquired *Endpoint*. Note that in Figure 7-32 and the related figures following, the Osnap Marker (*Endpoint* marker in this case) is anchored to the acquired point. The alignment vector always rotates about, and passes through, the acquired point (at the *Endpoint* marker). The current *Line* being constructed is at the cursor location.

Figure 7-32

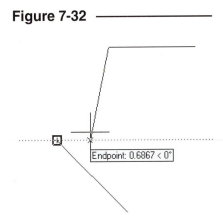

Moving the cursor to another location causes a different alignment vector to appear; this one displays vertical alignment with the same *Endpoint*.

Figure 7-33

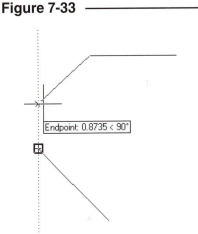

In addition, moving the cursor from an acquired point causes an array of alignment vectors to appear based on the current angular increment set in the *Polar Tacking* tab of the *Drafting Settings* dialog box (see Chapter 3, Figure 3-17, and related discussion). In Figure 7-34, alignment vectors are generated from the acquired point in 30-degree increments.

Figure 7-34

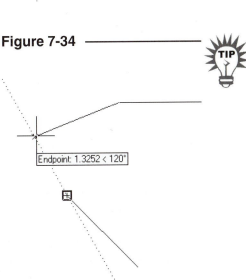

To acquire a point to use for Object Snap Tracking (when AutoCAD prompts to specify a point, move the cursor over the object point and pause briefly when the Osnap Marker appears, but do not pick the point. A small plus sign (+) is displayed when AutoCAD acquires the point (Fig. 7-35). The alignment vector appears as you move the cursor away from the acquired point. <u>You can acquire multiple points to generate multiple alignment vectors</u>.

Figure 7-35 —

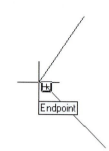

In Figure 7-35, the *Endpoint* object snap is on. Start a *Line* by picking its start point, move the cursor over another line's endpoint to acquire it, and then move the cursor along the horizontal, vertical, or polar alignment vector that appears (not shown in Figure 7-35) to locate the endpoint you want for the line you are drawing.

To clear an acquired point (in case you decide not to use an alignment vector from that point), move the cursor back over the point's acquisition marker until the plus sign (+) disappears. Acquired points also clear automatically when another command is issued. You can also toggle the word *OTRACK* on the status bar to clear acquired points.

To Use Object Snap Tracking:

1. Turn on Object Snap and Object Snap Tracking (press F3 and F11 or togle *OSNAP* and *OTRACK* on the Status Bar). You can use *OSNAP* single point selection when prompted to specify a point instead of using Running Osnap.

2. Start a *Draw* or *Modify* command that prompts you to specify a point.

3. Move the cursor over an Object Snap point to temporarily acquire it. Do not PICK the point but only pause over the point briefly to acquire it.

4. Move the cursor away from the acquired point until the desired vertical, horizontal, or polar alignment vector appears, then PICK the desired location for the line along the alignment vector.

Remember, you must set an Object Snap (single or running) before you can track from an object's snap point. <u>Object Snap and Object Snap Tracking must both be turned on to use Object Snap Tracking</u>. Polar tracking can also be turned on, but it is not necessary for Object Snap Tracking to operate.

Object Snap Tracking with Polar Tracking

For some cases you may want to use Object Snap Tracking in conjunction with Polar Tracking. This combination allows you to track from the last point specified (on the current *Line* or other operation) as well as to connect to an alignment vector from an existing object.

Figure 7-36 ────────

Figure 7-36 displays both Polar Tracking and Object Snap Tracking in use. The current *Line* (dashed vertical line) is Polar Tracking along a vertical vector from the last point specified (on the horizontal line above). The new *Line's* endpoint falls on the horizontal alignment vector (Object Snap Tracking) acquired from the *Endpoint* of the existing diagonal *Line*. Note that the tool tip displays the Polar Tracking angle ("Polar: <270") and the Object Snap Tracking mode and angle ("Endpoint: <0").

Endpoint: < 0°, Polar: < 270°

If the cursor is moved from the previous location (in the previous figure), additional Polar Tracking options and Object Snap Tracking options appear. For example in Figure 7-37, the current *Line* endpoint is tracking at 210 degrees (Polar Tracking) and falls on a vertical alignment vector from the *Endpoint* of the diagonal line (Object Snap Tracking).

Figure 7-37 ——————

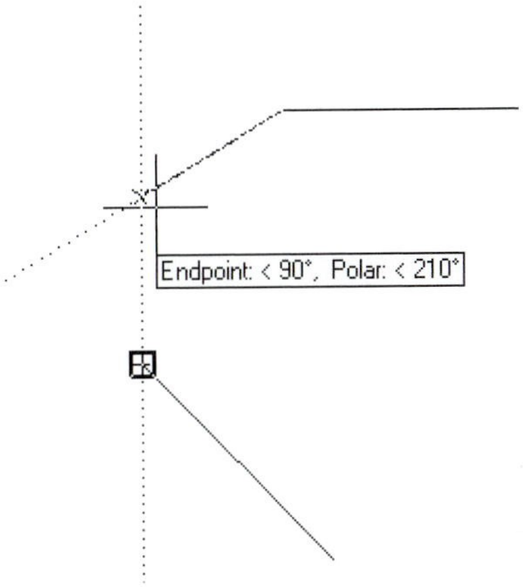

Endpoint: < 90°, Polar: < 210°

You can accomplish the same capabilities available with the combination of Polar Tracking and Object Snap Tracking by using only Object Snap Tracking and multiple acquired points. For example, Figure 7-38 illustrates the same situation as in Figure 7-36 but only Object Snap tracking is on (Polar Tracking is off). Notice that two *Endpoints* have been acquired to generate the desired vertical and horizontal alignment vectors.

Figure 7-38 ——————

Endpoint: < 0°, Endpoint: < 270°

Object Snap Tracking Settings

You can set the *Object Snap Tracking Settings* in the *Drafting Settings* dialog box (Fig. 7-39). The only options are to *Track orthogonally only* or to *Track using all polar angle settings*. The Object Snap Tracking vectors are determined by the *Increment angle* and *Additional angles* set for Polar Tracking (left side of dialog box). If you want to track using these angles, select *Track using all polar angle settings*.

Figure 7-39 ——————

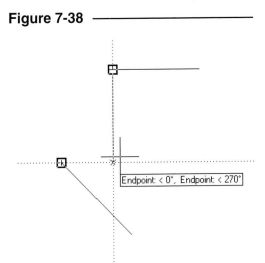

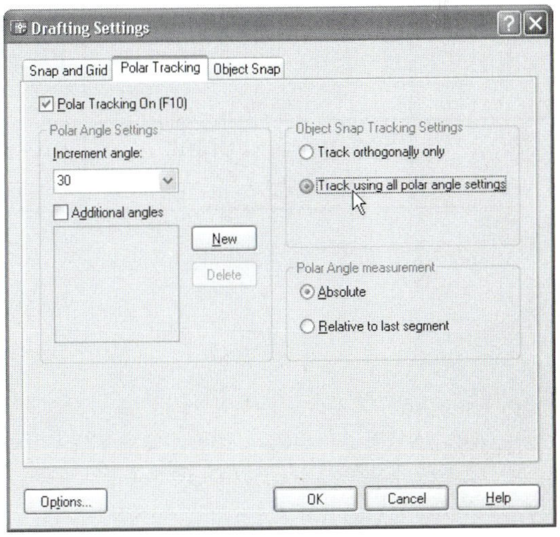

OSNAP APPLICATIONS

OSNAP can be used any time AutoCAD prompts you for a point. This means that you can invoke an *OSNAP* mode during any draw or modify command as well as during many other commands. *OSNAP* provides you with the potential to create 100% accurate drawings with AutoCAD. Take advantage of this feature whenever it will improve your drawing precision. Remember, any time you are prompted for a point, use *OSNAP* if it can improve your accuracy.

OSNAP PRACTICE

Single Point Selection Mode

1. Turn off *SNAP*, *POLAR*, *OSNAP*, and *OTRACK*. Draw two (approximately) vertical **Lines**. Follow these steps to draw another *Line* between *Endpoint*s.

STEPS	COMMAND PROMPT	PERFORM ACTION	COMMENTS
1.	Command:	select *Line* by any method	
2.	LINE Specify first point:	type *END* and press **Enter** (or Spacebar)	
3.	endp of	move the cursor near the end of one of the *Lines*	endpoint Marker appears (square box at *Line* end), "Endpoint" Snap Tip may appear
4.		**PICK** while the Marker is visible	rubberband line appears
5.	Specify next point or [Undo]:	type *END* and press **Enter** (or Spacebar)	
6.	endp of	move the cursor near the end of the second *Line*	endpoint Marker appears (square box at *Line* end), "Endpoint" Snap Tip may appear
7.		**PICK** while the Marker is visible	*Line* is created between endpoints
8.	Specify next point or [Undo]:	press **Enter**	completes command

2. Draw two *Circles*. Follow these steps to draw a *Line* between the *Centers*.

STEPS	COMMAND PROMPT	PERFORM ACTION	COMMENTS
1.	Command:	select *Line* by any method	
2.	LINE Specify first point:	invoke the cursor menu (press **Shift+#2** button) and select *Center*	
3.	cen of	move the cursor near a *Circle* object (or where you think the center is)	the AutoSnap Marker appears (small circle at center), "Center" Snap Tip may appear
4.		**PICK** while the Marker is visible	rubberband line appears
5.	Specify next point or [Undo]:	invoke the cursor menu (press **Shift+#2** button) and select *Center*	
6.	cen of	move the cursor near the second *Circle* object (or where you think the center is)	the AutoSnap Marker appears (small circle at center), "Center" Snap Tip may appear
7.		**PICK** the other *Circle* while the Marker is visible	*Line* is created between *Circle* centers
8.	Specify next point or [Undo]:	press **Enter**	completes command

3. *Erase* the *Line* only from the previous exercise. Draw another *Line* anywhere, but <u>not</u> attached to the *Circles*. Follow the steps to *Move* the *Line* endpoint to the *Circle* center.

STEPS	COMMAND PROMPT	PERFORM ACTION	COMMENTS
1.	Command:	select *Move* by any method	
2.	MOVE Select objects:	**PICK** the *Line*	the *Line* becomes highlighted
3.	Select objects:	press **Enter**	completes selection set
4.	Specify base point or displacement:	Select the **Endpoint** icon button from the *Object Snap* toolbar	
5.	endp of	move the cursor near a *Line* endpoint	AutoSnap Marker appears (square box), "Endpoint" Snap Tip may appear
6.		**PICK** while Marker is visible	endpoint becomes the "handle" for *Move*
7.	Specify second point of displacement:	select **Center** from *Object Snap* toolbar	
8.	cen of	move the cursor near a *Circle*	AutoSnap Marker appears (small circle), "Center" Snap Tip may appear
9.		**PICK** the *Circle* object while Marker is visible	selected *Line* is moved to *Circle* center

Running Object Snap Mode

4. Draw several **Lines** and **Circles** at random. To draw several *Lines* to *Endpoints* and *Tangent* to the *Circles*, follow these steps.

STEPS	COMMAND PROMPT	PERFORM ACTION	COMMENTS
1.	Command:	type *OSNAP* or *OS*	*Drafting Settings* dialog box appears
2.		select **Endpoint** and **Tangent** then press **OK**	turns on the Running Osnap modes
3.	Command:	invoke the **Line** command	use any method
4.	LINE Specify first point:	move the cursor <u>near</u> one *Line* endpoint	AutoSnap Marker appears (square box), "Endpoint" Snap Tip may appear
5.		**PICK** while Marker is visible	rubberband line appears, connected to endpoint
6.	Specify next point or [Undo]:	move cursor near endpoint of another *Line*	AutoSnap Marker appears (square box), "Endpoint" Snap Tip may appear
7.		**PICK** while Marker is visible	a *Line* is created between endpoints
8.	Specify next point or [Undo]:	move the cursor near a *Circle* object	AutoSnap Marker appears (small circle with tangent line segment), "Tangent" Snap Tip may appear
9.		**PICK** *Circle* while Marker is visible	a *Line* is created *Tangent* to the *Circle*
10.	Specify next point or [Undo]:	press **Enter**	ends *Line* command
11.	Command:	invoke the **Line** command	use any method
12.	LINE Specify first point:	move cursor near a *Circle*	AutoSnap Marker appears (small circle with tangent line segment), "Deferred Tangent" Snap Tip may appear
13.		**PICK** *Circle* while Marker is visible	rubberband line does NOT appear
14.	Specify next point or [Undo]:	move the cursor near a second *Circle* object	AutoSnap Marker appears (small circle with tangent line segment), "Deferred Tangent" Snap Tip may appear
15.		**PICK** *Circle* while Marker is visible	a *Line* is created tangent to the two *Circles*
16.	Specify next point or [Undo]:	click the word *OSNAP* on the Status Bar, press **F3** or**Ctrl+F**	temporarily toggles Running Osnaps off

(Continued)

17.		**PICK** a point near a *Line* end	*Line* is created, but does not snap to *Line* endpoint
18.	Specify next point or [Undo]:	click the word *OSNAP* on the Status Bar, press **F3** or **Ctrl+F**	toggles Running Osnaps back on
19.		**PICK** near a *Circle* (when Marker is visible	*Line* is created that snaps tangent to *Circle*
20.	Specify next point or [Undo]:	**Enter**	ends *Line* command
21.	Command:	invoke the cursor menu (press **Shift+#2**) and select *Osnap Settings*…	*Osnap Settings* dialog box appears
22.		Press the **Clear all** button, then **OK**	running *OSNAPS* are turned off

Object Snap Tracking

5. Begin a *New* drawing. Create an "L" shape by drawing one vertical and one horizontal *Line*, each 5 units long. To draw several *Lines* that track from *Endpoints* and *Midpoints*, follow these steps.

STEPS	COMMAND PROMPT	PERFORM ACTION	COMMENTS
1.	Command:	type *OSNAP* or *OS*	*Drafting Settings* dialog box appears, *Object Snap* tab
2.		select *Endpoint* and *Midpoint* options	turns on the Running Osnap modes
3.		select *Polar Tracking* tab and set *Increment Angle* to **45**, then select *OK*	sets tracking angle for Object Snap Tracking
4.		toggle <u>on</u> *OSNAP* and *OTRACK* and toggle <u>off</u> *SNAP*, *ORTHO*, and *POLAR*	select the words on the Status Bar (recessed is on and protruding is off
5.	Command:	invoke the *Line* command	
6.	LINE Specify first point:	move the cursor to the right horizontal *Line* endpoint and rest the cursor until the point is aquired	AutoSnap Marker appears (square box) and acquisition marker (plus) appears
7.		**PICK** while Marker is visible	rubberband line appears connected to endpoint and tracking vectors appear at 45 and 90 positions
8.	Specify next point	move the cursor to endpoint of vertical *Line* and rest the cursor until the point is aquired but <u>do</u> not PICK the point	AutoSnap Marker appears (square box) and acquistion marker (plus) appears

(Continued)

9.		move the cursor so two 90-degree alignment vectors appear (forming a square), then **PICK**	*Line* is drawn from first endpoint at 90-degree alignment to both endpoints
10.	Specify next point:	move cursor near top endpoint of vertical *Line* and PICK the point	*Line* is draw from last endpoint at 90-degree alignment to endpoint of vertical *Line* to form a square
11.	Specify next point:	press **Enter**	ends *Line* command
12.	Command:	invoke the *Line* command	use any method
13.	LINE Specify next point:	move the cursor to the vertical *Line* endpoint (top) and rest the cursor until the point is acquired	AutoSnap Marker appears (square box) and acquistion marker (plus) appears
14.		**PICK** *Line* while Marker is visible	rubberband line appears connected to endpoint and tracking vectors appear at 45 and 90 positions
15.	Specify next point:	move cursor to midpoint of original horizontal *Line* and rest the cursor until the point is aquired but <u>do not</u> PICK the point	AutoSnap Marker appears (triangle) and acquistion marker (plus) appears
16.		move the cursor upward so vertical alignment vector from horizontal line and 45-degree vector from vertical line appears, then PICK	diagonal *Line* is drawn above midpoint and aligned (45 degrees) to vertical line endpoint (to center of square)
17.	Specify next point:	press **Enter**	ends *Line* command

CHAPTER EXERCISES

1. *OSNAP* **Single Point Selection**

    Open **the CH6EX1A** drawing and begin constructing the sheet metal part. Each unit in the drawing represents one inch.

    A. Create four *Circles*. All *Circles* have a radius of **1.685**. The *Circles'* centers are located at **5,5**, **5,13**, **19,5**, and **19,13**.

    B. Draw four *Lines*. The *Lines* (highlighted in Figure 7-40) should be drawn on the outside of the *Circles* by using the **Quadrant** *OSNAP* mode as shown for each *Line* endpoint.

Figure 7-40 ————————

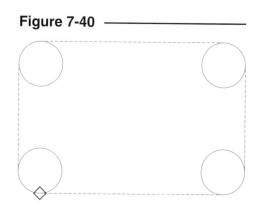

C. Draw two *Lines* from the *Center* of the existing *Circles* to form two diagonals as shown in Figure 7-41.

D. At the *Intersection* of the diagonals create a *Circle* with a **3** unit radius.

Figure 7-41

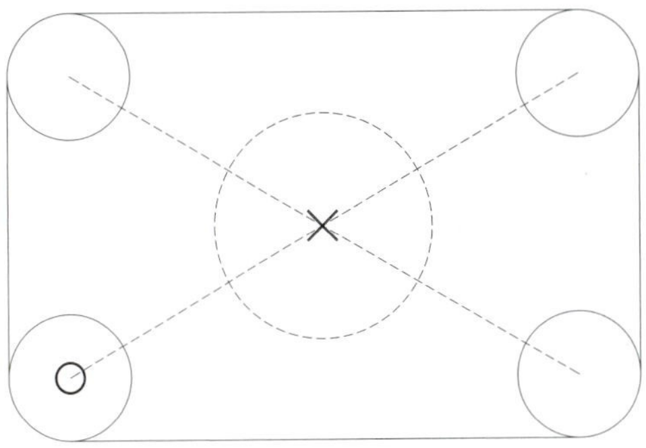

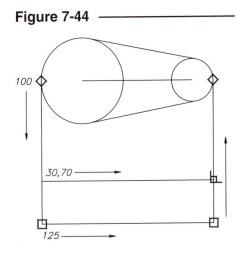

E. Draw two *Lines*, each from the *Intersection* of the diagonals to the *Midpoint* of the vertical *Lines* on each side. Finally, construct four new *Circles* with a radius of **.25,** each at the *Center* of the existing ones (Fig. 7-42).

F. *SaveAs* **CH7EX1**.

Figure 7-42

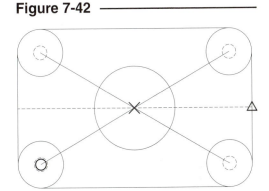

2. *OSNAP* **Single Point Selection**

A multiview drawing of a mechanical part is to be constructed using the A-METRIC drawing. All dimensions are in millimeters, so each unit equals one millimeter.

A. *Open* **A-METRIC** from Chapter 6 Exercises. Draw a *Line* from **60,140** to **140,140**. Create two *Circles* with the centers at the *Endpoints* of the *Line,* one *Circle* having a <u>diameter</u> of **60** and the second *Circle* having a diameter of **30**. Draw two *Lines Tangent* to the *Circles* as shown in Figure 7-43. *SaveAs* **PIVOTARM CH7**.

Figure 7-43

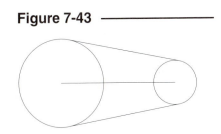

B. Draw a vertical *Line* down from the far left *Quadrant* of the *Circle* on the left. Specify relative polar (**@100<270**) or direct distance entry coordinates to make the *Line* 100 units. Draw a horizontal *Line* **125** units from the last *Endpoint* using relative polar or direct distance entry coordinates. Draw another *Line* between that *Endpoint* and the *Quadrant* of the *Circle* on the right. Finally, draw a horizontal *Line* from point **30,70** and *Perpendicular* to the vertical *Line* on the right.

Figure 7-44

C. Draw two vertical *Lines* from the **Intersections** of the horizontal *Line* and *Circles* and **Perpendicular** to the *Line* at the bottom. Next, draw two **Circles** concentric to the previous two and with diameters of **20** and **10** as shown in Figure 7-45.

Figure 7-45

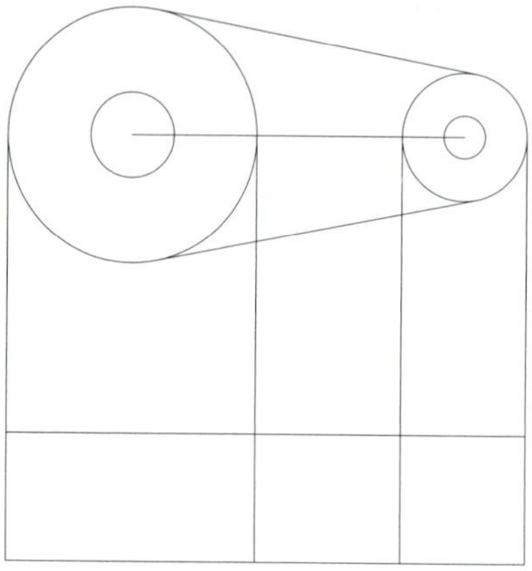

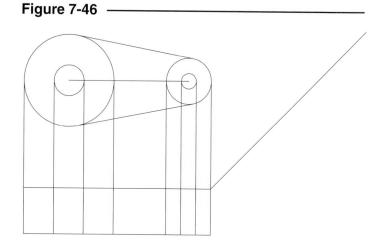

D. Draw four more vertical *Lines* as shown in Figure 7-46. Each *Line* is drawn from the new *Circles'* **Quadrant** and **Perpendicular** to the bottom line. Next, draw a miter *Line* from the **Intersection** of the corner shown to **@150<45**. *Save* the drawing for completion at a later time as another chapter exercise.

Figure 7-46

3. **Running Osnap**

Create a cross-sectional view of a door header composed of two 2 x 6 wooden boards and a piece of 1/2" plywood. (The dimensions of a 2 x 6 are actually 1-1/2" x 5-3/8".)

A. Begin a *New* drawing and assign the name **HEADER**. Draw four vertical lines as shown in Figure 7-47.

B. Use the *OSNAP* command or select *Drafting Settings...* from the *Tools* pull-down menu and turn on the *Endpoint* and *Intersection* modes.

Figure 7-47

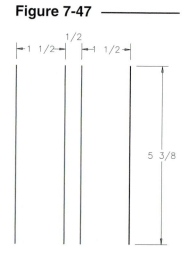

C. Draw the remaining lines as shown in Figure 7-48 to complete the header cross-section. *Save* the drawing.

Figure 7-48 ——

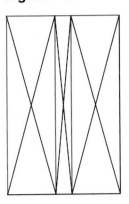

4. **Running Object Snap and *OSNAP* toggle**

Assume you are commissioned by the local parks and recreation department to provide a layout drawing for a major league sized baseball field. Lay out the location of bases, infield, and outfield as follows.

A. ***Open*** the **BALL FIELD CH6.DWG** that you set up in Chapter 6. Make sure *Limits* are set to 512',384', *Snap* is set to 10', and *Grid* is set to 20'. Ensure *SNAP* and *GRID* are on. Use *SaveAs* to save and rename the drawing to **BALL FIELD CH7**.

B. Begin drawing the baseball diamond by using the *Line* command and using direct distance coordinate entry. **PICK** the "Specify first point:" at **20',20'** (watch *Coords*). Draw the foul line to first base by turning on **ORTHO** or **POLAR**, move the cursor to the right (along the X direction) and enter a value of **90'** (don't forget the apostrophe to indicate feet). At the "Specify next point or [Undo]:" prompt, continue by drawing a vertical *Line* of **90'**. Continue drawing a square with **90'** between bases (Fig. 7-49).

Figure 7-49 ——

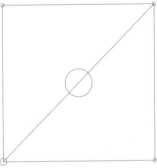

C. Invoke the *Drafting Settings* dialog box by any method. Turn on the *Endpoint*, *Midpoint*, and *Center* object snaps.

D. Use the *Circle* command to create a "base" at the lower-right corner of the square. At the "Specify center point for circle or [3P/2P/Ttr]:" prompt, **PICK** the *Line Endpoint* at the lower-right corner of the square. Enter a *Radius* of **1'** (don't forget the apostrophe). Since Running Object Snaps are on, AutoCAD should display the Marker (square box) at each of the *Line Endpoints* as you move the cursor near; therefore, you can easily "snap" the center of the bases (*Circles*) to the corners of the square. Draw *Circles* of the same *Radius* at second base (upper-right corner), and third base (upper-left corner of the square). Create home plate with a *Circle* of a **2'** *Radius* by the same method (see Figure 7-49).

E. Draw the pitcher's mound by first drawing a *Line* between home plate and second base. **PICK** the "Specify first point:" at home plate (*Center* or *Endpoint*), then at the "Specify next point or [Undo]:" prompt, PICK the *Center* or the *Endpoint* at second base. Construct a *Circle* of **8'** *Radius* at the *Midpoint* of the newly constructed diagonal line to represent the pitcher's mound (see Figure 7-49).

F. *Erase* the diagonal *Line* between home plate and second base. Draw the foul lines from first and third base to the outfield. For the first base foul line, construct a *Line* with the "Specify first point:" at the *Endpoint* of the existing first base line or *Center* of the base. Move the cursor (with *ORTHO* on) to the right (X positive) and enter a value of **240'** (don't forget the apostrophe). Press **Enter** to complete the *Line* command. Draw the third base foul line at the same length (in the positive Y direction) by the same method. (see Fig. 7-50).

Figure 7-50 ─────────

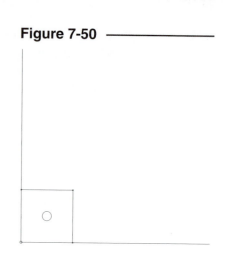

G. Draw the home run fence by using the *Arc* command from the *Draw* pull-down menu. Select the *Start, Center, End* method. **PICK** the end of the first base line (*Endpoint* object snap) for the *Start* of the *Arc* (Fig. 7-51, point 1), **PICK** the pitcher's mound (*Center* object snap) for the *Center* of the *Arc* (point 2), and the end of the third base line (*Endpoint* object snap) for the *End* of the *Arc* (point 3). Don't worry about *ORTHO* in this case because *OSNAP* overrides *ORTHO*.

Figure 7-51 ─────────

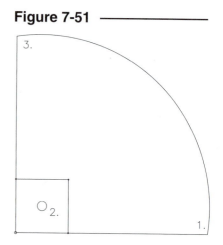

H. Now turn off **ORTHO**. Construct an *Arc* to represent the end of the infield. Select the *Arc Start, Center, End* method from the *Draw* pull-down menu. For the *Start* point of the *Arc*, toggle Running Osnaps <u>off</u> by pressing **F3, Ctrl+F,** or clicking the word **OSNAP** on the Status Bar and **PICK** location **140', 20', 0'** on the first base line (watch *Coords* and ensure *SNAP* and *GRID* are on). (See Figure 7-52, point 1) Next, toggle Running Osnaps back on, and **PICK** the *Center* of the pitcher's mound as the *Center* of the *Arc* (point 2). Third, toggle Running Osnaps off again and **PICK** the *End* point of the *Arc* on the third base line (point 3).

Figure 7-52 ─────────

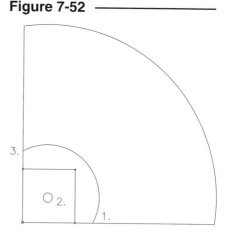

I. Lastly, the pitcher's mound should be moved to the correct distance from home plate. Type **M** (the command alias for *Move*) or select **Move** from the *Modify* pull-down menu. When prompted to "Select objects:" **PICK** the *Circle* representing pitcher's mound. At the "Specify base point or displacement:" prompt, toggle Running Osnaps <u>on</u> and **PICK** the *Center* of the mound. At the "Specify second point of displacement:" prompt, enter **@3'2"<225**. This should reposition the pitcher's mound to the regulation distance from home plate (60'-6"). Compare your drawing to Figure 7-52. *Save* the drawing (as BALL FIELD CH7).

J. Activate a *Layout* tab. Assuming your system is configured for a printer using an 11 x 8.5 sheet and is configured to automatically create a viewport (see "Setting Layout Options and Plot Options" in Chapter 6), a viewport should appear. (If the *Page Setup* dialog box automatically appears, select *OK*.) If a viewport appears, proceed to step L. If no viewport appears, complete step K.

K. To make a viewport, type *–Vports* at the Command prompt (don't forget the hyphen) and press **Enter**. Accept the default (*Fit*) option by pressing **Enter**. A viewport appears to fit the printable area.

L. Double-click inside the viewport, then type *Z* for *Zoom* and press **Enter**, and type *A* for *All* and press **Enter**. Activate the *Viewports* drop-down list by right-clicking on any icon button and selecting *Viewports* from the list of available toolbars that appears. Select *1/64"=1'* from the list. Make a print of the drawing on an 11 x 8.5 inch sheet.

M. Activate the *Model* tab. *Save* the drawing as **BALL FIELD CH7**.

5. **Polar Tracking and Object Snap Tracking**

A. In this exercise, you will create a table base using Polar Tracking and Object Snap Tracking to locate points for construction of holes and other geometry. Begin a *New* drawing and use the **ACAD.DWT** template. *Save* the drawing and assign the name **TABLE-BASE**. Set the drawing limits to **48 x 32**.

B. Invoke the *Drafting Settings* dialog box. In the *Snap and Grid* tab, set *Polar Distance* to **1.00** and make *Polar Snap* current. In the *Polar Tracking* tab, set the *Increment Angle* to **45**. In the *Object Snap* tab, turn on the *Endpoint*, *Midpoint* and *Center Osnap* options. Select *OK*. On the Status Bar ensure *SNAP*, *POLAR*, *OSNAP*, and *OTRACK* are on.

Figure 7-53 ————————

C. Use a *Line* and create the three line segments representing the first leg as shown in Figure 7-53. Begin at the indicated location. Use Polar Tracking and Polar Snap to assist drawing the diagonal lines segments. (Do not create the dimensions in your drawing.)

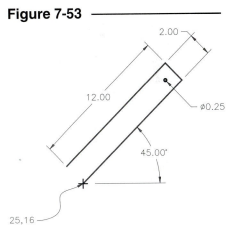

Figure 7-54 ————————

D. Place the first drill hole at the end of the leg using *Circle* with the *Center, Radius* option. Use Object Snap Tracking to indicate the center for the hole (*Circle*). Track vertically and horizontally from the indicated corners of the leg in Figure 7-54. Use *Endpoint Osnap* to snap to the indicated corners. (See Figure 7-53 for hole dimension.)

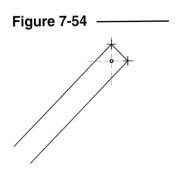

E. Create the other 3 legs in a similar fashion. You can track from the *Endpoints* on the bottom of the existing leg to identify the "first point" of the next *Line* segment as shown in Figure 7-55.

Figure 7-55

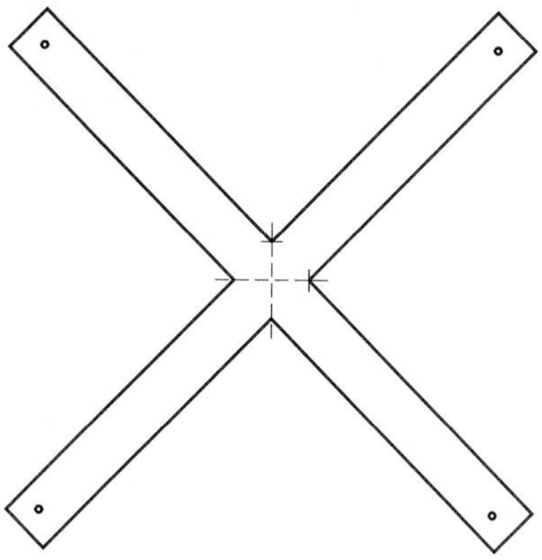

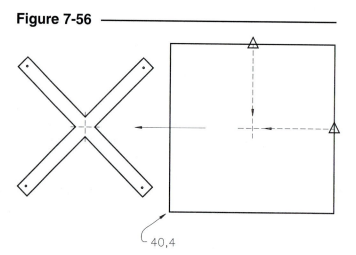

F. Use *Line* to create a 24" x 24" square table top on the right side of the drawing as shown in Figure 7-56. Use Object Snap Tracking and Polar Tracking with the *Move* command to move the square's center point to the center point of the 4 legs (HINT: At the "Specify base point or displacement:" prompt, Osnap Track to the square's *Midpoints*. At the "Specify second point of displacement:" prompt, use *Endpoint* Osnaps to locate the legs' center.

Figure 7-56

40,4

G. Create a smaller square (*Line*) in the center of the table which will act as a support plate for the legs. Track from the *Midpoint* of the legs to create the lines as indicated (highlighted) in Figure 7-57.

Figure 7-57

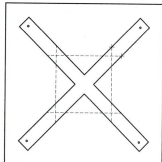

H. *Save* the drawing. You will finish the table base in Chapter 9.

8

DRAW
COMMANDS I

Chapter Objectives

After completing this chapter you should:

1. know where to locate and how to invoke the draw commands;

2. be able to draw *Lines*;

3. be able to draw *Circles* by each of the five options;

4. be able to draw *Arcs* by each of the eleven options;

5. be able to create *Point* objects and specify the *Point Style*;

6. be able to create *Plines* with width and combined of line and arc segments.

CONCEPTS

Draw Commands—Simple and Complex

Draw commands create objects. An object is the smallest component of a drawing. The draw commands listed immediately below create simple objects and are discussed in this chapter. Simple objects appear as one entity.

Figure 8-1 —

Line, Circle, Arc, and *Point*

Other draw commands create more complex shapes. Complex shapes appear to be composed of several components, but each shape is usually one object. An example of an object that is one entity but usually appears as several segments is listed below and is also covered in this chapter:

Pline

Other draw commands discussed in Chapter 15 (listed below) are a combination of simple and complex shapes:

Xline, Ray, Polygon, Rectangle, Donut, Spline, Ellipse, Divide, Mline, Measure, Sketch, Solid, Region, and *Boundary*

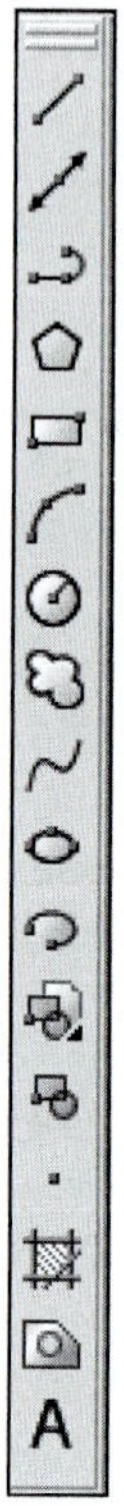

Draw Command Access

As a review from Chapter 3, Draw Command Concepts, remember that any of the five methods can be used to access the draw commands: *Draw* toolbar (Fig. 8-1), *Draw* pull-down menu (Fig. 8-2), *DRAW1* and *DRAW2* screen menus, keyboard entry of the command or alias, and digitizing tablet icons.

Figure 8-2 —————————————————

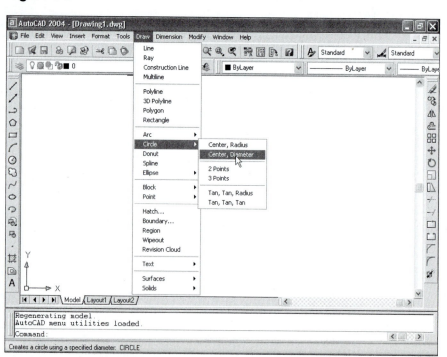

Coordinate Entry

When creating objects with draw commands, AutoCAD always prompts you to indicate points (such as endpoints, centers, radii) to describe the size and location of the objects to be drawn. An example you are familiar with is the *Line* command, where AutoCAD prompts for the "Specify first point:". Indication of these points, called <u>coordinate entry</u> can be accomplished by five formats (for 2D drawings):

1.	**Interactive**	**PICK** points on screen with input device
2.	**Absolute coordinates**	**X,Y**
3.	**Relative rectangular coordinates**	**@X,Y**
4.	**Relative polar coordinates**	**@distance<angle**
5.	**Direct distance entry**	**dist,direction** (Type a distance value relative to the last point, indicate direction with the cursor, then press Enter.)

Any of these methods can be used <u>whenever</u> AutoCAD prompts you to specify points. (For practice with these methods, see Chapter 3, Draw Command Concepts.)

Interactive Entry Tools

Also keep in mind that you can specify points interactively using the following AutoCAD features individually or in combination: Grid Snap, Polar Snap, Ortho, Polar Tracking, Object Snap, and Object Snap Tracking. Use these drawing tools <u>whenever</u> AutoCAD prompts you to select points. (See Chapters 3 and 7 for details on these tools.)

COMMANDS

LINE

Pull-down Menu	COMMAND (TYPE)	ALIAS (TYPE)	Short-cut	Screen (side) Menu	Tablet Menu
Draw *Line*	LINE	L	...	*DRAW 1* *Line*	*J,10*

This is the fundamental drawing command. The *Line* command creates straight line segments; each segment is an object. One or several line segments can be drawn with the *Line* command.

> Command: **Line**
> Specify first point: **PICK** or (**coordinates**) (A point can be designated by interactively selecting with the input device or by entering coordinates. Use any of the drawing tools to assist with interactive entry.)
> Specify next point or [Undo]: **PICK** or (**coordinates**) (Again, device input or keyboard input can be used.)
> Specify next point or [Undo]: **PICK** or (**coordinates**)
> Specify next point or [Close/Undo]: **PICK** or (**coordinates**) or *C*
> Specify next point or [Close/Undo]: press **Enter** to finish command
> Command:

Figure 8-3

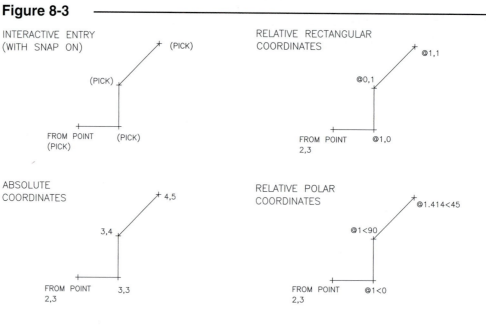

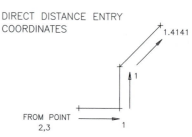

Figure 8-3 shows five examples of creating the same *Line* segments using different methods of coordinate entry.

Refer to Chapter 3, Draw Command Concepts, for examples of drawing vertical, horizontal, and inclined lines using the five formats for coordinate entry.

CIRCLE

Pull-down Menu	COMMAND (TYPE)	ALIAS (TYPE)	Short-cut	Screen (side) Menu	Tablet Menu
Draw *Circle >*	*CIRCLE*	*C*	...	DRAW 1 *Circle*	J,9

The *Circle* command creates one object. Depending on the option selected, you can provide two or three points to define a *Circle*. As with all commands, the Command line prompt displays the possible options:

Command: **Circle**
Specify center point for circle or [3P/2P/Ttr (tan tan radius)]: **PICK** or (**coordinates**) or (**option**).
(PICKing or entering coordinates designates the center point for the circle. You can enter 3P, 2P, or T for another option.)

As with most commands, the default and other options are displayed on the Command line. The default option always appears first. The other options can be invoked by typing the numbers and/or uppercase letter(s) that appear in brackets [ ].

All of the options for creating *Circles* are available from the *Draw* pull-down menu and from the *DRAW 1* screen (side) menu. The tool (icon button) for only the *Center, Radius* method is included in the *Draw* toolbar. Although the *Circle* options are not available on the *Draw* toolbar, you can use the *Circle* button, then right-click for a shortcut menu showing all the options.

The options, or methods, for drawing *Circles* are listed below. Each figure gives several possibilities for each option, with and without *OSNAP*s.

Center, Radius
Specify a center point, then a radius (Fig. 8-4).

The *Radius* (or *Diameter*) can be specified by entering values or by indicating a length inter-actively (PICK two points to specify a length when prompted). As always, points can be specified by PICKing or entering coordinates. Watch *Coords* for coordinate or distance (polar format) display. Grid Snap, Polar Snap, Polar Tracking, Object Snap, and Object Snap Tracking can be used for interactive point specification.

Figure 8-4

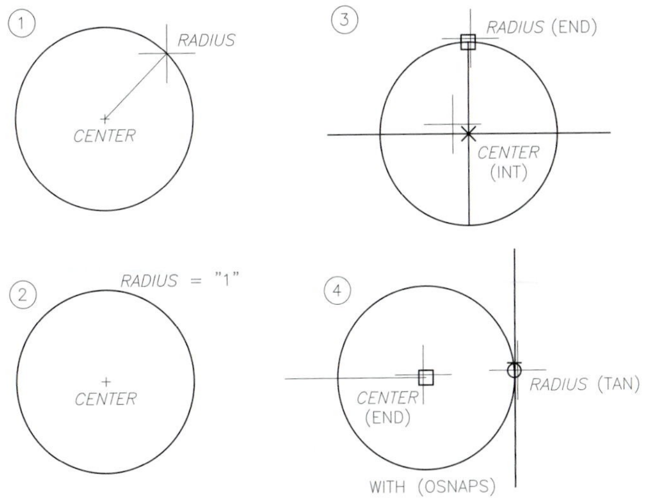

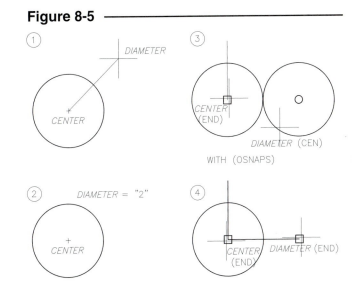

Center, Diameter
Specify the center point, then the diameter (Fig. 8-5).

Figure 8-5

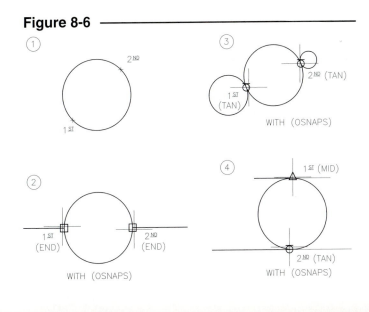

2 Points
The two points specify the location and diameter.

The *Tangent OSNAP*s can be used when selecting points with the *2 Point* and *3 Point* options, as shown in Figures 8-6 and 8-7.

Figure 8-6

3 Points

The *Circle* passes through all three points specified.

Figure 8-7 ──────────────

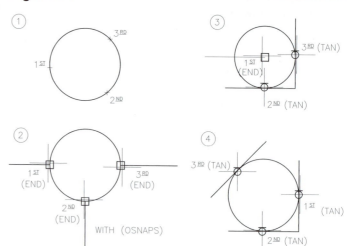

Tangent, Tangent, Radius

Specify two objects for the *Circle* to be tangent to; then specify the radius.

The *TTR* (Tangent, Tangent, Radius) method is extremely efficient and productive. The *OSNAP Tangent* modes are automatically invoked. This is the <u>only</u> draw command option that automatically calls *OSNAP*s.

Figure 8-8 ──────────────

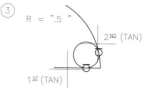

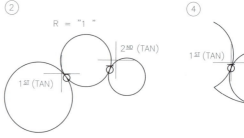

ARC

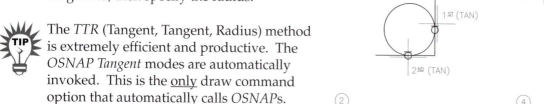

	Pull-down Menu	COMMAND (TYPE)	ALIAS (TYPE)	Short-cut	Screen (side) Menu	Tablet Menu
	Draw Arc >	ARC	A	...	DRAW 1 Arc	R,10

An arc is part of a circle; it is a regular curve of <u>less</u> than 360 degrees. The *Arc* command in AutoCAD provides eleven options for creating arcs. An *Arc* is one object. *Arcs* are always drawn by default in a <u>counterclockwise</u> direction. This occurrence forces you to decide in advance which points should be designated as *Start* and *End* points (for options requesting those points). For this reason, it is often easier to create arcs by another method, such as drawing a *Circle* and then using *Trim* or using the *Fillet* command. (See "Use *Arcs* or *Circles*?" at the end of this section on *Arcs*.) The *Arc* command prompt is:

Command: **Arc**
Specify start point of arc or [Center]: **PICK** or (**coordinates**) or **C** (Interactively select or enter coordinates in any format for the start point. Type C to use the Center option.)

The prompts displayed by AutoCAD are different depending on which option is selected. At any time while using the command, you can select from the options listed on the Command line by typing in the capitalized letter(s) for the desired option.

Alternately, to use a particular option of the *Arc* command, you can select from the *Draw* pull-down menu or from the *DRAW 1* screen (side) menu. The tool (icon button) for only the *3 Points* method is included in the *Draw* toolbar. However, you can select the *Arc* button, then right-click for a shortcut menu at any time during the *Arc* command to show other possible options at that point. These options require coordinate entry of <u>points in a specific order</u>.

Figure 8-9 ————————————————

3Points

Specify three points through which the *Arc* passes (Fig. 8-9).

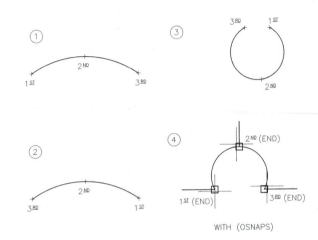

Start, Center, End

The radius is defined by the first two points that you specify (Fig. 8-10).

Figure 8-10 ————————————————

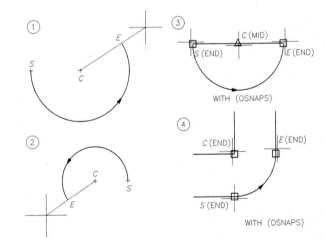

Start, Center, Angle

The angle is the <u>included</u> angle between the sides from the center to the endpoints. A <u>negative</u> angle can be entered to generate an *Arc* in a <u>clockwise</u> direction.

Figure 8-11 ————————————————

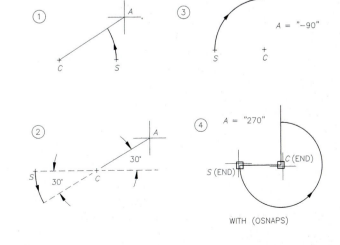

Start, Center, Length

Length means length of chord. The length of chord is between the start and the other point specified. A negative chord length can be entered to generate an *Arc* of 180+ degrees.

Figure 8-12

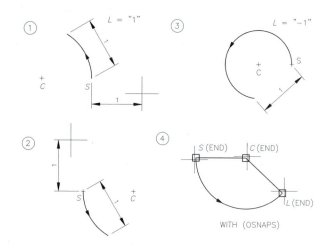

Start, End, Angle

The included angle is between the sides from the center to the endpoints. Negative angles generate clockwise *Arcs*.

Figure 8-13

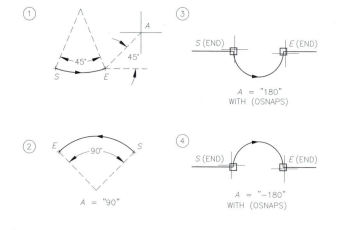

Start, End, Radius

The radius can be PICKed or entered as a value. A negative radius value generates an *Arc* of 180+ degrees.

Figure 8-14

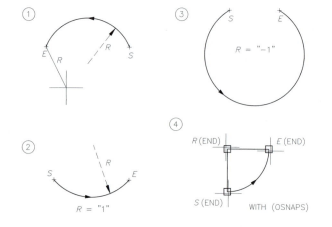

Start, End, Direction
The direction is tangent to the start point.

Figure 8-15

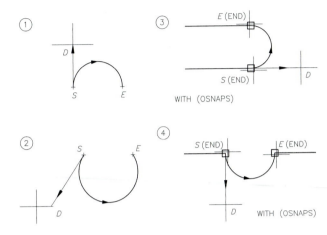

Center, Start, End
This option is like *Start, Center, End* but in a different order.

Figure 8-16

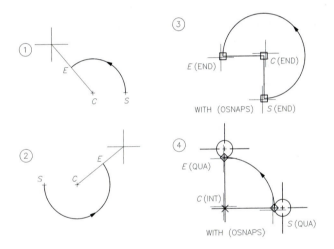

Center, Start, Angle
This option is like *Start, Center, Angle* but in a different order.

Figure 8-17

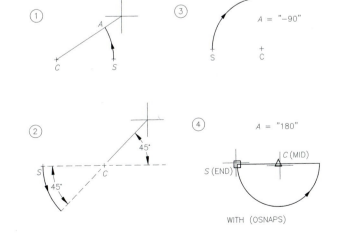

Center, Start, Length

This is similar to the *Start, Center, Length* option but in a different order. *Length* means length of chord.

Figure 8-18

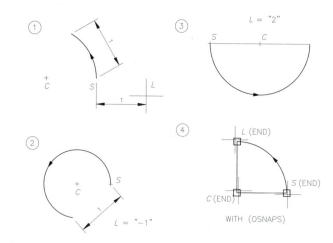

Continue

The new *Arc* continues from and is tangent to the last point. The only other point required is the endpoint of the *Arc*. This method allows drawing *Arcs* tangent to the preceding *Line* or *Arc*.

Arcs are always created in a <u>counterclockwise</u> direction. This fact must be taken into consideration when using any method <u>except</u> the *3-Point*, the *Start, End, Direction,* and the *Continue* options. The direction is explicitly specified with *Start, End, Direction* and *Continue* methods, and direction is irrelevant for *3-Point* method.

Figure 8-19

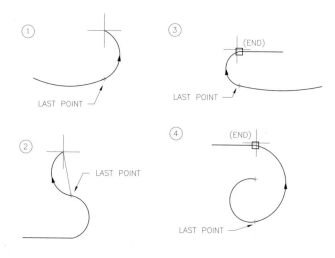

As usual, points can be specified by PICKing or entering coordinates. Watch *Coords* to display coordinate values or distances. Grid Snap, Polar Snap, Polar Tracking, Object Snap, and Object Snap Tracking can be used when PICKing. The *Endpoint, Intersection, Center, Midpoint,* and *Quadrant OSNAP* options can be used with great effectiveness. The *Tangent OSNAP* option <u>cannot</u> be used effectively with most of the *Arc* options. The *Radius, Direction, Length,* and *Angle* specifications can be given by entering values or by PICKing with or without *OSNAPs*.

Use *Arcs* or *Circles*?

Although there are sufficient options for drawing *Arcs,* <u>usually it is easier to use the *Circle* command</u> followed by *Trim* to achieve the desired arc. Creating a *Circle* is generally an easier operation than using *Arc* because the counterclockwise direction does not have to be considered. The unwanted portion of the circle can be *Trimmed* at the *Intersection* of or *Tangent* to the connecting objects using *OSNAP*. The *Fillet* command can also be used instead of the *Arc* command to add a fillet (arc) between two existing objects (see Chapter 9, Modify Commands I).

POINT

	Pull-down Menu	COMMAND (TYPE)	ALIAS (TYPE)	Short-cut	Screen (side) Menu	Tablet Menu
	Draw *Point >* *Single Point* or *Multiple Point*	*POINT*	*PO*	...	DRAW 2 *Point*	*O,9*

A *Point* is an object that has no dimension; it only has location. A *Point* is specified by giving only one coordinate value or by PICKing a location on the screen.

Figure 8-20 compares *Points* to *Line* and *Circle* objects.

> Command: **point**
> Current point modes: PDMODE=0 PDSIZE=0.0000
> Specify a point: **PICK** or (**coordinates**)
> (Select a location for the *Point* object.)

Points are useful in construction of drawings to locate points of reference for subsequent construction or locational verification. The *Node OSNAP* option is used to snap to *Point* objects.

Figure 8-20

LINE

POINT OBJECTS   CIRCLE

Points are drawing objects and therefore appear in prints and plots. The default "style" for points is a tiny dot. The *Point Style* dialog box can be used to define the format you choose for *Point* objects (the *Point* type [PDMODE] and size [PDSIZE]).

The *Draw* pull-down menu offers the *Single Point* and the *Multiple Point* options. The *Single Point* option creates one *Point,* then returns to the command prompt. This option is the same as using the *Point* command by any other method. Selecting *Multiple Point* continues the *Point* command until you press the Escape key.

DDPTYPE

	Pull-down Menu	COMMAND (TYPE)	ALIAS (TYPE)	Short-cut	Screen (side) Menu	Tablet Menu
	Format *Point Style...*	*DDPTYPE*	...	...	DRAW 2 *Point* *Ddptype:*	*U,1*

The *Point Style* dialog box (Fig. 8-21) is available only through the methods listed in the command table above. This dialog box allows you to define the format for the display of *Point* objects. The selected style is applied immediately to all newly created *Point* objects. The *Point Style* controls the format of *Points* for printing and plotting as well as for the computer display.

You can set the *Point Size* in *Absolute* units or *Relative to Screen* (default option). The *Relative to Screen* option keeps the *Points* the same size on the display when you *Zoom* in and out, whereas setting *Point Size* in *Absolute Units* gives you control over the size of *Points* for prints and plots. The *Point Size* is stored in the *PDSIZE* system variable. The selected *Point Style* is stored in the *PDMODE* variable.

Figure 8-21

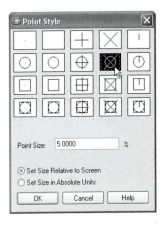

PLINE

	Pull-down Menu	COMMAND (TYPE)	ALIAS (TYPE)	Short-cut	Screen (side) Menu	Tablet Menu
	Draw Polyline	PLINE	PL	...	DRAW 1 *Pline*	N,10

A *Pline* (or *Polyline*) has special features that make this object more versatile than a *Line*. Three features are most noticeable when first using *Plines*:

1. A *Pline* can have a specified *width,* whereas a *Line* has no width.
2. Several *Pline* segments created with one *Pline* command are treated by AutoCAD as <u>one</u> object, whereas individual line segments created with one use of the *Line* command are individual objects.
3. A *Pline* can contain arc segments.

Figure 8-22 illustrates *Pline* versus *Line* and *Arc* comparisons.

The *Pline* command begins with the same prompt as *Line;* however, <u>after</u> the "start point:" is established, the *Pline* options are accessible.

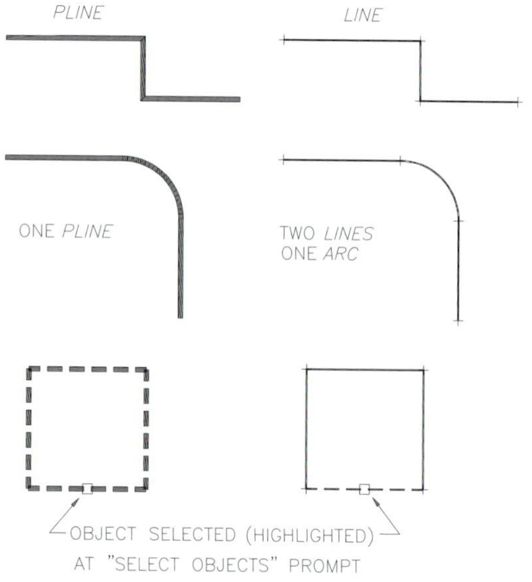

```
Command: Pline
Specify start point: PICK or (coordinates)
Current line-width is 0.0000
Specify next point or [Arc/Close/Halfwidth/Length/Undo/Width]:
```

Similar to the other drawing commands, you can invoke the *Pline* command by any method, then right-click for a shortcut menu at any time during the command to show other possible options at that point.

The options and descriptions follow.

Width
You can use this option to specify starting and ending widths. Width is measured perpendicular to the centerline of the *Pline* segment (Fig. 8-23). *Plines* can be tapered by specifying different starting and ending widths. See NOTE at the end of this section.

Halfwidth
This option allows specifying half of the *Pline* width. *Plines* can be tapered by specifying different starting and ending widths (Fig. 8-23).

Figure 8-22

PLINE   LINE

ONE *PLINE*   TWO *LINES* ONE *ARC*

OBJECT SELECTED (HIGHLIGHTED) AT "SELECT OBJECTS" PROMPT

Figure 8-23

WIDTH   HALF WIDTH

STARTING WIDTH   ENDING WIDTH   STARTING HALFWIDTH   ENDING HALFWIDTH

STARTING WIDTH   ENDING WIDTH   STARTING HALFWIDTH   ENDING HALFWIDTH

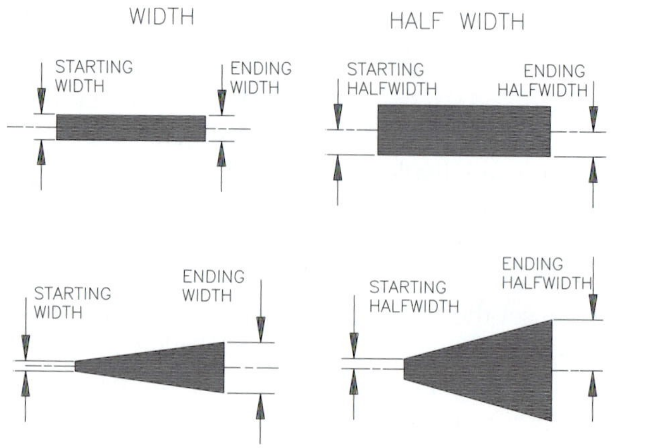

Arc

This option (by default) creates an arc segment in a manner similar to the *Arc Continue* method (Fig. 8-24). Any of several other methods are possible (see "*Pline Arc* Segments").

Figure 8-24

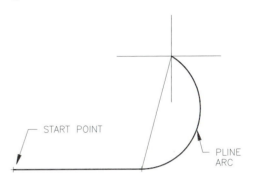

START POINT

PLINE ARC

Close

The *Close* option creates the closing segment connecting the first and last points specified with the current *Pline* command as shown in Figure 8-25.

This option can also be used to close a group of connected *Pline* segments into one continuous *Pline*. (A *Pline* closed by PICKing points has a specific start and endpoint.) A *Pline Closed* by this method has special properties if you use *Pedit* for *Pline* editing or if you use the *Fillet* command with the *Pline* option (see *Fillet* in Chapter 9 and *Pedit* in Chapter 16).

Figure 8-25

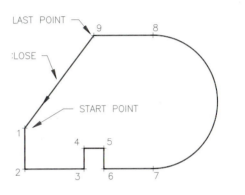

LAST POINT

CLOSE

START POINT

Length

Length draws a *Pline* segment at the same angle as and connected to the previous segment and uses a length that you specify. If the previous segment was an arc, *Length* makes the current segment tangent to the ending direction (Fig. 8-26).

Undo

Use this option to *Undo* the last *Pline* segment. It can be used repeatedly to undo multiple segments.

Figure 8-26

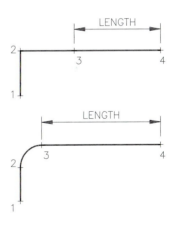

LENGTH

LENGTH

NOTE: If you change the *Width* of a *Pline*, be sure to respond to <u>both prompts</u> for width ("Specify starting width:" and "Specify ending width:") before you draw the first *Pline* segment. It is easy to hastily PICK the endpoint of the *Pline* segment after specifying the "Specify starting width:" instead of responding with a value (or Enter) for the "Specify ending width:". In this case (if you PICK at the "Specify ending width:" prompt), AutoCAD understands the line length that you interactively specified to be the ending width you want for the next line segment (you can PICK two points in response to the "Specify starting width:" or "Specify ending width:" prompts).

```
Command: Pline
Specify start point: PICK
Current line-width is 0.0000
Specify next point or [Arc/Close/Halfwidth/Length/Undo/Width]: w
Specify starting width <0.0000>: .2
Specify ending width <0.2000>: Enter a value or press Enter—do not PICK the "next point."
```

Pline Arc **Segments**

When the *Arc* option of *Pline* is selected, the prompt changes to provide the various methods for construction of arcs:

Specify endpoint of arc or [Angle/CEnter/CLose/Direction/Halfwidth/Line/Radius/Second pt/Undo/Width]:

Angle
You can draw an arc segment by specifying the included angle (a negative value indicates a clockwise direction for arc generation).

CEnter
This option allows you to specify a specific center point for the arc segment.

CLose
This option closes the *Pline* group with an arc segment.

Direction
Direction allows you to specify an explicit starting direction rather than using the ending direction of the previous segment as a default.

Line
This switches back to the line options of the *Pline* command.

Radius
You can specify an arc radius using this option.

Second pt
Using this option allows specification of a 3-point arc.

Because a shape created with one *Pline* command is <u>one object</u>, manipulation of the shape is generally easier than with several objects. For some applications, one *Pline* shape can have advantages over shapes composed of several objects (see *"Offset,"* Chapter 9). Editing *Plines* is accomplished by using the *Pedit* command. As an alternative, *Plines* can be "broken" back down into individual objects with *Explode*.

Drawing and editing *Plines* can be somewhat involved. As an alternative, you can draw a shape as you would normally with *Line, Circle, Arc, Trim,* etc., and then <u>convert</u> the shape to one *Pline* object using *Pedit*. (See Chapter 16 for details on converting *Lines* and *Arcs* to *Plines*.)

CHAPTER EXERCISES

Create a *New* drawing. When the *Create New Drawing* dialog box appears, select **Start from Scratch** and select **Imperial** settings. Set **Polar Snap On** and set the **Polar Distance** to **.25**. Turn on *SNAP, POLAR, OSNAP,* and *OTRACK*. *Save* the drawing as **CH8EX**. For each of the following problems, *Open* **CH8EX**, complete one problem, then use *SaveAs* to give the drawing a new name.

1. *Open* **CH8EX**. Create the geometry shown in Figure 8-27. Start the first *Circle* center at point **4,4.5** as shown. Do not copy the dimensions. *SaveAs* **LINK**. (HINT: Locate and draw the two small *Circles* first. Use *Arc, Start, Center, End* or *Center, Start, End* for the rounded ends.)

Figure 8-27

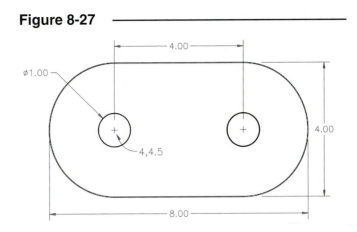

2. *Open* **CH8EX**. Create the geometry as shown in Figure 8-28. Do not copy the dimensions. Assume symmetry about the vertical axis. *SaveAs* **SLOTPLATE CH8**.

Figure 8-28

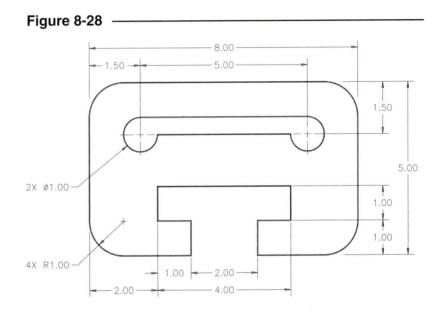

3. *Open* **CH8EX**. Create the shapes shown in Figure 8-29. Do not copy the dimensions. *SaveAs* **CH8EX3**.

   Draw the *Lines* at the bottom first, starting at coordinate **1,3**. Then create *Point* objects at **5,7**, **5.4,7**, **5.8,7**, etc. Change the *Point Style* to an X and *Regen*. Use the *NODe OSNAP* mode to draw the inclined *Lines*. Create the *Arc* on top with the *Start, End, Direction* option.

Figure 8-29

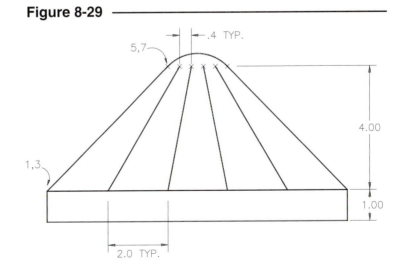

4. ***Open* CH8EX**. Create the shape shown in Figure 8-30. Draw the two horizontal *Lines* and the vertical *Line* first by specifying the endpoints as given. Then create the *Circle* and *Arcs*. *SaveAs* **CH8EX4**.

   HINT: Use the *Circle 2P* method with *Endpoint OSNAP*s. The two upper *Arcs* can be drawn by the *Start, End, Radius* method.

Figure 8-30 ————

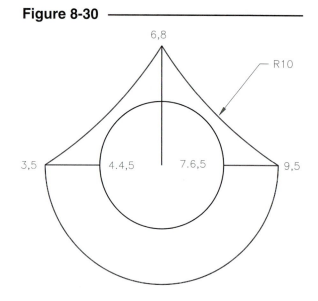

5. ***Open* CH8EX**. Draw the shape shown in Figure 8-31. Assume symmetry along a vertical axis. Start by drawing the two horizontal *Lines* at the base.

   Next, construct the side *Arcs* by the *Start, Center, Angle* method (you can specify a negative angle). The small *Arc* can be drawn by the *3P* method. Use *OSNAP*s when needed (especially for the horizontal *Line* on top and the *Line* along the vertical axis). *SaveAs* **CH8EX5**.

Figure 8-31 ————

6. ***Open* CH8EX**. Complete the geometry in Figure 8-32. Use the coordinates to establish the *Lines*. Draw the *Circles* using the *Tangent, Tangent, Radius* method. *SaveAs* **CH8EX6**.

Figure 8-32 ————

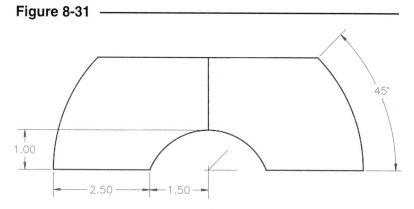

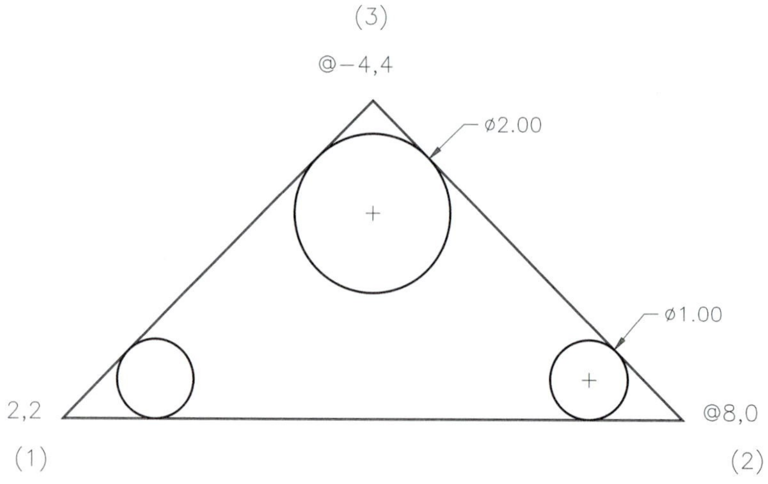

7. **Retaining Wall**

Figure 8-33

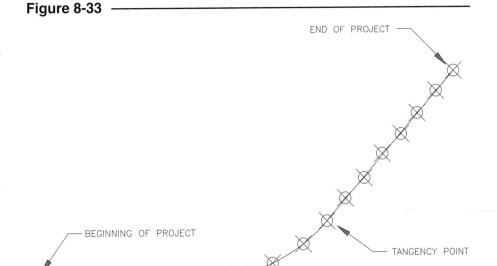

A contractor plans to stake out the edge of a retaining wall located at the bottom of a hill. The following table lists coordinate values based on a survey at the site. Use the drawing you created in Chapter 6 Exercises named **CH6EX2**. *Save* the drawing as **RET-WALL.**

Place *Points* at each coordinate value in order to create a set of data points. Determine the location of one *Arc* and two *Lines* representing the centerline of the retaining wall edge. The centerline of the retaining wall should match the data points as accurately as possible (Fig. 8-33). The data points given in the following table are in inch units.

Pt #	X	Y	
1.	60.0000	108.0000	
2.	81.5118	108.5120	
3.	101.4870	108.2560	
4.	122.2305	107.4880	
5.	141.4375	108.5120	
6.	158.1949	108.0000	
7.	192.0000	108.0000	Recommended tangency point
8.	215.3914	110.5185	
9.	242.3905	119.2603	
10.	266.5006	133.6341	
11.	285.1499	151.0284	Recommended tangency point
12.	299.5069	168.1924	
13.	314.2060	183.4111	
14.	328.3813	201.7784	
15.	343.0811	216.4723	
16.	355.6808	232.2157	
17.	370.3805	249.0087	
18.	384.0000	264.0000	

(Hint: Change the *Point Style* to an easily visible format.)

8. *Pline*

    Create the shape shown in
    Figure 8-34. Draw the outside
    shape with <u>one continuous</u>
    Pline (with 0.00 width).
    When finished, *SaveAs*
    PLINE1.

Figure 8-34

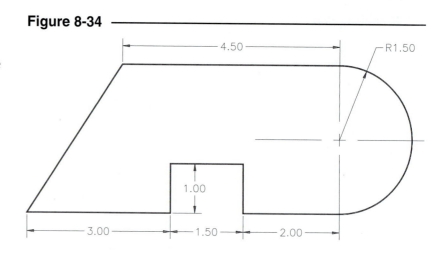

MODIFY COMMANDS I

Chapter Objectives

After completing this chapter you should:

1. be able to *Erase* objects;

2. be able to *Move* objects from a base point to a second point;

3. know how to *Rotate* objects about a base point;

4. be able to enlarge or reduce objects with *Scale*;

5. be able to *Stretch* objects and change the length of *Lines* and *Arcs* with *Lengthen*;

6. be able to *Trim* away parts of objects at cutting edges and *Extend* objects to boundary edges;

7. know how to use the four *Break* options;

8. be able to *Copy* objects and make *Mirror* images of selected objects;

9. know how to create parallel copies of objects with *Offset*;

10. be able to make *rectangular* and *polar Arrays* of objects;

11. be able to create a *Fillet* and a *Chamfer* between two objects.

CONCEPTS

Draw commands are used to create new objects. *Modify* commands are used to change existing objects or to use existing objects to create new and similar objects. The *Modify* commands covered first in this chapter (listed below) only <u>change existing objects</u>.

Erase, Move, Rotate, Scale, Stretch, Lengthen, Trim, Extend, and *Break*   **Figure 9-1**

The *Modify* commands covered near the end of this chapter (listed below) <u>use existing objects to create new and similar objects</u>. For example, the *Copy* command prompts you to select an object (or set of objects), then creates an identical object (or set).

Copy, Mirror, Offset, Array, Fillet and *Chamfer*

Modify commands can be invoked by any of the five command entry methods: toolbar icons, pull-down menus, screen menus, keyboard entry, and digitizing tablet menu. The *Modify* toolbar is visible and docked to the right side of the screen by default in AutoCAD 2004 (Fig. 9-1).

Several commands are found in the *Edit* pull-down menu such as *Cut*, *Copy*, and *Paste*. Although these command names appear similar (or the same in the case of *Copy*) to some of those in the *Modify* menu, these AutoCAD command names are actually *Cutclip*, *Copyclip*, and *Pasteclip*. These commands are for OLE operations (cutting and pasting) between two different AutoCAD drawings or other software applications. See Chapter 31, Object Linking and Embedding, for more information.

The *Modify* pull-down menu (Fig. 9-2) contains commands that change existing geometry and that use existing geometry to create new but similar objects. All of the *Modify* commands are contained in this single pull-down menu.

Figure 9-2

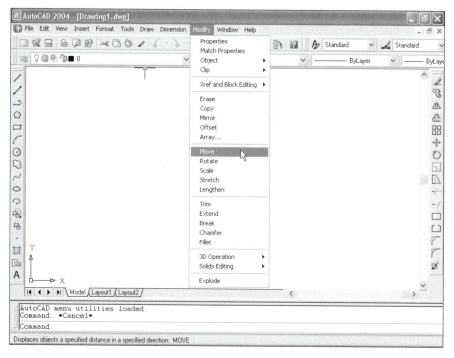

The *MODIFY1* and *MODIFY2* screen menus branch to the individual commands and options (Fig. 9-3).

Since all *Modify* commands affect or use existing geometry, the first step in using any *Modify* command is to construct a selection set (see Chapter 4). This can be done by one of two methods:

1. Invoking the desired command and then creating the selection set in response to the "Select Objects:" prompt (Verb/Noun syntax order) using any of the select object options;

2. Selecting the desired set of objects with the pickbox or *Auto Window* or *Crossing Window* <u>before</u> invoking the edit command (Noun/Verb syntax order).

Figure 9-3

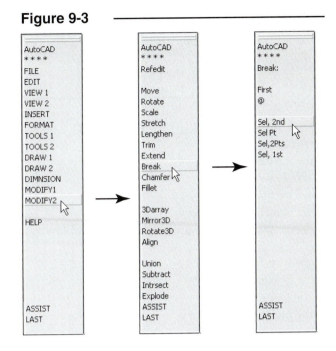

The first method allows use of any of the selection options (*Last, All, WPolygon, Fence,* etc.), while the latter method allows <u>only</u> the use of the pickbox and *Auto Window* and *Crossing Window*.

COMMANDS

ERASE

Pull-down Menu	COMMAND (TYPE)	ALIAS (TYPE)	Short-cut	Screen (side) Menu	Tablet Menu
Modify *Erase*	*ERASE*	*E*	**(Edit Mode)** *Erase*	*MODIFY1* *Erase*	*V,14*

The *Erase* command deletes the objects you select from the drawing. Any of the object selection methods can be used to highlight the objects to *Erase*. The only other required action is for you to press *Enter* to cause the erase to take effect.

```
Command: erase
Select Objects: PICK (Use any object selection method.)
Select Objects: PICK (Continue to select desired objects.)
Select Objects: Enter (Confirms the object selection process and causes Erase to take effect.)
Command:
```

If objects are erased accidentally, *U* can be used immediately following the mistake to undo one step, or *Oops* can be used to bring back into the drawing whatever was *Erased* the last time *Erase* was used (see Chapter 5). If only part of an object should be erased, use *Trim* or *Break*.

MOVE

Pull-down Menu	COMMAND (TYPE)	ALIAS (TYPE)	Short-cut	Screen (side) Menu	Tablet Menu
Modify *Move*	*MOVE*	*M*	(Edit Mode) *Move*	*MODIFY2* *Move*	*V,19*

Move allows you to relocate one or more objects from the existing position in the drawing to any other position you specify. After selecting the objects to *Move*, you must specify the "base point" and "second point of displacement." You can use any of the five coordinate entry methods to specify these points. Examples are shown in Figure 9-4.

Figure 9-4

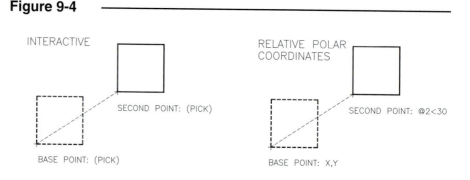
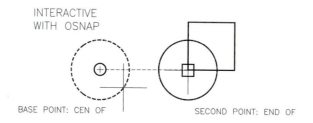

```
Command: move
Select Objects: PICK   (Use any of the object selection methods.)
Select Objects: PICK   (Continue to select other desired objects.)
Select Objects: Enter   (Press Enter to indicate selection of objects is complete.)
Specify base point or displacement: PICK or (coordinates) (This is the point to move from.  Select a
point to use as a "handle."  An Endpoint or Center, etc., can be used.)
Specify second point of displacement or <use first point as displacement>: PICK or (coordinates)
(This is the point to move to.  OSNAPs can also be used here.)
Command:
```

Keep in mind that *OSNAPs* can be used when PICKing any point. It is often helpful to toggle *ORTHO* or *POLAR ON* to force the *Move* in a horizontal or vertical direction.

If you know a specific distance and an angle that the set of objects should be moved, Polar Tracking or relative polar coordinates can be used. In the following sequence, relative polar coordinates are used to move objects 2 units in a 30 degree direction (see Figure 9-4, coordinate entry).

```
Command: move
Select Objects: PICK
Select Objects: PICK
Select Objects: Enter
Specify base point or displacement: X,Y (coordinates)
Specify second point of displacement or <use first point as displacement>:  @2<30
Command:
```

Direct distance coordinate entry can be used effectively with *Move*. For example, assume you wanted to move the right side view of a multiview drawing 20 units to the right using direct distance entry (Fig. 9-5). First, invoke the *Move* command and select all of the objects comprising the right side view in response to the "Select Objects:" prompt. The command sequence is as follows:

Figure 9-5

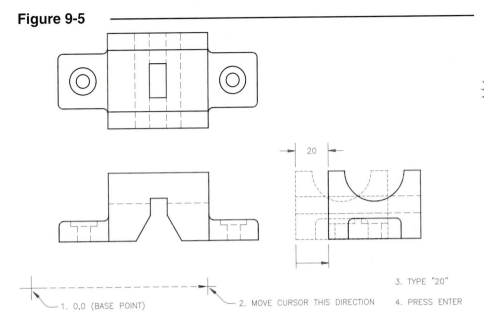

3. TYPE "20"

4. PRESS ENTER

1. 0,0 (BASE POINT)

2. MOVE CURSOR THIS DIRECTION

Command: **move**
Select Objects: **PICK**
Select Objects: **Enter**
Specify base point or displacement: **0,0** (or any value)
Specify second point of displacement or <use first point as displacement>: **20** (with *ORTHO* or *POLAR* on, move cursor to right), then press **Enter**

In response to the "Specify base point or displacement:" prompt, PICK a point or enter any coordinate pairs or single value. At the "Specify second point of displacement :" prompt, move the cursor to the right any distance (using Polar Tracking or *ORTHO* on), then type "20" and press Enter. The value specifies the distance, and the cursor location from the last point indicates the direction of movement.

In the previous example, note that <u>any value</u> can be entered in response to the "Specify base point or displacement:" prompt. If a single value is entered ("3," for example), AutoCAD recognizes it as direct distance entry. The point designated is 3 units from the last PICK point in the direction specified by wherever the cursor is at the time of entry.

ROTATE

	COMMAND (TYPE)	**ALIAS (TYPE)**	**Short-cut**	**Screen (side) Menu**	**Tablet Menu**
Pull-down Menu					
Modify *Rotate*	*ROTATE*	*RO*	(Edit Mode) *Rotate*	*MODIFY2* *Rotate*	*V,20*

Selected objects can be rotated to any position with this command. After selecting objects to *Rotate*, you select a "base point" (a point to rotate about) then specify an angle for rotation. AutoCAD rotates the selected objects by the increment specified from the original position.

For example, specifying a value of **45** would *Rotate* the selected objects 45 degrees counterclockwise from their current position; a value of **-45** would *Rotate* the objects 45 degrees in a clockwise direction (Fig. 9-6).

Figure 9-6

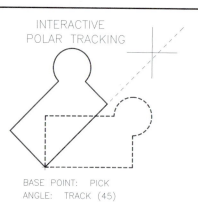

```
Command: rotate
Current positive angle in UCS: ANGDIR=counterclockwise  ANGBASE=0
Select objects: PICK
Select objects: PICK
Select objects: Enter (Indicates completion of object selection.)
Specify base point: PICK or (coordinates) (Select a point to rotate about.)
Specify rotation angle or [Reference]: PICK or (coordinates) (Interactively rotate the set or enter a
value for the number of degrees to rotate the object set.)
Command:
```

The base point is often selected interactively with *OSNAPs*. When specifying an angle for rotation, you can enter an angular value, use Polar Tracking, or turn on *ORTHO* for 90-degree rotation. Note the status of the two related system variables is given: *ANGDIR* (counterclockwise or clockwise rotation) and *ANGBASE* (base angle used for rotation).

The **Reference** option can be used to specify a vector as the original angle before rotation (Fig. 9-7). This vector can be indicated interactively (*OSNAPs* can be used) or entered as an angle using keyboard entry. Angular values that you enter in response to the "New angle:" prompt are understood by AutoCAD as <u>absolute</u> angles for the *Reference* option only.

Figure 9-7

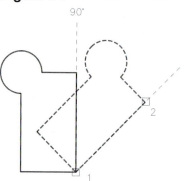

```
Command: rotate
Current positive angle in UCS: ANGDIR=counterclockwise
ANGBASE=0
Select objects: PICK
Select objects: Enter (Indicates completion of object selection.)
Specify base point: PICK or (coordinates) (Select a point to
rotate about.)
Specify rotation angle or [Reference]: R (Indicates the Reference
option.)
Specify the reference angle <0>: PICK or (value)  (PICK the first
point of the vector.)
Specify second point: PICK (Indicates the second point of the vector.)
Specify the new angle: (value) (Indicates the new angle.)
Command:
```

SCALE

Pull-down Menu	COMMAND (TYPE)	ALIAS (TYPE)	Short-cut	Screen (side) Menu	Tablet Menu
Modify *Scale*	*SCALE*	*SC*	(Edit Mode) *Scale*	*MODIFY2* *Scale*	*V,21*

The *Scale* command is used to increase or decrease the size of objects in a drawing. The *Scale* command does not normally have any relation to plotting a drawing to scale.

After selecting objects to *Scale*, AutoCAD prompts you to select a "Base point:", which is the <u>stationary point</u>. You can then scale the size of the selected objects interactively or enter a scale factor (Fig. 9-8).

Figure 9-8

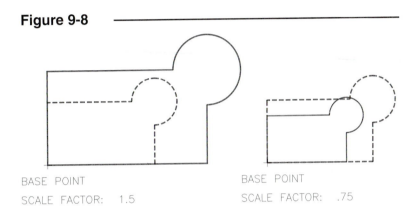

BASE POINT

SCALE FACTOR: 1.5

BASE POINT

SCALE FACTOR: .75

Using interactive input, you are presented with a rubberband line connected to the base point. Making the rubberband line longer or shorter than 1 unit increases or decreases the scale of the selected objects by that proportion; for example, pulling the rubberband line to two units length increases the scale by a factor of two.

        Command: *scale*
        Select objects: **PICK** or (**coordinates**) (Select the objects to scale.)
        Select objects: **Enter** (Indicates completion of the object selection.)
        Specify base point: **PICK** or (**coordinates**) (Select the stationary point.)
        Specify scale factor or [Reference]: **PICK** or (**value**) or **R** (Interactively scale the set of objects or enter a value for the scale factor.)
        Command:

It may be desirable in some cases to use the ***Reference*** option to specify a value or two points to use as the reference length. This length can be indicated interactively (*OSNAP*s can be used) or entered as a value. This length is used for the subsequent reference length that the rubberband uses when interactively scaling. For example, if the reference distance is 2, then the "rubberband" line must be stretched to a length greater than 2 to increase the scale of the selected objects.

Scale normally should <u>not</u> be used to change the scale of an entire drawing in order to plot on a specific size sheet. CAD drawings should be created <u>full size</u> in <u>actual units</u>.

STRETCH

Pull-down Menu	COMMAND (TYPE)	ALIAS (TYPE)	Short-cut	Screen (side) Menu	Tablet Menu
Modify *Stretch*	*STRETCH*	*S*	...	MODIFY2 *Stretch*	*V,22*

Objects can be made longer or shorter with *Stretch*. The power of this command lies in the ability to *Stretch* groups of objects while retaining the connectivity of the group (Fig. 9-9). When *Stretched*, *Lines* and *Plines* become longer or shorter and *Arcs* change radius to become longer or shorter. *Circles* do not stretch; rather, they move if the center is selected within the Crossing Window.

Objects to *Stretch* are not selected by the typical object selection methods but are indicated by a <u>Crossing Window or Crossing Polygon</u> only. The *Crossing Window or Polygon* should be created so the objects to *Stretch* <u>cross</u> through the window. *Stretch* actually moves the object endpoints that are located within the *Crossing Window*.

Figure 9-9 ————————————

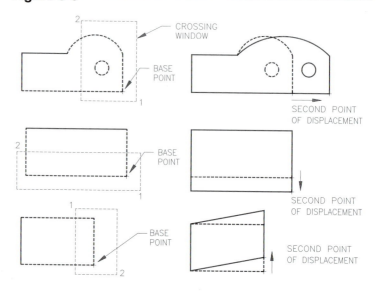

Following is the command sequence for *Stretch*.

```
Command: stretch
Select objects to stretch by crossing-window or crossing-polygon...
Select objects: PICK
Specify opposite corner: PICK
Select objects: Enter
Specify base point or displacement: PICK or (coordinates) (Select a point to stretch from.)
Specify second point of displacement: PICK or (coordinates) (Select a point to stretch to.)
Command:
```

Stretch can be used to lengthen one object while shortening another. Application of this ability would be repositioning a door or window on a wall (Fig. 9-10).

In summary, the *Stretch* command <u>stretches objects that cross</u> the selection Window and <u>moves objects that are completely within</u> the selection Window, as shown in Figure 9-10.

Figure 9-10 ————————————

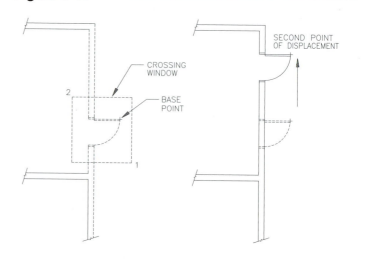

LENGTHEN

	Pull-down Menu	COMMAND (TYPE)	ALIAS (TYPE)	Short-cut	Screen (side) Menu	Tablet Menu
	Modify *Lengthen*	*LENGTHEN*	*LEN*	...	MODIFY2 *Lengthen*	*W,14*

Lengthen changes the length (longer or shorter) of linear objects and arcs. No additional objects are required (as with *Trim* and *Extend*) to make the change in length. Many methods are provided as displayed in the command prompt.

    Command: **lengthen**
    Select an object or [DElta/Percent/Total/DYnamic]:

Select an object
Selecting an object causes AutoCAD to report the current length of that object. If an *Arc* is selected, the included angle is also given.

DElta
Using this option returns the prompt shown next. You can change the current length of an object (including an arc) by any increment that you specify. Entering a positive value increases the length by that amount, while a negative value decreases the current length. The end of the object that you select changes while the other end retains its current endpoint (Fig. 9-11).

Figure 9-11

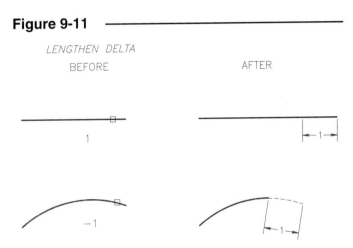

LENGTHEN DELTA
BEFORE
AFTER

1

−1

    Enter delta length or [Angle] <0.0000>: (**value**) or **a**

The *Angle* option allows you to change the <u>included angle</u> of an arc (the length along the curvature of the arc can be changed with the *Delta* option). Enter a positive or negative value (degrees) to add or subtract to the current included angle, then select the end of the object to change (Fig. 9-12).

Figure 9-12

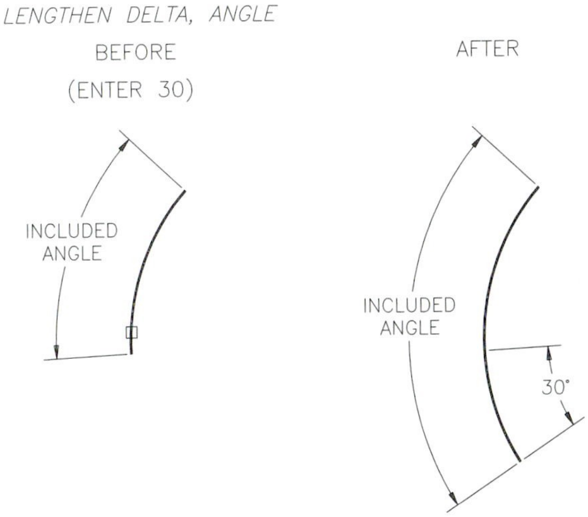

LENGTHEN DELTA, ANGLE
BEFORE
(ENTER 30)
AFTER

INCLUDED ANGLE

INCLUDED ANGLE

30°

Percent

Use this option if you want to change the length by a percentage of the current total length. For arcs, the percentage applied affects the length and the included angle equally, so there is no *Angle* option. A value of greater than 100 increases the current length, and a value of less than 100 decreases the current length. Negative values are not allowed. The end of the object that you select changes.

Figure 9-13

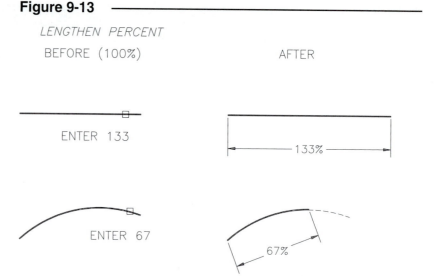

Total

This option lets you specify a value for the new total length. Simply enter the value and select the end of the object to change. The angle option is used to change the total included angle of a selected arc.

Specify total length or [Angle]
<1.0000)>: (**value**) or **A**

Figure 9-14

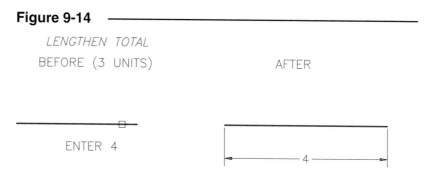

DYnamic

This option allows you to change the length of an object by dynamic dragging. Select the end of the object that you want to change. Object Snaps and Polar Tracking can be used.

Figure 9-15

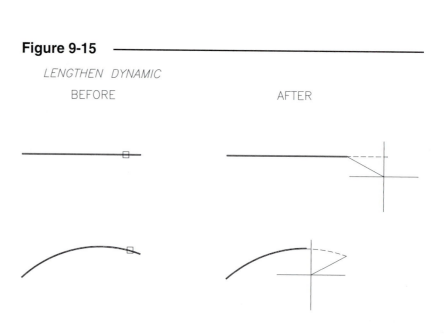

TRIM

	Pull-down Menu	COMMAND (TYPE)	ALIAS (TYPE)	Short-cut	Screen (side) Menu	Tablet Menu
	Modify *Trim*	*TRIM*	*TR*	...	MODIFY2 *Trim*	*W,15*

The *Trim* command allows you to trim (shorten) the end of an object back to the intersection of another object (Fig. 9-16). The middle section of an object can also be *Trimmed* between two intersecting objects. There are two steps to this command: first, PICK one or more "cutting edges" (existing objects); then PICK the object or objects to *Trim* (portion to remove). The cutting edges are highlighted after selection. Cutting edges themselves can be trimmed if they intersect other cutting edges, but lose their highlight when trimmed.

Figure 9-16

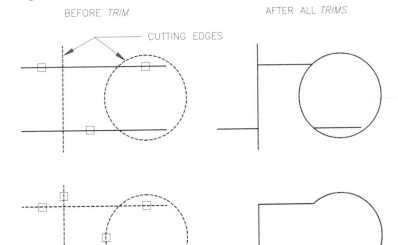

```
Command: Trim
Current settings: Projection=UCS
Edge=None
Select cutting edges ...
Select objects: PICK (Select an
object to use as a cutting edge.)
Select objects: Enter
Select object to trim or shift-select to extend or [Project/Edge/Undo]: PICK (Select the end of an
     object to trim.)
Select object to trim or shift-select to extend or [Project/Edge/Undo]: PICK
Select object to trim or shift-select to extend or [Project/Edge/Undo]: Enter
Command:
```

Edge

The *Edge* option can be set to *Extend* or *No extend*. In the *Extend* mode, objects that are selected as trimming edges will be <u>imaginarily extended</u> to serve as a cutting edge. In other words, lines used for trimming edges are treated as having infinite length (Fig. 9-17). The *No extend* mode considers only the actual length of the object selected as trimming edges.

Figure 9-17

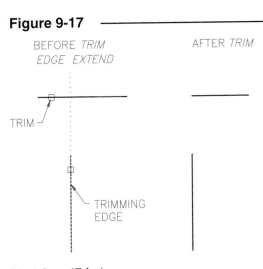

```
Command: Trim
Current settings: Projection=UCS Edge=None
Select cutting edges ...
Select objects: PICK (Select an object to use as a
cutting edge.)
Select objects: Enter
Select object to trim or shift-select to extend or [Project/Edge/Undo]: e  (Edge)
Enter an implied edge extension mode [Extend/No extend] <No extend>: e  (Extend)
Select object to trim or shift-select to extend or [Project/Edge/Undo]: PICK  (Select an object to trim.)
Select object to trim or shift-select to extend or [Project/Edge/Undo]: Enter
Command:
```

Projection

The *Projection* switch controls how *Trim* and *Extend* operate in 3D space. *Projection* affects the projection of the cutting edge and boundary edge. The three options are described here:

> Select object to trim or shift-select to extend or [Project/Edge/Undo]: **p** (*Project*)
> Enter a projection option [None/Ucs/View] <Ucs>:

None

The *None* option does not project the cutting edge. This mode is used for normal 2D drawing when all objects (cutting edges and objects to trim) lie in the current drawing plane. You can use this mode to *Trim* in 3D space when all objects lie in the plane of the current UCS or in planes parallel to it and the objects physically intersect. You can also *Trim* objects not in the UCS if the objects physically intersect, are in the same plane, and are not perpendicular to the UCS. In any case, the <u>objects must physically intersect</u> for trimming to occur.

The other two options (*UCS* and *View*) allow you to trim objects that do not physically intersect in 3D space.

UCS

This option projects the cutting edge perpendicular to the UCS. Any objects crossing the "projected" cutting edge in 3D space can be trimmed. For example, selecting a *Circle* as the cutting edge creates a projected cylinder used for cutting (Fig. 9-18).

Figure 9-18

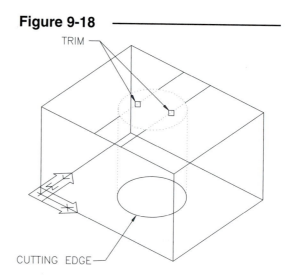

View

The *View* option allows you to trim objects that <u>appear</u> to intersect from the current viewpoint. The objects do not have to physically intersect in 3D space. The cutting edge is projected perpendicularly to the screen (parallel to the line of sight). This mode is useful for trimming "hidden edges" of a wireframe model to make the surfaces appear opaque (Fig. 9-19).

Figure 9-19

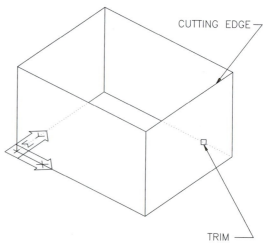

Undo

The *Undo* option allows you to undo the last *Trim* in case of an accidental trim.

Shift-Select

See "*Trim* and *Extend* Shift-Select Option."

EXTEND

	COMMAND (TYPE)	ALIAS (TYPE)	Short-cut	Screen (side) Menu	Tablet Menu
Pull-down Menu					
Modify *Extend*	*EXTEND*	*EX*	...	*MODIFY2* *Extend*	*W,16*

Extend can be thought of as the opposite of *Trim*. Objects such as *Lines, Arcs,* and *Plines* can be *Extended* until intersecting another object called a boundary edge (Fig. 9-20). The command first requires selection of <u>existing</u> objects to serve as boundary edge(s) which become high-lighted; then the objects to extend are selected. Objects extend until, and only if, they eventually intersect a boundary edge. An *Extended* object acquires a new endpoint at the boundary edge intersection.

Figure 9-20

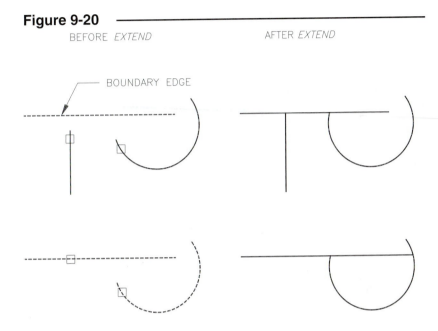

BEFORE *EXTEND*   AFTER *EXTEND*

BOUNDARY EDGE

Command: **extend**
Current settings: Projection=UCS Edge=None
Select boundary edges ...
Select objects: **PICK** (Select boundary edge.)
Select objects: **Enter**
Select object to extend or shift-select to trim or [Project/Edge/Undo]: **PICK** (Select object to extend.)
Select object to extend or shift-select to trim or [Project/Edge/Undo]: **PICK** (Select object to extend.)
Select object to extend or shift-select to trim or [Project/Edge/Undo]: **Enter**
Command:

Edge/Projection
The *Edge* and *Projection* switches operate identically to their function with the *Trim* command. Use *Edge* with the *Extend* option if you want a boundary edge object to be imaginarily extended (Fig. 9-21).

Shift-Select
See "*Trim* and *Extend* Shift-Select Option."

Figure 9-21

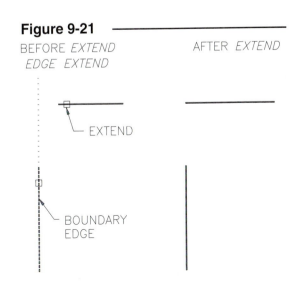

BEFORE *EXTEND* EDGE EXTEND   AFTER *EXTEND*

EXTEND

BOUNDARY EDGE

Trim and Extend Shift-Select Option

An enhancement to the *Trim* and *Extend* commands (new in AutoCAD 2002) is the shift-select option. With this feature, you can toggle between *Trim* and *Extend* by holding down the Shift key. For example, if you invoked *Trim* and are in the process of trimming lines but decide to then *Extend* a line, you can simply hold down the Shift key to change to the *Extend* command without leaving *Trim*.

```
Command: trim
Current settings: Projection=UCS, Edge=None
Select cutting edges ...
Select objects: PICK
Select objects: Enter
Select object to trim or shift-select to extend or [Project/Edge/Undo]: PICK (to trim)
Select object to trim or shift-select to extend or [Project/Edge/Undo]: Shift, then PICK (to extend)
```

Not only does holding down the Shift key toggle between *Trim* and *Extend*, but objects you selected as *cutting edges* become *boundary edges* and vice versa. Therefore, if you want to use this feature effectively, during the first step of either command (*Trim* or *Extend*), you must anticipate and select edges you might potentially use as both cutting edges and boundary edges.

BREAK

Pull-down Menu	COMMAND (TYPE)	ALIAS (TYPE)	Short-cut	Screen (side) Menu	Tablet Menu
Modify *Break*	*BREAK*	*BR*	...	*MODIFY2* *Break*	*W,17*

Break allows you to break a space in an object or break the end off an object. You can think of *Break* as a partial erase. If you choose to break a space in an object, the space is created between two points that you specify (Fig. 9-22). In this case, the *Break* creates two objects from one.

Figure 9-22

BEFORE *BREAK*   AFTER *BREAK*

If *Break*ing a circle (Fig. 9-23), the break is in a <u>counterclockwise</u> direction from the first to the second point specified.

Figure 9-23

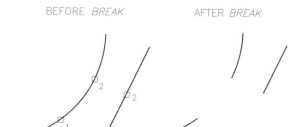

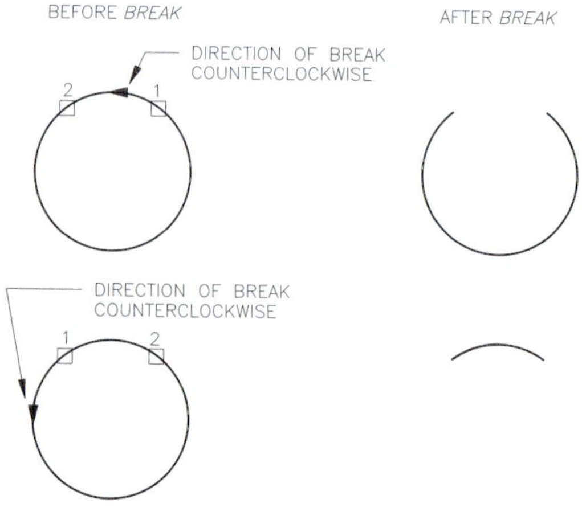

BEFORE *BREAK*   AFTER *BREAK*

DIRECTION OF BREAK
COUNTERCLOCKWISE

DIRECTION OF BREAK
COUNTERCLOCKWISE

If you want to *Break* the end off a *Line* or *Arc,* the first point should be specified at the point of the break and the second point should be just off the end of the *Line* or *Arc* (Fig. 9-24).

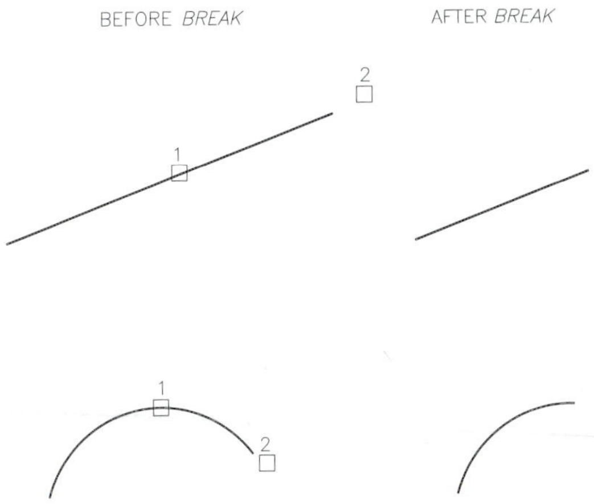

There are four options for *Break.* The *Modify* toolbar includes icons for only the default option (see *Select, Second*) and the *Break at Point* option. All four methods are available only from the screen menu. If you are typing, the desired option is selected by keying the letter *F* (for *First point*) or symbol @ (for "last point"). The four options are explained next in detail.

Figure 9-24

Select, Second
This method (the default method) has two steps: select the object to break; then select the second point of the break. The first point used to select the object is also the first point of the break (see Figures 9-22, 9-23, 9-24).

    Command: **break**
    Select object: **PICK** (This is the first point of the break.)
    Specify second break point or [First point]: **PICK** (This is the second point of the break.)
    Command:

Select, 2 Points
This method uses the first selection only to indicate the object to *Break.* You then specify the point that is the first point of the *Break,* and the next point specified is the second point of the *Break.* This option can be used with *OSNAP Intersection* to achieve the same results as *Trim.*

Selecting this option from the screen menu automatically sequences through the correct prompts (Fig. 9-25).

Figure 9-25

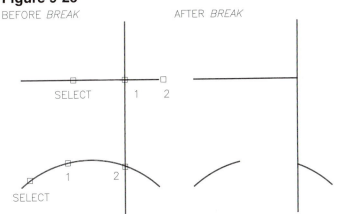

If you are typing, the command sequence is as follows:

    Command: **break**
    Select object: **PICK** (This selects only the object to break.)
    Specify second break point or [First point]: **f** (Indicates respecification of the first point.)
    Specify first break point: **PICK** (This is the first point of the break.)
    Specify second break point: **PICK** (This is the second point of the break.)
    Command:

Select Point

This option breaks one object into two separate objects with <u>no space</u> between. You specify only <u>one</u> point with this method. When you select the object, the point indicates the object <u>and</u> the point of the break. When prompted for the second point, the @ symbol (translated as "last point") is specified. This second step is done automatically only if selected from the screen menu. If you are typing the command, the @ symbol (last point) must be typed.

> Command: **break**
> Select object: **PICK** (This is the first point of the break.)
> Specify second break point or [First point]: **@** (Indicates the break to be at the last point.)
> Command:

The resulting object should <u>appear</u> as before; however, it has been transformed into <u>two</u> objects with matching endpoints (Fig. 9-26).

Figure 9-26

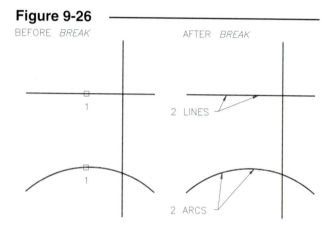

Break at Point

This option creates a break with <u>no space,</u> like the *1 Point* option; however, you can select the object you want to *Break* first and the point of the *Break* next (Fig. 9-27).

> Command: **break**
> Select object: **PICK** (This selects only the object to break.)
> Specify second break point or [First point]: **f** (Indicates respecification of the first point.)
> Specify first break point: **PICK** (This is the first point of the break.)
> Specify second break point: **@** (The second point of the break is the last point.)

Figure 9-27

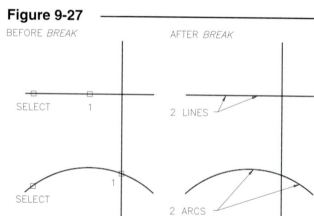

COPY

	Pull-down Menu	COMMAND (TYPE)	ALIAS (TYPE)	Short-cut	Screen (side) Menu	Tablet Menu
	Modify *Copy*	*COPY*	*CO* or *CP*	(Edit Mode) *Copy Selection*	*MODIFY1* *Copy*	*V,15*

Copy creates a duplicate set of the selected objects and allows placement of those copies. The *Copy* operation is like the *Move* command, except with *Copy* the original set of objects remains in its original location. You specify a "base point:" (point to copy <u>from</u>) and a "specify second point of displacement or <use first point as displacement>:" (point to copy <u>to</u>). See Figure 9-28, on the next page.

The command syntax for *Copy* is as follows.

Command: **copy**
Select Objects: **PICK** (Select objects to be copied.)
Select Objects: **Enter** (Indicates completion of the selection set.)
Specify base point or displacement, or [Multiple]: **PICK** or (**coordinates**) (This is the point to copy <u>from</u>. Select a point, usually on the object, to use as a "handle" or reference point. *OSNAPs* can be used. Coordinates in any format can also be entered.)
Specify second point of displacement or <use first point as displacement>: **PICK** or (**coordinates**) (This is the point to copy <u>to</u>. Select a point. *OSNAPs* can be used. Coordinates in any format can be entered.)
Command:

Figure 9-28

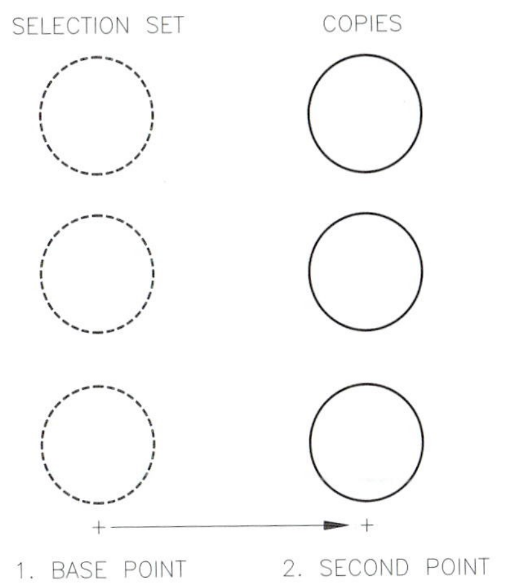

SELECTION SET   COPIES

1. BASE POINT   2. SECOND POINT

In many applications it is desirable to use *OSNAP* options to PICK the "base point:" and the "second point of displacement (Fig. 9-29).

Figure 9-29

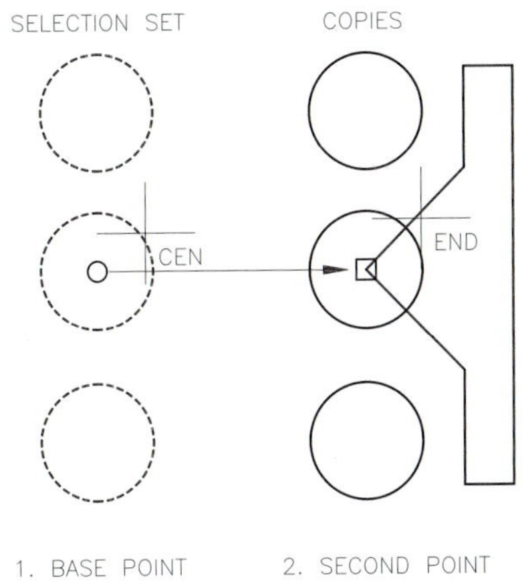

SELECTION SET   COPIES

CEN   END

1. BASE POINT   2. SECOND POINT

Alternately, you can PICK the "base point:" and use Polar Tracking or enter relative rectangular coordinates, relative polar coordinates, or direct distance entry to specify the "second point of displacement:".

Figure 9-30

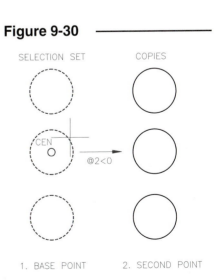

SELECTION SET   COPIES

CEN   @2<0

1. BASE POINT   2. SECOND POINT

The *Copy* command has a *Multiple* option. The *Multiple* option allows creating and placing multiple copies of the selection set.

Command: *copy*
Select Objects: **PICK**
Select Objects: **Enter**
Specify base point or displacement, or [Multiple]: *m*
Specify base point: **PICK**
Specify second point of displacement or <use first point as displacement>: **PICK**
Specify second point of displacement or <use first point as displacement>: **PICK**
Specify second point of displacement or <use first point as displacement>: **Enter**
Command:

Figure 9-31

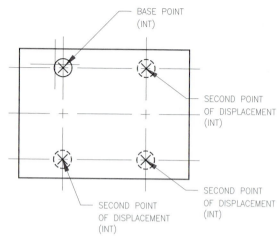

MIRROR

Pull-down Menu	COMMAND (TYPE)	ALIAS (TYPE)	Short-cut	Screen (side) Menu	Tablet Menu
Modify *Mirror*	*MIRROR*	*MI*	...	*MODIFY1* *Mirror*	*V,16*

This command creates a mirror image of selected existing objects. You can retain or delete the original objects ("old objects"). After selecting objects, you create two points specifying a "rubberband line," or "mirror line," about which to *Mirror*.

The length of the mirror line is unimportant since it represents a vector or axis (Fig. 9-32).

Figure 9-32

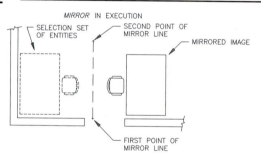

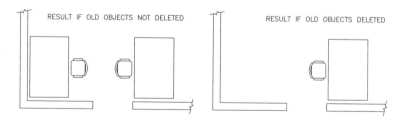

Command: *mirror*
Select Objects: **PICK** (Select object or group of objects to mirror.)
Select Objects: **Enter** (Press Enter to indicate completion of object selection.)
Specify first point of mirror line: **PICK** or (**coordinates**) (Draw first endpoint of line to represent mirror axis by PICKing or entering coordinates.)
Specify second point of mirror line: **PICK** or (**coordinates**) (Select second point of mirror line by PICKing or entering coordinates.)
Delete source objects? <N> **Enter** or **Y** (Press Enter to yield both sets of objects or enter Y to keep only the mirrored set.)
Command:

If you want to *Mirror* only in a vertical or horizontal direction, toggle *ORTHO* or *POLAR On* before selecting the "second point of mirror line."

Mirror can be used to draw the other half of a symmetrical object, thus saving some drawing time (Fig. 9-33).

Figure 9-33

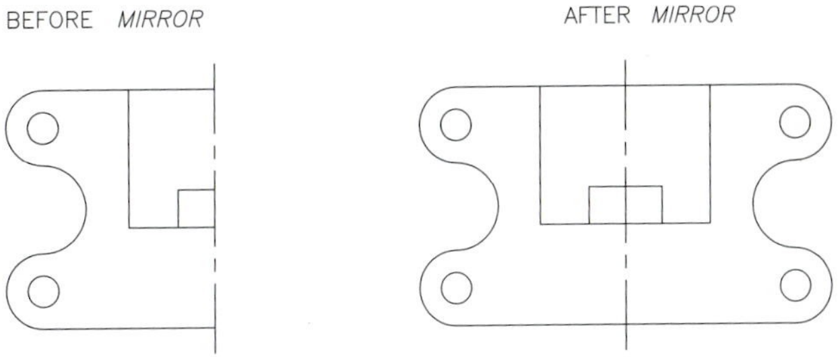

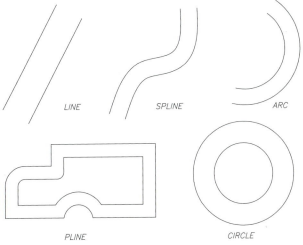

BEFORE *MIRROR*   AFTER *MIRROR*

You can control whether <u>text</u> is mirrored by using the *MIRRTEXT* variable. Type *MIRRTEXT* at the Command prompt and change the value to 1 if you want the text to be reflected; otherwise, the default setting of 0 <u>does not</u> mirror text along with other selected objects (Fig. 9-34). Dimensions are <u>not</u> affected by the *MIRRTEXT* variable and therefore are not reflected (that is, if the dimensions are associative). (See Chapter 28 for details on dimensions.)

Figure 9-34

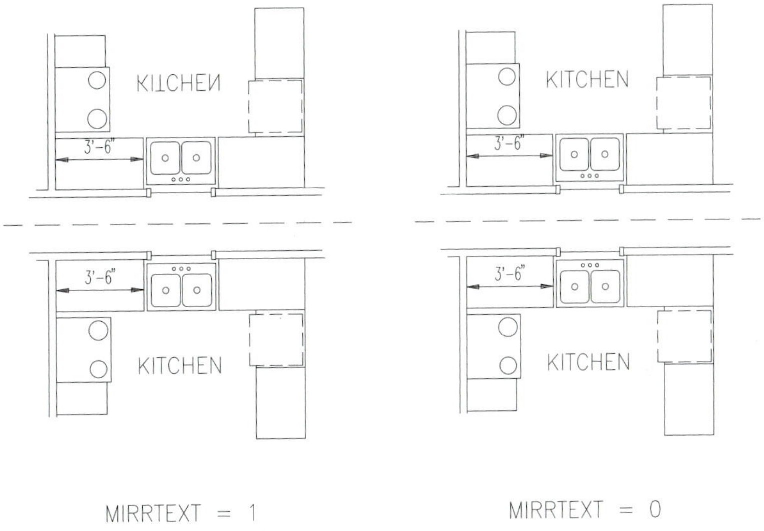

KITCHEN   3'-6"   KITCHEN   3'-6"

3'-6"   KITCHEN   3'-6"   KITCHEN

MIRRTEXT = 1   MIRRTEXT = 0

2004

TIP

OFFSET

Pull-down Menu	COMMAND (TYPE)	ALIAS (TYPE)	Short-cut	Screen (side) Menu	Tablet Menu
Modify *Offset*	*OFFSET*	*O*	...	*MODIFY1* *Offset*	V,17

Offset creates a <u>parallel copy</u> of selected objects. Selected objects can be *Line*s, *Arc*s, *Circle*s, *Pline*s, or other objects. *Offset* is a very useful command that can increase productivity greatly, particularly with *Pline*s.

Depending on the object selected, the resulting *Offset* is drawn differently (Fig. 9-35). *Offset* creates a parallel copy of a *Line* equal in length and perpendicular to the original. *Arc*s and *Circle*s have a concentric *Offset*. *Offset*ting closed *Pline*s or *Spline*s results in a complete parallel shape.

Figure 9-35

LINE   SPLINE   ARC

PLINE   CIRCLE

Two options are available with *Offset*: (1) *Offset* a specified *distance* and (2) *Offset through* a specified point (Fig. 9-36).

Figure 9-36

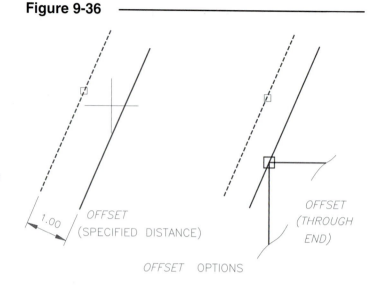

Distance

The *Distance* option command sequence is as follows:

> Command: ***offset***
> Specify offset distance or [Through] <1.0000>: (**value**) or **PICK** (Indicate distance to offset by entering a value or PICKing two points.)
> Select object to offset or <exit>: **PICK** (Only one object can be selected.)
> Specify point on side to offset: **PICK** (Select which side of the selected object for the offset to be drawn.)
> Select object to offset or <exit>: **PICK** or **Enter** (*Offset* can be used repeatedly to offset at the same distance or press Enter to exit.)
> Command:

Through

The command sequence for *Offset through* is as follows:

> Command: ***offset***
> Specify offset distance or [Through] <1.0000>: ***t***
> Select object to offset or <exit>: **PICK** (Only one object can be selected.)
> Specify through point: **PICK** or (**coordinates**) (Select a point for the object to be drawn through. You can use *OSNAPs*, Polar Tracking, and other tools or enter coordinates in any format.)
> Select object to offset or <exit>: **PICK** or **Enter** (*Offset* can be used repeatedly by PICKing a new through point or press Enter to exit.)
> Command:

Notice the power of using *Offset* with closed *Pline* or *Spline* shapes (Fig. 9-37). Because a *Pline* or a *Spline* is one object, *Offset* creates one complete smaller or larger "parallel" object. Any closed shape composed of *Lines* and *Arcs* can be converted to one *Pline* object (see "*Pedit*," Chapter 16).

Figure 9-37

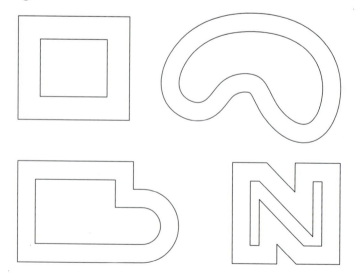

ARRAY

	Pull-down Menu	COMMAND (TYPE)	ALIAS (TYPE)	Short-cut	Screen (side) Menu	Tablet Menu
	Modify *Array*	*ARRAY*	AR	...	*MODIFY1* *Array*	*V,18*

The *Array* command creates either a *Rectangular* or a *Polar* (circular) pattern of existing objects that you select. The pattern could be created from a single object or from a group of objects. *Array* copies a duplicate set of objects for each "item" in the array.

The *Array* command produces a dialog box to enter the desired array parameters and to preview the pattern (Fig. 9-38). The first step is to select the type of array you want (at the top of the box): *Rectangular Array* or *Polar Array*.

Rectangular Array

A rectangular array is a pattern of objects generated into rows and columns. Selecting a *Rectangular Array* produces the central area of the box and preview image as shown in Figure 9-38. Follow these steps to specify the pattern for the *Rectangular Array*.

1. Pick the *Select objects* button in the upper-right corner of the dialog box to select the objects you want to array.
2. Enter values in the *Rows:* and *Columns:* edit boxes to indicate how many rows and columns you want. The image area indicates the number of rows and columns you request, but does not indicate the shape of the objects you selected.
3. Enter values in the *Row offset:* and *Column offset:* edit boxes to indicate the distance between the rows and columns (see Figure 9-39). Enter positive values to create an array to the right (+X) and upward (+Y) from the selected set, or enter negative values to create the array in the –X and –Y directions. You can also use the small select buttons to the far right of these edit boxes to interactively PICK two points for the *Row offset* and *Column offset*. Alternately, select the large button to the immediate right of the *Row offset* and *Column offset* edit boxes to use the "unit cell" method. With this method, you select diagonal corners of a window to specify both directions and both distances for the array (see Figure 9-40).
4. If you want to create an array at an angle (instead of generating the rows vertically and columns horizontally), enter a value in the *Angle of array* edit box or use the select button to PICK two points to specify the angle of the array.

Figure 9-38

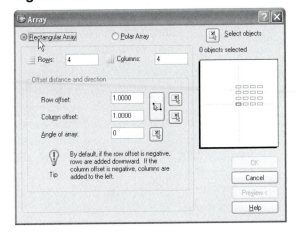

Figure 9-39

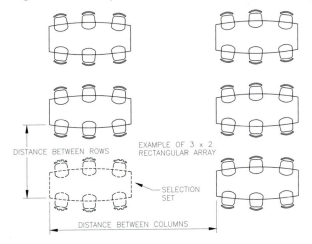

Figure 9-40

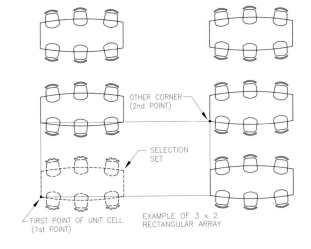

5. Select the *Preview* button to temporarily close the dialog box and return to the drawing where you can view the array and chose to *Accept* or *Modify* the array. Selecting *Modify* returns you to the *Array* dialog box.

Polar Array

The *Polar Array* option generates a circular pattern of the selection set. Selecting the *Polar Array* button produces the central area of the box and preview image similar to that shown in Figure 9-41. Typical steps are as follows.

Figure 9-41

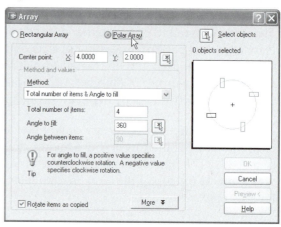

1. Pick the *Select objects* button in the upper-right corner of the dialog box to select the objects you want to array.
2. Enter values for the *Center point* of the circular pattern in the X and Y edit boxes, or use the select button to PICK a center point interactively.
3. Select the *Method* from the drop-down list. The three possible methods are:
    Total number of items & Angle to fill
    Total number of items & Angle between items
    Angle to fill & Angle between items
    Whichever option you choose enables the two applicable edit boxes below.
4. You must supply two of the three possible parameters following to specify the configuration of the array based on your selection in the *Method* drop-down list.
    Total number of items (the total number <u>includes</u> the selected object set)
    Angle to fill
    Angle between items
    Arrays are generated counterclockwise by default. To produce a clockwise array, enter a negative value in *Angle to fill*; or enter a negative value in *Angle to fill*, then switch to the *Total number of items & Angle between items*. The image area indicates the configuration of the array according to the options you selected and values you specified in the *Method and values* section. The image area does not indicate the shape of the object(s) you selected to array.
5. If you want the set of objects to be rotated but remain in the same orientation, remove the check in the *Rotate items as copied* box.
6. Select the *Preview* button to temporarily close the dialog box and return to the drawing where you can view the array and chose to *Accept* or *Modify* the array. Selecting *Modify* returns you to the *Array* dialog box.

Figure 9-42 illustrates a *Polar Array* created using 8 as the *Total number of items*, 360 as the *Angle to fill*, and selecting *Rotate items as copied*.

Figure 9-42

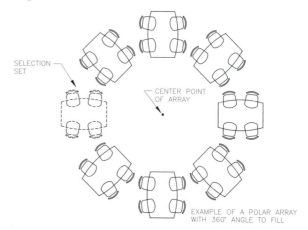

SELECTION SET

CENTER POINT OF ARRAY

EXAMPLE OF A POLAR ARRAY WITH 360° ANGLE TO FILL

2002

Figure 9-43 illustrates a *Polar Array* created using 5 as the *Total number of items*, 180 as the *Angle to fill*, and selecting *Rotate items as copied*.

Figure 9-43

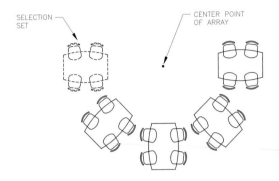

EXAMPLE OF A POLAR ARRAY
WITH 180° ANGLE TO FILL
AND 5 ITEMS TO ARRAY

SELECTION SET

CENTER POINT OF ARRAY

Occasionally an unexpected pattern may result if you remove the check from *Rotate items as copied*. In such a case, the objects may not seem to rotate about the specified center, as shown in Figure 9-44. The reason is that AutoCAD must select a single base point from the set of objects to use for the array such as the end of a line. This base point is not necessarily at the center of the set of objects.

Figure 9-44

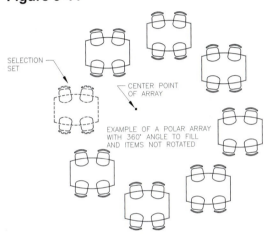

SELECTION SET

CENTER POINT OF ARRAY

EXAMPLE OF A POLAR ARRAY
WITH 360° ANGLE TO FILL
AND ITEMS NOT ROTATED

The solution to this problem is to use the *More* button near the bottom of the *Array* dialog box to open the lower section of the dialog box shown in Figure 9-45. In this area, remove the check for *Set to object's default*, then specify a *Base point* at the center of the selected set of objects by either entering values or PICKing a point interactively.

Figure 9-45

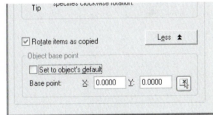

As an alternative to using the *Array* command to produce the *Array* dialog box, you can type –*Array* to use the Command line version. After selecting objects, enter the letter "r" to produce a *Rectangular Array* as shown in the following sequence.

```
Command: -array
Select objects: PICK
Select objects: Enter
Enter the type of array [Rectangular/Polar] <R>: r
Enter the number of rows (—-) <1>: (value)
Enter the number of columns (|||) <1> (value)
Enter the distance between rows or specify unit cell (—-): PICK or (value)
Specify opposite corner: PICK
Command:
```

To create a *Polar Array*, enter the letter "p" at the "Enter type of array" prompt as shown.

```
Command: -array
Select objects: PICK
Select objects: Enter
Enter the type of array [Rectangular/Polar] <R>: p
Specify center point of array or [Base]: PICK or (coordinates)
Enter the number of items in the array: (value)
Specify the angle to fill (+=ccw, -=cw) <360>: (value)
Rotate arrayed objects? [Yes/No] <Y>: (option)
Command:
```

FILLET

Pull-down Menu	COMMAND (TYPE)	ALIAS (TYPE)	Short-cut	Screen (side) Menu	Tablet Menu
Modify *Fillet*	*FILLET*	*F*	...	*MODIFY2* *Fillet*	*W,19*

The *Fillet* command automatically rounds a sharp corner (intersection of two *Lines*, *Arcs*, *Circles*, or *Pline* vertices) with a radius. You specify only the radius and select the objects' ends to be *Filleted*. The objects to fillet do <u>not</u> have to completely intersect but can overlap. You can specify whether or not the objects are automatically extended or trimmed as necessary (Fig. 9-46).

Figure 9-46

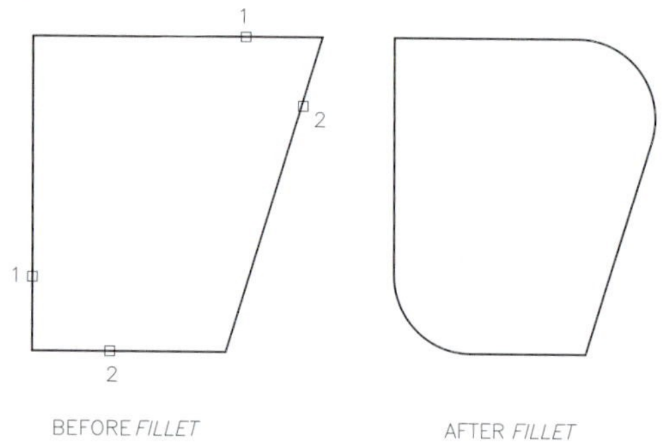

BEFORE *FILLET*   AFTER *FILLET*

```
Command: fillet
Current settings: Mode = TRIM, Radius = 0.5000
Select first object or [Polyline/Radius/Trim]: PICK (Select
one Line, Arc, or Circle near the point where the fillet
should be created.)
Select second object: PICK (Select second object to fillet near fillet location.)
Command:
```

The fillet is created at the corner selected.

Treatment of *Arcs* and *Circles* with *Fillet* is shown in Figure 9-47. Note that the objects to *Fillet* do not have to intersect but can overlap.

Figure 9-47

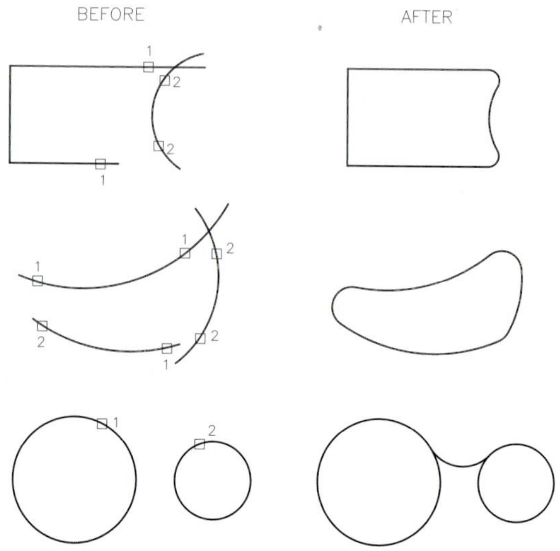

BEFORE   AFTER

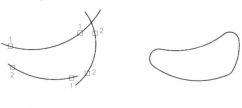

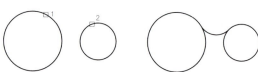

Radius
Use the *Radius* option to specify the desired radius for the fillets. *Fillet* uses the specified radius value for all new fillets until the value is changed.

> Command: *fillet*
> Current settings: Mode = TRIM, Radius = 0.5000
> Select first object or [Polyline/Radius/Trim]: *r*
> Specify fillet radius <0.5000>: *.75*
> Select first object or [Polyline/Radius/Trim]:

Since trimming and extending are done automatically by *Fillet* when necessary, using *Fillet* with a *radius* of *0* has particular usefulness. Using *Fillet* with a *0 radius* creates clean, sharp corners even if the original objects overlap or do not intersect (Fig. 9-48).

Figure 9-48

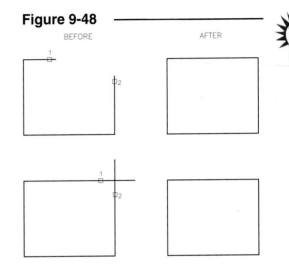

If parallel objects are selected, the *Fillet* is automatically created to the correct radius (Fig. 9-49). Therefore, parallel objects can be filleted at any time <u>without specifying a radius value</u>.

Figure 9-49

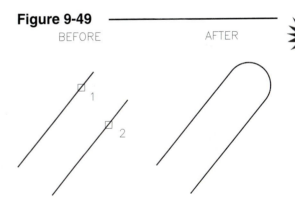

Polyline
Fillets can be created on *Polylines* in the same manner as two *Line* objects. Use *Fillet* as you normally would for *Lines*. However, if you want a fillet equal to the specified radius to be added to <u>each vertex</u> of the *Pline* (except the endpoint vertices), use the *Polyline* option; then select anywhere on the *Pline* (Fig. 9-50).

> Command: *fillet*
> Current settings: Mode = TRIM, Radius = 0.5000
> Select first object or [Polyline/Radius/Trim]: *P*
> (Indicates *Polyline* option.)
> Select 2D polyline: **PICK** (Select the desired polyline.)

Figure 9-50

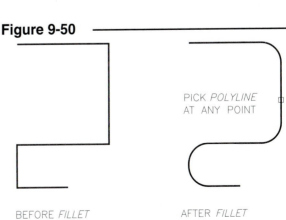

Closed Plines created by the *Close* option of *Pedit* react differently with *Fillet* than *Plines* connected by PICKing matching endpoints. Figure 9-51 illustrates the effect of *Fillet Polyline* on a *Closed Pline* and on a connected *Pline*.

Figure 9-51

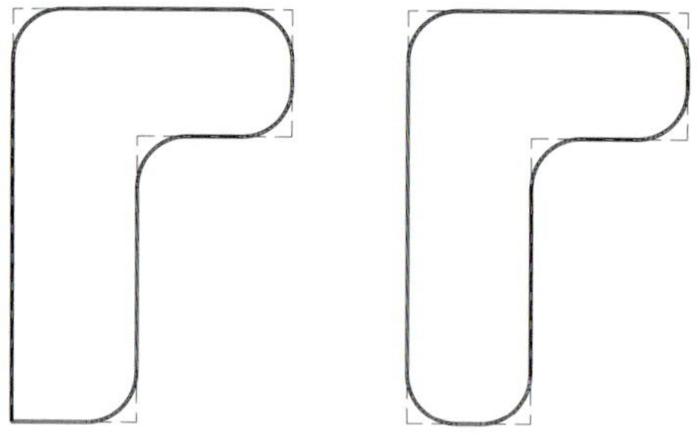

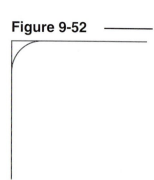

PLINE CONNECTED WITH PICK
AFTER FILLET

CLOSED PLINE
AFTER FILLET

Trim/Notrim

In the previous figures, the objects are shown having been filleted in the *Trim* mode—that is, with automatically trimmed or extended objects to meet the end of the new fillet radius. The *Notrim* mode creates the fillet without any extending or trimming of the involved objects (Fig. 9-52). Note that the command prompt indicates the current mode (as well as the current radius) when *Fillet* is invoked.

Figure 9-52

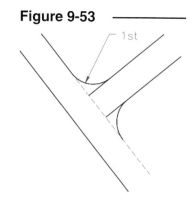

FILLET   NOTRIM

    Command: **fillet**
    Current settings: Mode = TRIM, Radius = 0.5000
    Select first object or [Polyline/Radius/Trim]: **t**
    (Invokes the *Trim/No trim* option.)
    Enter Trim mode option [Trim/No trim] <Trim>: **n** (Sets fillet to *No trim*.)
    Select first object or [Polyline/Radius/Trim]:

The *Notrim* mode is helpful in situations similar to that in Figure 9-53. In this case, if the intersecting *Lines* were trimmed when the first fillet was created, the longer *Line* (highlighted) would have to be redrawn in order to create the second fillet.

Figure 9-53

1st

Changing the *Trim/Notrim* option sets the *TRIMMODE* system variable to 0 (*Notrim*) or 1 (*Trim*). This variable controls trimming for both *Fillet* and *Chamfer*, so if you set *Fillet* to *Notrim*, *Chamfer* is also set to *Notrim*.

FILLET   NOTRIM

CHAMFER

	COMMAND (TYPE)	ALIAS (TYPE)	Short-cut	Screen (side) Menu	Tablet Menu
Pull-down Menu					
Modify *Chamfer*	*CHAMFER*	*CHA*	...	*MODIFY2* *Chamfer*	*W,18*

Chamfering is a manufacturing process used to replace a sharp corner with an angled surface. In AutoCAD, *Chamfer* is commonly used to change the intersection of two *Lines* or *Plines* by adding an angled line. The *Chamfer* command is similar to *Fillet*, but rather than rounding with a radius or "fillet," an angled line is automatically drawn at the distances (from the existing corner) that you specify.

Chamfers can be created by <u>two methods</u>: *Distance* (specify <u>two distances</u>) or *Angle* (specify a <u>distance and an angle</u>). The current method and the previously specified values are displayed at the Command prompt along with the options:

    Command: **chamfer**
    (TRIM mode) Current chamfer Dist1 = 0.5000, Dist2 = 0.5000
    Select first line or [Polyline/Distance/Angle/Trim/Method]:

Method
Use this option to indicate which of the two methods you want to use: *Distance* (specify 2 distances) or *Angle* (specify a distance and an angle).

Distance

Figure 9-54

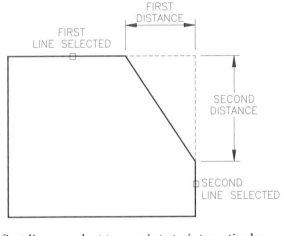

The *Distance* option is used to specify the two values applied <u>when the *Distance Method* is used</u> to create the chamfer. The values indicate the distances from the corner (intersection of two lines) to each chamfer endpoint (Fig. 9-54).

    Command: **chamfer**
    (TRIM mode) Current chamfer Dist1 = 0.5000, Dist2 = 0.5000
    Select first line or
    [Polyline/Distance/Angle/Trim/Method]: **d** (Indicates the *Distance* option.)
    Specify first chamfer distance <0.5000>: (**value**) or
    PICK (Enter a value for the distance from the
existing corner to the endpoint of the chamfer on the first line or select two points to interactively specify the distance.)
    Specify second chamfer distance <0.5000>: **Enter**, (**value**) or **PICK** (Press Enter to use the same distance as the first, enter another value, or PICK as before.)
    Select first line or [Polyline/Distance/Angle/Trim/Method]:

Chamfer with distances of 0 reacts like *Fillet* with a radius of 0; that is, overlapping corners can be automatically trimmed and non-intersecting corners can be automatically extended (using the *Trim* mode).

Angle

Figure 9-55

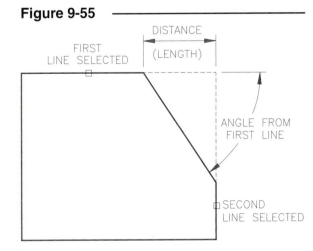

The *Angle* option allows you to specify the values that are <u>used for the *Angle Method*</u>. The values specify a distance <u>along the first line</u> and an <u>angle from the first line</u> (Fig. 9-55).

    Select first line or [Polyline/Distance/Angle/Trim/Method]: **a**
    Specify chamfer length on the first line <1.0000>: (**value**)
    Specify chamfer angle from the first line <0>: (**value**)

Trim/Notrim

The two lines selected for chamfering do not have to intersect but can overlap or not connect. The *Trim* setting automatically trims or extends the lines selected for the chamfer (Fig. 9-56), while the *Notrim* setting adds the chamfer without changing the length of the selected lines. These two options are the same as those in the *Fillet* command (see "*Fillet*"). Changing these options sets the *TRIMMODE* system variable to 0 (*Notrim*) or 1 (*Trim*). The variable controls trimming for both *Fillet* and *Chamfer*.

Figure 9-56

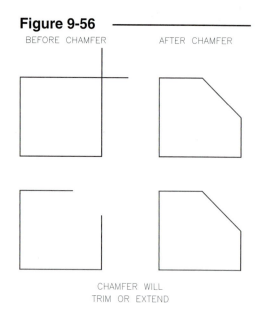

BEFORE CHAMFER   AFTER CHAMFER

CHAMFER WILL
TRIM OR EXTEND

Polyline

The *Polyline* option of *Chamfer* creates chamfers on all vertices of a *Pline*. All vertices of the *Pline* are chamfered with the supplied distances (Fig. 9-57). The first end of the *Pline* that was <u>drawn</u> takes the first distance. Use *Chamfer* without this option if you want to chamfer only one corner of a *Pline*. This is similar to the same option of *Fillet* (see "*Fillet*").

Figure 9-57

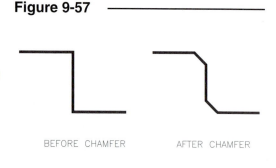

BEFORE CHAMFER   AFTER CHAMFER

A POLYLINE CHAMFER EXECUTES
A CHAMFER AT EACH VERTEX

CHAPTER EXERCISES

1. *Move*

   Begin a *New* drawing and create the geometry in Figure 9-58A using *Lines* and *Circles*. If desired, set *Polar Distance* to .25 to make drawing easy and accurate.

   For practice, <u>turn *SNAP OFF*</u> (**F9**). Use the *Move* command to move the *Circles* and *Lines* into the positions shown in Fig. 9-58B. *OSNAP*s are required to *Move* the geometry accurately (since *SNAP* is off). *Save* the drawing as **MOVE1**.

Figure 9-58

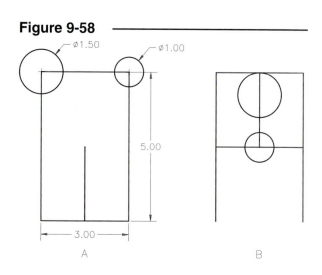

A   B

2. *Rotate*

Begin a *New* drawing and create the geometry in Figure 9-59A.

Rotate the shape into position shown in step B. *SaveAs* **ROTATE1**.

Use the *Reference* option to *Rotate* the box to align with the diagonal *Line* as shown in C. *SaveAs* **ROTATE2**.

Figure 9-59 ─────────────

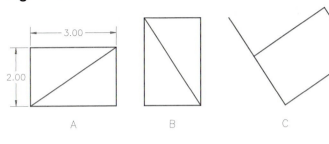

3. *Scale*

Open **ROTATE1** to again use the shape shown in Figure 9-60B and *SaveAs* **SCALE1**. *Scale* the shape by a factor of **1.5**.

Open **ROTATE2** to again use the shape shown in C. Use the *Reference* option of *Scale* to increase the scale of the three other *Lines* to equal the length of the original diagonal *Line* as shown. (HINT: *OSNAP*s are required to specify the *Reference length* and *New length*.) *SaveAs* **SCALE2.**

Figure 9-60 ─────────────

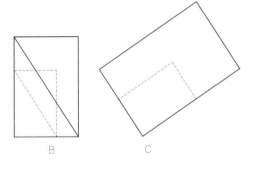

4. *Stretch*

A design change has been requested. *Open* the **SLOTPLATE CH8** drawing and make the following changes.

A. The top of the plate (including the slot) must be moved upward. This design change will add 1" to the total height of the Slot Plate. Use *Stretch* to accomplish the change, as shown in Figure 9-61. Draw the *Crossing Window* as shown.

Figure 9-61 ─────────────

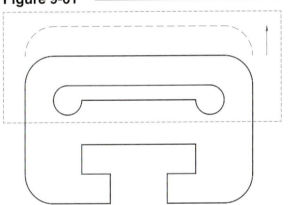

B. The notch at the bottom of the plate must be adjusted slightly by relocating it .50 units to the right, as shown in Figure 9-62. Draw the *Crossing Window* as shown. Use *SaveAs* to reassign the name to **SLOTPLATE 2**.

Figure 9-62 ─────────────

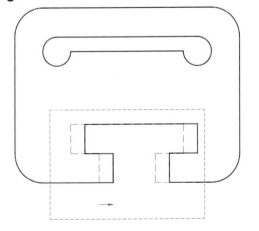

5. *Trim*

A. Create the shape shown in Figure 9-63A. *SaveAs* **TRIM-EX**.

B. Use *Trim* to alter the shape as shown in B. *SaveAs* **TRIM1**.

C. *Open* **TRIM-EX** to create the shapes shown in C and D using *Trim*. *SaveAs* **TRIM2** and **TRIM3**.

6. *Extend*

Open each of the drawings created as solutions for Figure 9-63 (**TRIM1**, **TRIM2**, and **TRIM3**). Use *Extend* to return each of the drawings to the original form shown in Figure 9-63A. *SaveAs* **EXTEND1**, **EXTEND2**, and **EXTEND3**.

Figure 9-63 —————————

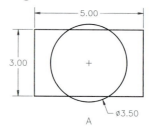

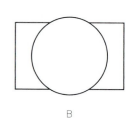

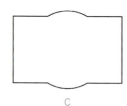

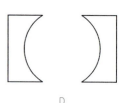

7. *Break*

A. Create the shape shown in Figure 9-64A. *SaveAs* **BREAK-EX**.

B. Use *Break* to make the two breaks as shown in B. *SaveAs* **BREAK1**. (HINT: You may have to use *OSNAP*s to create the breaks at the *Intersections* or *Quadrants* as shown.)

C. Open **BREAK-EX** each time to create the shapes shown in C and D with the *Break* command. *SaveAs* **BREAK2** and **BREAK3**.

Figure 9-64 —————————

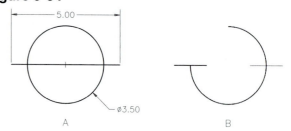

8. **Complete the Table Base**

A. Open the **TABLE-BASE** drawing you created in the Chapter 7 Exercises. The last step in the previous exercise was the creation of the square at the center of the table base. Now, use *Trim* to edit the legs and the support plate to look like Figure 9-65.

Figure 9-65 —————————

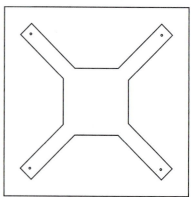

B. Use *Circle* with the *Center, Radius* option to create a 4″ diameter circle to represent the welded support for the center post for this table. Finally, add the other 4 drill holes (.25″ diameter) on the legs at the intersection of the vertical and horizontal lines representing the edges of the base. Use **Polar Tracking** or *Intersection* OSNAP for this step. The table base should appear as shown in Figure 9-66. *Save* the drawing.

Figure 9-66

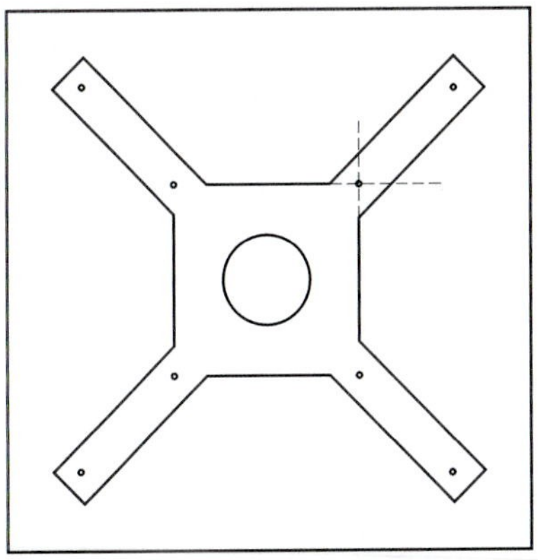

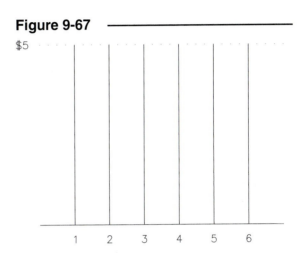

9. *Lengthen*

Five of your friends went to the horse races, each with $5.00 to bet. Construct a simple bar graph (similar to Fig. 9-67) to illustrate how your friends' wealth compared at the beginning of the day.

Modify the graph with *Lengthen* to report the results of their winnings and losses at the end of the day. The reports were as follows: friend 1 made $1.33 while friend 2 lost $2.40; friend 3 reported a 150% increase and friend 4 brought home 75% of the money; friend 5 ended the day with $7.80 and friend 6 came home with $5.60. Use the *Lengthen* command with the appropriate options to change the line lengths accordingly.

Who won the most? Who lost the most? Who was closest to even? Enhance the graph by adding width to the bars and other improvements as you wish (similar to Fig. 9-68). *SaveAs* **FRIENDS GRAPH.**

Figure 9-67

$5

1   2   3   4   5   6

Figure 9-68

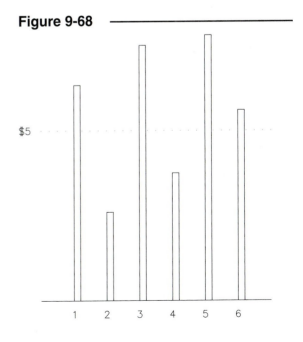

$5

1   2   3   4   5   6

10. **Trim, Extend**

A. **Open** the **PIVOTARM CH7** drawing from the Chapter 7 Exercises. Use **Trim** to remove the upper sections of the vertical *Lines* connecting the top and front views. Compare your work to Figure 9-69.

Figure 9-69

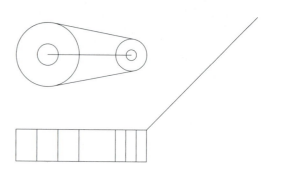

B. Next, draw a horizontal *Line* in the front view between the **Midpoints** of the vertical *Line* on each end of the view, as shown in Figure 9-70. **Erase** the horizontal *Line* in the top view between the *Circle* centers.

Figure 9-70

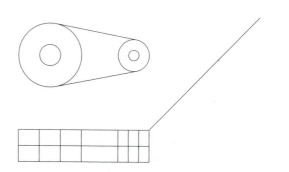

C. Draw vertical *Lines* from the **Endpoints** of the inclined *Line* (side of the object) in the top view down to the bottom *Line* in the front view, as shown in Figure 9-71. Use the two vertical lines as **Cutting edges** for **Trimming** the *Line* in the middle of the front view, as shown highlighted.

Figure 9-71

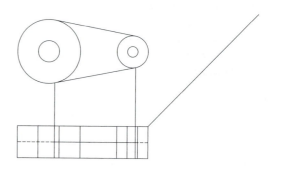

D. Finally, **Erase** the vertical lines used for **Cutting edges** and then use **Trim** to achieve the object as shown in Figure 9-72. Use *SaveAs* and name the drawing **PIVOTARM CH9**.

Figure 9-72

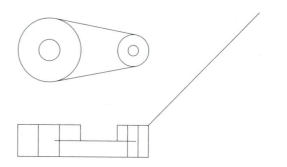

11. **GASKETA**

Begin a *New* drawing. Set *Limits* to **0,0** and **8,6**. Set *SNAP* and *GRID* values appropriately. Create the Gasket as shown in Figure 9-73. Construct only the gasket shape, not the dimensions or centerlines. (HINT: Locate and draw the four .5" diameter *Circles,* then create the concentric .5" radius arcs as full *Circles,* then *Trim.* Make use of *OSNAPs* and *Trim* whenever applicable.) *Save* the drawing and assign the name **GASKETA.**

Figure 9-73

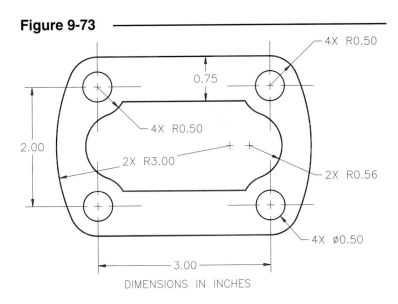

DIMENSIONS IN INCHES

12. **Chemical Process Flow Diagram**

Begin a *New* drawing and re-create the chemical process flow diagram shown in Figure 9-74. Use *Line, Circle, Arc, Trim, Extend, Scale, Break,* and other commands you feel necessary to complete the diagram. Because this is diagrammatic and will not be used for manufacturing, dimensions are not critical, but try to construct the shapes with proportional accuracy. *Save* the drawing as **FLOWDIAG.**

Figure 9-74

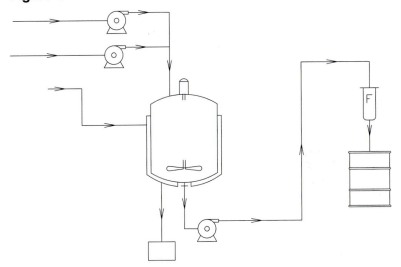

13. *Copy*

Begin a *New* drawing. Set the *Limits* to **24,18**. *Set Polar Snap* to **.5** and set your desired running *OSNAPs.* Turn On *SNAP, GRID, OSNAP,* and *POLAR.* Create the sheet metal part composed of four *Lines* and one **Circle** as shown in Figure 9-75. The lower-left corner of the part is at coordinate **2,3**.

Figure 9-75

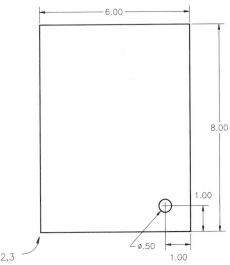

Use *Copy* to create two copies of the rectangle and hole in a side-by-side fashion as shown in Figure 9-76. Allow 2 units between the sheet metal layouts. *SaveAs* **PLATES**.

Figure 9-76

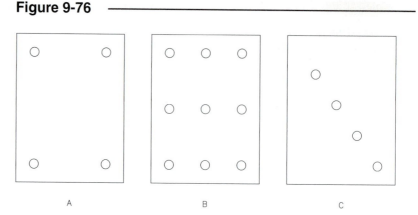

A   B   C

A. Use *Copy* to create the additional 3 holes equally spaced near the other corners of the part as shown in Figure 9-76A. *Save* the drawing.

B. Use *Copy Multiple* to create the hole configuration as shown in Figure 9-76B. *Save*.

C. Use *Copy* to create the hole placements as shown in C. Each hole <u>center</u> is **2** units at **125** degrees from the previous one (use relative polar coordinates or set an appropriate *Polar Angle*). *Save* the drawing.

14. *Mirror*

A manufacturing cell is displayed in Figure 9-77. The view is from above, showing a robot <u>centered</u> in a work station. The production line requires 4 cells. Begin by starting a *New* drawing, setting *Units* to *Engineering* and *Limits* to **40′ x 30′**. It may be helpful to set *Polar Snap* to **6″**. Draw one cell to the dimensions indicated. Begin at the indicated coordinates of the lower-left corner of the cell. *SaveAs* **MANFCELL**.

Figure 9-77

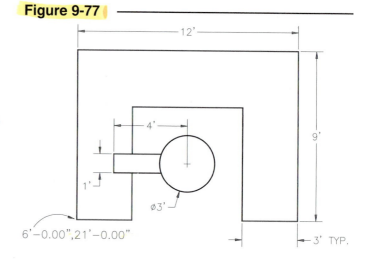

Use *Mirror* to create the other three manufacturing cells as shown in Figure 9-78. Ensure that there is sufficient space between the cells as indicated. Draw the two horizontal *Lines* representing the walkway as shown. *Save*.

Figure 9-78

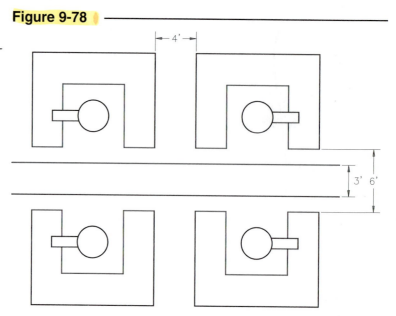

15. *Offset*

Create the electrical schematic as shown in Figure 9-79. Draw *Lines* and *Plines*, then *Offset* as needed. Use *Point* objects with an appropriate (circular) *Point Style* at each of the connections as shown. Check the *Set Size to Absolute Units* radio button (in the *Point Style* dialog box) and find an appropriate size for your drawing. Because this is a schematic, you can create the symbols by approximating the dimensions. Omit the text. *Save* the drawing as **SCHEMATIC1**. Create a *Plot* and check *Scaled to Fit*.

Figure 9-79

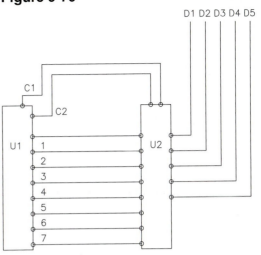

16. *Array, Polar*

Begin a *New* drawing. Create the starting geometry for a Flange Plate as shown in Figure 9-80. *SaveAs* **ARRAY**.

Figure 9-80

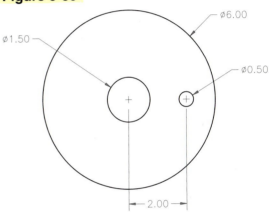

A. Create the *Polar Array* as shown in Figure 9-81A. *SaveAs* **ARRAY1**.

B. *Open* **ARRAY**. Create the *Polar Array* as shown in Figure 9-81B. *SaveAs* **ARRAY2**. (HINT: Use a negative angle to generate the *Array* in a clockwise direction.)

Figure 9-81

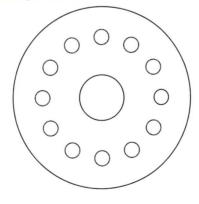

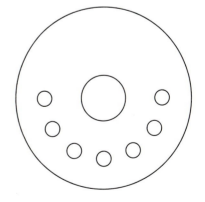

A

B

17. *Array, Rectangular*

Begin a *New* drawing. Select *Start from Scratch, Imperial* defaults. Use *Save* and assign the name **LIBRARY DESKS**. Create the *Array* of study carrels (desks) for the library as shown in Figure 9-82. The room size is 36' x 27' (set *Units* and *Limits* accordingly). Each carrel is **30" x 42"**. Design your own chair. Draw the first carrel (highlighted) at the indicated coordinates. Create the *Rectangular Array* so that the carrels touch side to side and allow a 6' aisle for walking between carrels (not including chairs).

Figure 9-82

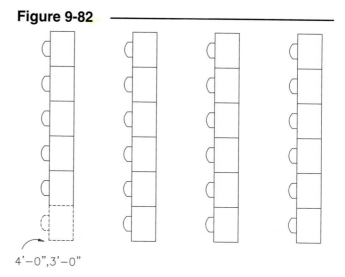

4'-0",3'-0"

18. *Array, Rectangular*

Create the bolt head with one thread as shown in Figure 9-83A. Create the small line segment at the end (0.40 in length) as a *Pline* with a *Width* of .02. *Array* the first thread (both crest and root lines, as indicated in B) to create the schematic thread representation. There is **1** row and **10** columns with **.2** units between each. Add the *Lines* around the outside to complete the fastener. *Erase* the *Pline* at the small end of the fastener (Fig. 9-83B) and replace it with a *Line*. *Save* as **BOLT**.

Figure 9-83

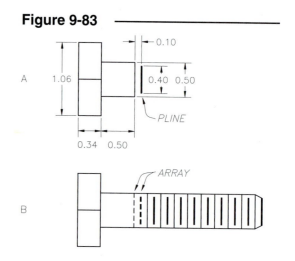

19. *Fillet*

Create the "T" Plate shown in Figure 9-84. Use *Fillet* to create all the fillets and rounds as the last step. When finished, *Save* the drawing as **T-PLATE** and make a plot.

Figure 9-84

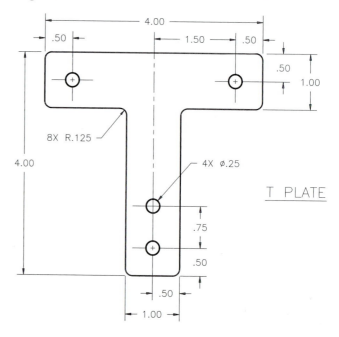

20. *Copy, Fillet*

Create the Gasket shown in Figure 9-85. Use the *Start from Scratch, Imperial* settings when you begin. Include center-lines in your drawing. *Save* the drawing as **GASKETB**.

Figure 9-85

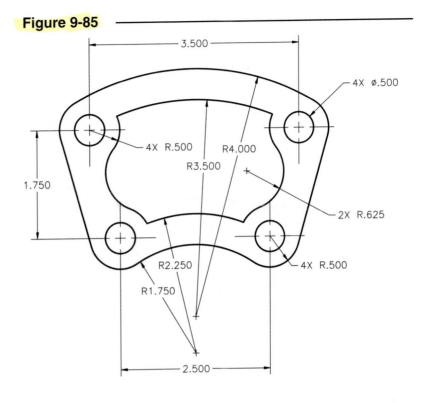

21. *Array*

Create the perforated plate as shown Figure 9-86. *Save* the drawing as **PERFPLAT**. (HINT: For the *Rectangular Array* in the center, create 100 holes, then *Erase* four holes, one at each corner for a total of 96. The two circular hole patterns contain a total of 40 holes each.)

Figure 9-86

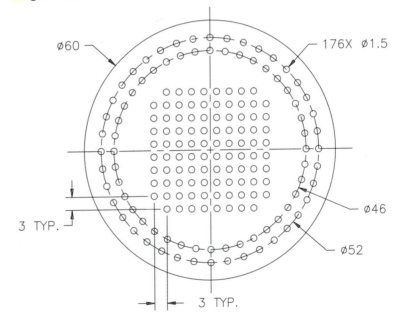

22. *Chamfer*

Begin a *New* drawing and select *Start from Scratch, Imperial* defaults. Set the *Limits* to **279,216**.
Next, type *Zoom* and use the *All* option. Set *Snap* to **2** and *Grid* to **10**. Create the Catch Bracket
shown in Figure 9-87. Draw the shape with all vertical and horizontal *Lines*, then use *Chamfer* to
create the six chamfers. *Save* the drawing as **CBRACKET** and create a plot.

Figure 9-87

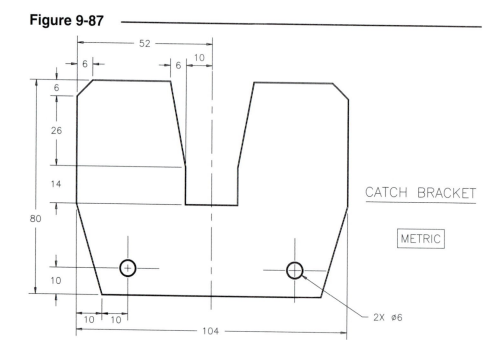

10

VIEWING COMMANDS

Chapter Objectives

After completing this chapter you should:

1. understand the relationship between the drawing objects and the display of those objects;

2. be able to use all of the *Zoom* options to view areas of a drawing;

3. be able to *Pan* the display about your screen;

4. be able to save and restore *Views*;

5. know how to use *Viewres* to change the display resolution for curved shapes;

6. be able to use *Aerial View* to *Pan* and *Zoom*;

7. be able to turn on and off the *UCS Icon*;

8. be able to create, save, and restore model space viewports using the *Vports* command.

CONCEPTS

The accepted CAD practice is to draw full size using actual units. Since the drawing is a virtual dimensional replica of the actual object, a drawing could represent a vast area (several hundred feet or even miles) or a small area (only millimeters). The drawing is created full size with the actual units, but it can be displayed at any size on the screen. Consider also that CAD systems provide for a very high degree of dimensional precision, which permits the generation of drawings with great detail and accuracy.

Combining those two CAD capabilities (great precision and drawings representing various areas), a method is needed to view different and detailed segments of the overall drawing area. In AutoCAD the commands that facilitate viewing different areas of a drawing are *Zoom*, *Pan*, and *View*.

The viewing commands are found in the *View* pull-down menu (Fig. 10-1).

Figure 10-1

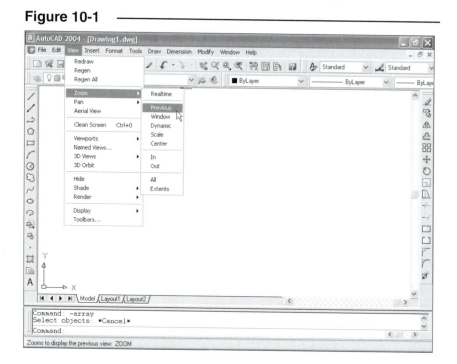

The Standard toolbar contains a group of tools (icon buttons) for the viewing commands located near the right end of the toolbar (Fig. 10-2). The *Realtime* options of *Pan* and *Zoom*, and *Zoom Previous* have icons permanently displayed on the toolbar, whereas the other *Zoom* options are located on flyouts.

Like other commands, the viewing commands can also be invoked by typing the command or command alias, using the screen menu (*VIEW1*), or using the digitizing tablet menu (if available).

The commands discussed in this chapter are:

> *Zoom, Pan, View, Dsviewer, Viewres, Ucsicon, Syswindows,* and *Vports*

The *Realtime* options are the default options for both the *Pan* and *Zoom* commands. *Realtime* options of *Pan* and *Zoom* allow you to <u>interactively</u> change the drawing display by moving the cursor in the drawing area.

Figure 10-2

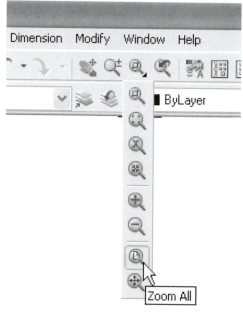

ZOOM and *PAN* with the Mouse Wheel

Zoom and *Pan*

To zoom in means to magnify a small segment of a drawing and to zoom out means to display a larger area of the drawing. Zooming does not change the size of the drawing objects; zooming changes only display of those objects—the area that is displayed on the screen. All objects in the drawing have the same dimensions before and after zooming. Only your display of the objects changes.

To pan means to move the display area slightly without changing the size of the current view window. Using the pan function, you "drag" the drawing across the screen to display an area outside of the current view in the Drawing Editor.

Although there are many ways to change the area of the drawing you want to view using either the *Zoom* or *Pan* commands, the fastest and easiest method for simple zoom and pan operations is to use the mouse wheel (the small wheel between the two mouse buttons) if you have one. Using the mouse wheel to zoom or pan does not require you to invoke the *Zoom* or *Pan* commands. Additionally, the mouse wheel zoom and pan functions are transparent, meaning that you can use this method while another command is in use. So, if you have a mouse wheel, you can zoom and pan at any time without using any commands.

Figure 10-3

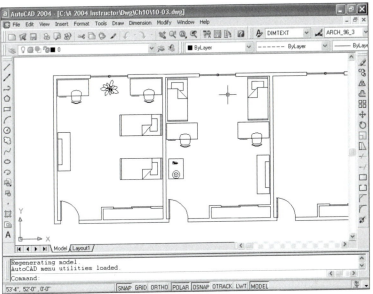

Zoom with the Mouse Wheel

If you have a mouse wheel, you can zoom by simply turning the wheel. Turn the wheel forward to zoom in and turn the wheel backward to zoom out. The current location of the cursor is the center for zooming. In other words, if you want to zoom in to an area, simply locate your cursor on the spot and turn the wheel forward. This type of zooming can be done transparently (during another command operation). You can also pan using the wheel (see "*Pan* with the Mouse Wheel or Third Button").

For example, Figure 10-3 displays a floor plan of a dorm room. Notice the location of the cursor in the upper-right corner of the drawing area. By turning the mouse wheel forward, the display changes by zooming in (enlarging) to the area designated by the cursor location. The resulting display (after turning the mouse wheel forward) is shown in Figure 10-4.

Figure 10-4

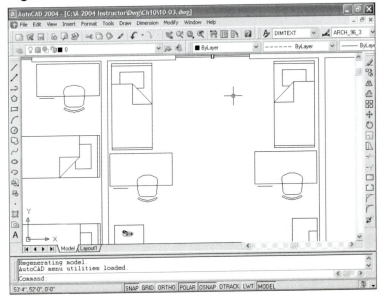

Using the same method, you could zoom out, or change the display from Figure 10-4 to Figure 10-3, by turning the wheel backward. When zooming out, the cursor location also controls the center for zooming.

The amount of zooming in or out that occurs with each turn of the mouse wheel is controlled by the *ZOOMFACTOR* system variable. The default value is 60. Increasing this value causes a greater degree of zooming with each increment of the mouse wheel's forward or backward movement; whereas, decreasing the value results in smaller changes in zooming with the wheel movement. The setting for *ZOOMFACTOR* can be between 3 and 100 and is stored with the individual computer (in the registry).

Pan with the Mouse Wheel or Third Button

If you have a mouse wheel or third mouse button, you can pan by <u>holding down</u> the wheel or button so the "hand" appears (Fig. 10-5), then move the hand cursor in any direction to "drag" the drawing around on the screen. Panning is typically used when you have previously zoomed in to an area but then want to view a different area that is slightly out of the current viewing area (off the screen). Panning with the mouse wheel or third button, similar to zooming with the mouse wheel, does not require you to invoke any commands. In addition, the pan feature is transparent, so you can pan with the wheel or third button while another command is in operation. Using the mouse wheel or third button to pan is essentially the same as using

Figure 10-5

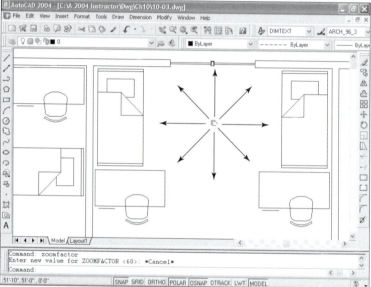

the *Pan* command with the *Realtime* option, except you do not have to invoke the *Pan* command.

Your ability to pan with the wheel or third button is based on the *MBUTTONPAN* system variable setting. If *MBUTTONPAN* is set to 1, then pressing the wheel or third button activates the *Realtime Pan* feature. If *MBUTTONPAN* is set to 0, pressing the wheel or button triggers the action defined in the ACAD.MNU file, normally set to activate the *Osnap* shortcut menu. The *MBUTTONPAN* setting is saved on the individual computer (in the registry).

COMMANDS

ZOOM

Pull-down Menu	COMMAND (TYPE)	ALIAS (TYPE)	Short-cut	Screen (side) Menu	Tablet Menu
View *Zoom >*	*ZOOM*	*Z*	(Default Menu) *Zoom*	*VIEW 1* *Zoom*	*J2 - J,5 or* *K,3 - K,5*

The *Zoom* options described next can be typed or selected by any method. Unlike most commands, AutoCAD provides a tool (icon button) for each <u>option</u> of the *Zoom* command. If you type *Zoom*, the options can be invoked by typing the first letter of the desired option or pressing Enter for the *Realtime* option.

The Command prompt appears as follows.

Command: **zoom**
Specify corner of window, enter a scale factor (nX or nXP), or
[All/Center/Dynamic/Extents/Previous/Scale/Window] <real time>:

Realtime (RTZOOM)

Realtime is the default option of *Zoom*. If you type *Zoom*, just press Enter to activate the *Realtime* option. Alternately, you can type *Rtzoom* to invoke this option directly.

With the *Realtime* option you can inter- actively zoom in and out with vertical cursor motion. When you activate the *Realtime* option, the cursor changes to a magnifying glass with a plus (+) and a minus (-) symbol. Move the cursor to any location on the drawing area and hold down the PICK (left) mouse button. Move the cursor up to zoom in and down to zoom out (Fig. 10-6). Horizontal movement has no effect. You can zoom in or out repetitively. Pressing Esc or Enter exits *Realtime Zoom* and returns to the Command prompt.

Figure 10-6

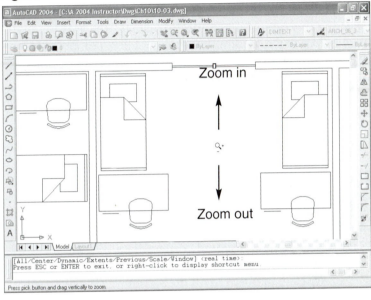

The current drawing window is used to determine the zooming factor. Moving the cursor half the window height (from the center to the top or bottom) zooms in or out to a zoom factor of 100%. Starting at the bottom or top allows zooming in or out (respectively) at a factor of 200%.

If you have zoomed in or out repeatedly, you can reach a zoom limit. In this case the plus (+) or minus (-) symbol disappears when you press the left mouse button. To zoom further with *Rtzoom*, first type *Regen*.

Pressing the right mouse button (or button #2 on digitizing pucks) produces a small cursor menu with other viewing options (Fig. 10-7). This same menu and its options can also be displayed by right-clicking during *Realtime Pan*. The options of the cursor menu are described below.

Figure 10-7

Exit

Select *Exit* to exit *Realtime Pan* or *Zoom* and return to the Command prompt. The Escape or Enter key can be used to accomplish the same action.

Pan

This selection switches to the *Realtime* option of *Pan*. A check mark appears here if you are currently using *Realtime Pan*. See *"Pan"* next in this chapter.

Zoom

A check mark appears here if you are currently using *Realtime Zoom*. If you are using the *Realtime* option of *Pan*, check this option to switch to *Realtime Zoom*.

Zoom Window
Select this option if you want to display an area to zoom in to by specifying a *Window*. This feature operates differently but accomplishes the same action as the *Window* option of *Zoom*. With this option, window selection is made with one PICK. In other words, PICK for the first corner, hold down the left button and drag, then release to establish the other corner. With the *Window* option of the *Zoom* command, PICK once for each of two corners. See "*Window.*"

Zoom Original
Selecting this option automatically displays the area of the drawing that appeared on the screen immediately before using the *Realtime* option of *Zoom* or *Pan*. Using this option successively has no effect on the display. This feature is different from the *Previous* option of the *Zoom* command, in which ten successive previous views can be displayed.

Zoom Extents
Use this option to display the entire drawing in its largest possible form in the drawing area. This option is identical to the *Extents* option of the *Zoom* command. See "*Extents.*"

Window

To *Zoom* with a *Window* is to draw a rectangular window around the desired viewing area. You PICK a first and a second corner (diagonally) to form the rectangle. The windowed area is magnified to fill the screen (Fig. 10-8). It is suggested that you draw the window with a 4 x 3 (approximate) proportion to match the screen proportion. If you type *Zoom* or *Z*, *Window* is an automatic option so you can begin selecting the first corner of the window after issuing the *Zoom* command without indicating the *Window* option as a separate step.

Figure 10-8

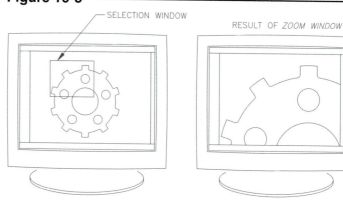

All

This option displays all of the objects in the drawing and all of the *Limits*. In Figure 10-9, notice the effects of *Zoom All*, based on the drawing objects and the drawing *Limits*.

Extents

This option results in the largest possible display of all of the objects, disregarding the *Limits*. (*Zoom All* includes the *Limits*.) See Figure 10-9.

Invoking *Zoom Extents* causes AutoCAD to execute two steps: (1) a true *Zoom Extents* is performed which causes the geometry to be maximized so it is "pressed" against the edge of

Figure 10-9

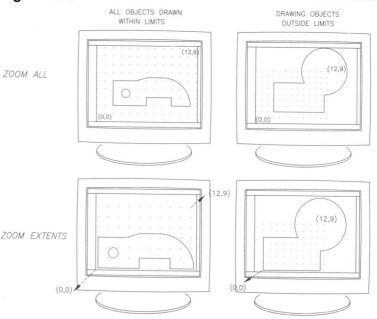

the drawing area window, and (2) a *Zoom Scale .95X* is performed which causes the geometry to be scaled down slightly to fit just within the window (see the *Scale* option).

Scale (X/XP)

This option allows you to enter a scale factor for the desired display. The value that is entered can be relative to the full view (*Limits*) or to the current display (Fig. 10-10). A value of **1, 2,** or **.5** causes a display that is 1, 2, or .5 times the size of the *Limits,* centered on the current display. A value of **1X, 2X,** or **.5X** yields a display 1, 2, or .5 times the size of the <u>current display.</u> If you are using paper space and viewing model space, a value of **1XP, 2XP,** or **.5XP** yields a model space display scaled 1, 2, or .5 times paper space units.

If you type *Zoom* or *Z,* you can enter a scale factor at the *Zoom* command prompt without having to type the letter *S.*

Figure 10-10

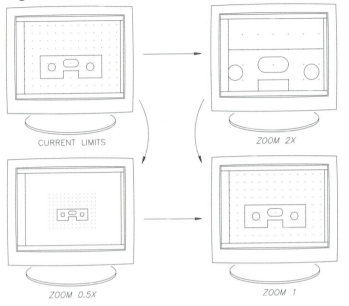

CURRENT LIMITS

ZOOM 2X

ZOOM 0.5X

ZOOM 1

In

Zoom In magnifies the current display by a factor of 2X (2 times the current display). Using this option is the same as entering a *Zoom Scale* of 2X.

Out

Zoom Out makes the current display smaller by a factor of .5X (.5 times the current display). Using this option is the same as entering a *Zoom Scale* of .5X.

Center

First, specify a location as the center of the zoomed area; then specify either a *Magnification factor* (see *Scale X/XP*), a *Height* value for the resulting display, or PICK two points forming a vertical to indicate the height for the resulting display (Fig. 10-11).

Previous

Selecting this option automatically changes to the previous display. AutoCAD saves the previous ten displays changed by *Zoom, Pan,* and *View.* You can successively change back through the previous ten displays with this option.

Figure 10-11

SELECTION OF CENTER

RESULT OF *ZOOM CENTER — 2X*

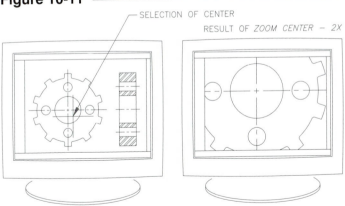

Dynamic

With this option you can change the display from one windowed area in a drawing to another without using *Zoom All* (to see the entire drawing) as an intermediate step. *Zoom Dynamic* causes the screen to display the drawing *Extents* bounded by a box. The current view or window is bounded by a box in a broken-line pattern. The view box is the box with an "X" in the center which can be moved to the desired location (Fig. 10-12). The desired location is selected by pressing **Enter** (not the PICK button as you might expect). See NOTE below.

Pressing the left button allows you to resize the view box (displaying an arrow instead of the "X") to the desired size. Move the mouse or puck left and right to increase and decrease the window size. Press the left button again to set the size and make the "X" reappear.

Figure 10-12

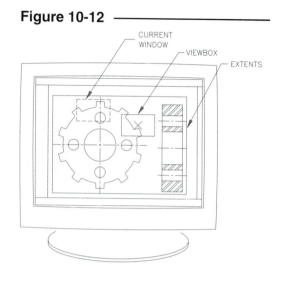

NOTE: With the normal setting for *Shortcut menus in the drawing area* turned on (in the *User Preferences* tab of the *Options* dialog box), right-clicking during *Zoom Dynamic* produces a shortcut menu. Select *Enter* to display the desired zoomed area. If you turn off shortcut menus, right-clicking during *Zoom Dynamic* automatically displays the zoomed area.

Zoom **Is Transparent**

Zoom is a transparent command, meaning it can be invoked while another command is in operation. You can, for example, begin the *Line* command and at the "Specify next point or [Undo]:" prompt *'Zoom* with a *Window* to better display the area for selecting the endpoint. This transparent feature is automatically entered if *Zoom* is invoked by the screen, pull-down, or tablet menus or toolbar icons, but if typed it must be prefixed by the apostrophe (') symbol, e.g., "Specify next point or [Undo]: *'Zoom.*" If *Zoom* has been invoked transparently, the >> symbols appear at the command prompt before the listed options as follows:

```
Command: line
Specify first point: 'zoom
>>Specify corner of window, enter a scale factor (nX or nXP), or
[All/Center/Dynamic/Extents/Previous/Scale/Window] <real time>:
```

PAN

Pull-down Menu	COMMAND (TYPE)	ALIAS (TYPE)	Short-cut	Screen (side) Menu	Tablet Menu
View *Pan*	*PAN* or -PAN	*P* or -P	(Default Menu) *Pan*	VIEW 1 *Pan*	N,10-P,11

Using the *Pan* command by typing or selecting from the tool, screen, or digitizing tablet menu produces the *Realtime* version of *Pan*. The *Point* option is available from the pull-down menu or by typing -*Pan*. The other "automatic" *Pan* options (*Left, Right, Up, Down*) can only be selected from the pull-down menu.

The *Pan* command is useful if you want to move (pan) the display area slightly without changing the size of the current view window. With *Pan*, you "drag" the drawing across the screen to display an area outside of the current view in the drawing area.

Realtime (RTPAN)

Realtime is the default option of *Pan*. It is invoked by selecting *Pan* by any method, including typing *Pan* or *Rtpan*. *Realtime Pan* is essentially the same as using the mouse wheel or third button to pan. That is, *Realtime Pan* allows you to interactively pan by "pulling" the drawing across the screen with the cursor motion. After activating the command, the cursor changes to a hand cursor. Move the hand to any location on the drawing, then hold the PICK (left) mouse button down and drag the drawing around on the screen to achieve the desired view (see Figure 10-5). When you release the mouse button, panning is discontinued. You can move the hand to another location and pan again without exiting the command.

You must press Escape or Enter or use *Exit* from the pop-up cursor menu to exit *Realtime Pan* and return to the Command prompt. The following Command prompt appears during *Realtime Pan*.

> Command: **pan**
> Press Esc or Enter to exit, or right-click to display shortcut menu.

If you press the right mouse button (#2 button), a small cursor menu pops up. This is the same menu that appears during *Realtime Zoom* (see Figure 10-7). This menu provides options for you to *Exit*, *Zoom* or *Pan* (realtime), *Zoom Window*, *Zoom Original*, or *Zoom Extents*. See "*Zoom*" earlier in this chapter.

Point

The *Point* option is available only through the pull-down menu or by typing "*-Pan*" (some commands can be typed with a hyphen [-] prefix to invoke the command line version of the command without dialog boxes, etc.). Using this option produces the following Command prompt.

> Command: **-pan**
> Specify base point or displacement: **PICK**
> or (**value**)
> Specify second point: **PICK** or
> (**value**)
> Command:

Figure 10-13

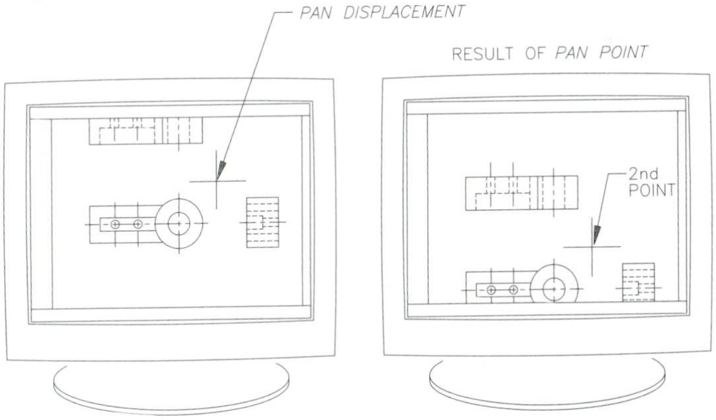

PAN DISPLACEMENT

RESULT OF PAN POINT

2nd POINT

The "base point:" can be thought of as a "handle," or point to move <u>from</u>, and the "second point:" as the new location to move <u>to</u> (Fig. 10-13). You can PICK each of these points with the cursor.

You can also enter coordinate values rather than interactively PICKing points with the cursor. The command syntax is as follows:

> Command: **-pan**
> Specify base point or displacement: **0 , -2**
> Specify second point: **Enter**
> Command:

Entering coordinate values allows you to *Pan* to a location outside of the current display. If you use the interactive method (PICK points), you can *Pan* only within the range of whatever is visible on the screen.

The following *Pan* options are available from the *View* pull-down menu only.

Left
Automatically pans to the left, equal to about 1/8 of the current drawing area width.

Right
Pans to the right—similar to but opposite of *Left*.

Up
Automatically pans up, equal to about 1/8 of the current drawing area height.

Down
Pans down—similar to but opposite of *Up*.

Pan Is Transparent
Pan, like *Zoom,* can be used as a transparent command. The transparent feature is automatically entered if *Pan* is invoked by the screen, pull-down, or tablet menus or icons. However, if you <u>type</u> *Pan* during another command operation, it must be prefixed by the apostrophe (') symbol.

```
Command: line
Specify first point: 'pan
>>Press ESC or ENTER to exit, or right-click to display shortcut menu. (Pan, then) Enter
Resuming LINE command.
Specify next point or [Undo]:
```

Scroll Bars

You can also use the horizontal and vertical scroll bars directly below and to the right of the graphics area to pan the drawing (Fig. 10-14). These scroll bars pan the drawing in one of three ways: (1) click on the arrows at the ends of the scroll bar, (2) move the thumb wheel, or (3) click inside the scroll bar. The scroll bars can be turned off (removed from the screen) by using the *Display* tab in the *Options* dialog box.

Figure 10-14 ——

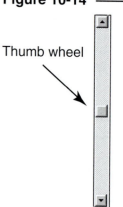

Thumb wheel

VIEW

Pull-down Menu	COMMAND (TYPE)	ALIAS (TYPE)	Short-cut	Screen (side) Menu	Tablet Menu
View *Named Views...*	VIEW or -VIEW	V or -V	...	VIEW 1 *Ddview*	M,5

The *View* command provides a dialog box for you to create *New* views of a specified display window and restore them (make *Current*) at a later time. For typical applications you would first *Zoom* in to achieve a desired display, then use *View, New* to save that display under an assigned name. Later in the drawing session, the named *View* can be made *Current* any number of times. This method is preferred to continually *Zooming* in and out to display several of the same areas repeatedly. Making a named *View* the *Current* view requires no regeneration time.

A *View* can be sent to the plotter as a separate display as long as the view is named and saved. Do this by toggling the *View* option in the *Plot Area* section of the *Plot* dialog box. Plotting is discussed in detail in Chapter 14.

Using *View* invokes the *Named Views* tab of the *View* dialog box (Fig. 10-15) and *–View* produces the Command line format. The options of the *View* command are described below.

Figure 10-15 ───────────

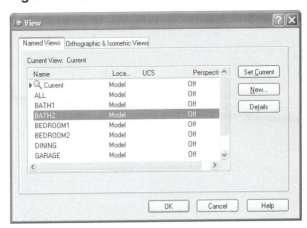

To save a new *View*, first use *Zoom* or *Pan* to achieve the desired display; then invoke the *View* dialog box and select *New*. In the *New View* dialog box that appears (Fig. 10-16), enter the desired name in the *View Name:* edit box. Ensure the *Current Display* button is selected.

If you do not *Zoom* into the desired display before using *View*, you can select the *Define Window* button, then press the *Define View Window* button on the right (see the pointer in Figure 10-16). This action temporarily removes the dialog box so you can *Zoom* with a window to define the new view area.

Figure 10-16 ───────────

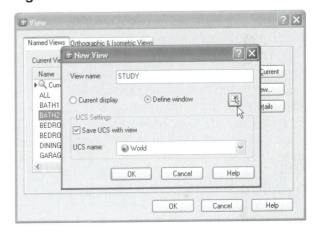

To *Restore* a named *View*, double-click on the name from the list in the *Named View* tab, or select the name (one click) and select *Set Current*. This action makes the selected view current in the drawing area.

In the *UCS Settings* section of the *New View* dialog box, you can decide whether or not to save the UCS (User Coordinate System) with the view. This feature is used primarily with 3D drawings. See Chapter 36, User Coordinate Systems, for more information on this feature and the *UCSVIEW* system variable.

The *Orthographic and Isometric Views* tab of the *View* dialog box is also used for 3D models. This tab and related options are discussed in Chapter 35, 3D Display and Viewing.

If you prefer the Command line version of the command, type *–View* to produce the following prompt.

```
Command: -view
Enter an option [?/Orthographic/Delete/Restore/Save/Ucs/Window]:
```

The options here are generally the same as those available in the dialog box.

?	Displays the list of named views.
Orthographic	Provides orthographic and isometric viewpoints for 3D models.
Delete	Deletes one or more saved views that you enter.
Restore	Displays the named view you request.
Save	Saves the current display as a *View* with a name you assign.
UCS	Allows you to save the UCS with the view.
Window	Allows you to specify a window in the current display and save it as a named *View*.

DSVIEWER

Pull-down Menu	COMMAND (TYPE)	ALIAS (TYPE)	Short-cut	Screen (side) Menu	Tablet Menu
View *Aerial View*	*DSVIEWER*	*AV*	...	VIEW 1 *Dsviewer*	*K,2*

Aerial View is a tool for navigating in a drawing. It can be used to *Zoom* and *Pan* while in another command. Invoking the *Dsviewer* command displays the window shown in Figure 10-17. The options and operations are described next.

Figure 10-17 ——————

Realtime Zoom

When Aerial View is used in *Realtime Zoom* mode, you can *Zoom* and *Pan* inside the viewer which causes the <u>windowed area to be dynamically displayed</u> in the main drawing area. *Pan* by clicking inside the viewer to *Pan* the current view box inside the viewer. Click a second time to resize the view box by moving left and right. This process is similar to using *Zoom Dynamic*. Right-click to release the cursor from the Aerial Viewer and fix the view box. When *Realtime Zoom* is disabled, all these features are available except the dynamic display (tracking the view box) in the graphics area.

This option is the most unique and useful feature of Aerial View. It allows you to dynamically move the view box around in the viewer while the magnified area is displayed <u>instantaneously and dynamically</u> in the main drawing area (Fig. 10-18). This feature is especially helpful for locating small details in the drawing, like reading text.

Figure 10-18 ——————

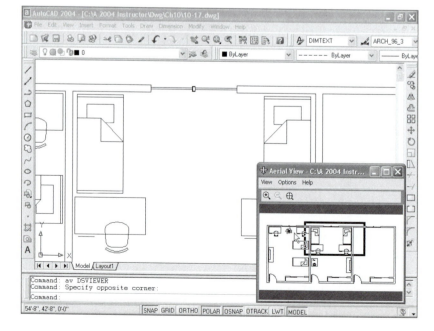

Dynamic Update

When this option is enabled, the Aerial View window is updated while you edit the drawing. When *Dynamic Update* is off, AutoCAD does not update the Aerial View window until you click in the Aerial View window.

Auto Viewport

Use this option when you are using multiple model space (tiled) viewports (see the *Vports* command later in this chapter). When *AutoViewport* is active, the view of the active viewport is automatically displayed in the Aerial Viewer.

Zoom In

PICKing this option causes the <u>display in the viewer</u> to be magnified 2X. This does <u>not</u> affect the display in the main graphics area.

Zoom Out

This option causes the <u>display in the viewer</u> to be reduced to .5X. This does <u>not</u> affect the display in the main graphics area.

Global

Selecting this option will resize the <u>drawing display in the viewer</u> to its maximum size.

NOTE: The *Zoom Out* and *Zoom In* options may be disabled. For example, when the entire drawing is displayed in the Aerial View, *Zoom Out* is disabled. When the current view nearly fills the Aerial View window, *Zoom In* is disabled.

VIEWRES

Pull-down Menu	COMMAND (TYPE)	ALIAS (TYPE)	Short-cut	Screen (side) MenuMenu	Tablet
Tools *Options...* *Display* *Display Resolution*	VIEWRES	...	...	*TOOLS2* *Options* *Display ...* *Display Resolution*	...

Viewres controls the resolution of curved shapes for the <u>screen display</u> only. Its purpose is to speed regeneration time by displaying curved shapes as linear approximations of curves; that is, a curved shape (such as an *Arc, Circle,* or *Ellipse*) appears as several short, straight line segments (Fig. 10-19). The drawing database and the plotted drawing, however, always define a true curve. The range of *Viewres* is from 1 to 20,000 with 1000 being the default. The higher the value (called "circle zoom percentage"), the more accurate the display of curves and the slower the regeneration time.

Figure 10-19

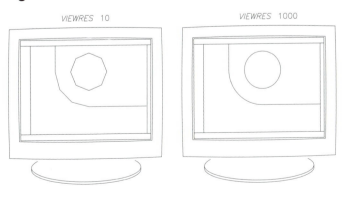

The lower the *Viewres* value, the faster is the regeneration time but the more "jagged" the curves. A value of 1000 to 5,000 is suggested for most applications. The command syntax is shown here.

```
Command: viewres
Do you want fast zooms? [Yes/No] <Y>: Y
Enter circle zoom percent (1-20000) <1000>: 5000
Command:
```

2002

Fast zooms are no longer a functioning part of this command, so your response to "Do you want fast zooms?" is irrelevant. The prompt is kept only for script compatibility.

UCSICON

Pull-down Menu	COMMAND (TYPE)	ALIAS (TYPE)	Short-cut	Screen (side) Menu	Tablet Menu
View *Display >* *UCS Icon >*	*UCSICON*	...	...	VIEW 2 *UCSicon*	L,2

2002

The icon that appears in the lower-left corner of the AutoCAD Drawing Editor is the Coordinate System Icon (Fig. 10-20). The icon is sometimes called the "UCS icon" because it automatically orients itself to the new location of a coordinate system that you create, called a "UCS" (User Coordinate System).

The *Ucsicon* command controls the appearance and positioning of the Coordinate System Icon. It is important to display the icon when working with 3D drawings so that you can more easily visualize the orientation of the X, Y, and Z coordinate system. However, when you are working with 2D drawings, it is not necessary to display this icon. (By now you are accustomed to normal orientation of the positive X and Y axes—X positive is to the right and Y positive is up.)

Figure 10-20

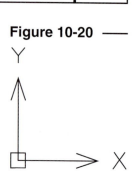

You can use the *Off* option of the *Ucsicon* command to remove the icon from the display. Use *Ucsicon* again with the *On* option to display the icon again. The Command prompt is as follows:

```
Command: ucsicon
Enter an option [ON/OFF/All/Noorigin/ORigin/Properties] <OFF>:
```

The *Properties* option can be used to change the type and properties of icon that are displayed. Using *Properties* produces the *UCS Icon* dialog box displayed in Figure 10-21. Here you can select from the *3D* (3-pole) icon or the *2D* (flat, or planar) icon. Either icon is suitable for 2D and 3D work. In fact, even though the older 2D-style icon only shows the direction for the X and Y axes, many AutoCAD users prefer it for 3D work because it makes the <u>orientation of the XY plane</u> much more apparent than the newer "3D" style icon.

Figure 10-21

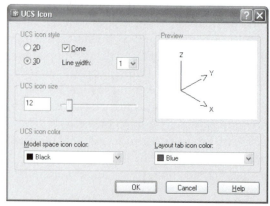

You can also control the visibility of the icon using the *View* pull-down menu. Select *Display*, then *UCS Icon*, then remove the check by the word *On*. Make sure you turn on the Coordinate System Icon when you begin working in 3D.

SYSWINDOWS

Pull-down Menu	COMMAND (TYPE)	ALIAS (TYPE)	Short-cut	Screen (side) Menu	Tablet Menu
Window (option)	*SYSWINDOWS*	...	...	...	...

Since AutoCAD 2000 you can open <u>multiple drawings simultaneously</u>. The *Syswindows* command gives you control of how you want multiple drawings to appear on the screen. When several drawings are open simultaneously, you can display each one as a full screen display, then use Ctrl+Tab to toggle between drawings.

Issuing *Syswindows* at the command line provides the same options as available from the *Window* pull-down menu.

```
Command: syswindows
Enter an option [Cascade/tile Horizontal/tile Vertical/Arrange icons]:
```

Cascade

The *Cascade* option arranges the windows one on another with only the title bar (including file name) appearing (Fig. 10-22). Make a drawing current and "on top" by selecting the title bar or any part of the window or by selecting its name in the list of files open at the bottom of the Window pull-down menu.

Figure 10-22

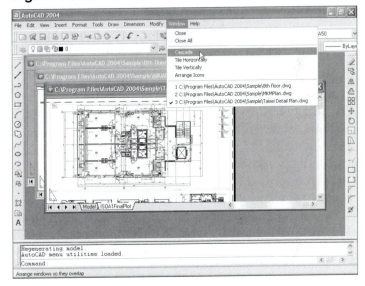

Tile Horizontal

This selection automatically arranges the display so all windows (drawings) are visible and oriented as shown in Figure 10-23. Make a drawing current by picking anywhere in the window or by selecting its name from the *Window* pull-down menu.

Figure 10-23

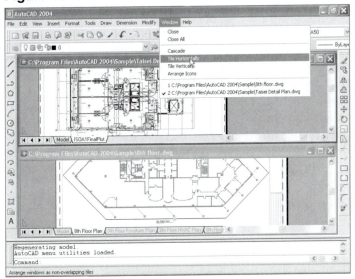

Tile Vertical

Use this option to automatically arrange the display so all windows are arranged side by side as shown in Figure 10-24.

Arrange Icons

This option is useful only when each drawing is reduced to an icon.

Figure 10-24

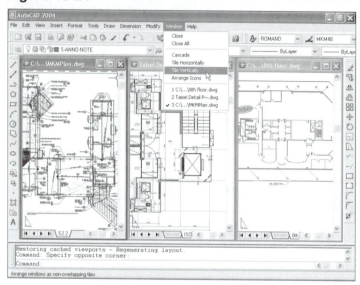

VPORTS

	Pull-down Menu	COMMAND (TYPE)	ALIAS (TYPE)	Short-cut	Screen (side) Menu	Tablet Menu
	View Viewports>	*VPORTS or -VPORTS*	...	...	VIEW 1 *Vports*	M,3 and M,4

A "viewport" is one of several simultaneous views of a drawing on the screen. The *Vports* command invokes the *Viewports* dialog box. With this dialog box you can create model space viewports or paper space viewports, depending on which space is current when you invoke *Vports*. *Model* tab (model space) viewports are sometimes called "tiled" viewports because they fit together like tiles with no space between. *Layout* tab (paper space) viewports can be any shape and configuration and are used mainly for setting up several views on a sheet for plotting. Only an introduction to *Model* tab viewports is given in this chapter. Using model space viewports for constructing 3D models is discussed in Chapter 35. *Layout* tab (paper space) viewports are discussed in Chapters 13 and 33, and using paper space viewports for 3D applications is discussed in Chapter 42.

If the *Model* tab is active and you use the *Vports* command, you will create model space viewports. Several viewport configurations are available. *Vports* in this case simply divides the <u>screen</u> into multiple sections, allowing you to view different parts of a drawing. Seeing several views of the same drawing simultaneously can make construction and editing of complex drawings more efficient than repeatedly using other display commands to view detailed areas. Model space *Vports* affect <u>only the screen display</u>. The viewport configuration <u>cannot be plotted</u>. If the *Plot* command is used from Model space, only the <u>current</u> viewport is plotted.

Figure 10-25 displays the AutoCAD drawing editor after the *Vports* command

Figure 10-25

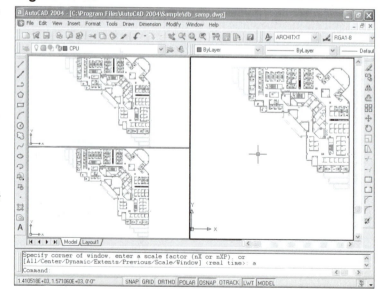

TIP

was used to divide the screen into viewports. Tiled viewports always fit together like tiles with no space between. The shape and location of the viewports are not variable as with paper space viewports.

The *View* pull-down menu (see Figure 10-1) can be used to display the *Viewports* dialog box (Fig. 10-26). The dialog box allows you to PICK the configuration of the viewports that you want. If you are working on a 2D drawing, make sure you select *2D* in the *Setup* box.

After you select the desired layout, the previous display appears in <u>each</u> of the viewports. For example, if a full view of the office layout is displayed when you use the *Vports* command, the resulting display in <u>each</u> viewport would be the same full view of the office (see Figure 10-25). It is up to you then to use viewing commands (*Zoom*, *Pan*, etc.) in each viewport to specify what areas of the drawing you want to see in each viewport. There is no automatic viewpoint configuration option for *Vports* for 2D drawings.

Figure 10-26 ─────────────

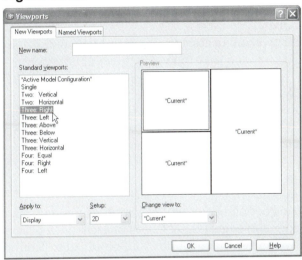

A popular arrangement of viewpoints for construction and editing of 2D drawings is a combination of an overall view and one or two *Zoomed* views (Fig. 10-27). You cannot draw or project from one viewport to another. Keep in mind that there is only <u>one model</u> (drawing) but several views of it on the screen. Notice that the active or current viewport displays the cursor, while moving the pointing device to another viewport displays only the pointer (small arrow) in that viewport.

A viewport is made active by PICKing in it. Any display commands (*Zoom*, *Pan*, *Redraw*, etc.) and some drawing aids (*SNAP*, *GRID*) used affect only the <u>current viewport</u>. Draw and modify commands that affect the model are potentially apparent in all viewports (for every display of the affected part of the model). *Redrawall* and *Regenall* can be used to redraw and regenerate all viewports.

Figure 10-27 ─────────────

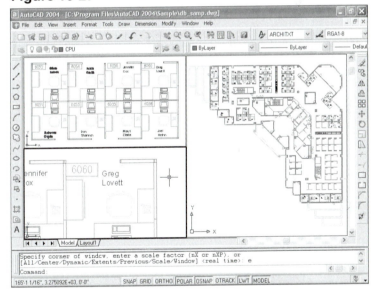

You can begin a drawing command in one viewport and finish in another. In other words, you can toggle viewports within a command. For example, you can use the *Line* command to PICK the "first point:" in one viewport, then make another viewport current to PICK the "next point:".

You can save a viewport configuration (including the views you prepare) by entering a name in the *New Name* box. If you save several viewport configurations for a drawing, you can then select from the list in the *Named Viewports* tab.

You can type –*Vports* to use the Command line version. The format is as follows:

> Command: **-vports**
> Enter an option [Save/Restore/Delete/Join/SIngle/?/2/3/4] <3>:

Save

Allows you to assign a name and save the current viewport configuration. The configuration can be *Restored* at a later time.

Restore

Redisplays a previously *Saved* viewport configuration.

Delete

Deletes a named viewport configuration.

Join

This option allows you to combine (join) two adjacent viewports. The viewports to join must share a common edge the full length of each viewport. For example, if four equal viewports were displayed, two adjacent viewports could be joined to produce a total of three viewports. You must select a *dominant* viewport. The *dominant* viewport determines the display to be used for the new viewport.

SIngle

Changes back to a single screen display using the current viewport's display.

?

Displays the identification numbers and screen positions of named (saved) and active viewport configurations. The screen positions are relative to the lower-left corner of the screen (0,0) and the upper-right corner (1,1).

2, 3, 4

Use these options to create 2, 3, or 4 viewports. You can choose the configuration. The possibilities are illustrated if you use the *Viewports* dialog box.

CHAPTER EXERCISES

Figure 10-28

1. *Zoom Extents, Realtime, Window, Previous, Center*

Open the sample drawing supplied with AutoCAD called **DB_SAMP.DWG**. If your system has the standard installation of AutoCAD, the drawing is located in the **C:\Program Files\AutoCAD 2004\Sample** directory. The drawing shows an office building layout (Fig. 10-28). Use *SaveAs* and rename the drawing to **ZOOM TEST** and locate it in your working folder (directory). (When you view sample drawings, it is a good idea to copy them to another name so you do not accidentally change the original drawings.)

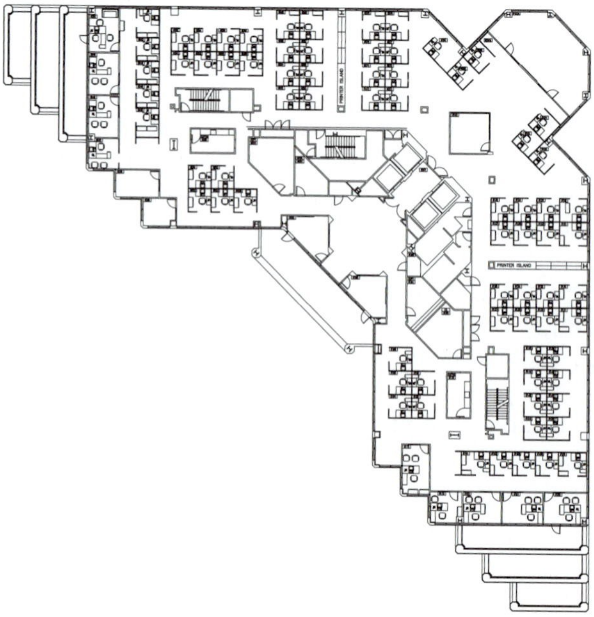

A. First, use *Zoom Extents* to display the office drawing as large as possible on your screen. Next, use *Zoom Realtime* and zoom in to the center of the layout (the center of your screen). You should be able to see rooms 6002 and 6198 located against the exterior wall (Fig. 10-29). Right-click and use *Zoom Extents* from the cursor pop-up menu to display the entire layout again. Press **Esc**, **Enter**, or right-click and select *Exit* to exit *Realtime*.

Figure 10-29

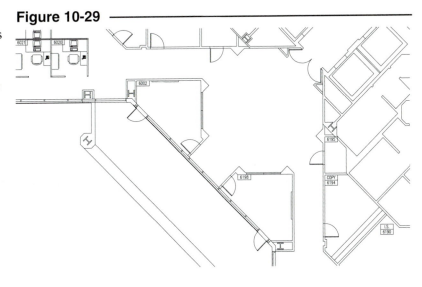

B. Use *Zoom Window* to closely examine room 6050 at the top left of the building. Whose office is it? What items on the desk are magenta in color? What item is cyan (light blue)?

C. Use *Zoom Center*. **PICK** the telephone on the desk next to the computer as the center. Specify a height of **24** (2'). You should see a display showing only the telephone. How many keys are on the telephone's number pad (*Point* objects)?

D. Next use *Zoom Previous* repeatedly until you see the same display of rooms 6002 and 6198 as before (Fig. 10-29). Use *Zoom Previous* repeatedly until you see the office drawing *Extents*.

2. *Pan Realtime*

A. Using the same drawing as exercise 1 (ZOOM TEST.DWG, originally DB_SAMP.DWG), use *Zoom Extents* to ensure you can view the entire drawing. Next *Zoom* with a *Window* to an area about 1/4 the size of the building.

B. Invoke the *Real Time* option of *Pan*. *Pan* about the drawing to find a coffee room. Can you find another coffee room? How many coffee rooms are there?

3. *Pan* and *Zoom* with the Mouse Wheel

In the following exercise, *Zoom* and *Pan* using the wheel on your mouse (turn the wheel to *Zoom* and press and drag the wheel to *Pan*). If you do not have this capability, use *Realtime Pan* and *Zoom* in concert to examine specific details of the office layout.

A. Find your new office. It is room 6100. (HINT: It is centrally located in the building.) Your assistant is in the office just to the right. What is your assistant's name and room number?

B. Although you have a nice office, there are some disadvantages. How far is it from your office door to the nearest coffee room (nearest corner of the coffee room)? HINT: Use the *Dist* command with *Endpoint OSNAP* to select the two nearest doors.

C. *Pan* and/or *Zoom* to find the copy room 6006. How far is it to the copy room (direct distance, door to door)?

D. Perform a *Zoom Extents*. Use the wheel to zoom in to Kathy Ragerie's office (room 6150) in the lower-right corner of the building. Remember to locate the cursor at the spot in the drawing that you want to zoom in to. If you need to pan, hold down the wheel and "drag" the drawing across the screen until you can view all of room 6150 clearly. How many chairs are in Kathy's office? Finally use *Zoom Extents* to size the drawing to the screen.

4. *Pan Point*

A. *Zoom* in to your new office (room 6100) so that you can see the entire room and room number. Assume you were listening to music on your computer with the door open and calculated it could be heard about 100' away. Naturally, you are concerned about not bothering the company CEO who has an office in room 6048. Use *Pan Point* to see if the CEO's office is within listening range. (HINT: enter **100'<0** at the "Specify base point or displacement:" prompt.) Should you turn down the system?

B. Do not *Exit* the **ZOOM TEST** drawing.

5. *Syswindows*

A. *Open* the **TABLET** drawing from the **Sample** folder (C:\Program Files\AutoCAD 2004\Sample). Use *SaveAs*, name the new drawing **TABLET TEST**, and locate it in your working folder.

B. You should now have two drawings open. Use the *Window* pull-down menu and select each of the options: *Cascade*, *Tile Horizontally*, and *Tile Vertically*. Which of these options would you generally use to display two drawings so they are optimally sized and arranged? Experiment with *Realtime Pan* and *Zoom* in each window.

C. Next, maximize one of the drawings by clicking the *Maximize* button in the drawing window's upper right corner (the center of the three small buttons). The drawing should fill the entire screen and the other drawing should disappear. Next toggle between drawings by holding down the **Ctrl** key and pressing the **Tab** key a couple of times.

D. Make the **TABLET TEST** drawing the current drawing.

6. *Viewres, Zoom Dynamic*

A. Make sure the **TABLET TEST** drawing is current. Use *Zoom* with the *Window* option (or you can try *Realtime Zoom*). Zoom in to the first several command icons located in the middle left section of the tablet menu. Your display should reveal icons for *Zoom* and other viewing commands. *Pan* or *Zoom* in or out until you see the *Zoom* icons clearly.

B. Notice how the *Zoom* command icons are shown as polygons instead of circles? Use the *Viewres* command and change the value to **5000**. Do the icons now appear as circles? Use *Zoom Realtime* again. Does the new setting change the speed of zooming?

C. Use *Zoom Dynamic*, then try moving the view box immediately. Can you move the box before the drawing is completely regenerated? Change the view box size so it is approximately equal to the size of one icon. Zoom in to the command in the lower-right corner of the tablet menu. What is the command? What is the command in the lower-left corner? For a drawing of this complexity and file size, which option of *Zoom* is faster and easier for your system—*Realtime* or *Dynamic*?

D. If you wish, you can *Exit* the **TABLET TEST** drawing.

7. *View*

     A. Make the **ZOOM TEST** drawing that you used in previous exercises the current drawing. *Zoom* in to the office that you will be moving into (room 6100). Make sure you can see the entire office. Use the *View* command and create a *New* view named **6100**. Next, *Zoom* or *Pan* to room 6048. *Save* the display as a *View* named **6048**.

     B. Use the *View* dialog box again and make view **6100** *Current*.

     C. In order for the CEO to access information on your new computer, a cable must be stretched from your office to room 6048. Find out what length of cable is needed to connect the upper left corner of office 6100 to the upper right corner of office 6048. (HINT: Use *Dist* to determine the distance. Type the '*–View* command <u>transparently</u> [prefix with an apostrophe and hyphen] during the *Dist* command and use *Running Endpoint OSNAP* to select the two indicated office corners. The *View* dialog box <u>is not transparent</u>.) What length of cable is needed? Do not *Exit* the drawing.

8. *Aerial View*

     A. It has been decided to connect several other computers to your office computer. Using a similar technique as in the previous exercise (using *Distance*), determine cable lengths to other locations. Rather than saving a *View* for each location, activate *Aerial View*.

     B. Use the *Global* option in *Aerial View* to display the entire office. Next, click inside the viewer and make the view box about the size of your office and right-click to zoom in to it (inside *Aerial View*). Your office should appear in the main drawing area.

     C. Ensure that *Endpoint* is checked in the *Running Osnap* dialog box. Activate *Dist* and **PICK** the lower-left corner of your office (in the main graphics display). Next in *Aerial View*, select a *Zoom Window* box around the room 6105 (just to the right of your office). What length of cable is needed to connect to the nearest left corner of the room?

     D. Use the same technique to connect the lower corner of office space 6092 to your office (upper-right corner), the lower corner of room 6093 to your office. What lengths of cable are needed?

     E. If you want, *Exit* the **ZOOM TEST** drawing for the next exercise.

     NOTE: If you view and experiment with other sample drawings, remember to immediately use *Saveas* to create new drawings so the originals are not accidentally changed.

9. *Zoom All, Extents*

Figure 10-30

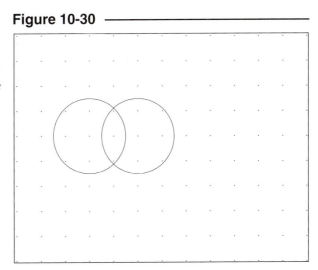

     A. Begin a *New* drawing. Turn on the *SNAP* (**F9**) and *GRID* (**F7**). Draw two **Circles**, each with a **1.5** unit *radius*. The *Circle* centers are at **3,5** and at **5,5**. See Figure 10-30.

     B. Use *Zoom All*. Does the display change? Now use *Zoom Extents*. What happens? Now use *Zoom All* again. Which option <u>always</u> shows all of the *Limits*?

C. Draw a *Circle* with the center at **10,10** and with a *radius* of **5**. Now use *Zoom All*. Notice the *GRID* appears only on the area defined by the *Limits*. Can you move the cursor to 0,0? Now use *Zoom Extents*. What happens? Can you move the cursor to 0,0?

D. *Erase* the large *Circle*. Use *Zoom All*. Can you move the cursor to 0,0? Use *Zoom Extents*. Can you find point 0,0?

E. *Exit* the drawing and discard changes.

10. *Vports*

A. *Open* the **ZOOM TEST** drawing again. Perform a *Zoom Extents*. Invoke the *Viewports* dialog box by any method. When the dialog box appears, choose the *Three: Right* option. The resulting viewport configuration should appear as Figure 10-25, shown earlier in the chapter.

B. Using *Zoom* and *Pan* in the individual viewports, produce a display with an overall view on the top and two detailed views as shown in Figure 10-27.

C. Type the *Vports* command. Use the *New* option to save the viewport configuration as **3R**.

D. Click in the bottom-left viewport to make it the current viewport. Use *Vports* again and change the display to a *Single* screen. Now use *Realtime Zoom* and *Pan* to locate and zoom to room 6100.

E. Use the *Viewports* dialog box again. Select the *Named Viewports* tab and select **3R** as the viewport configuration to restore. Ensure the **3R** viewport configuration was saved as expected, including the detail views you prepared.

F. Invoke a *Single* viewport again. Finally, use the *Vports* dialog box to create another viewport configuration of your choosing. *Save* the viewport configuration and assign an appropriate name. *Save* the **ZOOM TEST** drawing.

11
LAYERS AND OBJECT PROPERTIES

Chapter Objectives

After completing this chapter you should:

1. understand the strategy of grouping related geometry with *Layers*;

2. be able to create *Layers*;

3. be able to assign *Color, Linetype,* and *Lineweight* to *Layers*;

4. be able to control a layer's properties and visibility settings (*On, Off, Freeze, Thaw, Lock, Unlock*);

5. be able to select specific layers from a long list using the *Named Layer Filters* dialog box;

6. be able to assign *Color, Linetype,* and *Lineweight* to objects;

7. be able to set *LTSCALE* to adjust the scale of linetypes globally;

8. understand the concept of object properties;

9. be able to change an object's properties (layer, color, linetype, linetype scale, and lineweight) with the *Object Properties* toolbar, the *Properties* window, and with *Match Properties.*

CONCEPTS

In a CAD drawing, layers are used to group related objects in a drawing. Objects (*Lines, Circles, Arcs,* etc.) that are created to describe one component, function, or process of a drawing are perceived as related information and, therefore, are typically drawn on one layer. A single CAD drawing is generally composed of several components and, therefore, several layers. Use of layers provides you with a method to control visible features of the components of a drawing. For each layer, you can control its color on the screen, the linetype and lineweight it will be displayed with, and its visibility setting (on or off). You can also control if the layer is plotted or not.

Layers in a CAD drawing can be compared to clear overlay sheets on a manual drawing. For example, in a CAD architectural drawing, the floor plan can be drawn on one layer, electrical layout on another, plumbing on a third layer, and HVAC (heating, ventilating, and air conditioning) on a fourth layer (Fig. 11-1). Each layer of a CAD drawing can be assigned a different color, linetype, lineweight, and visibility setting similar to the way clear overlay sheets on a manual drawing can be used. Layers can be temporarily turned *Off* or *On* to simplify drawing and

Figure 11-1

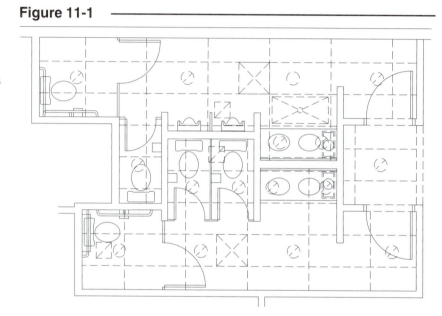

editing, like overlaying or removing the clear sheets. For example, in the architectural CAD drawing, only the floor plan layer can be made visible while creating the electrical layout, but can later be cross-referenced with the HVAC layout by turning its layer on. Layers can also be made "non-plottable." Final plots can be made of specific layers for the subcontractors and one plot of all layers for the general contractor by controlling the layers' *Plot/No Plot* icon before plotting.

AutoCAD allows you to create a practically unlimited number of layers. You should assign a name to each layer when you create it. The layer names should be descriptive of the information on the layer.

Assigning Colors, Linetypes, and Lineweights

There are two strategies for assigning colors, linetypes, and lineweights in a drawing: assign these properties to layers or assign them to objects.

Assign colors, linetypes, and lineweights to layers
Usually, layers are assigned a color, linetype, and lineweight so that all objects drawn on a single layer have the same color, linetype, and lineweight. Assigning colors, linetypes, and lineweights to layers is called *ByLayer* color, linetype, and lineweight setting. Using the *ByLayer* method makes it visually apparent which objects are related (on the same layer). All objects on the same layer have the same linetype and color.

Assign colors, linetypes, and lineweights to individual objects

Alternately, you can assign colors, linetypes, and lineweights to specific objects, overriding the layer's color, linetype, and lineweight setting. This method is fast and easy for small drawings and works well when layering schemes are not used or for particular applications. However, using this method makes it difficult to see which layers the objects are located on.

Object Properties

Another way to describe the assignment of color, linetype, and lineweight properties is to consider the concept of Object Properties. Each object has properties such as a layer, a color, a linetype, and a lineweight. The color, linetype, and lineweight for each object can be designated as *ByLayer* or as a specific color, linetype, or lineweight. Object properties for the two drawing strategies (schemes) are described as follows.

Object Properties

	1. *ByLayer* Drawing Scheme	2. Object-Specific Drawing Scheme
Layer Assignment	Layer name descriptive of geometry on the layer	Layer name descriptive of geometry on the layer
Color Assignment	*ByLayer*	*Red*, *Green*, *Blue*, *Yellow*, or other specific color setting
Linetype Assignment	*ByLayer*	*Continuous*, *Hidden*, *Center*, or other specific linetype setting
Lineweight Assignment	*ByLayer*	0.05 mm, 0.15 mm, 0.010", or other specific lineweight setting

It is recommended that beginners use only one method for assigning colors, linetypes, and lineweights. After gaining some experience, it may be desirable to combine the two methods only for specific applications. Usually, the *ByLayer* method is learned first, and the object-specific color, linetype, and lineweight assignment method is used only when layers are not needed or when complex applications are needed. (The *ByBlock* color, linetype, and lineweight assignment has special applications for *Blocks* and is discussed in Chapter 21.)

Plot Styles

Plot style is an object property that controls how an object appears in a plot. A plot style can be assigned to any object, but can also be assigned to a layer (*ByLayer*). A plot style can control such appearances as (plotted) color, lineweight, screen percentage, line end styles, line join styles, and pen number. A plot style is actually another object property, just like color, linetype, and lineweight, except that assigning it is optional. You should assign plot styles only if you want the plot to have features other than the way the drawing appears on the screen.

There are two types of plot styles: color-dependent and named plot styles. The color-dependent plot style is the simpler of the two because it is based on color. Since colors are usually assigned to layers, each layer's color on the screen controls how the drawing is plotted. Each screen color can be assigned to use a separate pen, screening percentage, line end style, and so on. In this way, all objects that appear in one color will be plotted with the same appearance. On the other hand, named plot styles are not based on color, so they can be assigned to any object. With named plot styles, objects that are drawn in one color can be plotted with different appearances. Plot styles are discussed in detail in Chapter 33, Advanced Layouts and Plotting.

LAYERS AND LAYER PROPERTIES CONTROLS

Layer Control Drop-Down List

The Object Properties toolbar contains a drop-down list for making layer control quick and easy (Fig. 11-2). The window normally displays the current layer's name, visibility setting, and properties. When you pull down the list, all the layers (unless otherwise specified in the *Named Layer Filters* dialog box) and their settings are displayed. Selecting any layer <u>name</u> makes it current. Clicking on any of the visibility/properties icons changes the layers' setting as described in the following section. Several layers can be changed in one "drop." You cannot change a layer's color, linetype, or lineweight, nor can you create new layers using this drop-down list.

Figure 11-2

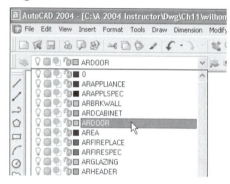

LAYER

Pull-down Menu	COMMAND (TYPE)	ALIAS (TYPE)	Short-cut	Screen (side) Menu	Tablet Menu
Format *Layer*	*LAYER* or -LAYER	*LA* or -LA	...	*FORMAT* *Layer*	U,5

The way to gain complete layer control is through the *Layer Properties Manager* (Fig. 11-3). The *Layer Properties Manager* is invoked by using the icon button (shown above), typing the *Layer* command or *LA* command alias, or selecting *Layer* from the *Format* pull-down or screen menu.

This dialog box allows full control for all layers in a drawing. Layers existing in the drawing appear in the list at the central area (only Layer 0 exists in new drawings and in those created from standard templates such as ACAD.DWT). New layers can be created by selecting the *New* button near the upper right corner of the dialog box.

All properties and visibility settings of layers can be controlled by highlighting the layer name and then selecting one of the icons for the layer such as the light bulb icon (*On, Off*),

Figure 11-3

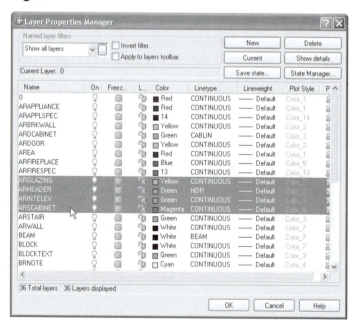

sun/snowflake icon (*Thaw/Freeze*), padlock icon (*Lock/Unlock*), *Color* tile, *Linetype*, *Lineweight*, or *Plot/No plot* icon.

Typical Windows dialog box features apply as explained in this paragraph. Multiple layers can be highlighted (see Figure 11-3) by holding down the Ctrl key while PICKing (to highlight one at a time) or holding down the Shift key while PICKing (to select a range of layers between and including two selected names). Right-clicking in the list area displays a cursor menu allowing you to *Select All* or *Clear All* names in the list and other options (see Figure 11-5). You can rename a layer by clicking twice <u>slowly</u> on the name (this is the same as PICKing an already highlighted name).

The column widths can be changed by moving the pointer to the "crack" between column headings until double arrows appear (Fig. 11-4). You can also resize the entire dialog box by placing your pointer on the extreme border or corner until double arrows appear, then clicking and dragging.

Figure 11-4

A particularly useful feature is the ability to sort the layers in the list by any one of the headings (*Name, On, Freeze, Linetype*, etc.) by clicking on the heading tile. For example, you can sort the list of names in alphabetical order (or reverse order) by clicking once (or twice) on the *Name* heading above the list of names. Or you may want to sort the *Frozen* and *Thawed* layers or sort layers by *Color* by clicking on the column heading.

The *Show details* tile near the top-right corner of the dialog box displays the details (properties and visibility settings) of the highlighted layer in the list (Fig. 11-5). This area is basically an <u>alternative</u> method of controlling layer properties and visibility settings (instead of using the icons in the list area).

Figure 11-5

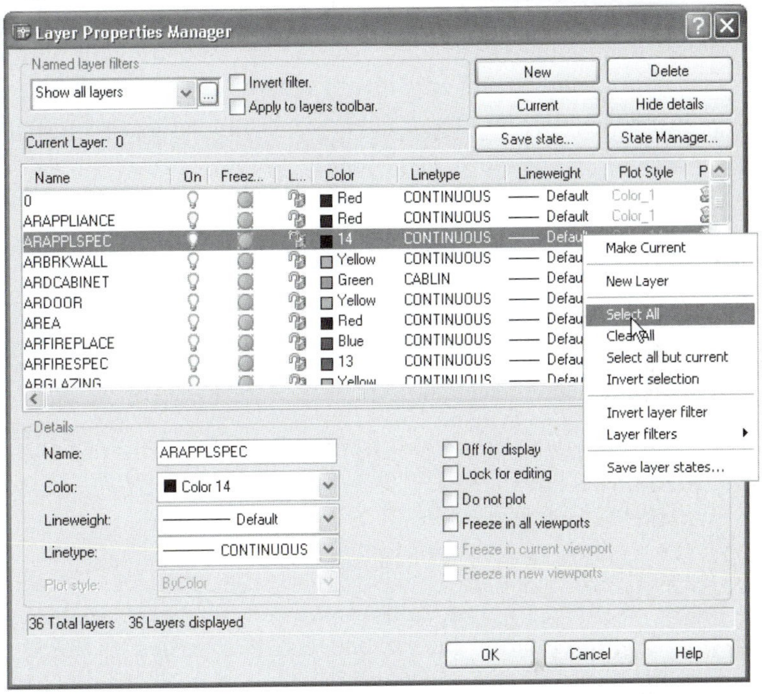

As an alternative to the dialog box, the *Layer* command can be used in command line format by typing *Layer* with a hyphen (-) prefix. The Command line format of *Layer* shows all of the available options.

```
Command: -layer
Current layer: "0"
Enter an option
[?/Make/Set/New/ON/OFF/Color/Ltype/LWeight/Plot/PStyle/Freeze/Thaw/LOck/Unlock/stAte]:
```

Detailed explanations of the options for controlling layer properties and visibility settings, whether using the *Layer Properties Manager* or using -*Layer* in Command line format, are listed next.

Current or *Set*

To *Set* a layer as the *Current* layer is to make it the active drawing layer. Any objects created with draw commands are created on the *Current* layer. You can, however, edit objects on any layer, but draw only on the current layer. Therefore, if you want to draw on the FLOORPLAN layer (for example), use the *Set* or *Current* option or double-click on the layer name. If you want to draw with a certain *Color* or *Linetype*, set the layer with the desired *Color, Linetype,* and *Lineweight* as the *Current* layer. Any layer can be made current, but only <u>one layer at a time</u> can be current.

To set the current layer with the *Layer Properties Manager* (see Figures 11-3 and 11-5), select the desired layer from the list and then select the *Current* tile. You can instead double-click on the desired layer name. Alternately, if you are typing, use the *Set* option of the -*Layer* command to make a layer the current layer.

On, Off

If a layer is *On*, it is visible. Objects on visible layers can be edited or plotted. Layers that are *Off* are not visible. Objects on layers that are *Off* will not plot and cannot be edited (unless the *ALL* selection option is used, such as *Erase, All*). It is not advisable to turn the current layer *Off*.

Freeze, Thaw

Freeze and *Thaw* override *On* and *Off*. *Freeze* is a more protected state than *Off*. Like being *Off*, a frozen layer is not visible, nor can its objects be edited or plotted. Objects on a frozen layer cannot be accidentally *Erase*d with the *ALL* option. *Freezing* also prevents the layer from being considered when *Regen*s occur. *Freezing* unused layers speeds up computing time when working with large and complex drawings. *Thawing* reverses the *Freezing* state. Layers can be *Thawed* and also turned *Off*. Frozen layers are not visible even though the light bulb icon is on.

Lock, Unlock

Layers that are *Locked* are protected from being edited but are still visible and can be plotted. *Locking* a layer prevents its objects from being changed even though they are visible. Objects on *Locked* layers cannot be selected with the *ALL* selection option (such as *Erase, All*). Layers can be *Locked* and *Off*.

Color, Linetype, Lineweight, and Other Properties

Layers have properties of *color, linetype,* and *lineweight* such that (generally) an object that is drawn on, or changed to, a specific layer assumes the layer's linetype and color. Using this scheme (*ByLayer*) enhances your ability to see what geometry is related by layer. It is also possible, however, to assign specific color, linetype, and lineweight to objects which will override the layer's color, linetype, and lineweight (see "*Color, Linetype,* and *Lineweight* Commands" and "Changing Object Properties").

Figure 11-6

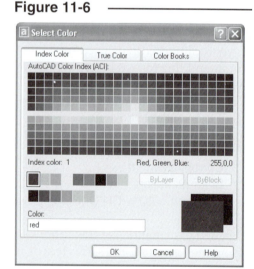

Color

Selecting one of the small color boxes in the list area of the *Layer Properties Manager* causes the *Select Color* dialog box to pop up (Fig. 11-6). The desired color can then be selected or the name or color number (called the ACI—AutoCAD Color Index) can be typed in the edit box. This action retroactively changes the color assigned to a layer. Since the color setting is assigned to the layer, all objects on the layer that have the *ByLayer* setting change to the new layer color. Objects with specific color assigned (not *ByLayer*) are not affected. Alternately, the *Color* option of the *–Layer* command (hyphen prefix) can be typed to enter the color name or ACI number.

2004

In AutoCAD 2004, the *Select Color* dialog box contains tabs for the *Index Color* (ACI), *True Color*, and *Color Books*. See "True Color and Color Books" later in this chapter for information.

Linetype

To set a layer's linetype, select the *Linetype* (word such as *Continuous* or *Hidden*) in the layer list, which in turn invokes the *Select Linetype* dialog box (Fig. 11-7). Select the desired linetype from the list.

Figure 11-7

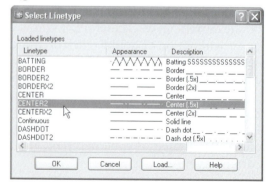

Alternately, the *Linetype* option of the *–Layer* command (hyphen prefix) can be typed. Similar to changing a layer's color, all objects on the layer with *ByLayer* linetype assignment are retroactively displayed in the selected layer linetype while non-*ByLayer* objects remain unchanged.

The ACAD.DWT and ACADISO.DWT template drawings as supplied by Autodesk have only one linetype available (*Continuous*). Before you can use other linetypes in a drawing, you must load the linetypes by selecting the *Load* tile (see the *Linetype* command) or by using a template drawing that has the desired linetypes already loaded.

Lineweight

The lineweight for a layer can be set by selecting the *Lineweight* (word such as *Default* or *0.20 mm, 0.50 mm, 0.010"* or *0.020"*, etc.) in the layer list. This action produces the *Lineweight* dialog box (Fig. 11-8). Select the desired lineweight from the list. Alternately, the *Lineweight* option of the *–Layer* command (hyphen prefix) can be typed. Like the *Color* and *Linetype* properties, all objects on the layer with *ByLayer* lineweight assignment are retroactively displayed in the lineweight assigned to the layer while non-*ByLayer* objects remain unchanged. You can set the units for lineweights (mm or inches) using the *Lineweight* command.

Figure 11-8 ——————

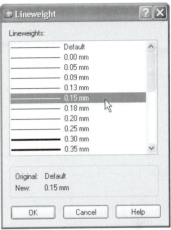

Plot Style

This column of the *Layer Properties Manager* designates the plot style assigned to the layer. If the drawing has a color-dependent Plot Style Table attached, this section is disabled since the plot styles are automatically assigned to colors. If a named Plot Style Table is attached to the drawing you can select this section to assign plot styles from the table to layers. Plot styles are discussed in Chapter 33, Advanced Layouts and Plotting.

Plot/No Plot

You can prevent a layer from plotting by clicking the printer icon so a red circle appears over the printer symbol. There are actually three methods for preventing a layer from appearing on the plot: *Freeze* it, turn it *Off*, or change the *Plot/No Plot* icon. The *Plot/No Plot* icon is generally preferred since the other two options prevent the layer from appearing in the drawing (screen) as well as in the plot.

Current Viewport Freeze, New Viewport Freeze

These options are used and are displayed in the *Layer Properties Manager* only when paper space viewports exist in the drawing. Using these options, you can control what geometry (layers) appears in specific viewports. See Chapters 13 and 33 for more information on these options.

New

The *New* option allows you to make new layers. There is only one layer in the AutoCAD templates (ACAD.DWT and ACADISO.DWT) as they are provided to you "out of the box." That layer is Layer 0. Layer 0 is a part of every AutoCAD drawing because it cannot be deleted. You can, however, change the *Color, Linetype,* and *Lineweight* of Layer 0 from the defaults (*Continuous* linetype, *Default* lineweight, and color #7 *White*). Layer 0 is generally used as a construction layer or for geometry not intended to be included in the final draft of the drawing. Layer 0 has special properties when creating *Blocks* (see Chapter 21).

You should create layers for each group of related objects and assign appropriate layer names for that geometry. You can use up to 256 characters (including spaces) for layer names. There is practically no limit to the number of layers that can be created (although 32,767 has been found to be the actual limit). To create layers in the *Layer Properties Manager* box, select the *New* tile.

A new layer named "Layer1" (or other number) then appears in the list with the default color (*White*), linetype (*Continuous*), and lineweight (*Default*). If an existing layer name is highlighted when you make a new layer, the highlighted layer's properties (*color, linetype, lineweight,* and *plot style*) are used as a template for the new layer. The new layer name initially appears in the rename mode so you can immediately assign a more appropriate and descriptive name for the layer. You can create many new names quickly by typing (renaming) the first layer name, then typing a comma before other names. A comma forces a new "blank" layer name to appear. Colors, linetypes, and lineweights should be assigned as the next step.

If you want to create layers by typing, the *New* and *Make* options of the *Layer* command can be used. *New* allows creation of one or more new layers. *Make* allows creation of one layer (at a time) and sets it as the current layer.

Delete

The *Delete* tile allows you to delete layers. <u>Only layers with no geometry can be deleted</u>. You cannot delete a layer that has objects on it, nor can you delete Layer 0, the current layer, or layers that are part of externally referenced (*Xref*) drawings. If you attempt to *Delete* such a layer, accidentally or intentionally, a warning appears.

Figure 11-9

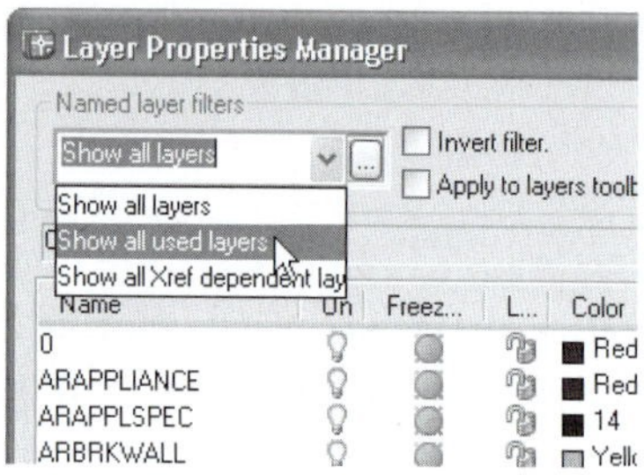

Save State, Restore State
See "Layer States."

Named Layer Filters

The *Named Layer Filters* drop-down list determines what names appear in the list of layers (Fig. 11-9). The choices are as follows.

Show all layers	shows all layers in the drawing
Show all used layers	shows all layers that have objects on them
Show all Xref dependent layers	shows all layers that are part of externally referenced drawings

When working on drawings with a large number of layers, it is time-consuming to scroll through the list of layer names in the *Layer Properties Manager* or in the Layer Control drop-down list on the Object Properties toolbar. The *Named Layer Filters* drop-down list in the upper left corner of the *Layer Properties Manager* controls names that appear in the list in that dialog box. Select the *Apply to Layers Toolbar* checkbox to apply the filter to the Layer Control drop-down list.

Figure 11-10

Named Layer Filters Dialog Box

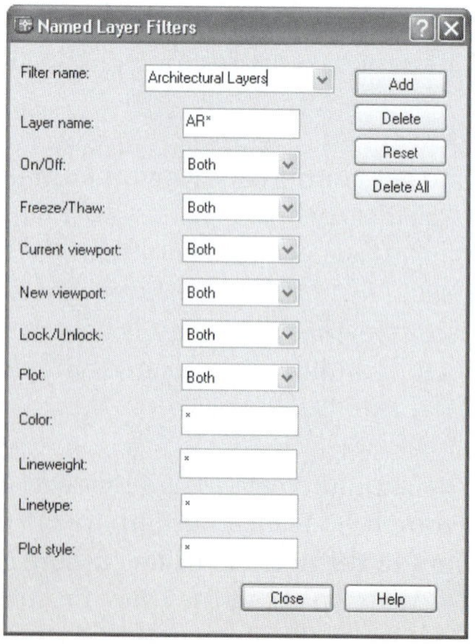

Selecting the small button just to the right of the *Named Layer Filters* drop-down list produces the *Named Layer Filters* dialog box (Fig. 11-10). The *Named Layer Filters* dialog box enables you to specify criteria for displaying a selected set of layers in both lists. For example, you may want to list only layer names that are *Thawed*. By using layer filters, you can shorten the layer list by filtering out the layers that do not meet this criterion. There are eleven criteria that can be specified in the *Named Layer Filters* dialog box. The criteria define <u>what you want to see</u> in the layer listing. When the criteria are set to *Both* or *, all layer names are displayed.

To create a layer filter, assign a *Filter Name*, specify the criteria from the boxes below, then select *Add*. The named filter then appears in the *Named Layer Filters* drop-down list in the *Layer Properties*

Manager along with *Show all layers*, *Show all used layers*, etc. To apply the new filter, you must then select it from the drop-down list.

The boxes displaying an asterisk (*) symbol enable you to enter specific values. These edit boxes allow you to list only layers that match particular names, colors, linetypes, lineweights, or plot styles. As an example, if you set the *Layer Names* filter to **AR***, only those layers whose names begin with "AR" will be displayed in the layer listing (see Figure 11-10).

The other criteria are in pairs, enabling you to choose one or "*Both*" of the settings. When a criteria is set to *Both*, you will see layers listed which match both items in the pair, such as layers that are *On* <u>and</u> *Off*. Selecting one of the pair displays only layer names meeting that condition. Selecting *Thawed* from the *Freeze/Thaw* box, for example, causes a display of only layers that are thawed.

When a named layer filter is applied, <u>only layers that match the criteria</u> are displayed in the list. Selecting the *Invert Filter* checkbox causes <u>all layers that do not match the filter</u> to be displayed. Only one layer filter configuration can be applied at a time.

Displaying an Object's Properties and Visibility Settings

When the Layer Control drop-down list in the Object Properties toolbar is in the normal position (not "dropped down"), it can be used to display an <u>object's</u> layer properties and visibility settings. Do this by selecting an object (with the pickbox, window, or crossing window) when no commands are in use.

When an object is selected, the Object Properties toolbar displays the layer, color, linetype, lineweight, and plot style <u>of the selected object</u> (Fig. 11-11). The Layer Control box displays the selected object's layer name and layer settings rather than that of the current layer. The Color Control, Linetype Control, Lineweight Control, and Plot Style Control boxes in the Object Properties toolbar (just to the right) also temporarily reflect the color, linetype, lineweight, and plot style properties of the selected object. Pressing the Escape key causes the list boxes to display the current layer name and settings again, as would normally be displayed. If more than one object is selected, and the objects are on different layers or have different color, linetype, and lineweight properties, the list boxes go blank until the objects become unhighlighted or a command is used.

Figure 11-11

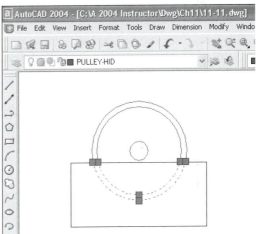

These sections of the Object Properties toolbar have another important new feature that allows you to change a highlighted object's (or set of objects) properties. See "Changing Object Properties" near the end of this chapter.

Make Object's Layer Current

This productive feature can be used to make a desired layer current simply by selecting <u>any object on the layer</u>. This option is generally faster than using the *Layer Properties Manager* or -*Layer* drop-down list to set a layer current. The *Make Object's Layer Current* feature is particularly useful when you want to draw objects on the same layer as other objects you see, but are not sure of the layer name. This feature (not a formal command) is available only by selecting the icon just to the right of the layer drop-down list (see Figure 11-11). It cannot be invoked by any other method.

There are two steps in the procedure to make an object's layer current: (1) select the icon and (2) select any object on the desired layer. The selected object's layer becomes current and the new current layer name immediately appears in the Layer Control list box in the Object Properties toolbar.

LAYERP

	Pull-down Menu	COMMAND (TYPE)	ALIAS (TYPE)	Short-cut	Screen (side) Menu	Tablet Menu
	...	*LAYERP*	...	...	...	...

Layerp (*Layer Previous*) is an *Undo* command only for layers. In other words, when you use *Layerp*, all the changes you made the last time you used the *Layer Properties Manager*, the *Layer Control* drop-down list, or the *–Layer* command are undone. Using *Layerp* does not affect any other activities that occurred to the drawing since the last layer settings, such as creating or editing geometry or viewing controls like *Pan* or *Zoom*.

Use *Layerp* to return the drawing to the previous layer settings. For example, if you froze several layers and changed some of the geometry in a drawing, but now want to thaw those frozen layers again without affecting the geometry changes, use *Layerp*. Or, if you changed the color and linetype properties of several layers but later decide you prefer the previous property settings, use *Layerp* to undo the changes and restore the original layer settings.

Layerp affects only layer-related activities; however, *Layerp* does not undo the following changes:

> Renamed layers: If you rename a layer and change its properties, *Layerp* restores the
> original properties but not the original layer name.
> Deleted layers: If you delete or purge a layer, using *Layerp* does not restore it.
> New layers: If you create a new layer in a drawing, using *Layerp* does not remove it.

The *LAYERPMODE* system variable setting (*On* or *Off*) enables or disables the *Layerp* command and the related layer tracking function. Since there is a modest performance loss for *Layerp* tracking, you can suspend layer tracking when you don't need it, such as when you run large scripts.

LAYER STATES

	Pull-down Menu	COMMAND (TYPE)	ALIAS (TYPE)	Short-cut	Screen (side) Menu	Tablet Menu
	Format Layer... Save state	...	...	...	*FORMAT Layer Save state*	...

Use the *Layer Properties Manager* to save and restore layer settings. In the upper-right corner of the *Layer Properties Manager*, two new buttons are available (Fig. 11-12): *Save state* and *State Manager*. The layer states functions can be invoked <u>only</u> through the *Layer Properties Manager*.

Figure 11-12 ————

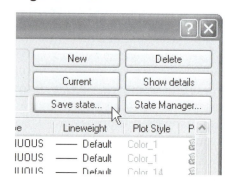

Why use Layer States?

Layer states allow you to manipulate layer settings—combinations of *On/Off, Freeze/Thaw, Lock/Unlock,* etc.—for groups of layers at one time. With many drawings you may want certain combinations of layers that are frozen, thawed, on, or off for particular operations. For example, you may require a certain combination of layers to be *On* for editing a particular aspect of a drawing—say, the floor plan, the HVAC layout, and the HVAC dimensions—while a different set of layers may be *On* for working with another aspect of the drawing—say, the floor plan, plumbing layout and fixtures. Instead of manipulating the list of all layers in the *Layer Properties Manager* each time you work with the floor plan, plumbing layout and fixtures and each time you create and edit the HVAC and dimensions, you can *Save* these settings to be *Restored* at any later time. It is suggested that you use layer states as a drawing construction feature primarily for the *Model* tab. Layer states help you change layer settings faster during drawing and editing operations.

Keep in mind that without layer states you can create and save layer settings (*Frozen, Thawed, On, Off*, etc.) with each viewport in each *Layout* tab in any drawing (see Chapters 13 and 33). Using this feature—creating *Layouts* with distinct layer settings—is a relatively simple method for saving several "looks" in one drawing. However, layouts are generally intended for maintaining different presentation plot setups (with different scales, print devices, and/or areas of a drawing as well as layer settings) in one drawing because you can save all information needed for printing or plotting (including layer settings) in a layout.

Save Layer States **Dialog Box**

The *Save state…* button (see Figure 11-12) produces the *Save Layer States* dialog box (Fig. 11-13). Since this box is used to save the current layer status of the drawing, you should specify those settings in the *Layer Properties Manager* before invoking this dialog box. In other words, to save the current combination of settings (*On, Off, Freeze, Thaw*, etc.), save the layer state. Assign a name to the current layer settings in the *New layer state name:* edit box. Layer states are saved in the current drawing.

Figure 11-13

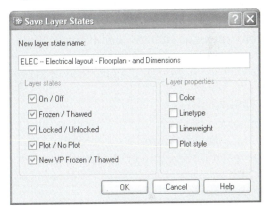

The *Layer states* checkboxes allow you to specify which settings (visibility, lock, and plot settings) you want to save with the layer state. In other words, with all the boxes checked you save all the settings as they appear for each layer in the *Layer Properties Manager*. If on the other hand, only the *On/Off* box is checked, only the *On* or *Off* setting of each layer as it is listed in the *Layer Properties Manager* is saved with the layer state.

The *Layer properties* checkboxes are generally not changed often during a drawing and editing session, unlike the frozen, thawed, on, or off status of layers in a drawing. However, this feature is very useful. Since properties of a layer such as color, lineweight, and plot style usually affect printing and plotting, you can create and restore layer states to prepare for different plot setups. Or, you may prefer a certain color and lineweight "look" on screen for drawing and editing, then restore a layer state that complies with the company standard before saving the drawing or for creating a print or plot. For example, to prepare for printing on a black and white printer, you could create and restore a layer state that quickly sets all layer colors to black. The *Layer properties* feature is also helpful if you intend to *Export* a layer state from one drawing then *Import* it to another. In this case, you can clone one drawing's color, linetype, lineweight, and/or plot characteristics to another drawing, assuming each drawing has the same layer names.

2002

Layer States Manager

Selecting the *State Manager…* button in the *Layer Properties Manager* (see Figure 11-12) produces the *Layer States Manager* dialog box (Fig. 11-14). The drawing's saved layer states are listed in the central area.

Restore

Highlight a name from the list, then use this option to restore the selected saved layer state. AutoCAD automatically sets the current drawing's global *On, Off, Freeze, Thaw,* etc. settings to those saved under the assigned name.

Figure 11-14

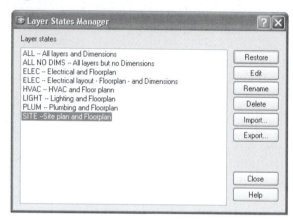

Edit…

This option produces the *Edit Layer States* dialog box (not shown), which appears to be essentially the same as the *Save Layer States* dialog box (see Figure 11-13). However, this feature allows you to select what layer characteristics are <u>restored</u> when you select *Restore*.

This feature does not actually allow you to change the individual layers' settings <u>saved</u> with a particular layer state—you must create a new state with a new name to do this. This option instead allows you to change which characteristics of the set of layers you want to <u>restore</u> (*On/Off, Freeze/Thaw, Color, Lineweight,* etc.) with the selected layer state. Surprisingly, even though you may not have specified that a particular characteristic (*On/Off, Freeze/Thaw, Color,* etc.) was checked when you saved the layer state, you can restore it here. For example, if you changed the color of several layers after you created a layer state, you can restore the original colors using the *Edit Layer States* dialog box, selecting *Color,* then *Restore,* even though the *Color* box was not checked when you saved the layer state!

Export…

Once you have created a layer state, you can save it to a file (with a .LAS extension). Each named layer state is saved as a separate .LAS file; you cannot save the entire set of layer states to a file. A layer state can be imported into other drawings with the *Import…* option.

Import…

You can import previously saved layer states from a .LAS file. Exporting and importing layer states is intended to be used primarily when you have several drawings with the same set of layer names (as when similar drawings are created from the same template drawing). Layer states can be saved in one drawing, then *Exported* to a file and *Imported* to the other similar drawings.

Layer States with *Blocks* and *Xrefs*

Layer states saved in a drawing that is inserted in your current drawing as a *Block* are also added to the current drawing. The saved layer states of *Xref* drawings are not accessible from the current drawing. (See Chapter 21, Blocks and DesignCenter, and Chapter 30, Xreferences.)

OBJECT-SPECIFIC PROPERTIES CONTROLS

The commands in this section are used to control the *color, linetype,* and *lineweight* properties of individual objects. This method of object property assignment is used only in special cases—when you want the objects' *color, linetype,* and *lineweight* to <u>override those properties assigned to the layers</u> on which the objects reside. Using this method makes it difficult to see which objects are on which layers. In most cases, the *Bylayer* property is assigned instead to individual objects so the objects assume the properties of their layers.

LINETYPE

	Pull-down Menu	COMMAND (TYPE)	ALIAS (TYPE)	Short-cut	Screen (side) Menu	Tablet Menu
	Format Linetype...	*LINETYPE* or *-LINETYPE*	*LT* or *-LT*	...	*FORMAT Linetype*	*U,3*

Invoking this command presents the *Linetype Manager* (Fig. 11-15). Even though this looks similar to the dialog box used for assigning linetypes to layers (shown earlier in Figure 11-7), beware!

When linetypes are selected and made *Current* using the *Linetype Manager*, they are <u>assigned to objects</u>—not to layers. That is, selecting a linetype by this manner (making it *Current*) causes all objects <u>from that time on</u> to be drawn using that linetype, regardless of the layer that they are on (unless the *ByLayer* type is selected). In contrast, selecting linetypes during the *Layer* command (using the *Layer Properties Manager*) results in assignment of linetypes to <u>layers</u>. Remember that using both of these methods for color, linetype, and lineweight assignment in one drawing can be very confusing until you have some experience using both methods.

Figure 11-15

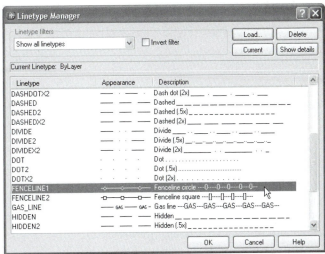

To draw an object in a specific linetype, simply select the linetype from the *Linetype Manager* list, select *Current* (or double-click the linetype), and select the *OK* tile. That linetype stays in effect (all objects are drawn with that linetype) until another linetype is selected. If you want to draw objects in the <u>layer's</u> assigned linetype, select *ByLayer*. *ByBlock* linetype assignment is discussed in Chapter 21.

Load

The *Linetype Manager* also lets you *Load* linetypes that are not already in the drawing. Just select the *Load* button to view and select from the list of available linetypes in the *Load or Reload Linetypes* dialog box (Fig. 11-16). The linetypes are loaded from the ACAD.LIN or ACADISO.LIN file. If you want to load all linetypes into the drawing, right-click to display a small cursor menu, then choose *Select All* to highlight all linetypes (see Figure 11-16). You can also select multiple linetypes by holding down the Ctrl key or select a range by holding down the Shift key. The loaded linetypes are then available for assignment from both the *Linetype Manager* and the *Layer Properties Manager*.

Figure 11-16

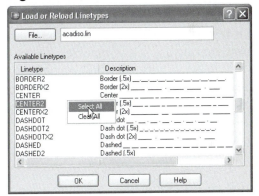

AutoCAD supplies numerous linetypes that can be viewed in the *Load or Reload Linetypes* dialog box (Figure 11-16). The ACAD_ISO*n*W100 linetypes are intended to be used with metric drawings. For the ACAD_ISO*n*W100 linetypes, *Limits* should be set to metric sheet sizes to accommodate the relatively large linetype spacing. Avoid using both ACAD_ISO*n*W100 linetypes and non-ISO linetypes in one drawing due to the difficulty managing the two sets of linetype scales.

Delete
The *Delete* button is useful for deleting any unused linetypes from the drawing. Unused linetypes are those which have been loaded into the drawing but have not been assigned to layers or to objects. Freeing the drawing of unused linetypes can reduce file size slightly.

The *Linetype Manager* has several features that are similar to the *Layer Properties Manager* such as a *Linetype filters* drop-down list and *Details/Hide Details* button. If details are visible, you can change the *Name* or *Description*, or set the *Global Scale Factor* (see *"LTSCALE"*) or *Current Object Scale* (see *"CELTSCALE"*).

Typing *-Linetype* (use the hyphen prefix) produces the command line format of *Linetype*. Although this format of the command does not offer the capabilities of the dialog box version, you can assign linetypes to objects or assign the *ByLayer* linetype setting.

> Command: ***-linetype***
> Current line type: "ByLayer"
> Enter an option [?/Create/Load/Set]:

You can list (?), *Load*, *Set*, or *Create* linetypes with the typed form of the command. Type a question mark (?) to display the list of available linetypes. Type *L* to *Load* any linetypes. The *Set* option of *Linetype* accomplishes the same action as selecting a linetype from the list in the dialog box. That is, you can set a specific linetype for all subsequent objects regardless of the layer's linetype setting, or you can use *Set* to assign the new objects to use the layer's (*ByLayer*) linetype.

You can create your own custom linetypes with the *Create* option or by using a text editor. Both simple linetypes (line, dash, and dot combinations) and complex linetypes (including text or other shapes) are possible. See Chapter 44, Basic Customization, for more information on creating linetypes.

Linetype Control Drop-Down List

The Object Properties toolbar contains a drop-down list for selecting linetypes (Fig. 11-17). Although this appears to be quick and easy, you can only assign linetypes to objects by this method unless *ByLayer* is selected to use the layers' assigned linetypes. Any linetype you select from this list becomes the current object linetype. If you want to select linetypes for layers, make sure this list displays the *ByLayer* setting, then use the *Layer Properties Manager* to select linetypes for layers.

Figure 11-17

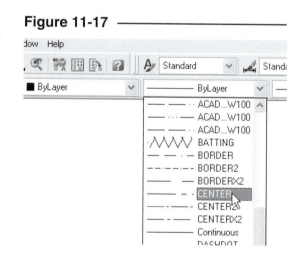

Keep in mind that the Linetype Control drop-down list as well as the Layer Control, Color Control, and Lineweight Control drop-down lists display the current settings for selected objects (objects selected when no commands are in use). When linetypes are assigned to layers rather than objects, the layer drop-down list reports a *ByLayer* setting.

LWEIGHT

	COMMAND (TYPE)	ALIAS (TYPE)	Short-cut	Screen (side) Menu	Tablet Menu
Pull-down Menu					
Format *Lineweight...*	*LWEIGHT*	*LW*	...	...	...

The *Lweight* command (short for lineweight) produces the *Lineweight Settings* dialog box (Fig. 11-18).
This dialog box is the lineweight <u>equivalent</u> of the
Linetype Manager—that is, this dialog box assigns
lineweights <u>to objects, not layers</u> (unless the *ByLayer*
or *Default* lineweight is selected). See the previous
discussion under "*Linetype.*" The *ByBlock* setting is
discussed in Chapter 21.

Figure 11-18

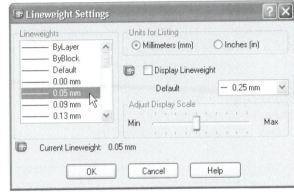

Any lineweight selected in the *Lineweight Settings*
dialog box automatically becomes current (without
having to select a *Current* button or double-click). The
current lineweight is assigned to all subsequently
drawn objects. That is, selecting a lineweight by this
manner (making it current) causes all objects <u>from that time on</u> to be drawn using that lineweight,
regardless of the layer that they are on (unless *ByLayer* or *Default* is selected). In contrast, selecting
lineweights during the *Layer* command (using the *Layer Properties Manager*) results in assignment of
lineweights to <u>layers</u>.

As a reminder, this type of drawing method (object-specific property assignment) can be difficult for
beginning drawings and for complex drawings. If you want to draw objects in the <u>layer's</u> assigned
lineweight, select *ByLayer*. If you want all layers to have the same lineweight, select *Default*.

Units for Listing
You can choose *Millimeters (mm)* or *Inches (in)* as the units for the lineweight list. The values reported in
the list (0.05 mm, 0.15 mm, 0.010", 0.015", etc.) indicate the thickness of the selected line. The setting is
stored in the *LWUNITS* system variable.

Display Lineweight
Checking this option causes the assigned lineweight thickness to be displayed anywhere in the
drawing—in model space (the *Model* tab), for model space geometry that appears in paper space viewports (viewports in a *Layout* tab), and in paper space (in a *Layout* tab, but not in a viewport).

When lineweights are displayed in model space (the *Model* tab), the line thickness is relative to the screen
size, so zooming in or out does not make the lines appear wider or narrower. However, when
lineweights are displayed in paper space (a *Layout* tab), the line thickness is absolute, so zooming
changes the lineweight appearance.

The *Display Lineweight* checkbox has the same function as
toggling the word *LWT* on the Status Bar (Fig. 11-19). The
LWT toggle is somewhat more accessible to use. The status
of the lineweight display for the drawing is stored in the
LWDISPLAY system variable. Using either the *Display
Lineweight* checkbox or the *LWT* toggle changes the setting
of the *LWDISPLAY* variable, and vice versa.

Figure 11-19

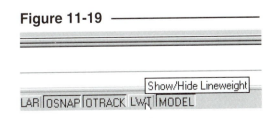

Default

Although the *Current* lineweight for new drawings is *ByLayer*, when new layers are created, all layers have the *Default* lineweight. This feature (*Default* rather than *ByLayer*) is exclusive for lineweights. Assigning the *Default* setting ensures that all layers (or objects) with this setting have the same lineweight. The advantage is that the <u>lineweight can be changed globally</u> for all layers (or objects) that have the *Default* setting.

Selecting a new lineweight in the *Default* drop-down list of the *Lineweight Settings* dialog box changes the lineweight thickness for existing layers (or objects) with a *Default* setting and assigns that setting for new layers (or objects) that are subsequently created. In contrast, to change the lineweight for existing layers with a *ByLayer* setting, <u>each layer's lineweight</u> must be changed individually. The *Default* lineweight setting is stored in the *LWDEFAULT* system variable.

Adjust Display Scale

This adjustment affects only the display of lineweights in the *Lineweight Settings* dialog box and the Lineweight Control drop-down list.

Typing *-Lweight* produces the Command line format of *Lweight*. This command line version only allows you to set the current lineweight for new objects and does not offer the capabilities of the dialog box version. Typing *-Lweight* (use the hyphen prefix) produces the following prompt:

        Command: **-lweight**
        Current lineweight: ByLayer
        Enter default lineweight for new objects or [?]:

Lineweight Control Drop-Down List

The Object Properties toolbar contains a drop-down list for selecting lineweights (Fig. 11-20). Similar to the function of the Linetype Control drop-down list, you can <u>only assign lineweights to objects</u> by this method unless *ByLayer* is selected to use the layers' assigned lineweights. Any lineweight you select from this list becomes the current object lineweight. If you want to select lineweights for layers, make sure this list displays the *ByLayer* setting, then use the *Layer Properties Manager* to select lineweights for layers.

Remember that the Lineweight Control drop-down list as well as the Layer Control, Color Control, and Linetype Control drop-down lists display the current settings for selected objects (objects selected when no commands are in use). When lineweights are assigned to layers rather than objects, the layer drop-down list reports a *ByLayer* setting.

Figure 11-20

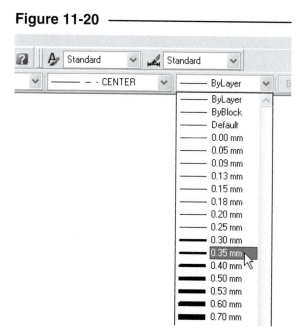

COLOR

Pull-down Menu	COMMAND (TYPE)	ALIAS (TYPE)	Short-cut	Screen (side) Menu	Tablet Menu
Format Color...	*COLOR*	*COL*	...	*FORMAT Color*	*U,4*

Similar to linetypes and lineweights, colors can be assigned to layers or to objects. Using the *Color* command assigns a color for all newly created <u>objects</u>, regardless of the layer's color designation (unless the *ByLayer* color is selected). This color setting <u>overrides</u> the layer color for any newly created objects so that all new objects are drawn with the specified color no matter what layer they are on. This type of color designation prohibits your ability to see which objects are on which layers by their color; however, for some applications object color setting may be desirable. Use the *Layer Properties Manager* to set colors for layers.

Invoking this command by the menus or by typing *Color* presents the *Select Color* dialog box shown in Figure 11-21. This is essentially the same dialog box used for assigning colors to layers; however, the *ByLayer* and *ByBlock* tiles are accessible. (The buttons are grayed-out when this dialog is invoked from the *Layer Properties Manager* because, in that case, any setting is a *ByLayer* setting.) *ByBlock* color assignment is discussed in Chapter 21.

Figure 11-21

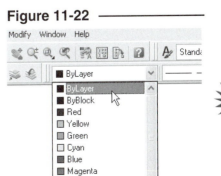

If you type *–Color* (include the hyphen prefix), the Command line format is displayed as follows:

```
Command: -color
Enter default object color <BYLAYER>: (Specify a color
name, number, ByLayer or ByBlock setting.)
Command:
```

The current color (whether *ByLayer* or specific object color) is saved in the *CECOLOR* (Current Entity Color) variable. The current color can be set by changing the value in the *CECOLOR* variable directly at the command prompt (by typing *CECOLOR*), by using the *Color* command (*Select Color* dialog box), or by using the Color Control drop-down list in the Object Properties toolbar. The *CECOLOR* variable accepts a string value (such as "red" or "bylayer") or accepts the ACI number (0 through 255). Valid values for true colors are a string of integers each from 1 to 255 separated by commas and preceded by RGB. The true color setting is entered as follows: RGB:000,000,000.

Color Control Drop-Down List

When an object-specific color has been set, the top item in the Object Properties toolbar Color Control drop-down list displays the current color (Fig. 11-22). Beware—using this list to select a color assigns an <u>object-specific color unless *ByLayer* is selected</u>. Any color you select from this list becomes the current object color. Choosing *Select Other...* from the bottom of the list invokes the *Select Color* dialog box (see Figure 11-21). If you want to assign colors to layers, make sure this list displays the *ByLayer* setting, then use the *Layer Properties Manager* to select colors for layers.

Figure 11-22

The *Color Control* drop-down list, as well as the others in the Object Properties toolbar, displays the current settings for selected objects (when no commands are in use).

True Color and Color Books

Figure 11-23 ————————

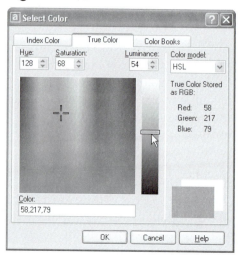

In AutoCAD 2004, support was added for true color and industry-standard color books. Three tabs are available in the *Select Color* dialog box—*Index Color*, *True Color*, and *Color Books*. The *Index Color* tab, discussed earlier in the chapter (see Figure 11-21), allows you to specify color using the 256-color ACI (AutoCAD Color Index).

True Color

The *True Color* tab in the *Select Color* dialog box provides a tool for you to specify colors by the *HLS* or the *RGB* color systems. Select the *Color Model* drop-down list to select which system you prefer (Fig. 11-23, upper-right corner). Watch the color swatch in the lower-right corner to dynamically display the color you specify.

HLS

This system specifies the color by *Hue*, *Luminance*, and *Saturation* (Fig. 11-23). *Hue* controls the pure color (red, yellow, blue), *Luminance* controls the color "value" (white to black—100 to 0), and *Saturation* controls the purity of the color (mix of *Hue* and *Luminance*—100 is pure color, 0 is no color). Use the pointer to specify *Hue* and *Saturation* and the slider for *Luminance* or enter values in the edit boxes.

Figure 11-24 ————————

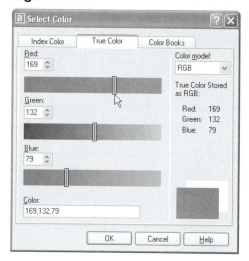

RGB

This system determines the color by the amount of red, green, and blue components (Fig. 11-24). Each of these colors is specified on a scale of 0 (no color) to 255 (maximum color). You can use the sliders or enter in the RGB values in the edit box.

Color Books

The *Color Books* tab in the *Select Color* dialog box allows you to specify colors using the Pantone and RAL commercial color systems. These systems are used largely in the architectural and interior design industries as a standard for specifying colors for paint and interior furnishings. These standard colors have been traditionally supplied to professionals in books displaying color samples for design and color matching.

Figure 11-25 ————————

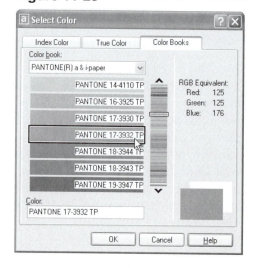

Select the desired Pantone or RAL "book" from the *Color book* drop-down list (Fig. 11-25). Use the slider and up/down arrows to display the color samples. Specify a particular color by clicking on it so the color appears in the color swatch (lower-right corner) and the index number appears in the *Color* edit box. If you know the desired index number, you can enter it directly in the edit box.

CONTROLLING LINETYPE SCALE

LTSCALE	Pull-down Menu	COMMAND (TYPE)	ALIAS (TYPE)	Short-cut	Screen (side) Menu	Tablet Menu
	Format Linetype... Show details>> Global scale factor	*LTSCALE*	*LTS*	...	*FORMAT Linetype Show details>> Global scale factor*	...

Hidden, dashed, dotted, and other linetypes that have spaces are called <u>non-continuous</u> linetypes. When drawing objects that have non-continuous linetypes (either *ByLayer* or object-specific linetype designations), the linetype's dashes or dots are automatically created and spaced. The *LTSCALE* (Linetype Scale) system variable controls the length and spacing of the dashes and/or dots. The value that is specified for *LTSCALE* affects the drawing <u>globally and retroactively</u>. That is, all existing non-continuous lines in the drawing as well as new lines are affected by *LTSCALE*. You can therefore adjust the drawing's linetype scale for all lines at any time with this one command.

If you choose to make the dashes of non-continuous lines smaller and closer together, reduce *LTSCALE*; if you desire larger dashes, increase *LTSCALE*. The *Hidden* linetype is shown in Figure 11-26 at various *LTSCALE* settings. Any positive value can be specified.

Figure 11-26

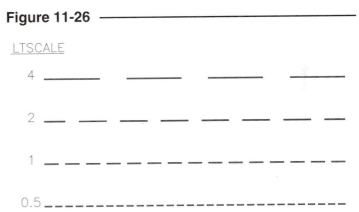

LTSCALE can be set in the *Details* section of the *Linetype Manager* or in Command line format. In the *Linetype Manager*, select the *Details* button to allow access to the *Global scale factor* edit box (Fig. 11-27). Changing the value in this edit box sets the *LTSCALE* variable. Changing the value in the *Current object scale* edit box sets the linetype scale for the current object only (see "*CELTSCALE*"), but does not affect the global linetype scale (*LTSCALE*).

LTSCALE can also be used in the Command line format.

Figure 11-27

```
Command: ltscale
Enter new linetype scale factor <1.0000>:
(value) (Enter any positive value.)
Command:
```

The *LTSCALE* for the default template drawing (ACAD.DWT) is 1. This value represents an appropriate *LTSCALE* for objects drawn within the default *Limits* of 12 x 9.

As a general rule, you should change the *LTSCALE* <u>proportionally</u> when *Limits* are changed (more specifically, when *Limits* are changed to other than the intended plot sheet size). For example, if you increase the drawing area defined by *Limits* by a factor of 2 (to 24 x 18) from the default (12 x 9), you might also change *LTSCALE* proportionally to a value of 2. Since *LTSCALE* is retroactive, it can be changed at a later time or repeatedly adjusted to display the desired spacing of linetypes.

If you load the *Hidden* linetype, it displays dashes (when plotted 1:1) of 1/4" with a *LTSCALE* of 1. The ANSI and ISO standards state that hidden lines should be displayed on drawings with dashes of approximately 1/8" or 3mm. To accomplish this, you can use the *Hidden* linetype and change *LTSCALE* to .5. It is recommended, however, that you use the *Hidden2* linetype that has dashes of 1/8" when plotted 1:1. Using this strategy, *LTSCALE* remains at a value of 1 to create the standard linetype sizes for *Limits* of 12 x 9 and can easily be changed in proportion to the *Limits*. (For more information on *LTSCALE*, see Chapter 12 and Chapter 13.)

The ACAD_ISO*n*W100 linetypes are intended to be used with metric drawings. Changing *Limits* to metric sheet sizes automatically displays these linetypes with appropriate linetype spacing. For these linetypes only, *LTSCALE* is changed automatically to the value selected in the *ISO Pen Width* box of the *Linetype* tab (see Figure 11-27). Using both ACAD_ISO*n*W100 linetypes and other linetypes in one drawing is discouraged due to the difficulty managing two sets of linetype scales.

Even though you have some control over the size of spacing for non-continuous lines, you have almost <u>no control</u> over the <u>placement</u> of the dashes for non-continuous lines. For example, the short dashes of center lines <u>cannot</u> always be controlled to intersect at the centers of a series of circles. The spacing can only be adjusted globally (all lines in the drawing) to reach a compromise. You can also adjust individual objects' linetype scale (*CELTSCALE*) to achieve the desired effect. See "*CELTSCALE*" (next) and Chapter 24, Multiview Drawing, for further discussion and suggestions on this subject.

CELTSCALE

Pull-down Menu	COMMAND (TYPE)	ALIAS (TYPE)	Short-cut	Screen (side) Menu	Tablet Menu
Format *Linetype...* *Show details>>* *Current object* *scale*	*CELTSCALE*	...	...	*FORMAT* *Linetype* *Show details>>* *Current object* *scale*	...

CELTSCALE stands for Current Entity Linetype Scale. *CELTSCALE* is actually a system variable for linetype scale stored with each object—an object property. This setting changes the <u>object-specific linetype scale proportional</u> to the *LTSCALE*. The *LTSCALE* value is global and retroactive, whereas *CELTSCALE* sets the linetype scale for all <u>newly created</u> objects and is <u>not retroactive</u>. *CELTSCALE* is object-specific. Using *CELTSCALE* to set an object linetype scale is similar to setting an object color, linetype, and lineweight in that the properties are <u>assigned to the specific object</u>.

For example, if you wanted all non-continuous lines (dashes and spaces) in the drawing to be two times the default size, set *LTSCALE* to 2 (*LTSCALE* is global and retroactive). If you then wanted only, say, a select two or three lines to have smaller spacing, change *CELTSCALE* to .5, draw the new lines, and then change *CELTSCALE* back to 1.

You can also use the *Linetype Manager* to set the *CELTSCALE* (see Figure 11-27). Set the desired *CELTSCALE* by listing the *Details* and changing the value in the *Current object scale* edit box.

This setting affects all linetypes for all <u>newly created</u> objects. The linetype scale for individual objects can also be changed retroactively using the method explained in the following paragraph.

Using <u>*CELTSCALE* is not the recommended method</u> for adjusting individual lines' linetype scale. Setting this variable each time you wanted to create a new entity with a different linetype scale would be too time consuming and confusing. The <u>recommended</u> method for adjusting linetypes in the drawing to different scales is <u>not to use *CELTSCALE*</u>, but to use the following strategy.

1. Set *LTSCALE* to an appropriate value for the drawing.
2. Create all objects in the drawing with the desired linetypes.
3. Adjust the *LTSCALE* again if necessary to globally alter the linetype scale.
4. Use *Properties* and *Matchprop* to <u>retroactively</u> change the linetype scale (*CELTSCALE*) <u>for selected objects</u> (see "*Properties*" and "*Matchprop*").

CHANGING OBJECT PROPERTIES

Often it is desirable to change the properties of an object after it has been created. For example, an object's *Layer* property could be changed. This can be thought of as "moving" the object from one layer to another. When an object is changed from one layer to another, it assumes the new layer's color, linetype, and lineweight, provided the object was originally created with color, linetype, and lineweight assigned *ByLayer*, as is generally the case. In other words, if an object was created on the wrong layer (possibly with the wrong linetype or color), it could be "moved" to the desired layer, therefore assuming the new layer's linetype and color.

Another example is to change an individual object's linetype scale. In some cases an individual object's linetype scale requires an adjustment to other than the global linetype scale (*LTSCALE*) setting. One of several methods can be used to adjust the individual object's linetype scale (*CELTSCALE*) to an appropriate value.

Several methods can be used to retroactively change properties of selected objects. The Object Properties toolbar, the *Properties* window, and the *Match Properties* command can be used to <u>retroactively</u> change the properties of individual objects. Properties that can be changed by these three methods are *Layer*, *Linetype, Lineweight, Color*, (object-specific) *Linetype scale, Plot Style*, and other properties.

Although some of the commands discussed in this section have additional capabilities, the discussion is limited to changing the object properties covered in this chapter—specifically layer, color, linetype, and lineweight. For this reason, these commands and features are also discussed in Chapter 16, Modify Commands II, and in other chapters.

Object Properties Toolbar

The five drop-down lists in the Object Properties toolbar (when not "dropped down") generally show the current layer, color, linetype, lineweight, and plot style. However, if an object or set of objects is selected, the information in these lists changes to display the current objects' settings. <u>You can change the selected objects' settings</u> by "dropping down" any of the lists and making another selection.

Figure 11-28

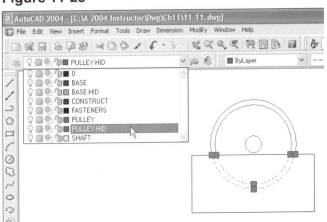

First, select (highlight) an object when no commands are in use. Use the pickbox (that appears on the crosshairs), window or crossing window to select the desired object or set of objects. The entries in the five lists (Layer Control, Color Control, Linetype Control, Lineweight Control, and Plot Style Control) then change to display the settings for the <u>selected object or objects</u>. If several objects are selected that have different properties, the boxes display no information (go "blank"). Next, use any of the drop-down lists to make another selection (Fig. 11-28). The highlighted object's properties are changed to those selected in the lists. Press the Escape key to complete the process.

Remember that in most cases color, linetype, and lineweight settings are assigned *ByLayer*. In this type of drawing scheme, to change the linetype or color properties of an object, you would change the object's <u>layer</u> (see Figure 11-28). If you are using this type of drawing scheme, refrain from using the Color Control, Linetype Control, and Lineweight Control drop-down lists for changing properties.

PROPERTIES

Pull-down Menu	COMMAND (TYPE)	ALIAS (TYPE)	Short-cut	Screen (side) Menu	Tablet Menu
Modify Properties	PROPERTIES	*PR* or *CH*	(Edit Mode) *Properties* or *Ctrl+1*	MODIFY1 *Property*	*Y,12* to *Y,13*

The *Properties* palette (Fig. 11-29) gives you complete access to one or more objects' properties. The contents of the palette change based on what type and how many objects are selected.

You can use the palette two ways: you can invoke the palette, then select (highlight) one or more objects, or you can select objects first and then invoke the palette. Once opened, this palette remains on the screen until dismissed by clicking the "X" in the upper corner. You can even toggle the palette on and off with Ctrl+1 (to appear and disappear). When multiple objects are selected, a drop-down list in the top of the palette appears. Select the object(s) whose properties you want to change.

To change an object's layer, linetype, lineweight, or color properties, highlight the desired objects in the drawing and select the properties you want to change from the right side of the palette. In the *General* list, the layer, linetype, lineweight, and color properties are located in the top half of the list.

If you use the *ByLayer* strategy of linetype, lineweight, and color properties assignment, you can change an object's linetype, lineweight, and color simply by changing its layer (see Figure 11-29). This method is recommended for most applications.

Figure 11-29

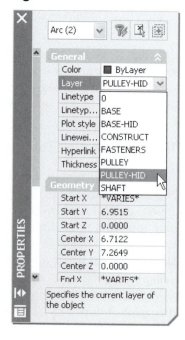

MATCHPROP

	Pull-down Menu	COMMAND (TYPE)	ALIAS (TYPE)	Short-cut	Screen (side) Menu	Tablet Menu
	Modify *Match* *Properties*	*MATCHPROP*	*MA*	...	*MODIFY1* *Matchprp*	**Y,14** **and** **Y,15**

Matchprop is used to "paint" the properties of one object to another. The process is simple. After invoking the command, select the object that has the desired properties (source object), then select the object you want to "paint" the properties to (destination object). The command prompt is as follows.

```
Command: matchprop
Select source object: PICK
Current active settings:  Color Layer Ltype Ltscale Lineweight Thickness PlotStyle Text Dim Hatch
     Polyline Viewport
Select destination object(s) or [Settings]: PICK
Select destination object(s) or [Settings]: Enter
Command:
```

Only one "source object" can be selected, but its properties can be painted to several "destination objects." The "destination object(s)" assume all of the properties of the "source object" (listed as "Current active settings").

Use the *Settings* option to control which of several possible properties and other settings are "painted" to the destination objects. At the "Select destination object(s) or [Settings]:" prompt, type *S* to display the *Property Settings* dialog box (Fig. 11-30). In the dialog box, designate which of the *Basic Properties* or *Special Properties* are to be painted to the "destination objects." The following *Basic Properties* correspond to properties discussed in this chapter.

Figure 11-30

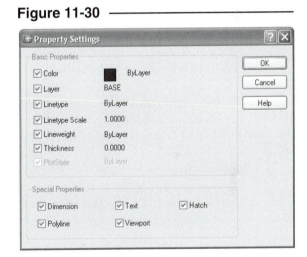

Color	paints the object-specific or *ByLayer* color
Layer	moves selected objects to Source Object layer
Linetype	paints the object-specific or *ByLayer* linetype
Lineweight	paints the object-specific or *ByLayer* lineweight
Linetype Scale	changes the individual object's linetype scale (*CELTSCALE*), not global (*LTSCALE*)

The *Special Properties* of the *Property Settings* dialog box are discussed in Chapter 16, Modify Commands II. Also see Chapter 16 for a full explanation of the *Properties* palette (*Properties*).

CHAPTER EXERCISES

1. *Layer Properties Manager,* **Layer Control drop-down list,** and *Make Object's Layer Current*

   Open the **WILHOME.DWG** sample drawing located in the C:\Program Files\AutoCAD 2004\Sample\ directory. Activate model space by selecting the *Model* tab. Next, use *Saveas* to save the drawing <u>in your working directory</u> as **WILHOME2**.

   A. Invoke the *Layer* command in Command line format by typing **-LAYER** or **-LA**. Use the **?** option to yield the list of layers in the drawing. Notice that all the layers have the default line-type except three. Which layers have a different linetype?

   B. Use the **Object Properties** toolbar to indicate the layers of selected items. On the right side of the drawing, **PICK** the yellow lines around the floor plan. What layer are the dimensions on? Press the **Esc** key twice to cancel the selection. Next, **PICK** the cyan (light blue) text around the semi-circular room near the top of the floor plan on the right. What layer is the text on? Press the **Esc** key twice to cancel the selection.

   C. Invoke the *Layer Properties Manager* by selecting the icon button or using the *Format* pull-down menu. Turn *Off* the **Dimensions** layer (click the light bulb icon). Select *OK* to return to the drawing and check the new setting. Are the dimensions displayed? Next, use the *Layerp* command to turn the **Dimensions** layer back *On*.

   D. Use the *Layer Properties Manager* and **PICK** the *Select All* option (right-click for menu). *Freeze* all layers. Which layer will not *Freeze*? Notice the light bulb icon indicates the layers are still *On* although the layers do not appear in the drawing (*Freeze* overrides *On*). Use the same procedure to select all layers and *Thaw* them.

   E. Use the *Layer Properties Manager* again and drop down the *Named Layer Filters* list. Are there any layers in the drawing that are unused (select *Show All Used Layers*, then *Invert Filter*)? Use the *Named Layer Filters* list again and list *All* layers.

   F. Now we want to *Freeze* all the *Yellow* layers. Using the *Layer Properties Manager* again, sort the list by color. **PICK** the first layer name in the list of the four yellow ones. Now hold down the **Shift** key and **PICK** the last name in the list. All four layers should be selected. *Freeze* all four by selecting any one of the sun/snowflake icons. Select *OK* to return to the drawing.

   G. Assuming you wanted to work only with the layers beginning with "AR," you can set the filter to display only those layers. First, activate the *Named Layer Filters* dialog by selecting the small button to the right of the drop-down list of the *Layer Properties Manager*. In the *Named Layer Filters* dialog box, enter **AR*** in the *Layer Names* edit box, then assign the name **AR Layers** in the *Filter Name* edit box and select *Add*. *Close* the dialog box. In the *Layer Properties Manager*, check *Apply to Layer toolbar*. Find **AR Layers** in the *Named layer filters* drop-down list and select it. Examine the list. Now select *OK* in the *Layer Properties Manager* to return to the drawing. Examine the **Layer Control** drop-down list. Does it display only layer names beginning with AR? Return to the *Layer Properties Manager* and select *Show All Layers* in the *Named layer filters* list. Select *OK* and examine the **Layer Control** drop-down list again. It should also show all layer names in the list.

H. Ensure all layers are *Thawed*. Sort the list alphabetically by *Name*. *Freeze* all layers beginning with **A** and **B**. Select *OK* and then use the drop-down list to display the names. Assuming you were working only with the thawed layers, it would be convenient to display <u>only</u> those names in the drop-down list. Use the *Named Layer Filters* dialog box to set a filter named **Thawed Layers**. In the *Layer Properties Manager*, find **Thawed Layers** in the *Named layer filters* drop-down list and select it. Ensure the *Apply to Object Properties toolbar* checkbox is checked. Return to the drawing and examine the drop-down list. Do only names of *Thawed* layers appear?

I. Next, you will "move" objects from one layer to another using the **Object Properties** toolbar. Make layer **0** the current layer by using the drop-down list and selecting the name. Use the *Layer Properties Manager* and right-click to *Select All* names. **PICK** one icon to *Freeze* all layers (except the current layer). Then select *Clear All* and *Thaw* layer **General Note**. Select *OK* and return to the drawing. Use a crossing window to select all of the objects in the drawing (the small blue Grips may appear). The **Object Properties** toolbar should indicate the layer on which the objects reside. Next, list the layers using the drop-down list and **PICK** layer **0**. Wait several seconds. Press **Esc** twice when the drawing regenerates. The objects should change color and be on layer 0. You actually "moved" the objects to layer 0. Finally, type *U* to undo the last action so the notes are on layer Gennote (cyan). Also, *Thaw* all layers.

J. Let's assume you want to draw some additional objects on the text layer and on the dimensions layer. First, *Zoom* in to the upper half of the floor plan on the right. Now select the *Make Object's Layer Current* icon button (just to the right of the layer drop-down list). At the "Select object whose layer will become current" prompt, select a text object (cyan color). The **General Note** layer name should appear in the Layer Control box as the current layer. Draw a *Line* and it should be on layer General Note and cyan in color. Next, use *Make Object's Layer Current* again and select a dimension object. The **Dimensions** layer should appear as the current layer. Draw another *Line*. It should be on layer Dimensions and yellow in color. Keep in mind that this method of setting a layer current works especially well when you do not know the layer names or know what objects are on which layers.

K. *Close* the drawing. Do not save the changes.

2. *Layer Properties Manager* **and** *Linetype Manager*

A. Begin a *New* drawing and use *Save* to assign the name **CH11EX2**. Set *Limits* to **11 x 8.5**, then *Zoom All*. Use the *Linetype Manager* to list the loaded linetypes. Are any linetypes already loaded? **PICK** the *Load* button then right-click to *Select All* linetypes. Select *OK*, then close the *Linetype Manager*.

B. Use the *Layer Properties Manager* to create 3 new layers named **OBJ, HID,** and **CEN**. Assign the following colors, linetypes, and lineweights to the layers by clicking on the *Color, Linetype,* and *Lineweight* columns.

OBJ	*Red*	*Continuous*	*Default*
HID	*Yellow*	*Hidden2*	*Default*
CEN	*Green*	*Center2*	*Default*

C. Make the **OBJ** layer *Current* and PICK the *OK* tile. Verify the current layer by looking at the *Layer Control* drop-down list, then draw the visible object *Lines* and the *Circle* only (not the dimensions) as shown in Figure 11-31.

Figure 11-31

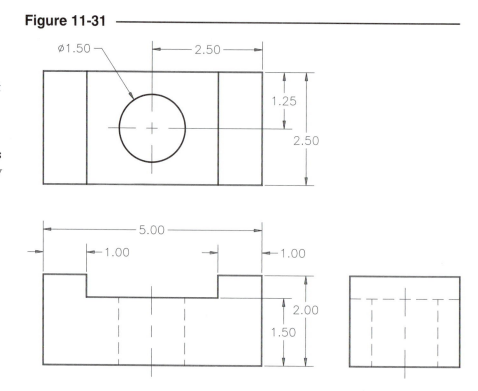

D. When you are finished drawing the visible object lines, create the necessary hidden lines by making layer **HID** the *Current* layer, then drawing *Lines*. Notice that you only specify the *Line* endpoints as usual and AutoCAD creates the dashes.

E. Next, create the center lines for the holes by making layer **CEN** the *Current* layer and drawing *Lines*. Make sure the center lines extend slightly beyond the *Circle* and beyond the horizontal *Lines* defining the hole. *Save* the drawing.

F. Now create a *New* layer named **BORDER** and draw a border and title block of your design on that layer. The final drawing should appear as that in Figure 11-32.

Figure 11-32

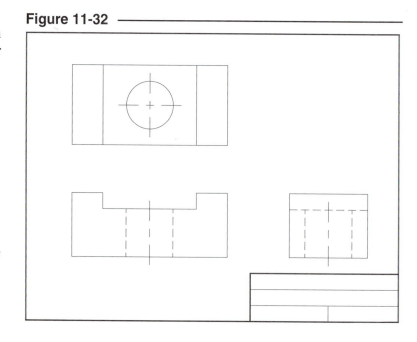

G. Open the *Linetype Manager* again and right-click to *Select All*, then select the *Delete* button. All unused linetypes should be removed from the drawing.

H. *Save* the drawing, then *Close* the drawing.

3. *LTSCALE, Properties* **palette,** *Matchprop*

A. Begin a *New* drawing and use *Save* to assign the name **CH11EX3**. Create the same four *New* layers that you made for the previous exercise (**OBJ**, **HID**, **CEN**, and **BORDER**) and assign the same linetypes, lineweights, and colors. Set the *Limits* equal to a "C" size sheet, **22 x 17**, then *Zoom All*.

B. Create the part shown in Figure 11-33. Draw on the appropriate layers to achieve the desired linetypes.

Figure 11-33

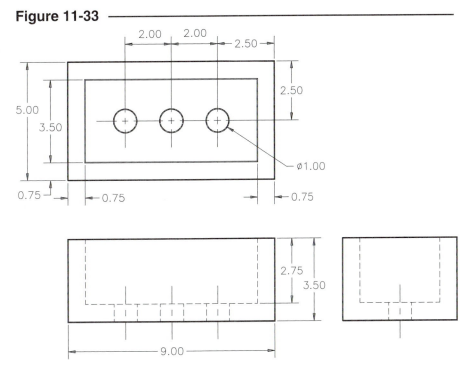

C. Notice that the *Hidden* and *Center* linetype dashes are very small. Use *LTSCALE* to adjust the scale of the non-continuous lines. Since you changed the *Limits* by a factor of slightly less than 2, try using **2** as the *LTSCALE* factor. Notice that all the lines are affected (globally) and that the new *LTSCALE* is retroactive for existing lines as well as for new lines. If the dashes appear too long or short because of the small hidden line segments, *LTSCALE* can be adjusted. Remember that you cannot control where the multiple <u>short</u> centerline dashes appear for one line segment. Try to reach a compromise between the hidden and center line dashes by adjusting the *LTSCALE*.

D. The six short vertical hidden lines in the front view and two in the side view should be adjusted to a smaller linetype scale. Invoke the *Properties* palette by selecting *Properties* from the *Modify* pull-down menu or selecting the *Properties* icon button. Select <u>only the two short lines in the side view</u>. Use the palette to retroactively adjust the two individual objects' *Linetype Scale* to **0.6000**.

E. When the new *Linetype Scale* (actually, the *CELTSCALE*) for the two selected lines is adjusted correctly, use *Matchprop* to "paint" the *Linetype Scale* to the several other short lines in the front view. Select **Match Properties** from the **Modify** pull-down menu or select the "paint brush" icon. When prompted for the "source object," select one of the two short vertical lines in the side view. When prompted for the "destination object(s)," select <u>each</u> of the short lines in the front view. This action should "paint" the *Linetype Scale* to the lines in the front view.

F. When you have the desired linetype scales, *Exit* AutoCAD and *Save Changes*. Keep in mind that the drawing size may appear differently on the screen than it will on a print or plot. Both the plot scale and the paper size should be considered when you set *LTSCALE*. This topic will be discussed further in Chapter 12, Advanced Drawing Setup.

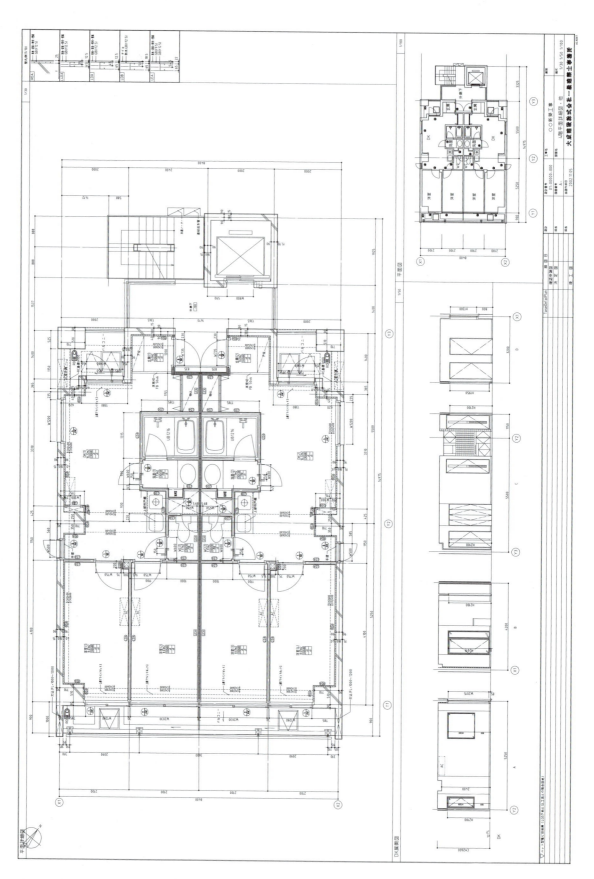

TAISEI DETAIL PLAN.DWG, Courtesy of Autodesk, Inc.

ADVANCED DRAWING SETUP

Chapter Objectives

After completing this chapter you should:

1. know the steps for setting up a drawing;

2. be able to determine an appropriate *Limits* setting for the drawing;

3. be able to calculate and apply the "drawing scale factor";

4. be able to access existing and create new template drawings;

5. know what setup steps can be considered for creating template drawings.

CONCEPTS

When you begin a drawing, there are several steps that are typically performed in preparation for creating geometry, such as setting *Units, Limits,* and creating *Layers* with *linetypes, lineweights,* and *colors.* Some of these basic concepts were discussed in Chapters 6 and 11. This chapter discusses setting *Limits* for correct plotting as well as other procedures, such as layer creation and variables settings, that help prepare a drawing for geometry creation.

To correctly set up a drawing for printing or plotting to scale, <u>any two</u> of the following three variables must be known. The third can be determined from the other two.

> Drawing *Limits*
> Print or plot sheet (paper) size
> Print or plot scale

The method given in this text for setting up a drawing (Chapters 6 and 12) is intended for plotting from both the *Model* tab (model space) and from *Layout* tabs (paper space). The method applies when you plot from the *Model* tab or plot with <u>one viewport</u> in a *Layout* tab. If multiple viewports are created in one layout, the same general steps would be taken; however, the plot scale might vary for each viewport based on the size and number of viewports. (See Chapters 13 and 33 for more information on layouts, viewports, and plotting layouts.)

Rather than performing the steps for drawing setup each time you begin, you can use "template drawings." Template drawings have many of the setup steps performed but contain no geometry. AutoCAD provides several template drawings and you can create your own template drawings. A typical engineering, architectural, design, or construction office generally produces drawings that are similar in format and can benefit from creation and use of individualized template drawings. The drawing similarities may be subject of the drawing (geometry), plot scale, sheet size, layering schemes, dimensioning styles, and/or text styles. Template drawing creation and use are discussed in this chapter.

STEPS FOR DRAWING SETUP

Assuming that you have in mind the general dimensions and proportions of the drawing you want to create, and the drawing will involve using layers, dimensions, and text, the following steps are suggested for setting up a drawing:

1. Determine and set the *Units* that are to be used.
2. Determine and set the drawing *Limits;* then *Zoom All.*
3. Set an appropriate *Snap Type* (*Polar Snap* or *Grid Snap*), *Snap* spacing, and *Polar* spacing if useful.
4. Set an appropriate *Grid* value if useful.
5. Change the *LTSCALE* value based on the new *Limits.* Set *PSLTSCALE* to 0.
6. Create the desired *Layers* and assign appropriate *linetype, lineweight,* and *color* settings.
7. Create desired *Text Styles* (optional, discussed in Chapter 18).
8. Create desired *Dimension Styles* (optional, discussed in Chapter 29).
9. Activate a *Layout* tab, set it up for the plot or print device and paper size, and create a viewport (if not already existing).
10. Create or insert a title block and border in the layout.

Each of the steps for drawing setup is explained in detail here.

1. Set *Units*

This task is accomplished by using the *Units* command or the *Quick Setup* or *Advanced Setup* wizard. Set the linear units and precision desired. Set angular units and precision if needed. (See Chapter 6 for details on the *Units* command and setting *Units* using the wizards.)

2. Set *Limits*

Before beginning to create an AutoCAD drawing, determine the size of the drawing area needed for the intended geometry. Using the actual *Units,* appropriate *Limits* should be set in order to draw the object or geometry to the <u>real-world size</u>. *Limits* are set with the *Limits* command by specifying the lower-left and upper-right corners of the drawing area. Always *Zoom All* after changing *Limits.* *Limits* can also be set using the *Quick Setup* or *Advanced Setup* wizard. (See Chapter 6 for details on the *Limits* command and the setup wizards.)

If you are planning to plot the drawing to scale, *Limits* should be set to a proportion of the sheet size you plan to plot on. For example, if the sheet size is 11" x 8.5", set *Limits* to 11 x 8.5 if you want to plot full size (1"=1"). Setting *Limits* to 22 x 17 (2 times 11 x 8.5) provides 2 times the drawing area and allows plotting at 1/2 size (1/2"=1") on the 11" x 8.5" sheet. Simply stated, set *Limits* to a <u>proportion</u> of the paper size.

Setting *Limits* to the <u>paper size</u> allows plotting at 1=1 scale. Setting *Limits* to a <u>proportion</u> of the sheet size allows plotting at the <u>reciprocal</u> of that proportion. For example, setting *Limits* to 2 times an 11" x 8.5" sheet allows you to plot 1/2 size on that sheet. Or setting *Limits* to 4 times an 11" x 8.5" sheet allows you to plot 1/4 size on that sheet. (Standard paper sizes are given in Chapter 14.)

Even if you plan to use a layout tab (paper space) and create one viewport, setting model space *Limits* to a proportion of the paper size is recommended. In this way you can easily calculate the <u>viewport scale</u> (the proportion of paper space units to model space units); therefore, the model space geometry can be plotted to a standard scale.

If you plan to plot from a layout, *Limits* in <u>paper space</u> should be set to the actual paper size; however, setting the paper space *Limits* values is automatically done for you when you select a *Plot Device* and *Paper Size* from the *Page Setup* or *Plot* dialog box (see Chapter 13, Layouts and Viewports, and Chapter 14, Printing and Plotting).

To set model space *Limits,* you can use the individual commands for setting *Units* and *Limits* (*Units* command and *Limits* command) or use the *Quick Setup* wizard or *Advanced Setup* wizard.

Drawing Scale Factor

<u>The proportion of the *Limits* to the intended print or plot sheet size is the "drawing scale factor."</u> This factor can be used as a general scale factor for other size-related drawing variables such as *LTSCALE,* dimension *Overall Scale (DIMSCALE),* and *Hatch* pattern scale. The drawing scale factor can also be used to determine the viewport scales.

Most size-related AutoCAD drawing variables are set to 1 by default. This means that variables (such as *LTSCALE*) that control sizing and spacing of objects are set appropriately for creating a drawing plotted full size (1=1).

Figure 12-1

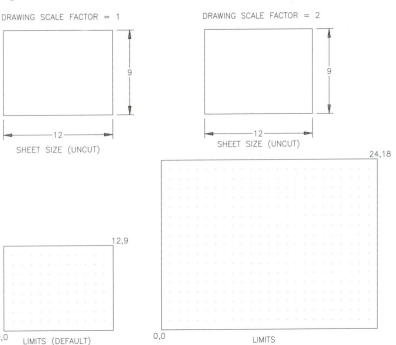

Therefore, when you set *Limits* to the sheet size and print or plot at 1=1, the sizing and spacing of linetypes and other variable-controlled objects are correct. When *Limits* are changed to some proportion of the sheet size, the size-related variables should also be changed proportionally. For example, if you intend to plot on a 12 x 9 sheet and the default *Limits* (12 x 9) are changed by a factor of 2 (to 24 x 18), then 2 becomes the drawing scale factor. Then, as a general rule, the values of variables such as *LTSCALE, DIMSCALE* (*Overall Scale*), and other scales should be multiplied by a factor of 2 (Fig. 12-1, on the previous page). When you then print the 24 x 18 area onto a 12" x 9" sheet, the plot scale would be 1/2, and all sized features would appear correct.

Limits should be set to the paper size or to a proportion of the paper size used for plotting or printing. In many cases, "cut" paper sizes are used based on the 11" x 8.5" module (as opposed to the 12" x 9" module called "uncut sizes"). Assume you plan to print on an 11" x 8.5" sheet. Setting *Limits* to the sheet size provides plotting or printing at 1=1 scale. Changing the *Limits* to a proportion of the sheet size <u>by some multiplier</u> makes that value the drawing scale factor. The <u>reciprocal</u> of the drawing scale factor is the scale for plotting on that sheet (Fig. 12-2).

Figure 12-2

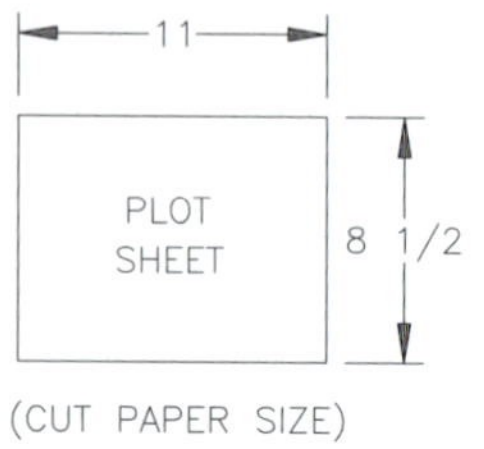

LIMITS	SCALE FACTOR	PLOT SCALE	
11 x 8.5	1	1=1	(FULL SIZE)
22 x 17	2	1=2	(HALF SIZE)
44 x 34	4	1=4	(1/4 SIZE)
110 x 85	10	1=10	(1/10 SIZE)

Since the drawing scale factor (DSF) is the proportion of *Limits* to the sheet size, you can use this formula:

$$DSF = \frac{Limits}{Sheet\ size}$$

Because plot/print scale is the reciprocal of the drawing scale factor:

$$Plot\ scale = \frac{1}{DSF}$$

The term "drawing scale factor" is not a variable or command that can be found in AutoCAD or its official documentation. This concept has been developed by AutoCAD users to set system variables appropriately for printing and plotting drawings to standard scales.

3. Set *Snap*

Use the *Snap* command or *Snap and Grid* tab of the *Drafting Settings* dialog box to set the *Snap* type (*Grid Snap* and/or *Polar Snap*) and appropriate *Snap Spacing* and/or *Polar Spacing* values. The *Snap Spacing* and *Polar Spacing* values are dependent on the <u>interactive</u> drawing accuracy that is desired.

The accuracy of the drawing and the size of the *Limits* should be considered when setting the *Snap Spacing* and *Polar Spacing* values. On one hand, to achieve accuracy and detail, you want the values to be the smallest dimensional increments that would commonly be used in the drawing. On the other hand, depending on the size of the *Limits*, the *Snap* value should be large enough to make interactive selection (PICKing) fast and easy. As a starting point for determining appropriate *Snap Spacing* and *Polar Spacing* values, the default values can be multiplied by the "drawing scale factor."

If you are creating a template drawing (.DWT file), it is of no help to set the *Snap Type* (Grid Snap or Polar Snap), *Polar Spacing* value, or *Polar Tracking* angles since these values are stored in the system registry, not in the drawing. The *Grid Snap Spacing* value, however, is stored in the drawing file.

4. Set *Grid*

The *Grid* value setting is usually set equal to or proportionally larger than that of the *Grid Snap* value. *Grid* should be set to a proportion that is <u>easily visible</u> and to some value representing a regular increment (such as .5, 1, 2, 5, 10, or 20). Setting the value to a proportion of *Grid Snap* gives visual indication of the *Grid Snap* increment. For example, a *Grid* value of 1X, 2X, or 5X would give visual display of every 1, 2, or 5 *Grid Snap* increments, respectively. If you are not using *Snap* because only extremely small or irregular interval lengths are needed, you may want to turn *Grid* off. (See Chapter 6 for details on the *Grid* command.)

5. Set the *LTSCALE* and *PSLTSCALE*

A change in *Limits* (then *Zoom All*) affects the display of non-continuous (hidden, dashed, dotted, etc.) lines. The *LTSCALE* variable controls the spacing of non-continuous lines. The drawing scale factor can be used to determine the *LTSCALE* setting. For example, if the default *Limits* (based on sheet size) have been changed by a factor of 2, the drawing scale factor=2; so set the *LTSCALE* to 2. In other words, if the DSF=2 and plot scale=1/2, you would want to double the linetype dashes to appear the correct size in the plot, so *LTSCALE*=2. If you prefer a *LTSCALE* setting of other than 1 for a drawing in the default *Limits*, multiply that value by the drawing scale factor.

American National Standards Institute (ANSI) requires that <u>hidden lines be plotted with 1/8"</u> <u>dashes</u>. An AutoCAD drawing plotted full size (1 unit=1") with the default *LTSCALE* value of 1 plots the *Hidden* linetype with 1/4" dashes. Therefore, it is recommended for English drawings that you use the *Hidden2*, *Center2*, and other *2 linetypes with dash lengths .5 times as long so an *LTSCALE* of 1 produces the ANSI standard dash lengths. Multiply your *LTSCALE* value by the drawing scale factor when plotting to scale. If individual lines need to be adjusted for linetype scale, use *Properties* when the drawing is nearly complete. (See Chapter 11 for details on the *LTSCALE* and *CELTSCALE* variables.)

Assuming you are planning to create only <u>one viewport</u> in a *Layout* tab, set the *PSLTSCALE* value (paper space *LTSCALE*) to 0. AutoCAD's default setting is 1. A setting of 0 forces the linetype scale to appear the same in *Layout* tabs as it does in the *Model* tab. To set the *PSLTSCALE* value, similar to setting the *LTSCALE*, simply type *PSLTSCALE* at the Command prompt. (See Chapter 13 for details on *PSLTSCALE*.)

6. Create *Layers*; Assign *Linetypes*, *Lineweights*, and *Colors*

Use the *Layer Properties Manager* to create the layers that you anticipate needing. Multiple layers can be created with the *New* option. You can type in several new names separated by commas. Assign a descriptive name for each layer, indicating its type of geometry, part name, or function. Include a *linetype* designator in the layer name if appropriate; for example, PART1-H and PART1-V indicate hidden and visible line layers for PART1.

Once the *Layers* have been created, assign a *Color*, *Linetype*, and *Lineweight* to each layer. *Colors* can be used to give visual relationships to parts of the drawing. Geometry that is intended to be plotted with different pens (pen size or color) or printed with different plotted appearances should be drawn in different screen colors (especially when color-dependent plot styles are used). Use the *Linetype* command to load the desired linetypes. You should also select *Lineweights* for each layer (if desired) using the *Layer Properties Manager*. (See Chapter 11 for details on creating layers and assigning linetypes, lineweights, and colors.)

7. Create Text Styles

AutoCAD has only one text style as part of the standard template drawings (ACAD.DWT and ACADISO.DWT) and one or two text styles for most other template drawings. If you desire other *Text Styles*, they are created using the *Style* command or the *Text Style…* option from the *Format* pull-down menu. (See Chapter 18, Creating and Editing Text.) If you desire engineering standard text, create a *Text Style* using the ROMANS.SHX or .TTF font file.

8. Create Dimension Styles

If you plan to dimension your drawing, Dimension Styles can be created at this point; however, they are generally created during the dimensioning process. Dimension Styles are names given to groups of dimension variable settings. (See Chapter 29 for information on creating Dimension Styles.)

Although you do not have to create Dimension Styles until you are ready to dimension the geometry, it is helpful to create Dimension Styles as part of a template drawing. If you produce similar drawings repeatedly, your dimensioning techniques are probably similar. Much time can be saved by using a template drawing with previously created Dimension Styles. Several of the AutoCAD-supplied template drawings have prepared Dimension Styles.

9. Activate a *Layout* Tab, Set the Plot Device, and Create a Viewport

This step can be done just before making a print or plot, or you can complete this and the next step before creating your drawing geometry. Completing these steps early allows you to see how your printed drawing might appear and how much area is occupied by the titleblock and border.

Activate a *Layout* tab. If your *Options* are set as explained in Chapter 6, Basic Drawing Setup, the layout is automatically set up for your default print/plot device and an appropriate-sized viewport is created. Otherwise, use the *Page Setup* dialog box (*Pagesetup* command) to set the desired print/plot device and paper size. Then use the *Vports* command to create one or more viewports. This topic is discussed briefly in Chapter 6, Basic Drawing Setup, and is discussed in more detail in Chapter 13, Layouts and Viewports.

10. Create a Titleblock and Border

For 2D drawings, it is helpful to insert a titleblock and border early in the drawing process. This action gives a visual drawing boundary as well as reserves the space needed by the titleblock.

Rather than create a new titleblock and border for each new drawing, a common practice is to use the *Insert* command to insert an existing titleblock and border as a *Block*.

You can draw or *Insert* a titleblock and border in the *Model* tab (model space) or into a layout (paper space), depending on how you want to plot the drawing. If you plan to plot from the *Model* tab, multiply the actual size of the titleblock and border by the drawing scale factor to determine their sizes for the drawing. If you want to plot from a layout, draw or insert the titleblock and border at the actual size (1:1) since a layout represents the plot sheet. (See Chapters 21 and 22 for information on *Block* creation and *Insertion*.)

USING AND CREATING TEMPLATE DRAWINGS

Instead of going through the steps for setup each time you begin a new drawing, you can create one or more template drawings or use one of the AutoCAD-supplied template drawings. AutoCAD template drawings are saved as a .DWT file format. A template drawing has the initial setup steps (*Units*, *Limits*, *Layers*, *Linetypes*, *Lineweights*, *Colors*, etc.) completed and saved, but no geometry has been created yet. For some templates, *Layout* tabs have been set up with titleblocks, viewports, and plot settings. Template drawings are used as a template or starting point each time you begin a new drawing. AutoCAD actually makes a copy of the template you select to begin the drawing. The creation and use of template drawings can save hours of preparation.

Using Template Drawings

To use a template drawing, select the *Template* option in the *Startup* or *Create New Drawing* dialog box (Fig. 12-3, on the next page) or in the *Select Template* dialog box. This option allows you to select the desired template (.DWT) drawing from a list including all templates in the Template folder.

Any template drawings that you create using the *SaveAs* command (as described in the next section) appear in the list of available templates. Select the *Browse...* option to browse other folders.

Several template drawings are supplied with AutoCAD (see the selections in Figure 12-3) and are described in the following table. Generally, the template drawing files include the titleblocks. In some cases, Dimension Styles have been created but other changes have not been made to the template drawings such as layer setups or loading linetypes.

Figure 12-3

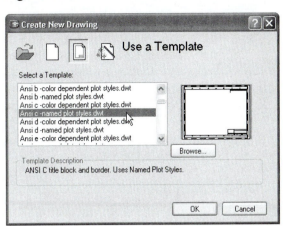

Table of AutoCAD-Supplied Template Drawing Settings

Template	Intended Sheet Size	Limits (Model)	Title Block	Orientation	Layers	Dimen. Styles	Plot Style
ACAD - Named Plot Styles.dwt	---	12,9	No	Landscape	0	Standard	Named
acad.dwt	---	12,9	No	Landscape	0	Standard	Color Dependent
ACADISO - Named Plot Styles.dwt	---	420,297	No	Landscape	0	Yes[8]	Named
acadiso.dwt	---	420,297	No	Landscape	0	Yes[8]	Color Dependent
ANSI A (portrait) - Color Dependent Plot Styles.dwt	8.5",11"	12,9	Yes	Portrait	Yes[1]	Standard	Color Dependent
ANSI A (portrait) - Named Plot Styles.dwt	8.5",11"	12,9	Yes	Portrait	Yes[1]	Standard	Named
ANSI A - Color Dependent Plot Styles.dwt	11",8.5"	12,9	Yes	Landscape	Yes[1]	Standard	Color Dependent
ANSI A - Named Plot Styles.dwt	11",8.5"	12,9	Yes	Landscape	Yes[1]	Standard	Named
ANSI B - Color Dependent Plot Styles.dwt	17"x11"	12,9	Yes	Landscape	Yes[1]	Standard	Color Dependent
ANSI B - Named Plot Styles.dwt	17"x11"	12,9	Yes	Landscape	Yes[1]	Standard	Named
ANSI C - Color Dependent Plot Styles.dwt	22"x17"	12,9	Yes	Landscape	Yes[1]	Standard	Color Dependent
ANSI C - Named Plot Styles.dwt	22"x17"	12,9	Yes	Landscape	Yes[1]	Standard	Named
ANSI D - Color Dependent Plot Styles.dwt	22"x34"	12,9	Yes	Landscape	Yes[1]	Standard	Color Dependent
ANSI D - Named Plot Styles.dwt	22"x34"	12,9	Yes	Landscape	Yes[1]	Standard	Named

Template	Intended Sheet Size	Limits (MS)	Title Block	Orientation	Layers	Dimen. Styles	Plot Style
ANSI E - Color Dependent Plot Styles.dwt	34"x44"	12,9	Yes	Landscape	Yes[1]	Standard	Color Dependent
ANSI E - Named Plot Styles.dwt	34"x44"	12,9	Yes	Landscape	Yes[1]	Standard	Named
ANSI Layout templates.dwt	---	12,9	No	---	Yes[1]	Standard	Color Dependent
Architectural, English units -Color Dependent Plot Styles.dwt	24"x36"	1'-0"x9"	Yes	Landscape	Yes[1]	Standard	Color Dependent
Architectural, English units -Named Plot Styles.dwt	24"x36"	1'-0"x9"	Yes	Landscape	Yes[1]	Standard	Named
DIN A0 - Color Dependent Plot Styles.dwt	1189mmx841mm	420x297	Yes	Landscape	Yes[2]	Yes[9]	Color Dependent
DIN A0 - Named Plot Styles.dwt	1189mmx841mm	420x297	Yes	Landscape	Yes[2]	Yes[9]	Named
DIN A1 - Color Dependent Plot Styles.dwt	841mmx594mm	420x297	Yes	Landscape	Yes[2]	Yes[9]	Color Dependent
DIN A1 - Named Plot Styles.dwt	841mmx594mm	420x297	Yes	Landscape	Yes[2]	Yes[9]	Named
DIN A2 -Color Dependent Plot Styles.dwt	594mmx420mm	420x297	Yes	Landscape	Yes[2]	Yes[9]	Color Dependent
DIN A2 - Named Plot Styles.dwt	594mmx420mm	420x297	Yes	Landscape	Yes[2]	Yes[9]	Named
DIN A3 -Color Dependent Plot Styles.dwt	420mmx297mm	420x297	Yes	Landscape	Yes[2]	Yes[9]	Color Dependent
DIN A3 -Named Plot Styles.dwt	420mmx297mm	420x297	Yes	Landscape	Yes[2]	Yes[9]	Named
DIN A4 -Color Dependent Plot Styles.dwt	210mmx297mm	420x297	Yes	Portrait	Yes[2]	Yes[9]	Color Dependent
DIN A4 -Named Plot Styles.dwt	210mmx297mm	420x297	Yes	Portrait	Yes[2]	Yes[9]	Named
Gb –Color Dependent Plot Styles.dwt	---	256.8069 x 149.3068	No	Landscape	0	Yes[11]	Color Dependent
Gb –Named Plot Styles.dwt	---	420x297	No	Landscape	0	Yes[11]	Named
Gb_a0 –Color Dependent Plot Styles.dwt	1306mmx924mm	420x 297	Yes	Landscape	Yes[6]	Yes[11]	Color Dependent
Gb_a0 –Named Plot Styles.dwt	1306mmx924mm	1154x821	Yes	Landscape	Yes[6]	Yes[11]	Named
Gb_a1 –Color Dependent Plot Styles.dwt	924mmx653mm	806x574	Yes	Landscape	Yes[6]	Yes[11]	Color Dependent
Gb_a1 –Named Plot Styles.dwt	924mmx653mm	420x297	Yes	Landscape	Yes[6]	Yes[11]	Named
Gb_a2 –Color Dependent Plot Styles.dwt	653mmx462mm	420x297	Yes	Landscape	Yes[6]	Yes[11]	Color Dependent

Template	Intended Sheet Size	Limits (MS)	Title Block	Orientation	Layers	Dimen. Styles	Plot Style
Gb_a2 –Named Plot Styles.dwt	653mmx462mm	559x400	Yes	Landscape	Yes[6]	Yes[11]	Named
Gb_a3 –Color Dependent Plot Styles.dwt	653mmx462mm	390x287	Yes	Landscape	Yes[6]	Yes[11]	Color Dependent
Gb_a3 –Named Plot Styles.dwt	462mmx326mm	420x297	Yes	Landscape	Yes[6]	Yes[11]	Named
Gb_a4 –Color Dependent Plot Styles.dwt	435mmx307mm	420x297	Yes	Landscape	Yes[6]	Yes[11]	Color Dependent
Gb_a4 –Named Plot Styles.dwt	435mmx307mm	180x287	Yes	Landscape	Yes[6]	Yes[11]	Named
Generic 24in x 32in Title Block -Color Dependent Plot Styles.dwt	36"x24"	12x9	Yes	Landscape	Yes[1]	Standard	Color Dependent
Generic 24in x 32in Title Block -Named Plot Styles.dwt	36"x24"	12x9	Yes	Landscape	Yes[1]	Standard	Named
ISO A0 -Color Dependent Plot Styles.dwt	1189mmx841mm	420x297	Yes	Landscape	Yes[3]	Yes[8]	Color Dependent
ISO A0 -Named Plot Styles.dwt	1189mmx841mm	420x297	Yes	Landscape	Yes[4]	Yes[8]	Named
ISO A1 -Color Dependent Plot Styles.dwt	841mmx594mm	420x297	Yes	Landscape	Yes[4]	Yes[8]	Color Dependent
ISO A1 -Named Plot Styles.dwt	841mmx594mm	420x297	Yes	Landscape	Yes[4]	Yes[8]	Named
ISO A2 -Color Dependent Plot Styles.dwt	594mmx420mm	420x297	Yes	Landscape	Yes[4]	Yes[8]	Color Dependent
ISO A2 -Named Plot Styles.dwt	594mmx420mm	420x297	Yes	Landscape	Yes[4]	Yes[8]	Named
ISO A3 -Color Dependent Plot Styles.dwt	420mmx297mm	420x297	Yes	Landscape	Yes[4]	Yes[8]	Color Dependent
ISO A3 -Named Plot Styles.dwt	420mmx297mm	420x297	Yes	Landscape	Yes[4]	Yes[8]	Named
ISO A4 -Color Dependent Plot Styles.dwt	210mmx297mm	420x297	Yes	Portrait	Yes[4]	Yes[8]	Color Dependent
ISO A4 -Named Plot Styles.dwt	210mmx297mm	420x297	Yes	Portrait	Yes[4]	Yes[8]	Named
JIS A0 -Color Dependent Plot Styles.dwt	1189mmx841mm	420x297	Yes	Landscape	Yes[5]	Yes[10]	Color Dependent
JIS A0 -Named Plot Styles.dwt	1189mmx841mm	420x297	Yes	Landscape	Yes[5]	Yes[10]	Named
JIS A1 -Color Dependent Plot Styles.dwt	841mmx594mm	420x297	Yes	Landscape	Yes[5]	Yes[10]	Color Dependent
JIS A1 -Named Plot Styles.dwt	841mmx594mm	420x297	Yes	Landscape	Yes[5]	Yes[10]	Named
JIS A2 -Color Dependent Plot Styles.dwt	594mmx420mm	420x297	Yes	Landscape	Yes[5]	Yes[10]	Color Dependent
JIS A2 -Named Plot Styles.dwt	594mmx420mm	420x297	Yes	Landscape	Yes[5]	Yes[10]	Named
JIS A3 -Color Dependent Plot Styles.dwt	420mmx297mm	420x297	Yes	Landscape	Yes[5]	Yes[10]	Color Dependent

2002

Template	Intended Sheet Size	Limits (MS)	Title Block	Orientation	Layers	Dimen. Styles	Plot Style
JIS A3 -Named Plot Styles.dwt	420mmx297mm	420x297	Yes	Landscape	Yes[5]	Yes[10]	Named
JIS A4 (landscape) - Color Dependent Plot Styles.dwt	297mmx210mm	420x297	Yes	Landscape	Yes[5]	Yes[10]	Color Dependent
JIS A4 (landscape) - Named Plot Styles.dwt	297mmx210mm	420x297	Yes	Landscape	Yes[5]	Yes[10]	Named
JIS A4 (portrait) -Color Dependent Plot Styles.dwt	297mmx210mm	420x297	Yes	Portrait	Yes[5]	Yes[10]	Color Dependent
JIS A4 (portrait) - Named Plot Styles.dwt	297mmx210mm	420x297	Yes	Portrait	Yes[5]	Yes[10]	Named
Metric Layout templates.dwt	---	420x297	Yes	---	Yes[7]	Yes[8]	Color Dependent

[1] Layers : 0, Title_block and Viewport.

[2] Layers : 0, 025, 035, 05, 07

[3] Layers : 0, FRAME 025, FRAME 050, FRAME 070, Title Block, and Viewport.

[4] Layers : 0, FRAME 025, FRAME 050, FRAME 070, FRAME 200, Title Block, and Viewport.

[5] Layers : 0, Object, Title Block, and Viewport.

[6] Layers : \M+5CDBC\M+5BFF2_\M+5B1EA\M+5CCE2\M+5C0B8\M+5CAF4\M+5D0D4
\M+5CDBC\M+5BFF2_\M+5B1EA\M+5CCE2\M+5C0B8\M+5CEC4\M+5D7D6
\M+5CDBC\M+5BFF2_\M+5BDC7\M+5CFDF
\M+5CDBC\M+5BFF2_\M+5C4DA\M+5BFF2\M+5CFDF
\M+5CDBC\M+5BFF2_\M+5CAD3\M+5BFDA
\M+5CDBC\M+5BFF2_\M+5CDE2\M+5BFF2\M+5CFDF, 0

[7] Layers: FRAME 025, FRAME 050, FRAME 070, FRAME 200, Object, RAHMEN025, RAHMEN035, RAHMEN05, RAHMEN07, Title block, Viewport

[8] Dimension Styles: ISO-25.

[9] Dimension Styles: DIN.

[10] Dimension Styles: JIS.

[11] Dimension Styles: GB-5

NOTE: The initial *PSLTSCALE* value for all templates is 1. (See Chapter 13 for details on *PSLTSCALE*.)

See the Tables of *Limits* Settings in Chapter 14 for standard plotting scales and intended sheet sizes for the *Limits* settings in these templates.

Creating Template Drawings

To make a template drawing, begin a *New* drawing using a default template drawing (ACAD.DWT) or other template drawing (*.DWT), then make the initial drawing setups. Alternately, *Open* an existing drawing that is set up as you need (layers created, linetypes loaded, *Limits* and other settings for plotting or printing to scale, etc.), and then *Erase* all the geometry. Next, use *SaveAs* to save the drawing under a different descriptive name with a .DWT file extension. Do this by selecting the *AutoCAD Drawing Template File (*.dwt)* option in the *Save Drawing As* dialog box (Fig. 12-4, on the next page).

2002

AutoCAD automatically (by default) saves the drawing in the folder where other template (*.DWT) files are found. Use the *Save in:* drop-down list at the top of the *Save Drawing As* dialog box to specify another location for saving the .DWT file.

Location of the Template folder in AutoCAD 2004 has been changed to align with the Microsoft Windows 2000 and XP conventions. That is, the templates are stored by default (for a stand-alone installation) in a folder multiple levels under the user's profile (C:\Documents and Settings\user's name\Local Settings\ Application Data\AutoCAD 2004\... and so on). To find the complete path, use the *Options* dialog box, *Files* tab, and expand *Drawing Template Settings*.

Figure 12-4 ————

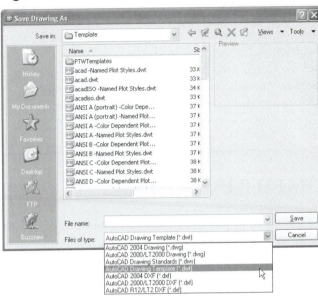

After you assign the file name in the *Save Drawing As* dialog box, you can enter a description of the template drawing in the *Template Description* dialog box that appears (Fig. 12-5). The description is saved with the template drawing and appears in the *Startup* and *Create New Drawing* dialog boxes when you highlight the template drawing name (see Figure 12-3).

Figure 12-5 ————

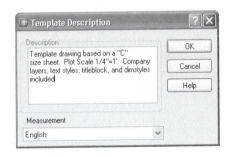

Multiple template drawings can be created, each for a particular situation. For example, you may want to create several templates, each having the setup steps completed but with different *Limits* and with the intention to plot or print each in a different scale or on a different size sheet. Another possibility is to create templates with different layout and layering schemes. There are many possibilities, but the specific settings used depend on your applications (geometry), typical scales, or your plot/print devices.

Typical drawing steps that can be considered for developing template drawings are listed below:

Set *Units*
Set *Limits*
Set *Snap*
Set *Grid*
Set *LTSCALE* and *PSLTSCALE*
Create *Layers* with color and linetypes assigned
Create *Text Styles* (see Chapter 18)
Create Dimension Styles (see Chapter 29)
Create *Layout* tabs, each with titleblock, border and plot settings saved (plot device, paper size, plot scale, etc.). (See Chapter 13, Layouts and Viewports.)

You can create template drawings for different sheet sizes and different plot scales. If you plot from the *Model* tab, a template drawing should be created for each case (scale and sheet size). When you learn to create *Layouts*, you can save one template drawing that contains several layouts, each layout for a different sheet size and/or scale.

Additional Advanced Drawing Setup Concepts

The drawing setup described in this chapter is appropriate for learning the basic concepts of setting up a drawing in preparation for creating layouts and viewports and plotting to scale. In Chapter 13, Layouts and Viewports, you will learn the basics of creating layouts and viewports—particularly for drawings that use one large viewport in a layout. In Chapter 33, Advanced Layouts and Plotting, other advanced ideas are introduced including the "reverse method" for drawing setup, which is a strategy where the *Limits* are not set and the drawing scale factor is not calculated initially, but size-related settings are based on the viewport scale.

CHAPTER EXERCISES

For the first three exercises, you will set up several drawings that will be used in other chapters for creating geometry. Follow the typical steps for setting up a drawing given in this chapter.

1. **Create a Metric Template Drawing**
   You are to create a template drawing for use with metric units. The drawing will be printed full size on an "A" size sheet (not on an A4 metric sheet), and the dimensions are in millimeters. Start a *New* drawing, select *Start from Scratch*, and select the <u>*Imperial*</u> default settings. Follow these steps.

   A. Set *Units* to *Decimal* and *Precision* to **0.00**.
   B. Set *Limits* to a metric sheet size, **279.4 x 215.9**; then *Zoom All* (scale factor is approximately 25).
   C. Set *Snap Spacing* (for *Grid Snap*) to **1**. (*Polar Tracking* and *Polar Snap* settings do not have to be preset since they are saved in the system registry.)
   D. Set *Grid* to **10**.
   E. Set *LTSCALE* to **25** (drawing scale factor of 25.4). Set *PSLTSCALE* to **0**.
   F. *Load* the *Center2* and *Hidden2 Linetypes*.
   G. Create the following *Layers* and assign the *Colors*, *Linetypes*, and *Lineweights* as shown:

OBJECT	red	continuous	0.40 mm
CONSTR	white	continuous	0.25 mm
CENTER	green	center2	0.25 mm
HIDDEN	yellow	hidden2	0.25 mm
DIM	cyan	continuous	0.25 mm
TITLE	white	continuous	0.25 mm
VORTS	white	continuous	0.25 mm

   H. Use the *SaveAs* command. In the dialog box under *Save as type:*, select *AutoCAD Drawing Template File (.dwt)*. Assign the name **A-METRIC**. Enter **Metric A Size Drawing** in the *Template Description* dialog box.

2. *Quick Setup* **wizard**
   A drawing of a mechanical part is to be made and plotted full size (1"=1") on an 11" x 8.5" sheet. Use the *Quick Setup* wizard to assist with the setup.

   A. Begin a *New* drawing. When the *Startup* or *Create New Drawing* dialog box appears, select *Wizard*. Select *Quick Setup*.
   B. In the *Quick Setup* dialog box, select *Decimal* in the *Units* step.
   C. Press *Next* to proceed to the *Area* step. Enter **11** and **8.5** in the two edit boxes so 11 appears as the *Width* (X direction) and 8.5 appears as the *Length* (in the Y direction).

D. Select the *Finish* button.

E. Check to ensure the *Limits* are set to 11,8.5 and *Grid* is set to .5.

F. Change the *Snap* to **.250**.

G. Set *LTSCALE* to **1**. Set *PSLTSCALE* to **0**.

H. *Load* the *Center2* and *Hidden2 Linetypes*.

I. Create the following *Layers* and assign the given *Colors, Linetypes,* and *Lineweights*:

OBJECT	*red*	*continuous*	*0.40 mm*
CONSTR	*white*	*continuous*	*0.25 mm*
CENTER	*green*	*center2*	*0.25 mm*
HIDDEN	*yellow*	*hidden2*	*0.25 mm*
DIM	*cyan*	*continuous*	*0.25 mm*
TITLE	*white*	*continuous*	*0.25 mm*
VPORTS	*white*	*continuous*	*0.25 mm*

J. Save the drawing as **BARGUIDE** for use in another chapter exercise.

3. *Quick Setup* **wizard**

A floor plan of an apartment has been requested. Dimensions are in feet and inches. The drawing will be plotted at 1/2"=1' scale on an 24" x 18" sheet.

A. Begin a *New* drawing and select the *Quick Setup* wizard.

B. Select *Architectural* in the *Units* step.

C. In the *Area* step, set *Width* and *Length* to **48'** x **36'** (576" x 432"), respectively (use the apostrophe symbol for feet). Select *Finish*.

D. Set the *Snap* value to **1** (inch).

E. Set the *Grid* value to **12** (inches).

F. Set the *LTSCALE* to **12**. Set *PSLTSCALE* to **0**.

G. Create the following *Layers* and assign the *Colors* and *Lineweights:*

FLOORPLN	*red*	*0.016"*
CONSTR	*white*	*0.010"*
TEXT	*green*	*0.010"*
TITLE	*yellow*	*0.010"*
DIM	*cyan*	*0.010"*
VPORTS	*white*	*0.010"*

H. *Save* the drawing and assign the name **APARTMENT** to be used later.

4. *Quick Setup* **wizard**

A drawing is to be created and printed in 1/4"=1" scale on an A size (11" x 8.5") sheet. The *Quick Setup* wizard can be used to automate the setup.

A. Begin a *New* drawing. Select *Quick Setup* wizard.

B. In the *Quick Setup* dialog box, select *Decimal* in the *Units* step.

C. Press *Next* to proceed to the *Area* step. Enter **44** and **34** in the two edit boxes so 44 appears in the *Width* edit box (X direction) and 34 appears in the *Length* edit box (Y direction).

D. Select the *Finish* button.

E. Check to ensure the *Limits* are set to 44,34 and *Snap* and *Grid* are set to 1.

F. Save the drawing as **CH12EX4**.

5. **Create a Template Drawing**
    In this exercise, you will create a "generic" template drawing that can be used at a later time. Creating templates now will save you time later when you begin new drawings.

    A. Create a template drawing for use with decimal dimensions and using standard paper "A" size format. Begin a *New* drawing and select *Start from Scratch*. Select the *Imperial* default settings.
    B. Set *Units* to *Decimal* and *Precision* to **0.00**.
    C. Set *Limits* to **11 x 8.5**.
    D. Set *Snap* to **.25**.
    E. Set *Grid* to **1**.
    F. Set *LTSCALE* to **1**. Set *PSLTSCALE* to **0**.
    G. Create the following *Layers*, assign the *Linetypes* and *Lineweights* as shown, and assign your choice of *Colors*. Create any other layers you think you may need or any assigned by your instructor.

OBJECT	*continuous*	*0.016"*
CONSTR	*continuous*	*0.010"*
TEXT	*continuous*	*0.010"*
TITLE	*continuous*	*0.010"*
VPORTS	*continuous*	*0.010"*
DIM	*continuous*	*0.010"*
HIDDEN	*hidden2*	*0.010"*
CENTER	*center2*	*0.010"*
DASHED	*dashed2*	*0.010"*

    H. Use *SaveAs* and name the template drawing **ASHEET**. Make sure you select *AutoCAD Drawing Template File (*.dwt)* from the *Save as Type:* drop-down list in the *Save Drawing As* dialog box. Also, from the *Save In:* drop-down list on top, select <u>your working directory</u> as the location to save the template.

13

LAYOUTS AND VIEWPORTS

Chapter Objectives

After completing this chapter you should:

1. know the difference between paper space and model space and between tiled viewports and paper space viewports;

2. know that the purpose of a layout is to prepare model space geometry for plotting;

3. know the "Guidelines for Using Layouts and Viewports";

4. be able to create layouts using the *Layout Wizard* and the *Layout* command;

5. know how to set up a layout for plotting using the *Page Setup* dialog box;

6. be able to use the options of *Vports* and *-Vports* to create and control viewports in paper space;

7. and know how to scale the model geometry displayed in paper space viewports.

CONCEPTS

You are already somewhat familiar with model space (the *Model* tab) and paper space (*Layout* tabs). Model space (the *Model* tab) is used for construction of geometry, whereas paper space (*Layout* tabs) is used for preparing to print or plot the model geometry. In order to view the model geometry in a layout, one or more viewports must be created. This can be done automatically by AutoCAD when you activate a *Layout* tab for the first time, or you can use the *Vports* command to create viewports in a layout. An introduction to these concepts is given in Chapter 6, Basic Drawing Setup, "Introduction to Layouts and Printing." However, this chapter (Chapter 13) gives a full explanation of paper space, layouts, and creating paper space viewports. Advanced features of layouts and plotting are discussed in Chapter 33 and using layouts for 3D applications is discussed in Chapter 42.

Paper Space and Model Space

The two drawing spaces that AutoCAD provides are model space and paper space. Model space is activated by selecting the *Model* tab and paper space is activated by selecting the *Layout1*, *Layout2*, or other layout tab. When you start AutoCAD and begin a drawing, model space is active by default. Objects that represent the subject of the drawing (model geometry) are normally drawn in model space. Dimensioning is traditionally performed in model space because it is associative—directly associated to the model geometry. The model geometry is usually completed before using paper space. (Since AutoCAD 2002, it is possible to create dimensions in paper space that are attached to objects in model space; however, this practice is recommended only for certain applications. See "Dimensioning in Paper Space Layouts" in Chapter 29.)

Paper space represents the <u>paper that you plot or print on</u>. When you enter paper space (by activating a *Layout* tab) for the first time, you may see only a blank "sheet," unless your system is configured to automatically generate a viewport as explained in Chapter 6. Normally the only geometry you would create in paper space is the title block, drawing border, and possibly some other annotation (text). In order to see any model geometry from paper space, viewports must be created (like cutting rectangular holes) so you can "see" into model space. Viewports are created either automatically by AutoCAD or by using the *Vports* command. Any number or size of rectangular or non-rectangular viewports can be created in paper space. Since there is only <u>one model space</u> in a drawing, you see the same model space geometry in each viewport initially. You can, however, control the size and area of the geometry displayed and which layers are *Frozen* and *Thawed* <u>in each viewport</u>. Since paper space represents the actual paper used for plotting, you should plot from paper space at a scale of 1:1.

The *TILEMODE* system variable controls which space is active—paper space (*TILEMODE*=0) or model space (*TILEMODE*=1). *TILEMODE* is automatically set when you activate the *Model* tab or a *Layout* tab. The *TILEMODE* system variable also controls which type of viewports can be created with the *Vports* command—paper space viewports (when *TILEMODE*=0) or model space (tiled) viewports (when *TILEMODE*=1).

Layouts

In AutoCAD Release 14 and older versions, there was only one paper space and one model space. Although there is still only one model space, you can now have multiple paper spaces, currently known as layouts. By default there is a *Layout1* tab and a *Layout2* tab (no *Layout2* tab exists when opening drawings created with Release 14 or earlier releases). You can produce any configuration of layouts by creating new layouts or copying, renaming, deleting, or moving existing layouts.

A shortcut menu is available (right-click while your pointer is on a layout tab) that allows you to *Rename* the selected layout as well as create a *New Layout* from scratch or *From Template*, *Delete*, *Move or Copy* a layout, *Select all Layouts*, or set up for plotting (Fig. 13-1).

Figure 13-1

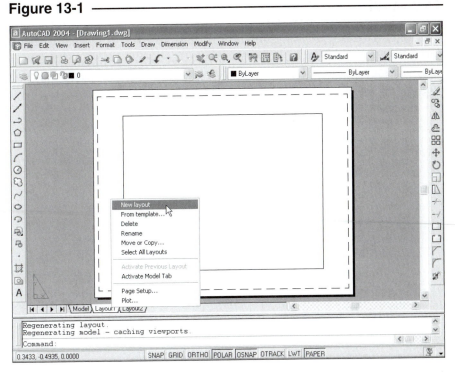

The primary function of a layout is to prepare the model geometry for plotting. A layout simulates the sheet of paper you will plot on and allows you to specify plotting parameters and see the changes in settings that you make, similar to a plot preview. Plot specifications are made using the *Page Setup* dialog box to select such parameters as plot device, paper size and orientation, scale, and Plot Style Tables to attach.

Since you can have multiple layouts in AutoCAD, you can set up multiple plot schemes. For example, assume you have a drawing (in model space) of an office complex. You can set up one layout to plot the office floor plan on a "C" size sheet with an electrostatic plotter in 1/4"=1' scale and set up a second layout to plot the same floor plan on a "A" size sheet with a laser jet printer at 1/8"=1' scale. Since the plotting setup is saved with each layout, you can plot any layout again without additional setup.

Viewports

When you activate a layout tab (enter paper space) for the first time, you may see a "blank sheet." However, if *Create Viewport in New Layouts* is checked in the *Display* tab of the *Options* dialog box (see Figure 13-2, lower-left corner), you may see a viewport that is automatically created. If no viewport exists, you can create one or more with the *Vports* command. A viewport in paper space is a window into model space. Without a viewport, no model geometry is visible in a layout.

Figure 13-2

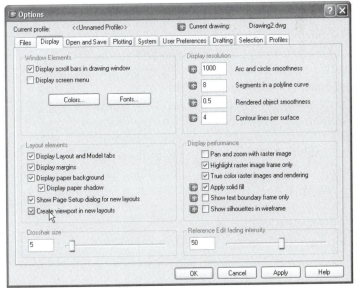

There are two types of viewports in AutoCAD, (1) viewports in model space that divide the screen like "tiles" described in Chapter 10 (known as model space viewports or tiled viewports) and (2) viewports in paper space described in this chapter (known as paper space viewports or floating viewports). The term "floating viewports" is used because these viewports can be moved or resized and can take on any shape. Floating viewports are objects that can be *Erased*, *Moved*, *Copied*, stretched with grips, or the viewport's layer can be turned *Off* or *Frozen* so no viewport "border" appears in the plot.

The command generally used to create viewports, *Vports*, can create either tiled viewports or floating viewports, but only one type at a time depending on the active space—*Model* tab or *Layout* tab. When the *Model* tab is active (*TILEMODE=1*), *Vports* creates tiled viewports. When a *Layout* tab is active (*TILE-MODE=0*), *Vports* creates paper space viewports.

Figure 13-3

Typically there is only one viewport per layout; however, you can create multiple viewports in one layout. This capability gives you the flexibility to display two or more areas of the same model (model space) in one layout. For example, you may want to show the entire floor plan in one viewport and a detail (enlarged view) in a second viewport, both in the same layout (Fig. 13-3).

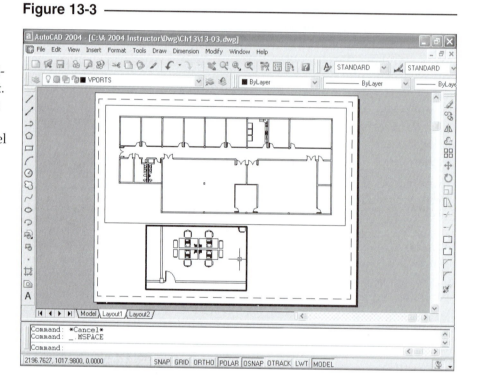

You can control the layer visibility in each viewport. In other words, you can control which layers are visible in which viewports. Do this by using the *Current VP Freeze* button and *New VP Freeze* button in the *Layer Properties Manager* (Fig. 13-4, last two columns). For example, notice in Figure 13-3 that the furniture layer is on in the detail viewport, but not in the overall view of the office floor plan above. (Using multiple viewports and setting viewport-specific layer visibility is discussed in detail in Chapter 33, Advanced Layouts and Plotting.)

Figure 13-4

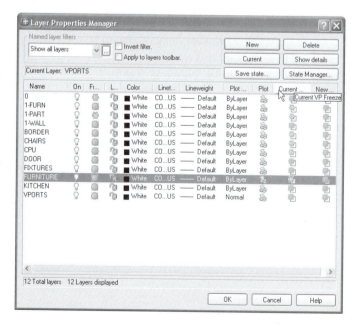

You can also scale the display of the geometry that appears in a viewport. You can set the viewport scale by using the *Viewports* toolbar drop-down list (Fig. 13-5), using the *Properties* dialog box, or entering a *Zoom XP* factor. This action sets the proportion (or scale) of paper space units to model space units.

Figure 13-5

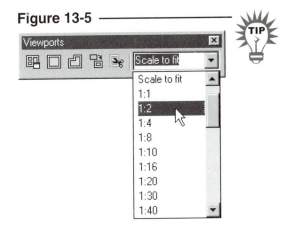

Since a layout represents the plotted sheet, you normally plot the layout at 1:1 scale. However, the geometry in each viewport is scaled to achieve the scale for the drawing objects. For example, you could have two or more viewports in one layout, each displaying the geometry at different scales, as is the case for Figure 13-6. Or, you can set up one layout to display model geometry in one scale and set up a similar layout to display the same geometry at a different scale.

Figure 13-6

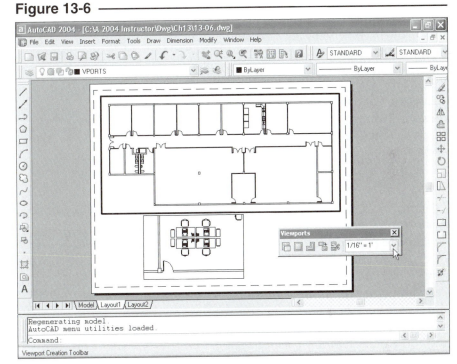

An important concept to remember when using layouts is that the *Plot Scale* that you select in the *Page Setup* and *Plot* dialog boxes should almost always be 1:1. This is because the layout is the actual paper size. Therefore, the geometry displayed in the viewports must be scaled to the desired plot scale (the reciprocal of the "drawing scale factor," which can be set by using the *Viewports* toolbar, the *Properties* dialog box for the viewport, or a *Zoom XP* factor.

Layouts and Viewports Example

Consider this brief example to explain the basics of using layouts and viewports. In order to keep this example simple, only one viewport is created in the layout to set up a drawing for plotting. This example assumes that no automatic viewports or page setups are created. Your system may be configured to automatically create viewports when a *Layout* tab is activated (as recommended in Chapter 6 to simplify viewport creation). Disabling automatic setups is accomplished by ensuring *Show Page Setup Dialog for New Layouts* and *Create Viewport in New Layouts* is unchecked in the *Display* tab in the *Options* dialog box.

First, the part geometry is created in the *Model* tab as usual (Fig. 13-7). Associative dimensions are also normally created in the *Model* tab. This step is the same method that you would have normally used to create a drawing.

Figure 13-7

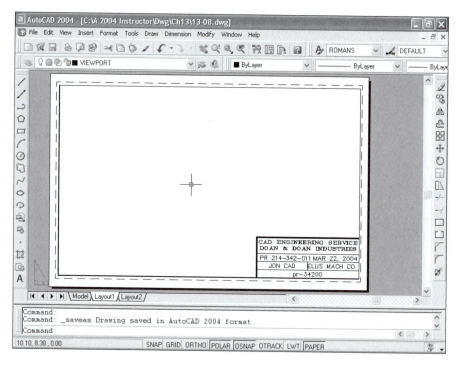

When the part geometry is complete, enable paper space by double-clicking a *Layout* tab. AutoCAD automatically changes the setting of the *TILEMODE* variable to 0. When you enable paper space for the first time in a drawing, a "blank sheet" appears (assuming no automatic viewports or page setups are created as a result of settings in the *Options* dialog box).

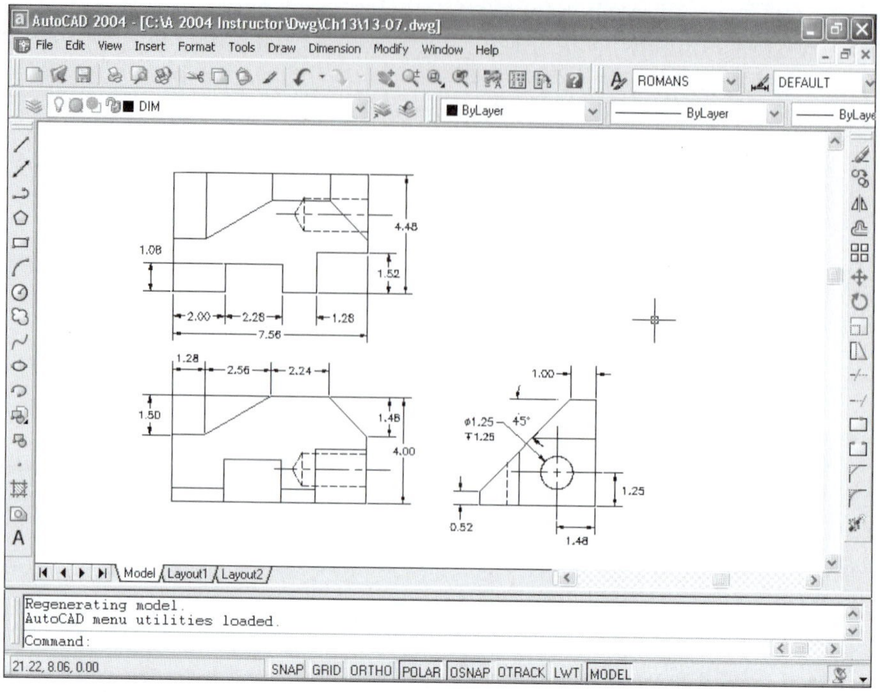

By activating a *Layout* tab, you are automatically switched to paper space. When that occurs, the paper space *Limits* are set automatically based on the selected plot device and sheet size previously selected for plotting. (The default plot device is set in the *Plotting* tab of the *Options* dialog box.) Objects such as a title block and border should be created in the paper space layout (Fig. 13-8). Normally, only objects that are annotations for the drawing (title blocks, tables, border, company logo, etc.) are drawn in the layout.

Figure 13-8

Next, create a viewport in the "paper" with the *Vports* command in order to "look" into model space. All of the model geometry in the drawing initially appears in the viewport. In this example, the *Vports* command is selected from the *Viewports* toolbar. At this point, there is no specific scale relation between paper space units and the size of the geometry in the viewport (Fig. 13-9).

You can control what part of model space geometry you see in a viewport and control the scale of model space to paper space. Two methods are used to control the geometry that is visible in a particular viewport.

Figure 13-9

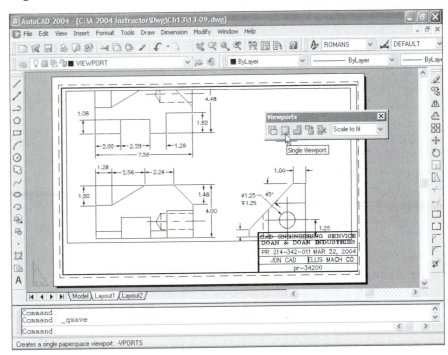

1. **Display commands and *Viewport Scale***

   The *Zoom*, *Pan*, *View*, *3Dorbit* and other display commands allow you to specify what area of model space you want to see in a viewport. In addition, the scale of the model geometry in the viewport can be set by using the *Viewports* toolbar, the *Properties* dialog box for the viewport, or a *Zoom XP* factor.

2. **Viewport-specific layer visibility control**

   You can control what <u>layers</u> are visible in specific viewports. This function is often used for displaying different model space geometry in separate viewports. You can control which layers appear in which viewports by using the *Layer Properties Manager*. (See Chapter 33 for information on viewport-specific layer visibility control.)

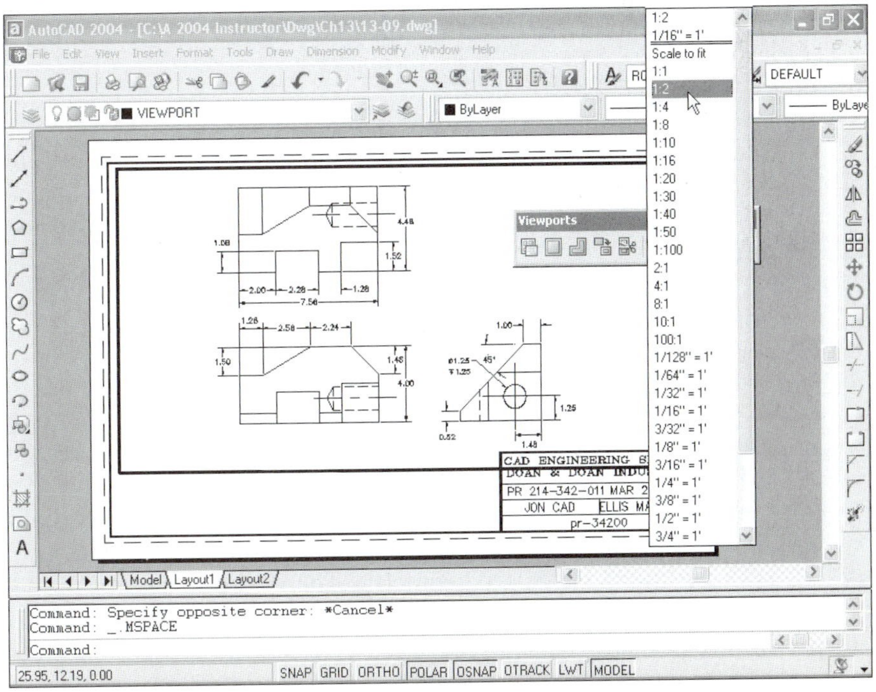

For example, the *Viewports* toolbar could be used to scale the model geometry in a viewport. In this example, double-click inside the viewport and select *1:2* to scale the model geometry to 1/2 size (model space units equal 1/2 times paper space units).

Figure 13-10

While in a layout with a viewport created to display model space geometry, you can double-click inside or outside the viewport. This action allows you to switch between model space (inside a viewport) and paper space (outside a viewport) so you can draw or edit in either space. You could instead use the *Mspace* (Model Space) and *Pspace* (Paper Space) commands or single-click the words *MODEL* or *PAPER* on the Status Bar to switch between inside and outside the viewport. The "crosshairs" can move completely across the screen when in paper space, but only within the viewport when in model space (model space inside a viewport).

An object <u>cannot be in both spaces</u>. You can, however, draw in paper space and <u>*OSNAP* to objects in model space</u>. Commands that are used will affect the objects or display of the current space.

When drawing is completed, activate paper space and plot at 1:1 since the layout size (paper space *Limits*) is set to the <u>actual paper size</u>. Typically, the *Page Setup* dialog box or the *Plot* dialog box is used to prepare the layout for the plot. The plot scale to use for the layout is almost always 1:1 since the layout represents the plotted sheet and the model geometry is already scaled.

Guidelines for Using Layouts and Viewports

Although there are other alternatives, the typical steps used to set up a layout and viewports for plotting are listed here. (These guidelines assume that no automatic viewports or page setups are created, as described in Chapter 6, "Introduction to Layouts and Printing.")

1. Create the part geometry in model space. Associative dimensions are typically created in model space.

2. Click on a *Layout* tab. When this is done, AutoCAD automatically sets up the size of the layout (*Limits*) for the default plot device, paper size, and orientation. Use the *Page Setup* dialog box to ensure the correct device, paper size, and orientation are selected. If not, make the desired selections. (If the layout does not automatically reflect the desired settings, check the *Plotting* tab of the *Options* dialog box to ensure that *Use the Plot Device Paper Size* is checked.)

3. Set up a title block and border as indicated.

    A. Make a layer named BORDER or TITLE and *Set* that layer as current.
    B. Draw, *Insert*, or *Xref* a border and a title block.

4. Make viewports in paper space.

    A. Make a layer named VIEWPORT (or other descriptive name) and set it as the *Current* layer (it can be turned *Off* later if you do not want the viewport objects to appear in the plot).
    B. Use the *Vports* command to make viewports. Each viewport contains a view of model space geometry.

5. Control the display of model space geometry in each viewport. Complete these steps for each viewport.

    A. Double-click inside the viewport or use the *MODEL/PAPER* toggle to "go into" model space (in the viewport). Move the cursor and PICK to activate the desired viewport if several viewports exist.
    B. Use the *Viewport Scale Control* drop-down list, the *Properties* dialog box for the viewport, or *Zoom XP* to scale the display of the paper space units to model units. This action dictates the plot scale for the model space geometry. The viewport scale is the same as the plot scale factor that would otherwise be used (reciprocal of the "drawing scale factor").
    C. Control the layer visibility for each viewport by using the icons in the *Layer Properties Manager*.

6. Plot from paper space at a scale of 1:1.

    A. Use the *MODEL/PAPER* toggle or double-click outside the viewport to switch to paper space.

    B. If desired, turn *Off* the VIEWPORT layer so the viewport borders do not plot.

    C. Invoke the *Plot* dialog box or the *Page Setup* dialog box to set the plot scale and other options. Normally, set the plot scale for the layout to *1:1* since the layout is set to the actual paper size. The paper space geometry will plot full size, and the resulting model space geometry will plot to the <u>scale</u> selected in the *Viewport Scale* drop-down list, *Properties* window, or by the designated *Zoom XP* factor.

    D. Make a *Full Preview* to check that all settings are correct. Make changes if necessary. Finally, make the plot.

CREATING AND SETTING UP LAYOUTS

Listed below are several ways to create layouts.

Shortcut menu	Right-click on a *Layout* tab and select *New Layout*, *From Template*, or *Move or Copy* (see Figure 13-1).
Layout command	Type or select from the *Insert* pull-down menu. Use the *New*, *Copy*, or *Template* option.
Layoutwizard command	Type or select from the *Insert* pull-down menu.

There are several steps involved in correctly setting up a layout, as listed previously in the "Guidelines for Using Layouts and Viewports." The steps involve specifying a plot device, setting paper size and orientation, setting up a new *Layout* tab, and creating the viewports and scaling the model geometry. One of the best methods for accomplishing all these tasks is to use the *Layout Wizard*.

LAYOUTWIZARD	Pull-down Menu	COMMAND (TYPE)	ALIAS (TYPE)	Short-cut	Screen (side) Menu	Tablet Menu
	Insert *Layout>* *Layout Wizard*	LAYOUTWIZARD	...	...	...	...

The *Layoutwizard* command produces a wizard to automatically lead you through eight steps to correctly set up a layout tab for plotting. The steps are essentially the same as those listed earlier in the "Guidelines for Using Layouts and Viewports."

Begin
The first step is intended for you to enter a name for the layout. The default name is whatever layout number is next in the sequence, for example, *Layout3*. Enter any name that has not yet been used in this drawing. You can change the name later if you want using the *Rename* option of the *Layout* command.

Figure 13-11

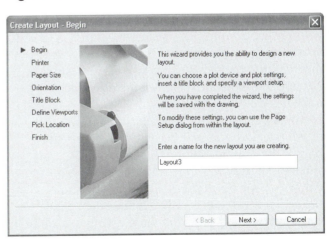

Printer
Because the *Limits* in paper space are automatically set based on the size and orientation of the selected paper, choosing a print or plot device is essential at this step. The list includes all previously configured printer and plotter devices. If you intend to use a different device, it may be wise to *Cancel*, use *Plotter Manager* to configure the new device, then begin *Layoutwizard* again (see Chapter 14, "Configuring Plotters and Printers").

Figure 13-12

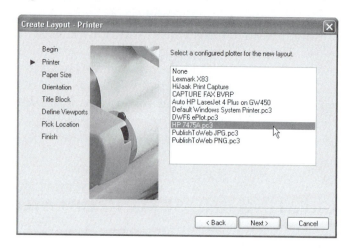

Paper Size
Select the paper size you expect to use for the layout (Fig. 13-13). Make sure you also select the units for the paper (*Drawing Units*) since the layout size (*Limits* in paper space) is automatically set in the selected units.

Orientation
Select either *Portrait* or *Landscape* (not shown). *Landscape* plots horizontal lines in the drawing along the long axis of the paper, whereas *Portrait* plots horizontal lines in the drawing along the short axis of the paper.

Figure 13-13

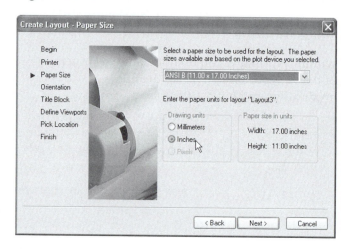

Title Block
In this step, you can select *None* or select from several AutoCAD-supplied ANSI, DIN, ISO and JIS standard title blocks. See "Table of AutoCAD-Supplied Template Drawing Settings" in Chapter 12 for exact sizes and specifications. The selected title block is inserted into the paper space layout. You can also specify whether you want the title block to be inserted as a *Block* or as an *Xref*. Normally inserting the title block as a *Block* object is safer because it becomes a permanent part of the drawing although it occupies a small bit of drawing space (increased file size). An *Xrefed* title block saves on file size because the actual title block is only referenced, but can

Figure 13-14

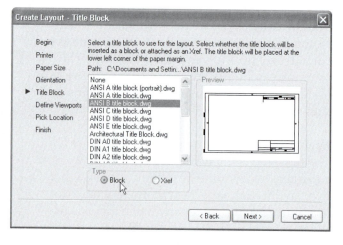

cause problems if you send the drawing to a client who does not have access to the same referenced drawing. (See Chapter 21 Blocks, DesignCenter, and Tool Palettes and Chapter 30, Xreferences, for more information on these subjects.) You can also create your own title block drawings and save them in the Template folder so they will appear in this list.

Define Viewports

This step has two parts. First select the type and number of viewports to set up. Use *None* if you want to set up viewports at a later time or want to create a non-rectangular viewport. Normally, use *Single* if you need only one rectangular viewport. Use *Std. 3D Engineering Views* if you have a 3D model. This option sets up four viewports with a top, front, side, and isometric view. The *Array* option creates multiple viewports with the number of *Rows* and *Columns* you enter in the edit boxes below. If you select *Std. 3D Engineering Views* or *Array*, you can then specify the *Spacing between rows* and *Spacing between columns*.

Figure 13-15

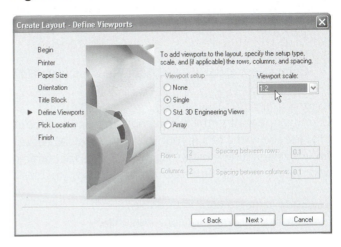

Pick Location

In this step, pick *Select Location* to temporarily exit the wizard and return to the layout to pick diagonal corners for the viewport(s) to fit within. Normally, you do not want the corners of the viewport(s) to overlap the title block or border. You can bypass this step to have AutoCAD automatically draw the viewport border at the extents of the printable area.

Finish

This is not really a step, rather a confirmation (not shown). In all other steps, you can use *Back* to go backward in the process to change some of the specifications. In this step, *Back* is disabled, so you can only select *Finish* or *Cancel*.

Figure 13-16

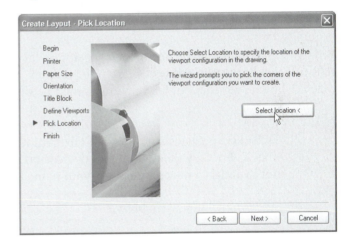

At this time, the new layout is activated so you can see the setup you specified. If you want to make changes to the plot device, paper size, orientation, or viewport scale, you can do so by using the *Page Setup* dialog box. You can also change the viewport sizes or configuration using *Vports*, or grips (explained later in this chapter).

LAYOUT

Pull-down Menu	COMMAND (TYPE)	ALIAS (TYPE)	Short-cut	Screen (side) Menu	Tablet Menu
Insert *Layout>* *New Layout*	*LAYOUT*	*LO* or *-LO*	...	...	...

If you prefer to create a layout without using the *Layout Wizard*, use the *Layout* command by typing it or selecting from the *Insert* pull-down menu or *Layout* toolbar. Using the wizard leads you through all the steps required to create a new layout including selecting plot device, paper size and orientation, and creating viewports. Use the *Layout* command to create viewports "manually" (if your system is not set to automatically create viewports). You will then have to use *Vports* to create viewports in the layout.

It is a good idea to ensure the correct plot device is specified before you use *Layout* to create a new layout. See the *New* option for more information. Besides creating a new layout, using a template, and copying an existing layout, you can rename, save, and delete layout.

> Command: **layout**
> Enter layout option [Copy/Delete/New/Template/Rename/SAveas/Set/?] <set>:

Copy

Use this option if you want to copy an existing layout including its plot specifications. If you do not provide the name of a layout to copy, AutoCAD assumes you want to copy the active layout tab. If you do not assign a new name, AutoCAD uses the name of the copied layout (*Floor Plan*, for example) and adds an incremental number in parentheses, such as *Floor Plan (2)*.

> Enter layout to copy <current>:
> Enter layout name for copy <default>:

Delete

Use this option to *Delete* a layout. If you do not enter a name, the most current layout is deleted by default.

> Enter name of layout to delete <current>:

Wildcard characters can be used when specifying layout names to delete. You can select several *Layout* tabs to delete by holding down Shift while you pick. If you select all layouts to delete, all the layouts are deleted, and a single layout tab remains named *Layout1*. The *Model* tab cannot be deleted.

New

This option creates a new layout tab. AutoCAD creates a new layout from scratch based on settings defined by the default plot device and paper size (in the *Options* dialog box). If you choose not to assign a new name, the layout name is automatically generated (*Layout3*, for example).

> Enter new layout name <Layout#>:

Keep in mind that when the new layout is created, its size (*Limits*) and shape are determined by the *Use as Default Output Device* setting specified in the *Plotting* tab of the *Options* dialog. Therefore, it is a good idea to ensure the correct plot device is specified before you use the *New* option. You can, however, change the layout at a later time using the *Page Setup* dialog box, assuming that *Use Plot Device Paper Size* is checked in the *Plotting* tab of the *Options* dialog box (see *Options* in this chapter).

Template

You can create a new template based on an existing layout in a template file (.DWT) or drawing file (.DWG) with this option. The layout, viewports, associated layers, and all the paper space geometry in the layout (from the specified template or drawing file) are inserted into the current drawing. No dimension styles or other objects are imported. The standard file navigation dialog box is displayed for you to select a file to use as a template. Next, the *Insert Layout* dialog box appears for you to select the desired layout from the drawing (Fig. 13-17). See "Table of AutoCAD-Supplied Template Drawings Settings" in Chapter 12.

Figure 13-17

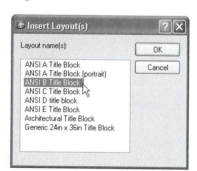

Rename
Use *Rename* to change the name of an existing layout. The last current layout is used as the default for the layout to rename.

> Enter layout to rename <current>:
> Enter new layout name:

Layout names can contain up to 255 characters and are not case sensitive. Only the first 32 characters are displayed in the tab.

Save
This option creates a .DWT file. The file contains all of the (paper space) geometry in the layout and all of the plot settings for the layout, but no model space geometry. All *Block* definitions appearing in the layout, such as a titleblock, are also saved to the .DWT file, but not unused *Block* definitions. All layouts are stored in the template folder as defined in the *Options* dialog box. The last current layout is used as the default for the layout to save.

> Enter layout to save to template <current>:

The standard file selection dialog box is displayed in which you can specify a file name for the .DWT file. When you later create new layout from a template (using the *Template* option of *Layout*), all the saved information (plot settings, layout geometry and *Block* definitions) are imported into the new layout.

Set
This option simply makes a layout current. This option has identical results as picking a layout tab.

? (List Layouts)
Use this option to list all the layouts when you have turned off the layout tabs in the *Display* tab of the *Options* dialog box. Otherwise, layout tabs are visible at the bottom of the drawing area.

Inserting Layouts with AutoCAD DesignCenter

A feature in AutoCAD, called DesignCenter, allows you to insert content from any drawing into another drawing. Content that can be inserted includes drawings, *Blocks*, *Dimstyles*, *Layers*, *Layouts*, *Linetypes*, *Textstyles*, *Xrefs*, raster images, and URLs (web site addresses). If you have created a layout in a drawing and want to insert it into the current drawing, you can use the *Template* option of *Layout* (as previously explained) or use DesignCenter to drag and drop the layout name into the drawing. With DesignCenter, only layouts from .DWG files (not .DWT files) can be imported. See Chapter 21 Blocks, DesignCenter, and Tool Palettes for more information.

Setting Up the Layout

When you do not use the *Layout Wizard*, the steps involved in setting up the layout must be done individually. In cases when you need flexibility, such as creating non-rectangular viewports or when some factors are not yet known, it is desirable to take each step individually. These steps are listed in "Guidelines for Using Layouts and Viewports."

OPTIONS

Pull-down Menu	COMMAND (TYPE)	ALIAS (TYPE)	Short-cut	Screen (side) Menu	Tablet Menu
Tools *Options...*	*OPTIONS*	*OP*	(Default) *Options....*	*TOOLS2* *Options*	*Y,10*

When you create *New* layouts using the *Layout* command, the size and shape of the layout is determined by the selected plot device, paper size, and paper size. Default settings are made in the *Plotting* tab of the *Options* dialog box (Fig. 13-18). Although most settings concerned with plotting a layout can later be changed using the *Page Setup* dialog box, it may be efficient to make the settings in the *Options* dialog box <u>before</u> creating layouts from scratch. Four settings in this dialog box affect layouts.

Figure 13-18

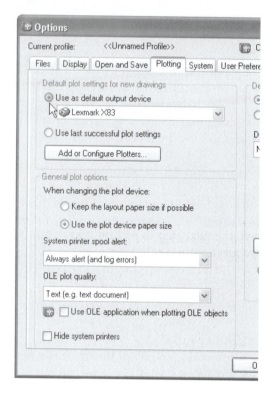

Default Plot Settings for New Drawings
Your choice in this section determines what plot device is used to automatically set the paper size (paper space *Limits*) when new layouts are created. Your setting here applies to <u>new drawings and the current drawing</u>. (Paper size information for each device is stored either in the plotter configuration file [.PC3] or in the default system settings if the output device is a system printer.)

Use as Default Output Device
The selected device is used to determine the paper size (paper space *Limits*) that are set automatically when a new layout is created in the current drawing and for new drawings. This drop-down list contains all configured plot or print devices (any plotter configuration files [PC3] and any system printers that are configured in the system).

Use Last Successful Plot Settings
This button uses the plotting settings (device and paper size) according to those of the last successful plot that was made on the system (a plot must be made from the layout for this option to be valid) and applies them to new layouts that are created. This setting affects new layouts created for the current drawing as well as new drawings. This option overrides the output device appearing in the *Use as Default Output Device* drop-down list.

General Plot Options
These options apply only when plot devices are changed for existing layouts <u>and</u> the layout has a viewport created. Plot devices for layouts can be changed using the *Page Setup* or *Plot* dialog boxes. When you change the plot device for an existing layout, the paper size (*Limits* setting) of the layout is determined by your choice in this section.

Keep the Layout Paper Size if Possible
If this button is pressed, AutoCAD attempts to use the paper size initially specified for the layout if you decide later to change the plot device for the layout. This option is useful if you have two devices that can use the same size paper as specified in the layout, so you can select either device to use without affecting the layout. However, if the selected output device cannot plot to the previously specified

paper size, AutoCAD displays a warning message and uses the paper size specified by the plot device you select in the *Page Setup* or *Plot* dialog box. If no viewports have been created in the layout, you can change the plot device and automatically reset the layout limits without consequence. This button sets the *PAPERUPDATE* system variable to 0 (layout paper size is not updated).

Use the Plot Device Paper Size
As the alternative to the previous button, this one sets the paper size (*Limits*) for an existing layout to the paper size of whichever device is selected for that layout in the *Page Setup* or *Plot* dialog box. This option sets *PAPERUPDATE* to 1.

PAGESETUP

Pull-down Menu	COMMAND (TYPE)	ALIAS (TYPE)	Short-cut	Screen (side) Menu	Tablet Menu
File *Page Setup...*	*PAGESETUP*	...	...	...	V,25

After ensuring that the desired settings have been made in the *Plotting* tab of the *Options* dialog box, the next step in setting up a layout (assuming that you are not using the *Layout Wizard*) is to use the *Page Setup* dialog box. Here you select or change the selected plot device, paper size, and orientations for the layout. All settings made in the *Page Setup* dialog box are saved with the layout. In this way, plot settings are already prepared whenever you want to make a plot.

Figure 13-19

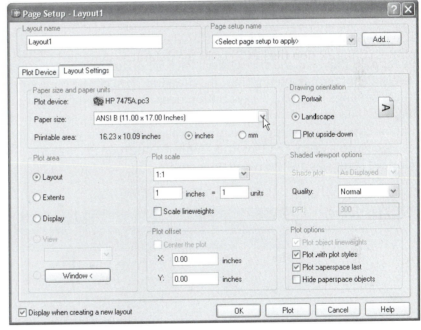

Using the *Page Setup* dialog box (Fig. 13-19) at this time is highly recommended after you create a *New* layout. If you are using a *Template* layout, this step is also suggested to ensure the plot specifications are set as you expect.

The primary selections to make are as follows:

1. Select a plotter or printer in the *Plot Device* tab.
2. Select *Paper Size* and *Drawing Orientation* in the *Layout Settings* tab.
3. Specify the *Plot area* and *Plot options*.
4. Ensure that *Plot Scale* is set to 1:1.

Remember that the size of the layout (paper space *Limits*) is automatically set based on *Paper Size* and *Drawing Orientation* you select here. Depending on your selection of *Keep the Layout Size if Possible* or *Use the Plot Device Paper Size* in the *Plotting* tab of the *Options* dialog box (see previous section), the layout size and shape may or may not change to the new plot device settings. You can, however, select a new plot device, paper size, or orientation <u>after creating the layout, but before creating viewports</u>.

Since the layout size and shape may change with a new device, viewports that were previously created may no longer fit on the page. In that case, the viewports must be changed to a new size or shape. Alternately, you can *Erase* the viewports and create new ones. Either of these choices is not recommended. Instead, ensure you have the desired plot device and paper settings <u>before creating viewports</u>. (See the "Guidelines for Using Layouts and Viewports.")

Plot Scale is almost always set to *1:1* since the layout represents the sheet you will be plotting on. The model geometry that appears in the layout (in a viewport) will be scaled by setting a viewport scale (see "Scaling Viewport Geometry").

Layout Name

The *Layout Name* at the top of the *Page Setup* dialog box generally displays the name appearing on the active layout tab. You can change the name of the tab here or use the *Rename* option of *Layout* or right-click on the tab.

Page Setup Name

Page setups are automatically saved using the name appearing in the *Layout Name* edit box. If you want to use a setup previously created for an existing layout, select it from the list. For example, if you wanted to copy the setup you made for *Layout1* to the active layout (*Layout2*), select **Layout1** from the drop-down list.

Add

Use the *Add* button to specify a unique name for the layout (a name other than the name appearing on the tab). In this case, the *User Defined Page Setup* dialog box appears allowing you to specify a new name or select another page setup from the drawing. You can also import user-defined page setups from other drawings using the *Import* button (also see "*Psetupin*"). If you define a user-defined page setup, make all the settings before assigning the name. Once a user-defined page setup name is defined, you cannot edit it. This feature is helpful in cases where you want to plot the same drawing with different settings (such as attached Plot Style Tables). Additionally, this feature could be used to import company standard setups as an alternative to including the standards in layout templates.

PSETUPIN

Pull-down Menu	COMMAND (TYPE)	ALIAS (TYPE)	Short-cut	Screen (side) Menu	Tablet Menu
...	PSETUPIN or -PSETUPIN	...	...	...	...

Psetupin imports a user-defined page setup into a new drawing layout. This feature provides the ability to import a saved, named page setup from another drawing into the current drawing. User-defined page setups are created using the *Page Setup* dialog box with the *Add* option (see previous section).

Psetupin causes AutoCAD to display the standard file navigation dialog box (not shown) in which you can select the drawing file whose page setups you want to import. The selected .DWG file must contain a user-defined page setup. After you select the drawing file you want to use, the *Import User Defined Page Setups* dialog box is displayed (not shown) listing all user-defined page setups contained in the selected .DWG file. When the setup is imported, the settings that are saved in the named page setup can then be applied to any layout in the new drawing.

If you enter *–Psetupin* at the Command line, the following prompts are displayed.

        Command: *-psetupin*
        (The file navigation dialog box appears. Select the desired .DWG file.)
        Enter user defined page setup(s) to import or [?]:

USING VIEWPORTS IN PAPER SPACE

If you use the *Layout Wizard*, viewports can be created during that process. If you do not use the wizard, viewports should be created for a layout only after specifying the plot device, paper size, and orientation in the *Page Setup* dialog box or through the default plot options set in the *Options* dialog box. You can create viewports with the *Vport* or *-Vport* commands. The Command line version (*-Vports*) produces several options not available in the dialog box, including options to create non-rectangular viewports.

VPORTS

Pull-down Menu	COMMAND (TYPE)	ALIAS (TYPE)	Short-cut	Screen (side) Menu	Tablet Menu
View *Viewports>*	*VPORTS or* *-VPORTS*	...	...	VIEW 1 Vports	M,3 and M,4

The *Vports* command produces the *Viewports* dialog box. Here you can select from several configurations to create rectangular viewports as shown in Figure 13-20. Making a selection in the *Standard Viewports* list on the left of the dialog box in turn displays the selected *Preview* on the right.

After making the desired selection from the dialog box, AutoCAD issues the following prompt.

> Command: **vports**
> Specify first corner or [Fit] <Fit>:

You can pick two diagonal corners to specify the area for the viewports to fill. Generally, pick just inside the titleblock and border if one has been inserted. If you use the *Fit* option, AutoCAD automatically fills the printable area with the specified number of viewports.

The *Setup: 3D* and *Change View to:* options are intended for use with 3D models. (See Chapter 42 for more information on these options.)

You may notice that the *Viewports* dialog box is the same dialog box that appears when you use *Vports* while in the *Model* tab. However, when *Vports* dialog is used in the *Model* tab, <u>tiled</u> viewports are created, whereas when the *Vports* is used in a layout (paper space), paper space viewports are created.

Figure 13-20

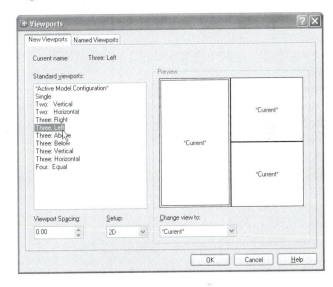

Figure 13-21

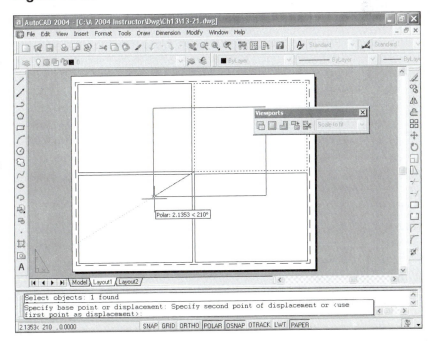

Paper space viewports differ from tiled viewports. AutoCAD treats paper space viewports created with *Vports* as <u>objects</u>. Like other objects, paper space viewports can be affected by most editing commands. For example, you could use *Vports* to create one viewport, then use *Copy* or *Array* to create other viewports. You could edit the size of viewports with *Stretch* or *Scale*. You can use Grips to change paper space viewports. Additionally, you can use *Move* (Fig. 13-21, previous page) to relocate the position of viewports. Delete a viewport using *Erase*. To edit a paper space viewport, you must be in paper space to PICK the viewport objects (borders).

Paper Space viewports can also overlap. This feature makes it possible for geometry appearing in different viewports to occupy the same area on the screen or on a plot.

NOTE: Avoid creating one viewport completely within another's border because visibility and selection problems may result.

Figure 13-22

Another difference between tiled (*Model* tab) viewports and paper space viewports is that viewports in layouts can have space between them. The option in the bottom left of the *Viewports* dialog box called *Viewport Spacing* allows you to create space between the viewports (if you choose any option other than *Single*) by entering a value in the edit box. Figure 13-22 illustrates a layout with three viewports after entering .5 in the *Viewport Spacing* edit box. Also note several of the –*Vports* command options are available in the right-click shortcut menu that appears when a viewport object is selected.

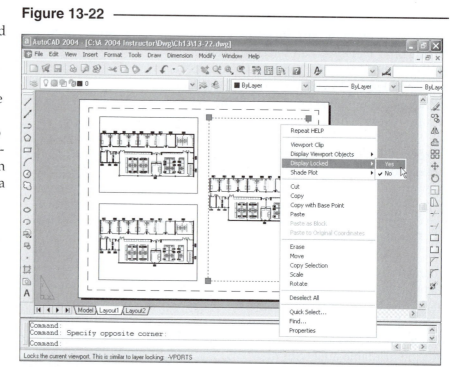

Using the –*Vports* command (with a hyphen prefix) invokes the Command line version and produces the following options.

    Command: **-vports**
    Specify corner of viewport or [ON/OFF/Fit/Shadeplot/Lock/Object/Polygonal/Restore/2/3/4] <Fit>:

The default option (Specify corner of viewport) allows you to pick two diagonal corners for one viewport. The other options are as follows. Remember to select the viewport objects (borders) while in paper space when AutoCAD prompts you to select a viewport.

ON
On turns on the display of model space geometry in the selected viewport. Select the desired viewport to turn on.

OFF
Turn off the display of model space geometry in the selected viewport with this option. Select the desired viewport to turn off.

Fit

Fit creates a new viewport and fits it to the size of the printable area. The new viewport becomes the current viewport.

Shadeplot

If you are displaying a 3D surface or solid model, *Shadeplot* allows you to select the type of display to print the current viewport such as *Wireframe*, *Hidden*, *Rendered*, or *As Displayed*. This feature is new with AutoCAD 2004. (See Chapter 35, 3D Display and Viewing, for more information.)

Lock

Use this option to <u>lock the scale</u> of the current viewport. Turn viewport locking *On* or *Off*. With paper space viewports, you set the scale of each viewport individually using *Zoom* with an *XP* scale factor or by using the *Viewports* toolbar scale drop-down list (see "Scaling Viewport Geometry"). Double-clicking inside a viewport and then using *Zoom* normally changes the scale of a viewport. If *Lock* is used on the viewport, the specified scale of the viewport cannot be changed (unless viewport locking is turned *Off*).

If you attempt to *Zoom* inside a locked viewport, the display of the entire layout (paper space) is zoomed.

Figure 13-23

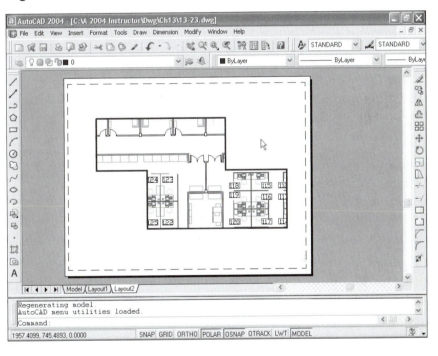

Object

You can use *Object* to convert a closed *Polyline*, *Ellipse*, *Spline*, *Region*, or *Circle* into a viewport (border). The polyline you specify must be closed and contain at least three vertices. It can be self-intersecting and can contain arcs as well as line segments. Figure 13-23 displays a closed *Pline* converted to a viewport.

Polygonal

Use this option to create an irregularly shaped viewport defined by specifying points. You can define the viewport by straight line or arc segments.

```
Command: -vports
Specify corner of viewport or [ON/OFF/Fit/Shadeplot/Lock/Object/Polygonal/Restore/2/3/4] <Fit>: p
Specify start point: PICK
Specify next point or [Arc/Close/Length/Undo]: PICK
Specify next point or [Arc/Close/Length/Undo]: PICK
Specify next point or [Arc/Close/Length/Undo]: c
Regenerating model.
```

If you choose the *Arc* option, the following prompts allow all arc creation options, identical to those found in the full *Arc* command.

```
Enter an arc boundary option [Angle/CEnter/CLose/Direction/Line/Radius/Second pt/Undo/Endpoint of arc]
<Endpoint>:
```

Figure 13-24 displays a viewport created with the *Polygonal* option of *–Vports*. Notice the arc segment. Keep in mind both the *Object* and *Polygonal* options can be used to create non-rectangular and curved viewports.

Figure 13-24 ────────────────

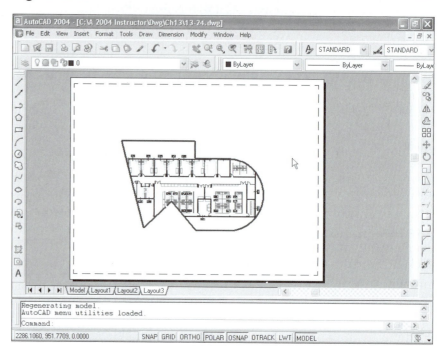

Restore

Use this option to restore a previously created named viewport configuration.

> Enter viewport configuration name or [?] <*Active>: Enter named viewport configuration.
> Specify first corner or [Fit] <Fit>: **PICK** or *F*
> Regenerating model.

2/3/4

These options create a number of viewports within a rectangular area that you specify. The *2* option allows you to arrange 2 viewports either *Vertically* or *Horizontally*. The *4* option automatically divides the specified area into 4 equal viewports. The possible configurations using the *3* option are displayed if you use the *Viewports* dialog box. Using the 3 option yields the following prompt:

> Horizontal/Vertical/Above/Below/Left/<Right>:

PSPACE

Pull-down Menu	COMMAND (TYPE)	ALIAS (TYPE)	Short-cut	Screen (side) Menu	Tablet Menu
...	*PSPACE*	*PS*	...	*VIEW1* *Pspace*	*L,5*

The *Pspace* command switches from model space (inside a viewport) to paper space (outside a viewport). The cursor is displayed in paper space and can move <u>across the entire screen</u> when in paper space, while the cursor appears only inside the current viewport if the *Mspace* command is used. This command accomplishes the same action as double-clicking outside the viewport in a layout or using the *MODEL/PAPER* toggle on the Status Bar.

If you type this command while in the *Model* tab, the following message appears:

> Command: *pspace*
> PSPACE
> ** Command not allowed in Model Tab **

MSPACE

Pull-down Menu	COMMAND (TYPE)	ALIAS (TYPE)	Short-cut	Screen (side) Menu	Tablet Menu
...	*MSPACE*	*MS*	...	*VIEW* *Mspace*	*L,4*

The *Mspace* command switches from paper space to <u>model space inside a viewport</u>, similar to double-clicking inside a viewport. If several viewports exist in the layout when you use the command, you are switched to the last active viewport. The cursor appears only in that viewport. The current viewport displays a heavy border. You can switch to another viewport (make another current) by PICKing in it. To use *Mspace*, there must be at least one viewport in the layout, and it must be *On* (not turned *Off* by the *-Vports* command) or AutoCAD issues a message and cancels the command.

MODEL

Pull-down Menu	COMMAND (TYPE)	ALIAS (TYPE)	Short-cut	Screen (side) Menu	Tablet Menu
...	*MODEL*	...	...	...	...

The *Model* command accomplishes the same action as selecting the *Model* tab; that is, *Model* makes the *Model* tab active. If you issue this command while you are in the *Model* tab, nothing happens.

Scaling the Display of Viewport Geometry

Paper space (layout) objects such as title block and border should correspond to the paper at a 1:1 scale. A paper space layout is intended to represent the print or plot sheet. *Limits* in a <u>layout</u> are automatically set to the exact paper size, and the finished drawing is plotted from the layout to a scale of 1:1. The model space geometry, however, should be true scale in real-world units, and *Limits* in <u>model space</u> are generally set to accommodate that geometry.

When model space geometry appears in a viewport in a layout, the size of the <u>displayed</u> geometry can be controlled so that it appears and plots in the correct scale. There are three ways to scale the display of model space geometry in a paper space viewport. You can (1) use the *Viewport Scale Control* drop-down list, (2) use the *Properties* palette, or (3) use the *Zoom* command and enter an *XP* factor. Once you set the scale for the viewport, use the *Lock* option of *–Vports* or the *Properties* dialog box to lock the scale of the viewport so it is not accidentally changed when you *Zoom* or *Pan* inside the viewport.

When you use layouts, the <u>*Plot Scale* specified in the *Plot* dialog box and *Page Setup* dialog box</u> dictates the scale for the layout, which <u>should normally be set to 1:1</u>. Therefore, <u>the viewport scale actually determines the plot scale of the geometry in a finished plot</u>. Set the viewport scale to the reciprocal of the "drawing scale factor," as explained in Chapter 12, Advanced Drawing Setup.

Viewport Scale Control Drop-Down List
The *Viewport Scale Control* drop-down list is located in the *Viewports* toolbar (see Figures 13-5 and 13-10). Invoke the *Viewports* toolbar using the *Toolbars...* option at the bottom of the *View* pull-down menu.

To set the viewport scale (scale of the model geometry in the viewport), first make the viewport current by double-clicking in it or typing *Mspace*. When the crosshairs appear in the viewport and the viewport border is highlighted (a wide border appears), select the desired scale from the list.

The *Viewport Scale Control* drop-down list provides only standard scales for decimal, metric, and architectural scales. <u>You can select from the list or enter a value</u>. Values can be entered as a decimal, a fraction (proper or improper), a proportion using a colon symbol (:), or an equation using an equal symbol (=). Feet (') and inch (") unit symbols can also be entered. For example, if you wanted to scale the viewport geometry at 1/2"=1', you could enter any of the following, as well as other, values.

    1/2"=1'
    .5"=1'
    1/2=12
    1/24
    1:24
    1=24
    .0417

Remember that the viewport scale is actually the desired plot scale for the geometry in the viewport, which is the reciprocal of the "drawing scale factor" (see Chapter 12).

Properties

The *Properties* palette can also be used to specify the scale for the display of model space geometry in a viewport. Do this by first double-clicking <u>in paper space</u> or issuing the *Pspace* command, then invoking *Properties*. When (or before) the dialog box appears, select the viewport object (border). Ensure the word "Viewport" appears in the top of the box (Fig. 13-25).

Under *Standard Scale*, select the desired scale from the drop-down list. This method offers only standard scales. Alternately, you can enter any values in the *Custom Scale* box. (See *Properties*, Chapter 16.) Similar to using the *Viewport Scale Control* edit box, you can enter values as a decimal or fraction and use a colon symbol or an equal symbol. Feet and inch unit symbols can also be entered. You can also *Lock* the viewport scale using the *Display Locked* option in the palette. (See *Properties*, Chapter 16.)

Figure 13-25 ——————————

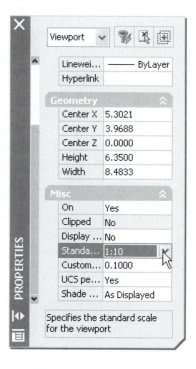

ZOOM XP Factors

To scale the display of model space geometry relative to paper space (in a viewport), you can also use the *XP* option of the *Zoom* command. *XP* means "times paper space." Thus, model space geometry is *Zoomed* to some factor "times paper space."

Since paper space is set to the actual size of the paper, <u>the *Zoom XP* factor that should be used for a viewport is equivalent to the plot scale</u> that would otherwise be used for plotting that geometry in model space. The *Zoom XP* factor is the reciprocal of the "drawing scale factor." *Zoom XP* <u>only</u> while you are "in" the desired model space viewport. Fractions or decimals are accepted.

For example, if the model space geometry would normally be plotted at 1/2"=1" or 1:2, the *Zoom* factor would be .5*XP* or 1/2*XP*. If a drawing would normally be plotted at 1/4"=1', the *Zoom* factor would be 1/48*XP*. Other examples are given in the following list.

1:5	*Zoom* .2*XP* or 1/5 *XP*
1:10	*Zoom* .1*XP* or 1/10*XP*
1:20	*Zoom* .05*XP* or 1/20*XP*
1/2"=1"	*Zoom* 1/2*XP*
3/8"=1"	*Zoom* 3/8*XP*
1/4"=1"	*Zoom* 1/4*XP*
1/8"=1"	*Zoom* 1/8*XP*
3"=1'	*Zoom* 1/4*XP*
1"=1'	*Zoom* 1/12*XP*
3/4'=1'	*Zoom* 1/16*XP*
1/2"=1'	*Zoom* 1/24*XP*
3/8"=1'	*Zoom* 1/32*XP*
1/4"=1'	*Zoom* 1/48*XP*
1/8"=1'	*Zoom* 1/96*XP*

Refer to the "Tables of Limits Settings," Chapter 14, for other plot scale factors.

Locking Viewport Geometry

Once the viewport scale has been set to yield the correct plot scale, you should lock the scale of the viewport so it is not accidentally changed when you *Zoom* or *Pan* inside the viewport. If viewport lock is on and you use *Zoom* or *Pan* when the viewport is active, AutoCAD automatically switches to paper space (*Pspace*) for zooming and panning.

You can lock the viewport scale by the following methods.

1. Use the *Lock* option of the *–Vports* command (see *–Vports*).
2. Use the *Display Locked* setting in the *Properties* palette (see Figure 13-25).
3. Use the *Display Locked* option in the right-click shortcut edit menu that appears when a viewport object is selected (see Figure 13-22).

Linetype Scale in Viewports—*PSLTSCALE*

The *LTSCALE* (linetype scale) setting controls how hidden, center, dashed, and other non-continuous lines appear in the *Model* tab. *LTSCALE* can be set to any value. The *PSLTSCALE* (paper space linetype scale) setting determines if those lines appear and plot the same in *Layout* tabs as they do in the *Model* tab. *PSLTSCALE* can be set to 1 or 0 (on or off).

If *PSLTSCALE* is set to 0, the non-continuous line scaling that appears in *Layout* tabs (and in viewports) is the same as that in the *Model* tab. In this way, the linetype spacing always looks the same relative to model space units, no matter whether you view it from the *Model* tab or from a viewport in a *Layout*. For example, if in the *Model* tab a particular center line shows one short dash and two long dashes, it will look the same when viewed in a viewport in a layout—one short dash and two long dashes. When *PSLTSCALE* is 0, linetype scaling is controlled exclusively by the *LTSCALE* setting. A *PSLTSCALE* setting of 0 is recommended for most drawings unless they contain multiple viewports or layouts.

If *PSLTSCALE* is set to 1 (the default setting for all AutoCAD template drawings), the linetype scale for non-continuous lines in viewports is automatically changed relative to the viewport scale; therefore, the scale of those lines in *Layout* tabs can appear differently than they do in the *Model* tab. A *PSLTSCALE* setting of 1 is recommended for drawings that contain multiple viewports or layouts, especially when they are at different scales. In such a situation, the line dashes would appear the same size in different viewports relative to paper space, even though the viewport scales were different. Technically speaking, if *PSLTSCALE* is set to 1, *LTSCALE* controls linetype scaling globally for the drawing, but lines that appear in viewports are automatically scaled to the *LTSCALE* times the viewport scale.

Until you gain more experience with layouts and viewports, and in particular creating multiple viewports, a *PSLTSCALE* setting of 0 is the simplest strategy. This setting also agrees with the strategy used in this text discussed earlier (in Chapter 6 and Chapter 12) for drawing setup. You can change the default *PSLTSCALE* setting of 1 to 0 by typing *PSLTSCALE* at the Command prompt.

Chapter 33, Advanced Layouts and Plotting, gives a more detailed discussion of *PSLTSCALE*, creating multiple viewports and layouts, and other drawing setup strategies.

Advanced Applications of Paper Space Viewports

In this chapter the basic applications of layouts and viewports are discussed. The basic procedures include creating a layout, setting up the layout for printing or plotting, creating viewports, and scaling the geometry in the viewports to plot to scale.

Layouts can be used for many other advanced applications such as setting up multiple viewports to display views of one drawing at different scales or for different plot devices, controlling what layers appear in each of several viewports, applying Plot Style Tables, displaying multiple drawings (*Xrefs*) in one drawing, and displaying and plotting several views of a 3D model. These topics are discussed in Chapter 33, Advanced Layouts and Plotting; Chapter 30, Xreferences; and Chapter 42, Creating 2D Drawings from 3D Models.

CHAPTER EXERCISES

1. *Pagesetup, Vports*

   A. *Open* the **A-METRIC.DWT** template drawing that you created in the Chapter 12 Exercises. Activate a *Layout* tab. If your *Options* are set as explained in Chapter 6, the layout is automatically set up for your default print/plot device. Otherwise, use the *Page Setup* dialog box (*Pagesetup* command) to set the desired print/plot device and to select an **11 x 8.5** paper size.

   B. If a viewport already exists in the layout, *Erase* it. Make **VPORTS** the *Current* layer. Then use the *Vports* command and select a *Single* viewport and accept the default (*Fit*) option to create one viewport at the maximum size for the printable area. Ensure that *PSLTSCALE* is set to **0** by typing it at the Command prompt. Finally, make the *Model* tab active. *Save* and *Close* the drawing.

2. *Pagesetup, Vports*

    A. *Open* the **BARGUIDE** drawing that you set up in the Chapter 12 Exercises. Activate a *Layout* tab. If your *Options* are set as explained in Chapter 6, the layout is automatically set up for your default print/plot device. Otherwise, use *Pagesetup* to set the desired print/plot device and to select an **11 x 8.5** paper size.

    B. If a viewport already exists in the layout, *Erase* it. Make **VPORTS** the *Current* layer. Then use the *Vports* command and select a *Single* viewport and accept the default (*Fit*) option to create one viewport at the maximum size for the printable area. Type *PSLTSCALE* and set the value to **0**. Finally, make the *Model* tab active. *Save* and *Close* the drawing.

3. **Create a Layout and Viewport for the ASHEET Template Drawing**

    A. *Open* the **ASHEET.DWT** template drawing you created in the Chapter 12 Exercises. Activate a *Layout* tab. Unless already set up, use *Pagesetup* to set the desired print/plot device and to select an **11 x 8.5** paper size.

    B. If a viewport already exists in the layout, *Erase* it. Make **VPORTS** the *Current* layer. Then use the *Vports* command and select a *Single* viewport and accept the default (*Fit*) option to create one viewport at the maximum size for the printable area. Ensure that *PSLTSCALE* is set to **0**. Finally, make the *Model* tab active. *Save* and *Close* the ASHEET template drawing.

4. **Create a New BSHEET Template Drawing**
Complete this exercise if your system is configured with a "B" size printer or plotter.

    A. Using the template drawing in the previous exercise, create a template for a standard engineering "B" size sheet. First, use the *New* command and select the *Template* option from the *Create New Drawing* dialog box. Select the *Browse* button. When the *Select a Template File* dialog box appears, select the **ASHEET.DWT**. When the drawing opens, set the model space *Limits* to **17 x 11** and *Zoom All*.

    B. Activate the *Layout 1* tab. *Erase* the existing viewport. Activate the *Page Setup* dialog box. In the **Plot Device** tab, select a print or plot device that can use a B size sheet (17 x 11). In the **Layout Settings** tab, select *ANSI B (11 x 17 Inches)* in the *Paper Size* drop-down list. Select *OK* to dismiss the *Page Setup* dialog box.

    C. Make **VPORTS** the *Current* layer. Then use the *Vports* command and select a *Single* viewport and accept the default (*Fit*) option to create one viewport at the maximum size for the printable area. Finally, make the *Model* tab active. All other settings and layers are okay as they are. Use *Saveas* and save the new drawing as a template (.**DWT**) drawing in your working directory. Assign the name **BSHEET**.

5. **Create Multiple Layouts for Plotting on C and D Size Sheets**
   Complete this exercise if your system is configured with a "C" and "D" sized plotter.

   A. Using the ASHEET.DWT template drawing from an earlier exercise, create a template for standard engineering "C" and "D" size sheets. First, use the *New* command and select the *Template* option from the *Create New Drawing* dialog box. Select the *Browse* button. When the *Select a Template File* dialog box appears, select the **ASHEET.DWT**. When the drawing opens, set the model space *Limits* to **34** x **22** and *Zoom All*.

   B. Activate the *Layout1* tab. *Erase* the existing viewport. Activate the *Page Setup* dialog box. In the *Plot Device* tab, select a plotter that can use a C size sheet (22 x 17). In the *Layout Settings* tab, select *ANSI C (22 x 17 Inches)* in the *Paper Size* drop-down list. Select *OK* to dismiss the *Page Setup* dialog box.

   C. Make **VPORTS** the *Current* layer. Then use the *Vports* command and select a *Single* viewport and use the *Fit* option to create one viewport at the maximum size for the printable area. Right-click on the *Layout 1* tab and select *Rename* from the shortcut menu. Rename the layout to "**C Sheet**."

   D. Activate the *Layout2* tab. *Erase* the existing viewport if one exists. Activate the *Page Setup* dialog box. In the *Plot Device* tab, select a print or plot device that can use a D size sheet (34 x 22). (This can be the same device you selected in step B, as long as it supports a D size sheet.) In the *Layout Settings* tab, select *ANSI D (34 x 22 Inches)* in the *Paper Size* drop-down list. Select *OK* to dismiss the *Page Setup* dialog box.

   E. Make **VPORTS** the *Current* layer. Then use the *Vports* command and select *Single* to create one viewport. *Rename* the layout "**D Sheet**."

   F. Finally, make the *Model* tab active. All other settings and layers are okay as they are. Use *Saveas* and save the new drawing as a template (.**DWT**) drawing in your working directory. Assign the name **C-D-SHEET**.

6. *Layout Wizard*
   In this exercise, you will open an existing drawing and use the *Layout Wizard* to set up a layout for plotting.

   A. *Open* the **DB_SAMP** drawing that is located in the AutoCAD 2004\Sample folder. Use the *Saveas* command and assign the name **VP_SAMP** and specify your working folder as the location to save.

   B. Create a new layer named **VPORTS** and make it the *Current* layer. Invoke the *Layout Wizard*. In the *Begin* step, enter a name for the new layout, such as **Layout 3**. Select *Next*.

   C. In the *Printer* step, the list of available printers in your lab or office appears. The default printer should be highlighted. Choose any device that you know will operate for your lab or office and select *Next*. If you are unsure of which devices are usable, keep the default selection and select *Next*.

   D. For *Paper Size*, select the largest paper size supported by your device. Ensure *Inches* is selected unless you are using a standard metric sheet. Select *Next*.

   E. Select *Landscape* for the *Orientation*.

F. Select a *Title Block* that matches the paper size you specified. If you are unsure of which to choose, refer to the Layout Templates table in the chapter. Insert the title block as a **Block**. Select **Next**.

G. In the *Define Viewports* step, select a **Single** viewport. Under *Viewport Scale*, select **Scaled to Fit**. Proceed to the next step.

H. Bypass the next step in the wizard, *Pick Location*, by selecting **Next**. Also, when the *Finish* step appears, select **Next**. This action causes AutoCAD to create a viewport that fits the extents of the printable area. The resulting layout should look similar to that shown in Figure 13-26. As you can see, the model space geometry extends beyond the title block. Your layout may appear slightly different depending on the sheet size and title block you selected. **Save** the drawing.

Figure 13-26

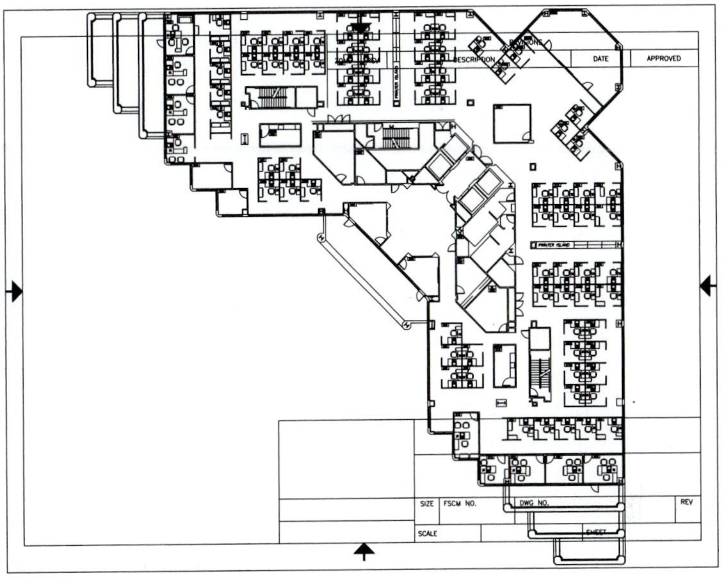

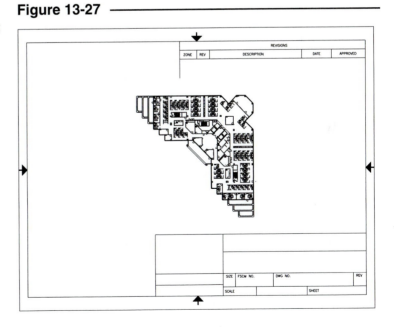

I. Use the *Mspace* command or double-click inside the viewport to activate model space. To specify a standard scale for the geometry in the viewport, invoke the **Viewports** toolbar. In the **Viewport Scale Control** drop-down list, select a scale (such as **1/64"=1'**) until the model geometry appears completely within the title block border (Fig. 13-27).

Figure 13-27

J. *Freeze* layer **VPORTS** so the viewport border does not appear. **Save** the drawing. With the layout active, access the **Plot** dialog box and ensure the *Plot Scale* is set to **1:1**. Select **Plot**.

7. *Page Setup,* **Vports,** and *Zoom XP*

In this exercise, you will use an existing metric drawing, access *Layout1*, draw a border and title block, create one viewport, and use *Zoom XP* to scale the model geometry to paper space. Finally, you will *Plot* the drawing to scale on an "A" size sheet.

A. *Open* the **CBRACKET** drawing you worked on in Chapter 9 Exercises. If you have already created a title block and border, *Erase* them.

B. Invoke the *Options* dialog box and access the *Display* tab. In the *Layout Elements* section (lower left), ensure that *Show Page Setup Dialog for New Layouts* is checked and *Create Viewport in New Layouts* is not checked.

C. Select the *Layout1* tab. The *Page Setup* dialog box should appear. In the *Plot Device* tab, select a plot device from the list that will allow you to plot on an A size sheet (11" x 8.5"). In the *Layout Settings* tab, select the correct sheet size and select *Landscape* orientation. Pick *OK* to finish the setup.

D. Set layer **TITLE** *Current*. Draw a title block and border (in paper space). HINT: Use *Pline* with a *width* of .02, and draw the border with a .5 unit margin within the edge of the *Limits* (border size of 10 x 7.5). Provide spaces in the title block for your school or company name, your name, part name, date, and scale as shown in the following figure. Use *Saveas* and assign the name **CBRACKET-PS**.

E. Create layer **VPORTS** and make it *Current*. Use the *Vports* command to create a *Single* viewport. Pick diagonal corners for the viewport at **.5,1.5** and **9.5,7**. The CBRACKET drawing should appear in the viewport at no particular scale, similar to Figure 13-28.

Figure 13-28

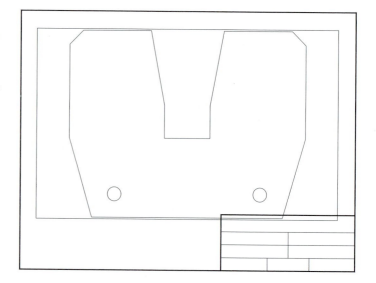

F. Next, use the *Mspace* command, the **MODEL/PAPER** toggle on the Status Bar, or double-click inside the viewport to bring the cursor into model space. Use *Zoom* with a **.03937XP** factor to scale the model space geometry to paper space. (The conversion from inches to millimeters is 25.4 and from millimeters to inches is .03937. Model space units are millimeters and paper space units are inches; therefore, enter a *Zoom XP* factor of .03937.)

G. Finally, activate paper space by using the *Pspace* command, the *MODEL/PAPER* toggle, or double-clicking outside the viewport. Use the *Layer Properties Manager* to make layer **VPORTS** *non-plottable*. Your completed drawing should look like that in Figure 13-29. *Save* the drawing. *Plot* the drawing from the layout at **1:1** scale.

Figure 13-29

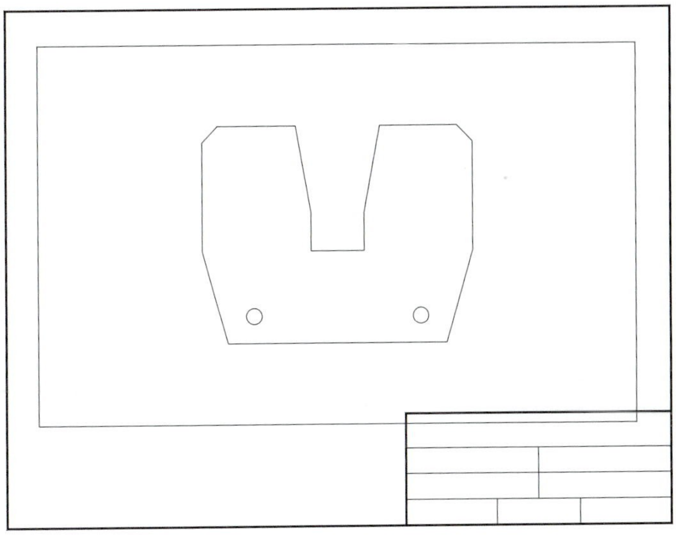

8. **Using a Layout Template**

This exercise gives you experience using a layout template already set up with a title block and border. You will draw the hammer (see Figure 13-30) and use a layout template to plot on a B size sheet. It is necessary to have a plot or print device configured for your system that can use a "B" size sheet.

A. Start AutoCAD. In the **Options** dialog box, select the **System** tab. In the *General Options* section, select **Show Startup Dialog Box**.

B. Now access the **Display** tab in the *Options* dialog box. In the *Layout Elements* section (lower left), remove the check for both *Show Page Setup Dialog for New Layouts* and *Create Viewport in New Layouts*. Select **OK**.

C. Begin a *New* drawing. In the *Create New Drawing* dialog box that appears, select **Use a Wizard** and select the **Advanced Setup Wizard**. In the first step, *Units*, select **Decimal** units and a **Precision** of **0.00**. Accept the defaults for the next three steps. In the *Area* step, enter a **Width** of **17** and a **Length** of **11**.

D. Right-click on an existing layout tab and use the **From Template** option from the shortcut menu. Select the **ANSI B -Color Dependent Plot Styles** template drawing (.DWT) in the *Select Template From File* dialog box, then select the **ANSI B Title Block** in the *Insert Layout(s)* dialog box. Access the new tab. Note that the template loads a layout with a title block and border in paper space and has one viewport created. Access the **Page Setup** dialog box and select the matching plot device and sheet size for the selected ANSI B title block.

E. Pick the **Model** tab. Set up appropriate *Snap* and *Grid* (for the *Limits* of 17 x 11) if desired. Also set *LTSCALE* and any running *Osnaps* you want to use. Type **PSLTSCALE** and set the value to **0**.

F. Create the following layers and assign linetypes and line weights. Assign colors of your choice.

       GEOMETRY   *Continuous*
       CENTER   *Center2*

Notice the *Title Block* and *Viewport* layers already exist as part of the template drawing.

G. Draw the hammer according to the dimensions given in Figure 13-30. *Save* the drawing as **HAMMER**.

Figure 13-30

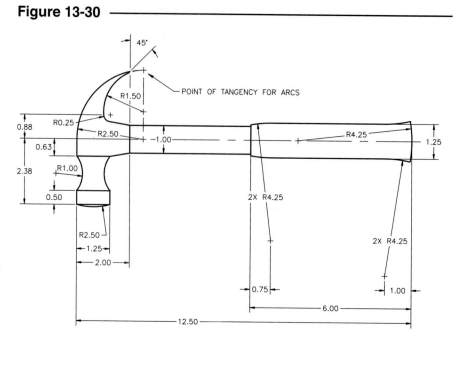

H. When you are finished with the drawing, select the *ANSI B Title Block* tab. Now, scale the hammer to plot to a standard scale. You can use either the *Viewports* toolbar or the *Properties* window to set the viewport scale to **1:1**. Alternately, use *Zoom 1XP* to correctly size model space units to paper space units.

I. Your completed drawing should look like that in Figure 13-31. Use the *Layer Properties Manager* to make layer **Viewports** *non-plottable*. *Save* the drawing. Make a *Plot* from the layout at a scale of **1:1**.

Figure 13-31

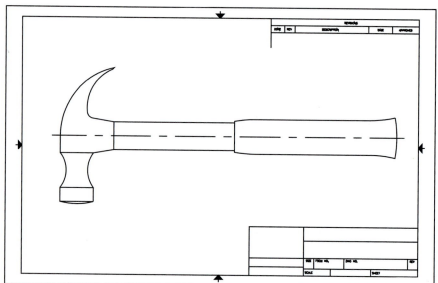

9. **Using a Template Drawing**

This exercise gives you experience using a template drawing that includes a title block and border. You will construct the wedge block in Figure 13-32 and plot on a B size sheet. It is necessary to have a plot or print device configured for your system that can use a "B" size sheet.

A. Complete steps 8.A. and 8.B. from the previous exercise.

B. Begin a *New* drawing. In the *Create New Drawing* dialog box that appears, select *Use a Template*. Select *ANSI B -Color Dependent Plot Styles.Dwt.*

C. Select the *ANSI B Title Block* tab. Then invoke the *Page Setup* dialog box and select a matching plot or print device and sheet size for the ANSI B title block.

D. Pick the *Model* tab. Set up model space with *Limits* of **17** x **11**. Set up appropriate *Snap* and *Grid* if desired. Also set *LTSCALE* and any running *Osnaps* you want to use. Type *PSLTSCALE* and set the value to **0**.

E. Create the following layers and assign linetypes and line weights. Assign colors of your choice.

GEOMETRY *Continuous*
CENTER *Center2*
HIDDEN *Hidden2*

Notice the *Title Block* and *Viewport* layers already exist as part of the template drawing.

F. Draw the wedge block according to the dimensions given (Fig. 13-32). *Save* the drawing as **WEDGEBK**.

G. When you are finished with the views, select the *ANSI B Title Block* tab. To scale the model space geometry to paper space, you can use either the *Viewports* toolbar or the *Properties* window to set the viewport scale to **1:2.** You may also need to use *Pan* to center the geometry.

H. Next, use the *Layer Properties Manager* to make layer **Viewports** *non-plottable*. Your completed drawing should look like that in Figure 13-33. *Save* the drawing. From paper space, plot the drawing on a B size sheet and enter a scale of **1:1**

Figure 13-32

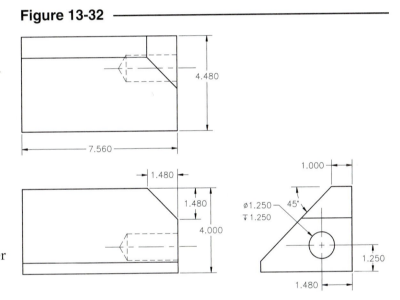

Figure 13-33

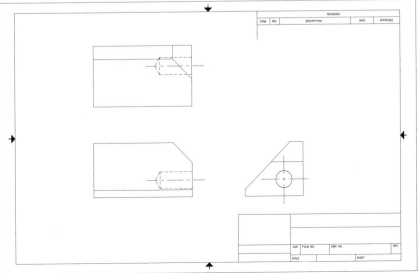

10. **Multiple Layouts**

Assume a client requests that a copy of the HAMMER drawing be faxed to him immediately; however, your fax machine cannot accommodate the B size sheet that contains the plotted drawing. In this exercise you will create a second layout using a layout template to plot the same model geometry on an A size sheet.

A. Open the drawing you created in a previous exercise named **HAMMER**. Access the *Display* tab in the *Options* dialog box. In the *Layout Elements* section (lower left), <u>remove the check for both</u> *Show Page Setup Dialog Box for New Layouts* and *Create Viewport in New Layouts*. Select **OK**.

B. Right-click on an existing layout tab and use the *From Template* option from the shortcut menu. Select the *ANSI A -Color Dependent Plot Styles* template drawing (.DWT) to import. When the layout is imported, click the new *ANSI A Title Block* tab. Note that the template loads a layout with a title block and border in paper space and has one viewport created. Access the *Page Setup* dialog box and select the matching print or plot device and sheet size for the selected ANSI A title block.

C. Double-click inside the viewport or use *Mspace* or the *MODEL/PAPER* toggle. Now scale the hammer to plot to a standard scale by using *Zoom 3/4XP* or enter .75 in the appropriate edit box in the *Viewports* toolbar or *Properties* window to correctly scale the model space geometry. Use *Pan* in the viewport to center the geometry within the viewport.

D. Your new layout should look like that in Figure 13-34. Access the *ANSI B Title Block* tab you created earlier. Note that you now have two layouts, each layout has settings saved to plot the same geometry using different plot devices and at different scales.

E. Access the new layout tab. *Save* the drawing. Make a *Plot* from the layout at a scale of **1:1** and send the fax.

Figure 13-34

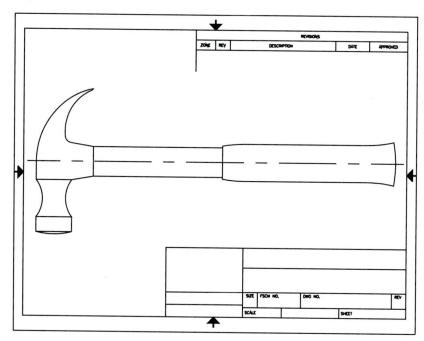

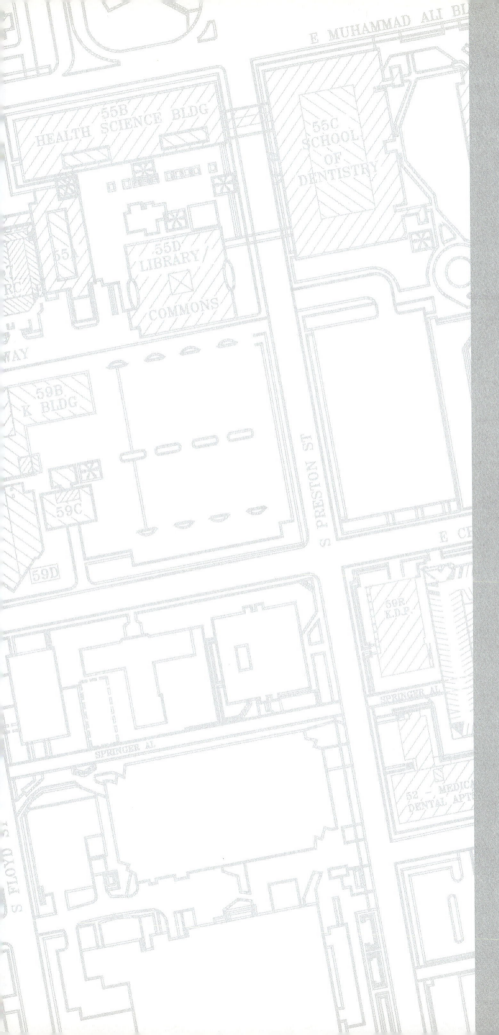

14

PRINTING AND PLOTTING

Chapter Objectives

After completing this chapter you should:

1. know the typical steps for printing or plotting;

2. be able to invoke and use the *Plot* and *Page Setup* dialog boxes;

3. be able to select from available plotting devices and set the paper size and orientation;

4. be able to specify what area of the drawing you want to plot;

5. be able to preview the plot before creating a plotted drawing;

6. be able to specify a scale for plotting a drawing;

7. know how to set up a drawing for plotting to a standard scale on a standard size sheet;

8. be able to use the Tables of *Limits* Settings to determine *Limits*, scale, and paper size settings;

9. know how to configure plot and print devices in AutoCAD.

CONCEPTS

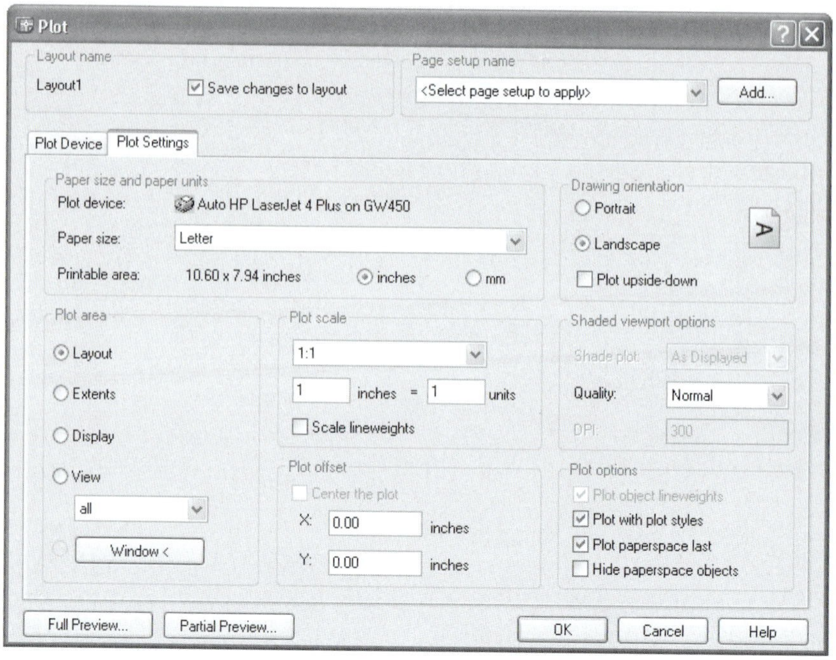

There are several concepts, procedures, and tools in AutoCAD related to printing and plotting. Many of these topics have already been discussed, such as setting up a drawing to draw true size, creating layouts and viewports, specifying the viewport scale, and using *Page Setup* to specify plot or print options. This chapter explains the typical steps to plotting from either the *Model* tab or *Layout* tabs, using both the *Page Setup* and *Plot* dialog boxes, plotting to scale, and configuring print and plot devices and options. Other advanced concepts of plotting, such as using plot styles and plot style tables, plot stamping, and advanced features of layouts, are explained in Chapter 33, Advanced Layouts and Plotting.

Figure 14-1 ——————————————

In AutoCAD, the term "plotting" can refer to plotting on a plot device (such as a pen plotter or electrostatic plotter) or printing with a printer (such as a laser jet printer). The *Plot* command is used to create a plot or print by producing the *Plot* dialog box. However, creating the plot or print can be accomplished using either the *Plot* dialog box or the *Page Setup* dialog box.

The *Plot* dialog box and the *Page Setup* dialog box have almost the same features and functions. The basic difference is that the *Page Setup* dialog box is used to specify and save plot and print specifications for each *Layout* or *Model* tab, whereas the *Plot* dialog box does not save the settings unless you plot, but has a few more functions including a plot preview. This chapter explains all features and distinctions of both dialog boxes.

The *Plot* dialog box, similar to the *Page Setup* dialog box, has two tabs, the *Plot Settings* tab (Fig. 14-1) and the *Plot Device* tab (see Figure 14-2). Use the *Plot Settings* tab to specify parameters of how the drawing will appear on the plot, such as orientation on the paper and paper size, what area of the drawing to plot, plot scale, and other options. Use the *Plot Device* tab to specify what type of printer or plotter you want to use, what plot style tables are attached, and other options.

TYPICAL STEPS TO PLOTTING

Assuming your CAD system and plotting devices have been properly configured, the typical basic steps to plotting the *Model* tab or a *Layout* tab are listed below.

1. Use *Save* to ensure the drawing has been saved in its most recent form before plotting (just in case some problem arises while plotting).

2. Make sure the plotter or printer is turned on, has paper (and pens for some devices) loaded, and is ready to accept the plot information from the computer.

3. Invoke the *Plot* dialog box.

4. Use the *Plot Device* tab to select the intended plot device from the drop-down list. The list includes all devices currently configured for your system.

5. In the *Plot Settings* tab, ensure the desired *Drawing Orientation* and *Paper Size* are selected.

6. Determine and select the desired *Plot Area* for the drawing: *Layout, Limits, Extents, Display, Window,* or *View.*

7. Select the desired scale from the *Scale* drop-down list or enter a *Custom* scale. If no standard scale is needed, select *Scaled to Fit.* If you are plotting from a *Layout,* normally set the scale to 1:1.

8. If necessary, specify a *Plot Offset* or *Center the Plot* on the sheet.

9. If necessary, specify the *Plot Options,* such as plotting with lineweights or plot styles.

10. Always preview the plot to ensure the drawing will plot as you expect. Select a *Full Preview* to view the drawing objects as they will plot. If the preview does not display the plot as you intend, make the appropriate changes. Otherwise, needless time and media could be wasted.

11. If all settings are acceptable, selecting the *OK* tile causes the drawing to be sent to the plotter or printer. All settings are saved in the drawing file when you plot, but are not saved if you *Cancel.*

Alternately, if you want to set up all the parameters for plotting and save them, but do not actually want to make a plot yet, you can use the *Page Setup* dialog box to do so. The *Page Setup* dialog box is almost the same as the *Plot* dialog box and can be used to set up and save plotting parameters.

USING THE *PLOT* AND *PAGE SETUP* DIALOG BOXES

PLOT

	Pull-down Menu	COMMAND (TYPE)	ALIAS (TYPE)	Short-cut	Screen (side) Menu	Tablet Menu
	File *Plot...*	PLOT or -PLOT	PRINT	Ctrl+P	FILE *Plot*	W,25

Using *Plot* invokes the *Plot* dialog box. The *Plot* dialog box has two tabs, the *Plot Settings* tab (see Figures 14-1 and 14-3) and the *Plot Device* tab (Figure 14-2).

The *Plot* dialog box allows you to set plotting parameters such as scale, orientation, offset, plot style, paper size, and plotting device. Settings made to plotting options can be saved for the *Model* tab and each *Layout* tab in the drawing so they do not have to be re-entered the next time, but only if you create the plot (see *"Save Changes to Layout"*).

Layout Name
At the top-left corner of the *Plot* dialog box, the *Layout Name* corresponds to the *Model* or *Layout* tab that is currently active (when the *Plot* command is used).

Save Changes to Layout
Check this box if you want settings you make to be saved with the drawing for the selected *Model* tab or *Layout* tab. You can save the settings for each layout. This is a great advantage since you can set up

several tabs, each showing a different aspect of the drawing and specifying different plot options such as scales, plot devices, layer combinations, plot styles, and so on. Using this feature allows you to load the drawing at a later time and reproduce plots with all plot specifications pre-set.

Page Setup Name

Typically page setups (plot specifications made in this dialog box) are automatically saved for each layout tab (including the *Model* tab) using the name of the tab as the *Page Setup Name*. For some situations, you may want to apply a previously saved *Page Setup* to a new layout by selecting a name from the list.

Plot Device **Tab**

Plotter Configuration

Many devices (printers and plotters) can be configured for use with AutoCAD, and all configurations are saved for your selection. For example, you can have both an "A" size and a "D" size plotter as well as a laser printer; any one or more could be used to plot the current drawing.

When you install AutoCAD, it automatically configures itself to use the Windows system printer. Additional devices for use specifically with AutoCAD can be configured within AutoCAD by using the *Plotter Manager*. The *Plotter Manager* can be accessed several ways. See "Configuring Plotters and Printers" near the end of this chapter.

Figure 14-2

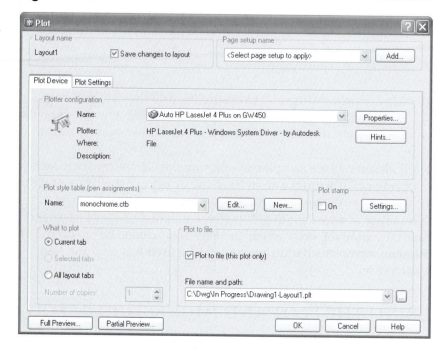

Name:
Select the desired device from the list. The list is composed of all previously configured devices for your system.

Properties…
This button invokes the *Plotter Configuration Editor*. In the editor, you can specify a variety of characteristics for the specific plot device shown in the *Name*: drop-down list. The *Plotter Configuration Editor* is also accessible through the *Plotter Manager* (see "*Plotter Configuration Editor*" near the end of this chapter).

Hints…
Use this button to invoke help on the device configuration.

Plot Style Table (pen assignments)

If desired, select a plot style table to use with your plot device. A plot style table can contain several plot styles. Each plot style can assign specific line characteristics (lineweight, screening, dithering, end and joint types, fill patterns, etc.). Normally, a plot style table can be assigned to the *Model* tab, or to *Layout* tabs. Plot styles are explained in Chapter 33.

Plot Stamp

You have the capability of adding an informational text "stamp" on each plot. For example, the drawing name, date, and time of plot can be added to each drawing that you plot for tracking and verification purposes. To specify the plot stamp, you can select the *Settings* button or use the *Plotstamp* command. Select the *On* button to apply the stamp you specified to the plotted drawing. For the full explanation of this feature, see *"Plotstamp"* in Chapter 33.

What to Plot

In this section you can specify which of the *Layout* or *Model* tabs you want to plot. You can plot the *Current tab, Selected tabs* or *All Layout tabs*. Multiple tabs can be selected with your pointer by holding down the Shift key, selecting the tabs, then using *Plot*. Multiple copies can also be generated.

Plot to File

Choosing this option writes the plot to a file instead of making a plot. This action generally creates a .PLT file type. The format of the file (for example, PCL or HP/GL language) depends on the device brand and model that is configured. The plot file can be printed or plotted later <u>without</u> AutoCAD, assuming the correct interpreter for the device is available. Specify the *File Name and Path* or use the browse button (…) to specify the name and location for the plot file.

Plot Settings Tab

Paper Size and Paper Units

Depending in the *Plot Device* selected, several paper sizes may be available. Select the desired *Paper Size* from the drop-down list, then select which units you want to use.

Inches or *mm* should be selected to correspond to the units used in the drawing. The selected units affect the *Scale* and *Plot Offset* units. For example, assuming the drawing was created using the actual units, a scale for plotting can be calculated without millimeter to inch conversion or vice versa.

Figure 14-3

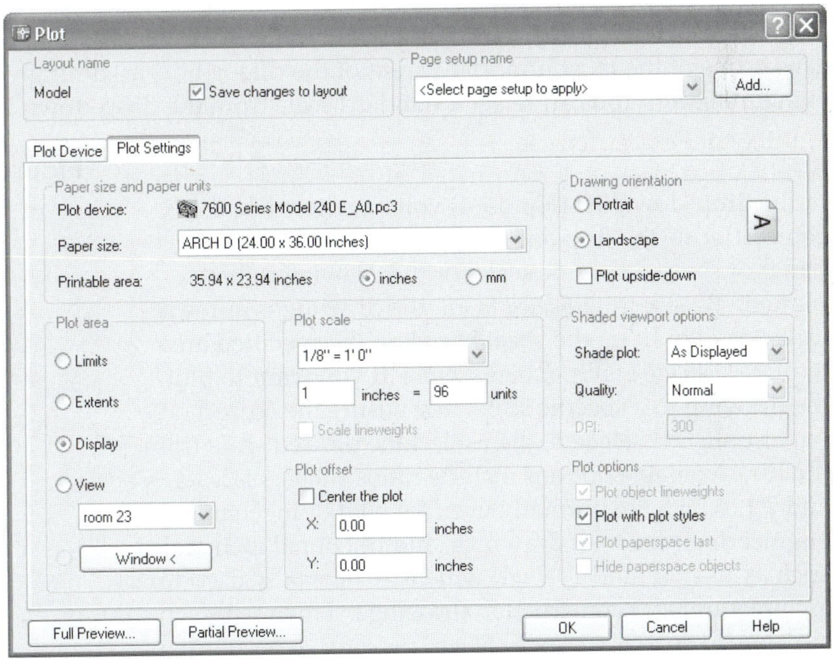

Drawing Orientation

Normally, *Landscape* is selected to match the orientation of the AutoCAD drawing area—that is, horizontal lines in the drawing plot lengthwise on the paper. *Portrait* positions horizontal lines in the drawing along the short axis of the paper.

Plot Area

Specify what part of the drawing you want to plot by selecting the desired button from the lower-left corner of the dialog box (see Figure 14-3). The choices are listed next.

Limits or *Layout*

This button name changes depending on whether you invoke the *Plot* or *Page Setup* dialog box while the *Model* tab or a *Layout* tab is current. If the *Model* tab is current, this selection plots the area defined by the *Limits* command (unless plotting a 3D object from other than the plan view). If a *Layout* tab is current, the layout is plotted as defined by the selected paper size.

Extents

Plotting *Extents* is similar to *Zoom Extents*. This option plots the entire drawing (all objects), disregarding the *Limits*. Use the *Zoom Extents* command in the drawing to make sure the extents have been updated before using this plot option.

Display

This option plots the current display on the screen. If using viewports, it plots the current viewport.

View

With this option you can plot a view previously saved using the *View* command. Creating *Views* of specific areas of a drawing is very helpful when you plan to make repetitive plots of the same area or plots of multiple areas of one drawing.

Window

Window allows you to plot any portion of the drawing. You must specify the window by picking or supplying coordinates for the lower-left and upper-right corners.

Plot Scale

This is one of the most important areas of the dialog box, assuming you want to plot the drawing to a standard scale. You can select a standard scale from the drop-down list or enter a custom scale.

Scale

In the drop-down list (Fig. 14-4) you have two basic choices: scale the drawing to automatically fit on the paper or select a specific scale for the drawing to be plotted. By selecting *Scaled to Fit*, the drawing is automatically sized to fit on the sheet based on the specified area to plot (*Display, Limits, Extents*, etc.). If you want to plot the drawing to a specific scale, first ensure the correct paper units are selected, then select the desired scale from the list. Available options include ratios such as 1:2, 1:4, and 2:1, which are normally used for civil or mechanical engineering (decimal) drawings, and fractional inch scales such as 1/4"=1' and 1/8"=1' which are normally used for architectural and construction drawings. Fractional inch scales such as 3/8"=1" or 5/8"=1" used for older mechanical drawings (pre-ANSI 1994 standards) must be specified in the *Custom* edit boxes.

Figure 14-4 ⎯⎯⎯⎯⎯⎯

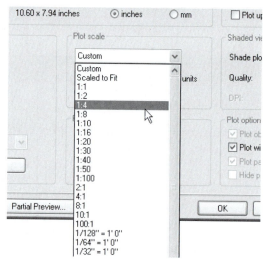

 If you are plotting the *Model* tab, select an option directly from the *Scale* drop-down list to specify the desired plot scale for the drawing. If you are plotting from a *Layout* tab, first set the scale for the viewport using the *Viewport* toolbar, *Properties* palette, or using *Zoom* with an *XP* value, then select *1:1* from the *Scale* drop-down list in the *Plot* or *Page Setup* dialog box. In either case, the plot scale for the drawing is equal to the reciprocal of the drawing scale factor and can be determined by referencing the Tables of *Limits* Settings. See "Plotting to Scale."

Custom
First select *Custom* from the *Scale* drop-down list, then enter the desired ratio in the unit boxes. Decimals or fractions can be entered. For example, to prepare for a plot of 3/8 size (3/8"=1"), the following ratios can be entered to achieve a plot at the desired scale: 3=8, 3/8=1, or .375=1 (inches=drawing units). For guidelines on plotting a drawing to scale, see "Plotting to Scale" in this chapter.

Scale Lineweights
This option scales the lineweights for layouts only. When checked, lineweights are scaled proportionally with the plot scale, so a 1mm lineweight, for example, would plot at .5mm when the drawing is plotted at 1:2 scale. Normally (no check) lineweights are plotted at their absolute value (0.010", 0.016", 0.25 mm, etc.) at any scale the drawing is plotted.

Plot Offset
Plotting and printing devices cannot plot to the edge of the paper, as noted by the difference in the *Paper Size* and *Printable Area* given in the *Paper Size and Paper Units* section of the dialog box. Therefore, the lower-left corner of your drawing will not plot at exactly the lower-left corner of the paper. You can choose to reposition, or offset, the drawing on the paper by entering positive or negative values in the X and Y edit boxes. Home position for plotters is the <u>lower-left</u> corner (landscape orientation) and for printers it is the <u>upper-left</u> corner of the paper (portrait orientation). If plotting the *Model* tab, you can choose to *Center the Plot* on the paper.

Shaded Viewport Options
This section of the *Plot* dialog box is new for AutoCAD 2004. These options are <u>useful only if you are printing from the *Model* tab</u> and you are printing a 3D surface or solid model. These options <u>do not affect the display of 3D objects in viewports</u> of a *Layout* tab. You should have a laserjet, inkjet, or electro-static type printer configured as the plot device (one that can print solid areas) since a pen plotter creates only line plots.

Shade Plot
Use these options if you printing from the *Model* tab. You can select from *As Displayed*, *Wireframe*, *Hidden*, or *Rendered*. Or, you can shade the 3D object on the screen using one of the *Shademodes*, then select *As Displayed* in the *Shade Plot* section to make a shaded print, or you can select *Wireframe*, *Hidden*, or *Rendered* from the drop-down list for printing the drawing no matter which display is visible on screen.

Quality, DPI
You can also select from *Draft* (wireframe), *Preview* (shaded at 150dpi), *Normal* (shaded at 300dpi), *Presentation* (shaded at 600dpi), or *Maximum* (for your printer) to specify the resolution quality of the print. Select *Custom* if you want to supply a different value in the *DPI* (dots per inch) edit box.

The label for this section is misleading since it has nothing to do with printing viewports. If you are printing from a *Layout* tab and want to print shaded images of 3D objects in viewports, you must control *Shade Plot* for each viewport. Use the *Shade Plot* option of the *-Vports* command or select the viewport and use the *Shade Plot* options in the *Properties* palette. See "Printing Shaded 3D Models" in Chapter 35 for more information concerning printing shaded and rendered drawings from the *Model* tab and from a *Layout* tab.

Plot Options
These checkboxes control the following options.

Plot Object Lineweights
Lineweights on a plot can be assigned by one of two ways: use lineweights in your drawing and plot them as they appear on the screen, or use plot styles to override lineweights assigned in the drawing. If you have assigned lineweights in your drawing, checking this option plots the lineweights as they appear in the drawing. Checking *Plot With Plot Styles* disables this option.

Plot with Plot Styles

You must have a plot style table attached (see *Plot Device* Tab, *Plot Style Table*) for this option to have effect. If you are using lineweights in the drawing and also have lineweights assigned in the attached plot style table, the plot style table lineweights have precedence for plots.

Plot Paperspace Last

Checking this option plots paper space geometry last if you are plotting a layout. Normally, paper space geometry is plotted before model space geometry.

Hide Paperspace Objects

This option is useful <u>only if you have 3D surface or solid objects in paperspace</u> and you want to remove hidden lines (edges obscured from view). If you are printing from the *Model* tab and want to remove hidden lines from 3D objects, use the *Shade Plot* options (see "Shaded Viewport Options"). If you are printing from a *Layout* tab, use the *Shade Plot* option of the -*Vports* command or select the viewport and use the *Shade Plot* options in the *Properties* palette.

Full Preview

Selecting the *Full Preview* button displays the complete drawing as it will plot on the sheet (Fig. 14-5). This function is particularly helpful to ensure the drawing is plotted as you expect. When you invoke a *Full Preview*, the drawing editor displays a simulated sheet of paper with the completed print or plot. This is helpful particularly when plot style tables are attached that create plots that appear differently than the *Model* tab or *Layout* tabs display. The *Zoom* feature is on by default so you can check specific areas of the drawing before making the plot. Right-clicking produces the shortcut, allowing you to use *Pan* and other *Zoom* options. Changing the display during a *Full Preview* <u>does not change</u> the area of the drawing to be plotted or printed.

Figure 14-5

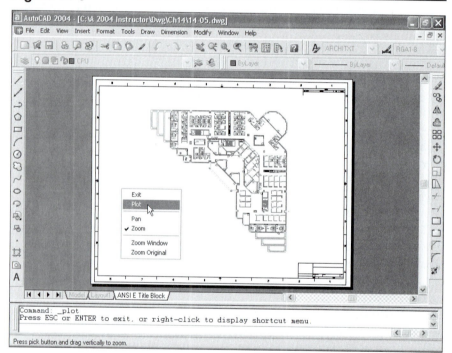

Partial Preview

Selecting the *Partial Preview* tile displays the effective plotting area as shown in Figure 14-6. Use the *Partial Preview* option to get a quick check showing how the drawing will fit on the sheet. The dotted border indicates the Printable Area and the blue filled Effective Area represents the size of the drawing as plotted on the sheet. The Effective Area is based on the selection made in the *Plot Area* section of the *Plot* dialog box (*Display, Extents, Limits,* etc.) and the selected *Plot Scale*. The red triangle designates the home position (location for the drawing's 0,0 point) for the plotting device.

Figure 14-6

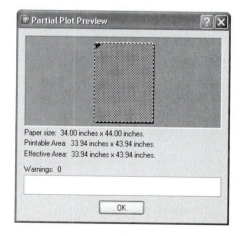

-PLOT

Entering –*Plot* at the Command line invokes the Command line version of *Plot*. This is an alternate method for creating a plot if you want to use the current settings or if you enter a page setup name to define the plot settings.

```
Command: -plot
Detailed plot configuration? [Yes/No] <No>:
Enter a layout name or [?] <ANSI E Title Block>:
Enter a page setup name <>:
Enter an output device name or [?] <HP 7586B.pc3>:
Write the plot to a file [Yes/No] <N>:
Save changes to layout [Yes/No]? <N>
Proceed with plot [Yes/No] <Y>:
Effective plotting area:  33.54 wide by 42.60 high
Plotting viewport 2.
Command:
```

PAGESETUP

Pull-down Menu	COMMAND (TYPE)	ALIAS (TYPE)	Short-cut	Screen (side) Menu	Tablet Menu
File *Page Setup...*	PAGESETUP	...	...	...	V,25

Produce the *Page Setup* dialog box by any method shown in the Command table above or right-click on a *Layout* or *Model* tab and select *Page Setup*. To cause the *Page Setup* dialog box to appear automatically when you create a new layout, use the *Display* tab of the *Options* dialog box. However, you can toggle this feature off by removing the check at the lower-left corner of the dialog box (Fig. 14-7).

The *Page Setup* dialog box is similar and has the same functions as the *Plot* dialog box with a few exceptions. The name of the second tab in this dialog box is *Layout Settings* instead of *Plot Settings*, although options in the tab are almost identical to the *Plot Settings* tab of the *Plot* dialog box.

Figure 14-7

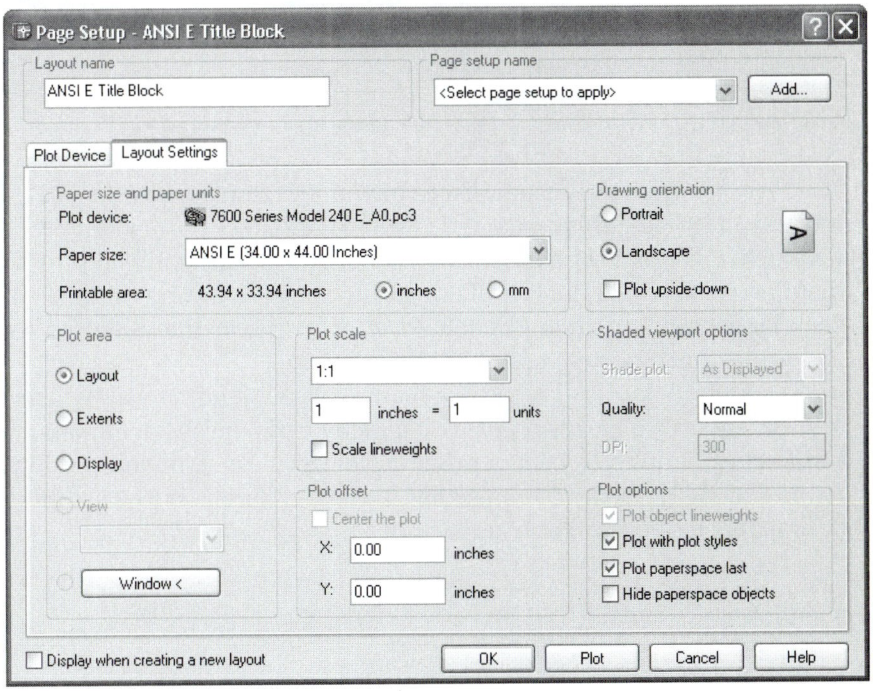

Page Setup - ANSI E Title Block

Layout name
ANSI E Title Block

Page setup name
<Select page setup to apply>   Add...

Plot Device | Layout Settings

Paper size and paper units
Plot device: 7600 Series Model 240 E_A0.pc3
Paper size: ANSI E (34.00 x 44.00 Inches)
Printable area: 43.94 x 33.94 inches   ⦿ inches   ○ mm

Drawing orientation
○ Portrait
⦿ Landscape
☐ Plot upside-down

Plot area
⦿ Layout
○ Extents
○ Display
○ View
Window <

Plot scale
1:1
1   inches = 1   units
☐ Scale lineweights

Plot offset
☐ Center the plot
X: 0.00   inches
Y: 0.00   inches

Shaded viewport options
Shade plot: As Displayed
Quality: Normal
DPI: 300

Plot options
☑ Plot object lineweights
☑ Plot with plot styles
☑ Plot paperspace last
☐ Hide paperspace objects

☐ Display when creating a new layout   OK   Plot   Cancel   Help

Should You Use the *Plot* Dialog Box or *Page Setup* Dialog Box?

You will notice that the *Plot* dialog box and *Page Setup* dialog box are almost identical (see previous Figure 14-1 and Figure 14-7). Both dialog boxes have essentially the same two tabs and features, and you can create a <u>plot from either</u> dialog box. There are four main differences described below.

1. Settings made in the *Page Setup* dialog box can be <u>saved for the *Layout* or *Model* tab without making a plot</u>, whereas settings made in the *Plot* dialog box <u>cannot be saved unless you plot</u>. You can exit the *Plot* dialog box only by *OK*, which makes a plot, or *Cancel,* which cancels everything. You can exit the *Page Setup* dialog box by *OK*, which automatically saves the settings, *Plot*, which saves settings and makes a plot, or *Cancel*.
2. The *Preview* button only appears in the *Plot* dialog box. Although you cannot preview from within the *Page Setup* dialog, you can exit and use the *Preview* button on the Standard toolbar.
3. Only the *Plot* dialog box allows you to *Plot to File*, plot multiple layouts, and apply a plot stamp.
4. Only with the *Page Setup* dialog box can you *Display Plot Styles* in layout tabs. When plot styles are attached, this option causes the display in the *Layout* tab to look like a plot or a *Full Preview*.

So which dialog box should you use? If you want to make settings for plotting but are <u>not</u> ready to plot, use the *Page Setup* dialog box to make the settings and save the changes (select *OK*), then select *Preview* from the standard toolbar. When you are ready to plot, use the *Plot* dialog box where you can change settings and *Preview* them immediately before you plot. Don't expect to save the changes in the *Plot* dialog box unless you plot.

PREVIEW

Pull-down Menu	COMMAND (TYPE)	ALIAS (TYPE)	Short-cut	Screen (side) Menu	Tablet Menu
File *Plot Preview*	*PREVIEW*	*PRE*	...	...	X,24

The *Preview* command accomplishes the same action as selecting a *Full Preview* in the *Plot* dialog box. The advantage to using this command is being able to see a full print preview directly without having to invoke the *Plot* dialog box first. This command is especially helpful if you prefer to use the *Page Setup* dialog box instead of the *Plot* dialog box to specify plotting options. All functions of this preview (right-click for shortcut menu to *Pan* and *Zoom*, etc.) operate the same as a full preview from the *Plot* dialog box including one important option, *Plot*.

You can print or plot during *Preview* directly from the shortcut menu (Fig. 14-8). Using the *Plot* option creates a print or plot based on your settings in the *Plot* or *Page Setup* dialog box. If you use this feature, ensure you first set the plotting parameters in the *Page Setup* dialog box.

Figure 14-8

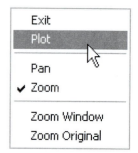

PLOTTING TO SCALE

When you create a <u>manual</u> drawing, a scale is determined before you can begin drawing. The scale is determined by the proportion between the size of the object on the paper and the actual size of the object. You then complete the drawing in that scale so the actual object is proportionally reduced or enlarged to fit on the paper.

With a <u>CAD</u> drawing you are not restricted to drawing on a sheet of paper, so the geometry can be created full size. Set *Limits* to provide an appropriate amount of drawing space; then the geometry can be drawn using the <u>actual dimensions</u> of the object. The resulting drawing on the CAD system is a virtual full-size replica of the actual object. Not until the CAD drawing is <u>plotted</u> on a fixed size sheet of paper, however, is it <u>scaled</u> to fit on the sheet.

You can specify the plot scale of an AutoCAD drawing in one of two ways. If you are plotting a *Layout* tab, specify the scale of the geometry that appears in the viewport by using the *Viewport* toolbar, the *Properties* palette, or the *Zoom XP* factor, and then select 1:1 in the *Plot* or *Page Setup* dialog box. If you plot the *Model* tab, select the desired scale directly in the *Plot* or *Page Setup* dialog box. In either case, the value or ratio that you specify is the proportion of paper space units to model space units—the same as the proportion of the paper size to the *Limits*. This ratio is also equal to the reciprocal of the "drawing scale factor," as explained in Chapters 12 and 13. Exceptions to this rule are cases in which multiple viewports are used.

For example, if you want to draw an object 15" long and plot it on an 11" x 8.5" sheet, you can set the model space *Limits* to 22 x 17 (2 x the sheet size). The value of **2** is then the drawing scale factor, and the plot scale is **1:2** (or **1/2,** the reciprocal of the drawing scale factor). If, in another case, the calculated drawing scale factor is **4**, then **1:4** (or **1/4**) would be the plot scale selected to achieve a drawing plotted at 1/4 actual size. In order to calculate *Limits* and drawing scale factors accurately, it is helpful to know the standard paper sizes.

Standard Paper Sizes

Size	Engineering (") (ANSI)	Architectural (")
A	8.5 x 11	9 x 12
B	11 x 17	12 x 18
C	17 x 22	18 x 24
D	22 x 34	24 x 36
E	34 x 44	36 x 48

Size	Metric (mm) (ISO)
A4	210 x 297
A3	297 x 420
A2	420 x 594
A1	594 x 841
A0	841 x 1189

Calculating the Drawing Scale Factor

 Assuming you are planning to plot from the *Model* tab or from a *Layout* tab using one viewport that fills almost all of the printable area, use the following method to determine the "drawing scale factor" and, therefore, the plot or print scale to specify.

1. In order to provide adequate space to create the drawing geometry full size, it is recommended to set the model space *Limits*, then *Zoom All*. *Limits* should be set to the intended sheet size used for plotting times a factor, if necessary, that provides enough area for drawing. This factor, <u>the proportion of the *Limits* to the sheet size, is the "drawing scale factor."</u>

$$DSF(mm) = \frac{Limits}{sheet\ size} \times 25.4$$

    You should also use a proportion that will yield a standard drawing scale (1/2"=1", 1/8"=1', 1:50, etc.), instead of a scale that is not a standard (1/3"=1", 3/5"=1', 1:23, etc.). See the "Tables of *Limits* Settings" for common standard drawing scales.

2. The "drawing scale factor" is used as the value (at least a starting point) for changing all size-related variables (*LTSCALE*, *Overall Scale* for dimensions, text height, *Hatch* pattern scale, etc.). The <u>reciprocal of the drawing scale factor is the plot scale</u> to use for plotting or printing the drawing.

$$Plot\ Scale = \frac{1}{DSF}$$

3. If the drawing is to be created using millimeter dimensions, set *Units* to *Decimal* and use metric values for the sheet size. In this way, the reciprocal of the drawing scale factor is the plot scale, the same as feet and inch drawings. However, multiply the drawing's scale factor by 25.4 (25.4mm = 1") to determine the factor for changing all size-related variables (*LTSCALE*, *Overall Scale* for dimensions, etc.).

$$DSF = \frac{Limits}{sheet\ size}$$

If drawing or inserting a border on the sheet, its maximum size cannot exceed the *Printable Area*. The *Printable Area* is given near the upper-left corner of the *Plot* or *Page Setup* dialog box, but the values can be different for different devices. Since plotters or printers do not draw or print all the way to the edge of the paper, the border should not be drawn outside of the *Printable Area*. Generally, approximately 1/4" to 1/2" (6mm to 12mm) offset from each edge of the paper (margin) is required. When plotting a *Layout*, the printable area is denoted by a dashed border around the sheet. If plotting the *Model* tab, you should multiply the margin (1/2 of the difference between the *Paper Size* and *Printable Area*) by the "drawing scale factor" to determine the margin size in model space drawing units.

Guidelines for Plotting the *Model* Tab to Scale

 Even though model space *Limits* can be changed and plot scale can be reset at <u>any time</u> in the drawing process, it is helpful to begin the process during the initial drawing setup (also see Chapter 12, Advanced Drawing Setup).

1. Set *Units* (*Decimal, Architectural, Engineering,* etc.) and *Precision* to be used in the drawing.
2. Set model space *Limits.* Set the *Limits* values to a proportion of the desired sheet size that provides enough area for drawing geometry as described in "Calculating the Drawing Scale Factor." The resulting value is the drawing scale factor.
3. Create the drawing geometry in the *Model* tab. Use the DSF as the initial value (or multiplier) for linetype scale, text size, dimension scale, hatch scale, etc.
4. Use the DSF to determine the scale factor to create, *Insert* or *Xref* the title block and border in model space, if one is needed.
5. Access the *Page Setup* or *Plot* dialog box. Select the desired scale in the *Scale* drop-down list or enter a value in the *Custom* edit boxes. The value to enter is 1/DSF. (Values are also given in the "Tables of *Limits* Settings.") Make a plot *Preview*, then make the plot or make the needed adjustments.

Guidelines for Plotting *Layouts* to Scale

Assuming you are using one viewport that fills almost all of the printable area, plot from the *Layout* tab using this method.

1. Set *Units* (*Decimal, Architectural, Engineering,* etc.) and *Precision* to be used in the drawing.
2. Set model space *Limits* and create the geometry in the *Model* tab. Use the DSF to determine values for linetype scale, text size, dimension scale, hatch scale, etc. Complete the drawing geometry in model space.
3. Click a *Layout* tab to begin setting up the layout. You can also use the *Layout Wizard* to create a new layout and complete steps 2 through 5.
4. Use *Page Setup* to select the desired *Plot Device* and *Paper Size.*
5. Create, *Insert,* or *Xref* the title block and border in the layout. If you are using one of AutoCAD's template drawings or another template, the title block and border may already exist.
6. Create a viewport using *Vports.*
7. Click inside the viewport to set the viewport scale. You can set the scale using the *Viewports* toolbar, the *Properties* palette, or use *Zoom* and enter an *XP* factor. Use 1/DSF as the viewport scale.
8. Activate the *Plot* or *Page Setup* dialog box and ensure the scale in the *Scale* drop-down list is 1:1. Make a plot *Preview*, then make the plot or make the needed adjustments.

If you want to print the same drawing using different devices and different scales, you can create multiple layouts. With each layout you can save the specific plot settings. Beware—a drawing printed at different scales may require changing or setting different values for linetype scale, text size, dimension variables, hatch scales, etc. for each layout. See Chapter 33, Advanced Layouts and Plotting, for more information on this topic.

Simplifying the process of plotting to scale and calculating model space *Limits* and the "drawing scale factor" can be accomplished by preparing template drawings. One method is to create a template for each sheet size that is used in the lab or office. In this way, the CAD operator begins the session by selecting the template drawing representing the sheet size and then multiplies those *Limits* by some factor to achieve the desired *Limits* and drawing scale factor. Another method is to create multiple layouts in the template drawing, one for each device and sheet size that you have available. In this way, a template drawing can be selected with the final layouts, title blocks, plot devices, and sheet sizes already specified, then you need only set the model space *Limits* and plot scale(s). (See Chapter 12 for help creating template drawings.)

TABLES OF *LIMITS* SETTINGS

Rather than making calculations of *Limits*, drawing scale factor, and plot scale for each drawing, the Tables of *Limits* Settings on the following pages can be used to make calculating easier. There are five tables, one for each of the following applications:

> Mechanical Engineering
> Architectural
> Metric (using ISO standard metric sheets)
> Metric (using ANSI standard engineering sheets)
> Civil Engineering

To use the tables correctly, you must know <u>any two</u> of the following three variables. The third variable can be determined from the other two.

> Approximate drawing <u>*Limits*</u>
> Desired print or plot <u>paper size</u>
> Desired print or plot <u>scale</u>

1. If you know the *Scale* and **paper size**:
    Assuming you know the scale you want to use, look along the top row to find the desired <u>scale</u> that you eventually want to print or plot. Find the desired <u>paper size</u> by looking down the left column. The intersection of the row and column yields the model space <u>*Limits*</u> settings to use to achieve the desired plot scale.

2. If you know the approximate model space **Limits** and **paper size**:
    Calculate how much space (minimum *Limits*) you require to create the drawing actual size. Look down the left column of the appropriate table to find the <u>paper size</u> you want to use for plotting. Look along that row to find the next larger <u>*Limits*</u> settings than your required area. Use these values to set the drawing *Limits*. The <u>scale</u> to use for the plot is located on top of that column.

3. If you know the *Scale* and approximate model space **Limits**:
    Calculate how much space (minimum *Limits*) you need to create the drawing objects actual size. Look along the top row of the appropriate table to find the desired <u>scale</u> that you eventually want to plot or print. Look down that column to find the next larger <u>*Limits*</u> than your required minimum area. Set the model space *Limits* to these values. Look to the left end of that row to give the <u>paper size</u> you need to print or plot the *Limits* in the desired scale.

NOTE: Common scales are given in the Tables of *Limits* Settings. If you need to create a print or plot in a scale that is not listed in the tables, you may be able to use the tables as a guide by finding the nearest table and standard scale to match your needs, then calculating <u>proportional</u> settings.

MECHANICAL TABLE OF *LIMITS* SETTINGS
For ANSI Sheet Sizes
(X axis x Y axis)

Paper Size (Inches)	Drawing Scale Factor 1, Scale 1" = 1", Proportion 1:1	1.33, 3/4" = 1", 3:4	2, 1/2" = 1", 1:2	2.67, 3/8" = 1", 3:8	4, 1/4" = 1", 1:4	5.33, 3/16" = 1", 3:16	8, 1/8" = 1", 1:8
A 11 x 8.5 In.	11.0 x 8.5	14.7 x 11.3	22.0 x 17.0	29.3 x 22.7	44.0 x 34.0	58.7 x 45.3	88.0 x 68.0
B 17 x 11 In.	17.0 x 11.0	22.7 x 14.7	34.0 x 22.0	45.3 x 29.3	68.0 x 44.0	90.7 x 58.7	136.0 x 88.0
C 22 x 17 In.	22.0 x 17.0	29.3 x 22.7	44.0 x 34.0	58.7 x 45.3	88.0 x 68.0	117.0 x 90.7	176.0 x 136.0
D 34 x 22 In.	34.0 x 22.0	45.3 x 29.3	68.0 x 44.0	90.7 x 58.7	136.0 x 88.0	181.3 x 117.3	272.0 x 176.0
E 44 x 34 In.	44.0 x 34.0	58.7 x 45.3	88.0 x 68.0	117.3 x 90.7	176.0 x 136.0	235.7 x 181.3	352.0 x 272.0

ARCHITECTURAL TABLE OF *LIMITS* SETTINGS
For Architectural Sheet Sizes
(X axis x Y axis)

Paper Size (Inches)		Drawing Scale Factor 12 / Scale 1" = 1' / Proportion 1:12	16 / 3/4" = 1' / 1:16	24 / 1/2" = 1' / 1:24	32 / 3/8" = 1' / 1:32	48 / 1/4" = 1' / 1:48	64 / 3/16" = 1' / 1:64	96 / 1/8" = 1' / 1:96
A 12 x 9	Ft	12 x 9	16 x 12	24 x 18	32 x 24	48 x 36	64 x 48	96 x 72
	In.	144 x 108	192 x 144	288 x 216	384 x 288	576 x 432	768 x 576	1152 x 864
B 18 x 12	Ft	18 x 12	24 x 16	36 x 24	48 x 32	64 x 48	96 x 64	128 x 96
	In.	216 x 144	288 x 192	432 x 288	576 x 384	768 x 576	1152 x 768	1536 x 1152
C 24 x 18	Ft	24 x 18	32 x 24	48 x 36	64 x 48	96 x 72	128 x 96	192 x 144
	In.	288 x 216	384 x 288	576 x 432	768 x 576	1152 x 864	1536 x 1152	2304 x 1728
D 36 x 24	Ft	36 x 24	48 x 32	72 x 48	96 x 64	144 x 96	192 x 128	288 x 192
	In.	432 x 288	576 x 384	864 x 576	1152 x 768	1728 x 1152	2304 x 1536	3456 x 2304
E 48 x 36	Ft	48 x 36	64 x 48	96 x 72	128 x 96	192 x 144	256 x 192	384 x 288
	In.	576 x 432	768 x 576	1152 x 864	1536 x 1152	2304 x 1728	3072 x 2304	4608 x 3456

METRIC TABLE OF *LIMITS* SETTINGS
For Metric (ISO) Sheet Sizes
(X axis x Y axis)

Paper Size (mm)		Drawing Scale Factor 25.4 / Scale 1:1 / Proportion 1:1	50.8 / 1:2 / 1:2	127 / 1:5 / 1:5	254 / 1:10 / 1:10	508 / 1:20 / 1:20	1270 / 1:50 / 1:50	2540 / 1:100 / 1:100
A4 297 x 210	mm	297 x 210	594 x 420	1485 x 1050	2970 x 2100	5940 x 4200	14,850 x 10,500	29,700 x 21,000
	m	.297 x .210	.594 x .420	1.485 x 1.050	2.97 x 2.10	5.94 x 4.20	14.85 x 10.50	29.70 x 21.00
A3 420 x 297	mm	420 x 297	840 x 594	2100 x 1485	4200 x 2970	8400 x 5940	21,000 x 14,850	42,000 x 29,700
	m	.420 x .297	.840 x .594	2.100 x 1.485	4.20 x 2.97	8.40 x 5.94	21.00 x 14.85	42.00 x 29.70
A2 594 x 420	mm	594 x 420	1188 x 840	2970 x 2100	5940 x 4200	11,880 x 8400	29,700 x 21,000	59,400 x 42,000
	m	.594 x .420	1.188 x .840	2.97 x 2.10	5.94 x 4.20	11.88 x 8.40	29.70 x 21.00	59.40 x 42.00
A1 841 x 594	mm	841 x 594	1682 x 1188	4205 x 2970	8410 x 5940	16,820 x 11,880	42,050 x 29,700	84,100 x 59,400
	m	.841 x .594	1.682 x 1.188	4.205 x 2.970	8.41 x 5.94	16.82 x 11.88	42.05 x 29.70	84.10 x 59.40
A0 1189 x 841	mm	1189 x 841	2378 x 1682	5945 x 4205	11,890 x 8410	23,780 x 16,820	59,450 x 42,050	118,900 x 84,100
	m	1.189 x .841	2.378 x 1.682	5.945 x 4.205	11.89 x 8.41	23.78 x 16.82	59.45 x 42.05	118.90 x 84.10

METRIC TABLE OF *LIMITS* SETTINGS
For ANSI Sheet Sizes
(X axis x Y axis)

Paper Size (mm)		Drawing Scale Factor 25.4 / Scale 1:1 / Proportion 1:1	50.8 / 1:2 / 1:2	127 / 1:5 / 1:5	254 / 1:10 / 1:10	508 / 1:20 / 1:20	1270 / 1:50 / 1:50	2540 / 1:100 / 1:100
A 279.4 x 215.9	mm	279.4 x 215.9	558.8 x 431.8	1397 x 1079.5	2794 x 2159	5588 x 4318	13,970 x 10,795	27,940 x 21,590
	m	0.2794 x 0.2159	0.5588 x 0.4318	1.397 x 1.0795	2.794 x 2.159	5.588 x 4.318	13.97 x 10.795	27.94 x 21.59
B 431.8 x 279.4	mm	431.8 x 279.4	863.6 x 558.8	2159 x 1397	4318 x 2794	8636 x 5588	21,590 x 13,970	43,180 x 27,940
	m	0.4318 x 0.2794	0.8636 x 0.5588	2.159 x 1.397	4.318 x 2.794	8.636 x 5.588	21.59 x 13.97	43.18 x 27.94
C 558.8 x 431.8	mm	558.8 x 431.8	1117.6 x 863.6	2794 x 2159	5588 x 4318	11,176 x 8636	27,940 x 21,590	55,880 x 43,180
	m	0.5588 x 0.4318	1.1176 x 0.8636	2.794 x 2.159	5.588 x 4.318	11.176 x 86.36	27.94 x 21.59	55.88 x 43.18
D 863.6 x 558.8	mm	863.6 x 558.8	1727.2 x 1117.6	4318 x 2794	8636 x 5588	17,272 x 11,176	43,180 x 27,940	86,360 x 55,880
	m	0.8636 x 0.5588	1.7272 x 1.1176	4.318 x 2.794	8.636 x 5.588	17.272 x 11.176	43.18 x 27.94	86.36 x 55.88
E 1117.6 x 863.6	mm	1117.6 x 863.6	2235.2 x 1727.2	5588 x 4318	11,176 x 8636	22,352 x 17,272	55,880 x 43,180	111,760 x 86,360
	m	1.1176 x 0.8636	2.2352 x 1.7272	5.588 x 4.318	11.176 x 8.636	22.352 x 17.272	55.88 x 43.18	111.76 x 86.36

CIVIL TABLE OF *LIMITS* SETTINGS
For ANSI Sheet Sizes
(X axis x Y axis)

Paper Size (Inches)	Drawing Scale Factor 120 Scale Tab 1″ = 10′ Proportion 1:120	240 1″ = 20′ 1:240	360 1″ = 30′ 1:360	480 1″ = 40′ 1:480	600 1″ = 50′ 1:600
A. 11 x 8.5 In. Ft	1320 x 1020 110 x 85	2640 x 2040 220 x 170	3960 x 3060 330 x 255	5280 x 4080 440 x 340	6600 x 5100 550 x 425
B. 17 x 11 In. Ft	2040 x 1320 170 x 110	4080 x 2640 340 x 220	6120 x 3960 510 x 330	8160 x 5280 680 x 440	10,200 x 6600 850 x 550
C. 22 x 17 In. Ft	2640 x 2040 220 x 170	5280 x 4080 440 x 340	7920 x 6120 660 x 510	10,560 x 8160 880 x 680	13,200 x 10,200 1100 x 850
D. 34 x 22 In. Ft	4080 x 2640 340 x 220	8160 x 5280 680 x 440	12,240 x 7920 1020 x 660	16,320 x 10,560 1360 x 880	20,400 x 13,200 1700 x 1100
E. 44 x 34 In. Ft	5280 x 4080 440 x 340	10,560 x 8160 880 x 680	15,840 x 12,240 1320 x 1020	21,120 x 16,320 1760 x 1360	26,400 x 20,400 2200 x 1700

Examples for Plotting to Scale

Following are several hypothetical examples of drawings that can be created using the "Guidelines for Plotting to Scale." As you read the examples, try to follow the logic and check the "Tables of *Limits* Settings."

A. A one-view drawing of a mechanical part that is 40" in length is to be drawn requiring an area of approximately 40" x 30", and the drawing is to be plotted on an 8.5" x 11" sheet. The template drawing *Limits* are preset to 0,0 and 11,8.5. (AutoCAD's default *Limits* are set to 0,0 and 12,9, which represents an uncut sheet size. Template *Limits* of 11 x 8.5 are more practical in this case.) The expected plot scale is 1/4"=1". The following steps are used to calculate the new *Limits*.

1. *Units* are set to *decimal*. Each unit represents 1.00 inches.
2. Multiplying the *Limits* of 11 x 8.5 by a factor of 4, the new **Limits** should be set to **44 x 34**, allowing adequate space for the drawing. The "drawing scale factor" is **4**.
3. All size-related variable default values (**LTSCALE, Overall Scale** for dimensions, etc.) are multiplied by **4**. The plot scale (for *Layout* tabs, entered in the *Viewport* toolbar, etc., or for the *Model* tab, entered in the *Plot* or *Page Setup* dialog box) is **1:4** to achieve a plotted drawing of 1/4"=1".

B. A floorplan of a residence will occupy 60' x 40'. The drawing is to be plotted on a "D" size architectural sheet (24" x 36"). No prepared template drawing exists, so the standard AutoCAD default drawing (12 x 9) *Limits* are used. The expected plot scale is 1/2"=1'.

1. *Units* are set to *Architectural*. Each unit represents 1".
2. The floorplan size is converted to inches (60' x 40' = 720" x 480"). The sheet size of 36" x 24" (or 3' x 2') is multiplied by **24** to arrive at **Limits** of **864" x 576"** (72' x 48'), allowing adequate area for the floor plan. The *Limits* are changed to those values.
3. All default values of size-related variables (**LTSCALE, Overall Scale** for dimensions, etc.) are multiplied by **24**, the drawing scale factor. The plot scale (for *Layout* tabs, entered in the *Viewport* toolbar, etc., or for the *Model* tab, entered in the *Plot* or *Page Setup* dialog box) is **1/2"=1'** (or 1/24, which is the reciprocal of the drawing scale factor).

C. A roadway cloverleaf is to be laid out to fit in an acquired plot of land measuring 1500' x 1000'. The drawing will be plotted on "D" size engineering sheet (34" x 22"). A template drawing with *Limits* set equal to the sheet size is used. The expected plot scale is 1"=50'.

1. *Units* are set to *Engineering* (feet and decimal inches). Each unit represents 1.00".
2. The sheet size of 34" x 22" (or 2.833' x 1.833') is multiplied by **600** to arrive at **Limits** of **20400" x 13200"** (1700' x 1100'), allowing enough drawing area for the site. The *Limits* are changed to **1700' x 1100'**.
3. All default values of size-related variables (**LTSCALE, Overall Scale** for dimensions, etc.) are multiplied by **600**, the drawing scale factor. The plot scale (for *Layout* tabs, entered in the *Viewport* toolbar, etc., or for the *Model* tab, entered in the *Plot* or *Page Setup* dialog box) is **1/600** (*inches = drawing units*) to achieve a drawing of 1"=50' scale.

NOTE: Many civil engineering firms use one unit in AutoCAD to represent one foot. This simplifies the problem of using decimal feet (10 parts/foot rather than 12 parts/foot); however, problems occur if architectural layouts are inserted or otherwise combined with civil work.

D. Three views of a small machine part are to be drawn and dimensioned in millimeters. An area of 480mm x 360mm is needed for the views and dimensions. The part is to be plotted on an A4 size sheet. The expected plot scale is 1:2.

1. *Units* are set to *decimal*. Each unit represents 1.00 millimeters.
2. Setting the *Limits* exactly to the sheet size (297 x 210) would not allow enough area for the drawing. The sheet size is multiplied by a factor of **2** to yield *Limits* of **594 x 420**, providing the necessary 480 x 360 area.
3. The plot scale (for *Layout* tabs, entered in the *Viewport* toolbar, etc., or for the *Model* tab, entered in the *Plot* or *Page Setup* dialog box) is **1:2** scale. Since this drawing is metric, the drawing scale factor for changing all size-related variable default values (*LTSCALE, Overall Scale* for dimensions, etc.) is multiplied by 25.4, or 2 x 25.4 = approximately **50**.

CONFIGURING PLOTTERS AND PRINTERS

Plotting Components

When a drawing is plotted, three main components are referenced to develop the finished plot.

1. The plot and page settings you set up in the *Plot* dialog box and *Page Setup* dialog box (saved in the .DWG file)
2. The plotter configuration information you specify in the *Add-A-Plotter Wizard* or *Plotter Configuration Editor* (saved in a .PC3 file)
3. A plot style table you create using the *Add-A-Plot Style Table Wizard* or *New* button in the *Plot* dialog box (saved in a .CTB or .STB file)

The three components are stored separately so they can be applied in different combinations. For example, any plot style table can be used with any plot device, or any page setup can use any plot style table.

The *Plot* and *Page Setup* dialog boxes were explained earlier in this chapter. The following section explains how to configure plotters and create .PC3 files using the *Plotter Manager*, *Add-A-Plotter Wizard*, and *Plotter Configuration Editor*. Plot style tables (.CTB and .STB files) are discussed in Chapter 33.

If you are using AutoCAD in a lab or office, the devices used for making plots and prints are most likely already configured for your use. However, no matter if you are a student or a professional, you may be required to configure a new device or change configuration settings on existing devices. This section explains those possibilities. In order to accomplish either of the tasks mentioned here, use the *Plottermanager* command to invoke the AutoCAD *Plotter Manager*.

PLOTTER- MANAGER	Pull-down Menu	COMMAND (TYPE)	ALIAS (TYPE)	Short-cut	Screen (side) Menu	Tablet Menu
	File *Plotter Manager...*	*PLOTTERMANAGER*	...	...	...	...

The AutoCAD *Plotter Manager* can be invoked by several methods as shown in the command table above. In addition to those methods, you can select the *Add or Configure Plotters* button in the *Plotting* tab of the *Options* dialog box.

In any case, the *Plotter Manager* is a separate system window which can run as an independent application to AutoCAD (Fig. 14-9). By default, the window displays icons of all plot devices configured specifically for AutoCAD. The default Windows configured device is included in this group. (When you install AutoCAD, the Windows configured system printer is automatically configured if you have a previously installed Windows printer.) You can add or configure a plotter with the *Plotter Manager*.

Figure 14-9

The display of the *Plotter Manager* can also be changed to list *Details* instead of *Icons*, as shown in Figure 14-10 (use the *View* pull-down menu and select *Details*). Note that the defined plotters are actually saved as .PC3 files. If you want to add a new plotter for use with AutoCAD, select the *Add-A-Plotter Wizard*. To change the settings on an existing plotter using the *Plotter Configuration Editor*, double-click on its icon or file name.

Figure 14-10

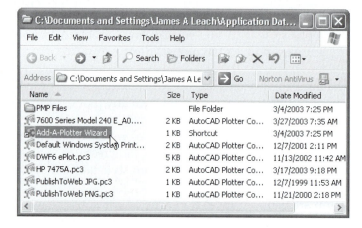

Add-A-Plotter Wizard

The *Add-A-Plotter Wizard* makes the process of configuring a new device very simple. Several choices are available (Fig. 14-11). You can configure a new device to be connected directly to and managed by your system or to be shared and managed by a network server. You can also reconfigure printers that were previously set up as Windows system printers. In this case, icons for the printers will appear in the *Plotter Manager* and you can access and control configuration settings to use specifically for AutoCAD.

Figure 14-11

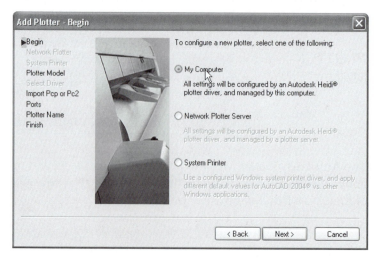

Start the *Add-A-Plotter Wizard* by double-clicking its icon in the *Plotter Manager* or selecting *Add Plotter...* from the *Wizards* cascading menu in the *Tools* pull-down menu.

After you have selected the device (*Plotter Model*) to configure (not shown), you can choose to import information about the plotter if it was previously configured for use with an earlier release of AutoCAD. Information such as paper size, optimization, port names, etc. were previously stored in .PCP and .PC2 files and can be converted to the .PC3 format (Fig. 14-12).

If you selected *My Computer* in the *Begin* step, you are presented with the *Ports* page next (not shown). You can select the port to which the device is connected or specify that the device will always *Plot to a File* or *Autospool*. If you specify a port, you can later select *Plot to File* in the *Plot* dialog box.

If you are using a device previously used as a Windows system printer, make sure you supply a different name for the device than the name Windows uses. This action ensures the new name appears in the *Plot* and *Page Setup* dialog boxes (Fig. 14-13).

Figure 14-12

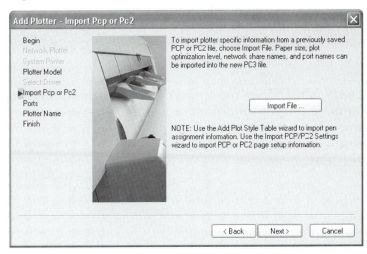

Figure 14-13

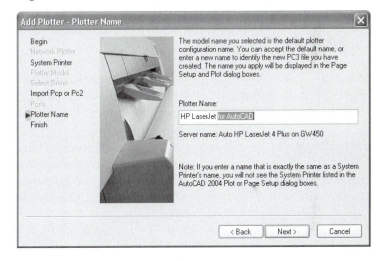

Once the new or previously used Windows system device has been configured for AutoCAD, its icon is displayed in the *Plotter Manager* along with any other configured devices.

Plotter Configuration Editor

After you create a configured plotter (.PC3) file using the *Add-A-Plotter Wizard*, you can edit the file using the *Plotter Configuration Editor*. The *Plotter Configuration Editor* allows you to specify a number of the plotter's output settings. You can start the *Plotter Configuration Editor* by one of the following methods.

1. Double-click the plotter's icon or file name in the *Plotter Manager*.
2. Choose *Edit Plotter Configuration* from the *Finish* step of the *Add-A-Plotter Wizard*.
3. Select *Properties* from the *Page Setup* dialog box or *Plot* dialog box.
4. Double-click a .PC3 file from Windows Explorer (by default, .PC3 files are stored in the AutoCAD 2002\Plotters folder).

The *Plotter Configuration Editor* contains three tabs. The *General* tab contains basic information about the plotter. Only the *Description* area is not read-only and can be edited. Information entered in the *Description* is saved with the .PC3 file.

The *Ports* tab allows you to specify ports to connect with the device. If you are configuring a local, non-Windows system plotter, you can respecify the port to which the device is connected. You can choose a serial (LPT), parallel (COM), or network port. If you are using a serial port, the settings within AutoCAD must match the plotter settings. For additional technical information, check AutoCAD 2004 *Help, Driver and Peripheral Guide*.

The *Device and Document Settings* tab contains the plotting options. These choices are different depending on your configured plotting device. For example, when you configure a pen plotter you have the option to modify the pen characteristics (Fig. 14-14). The *Device and Document Settings* tab can contain some or all of the following general option areas.

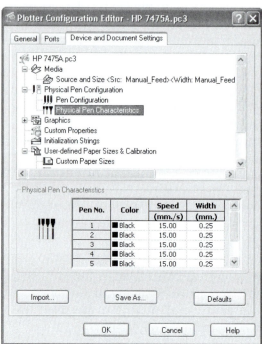

Figure 14-14

Media	Specifies a paper source, size, type, and destination
Physical Pen Configuration	Specifies settings for pen plotters
Graphics	Specifies settings for printing vector and raster graphics and True Type fonts
Custom Properties	Displays settings related to the device driver
Initialization Strings	Sets pre-initialization, post-initialization, and termination printer strings
User-Defined Paper Sizes and Calibration	Attaches a plot model parameter (.PMP) file to the .PC3 file, calibrates plotter, and adds, deletes, or revises custom paper sizes

The related .PC3 file contains these same categories of settings. Double-click any of the categories to view and change the specific settings.

You can also filter paper sizes that you do not use so they do not appear as a choice in the *Page Setup* and *Plot* dialog boxes. Expand the *User-defined Paper Sizes & Calibration* section, then select *Filter Paper Sizes*.

Because there are so many possible options for all the possible plotting devices it is not appropriate to include the information in this text. Additional technical information on this subject can be found in AutoCAD 2004 *Help, Driver and Peripheral Guide*.

.PCP, .PC2, and .PC3 Files

Plotting information for pre-AutoCAD 2000 releases was stored in .PCP and .PC2 files. These file types cannot be used in AutoCAD 2004, but the information in these files can be converted for use with AutoCAD 2004.

.PCP files, used with Release 12 and earlier versions of AutoCAD, contain information about plot settings, specifically pen settings. These files are device-independent, so one .PCP file could be used to define pen settings for many plotters. Although these were optional, pen assignments could be saved to a .PCP file and later imported rather than specifying the settings again when a different device was configured.

2002

In Releases 13 and 14, .PC2 files were used. These files contain pen assignment information as well as the device (plotter or printer) configuration information. In this way, when a .PC2 file was loaded, it automatically installed and configured the device and made it the current plotter. This feature was helpful for batch plotting because different plot devices could be operated while unattended.

AutoCAD 2000 through 2004 store plotter configuration information in .PC3 files. .PC3 files store the complete settings for particular devices, including media, paper sizes, printable area, and pen information. .PC3 files are drawing independent, portable, and sharable, so they can be used for any drawing or by any AutoCAD session.

.PCP and .PC2 can be imported and converted to the .PC3 format using the *Pcinwizard*. Therefore, information you have already saved about your devices can be automatically updated to use with AutoCAD 2004.

PCINWIZARD

Pull-down Menu	COMMAND (TYPE)	ALIAS (TYPE)	Short-cut	Screen (side) Menu	Tablet Menu
Tools *Wizards>* *Import* *Plot Settings...*	*PCINWIZARD*	...	...	...	...

Pcinwizard displays the *Import PCP or PC2 Plot Settings* wizard (Fig. 14-15). The wizard imports and updates plotter information previously stored in .PCP and .PC2 files (for Release 14 and earlier releases of AutoCAD) to the AutoCAD 2004 format, .PC3. Information that can be imported from .PCP or .PC2 files includes plot area, rotation, plot offset, plot optimization, plot to file, paper size, plot scale, and pen mapping.

The wizard prompts you for the name of the .PCP or .PC2 configuration file from which you want to import settings. You can view and modify the plot settings prior to importing them.

Figure 14-15

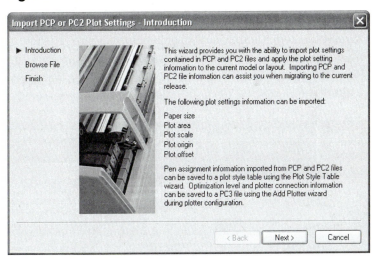

Like all the wizards, explanatory information is given in all steps. Upon completion, plot settings such as paper size, plot area, plot scale, plot origin, and plot offset are stored for each device and are ready for use with AutoCAD 2004.

CHAPTER EXERCISES

1. **Print GASKETA from the *Model* Tab**

   A. *Open* the **GASKETA** drawing that you created in Chapter 9 Exercises. Activate the *Model* tab. What are the model space *Limits*?

   B. From the *Model* tab, **Plot** the drawing **Extents** on an 11" x 8.5" sheet. Select the **Scale to Fit** option from the *Scale* drop-down list in the *Plot* dialog box.

   C. Next, **Plot** the drawing from the *Model* tab again. Plot the **Limits** on the same size sheet. **Scale to Fit.**

   D. Now, **Plot** the drawing **Limits** as before but plot the drawing at **1:1.** Measure the drawing and compare the accuracy with the dimensions given in the exercise in Chapter 9.

   E. Compare the three plots. What are the differences and why did they occur? (When you finish, there is no need to *Save* the changes.)

2. **Print GASKETA from a *Layout* Tab in Two Scales**

   A. Using the **GASKETA** drawing again, activate the *Layout1* tab. If a viewport exists in the layout, **Erase** it. Activate the *Page Setup* dialog box and configure the layout to use a printer to print on a *Letter* (or "A" size) sheet.

   B. Next, make layer **0 Current.** Use *Vports* to create a *Single* viewport using the **Fit** option. Double-click inside the viewport and **Zoom All** to see all of the model space *Limits*. Next, set the viewport scale to **1:1** using any method (*Zoom XP*, the *Viewport* toolbar, or the *Properties* palette). You may need to **Pan** to center the geometry, but do <u>not</u> *Zoom*.

   C. (Optional) Create a title block and border for the drawing in paper space.

   D. Activate the *Plot* dialog box. Ensure the *Scale* edit box is set at **1:1.** Plot the **Layout.** When complete, measure the plot for accuracy by comparing with the dimensions given for the exercise in Chapter 9. *Save* the drawing.

   E. Reset the viewport scale to **2:1** using any method (*Zoom XP*, the *Viewport* toolbar, or the *Properties* palette). Activate the *Plot* dialog box and make a second plot at the new scale. *Close* the drawing, but <u>do not save</u> it.

3. **Print the MANFCELL Drawing in Two Standard Scales**

   A. *Open* the **MANFCELL** drawing from Chapter 9. Check to make sure that the *Limits* settings are at **0,0** and **40',30'.**

   B. Activate the *Layout1* tab. If a viewport exists, **Erase** it. Use *Pagesetup* to select a printer to print on a letter ("A") sheet size. Next, create a *Single* viewport with the **Fit** option on layer **0.**

C. In this step you will make one print of the drawing at the largest possible size using a standard architectural scale. Use any method to set the viewport scale. Use the "Architectural Table of *Limits* Settings" in the chapter for guidance on an appropriate scale to set. (HINT: Look in the "A" size sheet row and find the next larger *Limits* settings that will accommodate your current *Limits*, then use the scale shown above at the top of the column.) *Save* the drawing. Make a print using the ***Plot*** or ***Page Setup*** dialog box.

D. Now make a plot of the drawing using the next smaller standard architectural scale. This time determine the appropriate scale to set using either the "Architectural Table of *Limits* Settings" or by selecting it from the *Properties* palette or *Viewport Scale Control* drop-down list.

4. **Use *Plotter Manager* to Configure a "D" Size Plot Device**
    Even if your system currently has a "D" size plot device configured, this exercise will give you practice with *Plotter Manager*. This exercise does not require that the device actually exist in your lab or office or that it be physically attached to your system.

    A. Invoke the ***Plotter Manager***. In *Plotter Manager*, double-click on ***Add-A-Plotter Wizard***.

    B. Select the ***Next*** button in the *Introduction* page. In the *Begin* page, select ***My Computer***. In the *Plotter Model* page, select any *Manufacturer* and *Model* that can plot on a "D" size sheet, such as ***CalComp 68444 Color Electrostatic, Hewlett-Packard Draft-Pro EXL (7576A)***, or ***Oce 9800 FBBS R3.x***.

    C. Select ***Next*** in both the *Import PCP or PC2* and *Ports* pages. In the *Plotter Name* step, select the default name but insert the manufacturer's name as a prefix, such as "HP" or "CalComp."

    D. Finally, activate the ***Plot*** or ***Page Setup*** dialog box, select the ***Plot Device*** tab, and ensure your new device is listed in the *Name* drop-down list.

5. **Create Multiple Layouts for Different Plot Devices, Set *PSLTSCALE***
    In this exercise you will create two layouts—one for plotting on an engineering "A" size sheet and one on a "C" size sheet. Check the table of "Standard Paper Sizes" in the chapter and find the correct size in inches for both engineering "A" and "C" size sheets. You will also adjust *LTSCALE* and *PSLTSCALE* appropriately for the layouts.

    A. ***Open*** drawing **CH11EX3**. Check to ensure the model space *Limits* are set at **22 x 17**. If you plan to print on a "A" size sheet, what is the drawing scale factor? Is the ***LTSCALE*** set appropriately?

    B. Activate the ***Layout1*** tab. Right-click and use ***Rename*** to rename the tab to **A Sheet**. If a viewport exists in the layout, ***Erase*** it. Activate the ***Page Setup*** dialog box and configure the layout to use a printer to print on a ***Letter*** ("A" size) sheet. Next, use ***Vports*** to create a viewport on layer **0** using the ***Fit*** option. Double-click inside the viewport and set the viewport scale to **1:2**.

    C. Toggle between ***Model*** tab and the ***A Sheet*** tab to examine the hidden and center lines. Do the dashes appear the same in both tabs? If the dashes appear differently in the two tabs, type ***PSLTSCALE*** and ensure the value is set to **0**. (Make sure you always use ***Regenall*** after using ***PSLTSCALE***.)

    D. Use the ***Page Setup*** or ***Plot*** dialog box to make the plot. Measure the plot for accuracy by comparing it with the dimensions given for the exercise in Chapter 11. To refresh your memory, that exercise involved adjusting the *LTSCALE* factor. Does the *LTSCALE* in the drawing yield hidden line dashes of 1/8" (for the long lines) on the plot? If not, make the ***LTSCALE*** adjustment and plot again. Use *SaveAs* to save and rename the drawing as **CH14EX5**.

E. Next, activate the *Layout2* tab. Right-click and use *Rename* to rename the tab to **C Sheet**. If a viewport exists in the layout, *Erase* it. Activate the *Page Setup* dialog box and configure the layout to use a plotter for a *ANSI C (22 x 17 Inches)* sheet. Next, on layer **0**, use *Vports* to create a viewport using the *Fit* option. Double-click inside the viewport and set the viewport scale. Since the model space limits are set to 22 x 17 (same as the sheet size), set the viewport scale to **1:1**.

F. Do the hidden lines appear the same in all tabs? *PSLTSCALE* set to 0 ensures that the hidden and center lines appear with the same *LTSCALE* in the viewports as in the *Model* tab. If everything looks okay, make the plot. Does the plot show hidden line dashes of 1/8"? *Save* the drawing.

6. **Create an Architectural Template Drawing**
    In this exercise you will create a new template for architectural applications to plot at 1/8"=1' scale on a "D" size sheet.

    A. Begin a *New* drawing using the template **ASHEET.DWT** you worked on in Chapter 13 Exercises. Use *Save* to create a new template (**.DWT**) named **D-8-AR** in <u>your working folder</u> (<u>not</u> where the AutoCAD-supplied templates are stored).

    B. Set *Units* to *Architectural*. Use the "Architectural Table of *Limits* Settings" to determine and set the new model space *Limits* for an architectural "**D**" size sheet to plot at **1/8"=1'**. Multiply the existing *LTSCALE* (**1.0**) times the drawing scale factor shown in the table. Turn *Snap* to *Polar Snap* and turn *Grid* off.

    C. Activate the *Layout2* tab. Right-click to produce the shortcut menu and select *Rename*. Rename the tab to **D Sheet**. *Erase* the viewport if one exists.

    D. Activate the *Page Setup* dialog box. In the *Plot Device* tab, select a device that can use an architectural D size sheet (36 x 24), such as the device you configured for your system in Exercise 4. In the *Layout Settings* tab, select *ARCH D (24 x 36 Inches)* in the *Paper Size* drop-down list. Select *OK* to dismiss the *Page Setup* dialog box and save the settings.

    E. Make **VPORTS** the current layer. Use the *Vports* command to create a *Single* viewport using the *Fit* option. Activate the *Viewports* toolbar. Double-click inside the viewport and set the viewport scale to **1/8"=1'**.

    F. Make the *Model* tab active and set the **OBJECT** layer current. Finally, *Save* as a template (.DWT) drawing in your working folder.

7. **Create a Civil Engineering Template Drawing**
    Create a new template for civil engineering applications to plot at 1"=20' scale on a "C" size sheet.

    A. Begin a *New* drawing using the template **C-D-SHEET**. Use *Save* to create a new template (.DWT) named **C-CIVIL-20** in <u>your working folder</u> (<u>not</u> where the AutoCAD-supplied templates are stored).

    B. Set *Units* to *Engineering*. Use the "Civil Table of *Limits* Settings" to determine and set the new *Limits* for a "C" size sheet to plot at **1"=20'**. Multiply the existing *LTSCALE* (.5) times the scale factor shown. Turn *Snap* to *Polar* and *Grid* Off.

    C. Activate the *D Sheet* layout tab. Use any method to set the viewport scale to **1:240**. *Save* the drawing as a template (.DWT) drawing in your working folder.

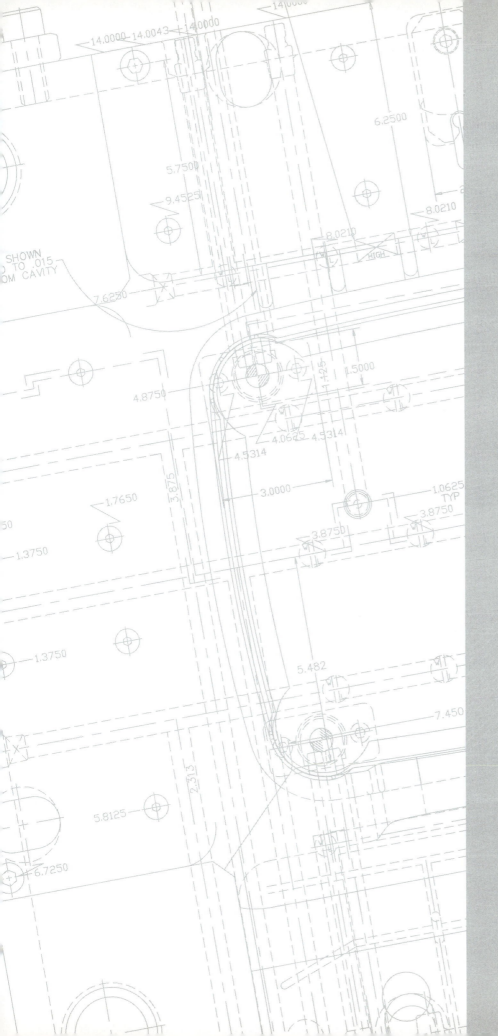

15

DRAW
COMMANDS II

Chapter Objectives

After completing this chapter you should:

1. be able to create construction lines using the *Xline* and *Ray* commands;

2. be able to create *Polygons* by the *Circumscribe*, *Inscribe*, and *Edge* methods;

3. be able to create *Rectangles*;

4. be able to use *Donut* to create circles with width;

5. be able to create *Spline* curves passing exactly through the selected points;

6. be able to create *Ellipses* using the *Axis End* method, the *Center* method, and the *Arc* method;

7. be able to use *Divide* to add points at equal parts of an object and *Measure* to add points at specified segment lengths along an object;

8. be able to draw multiple parallel lines with *Mline* and to create multi-line styles with *Mstyle*;

9. be able to use the *Sketch* command to create "freehand" sketch lines;

10. be able to use the *Solid* command to create a filled 2D shape with three or four straight edges;

11. be able to create a *Boundary* by PICKing inside a closed area;

12. know that objects forming a closed shape can be combined into one *Region* object.

CONCEPTS

Remember that *Draw* commands create AutoCAD objects. The draw commands addressed in this chapter create more complex objects than those discussed in Chapter 8, Draw Commands I. The draw commands covered previously (*Line, Circle, Arc, Point,* and *Pline*) create simple objects composed of one object. The shapes created by the commands covered in this chapter are more complex. Most of the objects <u>appear</u> to be composed of several simple objects, but each shape is actually treated by AutoCAD as <u>one object</u>. Only the *Divide, Measure* and *Sketch* commands create multiple objects. The following commands are explained in this chapter.

Xline, Ray, Polygon, Rectangle, Donut, Spline, Ellipse, Divide, Measure, Mline, Sketch, Solid, Boundary, and *Region*

These draw commands can be accessed using any of the command entry methods, including the menus and icon buttons, as illustrated in Chapter 8. Buttons that appear by the Command tables in this chapter are available in the default *Draw* toolbar or can be activated by creating your own custom toolbar (see Chapter 44 for information on customizing toolbars).

COMMANDS

XLINE

Pull-down Menu	COMMAND (TYPE)	ALIAS (TYPE)	Short-cut	Screen (side) Menu	Tablet Menu
Draw *Construction Line*	XLINE	XL	...	*DRAW 1* *Xline*	L,10

When you draft with pencil and paper, light "construction" lines are used to lay out a drawing. These construction lines are not intended to be part of the finished object lines but are helpful for preliminary layout such as locating intersections, center points, and projecting between views.

There are two types of construction lines—*Xline* and *Ray*. An *Xline* is a line with infinite length, therefore having no endpoints. A *Ray* has one "anchored" endpoint and the other end extends to infinity. Even though these lines extend to infinity, they do not affect the drawing *Limits* or *Extents* or change the display or plot area in any way.

An *Xline* has no endpoints (*Endpoint Osnap* cannot be used) but does have a <u>root</u>, which is the theoretical <u>midpoint</u> (*Midpoint Osnap* can be used). If *Trim* or *Break* is used with *Xlines* or *Rays* such that two endpoints are created, the construction lines become *Line* objects. *Xlines* and *Rays* are drawn on the current layer and assume the current linetype and color (object-specific or *ByLayer*).

There are many ways that you can create *Xlines* as shown by the options appearing at the command prompt. All options of *Xline* automatically repeat so you can easily draw multiple lines.

```
Command: xline
Specify a point or [Hor/Ver/Ang/Bisect/Offset]:
```

Specify a point

The default option only requires that you specify two points to construct the *Xline* (Fig. 15-1). The first point becomes the root and anchors the line for the next point specification. The second point, or "through point," can be PICKed at any location and can pass through any point (*Polar Snap*, *Polar Tracking*, *Osnaps*, and *Objects Snap Tracking* can be used). If horizontal or vertical *Xlines* are needed, *ORTHO* can be used in conjunction with the "Specify a point:" option.

Figure 15-1

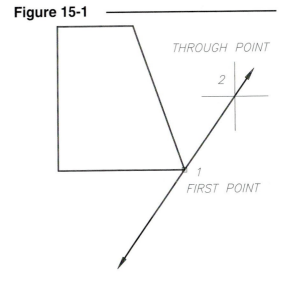
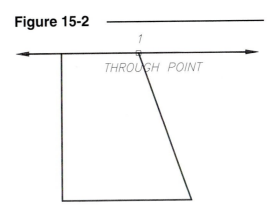

```
Command: xline
Specify a point or [Hor/Ver/Ang/Bisect/Offset]:
Specify through point:
```

Hor

This option creates a horizontal construction line. Type the letter "H" at the first prompt. You only specify one point, the through point or root (Fig. 15-2).

Figure 15-2

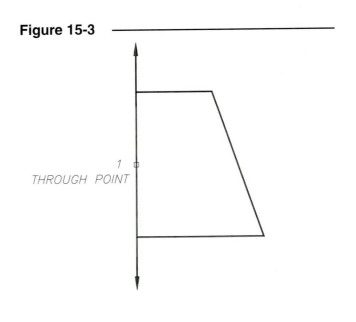

Ver

Ver creates a vertical construction line. You only specify one point, the through point (root) (Fig. 15-3).

Figure 15-3

Ang

The *Ang* option provides two ways to specify the desired angle. You can (1) *Enter angle* or (2) select a *Reference* line (*Line, Xline, Ray,* or *Pline*) as the starting angle, then specify an angle from the selected line (in a counterclockwise direction) for the *Xline* to be drawn (Fig. 15-4).

Command: *xline*
Specify a point or [Hor/Ver/Ang/Bisect/Offset]: *a*
Enter angle of xline (0) or [Reference]:

Figure 15-4

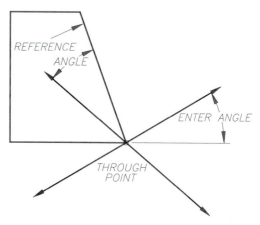

Bisect

This option draws the *Xline* at an angle between two selected points. First, select the angle vertex, then two points to define the angle (Fig. 15-5).

Command: *xline*
Specify a point or [Hor/Ver/Ang/Bisect/Offset]: *b*
Specify angle vertex point: **PICK**
Specify angle start point: **PICK**
Specify angle end point: **PICK**
Specify angle end point: **Enter**
Command:

Figure 15-5

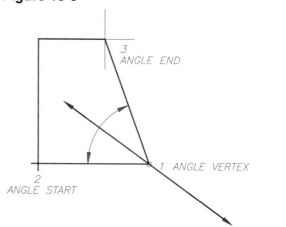

Offset

Offset creates an *Xline* parallel to another line. This option operates similarly to the *Offset* command. You can (1) specify a *Distance* from the selected line or (2) PICK a point to create the *Xline Through*. With the *Distance* option, enter the distance value, select a line (*Line, Xline, Ray,* or *Pline*), and specify on which side to create the offset *Xline*.

Command: *xline*
Specify a point or [Hor/Ver/Ang/Bisect/Offset]: *o*
Specify offset distance or [Through] <1.0000>:
(**value**) or **PICK**
Select a line object: **PICK**
Specify side to offset: **PICK**
Select a line object: **Enter**
Command:

Figure 15-6

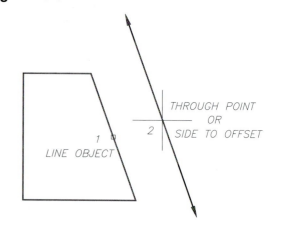

Using the *Through* option, select a line (*Line, Xline, Ray,* or *Pline*); then specify a point for the *Xline* to pass through. In each case, the anchor point of the *Xline* is the "root." (See "*Offset,*" Chapter 9.)

Command: *xline*
Specify a point or [Hor/Ver/Ang/Bisect/Offset]: *o*
Specify offset distance or [Through] <1.0000>: *t*
Select a line object: **PICK**
Specify through point: **PICK**
Select a line object: **Enter**
Command:

RAY

	Pull-down Menu	COMMAND (TYPE)	ALIAS (TYPE)	Short-cut	Screen (side) Menu	Tablet Menu
	Draw Ray	*RAY*	...	...	DRAW 1 *Ray*	*K,10*

A *Ray* is also a construction line (see *Xline*), but it extends to infinity in only <u>one direction</u> and has one "anchored" endpoint. Like an *Xline,* a *Ray* extends past the drawing area but does not affect the drawing *Limits* or *Extents.* The construction process for a *Ray* is simpler than for an *Xline,* only requiring you to establish a "start point" (endpoint) and a "through point" (Fig. 15-7). Multiple *Rays* can be created in one command.

Figure 15-7

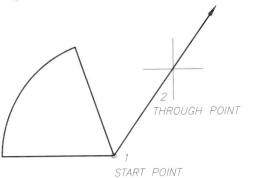

Command: *ray*
Specify start point: **PICK** or (**coordinates**)
Specify through point: **PICK** or (**coordinates**)
Specify through point: **PICK** or (**coordinates**)
Specify through point: **Enter**
Command:

Rays are especially helpful for construction of geometry about a central reference point or for construction of angular features. In each case, the geometry is usually constructed in only one direction from the center or vertex. If horizontal or vertical *Rays* are needed, just toggle on *ORTHO.* To draw a *Ray* at a specific angle, use relative polar coordinates or Polar Tracking. *Endpoint* and other appropriate *Osnaps* can be used with *Rays.* A *Ray* has one *Endpoint* but no *Midpoint.*

POLYGON

	Pull-down Menu	COMMAND (TYPE)	ALIAS (TYPE)	Short-cut	Screen (side) Menu	Tablet Menu
	Draw Polygon	*POLYGON*	*POL*	...	DRAW 1 *Polygon*	*P,10*

The *Polygon* command creates a regular polygon (all angles are equal and all sides have equal length). A *Polygon* object appears to be several individual objects but, like a *Pline,* is actually <u>one</u> object. In fact, AutoCAD uses *Pline* to create a *Polygon.* There are two basic options for creating *Polygons:* you can specify an *Edge* (length of one side) or specify the size of an imaginary circle for the *Polygon* to *Inscribe* or *Circumscribe.*

Inscribe/Circumscribe
The command sequence for this default method follows:

> Command: **polygon**
> Enter number of sides <4>: (**value**)
> Specify center of polygon or [Edge]: PICK or (**coordinates**)
> Enter an option [Inscribed in circle/Circumscribed about circle] <I>: *I* or *C*
> Specify radius of circle: PICK or (**coordinates**)
> Command:

The orientation of the *Polygon* and the imaginary circle are shown in Figure 15-8. Note that the *Inscribed* option allows control of one-half of the distance <u>across the corners</u>, and the *circumscribed* option allows control of one-half of the distance <u>across the flats</u>.

Using *ORTHO ON* with specification of the *radius of circle* forces the *Polygon* to a 90 degree orientation.

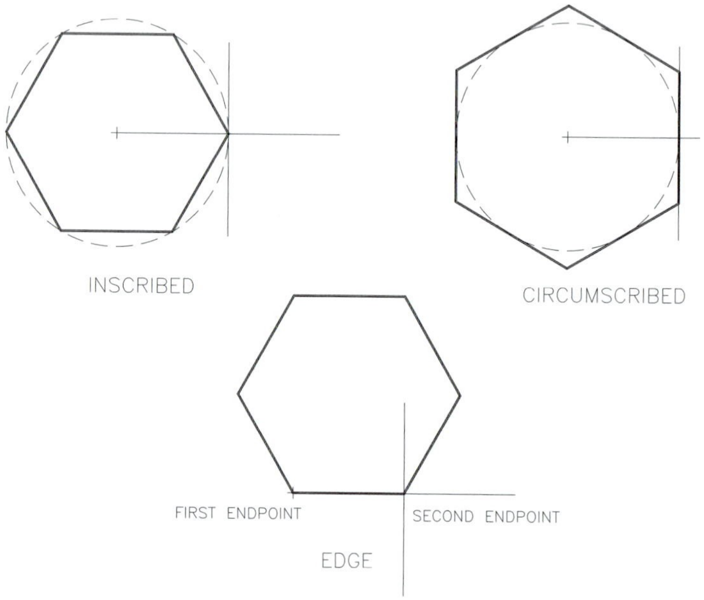

Edge
The *Edge* option only requires you to indicate the number of sides desired and to specify the two endpoints of one edge (Fig. 15-8).

> Command: **polygon**
> Enter number of sides <4>: (**value**)
> Specify center of polygon or [Edge]: *e*
> Specify first endpoint of edge: PICK or (**coordinates**)
> Specify second endpoint of edge: PICK or (**coordinates**)
> Command:

Figure 15-8

INSCRIBED

CIRCUMSCRIBED

FIRST ENDPOINT   SECOND ENDPOINT

EDGE

Because *Polygons* are created as *Plines*, *Pedit* can be used to change the line width or edit the shape in some way (see "*Pedit*," Chapter 16). *Polygons* can also be *Exploded* into individual objects similar to the way other *Plines* can be broken down into component objects (see "*Explode*," Chapter 16).

RECTANG

Pull-down Menu	COMMAND (TYPE)	ALIAS (TYPE)	Short-cut	Screen (side) Menu	Tablet Menu
Draw *Rectangle*	*RECTANG*	*REC*	...	DRAW 1 *Rectang*	Q,10

The *Rectang* command only requires the specification of two diagonal corners for construction of a rectangle, identical to making a selection window (Fig. 15-9, on the next page). The corners can be PICKed or dimensions can be entered. The rectangle can be any proportion, but the sides are always horizontal and vertical. The completed rectangle is <u>one AutoCAD object</u>, not four separate objects.

> Command: **rectang**
> Specify first corner point or [Chamfer/Elevation/Fillet/Thickness/Width]: PICK or (**coordinates**)
> Specify other corner point or [Dimensions]: *D* or PICK or (**coordinates**)

Several options of the *Rectangle* command affect the shape as if it were a *Pline* (see *Pline* and *Pedit*). For example, a *Rectangle* can have *width*:

> Specify first corner point or
> [Chamfer/Elevation/Fillet/Thickness/Width]: **w**
> Specify line width for rectangles <0.0000>: (**value**)

The *Fillet* and *Chamfer* options allow you to specify values for the fillet radius or chamfer distances:

> Specify first corner point or
> [Chamfer/Elevation/Fillet/Thickness/Width]: **f**
> Specify fillet radius for rectangles <0.0000>: (**value**)

> Specify first corner point or
> [Chamfer/Elevation/Fillet/Thickness/Width]: **c**
> Specify first chamfer distance for rectangles <0.0000>: (**value**)
> Specify second chamfer distance for rectangles <0.5000>: (**value**)

The *Dimensions* option prompts you for *length* and *width* dimensions for rectangles. (*Thickness* and *Elevation* are 3D properties. See Chapter 40, Surface Modeling, for more information.)

Figure 15-9

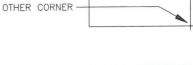

FIRST CORNER
OTHER CORNER

WIDTH=.05

FILLET
RADIUS=.5

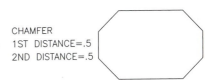

CHAMFER
1ST DISTANCE=.5
2ND DISTANCE=.5

DONUT

Pull-down Menu	COMMAND (TYPE)	ALIAS (TYPE)	Short-cut	Screen (side) Menu	Tablet Menu
Draw *Donut*	*DONUT*	*DO*	...	DRAW 1 *Donut*	*K,9*

A *Donut* is a circle with width (Fig. 15-10). Invoking the command allows changing the inside and outside diameters and creating multiple *Donuts*:

> Command: **donut**
> Specify inside diameter of donut <0.5000>: (**value**)
> Specify outside diameter of donut <1.0000>: (**value**)
> Specify center of donut or <exit>: **PICK**
> Specify center of donut or <exit>: **Enter**

Figure 15-10

FILL ON

INSIDE DIA = 0.5   INSIDE DIA = 0   INSIDE DIA = 1.8
OUTSIDE DIA = 1   OUTSIDE DIA = 1   OUTSIDE DIA = 2

FILL OFF

 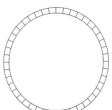

*Donut*s are actually solid filled circular *Pline*s with width. The solid fill for *Donut*s, *Pline*s, and other "solid" objects can be turned off with the *Fill* command or *FILLMODE* system variable.

SPLINE

	Pull-down Menu	COMMAND (TYPE)	ALIAS (TYPE)	Short-cut	Screen (side) Menu	Tablet Menu
	Draw Spline	SPLINE	SPL	...	DRAW 1 Spline	L,9

The *Spline* command creates a NURBS (non-uniform rational bezier spline) curve. The non-uniform feature allows irregular spacing of selected points to achieve sharp corners, for example. A *Spline* can also be used to create regular (rational) shapes such as arcs, circles, and ellipses. Irregular shapes can be combined with regular curves, all in one spline curve definition.

Spline is the newer and more functional version of a *Spline*-fit *Pline* (see "*Pedit*," Chapter 16). The main difference between a *Spline*-fit *Pline* and a *Spline* is that a *Spline* curve passes through the points selected, while the points selected for construction of a *Spline*-fit *Pline* only have a "pull" on the curve. Therefore, *Splines* are more suited to accurate design because the curve passes exactly through the points used to define the curve (data points) (Fig. 15-11).

Figure 15-11

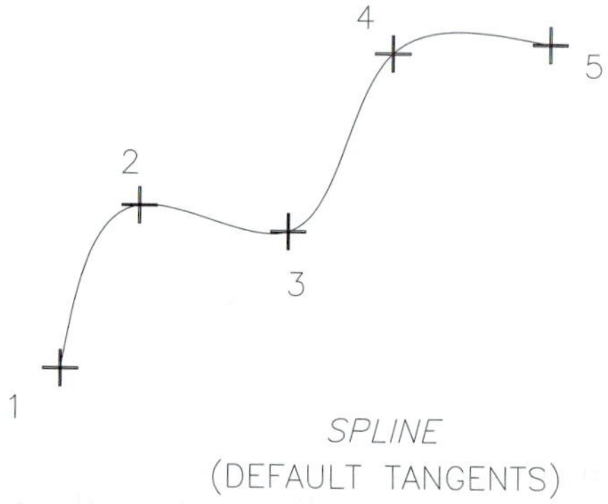

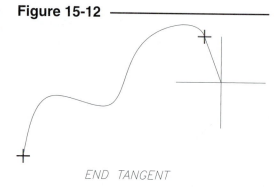

SPLINE
(DEFAULT TANGENTS)

The construction process involves specifying points that the curve will pass through and determining tangent directions for the two ends (for non-closed *Splines*) (Fig. 15-12).

Figure 15-12

END TANGENT

The *Close* option allows creation of closed *Splines* (these can be regular curves if the selected points are symmetrically arranged) (Fig. 15-13).

Figure 15-13

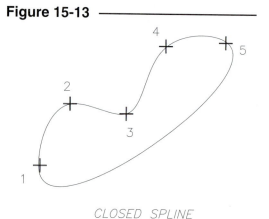

CLOSED SPLINE

```
Command: spline
Specify first point or [Object]: PICK or (coordinates)
Specify next point: PICK or (coordinates)
Specify next point or [Close/Fit tolerance] <start tangent>: PICK or (coordinates)
Specify next point or [Close/Fit tolerance] <start tangent>: PICK or (coordinates)
Specify next point or [Close/Fit tolerance] <start tangent>: Enter
Specify start tangent: PICK or Enter (Select direction for tangent or Enter for default)
Specify end tangent: PICK or Enter (Select direction for tangent or Enter for default)
Command:
```

The *Object* option allows you to convert *Spline*-fit *Plines* into NURBS *Splines*. Only *Spline*-fit *Plines* can be converted (see "*Pedit*," Chapter 16).

```
Command: spline
Specify first point or [Object]: o
Select objects to convert to splines ...
Select objects: PICK
Select objects: Enter
Command:
```

A *Fit Tolerance* applied to the *Spline* "loosens" the fit of the curve. A tolerance of 0 (default) causes the *Spline* to pass exactly through the data points. Entering a positive value allows the curve to fall away from the points to form a smoother curve (Fig. 15-14).

Figure 15-14

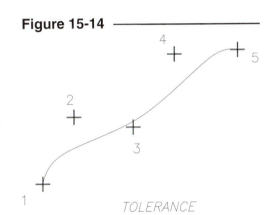

TOLERANCE

```
Specify next point or [Close/Fit tolerance] <start tangent>: f
Specify fit tolerance <0.0000>: (Enter a positive value)
```

ELLIPSE

	COMMAND (TYPE)	ALIAS (TYPE)	Short-cut	Screen (side) Menu	Tablet Menu
Pull-down Menu					
Draw *Ellipse*	*ELLIPSE*	*EL*	...	DRAW 1 *Ellipse*	*M,9*

An *Ellipse* is one object. AutoCAD *Ellipses* are (by default) NURBS curves (see *Spline*). There are three methods of creating *Ellipses* in AutoCAD: (1) specify one <u>axis</u> and the <u>end</u> of the second, (2) specify the <u>center</u> and the ends of each axis, and (3) create an elliptical <u>arc</u>. Each option also permits supplying a rotation angle rather than the second axis length.

```
Command: ellipse
Specify axis endpoint of ellipse or [Arc/Center]: PICK or (coordinates)
(This is the first endpoint of either the major or minor axis.)
Specify other endpoint of axis: PICK or (coordinates)
(Select a point for the other endpoint of the first axis.)
Specify distance to other axis or [Rotation]: PICK or (coordinates)
(This is the distance measured perpendicularly from the established axis.)
Command:
```

Axis End

This default option requires PICKing three points as indicated in the command sequence above (Fig. 15-15).

Figure 15-15

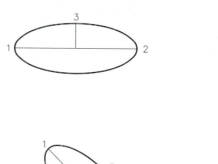
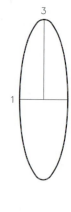
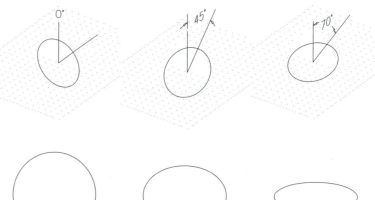

Rotation

If the *Rotation* option is used with the *Axis End* method, the following syntax is used:

Figure 15-16

Specify distance to other axis or [Rotation]: **r**
Specify rotation around major axis: **PICK** or **(value)**

The specified angle is the number of degrees the shape is rotated from the circular position (Fig. 15-16).

ROTATION = 0   ROTATION = 45   ROTATION = 70

Center

With many practical applications, the center point of the ellipse is known, and therefore the *Center* option should be used (Fig. 15-17):

Figure 15-17

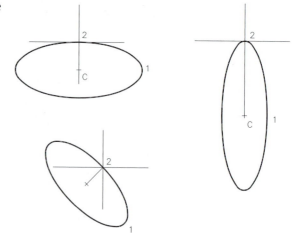

Command: *ellipse*
Specify axis endpoint of ellipse or [Arc/Center]: *c*
Specify center of ellipse: **PICK** or **(coordinates)**
Specify endpoint of axis: **PICK** or **(coordinates)**
(Select a point for the other endpoint of the first axis.)
Specify distance to other axis or [Rotation]: **PICK** or **(coordinates)** (This is the distance measured
perpendicularly from the established axis.)
Command:

The *Rotation* option appears and can be invoked after specifying the *Center* and first *Axis endpoint*.

Arc

Use this option to construct an elliptical arc
(partial ellipse). The procedure is identical
to the *Center* option with the addition of
specifying the start- and endpoints for the
arc (Fig. 15-18):

Figure 15-18

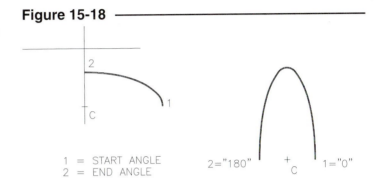

1 = START ANGLE   2="180"   1="0"
2 = END ANGLE

Command: *ellipse*
Specify axis endpoint of ellipse or [Arc/Center]: *a*
Specify axis endpoint of elliptical arc or [Center]: **PICK** or **(coordinates)**
Specify other endpoint of axis: **PICK** or **(coordinates)**
Specify distance to other axis or [Rotation]: **PICK** or **(coordinates)**
Specify start angle or [Parameter]: **PICK** or **(angular value)**
Specify end angle or [Parameter/Included angle]: **PICK** or **(angular value)**
Command:

The *Parameter* option allows you to specify the start point and endpoint for the elliptical arc. The parameters are based on the parametric vector equation: $p(u)=c+a*cos(u)+b*sin(u)$, where c is the center of the
ellipse and a and b are the major and minor axes.

DIVIDE

Pull-down Menu	COMMAND (TYPE)	ALIAS (TYPE)	Short-cut	Screen (side) Menu	Tablet Menu
Draw *Point>* *Divide*	DIVIDE	DIV	...	DRAW 2 *Divide*	V,13

The *Divide* and *Measure* commands add *Point* objects to existing objects. Both commands are found in
the *Draw* pull-down menu under *Point* because they create *Point* objects.

The *Divide* command finds equal intervals along an object such as a *Line, Pline, Spline,* or *Arc* and adds a
Point object at each interval. The object being divided is not actually broken into parts—it remains as
one object. *Point* objects are automatically added to display the "divisions."

The point objects that are added to the object can be used for subsequent construction by allowing you to
OSNAP to equally spaced intervals (*Nodes*).

The command sequence for the *Divide* command is as follows:

Command: **divide**
Select object to divide: **PICK** (Only one object can be selected.)
Enter the number of segments or [Block]: (**value**)
Command:

Point objects are added to divide the object selected into the desired number of parts. Therefore, there is <u>one less</u> *Point* added than the number of segments specified.

After using the *Divide* command, the *Point* objects may not be visible unless the point style is changed with the *Point Style* dialog box (*Format* pull-down menu) or by changing the *PDMODE* variable by command line format (see Chapter 8). Figure 15-19 shows *Points* displayed using a *PDMODE* of 3.

Figure 15-19

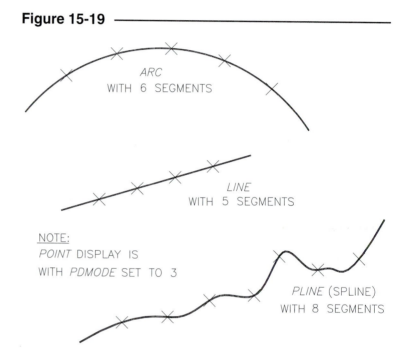

ARC
WITH 6 SEGMENTS

LINE
WITH 5 SEGMENTS

NOTE:
POINT DISPLAY IS
WITH PDMODE SET TO 3

PLINE (SPLINE)
WITH 8 SEGMENTS

You can request that *Block*s be inserted rather than *Point* objects along equal divisions of the selected object. Figure 15-20 displays a generic rectangular-shaped block inserted with *Divide*, both aligned and not aligned with a *Line, Arc,* and *Pline.* In order to insert a *Block* using the *Divide* command, the name of an <u>existing</u> *Block* must be given. (See Chapter 21, Blocks and DesignCenter.)

Figure 15-20

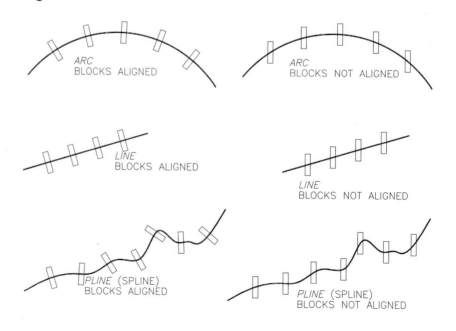

ARC
BLOCKS ALIGNED

ARC
BLOCKS NOT ALIGNED

LINE
BLOCKS ALIGNED

LINE
BLOCKS NOT ALIGNED

PLINE (SPLINE)
BLOCKS ALIGNED

PLINE (SPLINE)
BLOCKS NOT ALIGNED

MEASURE

	COMMAND (TYPE)	ALIAS (TYPE)	Short-cut	Screen (side) Menu	Tablet Menu
Pull-down Menu					
Draw *Point >* *Measure*	*MEASURE*	*ME*	...	DRAW 2 Measure	*V,12*

The *Measure* command is similar to the *Divide* command in that *Point* objects (or *Blocks*) are inserted along the selected object. The *Measure* command, however, allows you to designate the <u>length</u> of segments rather than the <u>number</u> of segments as with the *Divide* command.

> Command: **measure**
> Select object to measure: **PICK** (Only one object can be selected.)
> Specify length of segment or [Block]: (**value**) (Enter a value for length of one segment.)
> Command:

Point objects are added to the selected object at the designated intervals (lengths). One *Point* is added for each interval <u>beginning at the end nearest</u> the end used for object selection (Fig. 15-21). The intervals are of equal length except possibly the last segment, which is whatever length is remaining.

You can request that *Blocks* be inserted rather than *Point* objects at the designated intervals of the selected object. Inserting *Blocks* with *Measure* requires that an <u>existing</u> *Block* be used.

Figure 15-21

MLINE

	COMMAND (TYPE)	ALIAS (TYPE)	Short-cut	Screen (side) Menu	Tablet Menu
Pull-down Menu					
Draw *Multiline*	*MLINE*	*ML*	...	DRAW 1 Mline	*M,10*

Mline is short for multiline. A multiline is a set of <u>parallel lines</u> that behave as <u>one object</u>. The individual lines in the set are called line <u>elements</u> and are defined to your specifications. The set can contain up to 16 individual lines, each having its own offset, linetype, or color.

Mlines are useful for creating drawings composed of many parallel lines. An architectural floor plan, for example, contains many parallel lines representing walls. Instead of constructing the parallel lines individually (symbolizing the wall thickness), only <u>one</u> *Mline* need be drawn to depict the wall. The thickness and other details of the wall are dictated by the definition of the *Mline*.

<u>*Mline*</u> is the command to <u>draw</u> multilines. To <u>define</u> the individual line elements of a logical set (offset, linetype, color), use the <u>*Mlstyle*</u> command. An *Mlstyle* (multiline style) is a particular combination of defined line elements. Each *Mlstyle* can be saved in an external file with a .MLN file type. In this way, multilines can be used in different drawings without having to redefine the set. See *"Mlstyle"* next.

The *Mline* command operates similar to the *Line* command, asking you to specify points:

```
Command: mline
Current settings: Justification = Top, Scale = 1.00, Style = STANDARD
Specify start point or [Justification/Scale/STyle]: PICK or (coordinates)
Specify next point: PICK or (coordinates)
Specify next point or [Undo]: PICK or (coordinates)
Specify next point or [Close/Undo]: PICK or (coordinates)
Specify next point or [Close/Undo]: Enter
Command:
```

Mline provides the following three options for drawing multilines.

Justification

Justification determines how the multiline is drawn between the points you specify. The choices are *Top, Zero,* or *Bottom* (Fig. 15-22). *Top* aligns the top of the set of line elements along the points you PICK as you draw. A *Zero* setting uses the center of the set of line elements as the PICK point, and *Bottom* aligns the bottom line element between PICK points.

NOTE: The *Justification* methods are defined for drawing from <u>left to right</u> (positive X direction). In other words, using the *Top* justification, the top of the multiline (as it is defined in the *Mlstyle* dialog box) is between PICK points when you draw from left to right. PICKing points from right to left draws the *Mline* upside down.

Figure 15-22

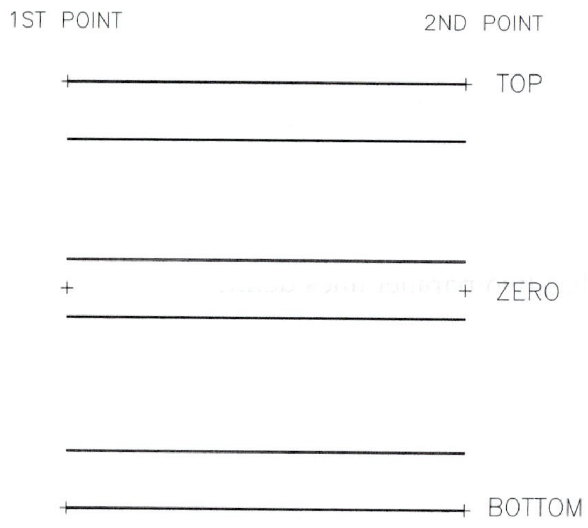

1ST POINT   2ND POINT

TOP

ZERO

BOTTOM

Scale

The *Scale* option controls the overall <u>width</u> of the multiline. The value you specify for the scale factor increases or decreases the defined multiline width proportionately. In Figure 15-23, a scale factor of 2.0 provides a multiple line pattern that is twice as wide as the *Mlstyle* definition; a scale of 3.0 provides a pattern three times the width.

Figure 15-23

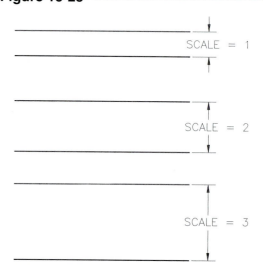

SCALE = 1

SCALE = 2

SCALE = 3

You can also enter negative scale values or a value of 0. A negative scale value flips the order of the multiple line pattern (Fig. 15-24). A value of 0 collapses the multiline into a single line. The utility of the *Scale* option allows you to vary a multiline set proportionally without the need for a complete new multiline definition.

Figure 15-24

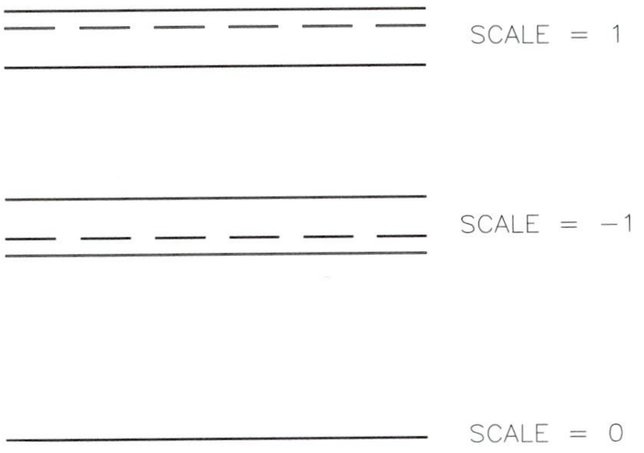

SCALE = 1

SCALE = −1

SCALE = 0

Figure 15-25

VARIOUS MULTILINE STYLES

Style
The *Style* option enables you to load different multiline styles provided they have been previously created. Some examples of multiline styles are shown in Figure 15-25. Only one multiline *Style* called STANDARD is provided by AutoCAD. The STANDARD multiline style has two parallel lines defined at 1 unit width.

STANDARD (DEFAULT)

SAMPLE 1

Using the *Style* option of the *Mline* command provides only the Command line format for setting a current style. Instead, you can use the *Mlstyle* dialog box to select a current multiline style by a more visible method.

SAMPLE 2

MLSTYLE

Pull-down Menu	COMMAND (TYPE)	ALIAS (TYPE)	Short-cut	Screen (side) Menu	Tablet Menu
Format Multiline Style...	*MLSTYLE*	...	...	DRAW 1 Mline Mlstyle:	V,5

The *Mline* command draws multilines, whereas the *Mlstyle* controls the configuration of the multilines that are drawn. Use the *Mlstyle* command to create new multiline definitions or to load, select, and edit previously created multiline styles. *Mlstyle* is not a draw command and is therefore found not in the *Draw* menus but in the *Format* pull-down and screen menus.

Using the *Mlstyle* command by any method invokes the *Multiline Styles* dialog box (Fig. 15-26). The areas of the three main dialog boxes are described next. Creating a new multiline style requires several steps that should follow a specific sequence. The "Steps for Creating New Multiline Styles" and "Steps for Editing Multiline Styles" are listed after the sections describing the *Multiline Styles* dialog box and nested dialog boxes.

Figure 15-26

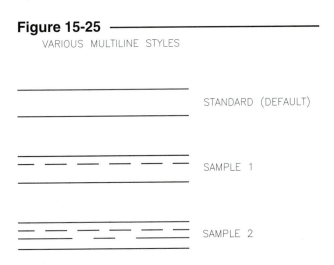

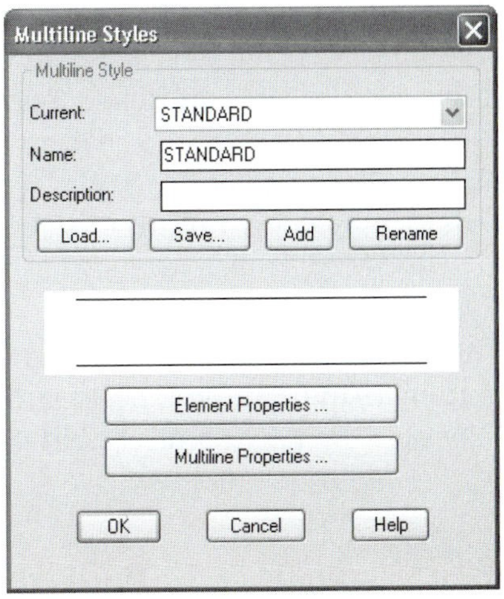

The *Multiline Styles* Dialog Box

Element Properties... and *Multiline Properties...*

These two tiles (see Figure 15-26) produce dialog boxes for defining new or changing existing multiline styles. When creating a new style, use these boxes first to define the line elements and their properties before *Naming, Adding,* and *Saving* the style. Features of these two dialog boxes are described later.

Current

This drop-down list provides you with the names of the currently loaded multiline styles. The image tile in the center of the dialog box provides a representation of the selected *Current* multiline style definition. Therefore, you can select names in the drop-down list to browse through the loaded multiline styles. The *Style* option of the *Mline* command serves the same purpose but operates only in Command line format and requires you to know the specific *Mlstyle* name and definition.

Name

When you create new multiline styles, use the *Name* edit box to enter the name of the style you have created. The *Mlstyle* can be *Named* after assigning *Element Properties* and *Multiline Properties*. Short names are desirable since a *Description* of the multiline configuration can be entered in the edit box just below *Name*.

Description

The *Description* edit box is an optional feature that allows you to provide a description for a new or existing multiline style definition. The description can contain up to 256 characters.

Load...

Figure 15-27

The *Load* tile invokes the *Load Multiline Styles* dialog box so that you can select an existing multiline definition from an external file (Fig. 15-27). The default multiline definition file is ACAD.MLN, which contains the STANDARD multiline style. This file can be appended or a new .MLN file can be created. An .MLN file can contain <u>more than one</u> multiline style.

Save...

Selecting the *Save* button invokes the typical save file dialog box (titled *Save Multiline Style*) so that you can save a multiline definition to an external file. Two or more multiline styles can be saved to one external file. Use any descriptive file name and an .MLN file extension is automatically appended. Saving a multiline style to an external file allows you to use the multiline definition in other drawings without having to recreate the style.

Add

Once you have defined or edited a new style using the *Element Properties* and *Multiline Properties* dialog boxes, use the *Add* tile. The *Add* tile adds the new multiline definition to the current multiline style drop-down list. Several multiline styles can be created in this manner.

NOTE: New and edited styles require a new *Name* before using *Add*. You must choose the *OK* button in order to save new multiline definitions to the drawing file.

Rename

The *Rename* tile allows you to rename the current multiline style. First, select the style to be renamed so it appears in both the *Current* and *Name* boxes. Next, change the name in the *Name* box and select *Rename*. The new name should appear in both boxes.

The *Element Properties* Dialog Box

The *Element Properties…* button (in the *Multiline Styles* dialog box) invokes the *Element Properties* dialog box (Fig. 15-28). This dialog box provides controls for you to define the offset (spacing), linetype, and color properties of <u>individual line elements</u> of a new or existing multiline style.

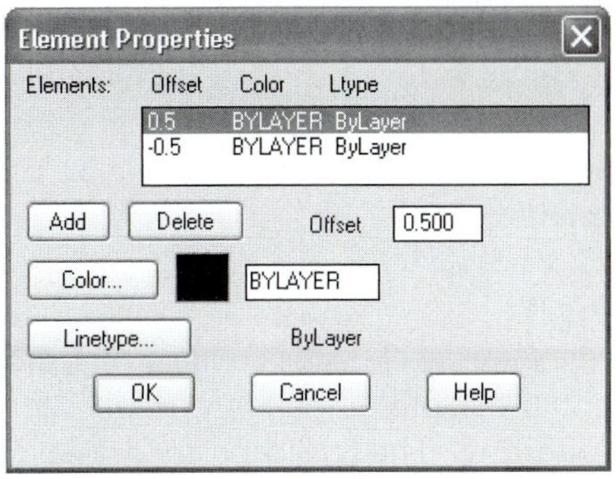

The *Elements:* list displays the <u>current</u> settings for the multiline style. Each line entry in a style has an offset, color, and linetype associated with it. A multiline definition can have as many as 16 individual line elements defining it. Select (highlight) any element in the list before changing its properties with the options below.

Figure 15-28

Add
New line elements are added to the list by selecting the *Add* button; then the *Offset, Color,* and *Linetype* properties can be assigned.

Delete
The current line element (the highlighted entry in the *Elements:* list) is deleted by selecting the *Delete* button.

Color
Selecting this tile produces the standard *Select Color* dialog box. *ByLayer* and *ByBlock* settings are valid. Any color selected is assigned to the highlighted line element in the list.

Linetype
Use this button to assign a linetype to the highlighted line element. The standard *Select Linetype* dialog box appears. Linetypes can be *Loaded* from this device.

Offset
Enter a value in the *Offset* edit box for each line element. The offset value is the distance measured perpendicular to the theoretical zero (center) position along the axis of the *Mline*.

The *Multiline Properties* Dialog Box

The *Multiline Properties* dialog box (Fig. 15-29) is also nested and invoked through the *Multiline Styles* dialog box. The options here enable you to control end capping, fill, and joint display properties of the multiline style. This device controls features of the <u>multiline style as a whole</u>, whereas the *Element properties* dialog treats the <u>individual</u> line element properties.

When multiline properties are assigned in this dialog box, the changes are reflected in the image tile upon returning to the *Multiline Styles* dialog box. You can cycle back and forth between these two boxes to "see" the changes as you make them. Figure 15-30 illustrates some examples of the options described on the next page.

Figure 15-29

Display Joints

When several segments of an *Mline* are drawn, AutoCAD automatically miters the "joints" so the segments connect with sharp, even corners. The joint is the bisector of the angle created by two adjoining segments. The *Display Joints* checkbox toggles on and off this miter line running across all the line elements.

Figure 15-30

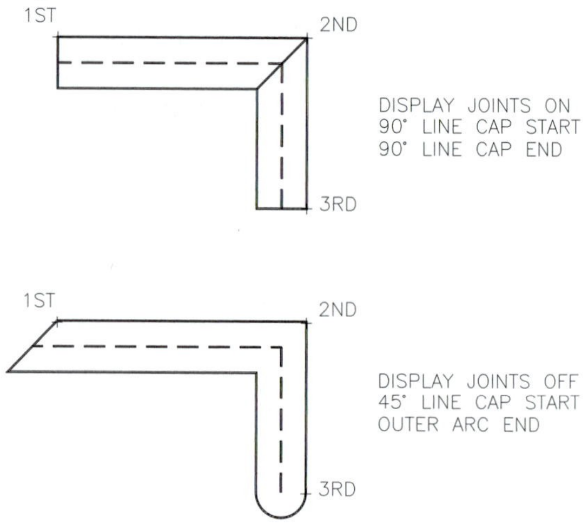

DISPLAY JOINTS ON
90° LINE CAP START
90° LINE CAP END

Line

The *Line* checkboxes create <u>straight</u> line caps across either or both of the *Start* or *End* segments of the multiline. The *Line* caps are drawn across all line elements.

DISPLAY JOINTS OFF
45° LINE CAP START
OUTER ARC END

Outer Arcs

This option creates an arc connecting and tangent to the two outermost line elements. You can control both the *Start* and *End* segments.

Inner Arcs

The inner arc option creates an arc between interior line segments. This option works with four or more line elements. In the case of an uneven number of line elements (five or more), the middle line is not capped.

Angle

The angle option controls the angle of start and end lines or arcs. Valid values range from 10 to 170 degrees.

Fill

The *Fill* toggle turns off or on the background fill of a multiline. The "X" must appear in the box before the *Color* tile is enabled. Selecting *Color* produces the standard *Select Color* dialog box. The image tile does not display the multiline fill, but fill appears in the drawing.

Steps for Creating New Multiline Styles

1. Use the *Mlstyle* command to invoke the *Multiline Styles* dialog box.

2. Use the *Element Properties* dialog box to define the line elements' offset, linetype, and color properties.

3. Define the multiline end capping, fill, and joint display properties using the *Multiline Properties* dialog box.

4. Return to the *Multiline Styles* dialog box and assign a *Name* to the new style.

5. Use the *Add* button to include the new style to the list of potential current styles and make it the current style.

6. Select *OK* to exit the *Multiline Styles* dialog box and begin drawing *Mlines* with the new style.

7. If you want to use the new style with other drawings, use *Save* in the *Multiline Styles* dialog box to save the style to an external .MLN file.

Steps for Editing Existing Multiline Styles

1. Use the *Mlstyle* command to invoke the *Multiline Styles* dialog box.

2. *Load* the desired multiline style from an external file or select it from the *Current* list. The style name must appear in the *Name* box.

3. If you want to edit the line elements' offset, linetype, and color properties, use the *Element Properties* dialog box.

4. If you want to edit the multiline end capping, fill, and joint display properties, use the *Multiline Properties* dialog box.

5. You must assign a <u>new *Name*</u> to the changed style in the *Multiline Styles* dialog box by editing the old name. You cannot use the same name. An "Invalid name" message appears if you attempt to *Add* using the old name.

6. Then you must *Add* the new (edited) style to the list of styles. This action also makes it the current style.

7. Select *OK* to exit the *Multiline Styles* dialog box and begin drawing *Mlines* with the new style.

8. If you want to use the new (edited) style with other drawings, use *Save* in the *Multiline Styles* dialog box to save the style to an external .MLN file.

Multiline Styles Example

Because the process of creating and editing multiline styles is somewhat involved, here is an example that you can follow to create a new multiline style. The specifications for two multilines are given in Figure 15-31. Follow these steps to create SAMPLE1.

Figure 15-31

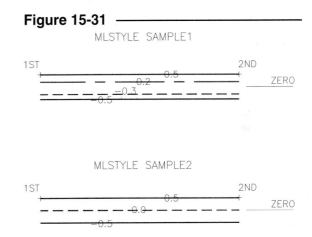

1. Invoke the *Multiline Styles* dialog box by any menu or icon or by typing *Mlstyle*.

2. Select the *Element Properties...* tile to invoke the *Element Properties* dialog box.

3. Select the *Add* button twice. Four line entries should appear in the *Elements:* list box. The two new entries have a 0.0 offset.

4. Highlight one of the new entries and change its offset by entering 0.2 in the *Offset* edit box. Next, change the linetype by selecting the *Linetype...* button and choosing the *Center2* linetype. (If the *Center2* linetype is not listed, you should *Load* it.)

5. Highlight the other new entry and change its *Offset* to -0.3. Next, change its *Linetype* to *Hidden2*.

6. Select the *OK* button to return to the *Multiline Styles* dialog box. At this point, enter the name SAMPLE1 in the *Name* edit box. Choose the *Add* button to save the new line definition and make it current. Last, choose the *OK* button to save the multiline style to the drawing database.

7. Use *Mline* to draw several segments and to test your new multiline style.

8. Use *Mlstyle* again and enter *Multiline Properties* to create *Line* end caps and a *Color Fill* to the style. Don't forget to change the name (to SAMPLE1-FILL, for example) and *Add* it to the list.

Now try to create SAMPLE2 multiline style on your own (see Figure 15-31).

Editing Multilines

Because a multiline is treated as one object and its complex configuration is defined by the multiline style, traditional editing commands such as *Trim* and *Extend*, cannot be used for typical editing functions. Instead, a special set of editing functions are designed for *Mlines*. The *Mledit* (multiline edit) command provides the editing functions for existing multilines and is discussed in Chapter 16. Your use for *Mlines* is limited until you can edit them with *Mledit*.

SKETCH

Pull-down Menu	COMMAND (TYPE)	ALIAS (TYPE)	Short-cut	Screen (side) Menu	Tablet Menu
...	*SKETCH*	...	...	...	...

The *Sketch* command is unlike other draw commands. *Sketch* quickly creates many short line segments (individual objects) by following the motions of the cursor. *Sketch* is used to give the appearance of a freehand "sketched" line, as used in Figure 15-32 for the tree and bushes. You do not specify individual endpoints, but rather draw a freehand line by placing the pen down, moving the cursor, and then picking up the pen. This action creates a large number of short *Line* segments. *Sketch* line segments are temporary (displayed in a different color) until *Recorded* or until *eXiting Sketch*.

Figure 15-32

 CAUTION: *Sketch* can increase the drawing file size greatly due to the relatively large number of line segment endpoints required to define the *Sketch* line.

> Command: **sketch**
> Record increment <0.1000>: (**value**) or **Enter** (Enter a value to specify the segment increment length or press Enter to accept the default increment.)
> Sketch. Pen eXit Quit Record Erase Connect. (**letter**) (Enter "p" or press button #1 to put the pen down and begin drawing or enter another letter for another option. After drawing a sketch line, enter "p" or press button #1 again to pick up the pen.)

It is important to specify an *increment* length for the short line segments that are created. This *increment* controls the "resolution" of the *Sketch* line (Fig. 15-33).

Figure 15-33

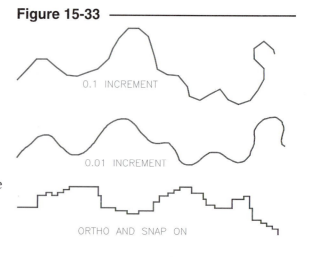

0.1 INCREMENT

0.01 INCREMENT

ORTHO AND SNAP ON

 Too large of an *increment* makes the straight line segments apparent, while too small of an *increment* unnecessarily increases file size. The default *increment* is 0.1 (appropriate for default *Limits* of 12 x 9) and should be changed proportionally with a change in *Limits*. Generally, multiply the default *increment* of 0.1 times the drawing scale factor.

Another important rule to consider whenever using *Sketch* is to turn *SNAP Off and ORTHO Off*, unless a "stair-step" effect is desired (see Figure 15-33). *Polar Snap* and *Polar Tracking* do not affect *Sketch* lines.

Sketch Options

Options of the *Sketch* command can be activated either by entering the corresponding letter(s) shown in uppercase at the *Sketch* command prompt or by pressing the desired mouse or puck button. For example, putting the pen up or down can be done by entering *P* at the Command line or by pressing button #1. The options of *Sketch* are as follows.

Pen	Lifts and lowers the pen. Position the cursor at the desired location to begin the line. Lower the pen and draw. Raise the pen when finished with the line.
Record	Records all temporary lines sketched so far without changing the pen position. After recording, the lines cannot be *Erased* with the *Erase* option of *Sketch* (although the normal *Erase* command can be used).
eXit	Records all temporary lines entered and returns to the Command: prompt.
Quit	Discards all temporary lines and returns to the Command: prompt.
Erase	Allows selective erasing of temporary lines (before recording). To erase, move backward from last sketch line toward the first. Press "p" to indicate the end of an erased area. This method works easily for relatively straight sections. To erase complex sketch lines, *eXit* and use the normal *Erase* command with window, crossing, or pickbox object selection.
Connect	Allows connection to the end of the last temporary sketch line (before recording). Move the cursor to the last sketch line and the pen is automatically lowered.
(period)	Draws a straight line (using *Line*) from the last sketched line to the cursor. After adding the straight lines, the pen returns to the up position.

Several options are illustrated in Figure 15-34. The *Pen* option lifts the pen up and down. A *period* (.) causes a straight line segment to be drawn from the last segment to the cursor location. *Erase* is accomplished by entering *E* and making a reverse motion.

As an alternative to the *Erase* option of *Sketch*, the *Erase* command can be used to erase all or part of the *Sketch* lines. Using a *Window* or *Crossing Window* is suggested to make selection of all the objects easier.

Figure 15-34 ──────────────

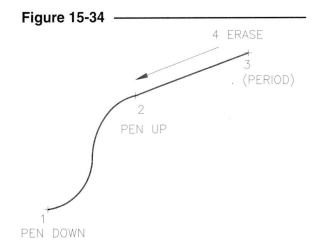

The *SKPOLY* Variable

The *SKPOLY* system variable controls whether AutoCAD creates connected *Sketch* line segments as one *Pline* (one object) or as multiple *Line* segments (separate objects). *SKPOLY* affects newly created *Sketch* lines only (Fig. 15-35).

Figure 15-35 ──────────────

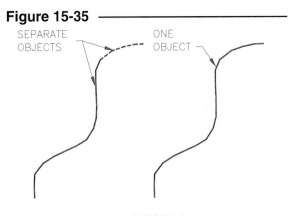

SKPOLY=0   This setting generates *Sketch* segments as individual *Line* objects. This is the <u>default</u> setting.

SKPOLY=1   This setting generates connected *Sketch* segments as <u>one *Pline* object</u>.

Using editing commands with *Sketch* lines can normally be tedious (when *SKPOLY* is set to the default value of 0). In this case, editing *Sketch* lines usually requires *Zooming* in since the line segments are relatively small. However, changing *SKPOLY* to 1 simplifies operations such as using *Erase* or *Trim*. For example, *Sketch* lines are sometimes used to draw break lines (Fig. 15-36) or to represent broken sections of mechanical parts, in which case use of *Trim* is helpful. If you expect to use *Trim* or other editing commands with *Sketch* lines, change *SKPOLY* to 1 before creating the *Sketch* lines.

Figure 15-36 ─────────────

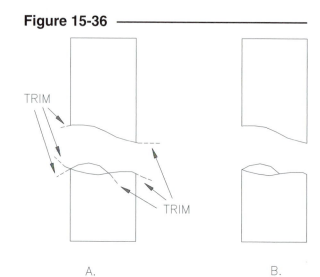

A.   B.

SOLID

Pull-down Menu	COMMAND (TYPE)	ALIAS (TYPE)	Short-cut	Screen (side) Menu	Tablet Menu
Draw *Surfaces >* *2D Solid*	*SOLID*	*SO*	...	*DRAW 2* *SURFACES* *Solid:*	*L,8*

The *Solid* command creates a 2D shape that is <u>filled solid</u> with the current color (object or *ByLayer*). The shape is <u>not</u> a 3D solid. Each 2D solid has either <u>three or four straight edges</u>. The construction process is relatively simple, but the order of PICKing corners is critical. In order to construct a square or rectangle, draw a bow-tie configuration (Fig. 15-37). Several *Solids* can be connected to produce more complex shapes.

Figure 15-37 ─────────────

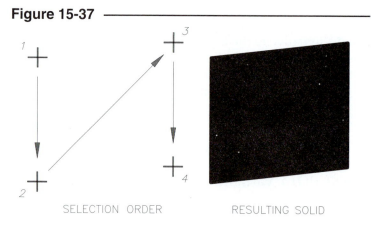

SELECTION ORDER   RESULTING SOLID

A *Solid* object is useful when you need shapes filled with solid color; for example, three-sided *Solids* can be used for large arrowheads. Keep in mind that solid filled areas may cause problems for some output devices such as plotters with felt pens and thin paper. As an alternative to using *Solid*, enclosed areas can be filled with hatch patterns or solid color (see "*Bhatch*," Chapter 26).

BOUNDARY

		COMMAND (TYPE)	ALIAS (TYPE)	Short-cut	Screen (side) Menu	Tablet Menu
	Pull-down Menu					
	Draw Boundary…	*BOUNDARY* or *-BOUNDARY*	*BO* or *-BO*	…	DRAW 2 *Boundary*	Q,9

Boundary finds and draws a boundary from a group of connected or overlapping shapes forming an enclosed area. The shapes can be any AutoCAD objects and can be in any configuration, as long as they form a totally enclosed area. *Boundary* creates either a *Polyline* or *Region* object forming the boundary shape (see *Region* next). The resulting *Boundary* does not affect the existing geometry in any way. *Boundary* finds and includes internal closed areas (called "islands") such as circles (holes) and includes them as part of the *Boundary*.

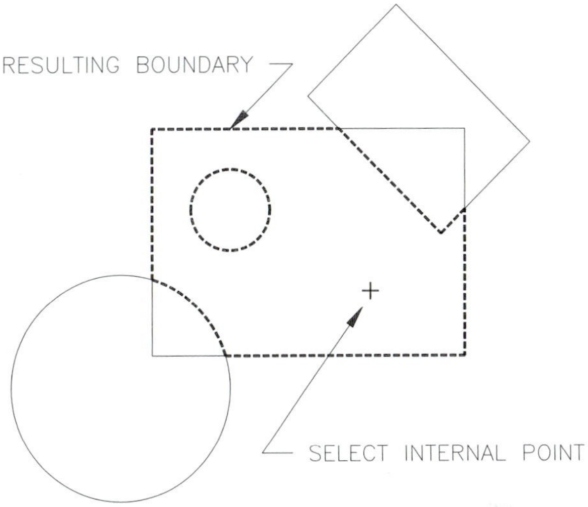

To create a *Boundary,* select an internal point in any enclosed area (Fig. 15-38). *Boundary* finds the boundary (complete enclosed shape) surrounding the internal point selected. You can use the resulting *Boundary* to construct other geometry or use the generated *Boundary* to determine the *Area* for the shape (such as a room in a floor plan). The same technique of selecting an internal point is also used to determine boundaries for sectioning using *Bhatch* (Chapter 26).

Figure 15-38

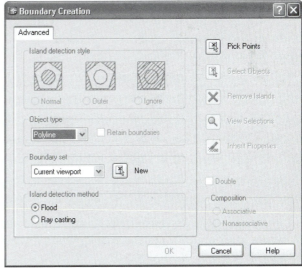

RESULTING BOUNDARY

SELECT INTERNAL POINT

Boundary operates in dialog box mode (Fig. 15-39), or type *-Boundary* for Command line mode. After setting the desired options, select the *Pick Points* tile to select the desired internal point (as shown in Fig. 15-38). The *Boundary Creation* dialog box is a subset of the *Advanced* tab of the *Boundary Hatch* dialog box; therefore, many options are disabled. (See "*Bhatch*" in Chapter 26.) The options are as follows.

Figure 15-39

Object Type
Construct either a *Region* or *Polyline* boundary (see *Region* next). If islands are found, two or more *Plines* are formed but only one *Region*.

Boundary Set
The *Current Viewport* option is sufficient for most applications; however, for large drawings you can make a new boundary set by selecting a smaller set of objects to be considered for the possible boundary.

Island Detection
This options tells AutoCAD whether or not to include interior enclosed objects in the Boundary. Flood automatically includes islands as boundary objects. Ray Casting sends a line out from the point you pick to the nearest object and then traces the boundary in a counterclockwise direction; therefore, islands are excluded as boundary objects.

Pick Points
Use this option to PICK an enclosed area in the drawing that is anywhere <u>inside</u> the desired boundary.

REGION

Pull-down Menu	COMMAND (TYPE)	ALIAS (TYPE)	Short-cut	Screen (side) Menu	Tablet Menu
Draw Region...	*REGION*	REG	...	DRAW 2 *Region*	R,9

The *Region* command converts one object or a set of objects forming a closed shape into one object called a *Region*. This is similar to the way in which a set of objects (*Lines, Arcs, Plines,* etc.) forming a closed shape and having matching endpoints can be converted into a closed *Pline*. A *Region*, however, has special properties:

1. Several *Regions* can be combined with Boolean operations known as *Union, Subtract,* and *Intersect* to form a "composite" *Region*. This process can be repeated until the desired shape is achieved.

2. A *Region* is considered a planar surface. The surface is defined by the edges of the *Region* and no edges can exist within the *Region* perimeter. *Regions* can be used with surface modeling.

In order to create a *Region,* a closed shape must exist. The shape can be composed of one or more objects such as a *Line, Arc, Pline, Circle, Ellipse,* or anything composed of a *Pline (Polygon, Rectangle, Boundary)*. If more than one object is involved, the <u>endpoints must match (having no gaps or overlaps)</u>. Simply invoke the *Region* command and select all objects to be converted to the *Region*.

```
Command: region
Select Objects: PICK
Select Objects: PICK
Select Objects: PICK
Select Objects: Enter
1 loop extracted.
1 Region created.
Command:
```

Consider the shape shown in Figure 15-40 composed of four connecting *Lines*. Using *Region* combines the shape into one object, a *Region*. The appearance of the object does not change after the conversion, even though the resulting shape is one object.

Although the *Region* appears to be no different than a closed *Pline,* it is more powerful because several *Regions* can be combined to form complex shapes (composite *Regions*) using the three Boolean operations explained in Chapter 16.

Figure 15-40 ————————————————

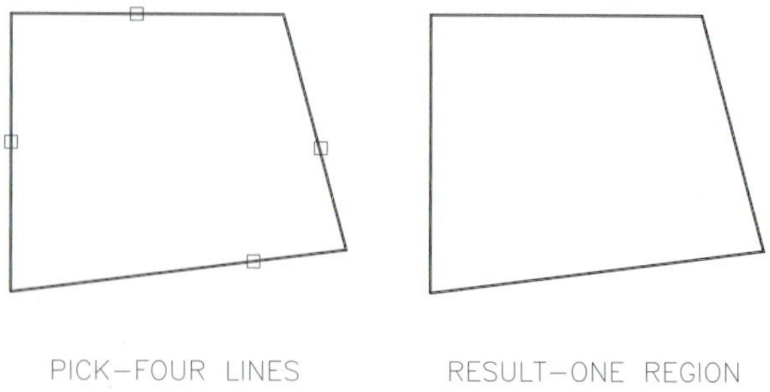

PICK—FOUR LINES   RESULT—ONE REGION

As an example, a set of *Regions* (converted *Circles*) can be combined to form the sprocket with only <u>one</u> *Subtract* operation (Fig. 15-41). Compare the simplicity of this operation to the process of using *Trim* to delete <u>each</u> of the unwanted portions of the small circles.

The Boolean operators, *Union*, *Subtraction*, and *Intersection*, can be used with *Regions* as well as solids. Any number of these commands can be used with *Regions* to form complex geometry (see Chapter 16, Modify Commands II).

Figure 15-41

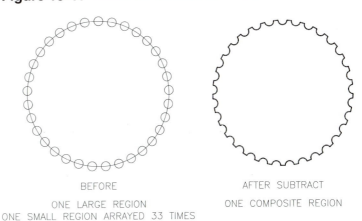

BEFORE
ONE LARGE REGION
ONE SMALL REGION ARRAYED 33 TIMES

AFTER SUBTRACT
ONE COMPOSITE REGION

WIPEOUT

Pull-down Menu	COMMAND (TYPE)	ALIAS (TYPE)	Short-cut	Screen (side) Menu	Tablet Menu
Draw *Wipeout*	*WIPEOUT*	...	...	...	...

Available only in the Express (or bonus) tools in previous releases of AutoCAD, this command has been included in the core AutoCAD 2004 product. *Wipeout* creates an "opaque" object that is used to "hide" other objects. The *Wipeout* command creates a closed *Pline* or uses an existing closed *Pline*. All objects behind the *Wipeout* become "invisible." For example, assume a *Pline* shape was created so as to cross other objects. You could use *Wipeout* to make the shape appear to be "on top of" the other objects (Fig. 15-42, left).

Figure 15-42

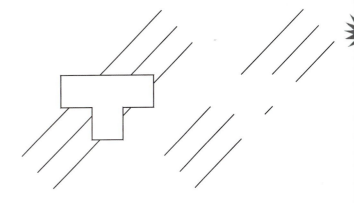

```
Command: wipeout
Specify first point or [Frames/Polyline] <Polyline>: PICK
Specify next point: PICK
Specify next point or [Undo]: PICK
Specify next point or [Close/Undo]: PICK
Specify next point or [Close/Undo]: Enter
Command:
```

You can use the *Frames* option to make the *Wipeout* object disappear. Doing so would result in a display similar to Figure 15-42, right.

```
Specify first point or [Frames/Polyline] <Polyline>: f
Enter mode [ON/OFF] <ON>: off
```

2004

If you have an existing *Pline* and want to use it to create a *Wipeout* of the same shape, use the *Polyline* option.

> Command: **wipeout**
> Specify first point or [Frames/Polyline] <Polyline>: **p**
> Select a closed polyline: **PICK**
> Erase polyline? [Yes/No] <No>:

Selecting *No* to the "Erase polyline?" prompt results in two objects that occupy the same space—the existing *Pline* and the new *Wipeout*. To create an "invisible" *Wipeout* you must specify the *Erase Polyline* option when you create the *Wipeout*, then use the *Frame Off* option to make the *Wipeout* object disappear. A *Wipeout* is an AutoCAD object. The *List* command reports the object as a *Wipeout* object.

REVCLOUD

Pull-down Menu	COMMAND (TYPE)	ALIAS (TYPE)	Short-cut	Screen (side) Menu	Tablet Menu
Draw *Revision Cloud*	*REVCLOUD*	...	...	...	...

New in AutoCAD 2004, this command was an Express (bonus) tool in AutoCAD 2000 and 2002. Use this command to draw a revision cloud on the current layer. Revision clouds are the customary method of indicating an area of a drawing (usually an architectural application) that contains a revision to the original design. Revision clouds should be created on a separate layer so they can be controlled for plotting.

Revcloud is an automated routine that simplifies drawing revision clouds. All you have to do is pick a point and move the cursor in a circular direction (Fig. 15-43). You do not have to "pick" an endpoint. The cloud automatically closes when your cursor approaches the start point and the command ends. A *Revcloud* is a series of *Pline* arc segments.

Figure 15-43

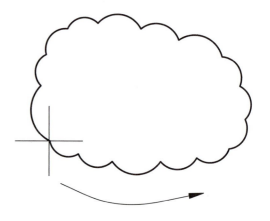

> Command: **revcloud**
> Minimum arc length: 0.5000   Maximum arc length: 0.5000
> Specify start point or [Arc length/Object] <Object>: **PICK**
> Guide crosshairs along cloud path... (move the cursor in a circular direction)
> Cloud finished.
> Command:

Use the *Arc length* option to specify a value in drawing units for the arc segments. Entering different values for the minimum and maximum arc lengths draws *Revclouds* with different size arcs, similar to that shown in Figure 15-43. You can convert some closed objects into *Revclouds* using the *Object* option. *Rectangles*, *Polygons*, *Circles*, and *Ellipses* will convert into *Revclouds*.

CHAPTER EXERCISES

1. *Polygon*

   Open the **PLINE1** drawing you created in Chapter 8 Exercises. Use *Polygon* to construct the polygon located at the *Center* of the existing arc, as shown in Figure 15-44. When finished, *SaveAs* **POLYGON1**.

Figure 15-44 ————

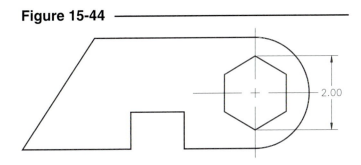

2. *Xline, Ellipse, Rectangle*

   Open the **APARTMENT** drawing that you created in Chapter 12. Draw the floor plan of the efficiency apartment on layer FLOORPLAN as shown in Figure 15-45. Use *SaveAs* to assign the name **EFF-APT**. Use *Xline* and *Line* to construct the floor plan with 8" width for the exterior walls and 5" for interior walls. Use the *Ellipse* and *Rectangle* commands to design and construct the kitchen sink, tub, wash basin, and toilet. *Save* the drawing but do not exit AutoCAD. Continue to Exercise 3.

Figure 15-45 ————

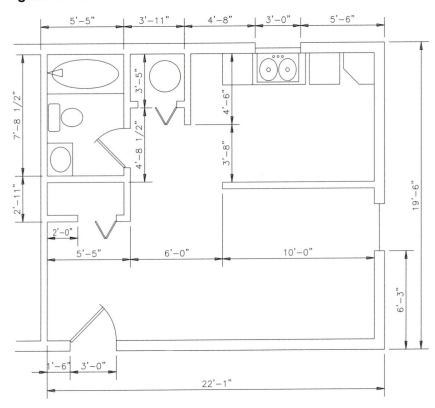

3. *Polygon, Sketch*

   Create a plant for the efficiency apartment as shown in Figure 15-46. Locate the plant near the entry. The plant is in a hexagonal pot (use *Polygon*) measuring 18" across the corners. Use *Sketch* to create 2 leaves as shown in figure A. Create a *Polar Array* to develop the other leaves similar to Figure B. *Save* the **EFF-APT** drawing.

Figure 15-46 ————

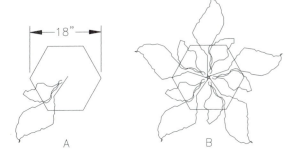

4. *Pline, Page Setup, Vports,* **and** *Viewport Scale*
This exercise gives you experience creating a border and plotting the efficiency apartment to scale. To complete the exercise as instructed, you should have a device configured to plot on a "C" size sheet.

A. Invoke the **Options** dialog box and access the **Display** tab. In the *Layout Elements* section (lower left), ensure that *Show Page Setup Dialog Box for New Layouts* and *Create Viewport in New Layouts* <u>are not checked</u>.

B. Select the **Layout1** tab. Activate the **Page Setup** dialog box. In the **Plot Device** tab, select a plot device from the list that will allow you to plot on an architectural "C" size sheet (**24" x 18"**). In the *Layout Settings* tab, select the correct sheet size. Pick **OK** to finish the setup.

C. Make layer **TITLE** the *Current* layer. Draw a border (in paper space) using a **Pline** with a *width* of **.04**, and draw the border with a **.75** unit margin within the edge of the *Limits* (border size of 22.5 x 16.5). *Save* the drawing as **EFF-APT-PS**.

D. Make layer **VPORTS** *Current*. Use *Vports* to create a *Single* viewport. Pick diagonal corners for the viewport at **1,1** and **23,17**. The apartment drawing should appear in the viewport at no particular scale. Use any method to set the viewport scale to *1/2"=1'*.

E. Next, *Freeze* layer **VPORTS**. Your completed drawing should look like that in Figure 15-47. *Save* the drawing. *Plot* the drawing from the layout at **1:1** scale.

Figure 15-47

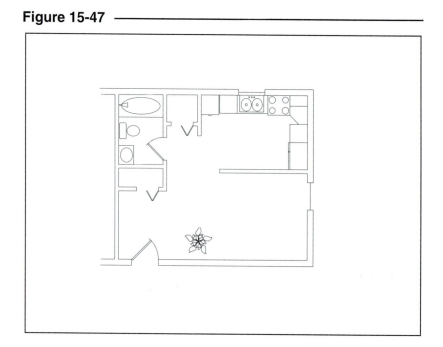

5. *Donut*

Open **SCHEMATIC1** drawing you created in Chapter 9 Exercises. Use the *Point Style* dialog box (*Format* pull-down menu) to change the style of the *Point* objects to dots instead of small circles. Turn on the *Node Running Osnap* mode. Use the *Donut* command and set the *Inside diameter* and the *Outside diameter* to an appropriate value for your drawing such that the *Inside diameter* is 1/2 the value of the *Outside diameter*. Create a *Donut* at each existing *Node*. *SaveAs* **SCHEMATIC1B**. Your finished drawing should look like that in Figure 15-48.

Figure 15-48

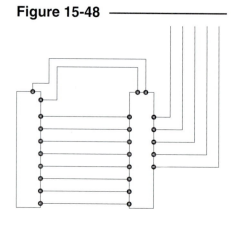

6. *Ellipse, Polygon*

Begin a *New* drawing using the template drawing that you created in Chapter 12 named **A-METRIC**. Use *SaveAs* and assign the name **WRENCH**.

Figure 15-49

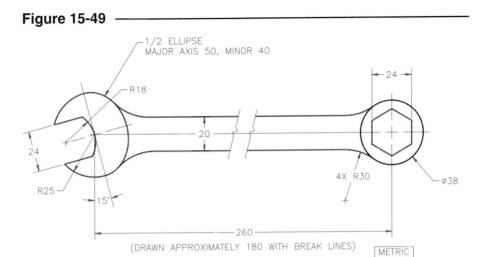

1/2 ELLIPSE
MAJOR AXIS 50, MINOR 40

R18

24

24

R25

15°

20

4X R30

Ø38

260

(DRAWN APPROXIMATELY 180 WITH BREAK LINES)

METRIC

Complete the construction of the wrench shown in Figure 15-49. Center the drawing within the *Limits*. Use *Line*, *Circle*, *Arc*, *Ellipse*, and *Polygon* to create the shape. For the "break lines," create one connected series of jagged *Lines*, then *Copy* for the matching set. Utilize *Trim*, *Rotate*, and other edit commands where necessary. HINT: Draw the wrench head in an orthogonal position; then rotate the entire shape 15 degrees. Notice the head is composed of 1/2 of a *Circle* (R25) 3 or plot device using *Pagesetup*, make one *Viewport*, and *Plot* the drawing at a standard scale.

7. *Xline, Spline*

Figure 15-50

Create the handle shown in Figure 15-50. Construct multiple *Horizontal* and *Vertical* *Xlines* at the given distances on a separate layer (shown as dashed lines in Figure 15-50). Construct two *Splines* that make up the handle sides by PICKing the points indicated at the *Xline* intersections. Note the **End Tangent** for one *Spline*. Connect the *Splines* at each end with horizontal **Lines.** *Freeze* the construction (*Xline*) layer. *Save* the drawing as **HANDLE**.

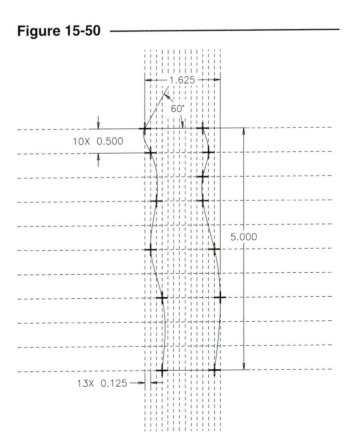

1.625

60°

10X 0.500

5.000

13X 0.125

8. *Divide, Measure*

Use the **ASHEET** template drawing, use
SaveAs, and assign the name **BILLMATL.**
Create the table in Figure 15-51 to be used
as a bill of materials. Draw the bottom
Line (as dimensioned) and a vertical *Line.*
Use *Divide* along the bottom *Line* and
Measure along the vertical *Line* to locate
Points as desired. Create *Offsets Through*
the *Points* using *Node OSNAP.* (*ORTHO*
and *Trim* may be of help.)

Figure 15-51 ──────────

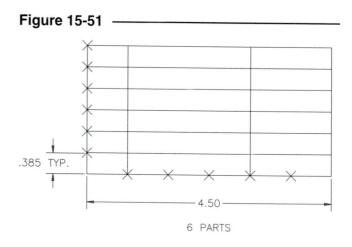

.385 TYP.

4.50

6 PARTS

9. *Mline, Mlstyle, Sketch*

Figure 15-52 ──────────

Using the **RET-WALL** drawing that
you created in Chapter 8, add the
concrete walls, island, and curbs (Fig.
15-52). Use *Mlstyle* to create a multiline style consisting of 2 line elements with *Offsets* of **1** and **-1.**
Assign *Start* and *End Caps* at **90**
degrees. *Name* the new style **WALL**
and *Add* it to the list to make it
current. Next, use *Mline* with the
Bottom Justification and draw one
Mline along the existing retaining
wall centerline. Use *Endpoint* to
snap to the ends of the existing 2
Lines and *Arc.* Construct a second
Mline above using the following coordinates:

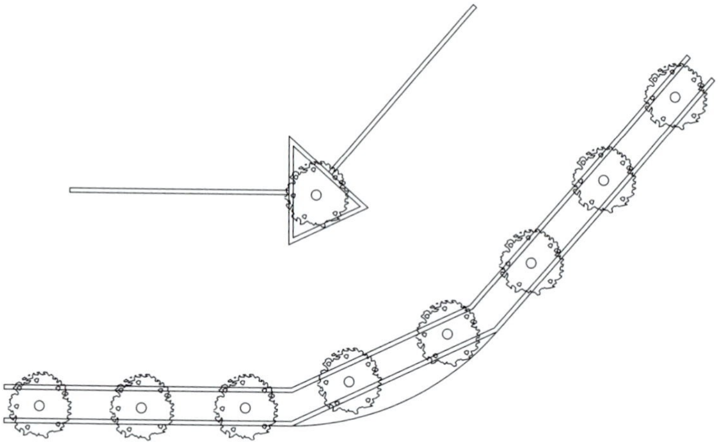

> 60,123
> 192,123
> 274,161
> 372.7,274

Next, construct the *Closed* triangular island with an *Mline* using *Top Justification.* The coordinates are:

> 192,192
> 222.58,206.26
> 192,233

For the parking curbs, draw two more *Mlines*; each *Mline* snaps to the *Midpoint* of the island and
is **100′** long. The curb on the right has an angle of **49** degrees (use polar coordinate entry).
Finally, create trees similar to the ones shown using *Sketch.* Remember to set the *SKPOLY* variable. Create one tree and *Copy* it to the other locations. Spacing between trees is approximately
48′. *Save* as **RET-WAL2.**

10. *Boundary*

Open the **GASKETA** drawing that you created in Chapter 9 Exercises. Use the *Boundary* command to create a boundary of the <u>outside shape only</u> (no islands). Use *Move Last* to displace the new shape to the right of the existing gasket. Use *SaveAs* and rename the drawing to **GASKET-AREA**. Keep this drawing to determine the *Area* of the shape after reading Chapter 17.

11. *Region*

Figure 15-53

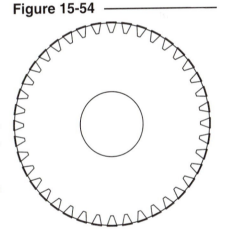

Create a gear, using *Regions*. Begin a *New* drawing, *Start from Scratch* with the *Imperial* defaults settings. Use *Save* and assign the name **GEAR-REGION**.

A. Set *Limits* at **8 x 6** and *Zoom All*. Create a *Circle* of **1** unit *diameter* with the center at **4,3**. Create a second concentric *Circle* with a *radius* of **1.557**. Create a closed *Pline* (to represent one gear tooth) by entering the following coordinates:

From point:	**5.571,2.9191**
To point:	**@.17<160**
To point:	**@.0228<94**
To point:	**@.0228<86**
To point:	**@.17<20**
To point:	**c**

The gear at this stage should look like that in Figure 15-53.

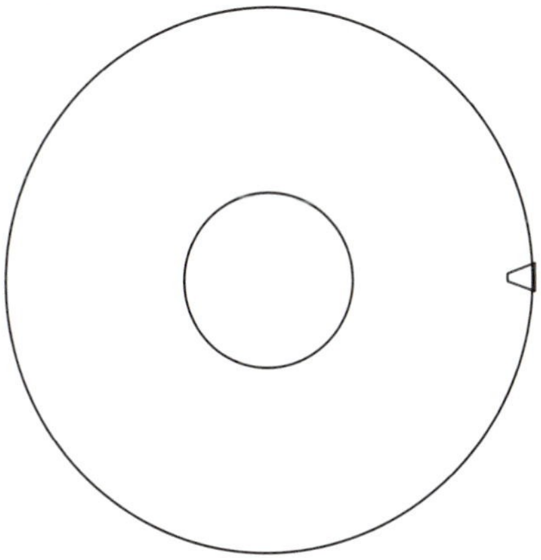

B. Use the *Region* command to convert the three shapes (two *Circles* and one "tooth") to three *Regions*.

C. *Array* the small *Region* (tooth) in a *Polar* array about the center of the gear. There are **40** items that are rotated as they are copied. This action should create all the teeth of the gear (Fig. 15-54).

D. *Save* the drawing. The gear will be completed in Chapter 16 Exercises by using the *Subtract* Boolean operation.

Figure 15-54

12. *Revcloud*

Figure 15-55

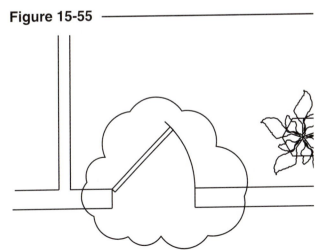

Assume you are working with the city engineering office checking the new apartment design submitted to your office for approval. *Open* **EFF-APT** drawing from Exercise 2. You notice that the location of the front door has some issues concerning passing the safety codes. Use *Revcloud* to denote the area for review. Use the *Arc length* option and specify a *Minimum arc length* of **1'** and a *Maximum arc length* of **2'**. Draw a revision cloud similar to that shown in Figure 15-55. Use *SaveAs* to save the drawing as **EFF-APP-REV**.

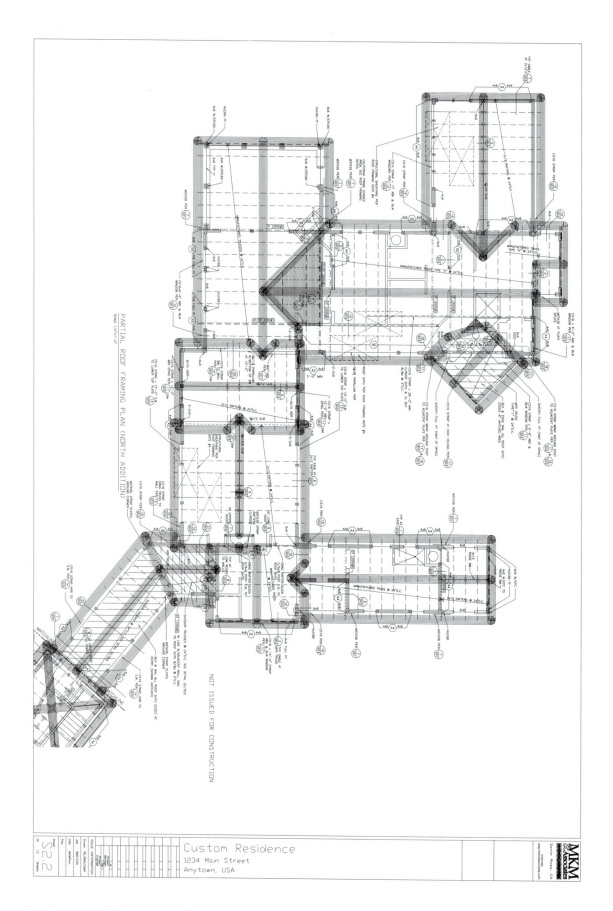

PARTIAL ROOF FRAMING PLAN (NORTH ADDITION)

SCALE 1/4"=1'-0"

NOT ISSUED FOR CONSTRUCTION

Custom Residence
1234 Main Street
Anytown, USA

MKN PLAN.DWG Courtesy, Autodesk, Inc.

MODIFY COMMANDS II

Chapter Objectives

After completing this chapter you should:

1. be able to use *Properties* to modify any type of object;

2. know that you can double-click on any object to produce the *Properties* window or a more specific editing tool;

3. be able to use *Matchprop* and the *Object Properties* toolbar to change properties of an object;

4. be able to use *Chprop* to change an object's properties;

5. know how to use *Change* to change points or properties;

6. be able to *Explode* a *Pline* into its component objects;

7. be able to *Align* objects with other objects;

8. be able to use all the *Pedit* options to modify *Plines* and to convert *Lines* and *Arcs* to *Plines*;

9. be able to modify *Splines* with *Splinedit*;

10. be able to edit existing *Mlines* using *Mledit*;

11. know that composite *Regions* can be created with *Union*, *Subtract*, and *Intersect*.

CONCEPTS

This chapter examines commands that are similar to, but generally more advanced and powerful than, those discussed in Chapter 9, Modify Commands I. None of the commands in this chapter create new duplicate objects from existing objects but instead modify the properties of the objects or convert objects from one type to another. Several commands are used to modify specific types of objects such as *Pedit* (modifies *Plines*), *Splinedit* (modifies *Splines*), *Mledit* (modifies *Mlines*), and *Union, Subtract,* and *Intersect* (modify *Regions*). Only one command in this chapter, *Align,* does not modify object properties but combines *Move* and *Rotate* into one operation. Several of the commands and features discussed in this chapter were mentioned in Chapter 11 (*Properties, MatchProp,* and *Object Properties* toolbar) but are explained completely here.

Only some commands in this chapter that modify general object properties have icon buttons that appear in the AutoCAD Drawing Editor by default. For example, you can access *Properties* and *Matchprop* from the *Object Properties* toolbar and *Explode* by using its icon from the *Modify* toolbar.

Other commands that modify specific objects such as *Pedit, Mledit,* and *Splinedit* appear in the *Modify II* toolbar (Fig. 16-1). The Boolean operators (*Union, Subtract,* and *Intersect*) appear in the *Solids Editing* toolbar. Activate a toolbar by right-clicking on any tool (icon button) and selecting from the list.

Figure 16-1

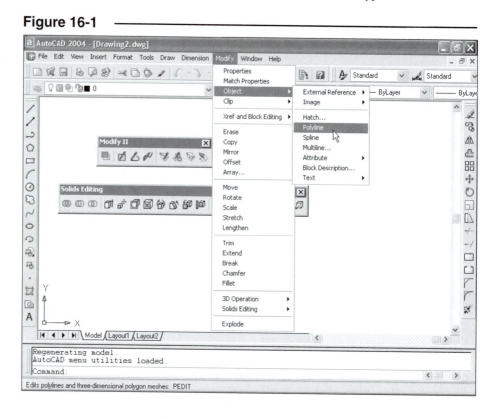

COMMANDS

You can use several methods to change properties of an object or of several objects. If you want to change an object's layer, color, or linetype only, the quickest method is by using the *Object Properties* toolbar (see "*Object Properties*" toolbar). If you want to change many properties of one or more objects (including layer, color, linetype, linetype scale, text style, dimension style, or hatch style) to match the properties of another existing object, use *Matchprop*. If you want to change any property (including coordinate data) for any object, use *Properties*.

In AutoCAD 2002 and 2004 you can double-click on any object (assuming the *Dblclkedit* command is set to *On*) to produce the *Properties* palette or a more specific editing tool, such as the *Hatch Edit* dialog box or the *Multiline Text Editor,* depending on the type of object you select.

Considerations for Changing Basic Properties of Objects

The *Properties* palette, the *Object Properties* toolbar, and the *Matchprop, Chprop,* and *Change* commands can be used to change the basic properties of objects. Here are several basic properties that can be changed and considerations when doing so.

Layer

By changing an object's *Layer,* the selected object is effectively <u>moved</u> to the designated layer. In doing so, if the object's *Color, Linetype,* and *Lineweight* are set to *BYLAYER,* the object assumes the color, linetype, and lineweight of the layer to which it is moved.

Color

It may be desirable in some cases to assign explicit *Color* to an object or objects independent of the layer on which they are drawn. The properties editing commands allow changing the color of an existing object from one object color to another, or from *BYLAYER* assignment to an object color. <u>An object drawn with an object-specific color can also be changed to *BYLAYER* with this option.</u>

Linetype

An object can assume the *Linetype* assigned *BYLAYER* or can be assigned an object-specific *Linetype.* The *Linetype* option is used to change an object's *Linetype* to that of its layer (*BYLAYER*) or to any object-specific linetype that has been loaded into the current drawing. <u>An object drawn with an object-specific linetype can also be changed to *BYLAYER* with this option.</u>

Linetype Scale

When an individual object is selected, the <u>object's individual linetype scale</u> can be changed with this option, but <u>not the global</u> linetype scale (*LTSCALE*). <u>This is the recommended method to alter an individual object's linetype scale.</u> First, draw all the objects in the current global *LTSCALE.* One of the properties editing commands could then be used with this option to retroactively adjust the linetype scale of <u>specific</u> objects. The result would be similar to setting the *CELTSCALE* before drawing the specific objects. Using this method to adjust a specific object's linetype scale <u>does not reset</u> the global *LTSCALE* or *CELTSCALE* variables.

Thickness

An object's *Thickness* can be changed by this option. *Thickness* is a three-dimensional quality (Z dimension) assigned to a two-dimensional object (see Chapter 40, Surface Modeling).

Lineweight

An object can assume the *Lineweight* assigned *ByLayer* or can be assigned an object-specific *Linetype.* The *Lineweight* option is used to change an object's *Lineweight* to that of its layer (*ByLayer*) or to any object-specific linetype that has been loaded into the current drawing. <u>An object drawn with an object-specific linetype can also be changed to *ByLayer* with this option.</u>

Plotstyle

Use this option to change an object's *Plotstyle.* Plot styles assigned as *ByLayer* or to individual objects can change the way the objects appear in plots, such as having certain screen patterns, line end joints, plotted colors, plotted lineweights, and so on. (See Chapter 33 for more information on plot styles.)

PROPERTIES

Pull-down Menu	COMMAND (TYPE)	ALIAS (TYPE)	Short-cut	Screen (side) Menu	Tablet Menu
Modify *Properties...*	*PROPERTIES*	*PR* or *CH*	**(Edit Mode)** *Properties* or *Ctrl+1*	*MODIFY1* *Property*	*Y12* to *Y,13*

The *Properties* palette (Fig. 16-2, right side) gives you complete access to one or more objects' properties. You can edit the properties simply by changing the entries in the right column.

Once opened, this window remains on the screen until dismissed by clicking the "X" in the upper-right corner. The *Properties* palette can be "docked" on the side of the screen (Fig. 16-2) or can be "floating." You can toggle the palette on and off with Ctrl+1 (to appear and disappear).

You can also set the palette to *Auto-hide* so only the title bar appears on your screen until you point to it to produce the full palette. If you click on the bottom icon on the title bar (*Properties*), a shortcut menu displays options to *Move, Size, Close, Allow Docking*, toggle *Auto-hide*, and toggle the *Description* at the bottom of the palette (Fig. 16-3). Changes you make to this palette are "persistent" (remain until changed).

The contents of the palette change based on what type and how many objects are selected. The power of this feature is apparent because the contents of the palette are specific to the type of object that you select. For example, if you select a *Line*, entries specific to that *Line* appear, allowing changes to any properties that the *Line* possesses (see Figure 16-2). Or, if you select a *Circle* or some *Text*, a window appears specific to the *Circle* or *Text* properties.

You can use the palette three ways: you can invoke the palette, then select (highlight) one or more objects, you can select objects first and then invoke the palette, or you can double-click on an object (for most objects) to produce the *Properties* palette (see *Dblclkedit*). When multiple objects are selected, a drop-down list in the top of the palette appears. Select the listed object(s) whose properties you want to change.

If you prefer to leave the palette on your screen, select the objects you want to change when no commands are in use with the pickbox or *Auto* window (see Figure 16-2, highlighted line). After changing the objects' properties, press Escape to deselect the objects. Pressing Escape does not dismiss the *Properties* palette, nor does it undo the changes as long as the cursor is outside the palette.

If no objects are selected, the *Properties* palette displays "No selection" in the drop-down list at the top (Fig. 16-3) and gives the current settings for the drawing. Drawing-wide settings can be changed such as the *LTSCALE* (see Figure 16-3).

Figure 16-2

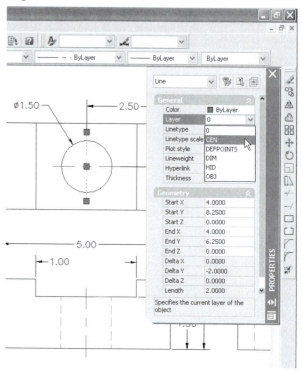

Figure 16-3

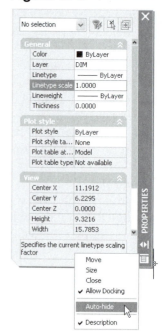

When objects are selected, the dialog box gives access to properties of the selected objects. Figure 16-2 displays the palette after selecting a *Line*. Notice the selection (*Line*) in the drop-down box at the top of the palette. The properties displayed in the central area of the window are specific to the object or group of objects highlighted in the drawing and selected from the drop-down list. In this case (see Figure 16-2), any aspect of the *Line* can be modified.

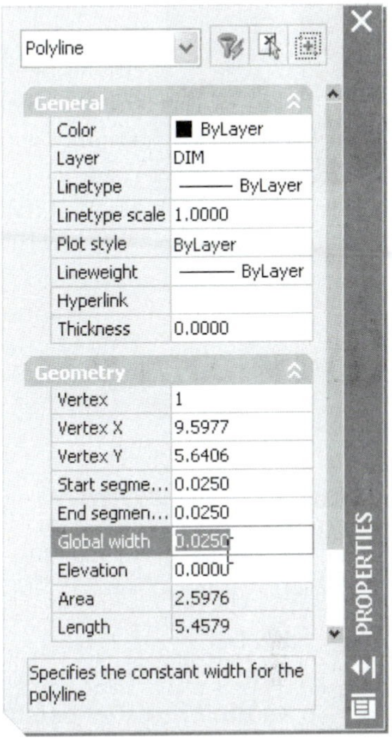

If a *Pline* is selected for example, any property of the *Pline* can be changed. For example, the *Width* of the *Pline* can be changed by entering a new value in the *Properties* palette (Fig. 16-4).

Figure 16-4

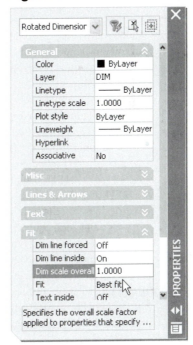

If a dimension is selected, for example, properties specific to a dimension, or to a group of dimensions if selected, can be changed. Note the list of variables that can be changed for a single dimension in Figure 16-5.

Figure 16-5

The *General* group lists basic properties for an object such as *Layer*, *Color*, or *Linetype*. To change an object's layer, linetype, lineweight, or color properties, highlight the desired objects in the drawing and select the properties you want to change from the right side of the palette. If you use the *ByLayer* strategy of linetype, lineweight, and color properties assignment, you can change an object's linetype, lineweight, and color simply by changing its layer (see Figure 16-2). This method is recommended for most applications and is described in Chapter 11.

The *Geometry* group lists and allows changing any values that control the selected object's geometry. For example, to change the diameter of a *Circle*, highlight the property and change the value in the right side of the palette (Fig. 16-6).

Figure 16-6

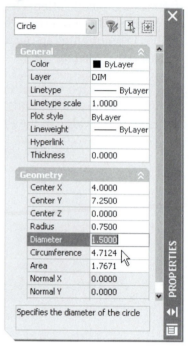

Quick Select

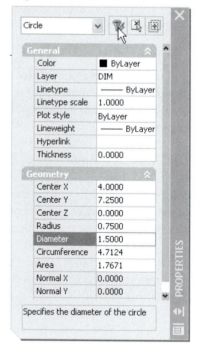

The *Quick Select* button appears near the upper-right corner of the *Properties* palette (Fig. 16-7). Selecting this button produces the *Quick Select* dialog box in which a selection set can be constructed based on criteria you choose from the dialog box. (See Chapter 20, Advanced Selection Sets, for information on *Quick Select*.)

Figure 16-7

Select Objects

Normally you can select objects by the typical methods any time the *Properties* palette is open. When objects are selected, AutoCAD searches its database and presents the properties of each object, one at a time, in the *Properties* palette. Instead, you can use the *Select Objects* button to save time posting information on each object individually to the window. Using *Select objects*, the properties of the total set of selected objects are not posted to the *Properties* palette until you press Enter.

Toggle Value of Pickadd Sysvar

This button toggles the *PICKADD* system variable on or off (1 or 0). This variable affects object selection at all times, not only for use with the *Properties* palette. *PICKADD* determines whether objects you PICK are added to the current selection set (normal setting, *PICKADD*=1) or replace the previous object or set (*PICKADD*=0). Changing the *PICKADD* variable to 0 works well with the *Properties* palette because selecting an object replaces the previous set of properties appearing in the palette with the new object's properties.

The button position of "+" (plus symbol) indicates a current setting of 1, or on, for *PICKADD* (PICKed objects are added). Confusing as it appears, a button position of "1" indicates a <u>current setting of 0</u>, or off, for *PICKADD* (PICKed objects replace the previous ones).

NOTE: Since the *PICKADD* setting affects object selection anytime, <u>make sure you set this variable back to your desired setting (usually on, or "+") before dismissing the *Properties* palette</u>. (See Chapter 20, Advanced Selection Sets, for information on *PICKADD*.)

DBLCLKEDIT

Pull-down Menu	COMMAND (TYPE)	ALIAS (TYPE)	Short-cut	Screen (side) Menu	Tablet Menu
...	*DBLCLKEDIT*	...	...	...	...

Since AutoCAD 2002 you can double-click on an object to produce the *Properties* palette or other similar dialog box to edit the object. The *Dblclkedit* command (double-click edit) controls whether double-click-ing an object produces a dialog box.

Command: **dblclkedit**
Enter double-click editing mode [ON/OFF] <ON>:

If double-click editing is turned on, the *Properties* palette or other dialog box is displayed when an object is double-clicked. When you double-click most objects, the *Properties* palette is displayed.

However, double-clicking some types of objects displays editing tools that are specific to the type of object. For example, double-clicking a line of *Text* produces the *Text Edit* dialog box. The object types (and result-ing editing tool) <u>that do not produce the *Properties* palette</u> when the object is double-clicked are listed below. These objects and the related editing tools are discussed fully in upcoming chapters of this text.

Attribute	Displays the *Edit Attribute Definition* dialog box (*Ddedit*).
Attribute within a block	Displays the *Enhanced Attribute Editor* (*Eattedit*).
Block	Displays the *Reference Edit* dialog box (*Refedit*).
Hatch	Displays the *Hatch Edit* dialog box (*Hatchedit*).
Leader text	Displays the *Multiline Text Editor* dialog box (*Ddedit*).
Mline	Displays the *Multiline Edit Tools* dialog box (*Mledit*).
Mtext	Displays the *Multiline Text Editor* dialog box (*Ddedit*).
Text	Displays the *Edit Text* dialog box (*Ddedit*).
Xref	Displays the *Reference Edit* dialog box (*Refedit*).

Dblclkedit is a command, not a system variable.

NOTE: The *PICKFIRST* system variable must be on (set to 1) for the *Properties* palette or other editing tool to appear when an object is double-clicked. (See Chapter 20, Advanced Selection Sets, for informa-tion on the *PICKFIRST* system variable.)

MATCHPROP

	Pull-down Menu	COMMAND (TYPE)	ALIAS (TYPE)	Short-cut	Screen (side) Menu	Tablet Menu
	Modify Match Properties	MATCHPROP	MA	...	MODIFY1 *Matchprp*	Y,14 and Y,15

Matchprop is explained briefly in Chapter 11 but is explained again in this chapter with the full details of the *Special Properties* palette.

Matchprop is used to "paint" the properties of one object to another. Simply invoke the command, select the object that has the desired properties ("source object"), then select the object you want to "paint" the properties to ("destination object"). The Command prompt is as follows:

```
Command: matchprop
Select source object: PICK
Current active settings:  Color Layer Ltype LTSCALE Lineweight Thickness PlotStyle Text Dim Hatch
    Polyline Viewport
Select destination object(s) or [Settings]: PICK
Select destination object(s) or [Settings]: PICK
Select destination object(s) or [Settings]: Enter
Command:
```

You can select several destination objects. The destination object(s) assume all of the "Current active settings" of the source object.

The *Property Settings* dialog box can be used to set which of the *Basic Properties* palette and *Special Properties* palette are to be painted to the destination objects (Fig. 16-8). At the "Select destination object(s) or [Settings]:" prompt, type *S* to display the dialog box. You can control the following *Basic Properties* palette. Only the checked properties are painted to the destination objects.

Figure 16-8

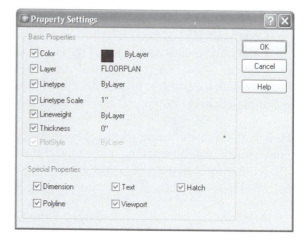

Color	This option paints the object-specific or *ByLayer* color.
Layer	Move selected objects to the source object layer with this option checked.
Linetype	Paint the object-specific or *ByLayer* linetype of the source object to the destination object.
Linetype Scale	This option changes the individual object's linetype scale (*CELTSCALE*) not global linetype scale (*LTSCALE*).
Thickness	Thickness is a 3-dimensional quality (see Chapter 40).
Lineweight	You can paint the object-specific or *ByLayer* lineweight of the source object to the destination object.
PlotStyle	The plot style of the source object is painted to the destination object when this setting is checked. (See Chapter 33 for information on plot styles.)

The *Special Properties* palette section allows you to specify features of dimensions, text, hatch patterns, viewports, and polylines to match, as explained next.

Dimension	This setting paints the *Dimension Style*. A *Dimension Style* defines the appearance of a dimension such as text style, size of arrows and text, tolerances if used, and many other features. (See Chapter 29.)
Text	This setting paints the source object's text *Style*. The text *Style* defines the text font and many other parameters that affect the appearance of the text. (See Chapter 18.)
Hatch	Checking this box paints the hatch properties of the source object to the destination object(s). The properties can include the hatch *Pattern, Angle, Scale,* and other characteristics. (See Chapter 26.)
Polyline	If you use *Matchprop* with *Plines*, the *Width* and *Linetype Generation* properties in addition to basic object properties of the first *Pline* are painted to the second. If the source polyline has variable *Width*, it is not transferred.
Viewport	If you match properties from one paper space viewport to another, the following properties are changed in addition to the basic object properties: *On/Off, Display Locking,* standard or custom *Scale, Shadeplot, Snap, Grid,* and *UCSicon* settings.

Matchprop is a simple and powerful command to use, especially if you want to convert dimensions, text or hatch patterns to look like other objects in the drawing. This method works only when you have existing objects in the drawing you want to "match." If you want to convert only layer, linetype, and color properties without changing the other properties or if you do not have existing objects to match, you can use the *Object Properties* palette toolbar.

Object Properties Toolbar

The five drop-down lists in the Object Properties toolbar (when not "dropped down") generally show the <u>current</u> layer, color, linetype, lineweight, and plot style. However, if an object or set of objects is selected, the information in these boxes changes to display the current <u>object's</u> settings. You can change an object's settings by picking an object (when no commands are in use), then dropping down any of the three lists and selecting a different layer, linetype, or color, etc.

Make sure you select (highlight) an object when no commands are in use. Use the pickbox (that appears on the cursor), *Window,* or *Crossing Window* to select the desired object or set of objects. The entries in the five boxes (Layer Control, Color Control, Linetype Control, etc.) then change to display the settings for the <u>selected object or objects</u>. If several objects are selected that have different properties, the boxes display no information (go "blank").

Figure 16-9

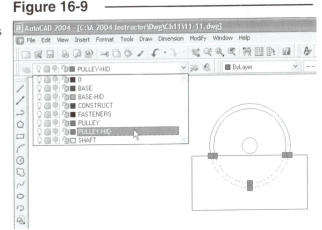

Next, use any of the drop-down lists to make another selection (Fig. 16-9). The highlighted object's properties are changed to those selected in the lists. Press the Escape key to complete the process.

Remember that in most cases color, linetype, lineweight, and plot style settings are assigned *ByLayer.* In this type of drawing scheme, to change the linetype or color properties of an object, you would change only the object's <u>layer</u> (see Figure 16-9). If you are using this type of drawing scheme, refrain from using the Color Control and Linetype Control drop-down lists for changing properties.

This method for changing object properties is about as quick and easy as *Matchprop*; however, only the layer, linetype, color, lineweight, or plot style can be changed with the *Object Properties* toolbar. The *Object Properties* toolbar method works well if you do not have other existing objects to match or if you want to change <u>only</u> layer, linetype, color, lineweight, and plot style <u>without</u> matching text, dimension, and hatch styles, and viewport or polyline properties.

CHPROP

Pull-down Menu	COMMAND (TYPE)	ALIAS (TYPE)	Short-cut	Screen (side) Menu	Tablet Menu
...	*CHPROP*	...	...	...	...

Chprop allows you to change basic properties of one or more objects using Command line format. If you want to change properties of several objects, pick several objects or select with a window or crossing window at the "Select objects:" prompt.

> Command: **chprop**
> Select objects: **PICK**
> Select objects: **Enter**
> Enter property to change [Color/LAyer/LType/ltScale/LWeight/Thickness]:

CHANGE

Pull-down Menu	COMMAND (TYPE)	ALIAS (TYPE)	Short-cut	Screen (side) Menu	Tablet Menu
...	*CHANGE*	-CH	...	...	...

The *Change* command allows changing three options: *Points*, *Properties*, or *Text*.

Point

This option allows changing the endpoint of an object or endpoints of several objects to one new position:

> Command: **change**
> Select objects: **PICK**
> Select objects: **Enter**
> Specify change point or [Properties]: **PICK** (Select a point to establish as new endpoint of all objects. *OSNAP*s can be used.)

The endpoint(s) of the selected object(s) <u>nearest</u> the new point selected at the "Specify change point or [Properties]:" prompt is changed to the new point (Fig. 16-10).

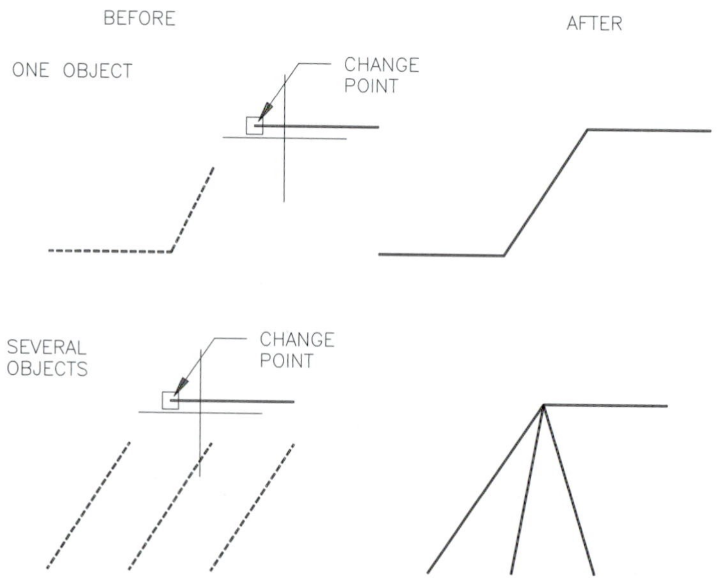

Properties

These options are discussed previously. The *Elevation* property (a 3D property) of an object can also be changed only with *Change* (see Chapter 40 for information on *Elevation*).

> Specify change point or [Properties]: **p**
> Enter property to change
> [Color/Elev/LAyer/LType/ltScale/
>   LWeight/Thickness]:

Figure 16-10

BEFORE   AFTER

ONE OBJECT   CHANGE POINT

SEVERAL OBJECTS   CHANGE POINT

Text

Although the word *"Text"* does <u>not</u> appear as an option, the *Change* command recognizes text created with *Dtext* if selected. <u>*Change* does not change *Mtext*</u>. (See Chapter 18 for information on *Dtext* and *Mtext*.)

You can change the following characteristics of *Dtext* objects:

> Text insertion point
> Text style
> Text height
> Text rotation angle
> Textual content

To change text, use the following command syntax:

> Command: **change**
> Select objects: **PICK** (Select one or several lines of text)
> Select objects: **Enter**
> Specify change point or [Properties]: **Enter**
> Specify new text insertion point <no change>: **PICK** or **Enter**
> Enter new text style <Standard>: (**text style name**) or **Enter**
> Specify new height <0.2000>: (**value**) or **Enter**
> Specify new rotation angle <0>: (**value**) or **Enter**
> Enter new text <text>: (**new text**) or **Enter** (Enter the complete new line of text)

EXPLODE

	Pull-down Menu	COMMAND (TYPE)	ALIAS (TYPE)	Short-cut	Screen (side) Menu	Tablet Menu
	Modify *Explode*	*EXPLODE*	X	...	*MODIFY2* *Explode*	*Y,22*

Many graphical shapes can be created in AutoCAD that are made of several elements but are treated as one object, such as *Plines, Polygons, Blocks, Hatch* patterns, and dimensions. The *Explode* command provides you with a means of breaking down or "exploding" the complex shape from one object into its many component segments (Fig. 16-11). Generally, *Explode* is used to allow subsequent editing of one or more of the component objects of a *Pline, Polygon,* or *Block,* etc., which would otherwise be impossible while the complex shape is considered one object.

Figure 16-11

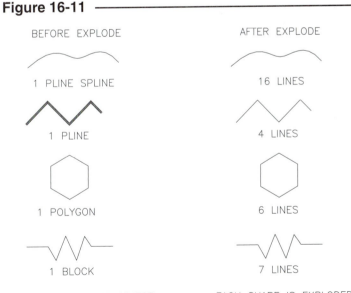

BEFORE EXPLODE — AFTER EXPLODE

1 PLINE SPLINE — 16 LINES

1 PLINE — 4 LINES

1 POLYGON — 6 LINES

1 BLOCK — 7 LINES

EACH SHAPE IS ONE OBJECT — EACH SHAPE IS EXPLODED INTO SEVERAL OBJECTS

The *Explode* command has no options and is simple to use. You only need to select the objects to *Explode*.

> Command: **explode**
> Select Objects: **PICK** (Select one or more *Plines, Blocks*, etc.)
> Select Objects: **Enter** (Indicates selection of objects is complete.)

When *Plines, Polygons, Blocks*, or hatch patterns are *Exploded*, they are transformed into *Line, Arc*, and *Circle* objects. Beware, *Plines* having *width* lose their width information when *Exploded* since *Line, Arc*, and *Circle* objects cannot have width. *Exploding* objects can have other consequences such as losing "associativity" of dimensions and hatch objects and increasing file sizes by *Exploding Blocks*.

ALIGN

Pull-down Menu	COMMAND (TYPE)	ALIAS (TYPE)	Short-cut	Screen (side) Menu	Tablet Menu
Modify *3D Operation>* *Align*	*ALIGN*	*AL*	...	*MODIFY2* *Align*	*X,14*

Align provides a means of aligning one shape (a simple object, group of objects, *Pline, Boundary, Region, Block*, or a 3D object) with another shape. *Align* provides a complex motion, usually a combined translation (like *Move*) and rotation (like *Rotate*), in one command.

The alignment is accomplished by connecting source points (on the shape to be moved) to destination points (on the stationary shape). You should use *OSNAP* modes to select the source and destination points to assure accurate alignment. Either a 2D or 3D alignment can be accomplished with this command. The command syntax for alignment in a <u>2D alignment</u> is as follows:

> Command: **align**
> Select objects: **PICK**
> Select objects: **Enter**
> Specify first source point: **PICK** (with *Osnap*)
> Specify first destination point: **PICK** (with *Osnap*)
> Specify second source point: **PICK** (with *Osnap*)
> Specify second destination point: **PICK** (with *Osnap*)
> Specify third source point or <continue>: **Enter**
> Scale objects based on alignment points? [Yes/No] <N>: **Enter** (or *Y* to scale the source object to match destination object)
> Command:

This command performs a translation and a rotation in one motion if needed to align the points as designated (Fig. 16-12).

First, the first source point is connected to (actually touches) the first destination point (causing a translation). Next, the vector defined by the first and second source points is aligned with the vector defined by the first and second destination points (causing rotation).

Figure 16-12

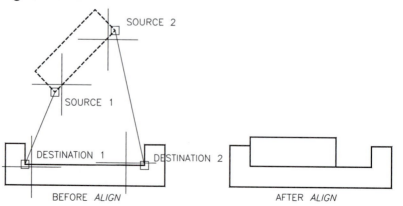

BEFORE *ALIGN*   AFTER *ALIGN*

If no third destination point is given (needed only for a 3D alignment), a 2D alignment is assumed and performed on the basis of the two sets of points.

Note that you can scale the source object based on the distance between the source points and the destination points. If you answer "Y" to the "Scale objects based on alignment points?" prompt, the source object is enlarged or reduced so the distance between its alignment points matches that of the destination points.

PEDIT

Pull-down Menu	COMMAND (TYPE)	ALIAS (TYPE)	Short-cut	Screen (side) Menu	Tablet Menu
Modify Object > Polyline	PEDIT	PE	...	MODIFY1 Pedit	Y,17

This command provides numerous options for editing *Polylines* (*Plines*). As an alternative, *Properties* can be used to change many of the *Pline*'s features in dialog box form (see Figure 16-4).

The list of options below emphasizes the great flexibility possible with *Polylines*. The first step after invoking *Pedit* is to select the *Pline* to edit.

```
Command: pedit
Select polyline or [Multiple]: PICK
Enter an option [Close/Join/Width/Edit vertex/Fit/Spline/Decurve/Ltype gen/Undo]:
```

Multiple
The *Multiple* option allows multiple *Plines* to be edited simultaneously. The selected *Plines* can be totally separate objects and do not have to be connected in any way. Once the *Multiple* option is invoked and the objects are selected, any *Pline* option, such as *Close, Open, Join, Width, Fit, Spline, Decurve,* or *Ltype gen,* operates on the selected *Plines*.

```
Command: Pedit
Select polyline or [Multiple]: m
Select objects: PICK
Select objects: PICK
Select objects: Enter
Enter an option [Close/Open/Join/Width/Fit/Spline/Decurve/Ltype gen/Undo]:
```

For example, you can change the width of all *Plines* in a drawing simultaneously using the *Multiple* option, selecting all *Plines*, and using the *Width* option.

Close
Close connects the last segment with the first segment of an existing "open" *Pline*, resulting in a "closed" *Pline* (Fig. 16-13). A closed *Pline* is one continuous object having no specific start or endpoint, as opposed to one closed by PICKing points. A *Closed Pline* reacts differently to the *Spline* option and to some commands such as *Fillet, Pline* option (see "*Fillet,*" Chapter 9).

Figure 16-13

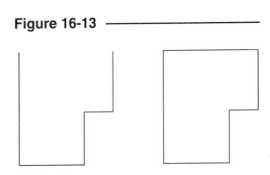

OPEN   CLOSE

Open
Open removes the closing segment if the *Close* option was used previously (Fig. 16-13).

Join

The *Join* option combines two or more objects into one *Pline*. In releases of AutoCAD prior to 2002, object endpoints had to match <u>exactly</u> for the objects to be joined into one *Pline*. With the enhanced version of *Pedit*, the line segments do not have to meet exactly for *Join* to work. You must first use the *Multiple* option, then *Join*.

```
Command: Pedit
Select polyline or [Multiple]: m
Select objects: PICK
Select objects: Enter
Enter an option [Close/Open/Join/Width/Fit/Spline/Decurve/Ltype gen/Undo]: j
Join Type = Extend
Enter fuzz distance or [Jointype] <0.0000>: .5
1 segments added to polyline
Enter an option [Close/Open/Join/Width/Fit/Spline/Decurve/Ltype gen/Undo]:
```

If the ends of the line segments do not touch but are within a distance that you can set, called the *fuzz distance*, the ends can be joined by *Pedit*. *Pedit* handles this automatically by either extending and trimming the line segments or by adding a new line segment based on your setting for *Jointype*.

```
Select objects: PICK
Enter an option [Close/Open/Join/Width/Fit/Spline/Decurve/Ltype gen/Undo]: j
Join Type = Extend
Enter fuzz distance or [Jointype] <2.0000>: j
Enter join type [Extend/Add/Both] <Extend>:
```

Extend

This option causes AutoCAD to join the selected polylines by extending or trimming the segments to the nearest endpoints (see Figure 16-14).

Add

Use this option to add a straight segment between the nearest endpoints (see Figure 16-14).

Both

If you use this option, the selected polylines are joined by extending or trimming if possible. If not, as in the case of near parallel lines when an extension would be outside the fuzz distance, a straight segment is added between the nearest endpoints.

Figure 16-14 ——————

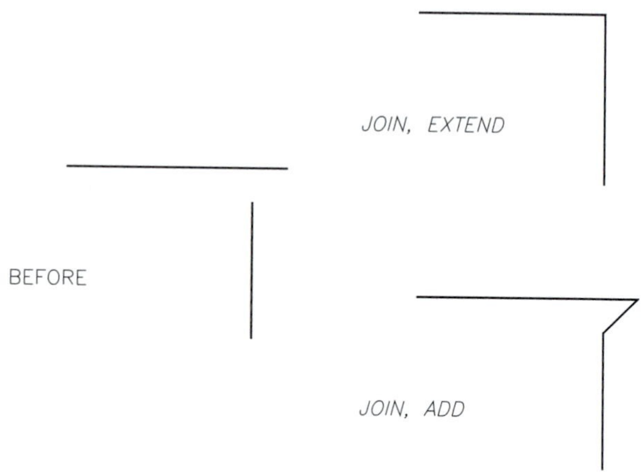

```
                          ┌────────
                          │
                JOIN, EXTEND

────────────   │
BEFORE          │

                          ┌────────
                          │
                JOIN, ADD
```

If the properties of several objects being joined into a polyline differ, the resulting polyline inherits the properties of the first object you select.

Width

Width allows specification of a uniform width for *Pline* segments (Fig. 16-15). Non-uniform width can be specified with the *Edit vertex* option.

Edit vertex

This option is covered in the next section.

Figure 16-15 ——————

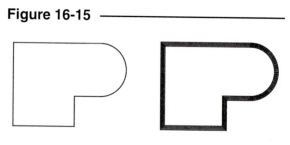

BEFORE WIDTH   AFTER WIDTH

2002

Fit

This option converts the *Pline* from straight line segments to arcs. The curve consists of two arcs for each pair of vertices. The resulting curve can be radical if the original *Pline* consists of sharp angles. The resulting curve passes <u>through all</u> vertices (Fig. 16-16).

Figure 16-16

BEFORE FIT   AFTER FIT

Spline

This option converts the *Pline* to a B-spline (Bezier spline) (Fig. 16-17). The *Pline* vertices act as "control points" affecting the shape of the curve. The resulting curve passes through <u>only the end</u> vertices. A *Spline*-fit *Pline* is <u>not the same as a spline curve created with the</u> <u>*Spline*</u> command. This option produces a less versatile version of the newer *Spline* object.

Figure 16-17

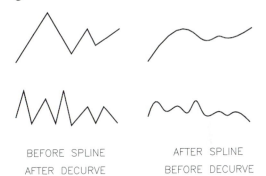

BEFORE SPLINE   AFTER SPLINE
AFTER DECURVE   BEFORE DECURVE

Decurve

Decurve removes the *Spline* or *Fit* curve and returns the *Pline* to its original straight line segments state (Fig. 16-17).

When you use the *Spline* option of *Pedit*, the amount of "pull" can be affected by setting the *SPLINETYPE* system variable to either 5 or 6 <u>before</u> using the *Spline* option. *SPLINETYPE* applies either a quadratic (5=more pull) or cubic (6=less pull) B-spline function (Fig. 16-18).

Figure 16-18

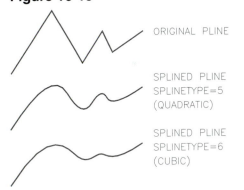

ORIGINAL PLINE

SPLINED PLINE
SPLINETYPE=5
(QUADRATIC)

SPLINED PLINE
SPLINETYPE=6
(CUBIC)

The *SPLINESEGS* system variable controls the number of line segments created when the *Spline* option is used. The variable should be set <u>before</u> using the option to any value (8=default): the higher the value, the more line segments. The actual number of segments in the resulting curve depends on the original number of *Pline* vertices and the value of the *SPLINETYPE* variable (Fig. 16-19).

Figure 16-19

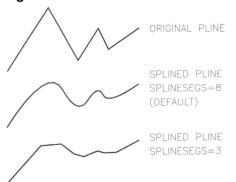

ORIGINAL PLINE

SPLINED PLINE
SPLINESEGS=8
(DEFAULT)

SPLINED PLINE
SPLINESEGS=3

TIP Changing the *SPLFRAME* variable to 1 causes the *Pline* frame (the original straight segments) to be displayed for *Splined* or *Fit Plines*. *Regen* must be used after changing the variable to display the original *Pline* "frame" (Fig. 16-20).

Figure 16-20

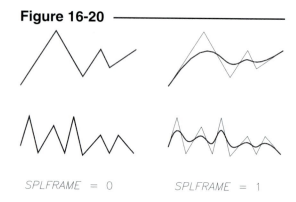

SPLFRAME = 0   SPLFRAME = 1

Ltype gen

This setting controls the generation of non-continuous linetypes for *Plines*. If *Off*, non-continuous linetype dashes start and stop at each vertex, as if the *Pline* segments were individual *Line* segments. For dashed linetypes, each line segment begins and ends with a full dashed segment (Fig. 16-21). If *On*, linetypes are drawn in a consistent pattern, disregarding vertices. In this case, it is possible for a vertex to have a space rather than a dash. Using the *Ltype gen* option <u>retroactively</u> changes *Plines* that have already been drawn. *Ltype gen* affects objects composed of *Plines* such as *Polygons*, *Rectangles*, and *Boundaries*.

Similarly, the *PLINEGEN* system variable controls how <u>new</u> non-continuous linetypes are drawn for *Plines*. A setting of 1 creates a consistent linetype pattern, disregarding vertices (like *Ltype gen On*). A *PLINEGEN* setting of 0 creates linetypes stopping and starting at each vertex (like *Ltype gen Off*). However, *PLINEGEN* is <u>not retroactive</u>—it affects only <u>new</u> *Plines*.

Figure 16-21

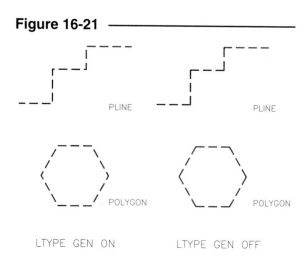

PLINE   PLINE

POLYGON   POLYGON

LTYPE GEN ON   LTYPE GEN OFF

Undo

Undo reverses the most recent *Pedit* operation.

eXit

This option exits the *Pedit* options, keeps the changes, and returns to the Command prompt.

Vertex Editing

Upon selecting the *Edit Vertex* option from the *Pedit* options list, the group of suboptions is displayed on the screen menu and Command line:

```
Command: pedit
Select polyline or [Multiple]: PICK
Enter an option [Close/Join/Width/Edit vertex/Fit/Spline/Decurve/Ltype gen/Undo]: e
Enter a vertex editing option
[Next/Previous/Break/Insert/Move/Regen/Straighten/Tangent/Width/eXit] <N>:
```

Next

AutoCAD places an **X** marker at the first endpoint of the *Pline*. The *Next* and *Previous* options allow you to sequence the marker to the desired vertex (Fig. 16-22).

Previous

See *Next* above.

Figure 16-22

VERTEX MARKER

PREVIOUS AND NEXT OPTIONS WILL MOVE THE MARKER TO THE DESIRED VERTEX

Break

This selection causes a break between the marked vertex and another vertex you then select using the *Next* or *Previous* option (Fig. 16-23).

        Enter an option [Next/Previous/Go/eXit] <N>:

Selecting *Go* causes the break. An endpoint vertex cannot be selected.

Figure 16-23

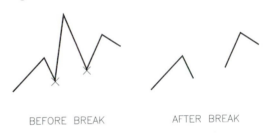

BEFORE BREAK   AFTER BREAK

Insert

Insert allows you to insert a new vertex at any location <u>after</u> the vertex that is marked with the **X** (Fig. 16-24). Place the marker before the intended new vertex, use *Insert*, then PICK the new vertex location.

Figure 16-24

BEFORE INSERT   AFTER INSERT

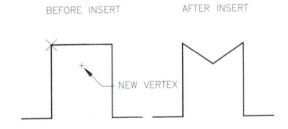

NEW VERTEX

Move

You are prompted to indicate a new location to *Move* the marked vertex (Fig. 16-25).

Regen

In older releases of AutoCAD, *Regen* should be used after the *Width* option to display the new changes.

Figure 16-25

BEFORE MOVE   AFTER MOVE

+ ◄— NEW LOCATION

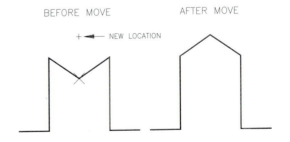

Straighten

You can *Straighten* the *Pline* segments between the current marker and the other marker that you then place by one of these options:

Enter an option [Next/Previous/Go/eXit] <N>:

Selecting *Go* causes the straightening to occur (Fig. 16-26).

Tangent

Tangent allows you to specify the direction of tangency of the current vertex for use with curve *Fitting*.

Width

This option allows changing the *Width* of the *Pline* segment immediately following the marker, thus achieving a specific width for one segment of the *Pline* (Fig. 16-27). *Width* can be specified with different starting and ending values.

eXit

This option exits from vertex editing, saves changes, and returns to the main *Pedit* prompt.

Figure 16-26

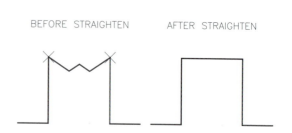

BEFORE STRAIGHTEN   AFTER STRAIGHTEN

Figure 16-27

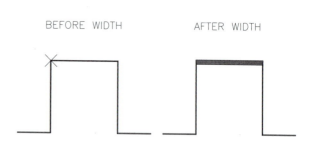

BEFORE WIDTH   AFTER WIDTH

Grips

Plines can also be edited easily using Grips (see Chapter 23). A Grip appears on each vertex of the *Pline*. Editing *Plines* with Grips is sometimes easier than using *Pedit* because Grips are more direct and less dependent on the command interface.

Converting *Lines* and *Arcs* to *Plines*

A very important and productive feature of *Pedit* is the ability to convert *Lines* and *Arcs* to *Plines* and closed *Pline* shapes. Potential uses of this option are converting a series of connected *Lines* and *Arcs* to a closed *Pline* for subsequent use with *Offset* or for inquiry of the area (*Area* command) or length (*List* command) of a single shape. The only requirement for conversion of *Lines* and *Arcs* to *Plines* is that the selected objects must have <u>exact</u> matching endpoints.

To accomplish the conversion of objects to *Plines,* simply select a *Line* or *Arc* object and request to turn it into one:

```
Command: pedit
Select polyline or [Multiple]: PICK  (Select only one Line or Arc)
Object selected is not a polyline
Do you want to turn it into one? <Y> Enter
Enter an option [Close/Join/Width/Edit vertex/Fit/Spline/Decurve/Ltype gen/Undo]: j  (Use the Join option)
Select objects: PICK
Select objects: Enter
1 segments added to polyline
Enter an option [Close/Join/Width/Edit vertex/Fit/Spline/Decurve/Ltype gen/Undo]: Enter
Command:
```

The resulting conversion is a closed *Polyline* shape.

	Pull-down Menu	COMMAND (TYPE)	ALIAS (TYPE)	Short-cut	Screen (side) Menu	Tablet Menu
SPLINEDIT	*Modify Object > Splinedit*	SPLINEDIT	SPE	...	MODIFY1 *Splinedt*	Y,18

Splinedit is an extremely powerful command for changing the configuration of existing *Splines*. You can use multiple methods to change *Splines*. All of the *Splinedit* methods fall under two sets of options.

The two groups of options that AutoCAD uses to edit *Splines* are based on two sets of points: data points and control points. Data points are the points that were specified when the *Spline* was created—the points that the *Spline* actually passes through (Fig. 16-28).

Figure 16-28

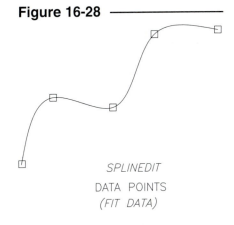

SPLINEDIT
DATA POINTS
(FIT DATA)

Control points are other points outside of the path of the *Spline* that only have a "pull" effect on the curve (Fig. 16-29).

Figure 16-29

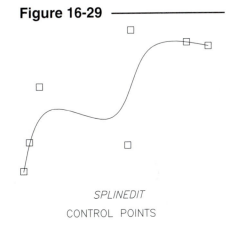

SPLINEDIT
CONTROL POINTS

Editing *Spline* Data Points

The command prompt displays several levels of options. The top level of options uses the control points method for editing. Select *Fit Data* to use data points for editing. The *Fit Data* methods are recommended for most applications. Since the curve passes directly through the data points, these options offer direct control of the curve path.

```
Command: splinedit
Select spline: PICK
Enter an option [Fit data/Close/Move vertex/Refine/rEverse/Undo]: f
Enter a fit data option [Add/Close/Delete/Move/Purge/Tangents/toLerance/eXit] <eXit>:
```

Add

You can add points to the *Spline*. The *Spline* curve changes to pass through the new points. First, PICK an existing point on the curve. That point and the next one in sequence (in the order of creation) become highlighted. The new point will change the curve between those two highlighted data points (Fig. 16-30).

> Specify control point: **PICK** (Select an existing point on the curve before the intended new point.)
> Specify new point: **PICK** (PICK a new point location between the two marked points.)

Figure 16-30 ───────────────

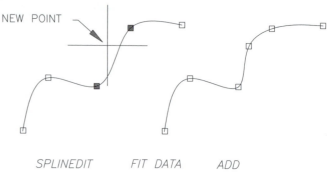

SPLINEDIT   FIT DATA   ADD

Close/Open

The *Close* option appears only if the existing curve is open, and the *Open* prompt appears only if the curve is closed. Selecting either option automatically forces the opposite change. *Close* causes the two ends to become tangent, forming one smooth curve (Fig. 16-31). This tangent continuity is characteristic of *Closed Splines* only. *Splines* that have matching endpoints do not have tangent continuity unless the *Close* option of *Spline* or *Splinedit* is used.

Figure 16-31 ───────────────

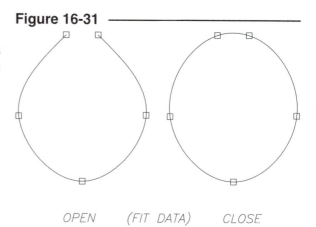

OPEN   (FIT DATA)   CLOSE

Move

You can move any data point to a new location with this option (Fig. 16-32). The beginning endpoint (in the order of creation) becomes highlighted. Type *N* for next or *S* to select the desired data point to move; then PICK the new location.

> Specify new location or [Next/Previous/Select point/eXit] <N>:

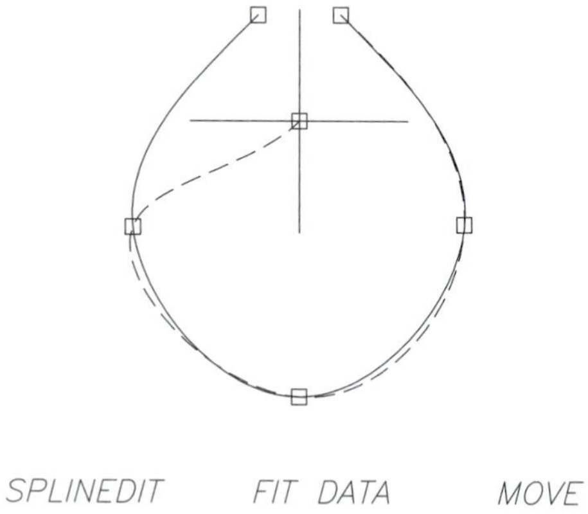

Purge

Purge deletes all data points and renders the *Fit Data* set of options unusable. You are returned to the control point options (top level). To reinstate the points, use *Undo*.

Figure 16-32 ───────────────

SPLINEDIT   FIT DATA   MOVE

Tangents

You can change the directions for the start and endpoint tangents with this option. This action gives the same control that exists with the "Enter start tangent" and "Enter end tangent" prompts of the *Spline* command used when the curves were created (see Figure 15-12, *Spline, End Tangent*).

toLerance

Use *toLerance* to specify a value, or tolerance, for the curve to "fall" away from the data points. Specifying a tolerance causes the curve to smooth out, or fall, from the data points. The higher the value, the more the curve "loosens." The *toLerance* option of *Splinedit* is identical to the *Fit Tolerance* option available with *Spline* (see Figure 15-14, *Spline, Tolerance*).

Editing *Spline* <u>Control Points</u>

Use the top level of command options (except *Fit Data*) to edit the *Spline's* control points. These options are similar to those used for editing the data points; however, the results are different since the curve does not pass through the control points.

```
Command: splinedit
Select spline: PICK
Enter an option [Fit data/Close/Move vertex/Refine/rEverse/Undo]:
```

Fit Data
Discussed previously.

Close/Open
These options operate similar to the *Fit Data* equivalents; however, the resulting curve falls away from the control points (Fig. 16-33; see also Fig. 16-31, *Fit Data, Close*).

Figure 16-33 ───────────

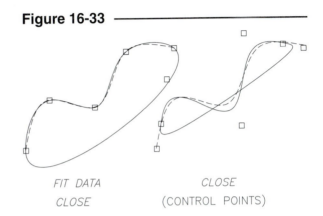

FIT DATA
CLOSE

CLOSE
(CONTROL POINTS)

Move Vertex
Move Vertex allows you to move the location of any control points. This is the control points' equivalent to the *Move* option of *Fit Data* (see Figure 16-32, *Fit Data, Move*). The method of selecting points (*Next/Previous/Select point/eXit/*) is the same as that used for other options.

Refine
Selecting the *Refine* option reveals another level of options.

```
Enter a refine option [Add control point/Elevate order/Weight/eXit] <eXit>:
```

Add control points is the control points' equivalent to *Fit Data Add* (see Figure 16-30). At the "Select a point on the Spline" prompt, simply PICK a point near the desired location for the new point to appear. Once the *Refine* option has been used, the *Fit Data* options are no longer available.

Elevate order allows you to <u>increase the number of control points</u> uniformly along the length of the *Spline*. Enter a value from n to 26, where n is the current number of points + one. Once a *Spline* is elevated, it cannot be reduced.

Weight is an option that you use to assign a value to the <u>amount of "pull"</u> that a <u>specific control point</u> has on the *Spline* curve. The higher the value, the more "pull," and the closer the curve moves toward the control point. The typical method of selecting points (*Next/Previous/Select point/eXit/*) is used.

rEverse
The *rEverse* option reverses the direction of the *Spline*. The first endpoint (when created) then becomes the last endpoint. Reversing the direction may be helpful for selection during the *Move* option.

Grips
Splines can also be edited easily using *Grips* (see Chapter 23). The *Grip* points that appear on the *Spline* are identical to the *Fit Data* points. Editing with *Grips* is a bit more direct and less dependent on the command interface.

MLEDIT

Pull-down Menu	COMMAND (TYPE)	ALIAS (TYPE)	Short-cut	Screen (side) Menu	Tablet Menu
Modify *Object >* *Multiline...*	MLEDIT or -MLEDIT	...	...	MODIFY1 *Mledit*	Y,19

The *Mledit* command provides tools for editing multilines created with the *Mline* command. Somewhat surprisingly, the typical line editing commands such as *Trim, Extend, Break, Fillet,* and *Chamfer* cannot be used with *Mlines* unless the *Mlines* are *Exploded* (see "Other Editing Possibilities for *Mlines*" at the end of this section). Instead, the *Mledit* command is required for *Mline* modification and contains several editing functions. These editing functions give you control of multiline intersections.

Figure 16-34

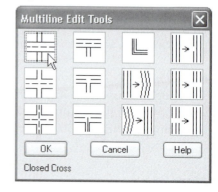

Invoking the *Mledit* command by any method produces the *Multiline Edit Tools* dialog box. The dialog box presents 12 image tiles that you PICK to activate the desired function. Single-clicking on an image tile displays the option name in the lower-left corner of the dialog box (Fig. 16-34). Double-clicking on an image tile activates that option. The dialog box then disappears, allowing you to select the desired multiline segments for the editing action.

Command: **mledit** (The *Multiline Edit Tools* dialog box appears. Select the desired option.)
Select first mline: **PICK**
Select second mline: **PICK**
Select first mline or [Undo]: **PICK** or **Enter**
Command:

The *Multiline Edit Tools* dialog box is organized in columns as follows:

Intersection—Cross	Intersection—Tee	Corner, Vertices	Lines
Closed Cross	Closed Tee	Corner Joint	Cut Single
Open Cross	Open Tee	Vertex Add	Cut All
Merged Cross	Merged Tee	Vertex Delete	Weld

The options are described and illustrated next.

Closed Cross

Use this option to trim one of two intersecting *Mlines*. The first *Mline* PICKed is trimmed to the outer edges of the second. The second *Mline* is "closed." All line elements in the first multiline are trimmed (Fig. 16-35).

Figure 16-35

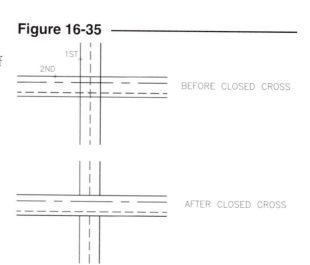

Open Cross

This option <u>trims both</u> intersecting *Mlines*. Both *Mlines* are "open." <u>All</u> line elements of the <u>first</u> *Mline* PICKed are trimmed to the outer edges of the second. Only the outer line elements of the second multiline are trimmed, while the inner lines continue through the intersection (Fig. 16-36).

Figure 16-36 ─────────

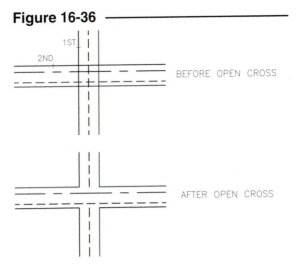

Merged Cross

With this option, the <u>outer</u> line elements of <u>both</u> intersecting *Mlines* are trimmed and the <u>inner line elements merge</u>. The inner line elements merge at the second *Mline's* next set (Fig. 16-37). A full merge (both inner lines continue through) occurs only if the second *Mline* PICKed has no or only one inner line element.

Figure 16-37 ─────────

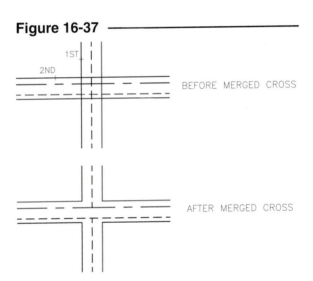

Closed Tee

As indicated by the image tile, a "T" intersection is created rather than a "crossing" intersection. The <u>first</u> *Mline* PICKed is <u>trimmed</u> to the <u>nearest</u> (to the PICK point) outer edge of the second. Only the side of the first *Mline* nearest the PICK point remains. The second *Mline* is not affected (Fig. 16-38).

Figure 16-38 ─────────

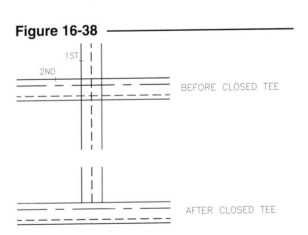

Open Tee

With this "T" intersection, both outer edges of two intersecting *Mlines* are "open." All line elements of the first *Mline* PICKed are trimmed at the outer edges of the second. Only the outer line element of the second is trimmed (Fig. 16-39).

Figure 16-39

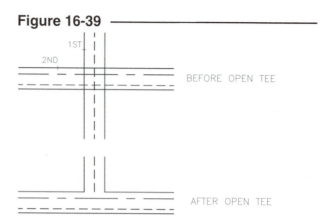

BEFORE OPEN TEE

AFTER OPEN TEE

Merged Tee

This "T" option allows the inner line elements of both intersecting *Mlines* to *merge*. The merge occurs at the second *Mline's* first inner element. Only the side of the first line nearest the PICK point remains (Fig. 16-40).

Figure 16-40

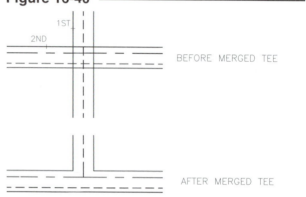

BEFORE MERGED TEE

AFTER MERGED TEE

Corner Joint

This option trims both *Mlines* to create a corner. Only the PICKed sides of the *Mlines* remain and the extending portions of both (if any) are trimmed. All of the inner line elements merge (Fig. 16-41).

Figure 16-41

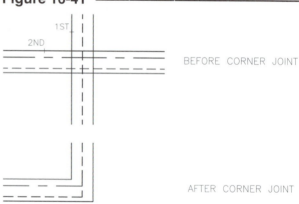

BEFORE CORNER JOINT

AFTER CORNER JOINT

Add Vertex

If you want to add a new corner (vertex) to an existing *Mline*, use this feature. A new vertex is added where you PICK the *Mline*. However, it is not readily apparent that a new vertex exists, nor does the *Mline* visibly change in any way. You must further edit the *Mline* with *Stretch* or Grips (see Chapter 23) to create a new "corner" (Fig. 16-42).

Figure 16-42

BEFORE ADD VERTEX

AFTER ADD VERTEX
THEN STRETCH

Delete Vertex

Use this feature to remove a corner (vertex) of an *Mline*. The vertex <u>nearest</u> the location you PICK is deleted. The resulting *Mline* contains only one straight segment between the adjacent two vertices. Unlike *Add Vertex*, the *Mline* immediately changes and no further editing is needed to see the effect (Fig. 16-43).

Figure 16-43

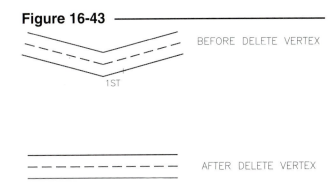

Cut Single

The *Cut* options are used to break (cut) a space in one *Mline*. *Cut Single* <u>breaks any line element</u> that is selected. Similar to *Break 2Points*, the break occurs <u>between the two PICK points</u> (Fig. 16-44). The break points can be on either side of a vertex.

Figure 16-44

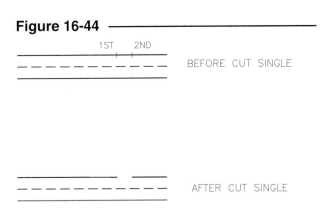

Cut All

This *Cut* option <u>breaks all line elements</u> of the selected *Mline*. Any line elements can be selected. The line elements are cut at the PICK points in a direction perpendicular to the axis of the *Mline* (Fig. 16-45). NOTE: Although the *Mline* appears to be cut into two separate *Mlines*, it <u>remains one single object</u>. Using *Stretch* or Grips to "move" the *Mline* causes the cut to close again.

Figure 16-45

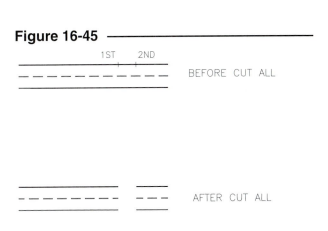

Weld

This option <u>reverses the action of a *Cut*</u>. The *Mline* is restored to its original continuous configuration. PICK on both sides of the "break" (Fig. 16-46).

Figure 16-46

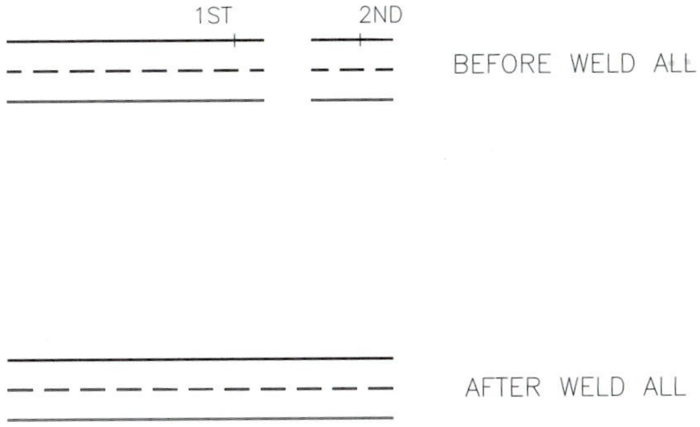

Like many other dialog box-based commands, *Mledit* can also be used in command line format by using the hyphen (-) symbol as a prefix to the command. Key in "*-MLEDIT*" to force the command line interface. The following prompt appears:

```
Command: -mledit
Enter mline editing option [CC/OC/MC/CT/OT/MT/CJ/AV/DV/CS/CA/WA]:
```

Entering the acronym activates the related option.

CC	*Closed Cross*
OC	*Open Cross*
MC	*Merged Cross*
CT	*Closed Tee*
OT	*Open Tee*
MT	*Merged Tee*
CJ	*Corner Joint*
AV	*Add Vertex*
DV	*Delete Vertex*
CS	*Cut Single*
CA	*Cut All*
WA	*Weld All*

NOTE: *Mledit* operates only with co-planar *Mlines*.

Other Editing Possibilities for *Mlines*

Although it appears that *Mledit* contains the necessary tools to handle all editing possibilities for *Mlines*, that is not necessarily the case. You may experience situations where *Mledit* cannot handle your desired editing request and you have to resort to traditional editing commands. Only in the case that *Mledit* cannot perform as you want, *Explode* the *Mline*, which converts each line element to an individual object. That action allows you to use Grips, *Trim, Extend, Break, Fillet,* or *Chamfer* on the individual line elements. Once an *Mline* is *Exploded*, it cannot be converted back to an *Mline* nor can *Mledit* be used with it.

Boolean Commands

Region combines one or several objects forming a closed shape into one object, a *Region*. The appearance of the object(s) does not change after the conversion, even though the resulting shape is one object (see "*Region*," Chapter 15).

Although the *Region* appears to be no different than a closed *Pline*, it is more powerful because several *Regions* can be combined to form complex shapes (called "composite *Regions*") using the three Boolean operations explained next. As an example, a set of *Regions* (converted *Circles*) can be combined to form the sprocket with only one *Subtract* operation, as shown in Chapter 15, Figure 15-41.

The Boolean operators, *Union, Subtract,* and *Intersect,* can be used with *Regions* as well as solids. Any number of these commands can be used with *Regions* to form complex geometry. To illustrate each of the Boolean commands, consider the shapes shown in Figure 16-47. The *Circle* and the closed *Pline* are first converted to *Regions*; then *Union, Subtract,* or *Intersection* can be used.

Figure 16-47

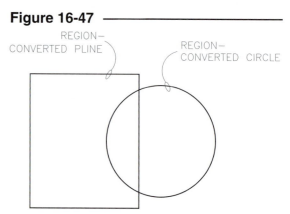

UNION

	Pull-down Menu	COMMAND (TYPE)	ALIAS (TYPE)	Short-cut	Screen (side) Menu	Tablet Menu
	Modify Solids Editing> Union	UNION	UNI	...	MODIFY2 *Union*	X,15

Union combines <u>two or more</u> *Regions* (or solids) into <u>one</u> *Region* (or solid). The resulting composite *Region* has the encompassing perimeter and area of the original *Regions*. Invoking *Union* causes AutoCAD to prompt you to select objects. You can select only existing *Regions* (or solids).

Command: **union**
Select Objects: **PICK** (*Region*)
Select Objects: **PICK** (*Region*)
Select Objects: **Enter**
Command:

The selected *Regions* are combined into one composite *Region* (Fig. 16-48). Any number of Boolean operations can be performed on the *Region*(s).

Figure 16-48

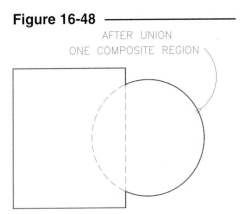

AFTER UNION
ONE COMPOSITE REGION

Several *Regions* can be selected in response to the "Select objects:" prompt. For example, a composite *Region* such as that in Figure 16-49 can be created with one *Union*.

Two or more *Regions* can be *Unioned* even if they do not overlap. They are simply combined into one object although they still appear as two.

Figure 16-49

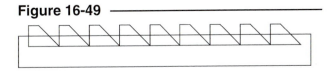

BEFORE— 10 REGIONS

AFTER UNION— ONE REGION

SUBTRACT

	Pull-down Menu	COMMAND (TYPE)	ALIAS (TYPE)	Short-cut	Screen (side) Menu	Tablet Menu
	Modify Solids Editing > Subtract	SUBTRACT	SU	...	MODIFY2 *Subtract*	X,16

Subtract enables you to remove one *Region* (or set of *Regions*) from another. The *Regions* must be created before using *Subtract* (or another Boolean operator). *Subtract* also works with solids (as do the other Boolean operations).

There are two steps to *Subtract*. First, you are prompted to select the *Region* or set of *Regions* to "subtract from" (those that you wish to <u>keep</u>), then to select the *Regions* "to subtract" (those you want to <u>remove</u>). The resulting shape is one composite *Region* comprising the perimeter of the first set minus the second (sometimes called "difference").

```
Command: subtract
Select solids and regions to subtract from...
Select Objects: PICK
Select Objects: Enter
Select solids and regions to subtract...
Select Objects: PICK
Select Objects: Enter
Command:
```

Consider the two shapes previously shown in Figure 16-47. The resulting *Region* shown in Figure 16-50 is the result of *Subtracting* the circular *Region* from the rectangular one.

Keep in mind that <u>multiple</u> *Regions* can be selected as the set to keep or as the set to remove. For example, the sprocket illustrated previously (Fig. 15-41) was created by subtracting several circular *Regions* in one operation. Another example is the removal of material to create holes or slots in sheet metal (discussed in Chapter 40, Surface Modeling).

Figure 16-50 ——————————

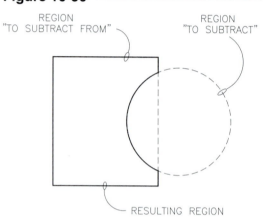

INTERSECT

Pull-down Menu	COMMAND (TYPE)	ALIAS)TYPE)	Short-cut	Screen (side) Menu	Tablet Menu
Modify *Solids Editing>* *Intersect*	*INTERSECT*	*IN*	...	MODIFY2 *Intrsect*	X,17

Intersect is the Boolean operator that finds the common area from two or more *Regions*.

Consider the rectangular and circular *Regions* previously shown (Fig. 16-47). Using the *Intersect* command and selecting both shapes results in a *Region* comprising only that area that is shared by both shapes (Fig. 16-51):

```
Command: intersect
Select Objects: PICK
Select Objects: PICK
Select Objects: Enter
Command:
```

Figure 16-51 ——————————

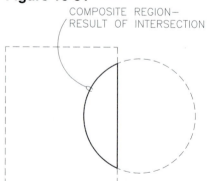

If more than two *Regions* are selected, the resulting *Intersection* is composed of only the common area from all shapes (Fig. 16-52). If <u>all</u> of the shapes selected do not overlap, a <u>null</u> *Region* is created (all shapes disappear because no area is common to <u>all</u>).

Intersect is a powerful operation when used with solid modeling. Techniques for saving time using Boolean operations are discussed in Chapter 38, Solid Modeling Construction.

Figure 16-52

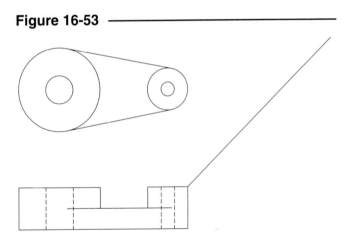

BEFORE INTERSECTION   RESULTING REGION
                             COMMON AREA

CHAPTER EXERCISES

1. *Chprop* **or** *Properties*

   Open the **PIVOTARM CH9** drawing that you worked on in Chapter 9 Exercises. *Load* the *Hidden2* and *Center2 Linetypes*. Make two *New Layers* named **HID** and **CEN** and assign the matching linetypes and yellow and green colors, respectively. Check the *Limits* of the drawing; then calculate and set an appropriate *LTSCALE.* Use *Chprop* or *Properties* palette to change the *Lines* representing the holes in the front view to the **HID** layer as shown in Figure 16-53. Use *SaveAs* and name the drawing **PIVOTARM CH16.**

Figure 16-53

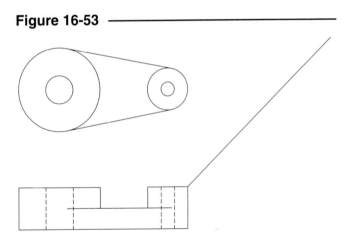

2. *Change*

   Open **CH8EX3.** *Erase* the *Arc* at the top of the object. *Erase* the *Points* with a window. Invoke the *Change* command. When prompted to *Select objects*, **PICK** all of the inclined *Lines* near the top. When prompted to "Specify change point," enter coordinate **6,8.** The object should appear as that in Figure 16-54. Use *SaveAs* and assign the name **CH16EX2.**

Figure 16-54

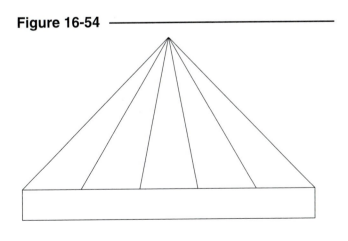

3. *Properties* **Palette**

Figure 16-55

A design change is required for the bolt holes in **GASKETA** (from Chapter 9 Exercises). *Open* **GASKETA** and invoke the *Properties* palette. Change each of the four bolt holes to *.375* diameter (Fig. 16-55). *Save* the drawing.

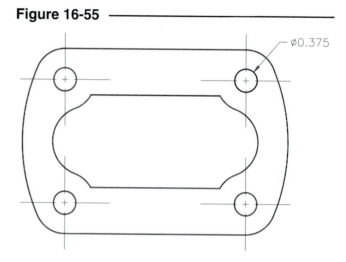
⌀0.375

4. *Align*

Open the **PLATES** drawing you created in Chapter 9. The three plates are to be stamped at one time on a single sheet of stock measuring 15″ x 12″. Place the three plates together to achieve optimum nesting on the sheet stock.

A. Use *Align* to move the plate in the center (with 9 holes). Select the *First source* and *destination points* (1S, 1D) and *Second source* and *destination points* (2S, 2D) as shown in Figure 16-56.

Figure 16-56

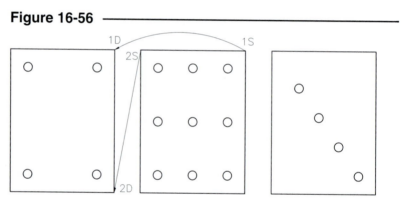

B. After the first alignment is complete, use *Align* to move the plate on the right (with 4 diagonal holes). The *source* and *destination points* are indicated in Figure 16-57.

Figure 16-57

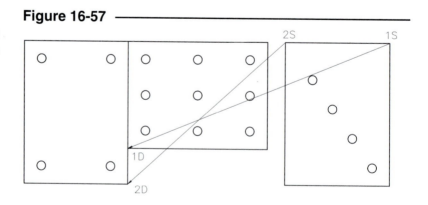

C. Finally, draw the sheet stock outline (15" x 12") using *Line* as shown in Figure 16-58. The plates are ready for production. Use *SaveAs* and assign the name **PLATENEST**.

Figure 16-58 ───────────

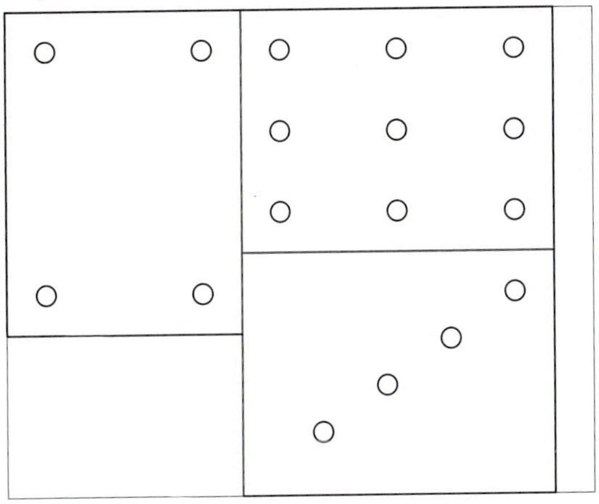

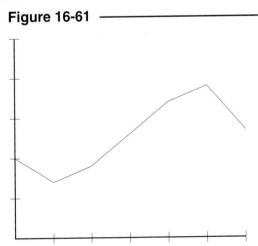

5. *Explode*

Figure 16-59 ─────────────────────

Open the **POLYGON1** drawing that you completed in Chapter 15 Exercises. You can quickly create the five-sided shape shown (continuous lines) in Figure 16-59 by *Exploding* the *Polygon*. First, *Explode* the *Polygon* and *Erase* the two *Lines* (highlighted). Draw the bottom *Line* from the two open *Endpoints*. Do not exit the drawing.

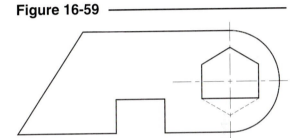

6. *Pedit*

Figure 16-60 ─────────────────

Use *Pedit* with the *Edit vertex* options to alter the shape as shown in Figure 16-60. For the bottom notch, use *Straighten*. For the top notch, use *Insert*. Use *SaveAs* and change the name to **PEDIT1**.

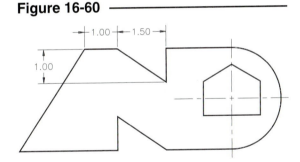

7. *Pline, Pedit*

A. Create a line graph as shown in Figure 16-61 to illustrate the low temperatures for a week. The temperatures are as follows:

X axis	Y axis
Sunday	20
Monday	14
Tuesday	18
Wednesday	26
Thursday	34
Friday	38
Saturday	27

Use equal intervals along each axis. Use a *Pline* for the graph line. *Save* the drawing as **TEMP-A**. (You will label the graph at a later time.)

Figure 16-61 ──────────────

B. Use *Pedit* to change the *Pline* to a *Spline*. Note that the graph line is no longer 100% accurate because it does not pass through the original vertices (see Fig. 16-62). Use the *SPLFRAME* variable to display the original "frame" (*Regen* must be used after).

Use the *Properties* palette and try the *Cubic* and *Quadratic* options. Which option causes the vertices to have more pull? Find the most accurate option. Set *SPLFRAME* to **0** and *SaveAs* **TEMP-B**.

C. Use *Pedit* to change the curve from *Pline* to *Fit Curve*. Does the graph line pass through the vertices? *Saveas* **TEMP-C**.

D. *Open* drawing **TEMP-A**. *Erase* the *Splined Pline* and construct the graph using a *Spline* instead (Fig. 16-63, see Exercise 7A for data). *SaveAs* **TEMP-D**. Compare the *Spline* with the variations of *Plines*. Which of the four drawings (A, B, C, or D) is smoothest? Which is the most accurate?

E. It was learned that there was a mistake in reporting the temperatures for that week. Thursday's low must be changed to 38 degrees and Friday's to 34 degrees. *Open* **TEMP-D** (if not already open) and use *Splinedit* to correct the mistake. Use the *Move* option of *Fit Data* so that the exact data points can be altered as shown in Figure 16-64. *SaveAs* **TEMP-E**.

Figure 16-62

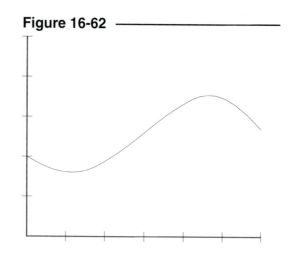

Figure 16-63

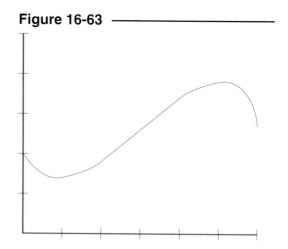

Figure 16-64

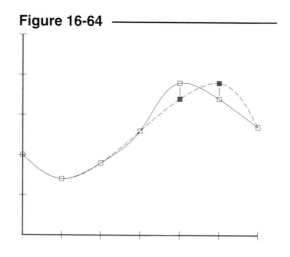

8. **Converting** *Lines, Circles,* **and** *Arcs* **to** *Plines*

    A. Begin a *New* drawing and use the **A-METRIC** template (that you worked on in Chapter 13 Exercises). Use *Save* and assign the name **GASKETC**. Change the *Limits* for plotting on an A sheet at 2:1 (refer to the Metric Table of *Limits* Settings and set *Limits* to 1/2 x *Limits* specified for 1:1 scale). Change the **LTSCALE** to **12**. First, draw only the <u>inside</u> shape using *Lines* and *Circles* (with *Trim*) or *Arcs* (Fig. 16-65). Then convert the *Lines* and *Arcs* to one closed *Pline* using *Pedit*. Finally, locate and draw the 3 bolt holes. *Save* the drawing.

Figure 16-65

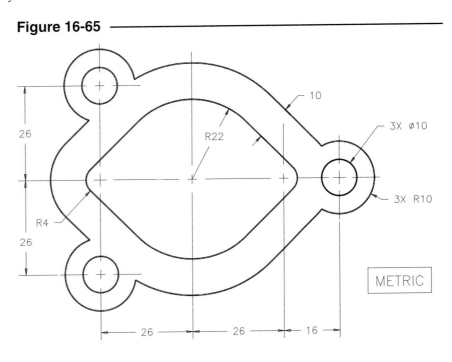

    B. *Offset* the existing inside shape to create the outside shape. Use *Offset* to create concentric circles around the bolt holes. Use *Trim* to complete the gasket. *Save* the drawing and create a plot at 2:1 scale.

9. *Splinedit*

    Open the **HANDLE** drawing that you created in Chapter 15 Exercises. During the testing and analysis process, it was discovered that the shape of the handle should have a more ergonomic design. The finger side (left) should be flatter to accommodate varying sizes of hands, and the thumb side (right) should have more of a protrusion on top to prevent slippage.

    First, add more control points uniformly along the length of the left side with *Spinedit, Refine. Elevate* the *Order* from 4 to **6**, then use *Move Vertex* to align the control points as shown in Figure 16-66.

    On the right side of the handle, *Add* two points under the *Fit Data* option to create the protrusion shown in Figure 16-67. You may have to *Reverse* the direction of the *Spline* to add the new points between the two highlighted ones. *SaveAs* **HANDLE2**.

Figure 16-66 —   **Figure 16-67 —**

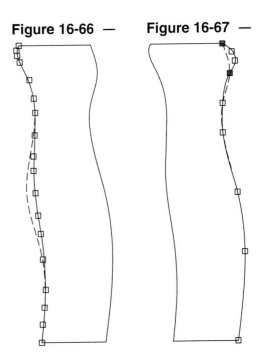

10. *Mline, Mledit, Stretch*

Figure 16-68

Draw the floor plan of the storage room shown in Figure 16-68. Use *Mline* objects for the walls and *Plines* for the windows and doors. Use *Mledit* to treat the intersections as shown. When your drawing is complete according to the given specifications, use *Stretch* to center the large window along the top wall. Save the drawing as **STORE ROOM**.

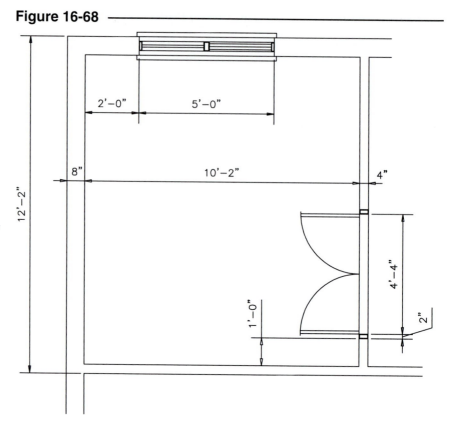

11. **Gear Drawing**

Figure 16-69

Complete the drawing of the gear you began in Chapter 15 Exercises called **GEAR-REGION**. If you remember, three shapes were created (two *Circles* and one "tooth") and each was converted to a *Region*. Finally, the "tooth" was *Arrayed* to create the total of 40 teeth.

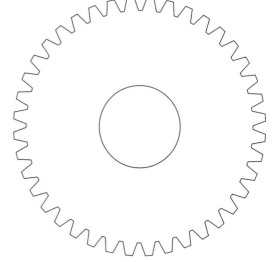

A. To complete the gear, *Subtract* the small circular *Region* and all of the teeth from the large circular *Region*. First, use *Subtract*. At the "Select solids and regions to subtract from..." prompt, PICK the large circular *Region*. At the "Select solids and regions to subtract..." prompt, use a window to select <u>everything</u> (the large circular *Region* is automatically filtered out). The resulting gear should resemble Figure 16-69. *Save* the drawing as **GEAR-REGION 2**.

B. Consider the steps involved if you were to create the gear (as an alternative) by using *Trim* to remove 40 small sections of the large *Circle* and all unwanted parts of the teeth. *Regions* are clearly easier in this case.

12. Wrench Drawing

Create the same wrench you created in Chapter 15 again, only this time use region modeling. Refer to Chapter 15, Exercise 6 for dimensions (exclude the break lines). Begin a *New* drawing and use the **A-METRIC** template. Use *SaveAs* and assign the name **WRENCH-REG**.

Figure 16-70

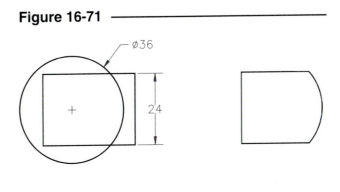

A. Set *Limits* to **372,288** to prepare the drawing for plotting at 3:4 (the drawing scale factor is 33.87). Set the *GRID* to **10**.

B. Draw a *Circle* and an *Ellipse* as shown on the left in Figure 16-70. The center of each shape is located at **60,150**. *Trim* half of each shape as shown (highlighted). Use *Region* to convert the two remaining halves into a *Region* as shown on the right of Figure 16-70.

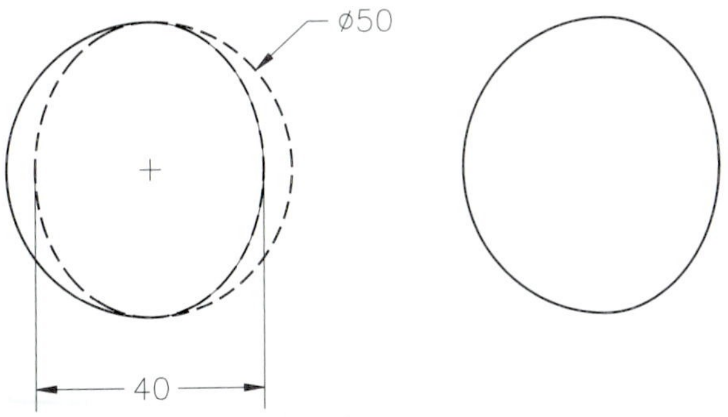

C. Next, create a *Circle* with the center at **110,150** and a diameter as shown in Figure 16-71. Then draw a closed *Pline* in a rectangular shape as shown. The height of the rectangle must be drawn as specified; however, the width of the rectangle can be drawn <u>approximately</u> as shown on the left. Convert each shape to a *Region*; then use *Intersect* to create the region as shown on the right side of the figure.

Figure 16-71

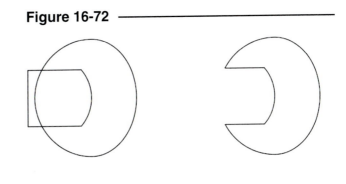

D. *Move* the rectangular-shaped region **68** units to the left to overlap the first region as shown in Figure 16-72. Use *Subtract* to create the composite region on the right representing the head of the wrench.

Figure 16-72

E. Complete the construction of the wrench in a manner similar to that used in the previous steps. Refer to Chapter 15 Exercises for dimensions of the wrench. Complete the wrench as one *Region*. *Save* the drawing as **WRENCH-REG**.

13. **Retaining Wall**

Figure 16-73

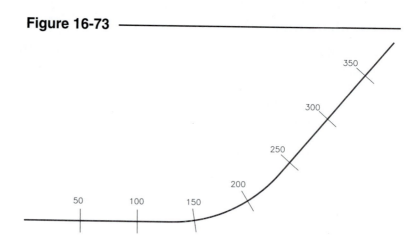

Open the **RET-WALL** drawing that you created in Chapter 8 Exercises. Annotate the wall with 50 unit stations as shown in Figure 16-73.

HINT: Convert the centerline of the retaining wall to a *Pline*; then use the *Measure* command to place *Point* objects at the 50 unit stations. *Offset* the wall on both sides to provide a construction aid in the creation of the perpendicular tick marks. Use *Node* and *Perpendicular* Osnaps. *Save* the drawing as **RET-WAL3**.

INQUIRY COMMANDS

Chapter Objectives

After completing this chapter you should:

1. be able to list the *Status* of a drawing;

2. be able to *List* the AutoCAD database information about an object;

3. know how to list the entire database of all objects with *Dblist*;

4. be able to calculate the *Area* of a closed shape with and without "islands";

5. be able to find the distance between two points with *Dist*;

6. be able to report the coordinate value of a selected point using the *ID* command;

7. know how to list the *Time* spent on a drawing or in the current drawing session;

8. be able to use *Setvar* to change system variable settings or list current settings.

CONCEPTS

AutoCAD provides several commands that allow you to find out information about the current drawing status and specific objects in a drawing. These commands as a group are known as "*Inquiry* commands" and are grouped together in the menu systems. The *Inquiry* commands are located in the *Inquiry* toolbar (Fig. 17-1). You can also use the *Tools* pull-down menu to access the *Inquiry* commands (Fig. 17-2).

Figure 17-1

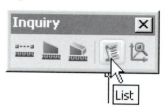

Using *Inquiry* commands, you can find out such information as the amount of time spent in the current drawing, the distance between two points, the area of a closed shape, the database listing of properties for specific objects (coordinates of endpoints, lengths, angles, etc.), and current settings for system variables as well as other information. The *Inquiry* commands are:

> *Status, List, Dblist, Area, Dist, ID, Time,* and *Setvar*

Figure 17-2

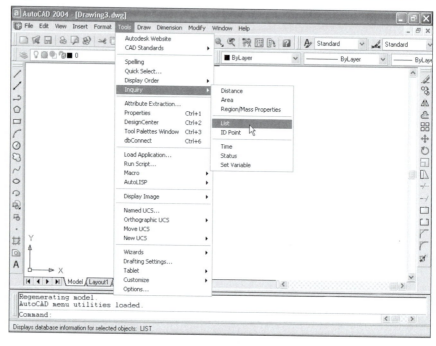

COMMANDS

STATUS

Pull-down Menu	COMMAND (TYPE)	ALIAS (TYPE)	Short-cut	Screen (side) Menu	Tablet Menu
Tools *Inquiry >* *Status*	*STATUS*	...	...	*TOOLS 1* *Status*	...

The *Status* command gives many pieces of information related to the current drawing. Typing or **PICK**ing the command from the icon or one of the menus causes a text screen to appear similar to that shown in Figure 17-3, on the next page. The information items are:

Total number of objects in the current drawing
Paper space limits: values set by the *Limits* command in Paper Space (*Limits* in Paper Space are set by selecting a *Paper size* in the *Page Setup* or *Plot* dialog box.)
Paper space uses: area used by the objects (drawing extents) in Paper Space
Model space limits: values set by the *Limits* command in Model Space
Model space uses: area used by the objects (drawing extents) in Model Space

Display shows: current display or windowed area
Insertion basepoint: point specified by the *Base* command or default (0,0)
Snap resolution: value specified by the *Snap* command
Grid spacing: value specified by the *Grid* command
Current space: Paper Space or Model Space
Current layer: name
Current color: current color assignment
Current linetype: current linetype assignment
Current lineweight: current lineweight assignment
Current plot style: current plot style assignment
Current elevation, thickness: 3D properties—current height above the XY plane and Z dimension
On or off status: *FILL, GRID, ORTHO, QTEXT, SNAP, TABLET*
Object Snap Modes: current *Running OSNAP* modes
Free dwg disk: space on the current drawing hard disk drive
Free temp disk: space on the current temporary files hard disk drive
Free physical memory: amount of free RAM (total RAM)
Free swap file space: amount of free swap file space (total allocated swap file)

Figure 17-3

```
AutoCAD Text Window - C:\Program Files\AutoCAD 2004\Sample\Wilhome...
Edit
Command: '_status 8817 objects in C:\Program Files\AutoCAD
2004\Sample\Wilhome.dwg
Model space limits are X:    -2'-0"   Y:    -2'-0"  (Off)
                       X: 142'-3 5/16"   Y: 95'-0 5/16"
Model space uses       X: 0'-7 15/16"   Y: 6'-1 1/4"
                       X: 127'-10 7/8"   Y: 89'-8 5/8"
Display shows          X: 127'-9 1/16"   Y:      0'-5"
                       X: 270'-9 1/8"   Y: 84'-10 3/8"
Insertion base is      X:      0'-0"   Y:      0'-0"   Z:      0'-0"
Snap resolution is     X: 0'-0 1/2"   Y: 0'-0 1/2"
Grid spacing is        X: 0'-0 1/2"   Y: 0'-0 1/2"

Current space:         Model space
Current layout:        Model
Current layer:         "0"
Current color:         BYLAYER -- 7 (white)
Current linetype:      BYLAYER -- "Continuous"
Current lineweight:    BYLAYER
Current elevation:      0'-0"  thickness:     0'-0"
Fill on  Grid off  Ortho off  Qtext off  Snap off  Tablet off
Object snap modes:     Center, Endpoint, Intersection, Midpoint, Ext
Free dwg disk (C:) space: 503.8 MBytes
Free temp disk (C:) space: 503.8 MBytes
Free physical memory: 295.4 Mbytes (out of 511.0M).
Free swap file space: 1079.5 Mbytes (out of 1250.2M).

Command:
```

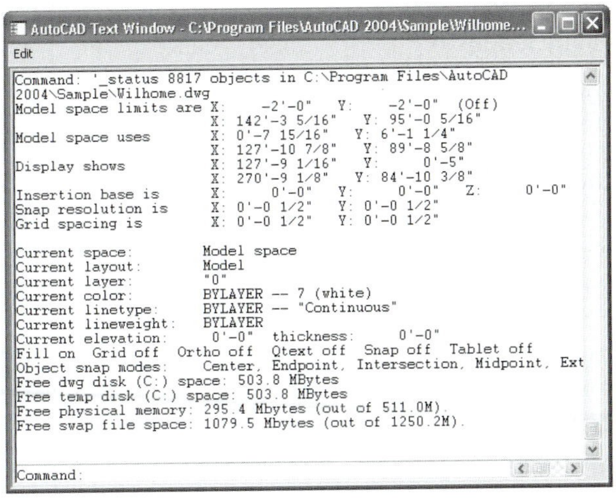

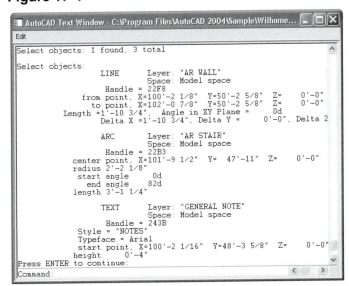

LIST

Pull-down Menu	COMMAND (TYPE)	ALIAS (TYPE)	Short-cut	Screen (side) Menu	Tablet Menu
Tools Inquiry > List	*LIST*	*LS or LI*	...	*TOOLS 1 List*	*U,8*

The *List* command displays the database list of information in text window format for one or more specified objects. The information displayed depends on the <u>type</u> of object selected. Invoking the *List* command causes a prompt for you to select objects. AutoCAD then displays the list for the selected objects (see Figs. 17-4 and 17-5, on the next page). A *List* of a *Line* and an *Arc* is given in Figure 17-4.

For a *Line,* coordinates for the endpoints, line length and angle, current layer, and other information are given.

For an *Arc,* the center coordinate, radius, start and end angles, and length are given.

Figure 17-4

```
AutoCAD Text Window - C:\Program Files\AutoCAD 2004\Sample\Wilhome...
Edit
Select objects: 1 found, 3 total

Select objects:
                 LINE      Layer: "AR WALL"
                           Space: Model space
                  Handle = 22F8
             from point, X=100'-2 1/8"  Y=50'-2 5/8"  Z=    0'-0"
               to point, X=102'-0 7/8"  Y=50'-2 5/8"  Z=    0'-0"
          Length =1'-10 3/4",  Angle in XY Plane =     0d
                 Delta X =1'-10 3/4", Delta Y =     0'-0", Delta Z

                 ARC       Layer: "AR STAIR"
                           Space: Model space
                  Handle = 22B3
            center point, X=101'-9 1/2"  Y=  47'-11"  Z=    0'-0"
            radius 2'-2 1/8"
             start angle     0d
               end angle    82d
            length 3'-1 1/4"

                 TEXT      Layer: "GENERAL NOTE"
                           Space: Model space
                  Handle = 243B
             Style = "NOTES"
             Typeface = Arial
             start point, X=100'-2 1/16"  Y=48'-3 5/8"  Z=    0'-0"
            height    0'-4"
Press ENTER to continue:
Command:
```

The *List* for a *Pline* is shown in Figure 17-5. The location of each vertex is given, as well as the length and perimeter of the entire *Pline*.

Since Release 14, *Plines* are created and listed as *Lwpolylines,* or "lightweight polylines." In previous releases, complete data for each *Pline* vertex (starting width, ending width, color, etc.) were stored along with the coordinate values of the vertex, then repeated for each vertex. In AutoCAD 2004, the data common to all vertices are stored only once, and only the coordinate data are stored for each vertex. Because this data structure saves file space, the new *Plines* are known as lightweight *Plines*.

Figure 17-5

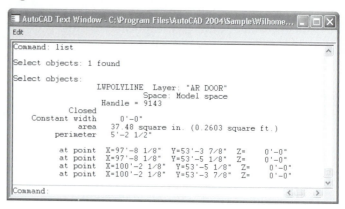

DBLIST

Pull-down Menu	COMMAND (TYPE)	ALIAS (TYPE)	Short-cut	Screen (side) Menu	Tablet Menu
...	DBLIST	...	...	...	...

The *Dblist* command is similar to the *List* command in that it displays the database listing of objects; however, *Dblist* gives information for <u>every</u> object in the current drawing! This command is generally used when you desire to send the list to a printer or when only a few objects are in the drawing. If you use this command in a complex drawing, be prepared to page through many screens of information. Press Escape to cancel *Dblist* and return to the Command: prompt. Press F2 to open and close the text window.

AREA

Pull-down Menu	COMMAND (TYPE)	ALIAS (TYPE)	Short-cut	Screen (side) Menu	Tablet Menu
Tools Inquiry > Area	AREA	AA	...	TOOLS 1 Area	T,7

The *Area* command is helpful for many applications. With this command AutoCAD calculates the area and the perimeter of any enclosed shape in a matter of milliseconds. You specify the area (shape) to consider for calculation by PICKing the *Object* (if it is a closed *Pline, Polygon, Circle, Boundary, Region* or other closed object) or by PICKing points (corners of the outline) to define the shape. The options are given below.

Specify first corner point
The command sequence for specifying the area by PICKing points is shown below. This method should be used only for shapes with <u>straight</u> sides. An example of the *Point* method (PICKing points to define the area) is shown in Figure 17-6.

Figure 17-6

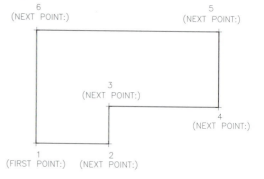

Command: ***area***
Specify first corner point or [Object/Add/Subtract]: **PICK** (Locate the first corner to define the shape.)
Specify next corner point or press ENTER for total: **PICK** (Locate the second corner on the shape.)
Specify next corner point or press ENTER for total: **PICK** (Locate the next corner.)
Specify next corner point or press ENTER for total: **PICK** (Continue selecting points until all corners have been defined.)
Specify next corner point or press ENTER for total: **Enter**
Area = *nn.nnn* perimeter = *nn.nnn*
Command:

Object

If the shape for which you want to find the area and perimeter is a *Circle, Polygon, Rectangle, Ellipse, Boundary, Region,* or closed *Pline,* the *Object* option of the *Area* command can be used. Select the shape with one PICK (since all of these shapes are considered as one object by AutoCAD).

The ability to find the area of a closed *Pline, Region,* or *Boundary* is extremely helpful. Remember that <u>any</u> closed shape, even if it includes *Arcs* and other curves, can be converted to a closed *Pline* with the *Pedit* command (as long as there are no gaps or overlaps) or can be used with the *Boundary* command. This method provides you with the ability to easily calculate the area of any shape, curved or straight. In short, convert the shape to a closed *Pline, Region,* or *Boundary* and find the *Area* with the *Object* option.

Add, Subtract

Add and *Subtract* provide you with the means to find the area of a closed shape that has islands, or negative spaces. For example, you may be required to find the surface area of a sheet of material that has several punched holes. In this case, the area of the holes is subtracted from the area defined by the perimeter shape. The *Add* and *Subtract* options are used specifically for that purpose. The following command sequence displays the process of calculating an area and subtracting the area occupied by the holes.

Command: ***area***
Specify first corner point or [Object/Add/Subtract]: **a** (Use the *Add* option to begin a running total.)
Specify first corner point or [Object/Subtract]: **o** (Use the *Object* option to select the outside shape.)
(ADD mode) Select objects: **PICK** (Select the closed object.)
Area = 13.31, Perimeter = 14.39
Total area = 13.31

(ADD mode) Select objects: **Enter** (Completion of *Add* mode.)
Specify first corner point or [Object/Subtract]: **s** (Switch to *Subtract* mode.)
Specify first corner point or [Object/Add]: **o** (Use *Object* mode.)
(SUBTRACT mode) Select objects: **PICK** (Select the first *Circle* to subtract.)
Area = 0.69, Length = 2.95
Total area = 12.62

(SUBTRACT mode) Select objects: **PICK** (Select the second *Circle* to subtract.)
Area = 0.69, Length = 2.95
Total area = 11.93

(SUBTRACT mode) Select objects: **Enter** (Completion of *Subtract* mode.)
Specify first corner point or [Object/Add]: **Enter** (Completion of *Area* command.)
Command:

Make sure that you press Enter between the *Add* and *Subtract* modes.

An example of the last command sequence used to find the area of a shape minus the holes is shown in Figure 17-7. Notice that the object selected in the first step is a closed *Pline* shape, including an *Arc*.

Figure 17-7

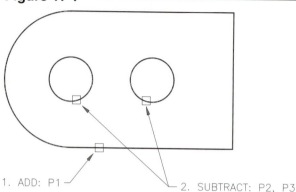

1. ADD: P1
2. SUBTRACT: P2, P3

DIST

Pull-down Menu	COMMAND (TYPE)	ALIAS (TYPE)	Short-cut	Screen (side) Menu	Tablet Menu
Tools Inquiry > Distance	DIST	DI	...	TOOLS 1 Dist	T,8

 The *Dist* command reports the distance between any two points you specify. *OSNAP*s can be used to snap to the existing points. This command is helpful in many engineering or architectural applications, such as finding the clearance between two mechanical parts, finding the distance between columns in a building, or finding the size of an opening in a part or doorway. The command is easy to use.

    Command: **dist**
    Specify first point: **PICK** (Use *Osnaps* if needed.)
    Specify second point: **PICK** (Use *Osnaps* if needed.)
    Distance = 3.63, Angle in XY Plane = 165, Angle from XY Plane = 0
    Delta X = -3.50, Delta Y = 0.97, Delta Z = 0.00
    Command:

AutoCAD reports the absolute and relative distances as well as the angle of the line between the points.

ID

Pull-down Menu	COMMAND (TYPE)	ALIAS (TYPE)	Short-cut	Screen (side) Menu	Tablet Menu
Tools Inquiry > ID Point	ID	...	...	TOOLS 1 ID	U,9

The *ID* command reports the coordinate value of any point you select with the cursor. If you require the location associated with a specific object, an *OSNAP* mode (*Endpoint, Midpoint, Center,* etc.) can be used.

    Command: **id**
    Specify point: **PICK** (Use *Osnaps* if needed.)
    X = 7.63   Y = 6.25   Z = 0.00
    Command:

 NOTE: *ID* also sets AutoCAD's "last point." The last point can be referenced in commands by using the @ (at) symbol with relative rectangular or relative polar coordinates.

TIME

Pull-down Menu	COMMAND (TYPE)	ALIAS (TYPE)	Short-cut	Screen (side) Menu	Tablet Menu
Tools Inquiry > Time	TIME	...	...	TOOLS 1 Time	...

This command is useful for keeping track of the time spent in the current drawing session or total time spent on a particular drawing. Knowing how much time is spent on a drawing can be useful in an office situation for bidding or billing jobs.

The *Time* command reports the information shown in Figure 17-8. The *Total editing time* is automatically kept, starting from when the drawing was first created until the current time. Plotting and printing time is not included in this total, nor is the time spent in a session when changes are discarded.

Figure 17-8

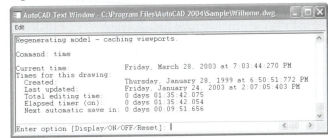

Display
The *Display* option causes *Time* to repeat the display with the updated times.

ON/OFF/Reset
The *Elapsed timer* is a separate compilation of time controlled by the user. The *Elapsed timer* can be turned *ON* or *OFF* or can be *Reset*.

Time also reports when the next automatic save will be made. The time interval of the Automatic Save feature is controlled by the *SAVETIME* system variable. To set the interval between automatic saves, type *SAVETIME* at the Command line and specify a value for time (in minutes). See also *SAVETIME* in Chapter 2.

SETVAR

Pull-down Menu	COMMAND (TYPE)	ALIAS (TYPE)	Short-cut	Screen (side) Menu	Tablet Menu
Tools Inquiry > Set Variable	SETVAR	SET	...	TOOLS 1 Setvar	U,10

The settings (values or on/off status) that you make for many commands, such as *Limits, Grid, Snap, Running Osnaps, Fillet* values, *Pline* width, etc. are saved in system variables. In AutoCAD 2004 there are 397 system variables. The variables store the settings that are used to create and edit the drawing. *Setvar* ("set variable") gives you access to the system variables. (See Appendix A for a complete list of the system variables, including an explanation, default setting, and possible settings for each.)

Setvar allows you to perform two functions: (1) change the setting for any system variable and (2) display the current setting for one or all system variables. To change a setting for a system variable using *Setvar*, just use the command and enter the name of the variable. For example, the following syntax lists values for the *GRID* setting and current *Fillet* radius:

Command: *setvar*
Enter variable name or [?]: *gridunit*
Enter new value for GRIDUNIT <1.00,1.00>:

Command: *setvar*
Enter variable name or [?]: *filletrad*
Enter new value for FILLETRAD <0.50>:

With recent releases of AutoCAD, *Setvar* is not needed to set system variables. You can enter the variable name directly at the Command prompt without using *Setvar* first.

Command: *filletrad*
Enter new value for FILLETRAD <0.50>:

To list the current settings for all system variables, use *Setvar* with the *?* (question mark) option. The complete list of system variables is given in a text window with the current setting for each variable (Fig. 17-9).

Figure 17-9

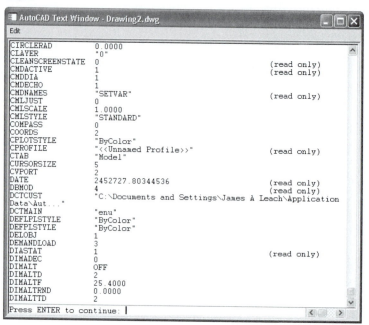

You can also use *Help* to list the system variables with a short explanation for each. In the *Help Topics* dialog box, select the *Contents* tab, select *Command Reference,* then *System Variables.*

MASSPROP

The *Massprop* command is used to give mass and volumetric properties for AutoCAD solids. See Chapter 39, Advanced Solids Features, for use of *Massprop.*

CHAPTER EXERCISES

1. *List*

    A. *Open* the **PLATENEST** drawing from Chapter 16 Exercises. Assume that a laser will be used to cut the plates and holes from the stock, and you must program the coordinates. Use the *List* command to give information on the *Line*s and *Circle*s for the one plate with four holes in a diagonal orientation. Determine and write down the coordinates for the 4 corners of the plate and the centers of the 4 holes.

    B. *Open* the **EFF-APT** drawing from Chapter 15 Exercises. Use *List* to determine the area of the inside of the tub. If the tub were filled with 10" of water, what would be the volume of water in the tub?

2. *Area*

    A. *Open* the **EFF-APT** drawing. The entry room is to be carpeted at a cost of $12.50 per square yard. Use the *Area* command (with the PICK points option) to determine the cost for carpeting the room, not including the closet.

    B. *Open* the **PLATENEST** drawing from Chapter 16 Exercises. Use *SaveAs* to create a file named **PLATE-AREA**. Using the *Area* command, calculate the wasted material (the two pieces of stock remaining after the 3 plates have been cut or stamped). HINT: Use *Boundary* to create objects from the waste areas for determining the *Area*.

    C. Create a *Boundary* (with islands) of the plate with four holes arranged diagonally. *Move* the new boundary objects **10** units to the right. Using the *Add* and *Subtract* options of *Area*, calculate the surface area for painting 100 pieces, both sides. (Remember to press Enter between the *Add* and *Subtract* operations.) *Save* the drawing.

3. *Dist*

    A. *Open* the **EFF-APT** drawing. Use the *Dist* command to determine the best location for installing a wall-mounted telephone in the apartment. Where should the telephone be located in order to provide the most equal access from all corners of the apartment? What is the farthest distance that you would have to walk to answer the phone?

    B. Using the *Dist* command, determine what length of pipe would be required to connect the kitchen sink drain (use the center of the far sink) to the tub drain (assume the drain is at the far end of the tub). Calculate only the direct distance (under the floor).

4. *ID*

    Open the **PLATENEST** drawing once again. You have now been assigned to program the laser to cut the plate with 4 holes in the corners. Use the *ID* command (with *OSNAP*s) to determine the coordinates for the 4 corners and the hole centers.

5. **Time**

    Using the *Time* command, what is the total amount of editing time you spent with the
    PLATENEST drawing? How much time have you spent in this session? How much time until
    the next automatic save?

6. **Setvar**

    Using the **PLATENEST** drawing again, use *Setvar* to list system variables. What are the current
    settings for *Fillet* radius (**FILLETRAD**), the last point used (**LASTPOINT**), *Linetype Scale*
    (**LTSCALE**), automatic file save interval (**SAVETIME**), and text *Style* (**TEXTSTYLE**).

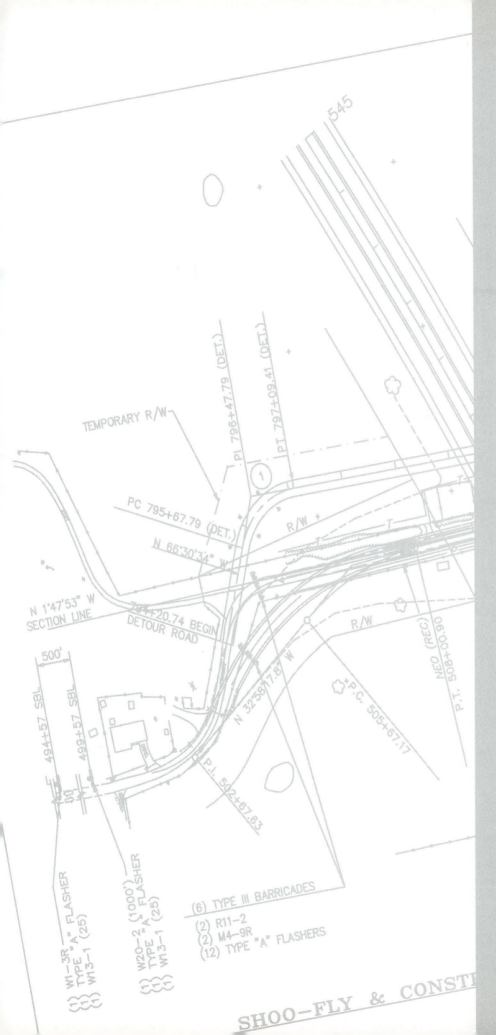

18

CREATING AND EDITING TEXT

Chapter Objectives

After completing this chapter you should:

1. be able to create lines of text in a drawing using *Dtext*;

2. be able to create and format paragraph text using *Mtext*;

3. be able to create text styles with the *Style* command;

4. know how to import external text into AutoCAD;

5. know that *Ddedit* can be used to edit the content of existing text and *Properties* can be used to modify any property of existing text;

6. be able to use *Spell* to check spelling and be able to *Find and Replace* text in a drawing;

7. know how to change the justification and scale of existing text using *Scaletext* and *Justifytext*;

8. know how features such as *Qtext* (quick text), *TEXTFILL*, and font mapping can be used to control the appearance of text in a drawing or print/plot.

CONCEPTS

The *Dtext* and *Mtext* commands provide you with a means of creating text in an AutoCAD drawing. "Text" in CAD drawings usually refers to sentences, words, or notes created from alphabetical or numerical characters that appear in the drawing. The numeric values that are part of specific dimensions are generally <u>not</u> considered "text," since dimensional values are a component of the dimension created automatically with the use of dimensioning commands.

Text in technical drawings is typically in the form of notes concerning information or descriptions of the objects contained in the drawing. For example, an architectural drawing might have written descriptions of rooms or spaces, special instructions for construction, or notes concerning materials or furnishings (Fig. 18-1). An engineering drawing may contain, in addition to the dimensions, manufacturing notes, bill of materials, schedules, or tables (Fig. 18-2). Technical illustrations may contain part numbers or assembly notes. Title blocks also contain text.

Figure 18-1

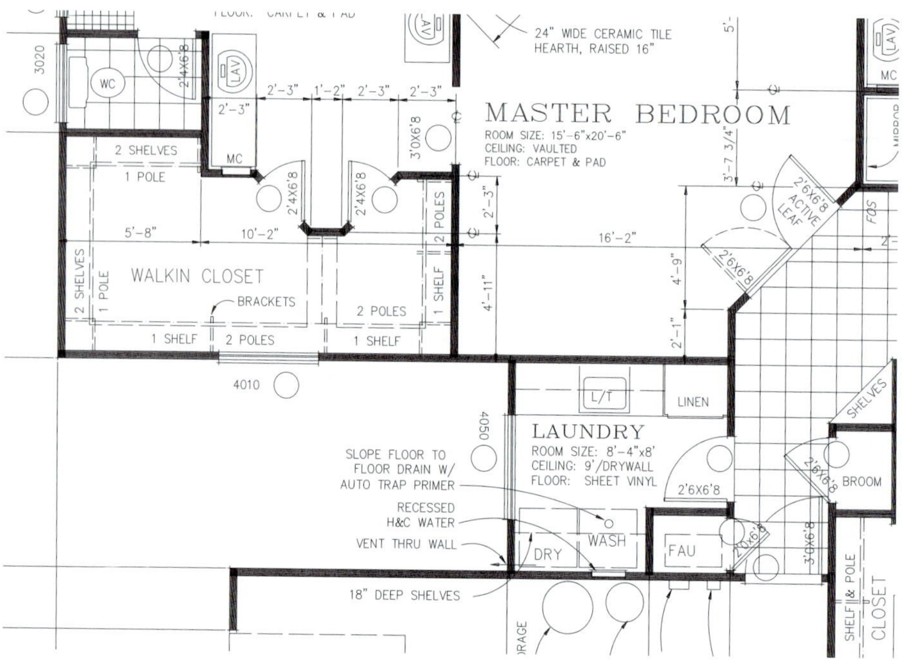

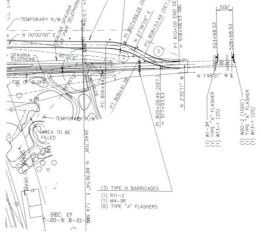

Figure 18-2

A line of text or paragraph of text in an AutoCAD drawing is treated as an object, just like a *Line* or a *Circle*. Each text object can be *Erased*, *Moved*, *Rotated*, or otherwise edited as any other graphical object. The letters themselves can be changed individually with special text editing commands. A spell checker is available by using the *Spell* command.

ROUTE	POINT	STATION	COORDINATES	
			NORTH	EAST
℄ BASE LINE	P.I.	502+67.63	452555.90	2855805.56
℄ BASE LINE	P.C.	505+67.16	453207.20	2855642.55
℄ BASE LINE	P.I.	506+86.85	453307.61	2855577.41
℄ BASE LINE	P.T.	508+00.90	453427.17	2855572.01
℄ BASE LINE	P.T.	510+00.00	453626.07	2855563.02
℄ BASE LINE	P.T.	516+50.00	454275.41	2855533.69
℄ BASE LINE	P.O.T.	519+00.00	454264.13	2855783.44
DETOUR @ STA 551+80				
℄ DETOUR	P.O.T.	794+20.74	453044.48	2855659.62
℄ DETOUR	P.C.	795+67.79	453103.09	2855524.75
℄ DETOUR	P.I	796+47.79	453134.98	2855451.38
℄ DETOUR	P.T.	797+09.41	453214.98	2855451.38
℄ DETOUR	P.C.	806+94.29	454199.87	2855451.38
℄ DETOUR	P.I.	807+69.29	454274.87	2855451.38
℄ DETOUR	P.T.	808+41.38	454341.30	2855486.12
℄ DETOUR	P.C.	808+53.49	454352.03	2855491.73
℄ DETOUR	P.I.	809+28.49	454418.50	2855526.48
℄ DETOUR	P.C.	810+00.00	454493.42	2855523.09

Since text is treated as a graphical element, the use of many lines of text in a drawing can slow regeneration time and increase plotting time significantly.

The *Dtext* and *Mtext* commands perform basically the same function; they create text in a drawing. *Mtext* is the newer and more sophisticated method of text entry. With *Mtext* (multiline text) you can create a paragraph of text that "wraps" within a text boundary (rectangle) that you specify. An *Mtext* paragraph is treated as one AutoCAD object. The *Dtext* command is intended to be used for creating single or multiple independent lines of text. The *Text* command, available in AutoCAD Release 14 and previous releases, has been converted to operate identically to the *Dtext* command.

Many options for text justification are available. *Justification* is the method of aligning multiple lines of text. For example, if text is right justified, the right ends of the lines of text are aligned. The form or shape of the individual letters is determined in AutoCAD by the text *Style*. Creating a *Style* begins with selecting a Windows standard TrueType or AutoCAD-supplied font file. Font files supplied with AutoCAD have file extensions of .TTF (TrueType) or .SHX (AutoCAD compiled shape files). The AutoCAD .SHX fonts are located in the C:\Program Files\AutoCAD 2004\Fonts directory (by the default installation). The TrueType fonts are installed with the other Windows fonts in the C:\Windows\Fonts directory. Additional fonts can be purchased or may already be on your computer (supplied with Windows, word processors, or other software).

After a font for the *Style* is selected, other parameters (such as width and obliquing angle) can be specified to customize the *style* to your needs. Only one *style*, called *Standard*, has been created as part of the traditional default template drawings (ACAD.DWT and ACADISO.DWT) and uses the TXT.SHX font file. Other template drawings may have two or more created *styles*.

If any other style of text is desired, it can be created with the *Style* command. When a new *style* is created, it becomes the current one used by the *Dtext* or *Mtext* command. If several *styles* have been created in a drawing, a particular one can be recalled or made current by using the *style* <u>option</u> of the *Dtext* or *Mtext* commands. Alternately, in AutoCAD 2004, you can select the current text style from the *Text Style Control* drop-down list just above the right end of the Object Properties toolbar. All styles that have been created for the drawing are included in this list.

In summary, the *Style* <u>command</u> allows you to design <u>new</u> styles with your choice of options, such as fonts, width factor, and obliquing angle, whereas the *style* <u>option</u> of *Dtext* and *Mtext* and the *Text Style Control* drop-down list allow you to select from <u>existing</u> styles in the drawing that you previously created.

Commands related to creating or editing text in an AutoCAD drawing include:

Dtext	Places individual lines of text in a drawing and allows you to see each letter as it is typed.
Mtext	Places text in paragraph form (with word wrap) within a text boundary and allows many methods of formatting the appearance of the text.
Style	Creates text styles for use with any of the text creation commands. You can select from font files, specify other parameters to design the appearance of the letters, and assign a name for each style.
Spell	Checks the spelling of existing text in a drawing.
Find	Used to find or replace text strings globally in the drawing. Text created with *Dtext* or *Mtext* can be located with *Find*.
Ddedit	Invokes a dialog box for editing text. If you select *Text* or *Dtext* for editing, you can change the text (characters) only; if you select *Mtext* objects, you can edit individual characters and change the appearance of individual characters or the entire paragraph(s).
Scaletext	Allows you to change the scale of multiple text objects without altering the text insertion points.
Justifytext	Allows you to change the insertion point and justification of existing text without changing the text position.
Qtext	Short for quick-text, temporarily displays a line of text as a box instead of individual characters in order to speed up regeneration time and plotting time.

TEXT CREATION COMMANDS

The commands for creating text are formally named *Dtext* and *Mtext* (these are the commands used for typing). The *Draw* pull-down and screen (side) menus provide access to the two commonly used text commands, *Multiline Text...* (*Mtext*) and *Single-Line Text* (*Dtext*) (Fig. 18-3). Only the *Mtext* command has an icon button (by default) near the bottom of the *Draw* toolbar (Fig. 18-4).

Figure 18-3 ——————— **Figure 18-4**

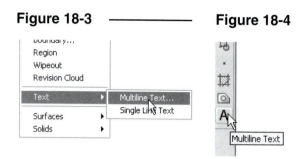

DTEXT

Pull-down Menu	COMMAND (TYPE)	ALIAS (TYPE)	Short-cut	Screen (side) Menu	Tablet Menu
Draw *Text >* *Single Line Text...*	DTEXT	DT	...	DRAW 2 *Dtext*	K,8

Dtext (dynamic text) lets you insert single lines of text into an AutoCAD drawing. *Dtext* displays each character in the drawing as it is typed. You can enter multiple lines of text without exiting the *Dtext* command. The lines of text do not wrap. The options are presented below:

```
Command: dtext
Current text style: "Standard" Text height: 0.2000
Specify start point of text or [Justify/Style]:
```

Start Point

The *Start point* for a line of text is the <u>left end</u> of the baseline for the text (Fig. 18-5). *Height* is the distance from the baseline to the top of upper case letters. Additional lines of text are automatically spaced below and left justified. The *rotation angle* is the angle of the baseline (Fig. 18-6).

The command sequence for this option is:

```
Command: dtext
Current text style: "Standard" Text height: 0.2000
Specify start point of text or [Justify/Style]: PICK or (coordinates)
Specify height <0.2000>: Enter or (value)
Specify rotation angle of text <0>: Enter or (value)
Enter text: (Type the desired line of text and press Enter)
Enter text: (Type another line of text and press Enter)
Enter text: Enter
Command:
```

Figure 18-5 ———————

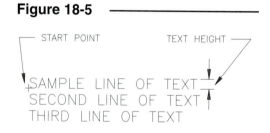

Figure 18-6 ———————

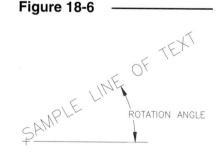

NOTE: When the "Enter text:" prompt appears, you can also PICK a new location for the next line of text anywhere in the drawing.

Justify
If you want to use one of the justification methods, invoking this option displays the choices at the prompt:

        Command: **dtext**
        Current text style: "Standard" Text height: 0.2000
        Specify start point of text or [Justify/Style]: **J** (*Justify* option)
        Enter an option [Align/Fit/Center/Middle/Right/TL/TC/TR/ML/MC/MR/BL/BC/BR]:

After specifying a justification option, you can enter the desired text in response to the "Enter text:" prompt. The text is not justified until <u>after</u> you press Enter.

Align
Aligns the line of text between the two points specified (P1, P2). The text height is adjusted automatically (Fig. 18-7).

Figure 18-7 ────────────

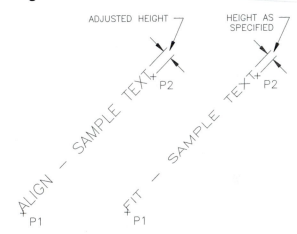

Fit
Fits (compresses or extends) the line of text between the two points specified (P1, P2). The text height does not change (Fig. 18-7).

Center
Centers the baseline of the first line of text at the specified point. Additional lines of text are centered below the first (Fig. 18-8).

Middle
Centers the first line of text both vertically and horizontally about the specified point. Additional lines of text are centered below it (Fig. 18-8).

Figure 18-8 ────────────

LEFT JUSTIFIED TEXT
(START POINT)
SAMPLE

RIGHT JUSTIFIED TEXT
(R OPTION)
SAMPLE

CENTER JUSTIFIED TEXT
(C OPTION)
SAMPLE

MIDDLE JUSTIFIED TEXT
(M OPTION)
SAMPLE

Right
Creates text that is right justified from the specified point (Fig. 18-8).

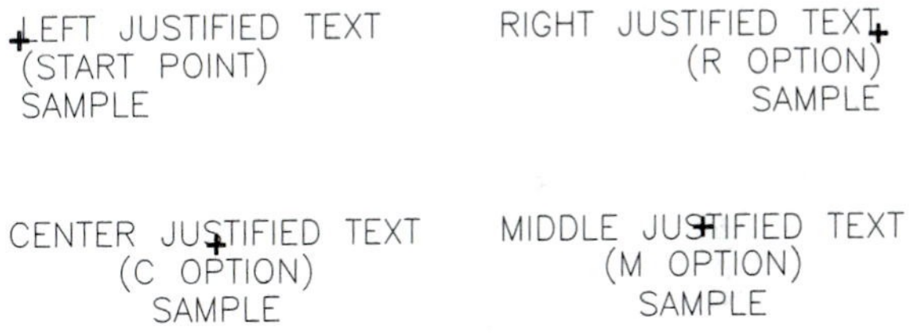

TL
Top Left. Places the text in the drawing so the top line (of the first line of text) is at the point specified and additional lines of text are left justified below the point. The top line is defined by the upper case and tall lower case letters (Fig. 18-9).

TC
Top Center. Places the text so the top line of text is at the point specified and the line(s) of text are centered below the point (Fig. 18-9).

Figure 18-9 ────────────

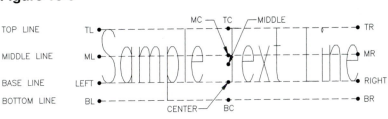

TR
Top Right. Places the text so the top right corner of the text is at the point specified and additional lines of text are right justified below that point (Fig. 18-9).

ML

Middle Left. Places text so it is left justified and the middle line of the first line of text aligns with the point specified. The middle line is half way between the top line and the baseline, not considering the bottom (extender) line (Fig. 18-9).

MC

Middle Center. Centers the first line of text both vertically and horizontally about the midpoint of the middle line. Additional lines of text are centered below that point (Fig. 18-9).

MR

Middle Right. Justifies the first line of text at the right end of the middle line. Additional lines of text are right justified (Fig. 18-9).

BL

Bottom Left. Attaches the bottom (extender) line of the first line of text to the specified point. The bottom line is determined by the lowest point of lower case extended letters such as y, p, q, j, and g. If only upper-case letters are used, the letters appear to be located above the specified point. Additional lines of text are left justified (Fig. 18-9).

BC

Bottom Center. Centers the first line of text horizontally about the bottom (extender) line (Fig. 18-9).

BR

Bottom Right. Aligns the bottom (extender) line of the first line of text at the specified point. Additional lines of text are right justified (Fig. 18-9).

NOTE: Because there is a separate baseline and bottom (extender) line, the *MC* and *Middle* points do not coincide and the *BL, BC, BR* and *Left, Center, Right* options differ. Also, because of this feature, when all uppercase letters are used, they ride above the bottom line. This can be helpful for placing text in a table because selecting a horizontal *Line* object for text alignment with *BL, BC,* or *BR* options automatically spaces the text visibly above the *Line*.

Style (option of *Dtext* or *Mtext*)

The *style* <u>option</u> of the *Dtext* or *Mtext* command allows you to select from the <u>existing</u> text styles that have been previously created as part of the current drawing. The style selected from the list becomes the current style and is used when placing text with *Dtext* or *Mtext*. Alternately, <u>before you use *Dtext* or *Mtext*</u>, you can select the current text style from the *Text Style Control* drop-down list (just above the right end of the Object Properties toolbar) to make it the current text style.

Since only one text *style, Standard,* is available in the traditional (inch) template drawing (ACAD.DWT) and the metric template drawing (ACADISO.DWT), other styles must be created before the *style* option of *Dtext* is of any use. Various text styles are created with the *Style* <u>command</u> (this topic is discussed later).

Use the *Style* option of *Dtext* or *Mtext* to list existing styles and specifications. An example listing is shown below:

```
Command: dtext
Current text style: "Standard" Text height: 0.2000
Specify start point of text or [Justify/Style]: S (Style option)
Enter style name or [?] <STANDARD>: ? (list option)
Enter text style(s) to list <*>: Enter
Text styles:
```

Style name: "ROMANS"   Font files: romans
   Height: 1.50 Width factor: 1.00 Obliquing angle: 0.000
   Generation: Normal

Style name: "STANDARD"   Font files: txt
   Height: 0.00 Width factor: 1.00 Obliquing angle: 0.000
   Generation: Normal

Current text style: "STANDARD"
Text height: 0.30
Specify start point of text or [Justify/Style]:

MTEXT

Pull-down Menu	COMMAND (TYPE)	ALIAS (TYPE)	Short-cut	Screen (side) Menu	Tablet Menu
Draw *Text >* *Multiline Text...*	*MTEXT or* *-MTEXT*	*T, -T or* *MT*	...	*DRAW 2* *Mtext*	*J,8*

Multiline Text (*Mtext*) has more editing options than other text commands. You can apply underlining, color, bold, italic, font, and height changes to individual characters or words within a paragraph or multiple paragraphs of text.

Mtext allows you to create paragraph text defined by a text boundary. The text boundary is a reference rectangle that specifies the paragraph width. The *Mtext* object that you create can be a line, one paragraph, or several paragraphs. AutoCAD references *Mtext* (created with one use of the *Mtext* command) as one object, regardless of the amount of text supplied. Like *Dtext*, several justification methods are possible.

Command: **mtext**
Current text style: "Standard" Text height: 0.2000
Specify first corner: **PICK**
Specify opposite corner or [Height/Justify/Line spacing/Rotation/Style/Width]: **PICK** or (**option**)

You can PICK two corners to invoke the Multiline Text Editor, or enter the first letter of one of these options: *Height, Justify, Rotation, Style, Line Spacing*, or *Width*. Other options can also be accessed within the Multiline Text Editor.

Using the default option the *Mtext* command, you supply a "first corner" and "opposite corner" to define the diagonal corners of the text boundary (like a window). Although this boundary confines the text on two or three sides, one or two arrows indicate the direction text flows if it "spills" out of the boundary (Fig. 18-10). (See "Text Flow and *Justification*.")

Figure 18-10

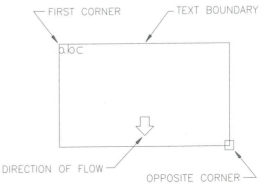

After you PICK the two points defining the text boundary, the Multiline Text Editor appears ready for you to enter the desired text (Fig. 18-11). The text wraps based on the width you defined for the text boundary. Select the *OK* button to have the text entered into the drawing.

This "frameless" text editor is new for AutoCAD 2004. The new editor is simpler to use than the previous editor and contains several new features.

Figure 18-11

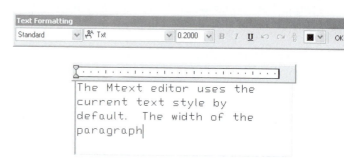

Two methods (in addition to the command line options) are available for you to format and edit text while you use the Multiline Text Editor: the *Text Formatting* toolbar (Fig. 18-11, top) and the right-click shortcut menu (described later). Using the options in these menus is <u>dynamic</u>—the text in the editor immediately reflects the changes made for your preferences.

Text Formatting Toolbar and Command Line Options

Style
This option is the first (far left) drop-down list on the *Text Formatting* toolbar. The box (not dropped down) displays the current text *Style* (*Standard* for most template drawings), and the list displays all existing text styles in the drawing. Select the style you want to use (make current) from this list. Alternately, <u>before you use *Dtext* or *Mtext*</u>, you can select the current text style from the *Text Style Control* drop-down list (just above the right end of the Object Properties toolbar) to make it the current text style. (See the *Style* command for information on creating new text styles.)

Font
Choose from any font in the drop-down list. The list includes all fonts registered on your system. Use this feature to change the appearance of selected words in the paragraph. Your selection here <u>overrides the font originally assigned to the text *Style*</u> for the selected or newly created words or paragraphs. Even though you can change the font for the entire *Mtext* object (paragraph) using this feature, it is recommended that you set the paragraph to the desired *Style* (in the first drop-down list), rather than change the characters in the *Style* to a different font for the current *Mtext* object. See following NOTE.

Text Color
Select individual text, then use this drop-down list (see Figure 18-11, far right) to select a color for the selected text. This selection overrides the layer color. See following NOTE.

NOTE: The *Font* and *Color* options in the *Text Formatting* toolbar override the properties of the text *Style* and layer color. For example, changing the font in the *Font* list overrides the text style's font used for the paragraph so it is possible to have one font used for the paragraph and others for individual characters within the paragraph. This is analogous to object-specific color and linetype assignment in that you can have a layer containing objects with different linetypes and colors than the linetype and color assigned to the layer. To avoid confusion, it is recommended for <u>global</u> formatting to create text *Styles* to be used for the paragraphs, layer color to determine color for the paragraphs, then if needed, use the *Font* and *Color* options in the *Text Formatting* toolbar to change fonts and colors for <u>selected</u> text rather than for the entire paragraph.

Height

Select from the list or enter a new value for the height to be used globally for the paragraph or for selected words or letters. If *Height* was defined when the selected text *Style* was created for the drawing, this option overrides the *Style's Height*.

Bold, Italic, Underline

Select (highlight) the desired letters or words then PICK the desired button. Only authentic TrueType fonts (not the AutoCAD-supplied .SHX equivalents) can be bolded or italicized.

Undo, Redo

Use *Undo* to reverse the action of the last formatting option used. You can *Redo* the last *Undo*, but only immediately after the *Undo*.

Stack/Unstack

This option is a toggle to *Stack* or *Unstack* specific text (see "Creating Stacked Text"). Highlight the fraction or text, then use this option to stack or unstack the fraction or text.

Width

Even though this option does not appear on the *Text Formatting* toolbar, you can interactively change the width of the existing text boundary. Do this by placing your pointer at the right border of the ruler above the text so the double arrows appear, then adjust the paragraph width (Fig. 18-12). The lines of text will automatically "wrap" to adjust to the new text boundary. You can also use the *Width* option from the Command line to set the width to a specific value. Also see "Text Flow and Justification" for changing the width of the text boundary with Grips.

Figure 18-12

[Width: 5.1915]

You can adjust the width of the paragraph by placing your pointer at the right end of the ruler, then click and drag.

Rotation

The *Rotation* option specifies the rotation angle of the entire *Mtext* object (paragraph) including the text boundary. This option is available only by command line; however, the Grips Rotate option can be used after the *Mtext* object has been created. Using the Command line method, you can see the rotated text boundary as you specify the corners (Fig. 18-13). The rotation angle is <u>not</u> reflected in the text appearing in the editor but only in the drawing.

Figure 18-13

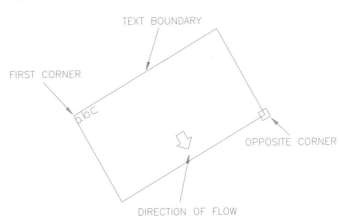

TEXT BOUNDARY

FIRST CORNER

abc

OPPOSITE CORNER

DIRECTION OF FLOW

Line Spacing

The *Line Spacing* option is available only by Command line. Line spacing sets the spacing between lines for the paragraph. The spacing increment is the vertical distance between the bottom (or baseline) of one line of text and the bottom of the next line of text.

Specify opposite corner or [Height/Justify/Line spacing/Rotation/Style/Width]: **L**
Enter line spacing type [At least/Exactly] <At least>:

Select *At Least* (the default setting) if you have different size characters in the paragraph. This option automatically adds space between lines based on the height of the largest character in the line so the spacing can vary depending on the size of the characters.

The *Exactly* option forces the line spacing to be the same for all lines of text in the *Mtext* object. Use this option to insert text into a table or to ensure that line spacing is identical in multiple *Mtext* objects. You can set the spacing increment to multiples of single-line spacing (1x, 1.5x, 2x, etc.) or to an absolute value such as 1 for spacing of exactly 1.0 units, regardless of the text height. Using *Exactly* can cause text in lines above or below lines with large font characters to overlap the larger characters.

Mtext **Right-Click Shortcut Menu**

Right-click in the Multiline Text Editor to produce the short-cut menu shown in Figure 18-14. If you want to change specific text, highlight the text before invoking the menu.

Figure 18-14 ——————————

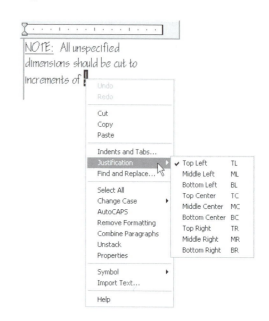

Undo, Redo
These options are described previously (see "*Text Formatting Toolbar*").

Cut, Copy, Paste
These selections allow you to *Cut* (erase) highlighted text from the paragraph, *Copy* highlighted text, and *Paste* (the *Cut* or *Copied*) text to the current cursor position. Text that you *Cut* or *Copy* is held in the Clipboard until *Pasted*.

Indents and Tabs...
Selecting this option produces the *Indents and Tabs* dialog box (Fig. 18-15). Simply key in values (in drawing units) for the desired indentations for *First Line* and/or for the entire *Paragraph*. Tab stop positions can be set by entering a value in the top edit box, then select *Set* to make the value apply to the *Mtext* object. *Clear* deletes the highlighted tab set values from the list.

Figure 18-15 ——————————

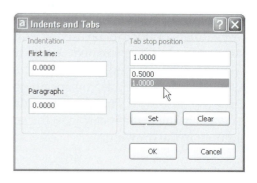

Alternately, you can set the first line indent and the paragraph indent interactively by using your pointer on the ruler (just above the text boundary). Move the top indent marker to set the indent for the *First line* (Fig. 18-16) and the bottom indent marker to set an indent for the entire *Paragraph*.

Figure 18-16 ——————————

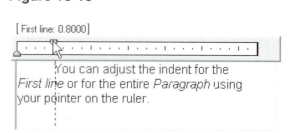

You can also set tab positions by clicking in the ruler appearing above the text box (Fig. 18-17). Pick the desired position to make a tab appear. Drag and drop tab stops out of the ruler to clear them. Default tabs are already set at each long ruler mark (0.8 inches or 10 mm).

Figure 18-17

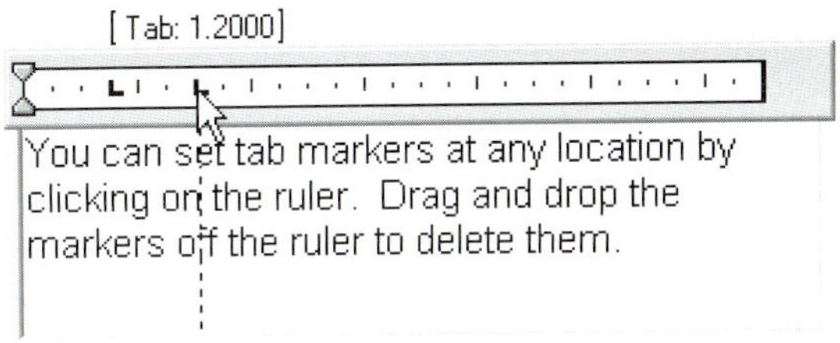

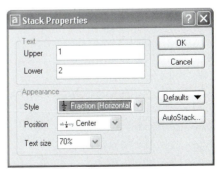

[Tab: 1.2000]

You can set tab markers at any location by clicking on the ruler. Drag and drop the markers off the ruler to delete them.

2004

Find and Replace
See "Editing Text" later in this chapter.

Justification
This property determines how the paragraph is located and direction of flow with respect to the text boundary (see "Text Flow and *Justification*").

Select All
Picking this option automatically selects all the text contained in the *Mtext* object (paragraph or paragraphs).

Change Case
First select the desired text, then change it to either *UPPERCASE* or *lowercase*. All letters are changed, so with these options you cannot have mixed upper and lower case, such as Title Case, unless you change letters individually.

AutoCAPS
This option affects only newly typed and imported text. When checked, *AutoCAPS* turns on Caps Lock on your keyboard so only uppercase letters appear as you type. If needed, use *Change Case* to convert the text back to lowercase.

Remove Formatting
Select the desired text, then select this option to remove bold, italic, or underline formatting.

Combine Paragraphs
If you have more than one paragraph in one *Mtext* object, you can select all the text in the paragraphs, then use this option. The selected paragraphs are converted into a single paragraph and each paragraph return is replaced with a space.

Stack/Unstack
Stack/Unstack appears in the shortcut menu if you highlight only stacked text, such as a fraction, then right-click to invoke the menu. This option is a toggle so that stacked characters become unstacked and vice versa. See "Creating Stacked Text."

Properties
This option appears in the shortcut menu if you highlight only stacked text such as a fraction, then right-click. The *Stack Properties* dialog box appears (Fig. 18-18). See also "Creating Stacked Text."

Figure 18-18

The *Text* section of the *Stack Properties* dialog box allows you to change the text characters (numbers or letters) contained in the stacked set. Although using a slash (/), pound (#), or carat (^) to specify a stack normally creates horizontal, diagonal, and tolerance format, respectively, that format can be changed with the *Style* drop-down list in this dialog box. Use the *Position* option to align the fraction with the *Top*, *Center*, or *Bottom* of the other normal characters in the line of text. You can also specify the *Text size* for the stack (which should normally be smaller than normal characters since the stacked set occupies more vertical space).

Symbol

Common symbols (*Degrees, Plus/Minus, Diameter*) can be inserted. Note that in the Multiline Text Editor, these symbols are displayed with a "%%" and a letter, and the nonbreaking space is displayed as a hollow rectangle, but will be displayed correctly in the drawing. Selecting *Other...* produces a character map to select symbols from (See *Other Symbols* next).

Other Symbols

The steps for inserting symbols from the *Character Map* dialog box (Fig. 18-19) are as follows:

1. Highlight the symbol.
2. Double-click or pick *Select* so the item appears in the *Characters to copy:* edit box.
3. Select the *Copy* button to copy the item(s) to the Windows Clipboard.
4. *Close* the dialog box.
5. In the Multiline Text Editor, move the cursor to the desired location to insert the symbol.
6. Finally, right-click and select *Paste* from the menu.

Figure 18-19

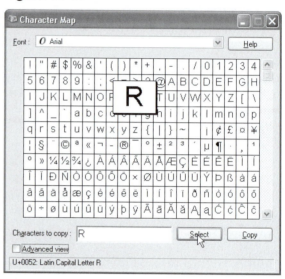

Import Text

See "Importing External Text into AutoCAD."

Text Flow and *Justification*

Figure 18-20

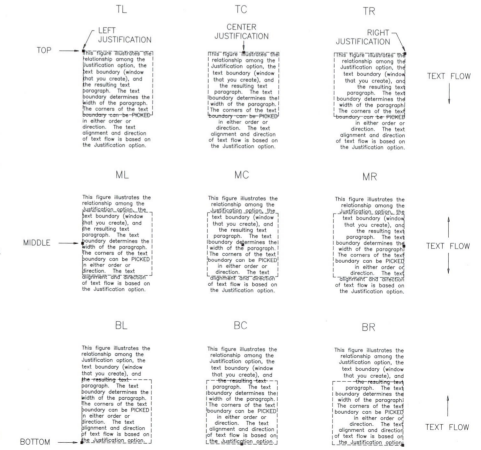

When you PICK two corners to define the text boundary, you determine the width of the paragraph and the direction that the text flows. You can draw the text boundary in any direction from the "first corner" to the "other corner." Text does not always fit within the boundary but is confined on two or three sides and "spills" out of the boundary (up, down, or both) in the direction of the arrow(s) displayed when you draw the boundary (see Figure 18-10).

Justification is the method of aligning the text with respect to the text boundary. The *Justification* options are *TL, TC, TR, ML, MC, MR, BL, BC,* and

BR (*Top Left, Top Centered, Top Right, Middle Left, Bottom Left,* etc.) The text paragraph (*Mtext* object) is effectively "attached" to the boundary based on the *Justification* option selected. *Justification* can be specified by two methods: in Command line format before specifying the text boundary or in the Multiline Text Editor shortcut menu.

The *Justification* methods are illustrated in Figure 18-20. The illustration shows the relationship among the *Justification* option, the text boundary, the direction of flow, and the resulting text paragraph.

If you want to adjust the text boundary after creating the *Mtext* object, you can use *Mtedit* (or double-click on the *Mtext* object), *Ddedit,* or *Properties* (see "Editing Text") or use Grips (Chapter 23) to stretch the text boundary. If you activate the Grips, four grips appear at the text boundary corners and one grip appears at the defined *Justification* point (Fig. 18-21).

Figure 18-21 ———————

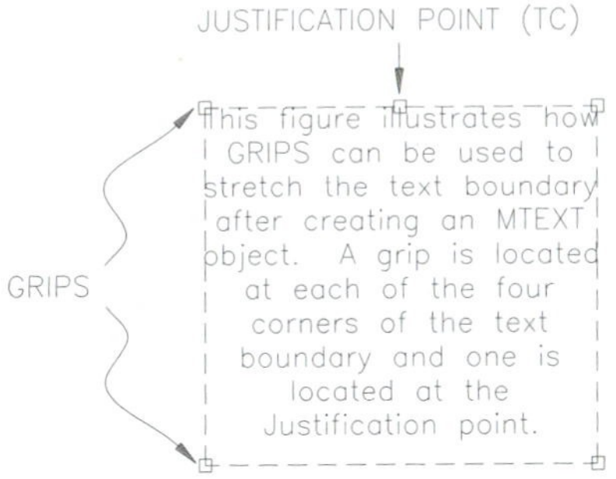

JUSTIFICATION POINT (TC)

This figure illustrates how GRIPS can be used to stretch the text boundary after creating an MTEXT object. A grip is located at each of the four corners of the text boundary and one is located at the Justification point.

GRIPS

Creating Stacked Text

"Stacked" text is a combination of two elements of text, one above the other, usually separated by a horizontal line. The most common type of stacked text is a fraction, although letters or words can also be stacked. If you want to create stacked text (numbers or letters), enter a slash (/), pound (#), or carat (^) as a separator between the upper and lower lines of text. Entering a slash, pound, or carat in the Multiline Text Editor automatically invokes the *AutoStack Properties* dialog box (Fig. 18-22). Although the slash, pound, and carat each specify a particular stacked format, <u>any of these formats can be changed using the *Properties* option in the *Mtext* shortcut menu</u> (see "*Properties*" in "*Mtext* Right-Click Shortcut Menu").

Figure 18-22 ———————

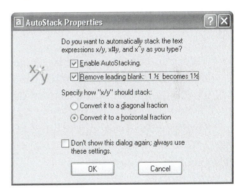

Enable AutoStacking
When this box is checked, fractions are automatically stacked as you type. Anytime you enter numeric characters before and after the carat, slash, or pound character, the numbers are automatically stacked. Each of these characters causes the following style of stacking:

slash (/) causes a vertical stack separated by a horizontal line
pound (#) causes a diagonal stack separated by a diagonal line
carat (^) causes a vertical stack without a horizontal separator line

Remove Leading Blank
This option removes blanks between a whole number and a fraction. For example, you would normally type one and one-half with a space (1 1/2) which would convert to a whole number plus a space before the fraction.

Convert it to a Diagonal Fraction and *Convert it to a Horizontal Fraction*
These options convert the slash character to a diagonal or horizontal fraction when AutoStack is on. Whether AutoStack is on or off, the pound character is always converted to a diagonal fraction, and the carat character is always converted to a tolerance format.

Don't Show This Dialog Again; Always Use These Settings
Place a check in this box to suppresses display of the *AutoStack Properties* dialog box using the current property settings for all stacked text. When this option is cleared, the AutoStack Properties dialog box is automatically displayed if you type two numbers separated by a slash, carat, or pound sign followed by a space or nonnumeric character.

Calculating Text Height for Scaled Drawings

To achieve a specific height in a drawing intended to be plotted to scale, multiply the desired text height for the plot by the drawing scale factor. (See Chapter 14, Printing and Plotting.) For example, if the drawing scale factor is 48 and the desired text height on the plotted drawing is 1/8", enter **6** (1/8 x 48) in response to the "Height:" prompt of the *Dtext* or *Mtext* command.

If you know the plot scale (for example, 1/4"= 1') but not the drawing scale factor (DSF), calculate the reciprocal of the plot scale to determine the DSF, then multiply the intended text height for the plot by the DSF, and enter the value in response to the "Height:" prompt. For example, if the plot scale is 1/4"=1', then the DSF = 48 (reciprocal of 1/48).

If *Limits* have already been set and you do not know the drawing scale factor, use the following steps to calculate a text height to enter in response to the "Height:" prompt to achieve a specific plotted text height.

1. Determine the sheet size to be used for plotting (for example, 36" x 24").
2. Decide on the text height for the finished plot (for example, .125").
3. Check the *Limits* of the current drawing (for example, 144' x 96' or 1728" x 1152").
4. Divide the *Limits* by the sheet size to determine the drawing scale factor (1728"/36" = 48).
5. Multiply the desired text height by the drawing scale factor (.125 x 48 = 6).

Also see "*Spacetrans*" below.

SPACETRANS

Pull-down Menu	COMMAND (TYPE)	ALIAS (TYPE)	Short-cut	Screen (side) Menu	Tablet Menu
...	SPACETRANS	...	...	...	...

The *Spacetrans* command, introduced in AutoCAD 2002, is used to convert a distance in one space (model space or paper space) to the equivalent distance in the other space. *Spacetrans* is intended to be used transparently (within a command) to pass the translated value to the current command. For example, using *Spacetrans* while drawing a *Line* in paper space allows you to draw the line length to an equivalent model space distance.

Spacetrans is especially helpful for creating text objects in model space that print at a specific size in paper space. Instead of having to calculate the text height in model space units times the drawing scale factor (text height x DSF) as described previously in "Calculating Text Height for Scaled Drawings," using *Spacetrans* allows you to enter only the desired plotted text height (assuming you are plotting from paper space at 1:1). Using *Spacetrans* in model space (inside a viewport) allows you to create model space text in paper space units.

As an example, assume you are creating *Dtext* inside a viewport (in model space) and want the text height to be equivalent to .125 units (1/8") in paper space. You would use a procedure similar to that shown below—that is, invoking *Spacetrans* transparently during the *Dtext* command to convert a paper space value (.125) to model space units and pass the value to the *Dtext* command.

```
Command: dtext
Current text style: "romans" Text height: 0.2000
Specify start point of text or [Justify/Style]: PICK
Specify height <0.2000>: 'spacetrans
>>Specify paper space distance <1.0000>: .125
Resuming DTEXT command.
Specify height <0.2000>: 0.500000000000000
Specify rotation angle of text <0>: Enter
Enter text:
```

When you are first prompted to "Specify height," invoke *Spacetrans* transparently. Enter the desired height in paper space units at the "Specify paper space distance" prompt. *Spacetrans* then converts the .125 units to the equivalent distance in model space (0.500000000000000) and passes the value to the second "Specify height" prompt, thus creating text in model space to print at .125 in paper space.

Technically, *Spacetrans* computes the entered paper space units times the reciprocal of the viewport scale (or viewport zoom factor). In the previous example, the viewport scale is 1:4; therefore, .125 x 4 = .5.

Spacetrans cannot be used in the *Model* tab—you must be in a layout with at least one viewport created, and you must be in the desired space when *Spacetrans* is invoked. If you are in paper space when *Spacetrans* is invoked and there is more than one viewport, you are prompted to "Select a viewport."

2002

TEXT STYLES, FONTS, EXTERNAL TEXT, AND EXTERNAL EDITORS

STYLE

	Pull-down Menu	COMMAND (TYPE)	ALIAS (TYPE)	Short-cut	Screen (side) Menu	Tablet Menu
	Format *Text Style...*	*STYLE* or *-STYLE*	*ST*	...	*FORMAT* *Style:*	*U,2*

Text styles can be created by using the *Style* command. A text *Style* is created by selecting a font file as a foundation and then specifying several other parameters to define the configuration of the letters.

Using the *Style* command invokes the *Text Style* dialog box (Fig. 18-23). All options for creating and modifying text styles are accessible from this device. Selecting the font file is the initial step in creating a style. The font file selected then becomes a foundation for "designing" the new style based on your choices for the other parameters (*Effects*).

Figure 18-23

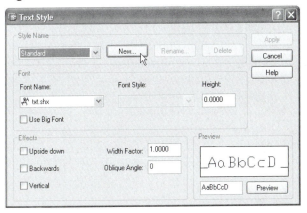

Recommended steps for creating text *Styles* are as follows:

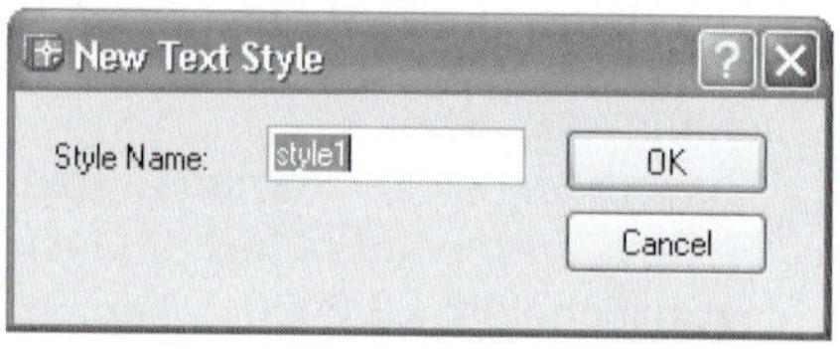

1. Use the *New* button to create a new text *Style*. This button opens the *New Text Style* dialog box (Fig. 18-24). By default, AutoCAD automatically assigns the name Style*n*, where *n* is a number that starts at 1. Enter a descriptive name into the edit box. Style names can be up to 256 characters long and can contain letters, numbers, and the special characters dollar sign ($), underscore (_), and hyphen (-).

Figure 18-24

2. Select a font to use for the style from the *Font Name* drop-down list (see Figure 18-23). TrueType equivalents of the traditional AutoCAD fonts files (.SHX) and authentic TrueType fonts are available.
3. Select parameters in the *Effects* section of the dialog box (Fig. 18-23, lower left). Changes to the *Upside down, Backwards, Width Factor,* or *Oblique Angle* are displayed immediately in the *Preview* tile.
4. You can enter specific words or characters in the edit box in the lower-right corner, then press *Preview* to view the characters in the new style.
5. Select *Apply* to save the changes you made in the *Effects* section to the new style.
6. Create other new *Styles* using the same procedure listed in steps 1 through 5.
7. Select *Close*. The new (or last created) text style that is created automatically becomes the current style inserted when *Dtext* or *Mtext* is used.

You can modify existing styles using this dialog box by selecting the existing *Style* from the list, making the changes, then selecting *Apply* to save the changes to the existing *Style*.

Rename

Select an existing *Style* from the drop-down list, then select *Rename*. Enter a new name in the edit box.

Delete

Select an existing *Style* from the drop-down list, then select *Delete*. You cannot delete the current *Style* or *Styles* that have been used for creating text in the drawing.

Alternately, you can enter -*Style* (use the hyphen prefix) at the command prompt to display the Command line format of *Style* as shown below:

```
Command: -style
Enter name of text style or [?] <Standard>: name  (Enter new style name.)
New style.
Specify full font name or font filename (TTF or SHX) <txt>: name  (Enter desired font file.)
Specify height of text <0.0000>: Enter or (value)
Specify width factor <1.0000>: Enter or (value)
Specify obliquing angle <0>: Enter or (value)
Display text backwards? [Yes/No] <N>: Enter or Y
Display text upside-down? [Yes/No] <N>: Enter or Y
Vertical? <N> Enter or Y
"(new style)" is now the current text style.
Command:
```

The options that appear in both the *Text Style* dialog box and in the Command line format are described in detail next.

Height <0.000>
The height should be 0.000 if you want to be prompted again for height each time the *Dtext* command is used. In this way, the height is variable for the style each time you create text in the drawing. If you want the height to be constant, enter a value other than 0. Then, *Dtext* will not prompt you for a height since it has already been specified. The *Mtext* command, however, allows you to change the height in the Multiline Text Editor, even if you specified a height when you created the text style. A specific height assignment with the *Style* command also overrides the *DIMTXT* setting (see Chapter 29).

Width factor <1.000>
A *width factor* of 1 keeps the characters proportioned normally. A value less than 1 compresses the width of the text (horizontal dimension) proportionally; a value of greater than 1 extends the text proportionally.

Obliquing angle <0>
An angle of 0 keeps the font file as vertical characters. Entering an angle of 15, for example, would slant the text forward from the existing position, or entering a negative angle would cause a back-slant on a vertically oriented font (Fig. 18-25).

Figure 18-25

0° OBLIQUING ANGLE

15° OBLIQUING ANGLE

−15° OBLIQUING ANGLE

Backwards
Backwards characters can be helpful for special applications, such as those in the printing industry.

Figure 18-26

UPSIDE DOWN TEXT

TEXT ROTATED 180°.

Upside-down
Each letter is created upside-down in the order as typed (Fig. 18-26). This is different than entering a rotation angle of 180 in the *Dtext* command. (Turn this book 180 degrees to read the figure.)

Figure 18-27

V T
E E
R X
T T
I
C
A
L

Vertical
Vertical letters are shown in Figure 18-27. The normal rotation angle for vertical text when using *Dtext* or *Mtext* is 270. Only .SHX fonts can be used for this option. *Vertical* text does not display in the *Preview* image tile.

Since specification of a font file is an initial step in creating a style, it seems logical that different styles could be created using one font file but changing the other parameters. It is possible, and in many cases desirable, to do so. For example, a common practice is to create styles that reference the same font file but have different obliquing angles (often used for lettering on isometric planes). This can be done by using the *Style* command to assign the same font file for each style, but assign different parameters (obliquing angle, width factor, etc.) and a unique name to each style. The relationship between fonts, styles, and resulting text is shown in Figure 18-28.

Figure 18-28

FONT FILE	STYLE NAME (EXAMPLE)	RESULTING TEXT (BASED ON OTHER PARAMETERS)
ROMAN SIMPLEX (ROMANS.SHX)	ROMANS–VERT	ABCDEFG 1234
	ROMANS–ITAL	ABCDEFG 1234
ROMAN DUPLEX (ROMAND.SHX)	ROMAND–EXT	ABC 1234
	ROMAND–COMP	ABCDEFG 1234

.SHX Fonts

The .SHX fonts are fonts created especially for AutoCAD drawings by Autodesk. They are generally smaller files and are efficient to use for AutoCAD drawings where file size and regeneration time is critical. These fonts, however, are composed of single line segments and are not solid-filled such as most TrueType fonts (use *Zoom* to closely examine and compare .SHX vs. TrueType fonts).

Because the Multiline Text Editor is a Windows-compliant dialog box, it can display only fonts that are recognized by Windows. Since AutoCAD .SHX fonts are not recognized by Windows, AutoCAD supplies a TrueType equivalent when you select an .SHX or any other non-TrueType font for display in the Multiline Text Editor only. When you then use a text creation command, AutoCAD creates the text in the drawing with the true .SHX font.

All AutoCAD .SHX shape fonts are Unicode fonts since Release 14. A Unicode font can contain 65,535 characters with shapes for many languages (see *Big Fonts*). Unicode fonts contain many more characters than are shown on your keyboard in your system; therefore, to use a character not directly available from the keyboard, you can enter the escape sequence $\backslash U+nnnn$ where *nnnn* represents the Unicode hexadecimal value for the character (see "Special Text Characters").

Big Fonts

Big Fonts are used for characters not found in the English language. Text files for some alphabets contain thousands of non-ASCII characters. For these applications, AutoCAD provides a special type of shape definition known as a *Big Font* file that contains many other characters.

When you check *Use Big Font*, the *Font Style* box changes to a *Big Font* box. You can select *Use Big Font* only when using .SHX fonts. Set a *Style* to use both regular and *Big Font* files for other than English languages. Otherwise, for most applications, select only the regular .SHX font files.

TrueType Fonts

AutoCAD uses the Windows operating system directly to display TrueType text on the screen for most applications. However, when text in AutoCAD has been transformed (mirrored, upside-down, backward, oblique, has a width factor not equal to 1, or is in an orientation that is not co-planar with the screen), AutoCAD must draw the TrueType text. Text that has been transformed might appear slightly more bold in some circumstances, especially at lower resolutions. This difference is only in the screen display of the font and does not affect the plotted drawing.

For TrueType fonts only, the value specified for text *Height* might not represent the actual height of uppercase letters. The height specified represents the height of a capital letter plus an "ascent" area reserved for accent marks and other marks used in non-English languages. TrueType fonts also have a "descent" area for portions of characters with extenders (such as the characters y, j, p, g, and q).

Proxy Fonts

For third-party or custom .SHX fonts that have no TrueType equivalent, AutoCAD supplies up to eight different TrueType fonts called proxy fonts. Proxy fonts appear different in the Multiline Text Editor from the font they represent to indicate that they are substitutions for the fonts used in the drawing.

Importing External Text into AutoCAD

You can import ASCII text files created in other text editors or word processors into an AutoCAD drawing. There are three methods: use the *Import Text* button in Multiline Text Editor, use *Copy* and *Paste* (using the Windows Clipboard), or "drag and drop" a file icon from the Windows Explorer.

Importing text in ASCII or RTF files can save you drawing time. For example, you can create a text file of standard notes that you include in many drawings or use information prepared for a report or other document. Instead of entering (typing) this information in the drawing, you can import the file. The imported text becomes an AutoCAD text object, which you can edit as if you created it in AutoCAD. Imported text retains its original formatting properties from the text editor in which it was created.

Importing with the Multiline Text Editor

To import text into the Multiline Text Editor, first use the *Mtext* command. PICK two points to define the text boundary as usual. When the Multiline Text Editor appears, right-click and select *Import Text* from the shortcut menu. The *Select File* dialog box appears for you to locate and select the desired file. The file to import must be in an ASCII or RTF format. (An ASCII text file often has a .TXT file extension.) The selected text appears in the editor and can be manipulated just as any text you entered.

You can also *Copy* text (to the Windows Clipboard) from any Windows document and *Paste* it into AutoCAD. The best method to use with *Copy* and *Paste* is to use the Multiline Text Editor. Open the editor, then use the Alt+Tab key combination to switch to the other document. Highlight the desired text and select *Copy* from the *Edit* menu or right-click menu. Switch back to the Multiline Text Editor, right-click for the pop-up menu, and select *Paste*. The imported text can be edited like any other AutoCAD *Mtext* object.

Using Drag and Drop to Import Text Files

You can also use the drag-and-drop feature to insert ASCII text into a drawing. Open the Windows Explorer (file manager) and size the window so AutoCAD is also visible on the screen. Drag the file name or icon and drop it into an AutoCAD drawing. The pasted text uses the formats and fonts defined by the current AutoCAD text style. The width of the lines or paragraph is determined by line breaks and carriage returns in the original document. You can drag only files with a file extension of .TXT into an AutoCAD drawing.

Using an External Text Editor for *Multiline Text*

You can specify a different text editor to use instead of the "internal" Multiline Text Editor. For example, you may prefer the Windows Notepad or other editor, although these editors do not offer the text formatting capabilities available in the Multiline Text Editor.

To specify an external text editor, you can use the *Files* tab of the *Options* dialog box from the *Tools* pull-down menu (Fig. 18-29) or use the *MTEXTED* system variable. Enter "internal" to use the internal AutoCAD Multiline Text Editor.

If you use an external text editor (other than the AutoCAD internal editor), you must enter the formatting codes as you enter the text using the text editor. Text features that can be controlled are underline, overline, height, color, spacing, stacked text, and other options. The following table lists the codes and the resulting text.

Figure 18-29

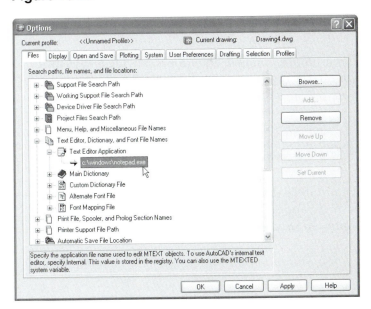

Format Codes for Creating *Multiline* Text in External Text Editors

Format code	Purpose	Type this. . .	To produce this
\O. . .\o	Turns overline on and off	You have \Omany\o choices	You have $\overline{\text{many}}$ choices
\L. . .\l	Turns underline on and off	You have \Lmany\l choices	You have <u>many</u> choices
\~	Inserts a nonbreaking space	Keep these\~words together	Keep these words together
\\	Inserts a backslash	slash\\backslash	slash\ backslash
\{. . .\}	Inserts an opening and closing brace	The \{bracketed\} word	The {bracketed} word
\C*value*;	Changes to the specified color	Change \C2;these colors	Change these colors
\F*filename*;	Changes to the specified font file	Change \Farial;these fonts	Change **these fonts**
\H*value*;	Changes to the specified text height	Change \H2;these sizes	Change **these sizes**
\S. . .^. . .	Stacks the subsequent text at the \ or ^ symbol	1.000\S+0.010^0.000;	$1.000^{+0.010}_{-0.000}$
\T*value*;	Adjusts the space between characters, from .75 to 4 times	\T2;TRACKING	T R A C K I N G
\Q*angle*;	Changes obliquing angle	\Q20;OBLIQUE	*OBLIQUE*
\W*value*;	Change width factor to produce wide text	\W2;Wide	**WIDE**
\A	Sets the alignment value to 0 (bottom), 1 (center), or 2 (top)	\A1;Center\S1/2	Cente$\frac{1}{2}$
\P	Ends paragraph	First paragraph\PSecond paragraph	First paragraph Second paragraph

For example, you may enter the following text in the Windows Notepad editor using the *Mtext* command.

```
\C1;This sample line of text is red in color.\P
\C7;\H.25;This text has a height of 0.25.\P
\H.2;\Fromanc;This line is an example of romanc.shx.\P
\Ftxt;This line contains stacked fractions such as \S1/2
```

Figure 18-30 displays the text as it might appear in the AutoCAD drawing editor.

Figure 18-30 ———————————————————————

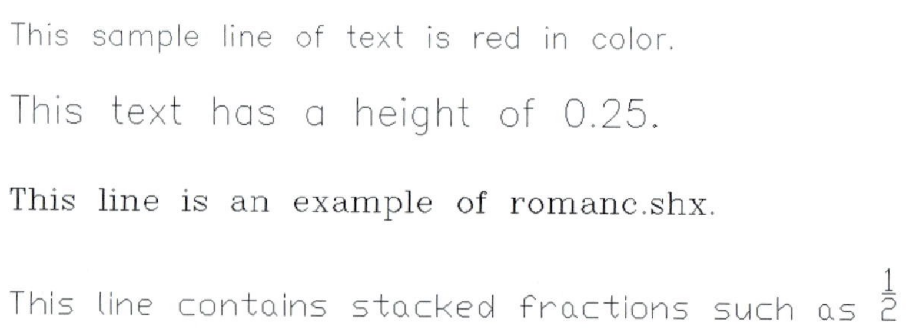

This sample line of text is red in color.

This text has a height of 0.25.

This line is an example of romanc.shx.

This line contains stacked fractions such as $\frac{1}{2}$

Special Text Characters

Special characters that are often used in drawings can be entered easily in the internal Multiline Text Editor by selecting *Symbol* from the shortcut menu (see Figure 18-14). In an external editor or with the *Dtext* command, you can code the text by using the "%%" symbols or by entering the Unicode values for high-ASCII (characters above the 128 range). The %% symbols can be used with the *Dtext* command, but the Unicode values must be used for creating *Mtext* with an external text editor. The following codes are typed in response to the "Enter text:" prompt in the *Dtext* command or entered in the external text editor.

External Editor	*Dtext*	Result	Description
\U+2205	%%c	ø	diameter (metric)
\U+00b0	%%d	°	degrees
	%%o	—	overscored text
	%%u	__	underscored text
\U+00b1	%%p	±	plus or minus
	%%*nnn*	varies	ASCII text character number
\U+*nnnn*		varies	Unicode text hexadecimal value

For example, entering "**Chamfer 45%%d**" at the "Enter text:" prompt of the *Dtext* command draws the following text: **Chamfer 45°**.

EDITING TEXT

SPELL

Pull-down Menu	COMMAND (TYPE)	ALIAS (TYPE)	Short-cut	Screen (side) Menu	Tablet Menu
Tools *Spelling*	*SPELL*	*SP*	...	*TOOLS 1* *Spell*	*T,10*

AutoCAD has an internal spell checker that can be used to spell check and correct existing text in a drawing after using *Dtext* or *Mtext*. You can also select *Block* attributes and *Blocks* containing text to spell check. The *Check Spelling* dialog box (Fig. 18-31) has options to *Ignore* the current word or *Change* to the suggested word. The *Ignore All* and *Change All* options treat every occurrence of the highlighted word.

AutoCAD matches the words in the drawing to the words in the current dictionary. If the speller indicates that a word is misspelled but it is a proper name or an acronym you use often, it can be added to a custom dictionary. Choose *Add* if you want to leave a word unchanged but add it to the current custom dictionary.

Figure 18-31

Selecting the *Change Dictionaries...* tile produces the dialog box shown in Figure 18-32. You can select from other *Main Dictionaries* that are provided with your version of AutoCAD. The current main dictionary can also be changed in the *Files* tab of the *Options* dialog box (see Figure 18-29), and the name is stored in the *DCTMAIN* system variable.

The AutoCAD default custom dictionary is SAMPLE.CUS (Fig. 18-32). If you use the *Add* function (in the *Check Spelling* dialog box), the selected word is added to the current custom dictionary. You can also create a custom dictionary "on the fly" by entering any name in the *Custom dictionary* edit box. A file extension of .CUS should be used with the name (although other extensions will work). A custom dictionary name must be specified before you can add words. Words can be added by entering the desired word in the *Custom dictionary words* edit box or by using the *Add* tile in the *Check Spelling* dialog box. Custom dictionaries can be changed during a spell check. The custom dictionary can also be changed in the *Files* tab of the *Options* dialog box (see Figure 18-29), and its name is stored in the *DCTCUST* system variable.

Figure 18-32

DDEDIT

Pull-down Menu	COMMAND (TYPE)	ALIAS (TYPE)	Short-cut	Screen (side) Menu	Tablet Menu
Modify *Object >* *Text>* *Edit...*	*DDEDIT*	*ED*	*Modify1*	*Ddedit*	*Y,21*

Ddedit invokes a dialog box for editing existing text in a drawing. You can edit <u>individual characters</u> or the entire line or paragraph. If the selected text was created by the *Dtext* command, the *Edit Text* dialog appears displaying one line of text (Fig. 18-33). If *Mtext* created the selected text, the Multiline Text Editor appears (or current text editor).

Figure 18-33

PROPERTIES

Pull-down Menu	COMMAND (TYPE)	ALIAS (TYPE)	Short-cut	Screen (side) Menu	Tablet Menu
Modify *Properties...*	*PROPERTIES*	*PR or* *CH*	(Default Menu) *Find or* *Ctrl + 1*	*MODIFY1* *Modify*	*Y,14*

The *Properties* command invokes the *Properties* palette (Fig. 18-34, on the next page). As described in Chapter 16, the palette that appears is <u>specific</u> to the type of object that is PICKed—kind of a "smart" properties palette. Remember that you can select objects and then invoke the palette, or you can keep the *Properties* palette on the screen and select objects in the drawing whose properties you want to change. (See Chapter 16 for more information on the *Properties* palette.)

If a line of *Dtext* is selected, the palette displays properties as shown in Figure 18-34. The *Properties* palette allows you to change almost any properties associated with the text, including the *Contents* (text), *Style*, *Justification*, *Height*, *Rotation*, *Width*, and *Obliquing*. Simply locate the property in the left column and change the value in the right column.

Figure 18-34

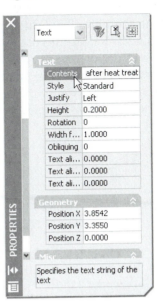

If you select an *Mtext* paragraph, the *Properties* palette appears allowing you to change the properties as for a *Dtext* object. However, if you select the *Contents* section, a small button on the right appears (Fig. 18-35). Pressing this button invokes the Multiline Text Editor with the *Mtext* contents and allows you to edit the text with the full capabilities of the editor.

Figure 18-35

FIND

	Pull-down Menu	COMMAND (TYPE)	ALIAS (TYPE)	Short-cut	Screen (side) Menu	Tablet Menu
	Edit *Find...*	*FIND*	...	(Default menu) *Find...*	...	X,10

Find is used to find or replace text strings <u>globally</u> in a drawing. It can find and replace any text in a drawing, whether it is *Dtext* or *Mtext*. *Find* produces the *Find and Replace* dialog box (see Figure 18-36, on the next page).

The *Find* command was introduced in AutoCAD 2000. The older *Find/Replace* tab in the Multiline Text Editor (introduced in AutoCAD Release 14) operates only for *Mtext* and only for one *Mtext* object at a time. The new *Find* command can search a drawing globally and operates with all types of text, including dimensions and *Block* attributes.

Figure 18-36

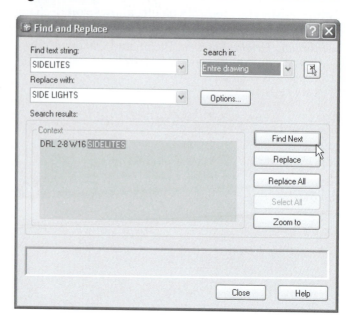

Find Text String, Find/Find Next
To find text only (not replace), enter the desired string in the *Find Text String* edit box and press *Find* (Fig. 18-36). When matching text is found, it appears in the context of the sentence or paragraph in the *Search Results* area. When one instance of the text string is located, the *Find* button changes to *Find Next*.

Replace With, Replace
You can find any text string in a drawing, or you can search for text and replace it with other text. If you want to search for a text string and replace it with another, enter the desired text strings in the *Find Text String* and *Replace With* edit boxes, then press the *Find/Find Next* button. You can verify and replace each instance of the text string found by alternately using *Find Next* and *Replace*, or you can select *Replace All* to globally replace all without verification. The status area confirms the replacements and indicates the number of replacements that were made.

Select Objects
The small button in the upper-right corner of the dialog box allows you to select objects with a pickbox or window to determine your selection set for the search. Press Enter when objects have been selected to return to the dialog box. Once a selection set has been specified, you can search the *Entire drawing* or limit the search to the *Current selection* by choosing these options in the *Search in:* drop-down list.

Select All
This options finds and selects (highlights) all objects in the current selection set containing instances of the text that you enter in *Find Text String*. This option is available only when you use *Select Objects* and set *Search In* to *Current Selection*. When you choose *Select All*, the dialog box closes and AutoCAD displays a message on the Command line indicating the number of objects that it found and selected. Note that *Select All* does not replace text; AutoCAD ignores any text in *Replace With*.

Zoom To

This feature is extremely helpful for finding text in a large drawing. Enter the desired text string to search for in the *Find Text String* edit box, select *Find/Find Next* button, then use *Zoom to*. AutoCAD automatically locates the desired text string and zooms in to display the text (Fig. 18-37). Although AutoCAD searches model space and all layouts defined for the drawing, you can zoom only to text in the current *Model* or *Layout* tab.

Figure 18-37

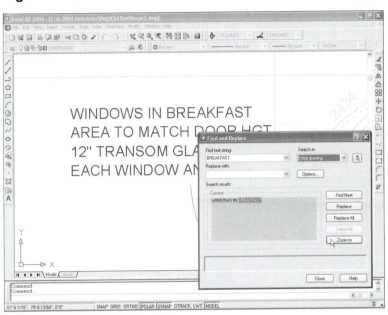

Options

Selecting the *Options* button in the *Find and Replace* dialog box produces the *Find and Replace Options* dialog box (Fig. 18-38). Note that the *Find* feature can locate and replace *Block Attribute Values, Dimension Annotation Text, Text (Mtext, Dtext, Text), Hyperlink Description,* and *Hyperlinks*. Removing any check disables the search for that text type.

If it is important that the search exactly match the case (upper and lowercase letters) you entered in the *Find Text String* and *Replace With* edit boxes, check the *Match Case* box. If you want to search for complete words only, check the *Find whole words only* box. For example, without a check in either of these boxes, entering "door" in the *Find Text String* edit box would yield a find of "Doors" in the *Search Results* area. With a check in either *Match Case* or *Find whole words only*, a search for "door" would not find "Doors."

Figure 18-38

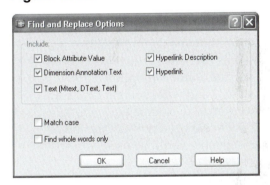

Find/Replace in the Multiline Text Editor

An older version of the *Find* command is the *Find and Replace* option available from the Multiline Text Editor shortcut menu (see previous Figure 18-14). With this feature you can find and replace text only within the Multiline Text Editor. *Find and Replace* operates only for *Mtext* objects, not for other text objects.

Selecting *Find and Replace* from the Multiline Text Editor shortcut menu produces the *Replace* dialog box (Fig. 18-39). The options here operate similarly to the same options in the *Find and Replace* dialog box of the *Find* command (see previous explanation of "*Find*").

Figure 18-39

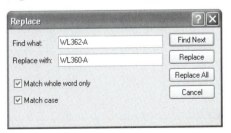

This feature is useful primarily if you are currently using the Multiline Text Editor or if you have located a specific *Mtext* object you want to deal with. In most cases, however, the *Find* command is preferred since it provides more features and options for finding and replacing text for all types of AutoCAD text objects.

SCALETEXT

Pull-down Menu	COMMAND (TYPE)	ALIAS (TYPE)	Short-cut	Screen (side) Menu	Tablet Menu
Modify Object > Text > Scale	*SCALETEXT*	...	...	...	...

You cannot effectively scale multiple lines of text (*Dtext* objects) or multiple paragraphs of text (*Mtext* objects) using the normal *Scale* command since all text objects become scaled with relation to only one base point; therefore, individual lines of text lose their original insertion point.

Since AutoCAD 2002 a new text editing command, *Scaletext*, allows you to scale multiple text objects using this one command, and each line or paragraph is scaled relative to its individual justification point.

For example, consider the three *Dtext* objects shown in Figure 18-40. Assume each line of text was created using the *Start point* option (left-justified) and with a height of .20. If you wanted to change the height of all three text objects to .25, the following sequence could be used.

Figure 18-40

SCALETEXT scales each text object individually.
Each line is scaled relative to the justification point.
These sample lines are created by DTEXT.

```
Command: scaletext
Select objects: PICK
Select objects: PICK
Select objects: PICK
Select objects: Enter
Enter a base point option for scaling
[Existing/Left/Center/Middle/Right/TL/TC/TR/ML/MC/MR/BL/BC/BR] <Existing>: Enter
Specify new height or [Match object/Scale factor] <0.2000>: .25
Command:
```

The resulting three lines of scaled text appear in Figure 18-41. Note that each line increased in height, which is evident since the space between lines has been reduced. Note also that each line of text has retained its insertion point; therefore, the original justification is unchanged.

Figure 18-41

SCALETEXT scales each text object individually.
Each line is scaled relative to the justification point.
These sample lines are created by DTEXT.

Scaletext also allows you to specify a new justification point. The justification option you select is applied to each text object individually. For example, selecting the *MC* option would scale each line of text in relation to <u>its own</u> middle center point.

> Enter a base point option for scaling
> [Existing/Left/Center/Middle/Right/TL/TC/TR/ML/MC/MR/BL/BC/BR] <Existing>: **mc**
> Specify new height or [Match object/Scale factor] <0.2000>: **.25**

Scale factor

This option allows you to specify a relative scale factor rather than a specific height. Entering a value of 1.5, for example, would scale each line of text 1.5 times the current size. This option is particularly helpful when the drawing contains lines of text at different heights and you want to change all lines proportionally.

> Specify new height or [Match object/Scale factor] <0.2500>: **s**
> Specify scale factor or [Reference] <0.2000>: **1.5**

Match object

You can select an existing text object that you want the text to match. After selecting the matching text, its height is listed and the text (that was selected in the first step) is scaled to the listed height.

> Specify new height or [Match object/Scale factor] <0.2500>: **m**
> Select a text object with the desired height: **PICK**
> Height=0.2400

JUSTIFYTEXT

Pull-down Menu	COMMAND (TYPE)	ALIAS (TYPE)	Short-cut	Screen (side) Menu	Tablet Menu
Modify Object > Text > Justify	*JUSTIFYTEXT*	...	...	...	...

The *Justifytext* command was introduced in AutoCAD 2002. *Justifytext* changes the justification point of selected text objects without changing the text locations. Technically, this command relocates the insertion point (sometimes called attachment point) for the text object (*Mtext* objects, *Dtext* objects, leader text objects, and block attributes) and then justifies the text to the new insertion point.

Consider the two *Dtext* objects and the single *Mtext* object shown in Figure 18-42. In each case, the default justification methods were used when creating the text, resulting in left-justified text and insertion points as shown by the small "blip" (the "blip" is not created as part of the text—it is shown here only for illustration).

Figure 18-42

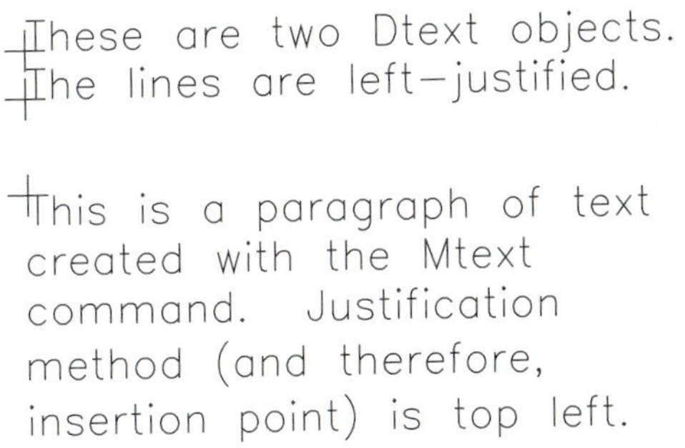

These are two Dtext objects.
The lines are left—justified.

This is a paragraph of text
created with the Mtext
command. Justification
method (and therefore,
insertion point) is top left.

Using previously available methods for changing the text justification yielded different results based on the command used or type of text object selected. For example, using the *Properties* palette to change the *Justify* field and selecting the *Bottom Right* option results in the changes shown in Figure 18-43. Note that for the *Dtext*

Figure 18-43 ───────────────────

These are two Dtext objects,
The lines are left—justified.

-+-   This is a paragraph of text
created with the Mtext
command. Justification
method (and therefore,
insertion point) is top left.

objects, the insertion point is not changed, but the text objects are moved to justify according to the original insertion point. On the other hand, the insertion point for the *Mtext* paragraph is changed from the original top-left insertion point (shown as dashed in the figure) to the bottom right and the text is justified to the new insertion point.

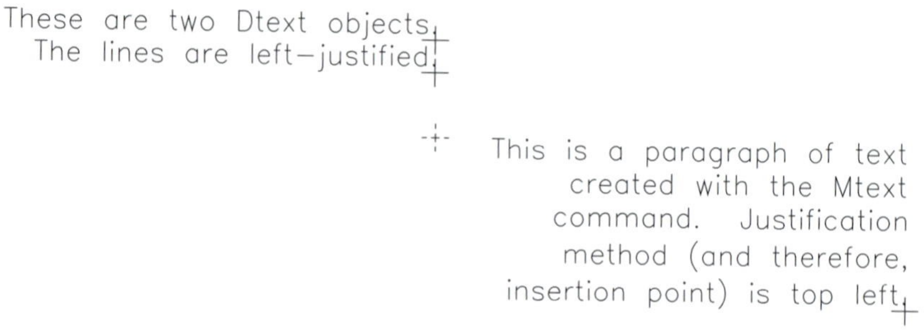

To avoid this problem, use the *Justifytext* command to change multiple text objects to produce similar results for all cases. For example, assume you used the *Justifytext* command to change the three text objects shown previously in Figure 18-42 to a new justification point of *Bottom right*.

```
Command: justifytext
Select objects: PICK
Select objects: PICK
Select objects: PICK
Select objects: Enter
Enter a justification option
[Left/Align/Fit/Center/Middle/Right/TL/TC/TR/ML/MC/MR/BL/BC/BR] <TL>: br
Command:
```

The resulting changes are shown in Figure 18-44. Compare these changes to the original text objects in Figure 18-42. Notice that the *Justifytext* command keeps the text in the same location, although the insertion points have been changed. Each text object has been justified in relation to its own new insertion point.

Figure 18-44 ───────────────────

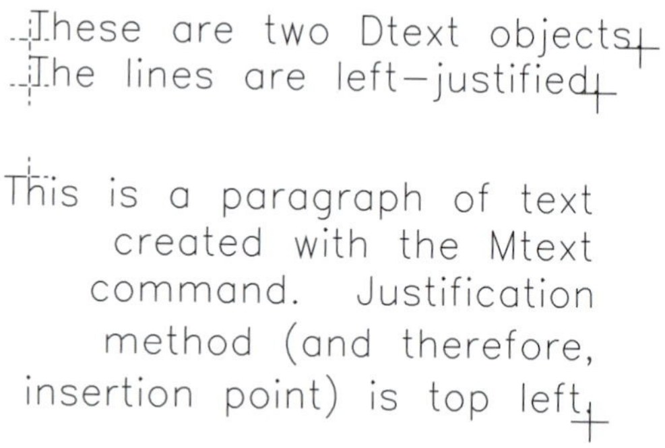

These are two Dtext objects
The lines are left—justified

This is a paragraph of text
created with the Mtext
command. Justification
method (and therefore,
insertion point) is top left.

QTEXT

Pull-down Menu	COMMAND (TYPE)	ALIAS (TYPE)	Short-cut	Screen (side) Menu	Tablet Menu
...	QTEXT	...	...	...	...

Qtext (quick text) allows you to display a line of text <u>as a box</u> in order to speed up drawing and plotting times. Because text objects are treated as graphical elements, a drawing with much text can be relatively slower to regenerate and take considerably more time plotting than the same drawing with little or no text.

When *Qtext* is turned *ON* and the drawing is regenerated, each text line is displayed as a rectangular box (Fig. 18-45, on the next page). Each box displayed represents one line of text and is approximately equal in size to the associated line of text.

Figure 18-45

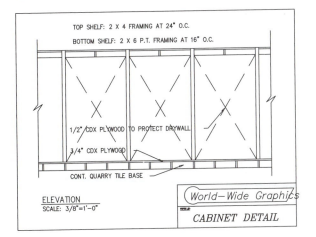

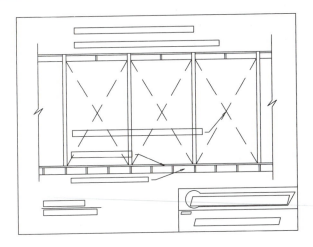

QTEXT OFF   QTEXT ON

For drawings with considerable amounts of text, *Qtext ON* noticeably reduces regeneration time. For check plots (plots made during the drawing or design process used for checking progress), the drawing can be plotted with *Qtext ON*, requiring considerably less plotting time. *Qtext* is then turned *OFF* and the drawing must be *Regenerated* to make the final plot.

When *Qtext* is turned *ON*, the text remains in a readable state until a *Regen* is invoked or caused. When *Qtext* is turned *OFF*, the drawing must be regenerated to read the text again.

TEXTFILL

Pull-down Menu	COMMAND (TYPE)	ALIAS (TYPE)	Short-cut	Screen (side) Menu	Tablet Menu
...	*TEXTFILL*	...	...	...	...

The *TEXTFILL* variable controls the display of TrueType fonts for <u>printing and plotting only</u>. *TEXTFILL* does not control the display of fonts for the screen—fonts always appear filled in the Drawing Editor. If *TEXTFILL* is set to 1 (on), these fonts print and plot with solid-filled characters (Fig. 18-46). If *TEXTFILL* is set to 0 (off), the fonts print and plot as out-lined text. The variable controls text display globally and retroactively.

Figure 18-46

Substituting Fonts

Alternate Fonts

Fonts in your drawing are specified by the font file referenced by the current text *Style* and by individual font formats specified for sections of *Mtext*. Text font files are not stored in the drawing; instead, text objects reference the font files located on the current computer system. Occasionally, a drawing is loaded that references a font file not available on your system. This occurrence is more common with the ability to create text styles using TrueType fonts.

By default, AutoCAD substitutes the SIMPLEX.SHX for unreferenced fonts. In other words, when AutoCAD loads a drawing that contains a font not on your system, the SIMPLEX.SHX font is automatically substituted. You can use the *FONTALT* system variable or the *Files* tab of the *Options* dialog box (Fig. 18-47) to specify other font files that you want to use as an alternate font whenever a drawing with unreferenced fonts is loaded. Typically, .SHX fonts are used as alternates.

Figure 18-47

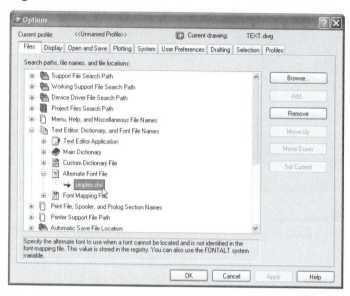

Font Mapping

You can also set up a <u>font mapping table</u> listing several fonts and substitutes to be used whenever a text object is encountered that references one of the fonts. The current table name is stored in the *FONTMAP* system variable (the default is ACAD.FMP). A font mapping table is a plain ASCII file with a .FMP file extension and contains one font mapping per line. Each line contains the base font followed by a semicolon and the substitute font.

You can use font mapping to simplify problems that may occur when exchanging drawings with clients. You can also use a font mapping table to substitute fast-drawing .SHX files while drawing and for test plots, then use another table to substitute back the more complex fonts for the final draft. For example, you might want to substitute the ROMANS.SHX font for the Times TrueType font and the TXT.SHX font for the Arial TrueType font. (The second font in each line is the new font that you want to appear in the drawing.)

```
times;romans.shx
arial;txt.shx
```

AutoCAD provides a sample font mapping table called ACAD.FMP. The table has 16 substitutions that are used to map the Release 13 PostScript fonts to the newer TrueType fonts. The ACAD.FMP contents are shown here.

```
cibt;CITYB___.TTF
cobt;COUNB___.TTF
eur;EURR_____.TTF
euro;EURRO___.TTF
par;PANROMAN.TTF
rom;ROMANTIC.TTF
romb;ROMAB___.TTF
romi;ROMAI___.TTF
sas;SANSS___.TTF
sasb;SANSSB__.TTF
sasbo;SANSSBO_.TTF
saso;SANSSO__.TTF
suf;SUPEF___.TTF
te;TECHNIC_.TTF
teb;TECHB___.TTF
tel;TECHL___.TTF
```

You can edit this table with an ASCII text editor or create a new font mapping table. You can specify a new font mapping file in the *Files* tab of the *Options* dialog box (see Figure 18-47) or use the *FONTMAP* system variable.

A font mapping table <u>forces</u> the listed substitutions, while an alternate font substitutes the specified font for <u>any</u> font but <u>only</u> when an unreferenced font is encountered.

When a drawing is opened and AutoCAD begins to locate the fonts referenced by text *Styles* in the drawing, AutoCAD follows the following priority:

1. AutoCAD uses the *FONTMAP* value(s) if defined.
2. If not defined, AutoCAD uses the font defined in the text *Style*.
3. If not found, Windows substitutes a similar font (if .TTF) or uses the *FONTALT* specified (if .SHX).
4. If an .SHX is not found, it prompts you for a new font.

Text Attributes

When you want text to be entered into a drawing and associated with some geometry, such as a label for a symbol or number for a part, *Block Attributes* can be used. *Attributes* are text objects attached to *Blocks*. When the *Blocks* are *Inserted* into a drawing, the text *Attributes* are also inserted; however, the content of the text can be entered at the time of the insertion. See Chapter 21, Blocks and DesignCenter, and Chapter 22, Block Attributes.

CHAPTER EXERCISES

1. *Dtext*

   Open the **EFF-APT-PS** drawing. Make *Layer* **TEXT** *Current* and use *Dtext* to label the three rooms: **KITCHEN**, **LIVING ROOM**, and **BATH**. Use the *Standard style* and the *Start point* justification option. When prompted for the *Height:*, enter a value to yield letters of 3/16″ on a 1/4″=1′ plot (3/16 x the drawing scale factor = text height). *Save* the drawing.

2. *Style, Dtext, Ddedit*

   Open the drawing of the temperature graph you created as **TEMP-D**. Use *Style* to create two styles named **ROMANS** and **ROMANC** based on the *romans.shx* and *romanc.shx* font files (accept all defaults). Use *Dtext* with the *Center* justification option to label the days of the week and the temperatures (and degree symbols) with the **ROMANS** style as shown in Figure 18-48. Use a *Height* of **1.6**. Label the axes as shown using the **ROMANC** style. Use *Ddedit* for editing any mistakes. Use *SaveAs* and name the drawing **TEMPGRPH**.

Figure 18-48

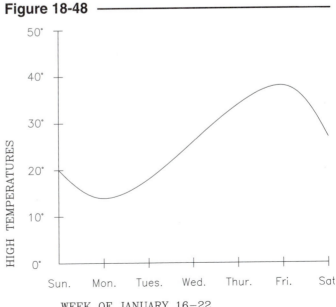

WEEK OF JANUARY 16—22

3. *Style, Dtext*

   Open the **BILLMATL** drawing created in the Chapter 15 Exercises. Use *Style* to create a new style using the *romans.shx* font. Use whatever justification methods you need to align the text information (not the titles) as shown in Figure 18-49. Next, type the *Style* command to create a new style that you name as **ROMANS-ITAL**. Use the *romans.shx* font file and specify a **15** degree *obliquing angle*. Use this style for the **NO.**, **PART NAME**, and **MATERIAL**. *SaveAs* **BILLMAT2**.

Figure 18-49

NO.	PART NAME	MATERIAL
1	Base	Cast Iron
2	Centering Screw	N/A
3	Slide Bracket	Mild Steel
4	Swivel Plate	Mild Steel
5	Top Plate	Cast Iron

4. *Edit Text*

   Open the **EFF-APT-PS** drawing. Create a new style named **ARCH1** using the *CityBlueprint* (.TTF) font file. Next, invoke the *Properties* command. Use this dialog box to modify the text style of each of the existing room names to the new style as shown in Figure 18-50. *SaveAs* **EFF-APT2**.

Figure 18-50

5. *Dtext, Mtext*

   Open the **CBRACKET** drawing from Chapter 9 Exercises. Using *romans.shx* font, use *Dtext* to place the part name and METRIC annotation (Fig. 18-51). Use a *Height* of **5** and **4**, respectively, and the *Center Justification* option. For the notes, use *Mtext* to create the boundary as shown. Use the default *Justify* method (**TL**)and a *Height* of **3**. Use *Ddedit* or *Properties* if necessary. *SaveAs* **CBRACTXT**.

Figure 18-51

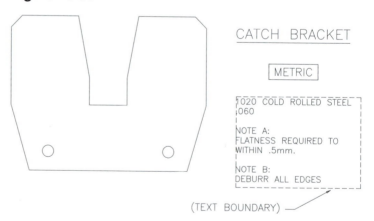

6. *Style*

   Create two new styles for each of your template drawings: **ASHEET**, **BSHEET**, and **C-D-SHEET**. Use the *romans.shx* style with the default options for engineering applications or *CityBlueprint* (.TTF) for architectural applications. Next, design a style of your choosing to use for larger text as in title blocks or large notes.

7. *Import Text, Ddedit, Properties*

Use a text editor such as Windows Wordpad or Notepad to create a text file containing words similar to "Temperatures were recorded at Sanderson Field by the National Weather Service." Then *Open* the **TEMPGRPH** drawing and use the *Import Text...* option of *Mtext* to bring the text into the drawing as a note in the graph as shown in Figure 18-52. Use *Ddedit* to edit the text if desired or use *Properties* to change the text style or height.

Figure 18-52

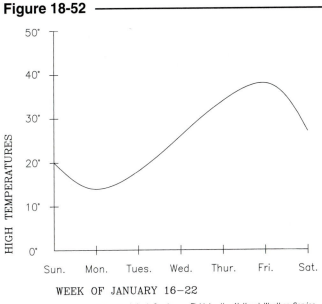

WEEK OF JANUARY 16−22

Temperatures were recorded at Sanderson Field by the National Weather Service.

8. *Create a Title Block*

A. Begin a *New* drawing and assign the name **TBLOCK**. Create the title block as shown in Figure 18-53 or design your own, allowing space for eight text entries. The dimensions are set for an A size sheet. Draw on *Layer* **0**. Use a *Pline* with **.02** *width* for the boundary and *Lines* for the interior divisions. (No *Lines* are needed on the right side and bottom because the title block will fit against the border lines.)

Figure 18-53

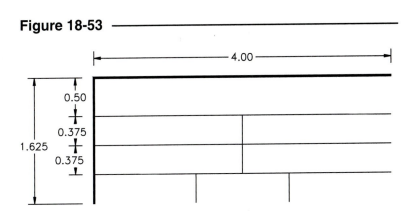

B. Create two text *Styles* using *romans.shx* and *romanc.shx* font files. Insert text similar to that shown in Figure 18-54. Examples of the fields to create are:

Company or School Name
Part Name or Project Title
Scale
Designer Name
Checker or Instructor Name
Completion Date
Check or Grade Date
Project or Part Number

Choose text items relevant to your school or office. *Save* the drawing.

Figure 18-54

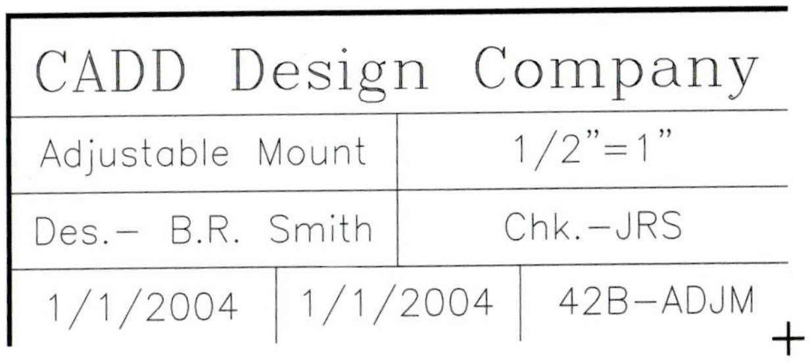

CADD Design Company		
Adjustable Mount	1/2"=1"	
Des.− B.R. Smith	Chk.−JRS	
1/1/2004	1/1/2004	42B−ADJM

9. *Mtext*

Open the **STORE ROOM** drawing that you created in Chapter 16 Exercises. Use *Mtext* to create text paragraphs giving specifications as shown in Figure 18-55. Format the text below as shown in the figure. Use *CityBlueprint* (.TTF) as the base font file and specify a base *Height* of **3.5"**. Use a *Color* of your choice and *CountryBlueprint* (.TTF) font to emphasize the first line of the Room paragraph. All paragraphs use the *TC Justify* methods except the Contractor Notes paragraph, which is *TL*. Save the drawing as **STORE ROOM2**.

Room: <u>STORAGE ROOM</u>
  11'-2" x 10'-2"
  Cedar Lined - 2 Walls

Doors: 2 - 2268 DOORS
  Fire Type A
  Andermax

Windows: 2 - 2640 CASEMENT WINDOWS
  Triple Pane Argon Filled
  Andermax

Notes: <u>Contractor Notes:</u>

  Contractor to verify all dimensions in field. Fill door and window roughouts after door and window placement.

Figure 18-55

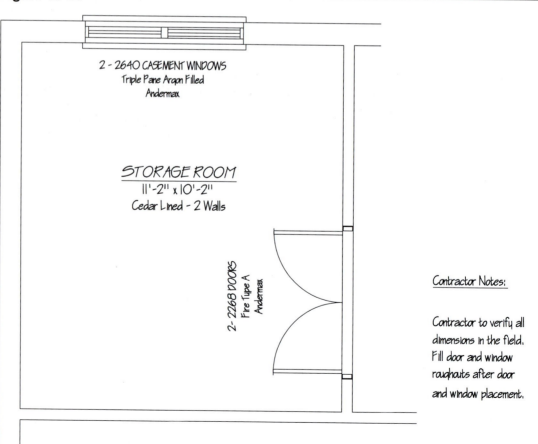

INTERNET TOOLS

Chapter Objectives

After completing this chapter you should:

1. be able to use the *Browser* command to launch your Web browser from within AutoCAD;

2. be able to use *Hyperlink* to link your drawings to other drawings, text documents, or spreadsheets on your computer, your network, or the Internet;

3. be able to use *Publishtoweb* to create Web pages displaying AutoCAD drawings;

4. know how to attach drawings to your email messages using *Etransmit*;

5. know how to use the *DWF6 ePlot* device driver to create .DWF files;

6. be able to use *Publish* to create drawing sets in .DWF format; and

7. and be able to use Autodesk Express Viewer to view .DWF files.

CONCEPTS

Through recent years Autodesk has created features to allow AutoCAD users to fully utilize the latest Internet-related technological advances to enhance their productivity. With AutoCAD 2000i, AutoCAD 2002, and now AutoCAD 2004, multiple tools have been added to enable you to collaborate and share your designs with colleagues and clients using Internet technologies.

AutoCAD 2004 includes features that allow you to be more connected to Web technology by enabling capabilities such as publishing Web pages that include drawings, emailing drawings, linking other remotely located drawings and documents to your drawing, creating single drawings or complete drawing sets that are viewable in standard Web browsers, and allowing automatic updating to new features through the Autodesk Web site. Two of the Internet-related features of AutoCAD 2000i and 2002 (the *Today* window and *MeetNow*) have been removed in AutoCAD 2004.

The commands explained in this chapter are listed below.

Browser	Launch your Web browser from within AutoCAD with the *Browser* command so you can have instant access to the Internet while in your drawing environment.
Communication Center	This new feature keeps you up to date by notifying you when new features or updates for AutoCAD are available through the Internet.
Hyperlink	Use the *Hyperlink* command to create pointers in your AutoCAD drawings that provide jumps to other files you want to associate to the current drawing. The other files can be located on your local computer or network or can link to Web pages on the Internet. In this way your AutoCAD drawings can "contain" additional graphical, numerical, or textual information.
PublishToWeb	With this powerful tool you can generate Web pages complete with embedded AutoCAD drawing images in .JPG, .PNG, and .DWF format. This tool is automatic so no experience with .HTML is required.
eTransmit	*eTransmit* opens your email system and allows you to include an AutoCAD drawing (and any associated files) as an email attachment.
Publish	This new feature allows you to create drawing sets that are readable, but not editable, from Autodesk Express Viewer or Microsoft Internet Explorer. The drawings are created in .DWF format.
Plot Drivers	You can create a Drawing Web Format (.DWF) file of the current drawing using the *DWF6 ePlot* device from the *Plot* dialog box. A .DWF file is a compact, vector-based file that is viewable without AutoCAD and easily transportable over the Internet. You can also create raster files of the current drawing using the *PublishToWeb PNG* or *PublishToWeb JPG* device.
Express Viewer	This new viewer can be used to view .DWF files. The Autodesk Express Viewer is free and downloadable, and can be installed by anyone—without a license for AutoCAD.

BROWSER

Pull-down Menu	COMMAND (TYPE)	ALIAS (TYPE)	Short-cut	Screen (side) Menu	Tablet Menu
...	*BROWSER*	...	...	...	...

AutoCAD 2004 includes this command specifically to launch your Web browser from within AutoCAD. An icon button on the *Web* toolbar (Fig. 19-1) and in the file navigation dialog boxes is available to invoke the *Browser* command.

Figure 19-1

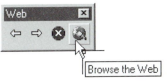

The information about your browser (location and name of the executable file) is stored in the system registry. Therefore, when you use *Browser*, AutoCAD automatically locates and launches your current (default) Web browser from within AutoCAD. This action makes it possible to view any Web site or select any viewable files located on your computer. If you have Autodesk Express Viewer, you can view .DWF files after you have created them, view other .DWF files on the Internet, and drag-and-drop related .DWG files from the Internet directly into your AutoCAD session.

When you use *Browser*, your registered Web browser is launched. By default (if you have not specified a different file or URL), the Autodesk home page is displayed (Fig. 19-2). You can locate another file or URL by entering it in the *Address* edit box of the browser.

Figure 19-2

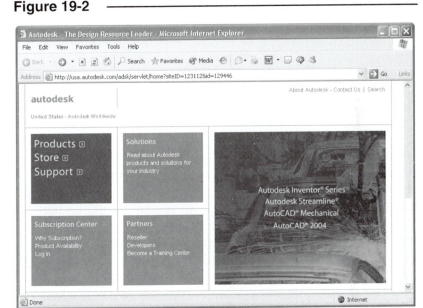

INETLOCATION

Pull-down Menu	COMMAND (TYPE)	ALIAS (TYPE)	Short-cut	Screen (side) Menu	Tablet Menu
Tool Options... Files	*INETLOCATION*	...	...	*TOOLS2 Options... Files*	...

The initial Web address that is located when *Browser* is used can be specified by using the *INETLOCATION* system variable or by using the *Options* dialog box. The default *INETLOCATION* for AutoCAD 2004 is:

> www.autodesk.com

You can specify any other viewable file or Internet address. Using the Command line method, the following prompts are issued:

```
Command: inetlocation
Enter new value for INETLOCATION <"http://www.autodesk.com">: (Enter desired URL)
Command:
```

If you prefer to use the *Options* dialog box (Fig. 19-3), expand the *Menu, Help, and Miscellaneous File Names* section. Expand *Default Internet Location*, highlight the address, and select the *Remove* button. Then enter the desired new address.

Figure 19-3

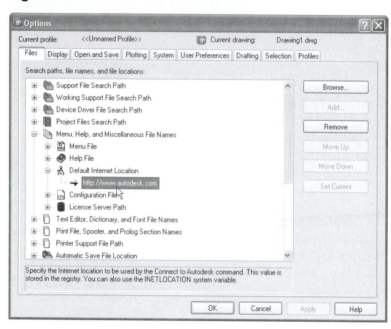

Communication Center

Communication Center is not a command, rather it is a feature that can be installed with AutoCAD 2004. Communication Center may or may not be installed on your system depending on options that were selected during installment. (See "Tray Settings" in Chapter 1 for information on enabling the Communication Center icon.) Communication Center is an interactive feature that connects your computer to the Autodesk Web site. Autodesk monitors information about your computer and AutoCAD installation and notifies you when new information and updates are available.

You must be connected to the Internet for Autodesk to deliver content and information. Each time Communication Center is connected, it sends information from your computer to Autodesk so that the correct resources can be returned to you. All information is sent to Autodesk anonymously to maintain your privacy. The following information is sent to Autodesk:

Product Name	the name of the product in which you are using Communication Center
Product Release Number	the version of the product
Product Language	the language version of your product
Country	the country that was specified in the Communication Center settings

Autodesk compiles statistics using the information sent from Communications Center to monitor how AutoCAD is being used and if it needs updating. Autodesk states that it will maintain your information in accordance with Autodesk's published privacy policy. The privacy policy is available at:

http://www.autodesk.com/privacy

If Communication Center is available on your computer, a small "satellite dish" icon appears in the extreme lower-right corner of the AutoCAD window. Left-click on the icon to produce the *Communication Center* dialog box. Right-click on the icon or select *Settings* from the *Communication Center* dialog box to produce the *Configurations Settings* dialog box.

2004

Communication Center Dialog Box

The *Communication Center* dialog box (Fig. 19-4) indicates whether your product is up to date and may give you a variety of other information depending on your configuration settings. Click on any of the items in the list to view the content. If you have not been connected to the Internet for some time, or you specified to update information *On Demand*, you can choose to *Refresh Content*. Selecting *Settings* produces the *Configuration Settings* dialog box.

Configurations Settings Dialog Box

Settings Tab
In the *Settings* tab (Fig. 19-5), specify the country you are in and how often you want to check for new content (*Daily*, *Weekly*, *Monthly*, or *On Demand*). Checking *Enable Balloon Notification for new announcements* causes a balloon message to appear in the lower-right corner of the AutoCAD window when new content is available.

Channels Tab
The *Channels* tab (Fig. 19-6) allows you to indicate what information should be sent to you. The following four content areas are available.

Live Update Maintenance Patches
You can receive notifications whenever new maintenance patches are released from Autodesk.

Subscription Info and Extension Announcements
This option provides announcements and subscription program news if you are an Autodesk subscription member.

Articles and Tips
Check this box to be notified when new articles and tips are available on Autodesk Web sites (see Figure 19-4).

Product Support Information
You can get breaking news from the Product Support team at Autodesk with this feature. Also stay informed about Autodesk company news and product announcements.

Controlling Communication Center on a Network
If you have a network installation of AutoCAD, you can use the CAD Manager Control utility to turn Communication Center on and off. For example, if you want to prevent Communication Center from sending information to Autodesk, you can turn it off. Information about how to use the utility is available by installing and running the utility, and then clicking *Help* in the *CAD Manager Control Utility* window. The utility can be installed from the AutoCAD 2004 CD under the *Network Deployment* tab.

Figure 19-4

Figure 19-5

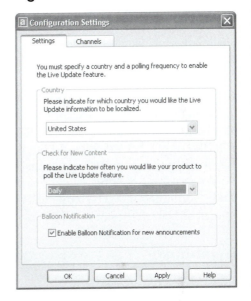

Figure 19-6

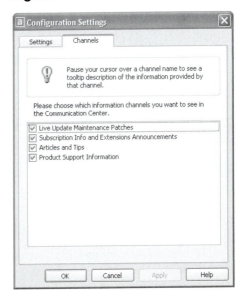

USING HYPERLINKS

The *Hyperlink* command allows you to attach a link to any AutoCAD graphical object. The link is actually a file name or Web address (URL) associated with the object. Therefore, *Hyperlink* can be used to link to files on the local computer or network (without Internet access) or to connect to files over the Internet, depending on what information is entered as the link.

Once a hyperlink is created, you can right-click on the object to display a shortcut menu allowing you to *Open* the link. The link automatically opens the related program (word processor, spreadsheet, or CAD program) and displays the file, or it opens your Web browser and displays the page or image. When the cursor passes over an AutoCAD object that has a hyperlink attached, the *Hyperlink* symbol and the related link information are displayed (Fig. 19-7).

Figure 19-7

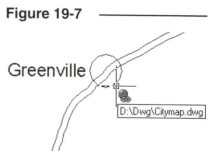

A *Hyperlink* can be used to create links within two types of AutoCAD files: (1) to create links in the current AutoCAD drawing (.DWG) and (2) to create links for a .DWF image that is created from the current drawing but is to be viewed from a Web browser. To create links to be used in .DWF images, use the *Hyperlink* command as you would normally to attach links to objects in the current drawing, then plot the drawing using the *DWF ePlot* device to create the related .DWF file (see "Creating .DWF Files with ePlot").

Keep in mind the power and potential usefulness of hyperlinks in AutoCAD drawings. Any hyperlink can call another file. For example, a drawing of a mechanical assembly could contain hyperlinks for each component of the assembly such that passing the cursor over each component and selecting *Open* from the right-click menu could open the related component drawing in AutoCAD. Or, as another example, opening another hyperlink for the entire assembly could in turn open a bill of materials in a word processing or spreadsheet program. The same capability hyperlinks add to an AutoCAD drawing can be contained in a .DWF image of the drawing viewed in a Web browser if those hyperlinks call other images or HTML files normally viewable in a browser.

HYPERLINK

Pull-down Menu	COMMAND (TYPE)	ALIAS (TYPE)	Short-cut	Screen (side) Menu	Tablet Menu
Insert Hyperlink...	HYPERLINK	...	Ctrl+K	...	...

When the *Hyperlink* command is invoked, you must first select the object you want to attach the link to. If you select graphical objects that do not already contain hyperlinks, AutoCAD displays the *Insert Hyperlink* dialog box (Fig. 19-8). If the graphical objects already contain hyperlinks, the *Edit Hyperlink* dialog box is displayed, which has the same options as the *Insert Hyperlink* dialog box. The Command prompt and the dialog box options are explained next.

Figure 19-8

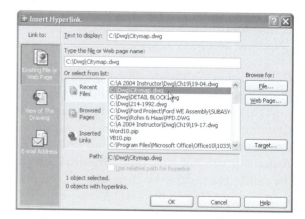

Command: **hyperlink**
Select objects: **PICK**
Select objects: **Enter**
(The *Insert Hyperlink* or *Edit Hyperlink* dialog box appears.)
Command:

When the *Insert Hyperlink* dialog box appears, first select (from the buttons on the left side of the dialog box) what kind of link you want to attach to the selected object—an *Existing File or Web Page*, a view of the current drawing (by selecting *View of this Drawing*), or an *Email Address*.

Existing File or Web Page

Selecting the *Existing File or Web Page* button allows you to type the information or select from *Recent Files*, *Browsed Pages*, or *Inserted Links*. In each case, files, pages, and URLs you recently used are displayed. There are also buttons to select another *File...* or *Web Page...* that is not displayed in the list. Selecting the *Target* button opens the *Select Place in Document* dialog box where you can select a named *View* in a drawing to link to (not the current drawing).

The text you enter in the *Text to display:* edit box is the text that appears in the drawing when the cursor passes over the object containing the hyperlink (see Figure 19-4).

The *Use Relative Path for Hyperlink* checkbox toggles the use of a relative path for the current drawing. If this option is selected, the full path to the linked file is not stored in the drawing with the hyperlink. AutoCAD sets the relative path to the value specified by the *HYPERLINKBASE* system variable or, if this variable is not set, the current drawing path. If this option is not selected or is disabled, the full path to the associated file is stored with the hyperlink. Edit the current path in the *Type the file or Web page name:* edit box to enable this checkbox. (See "*HYPERLINKBASE*" next.)

The *Remove Link* button appears only in the *Edit Hyperlink* dialog box. You can use this option to remove a previously attached hyperlink from the selected object.

View of This Drawing

If you want a specific view or layout of the <u>current drawing</u> to be associated to the selected object as a hyperlink, you can use this button to select from a list of *Views* that are contained in the current drawing (Fig. 19-9). In other words, this feature allows you to select an object in the current drawing and attach a hyperlink that causes a jump to a different layout or saved view of the drawing. AutoCAD displays that portion of the drawing (view or layout) when the hyperlink is opened.

Email Address

This feature allows you to attach a hyperlink to an object so that when this hyperlink is opened, it causes the computer's default email system to open ready to type a new message. The email address contained in the link (appearing in the *E-mail address:* edit box of the *Insert Hyperlink* dialog box, Figure 19-7) is automatically entered in the "Send to:" box of your new email message and the subject text in the link (contained in the *Subject:* edit box) is automatically entered in the "Subject:" box of your email message. You simply type the message and select *Send* in your email program. Use the *Text to display:* edit box to supply the text that appears in the drawing when the cursor passes over the object containing the hyperlink.

Figure 19-9 ──────────

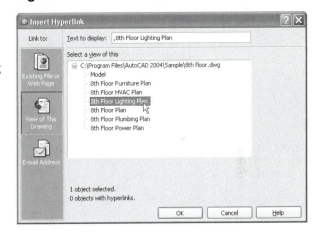

Figure 19-10 ──────────

HYPERLINKBASE

Pull-down Menu	COMMAND (TYPE)	ALIAS (TYPE)	Short-cut	Screen (side) Menu	Tablet Menu
...	*HYPERLINKBASE*	...	...	...	...

Use the *HYPERLINKBASE* system variable to specify the path you want to use for all relative hyperlinks in the drawing. If no value is specified, the current drawing path is used for all relative hyperlinks. Relative paths are used primarily for file-related hyperlinks rather than Web addresses.

> Command: `hyperlinkbase`
> Enter new value for HYPERLINKBASE, or . for none <"">:

Relative and Absolute Paths

With *HYPERLINKBASE* you can specify a relative path that is used by the hyperlinks you create in a drawing. Relative paths afford you greater flexibility and are easier to edit than absolute hyperlinks because with relative hyperlinks, you can update the relative path for all the hyperlinks in your drawing at the same time, rather than editing each hyperlink individually. Absolute hyperlinks work well in situations where the linked files always remain in the same location, which is more common with Web addresses. For example, if you subsequently move the files referenced by absolute hyperlinks to a different directory, all the hyperlink paths in the drawing must be updated.

SELECTURL

Pull-down Menu	COMMAND (TYPE)	ALIAS (TYPE)	Short-cut	Screen (side) Menu	Tablet Menu
...	*SELECTURL*	...	...	...	...

Use *Selecturl* to "select," or highlight, all objects in the drawing that have hyperlinks (URLs) attached. The highlighting action essentially enables grips on the objects. Press Escape to cancel the grips.

CREATING EMAIL AND WEB PAGES

In recent versions of AutoCAD, new features are included that allow you to create your own Web pages to display drawings and to send a drawing file as an email attachment. The *PublishToWeb* feature creates Web pages displaying selected drawings in .JPG, .PNG, or .DWF format. This feature automatically creates the necessary HTML code. The *eTransmit* feature prepares a drawing for emailing by compressing it into a .ZIP or .EXE file, then opens your default email system if desired.

PUBLISHTOWEB

Pull-down Menu	COMMAND (TYPE)	ALIAS (TYPE)	Short-cut	Screen (side) Menu	Tablet Menu
Files *Publish to Web*	*PUBLISHTOWEB*	*PTW*	...	...	...

The *PublishToWeb* feature in AutoCAD 2002 and 2004 is a surprisingly simple and quick method of generating a Web page (HTML document) complete with embedded AutoCAD drawing images (.JPG, .PNG, or .DWF files). You need no previous experience creating Web pages or HTML code—AutoCAD does it all for you automatically! All you need to begin are the AutoCAD drawings you want to make images of and, if you are ready to post the page to the Web, a location on a server to store the files.

An example Web page created by *PublishToWeb* is shown in Figure 19-11. The general structure of the page is determined by a template you select. You supply the page title and subtitle as well as the AutoCAD drawings to use. Each image and its accompanying description are hyperlinks to display a full-screen version of the image.

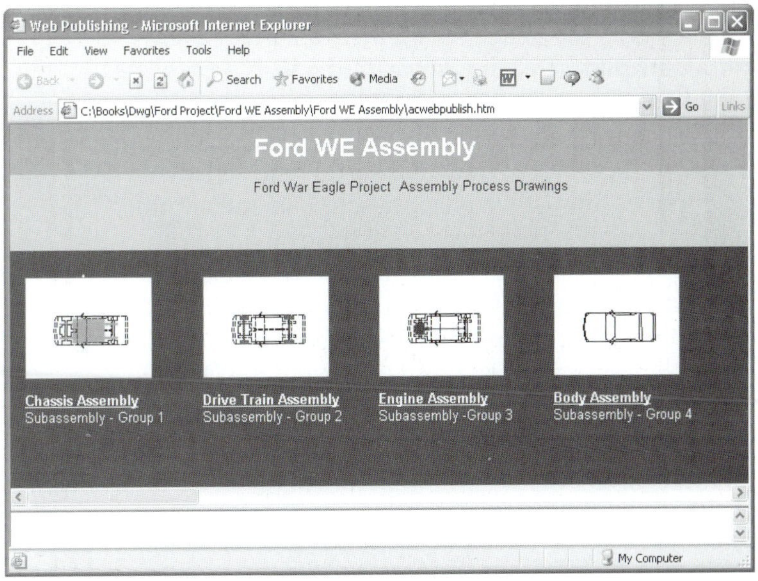

The *PublishToWeb* command produces a wizard that leads you through the process. The sequence is almost self-explanatory; however, here is some additional help you will need.

NOTE: You can publish only saved drawings. You can publish an open drawing if it has been previously saved. Because of the confusing AutoCAD alert that may appear, it is suggested that you *Save* the current drawing (even if it is a *New* drawing) before you use *PublishToWeb*.

Figure 19-11

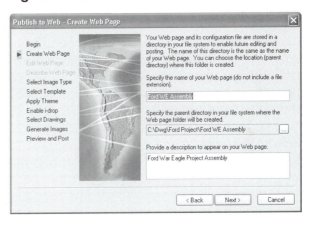

Begin
The first step in the *Publish to Web* wizard (not shown) is simply a choice of *Create a New Web Page* or *Edit Existing Web Page*. If this is your first run, select *Create Web Page*, then *Next*. If a confusing AutoCAD alert appears, select *Cancel* and see the NOTE above.

Create Web Page
In this step (Fig. 19-12) you supply the *name of your Web* page (in the first box), the *parent directory* where a folder will be created to store the files (use the browse button, right center), and a *description* (in the bottom box). The *name* appears on top of the final Web page (see previous Figure 19-11). The *name* you enter also determines the name of the folder (under the parent directory you select) that AutoCAD creates to store the files. The *description* appears on the Web page under the title (see previous Figure 19-11).

Select Image Type
Here you can select from a *DWF, JPEG,* or *PNG* format for the images that will be created (Fig. 19-13). A brief explanation for each image type is given <u>after</u> you make the selection. For *JPEG* and *PNG* images, you can select from *Small, Medium, Large,* or *Extra Large* image sizes.

Figure 19-12

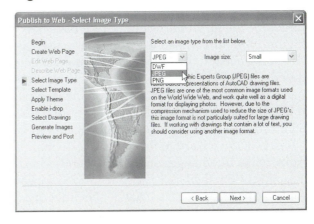

Figure 19-13

Select Template

Four basic templates are available to choose from (Fig. 19-14). The image tile on the right gives a simplified display of what your resulting Web page will look like. The two *Array* options display an array of images on the page so the viewer can click on any image to see a larger view of the drawing. The two *List* options allow the viewer to select an entry from the list to view a large version of the drawing in an image frame. Summary information (if selected) is the text that appears for each drawing in its *Summary* tab of the *Drawing Properties* dialog box. You can assign a description for a drawing using the *DWG-PROPS* command.

Figure 19-14

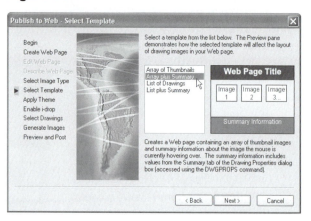

Apply Theme

AutoCAD provides several color schemes for you to choose for your Web page (Fig. 19-15). The colors are applied to different sections of the Web page, such as the title, description, and summary.

Figure 19-15

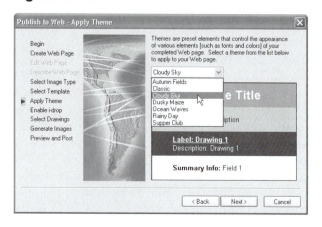

Enable i-Drop

The *Enable i-Drop* page (not shown) simply allows you to check whether you want to use this feature or not. If you enable i-Drop, copies of your drawing files are automatically included in the folder on the computer or server that contains your Web page. This allows viewers of the page to drag and drop the drawing files over the Internet directly into their AutoCAD session. This option is recommended only if you want to provide the page viewers access to your drawings. This option removes the paths (stored drive and directory location information) from each attached Xref or other file contained in the transmittal set (useful when *Preserve Directory Structure* is not checked).

Select Drawings

In this step (Fig. 19-16), you select the images that you want to use for the page and enter the text you want to appear under each image. First, select an AutoCAD drawing you want to use for an image (use the browse button). Next, select the *Model* (tab) or a *Layout* in the drawing you want to display. The text you enter in the *Label* box appears immediately under the image on the resulting Web page (see Figure 19-11, underlined text) and is also used as the label displayed in the *Image list*, so you must enter a descriptive label if you use multiple images on the page. The *Description* will appear on the Web page under the *Label* text. Select the *Add->* button to add all those entries and selections that appear on the left side of the wizard to the *Image list*. To make changes, highlight any item from the *Image list* on the right, make the changes on the left, then select *Update->*. The *Remove* button removes any highlighted selections from the *Image list*. The images appear on the page in the same order as the *Image list*. Use *Move Up* or *Move Down* to change the order of a selected item in the list. Select *Next* to proceed.

Figure 19-16

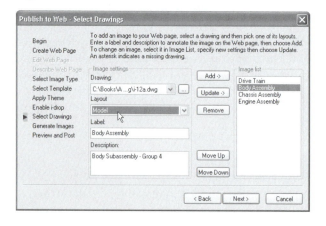

Generate Images

Selecting *Next* in this dialog box (not shown) causes AutoCAD to create the .PNG, .JPG, or .DWF images and generate the HTML document to display the images. No other input is required since the default option (*Regenerate images for drawings that have changed*) causes all new and all changed images to generate. Select *Regenerate all images* if you edit an existing Web page and want a duplicate set of images to be created in a second location.

Preview and Post

This last step (not shown) allows you to *Preview* the newly created Web page or to *Post Now*. The *Preview* option launches your Web browser with the page exactly as it will appear and operate when an outside viewer sees it. *Close* the browser to return to the *Publish to Web* wizard.

You can select *Back* at any step in the *Publish to Web* wizard, allowing you to change any aspect of the Web page that you specified in earlier steps. Use this feature to make corrections or changes before you post the Web page.

If you select *Post Now*, you are prompted for a location to store the files. Select the location on your Web server as specified by your system administrator.

The .PNG, .JPG, or .DWF files and the .HTM file that are generated by AutoCAD reside in a folder you specified as the *name of your Web page* in the *Create Web Page* step. A .PTW file is used by AutoCAD if you want to edit the Web page at a later time. This folder is located in the parent directory you also specified in that step. You can view the page and images at any time using your browser by selecting *Open* from the *File* menu, then using the *Browse* button to locate the .HTM file. If you select *Post Now* in the last step of the wizard, AutoCAD copies the files (not the .PTW file) to the location that you specify. The original files are not deleted.

ETRANSMIT

Pull-down Menu	COMMAND (TYPE)	ALIAS (TYPE)	Short-cut	Screen (side) Menu	Tablet Menu
File *eTransmit*	*ETRANSMIT*	...	...	...	...

eTransmit is a feature that greatly simplifies the process of sending AutoCAD drawings by email. The process is straightforward: *Open* the drawing you want to send, use *eTransmit*, enter any notes and select your preferences in the *Create Transmittal* dialog box. AutoCAD then creates a compressed .EXE or .ZIP file of the drawing package and opens your email software with the transmittal file attached. Just write your email note and send it!

eTransmit operates on the current drawing. First *Open* the drawing you want to transmit. If the drawing contains Xrefs, fonts, plot style tables, or compiled shapes, you can specify each item you want to include in the transmittal. The *eTransmit* command produces the *Create Transmittal* dialog box (Fig. 19-17).

Figure 19-17

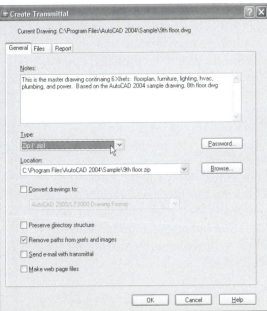

General Tab

Notes:

Enter any notes you want to include with the transmittal report (see *Report* tab). Alternately, you can specify a template of default notes to be included with all your transmittal sets by creating an ASCII text file called ETRANSMIT.TXT. Then, specify the location of the ETRANSMIT.TXT file in the *Support File Search Path* option on the *Files* tab of the *Options* dialog box.

Type

Here you can select from *Folder (set of files)*, *Self-extracting executable (*.exe)*, or *Zip (*.zip)*. The *Folder* option creates a new folder or uses an existing folder and creates the uncompressed transmittal files. The *Self-extracting executable (*.exe)* option creates one self-executable file that the recipient can double-click to decompress and restore the original files. The *Zip (*.zip)* option creates a transmittal set of files as one compressed .ZIP file.

Password

Here you can set a password that is required to decompress the files. Make sure you notify the recipient of the password that is required to decompress the files.

Location

Use this section to specify the location on your computer or network where the transmittal set is to be created. You can use the *Browse...* button to specify a new location.

Convert Drawings To

If you want to convert the drawing(s) to an earlier version of AutoCAD or AutoCAD LT, check this box and select from the drop-down list.

Preserve Directory Structure

Checking this box causes AutoCAD to preserve the directory structure of all files in the transmittal set during decompression and installation on another system. This feature may be helpful to the recipient if the set contains Xrefs, fonts, or shapes, etc. If this option is cleared, all files are installed to the target directory when the transmittal set is installed.

Remove Paths from Xrefs and Images

This option removes the paths (stored drive and directory location information) from each attached Xref or other file contained in the transmittal set (useful when *Preserve Directory Structure* is not checked).

Send E-mail with Transmittal

Check this box to launch your default email program when the transmittal set is created. When the program opens, AutoCAD automatically attaches the transmittal set to the email message and enters the drawing set title in the "Subject:" line of your note.

Make Web Page Files

This feature is especially powerful. Checking this box causes AutoCAD to generate a Web page (.HTM and .BMP files) that includes a bitmap image and a link to download the transmittal set (Fig. 19-18).

Figure 19-18

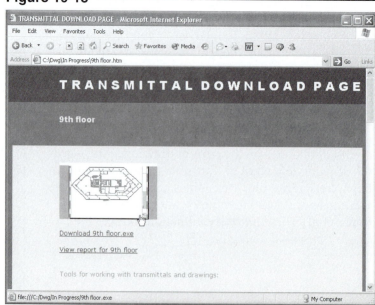

The page is created automatically using the *PublishToWeb* routine and templates. The .HTM and .BMP files are not included in the transmittal set, but are created separately and placed in the folder specified in the *Location:* edit box. Make sure you specify (in the *Location:* edit box) the final location on the server you intend to post these files since the download links (paths) are coded into the HTML document.

Files Tab

The *Files* tab lists the files to be included in the transmittal set. When you first access the tab, all files associated with the current drawing (such as related Xrefs, plot styles, and fonts) are listed. You can remove files from the transmittal set or use the *Add File* button to add additional files to the set.

List of Files

The list in the central area of the *Files* tab displays the names of all files to be included in the transmittal set (Fig. 19-19). You can choose either the *List View* or *Tree View* button above the list on the left to switch back and forth between these views. All files that are to be included in the set have a check appearing in the check box on the left of each file in the list. Remove a check if you do not want the related file included in the set to transmit.

Figure 19-19

List View

The left-most button above the list toggles the display of files to list view. In this format, files are listed in alphabetical order by default but can be sorted by *Filename*, *Type*, *Size*, or *Date* in normal or reverse order by clicking on the column heading.

Tree View

This arrangement displays a hierarchical listing—especially helpful for determining which Xrefs, fonts, and plot styles are related to which drawing files (see Figure 19-19).

Add File

Use this button to open the standard file selection dialog box where you can select additional files to include in the transmittal set.

Include Fonts

If you remove this check, all the current drawing's associated font files (.TXT and .SHX) are removed from the list display and the font files are not included in the transmittal set. Generally, most users already have the associated .TXT and .SHX fonts on their systems since they are included with AutoCAD. Because True Type fonts are proprietary, they are never included with the transmittal set. However, if any required True Type fonts are not registered on the computer that receives the transmittal set, the font specified by the *FONTALT* system variable is substituted.

Report Tab

This tab displays a report that is automatically generated and included with the transmittal set. The generated report is a .TXT file that has the same name as the current drawing. You can add additional notes in the report by entering the information in the *Notes* section of the *General* tab. The information automatically generated by AutoCAD explains what steps must be taken by the recipient for the transmittal set to work properly. For example, if AutoCAD detects .SHX fonts in one of the transmittal drawings, the report instructs the recipient where to copy these files so AutoCAD can detect the files when the transmittal drawing is opened.

2002

Save As
This button opens the *Save report file as* dialog box where you can specify a location to save a report (.TXT) file. A report file is always generated and included with all transmittal sets; however, you can save an additional copy of the report file for archival purposes using this option.

CREATING .DWF FILES

2002

CAD files contain very precise and detailed geometric information as well as an enormous amount of data beyond the drawing geometry that you see, such as coordinate data, named dependent objects, system variable settings, user preferences, and so on. Most CAD files are also proprietary—viewable only using the software product that created the CAD file. In today's collaborative environment, however, it is necessary for many people who are involved with various aspects of the design process to view drawings with some precision, but not edit the drawings or have access to the other data contained in the drawing file. An example of implementing this idea is AutoCAD's *Publish to Web* capability, which facilitates posting drawing images on a Web page in .JPG, .DWF, or .PNG format, so anyone with a Web browser can view the drawings.

2002

Autodesk's .DWF file format is also intended for this purpose—viewing a detailed drawing image without AutoCAD and without the ability to access unneeded data. Since the data is compressed, the file is easily transportable over the Internet. Originally .DWF images (created in releases previous to AutoCAD 2002) were viewable in standard Web browsers, but the *WHIP!* plug-in was required for the browser. With AutoCAD 2004, Autodesk Express Viewer is used to view .DWF files. Autodesk Express Viewer is included with AutoCAD 2004 and can also be downloaded free from the www.autodesk.com Web site.

2004

.DWF files are a good way to share AutoCAD drawing files with colleagues who don't have AutoCAD. Because the Autodesk Express Viewer is free and the interface is easy to use, even individuals with no CAD knowledge can easily view and navigate a .DWF file. You cannot change the .DWF image from within the viewer, although you can pan, zoom, print, manipulate layers and views, and activate hyperlinks. (See "Express Viewer—Viewing .DWF Files.")

2002

Raster Files and Vector Files

The common World Wide Web graphic standard uses a raster image format such as .GIF or .JPG. These files can be highly compressed, making the information displayed in the images quickly transportable over data transmission lines and devices used for the Internet. However, because raster images are pixel-based, or composed of tiny "dots," viewing detail is not possible—zooming in only enlarges the "dots." Figure 19-20 displays a raster image which has been enlarged.

Raster images, such as a "screen capture" for example, are defined in the data file as a map of pixels—rows and columns of picture elements either in black and white or in color. Raster images are sometimes referred to as "bitmaps," since the file retains the map of each dot (placement within the rows and columns) with its color information.

Figure 19-20

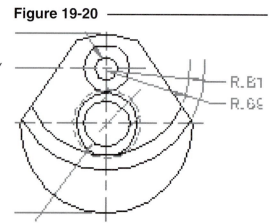

Examples of raster image file formats (file name extensions) are: .BMP, .CLP, .DIB, .GIF, .ICO, .IFF, .JPG, .MAC, .MSP, .NIF, .PBM, .PCX, .PSD, .RAS, .SGI, .TGA, .TIF, .XBM, .XPM, .XWD.

Raster images are of little use to professionals who require drawings with detailed information. Instead, the engineering, architecture, design, and construction industries use vector images to retain data with great precision and detail. CAD programs, which are the primary graphics creation and storage media for these industries, generate vector data. Figure 19-21 illustrates the same vector image (a CAD drawing) in a "zoomed in" display.

Figure 19-21

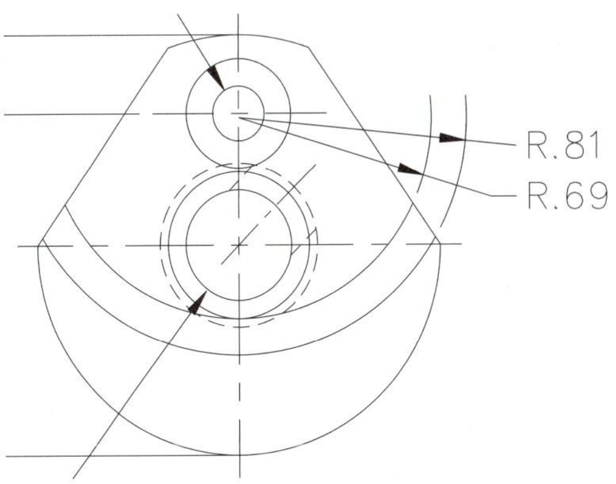

Because a vector file defines a line, for example, as a vector between two endpoints, only one dimension is retained—length. Normally, "zooming in" to a detailed section of the drawing does not create a second dimension (thickness or width) for the line. The thickness of the displayed line on a computer monitor is always one pixel, the size of which is determined by the display device.

NOTE: When creating a .DWF file, you can choose to display the drawings in .DWF format with line weight to give a representation of the printed image. In this case, zooming does change line thickness.

Examples of vector file formats include: .CDR, .CGM, .CMX, .DGN, .DWG, .DWF, .DXF, .GEM, .IGS, .MCS, .P10, .PCL, .PGL, .PIC, .PLT, .PRT, .WRL.

A .DWF image is a compressed vector file whose geometry can be viewed in detail by zooming in and out. Before the new .DWF standard was introduced, the World Wide Web provided inadequate usefulness for the engineering, architecture, design, and construction industries. Raster files are unable to retain the necessary detailed information, while most data-rich vector files generated by CAD programs are too large to be "viewable" over the Web.

Creating .DWF Files with ePlot

ePlot is a plot device designed to create a copy of the current drawing as a .DWF file. Several options are available for creating the .DWF, such as resolution, compression, font handling, background color, and inclusion of layer information.

ePlot is not a command, rather it appears as plot devices available through the *Plot Device* tab of the *Plot* dialog box (Fig. 19-22). When you use ePlot, AutoCAD makes a copy of the graphics geometry portion of the current drawing with other options you select and creates a .DWF file. If you plan to include hyperlinks in your .DWF or include layers or any other graphic information, that action must be performed in the original drawing before using ePlot to create the .DWF.

Figure 19-22

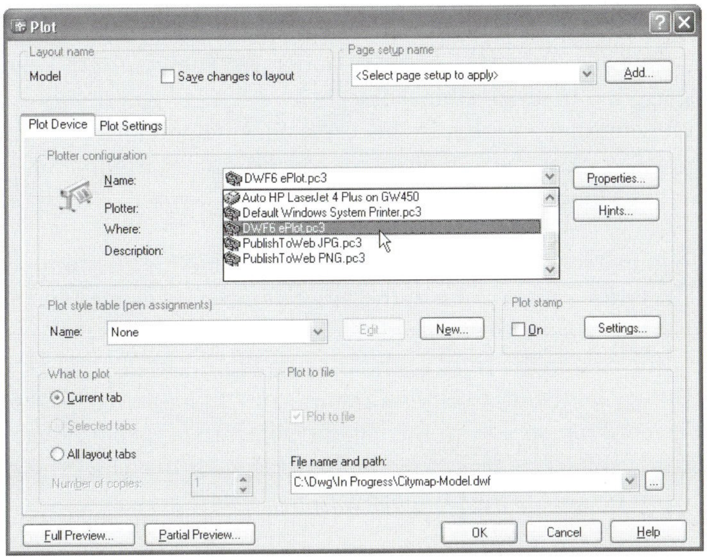

2004

DWF6 ePlot

To create a new (AutoCAD 2004) .DWF file, simply select the *DWF6 ePlot.pc3* device from the *Plot* dialog box (see Figure 19-22). Since you are creating a file rather than a print or plot, make sure you specify the *File name and path* in the *Plot to file* section. (Although you can use the *Page Setup* dialog box to make a .DWF, you cannot specify a file name and location; therefore, it is recommended that you use only the *Plot* dialog box to create .DWF images.) You can enter the location of a local computer or network folder to plot the file to or enter an Internet or intranet URL to plot the file to. (You can only plot .DWF files to the Internet using the FTP protocol.) Other options such as resolution, compression, and background color are accessible through the *Properties* button (see "ePlot Options"). Selecting *OK* in the *Plot* dialog box creates the .DWF file.

Older ePlot Devices

When you create .DWF files, you choose a plotter configuration file that uses a specific .DWF model. AutoCAD comes pre-configured with only one .PC3 (plotter configuration) file (*DWF6 ePlot*), although four older ePlot devices are available.

> *DWF Classic (R14 look)*
> *DWF ePlot (optimized for plotting)*
> *DWF ePlot (WHIP! 3.1 Compatible Version)*
> *DWF eView (optimized for viewing)*

Use the *Add-a-Plotter Wizard* to create these .DWF plotter configurations only if you want to create .DWF files compatible with the older viewing methods such as the *WHIP!* plug-in (see "Express Viewer—Viewing .DWF Files"). Use the *DWF eView (optimized for viewing), DWF Classic (R14 look),* or *DWF ePlot (WHIP! 3.1 Compatible Version)* to optimize the image for viewing. The resulting .DWF images have a higher resolution to allow for greater zooming while viewing. The resolution depends on the paper size and is not expressed in dots per inch (DPI). .DWF files produced by the *DWF ePlot (optimized for plotting)* model are optimized for printing and plotting. Their resolution is expressed in DPI, independent of paper size. The resulting .DWF images generally have less zoom depth but store higher quality raster images (if raster images are embedded in the drawing).

Specifying ePlot Options

You can create .DWF files with or without lineweights. By default, .DWF images are plotted with lineweights. This can cause areas of your .DWF files (when zoomed in the viewer) to look significantly different from how they appear as an AutoCAD drawing. If you have not specified lineweight values in the *Layer Properties Manager*, a default lineweight of .06 inches is applied to all graphical objects by ePlot. If you do not want lineweights to appear in the .DWF files when viewed in your browser, clear the *Plot object lineweights* option in the *Plot Settings* tab in the *Plot* dialog box.

To gain access to the other options available for creating your .DWF files, ensure one of the ePlot devices is selected in the *Plotter Configuration* drop-down list, then select the *Properties...* button just to the right of the list (see Figure 19-22). Next, in the *Plotter Configuration Editor* that appears (not shown), highlight *Custom Properties* from the list and select the *Custom Properties* button that appears below the list in the editor. The previous action produces the *DWF6 ePlot Properties* dialog box.

Figure 19-23

The options in the *DWF6 ePlot Properties* dialog box (Fig. 19-23) are explained on the next page.

2004

Resolution

Use the drop-down list to specify the resolution of the .DWF files that you create. The higher the resolution of the .DWF file, the greater its precision and the larger the file size. For most .DWF files, a low or medium resolution setting is sufficient.

Resolution for vector files (*Vector resolution*) does not affect the quality of the lines, rather it affects the level of detail in the drawing available when zooming in. When you create .DWF files of drawings that cover a large geometric extent, such as a map of a large topographical region, use a higher resolution setting. As an example, consider a map of the world that you want to output as a .DWF file. With a medium resolution setting, the maximum level of detail available with zoom may show only a state. With a high resolution setting, you might be able to zoom to a level of detail to display a city or a building. *Maximum raster resolution* affects only raster images that may be attached to the drawing.

Format

Compressed Binary (recommended)
The default setting, this option compresses the file without any loss of data. Use this for the output of most .DWF files.

Uncompressed Binary
You can also create uncompressed binary files, creating a larger file size than the compressed option but with no increase in data.

Zipped ASCII Encoded 2D Stream (advanced)
Use this option to create a readable (text) file, similar to the format of an ASCII .DXF file.

Font Handling

Capture None (all viewer supplied)
No fonts are saved with the drawing using this option. This option requires the viewer to use fonts native to the computer on which the drawings are being viewed. This option is acceptable if the drawing contains typical TrueType fonts such as Arial or Courier.

Capture Some (recommended)
This option is useful if you are using AutoCAD-supplied fonts (.TTF or .SHX) in the drawing such as GDT, TXT, GOTHICx, or ROMANx. AutoCAD supplied fonts are typically not registered on computers where AutoCAD has not been installed.

Capture All
Use this option if you are not sure what fonts are used in the drawing but want to ensure that fonts in the .DWF appear in the viewer just like the .DWG form of the drawing.

Edit List
Use the *Edit List* button to select which fonts you want to be captured with the *Capture some (recommended)* option.

Background Color Shown in Viewer
This option affects all .DWF files created for and viewed using the Express Viewer or the *WHIP!* Plug-in, regardless of the background color displayed in the AutoCAD drawing.

Include Layer Information
If this box is checked, you can control the *On/Off* visibility of the layers in the .DWF image when the resulting .DWF is viewed in the viewer, similar to controlling layer visibility in AutoCAD (see "Express Viewer—Viewing .DWF Files"). If this box is cleared, the resulting .DWF in the viewer offers no layer listing or visibility control.

Show Paper Boundaries

This option affects only .DWF files created for and viewed in a Web browser using the *WHIP!* Plug-in. If the .DWF image is viewed in the Express Viewer, the paper boundary display can be toggled on or off in the viewer.

Save Preview in DWF

Check this option to save a preview bitmap image when the .DWF is listed in a dialog box using the *Thumbnails* option.

Creating .DWF Files with *Publish*

The *Publish* command, also called *Design Publisher*, is a new feature available with AutoCAD 2004. *Publish* automates the process of creating .DWF files. *Publish* allows you to combine several drawings and/or layouts into one .DWF file.

PUBLISH

Pull-down Menu	COMMAND (TYPE)	ALIAS (TYPE)	Short-cut	Screen (side) Menu	Tablet Menu
File *Publish*	*PUBLISH*	...	...	...	...

The *Publish* command provides you with a utility to create drawing "sets" in .DWF format that can be viewed in Autodesk Express Viewer. *Publish* creates .DWF6 files automatically, rather than having to use the *DWF6 ePlot* device in the *Plot* dialog box. (Autodesk Express Viewer comes with AutoCAD 2004 and is free and downloadable from the www.autodesk.com Web site.)

With *Publish*, you can easily create a drawing "set" in .DWF format. A drawing set created by *Publish* can contain multiple "sheets" within one .DWF file. Therefore, you can publish a complete project composed of several .DWG files with multiple layouts. Each drawing or layout (or model space layout) you specify becomes a separate "sheet" within the .DWF file. Since many projects are composed of several drawings and/or layouts, *Publish* is especially helpful for creating one file that is easily transportable and viewable by other colleagues or clients—particularly when the clients or colleagues do not have AutoCAD.

When you view a multiple sheet .DWF file created by *Publish*, the list of sheets appears in the drop-down list at the top of the Autodesk Express Viewer (Fig. 19-24). Each sheet can be from a different layout of the same .DWG file, each can be from a separate .DWG, or the list can contain a combination of both. The sheets you want to appear are specified when you use *Publish* to create the .DWF.

Figure 19-24

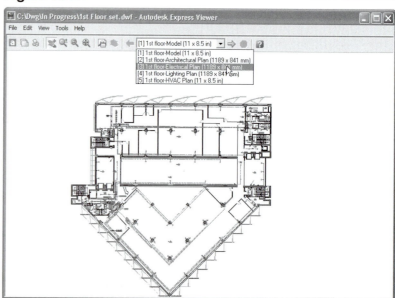

Using *Publish* by any method produces the *Publish Drawing Sheets* dialog box (Fig. 19-25). Normally, the steps for using this utility are:

Figure 19-25

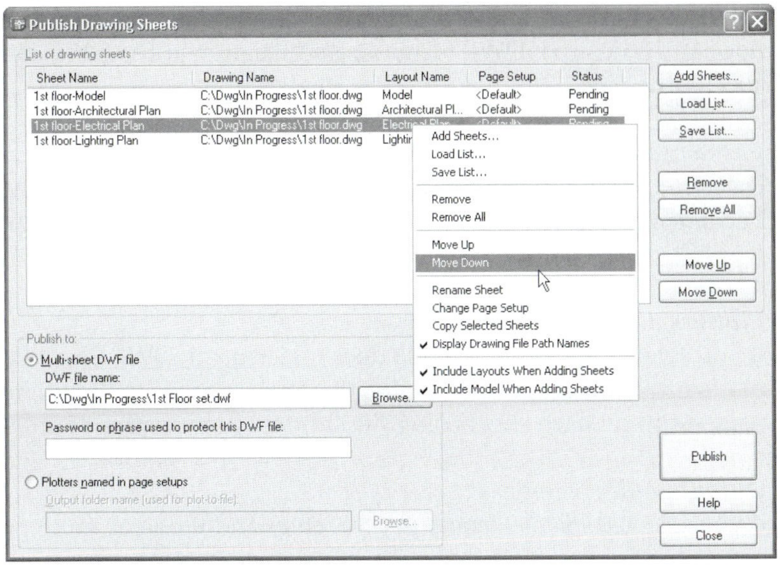

1. Use *Add Sheets…* to select the drawings and layouts you want to include in the .DWF file.
2. Change the order that the sheets appear when viewed using *Move Up* or *Move Down*.
3. Use *Save List* if you intend to publish this or a similar group of sheets again.
4. Select *Multi-sheet DWF File* if you want to create a .DWF file or select *Plotters named in page setups* if you want to plot each sheet to the device listed in its *Page setup* column.
5. Specify the name and location of the .DWF file you want to create.
6. Select *Publish* to create the .DWF file.

The options are described here.

Add Sheets…
This option produces the standard *Select Drawings* dialog box. Select the drawing files you want to include in the .DWF file to be created.

List of drawing sheets
The *List of drawing sheets* (in the *Publish Drawing Sheets* dialog box) lists by default every layout (including the *Model* tab) for each drawing selected with *Add Sheets*. Therefore, selection of one drawing in the *Select Drawings* dialog box will usually produce multiple entries in the *List of drawing sheets* since most drawings contain one model space layout and at least one other layout. In Figure 19-25 for example, all entries are different layouts of one drawing, 1ST FLOOR. Using the shortcut menu, you can specify whether or not layouts or the *Model* layout are automatically included when adding sheets to the list. You can also rename sheets (*Sheet Name*) and change the *Page Setup* for each sheet (see "*Plotters named in page setups*"). The *Status* field indicates whether the sheets have or have not yet been published (*Plotted* or *Pending*).

Load List…
If you saved a list of drawings previously using *Save List…*, you can load the same list again for publishing or modifying. You can load .DSD (drawing sheet description) or .BP3 (batch plot) files.

Save List…
This choice saves the current list of drawings and layouts in the *List of drawing sheets* as a .DSD file.

Remove and *Remove All*
Use these options to remove all entries or only highlighted entries from the *List of drawing sheets*.

Move Up and *Move Down*
These buttons allow you to specify the order of the drawing sheets when they are viewed. The order in the *List of drawing sheets* dictates the order of sheets that appear in the drop-down list of Express Viewer (see Figure 19-24).

Multi-sheet DWF file

Select this radio button to produce one .DWF file containing all the sheets in the *List of drawing sheets*. Specify the desired .DWF file name and location in the *DWF file name* box or use the *Browse* button to specify the path.

Password or phrase used to protect this DWF file

If desired, enter a password needed to view the .DWF file. .DWF passwords are case sensitive and can be made up of letters, numbers, punctuation, or non-ASCII characters. <u>If you lose or forget the password, it cannot be recovered</u>.

Plotters named in page setups

<u>If you select this option, *Publish* does not create a .DWF file</u>. Instead, each sheet is printed, plotted, or plotted to a file as specified by the device named in its *Page Setup* column. You can specify any named page setup for each sheet using the shortcut menu or by double-clicking in the *List of drawing sheets*.

Output Folder Name

If you selected *Plotters named in page setups* and if one or more page setups are set to *Plot to file*, use this box to specify the name and path for the output folder.

Publish

Select *Publish* only when you are ready to begin the publishing operation (all other options are specified). *Publish* creates a single- or multi-sheet .DWF file or plots to a device or file depending on the radio button selected in the *Publish To* area.

When publishing is finished, the *Publishing Complete* dialog box appears (Fig. 19-26) with two options. You can select to *View DWF File*, which opens Autodesk Express Viewer with the drawing set loaded (see Figure 19-24). You can also create a log file that contains detailed information about the sheets that were published. The log file has the same name and location as the .DWF file but with a .CSV (comma separated) file type (viewable in Microsoft Excel, Word, or other program). If a drawing sheet fails to plot for some reason, *Publisher* continues plotting the remaining sheets in the drawing set. In this case, a log file is automatically created that lists errors or warnings encountered during the publishing process.

Figure 19-26

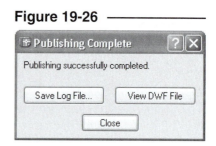

EXPRESS VIEWER—VIEWING .DWF FILES

Although .DWF files are created from AutoCAD drawings, you cannot view the .DWF files from within AutoCAD. To view .DWF files created with AutoCAD 2004 *DWF6 ePlot* you need Autodesk Express Viewer.

Autodesk Express Viewer comes with AutoCAD 2004, so everyone with AutoCAD 2004 can view .DWF images directly with Autodesk Express Viewer. Alternately, Autodesk Express Viewer can be downloaded free from www.autodesk.com and installed by anyone—without AutoCAD.

.DWF files created with AutoCAD 2002 or earlier releases or with AutoCAD 2004 *.DWF Classic (R14 Look)* or *.DWF ePlot (WHIP! 3.1 Compatible Version)* can be viewed using Autodesk Express Viewer or with Netscape Navigator or Microsoft Internet Explorer with the *WHIP!* Plug-in installed. The *WHIP!* Plug-in is no longer available. Autodesk Express Viewer is intended to replace the *WHIP!* Plug-in, so the newer (AutoCAD 2004 ePlot) .DWF images may not be displayed correctly with the *WHIP!* Plug-in.

Opening and Viewing .DWF Files

There are several methods that can be used to open .DWF files for viewing:

1. Use the *File* pull-down menu, then use *Open* in Autodesk Express Viewer to locate and open a .DWF file;
2. Drag-and-drop the .DWF file from Windows Explorer or My Computer into Autodesk Express Viewer;
3. Double-click on the .DWF file in Windows Explorer or My Computer to launch Express Viewer to view the file.

Once a .DWF file is loaded in Autodesk Express Viewer, you can right-click to produce a menu displaying zoom, pan, and other options (Fig. 19-27). Some menu options may be disabled when you are viewing a .DWF image based on the options specified when the .DWF was created. The right-click menu provides the following options.

Figure 19-27

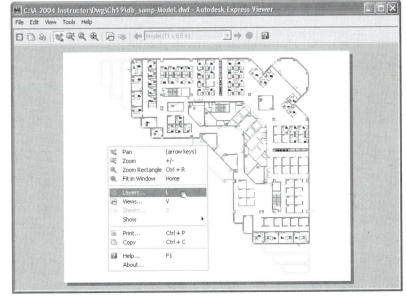

Pan

This option puts you in real-time *Pan* mode and the pointer becomes a "hand" icon. Press the left mouse button to drag the image around on the screen in a manner identical to the *Realtime* option of *Pan* in AutoCAD.

Zoom options

With *Zoom* capabilities of a .DWF image, genuine detail can be viewed, similar to capabilities in the original CAD program with the associated CAD drawing. By default, you can *Zoom* at any time (without selecting a menu option) by turning the mouse wheel if you have a wheel mouse. The *Zoom Rectangle* and *Fit in Window* options operate identically to the *Window* and *All* options, respectively, in AutoCAD (see Chapter 10, Viewing Commands).

Keep in mind that ePlot creates .DWF images with lineweights by default; therefore, if you want others to view your drawing with great zoom detail without the line thickness changing as zooming occurs, create the .DWF after clearing the *Plot object linetypes* option in the *Plot Settings* tab of the *Plot* dialog box.

Figure 19-28

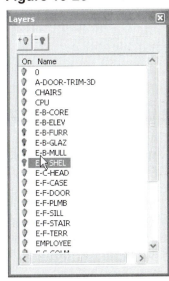

Layers

Using this option produces the *Layers* dialog box (Fig. 19-28). Here you can click the light bulb icon to turn a layer on or off in real time (assuming the *Custom Properties* were set to *Include Layer Information* when the .DWF file was created). You can select multiple layers by holding down Shift (to select a range) or Ctrl (to select individual multiple layers), then use the on or off buttons near the top of the dialog box.

Views

If the AutoCAD drawing from which the .DWF image was created contained named *Views*, those views are recorded in the .DWF image and can be viewed with this option. The *Views* dialog box appears with the list of named views for you to select from (Fig. 19-29). The selected view then appears in the viewer.

Show

A cascading menu appears with toggles for you to show *Hyperlinks* (if hyperlinks were included in the drawing), *Page Titles* (if the .DWF files were created using *Publish*), and to show the *Paper Background* and the *Toolbar* in the viewer.

Print

You can print the .DWF image using this option. Your system's *Print* dialog box is displayed with all printers configured for your computer to select from.

Copy

Use this option if you want to copy the .DWF image and *Paste* it into AutoCAD as an OLE object. (See Chapter 32, Raster Images and Vector Files.)

Figure 19-29

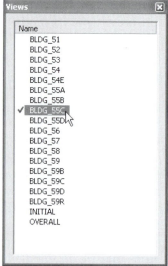

Viewing URLs (Hyperlinks) in a .DWF Image

When viewing a .DWF image in Express Viewer, passing the mouse pointer over a URL-linked object displays a small hand image. Clicking on the linked object automatically links to the new URL, which causes the text and/or images to display in Microsoft Internet Explorer or Express Viewer, depending on the file format (.HTM, .DWF, .JPG, etc.). For example, clicking on a door, window, or fixture in an architectural layout could, in turn, display a table of text information listing price, manufacturer, specifications, etc. Or, clicking on one component of a mechanical assembly could, in turn, display a detail drawing of the selected part. It would also be possible to view a map, select a building, and see written information about the business or link to the Web site of the business.

NOTE: Depending on the file format, clicking on a URL-linked object in Express Viewer may display the linked object in Microsoft Internet Explorer or another program, not in Express Viewer.

To create URLs in .DWF images, *Open* the original drawing in AutoCAD and use the *Hyperlink* command to specify a URL. Next, use the *Plot* command and select an ePlot device to create the related .DWF file with hyperlinks included (see "Using Hyperlinks" and "Creating .DWF Files with ePlot").

TRANSFERRING AutoCAD FILES OVER THE INTERNET

Opening and Saving .DWF and .DWG Files via the Internet

You can use AutoCAD to open and save files from the Internet using the typical *Open* and *Saveas* commands. If you use *Open* to open a drawing from a remote Internet site, the drawing file that you specify is downloaded to your computer and opened in the AutoCAD drawing area. You can then edit the drawing and save it, either locally or back to any Internet or intranet location assuming you have sufficient access privileges.

If you know the URL to the file you want to open or save to, you can enter it directly in the *Select File* dialog box that appears when you use *Open* or the *Save Drawing As* dialog box that appears when you use *Saveas*. You can also use the *Search the Web* button to produce the *Browse the Web* dialog box. Here you can navigate to find the Internet location for the file.

To <u>open</u> an AutoCAD file from the Internet using the *Select File* dialog box:

1. from the *File* menu, choose *Open*; then
2. enter the URL to the file in *File Name* and select *Open*.

You must enter the transfer protocol (for example, http:// or ftp://) and the extension (for example, .DWG or .DWT) of the file you want to open.

To <u>save</u> an AutoCAD file to an Internet location using the *Save Drawing As* dialog box:

1. from the *File* menu, select *Save As*;
2. enter the URL to the file in *File Name*; then
3. select a file format from the *Save As Type* list, and then choose *Save*.

In step 2 you must enter the File Transfer Protocol (ftp://) and the extension (for example, .DWG or .DWT) of the file you want to save. You can only save AutoCAD files to an Internet location using FTP.

Using i-drop to Drag-and-Drop Drawing Files into AutoCAD

Autodesk provides a utility called "i-drop" to allow users to drag and drop AutoCAD drawings from a Web page directly into AutoCAD. One component of this technology is the i-drop Indicator that modifies your browser behavior so that you can initiate the drag-and-drop action. If you have your own Web page, you can activate the i-drop capability so visitors to your page can drag and drop drawing files into an AutoCAD session. For information about using i-drop, downloading the i-drop Indicator, and creating an i-drop handle on your Web site, see the i-drop documentation on the Autodesk Web site at www.autodesk.com/idrop. If you use *PublishToWeb* in AutoCAD to create your Web pages, select the *Enable i-drop* option to make use of this capability (see *PublishToWeb*).

CHAPTER EXERCISES

1. ***Browser, Inetlocation*, Autodesk Express Viewer**

    A. Start AutoCAD if not already running. Type in the ***Browser*** command. Note that the default URL is http://www.autodesk.com. Press **Enter** to display your system's default browser and the Autodesk Web site.

    B. On the Autodesk site, locate **Products**, then click to follow the link. On the Products page, locate **Autodesk Express Viewer** and go to that page. If Autodesk Express Viewer is not installed on your computer, select the **Download Autodesk Express Viewer** button and follow the instructions.

    C. Once the Express Viewer is installed, start AutoCAD again if not already started. Enter ***Inetlocation*** at the command prompt. Change the default location to your favorite news source such as **http://www.usatoday.com**. Next, use ***Browser*** again, press **Enter** to accept the new URL, and validate that the new site appears in your Internet browser.

D. Now use the ***Tools*** pull-down menu and select ***Autodesk Website***. Does your browser launch and display the Autodesk Web site, or does the browser display the URL that you entered as the new *Inetlocation*?

2. ***Publish to Web***

A. Start AutoCAD if not already running. ***Save*** the current drawing. (It is recommended to always save the current drawing before using *PublishToWeb*.)

B. Invoke ***Publishtoweb***. Follow the necessary steps in the wizard to create a new Web page named **Project Drawings** and locate it in your working directory. Write an appropriate *description* when prompted. Specify *JPEG* image type and *Small* image size. For the template, specify ***Array plus Summary***. Select a *Theme* of your choice. In the *Select Drawings* step, specify four of your favorite drawings that you completed in the previous Chapter Exercises. Proceed to generate the images and *Preview* them. If the Web page needs improving, go *Back* and make the changes, or use *Publishtoweb* again and select ***Edit an Existing Web Page*** to generate new images. Finally, *Post* the Web page to the **Project Drawings** folder.

C. Start your browser, locate and open the **ACWEBPUB-LISH.HTM** file to view your Web page. Your new Web page should appear in the browser and look similar to Figure 19-30, depending on the drawings you selected to display. Click on any image to produce a larger view of the drawing. If you like, you can improve the page as you see fit by changing the drawings in AutoCAD and regenerating the Web page again with *Publishtoweb*.

Figure 19-30

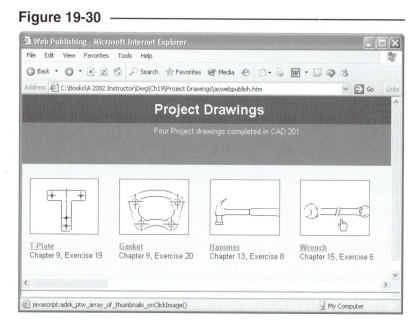

3. ***Etransmit***

A. Start AutoCAD if not already running. ***Open*** a drawing that you want to transmit to a friend. If you make any changes, *Save* the current drawing.

B. Invoke *eTransmit*. The *Create Transmittal* dialog box should appear. In the *Notes* section of the *General* tab, enter "Here is the drawing that I would like for you to review," or similar message. Also specify a ***Self-extracting executable (*exe)*** as the *Type* of file to create. Use the ***Browse*** button to specify your working directory as the *Location*. Clear the check boxes for the ***Preserve directory structure*** and ***Make web page files*** options. Check ***Remove paths from Xrefs and images***. In the ***Files*** tab, make sure all associated files are checked. Do not select *OK* yet.

C. If you have an email system available on the system, select ***Send e-mail with transmittal*** in the *General* tab. If not, clear the check for this option. Select ***OK***.

D. Depending on your selection in step C, your email system may open with the .EXE file as an attachment. If you chose not to have AutoCAD open your email program, copy the .EXE file to disk, and attach it to an email message at a later time. In either case, send an email to a friend and ask him or her to extract the file and view the drawing in AutoCAD. Request a confirmation and a reply email reporting on the success of the transmittal.

4. **Use *ePlot* to create .DWF files**

A. In AutoCAD, *Open* the **T-PLATE** drawing from Chapter 9 Exercises. Activate the *Model* tab. Use the *Plot* dialog box and *Plot Device* tab to select the **DWF6 ePlot.pc3** device. In the *Plot to file* section, accept the default name (T-PLATE-MODEL) and ensure your working directory is specified in the *Location* edit box. Select *OK* to make the .DWF file.

B. Start Autodesk Express Viewer. Use the *File* pull-down menu, then the *Open* command, and locate the **T-PLATE-MODEL.DWF**.

C. In AutoCAD, *Open* the **GASKETA** drawing from Chapter 9 Exercises. Follow step 1 to make a **.DWF** file. Use Autodesk Express Viewer to verify the .DWF was created correctly.

D. Create another **.DWF** file similar to the previous two, but this time activate a *Layout* tab that was set up previously with a viewport. If no layouts were set up, create one viewport but make sure the viewport border is slightly smaller than the printable area (dashed line), then make the .DWF file of the layout.

E. View the new .DWF image with Autodesk Express Viewer to verify it was created correctly. The image should appear similar to the Model tab .DWFs, except the viewport border should be visible.

5. **Use *Publish* to create a .DWF file**

A. In AutoCAD, use *Publish*. When the *Publish Drawing Sheets* dialog box appears, select *Add Sheets...*. Select four drawings of your choice. In the *List of drawing sheets*, specify the layouts you want to publish. Specify a *Multi-sheet DWF file*, then enter the desired *DWF file name* and location, such as **Exercise Drawings**. Your list and specifications should look similar to that in Figure 19-31.

Figure 19-31

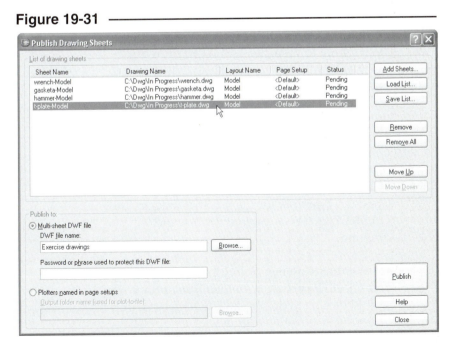

B.   Select **Publish** in the *Publish Drawing Sheets* dialog box. When the *Publishing Complete* dialog box appears, select **View DWF File**.

C.   Your new multi-sheet .DWF file should appear in the Autodesk Express Viewer. Use the pull-down list at the top of the viewer to inspect each of the drawings, similar to that in Figure 19-32. When you are finished, close the viewer, *Close* the *Publishing Complete* dialog box, and *Close* the *Publish Drawing Sheets* dialog box.

Figure 19-32

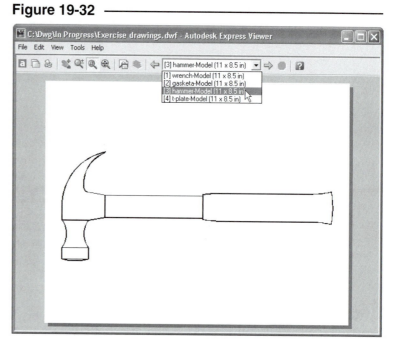

6.   **Insert** *Hyperlinks*

A.   In AutoCAD, **Open** the **GASKETA** drawing. Activate the *Model* tab. Use *SaveAs* to save the drawing as **GASKETA2**. *Close* the drawing.

B.   **Open** the **T-PLATE** drawing. Activate the *Model* tab. Use *SaveAs* to save the drawing as **T-PLATE2**. *Close* the drawing.

C.   **Open** the **GASKET2** drawing. Use *Dtext* to place text at the bottom of the drawing. Enter **"Click here to view the T-PLATE2 drawing."** The text should appear similar to that shown in Figure 19-33. Create a *Hyperlink* for the entire line of text. Associate the **T-PLATE2.DWG** as the file to link. *Save* and *Close* the drawing.

Figure 19-33

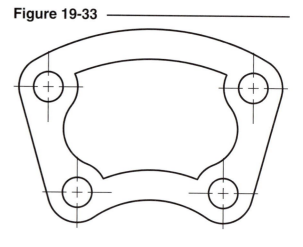

Click here to view the T—PLATE2 drawing

D.   **Open** the **T-PLATE2** drawing. Use *Dtext* to place text at the bottom of the drawing, stating **"Click here to view the GASKETA2 drawing."** Create a *Hyperlink* for the text that links to the **GASKETA2.DWG**. *Save* and *Close* the drawing.

E.   **Open** either drawing. Select the text, right-click, and select *Hyperlink*, then the *Open* option. You should be able to do the same in both drawings to "toggle" between the two drawings.

7.   **Create .DWF files with** *Hyperlinks*

    A.   *Open* the **GASKETA2** drawing. Activate the *Model* tab. Select the hyperlink, right-click to produce the shortcut menu, and select *Edit Hyperlink*. Edit the hyperlink to link to a .DWF, not the .DWG; for example, change C:\DWG\GASKETA2.DWG to **C:\DWG\GASKETA2.DWF**. Open the *Plot* dialog box to create a **.DWF** file. Use the *DWF6 ePlot.pc3* device. Make sure you specify **GASKETA2** for the name of the .DWF file, <u>not</u> GASKETA2-Model. Select *OK* to produce the .DWF file. *Save* and *Close* the drawing.

    B.   *Open* the **T-PLATE2** drawing. Follow the same procedure as in step A to create a .DWF file for the T-PLATE2 drawing. Make sure you convert the hyperlink extension to .DWF and save the .DWG with the correct name. *Save* and *Close* the drawing.

    C.   Use Express Viewer to view the .DWF images and test the hyperlinks. Make any necessary corrections.

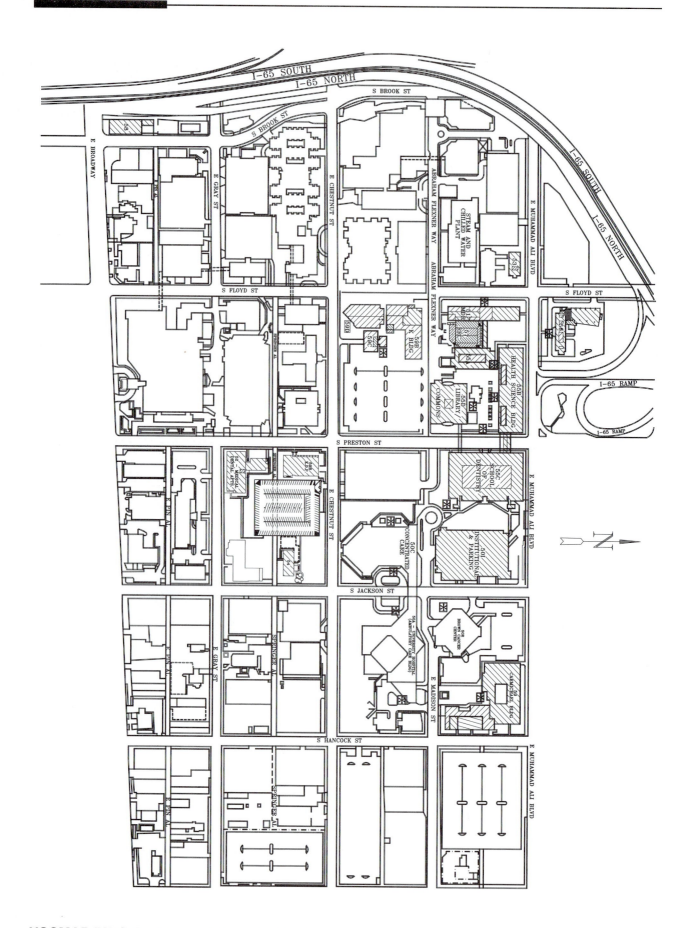

HSCMAP.DWG Courtesy, Michael Anderson

20

ADVANCED SELECTION SETS

Chapter Objectives

After completing this chapter you should:

1. be able to control the way objects are selected using the following dialog box options (or variables): *Noun/Verb Selection* (*PICKFIRST*), *Use Shift to Add to Selection* (*PICKADD*), *Implied Windowing* (*PICKAUTO*), and *Press and Drag* (*PICKDRAG*);

2. be able to use *Qselect* to specify criteria for creating a selection set to use with an editing command;

3. be able to use *Object Selection Filters* in a complex drawing to find a selection set for use with an editing command;

4. be able to create named selection sets (Object Groups) with the *Group* command and use options in the *Object Grouping* dialog box.

CONCEPTS

When you "Select Objects:", you determine which objects in the drawing are affected by the subsequent editing action by specifying a selection set. You can select the set of objects in several ways. The fundamental methods of object selection (such as PICKing with the pickbox, a *Window*, a *Crossing Window*, etc.) are explained in Chapter 4, Selection Sets. This chapter deals with advanced methods of specifying selection sets and the variables that control your preferences for how selection methods operate. Specifically, this chapter explains:

> Selection Set Variables
> Object Filters
> Object Groups

SELECTION SET VARIABLES

Four variables allow you to customize the way you select objects. Keep in mind that selecting objects occurs only for editing commands; therefore, the variables discussed in this chapter affect how you select objects when you edit AutoCAD objects. The variable names and the related action are briefly explained here:

Variable Name	Default Setting	Related Action
PICKFIRST	1	Enables and disables Noun/Verb command syntax. Noun/Verb means PICK objects (noun) FIRST and then use the edit command (verb).
PICKADD	1	Controls whether objects are ADDed to the selection set when PICKed or replace the selection set when picked. Also controls whether the Shift key + #1 button combination removes or adds selected objects to the selection set.
PICKAUTO	1	Enables or disables the PICKbox AUTOmatic window/crossing window feature for object selection.
PICKDRAG	0	Enables or disables single PICK window DRAGging. When PICKDRAG is set to 1, you start the window by pressing the PICK button, then draw the window by holding the button down and DRAGging to specify the diagonal corner, and close the window by releasing the button. In other words, windowing is done with one PICK and one release rather than with two PICKs.

Like many system variables, the selection set variables listed above hold an integer value of either 1 or 0. The 1 designates a setting of *ON* and 0 designates a setting of *OFF*. System variables that hold an integer value of either 1 or 0 "toggle" a feature *ON* or *OFF*.

Changing the Settings

Settings for the selection set variables can be changed by any of two ways:

1. You can type the variable name at the Command: prompt (just like an AutoCAD command) to change the setting.

2. Use the *Selection* tab of the *Options* dialog box (Fig. 20-1) to change the four selection set variables listed above. The four variables can be changed in this dialog box, but the syntax in the dialog box is not the same as the variable name. A check in the checkbox by each choice does not necessarily mean a setting of *ON* for the related variable.

When you change any of these four variables, the setting is recorded in the system registry, rather than in the current drawing file as with most variable settings. In this way, the change (which is generally the personal preference of the operator) is established at the workstation, not the drawing file.

OPTIONS

Pull-down Menu	COMMAND (TYPE)	ALIAS (TYPE)	Short-cut	Screen (side) Menu	Tablet Menu
Tools *Options...*	*OPTIONS*	*OP*	(Default menu) *Options...*	*TOOLS2* *Options*	*Y,10*

The *Selection* tab of the *Options* dialog box (Fig. 20-1) can be used to change the four selection set variables listed previously, namely, *PICKFIRST*, *PICKADD*, *PICKAUTO*, and *PICKDRAG*. Although the checkboxes on the left side of the *Selection* tab do not indicate the formal variable names, they are intended to be more descriptive.

The checkbox names and the respective variables are listed here:

Noun/Verb Selection	*PICKFIRST*
Use Shift to Add to Selection	*PICKADD*
Press and Drag	*PICKDRAG*
Implied Windowing	*PICKAUTO*

These selection set variables are explained in the sections next; each section is designated by the formal variable name.

Figure 20-1

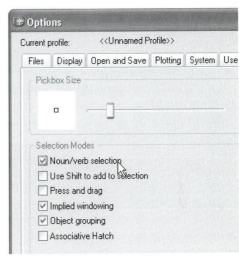

NOTE: It is advisable to turn off Grips when you are experimenting with these variables to avoid confusion. Turn off Grips by removing the check from *Enable Grips* on the right side of the dialog box or by typing *GRIPS* at the command prompt and changing the setting to 0 (off).

PICKFIRST

NOTE: If the *GRIPS* variable is set to 1, object grips are enabled. Object grips do not hinder your ability to use *PICKFIRST*, but they may distract your attention from the current topic. Setting the *GRIPS* variable to 0 disables object grips. You can type *GRIPS* at the command prompt to change the setting. See Chapter 23 for a full discussion of object grips.

Changing the setting of *PICKFIRST* is easily accomplished in command line mode (typing). However, the *Selection* tab of the *Options* dialog box can also be used to toggle the setting. The choice is titled *Noun/Verb Selection*, and a check appearing in the box means that *PICKFIRST* is set to 1 (*ON*).

The *PICKFIRST* system variable enables or disables the ability to select objects <u>before</u> using a command. If *PICKFIRST* is set to 1, or *ON*, you can select objects at the Command: prompt <u>before</u> a command is used. *PICKFIRST* means that you PICK the objects FIRST and then invoke the desired command. *PICKFIRST* set to 1 makes the small pickbox appear at the cursor when you are not using a command (at the open Command: prompt). *PICKFIRST* set to 1 enables you to select objects with the pickbox, *AUto* window, or *Crossing Window* methods when a command is not in use.

This order of editing is called "Noun/Verb"; the <u>objects are the nouns</u> and the <u>command is the verb</u>. Noun/Verb editing is preferred by some users because you can decide what objects need to be changed, then decide how (what command) you want to change them. Noun/Verb editing allows AutoCAD to operate like some other CAD systems; that is, the objects are PICKed FIRST and then the command is chosen.

The command syntax for Noun/Verb editing is given next using the *Move* command as an example:

> Command: **PICK** (Use the cursor pickbox or auto window/crossing window to select objects.)
> Command: **PICK** (Continue selecting desired objects.)
> Command: **Move** (Enter the desired command and AutoCAD responds with the number of objects selected.) 2 found
> Specify base point or displacement: **PICK** or **(coordinates)**
> Specify second point of displacement, or <use first point as displacement>: **PICK** or **(coordinates)**
> Command:

Notice that as soon as the edit command is invoked, AutoCAD reports the number of objects found and uses these as the selection set to act on. You do <u>not</u> get a chance to "Select objects:" within the command. The selection set PICKed <u>immediately</u> before the command is used for the editing action. The command then passes through the "Select objects:" step to the next prompt in the sequence. All editing commands operate the same as with Verb/Noun syntax order with the exception that the "Select objects:" step is bypassed.

<u>Only</u> the pickbox, auto window, and crossing window can be used for object selection with Noun/Verb editing. The other object selection methods (*ALL, Last, Previous, Fence, Window Polygon,* and *Crossing Polygon*) are only available when you are presented with the "Select objects:" prompt.

NOTE: To disable the cursor pickbox, *PICKFIRST* <u>and</u> *GRIPS* must be OFF.

NOTE: When *PICKFIRST* is set to 0, the Edit-mode shortcut menus do not appear (see "Shortcut Menus," Chapter 1).

PICKADD

The *PICKADD* variable controls whether objects are ADDed to the selection set when they are PICKed or whether selected objects replace the last selection set. This variable is *ON* (set to 1) by default. Most AutoCAD operators work in this mode.

Until you reached this chapter, it is probable that all PICKing you did was with *PICKADD* set to 1. In other words, every time you selected an object it was added to the selection set. In this way, the selection set is <u>cumulative</u>; that is, each object PICKed is added to the current set. This mode also allows you to use multiple selection methods to build the set. You can PICK with the pickbox, then with a window, then with any other method to continue selecting objects. The "Select objects:" process can only be ended by pressing Enter.

With the default option (when *PICKADD* is set to 1 or *ON*), the Shift+#1 key combination allows you to <u>deselect</u>, or remove, objects from the current selection set. This has the same result as using the *Remove* option. Deselecting is helpful if you accidentally select objects or if it is easier in some situations to select *ALL* and then deselect (Shift+#1) a few objects.

When the *PICKADD* variable is set to 0 (*OFF*), objects that you select <u>replace</u> the last selection set. Let's say you select five objects and they become highlighted. If you then select two other objects with a window, they would become highlighted and the other five objects automatically become deselected and unhighlighted.

The two new objects would replace the last five to define the new selection set. Figure 20-2 illustrates a similar scenario.

Figure 20-2

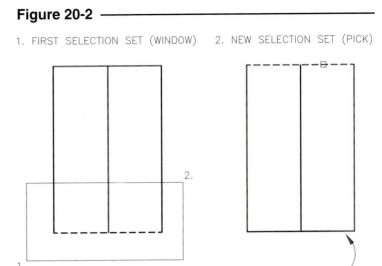

1. FIRST SELECTION SET (WINDOW)   2. NEW SELECTION SET (PICK)

AUTOMATICALLY DESELECTED

The *PICKADD* variable also controls whether the Shift+#1 button combination removes from or adds to the selection set. When *PICKADD* is set to 1 (*ON*), the Shift+#1 combination deselects (removes) objects from the selection set. When *PICKADD* is set to 0 (*OFF*), the Shift+#1 combination toggles objects in or out of the selection set, depending on the object's current state. In other words (when *PICKADD* is *OFF*), if the object is included in the set (highlighted), Shift+#1 deselects (unhighlights) it, or if the object is not in the selection set, Shift+#1 adds it.

You can change the variable by typing in *PICKADD* at the Command prompt. Changing the *PICKADD* variable in the *Options* dialog box is accomplished by making the desired choice in the checkbox. Changing the setting in this way is confusing because a check by *Use Shift to Add* means that *PICKADD* is *OFF*! Normally, a check means the related variable is *ON*. To avoid confusion, use only one method (either typing or dialog box) to change the setting until you are familiar with it.

You can also change the *PICKADD* variable using the *Toggle value of PICKADD Sysvar* button in the upper-right corner of the *Properties* palette. The symbol on this button changes (" + " or "1") to reflect the current setting for *PICKADD*; however, the symbols are confusing since changing *PICKADD* to 0 (*OFF*) changes the button to display a "1."

It may occur to you that you cannot imagine a practical application for using *PICKADD OFF*. It does make sense to operate AutoCAD for most applications with *PICKADD ON*. However, it can be useful to turn *PICKADD OFF* when using the *Properties* palette (so selecting one set of objects replaces the previous set in the *Properties* palette). For more information on *Properties*, see Chapter 16, Modify Commands II.

PICKAUTO

The *PICKAUTO* variable controls automatic windowing when the "Select objects:" prompt appears. Automatic windowing (*Implied Windowing*) is the feature that starts the first corner of a window or crossing window when you PICK in an open area. Once the first corner is established, if the cursor is moved to the right, a window is created, and if the cursor is moved to the left, a crossing window is started. See Chapter 4 for details of this feature.

When *PICKAUTO* is set to 1 (or *Implied Windowing* is checked in the dialog box, Fig. 20-1), the automatic window/crossing is available whenever you select objects. When *PICKAUTO* is set to 0 (or no check appears by *Implied Windowing*), the automatic windowing feature is disabled. However, the *PICKAUTO* variable is overridden if *GRIPS* are *ON* or if the *PICKFIRST* variable is *ON*. In either of these cases, the cursor pickbox and auto windowing are enabled so that objects can be selected at the open Command: prompt.

Auto windowing is a helpful feature and can be used to increase your drawing efficiency. The default setting of *PICKAUTO* (*ON*) is the typical setting for AutoCAD users.

PICKDRAG

Figure 20-3

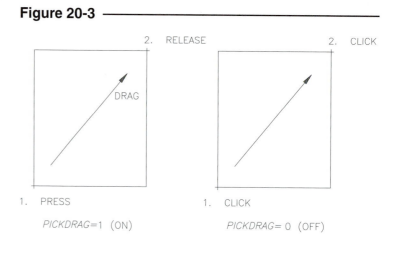

This variable controls the method of drawing a selection window. *PICKDRAG* set to 1 or *ON* allows you to draw the window or crossing window by clicking at the first corner, holding down the button, dragging to the other corner, and releasing the button at the other corner (Fig. 20-3). With *PICKDRAG ON* you can specify diagonal corners of the window with one press and one release rather than with two clicks. Many GUIs (graphical user interfaces) of other software use this method of mouse control.

The default setting of *PICKDRAG* is 0 or *OFF*. This method allows you to create a window by clicking at one corner and again at the other corner. Most AutoCAD users are accustomed to this style, which accounts for the default setting of *PICKDRAG* to 0 (*OFF*).

OBJECT SELECTION FILTERS

Large and complex AutoCAD drawings may require advanced methods of specifying a selection set. For example, consider working with a drawing of a manufacturing plant layout and having to select all of the doors on all floors, or having to select all metal washers of less than 1/2" diameter in a complex mechanical assembly drawing. Rather than spend several minutes PICKing objects, it may be more efficient to use one of the object filters that AutoCAD provides.

There are two selection filters provided in AutoCAD: *Qselect* (*Quick Select* dialog box) and *Filter* (*Object Selection Filter* dialog box). *Qselect* is simpler to use than *Filter*.

Both object selection filters allows you to specify a criteria based on object type and/or object property. In other words, you specify what type of object (*Line, Arc, Mtext,* etc.) and/or what property (*Layer, Color, Linetype,* etc.) and AutoCAD searches (applies the filter to) the drawing to find objects that match the criteria. When the objects are found, they can be used for some editing action, such as *Move, Copy, Erase,* or change properties.

The *Quick Select* dialog can be invoked by typing *Qselect* or is accessible through other dialog boxes, such as the *Properties* palette. *Qselect* should be used immediately before an editing command to locate the desired objects you want to edit. *Qselect* is <u>not</u> transparent (cannot be used during another command, except when it is accessible from other dialog boxes).

The *Object Selection Filter* dialog box can be invoked by typing *Filter*. *Filter* is more advanced than *Qselect* and can be used to specify a more complex filter criteria. Unlike *Qselect*, *Filter* is transparent, so it can be used during another command by typing *'Filter* (with an apostrophe prefix).

QSELECT

	Pull-down Menu	COMMAND (TYPE)	ALIAS (TYPE)	Short-cut	Screen (side) Menu	Tablet Menu
	Tools *Quick Select...*	*QSELECT*	...	(Default Menu) *Quick Select...*	...	X,9

With *Qselect*, you create a selection set of all objects that either match or do not match the object type and object property criteria that you specify. You can search the entire drawing or an existing selection set. If you use *Qselect* with a partially opened drawing, objects that are not loaded are not considered in the search. When the *OK* button is pressed in the *Quick Select* dialog box, AutoCAD finds and highlights the objects matching the criteria. The objects become the current selection set so any editing command used immediately after *Qselect* (by Noun/Verb selection) automatically finds the matching selection set.

Qselect is the simpler of the two available object selection filters. Although you cannot name and save selection sets with *Qselect*, you can search and find objects by *Object type* and *Properties* much easier than using the *Object Selection Filters* dialog box (explained in detail in the following section).

Qselect can be invoked by several methods listed in the command table above, including choosing *Quick Select* from the shortcut menu that appears when you right-click in the drawing area. The *Qselect* button also appears in the *Properties* palette (see Figure 16-7). *Qselect* produces the *Quick Select* dialog box (Fig. 20-4).

To use the *Quick Select* dialog box, follow these steps:

1. Invoke the *Quick Select* dialog box by any method described above.

2. Specify the *Entire Drawing* to search or use the *Select Object* button to specify a smaller search area.

3. If you want to find one *Object Type*, select it from the drop-down list; otherwise *Multiple* will be used.

4. Select the desired *Object Property* (only one) to search for (unless you need only to search by *Object Type*).

5. Specify the *Operator* and *Value*.

6. Specify if you want to *Include in New Selection Set* (search for the object type and property you specified) or *Exclude from New Selection Set* (search for everything but the object type and property you specified).

7. Press *OK*. AutoCAD highlights all objects that match the criteria.

8. Use an editing command to act on the highlighted objects (*Move, Erase*, change properties, etc.)

Figure 20-4 —————————

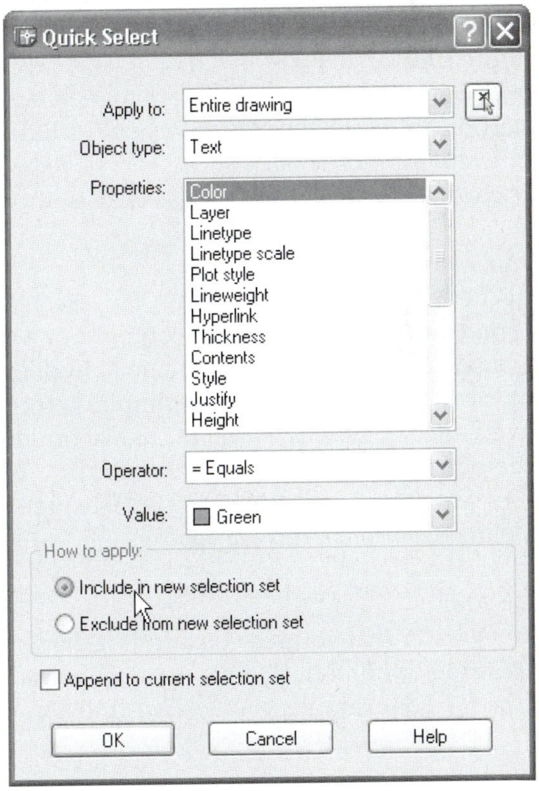

For example, assume the criteria shown in Figure 20-4 were applied to a process flow diagram, AutoCAD would find and highlight all *Text* objects in the drawing whose *Color Equals Green* (Fig. 20-5). Note the Grips (small squares) that appear on the highlighted objects when Grips are enabled.

Figure 20-5

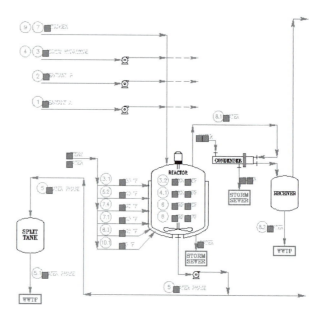

The specific areas of the dialog box are explained as follows.

Apply to

Specify whether you want to apply the filtering criteria to the entire drawing or the current selection set. If there is no current selection set when you invoke *Qselect*, *Entire Drawing* is the default value. Use the *Select Objects* button to return to the drawing and create a selection set.

Select Objects

The *Select Objects* button (upper right corner) temporarily returns to the drawing so you can select the objects on which you want to perform the *Quick Select* operation. AutoCAD uses the new selection set as the area to search for objects matching the specified criteria.

Object Type

Specifies the object type you want to search for (filter). The default is *Multiple*; therefore, if no previous selection set exists, the *Object Type* list includes all and only object types available in the <u>current drawing</u>. For example, if only *Lines* and *Circles* have been created in the drawing, only *Line* and *Circle* object types appear in the list. If a selection set exists, the list includes only the object types of the selected objects.

Properties

This field is specific to the *Object Type* selected in the list above. The *Properties* list <u>includes all searchable properties for the selected object type</u>. For some object types, the list of properties can be very long; for example, dimension objects contain a property field for each dimensioning variable. This feature gives you considerable power. For example, you could use this tool to highlight all *Linear Dimensions* that have a specific type of *Tolerance Display* or find all *Text* that matches a specific *Height* value. The sort order for the properties list (alphabetic or categorized) is based on the current sort order in the *Object Properties* palette (see "Properties" in Chapter 16). The property you select determines the options available in the *Operator* and *Value* boxes.

Operator

The *Operator* that you select controls if the search *Equals*, is *Not Equal To*, *Greater Than*, or *Less Than* the text string or numerical value in the *Value* field. Options in the *Operator* field vary depending on the *Property*. For example, *Greater Than* and *Less Than* are not available for some properties.

Value

Possibilities in this field depend on the selected object and property. If known values for the selected property are stored in the drawing, the *Value* box becomes a drop-down list from which you can choose an available value. For example, if *Color* is selected in the *Properties* list, *ByLayer*, *ByBlock*, *Red*, *Yellow*, and so on appear in the list. If no choices appear in the *Value* field, input a text string or numerical value.

How To Apply

Your choice here specifies whether you want the new selection set to include or exclude objects that match the specified filtering criteria. Choose *Include in new selection set* if you want to create a new selection set composed only of objects that <u>match</u> the filtering criteria. Choose *Exclude from new selection set* to create a new selection set composed of all objects that <u>do not match</u> the filtering criteria.

OK

After specifying the criteria for the objects to search in the fields described above, select *OK*. All objects matching (or not matching) the criteria set in the *Quick Select* dialog box become highlighted in the current drawing. These objects are treated as if you had selected them by the pickbox, window, or other selection method. Assuming Noun/Verb selection is enabled (*PICKFIRST*=1), the next editing command issued automatically finds the *Qselect* objects.

Append to Current Selection Set

Although you can only select one specific object type and one property at a time, you can use *Qselect* <u>repeatedly</u> to search for multiple object types and properties. Do this by specifying one set of criteria, then select *OK*. AutoCAD highlights all objects matching that criteria. Next, <u>while the objects are highlighted</u>, right-click to select *Qselect* again. Specify a second set of criteria and include *Append to current selection set*. When you press *OK*, AutoCAD highlights objects matching the new criteria and "adds" them to the original highlighted set.

FILTER

Pull-down Menu	COMMAND (TYPE)	ALIAS (TYPE)	Short-cut	Screen (side) Menu	Tablet Menu
...	*FILTER*	*FI*	...	*ASSIST* *Filters*	...

The *Object Selection Filters* dialog box is a more advanced version of *Qselect*. Here you can specify multiple selection criteria and save a criteria set under a name that you assign for use at a later time. You can also invoke *Filter* during a command (transparently) to select objects for use with the current command. Using the *Filter* command by any method produces the *Object Selection Filters* dialog box. Figure 20-6 displays the dialog box when it first appears. The *Select Filter* cluster near the left center is where you specify the search criteria. After you designate the filters and values, use *Add to List* to cause your choices to appear in the large area at the top of the dialog box. All filters appearing at the top of the box are applied to the drawing when *Apply* is selected.

Figure 20-6

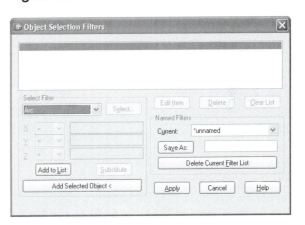

Select Filter (Drop-Down List)

Activating the drop-down list reveals the possible selection criteria (filters) that you can use, such as *Arc* or *Layer*. You can select one or more of the selection filters shown in the following list. You can also specify a set of values that applies to each filter, for example, *Arcs* with a *radius* of less than 1.00" or *Layers* that begin with "AR."

Arc	Attribute Position	Block Position
Arc Center	Attribute Tag	Block Rotation
Arc Radius	Block	Body
Attribute	Block Name	Circle

Circle Center	Line End	Spline
Circle Radius	Linetype	Text
Color	Linetype Scale	Text Height
Dimension	Multiline Style	Text Position
Dimension Style	Multiline	Text Value
Elevation	Normal Vector	Text Rotation
Ellipse	Point Position	Text Style Name
Ellipse Center	Point	Thickness
Hatch	Polyline	Tolerance
Hatch Pattern Name	Ray	Trace
Image	Region	3dface
Image Position	Shape Position	Viewport
Layer	Shape Name	Viewport Center
Leader	Shape	Xdata
Line	Solid	Xline
Line Start	Solid Body	

The list also provides a series of typical database grouping operators which include AND, OR, XOR, and NOT. For example, you may want to select all *Text* in a drawing but *NOT* the *Text* in a specific text style.

Select

If you select a filter that has multiple values, such as *Layer, Block Name,* and *Text Style Name,* the *Select* button is activated and enables you to select a value from the list (Fig. 20-7). If the *Select* button is grayed out, no existing values are available.

Figure 20-7

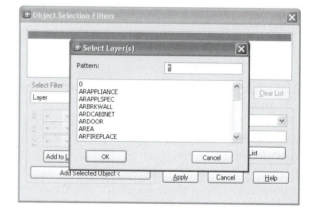

X

If you select a filter that requires alphanumeric input (numbers or text strings), the X field is used. For example, if you selected *Arc Radius* as a filter, you would select the operator (=, <, >, etc.) from the first drop-down list and key in a numeric value in the edit box (Fig. 20-8). Other examples requiring numeric input are *Circle Radius, Text Rotation,* and *Text Height.*

If you select a filter that requires a text string, the string is entered in the X field edit box. Examples are *Attribute Tag* and *Text Value.*

The X field is also used (in conjunction with the Y and Z fields) for filters that require coordinate input in an X,Y,Z format. Example filters include *Arc Center, Block Position,* and *Line Start.*

Figure 20-8

Y,Z
These fields are used for coordinate entry of the Y and Z coordinates when needed as one of the criteria for the filter (see "X" on the previous page). Additionally, you may use standard database relational operators such as equal, less than, or greater than, to more effectively specify the desired value.

Add to List
Use the *Add to List* button after selecting a filter to force the new selection to appear in the list at the top of the dialog box.

Substitute
After choosing a value from the *Select* listing or entering values for X, Y, and Z, *Substitute* replaces the value of the <u>highlighted</u> item in the list (at the top of the dialog box) with the new choice or value.

Add Selected Object
Use this button to select an object from the drawing that you want to include in the filter list.

Edit Item
Once an item is added to the list, you must use *Edit Item* to change it. First, highlight the item in the filter list (on top); then select *Edit Item*. The filter and values then appear in the *Select Filter* cluster ready for you to specify new values in the X,Y,Z fields or you can select a new filter by using the *Select* button.

Delete
This button simply deletes the highlighted item in the filter list.

Clear List
Use *Clear List* to delete all filters in the existing filter list.

Named Filters
The *Named Filters* cluster contains features that enable you to save the current filter configuration to a file for future use with the current drawing. This eliminates the need to rebuild a selection filter used previously.

Current
By default, the current filter list of criteria is *unnamed*, in much the same fashion as the default dimension style. Once a filter list has been saved by name, it appears as a selectable object filter configuration in the drop-down list of named filters.

Save As
Once object filter(s) have been specified and added to the list (on top), you can save the list by name. First, enter a name for the list; then PICK the *Save As* button. The list is automatically set as the current object filter configuration and can be recalled in the future for use in the drawing.

Delete Current Filter List
Deletes the named filter configuration displayed in the *Current* field.

Apply
The *Apply* button applies the current filter list to the current AutoCAD drawing. Using *Apply* causes the dialog box to disappear and the "Select objects:" prompt to appear. <u>You must specify a selection set (by any method) to be tested against the filter criteria.</u> If you want the filters to apply to the entire drawing (except *Locked* and *Frozen* layers), enter *ALL*. If you only need to search a smaller area of the drawing, PICK objects or use a window or other selection method.

The application of the current filters can be accomplished in two ways:

1. Within a command (transparently)
   At the "Select objects:" prompt, you can invoke *Filter* transparently by selecting from a menu or typing "*'filter*" (prefaced by an apostrophe). Then set the desired filters, select *Apply*, and specify the selection set to test against the criteria. The resulting (filtered) selection set is used for the current command.

2. At the Command prompt
   From the Command: prompt (when no command is in use), type or select *Filter*, specify the filter configuration, and then use *Apply*. You are prompted to "Select objects:". Specify *All* or use a selection method to specify the area of the drawing for the filter to be applied against. Press Enter to complete the operation. The filtered selection set is stored in the selection set <u>buffer</u> and can be recalled by using the *Previous* option in response to the next "Select objects:" prompt.

OBJECT GROUPS

Figure 20-9

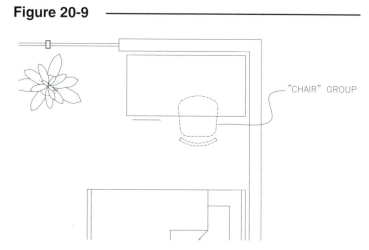

"CHAIR" GROUP

Often, a group of objects that are related in some way in a drawing may require an editing action. For example, all objects (*Lines* and *Arcs*) representing a chair in a floor plan may need to be selected for *Copying* or changing color or changing to another layer (Fig. 20-9). Or you may have to make several manipulations of all the fasteners in a mechanical drawing or all of the data points in a civil engineering drawing. In complex drawings containing a large number of objects, the process of building such a selection set (with the pickbox, window, or other methods) may take considerable effort. One disadvantage of the traditional selection process (without *Groups*) is that selection sets could not be saved and recalled for use at a later time. AutoCAD introduced the *Group* command to identify and organize named selection sets.

A <u>Group</u> is a <u>set of objects</u> that has an <u>assigned name</u> and description. The *Group* command allows you to determine the objects you want to include in the group and to assign a name and description. Usually the objects chosen to participate in a group relate to each other in some manner. Objects can be members of more than one group.

Once a group is created, it can be assigned a <u>*Selectable*</u> status. If a group is *Selectable*, the entire group can be automatically selected (highlighted) at any "Select objects:" prompt by PICKing any individual member, or by typing the word "group" or the letter "G," then giving the group name. For example, if all the objects in a mechanical drawing representing fasteners were assigned to a group named "fasten," the group could be selected during the *Move* command as follows:

```
Command: move
Select objects: g
Enter group name: fasten
18 found
Select objects: Enter
Specify base point or displacement:
        etc.
```

If the group is assigned a <u>nonselectable</u> status, the group <u>cannot</u> be selected as a whole by any method. Nonselectablility prevents accidentally *Copying* the entire group, for example. If a group is assigned a <u>selectable</u> status, the group can be manipulated as a whole <u>or</u> each member can be edited individually. The *PICKSTYLE* variable controls whether PICKing highlights the individual member or the entire group. In other words, you can PICK individual members if *PICKSTYLE*=0 <u>or</u> an entire selectable group if *PICKSTYLE*=1. The Ctrl+H key sequence toggles the *PICKSTYLE* variable.

A group can be thought of as a set of objects that has a level of distinction beyond that of a typical selection set (specified "on the fly" by the pickbox, window, or other method), but not as formal as a *Block*. A *Block* is a group of objects that is combined into <u>one object</u> (using the *Block* command), but individual entities <u>cannot</u> be edited separately. (Blocks are discussed in Chapter 21.) Using groups for effective drawing organization is similar to good layer control; however, with groups you have the versatility to allow members of a group to reside on different layers.

To summarize, using groups involves two basic activities: (1) use the *Group* command to define the objects and assign a name and (2) at any later time, locate (highlight) the group for editing action at the "Select objects:" prompt by PICKing a member or typing "G" and giving its name.

GROUP

	COMMAND (TYPE)	ALIAS (TYPE)	Short-cut	Screen (side) Menu	Tablet Menu
Pull-down Menu					
...	GROUP or -GROUP	G or -G	...	ASSIST Group	X,8

Invoking the *Group* command by any method produces the *Object Grouping* dialog box (Fig. 20-10). (The hyphen can be entered as a prefix, such as "-*group*," to present the command line equivalent.) The main purposes of this interface are to <u>create</u> groups and <u>change</u> the properties of existing groups.

The focus of the dialog box is the *Group Name* list (on top), which gives the list of existing groups and indicates each group's selectable status. If a group is <u>selectable</u>, the entire group can be highlighted by typing the word "group" or letter "G" at the "Select objects:" prompt, then entering the group name. Alternately, you can PICK one member to highlight the entire group, or use Ctrl+H to PICK only one member. A group that is <u>not selectable</u> cannot be selected as a whole.

Typically, your first activity with the *Object Grouping* dialog box is to create a new group.

Figure 20-10

Creating a Named Group

1. Enter the desired name in the *Group Name:* edit box. This must be done as the <u>first</u> step.

2. Enter a unique description for the group in the *Description* edit box. Use up to 256 characters to describe the relationship or characteristic features of the group.

3. Determine if the group is to be selectable or not and indicate so in the *Selectable* checkbox.

4. Use the *New* button to PICK the desired members (objects) to be included in the group. The dialog box is temporarily hidden until all desired objects are selected. Once the selection set is chosen (press Enter), the dialog box reappears with the new group added to the list.

All of the options and details of the *Group Identification* and *Create Group* clusters are explained here:

Group Name:

The *Group Name:* edit box (just below the list) is used to enter the name of a new group that you want to create or to display the name of an existing group selected from the list above that requires changing. When creating a new group, enter the desired name before choosing the set of objects (using the *New* tile).

Description

Including a *Description* for new groups is optional, but this feature is useful in organizing various selection sets in a drawing. A higher level of distinction is added to the groups if a description is included. The description may be up to 256 characters. If a group name is selected from the list, both the group name and description appear in the *Group Identification* cluster.

Find Name

Use this option to find the group name for any object in the drawing. (The name currently appearing in the *Group Name:* edit box does not have an effect on this option.) Using the *Find Name* tile temporarily removes the *Object Grouping* dialog box and prompts you to "Pick a member of a group." After doing so, AutoCAD responds with the names of the group(s) of which the selected object is a member (Fig. 20-11). If an object is PICKed that is not assigned to a group, AutoCAD responds with "Not a group member." This option can be used immediately upon entering the *Object Grouping* dialog box, with no other action required.

Figure 20-11 ───────────

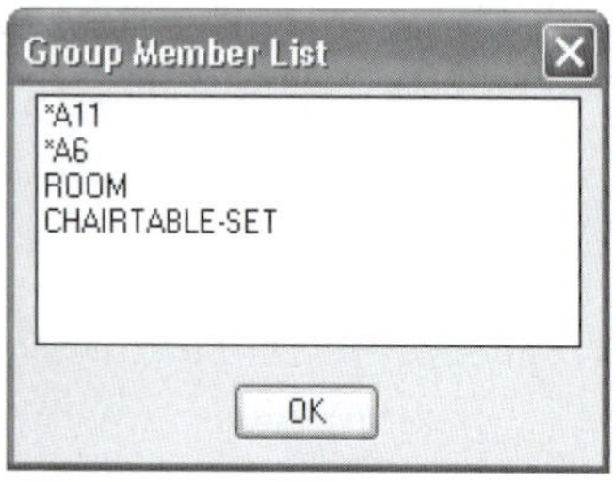

Highlight

Choosing this tile highlights all members of the group appearing in the *Group Name:* box. The dialog box temporarily disappears so the drawing can be viewed. Use this to verify which objects are members of the current group.

Include Unnamed

When a group is copied (with *Copy, Mirror*, etc.), the objects in the new group have a collective relationship like any other group; however, the group is considered an "unnamed group." Toggling the *Include Unnamed* checkbox forces any unnamed groups to be included in the list above (Fig. 20-12). In this way, you can select and manipulate copied groups like any other group but without having to go through the process to create the group from scratch. The copied group is given a default name such as *An, where n is the number of the group created since the beginning of the drawing. You can also create an unnamed group using the *Unnamed* toggle and the *New* tile. An unnamed group can be *Renamed*.

Figure 20-12 ───────────

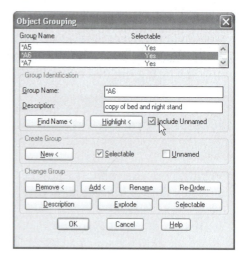

New

The *New* tile in the *Create Group* cluster is used to select the objects that you want to assign as group members. You are returned to the drawing and prompted to "Select objects:". (See "Creating a Named Group.")

Selectable

Use this toggle to specify the selectable status for the new group to create. If the check appears, the new group is created as a *Selectable* group. A selectable group can be selected as a whole by <u>two methods</u>. When prompted to "Select objects:" during an editing command, you can (1) PICK any individual member of a selectable group or (2) enter "G" and the group name.

When using the PICK method, the *PICKSTYLE* variable controls whether an individual member is selected or the entire group. When *PICKSTYLE* is set to 1, PICKing any member of a <u>selectable</u> group highlights (selects) the entire group. If *PICKSTYLE* is set to 0, PICKing any member of a selectable group highlights <u>only</u> that member. The Ctrl+H key sequence automatically changes the *PICKSTYLE* variable setting to 1 (*Group on*) or 0 (*Group off*). Therefore, you can PICK one member (of a selectable group) or the entire group by toggling Ctrl+H. *PICKSTYLE* has no effect on nonselectable groups.

NOTE: In releases of AutoCAD previous to 2004, Ctrl+A toggles the *PICKSTYLE* variable. In AutoCAD 2004, Ctrl+A selects all objects in the drawing.

Figure 20-13 ────────

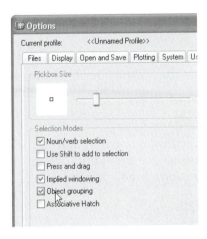

The *PICKSTYLE* variable can also be changed from the *Selections* tab of the *Options* dialog box. Checking the *Object Grouping* checkbox sets *PICKSTYLE* to 1 (Fig. 20-13). (*PICKSTYLE* also affects the selection of associative hatch objects. See Chapter 26 and Appendix A.) The *PICKSTYLE* variable is saved with the drawing file.

For drawings that contain objects that are members of multiple groups, normally the *Selectable* status of the groups (or all but one group) should be toggled off to avoid activating all groups when a member is selected. Groups or group members that are on *Frozen* layers cannot be selected.

Unnamed
If you want to create a group with a default name (*A*n, where *n* is the number of the group created since the beginning of the drawing), use this option. If the check appears in the checkbox, you cannot enter a name in the *Group Name* edit box. Use the *New* tile to select the objects to include in the unnamed group. If you keep unnamed groups, toggle *Include Unnamed* on to include these groups in the list. An unnamed group can be *Renamed*.

Typically, you should name groups that you create. When you make a copy of an existing group, assign a new name (with *Rename*) if changes are made to the members. However, if you keep an unnamed group that is an identical copy, a good strategy is to keep the default (*A*n) name as an identifier and add a description like "copy of fasten group." It is suggested that unnamed groups be <u>nonselectable</u> to avoid accidentally selecting the entire group by PICKing one object for a *Copy* or other operation. Remember, you can always change the selectable status when you want to access the entire unnamed group.

Once a group exists, it can be changed in a number of ways. The options for changing a group appear in the *Change Group* cluster of the *Object Grouping* dialog box.

Changing Properties of a Group

1. Select the group to change from the list on top of the *Object Grouping* dialog box (See Figure 20-12). Once a group is selected, the name and description appear in the related edit boxes. A group must be selected before the *Change Group* cluster options are enabled.

2. Select the desired option from the choices in the *Change Group* area.

Remove
The *Remove* tile allows you to <u>remove any member</u> of the specified group. The dialog box disappears temporarily and you are prompted to "Remove objects." If all members of a group are removed, the object group still remains defined in the list box.

Add

Use this option to add new members to the group appearing in the *Group Name:* edit box. The "Select objects:" prompt appears.

Rename

First, select the desired group to rename from the list. Next, type over the existing name in the *Group Name:* edit box and enter the desired new name. Then choose the *Rename* button.

Re-order

See "Re-ordering Group Members" at the end of this section.

Description

You can change the description for the specified object group with this option. First, select the desired group from the list. Next, edit the contents of the *Description* edit box. Then, select the *Description* tile.

Explode

Use this option to remove a group definition from the drawing. The *Explode* option breaks the current group into its original objects. The group definition is removed from the object group list box; however, the individual group members are not affected in any other way.

Selectable

This option changes the selectable status for a group. Highlight a group name from the list; then choose the *Selectable* tile. The *Yes* or *No* indicator in the Selectable column of the list toggles as you click the tile. See *"Selectable"* in the discussion of the *Create Group* cluster earlier.

Re-ordering Group Members

When you select objects to be included in new groups, AutoCAD assigns a number to each member. The number corresponds to the sequence that you select the objects. For special applications, the number associated with individual members is critical. For example, a group may be formed to generate a complete tool path where each member of the group represents one motion of the sequence.

The *Re-order* button (in the *Object Grouping* dialog box) invokes the *Order Group* dialog box (Fig. 20-14). This dialog box allows you to change the order of individual group members. To reverse the order, simply select *Reverse Order*. To re-order members of a group to another sequence, use these steps:

Figure 20-14 ——————————

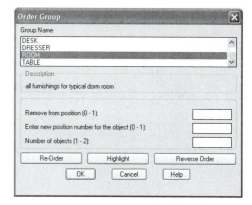

1. In the *Order Group* dialog box, select the group to re-order from the list on top.

2. Select the *Highlight* option to display each group member one at a time. In the small dialog box that appears, select *Next* repeatedly to go completely through the current sequence of group members. The sequence begins with 0, so the first object has position 0. Then select OK to return to the *Order Group* dialog box.

3. In the *Number of Objects* edit box, enter the number of objects that you want to reposition.

4. In the *Remove from Position* edit box, enter the current position of the object to re-order.

5. Enter the new position in the *Enter new position number* edit box.

6. Select *Re-order*.

7. Select *Highlight* again to verify the new order.

Operation of this dialog box is difficult and is not explained well in the documentation. It is possible to accidentally remove or duplicate members from the group using this device. A more straightforward and stable alternative for re-ordering is to *Explode* the group and make a new group composed of the same members, but PICK the members <u>in the desired order</u>.

Groups Examples

Imagine that you have the task to lay out the facilities for a college dormitory. Using the floor plan provided by the architect, you begin to go through the process of drawing furniture items to be included in a typical room, such as a desk, chair, and accessory furnishings. The drawing may look like that in Figure 20-15.

Figure 20-15

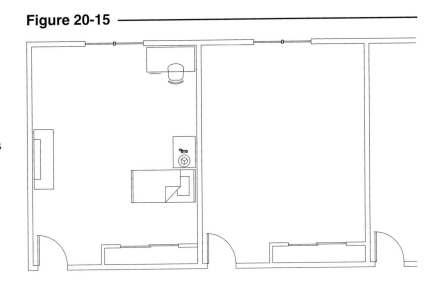

Since each room is to house two students, all objects representing each furniture item must be *Copied*, *Mirrored*, or otherwise manipulated several times to complete the layout for the entire dorm. To *Copy* the chair, for example, by the traditional method (without creating groups), each of the 12 objects (*Lines* and *Arcs*) comprising the chair must be selected <u>each time</u> the chair is manipulated. Instead, combining the individual objects into a group named "CHAIR" would make selection of these items for each operation much easier.

To make the CHAIR group, the *Group* command is used to activate the *Object Grouping* dialog box. The CHAIR group is created by selecting all *Lines* and *Arcs* as members (see previous Figure 20-9). The group is made *Selectable* as shown in Figure 20-16.

Now, any time you want to *Copy* the chair, at the "Select objects:" prompt enter the letter "G" and "CHAIR" to cause the entire group to be highlighted. Alternately, the CHAIR group can be selected by PICKing any *Line* or *Arc* on the chair since the group is selectable.

Figure 20-16

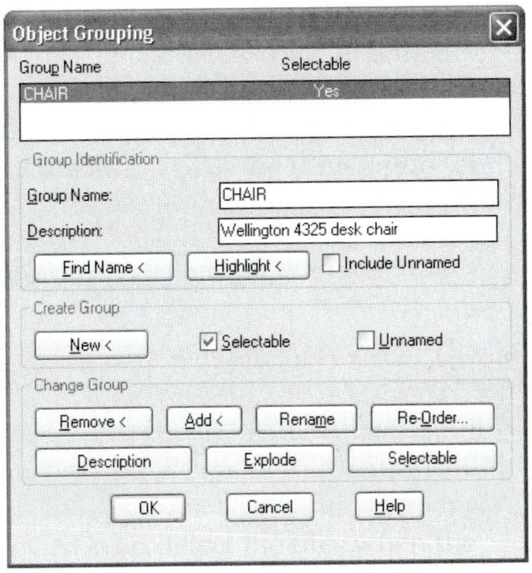

Object Grouping

Group Name	Selectable
CHAIR	Yes

Group Identification

Group Name: CHAIR

Description: Wellington 4325 desk chair

Find Name <   Highlight <   ☐ Include Unnamed

Create Group

New <   ☑ Selectable   ☐ Unnamed

Change Group

Remove <   Add <   Rename   Re-Order...

Description   Explode   Selectable

OK   Cancel   Help

The same procedure is used to combine several objects into a group for each furniture item. The complete suite of groups may appear as shown in the dialog box shown in Figure 20-17. Toggling on *Include Unnamed* forces the unnamed group to appear in the list. The unnamed group was automatically created when the CHAIR group was copied.

Using the named groups, completing the layout is simplified. Each group (CHAIR, DESK, etc.) is selected and *Copied* to complete the room layout.

Figure 20-17

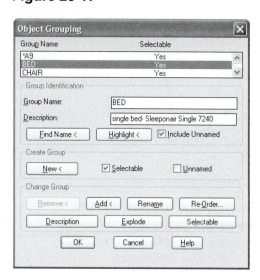

The next step in the drawing is to lay out several other rooms. Assuming that other rooms will have the same arrangement of furniture, another group could be made of the <u>entire room</u> layout. (Objects can be members of more than one group.) A group called ROOM is created in the same manner as before, and all objects that are part of furniture items are selected to be members of the group. If the ROOM group is made selectable, all items can be highlighted with one PICK and copied to the other rooms (Fig. 20-18).

Figure 20-18

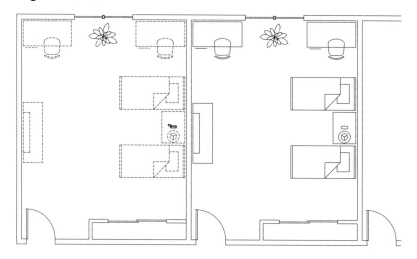

After submitting the layout to the client for review, a request is made for you to present alternative room layouts. Going back to the drawing, you discover that each time you select an individual furniture item to manipulate, the entire ROOM group is highlighted. To remedy the situation, the ROOM group is changed to <u>nonselectable</u> with the *GROUP* command. Now the individual furniture items can be selected since the original groups (CHAIR, DESK, BED, etc.) are still *Selectable* (Fig. 20-19).

Figure 20-19

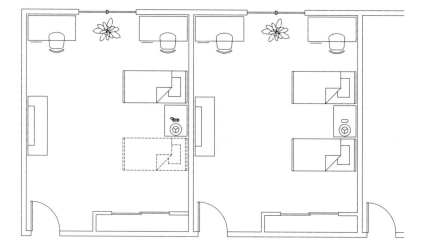

Another, more typical application is the use of *Groups* in conjunction with *Blocks*. A block is a set of objects combined into one object with the *Block* command. A block always behaves as <u>one object</u>. In our dorm room layout, each furniture item could be *Inserted* as a previously defined block. The entire set of blocks (bed, chair, desk, etc.) could be combined into a group called ROOM. In this case, the ROOM group would be set to *Selectable*. This would allow you to select <u>all</u> furniture items as a group by PICKing only one. However, if you wanted to select only <u>one</u> furniture item (Block), use Ctrl+H to change *PICKSTYLE* to 0. In this way, individual members of any group (blocks in this case) can be PICKed without having to change the *Selectable* status. See Chapter 21 for more information on *Blocks*.

CHAPTER EXERCISES

1. *PICKFIRST*

   Check to ensure that ***PICKFIRST*** is set to **1** (*ON*). Also make sure *GRIPS* are set to 0 (*OFF*) by typing ***GRIPS*** at the command prompt or invoking the *Selection* tab of the *Options* dialog box. For each of the following editing commands you use, make sure you PICK the objects FIRST, then invoke the editing command.

   A. *Open* the **PLATES** drawing that you created as an exercise in Chapter 10 (not the PLATNEST drawing). *Erase* 3 of the 4 holes from the plate on the left, leaving only the hole at the lower-left corner (**PICK** the *Circles*; then invoke *Erase*). Create a *Rectangular Array* with **7** rows and **5** columns and **1** unit between each hole (PICK the *Circle* before invoking *Array*). The new plate should have 35 holes as shown in Figure 20-20, plate A.

   Figure 20-20

   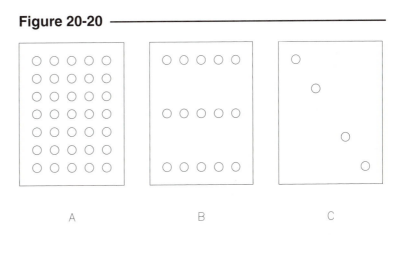

   A   B   C

   B. For the center plate, **PICK** the 3 holes on the <u>vertical</u> center; then invoke *Copy*. Use the *Multiple* option to create the additional 2 sets of 3 holes on each side of the center column. Your new plate should look like that in Figure 20-20, plate B. Use *SaveAs* to assign a new name, **PLATES2**.

   C. For the last plate (on the right), **PICK** the two top holes and *Move* them upward. Use the *Center* of the top hole as the "Specify base point or displacement, or [Multiple]:" and specify a "Specify second point of displacement or <use first point as displacement>:" as **1** unit from the top and left edges. Compare your results to Figure 20-20, plate C. *Save* the drawing (as **PLATES2**).

   Remember that you can leave the *PICKFIRST* setting *ON* always. This means that you can use both Noun/Verb or Verb/Noun editing at any time. You always have the choice of whether you want to PICK objects first or use the command first.

2. *PICKADD*

Figure 20-21

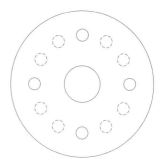

Change *PICKADD* to **0** (*OFF*). Remember that if you want to PICK objects and add them to the existing highlighted selection set, hold down the **SHIFT** key; then **PICK**.

A. *Open* the **ARRAY1** drawing. (The settings for *GRIPS* and *PICK-FIRST* have not changed by *Opening* a new drawing.) *Erase* the holes (highlighted *Circles* shown in Figure 20-21) leaving only 4 holes. Use either Noun/Verb or Verb/Noun editing. *SaveAs* **FLANGE1**.

B. *Open* drawing **ARRAY2**. Select all the holes and *Rotate* them 180 degrees to achieve the arrangement shown in Figure 20-22. Any selection method may be used. Try several selection methods to see how *PICKADD* reacts. (The *Fence* option can be used to select all holes without having to use the **Shift** key.) *SaveAs* **FLANGE2**. Change the *PICKADD* setting back to **1** (*ON*).

Figure 20-22

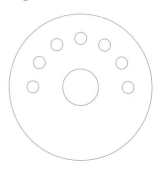

3. *PICKDRAG*

Figure 20-23

Open the **CH11EX3** drawing. Set *PICK-DRAG* to **1** (*ON*). Use *Stretch* to change the front and side views to achieve the new base thickness as indicated in Figure 20-23. (Remember to hold down the PICK button when dragging the mouse or puck.) When finished, *SaveAs* **HOLDR-CUP**. Change the *PICKDRAG* setting back to **0** (*OFF*).

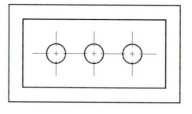

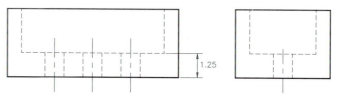

4. *Qselect*

This exercise involves using object selection filters to change the properties of objects in the drawing. *Open* the **DB_SAMP.DWG** from the AutoCAD 2004\Sample folder.

A. Activate the *Model* tab. Notice that all the doors in the drawing are green in color. Using *List*, verify that any door is on the E-F-Door layer.

B. Invoke *Qselect* by any method. Ensure that *Entire Drawing* appears in the *Apply to:* section. In the *Object Type* section, select *Block Reference* from the drop-down list. In *Properties, Operator* and *Value*, select *Layer, Equals,* and **E-F-DOOR**, respectively. Ensure *Include in new selection set* is checked and *Append to current selection set* is not checked. Then select the **OK** button. All the doors should be highlighted. The command line reports "39 items selected."

C. With the doors highlighted, use the Layer Control drop-down list and select layer **0**. All the doors are now "moved" to layer 0 and therefore change to black or white (the *ByLayer* color assignment) depending on your background color. Press **Escape** to clear the highlighted items and grips (if enabled).

Figure 20-24

D. Invoke the *Properties* palette (not shown). Use the *Quick Select* button to produce the *Quick Select* dialog box. In the dialog box (Fig. 20-24), ensure that *Entire Drawing* appears in the *Apply to:* section. In the *Object Type* section, select *Multiple* from the drop-down list. In *Properties*, *Operator* and *Value*, select *Layer*, *Equals*, and **E-F-STAIR**, respectively. Ensure *Include in new selection set* is checked and *Append to current selection set* is not checked. Then select the *OK* button. The stairs should be highlighted and the command line reports "217 items selected."

E. In the *Properties* palette, select the *Color* property and select *Green*. Click in the drawing area and press **Escape** to deselect the stair objects. The stairs should now be green.

F. *Close* the drawing and <u>do not save</u> the changes.

5. *Filter*

Figure 20-25

In this exercise, you will use the *Filter* command for several operations.

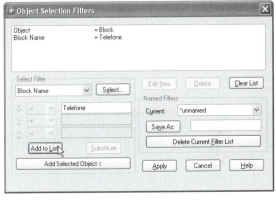

A. *Open* the **EXPO 98 maps.DWG** (download this drawing from the www.mhhe.com/leach Web site). Select the *Model* tab. The map indicates information about services such as telephone locations, transportation, restaurants, and so on.

B. Use the *Object Selection Filter* dialog box to count the number of telephones. Invoke *Filter*. When the dialog box appears, select *Block Name* from the *Select Filter* section drop-down list. Then, click on the *Select* button to select from the list of Blocks in the drawing. Select **TELEFONE** from the bottom of the list, then pick *OK*. In the *Object Selection Filter* dialog box, select *Add to List*. Two entries in the filter list above should appear (Fig. 20-25).

C. Select the *Apply* button. The dialog box disappears and the command line prompts to "Select objects:". Enter *All* so the filter is applied to the entire drawing. AutoCAD reports the number of (all) objects found and the number that were filtered out (not telephones). At the "Select objects:" prompt, press **Enter**. Although you have exited the *Filter* command, the objects remain highlighted.

D. Use the *Select* command to do the arithmetic. Type *Select* at the command prompt. AutoCAD reports "75 found."

E. Next, you will change the cafeteria symbols to layer 0. Do this by typing *Change* at the command line. At the "Select objects:" prompt, enter *'Filter* (with the apostrophe prefix). When the dialog box appears, highlight the existing items (telephone blocks) from the filter list and select *Delete* to remove them. Next, go through the process (as before) to create a filter for *Block Names* called **CAFETARIA**. Make sure you *Add to List*. Finally, select *Apply* to exit the dialog box. At the "Select objects:" prompt, type *All* so AutoCAD searches the entire drawing.

F. Press **Enter** at the next "Select objects:" prompt. Remember, you are still in the *Change* command, so AutoCAD prompts to "Specify change point or [Properties]:". Type *P* for the *Properties* option. Enter *LA* for *Layer* and specify layer **0** as the new layer name. Complete the *Change* command by pressing **Enter**.

G. Since the CAFETARIA objects have been moved to layer 0, use the *Layer* command to make layer **0** *Current*, then right-click to *Select All* and *Freeze* all layers. AutoCAD reports that the current layer (0) cannot be frozen. Pick *OK* to exit the *Layer Properties Manager*. The cafeteria symbols should be visible.

H. *Close* the drawing, but do not save the changes.

6. **Object Groups**

Figure 20-26 ───────────────

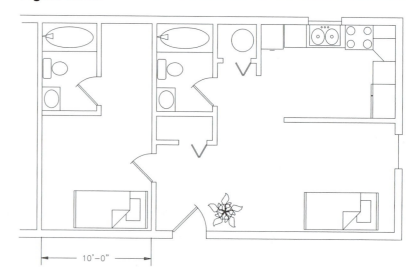

A. *Open* the **EFF-APT2** drawing that you last worked on in Chapter 18 exercises. Create a *Group* for each of the three bathroom fixtures. *Name* one group **WASHBASIN**, give it a *Description*, and include the *Ellipse* and the surrounding *Rectangle* as its members. Make two more groups named **WCLOSET** and **BATHTUB** and select the appropriate members. Next, combine the three fixture groups into one *Group* named **BATHROOM** and give it a description. Then, draw a bed of your own design and combine its members into a *Group* named **SINGLEBED**. Make all five groups *Selectable*. Use *SaveAs* to save and rename the drawing to **2BR-APT**.

B. A small bedroom and bathroom are to be added to the apartment, but a 10' maximum interior span is allowed. One possible design is shown in Figure 20-26. Other more efficient designs are possible. Draw the new walls for the addition, but design the bathroom and bedroom door locations to your personal specifications. Use *Copy* or *Mirror* where appropriate. Make the necessary *Trims* and other edits to complete the floor plan.

C. Next, *Copy* the bathroom fixture groups to the new bathroom. If your design is similar to that in the figure, the BATHROOM group can be copied as a whole. If you need to *Copy* only individual fixtures, take the appropriate action so you don't select the entire BATHROOM group. *Copy* or *Mirror* the **SINGLEBED**. Once completed, activate the *Object Grouping* dialog box and toggle on *Include Unnamed*. Can you explain what happened? Assign a *Description* to each of the unnamed groups and *Save* the drawing.

21

BLOCKS, DesignCenter, and Tool Palettes

Chapter Objectives

After completing this chapter you should:

1. understand the concept of creating and inserting symbols in AutoCAD drawings;

2. be able to use the *Block* command to transform a group of objects into one object that is stored in the current drawing's block definition table;

3. be able to use the *Insert* and *Minsert* commands to bring *Blocks* into drawings;

4. know that *color, linetype* and *lineweight* of *Blocks* are based on conditions when the *Block* is made;

5. be able to convert *Blocks* to individual objects with *Explode*;

6. be able to use *Wblock* to prepare .DWG files for insertion into other drawings;

7. be able to redefine and globally change previously inserted *Blocks*;

8. be able to use DesignCenter™ and Tool Palettes to drag and drop *Blocks* into the current drawing.

CONCEPTS

A *Block* is a <u>group</u> of objects that are combined into <u>one</u> object with the *Block* command. The typical application for *Blocks* is in the use of symbols. Many drawings contain symbols, such as doors and windows for architectural drawings, capacitors and resistors for electrical schematics, or pumps and valves for piping and instrumentation drawings. In AutoCAD, symbols are created first by constructing the desired geometry with objects like *Line*, *Arc*, and *Circle*, then transforming the set of objects comprising the symbol into a *Block*. A description of the objects comprising the *Block* is then stored in the drawing's "block definition table." The *Blocks* can then each be *Inserted* into a drawing many times and treated as a single object. Text can be attached to *Blocks* (called *Attributes*) and the text can be modified for each *Block* when inserted.

Figure 21-1 compares a shape composed of a set of objects and the same shape after it has been made into a *Block* and *Insert*ed back into the drawing. Notice that the original set of objects is selected (highlighted) individually for editing, whereas, the *Block* is only one object.

Figure 21-1 ───────────────────

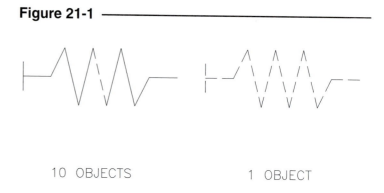

10 OBJECTS   1 OBJECT

Since an inserted *Block* is one object, it uses less file space than a set of objects that is copied with *Copy*. The *Copy* command creates a duplicate set of objects, so that if the original symbol were created with 10 objects, 3 copies would yield a total of 40 objects. If instead the original set of 10 were made into a *Block* and then *Insert*ed 3 times, the total objects would be 13 (the original 10 + 3).

Upon *Insert*ing a *Block*, its scale can be changed and rotational orientation specified without having to use the *Scale* or *Rotate* commands (Fig. 21-2). If a design change is desired in the *Blocks* that have already been *Insert*ed, the original *Block* can be redefined and the previously inserted *Blocks* are automatically updated. *Blocks* can be made to have explicit *Linetype*, *Lineweight* and *Color*, regardless of the layer they are inserted onto, or they can be made to assume the *Color*, *Linetype*, and *Lineweight* of the layer onto which they are *Insert*ed.

Figure 21-2 ───────────────────

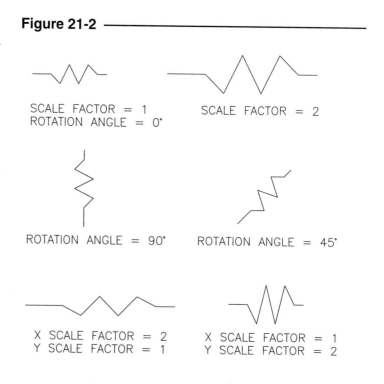

Blocks can be <u>nested</u>; that is, one *Block* can reference another *Block*. Practically, this means that the definition of *Block* "C" can contain *Block* "A" so that when *Block* "C" is inserted, *Block* "A" is also inserted as part of *Block* "C" (Fig. 21-3).

Figure 21-3 ───────────────────────────────────

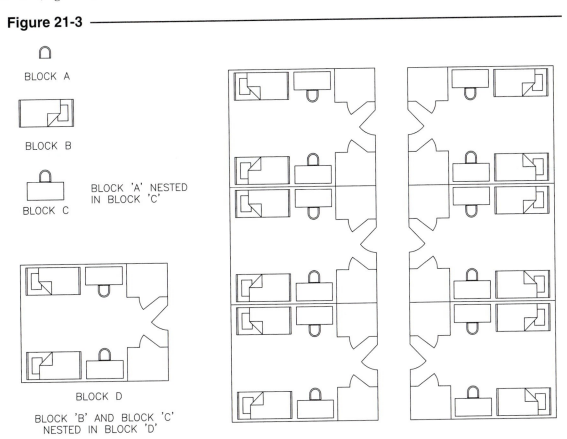

BLOCK A

BLOCK B

BLOCK C   BLOCK 'A' NESTED IN BLOCK 'C'

BLOCK D

BLOCK 'B' AND BLOCK 'C' NESTED IN BLOCK 'D'

Blocks created within the current drawing can be copied to disk as complete and separate drawing files (.DWG file) by using the *Wblock* command (Write Block). This action allows you to *Insert* the *Blocks* into other drawings.

Commands related to using *Blocks* are:

Block	Creates a *Block* from individual objects
Insert	Inserts a *Block* into a drawing
Minsert	Permits a multiple insert in a rectangular pattern
Explode	Breaks a *Block* into its original set of multiple objects
Wblock	Writes an existing *Block* or a set of objects to a file on disk
Base	Allows specification of an insertion base point
Purge	Deletes uninserted *Blocks* from the block definition table
Rename	Allows renaming *Blocks*
Adcenter	Invokes DesignCenter, which allows you to drag and drop *Blocks*, *Dimension Styles*, *Layers*, *Layouts*, *Linetypes*, *Text Styles*, and *Xrefs* into the current drawing (separate window)
Toolpalettes	Displays tool palettes that allow you to drag-and-drop *Blocks* and *Hatch* patterns into a drawing

Using Tool Palettes and DesignCenter are two easy methods for inserting *Blocks*. With DesignCenter, you can drag-and-drop *Blocks* <u>from other drawings into the current drawing</u>. These two features are described near the end of this chapter as well as in Chapters 26 and 30.

COMMANDS

BLOCK

	Pull-down Menu	COMMAND (TYPE)	ALIAS (TYPE)	Short-cut	Screen (side) Menu	Tablet Menu
	Draw Block > Make...	*BLOCK*, or *-BLOCK*	*B or -B*	...	*DRAW2 Bmake*	*N,9*

Selecting the icon button, using the pull-down or screen menu, or typing *Block* or *Bmake*, produces the *Block Definition* dialog box shown in Figure 21-4. This dialog box provides the same functions as using the *-Block* command (a hyphen prefix produces the Command line equivalent).

To make a *Block*, first create the *Lines, Circles, Arcs,* or other objects comprising the shapes to be combined into the *Block*. Next, use the *Block* command to transform the objects into one object—a *Block*.

In the *Block Definition* dialog box, enter the desired *Block* name in the *Name* edit box. Then use the *Select Objects* button (top center) to return to the drawing temporarily to select the objects you wish to comprise the *Block*. After selection of objects, the dialog box reappears. Use the *Specify Insertion Base Point* button (in the *Base Point* section of the dialog box) if you want to use a point other than the default 0,0,0 as the "insertion point" when the *Block* is later inserted. Usually select a point in the corner or center of the set of objects as the base point. When you select *OK*, the new *Block* is defined and stored in the drawing's block definition table awaiting future insertions.

If *Delete* is selected in the *Objects* section of the dialog box, the original set of "template" objects comprising the *Block* disappear even though the definition of the *Block* remains in the table. Checking *Retain* forces AutoCAD to retain the original objects (similar to using *Oops* after the *Block* command), or selecting *Convert to Block* keeps the original set of objects visible in the drawing but transforms them into a *Block*.

The bottom half of the dialog box is used to specify how *Blocks* are described when DesignCenter is used to drag and drop the *Blocks* into a drawing instead of using the *Insert* command. In DesignCenter, you can preview the *Blocks* and read the description of the *Blocks*. Generally the *Create Icon from Block Geometry* option is used to create the preview thumbnail sketch unless the *Block* geometry is very complex, in which case another icon file could be selected. The *Block Units* options are described later in this chapter. Use the *Hyperlink* button to produce the *Insert Hyperlink* dialog box for attaching a hyperlink to a *Block*. The *Names* drop-down list (Fig. 21-5) is used to select existing *Blocks* if you want to redefine a *Block* (see "Redefining Blocks" later in this chapter).

Figure 21-4 —————

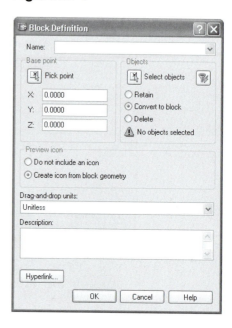

Figure 21-5 —————

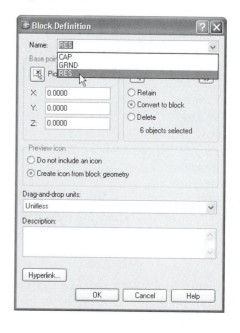

If you prefer to type, use *-Block* to produce the Command line equivalent of the *Block Definition* dialog box. The command syntax is as follows:

Command: ***-Block***
Block name (or ?): (**name**) (Enter a descriptive name for the *Block* up to 255 characters.)
Insertion base point: **PICK** or (**coordinates**) (Select a point to be used later for insertion.)
Select objects: **PICK**
Select objects: **PICK** (Continue selecting all desired objects.)
Select objects: **Enter**

The *Block* then <u>disappears</u> as it is stored in the current drawing's "block definition table." The *Oops* command can be used to restore the original set of "template" objects (they reappear), but the definition of the *Block* remains in the table. Using the *?* option of the *Block* command lists the *Blocks* stored in the block definition table.

Block Color, Linetype, **and** *Lineweight* **Settings**

The <u>color, linetype</u>, and <u>lineweight</u> of the *Block* are determined by one of the following settings when the *Block* is created:

1. When a *Block* is inserted, it is drawn on its original layer with its original *color, linetype* and *lineweight* (when the objects were <u>created</u>), regardless of the layer or *color, linetype* and *lineweight* settings that are current when the *Block* is inserted (unless conditions 2 or 3 exist).

2. If a *Block* is created on <u>Layer 0</u> (Layer 0 is current when the original objects comprising the *Block* are created), then the *Block* assumes the *color, linetype* and *lineweight* of any layer that is current when it is inserted (Fig. 21-6).

Figure 21-6

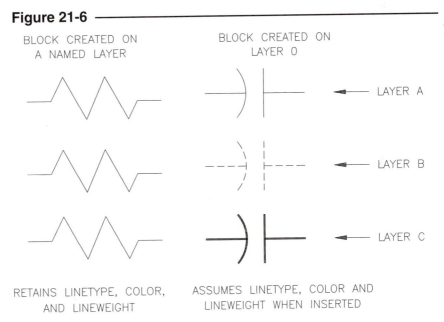

BLOCK CREATED ON A NAMED LAYER — RETAINS LINETYPE, COLOR, AND LINEWEIGHT

BLOCK CREATED ON LAYER 0 — ASSUMES LINETYPE, COLOR AND LINEWEIGHT WHEN INSERTED

LAYER A
LAYER B
LAYER C

3. If the *Block* is created with the special *BYBLOCK color, linetype* and *lineweight* setting, the *Block* is inserted with the *color, linetype* and *lineweight* settings that are <u>current during insertion</u>, whether the *BYLAYER* or explicit object *color, linetype* and *lineweight* settings are current.

INSERT

Pull-down Menu	COMMAND (TYPE)	ALIAS (TYPE)	Short-cut	Screen (side) Menu	Tablet Menu
Insert Block...	INSERT or -INSERT	I or -I	...	INSERT Ddinsert	T,5

Once the *Block* has been created, it is inserted back into the drawing at the desired location(s) with the *Insert* command. The *Insert* command produces the *Insert* dialog box (Fig. 21-7) which allows you to select which *Block* to insert and to specify the *Insertion Point*, *Scale,* and *Rotation*, either interactively (*On-screen*) or by specifying values.

First, select the *Block* you want to insert. All *Blocks* located in the drawing's block definition table are listed in the *Name* drop-down list. Next, determine the parameters for *Insertion Point*, *Scale*, and *Rotation*. You can enter values in the edit boxes if you have specific parameters in mind or check *Specify On-screen* to interactively supply the parameters. For example, with the settings shown in Figure 21-7, AutoCAD would allow you to preview the *Block* as you dragged it about the screen and picked the *Insertion Point*. You would not be prompted for a *Scale* or *Rotation* angle since they are specified in the dialog box as 1.0000 an 0 degrees, respectively. Entering any other values in the *Scale* or *Rotation* edit boxes causes AutoCAD to preview the *Block* at the specified scale and rotation angle as you drag it about the drawing to pick the insertion point. Remember that *Osnaps* can be used when specifying the parameters interactively. Check *Uniform Scale* to ensure the X, Y, and Z values are scaled proportionally. *Explode* can also be toggled, which would insert the *Block* as multiple objects (see "*Explode*").

Figure 21-7

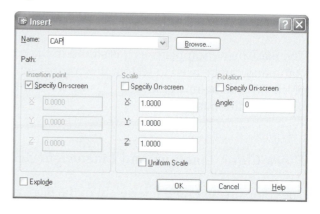

Inserting Other Drawings as *Blocks*

Selecting the *Browse* tile in the *Insert* dialog box produces the *Select Drawing* dialog box (Fig. 21-8). Here you can select <u>any</u> drawing (.DWG file) for insertion. When one drawing is *Inserted* into another, the entire drawing comes into the current drawing as a *Block*, or as <u>one object</u>. If you want to edit individual objects in the inserted drawing, you must *Explode* the object.

If you prefer the Command line equivalent, type *–Insert*.

Figure 21-8

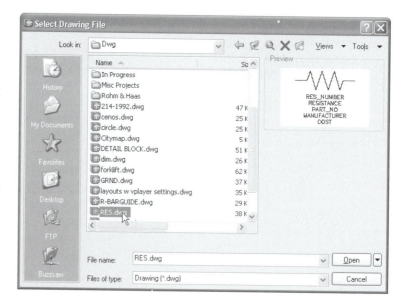

```
Command: -insert
Enter block name or [?] : name
(Type the name of an existing block
or .DWG file to insert.)
Specify insertion point or
[Scale/X/Y/Z/Rotate/PScale/PX/PY/PZ/PRotate]: PICK or option
Enter X scale factor, specify opposite corner, or [Corner/XYZ] <1>: value, PICK or option
Enter Y scale factor <use X scale factor>: value or PICK
Specify rotation angle <0>: value or PICK
```

Sometimes it is desirable to see the *Block* in the intended scale factor or rotation angle before you choose the insertion point. Presets ("PScale/PC/PY/PZ/PRotate") allow you to specify a rotation angle or scale factor <u>before</u> you dynamically drag the *Block* to pick the insertion point. (Normally, you would have to select the insertion point before the prompts for scale factor and rotation angle appear.)

MINSERT

Pull-down Menu	COMMAND (TYPE)	ALIAS (TYPE)	Short-cut	Screen (side) Menu	Tablet Menu
...	*MINSERT*	...	...	...	...

This command allows a <u>multiple insert</u> in a rectangular pattern (Fig. 21-9). *Minsert* is actually a combination of the *Insert* and the *Array Rectangular* commands. The *Blocks* inserted with *Minsert* are associated (the group is treated as one object) and cannot be edited independently (unless *Exploded*).

Examining the command syntax yields the similarity to a *Rectangular Array*.

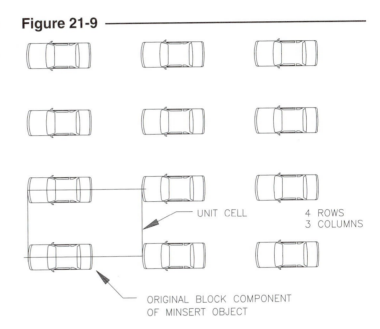

Figure 21-9

UNIT CELL   4 ROWS
3 COLUMNS

ORIGINAL BLOCK COMPONENT
OF MINSERT OBJECT

Command: **Minsert**
Enter block name [or ?] <current>: **name**
Specify insertion point or [Scale/X/Y/Z/Rotate/PScale/PX/PY/PZ/PRotate]: (**value**), **PICK** or option
Enter X scale factor, specify opposite corner, or [Corner/XYZ] <1>: (**value**) or **Enter**
Enter Y scale factor <use X scale factor>: (**value**) or **Enter**
Specify rotation angle <0>: (**value**) or **Enter**
Enter number of rows (—-) <1>: (**value**)
Enter number of columns (|||) <1>: (**value**)
Enter distance between rows or specify unit cell (—-): (**value**) or **PICK** (Value specifies Y distance
    from *Block* corner to *Block* corner; PICK allows drawing a unit cell rectangle.)
Distance between columns: (**value**) or **PICK** (Specifies X distance between *Block* corners.)
Command:

EXPLODE

Pull-down Menu	COMMAND (TYPE)	ALIAS (TYPE)	Short-cut	Screen (side) Menu	Tablet Menu
Modify *Explode*	*Explode*	X	...	MODIFY2 *Explode*	Y,22

Explode breaks a <u>previously</u> inserted *Block* back into its original set of objects (Fig. 21-10, on the next page), which allows you to edit individual objects comprising the shape. *Blocks* that have been *Minsert*ed cannot be *Explode*d. There are no options for this command.

Command: *explode*
Select objects: **PICK**
Select objects: **Enter**
Command:

Figure 21-10

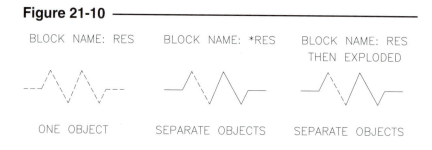

BLOCK NAME: RES
ONE OBJECT

BLOCK NAME: *RES
SEPARATE OBJECTS

BLOCK NAME: RES
THEN EXPLODED
SEPARATE OBJECTS

-INSERT with *

Using the *-Insert* command with the asterisk (*) allows you to insert a *Block*, not as one object, but as the original set of objects comprising the *Block*. In this way, you can edit individual objects in the *Block*, otherwise impossible if the *Block* is only one object (Fig. 21-10).

The normal *-Insert* command is used (type *-Insert* with a hyphen); however, when the desired <u>Block name</u> is entered, it is prefaced by the asterisk (*) symbol:

Command: *-insert*
Enter Block name (or ?): * (**name**) (Type the * symbol, then the name of an existing block or .DWG
   file to insert.)
Specify insertion point for block:
Scale factor <1>:
Specify rotation angle <0>:
Command:

This action accomplishes the same goal as using the *Insert* dialog box; then *Explode*.

XPLODE

Pull-down Menu	COMMAND (TYPE)	ALIAS (TYPE)	Short-cut	Screen (side) Menu	Tablet Menu
...	*XPLODE*	*XP*	...	...	...

When you *Insert* a *Block* into a drawing, all layers, linetypes, and colors contained in the *Block* are also inserted into the parent drawing if they do not already exist. If you use *Explode* to break down the *Block* into its component entities, those entities retain their native properties of layer, linetype, and color. For example, you may insert a *Block* that also inserts its layer named FLOORPLAN. If that *Block* is *Exploded*, its objects remain on layer FLOORPLAN. You may instead want to *Explode* the *Block* and have the objects reside on a layer existing in the current drawing, AR-WALL, for example.

The *Xplode* command is an expanded version of *Explode*. The *Xplode* command allows you to specify new properties for the objects that are exploded. When you use *Xplode*, you can also specify the new *Layer*, *Linetype*, *Color*, *Lineweight*, or choose to *Explode* the *Block* normally.

Command: *xplode*
Select objects to XPlode.
Select objects: **PICK**
Select objects: **Enter**
Enter an option [All/Color/LAyer/LType/Inherit from parent block/Explode] <Explode>:

The options are as follows.

Color
Use this option to specify a color for the exploded objects. Entering *ByLayer* causes the component objects to inherit the color of the exploded object's layer. Entering *ByBlock* causes the component objects to inherit the object-specific color of the exploded object.

> Enter an option [All/Color/LAyer/LType/Inherit from parent block/Explode] <Explode>: **c**
> Enter new color for exploded objects.
> [Red/Yellow/Green/Cyan/Blue/Magenta/White/BYLayer/BYBlock] <BYLAYER>:

Layer
With this option you can specify the layer of the component objects after you explode them. The default option is to inherit the current layer rather than the layer of the exploded object.

> Enter an option [All/Color/LAyer/LType/Inherit from parent block/Explode] <Explode>: **la**
> Enter new layer name for exploded objects <0>:

Linetype
You can enter the name of any linetype that is loaded in the drawing. The exploded objects assume the specified linetype. Entering *ByLayer* causes the component objects to inherit the linetype of the exploded object's layer. Entering *ByBlock* causes the component objects to inherit the object-specific linetype of the exploded object.

> Enter an option [All/Color/LAyer/LType/Inherit from parent block/Explode] <Explode>: **lt**
> Enter new linetype name for exploded objects <BYLAYER>:

Inherit from parent block
This option sets the color, linetype, lineweight, and layer of the exploded objects to that of the *Block* if the *Block* was created using *ByBlock* color, linetype, and lineweight and the objects were drawn on layer 0.

All
Use this option to specify a layer, color, linetype, <u>and lineweight</u> for the exploded objects. This is the only method for specifying a distinct lineweight.

> Enter an option [All/Color/LAyer/LType/Inherit from parent block/Explode] <Explode>: **all**
> Enter new color for exploded objects.
> [Red/Yellow/Green/Cyan/Blue/Magenta/White/BYLayer/BYBlock] <BYLAYER>:
> Enter new linetype name for exploded objects <BYLAYER> :
> Enter new lineweight <BYLayer>:
> Enter new layer name for exploded objects <0>:

WBLOCK

Pull-down Menu	COMMAND (TYPE)	ALIAS (TYPE)	Short-cut	Screen (side) Menu	Tablet Menu
File *Export...*	*WBLOCK*	*W*	...	*FILE* *Export*	*W,24*

The *Wblock* command writes a *Block* out to disk as a separate and complete drawing (.DWG) file. The *Block* used for writing to disk can exist in the current drawing's *Block* definition table or can be created by the *Wblock* command. Remember that the *Insert* command inserts *Blocks* (from the current drawing's block definition table) or finds and accepts .DWG files and treats them as *Blocks* upon insertion.

There are two ways to create a *Wblock* from the current drawing, (1) using an existing *Block* and (2) using a set of objects not previously defined in the current drawing as a *Block*. If you are using an existing *Block*, a copy of the *Block* is essentially transformed by the *Wblock* command to create a complete AutoCAD drawing (.DWG) file. The original block definition remains in the current drawing's block definition table. In this way, *Blocks* that were originally intended for insertion into the current drawing can be inserted into other drawings.

Figure 21-11 illustrates the relationship among a *Block*, the current drawing, and a *WBlock*. In the figure, SCHEM1.DWG contains several *Blocks*. The RES block is written out to a .DWG file using *Wblock* and named RESISTOR. RESISTOR is then *Inserted* into the SCHEM2 drawing.

Figure 21-11

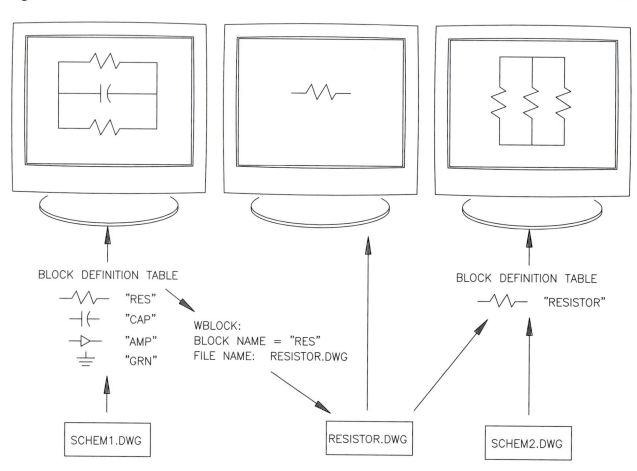

 If you want to transform a set of objects into a *Block* to be used in other drawings but not in the current one, you can use *Wblock* to transform (a copy of) the objects in the current drawing into a separate .DWG file. This action does not create a *Block* in the current drawing. As an alternative, if you want to create symbols specifically to be inserted into other drawings, each symbol could be created initially as a separate .DWG file.

The *Wblock* command produces the *Write Block* dialog box (Fig. 21-12). You should notice similarities to the *Block Definition* dialog box. Under *Source*, select *Block* if you want to write out an existing *Block* and select the *Block* name from the list, or select *Objects* if you want to transform a set of objects (not a previously defined *Block*) into a separate .DWG file.

Figure 21-12 ——————

The *Base Point* section allows you to specify a base point to use upon insertion of the *Block*. Enter coordinate values or use the *Specify Insert Base Point* button to pick a location in the drawing. The *Objects* section allows you to specify how you want to treat selected objects if you create a new .DWG file from objects (not from an existing *Block*). You can *Retain* the objects in their current state, *Convert to Block*, or *Delete from Drawing*.

The *Destination* section defines the desired *File Name and Path*, and *Insert Units*. Your choice for *Insert Units* is applicable only when you drag and drop a *Block*, as with DesignCenter. See "DesignCenter" later in this chapter for an explanation of this subject.

If you prefer the Command line equivalent to the *Write Block* dialog box, type -*Wblock* and follow the prompt sequence shown here to create *Wblocks* (.DWG files) from existing *Blocks*,

> Command: **-wblock**
> (At this point, the *Create Drawing File* dialog box appears, prompting you to supply a name for the .DWG file to be created. Typically, a new descriptive name would be typed in the edit box rather than selecting from the existing names.)
> Enter name of existing block or [= (block=output file)/* (whole drawing)] <define new drawing>:
> (Enter the name of the desired existing *Block*. If the file name given in the previous step is the same as the existing *Block* name, an "=" symbol can be entered, or enter an asterisk to write out the entire drawing.)
> Command:

A copy of the existing *Block* is then created in the selected directory as a *Wblock* (.DWG file).

An alternative method for creating a *Wblock* is using the *Export Data* dialog box accessed from the *File* pull-down menu. Make sure you select *Block* (*.DWG) as the type of file to export (bottom of the dialog box). You can create a *Wblock* (.DWG file) from an existing *Block* or from objects that you select from the screen. After specifying the .DWG name you want to create and the dialog box disappears, you are presented with the same command syntax as shown previously for the –*Wblock* command.

When *Wblocks* are *Inserted*, the *Color*, *Linetype*, and *Lineweight* settings of the *Wblock* are determined by the settings current when the original objects comprising the *Wblock* were created. The three possible settings are the same as those for *Blocks* (see "Block," *Color*, *Linetype*, and *Lineweight* Settings).

When a *Wblock* is *Inserted*, its parent (original) layer is also inserted into the current drawing. *Freezing* either the parent layer or the layer that was current during the insertion causes the *Wblock* to be frozen.

Redefining *Blocks*

If you want to change the configuration of a *Block*, even after it has been inserted, it can be accomplished by redefining the *Block*. In doing so, all of the previous *Block* insertions are automatically and globally updated (Fig. 21-13). AutoCAD stores two fundamental pieces of information for each *Block* insertion—the insertion point and the *Block* name. The actual block definition is stored in the block definition table. Redefining the *Block* involves changing that definition.

Figure 21-13

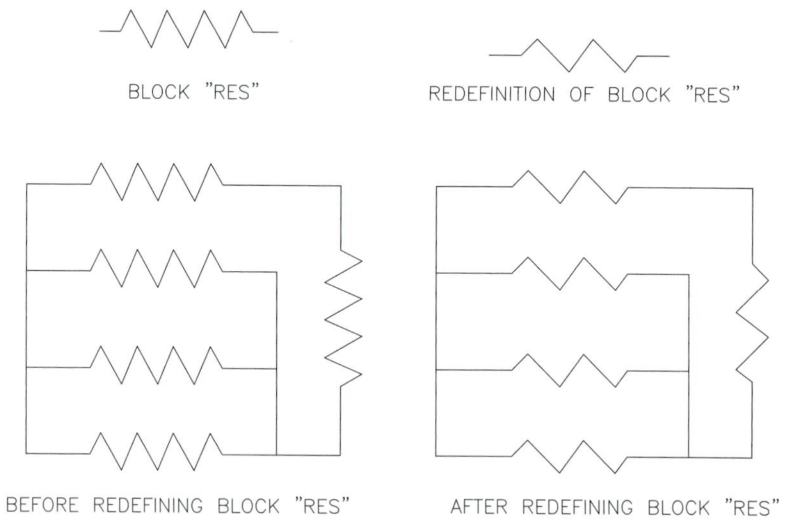

BLOCK "RES"   REDEFINITION OF BLOCK "RES"

BEFORE REDEFINING BLOCK "RES"   AFTER REDEFINING BLOCK "RES"

To redefine a *Block*, use the *Block* command. First, draw the new geometry or change the <u>original</u> "template" set of objects. (The change cannot be made using an inserted *Block* unless it is *Exploded* because a *Block* cannot reference itself.) Next, use the *Block* command and select the new or changed geometry. The old *Block* is redefined with the new geometry as long as the <u>original Block name</u> is used.

The *Refedit* command can also be used to redefine a *Block*. *Refedit* "opens" the *Block* for editing, then you make the necessary changes, and finally use *Refclose* to "close" the editing session and save the *Block*. This process has the same result as redefining the *Block*. See "*Refedit*" in Chapter 30, Xreferences.

BASE

Pull-down Menu	COMMAND (TYPE)	ALIAS (TYPE)	Short-cut	Screen (side) Menu	Tablet Menu
Draw *Block >* *Base*	*BASE*	...	...	DRAW2 *Base*	...

The *Base* command allows you to specify an "insertion base point" (see the *Block* command) in the current drawing for subsequent insertions. If the *Insert* command is used to bring a .DWG file into another drawing, the insertion base point of the .DWG is 0,0 by default. The *Base* command permits you to specify another location as the insertion base point. The *Base* command is used in the symbol drawing, that is, used in the drawing <u>to be inserted</u>. For example, while creating separate symbol drawings (.DWGs) for subsequent insertion into other drawings, the *Base* command is used to specify an appropriate point on the symbol geometry for the *Insert* command to use as a "handle" other than point 0,0 (Fig. 21-14).

Figure 21-14

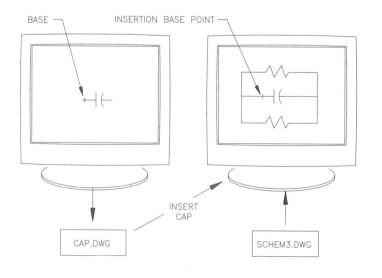

BASE   INSERTION BASE POINT

INSERT CAP

CAP.DWG   SCHEM3.DWG

Blocks Search Path

You can use the *Options* dialog box to specify the search path used when the *Insert* command attempts to locate .DWG files. As you remember, when *Insert* is used and a *Block* name is entered, AutoCAD searches the drawing's block definition table and, if no *Block* by the specified name is found, searches the specified path for a .DWG file by the name. (If you use the *Insert* dialog box, you can select the *Browse...* button to locate a .DWG file.) If the paths of the intended files are specified in the "Support File Search Path," you can type *Insert* and the name of the file, and AutoCAD will locate and insert the file. Otherwise, you have to locate the file using the *Browse* button each time you *Insert* a new file.

Figure 21-15

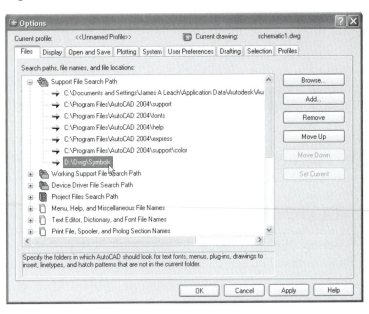

The *Files* tab of the *Options* dialog box allows you to list and modify the path that AutoCAD searches for drawings to *Insert* (Fig. 21-15). The search path is also used for finding font files, menus, plug-ins, line-types, and hatch patterns.

Suppose all of your symbols were stored in a folder called D:\Dwg\Symbols. Access the *Options* dialog box by typing *Options* or selecting from the *Tools* pull-down menu. In the *Files* tab, open the *Support File Search Path* section.

Select *Add..* or *Browse...* to enter the location of your symbols as shown in the figure. Use *Move Up* or *Move Down* to specify an order (priority) for searching. Then each time you type "*Insert,*" you could enter the name of a file without the path and AutoCAD can locate it.

PURGE	Pull-down Menu	COMMAND (TYPE)	ALIAS (TYPE)	Short-cut	Screen (side) Menu	Tablet Menu
	File *Drawing Utilities >* *Purge...*	*PURGE*	*PU*	...	*FILE* *Purge*	*X,25*

Purge allows you to selectively delete a *Block* that is not referenced in the drawing. In other words, if the drawing has a *Block* defined but not appearing in the drawing, it can be deleted with *Purge*. In fact, *Purge* can selectively delete <u>any named object</u> that is not referenced in the drawing. Examples of unreferenced named objects are:

Blocks that have been defined but not *Inserted*;
Layers that exist without objects residing on the layers;
Dimstyles that have been defined, yet no dimensions are created in the style (see Chapter 29);
Linetypes that were loaded but not used;
Shapes that were loaded but not used;
Text Styles that have been defined, yet no text has been created in the *Style*;
Mlstyles (multiline styles) that have been defined, yet no *Mlines* have been drawn in the style.

Since these named objects occupy a small amount of space in the drawing, using *Purge* to delete unused named objects can reduce the file size. This may be helpful when drawings are created using template drawings that contain many unused named objects or when other drawings are *Inserted* that contain many unused *Layers* or *Blocks*, etc.

Purge is especially useful for getting rid of unused *Blocks* because unused *Blocks* can occupy a huge amount of file space compared to other named objects (*Blocks* are the only geometry-based named objects). Although some other named objects can be deleted by other methods (such as selecting *Delete* in the *Text Style* dialog box), *Purge* is the only method for deleting *Blocks*.

The *Purge* command produces the *Purge* dialog box (Fig. 21-16). Here you can view all named objects in the drawing, but you can purge only those items that are not used in the current drawing.

Figure 21-16

View items you can purge
Generally this option is the default choice since you can remove unused named objects from your drawing only if this option is checked.

View items you cannot purge
This option is useful if you want to view all the named objects contained in the drawing. You can also get an explanation appearing at the bottom of the dialog box as to why a selected item cannot be removed.

Purge
Use the *Purge* button to remove only the items you select from the list. First, select the items, then select *Purge*. You can select more than one item using the Ctrl or Shift keys while selecting multiple items or a range, respectively.

Purge All
Use this button to remove all unused named objects from the drawing. You do not have to first select items from the list. To be safe, you should also select *Confirm each item to be purged* when using this option.

Confirm each item to be purged
If the *Confirm each item to be purged* box is checked, AutoCAD asks for confirmation before removing each object. If this box is not checked and you *Purge All*, all unused named objects in the drawing are removed at once without notification.

Purge nested items
You can remove unused nested *Blocks* and other nested named objects with this option. This option is particularly helpful if you have *Inserted* other drawings or *Xrefs* into the current drawing since all the named items contained in the inserted drawings are also inserted.

A Command line version can be invoked by typing *–Purge*.

```
Command: -purge
Enter type of unused objects to purge
[Blocks/Dimstyles/LAyers/LTypes/Plotstyles/SHapes/textSTyles/Mlinestyles/All]: b
Enter name(s) to purge <*>: Enter or (name)
Verify each name to be purged? [Yes/No] <Y>:
```

You can select one type of object to be purged or select *All* to have all named objects listed.

RENAME

	Pull-down Menu	COMMAND (TYPE)	ALIAS (TYPE)	Short-cut	Screen (side) Menu	Tablet Menu
	Format Rename...	*RENAME* or *-RENAME*	*REN* or *-REN*	...	*FORMAT Rename*	*V,1*

This utility command allows you to rename a *Block* or <u>any named object</u> that is part of the current drawing. *Rename* allows you to rename the named objects listed here:

Blocks, Dimension Styles, Layers, Linetypes, Plot Styles, Text Styles, User Coordinate Systems, Views, and *Viewport* configurations.

If you prefer to use dialog boxes, you can type *Rename* or select *Rename...* from the *Format* pull-down menu to access the dialog box shown in Figure 21-17.

You can select from the *Named Objects* list to display the related objects existing in the drawing. Then select or type the old name so it appears in the *Old Name:* edit box. Specify the new name in the *Rename To:* edit box. <u>You must then PICK *the Rename To:* tile</u> and the new names will appear in the list. Then select the *OK* tile to confirm.

Figure 21-17

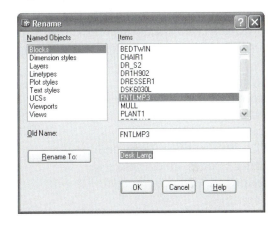

The *Rename* dialog box can be used with wildcard characters (see Chapter 43) to specify a list of named objects for renaming. For example, if you desired to rename all the "DIM-*" layers so that the letters "DIM" were replaced with only the letter "D", the following sequence would be used.

1. In the *Old Name* edit box, enter "DIM-*" and press Enter. All of the layers beginning with "DIM-" are highlighted.

2. In the *Rename To:* edit box, enter "D-*." Next, the *Rename To:* tile must be PICKed. Finally, PICK the *OK* tile to confirm the change.

The *-Rename* command (with a hyphen prefix) can be used to rename objects one at a time. For example, the command sequence might be as follows:

```
Command: -rename
Enter object type to rename [Block/Dimstyle/LAyer/LType/Plot style/Style/Ucs/VIew/VPort]: b
Enter old block name: chair7
Enter new block name: desk-chair
Command:
```

Wildcard characters are not allowed with the *-Rename* command but are allowed with the dialog box.

DESIGNCENTER

The DesignCenter window (Fig. 21-18) allows you to navigate, find, and preview a variety of content, including *Blocks*, located anywhere accessible to your workstation, then allows you to open or insert the content using drag-and-drop. "Content" that can be viewed and managed includes other drawings, *Blocks*, *Dimstyles*, *Layers*, *Layouts*, *Linetypes*, *Textstyles*, *Xrefs*, raster images, and URLs (Web site addresses). In addition, if you have multiple drawings open, you can streamline your drawing process by copying and pasting content, such as layer definitions, between drawings. (Dimension Styles, Xrefs, and Raster Images are discussed in Chapters 29, 30, 32, respectively.)

Figure 21-18

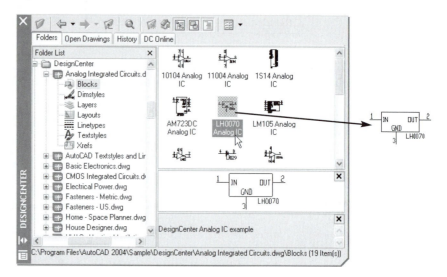

ADCENTER

Pull-down Menu	COMMAND (TYPE)	ALIAS (TYPE)	Short-cut	Screen (side) Menu	Tablet Menu
Tools *AutoCAD DesignCenter...*	*ADCENTER*	*ADC*	*Ctrl + 2*	...	...

Accessing *Adcenter* by any method produces the DesignCenter palette. There are two sections to the window. The left side is called the Tree View and displays a Windows Explorer-type hierarchical directory (folder) structure of the local system (Fig. 21-19). The right side is called the Content Area (formerly called the Palette) and displays lists, icons, or thumbnail sketches of the content selected in the Tree View. The Content Area can display *Blocks*, *Dimstyles*, *Layers*, *Layouts*, *Linetypes*, *Textstyles*, *Xrefs*, and a variety of other content.

Generally, the Content Area is used to drag and drop the icons or thumbnails from the content area into the current drawing (see Figure 21-18). However, you can also streamline a variety of tasks such as those listed here:

- Browse sources of drawing content including open drawings, other drawings, raster images, content within the drawings (*Blocks*, *Dimstyles*, *Layers*, and so on), content on network drives, or content on a Web page.
- Insert, attach, or copy and paste the content (drawings, images, *Blocks*, *Layers*, etc.) into the current drawing.

- Create shortcuts to drawings, folders, and Internet locations that you access frequently.

- Use a special search engine to find drawing content on your computer or network drives. You can specify criteria for the search based on key words, names of *Blocks*, *Dimstyles*, *Layers*, etc., or the date a drawing was last saved. Once you have found the content, you can load it into DesignCenter or drag it into the current drawing.

- Open drawings by dragging a drawing (.DWG) file from the Content Area into the drawing area.

These features of DesignCenter are explained in the following descriptions of the DesignCenter toolbar icons.

Figure 21-19

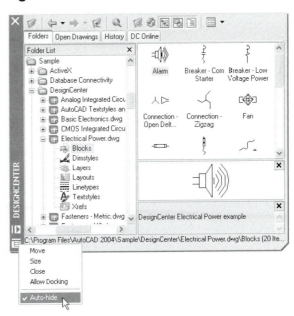

Resizing, Docking, and Hiding the DesignCenter Palette

You can change the width and height of DesignCenter by resting your pointer on one of the borders (not the title bar) until a double arrow appears, then dragging to the desired size. Alternately, rest your pointer on the lower corner (where the bevel appears) to resize both height and width. You can also move the bar between the Content Area and the Tree View area. Dock DesignCenter by clicking the title bar, then dragging it to either edge of the drawing window. You can hide the DesignCenter palette by using the *Auto-hide* option (click on the *Properties* button at the bottom of the title bar to produce the *Properties* menu as shown in Figure 21-19). When *Auto-hide* is on, the palette is normally hidden (only the title bar is visible) and the palette appears when you bring the pointer to the title bar.

DesignCenter Options

In addition to selecting the icons from the toolbar, you can right-click the Palette background to produce the shortcut menu and choose the desired option (Fig. 21-20).

Folders Tab

Folders is the default display for the Tree View side of DesignCenter. This choice displays a hierarchical structure of the desktop (local workstation). Because this arrangement is similar to Windows Explorer, you can navigate and locate content anywhere accessible to your system, including network drives. Figure 21-20 displays a typical hierarchical structure on the Tree View side. For example, you may want to import layer definitions (including color and linetype information) from a drawing into the current drawing by dragging and dropping.

Figure 21-20

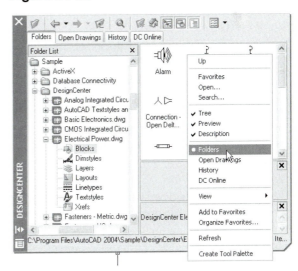

Open Drawings **Tab**

This option changes the Tree View to display all open drawings (Fig. 21-21). This feature is helpful when you have several drawings open and want to locate content from one drawing and import it into the current drawing. As shown in Figure 21-21, you may want to locate *Block* definitions from one drawing and *Insert* them into <u>another</u> drawing. Ensure you make the desired "target" drawing current in the drawing area before you drag and drop content from the Content Area.

Figure 21-21

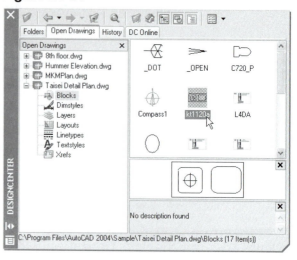

History **Tab**

This option displays a history (chronological list) of the last 20 file locations accessed through DesignCenter (Fig. 21-22). The purpose of this feature is simply to locate the file and load it into the Content Area. Load the file into the Content Area by double-clicking on it.

Figure 21-22

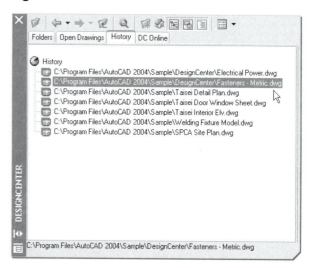

Tree View **Toggle**

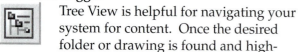
Tree View is helpful for navigating your system for content. Once the desired folder or drawing is found and highlighted in Tree View, you may want to toggle Tree View off so only the Content Area is displayed with the desired content. The desired content may be drawings, *Blocks*, images, or a variety of other content. For example, consider previous Figure 21-21 which displays Tree View on. Figure 21-23 illustrates Tree View toggled off and only the Content Area displayed. The resulting configuration displays only the *Blocks* contained in the selected drawing. Keep in mind that you can also change the Views of the Content Area to display *Large Icons*, *Small Icons*, a *List*, or *Details* (see "Views").

Figure 21-23

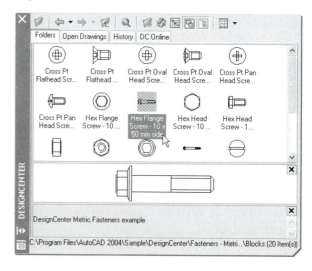

Favorites

This button displays the contents of the AutoCAD Favorites folder in the Content Area. The Tree View section displays the highlighted folder in the Desktop view.

You can add folders and files to *Favorites* by highlighting an item in Tree View or the Content Area, right-clicking on it, and selecting *Add to Favorites* from the shortcut menu (Fig. 21-24).

Figure 21-24

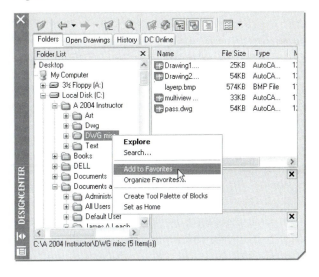

Load

Displays the *Load* dialog box (not shown), in which you can load the Content Area with content from anywhere accessible from your system. The *Load* dialog box is identical to the *Select File* dialog box (see "*Open*," Chapter 2, Working with Files). After selecting a file, DesignCenter automatically finds the file in Tree View and loads its content (*Blocks, Layers, Dimstyles,* etc.) into the Content Area. You can also load files into the Content Area using Windows Explorer (see "Loading the Content Area with Windows Explorer" at the end of this section).

Search

This button invokes the *Search* dialog box (Fig. 21-25), in which you can specify search criteria to locate drawing files, *Blocks, Layers, Dimstyles,* and other content within drawings. <u>Once the desired content is found in the list at the bottom of the dialog box, double-click on it to load it into the Content Area.</u>

This feature is extremely powerful and easy to use. The list of possible items to search for is displayed in the *Look For* drop-down list and includes the following choices:

> *Blocks*
> *Dimstyles*
> *Drawings*
> *Drawings and Blocks*
> *Layers*
> *Layouts*
> *Linetypes*
> *Text Styles*
> *Xrefs*

Figure 21-25

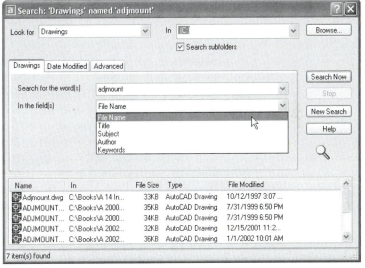

If *Drawings* is selected in the *Look for:* list, three tabs (shown in Figure 21-25) appear to allow you to refine the search criteria.

Drawings
Enter a text string in the *Search for words* edit box. Wildcards can be used (see Chapter 43 for valid wildcards in AutoCAD). From the *In the field(s)* drop-down list, select what field you want your text string to apply to:

> *File Name*
> *Title*
> *Subject*
> *Author*
> *Keywords*

Title, *Subject*, *Author*, and *Keywords* are searched for only in the *Drawing Properties* (description), so if you are searching for pre-AutoCAD 2000 drawings or have not specified drawing properties, these fields are not useful (see Chapter 2 for information on *Drawing Properties*).

Date Modified
You can search by *Date Modified* by specifying modification between two dates or during the previous X number of days or X number of months.

Advanced
The *Advanced* tab allows you to search for text strings in the *Drawing and Block description*, *Block Name*, *Attribute Tag*, or *Attribute Value*.

Up

Use the *Up* tool to display the next higher level in the Tree View hierarchy.

Preview

This button causes a preview image (thumbnail sketch) of the selected item to appear at the bottom of the Content Area. If there is no preview image saved with the selected item, the *Preview* area is empty. You can resize the preview image by dragging the bar between the Content Area and the preview pane.

To make a preview image appear, you must select an item (drawing, *Block*, or image file) <u>from the Content Area</u>. Selecting items in the Tree View does not cause a preview to appear. Figure 21-26 displays a drawing preview. Figure 21-19 displays a *Block* preview.

Figure 21-26

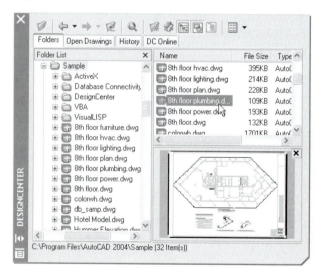

Description

This button displays a text description of the selected item at the bottom of the Content Area (Fig. 21-27). The description is displayed for *Blocks* (from the *Description* field when the *Block* is created) and for drawings (from *Drawing Properties*, see *DWGPROPS*, Chapter 2).

The description area can be resized. If you display both *Preview* and *Description* panes, the *Description* pane is displayed below the *Preview* pane, separated from it by a bar (see Figure 21-19).

You cannot edit the text description in DesignCenter; however, you can copy it to the Clipboard. Select the text you want to copy, right-click inside the pane, and then select *Copy* from the shortcut menu.

Views

Four possible *Views* of the Content Area are possible. Make the choice by selecting the down arrow to display the options.

Large Icons
Small Icons
List
Details

It is helpful to reset the configuration of the Content Area depending on the job you are performing. For example, if the *Preview* pane is displayed, you can set the Content Area to display only a *List* or *Details* (see Figure 21-26). Or if you prefer to "see" all the drawing files or *Blocks* in the Content Area, it would be better to set the View to *Large Icons* and toggle *Preview* off (see Figure 21-27).

Generally, items are sorted alphabetically by name in the Content Area. If you change the *View* to *Details*, you can sort <u>files</u> (not *Blocks*) by name, size, type, and other properties, depending on the type of content displayed in the Content Area (see Figure 21-26). Click on the column header for the column to sort by (click for ascending order, click again for descending order).

Back/Forward

The *Back* button contains a drop-down list displaying content that was previously loaded into the Content Area. Use the drop-down list to locate and load the desired item directly into the Content Area (Fig. 21-28). Once you use this feature (similar to the *Redo* drop-down list), the *Forward* drop-down list holds the list of items you previously used before going *Back*.

Figure 21-27

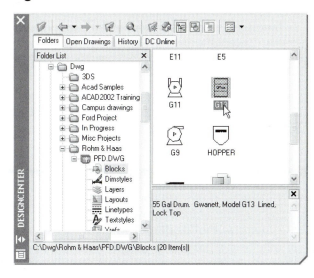

Figure 21-28

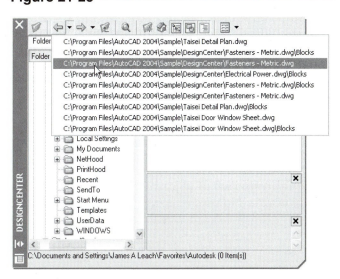

Loading the Content Area with Windows Explorer
You can use Windows Explorer to load content into the Content Area. For example, if you are browsing a network drive in Windows Explorer and locate a drawing file, you can drag the selected file directly into the Content Area. <u>The selected file must be dropped in to the Content Area area</u>, not the Tree View, Preview, or description areas. If you drop the file into any area other than the Content Area, the drawing is opened into the AutoCAD session.

DC Online

DesignCenter Online is a new feature in AutoCAD 2004. This very powerful section of DesignCenter connects you to the Internet and gives you access to thousands of symbols (drawings that can be inserted as *Blocks*) and manufacturer's product information (Fig. 21-29).

Figure 21-29

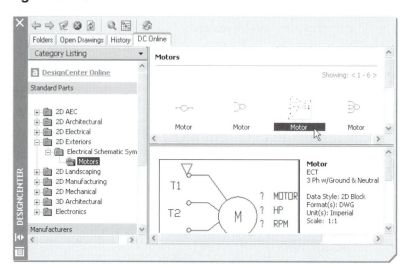

Using the *Category Listing* (see Figure 21-29, left panel) you can select from categories such as architectural, electrical, landscaping, mechanical, and manufacturing. Making one of these selections provides access to the specific drawings (in the Content Area) that you can select from to insert into AutoCAD. Simply drag and drop the desired drawing from the Content Area into your AutoCAD session.

You can select from the drop-down list at the top of the left panel or right-click to display options for *Category Listing*, *Search*, *Settings* (for searching) or *Collections* (updating the categories).

To Open a Drawing from DesignCenter

In addition to using DesignCenter to view and import content contained within drawing files, it is possible to use DesignCenter to *Open* drawings. <u>While holding down the Ctrl key</u>, drag the icon of the drawing file you want to open from the <u>Content Area</u> and drop it in the drawing area. Remember, you must drag it from the Content Area, not from the Tree View area.

Usually, you would want to drop the selected drawing into a blank (*New*) drawing. Unless the selected drawing is a titleblock and border, drop it into model space, not into a layout (paper space). Make sure the background (drawing area) is visible. You may need to resize the windows displaying any currently open drawings.

A drawing file that is dropped into AutoCAD is actually *Inserted* as a *Block*. The typical Command line prompts appear for insertion point, scale factors, and rotation angle. Generally, enter 0,0 as the insertion point and accept the defaults for X and Y scale factors and rotation angle. If you want to edit the geometry, you will have to *Explode* the drawing.

2004

To *Insert* a *Block* Using DesignCenter

One of the primary functions of DesignCenter is to insert *Block* definitions into a drawing. When you insert a *Block* into a drawing, the block definition is copied into the drawing database. Any instance of that *Block* that you *Insert* into the drawing from that time on references the original *Block* definition.

You cannot add *Blocks* to a drawing while another command is active. If you try to drop a *Block* into AutoCAD while a command is active at the Command line, the icon changes to a slash circle indicating the action is invalid.

There are two methods for inserting *Blocks* into a drawing using DesignCenter: (1) using drag-and-drop with Autoscaling and (2) using the *Insert* dialog box with explicit insertion point, scale, and rotation value entry.

Block Insertion with Drag-and-Drop

When you drag-and-drop a *Block* from DesignCenter into a drawing, you are not prompted for an insertion point, X and Y scale factors, or a rotation angle. Although you specify an insertion point interactively when you "drop" the *Block* icon, AutoCAD uses Autoscaling to determine the scaling parameters. Autoscaling is a process of comparing the specified units of the *Block* definition (when the *Block* was created) and the *Drawing Units for Block Inserts* set in the *Drawing Units* dialog box of the target drawing. The value options are *Unitless, Inches, Feet, Miles, Millimeters, Centimeters, Kilometers,* and many other choices.

The *Drawing Units for Block Inserts* set in the *Drawing Units* dialog box (Fig. 21-30) <u>should be set in the target drawing</u> and controls how the *Block* units are scaled when the *Block* is dropped (into the target drawing). It is helpful that the <u>setting here can be changed immediately before dropping a *Block* into the drawing</u>. (Also see "*Units*" in Chapter 6.)

In addition, you can specify the desired *Block* units in the *Block Definition* dialog box (Fig. 21-31). This setting is generally <u>assigned when each *Block* is created</u>. (See "*Block*" earlier in this chapter.)

Figure 21-30 ─────────────

Figure 21-31 ─────────────

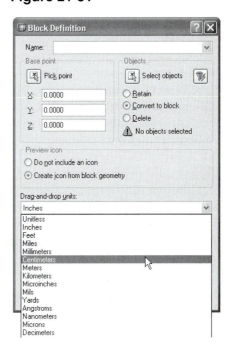

Resulting Scale Factors with Autoscale (Drag-and-Drop)

This table gives sample scale factors that Autoscaling uses when *Blocks* are inserted into other drawings by the drag-and-drop method. The first row (top) of the table lists the *Block* settings (made in the *Block Definition* dialog box) and the first column (left) lists the drawing units settings (made in the *Drawing Units* dialog box of the target drawing).

Drawing Units \ Block Units	Unitless	Inches	Feet	Millimeters
Unitless	1.000	1.000	1.000	1.000
Inches	1.000	1.000	12.000	0.039
Feet	1.000	0.083	1.000	0.003
Millimeters	1.000	25.400	304.800	1.000

Note that when the units in <u>either</u> case (defined in the *Block* or in the target *Drawing Units*) are set to *Unitless*, the scale factor is always 1.00. Also, when the two settings <u>match</u> (*Inches* and *Inches*, or *Feet* and *Feet*), the scale factor is always 1.00. The other cells of the table list a proportion of one unit to the other; for example, .0833 equals 1/12 (feet/inches).

To use Autoscaling without surprises, you should check the units set for the *Blocks* before you insert them and set the *Drawing Units for Block Inserts* in the target drawing according to your needs. You can check the settings for each *Block* by opening the drawing containing the *Blocks*, invoking the *Block* command, and selecting each *Block* from the *Name* drop-down list.

 Although a simple drag-and-drop appears to be a sufficient method for *Block* insertion, be advised that there can be drawbacks. For example, dimension values inside blocks may not be true if a *Block* (or another drawing) is scaled automatically when you drag it into the drawing from DesignCenter. If unsure, right-click from the DesignCenter Content Area and select *Insert Block*, then specify the scale factor at the Command prompt, as described next.

Figure 21-32

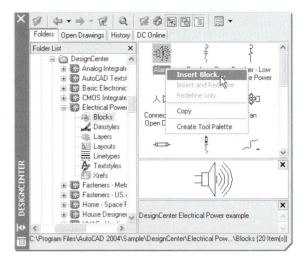

Block **Insertion with Specified Coordinates, Scale, and Rotation**

The second method of inserting *Blocks* from DesignCenter is to highlight the block name or icon, then right-click on it and select *Insert Block* from the shortcut menu (Fig. 21-32). Alternately, double-click on the block name or icon. This method produces the *Insert* dialog box (see "*Insert*" earlier this chapter). Here you can specify coordinates for insertion point, scale factor, and rotation angle. You have your choice (by using checkboxes) to specify these parameters in edit boxes or at the Command line prompts.

If you prefer this method but want to preview the block before inserting it, you can right-drag the block icon into the drawing area. The block appears in the drawing with the shortcut menu options. Select *Insert Block* and proceed to specify the parameters.

To Attach an Xref Using DesignCenter
See Chapter 30, Xreferences.

TOOL PALETTES

Tool palettes are new with AutoCAD 2004. Tool palettes introduce a faster and more visible process of inserting *Blocks* and *Hatch* patterns into a drawing. You can use tool palettes to insert a *Block* instead of using the *Insert* command or DesignCenter. You can also use tool palettes to insert hatch patterns (section lines or fill patterns) into a drawing. *Blocks* and hatches that reside on a tool palette are called tools, and several tool properties including scale, rotation, and layer can be set for each tool individually. The features of tool palettes, using tool palettes to insert *Blocks*, and creating tool palettes are discussed in this chapter. Using tool palettes to insert hatch patterns is discussed in Chapter 26, Section Views.

TOOLPALETTES

Pull-down Menu	COMMAND (TYPE)	ALIAS (TYPE)	Short-cut	Screen (side) Menu	Tablet Menu
Tools Tool Palettes Window	TOOLPALETTES	TP	Ctrl +3	...	...

Any of the methods shown in the Command Table above produce the *Tool Palettes* window (Fig. 21-33). Tool palettes are the individual tabbed areas (like pages) of the *Tool Palettes* window. When you need to add a *Block* to a drawing, you can drag it from the tool palette into your drawing instead of using the *Insert* command or using DesignCenter. You can create your own tool palettes by placing the *Blocks* that you use often on a tool palette.

Figure 21-33

Tool palettes are simple to use. To insert a *Block* (tool) from a tool palette, simply drag it from the palette into the drawing at the desired location (Fig. 21-34). You can use *OSNAPs* when dragging *Blocks* from a tool palette; however, Grid Snap is suppressed during dragging.

When a *Block* is dragged from a tool palette into a drawing, it is scaled automatically according to the ratio of units defined in the *Block* and units used in the current drawing. This action is similar to using DesignCenter, in that the scale and rotation angle may have to be changed after the *Block* is inserted. See "*Block* Insertion with Drag-and-Drop" and "Resulting Scale Factors with Autoscale (Drag-and-Drop)" earlier in this chapter.

Figure 21-34

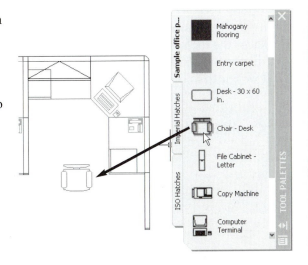

Using the *Tool Palette* window, you can preset the *Block* insertion properties or hatch pattern properties of any tool on a tool palette. For example, you can change the insertion scale of a *Block* or the angle of a hatch pattern. Do this by right-clicking on a tool and selecting *Properties...* (see "Tool *Properties*").

You can also change the scale and rotation angle after inserting the *Block* using Grips (see Chapter 23, Grip Editing).

Palette *Properties*

The options and settings for tool palettes are accessible from shortcut menus in different areas on the *Tool Palettes* window. Right-clicking inside the palette produces the menu shown in Figure 21-35. You can also click on the *Properties* button (bottom of the vertical title bar) to produce essentially the same menu with the addition of *Move, Size,* and *Close* options. Palette properties are saved with your AutoCAD profile.

Move, Size, Close, Allow Docking, Auto-hide
These options are identical to the same features in DesignCenter. See "Resizing, Docking, and Hiding the DesignCenter Palette" earlier in this chapter.

Figure 21-35 ────

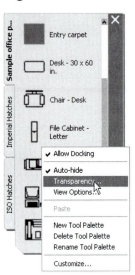

Transparency...
Use this option to make the *Tool Palettes* window transparent so you can "see through" the window to the drawing underneath. Set the desired level in the *Transparency* dialog box that appears (Fig. 21-36).

NOTE: In order to use transparency, you must select *Software Acceleration* in the *Options* dialog box. Select the *System* tab, then *Properties...* under *Current 3D Graphics Display*. Transparency is not available if you are using the Microsoft Windows NT operating system.

Figure 21-36 ────

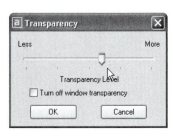

View Options...
This option produces the *View Options* dialog box (Fig. 21-37). You can specify the *Image size* and the *View style*. The *List View* (shown in Figure 21-35) places the tools with text in a vertical column regardless of the palette width, whereas the *Icon with text* option is best to use when you increase the window width to create multiple columns.

New Tool Palette, Delete Tool Palette, Rename Tool Palette, and *Customize*
See "Creating and Managing Tool Palettes" later in this section.

Figure 21-37 ────

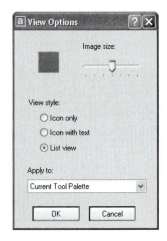

Tool *Properties*

If you right-click on a tool (*Block* or hatch pattern), a shortcut menu appears (Fig. 21-38) allowing you to change properties <u>for that specific tool</u>. You can *Cut* any tool and *Paste* it to another palette or *Copy* a tool to another palette. Selecting *Delete Tool* removes it from that palette only.

Figure 21-38

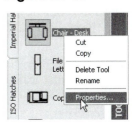

You can change the properties (such as scale and rotation angle) for any specific tool. Do this by right-clicking on the tool and selecting *Properties...* from the shortcut menu (see Figure 21-38) to produce the *Tool Properties* dialog box (Fig. 21-39).

The *Tool Properties* dialog box has two categories of properties—the *Insert* (or *Pattern*) properties category and the *General* properties category. You can control object-specific properties such as *Scale* and *Rotation* angle in the *Insert* list.

Use the *General* properties section <u>only</u> if you want to override the current drawing's property settings such as *Layer*, *Color*, and *Linetype*. (This action is sometimes referred to as setting a tool property <u>override</u>.) When a property override is set, special conditions may exist. For example, if a specified layer does not exist in the drawing, that layer is created automatically when the *Block* or hatch is inserted. If a *Block* or hatch is inserted on a layer that is turned off or frozen, the *Block* or hatch is created on the current layer instead.

Figure 21-39

Creating and Managing Tool Palettes

Creating Tool Palettes using DesignCenter

DesignCenter provides a simple method for creating new tool palettes. Begin by opening both the DesignCenter window and the *Tool Palettes* window. Locate the desired *Blocks* or hatch patterns you want to insert into a new (not yet created) palette and ensure they appear in the DesignCenter Content Area. Next, right-click on any tool you want to move into the new palette so a shortcut menu appears (Fig. 21-40, center). Select *Create Tool Palette* from the menu. A new palette appears with the selected *Block* shown at the top of the palette. Near the new *Block*, a small edit box (not shown) appears for you to enter in the desired tool palette name. After doing so, you are presented with the new palette containing the selected *Block* (as shown in Figure 21-40, right side).

Figure 21-40

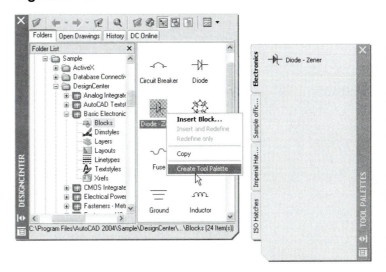

To add additional *Blocks* or hatch patterns into your new palette, drag and drop them from the DesignCenter Content Area into the new palette (Fig. 21-41). Next, use the tool *Properties* shortcut menu, if needed, to preset the scale and rotation angle for the tool, as described earlier.

Figure 21-41

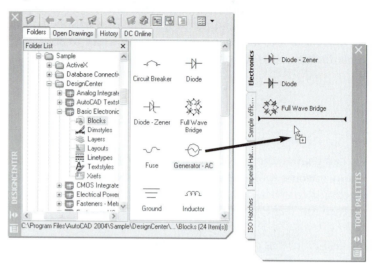

You can also create entire tool palettes fully populated with *Blocks* from all the *Blocks* contained in any drawing using DesignCenter. Do this by locating the drawing file in the Tree View area of DesignCenter, then right-clicking on the drawing file so the shortcut menu appears. Select *Create Tool Palette* (Fig. 21-42). A new tool palette is automatically added to the *Tool Palettes* window containing all the *Blocks* from the selected drawing. The new tool palette has the same name as the drawing.

Figure 21-42

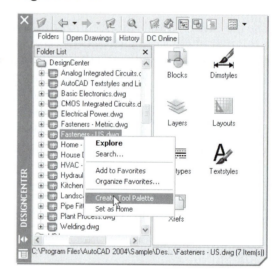

Creating Tool Palettes using Palette *Properties*
Selecting any blank area inside a tool palette and right-clicking or selecting the *Properties* button on the *Tool Palette* window produces the palette *Properties* menu (see Figure 21-35). Selecting *New Tool Palette* from this menu prompts you for a name and creates a new palette. You must then populate the palette with the desired *Blocks* or hatch patterns using DesignCenter as described previously.

Customize, Delete, and Rename Tool Palettes
Use the *Properties* button on the *Tool Palette* window or right-click inside a tool palette to produce the palette *Properties* menu (see previous Figure 21-35). Select *Delete Tool Palette* from the menu to delete an entire palette and all of its contents from the *Tool Palette* window. The *Confirm Tool Palette Deletion* dialog box (not shown) appears for you to *OK* or *Cancel* the deletion. The *Rename* option produces an edit box for you to specify a new name for the current palette.

Figure 21-43

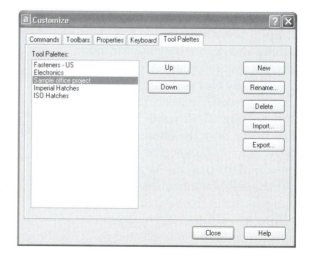

The *Customize* option from the shortcut menu invokes the *Customize* dialog box (Fig. 21-43) with the *Tool Palettes* tab activated. Here you can change the order of the palettes in the window using the *Up* and *Down* buttons. This dialog box also provides buttons for you to *Delete*, create *New*, and *Rename* existing palettes.

Import and *Export* **Tool Palettes**
With the *Export* and *Import* options, you can create several tool palettes to use, each for specific applications. You can also create and export tool palettes for use by others in your school or office.

The *Tool Palettes* tab of the *Customize* dialog box (see Figure 21-43) allows you to export tool palettes to a file. Selecting the *Export* button invokes the *Export Tool Palette* dialog box (not shown) where you can specify the location and name of the file. (The default path for tool palette files is set in the *Files* tab of the *Options* dialog box under *Tool Palettes File Locations*.) Tool palettes are automatically assigned a *.XTP file extension. Use the *Import* button to select and activate a specific tool palette file.

If a tool palette file is set with a read-only attribute, a "lock" icon displays in a lower corner of the tool palette. This indicates that the tool palette cannot be modified beyond changing its display settings and rearranging the icons.

NOTE: Tool palettes are not saved with the drawing file, but are saved as part of your AutoCAD profile. For example, when you open the *Tool Palettes* window from any drawing, your specific set of palettes appears, whereas another AutoCAD user on the same computer can have a different set of palettes based on his or her profile. Therefore, importing a tool palette file (*.XTP) adds the specific palettes to your profile for use in any drawing.

Updating the Icon for a Tool
If you change (redefine) a *Block* or hatch definition you must update its icon in the tool palette in which it appears. Right-click on the tool and select *Properties*. In the *Tool Properties* dialog box, change the entry in the *Source File* field for *Blocks* or the *Pattern name* field for hatches (to any path), and then change the entry back again to the correct location. This forces an update of the icon for that tool. Alternately, you can delete the tool and then replace it using DesignCenter.

2004

CHAPTER EXERCISES

1. *Block, Insert*
    In the next several exercises, you will create an office floor plan, then create pieces of furniture as *Blocks* and *Insert* them into the office. All of the block-related commands are used.

    A. Start a *New* drawing. Select **Start from Scratch** and use the **Imperial** defaults. Use *Save* and assign the name **OFF-ATT**. Set up the drawing as follows:

1. *Units*	Architectural	1/2" *Precision*		
2. *Limits*	48' x 36'	(1/4"=1' scale on an A size sheet), drawing scale factor = 48		
3. *Snap, Grid*	3			
4. *Grid*	12			
5. *Layers*	FLOORPLAN	continuous	.014	colors of your choice
	FURNITURE	continuous	.060	
	ELEC-HDWR	continuous	.060	
	ELEC-LINES	hidden 2	.060	
	DIM-FLOOR	continuous	.060	
	DIM-ELEC	continuous	.060	
	TEXT	continuous	.060	
	TITLE	continuous	.060	
6. *Text Style*	CityBlueprint	CityBlueprint (TrueType font)		
7. *Ltscale*	48			

B.   Create the floor plan shown in Figure 21-44. Center the geometry in the *Limits*. Draw on layer **FLOORPLAN**. Use any method you want for construction (e.g., *Line, Pline, Xline, Mline, Offset*).

Figure 21-44 ——————————————————————————

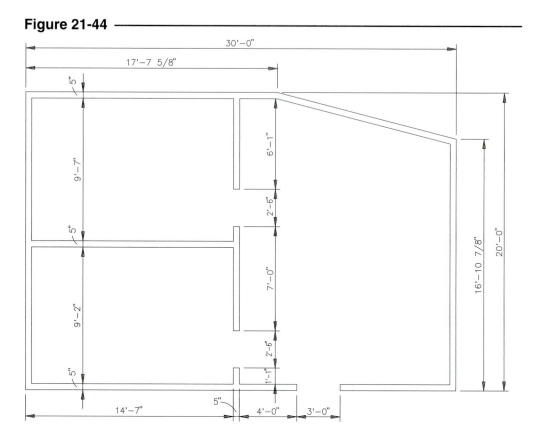

C.   Create the furniture shown in Figure 21-45. Draw on layer **FURNITURE**. Locate the pieces anywhere for now. Do <u>not</u> make each piece a *Block*. *Save* the drawing as **OFF-ATT**.

Now make each piece a *Block*. Use the *name* as indicated and the *insertion base point* as shown by the "blip." Next, use the *Block* command again but only to check the list of *Blocks*. Use *SaveAs* and rename the drawing **OFFICE**.

Figure 21-45 ——————————————————————————

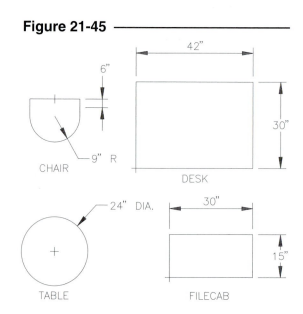

D. Use *Insert* to insert the furniture into the drawing, as shown in Figure 21-46. You may use your own arrangement for the furniture, but *Insert* the same number of each piece as shown. *Save* the drawing.

Figure 21-46

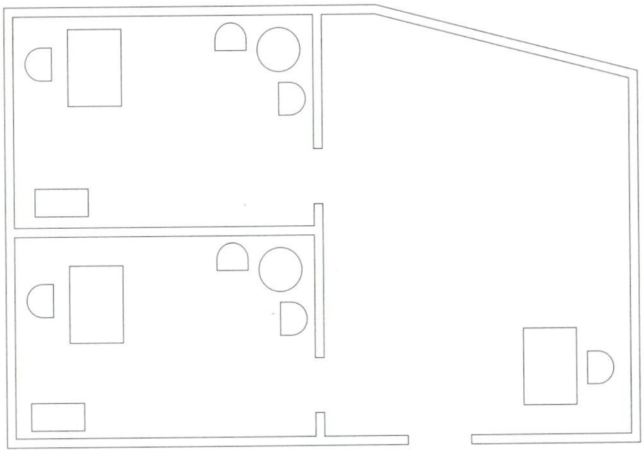

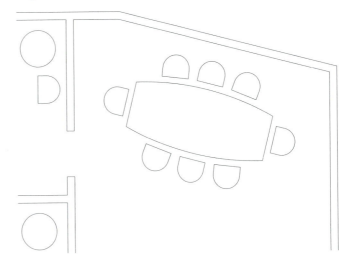

2. **Creating a .DWG file for *Insertion*, *Base***

Figure 21-47

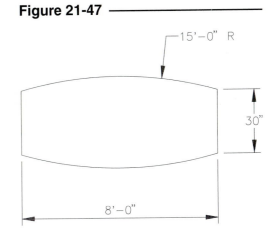

Begin a *New* drawing. Assign the name **CONFTABL**. Create the table as shown in Figure 21-47 on *Layer* 0. Since this drawing is intended for insertion into the **OFFICE** drawing, use the *Base* command to assign an insertion base point at the lower-left corner of the table.

When you are finished, *Save* the drawing.

3. *Insert, Explode, Divide*

Figure 21-48

A. *Open* the **OFFICE** drawing. Ensure that layer **FURNITURE** is current. Use **DesignCenter** to bring the **CONFTABL** drawing in as a *Block* in the placement shown in Figure 21-48.

Notice that the CONFTABL assumes the linetype and color of the current layer, since it was created on layer **0**.

B. *Explode* the CONFTABL. The *Exploded* CONFTABL returns to *Layer* 0, so use the *Properties* palette to change it back to *Layer* **FURNITURE**. Then use the *Divide* command (with the *Block* option) to insert the **CHAIR** block as shown in Figure 21-48. Also *Insert* a **CHAIR** at each end of the table. *Save* the drawing.

4. *Wblock, BYBLOCK setting*

Figure 21-49

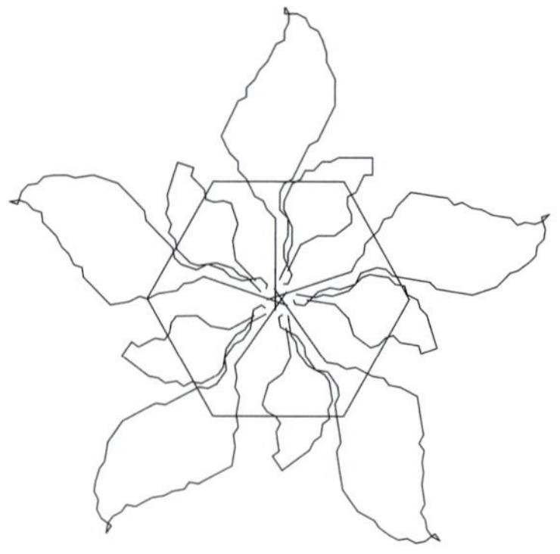

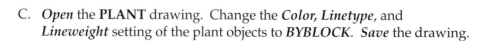

A. *Open* the **EFF-APT** drawing you worked on in Chapter 15 Exercises. Use the *Wblock* command to transform the plant into a .DWG file (Fig. 21-49). Use the name **PLANT** and specify the *Insertion base point* at the center. Do not save the EFF-APT drawing. *Close* **EFF-APT**.

B. *Open* the **OFFICE** drawing and use **DesignCenter** to *Insert* the **PLANT** into one of the three rooms. The plant probably appears in a different color than the current layer. Why? Check the *Layer* listing to see if any new layers came in with the PLANT block. *Erase* the PLANT block.

PLANT

C. *Open* the **PLANT** drawing. Change the *Color, Linetype*, and *Lineweight* setting of the plant objects to *BYBLOCK*. *Save* the drawing.

D. *Open* the **OFFICE** drawing again and *Insert* the **PLANT** onto the **FURNITURE** layer. It should appear now in the current layer's *color, linetype*, and *lineweight*. *Insert* a **PLANT** into each of the 3 rooms. *Save* the drawing.

5. **Redefining a** *Block*

Figure 21-50

After a successful meeting, the client accepts the proposed office design with one small change. The client requests a slightly larger chair than that specified. *Explode* one of the **CHAIR** blocks. Use the *Scale* command to increase the size slightly or otherwise redesign the chair in some way. Use the *Block* command to redefine the **CHAIR** block. All previous insertions of the CHAIR should reflect the design change. *Save* the drawing. Your design should look similar to that shown in Figure 21-50. *Plot* to a standard scale based on your plotting capabilities.

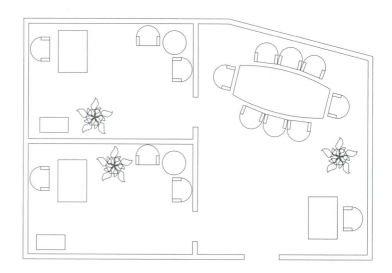

6. *Rename*

Open the **OFFICE** drawing. Enter *Rename* (or select from the *Format* pull-down menu) to produce the *Rename* dialog box. Rename the *Blocks* as follows:

New Name	Old Name
DSK	**DESK**
CHR	**CHAIR**
TBL	**TABLE**
FLC	**FILECAB**

Next, use the *Rename* dialog box again to rename the *Layers* as indicated below:

New Name	Old Name
FLOORPLN-DIM	**DIM-FLOOR**
ELEC-DIM	**DIM-ELEC**

Use the *Rename* dialog box again to change the *Text Style* name as follows:

New Name	Old Name
ARCH-FONT	**CITYBLUEPRINT**

Use *SaveAs* to save and rename the drawing to **OFFICE-REN**.

7. *Purge*

   Using the Windows Explorer, check the file size of **OFFICE-REN.DWG**. Now use *Purge* to remove any unreferenced named objects. *Exit AutoCAD* and *Save Changes*. Check the file size again. Is the file size slightly smaller?

8. **Tool Palettes**

   In this exercise you will change some of the *Blocks* in the OFFICE drawing using *Blocks* from the standard Tool Palettes window.

   A. *Open* the **OFFICE-REN** drawing again if not already open. Assume as a result of the meeting with the client, new desks and chairs are requested for the receptionist and the two offices. Use the *Toolpalettes* command or press **Ctrl+3** to produce the Tool Palettes window. Select the standard AutoCAD *Sample Office Project* palette (showing the office furniture).

   B. Right-click on the *Chair - Desk* tool and select *Properties*. In the *Tool Properties* dialog box that appears, note that the scale is 1.00 and the layer is set to *Bylayer*. Check the settings for the *Desk – 30 x 60 in*. With these settings, the new furniture should be usable without changes.

   C. Make **FURNITURE** the current layer. *Erase* the three desks and the three desk chairs. Drag and drop three chairs and three desks from the palette window and place them near the desired locations in the office. Use *Rotate* and *Move* to position the new *Blocks* into suitable locations and orientations. Your drawing should look similar to that in Figure 21-51. Use *SaveAs* and name the drawing **OFFICE-REV**.

Figure 21-51

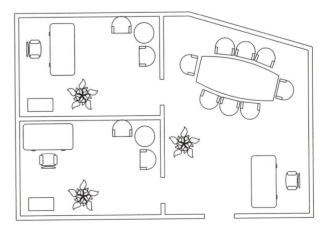

9. **Create a Tool Palette**

    A. *Close* all open drawings and begin a *New* drawing. Open **DesignCenter** and locate the **OFFICE** drawing in the Tree View area. Expand the plus (+) symbol next to OFFICE.DWG, then click on the word *Blocks* just below the drawing name. All of the *Blocks* contained in the drawing should appear in the Content Area.

    B. Open the **Tool Palettes**. Now, back in DesignCenter, locate the OFFICE drawing in the Tree View area and right-click on it. From the shortcut menu select *Create Tool Palette*.

    C. The previous action should have automatically created a new tool palette named OFFICE and populated it with all the *Blocks* contained in the drawing! Examine the new tool palette. Right-click on any tool and examine its *Properties*. Now you have a new tool palette ready to use for inserting the office furniture *Blocks* into any other drawings.

10. Create the process flow diagram shown in Figure 21-52. Create symbols (*Blocks*) for each of the valves and gates. Use the names indicated (for the *Blocks*) and include the text in your drawing. *Save* the drawing as **PFD**.

Figure 21-52

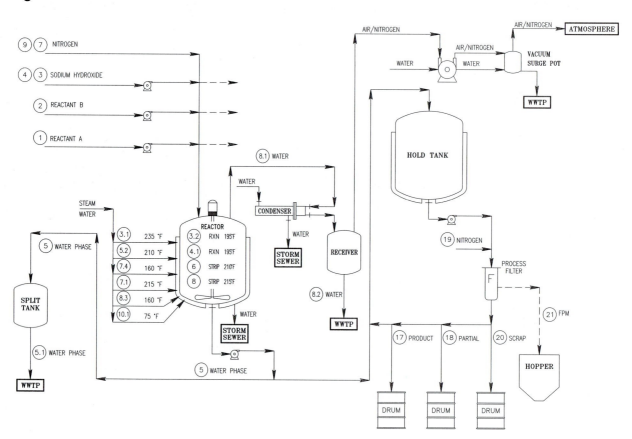

22

BLOCK ATTRIBUTES

Chapter Objectives

After completing this chapter you should:

1. be able to define block attributes using the *Attribute Definition* dialog box and the *–Attdef* command;

2. be able to control the display of attributes in the drawing with *Attdisp*;

3. know how to edit existing attributes with *Attedit*, the *Enhanced Attribute Editor*, and the *Find and Replace* dialog box;

4. be able to edit attributes globally and individually with *–Attedit*;

5. be able to redefine attributes for existing blocks using the *Block Attribute Manager* and the *Attredef* command;

6. be able to use the *Attribute Extraction* wizard to create extract files of the attribute information contained in one or more drawings.

CONCEPTS

A block <u>attribute</u> is a line of text (numbers or letters) associated with a block. An attribute can be thought of as a label or description for a block. A block can have multiple attributes. The attributes are included with the drawing objects when the block is defined (using the *Block* command). When the block is *Inserted*, its attributes are also *Inserted*.

Since *Blocks* are typically used as symbols in a drawing, attributes are text strings that label or describe each symbol. For example, you can have a series of symbols such as transistors, resistors, and capacitors prepared for creating electrical schematics. The associated attributes can give the related description of each block, such as ohms or wattage values, model number, part number, and cost. If your symbols are doors, windows, and fixtures for architectural applications, attached attributes can include size, cost, and manufacturer. A mechanical engineer can have a series of blocks representing fasteners with attached attributes to specify fastener type, major diameter, pitch, length, etc. Attributes could even be used to automate the process of entering text into a title block, assuming the title block was *Inserted* as a block or separate .DWG file.

Attributes can add another level of significance to an AutoCAD drawing. Not only can attributes automate the process of placing the text attached to a block, but the inserted attribute information can be <u>extracted</u> from a drawing to form a bill of materials or used for cost analysis, for example. Extracted text can be imported to a database or spreadsheet program for further processing and analysis.

Attributes are created with the *Attdef* (attribute define) command. *Attdef* operates similarly to *Dtext*, prompting you for text height, placement, and justification. During attribute definition, parameters can be adjusted that determine how the attributes will appear when they are inserted.

This chapter discusses defining attributes, inserting attributed blocks, displaying and editing attributes in a drawing, and extracting attributes from drawings.

CREATING ATTRIBUTES

Steps for Creating and Inserting Block Attributes

1. Create the objects that will comprise the block. Do not use the *Block* command yet; only draw the geometry.

2. Use the *Attdef* command to create and place the desired text strings associated with the geometry comprising the proposed block.

3. Use the *Block* command to convert the drawing geometry (objects) and the attributes (text) into a named block. When prompted to "Select objects:" for the block, select both the drawing objects and the text (attributes). The block and attributes disappear but are defined in the drawing's block definition table.

4. Use *Insert* to insert the attributed block into the drawing. Edit the text attributes as necessary (if parameters were set as such) during the insertion process.

ATTDEF

	COMMAND (TYPE)	ALIAS (TYPE)	Short-cut	Screen (side) Menu	Tablet Menu
Pull-down Menu					
Draw *Block >* *Define Attributes...*	*ATTDEF or* *-ATTDEF*	*ATT or* *-ATT*	...	DRAW2 *Attdef*	...

Attdef (attribute define) allows you to define attributes for future combination with a block. If you intend to associate the text attributes with drawing objects (*Line, Arc, Circle*, etc.) for the block, it is usually preferred to draw the objects before using *Attdef*. In this way, the text can be located in reference to the drawing objects.

Attdef allows you to create the text and provides justification, height, and rotation options similar to the *Dtext* command. You can also define parameters, called Attribute Modes, specifying how the text will be inserted—*Invisible, Constant, Verify,* or *Preset*.

Figure 22-1

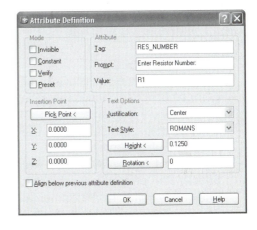

The *Attdef* command differs, depending on the method used to invoke it. If you select from any menus or the icon or type *Attdef*, the *Attribute Definition* dialog box appears (Fig. 22-1). If you type -*Attdef* (with a hyphen prefix), the command line format is used. The Command line format is presented below:

```
Command: -attdef
Current attribute modes: Invisible=N  Constant=N  Verify=N
Preset=N
Enter an option to change [Invisible/Constant/Verify/Preset] <done>: Enter or (option)
Enter attribute tag name: (Enter tag, no spaces)
Enter attribute prompt: (Enter desired prompt, spaces allowed)
Enter default attribute value: (Enter desired default text)
Current text style: "STANDARD"  Text height:  0.2000
Specify start point of text or [Justify/Style]: Enter or (option)
Specify height <0.2000>: Enter or (value)
Specify rotation angle of text <0>: Enter or (value)
Command:
```

The first option, "attribute modes," is described later. For this example, the default attribute modes have been accepted.

The *Attribute Definition* dialog box has identical options that are found in the Command line format, including justification options. The same entries have been made here to define the first attribute example shown in Figure 22-2.

Figure 22-2

Attribute Tag
This is the descriptor for the <u>type of text</u> to be entered, such as MODEL_NO, PART_NO, or NAME. <u>Spaces cannot be used</u> in the Attribute Tag.

Attribute Prompt
The prompt is what words (<u>prompt) you want to appear</u> when the block is *Inserted* and when the actual text (values) must be entered. Spaces and punctuation can be included.

Default Attribute Value

This is the <u>default text</u> that appears with the block when it is *Inserted*. Supply the typical or expected words or numbers.

Justify

Any of the typical text justification methods can be selected using this option. (See Chapter 18, Creating and Editing Text, if you need help with justification.)

Style

This option allows you to select from existing text *Styles* in the drawing. Otherwise, the attribute is drawn in the current *Style*.

As an example, assume that you are using the *Attdef* command to make attributes attached to a resistor symbol to be *Blocked* (Fig. 22-2). The scenario may appear as follows or as presented in the *Attribute Definition* dialog box in Figure 22-1:

```
Command: -attdef
Current attribute modes: Invisible=N  Constant=N  Verify=N  Preset=N
Enter an option to change [Invisible/Constant/Verify/Preset] <done>: Enter
Enter attribute tag name: RES_NUMBER
Enter attribute prompt: Enter Resistor Number:
Enter default attribute value: R1
Current text style: "ROMANS"  Text height:  0.2000
Specify start point of text or [Justify/Style]: J
Enter an option [Align/Fit/Center/Middle/Right/TL/TC/TR/ML/MC/MR/BL/BC/BR]: C
Specify center point of text: PICK
Specify height <0.2000>: .125
Specify rotation angle of text <0>: Enter
```

The attribute would then appear at the selected location, as shown in Figure 22-2. Keep in mind that the *Lines* comprising the resistor were created before creating attributes. Also remember that the resistor and the attribute are not yet *Blocked*. The *Attribute Tag* <u>only</u> is displayed until the *Block* command is used.

Two other attributes could be created and positioned beneath the first one by pressing **Enter** when prompted for the "<Start point>:" (Command line format) or selecting the "*Align below previous attribute definition*" checkbox (dialog box format). The command syntax may read like this:

```
Command: -attdef
Current attribute modes: Invisible=N  Constant=N  Verify=N  Preset=N
Enter an option to change [Invisible/Constant/Verify/Preset] <done>: Enter
Enter attribute tag name: RESISTANCE
Enter attribute prompt: Enter Resistor Number:
Enter default attribute value: 0 ohms
Current text style: "ROMANS"  Text height:  0.1250
Specify start point of text or [Justify/Style]: Enter
```

```
Command: -attdef
Current attribute modes: Invisible=N  Constant=N  Verify=N  Preset=N
Enter an option to change [Invisible/Constant/Verify/Preset] <done>: Enter
Enter attribute tag name: PART_NO
Enter attribute prompt: Enter Part Number:
Enter default attribute value: 0-0000
Current text style: "ROMANS"  Text height:  0.1250
Specify start point of text or [Justify/Style]: Enter
```

The resulting unblocked symbol and three attributes appear as shown in Figure 22-3. Only the *Attribute Tags* are displayed at this point. The *Block* command has not yet been used.

Figure 22-3

RES_NUMBER
RESISTANCE
PART_NO

The *Attributes Modes* define the appearance or the action required when you *Insert* the attributes. The options are as follows.

Invisible
An *Invisible* attribute is not displayed in the drawing after insertion. This option can be used to prevent unnecessary information from cluttering the drawing and slowing regeneration time. The *Attdisp* command can be used later to make these attributes visible.

Constant
This option gives the attribute a fixed value for all insertions of the block. In other words, the attribute always has the same text and cannot be changed. You are not prompted for the value upon insertion.

Verify
When attributes are entered in Command line format, this option forces you to verify that the attribute is correct when the block is inserted. This option has no effect when attributes are entered in dialog box format (when *ATTDIA*=1).

Preset
This option prevents you from having to enter a value for the attribute during insertion if you are using the Command line format. In this case, you are not prompted for the attribute value during insertion. Instead, the default value is used for the block, although it can be changed later with the *Attedit* command.

The *Attribute Modes* are toggles (Yes or No). In Command line format, the options are toggled by entering the appropriate letter, which reverses its current position (Y or N). For example, to create an *Invisible* attribute, type **"I"** at the prompt:

```
Command: -attdef
Current attribute modes: Invisible=N  Constant=N  Verify=N  Preset=N
Enter an option to change [Invisible/Constant/Verify/Preset] <done>: I
Current attribute modes: Invisible=Y  Constant=N  Verify=N  Preset=N
Enter an option to change [Invisible/Constant/Verify/Preset] <done>: Enter
```

Note that the second prompt (fourth line) reflects the new position of the previous request.

Alternately, attribute modes could be defined using the *Attribute Definition* dialog box (Fig. 22-4). The *Mode* cluster of checkboxes (upper-left corner) makes setting attribute modes simpler than in Command line format.

Using our previous example, two additional attributes can be created with specific *Invisible* and *Verify* attribute modes. Either the dialog box or Command line format can be used. The following command syntax shows the creation of an *Invisible* attribute:

Figure 22-4

```
Command: -attdef
Current attribute modes: Invisible=N  Constant=N  Verify=N  Preset=N
Enter an option to change [Invisible/Constant/Verify/Preset] <done>: I
Current attribute modes: Invisible=Y  Constant=N  Verify=N  Preset=N
Enter an option to change [Invisible/Constant/Verify/Preset] <done>: Enter
Enter attribute tag name: MANUFACTURER
Enter attribute prompt: Enter Manufacturer:
Enter default attribute value: Enter
Current text style:  "ROMANS"  Text height:  0.125000
Specify start point of text or [Justify/Style]: Enter
Command:
```

The first attribute has been defined. Next, a second attribute is defined having *Invisible* and *Verify* parameters. (Figure 22-4 also shows the correct entries for creation of this attribute.)

```
Command: -attdef
Current attribute modes: Invisible=Y  Constant=N  Verify=N  Preset=N
Enter an option to change [Invisible/Constant/Verify/Preset] <done>: V
Current attribute modes: Invisible=Y  Constant=N  Verify=Y  Preset=N
Enter an option to change [Invisible/Constant/Verify/Preset] <done>: Enter
Enter attribute tag name: COST
Enter attribute prompt: Enter Cost:
Enter default attribute value: 0.00
Current text style:  "ROMANS"  Text height:  0.125000
Specify start point of text or [Justify/Style]: Enter
Command:
```

Figure 22-5

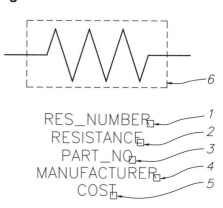

Tags for *Invisible* attributes appear as any other attribute tags. The *Values*, however, are invisible when the block is inserted.

Using the *Block* Command to Create an Attributed Block

Next, the *Block* command is used to transform the drawing objects and the text (attributes) into an attributed block. When you are prompted to "Select objects:" in the *Block* command, select the attributes in the order you desire their prompts to appear on insertion. In other words, the first attribute selected during the *Block* command (RES_NUMBER) is the first attribute prompted for editing during the *Insert* command. PICK the attributes one at a time in order. Then PICK the drawing objects (a crossing window can be used). Figure 22-5 shows this sequence.

Figure 22-6

If the attributes are selected with a window, they are inserted in reverse order of creation, so you would be prompted for COST first, MANUFACTURER second, and so on.

If you use the *Block* command (instead of *–Block*), the *Block Definition* dialog box appears (Fig. 22-6). Use the *Select objects* button to return to the geometry and select the attributes and the geometry as shown in Figure 22-5. Next, pick the *Base point* as the left *Endpoint* of the geometry.

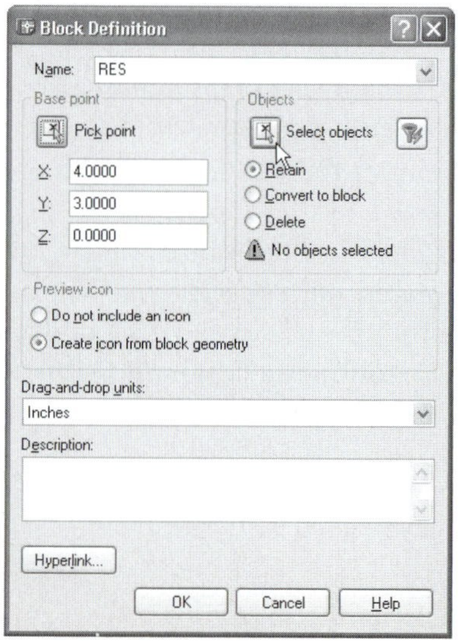

Using the *–Block* command, the following command sequence appears. Select objects as shown in Figure 22-5.

Command: **-block**
Enter block name or [?]: **res**
Specify insertion base point: **PICK**
Select objects: **PICK** (1)
Select objects: **PICK** (2)
Select objects: **PICK** (3)
Select objects: **PICK** (4)
Select objects: **PICK** (5)
Select objects: **PICK** (6)
Select objects: **Enter**
Command:

When the geometry to define the block has been defined and you select *OK* in the *Block Definition* dialog box, the *Edit Attributes* dialog box appears (see Figure 22-7). This is offered only to verify the desired default settings for the attribute values.

Inserting Attributed Blocks

Use the *Insert* command to bring attributed blocks into your drawing as you would to bring any block into your drawing. When the *Insert* dialog box appears (not shown), make the desired settings and select *OK*. Next, specify the insertion point (where you want to place the block in your drawing). At this point, the *Edit Attributes* dialog box appears (Fig. 22-7). Enter the desired attribute values for each block insertion.

Notice that the prompts appearing in the dialog box are those that you specified as the "attribute prompt" with the *Attdef* command. The order of the prompts matches the order of selection during the *Block* command.

Figure 22-7

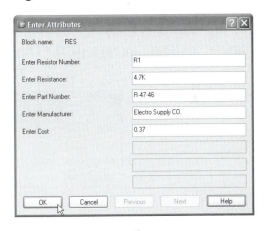

If you use the *–Insert* command, you are prompted at the command line for the attribute values. The Command line format is as follows:

Command: **-insert**
Enter block name or [?]: **res**
Specify insertion point or [Scale/X/Y/Z/Rotate/PScale/PX/PY/PZ/PRotate]: **PICK**
Enter X scale factor, specify opposite corner, or [Corner/XYZ] <1>: **Enter**
Enter Y scale factor <use X scale factor>: **Enter**
Specify rotation angle <0>: **Enter**
Enter attribute values
Enter Resistor Number: <0>: **R1**
Enter Resistance: <0 ohms>: **4.7K**
Enter Part Number: <0-0000>: **R4746**
Enter Manufacturer: **Electro Supply Co.**
Enter Cost <0.00>: **.37**
Verify attribute values
Enter Cost <.37>: **Enter**
Command:

Note the repeated prompt (*Verify*) for the COST attribute.

Entering the attribute values as indicated by either method above yields the block insertion shown in Figure 22-8. Notice the absence of the last two attribute values since they are *Invisible*.

Figure 22-8

The *ATTREQ* Variable

When *Inserting* blocks with attributes, you can force the attribute requests to be suppressed. In other words, you can disable the prompts asking for attribute values when the blocks are *Inserted*. To do this, use the *ATTREQ* (attribute request) system variable. A setting of **1** (the default) turns attribute requesting on, and a value of **0** disables the attribute value prompts.

The attribute prompts can be disabled when you want to *Insert* several blocks but do not want to enter the attribute values right away. The attribute values for each block can be entered at a later time using the *Attedit* command. These attribute editing commands are discussed on the following pages.

DISPLAYING AND EDITING ATTRIBUTES

ATTDISP

Pull-down Menu	COMMAND (TYPE)	ALIAS (TYPE)	Short-cut	Screen (side) Menu	Tablet Menu
View *Display >* *Attribute Display*	*ATTDISP*	...	...	VIEW 2 *Attdisp*	L,1

The *Attdisp* (attribute display) command allows you to control the visibility state of all inserted attributes contained in the drawing:

    Command: **attdisp**
    Normal/ON/OFF <current value>: (option)
    Command:

Attdisp has three positions. Changing the state forces a regeneration.

Figure 22-9

On
All attributes (normal and *Invisible*) in the drawing are displayed.

Off
No attributes in the drawing are displayed.

Normal
Normal attributes are displayed; *Invisible* attributes are not displayed.

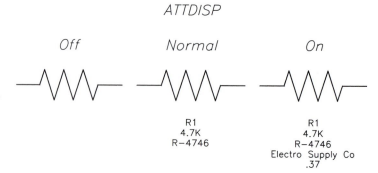

Figure 22-9 illustrates the RES block example with each of the three *Attdisp* options. The last two attributes were defined with *Invisible* modes, but are displayed with the *On* state of *Attdisp*. (Since *Attdisp* affects attributes globally, it is not possible to display the three options at one time in a drawing.)

When attributes are defined using the *Invisible* mode of the *Attdef* command, the attribute values are normally not displayed with the block after *Insertion*. However, the *Attdisp On* option allows you to override the *Invisible* mode. Turning all attributes *Off* with *Attdisp* can be useful for many applications when it is desirable to display or plot only the drawing geometry. The *Attdisp* command changes the *ATTMODE* system variable.

ATTEDIT

Pull-down Menu	COMMAND (TYPE)	ALIAS (TYPE)	Short-cut	Screen (side) Menu	Tablet Menu
...	*ATTEDIT*	*ATE*	...	MODIFY1 *Attedit*	...

The *Attedit* command invokes the *Edit Attributes* dialog box (Fig. 22-10). This dialog box allows you to edit only attribute <u>values of existing</u> blocks in the drawing. The configuration and operation of the box are identical to the *Edit Attributes* dialog box (Fig. 22-7).

When the *Attedit* command is entered, AutoCAD requests that you select a block. Any existing attributed block can be selected for attribute editing.

    Command: **attedit**
    Select block reference: **PICK**

At this point, the dialog box appears. After the desired changes have been made, selecting the *OK* tile updates the selected block with the indicated changes.

Figure 22-10

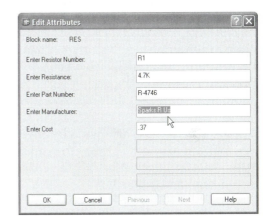

MULTIPLE ATTEDIT

Using *Attedit* with the *Multiple* modifier causes AutoCAD to repeat the *Attedit* command requesting multiple block selections over and over until you cancel the command with Escape. You must type *Attedit* (or any other command) for it to operate with the *Multiple* modifier. (See Chapter 43 for other uses of the *Multiple* modifier.)

    Command: **multiple**
    Enter command name to repeat: **attedit**
    Select block reference: **PICK**

This form of the command can be used to query blocks in an existing drawing. For example, instead of using the *Attdisp* command to display the *Invisible* attributes, *Multiple Attedit* can be used on selected blocks to display the *Invisible* attributes in dialog box form. Another example is the use of *Attdisp* to turn visibility of <u>all</u> attributes *Off* in a drawing, thus simplifying the drawing appearance and speeding regenerations. The *Multiple Attedit* command could then be used to query the blocks to display the attributes in dialog box form.

EATTEDIT

	Pull-down Menu	COMMAND (TYPE)	ALIAS (TYPE)	Short-cut	Screen (side) Menu	Tablet Menu
	Modify Object > Attribute > Single...	*EATTEDIT*	...	...	...	...

The *Eattedit* command produces the *Enhanced Attribute Editor* (Fig. 22-11). This tool is new with AutoCAD 2002 and is intended as an enhanced *Attedit* tool for existing block insertions. The *Enhanced Attribute Editor* allows you to modify attribute <u>values</u> similar to using *Attedit*; however, you can also modify attribute <u>text settings</u> and <u>properties</u> such as layer, linetype, and color. You can change multiple attributes, one at a time, in one session of the *Enhanced Attribute Editor* by using the *Select block* button in the editor's upper-right corner. Changes you make to the attributes are immediately visible in the drawing. If *Dblclkedit* is *On*, double-clicking on a *Block* with attributes automatically invokes the *Enhanced Attribute Editor*.

Figure 22-11 ——————————

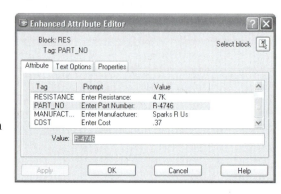

When you invoke the command by any method (shown in the previous command table), AutoCAD prompts you to select a block. You can select the *Block* geometry or any attribute of any *Block* in the drawing. After block selection, the *Enhanced Attribute Editor* appears and displays three tabs (*Attribute, Text Options,* and *Properties*).

    Command: **eattedit**
    Select a block:

Figure 22-12 ——————————

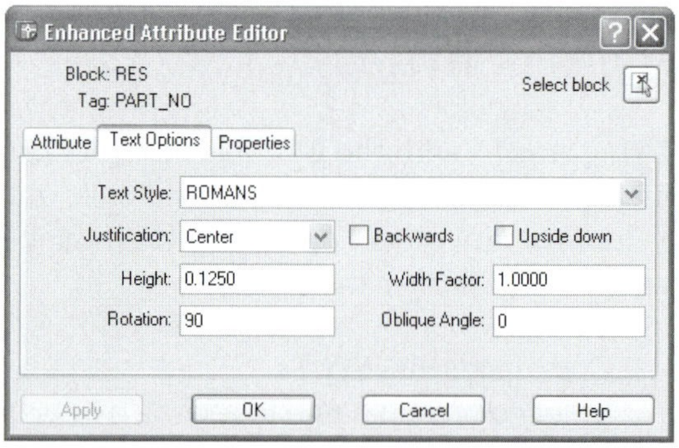

Attribute
When the *Enhanced Attribute Editor* appears, the *Attribute* tab is active by default (see Figure 22-11, above). If you PICKed an attribute during block selection, that specific attribute is highlighted in the editor and its text is placed in the *Value* edit box. You can change any attribute value of the *Block*, just as you can using the *Attedit* command.

Text Options
In this tab (Fig. 22-12), you can change any of the text properties for an attribute, such as the *Text Style, Justification, Height, Rotation, Width Factor, Oblique Angle,* and other properties, depending on the type of text font used (.SHX or .TTF). The changes apply only to the one attribute value you selected in the *Attribute* tab, so before using the *Text Options* tab, <u>select the desired attribute in the *Attribute* tab</u>.

Figure 22-13 ——————————

Properties
This tab gives you the ability to change the selected attribute's *Layer, Color, Linetype, Lineweight,* and *Plot style* properties (Fig. 22-13). Since the changes are applied to the highlighted attribute only, remember to <u>select the desired attribute in the *Attribute* tab</u> before using this tab.

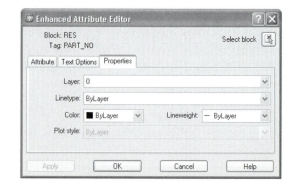

Apply, OK, and Cancel

Note that the *Enhanced Attribute Editor* has an *Apply* button in addition to the *OK* and *Cancel* buttons. Changes made in any tab of the *Enhanced Attribute Editor* are automatically and immediately displayed in the drawing temporarily until *Apply*, *OK*, or *Cancel* is used. The *Apply* button forces the change permanently and immediately to the drawing. Beware: once you use *Apply*, the changes to the drawing are not canceled if you close the dialog box using the *Cancel* button. Using the *OK* button will also apply all changes; however, unlike using the *Apply* button, you still have the option to use *Cancel* before ending the session. If you are making changes to only one block, use *OK* to apply the changes and close the *Enhanced Attribute Editor*. If you want to change attributes for several blocks, use the *Apply* button between block selection. If you want to cancel applied changes, you must use the *Undo* command immediately after using the *Enhanced Attribute Editor*.

PROPERTIES

Pull-down Menu	COMMAND (TYPE)	ALIAS (TYPE)	Short-cut	Screen (side) Menu	Tablet Menu
Modify *Properties*	*PROPERTIES*	*CH* or *PR*	*Ctrl +1*	*MODIFY1* *Property*	*Y, 12* or *Y, 13*

In AutoCAD 2004 you can edit attribute values from the new *Properties* palette (previous versions of AutoCAD do not display attributes of a *Block* in the *Properties* window). Simply open the *Properties* palette, select the desired attributed *Block*, and all of the attributes appear at the <u>bottom</u> of the palette (Fig. 22-14). You may have to scroll down to view the attributes section in the palette.

Remember, this method is useful only for editing the attribute <u>values</u>, similar to the function of using the *Attedit* command. However, one advantage to using *Properties* instead of *Attedit* is that you can edit the rotation angle and location of the *Block*. If you want to edit the properties of the individual attributes (text properties), use *Eattedit*.

Figure 22-14

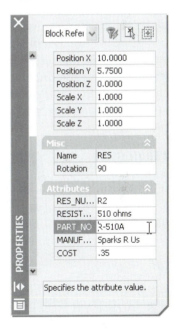

FIND

Pull-down Menu	COMMAND (TYPE)	ALIAS (TYPE)	Short-cut	Screen (side) Menu	Tablet Menu
Edit *Find...*	*FIND*	...	(Default Menu) *Find...*	...	*X,10*

Find is used to find or replace text strings <u>globally</u> in a drawing as explained previously in Chapter 18, Creating and Editing Text. *Find* can also be used to find and replace <u>attribute values</u> and is an alternative to using *Attedit*. *Attedit* requires that you actually PICK the attributes that you want to change, whereas *Find* searches the drawing for you based on the value you enter in the *Find text string* edit box.

Using the *Find* command produces the *Find and Replace* dialog box (Fig. 22-15). To use *Find* to edit attributes in a drawing, you must ensure that *Find* looks for block attributes when the search is made. To do this, use the *Options* button in the *Find and Replace* dialog box.

Figure 22-15

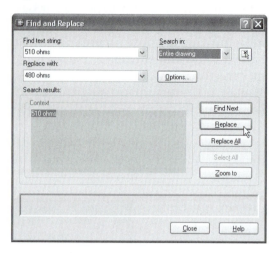

In the *Find and Replace Options* dialog box that appears, select *Block Attribute Value* to be included in the search (Fig. 22-16). Next, select *OK* to return to the main dialog box and to proceed with the search.

Figure 22-16

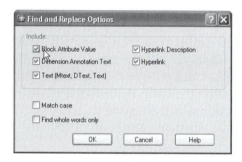

Specify the attribute value to find in the *Find text string* edit box (see Figure 22-15). Enter the replacement attribute value in the *Replace with* edit box. Select the *Find* button to begin the search, then use *Replace* and *Find Next*, or *Replace All* as needed. Refer to Chapter 18 for details on these and other features of the *Find and Replace* dialog box, such as the *Zoom to* feature.

-ATTEDIT

Pull-down Menu	COMMAND (TYPE)	ALIAS (TYPE)	Short-cut	Screen (side) Menu	Tablet Menu
Modify Object > Attribute > Global	-ATTEDIT	-ATE	...	MODIFY1 Attedit	...

-Attedit (Attribute Edit) is an older Command line attribute editing tool that has several features similar to those included in the newer dialog box-based commands. *–Attedit* is similar to *Attedit* and *Eattedit* in that you can edit the values of selected attributes in a drawing. *–Attedit* also has options similar to *Eattedit* that allow you to change the attribute text options and some properties. You can also use *–Attedit* in a manner similar to *Find* to globally edit attribute values in a drawing.

    Command: *-attedit*
    Edit attributes one at a time? [Yes/No] <Y>:

The following prompts depend on your response to the first prompt:

N (No)
Indicates that you want global editing. (In other words, all attributes are selected.) Attributes can then be further selected by block name, tag, or value. Only attribute values can be edited using this global mode.

Y (Yes)

Allows you to PICK each attribute you want to edit. You can further filter (restrict) the selected set by block name, tag, and value. You can then <u>edit any property or placement</u> of the attribute.

No matter which option you choose, the following prompts appear, which allow you to filter, or further restrict, the set of attributes for editing. You can specify the selection set to include only specific block names, tags, or values:

```
Enter block name specification <*>: Enter desired block name(s)
Enter attribute tag specification <*>: Enter desired tag(s)
Enter attribute value specification <*>: Enter desired value(s)
```

During this portion of the selection set specification, sets of names, tags, or values can be entered only to filter the set you choose. Commas and wildcard characters can be used (see Chapter 43 for information on using wildcards). Pressing Enter accepts the default option of all (*) selected.

Attribute values are <u>case sensitive</u> (upper- and lowercase letters must be specified exactly). If you have entered null values (for example, if *ATTREQ* was set to 0 during block insertion), you can select all null values to be included in the selection set by using the backslash symbol (\).

The subsequent prompts depend on your previous choice of global or individual editing.

No (Global Editing—Editing Values Only)

```
Command: -attedit
Edit attributes one at a time? [Yes/No] <Y>: n
Performing global editing of attribute values.
Edit only attributes visible on screen? [Yes/No] <Y>: (option)
Enter block name specification <*>: Enter or (desired block name)
Enter attribute tag specification <*>: Enter or (desired tag)
Enter attribute value specification <*>: Enter or (desired value)
Select Attributes: PICK
Select Attributes: PICK
Select Attributes: Enter
2 attributes selected.
Enter string to change: (Enter any existing string)
Enter new string: (Enter any new string)
Command:
```

The "only attributes visible on the screen" refers to those visible (Y) or those through the entire drawing (N). Answering "N" to this prompt and Enter to the next three would find all attributes in the drawing.

AutoCAD searches for the "string to change" and replaces all occurrences of the string with the new string. No change is made if the string is not found. Remember that only attribute values can be edited with the global mode. This feature has the same capability as using *Find*, described previously.

Yes (Individual Editing—Editing Any Property)

After filtering for the block name, attribute tag, and value, AutoCAD prompts:

```
Select Attributes: PICK
Select Attributes: PICK (continue selection)
Select Attributes: Enter
n attributes selected.
Enter an option [Value/Position/Height/Angle/Style/Layer/Color/Next] <N>:
```

A marker appears at one of the attributes. Use the *Next* option to move the marker to the attribute you wish to edit. You can then select any other option. After editing the marked attribute, the change is displayed immediately and you can position the marker at another attribute to edit.

You can change the attribute's *Value, Position, Height, Angle, Style, Layer,* or *Color* by selecting the appropriate option. These options are Command line versions of the *Text Options* and *Properties* tabs of the *Enhanced Attribute Editor* (see "*Eattedit*").

Global and Individual Editing Example

Assume that the RES block and a CAP block were inserted several times to create the partial schematic shown in Figure 22-17. The *Attdisp* command is set to *On* to display all the attributes, including the last two for each block, which are normally *Invisible*.

Figure 22-17

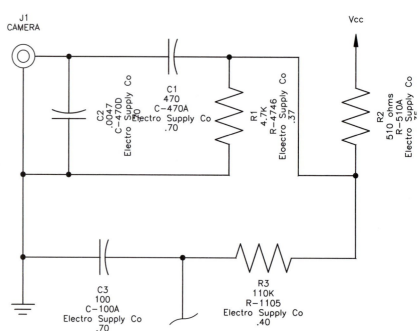

-*Attedit* is used to edit the attributes of RES and CAP. If you wish to change the <u>value</u> of one or several attributes, the global editing mode of -*Attedit* would generally be used, since <u>values only</u> can be edited in global mode. For this example, assuming that the supplier changed, the MANUFACTURER attribute for all blocks in the drawing is edited. The command syntax is as follows:

```
Command: -attedit
Edit attributes one at a time? [Yes/No] <Y>: n
Performing global editing of attribute values.
Edit only attributes visible on screen? [Yes/No] <Y>: Enter
Enter block name specification <*>: Enter
Enter attribute tag specification <*>: Manufacturer
Enter attribute value specification <*>: Enter
Select Attributes: PICK (Select specifically the 6 MANUFACTURER attributes)
Select Attributes: Enter
6 attributes selected.
Enter string to change: Electro Supply Co
Enter new string: Sparks R Us
Command:
```

All instances of the "Electro Supply Co" redisplay with the new string. Note that resistor R1 did not change because the string did not match (due to the mis-spelling, Fig. 22-18). *Attedit* or *Eattedit* can be used to edit the value of that attribute individually.

The individual editing mode of *-Attedit* or *Eattedit* can be used to change the <u>height</u> of text for the RES_NUMBER and CAP_NUMBER attribute for both the RES and CAP blocks. The command syntax is as follows:

Figure 22-18 ──────────────

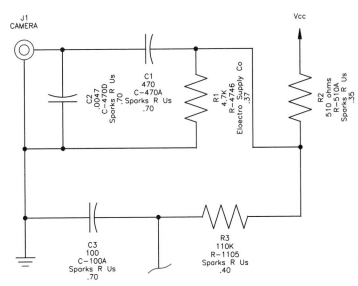

Command: *-attedit*
Edit attributes one at a time? [Yes/No] <Y>: **Enter**
Enter block name specification <*>: **Enter**
Enter attribute tag specification <*>: **Enter**
Enter attribute value specification <*>: **Enter**
Select Attributes: **PICK**
Select Attributes: **PICK** (Continue selecting all 6 attributes)
Select Attributes: **Enter**
6 attributes selected.
Enter an option [Value/Position/Height/Angle/Style/Layer/Color/Next] <N>: *H*
Specify new height <0.1250>: *.2*
Enter an option [Value/Position/Height/Angle/Style/Layer/Color/Next] <N>: *N*
Enter an option [Value/Position/Height/Angle/Style/Layer/Color/Next] <N>: *H*
Specify new height <0.1250>: *.2*
    etc.

The result of the new attribute text height for CAP and RES blocks is shown in Figure 22-19. *Attdisp* has been changed to *Normal* (*Invisible* attributes do not display).

Figure 22-19 ──────────────

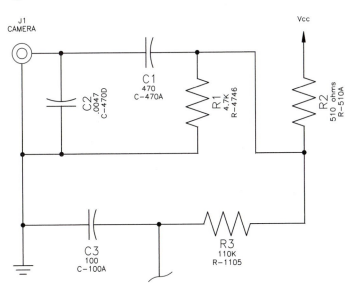

REDEFINING ATTRIBUTES

Redefining an attributed block is similar to redefining a block (discussed previously in Chapter 21); that is, the old block definition (only the attributes in this case) is replaced with a newer one. When a block definition is changed, all blocks in the drawing that reference that definition are updated and display the new changes; therefore, redefining a block is essentially a global update for all block insertions.

Using one of the tools described in this chapter to redefine an attributed block (but not the *Block* command) also results in a global update—all blocks in the drawing referencing the new block definition reflect the changed attributes. Therefore, the tools explained in this section are used to globally change attribute features such as the *Tag, Prompt, Default Value,* the order of the attributes that appear upon insertion, or the text properties of the values. If you want to change the attribute values for individual blocks, use *Eattedit, Attedit, Properties,* or *Find* (see "Displaying and Editing Attributes").

There are three methods you can use to redefine attributes: *Battman* (the *Block Attribute Manager*), *Attredef,* and *Refedit.*

The *Block Attribute Manager (Battman)* is a dialog box-based tool introduced in AutoCAD 2002. It allows you to change the *Tag, Prompt, Default Values, Mode,* the value text options and properties, and the order of the attribute prompting. Changes to attributes made using the *Block Attribute Manager* can be immediately reflected in the drawing.

Attredef is an older Command-line tool used exclusively to redefine blocks containing attributes (using the *Block* command to redefine an attributed block does not update existing attributes in the drawing that reference the new definition). *Attredef* requires *Exploding* an existing block, making the desired changes using additional commands, then using *Attredef* to create the new block definition.

Refedit allows you to select existing *Blocks* or *Xrefs* "in place." With this method, use *Refedit* to select the desired block, then use the *Properties* palette to change the attributes, then use *Refclose* to save the changes and redefine the block and attributes.

BATTMAN

Pull-down Menu	COMMAND (TYPE)	ALIAS (TYPE)	Short-cut	Screen (side) Menu	Tablet Menu
Model > *Object >* *Attribute >* *Block Attribute Manager...*	*BATTMAN*	...	...	...	...

The *Battman* command invokes the *Block Attribute Manager* (Fig. 22-20). With the *Block Attribute Manager,* you can modify attributes within a *Block* without having to *Explode* or redefine the entire *Block.* You can change almost any property of a *Block's* attributes such as the tag, prompt, default value, mode, text settings, and the order of attribute prompting. All changes made to an attributed *Block* are "pushed" out to all references of the *Block* in the drawing.

Figure 22-20

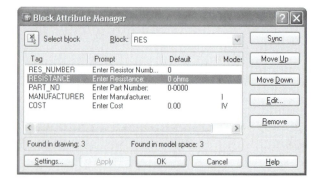

After producing the *Block Attribute Manager*, use the *Select block* button or the *Block:* drop-down list (containing all blocks defined in the drawing) to select the block you want to redefine. Once selected, all the attributes defined for the block appear in the attribute list in the central area. Next, select any attribute from the list you want to redefine. The buttons on the right and bottom of the *Block Attribute Manager* offer the options explained next. Most changes you make to attributes using the *Block Attribute Manager* can be immediately reflected in the drawing if the *Auto Preview Changes* box is checked in the *Edit Attribute* dialog box (see "*Edit Attribute* Dialog Box").

Sync

The *Sync* button synchronizes the highlighted block to change according to the definition as it exists in the block definition table before using the *Block Attribute Manager*. This action may be necessary only if attributes were redefined using methods other than the *Block Attribute Manager*. Using this button is essentially the same as using the *Attsync* command (see "ATTSYNC"). *Sync* does not update the existing attributes in the drawing to new changes you make in the *Block Attribute Manager*—use *Apply* or *OK* to accomplish that action.

Move Up, Move Down

These options change the order in which you are prompted to enter the attribute values upon insertion of the block. Changes to the order made by *Move Up* and *Move Down* <u>do not affect the order that the attributes appear in the inserted block</u>—only the prompting order in the *Edit Attributes* dialog box or at the Command line when using *Insert* or *Ddatte*.

Edit

This button produces the *Edit Attribute* dialog box (see Figure 22-21 and description under "*Edit Attribute* Dialog Box").

Remove

Highlight any attribute in the list, then select *Remove* to delete the attribute from the block definition and from existing insertions of the block if desired. If only one attribute exists, this option is disabled.

Settings

This button produces the *Settings* dialog box (see Figure 22-24 and description).

Apply

Use this button to immediately apply any changes you made in the *Block Attribute Manager* to the block definition table. The changes will also be pushed out to existing insertions of the block unless *Apply changes to existing references* is unchecked in the *Settings* dialog box. You cannot cancel the changes using *Cancel*. Using *Apply* is essentially the same as using *OK*, except the *Block Attribute Manager* does not close. If you want to use the *Block Attribute Manager* to change attributes from more than one block, use *Apply* between block selection.

OK

Use *OK* to apply all changes you made to the block definition tables and to close the *Block Attribute Manager*.

Edit Attribute Dialog Box

This dialog box is used to edit the properties of the selected attribute in the *Block Attribute Manager*. The *Edit Attribute* dialog box (see Figure 22-21) is available only from the *Block Attribute Manager*, although the name is almost identical to the *Edit Attributes* dialog box that appears using the *Insert* or *Ddatte* command. This dialog box has three tabs: *Attribute*, *Text Options*, and *Properties*. Although this dialog box looks similar to the *Enhanced Attribute Editor*, their functions are different. The *Enhanced Attribute Editor* is used to change <u>individual values</u> of attributes in existing block insertions, whereas the *Edit Attribute* dialog box changes the <u>block definition table</u> for attributed blocks as well as existing insertions if desired.

Auto preview changes
If this box is checked (and *Apply changes to existing references* in the *Settings* dialog box is checked), changes that you make in any of the tabs here and in the *Block Attribute Manager* are displayed in existing block insertions in the drawing. If *Auto preview changes* is not checked, changes are not visible in the drawing until you use *Apply* or *OK*. This option is disabled if *Apply changes to existing references* is unchecked in the *Settings* dialog box.

Figure 22-21 ————————

Attribute
The *Attribute* tab (Fig. 22-21) allows you to change the attribute *Tag, Prompt, Default value,* and *Mode* of the attribute highlighted in the *Block Attribute Manager*.

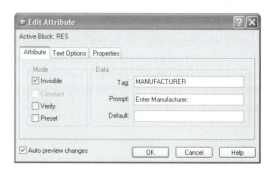

Text Options
In this tab (Fig. 22-22), you can change the *Text Style, Justification, Height, Rotation, Width Factor, Oblique Angle,* and other options depending on the text font used (.SHX or .TTF). Despite the similarity to the tab of the same name in the *Enhanced Attribute Editor*, this tab <u>changes the block definition, not individual attribute values</u>.

Figure 22-22 ————————

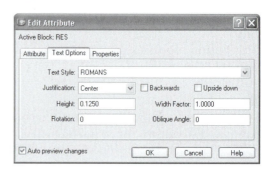

Properties
Use the *Properties* tab (Fig. 22-23) to change any attribute values for the block definition such as *Layer, Linetype, Color, Lineweight,* and *Plot style*. This tab is also similar to the tab of the same name in the *Enhanced Attribute Editor*; however, this tab <u>changes the block definition, not individual attribute values</u>.

Figure 22-23 ————————

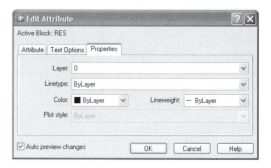

Settings Dialog Box
The *Settings* dialog box (Fig. 22-24, on the next page) is available through the *Settings* button in the *Block Attribute Manager*.

Display in list
The *Display in list* section controls which properties of the attributes appear in the attribute list of the *Block Attribute Manager*. Note the options selected in Figure 22-24 and compare the list in Figure 22-20. Selecting multiple properties is helpful if you need to compare attributes in the list based on specific properties.

Emphasize duplicate tags
It is possible for one block to contain two or more attributes with the same tag. Such a practice should be avoided to prevent confusion. You can use the *Emphasize duplicate tags* option to force attributes with the same tag to be displayed in the list in red rather than in black. This action makes it easier to locate and change tags that are duplicate.

Figure 22-24

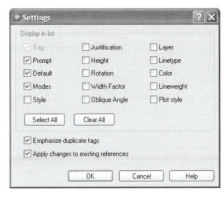

Apply changes to existing references
With this box checked, all changes that are made to attributes in the *Block Attribute Manager* (using *Apply* or *OK*) are applied to the existing block insertions in the drawing. Otherwise, the changes made redefine the block attributes and affect new block insertions but not existing block insertions. This option must be checked to use the *Auto preview changes* feature.

ATTREDEF

Pull-down Menu	COMMAND (TYPE)	ALIAS (TYPE)	Short-cut	Screen (side) Menu	Tablet Menu
...	*ATTREDEF*	...	...	...	...

The *Attredef* command (Attribute redefine) enables you to redefine an attributed block. To use the *Attredef* feature, the objects (to be *Blocked*) with new attributes defined must be in place <u>before</u> invoking the command. This method is older and tedious compared to redefining attributes using *Battman*.

Creating the new objects and attributes is accomplished by *Inserting* the old block, *Exploding* it, then making the desired changes to the attributes and objects before using *Attredef*. Figure 22-25 illustrates a new RES block definition (in the lower-right corner) before using *Attredef*.

After *Exploding* an inserted block, edit the attributes to reflect the desired change. You can use several methods to edit existing attributes (explained in "Other Attribute Editing Methods"). Once the attributes and geometry have been edited or created to your liking, invoke *Attredef*. The command operates only in Command line format as follows:

Command: **attredef**
Enter name of the block you wish to redefine: **res**
Select objects for new Block...
Select objects: **PICK** (Select attributes in order that you want them to appear; see Figure 22-5.)
Select objects: **PICK** (Continue selecting all attributes.) etc.
Select objects: **PICK** (Select geometry last.)
Select objects: **Enter**
Specify insertion base point of new Block: **PICK**
Command:

Figure 22-25

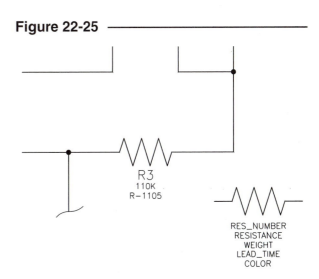

The block definition is updated, and all instances of the block automatically and globally change to display the new changes. All attributes that were changed reflect the default values. *Attedit* can be used later to make the desired corrections to the individual attributes.

Figure 22-26

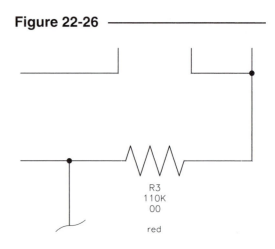

Figure 22-26 displays one of the RES blocks after redefining. Notice that the forth attribute is invisible as it was before it was *Exploded*. For this attribute, the mode was not changed. The new last attribute, however, was changed to a normal attribute mode and therefore appears after using *Attredef*.

REFEDIT, PROPERTIES, and *REFCLOSE*

Another method of redefining blocks is to use *Refedit*, the *Properties* palette, and *Refclose*. This technique is also known as "in-place editing" because the references (*Blocks* or *Xrefs*) can be edited after being inserted or attached. However, the introduction of the *Block Attribute Manager* in AutoCAD 2002 makes the *Refedit* method relatively involved and limited; therefore, this method for the purpose of redefining block attributes is explained briefly in this chapter only for reference. Refer to Chapter 30, Xreferences, for a full explanation of *Refedit* and other related commands and variables for use with *Blocks* and *Xrefs*.

To redefine one block attribute at a time using *Refedit*, *Properties*, and *Refclose*, follow these steps.

1. Invoke the *Properties* palette.
2. Invoke *Refedit*. Select the <u>Block</u> (not the attributes) containing the attributes you want to redefine.
3. In the *Reference Edit* dialog box that appears, select the *Display attribute definitions for editing* checkbox. Select *OK*.
4. At the "Select nested objects" prompt, select the *Block* again (not the attributes).
5. Use the *Properties* palette to change the *Tag* or *Prompt* property. When finished, press Escape twice to deselect the *Block*.
6. Use the *Refclose* command and *Save* option to save the changes and redefine the attributes. All insertions of the *Block* are redefined.

When *Blocks* are redefined using this method, the block definition is changed in the current drawing. However, when *Xrefs* are changed using *Refedit*, the original externally referenced drawing can be changed from within the current drawing (see Chapter 30, Xreferences).

Other Attribute Editing Methods

Keep in mind that if you want to modify attributes after *Exploding* the old block, there are three alternative methods: you can use *Ddedit*, *Properties*, or *Attdef*. *Ddedit*, the dialog box for editing text, recognizes the text as attributes but allows you to change only the attribute *Tags*. Figure 22-27 illustrates using the *Edit Attribute Definition* dialog box (*Ddedit*) for changing the RES_NUMBER tag.

Figure 22-27

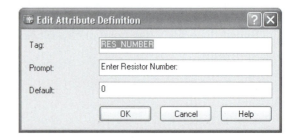

Optionally, the *Properties* palette can be invoked. Not only does *Properties* recognize the text as attributes, but the *Tag, Prompt,* default *Value,* and attribute modes can be modified. Figure 22-28 displays this dialog box for the same operation. Notice all of the possible changes that can be made.

Figure 22-28

Instead of using *Ddedit* or *Properties* to change <u>existing</u> attributes, new attributes can be created "from scratch" with *Attdef.* Remember that if you want to change the attributes when redefining a block, one of these three operations must take place before using *Attredef.*

Grip Editing Attribute Location

Grips can be used to "move" the location of the attributes after an attributed *Block* has been inserted. To do this (with Grips on), PICK the *Block* so grips appear at each attribute and at the block insertion point. Next, PICK the attribute you want to move so its grip becomes "hot." The attribute can be moved to any location and still retain its association with the block. (See Chapter 23, Grip Editing.)

ATTSYNC

Pull-down Menu	COMMAND (TYPE)	ALIAS (TYPE)	Short-cut	Screen (side) Menu	Tablet Menu
...	*ATTSYNC*	...	...	...	...

The *Attsync* command synchronizes the selected block attributes in the drawing to change according to the definition as it exists in the block definition table. *Attsync* does not affect individual attribute values, only changes made to the properties of the inserted attributes that are different from what is established in the definition table for the block. *Attsync* is usually not needed if you used the *Block Attribute Manager* or *Attredef* to redefine block attributes. This command may be helpful in the following cases.

Use *Attsync* if you used the *Block* command to redefine a block containing attributes, since redefining a block with the *Block* command does not update attributes in blocks inserted in the drawing; using *Block* to redefine attributes updates only the drawing geometry. For example, you may have *Exploded* a block, deleted an attribute, then used *Block* to redefine it. *Attsync* would be necessary to update all the existing blocks in the drawing to the new definition.

Use *Attsync* if you edited the format of the attributes by some method and want to change them back to the original definition. For example, if you previously used the *Enhanced Attribute Editor* to change specific properties of an attribute in the drawing such as text height or color, using *Attsync* would change the properties back to the original settings as established in the block's definition. *Attsync* does not affect any changes (text or numeric content) made to the attribute values themselves.

When *Attsync* is invoked, you are prompted for the names of blocks you want to update with the current attributes defined for the blocks.

    Command: **attsync**
    Enter an option [?/Name/Select] <Select>:

Enter the name of the block you want to update. Entering a question mark (?) displays a list of all block definitions in the drawing. Alternately, you can press Enter to select the block whose attributes you want to update. If a block you specify does not contain attributes or does not exist, an error message is displayed, and you are prompted to specify another block.

EXTRACTING ATTRIBUTES

By extracting attributes from AutoCAD drawings, you can create a report of the existing blocks and related text information contained in the drawings. In other words, a list of some or all of the blocks and attributes in drawings can be written to a separate file (extracted). During the attribute extraction process, you can choose to convert this text and/or numeric information into a spreadsheet, database, or text file format. The resulting output (or extract) file can be used as a report indicating a variety of data such as the name, number, location, layer, and scale factor of the blocks, as well as any attribute information such as prices, manufacturers, sizes, dates, or any other attribute data that may be defined in the block insertions. Because of this attribute extract feature, AutoCAD drawings contain the intelligence that can be used to compile or calculate bills of materials, supplier information, employee records, quotations, billing totals, maintenance schedules, catalogs, and so on.

There are two methods you can use to extract attributes: the *Attribute Extraction* wizard (new in AutoCAD 2002) and the older *Attext* command.

The *Eattext* command produces the *Attribute Extraction* wizard. This wizard makes the process of extracting attributes far simpler and more visual than the earlier methods in AutoCAD. The wizard leads you through the steps of selecting drawings, selecting *Blocks* and/or *Xrefs*, selecting the desired attribute information, selecting the output format, and saving templates for future use. The *Attribute Extraction* wizard provides you with capabilities not available by other means.

Alternately, you can use the *Attext* command to produce the older *Attribute Extraction* dialog box or type *-Attext* to invoke the Command line equivalent. Using this method is far more complex than the newer *Attribute Extraction* wizard because template files must be created externally to AutoCAD and the chances of error are great.

EATTEXT

Pull-down Menu	COMMAND (TYPE)	ALIAS (TYPE)	Short-cut	Screen (side) Menu	Tablet Menu
Tools *Attribute Extraction...*	*EATTEXT*	...	...	...	...

To produce the *Attribute Extraction* wizard, use the *Eattext* command or any option shown in the command table. The *Attribute Extraction* wizard (Fig. 22-29) leads you through the seven steps to extract block attribute (text and numeric) information from an AutoCAD drawing into an external file. The resulting extract file can be used to generate reports of information in spreadsheet, database, or text formats.

The *Attribute Extraction* wizard introduced in AutoCAD 2002 offers a number of features that are not available by other means.

Figure 22-29 ————————————————

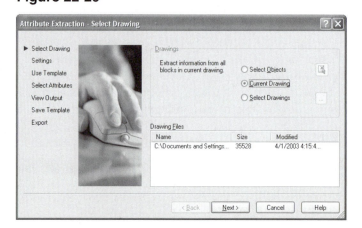

With the *Attribute Extraction* wizard, you can extract attribute data from multiple drawings in addition to the current drawing. You can also extract attribute data from nested *Blocks* and *Xrefs*. In addition to the ability to create the output files in Microsoft Excel, Microsoft Access, and ASCII text formats, you can view and edit the organization of the data before creating the extract file.

The seven steps required in the *Attribute Extraction* wizard are explained here. Use the *Next* or *Finish* buttons to proceed after each step and the *Back* button to go back and change an option at any time.

Select Drawing
When you first invoke the *Attribute Extraction* wizard, you are prompted to specify which drawing(s) or which blocks in the current drawing you want to include in the output file (see Figure 22-29). The *Select*

Objects option allows you to select only specific *Blocks* from the current drawing. Use the *Current Drawing* option to have all blocks considered for extraction. Use the *Select Drawings* option to locate and specify multiple drawings containing the blocks you want to extract information from. The list of drawings you have selected appears near the bottom of the dialog box. In the *Select Attributes* step, you can exclude any of the blocks you specified in this step.

Settings
In the second step (Fig. 22-30), you can choose to *Include xrefs* and to *Include nested blocks* if you wish. If these objects are not contained in the selected drawing(s) from the previous step, your selection here is irrelevant.

Use Template
Template files contain specifications indicating which blocks and attributes are to be extracted and the format of the output file. If this is your first time to use the *Attribute Extraction* wizard, select *No template* in this step (Fig. 22-31). If you have used the *Attribute Extraction* wizard previously and have saved a template file in .BLK format, or one exists for your project or company, you can select *Use template*, then the *Use Template* button to locate and select the file. The *Attribute Extraction* wizard cannot use template files in .TXT format created for use in previous versions of AutoCAD.

Select Attributes
The *Select Attributes* dialog box (Fig. 22-32) demonstrates the power and flexibility of the *Attribute Extraction* wizard. Here you can include or exclude which blocks and which attributes for each block you want to use in the output file by checking or unchecking the boxes. The *Blocks* list on the left contains the blocks individually selected (from the first step,

Figure 22-30

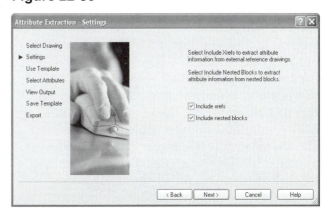

Figure 22-31

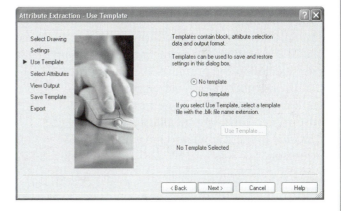

Figure 22-32

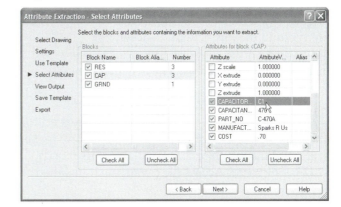

Select Objects), and all blocks contained in the specified drawings (*Current Drawing* or *Select Drawings*) including nested blocks and *Xrefs*, if previously selected. For each highlighted block in the *Blocks* list, the related attributes for that block are listed in the *Attributes for block* list on the left. For the highlighted block, check the attributes you want to include in the extract file. <u>This process must be repeated for each block</u> checked in the *Blocks* list.

Note that the *Attributes for block* list contains information in addition to the typical attributes that were specified for the block. Several other fields of information can be checked to include in the extract file: X, Y, and Z coordinates of the block insertion points; X, Y, and Z scale factors of the blocks; X, Y, and Z components of the blocks' extrusion direction; and the blocks' layer and orientation (rotation angle) upon insertion.

View Output

This feature (Fig. 22-33) allows you to choose between two possible orientations (views) for the output file. The default view (not shown) has only four columns: *Block Name*, *Count*, *Attribute Tag*, and *Attribute Value*. This orientation is useful when you have many blocks with different attribute tags; however, it is very difficult to count the number of blocks inserted. The alternate view (see Figure 22-33) provides a column for *Block Name*, *Count*, and a column for each attribute tag for all blocks. This view is useful for totaling the number of blocks inserted and is best in cases when all blocks have the same attribute tags.

Figure 22-33 ───────────────

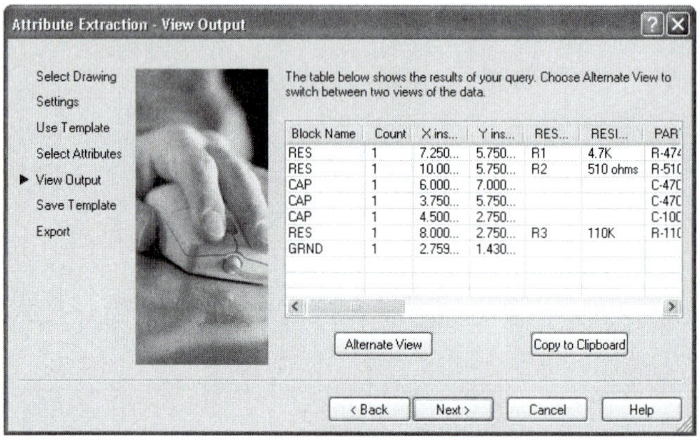

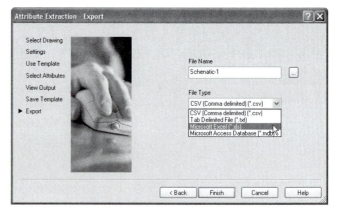

Save Template

The *Save Template* step (not shown) allows you to save, or not save, a template file. Template files are useful if you intend to extract attributes again at a later time. The template file keeps the information you selected in the previous two steps only— that is, the block names and attributes you want to include and the orientation of the output file. Template files saved using the *Attribute Extraction* wizard have a .BLK file extension.

Export

In the final step of the *Attribute Extraction* wizard, you supply the name you want for the output file and the type of file to be created (Fig. 22-34). You can select from two to four file types in the *File Type* drop-down list depending on the software programs that are installed on your system. The possible file types are:

Figure 22-34 ───────────────

.*TXT*   Tab delimited file readable in any text editor or word processor
.*CSV*   Comma delimited file readable in any text editor, word processor, or spreadsheet program
.*XLS*   Microsoft Excel file (available if installed)
.*MDB*   Microsoft Access file (available if installed)

Pressing the *Finish* button causes AutoCAD to create the output file. An example file for the options shown in the preceding figures (in Microsoft Excel format) is displayed in Figure 22-35. Note the orientation (view) gives only one *Block Name* for each insertion (3 RES blocks, 3 CAP blocks, and 1 GRND block). Also note that only the RES blocks have the RES_NUMBER and RESISTANCE attributes and only the CAP blocks have the CAPACITOR NUMBER and CAPACITANCE attributes. The PART_NUMBER, MANUFACTURER, and COST attributes are common to both blocks. The GRND block has no text attributes.

Figure 22-35

	A	B	C	D	E	F	G	H	I	J	K
1	Block Name	Count	X insertio	Y insertion	RES_N	RESISTAN	PART_NO	MANUFACTUR	COST	CAPA	CAPACITA
2	RES	1	7.25	5.75	R1	4.7K	R-4746	Sparks R Us	0.37		
3	RES	1	10.00	5.75	R2	510 ohms	R-510A	Sparks R Us	0.35		
4	CAP	1	6.00	7.00			C-470A	Sparks R Us	0.70	C1	470.00
5	CAP	1	3.75	5.75			C-470D	Sparks R Us	0.70	C2	0.00
6	CAP	1	4.50	2.75			C-100A	Sparks R Us	0.70	C3	100.00
7	RES	1	8.00	2.75	R3	110K	R-1105	Sparks R Us	0.40		
8	GRND	1	2.76	1.43							
9											

2002

ATTEXT

Pull-down Menu	COMMAND (TYPE)	ALIAS (TYPE)	Short-cut	Screen (side) Menu	Tablet Menu
...	*ATTEXT* or -*ATTEXT*	...	...	...	...

Before AutoCAD 2002, the *Attext* command was used to extract attributes. The *Attext* command produces the *Attribute Extraction* dialog box shown in Figure 22-36. This method is more complex and allows attribute extraction from the current drawing only; however, you can create a .DXF format output file with this method. Before invoking *Attext*, you must create the template file "manually" in a text editor or word processor. The steps are given below.

Steps for Creating Output Files Using *Attext*

1. In AutoCAD, *Insert* all desired *Blocks* and related attributes into the drawing. *Save* the drawing.

2. With a word processor or text editor, create a template file. The template file specifies what information should be included in the output file and how the output file should be structured. (A template file is not required if you specify a *DXF* format for the output file.)

3. In AutoCAD, use the *Attext* or -*Attext* command to specify the name of the template file to be used and to create the output file.

4. Examine the output file in the text editor or word processor or import the output file into a database or spreadsheet program.

Creating the Template File (Step 2)

The template file is created in a text editor (such as Windows Notepad) or word processor (such as Microsoft Word) in non-document mode. The template file must be in straight ASCII form (no internal word processing codes). The template file must be given a .TXT file extension.

Each line of the template file specifies a column of information, or field, to be written in the output file, including the name of the column, the width of the column (number of characters or numerals), and the numerical precision. Each *Block* that matches a line in the template file creates a column in the output file.

The possible fields that AutoCAD allows you to specify and the format that you must use for each are shown below. Use any of the fields you want when creating the template file. The first two columns below are included in the template file (not the comments in the third column):

BL:LEVEL	Nwwwddd	(*Block* nesting level)
BL:NAME	Cwww000	(*Block* name)
BL:X	Nwwwddd	(X coordinate of *Block* insertion point)
BL:Y	Nwwwddd	(Y coordinate of *Block* insertion point)
BL:Z	Nwwwddd	(Z coordinate of *Block* insertion point)
BL:NUMBER	Nwwwddd	(*Block* counter; same for all members of *MINSERT*)
BL:HANDLE	Cwww000	(*Block* handle; same for all members of *MINSERT*)
BL:LAYER	Cwww000	(*Block* insertion layer name)
BL:ORIENT	Nwwwddd	(*Block* rotation angle)
BL:XSCALE	Nwwwddd	(X scale factor of *Block*)
BL:YSCALE	Nwwwddd	(Y scale factor of *Block*)
BL:ZSCALE	Nwwwddd	(Z scale factor of *Block*)
BL:XEXTRUDE	Nwwwddd	(X component of *Block's* extrusion direction)
BL:YEXTRUDE	Nwwwddd	(Y component of *Block's* extrusion direction)
BL:ZEXTRUDE	Nwwwddd	(Z component of *Block's* extrusion direction)
BL:SPACE	Cwww000	(Space between fields)
Attribute Tag	Cwww000	(Attribute tag, character)
Attribute Tag	Nwwwddd	(Attribute tag, numeric)

All items in the first column (in Courier font) must be spelled exactly as shown, if used. Use spaces rather than tabs or indents to separate the second column entries. In the second column, the first letter must be N or C, representing a numerical or character field. The three "w"s indicate digits that specify the field width (number of characters or numbers). The last three digits (0 or d) specify the number of decimal places you desire for the field (0 indicates that no decimals can be specified for that field).

The template file can specify any or all of the possible fields in any order, but should be listed in the order you wish the extract file to display. Each template file must include at least one attribute field. If a block contains any of the specified attributes, it is listed in the extract file; otherwise it is skipped. If a block contains some, but not all, of the specified attributes, the blank fields are filled with spaces or zeros.

A sample template file for the schematic drawing example in Figure 22-19 is shown here:

BL:NAME	C004000
BL:X	N005002
BL:Y	N005002
BL:SPACE	C002000
RES_NUMBER	C004000
RESISTANCE	C009000
CAP_NUMBER	C004000
CAPACITANCE	C006000
PART_NO	C007000
MANUFACTURER	C012000
COST	N005002

Creating the Extract File (Step 3)
Once you have created a template file, you can use the *-Attext* command (for Command line format) or invoke the *Attribute Extraction* dialog box by the *Attext* command. The Command line format syntax is as follows:

    Command: **-attext**
    Enter extraction type or enable object selection [Cdf/Sdf/Dxf/Objects] <C>: **Enter** or **(option)**

Instead of using the *-Attext* command, you may prefer to use *Attext*, which invokes the *Attribute Extraction* dialog box (Fig. 22-36). This dialog box serves the same functions as the *Attext* command.

Figure 22-36 ——————————

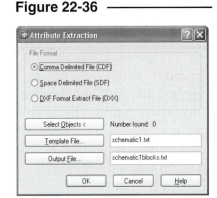

You can use the *Select Objects* option (or the *Objects* option in Command line format) if you want to PICK only certain block attributes to be included in the extract file.

File Format specifies the structure of the extract file. You can specify either a *SDF*, *CDF*, or *DXF* format for the extract file.

SDF
SDF (Space Delimited File) uses spaces to separate the fields. The *SDF* format is more readable because it appears in columnar format. Numerical fields are right justified, whereas character fields are left justified. Therefore, it may be necessary to include a space field (BL:SPACE or BL:DUMMY) after a numeric field that would otherwise be followed immediately by a character field. This method is used in the example template file shown previously.

CDF
CDF (Comma Delimited File) uses a character that you specify to separate the fields of the extract file. The default character for *CDF* is a comma (,). Some database packages require a *CDF* format for files to be imported.

DXF
The *DXF* format is a variation of the standard AutoCAD *DXF* (Data Interchange File format); however, it contains only block and attribute information. This format contains more information than the *SDF* and *CDF* files and is generally harder to interpret. *DXF* is a standard format and therefore does <u>not</u> request a template file. (Selecting *Export* from the *File* pull-down menu, then *DXX Extract*, allows you to create a *DXF* format extract file only. The *Attext* command must be typed if you want an *SDF* or *CDF* format extract file.)

After specifying the desired format, you must specify the *Template File* (name of previously created template file) and *Output File* (name of the extract file to create). The default extract file name is the same as the drawing name but with the .TXT extension.

Finally, AutoCAD then creates the extract file based on the specified parameters, displaying a message similar to the following:

    6 records in extract file.
    Command:

If you receive an error message, check the format of your template file. An error can occur if, for example, you have a BL:NAME field with a width of 10 characters, but a block in the drawing has a name 12 characters long.

Below is the sample extract file created using the electrical schematic drawing and the previous example template file.

Sample Extract File

```
RES  7.25 5.75  R1  4.7              R-4746 Sparks R Us  0.37
RES 10.00 5.75  R2  510 ohms         R-510A Sparks R Us  0.35
CAP  6.00 7.00          C1  470      C-470A Sparks R Us  0.70
CAP  3.75 5.75          C2  .0047    C-470D Sparks R Us  0.70
CAP  4.50 2.75          C3  100      C-100A Sparks R Us  0.70
RES  8.00 2.75  R3  110K             R-1105 Sparks R Us  0.40
```

This extract file lists the block name, X and Y location of the block resistor number, resistance value, capacitor number, capacitance value, manufacturer, and cost. Compare the resulting extract file above with the matching template file shown previously. Remember that each <u>line</u> in the template file creates a <u>column</u> in the extract file.

CHAPTER EXERCISES

1. **Create an Attributed Title Block**

    A. Open the **TBLOCK** drawing that you created in Chapter 18, Exercise 8. Because we need to create block attributes, *Erase* the existing text so only the lines remain. (Alternately, you can begin a *New* drawing and follow the instructions for Chapter 18, Exercise 8A. Do not complete 8B.)

    B. *SaveAs* **TBLOCKAT**. Check the drawing to ensure that 2 text *Styles* exist using *romans* and *romanc* font files (if not, create the two *Styles*). Next, define attributes similar to those described below using *Attdef*. Substitute your personalized *Tags*, *Prompts*, and *Values* where appropriate or as assigned (such as your name in place of B. R. Smith):

Tag	Prompt	Value	Mode
COMPANY		CADD Design Co.	const
PROJ_TITLE	Enter Project Title	PROJ-	
SCALE	Enter Scale	1"=1"	
DES_NAME		Des.-B.R.Smith	const
CHK_NAME	Enter Checker Name	Ch.-	
COMP_DATE	Enter Completion Date	1/1/04	
CHK_DATE	Enter Check Date	1/1/04	
PROJ_NO	Enter Project Number	PROJ	verif

The completed attributes should appear similar to those shown in Figure 22-37. *Save* the drawing.

C. Use the *Base* command to assign the *Insertion base point*. **PICK** the lower-right corner (shown in Figure 22-37 by the "blip") as the *Insertion base point*.

Since the entire TBLOCKAT drawing (.DWG file) will be inserted as a block, you do not have to use the *Block* command to define the attributed block. *Save* the drawing (as **TBLOCKAT**).

Figure 22-37 ———————

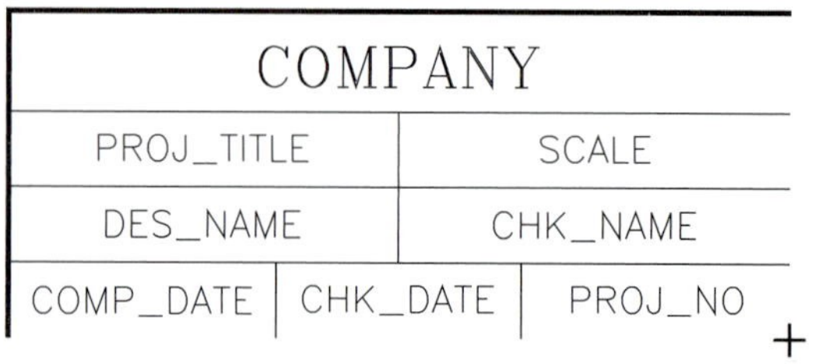

D. Begin a *New* drawing and use the **ASHEET** drawing as a *template*. Set the **TITLE** layer *current* and draw a border with a *Pline* of *.02 width*. Set the *ATTDIA* variable to **1** (*On*). Use the *Insert* command to bring the **TBLOCKAT** drawing in as a block. The *Enter Attributes* dialog box should appear. Enter the attributes to complete the title block similar to that shown in Figure 22-38. Do not *Save* the drawing.

Figure 22-38 ─────

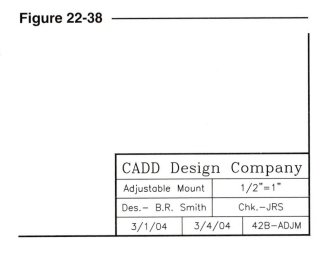

CADD Design Company	
Adjustable Mount	1/2"=1"
Des.— B.R. Smith	Chk.—JRS

3/1/04	3/4/04	42B—ADJM

2. **Create Attributed Blocks for the OFF-ATT Drawing**

Figure 22-39 ─────

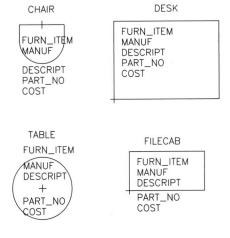

A. *Open* the **OFF-ATT** drawing (not OFFICE-REN) that you prepared in Chapter 21 Exercises. The drawing should have the office floor plan completed and the geometry drawn for the furniture blocks. The furniture has not yet been transformed to *Blocks*. You are ready to create attributes for the furniture, as shown in Figure 22-39.

Include the information below for the proposed blocks. Use the same *Tag*, *Prompt*, and *Mode* for each proposed block. The *Values* are different, depending on the furniture item.

Tag	Prompt	Mode
FURN_ITEM	Enter furniture item	
MANUF	Enter manufacturer	
DESCRIPT	Enter model or size	
PART_NO	Enter part number	Invisible
COST	Enter dealer cost	Invisible

Values

CHAIR	DESK	TABLE	FILE CABINET
WELLTON	KERSEY	KERSEY	STEELMAN
SWIVEL	50" x 30"	ROUND 2'	28 x 3
C-143	KER-29	KER-13	3-28-L
79.99	249.95	42.50	129.99

B. Next, use the **Block** command to make each attributed block Use the block *names* shown (Fig. 22-39). Select the attributes <u>in the order</u> you want them to appear upon insertion; then select the geometry. Use the *Insertion base points* as shown. *Save* the drawing.

C. Set the *ATTDIA* variable to **0** (*Off*). *Insert* the blocks into the office and accept the default values. Create an arrangement similar to that shown in Figure 22-40.

Figure 22-40

D. Use the *ATTDISP* variable and change the setting to *Off.* Do the attributes disappear?

E. A price change has been reported for all KERSEY furniture items. Your buyer has negotiated an additional 10% off the current dealer cost. Type in the *Multiple Attedit* command and select each block one at a time to view the attributes. Make the COST changes as necessary. Use *SaveAs* and assign the new name **OFF-ATT2**.

F. Change the *ATTDISP* setting back to *Normal.* It would be helpful to increase the *Height* of the FURN_ITEM attribute (CHAIR, TABLE, etc.). Use the *Enhanced Attribute Editor* to change the **FUNR_ITEM** tag for each block. You will have to select each block individually, then use the *Text Options* tab to change the height to **3"** for each block. If everything looks correct, *Save* the drawing.

3. **Extract Attributes**

A. A cost analysis of the office design is required to check against the $3,400.00 budget allocated for furnishings. This can be accomplished by extracting the attributes into a report. Use the *Attribute Extraction* wizard to extract the block attributes. Select only the X and Y insertion points and the text attributes to include in the output file as shown below. Specify an *Alternate View* such that each block insertion appears only once in the output file as shown below. Save a template file named **OFF-TMPL.BLK**. Create an output file named **OFF-EXT** and specify a file format (.TXT, .CVS, etc.) appropriate for your system. When finished examine the output file. The file should appear similar to that below.

```
DESK       165.00  348.00  DESK            KERSEY      224.95
DESK       165.00  228.00  DESK            KERSEY      224.95
DESK       405.00  174.00  DESK            KERSEY      224.95
CHAIR      156.00  327.00  CHAIR           WELLTON      79.99
CHAIR      153.00  204.00  CHAIR           WELLTON      79.99
CHAIR      249.00  336.00  CHAIR           WELLTON      79.99
CHAIR      249.00  216.00  CHAIR           WELLTON      79.99
CHAIR      447.00  153.00  CHAIR           WELLTON      79.99
CHAIR      387.00  189.00  CHAIR           WELLTON      79.99
TABLE      276.00  204.00  TABLE           KERSEY       38.25
TABLE      276.00  327.00  TABLE           KERSEY       38.25
FILECAB    450.00  225.00  FILE CABINET    STEELMAN    129.99
FILECAB    135.00  132.00  FILE CABINET    STEELMAN    129.99
FILECAB    135.00  249.00  FILE CABINET    STEELMAN    129.99
```

B. Total the COST column to acquire the total furnishings cost so far for the job. If you purchase the $880 conference table and 8 more chairs, how much money will be left in the budget to purchase plants and wall decorations?

4. Redefine the CHAIR block in the OFF-ATT drawing. Make a *Copy* of one chair and *Explode* it. Change the geometry in some way (add arms to the chair, for example). Use *Ddedit* or *Properties* to change the **COST** attribute. Increase the default cost value by 15%. Then use *Attredef* to define the new block and to redefine all the existing CHAIR blocks. Use *SaveAs* and assign the name **OFF-ATT3**. Finally, create a new extract file and calculate the total cost for the office (the original template file can be used again).

5. Create the electrical schematic illustrated in Figure 22-41. Use *Blocks* for the symbols. The text associated with the symbols should be created as attributes. Use the following block names and tags:

Block Names:	RES
	CAP
	GRD
	AMP
Attribute Tags:	PART_NUM
	MANUF_NUM
	RESISTANCE
	CAPACITANCE

(Only the RES and CAP blocks have values for resistance or capacitance.) When you are finished with the drawing, create an extract file reporting information for all attribute tags (coordinate data is not needed). Save the drawing as **SCHEM2**.

Figure 22-41

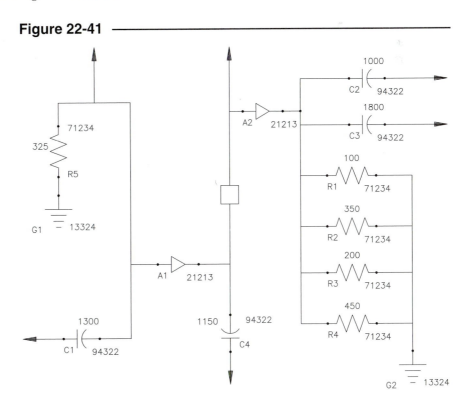

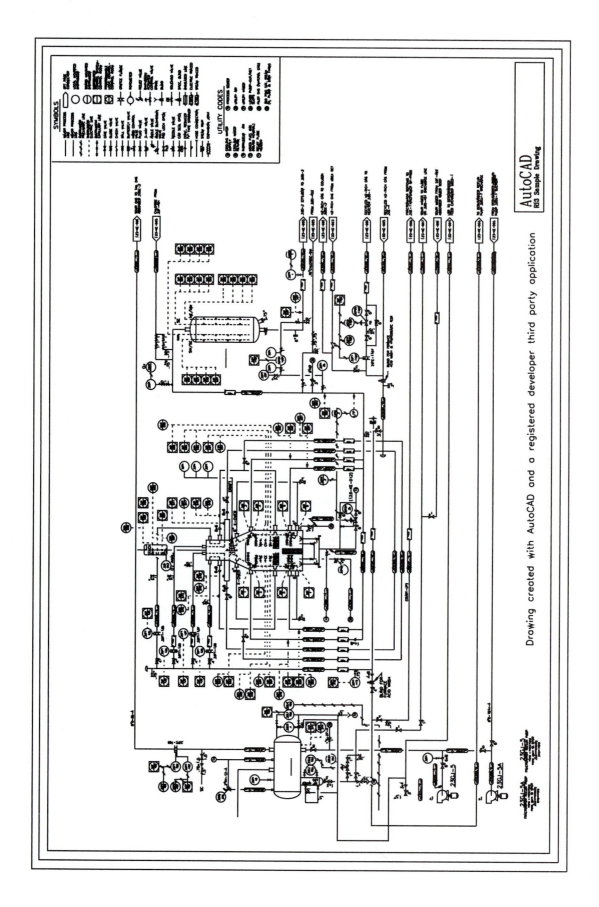

PNID.DWG, Courtesy of Autodesk, Inc.

23

GRIP EDITING

Chapter Objectives

After completing this chapter you should:

1. be able to use the *GRIPS* variable to enable or disable object grips;

2. be able to activate the grips on any object;

3. be able to make an object's grips hot or cold;

4. be able to use each of the grip editing options, namely, STRETCH, MOVE, ROTATE, SCALE, AND MIRROR;

5. be able to use the Copy and Base suboptions;

6. be able to use the auxiliary grid that is automatically created when the Copy suboption is used;

7. be able to change grip variable settings using the *Options* dialog box or the Command line format.

CONCEPTS

Grips provide an alternative method of editing AutoCAD objects. The object grips are available for use by setting the *GRIPS* variable to **1**. Object grips are small squares appearing on selected objects at end-points, midpoints, or centers, etc. The object grips are activated (made visible) by **PICK**ing objects with the cursor pickbox only <u>when no commands are in use</u> (at the open Command prompt). Grips are like small, magnetic *OSNAPs* (*Endpoint, Midpoint, Center, Quadrant*, etc.) that can be used for snapping one object to another, for example. If the cursor is moved within the small square, it is automatically "snapped" to the grip. Grips can replace the use of *OSNAP* for many applications. The grip option allows you to STRETCH, MOVE, ROTATE, SCALE, MIRROR, or COPY objects without invoking the normal editing commands or *OSNAP*s.

As an example, the endpoint of a *Line* could be "snapped" to the endpoint of an *Arc* (shown in Fig. 23-1) by the following steps:

Figure 23-1

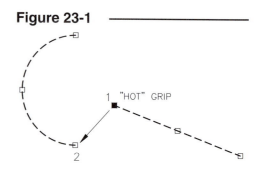

1. Activate the grips by selecting both objects. Selection is done when no commands are in use (during the open Command prompt).
2. Select the grip at the endpoint of the *Line* (1). The grip turns **hot** (red).
3. The ** STRETCH ** option appears in place of the Command prompt.
4. STRETCH the *Line* to the endpoint grip on the *Arc* (2). **PICK** when the cursor "snaps" to the grip.
5. The *Line* and the *Arc* should then have connecting endpoints. The Command prompt reappears. Press Escape to cancel (deactivate) the grips.

GRIPS FEATURES

GRIPS and
DDGRIPS

Pull-down Menu	COMMAND (TYPE)	ALIAS (TYPE)	Short-cut	Screen (side) Menu	Tablet Menu
Tools *Options...* *Selection*	*GRIPS* or *DDGRIPS*	*GR*	...	*TOOLS 2* *Options...* *Grips*	...

Grips are enabled or disabled by changing the setting of the system variable, *GRIPS*. A setting of 1 enables or turns *ON GRIPS* and a setting of **0** disables or turns *OFF GRIPS*. This variable can be typed at the Command prompt, or Grips can be invoked from the *Selection* tab of the *Options* dialog box, or by typing *Ddgrips* (Fig. 23-2). Using the dialog box, toggle *Enable Grips* to turn *GRIPS ON*. The default setting in AutoCAD for the *GRIPS* variable is **1** (*ON*). (See "Grip Settings" near the end of the chapter for explanations of the other options in the *Grips* section of the *Options* dialog box.)

Figure 23-2

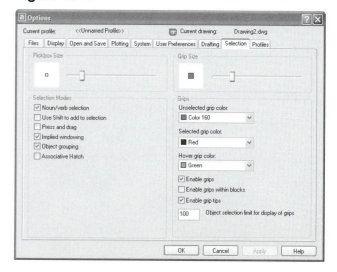

The *GRIPS* variable is saved in the system registry rather than in the current drawing as with most other system variables. Variables saved in the system registry are effective for any drawing session on that <u>particular computer</u>, no matter which drawing is current. The reasoning is that grip-related variables (and selection set-related variables) are a matter of personal preference and therefore should remain constant for a particular CAD station.

When *GRIPS* have been enabled, a small pickbox (3 pixels is the default size) appears at the center of the cursor crosshairs. (The pickbox also appears if the *PICKFIRST* system variable is set to **1**.) This pickbox operates in the same manner as the pickbox appearing during the "Select objects:" prompt. <u>Only</u> the pickbox, *AUTO window*, or *AUTO crossing window* methods can be used for selecting objects to activate the grips. (These three options are the only options available for Noun/Verb object selection as well.)

Activating Grips on Objects

Figure 23-3

The grips on objects are activated by selecting desired objects with the cursor pickbox, window, or crossing window. This action is done <u>when no commands are in use</u> (at the open Command prompt). When an object has been selected, two things happen: the grips appear and the object is highlighted. The grips are the small blue (default color) boxes appearing at the endpoints, midpoint, center, quadrants, vertices, insertion point, or other locations depending on the object type (Fig. 23-3). Highlighting indicates that the object is included in the selection set.

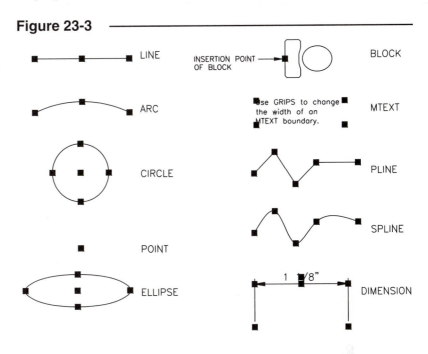

Cold and Hot Grips

In AutoCAD 2004, grips can have only two states: **cold** and **hot**. When *Grips* are turned on, and an object is selected (when no commands are in use), the object's grips become **cold**—its grips are displayed in blue (default color) and the object is highlighted.

Figure 23-4

One or more grips can then be made **hot**. A hot grip is created by "hovering" the cursor over a cold grip until the grip changes to a light green (default) color, then you PICK the grip so it changes to red (default color). The intermittent green grip is called a "hover" grip.

A **hot** grip is red (by default) and its object is almost always highlighted. Any grip can be changed to **hot** by selecting the grip itself. A <u>hot grip is the default base point</u> used for the editing action such as MOVE, ROTATE, SCALE, or MIRROR, or is the stretch point for STRETCH. When a **hot** grip exists, a new series of prompts appear in place of the Command prompt that displays the various grip editing options. The grip editing options are also available from a right-click cursor menu (Fig. 23-4). <u>A grip must be changed to **hot** before the editing options appear</u>. Two or more grips can be made **hot** simultaneously by pressing Shift while selecting <u>each</u> grip.

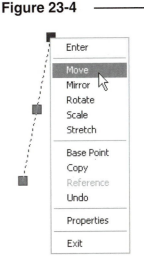

2004

If you have made a *Grip* **hot** and want to deactivate it, possibly to make another *Grip* **hot** instead, press Escape once. This returns the grip to a **cold** state. In effect, this is an undo only for the **hot** grip. Pressing Escape again cancels all *Grips*. Pressing Escape demotes hot grips or cancels all grips:

Grip State	Press Escape once	Press Escape twice
only **cold**	grips are deactivated	
hot and **cold**	**hot** demoted to **cold**	grips are deactivated

NOTE: Beware that there are <u>two right-click (shortcut) menus available when grips are activated</u>. The Grip menu (see previous Figure 23-4) is available only when a grip is <u>hot</u> (red). If you make any grip hot, then right-click, the <u>Grip menu</u> appears. The Grip menu contains grip editing options. However, if grips are <u>cold</u> and you right-click, the <u>Edit Mode menu</u> appears (Fig. 23-5). This menu appears any time one or more objects are selected, no commands are active, and you right-click (see Chapter 1, Edit Mode Menu). Although the commands displayed in the two menus appear to be the same, they are not. The Edit Mode menu contains the full commands. For example, *Move* in Figure 23-4 is the MOVE grip option, whereas *Move* in Figure 23-5 is the *Move* command.

NOTE: When *PICKFIRST* is set to 0, the Edit-mode shortcut menus do not appear—the Default menu appears instead.

Figure 23-5

Grips in Previous Versions of AutoCAD

In versions of AutoCAD previous to 2004, grips can have three states: **warm**, **cold**, and **hot** (Fig. 23-6). In these versions, when an object is selected and its grips appear, the grips (blue) are called "warm." If you then deselect the object using SHIFT+1 to remove the object from the selection set, the object becomes unhighlighted, yet the grips remain active, or "cold." This feature is useful for using these grips to "snap" to, but since the object is unhighlighted, it would not be included in the set to MOVE, ROTATE, SCALE, etc. However, in <u>AutoCAD 2004, grips are only cold or hot</u>. When an object whose grips are active (cold) is deselected in AutoCAD 2004, grips are canceled.

Figure 23-6

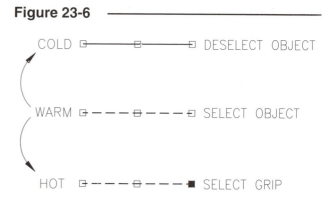

Grip Editing Options

When a **hot grip** has been activated, the grip editing options are available. The Command prompt is replaced by the STRETCH, MOVE, ROTATE, SCALE, or MIRROR grip editing options. You can sequentially cycle through the options by pressing the Space bar or Enter key. The editing options are displayed in Figures 23-7 through 23-11 on the next page.

Alternately, you can <u>right-click when a grip is **hot**</u> to activate the grip menu (see Figure 23-4). This menu has the same options available in Command line format with the addition of *Reference, Properties…*, and *Go to URL…*. The options are described in the following figures.

** STRETCH **
Specify stretch point or
[Basepoint/Copy/Undo/eXit]:

NOTE: Figures display a cold grip as unfilled
and a hot grip as filled solid.

Figure 23-7

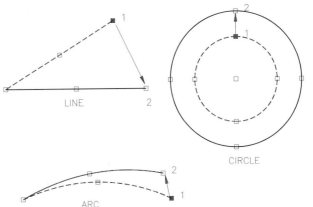

** MOVE **
Specify move point or
[Base point/Copy/Undo/eXit]:

Figure 23-8

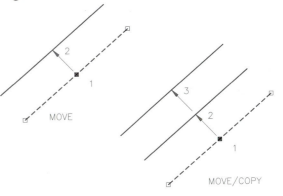

** ROTATE **
Specify rotation angle or
[Base point/Copy/Undo/Reference/eXit]:

Figure 23-9

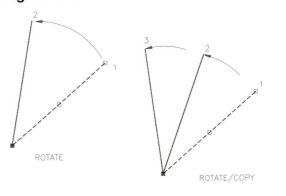

** SCALE **
Specify scale factor or
[Base point/Copy/Undo/Reference/eXit]:

Figure 23-10

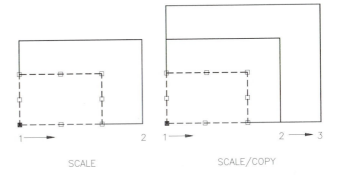

** MIRROR **
Specify second point or
[Base point/Copy/Undo/eXit]:

Figure 23-11

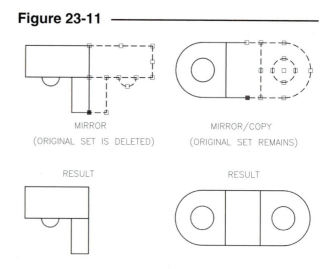

The *Grip* options are easy to understand and use. Each option operates like the full AutoCAD command by the same name. Generally, the editing option used (except for STRETCH) affects all highlighted objects. The **hot** grip is the base point for each operation. The suboptions, Base and Copy, are explained next.

NOTE: The STRETCH option differs from other options in that STRETCH affects only the object that is attached to the **hot** grip, rather than affecting all highlighted (**cold**) objects.

Base
The Base suboption appears with all of the main grip editing options (STRETCH, MOVE, etc.). Base allows using any other grip as the base point instead of the **hot** grip. Type the letter *B* or select from the right-click cursor menu to invoke this suboption.

Copy
Copy is a suboption of every main choice. Activating this suboption by typing the letter *C* or selecting from the right-click cursor menu invokes a Multiple copy mode, such that whatever set of objects is STRETCHed, MOVEd, ROTATEd, etc., becomes the first of an unlimited number of copies (see the previous five figures). The Multiple mode remains active until exiting back to the Command prompt.

Undo
The Undo option, invoked by typing the letter *U* or selecting from the right-click cursor menu will undo the last Copy or the last Base point selection. Undo functions only after a Base or Copy operation.

Reference
This option operates similarly to the reference option of the *Scale* and *Rotate* commands. Use *Reference* to enter or PICK a new reference length (SCALE) or angle (ROTATE). (See "*Scale* and *Rotate*," Chapter 9.) With grips, *Reference* is only enabled when the SCALE or ROTATE options are active.

Properties... **(Right-Click Menu Only)**
Selecting this option from the right-click grip menu (see Figure 23-4) activates the *Properties* palette. All highlighted objects become subjects of the palette. Any property of the selected objects can be changed with the palette. (See "*Properties*," Chapter 16.)

Auxiliary Grid

An auxiliary grid is <u>automatically</u> established on creating the first Copy (Fig. 23-12). The grid is activated by pressing Shift while placing the subsequent copies. The subsequent copies are then "snapped" to the grid in the same manner that *SNAP* functions. The spacing of this auxiliary grid is determined by the location of the first Copy, that is, the X and Y intervals between the base point and the second point.

Figure 23-12

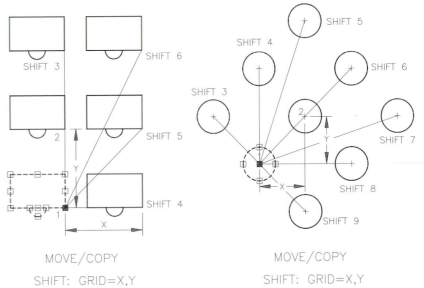

MOVE/COPY
SHIFT: GRID=X,Y

MOVE/COPY
SHIFT: GRID=X,Y

For example, a "polar array" can be simulated with grips by using ROTATE with the Copy suboption (Fig. 23-13). The "array" can be constructed by making one Copy, then using the auxiliary grid to achieve equal angular spacing. The steps for creating a "polar array" are as follows:

1. Select the object(s) to array.

2. Select a grip on the set of objects to use as the center of the array. Cycle to the ROTATE option by pressing Enter or selecting from the right-click cursor menu. Next, invoke the Copy suboption.

3. Make the first copy at any desired location.

4. After making the first copy, activate the auxiliary angular grid by holding down Shift while making the other copies.

5. Cancel the grips or select another command from the menus.

Figure 23-13

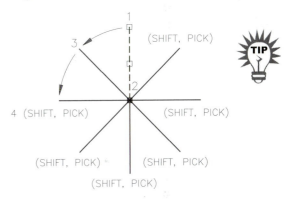

ROTATE/COPY
SHIFT — ANGLES EQUAL

Editing Dimensions

One of the most effective applications of grips is as a dimension editor. Because grips exist at the dimension's extension line origins, arrowhead endpoints, and dimensional value, a dimension can be changed in many ways and still retain its associativity (Fig. 23-14). See Chapter 28, Dimensioning, for further information about dimensions, associativity, and editing dimensions with *Grips*.

Figure 23-14

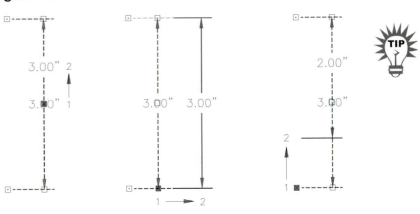

STRETCH TEXT   STRETCH EXTENSION LINE   CHANGE DIMENSION

GUIDELINES FOR USING GRIPS

Although there are many ways to use grips based on the existing objects and the desired application, a general set of guidelines for using grips is given here:

1. Create the objects to edit.

2. Select objects to make **cold** grips appear.

3. Select the desired **hot** grip(s). The *Grip* options should appear in place of the Command line.

4. Press Space or Enter to cycle to the desired editing option (STRETCH, MOVE, ROTATE, SCALE, MIRROR) or select from the right-click cursor menu.

5. Select the desired suboptions, if any. If the *Copy* suboption is needed or the base point needs to be re-specified, do so at this time. *Base* or *Copy* can be selected in either order.

6. Make the desired STRETCH, MOVE, ROTATE, SCALE, or MIRROR.

7. Cancel the grips by pressing Escape or selecting a command from a menu.

GRIPS SETTINGS

Several settings are available in the *Selection* tab of the *Options* dialog box (Fig. 23-15) that control the way grips appear or operate. The settings can also be changed by typing in the related variable name at the Command prompt.

Figure 23-15

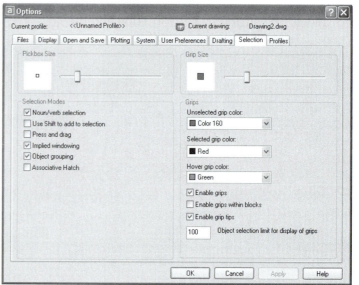

Grip Size
Use the slider bar to interactively increase or decrease the size of the grip "box." The *GRIPSIZE* variable could alternately be used. The default size is 5 pixels square (*GRIPSIZE*=5).

Unselected Grip Color
This setting enables you to change the color of **cold** grips. Select the desired color from the *Unselected Grip Color* drop-down list. PICKing the *Select Color...* tile produces the *Select Color* dialog box (identical to that used with other color settings). The default setting is blue (ACI number 5). Alternately, the *GRIPCOLOR* variable can be typed and any ACI number from 0 to 255 can be entered.

Selected Grip Color
The color of **hot** grips can be specified with this option. Select the desired color from the *Selected Grip Color* drop-down list. The *Select Color...* tile produces the *Select Color* dialog box. You can also type *GRIPHOT* and enter any ACI number to make the change. The default color is red (1).

Hover Grip Color
This setting specifies the color of grips when you "hover" your cursor over a **cold** (blue) grip, usually just before you PICK it to make it a hot grip. The drop-down list operates identically to the other grip color options.

Enable Grips

If a check appears in the checkbox, *Grips* are enabled for the workstation. A check sets the *GRIPS* variable to 1 (on). Removing the check disables grips and sets *GRIPS*=0 (off). The default setting is on.

Enable Grips within Blocks

Figure 23-16

When no check appears in this box, only one grip at the block's insertion point is visible (Fig. 23-16). This allows you to work with a block as one object. The related variable is *GRIPBLOCK*, with a setting of 0=off. This is the default setting (disabled).

When the box is checked, *GRIPBLOCK* is set to 1. All grips on all objects contained in the block are visible and functional (Fig. 23-16). This allows you to use any grip on any object within the block. This does <u>not</u> mean that the block is <u>*Exploded*</u>—the block retains its single-object integrity and individual entities in the block cannot be edited independently. This feature permits you to use the grips only on each of the block's components.

Notice in Figure 23-16 how the insertion point is not accessible when *GRIPBLOCK* is set to 1. Notice also that there are two grips on each of the *Normal* attributes and only one on the *Invisible* attribute and that these grips do not change with the two *GRIPBLOCK* settings. Making one of these grips hot allows you to "move" the attribute to any location and the attribute still retains its association with the block.

Enable Grip Tips

This option has no effect on standard AutoCAD objects. This option enables the grip tips only when custom (third-party) objects supporting grip tips are included in the drawing. (Alternately, set *GRIPTIPS*=1.)

Object Selection Limit for Display of Grips

This value sets the maximum number of selected objects that can display grips. When the initial selection set includes more than the specified number of objects, grips are not displayed. The valid range is 1 to 32,767. (The default setting is 20 for the *GRIPOBJLIMIT* system variable.)

NOTE: All *Grip*-related variable settings are saved in the system registry rather than in the current drawing file. This generally means that changes in these variables remain with the computer, not the drawing being used.

Point Specification Precedence

When you select a point with the input device, AutoCAD uses a point specification precedence to determine which location on the drawing to find. The hierarchy is listed here:

1. Object Snap (*OSNAP*)
2. Explicit coordinate entry (absolute, relative, or polar coordinates)
3. *ORTHO*
4. Point filters (XY filters)
5. *Grips* auxiliary grid (rectangular and circular)
6. *Grips* (on objects)
7. *SNAP* (F9 grid snap)
8. Digitizing a point location

Practically, this means that *OSNAP* has priority over any other point selection mode. As far as *Grips* are concerned, *ORTHO* overrides *Grips*, so turn off *ORTHO* if you want to snap to *Grips*. Although *Grips* override *SNAP* (F9), it is suggested that *SNAP be turned Off* while using *Grips* to simplify PICKing.

More Grips

Grips have a wide range of editing potential. For example, using grips for modifying dimensions is extremely efficient (see Chapter 28, Dimensioning). Grips in AutoCAD 2004 have a 3D feature, that is, they always appear parallel to the current UCS XY plane and operate in their respective planes (see Chapter 40, Surface Modeling).

CHAPTER EXERCISES

Figure 23-17 ———————————

1. *Open* the **TEMP-D** drawing from Chapter 16 Exercises. Use the grips on the existing *Spline* to generate a new graph displaying the temperatures for the following week. **PICK** the *Spline* to activate grips. Use the **STRETCH** option to stretch the first data point grip to the value indicated below for Sunday's temperature. **Cancel** the grips; then repeat the steps for each data point on the graph. *Save* the drawing as **TEMP-F**.

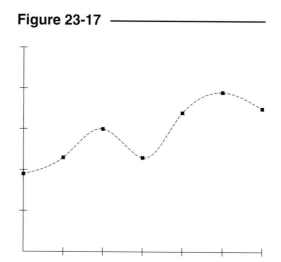

X axis	Y axis
Sunday	29
Monday	32
Tuesday	40
Wednesday	33
Thursday	44
Friday	49
Saturday	45

2. *Open* **CH16EX2** drawing. Activate **grips** on the *Line* to the far right. Make the top grip **hot**. Use the **STRETCH** option to stretch the top grip to the right to create a vertical *Line*. **Cancel** the grips.

Next, activate the **grips** for all the other *Lines* (not including the vertical one on the right); as shown in Figure 23-18. **STRETCH** the top of all inclined *Lines* to the top of the vertical *Line* by making the common top grips **hot**, then OSNAPing to the top **ENDpoint** of the vertical *Line*. *Save* the drawing as **CH23EX2**.

Figure 23-18 ———————————

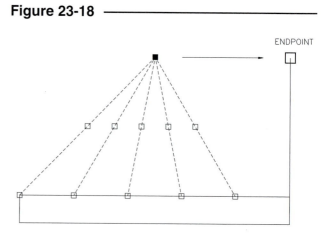

ENDPOINT

3. This exercise involves all the options of grip editing to create a Space Plate. Begin a *New* drawing or use the **ASHEET** *template* and use *SaveAs* to assign the name **SPACEPLT**.

A. Set the *Snap* value to **.125**. Set **Polar Snap** to **.125** and turn on **Polar Tracking**. Draw the geometry shown in Figure 23-19 using the *Line* and *Circle* commands.

Figure 23-19

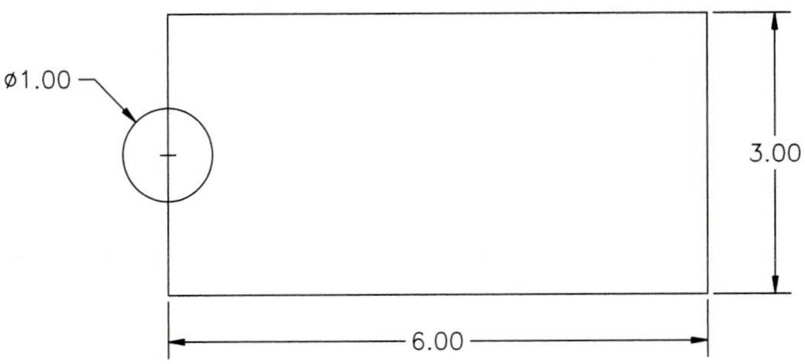

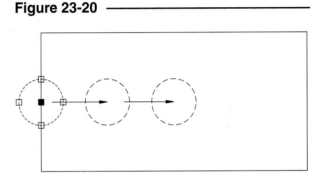

B. Activate the **grips** on the *Circle*. Make the center grip **hot**. Cycle to the **MOVE** option. Enter *C* for the **Copy** option. You should then get the prompt for **MOVE (multiple)**. Make two copies as shown in Figure 23-20. The new *Circles* should be spaced evenly, with the one at the far right in the center of the rectangle. If the spacing is off, use **grips** with the **STRETCH** option to make the correction.

Figure 23-20

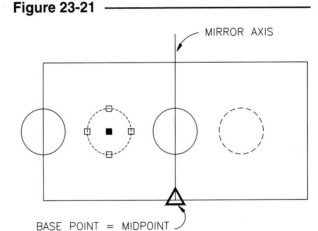

C. Activate the **grips** on the number 2 *Circle* (from the left). Make the center *Circle* grip **hot** and cycle to the **MIRROR** option. Enter *C* for the **Copy** option. (You'll see the **MIRROR (multiple)** prompt.) Then enter *B* to specify a new base point as indicated in Figure 23-21. Turn *On ORTHO* and specify the mirror axis as shown to create the new *Circle* (shown in Fig. 23-21 in hidden line-type).

Figure 23-21

MIRROR AXIS

BASE POINT = MIDPOINT

D. Use *Trim* to trim away the outer half of the *Circle* and the interior portion of the vertical *Line* on the left side of the Space Plate. Activate the **grips** on the two vertical *Lines* and the new *Arc*. Make the common grip **hot** (on the *Arc* and *Line* as shown in Fig. 23-22) and **STRETCH** it downward .5 units. (Note how you can affect multiple objects by selecting a common grip.) **Stretch** the upper end of the *Arc* upward .5 units.

Figure 23-22

E. *Erase* the *Line* on the right side of the Space Plate. Use the same method that you used in step C to **MIRROR** the *Lines* and *Arc* to the right side of the plate (as shown in Fig. 23-23).

(REMINDER: After you select the **hot** grip, use the **Copy** option <u>and</u> the **Base** point option. Use *ORTHO*.)

Figure 23-23

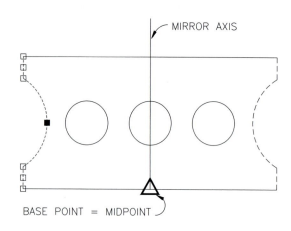

F. In this step, you will **STRETCH** the top edge upward one unit and the bottom edge downward one unit by selecting <u>multiple</u> **hot** grips.

Select the desired horizontal <u>and</u> attached vertical *Lines*. Hold down Shift while selecting <u>each</u> of the endpoint grips, as shown in Figure 23-24. Although they appear **hot**, you must select one of the two again to activate the **STRETCH** option.

Figure 23-24

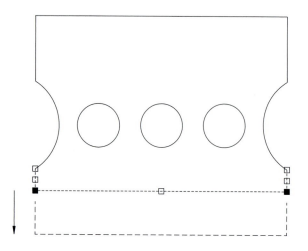

G. In this step, two more *Circles* are created by the **ROTATE** option (see Figure 23-25). Select the two outside *Circles* to make the grips appear. PICK the center grip to make it **hot**. Cycle to the **ROTATE** option. Enter *C* for the **Copy** option. Next, enter *B* for the **Base** option and *OSNAP* to the *CENter* as shown. Make sure *ORTHO* is *On* and create the new *Circles*.

Figure 23-25

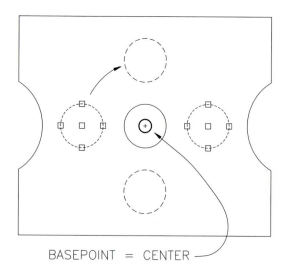

H. Select the center *Circle* and make the center grip **hot** (Fig. 23-26). Cycle to the **SCALE** option. The scale factor is **1.5**. Since the *Circle* is a 1 unit diameter, it can be interactively scaled (watch the *Coords* display), or you can enter the value. The drawing is complete. *Save* the drawing as **SPACEPLT**.

Figure 23-26

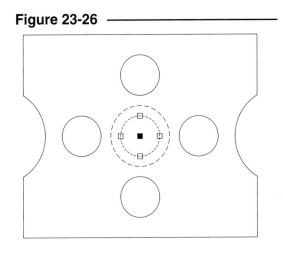

4. *Open* the **STORE ROOM2** drawing from Chapter 18 Exercises. Use the grips to **STRETCH, MOVE,** or **ROTATE** each text paragraph to achieve the results shown in Figure 23-27. *SaveAs* **STORE ROOM3**.

Figure 23-27

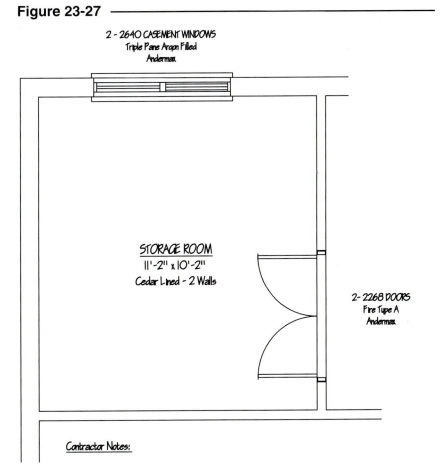

5. *Open* the **T-PLATE** drawing that you created in Chapter 9. An order has arrived for a modified version of the part. The new plate requires two new holes along the top, a 1″ increase in the height, and a .5″ increase from the vertical center to the hole on the left (Fig. 23-28). Use grips to **STRETCH** and **Copy** the necessary components of the existing part. Use *SaveAs* to rename the part to **TPLATEB**.

Figure 23-28

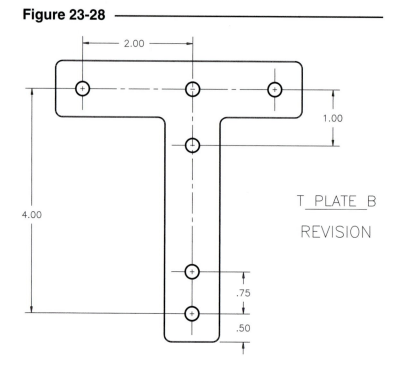

6. *Open* the **OFFICE** drawing that you worked on in Chapter 21 Exercises. Use grips to **STRETCH**, **MOVE**, **ROTATE**, **SCALE**, **MIRROR**, and **Copy** the furniture *Blocks*. Experiment with each option. Change the *GRIPBLOCK* variable to enable all the grips on the *Blocks* for some of the editing. *Save* any changes that you like.

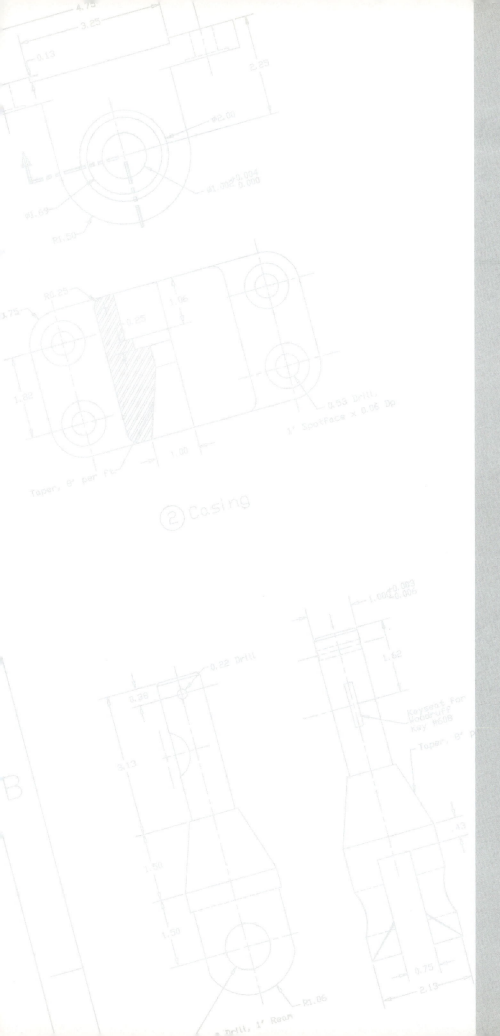

MULTIVIEW DRAWING

Chapter Objectives

After completing this chapter you should:

1. be able to draw projection lines using *ORTHO* and *SNAP*, and *Polar Tracking*;

2. be able to use *Xline* and *Ray* to create construction lines;

3. know how to use *Offset* for construction of views;

4. be able to use *Object Snap Tracking* for alignment of lines and views;

5. be able to use construction layers for managing construction lines and notes;

6. be able to use linetypes, lineweights, and layers to draw and manage ANSI standards;

7. know how to create fillets, rounds, and runouts;

8. know the typical guidelines for creating a three-view multiview drawing.

CONCEPTS

Multiview drawings are used to represent 3D objects on 2D media. The standards and conventions related to multiview drawings have been developed over years of using and optimizing a system of representing real objects on paper. Now that our technology has developed to a point that we can create 3D models, some of the methods we use to generate multiview drawings have changed, but the standards and conventions have been retained so that we can continue to have a universally understood method of communication.

This chapter illustrates methods of creating 2D multiview drawings with AutoCAD (without a 3D model) while complying with industry standards. (Creating 2D drawings from 3D models is addressed in Chapter 42.) Many techniques can be used to construct multiview drawings with AutoCAD because of its versatility. The methods shown in this chapter are the more common methods because they are derived from traditional manual techniques. Other methods are possible.

PROJECTION AND ALIGNMENT OF VIEWS

Projection theory and the conventions of multiview drawing dictate that the views be aligned with each other and oriented in a particular relationship. AutoCAD has particular features, such as *SNAP*, *ORTHO*, construction lines (*Xline, Ray*), Object Snap, Polar Tracking and Object Snap Tracking, that can be used effectively for facilitating projection and alignment of views.

Using *ORTHO* and *OSNAP* to Draw Projection Lines

ORTHO (F8) can be used effectively in concert with *OSNAP* to draw projection lines during construction of multiview drawings. For example, drawing a Line interactively with *ORTHO ON* forces the *Line* to be drawn in either a horizontal or vertical direction.

Figure 24-1 simulates this feature while drawing projection *Lines* from the top view over to a 45 degree miter line (intended for transfer of dimensions to the side view). The "first point:" of the *Line* originated from the *Endpoint* of the *Line* on the top view.

Figure 24-1 ──────────────

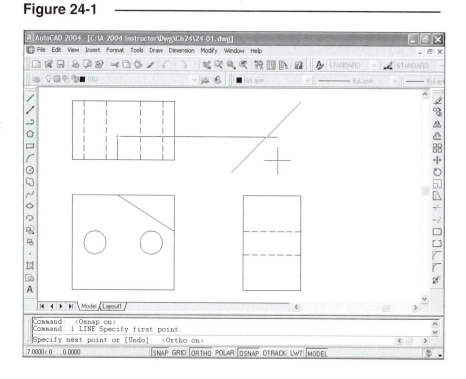

Figure 24-2 illustrates the next step. The vertical projection *Line* is drawn from the *Intersection* of the 45 degree line and the last projection line. *ORTHO* forces the *Line* to the correct vertical alignment with the side view.

Remember that any draw <u>or</u> edit command that requires PICKing is a candidate for *ORTHO* and/or *OSNAP*.

NOTE: *OSNAP* overrides *ORTHO*. If *ORTHO* is *ON* and you are using an *OSNAP* mode to PICK the "next point:" of a *Line*, the *OSNAP* mode has priority; and, therefore, the construction may not result in an orthogonal *Line*.

Figure 24-2

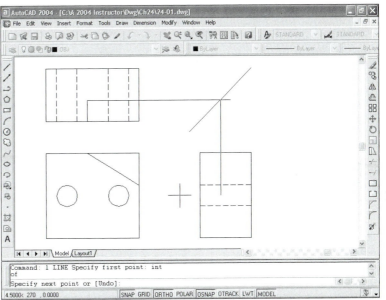

Using *Polar Tracking* to Draw Projection Lines

Similar to using *ORTHO* and *SNAP*, you can use *Polar Tracking* and *SNAP* options to draw projection lines at 90-degree increments. Using the previous example, *Polar Tracking* is used to project dimensions between the top and side views through the 45-degree miter line.

To use *Polar Tracking*, set the desired *Polar Angle Settings* in the *Polar Tracking* tab of the *Drafting Settings* dialog box. Ensure *POLAR* is appears recessed on the Status Bar. You can also set a *Polar Snap* increment or a *Grid Snap* increment to use with *Polar Tracking* (see Chapter 3 for more information on these settings).

For example, you could use the *Endpoint OSNAP* option to snap to the *Line* in the top view, then *Polar Tracking* forces the line to a previously set polar increment (0 degrees in this case).

Note that you can use *OSNAPs* in conjunction with *Polar Tracking*, as shown in Figure 24-3, such that the current horizontal line *OSNAPs* to its *Intersection* with the 45-degree miter line. (*OSNAPS* <u>cannot</u> be used effectively with *ORTHO*, since *OSNAPs* override *ORTHO*.)

Figure 24-3

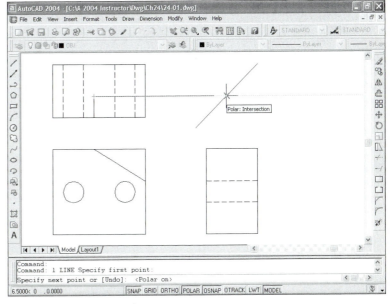

In the following step, use *Intersection OSNAP* option to snap to the intersection of the previous *Line* and the 45-degree miter line. Again *Polar Tracking* forces the *Line* to a vertical position (270 degrees).

Figure 24-4

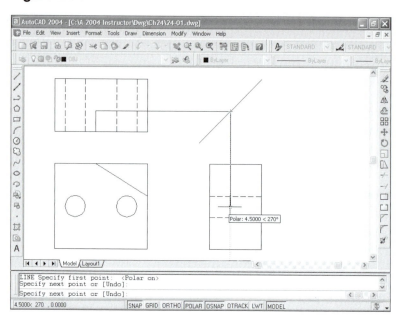

Using *Object Snap Tracking* to Draw Projection Lines Aligned with *OSNAP* Points

This feature, although the most complex, provides the greatest amount of assistance in constructing multiview drawings. The advantage is the availability of the features described in the previous method (*Polar Tracking* and *OSNAP* for alignment of vertical and horizontal *Lines*) in addition to the creation of *Lines* and other objects that align with *OSNAP* points (*Endpoint, Intersection, Midpoint,* etc.) of other objects.

To use *Object Snap Tracking*, first set the desired running *OSNAP* options (*Endpoint, Intersection,* etc.). Next, toggle on *OTRACK* at the Status Bar. Use *Polar Tracking* in conjunction with *Object Snap Tracking* by setting polar angle increments (see previous discussion) and toggling on *POLAR* on the Status Bar.

For example, to begin the *Line* in Figure 24-5, first "acquire" the *Endpoint* of the *Line* shown in the front view. Note that the "first point" of the new *Line* (in the top view) aligns vertically with the acquired *Endpoint*. At this point in time, the short vertical *Line* for the top view could be drawn from the intersection shown. *Object Snap Tracking* <u>prevents having to draw the vertical projection line</u> from the front up to the top view.

Figure 24-5

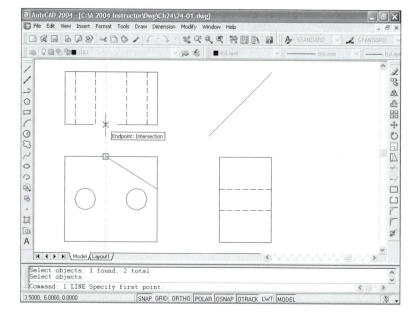

After constructing the two needed *Lines* for the top view, a tracking vector aligns with the acquired *Endpoint* of the indicated line in the top view and the *Intersection* of the 45-degree miter line. Note that a <u>horizontal projection line is not needed</u> since *Object Snap Tracking* ensures the "first point" aligns with the appropriate point from the top view.

Figure 24-6

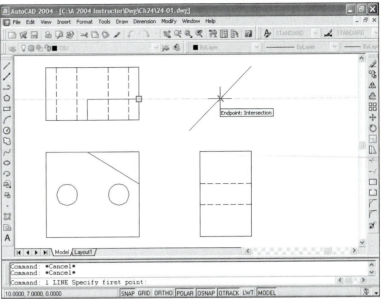

The next step is to draw the short horizontal visible line in the side view. Coming from the "first point" (on the 45-degree miter line found in the previous step), draw the vertical projection line down to the appropriate point in the side view that "tracks" with the related *Endpoint* in the front view (see horizontal tracking vector). Since this *Line* is constructed to the correct endpoint, <u>*Trimming* is unnecessary here</u>, but it would be needed in the previous two methods.

Figure 24-7

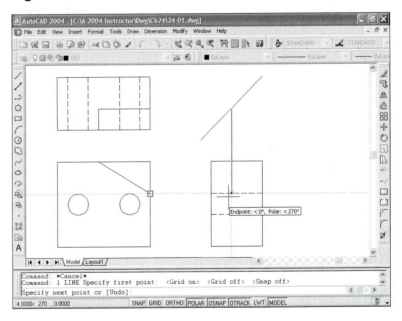

These AutoCAD features (*Object Snap Tracking* in conjunction with *Polar Tracking*) are probably the most helpful features for construction of multiview drawing since the introduction of AutoCAD in 1982.

Using *Xline* and *Ray* for Construction Lines

Another strategy for constructing multiview drawings is to make use of the AutoCAD construction line commands *Xline* and *Ray*.

Xlines can be created to "box in" the views and ensure proper alignment. *Ray* is suited for creating the 45 degree miter line for projection between the top and side views. The advantage to using this method is that horizontal and vertical *Xlines* can be created quickly.

These lines can be *Trimmed* to become part of the finished view, or other lines could be drawn "on top of" the construction lines to create the final lines of the views. In either case, **TIP** constructions lines should be drawn on a separate layer so that the layer can be frozen before plotting. If you intend to *Trim* the construction lines so that they become part of the final geometry, draw them originally on the view layers.

Figure 24-8 ———————————

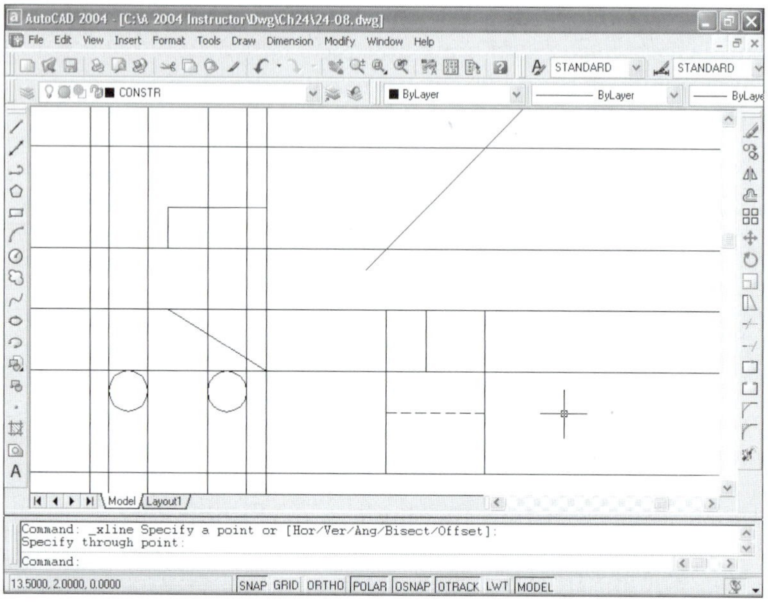

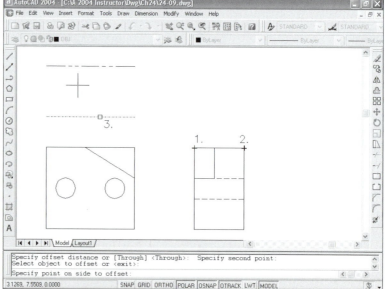

Using *Offset* for Construction of Views

An alternative to using the traditional miter line method for construction of a view by projection, the *Offset* command can be used to transfer distances from one view and to construct another. The *Distance* option of *Offset* provides this alternative.

For example, assume that the side view was completed and you need to construct a top view (Fig. 24-9). First, create a horizontal line as the inner edge of the top view (shown highlighted) by *Offset* or other method. To create the outer edge of the top view (shown in phantom linetype), use *Offset* and PICK points (1) and (2) to specify the *distance*. Select the existing line (3) as the "*Object to Offset:*", then PICK the "*side to offset*" at the current cursor position.

Figure 24-9 ———————————

Realignment of Views Using *Polar Snap* **and** *Polar Tracking*

Another application of *Polar Snap* and *Polar Tracking* is the use of *Move* to change the location of an entire view while retaining its orthogonal alignment.

For example, assume that the views of a multiview (Fig. 24-10) are complete and ready for dimensioning; however, there is not enough room between the front and side views. You can invoke the *Move* command, select the entire view with a window or other option, and "slide" the entire view outward. *Polar Tracking* ensures proper orthogonal alignment. *Polar Snap* forces the movement to a regular increment so the coordinate points of the geometry retain a relationship to the original points (for example, moving exactly 1 unit).

An alternative to *Mov*ing the view interactively is use of coordinate specification (absolute, relative rectangular, or relative polar or direct distance entry). (See Chapter 9 for an example of moving a view with direct distance entry.)

Figure 24-10

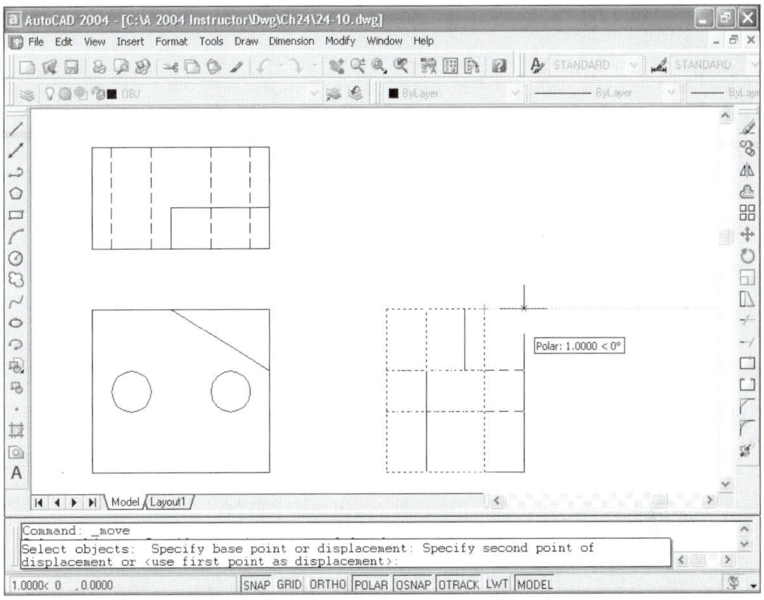

USING CONSTRUCTION LAYERS

The use of layers for isolating construction lines can make drawing and editing faster and easier. The creation of multiview drawings can involve construction lines, reference points, or notes that are not intended for the final plot. Rather than *Erasing* these construction lines, points, or notes before plotting, they can be created on a separate layer and turned *Off* or made *Frozen* before running the final plot. If design changes are required, as they often are, the construction layers can be turned *On*, rather than having to recreate the construction.

There are two strategies for creating construction objects on separate layers:

1. Use Layer 0 for construction lines, reference points, and notes. This method can be used for fast, simple drawings.

2. Create a new layer for construction lines, reference points, and notes. Use this method for more complex drawings or drawings involving use of *Blocks* on Layer 0.

For example, consider the drawing during construction in Figure 24-11. A separate layer has been created for the construction lines, notes, and reference points.

Figure 24-11 ──────────────

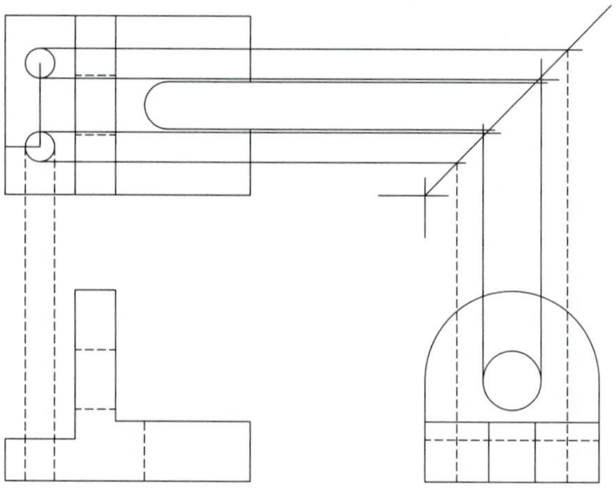

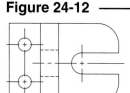

In Figure 24-12, the drawing is ready for making a print. Notice that the construction layer is *Frozen*.

Figure 24-12 ──────────────

If you are printing the *Limits*, the construction layer should be *Frozen*, rather than turned *Off*, since *Zoom All* and *Zoom Extents* are affected by geometry on layers turned *Off* and not *Frozen* (unless the objects are *Xlines* or *Rays*).

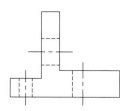

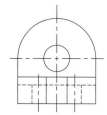

USING LINETYPES

Different types of lines are used to represent different features of a multiview drawing. Linetypes in AutoCAD are accessed by the *Linetype* command or through the *Layer Properties Manager*. *Linetypes* can be changed retroactively by the *Properties* window. *Linetypes* can be assigned to individual objects specifically or to layers (*ByLayer*). See Chapter 11 for a full discussion on this topic.

AutoCAD complies with the ANSI and ISO standards for linetypes. The principal AutoCAD linetypes used in multiview drawings and the associated names are shown in Figure 24-13.

Many other linetypes are provided in AutoCAD. Refer to Chapter 11 for the full list and illustration of the linetypes.

Figure 24-13 ──────────────

CONTINUOUS	————————————	DARK, WIDE
HIDDEN	– – – – – – – – –	MEDIUM
CENTER	—— – —— – ——	MEDIUM
PHANTOM	—— – – ——	VARIES
DASHED	– —— – —— –	VARIES

Other standard lines are created by AutoCAD automatically. For example, dimension lines can be automatically drawn when using dimensioning commands (Chapter 28), and section lines can be automatically drawn when using the *Hatch* command (Chapter 26).

Objects in AutoCAD can have lineweight. This is accomplished by using the *Lineweight* command or by assigning *Lineweight* in the *Layer Properties Manager* (see Chapter 11). Additionally, lineweights can be assigned by using plot styles or assigning plot device lineweights or pen thickness.

Drawing Hidden and Center Lines

Figure 24-14 ────────────

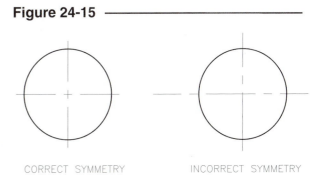

Figure 24-14 illustrates a typical application of AutoCAD *Hidden* and *Center* linetypes. Notice that the horizontal center line in the front view does not automatically locate the short dashes correctly, and the hidden lines in the right side view incorrectly intersect the center vertical line.

Although AutoCAD supplies ANSI standard linetypes, the application of those linetypes does not always follow ANSI standards. For example, you do <u>not</u> have control over the placement of the individual dashes of center lines and hidden lines. (You have control of only the endpoints of the lines and the *Ltscale*.) Therefore, the short dashes of center lines may not cross exactly at the circle centers, or the dashes of hidden lines may not always intersect as desired.

You do, however, have control of the endpoints of the lines. Draw lines with the *Center* linetype such that the endpoints are symmetric about the circle or group of circles. This action assures that the short dash occurs at the center of the circle (if an odd number of dashes are generated). Figure 24-15 illustrates correct and incorrect technique.

Figure 24-15 ────────────

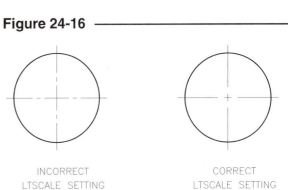

CORRECT SYMMETRY   INCORRECT SYMMETRY

You can also control the relative size of non-continuous linetypes with the *Ltscale* variable. <u>In some cases</u>, the variable can be adjusted to achieve the desired results.

For example, Figure 24-16 demonstrates the use of *Ltscale* to adjust the center line dashes to the correct spacing. Remember that *Ltscale* adjusts linetypes <u>globally</u> (all linetypes across the drawing).

Figure 24-16 ────────────

When the *Ltscale* has been optimally adjusted for the drawing <u>globally</u>, use the *Properties* palette to adjust the linetype scale of <u>individual objects</u>. In this way, the drawing lines can originally be created to the global linetype scale without regard to the *Celtscale*. The finished drawing linetype scale can be adjusted with *Ltscale* globally; then <u>only those objects</u> that need further adjusting can be fine-tuned retroactively with *Properties*.

INCORRECT
LTSCALE SETTING   CORRECT
LTSCALE SETTING

The *Center* command (a dimensioning command) can be used to draw center lines automatically with correct symmetry and spacing (see Chapter 28, Dimensioning).

ANSI Standards require that multiview drawings are created with object lines having a dark lineweight, and hidden, center, dimension, and other reference lines created in a medium lineweight (see Figure 24-13).

You can assign *Lineweight* to objects using the *Lineweight* command or assign *Lineweight* to layers in the *Layer Properties Manager*. As described previously in Chapter 11, *Lineweight* can be assigned to individual objects or to layers (*ByLayer*). Use the *Lineweight Settings* dialog box to assign *Lineweight* property to objects. Use the *Layer Properties Manager* to assign *Lineweight* to layers. Generally, the *ByLayer Lineweight* assignment is preferred, similar to the *ByLayer* method of assigning *Color* and *Linetype*.

Additionally, lineweights can be assigned by using plot styles or assigning plot device lineweights or pen thickness (see Chapter 33 for information on plot styles).

Managing Linetypes, Lineweights, and Colors

There are two strategies for assigning linetypes, lineweights, and colors: *ByLayer* and object-specific assignment. In either case, thoughtful layer utilization for linetypes will make your drawings more flexible and efficient.

BYLAYER Linetypes, Lineweights, and Colors

The *ByLayer* linetype, lineweight, and color settings are recommended when you want the most control over linetype visibility and plotting. This is accomplished by creating layers with the *Layer Properties Manager* and assigning linetype, lineweight, and color for each layer. After you assign *Linetypes* to specific layers, you simply set the layer (with the desired linetype) as the *Current* layer and draw on that layer in order to draw objects in a specific linetype.

A template drawing for creating typical multiview drawings could be set up with the layer and linetype assignments similar to that shown in Figure 24-17. Each layer has its own *color*, *linetype*, and *lineweight* setting, and the layer name indicates the type of lines used (hidden, center, object, etc.). This strategy is useful for simple, generic applications.

Figure 24-17 ─────────────

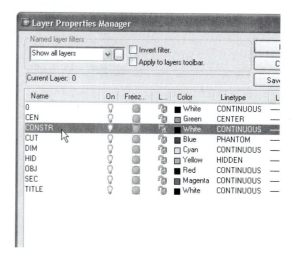

Using the *ByLayer* strategy (*ByLayer Linetype*, *Lineweight*, and *Color* assignment) gives you flexibility. You can control the linetype visibility by controlling the layer visibility (show only object layers or hidden line layers). You can also retroactively change the linetype, lineweight, and color of an existing object by changing the object's *Layer* property with the *Properties* palette or *Matchprop*. Objects changed to a different layer assume the *color*, *linetype*, and *lineweight* of the new layer.

Another strategy for multiview drawings involving several parts, such as an assembly, is to create layers for each line-type specific to each part, as shown in Figure 24-18. With this strategy, each part has the complete set of linetypes, but only one color per part in order to distinguish the part from others in the display.

Related layer groups can be selected within the dialog box using the *Named Layer Filters* dialog box. If the *-Layer* command is used instead, wildcards can be typed for layer selection. For example, entering "?????-HID" selects all of the layers with hidden lines, or entering "MOUNT*" would select all of the layers associated with the "MOUNT" part.

Object *Linetypes, Lineweights, and Colors*

Although this method can be complex, object-specific line-type, lineweight, and color assignment can also be managed by skillful utilization of layers. One method is to create one layer for each part or each group of related geometry (Fig. 24-19). The color, lineweight, and linetype settings should be left to the defaults. Then object-specific linetype settings can be assigned (for hidden, center, visible, etc.) using the *Linetype, Lineweight,* and *Color* commands.

For assemblies, all lines related to one part would be drawn on one layer. Remember that you can draw everything with one linetype and color setting, then use *Properties* to retroactively set the desired color and linetype for each set of objects. Visibility of parts can be controlled by *Freezing* or *Thawing* part layers. You cannot isolate and control visibility of linetypes or colors by this method.

Figure 24-18

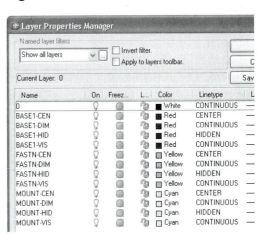

Figure 24-19

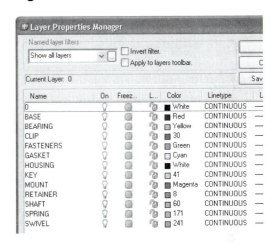

CREATING FILLETS, ROUNDS, AND RUNOUTS

Many mechanical metal or plastic parts manufactured from a molding process have slightly rounded corners. The otherwise sharp corners are rounded because of limitations in the molding process or for safety. A convex corner is called a round and a concave corner is called a fillet. These fillets and rounds are created easily in AutoCAD by using the *Fillet* command.

The example in Figure 24-20 shows a multiview drawing of a part with sharp corners before the fillets and rounds are drawn.

Figure 24-20

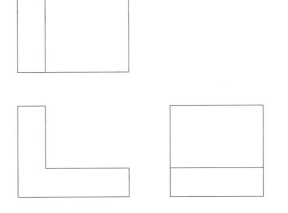

The corners are rounded using the *Fillet* command. First, use *Fillet* to specify the *Radius*. Once the *Radius* is specified, just select the desired lines to *Fillet* near the end to round.

If the *Fillet* is in the middle portion of a *Line* instead of the end, *Extend* can be used to reconnect the part of the *Line* automatically trimmed by *Fillet*, or *Fillet* can be used in the *Notrim* mode.

Figure 24-21

A <u>runout is a visual representation</u> of a complex fillet or round. For example, when two filleted edges intersect at less than a 90 degree angle, a runout should be drawn as shown in the top view of the multiview drawing (Fig. 24-22).

The finish marks (V-shaped symbols) indicate machined surfaces. Finished surfaces have sharp corners.

Figure 24-22

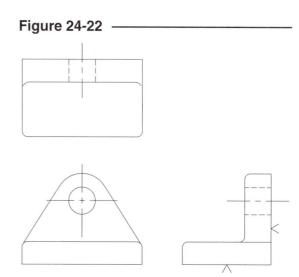

A close-up of the runouts is shown in Figure 24-23. No AutoCAD command is provided for this specific function. The *3point* option of the *Arc* command can be used to create the runouts, although other options can be used. Alternately, the *Circle TTR* option can be used with *Trim* to achieve the desired effect. As a general rule, use the same radius or slightly larger than that given for the fillets and rounds, but draw it less than 90 degrees.

Figure 24-23

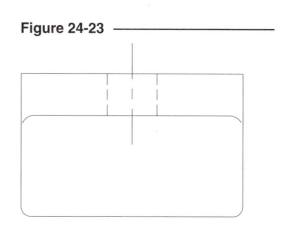

GUIDELINES FOR CREATING A TYPICAL THREE-VIEW DRAWING

Following are some guidelines for creating the three-view drawing in Figure 24-24. This object is used only as an example. The steps or particular construction procedure may vary, depending on the specific object drawn. Dimensions are shown in the figure so you can create the multiview drawing as an exercise.

Figure 24-24

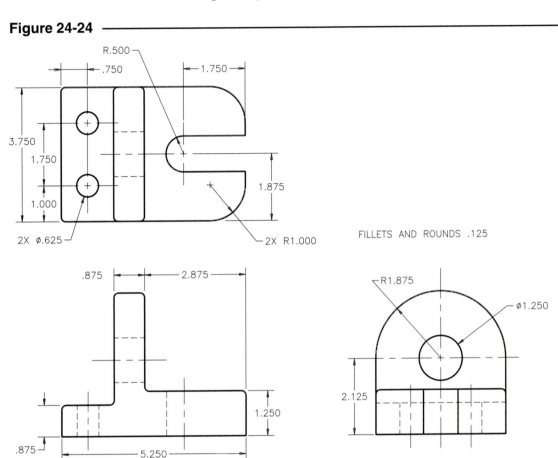

1. Drawing Setup

   Units are set to *Decimal* with 3 places of *Precision*. *Limits* of 22 x 17 are set to allow enough drawing space for both views. The finished drawing can be plotted on a B size sheet at a scale of 1"=1" or on an A size sheet at a scale of 1/2"=1". *Snap* is set to an increment of .125. *Grid* is set to an increment of .5. Although it is recommend that you begin drawing the views using *Grid Snap*, a *Polar Snap* increment of .125 is set, and *Polar Tracking* angles are set. Turn *POLAR* on. Set the desired running *OSNAPs*, such as *Endpoint, Midpoint, Intersection, Quadrant*, and *Center*. Turn on *OTRACK. Ltscale* is not changed from the default of 1. Layers are created (OBJ, HID, CEN, DIM, BORDER, and CONSTR) with appropriate *Linetypes, Lineweights*, and *Colors* assigned.

2. An outline of each view is "blocked in" by drawing the appropriate *Lines* and *Circles* on the OBJ layer similar to that shown in Figure 24-25. Ensure that *SNAP* is *ON*. *ORTHO* or *Polar Tracking* should be turned *ON* when appropriate. Use the cursor to ensure that the views align horizontally and vertically. Note that the top edge of the front view was determined by projecting from the *Circle* in the right side view.

Figure 24-25

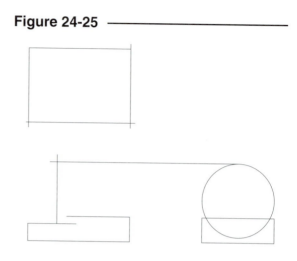

Another method for construction of a multiview drawing is shown in Figure 24-26. This method uses the *Xline* and *Ray* commands to create construction and projection lines. Here all the views are blocked in and some of the object lines have been formed. The construction lines should be kept on a separate layer, except in the case where *Xlines* and *Rays* can be trimmed and converted to the object lines. (The following illustrations do not display this method because of the difficulty in seeing which are object and construction lines.)

Figure 24-26

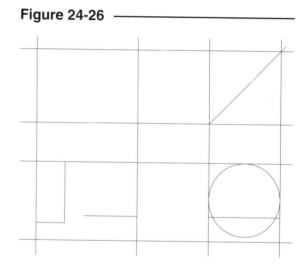

3. This drawing requires some projection between the top and side views (Fig. 24-27). The CONSTR layer is set as *Current*. Two *Lines* are drawn from the inside edges of the two views (using *OSNAP* and *ORTHO* for alignment). A 45 degree miter line is constructed for the projection lines to "make the turn." A *Ray* is suited for this purpose.

Figure 24-27

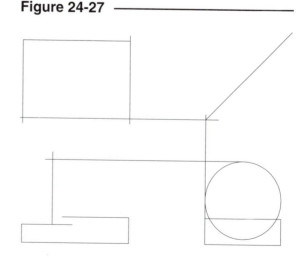

4. Details are added to the front and top views
    (Fig. 24-28). The projection line from the side view to
    the front view (previous figure) is *Trimmed*. A *Circle*
    representing the hole is drawn in the side view and
    projected up and over to the top view and to the front
    view. The object lines are drawn on layer OBJ and
    some projection lines are drawn on layer CONSTR.
    The horizontal projection lines from the 45 degree
    miter line are drawn on layer HID awaiting *Trimming*.
    Alternately, those two projection lines could be drawn
    on layer CONSTR and changed to the appropriate
    layer with *Properties* after *Trimming*. Keep in mind
    that *Object Snap Tracking* can be used here to ensure
    proper alignment with object features and to prevent
    having to actually draw some construction lines.

Figure 24-28

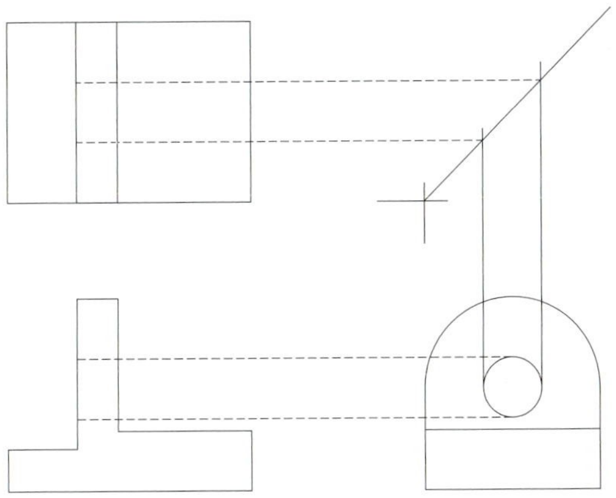

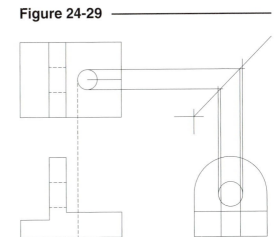

5. The hidden lines used for projection to the top view
    and front view (previous figure) are *Trimmed*. The
    slot is created in the top view with a *Circle* and pro-
    jected to the side and front views. It is usually faster
    and easier to draw round object features in their cir-
    cular view first, then project to the other views. Make
    sure you use the correct layers (OBJ, CONSTR, HID)
    for the appropriate features. If you do not, *Properties*
    or *Matchprop* can be used retroactively.

Figure 24-29

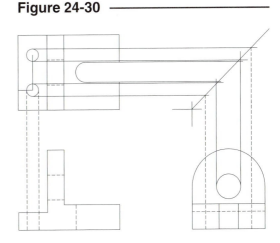

6. The lines shown in the previous figure as projection
    lines or construction lines for the slot are *Trimmed*
    or *Erased*. The holes in the top view are drawn on
    layer OBJ and projected to the other views. The
    projection lines and hidden lines are drawn on
    their respective layers.

Figure 24-30

7. *Trim* the appropriate hidden lines. *Freeze* layer CONSTR. On layer OBJ, use *Fillet* to create the rounded corners in the top view. Draw the correct center lines for the holes on layer CEN. The value for *Ltscale* should be adjusted to achieve the optimum center line spacing. The *Properties* palette can be used to adjust individual object linetype scale.

Figure 24-31

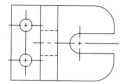

8. Fillets and rounds are added using the *Fillet* command. The runouts are created by drawing a *3point Arc* and *Trimming* or *Extending* the *Line* ends as necessary. Use *Zoom* for this detail work.

Figure 24-32

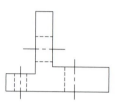
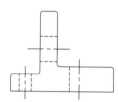
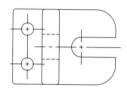

9. Activate a *Layout* tab, configure a print or plot device using *Pagesetup*, make one *Viewport*, and set the *Viewport scale* to a standard scale. Add a border and a title block using *Pline*. Include the part name, company, draftsperson, scale, date, and drawing file name in the title block. The drawing is ready for dimensioning and manufacturing notes.

Figure 24-33

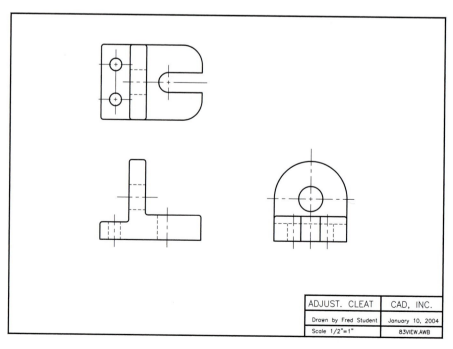

ADJUST. CLEAT	CAD, INC.
Drawn by Fred Student	January 10, 2004
Scale 1/2"=1"	83VIEW.AWB

CHAPTER EXERCISES

1. *Open* the **PIVOTARM CH16** drawing.

    A. Create the right side view. Use *OSNAP* and **ORTHO** or *Polar Tracking* to create *Lines* or *Rays* to the miter line and down to the right side view as shown in Figure 24-34. *Offset* may be used effectively for this purpose instead. Use *Extend, Offset,* or *Ray* to create the projection lines from the front view to the right side view. Use *Object Snap Tracking* when appropriate.

    Figure 24-34

    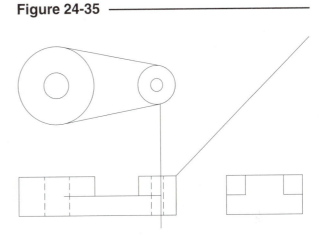

    B. *Trim* or *Erase* the unwanted projection lines, as shown in Figure 24-35. Draw a *Line* or *Ray* from the *Endpoint* of the diagonal *Line* in the top view down to the front to supply the boundary edge for *Trimming* the horizontal *Line* in the front view as shown.

    Figure 24-35

    C. Next, create the hidden lines for the holes by the same fashion as before. Use previously created *Layers* to achieve the desired *Linetypes*. Complete the side view by adding the horizontal hidden *Line* in the center of the view.

    Figure 24-36

    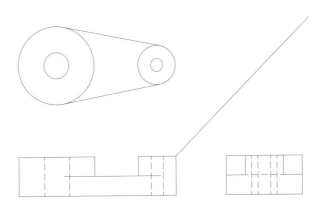

D. Another hole for a set screw must be added to the small end of the Pivot Arm. Construct a *Circle* of 4mm diameter with its center located 8mm from the top edge in the side view as shown in Figure 24-37. Project the set screw hole to the other views.

Figure 24-37

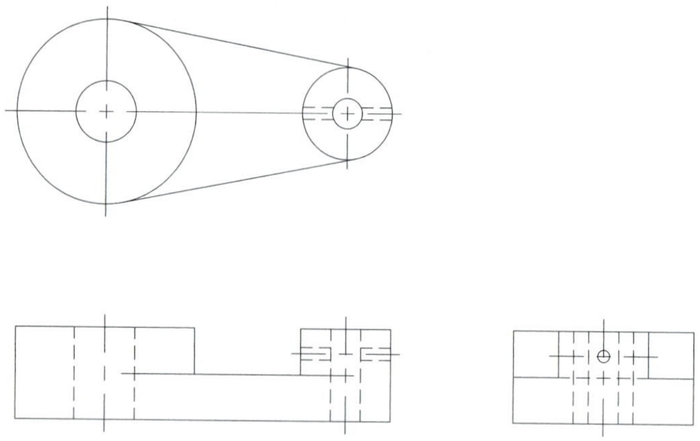

E. Make new layers **CONSTR**, **OBJ**, and **TITLE** and change objects to the appropriate layers with *Properties*. *Freeze* layer **CONSTR**. Add centerlines on the **CEN** layer as shown in Figure 24-37. Change the *Ltscale* to **18**. Activate a *Layout* tab, configure a print or plot device using *Pagesetup*, make one *Viewport*, and set the *Viewport scale* to a standard scale. To complete the PIVOTARM drawing, draw a *Pline* border (*width* **.02** x scale factor) in the layout and *Insert* the **TBLOCKAT** drawing that you created in Chapter 22 Exercises. *SaveAs* **PIVOTARM CH24**.

For exercises 2 through 5, construct and plot the multiview drawings as instructed. Use an appropriate *template* drawing for each exercise unless instructed otherwise. Use conventional practices for *layers*, *linetypes*, and *linetypes*. Draw a *Pline* border with the correct *width* and *Insert* your **TBLOCK** or **TBLOCKAT** drawing.

2. Make a two-view multiview drawing of the Clip. *Plot* the drawing full size (**1=1**). Use the **ASHEET** template drawing to achieve the desired plot scale. *Save* the drawing as **CLIP**.

Figure 24-38

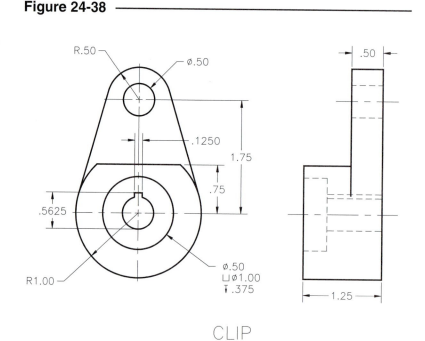

CLIP

3. Make a three-view multiview drawing of the Bar Guide and *Plot* it **1=1**. Use the **BARGUIDE** drawing you set up in Chapter 12 Exercises. Note that a partial *Ellipse* will appear in one of the views.

Figure 24-39

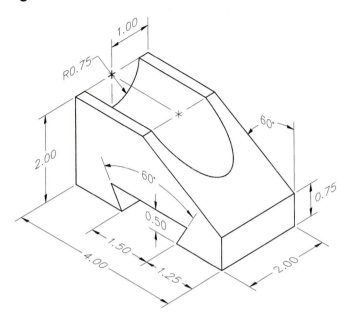

4. Construct a multi-view drawing of the Saddle. Three views are needed. The channel along the bottom of the part intersects with the saddle on top to create a slotted hole visible in the top view. *Plot* the drawing at **1=1**. *Save* as **SADDLE**.

Figure 24-40

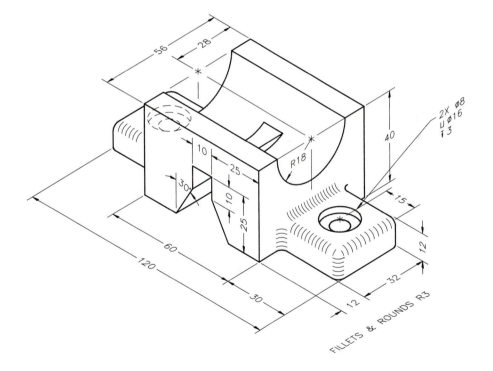

5. Draw a multiview of the Adjustable Mount shown in Figure 24-41. Determine an appropriate template drawing to use and scale for plotting based on your plotter capabilities. *Plot* the drawing to an accepted scale. *Save* the drawing as **ADJMOUNT**.

Figure 24-41

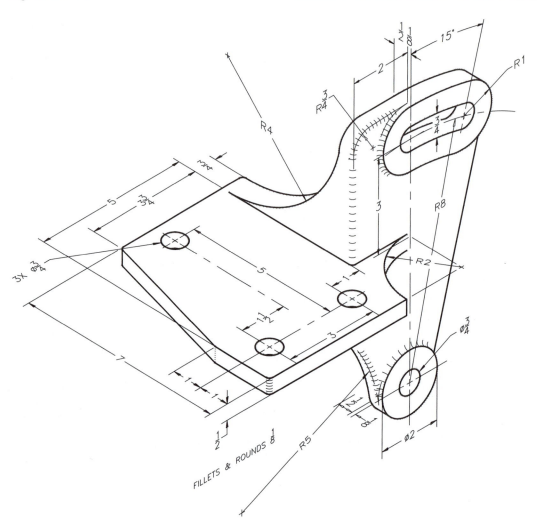

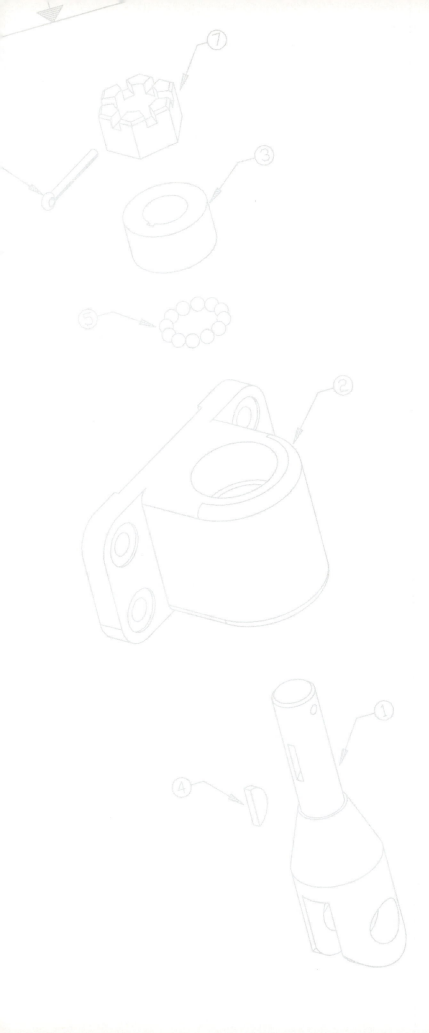

25

PICTORIAL DRAWINGS

Chapter Objectives

After completing this chapter you should be able to:

1. activate the *Isometric Style* of *Snap* for creating isometric drawings;

2. draw on the three isometric planes by toggling *Isoplane* using Ctrl+E;

3. create isometric ellipses with the *Isocircle* option of *Ellipse*;

4. construct an isometric drawing in AutoCAD;

5. create Oblique Cavalier and Cabinet drawings in AutoCAD.

CONCEPTS

Isometric drawings and oblique drawings are pictorial drawings. Pictorial drawings show three principal faces of the object in one view. A pictorial drawing is a drawing of a 3D object as if you were positioned to see (typically) some of the front, some of the top, and some of the side of the object. All three dimensions of the object (width, height, and depth) are visible in a pictorial drawing.

Figure 25-1

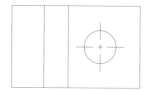

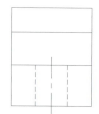

Multiview drawings differ from pictorial drawings because a multiview only shows two dimensions in each view, so two or more views are needed to see all three dimensions of the object. A pictorial drawing shows all dimensions in the one view. Pictorial drawings depict the object similar to the way you are accustomed to viewing objects in everyday life, that is, seeing all three dimensions. Figure 25-1 and Figure 25-2 show the same object in multiview and in pictorial representation, respectively. Notice that multiview drawings use hidden lines to indicate features that are normally obstructed from view, whereas <u>hidden lines are normally omitted</u> in isometric drawings (unless certain hidden features must be indicated for a particular function or purpose).

Figure 25-2

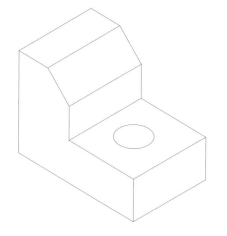

Types of Pictorial Drawings

Pictorial drawings are classified as follows:

1. Axonometric drawings
    a. Isometric drawings
    b. Dimetric drawings
    c. Trimetric drawings

2. Oblique drawings

Axonometric drawings are characterized by how the angle of the edges or axes (axon-) are measured (-metric) with respect to each other.

Isometric drawings are drawn so that each of the axes have equal angular measurement. ("Isometric" means equal measurement.) The isometric axes are always drawn at 120 degree increments (Fig. 25-3). All rectilinear lines on the object (representing horizontal and vertical edges—not inclined or oblique) are drawn on the isometric axes.

Figure 25-3

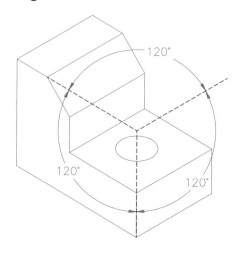

A 3D object seen "in isometric" is thought of as being oriented so that each of three perpendicular faces (such as the top, front, and side) are seen equally. In other words, the angles formed between the line of sight and each of the principal faces are equal.

Dimetric drawings are constructed so that the angle between any two of the three axes is equal. There are many possibilities for dimetric axes. A common orientation for dimetric drawings is shown in Figure 25-4. For 3D objects seen from a dimetric viewpoint, the angles formed between the line of sight and each of two principal faces are equal.

Figure 25-4

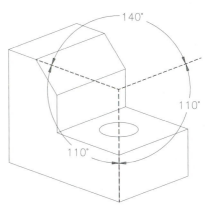

Trimetric drawings have three unequal angles between the axes. Numerous possibilities exist. A common orientation for trimetric drawings is shown in Figure 25-5.

Figure 25-5

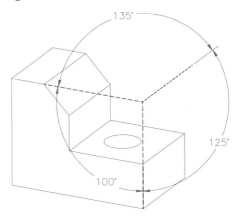

Oblique drawings are characterized by a vertical axis and horizontal axis for the two dimensions of the front face and a third (receding) axis of either 30, 45, or 60 degrees (Fig. 25-6). Oblique drawings depict the true size and shape of the front face, but add the depth to what would otherwise be a typical 2D view. This technique simplifies construction of drawings for objects that have contours in the profile view (front face) but relatively few features along the depth. Viewing a 3D object from an oblique viewpoint is not possible.

Figure 25-6

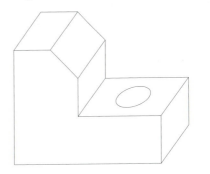

This chapter will explain the construction of isometric and oblique drawings in AutoCAD.

Pictorial Drawings Are 2D Drawings

Isometric, dimetric, trimetric, and oblique drawings are <u>2D drawings</u>, whether created with AutoCAD or otherwise. Pictorial drawing was invented before the existence of CAD and therefore was intended to simulate a 3D object on a 2D plane (the plane of the paper). If AutoCAD is used to create the pictorial, the geometry lies on a 2D plane—the XY plane. All coordinates defining objects have X and Y values with a Z value of 0. When the *Isometric* style of *Snap* is activated, an isometrically structured *SNAP* and *GRID* appear on the XY plane.

Figure 25-7 illustrates the 2D nature of an isometric drawing created in AutoCAD. Isometric lines are created on the XY plane. The *Isometric SNAP* and *GRID* are also on the 2D plane. (The *Vpoint* command was used to give other than a *Plan* view of the drawing in this figure.)

Although pictorial drawings are based on the theory of projecting 3D objects onto 2D planes, it is physically possible to achieve an axonometric (isometric, dimetric, or trimetric) viewpoint of a 3D object using a 3D CAD system. In AutoCAD, the *Vpoint* command is used to specify the observer's position in 3D space with respect to a 3D model. Chapter 35 discusses the specific commands and values needed to attain axonometric viewpoints of a 3D model.

Figure 25-7

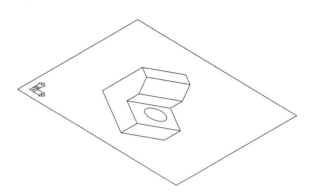

ISOMETRIC DRAWING IN AutoCAD

AutoCAD provides the capability to construct isometric drawings. An isometric *SNAP* and *GRID* are available, as well as a utility for creation of isometrically correct ellipses. Isometric lines are created with the *Line* command. There are no special options of *Line* for isometric drawing, but isometric *SNAP* and *GRID* can be used to force *Lines* to an isometric orientation. Begin creating an isometric drawing in AutoCAD by activating the *Isometric Style* option of the *Snap* command. This action can be done using any of the options listed in the following Command table.

SNAP

Pull-down Menu	COMMAND (TYPE)	ALIAS (TYPE)	Short-cut	Screen (side) Menu	Tablet Menu
Tools *Drafting Settings...* *Snap and Grid*	*SNAP*	*SN*	*F9 or Ctrl+B*	*TOOLS 2* *Grid*	*W,10*

Command: **snap**
Specify snap spacing or
[ON/OFF/Aspect/Rotate/Style/Type] <0.5000>: **s**
Enter snap grid style [Standard/Isometric] <S>: **i**
Specify vertical spacing <0.5000>: **Enter**
Command:

Alternately, toggling the indicated checkbox in the lower-right corner of the *Drafting Settings* dialog box activates the *Isometric Snap* and *Grid Snap* (Fig. 25-8).

Figure 25-8

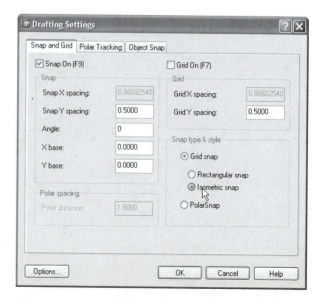

Figure 25-9 illustrates the effect of setting the *Isometric SNAP* and *GRID*. Notice the new position of the cursor.

Figure 25-9

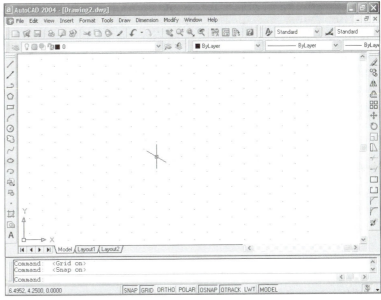

Using <u>Ctrl+E</u> (pressing the Ctrl key and the letter "E" simultaneously) toggles the cursor to one of three possible *Isoplanes* (AutoCAD's term for the three faces of the isometric pictorial). If *ORTHO* is *ON*, only <u>isometric</u> lines are drawn; that is, you can only draw *Lines* aligned with the isometric axes. *Lines* can be drawn on only two axes for each isoplane. Ctrl+E allows drawing on the two axes aligned with another face of the object. *ORTHO* is *OFF* in order to draw inclined or oblique lines (not on the isometric axis). The functions of *GRID* (F7) and *SNAP* (F9) remain unchanged.

With *SNAP ON*, toggle *Coords* (F6) several times and examine the readout as you move the cursor. The <u>absolute coordinate format is of no particular assistance</u> while drawing in isometric because of the configuration of the *GRID*. The <u>relative polar</u> format, however, is <u>very helpful</u>. Use relative polar format for *Coords* while drawing in isometric (Fig. 25-10).

Alternately, use *Polar Tracking* instead of *ORTHO*. Set the *Polar Angle Settings* to 30 degrees. The advantage of using *Polar Tracking* is that the current line length is given on the polar tracking tip (see Figure 25-10). The disadvantage is that it is possible to draw non-isoplane lines accidentally. Only one setting at a time can be used in AutoCAD—*Polar Tracking* or *ORTHO*. (The remainder of the figures illustrate the use of *ORTHO*.)

Figure 25-10

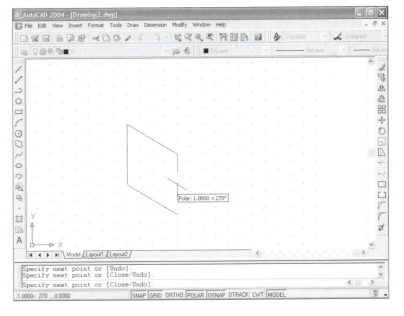

The effects of changing the *Isoplane* are shown in the following figures. Press Ctrl+E to change *Isoplane*.

With *ORTHO ON*, drawing a *Line* is limited to the two axes of the current *Isoplane*. Only one side of a cube, for example, can be drawn on the current *Isoplane*. Watch *Coords* (in a polar format) to give the length of the current *Line* as you draw (lower-left corner of the screen).

Toggling Ctrl+E switches the cursor and the effect of *ORTHO* to another *Isoplane*. One other side of a cube can be constructed on this *Isoplane* (Fig. 25-11).

Figure 25-11

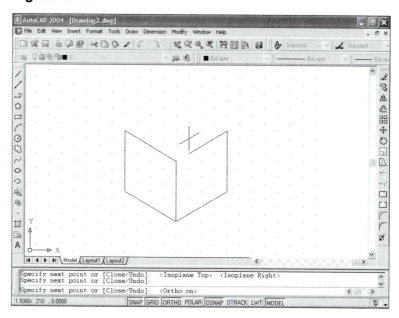

Direct Distance Entry can be of great help when drawing isometric lines. Use Ctrl+E and *ORTHO* to force the *Line* to the correct orientation, then enter the desired distance value at the Command line.

Isometric Ellipses

Isometric ellipses are easily drawn in AutoCAD by using the *Isocircle* option of the *Ellipse* command. This option appears <u>only</u> when the isometric *SNAP* is *ON*.

ELLIPSE

Pull-down Menu	COMMAND (TYPE)	ALIAS (TYPE)	Short-cut	Screen (side) Menu	Tablet Menu
Draw *Ellipse*	*ELLIPSE*	*EL*	...	*DRAW 1* *Ellipse*	*M,9*

Although the *Isocircle* option does not appear in the pull-down or digitizing tablet menus, it can be invoked as an option of the *Ellipse* command. The *Isocircle* option of *Ellipse* appears only when the *Snap Type* is set to *Isometric*. You <u>must type "I"</u> to use the *Isocircle* option. The command syntax is as follows:

    Command: ellipse
    Specify axis endpoint of ellipse or [Arc/Center/Isocircle]: i
    Specify center of isocircle: PICK or (coordinates)
    Specify radius of isocircle or [Diameter]: PICK or (coordinates)
    Command:

After selecting the center point of the *Isocircle*, the isometrically correct ellipse appears on the screen on the current *Isoplane*. Use Ctrl+E to toggle the ellipse to the correct orientation. When defining the radius interactively, use *ORTHO* to force the rubberband line to an isometric axis (Fig. 25-12, next page).

Since isometric angles are equal, all isometric ellipses have the same proportion (major to minor axis). The only differences in isometric ellipses are the size and the orientation (*Isoplane*).

Figure 25-12

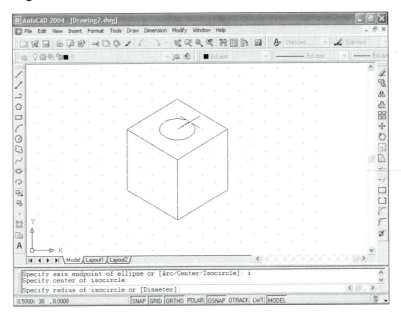

Figure 25-13 shows three ellipses correctly oriented on their respective faces. Use Ctrl+E to toggle the correct *Isoplane* orientation: *Isoplane Top*, *Isoplane Left*, or *Isoplane Right*.

Figure 25-13

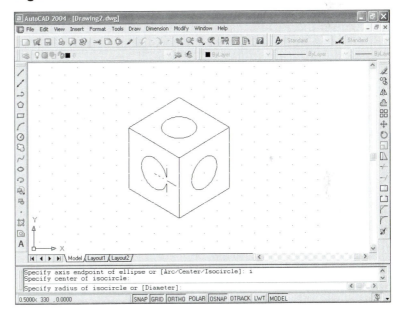

When defining the radius or diameter of an ellipse, it should always be measured in an isometric direction. In other words, an isometric ellipse is always measured on the two isometric axes (or center lines) parallel with the plane of the ellipse.

If you define the radius or diameter interactively, use *ORTHO ON*. If you enter a value, AutoCAD automatically applies the value to the correct isometric axes.

Creating an Isometric Drawing

In this exercise, the object in Figure 25-14 is drawn in isometric.

The initial steps to create an isometric drawing begin with the typical setup (see Chapter 6, Drawing Setup):

1. Set the desired *Units*.

2. Set appropriate *Limits*.

3. Set the *Isometric Style* of *Snap* and specify an appropriate value for spacing.

Figure 25-14

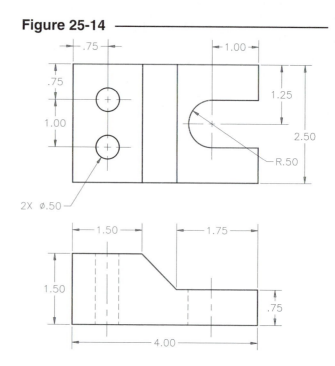

4. The next step involves creating an isometric framework of the desired object. In other words, draw an isometric box equal to the overall dimensions of the object. Using the dimensions given in Figure 25-14, create the encompassing isometric box with the *Line* command (Fig. 25-15).

   Use *ORTHO* to force isometric *Lines*. Watch the *Coords* display (in a relative polar format) to give the current lengths as you draw or use direct distance entry.

Figure 25-15

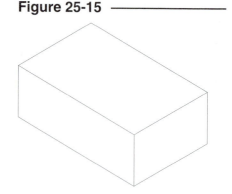

5. Add the lines defining the lower surface. Define the needed edge of the upper isometric surface as shown.

Figure 25-16

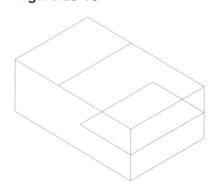

6. The <u>inclined</u> edges of the inclined surface can be drawn (with *Line*) only when *ORTHO* is *OFF*. <u>Inclined</u> lines in isometric cannot be drawn by transferring the lengths of the lines, but only by defining the <u>ends</u> of the inclined lines on <u>isometric</u> lines, then connecting the endpoints. Next, *Trim* or *Erase* the necessary *Lines*.

Figure 25-17 ——————

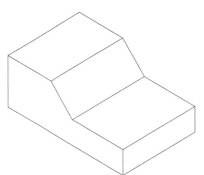

7. Draw the slot by constructing an *Ellipse* with the *Isocircle* option. Draw the two *Lines* connecting the circle to the right edge. *Trim* the unwanted part of the *Ellipse* (highlighted) using the *Lines* as cutting edges.

Figure 25-18 ——————

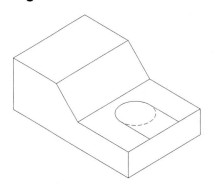

8. *Copy* the far *Line* and the *Ellipse* down to the bottom surface. Add two vertical *Lines* at the end of the slot.

Figure 25-19 ——————

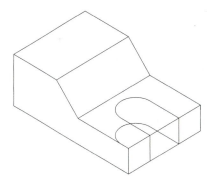

9. Use *Trim* to remove the part of the *Ellipse* that would normally be hidden from view. *Trim* the *Lines* along the right edge at the opening of the slot.

Figure 25-20 ——————

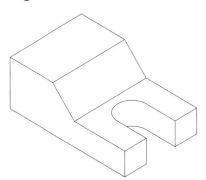

10. Add the two holes on the top with *Ellipse, Isocircle* option. Use *ORTHO ON* when defining the radius. *Copy* can also be used to create the second *Ellipse* from the first.

Figure 25-21 —————————

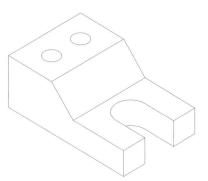

Dimensioning Isometric Drawings in AutoCAD

Refer to Chapter 29, Dimension Styles and Variables, for details on how to dimension isometric drawings.

OBLIQUE DRAWING IN AutoCAD

Figure 25-22 ——————

Oblique drawings are characterized by having two axes at a 90 degree orientation. Typically, you should locate the <u>front face</u> of the object along these two axes. Since the object's characteristic shape is seen in the front view, an oblique drawing allows you to create all shapes parallel to the front face true size and shape as you would in a multiview drawing. Circles on or parallel to the front face can be drawn as circles. The third axis, the receding axis, can be drawn at a choice of angles, 30, 45, or 60 degrees, depending on whether you want to show more of the top or the side of the object.

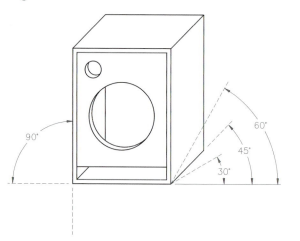

Figure 25-22 illustrates the axes orientation of an oblique drawing, including the choice of angles for the receding axis.

Figure 25-23 ————————

Another option allowed with oblique drawings is the measurement used for the receding axis. Using the full depth of the object along the receding axis is called <u>Cavalier</u> oblique drawing. This method depicts the object (a cube with a hole in this case) as having an elongated depth (Fig. 25-23).

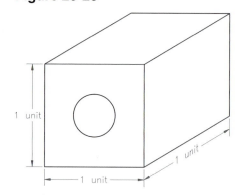

Using 1/2 or 3/4 of the true depth along the receding axis gives a more realistic pictorial representation of the object. This is called a <u>Cabinet</u> oblique (Fig. 25-24).

Figure 25-24 ————————

No functions or commands in AutoCAD are intended specifically for oblique drawing. However, *Polar Snap* and *Polar Tracking* can simplify the process of drawing lines on the front face of the object and along the receding axis. The steps for creating a typical oblique drawing are given next.

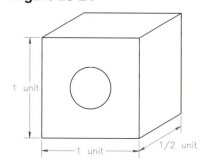

The object in Figure 25-25 is used for the example. From the dimensions given in the multiview, create a cabinet oblique with the receding axis at 45 degrees.

Figure 25-25

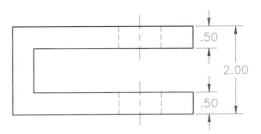

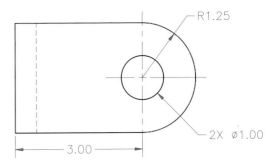

1. Create the characteristic shape of the front face of the object as shown in the front view.

Figure 25-26

2. Use *Copy* with the *Multiple* option to copy the front face back on the receding axis. *Polar Snap* and *Polar Tracking* can be used to specify the "second point of displacement" as shown in Figure 25-27. Notice that a distance of .25 along the receding axis is used (1/2 of the actual depth) for this cabinet oblique.

Figure 25-27

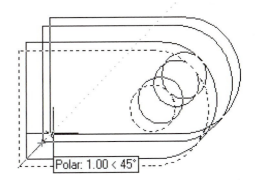

3. Draw the *Line* representing the edge on the upper-left of the object along the receding axis. Use *Endpoint OSNAP* to connect the *Lines*. Make a *Copy* of the *Line* or draw another *Line* .5 units to the right. Drop a vertical *Line* from the *Intersection* as shown.

Figure 25-28

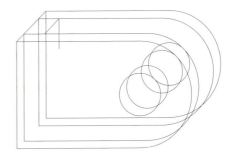

4. Use *Trim* and *Erase* to remove the unwanted parts of the *Lines* and *Circles* (those edges that are normally obscured).

Figure 25-29

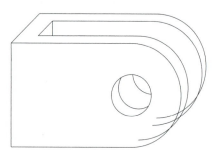

5. *Zoom* with a *window* to the lower-right corner of the drawing. Draw a *Line Tangent* to the edges of the arcs to define the limiting elements along the receding axis. *Trim* the unwanted segments of the arcs.

Figure 25-30

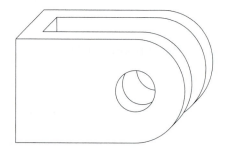

The resulting cabinet oblique drawing should appear like that in Figure 25-31.

Figure 25-31

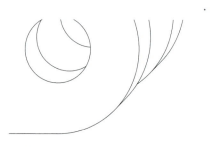

CHAPTER EXERCISES

Isometric Drawing

For exercises 1, 2, and 3 create isometric drawings as instructed. To begin, use an appropriate template drawing and draw a *Pline* border and insert the **TBLOCKAT**.

1. Create an isometric drawing of the cylinder shown in Figure 25-32. *Save* the drawing as **CYLINDER** and *Plot* so the drawing is *Scaled to Fit* on an A size sheet.

Figure 25-32

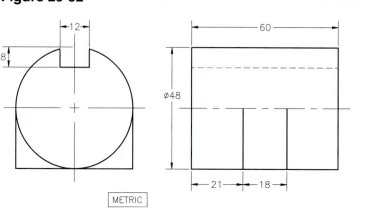

METRIC

2. Make an isometric drawing of the Corner Brace shown in Figure 25-33. *Save* the drawing as **CRNBRACE**. *Plot* at **1=1** scale on an A size sheet.

Figure 25-33

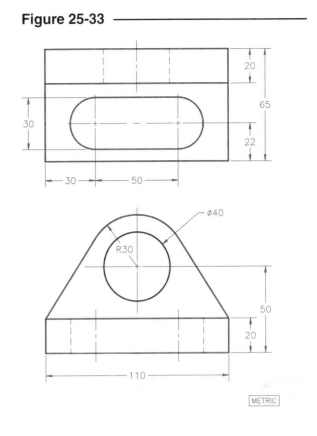

3. Draw the Support Bracket (Fig. 25-34) in isometric. The drawing can be *Plotted* at **1=1** scale on an A size sheet. *Save* the drawing and assign the name **SBRACKET**.

Figure 25-34

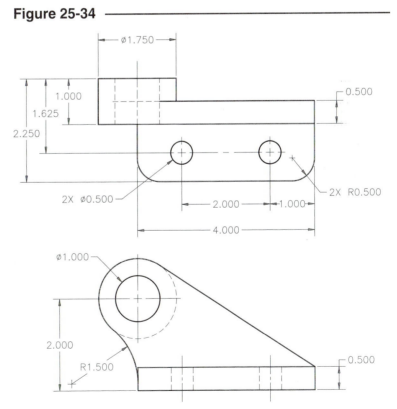

Oblique Drawing

For exercises 4 and 5, create oblique drawings as instructed. To begin, use an appropriate template drawing and draw a *Pline* border and *Insert* the **TBLOCKAT**.

4. Make an oblique cabinet projection of the Bearing shown in Figure 25-35. Construct all dimensions on the receding axis 1/2 of the actual length. Select the optimum angle for the receding axis to be able to view the 15 x 15 slot. *Plot* at **1=1** scale on an A size sheet. *Save* the drawing as **BEARING**.

Figure 25-35 ────────────────

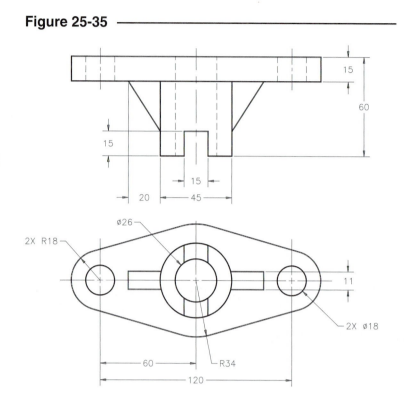

5. Construct a cavalier oblique drawing of the Pulley showing the circular view true size and shape. The illustration in Figure 25-36 gives only the side view. All vertical dimensions in the figure are diameters. *Save* the drawing as **PULLEY** and make a *Plot* on an A size sheet at **1=1**.

Figure 25-36 ────────────────

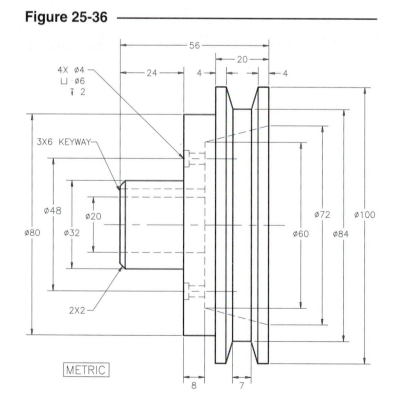

26

SECTION VIEWS

Chapter Objectives

After completing this chapter you should:

1. be able to use the *Bhatch* command to select associative hatch patterns;

2. be able to specify a *Scale* and *Angle* for hatch lines;

3. know how to define a boundary for hatching using the *Pick Points* and *Select Objects* methods;

4. be able to use *Hatch* to create non-associative hatch lines and discard the boundary;

5. be able to drag and drop hatch patterns from a Tool Palette and be able to create your own Tool Palettes;

6. know how to use *Hatchedit* to modify parameters of existing hatch patterns in the drawing;

7. be able to edit hatched areas using Grips, selection options, and *Draworder*;

8. know how to draw cutting plane lines for section views.

CONCEPTS

A section view is a view of the interior of an object after it has been imaginarily cut open to reveal the object's inner details. A section view is only one of two or more views of a multiview drawing describing the object. For example, a multiview drawing of a machine part may contain three views, one of which is a section view.

Hatch lines (also known as section lines) are drawn in the section view to indicate the solid material that has been cut through. Each combination of section lines is called a hatch <u>pattern</u>, and each pattern is used to represent a specific material. In full and half section views, hidden lines are omitted since the inside of the object is visible.

ANSI (American National Standards Institute) and ISO (International Organization of Standardization) have published standard configurations for section lines, and AutoCAD supports those standards. However, ANSI no longer specifies section line pattern standards.

A cutting plane line is drawn in an adjacent view to the section view to indicate the plane that imaginarily cuts through the object. Arrows on each end of the cutting plane line indicate the line of sight for the section view. ANSI dictates that a thick dashed or phantom line be used as the standard cutting plane line.

This chapter discusses the AutoCAD methods used to draw hatch lines for section views and related cutting plane lines. The *Bhatch* (boundary hatch) command allows you to select an enclosed area and select the hatch pattern and the parameters for the appearance of the hatch pattern; then AutoCAD automatically draws the hatch (section) lines. Existing hatch lines in the drawing can be modified using *Hatchedit*. Cutting plane lines are created in AutoCAD by using a dashed linetype. Cutting plane lines should be assigned a heavy *Lineweight* or should be drawn with a *Pline* with *Width*.

DEFINING HATCH PATTERNS AND HATCH BOUNDARIES

A hatch pattern is composed of many lines that have a particular linetype, spacing, and angle. Many standard hatch patterns are provided by AutoCAD for your selection. Rather than having to draw each section line individually, you are required only to specify the area to be hatched and AutoCAD fills the designated area with the selected hatch pattern. An AutoCAD hatch pattern is inserted as <u>one object</u>. For example, you can *Erase* the inserted hatch pattern by selecting only one line in the pattern, and the entire pattern in the area is *Erased*.

In a typical section view (Fig. 26-1), the hatch pattern completely fills the area representing the material that has been cut through. With the *Bhatch* command you can define the boundary of an area to be hatched simply by pointing inside of an enclosed area.

Figure 26-1

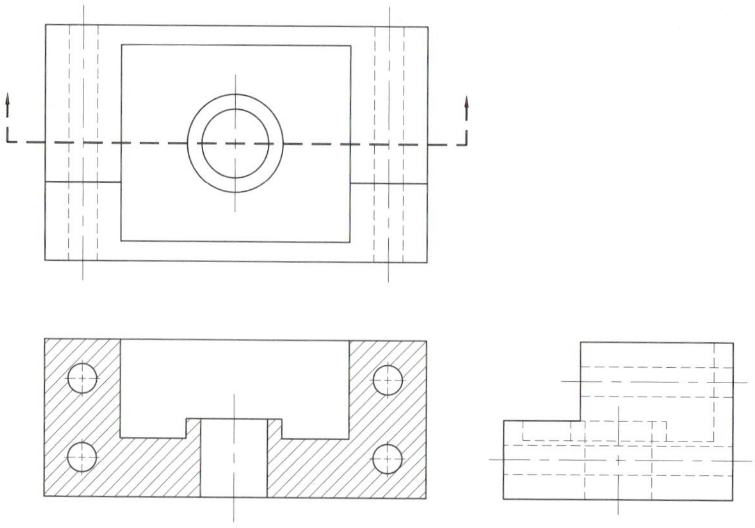

Both the *Hatch* and the *Bhatch* commands fill a specified area with a selected hatch pattern. *Hatch* requires that you select <u>each object</u> defining the boundary, whereas the *Bhatch* command finds the boundary automatically when you point inside it. Additionally, *Hatch* operates in command line format, whereas *Bhatch* operates in dialog box fashion. For these reasons, *Bhatch* is superior to *Hatch* and is recommended in most cases for drawing section views.

Hatch patterns created with *Bhatch* are <u>associative</u>. Associative hatch patterns are associated to the boundary geometry such that when the shape of the boundary changes (by *Stretch, Scale, Rotate, Move, Properties, Grips,* etc.), the hatch pattern automatically reforms itself to conform to the new shape (Fig. 26-2). For example, if a design change required a larger diameter for a hole, *Properties* could be used to change the diameter of the hole, and the surrounding section lines would automatically adapt to the new diameter.

Figure 26-2

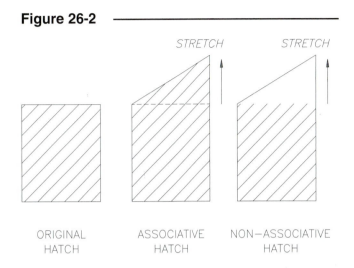

Once the hatch patterns have been drawn, any feature of the existing hatch pattern (created with *Bhatch*) can be changed retroactively using *Hatchedit*. The *Hatchedit* dialog box gives access to the same options that were used to create the hatch (in the *Boundary Hatch* dialog box). Changing the scale, angle, or pattern of any existing section view in the drawing is a simple process.

Steps for Creating a Section View Using the *Bhatch* Command

1. Create the view that contains the area to be hatched using typical draw commands such as *Line, Arc, Circle,* or *Pline*. If you intend to have text or dimensions inside the area to be hatched, add them before hatching.

2. Invoke the *Bhatch* command. The *Boundary Hatch and Fill* dialog box appears (see Figure 26-3).

3. Specify the *Type* to use. Select the desired pattern from the *Pattern* drop-down list or select the *Swatch* tile to allow you to select from the *Hatch Pattern Palette* image tiles.

4. Specify the *Scale* and *Angle* in the dialog box.

5. Define the area to be hatched by PICKing an internal point (*Pick Points* button) or by individually selecting the objects (*Select objects* button).

6. *Preview* the hatch to make sure everything is as expected. Adjust hatching parameters as necessary and *Preview* again.

7. Apply the hatch by selecting *OK*. The hatch pattern is automatically drawn and becomes an associated object in the drawing.

8. If other areas are to be hatched, additional internal points or objects can be selected to define the new area for hatching. The parameters used previously appear again in the *Boundary Hatch and Fill* dialog box by default. You can also *Inherit Properties* from a previously applied hatch.

9. For mechanical drawings, draw a cutting plane line in a view adjacent to the section view. A *Lineweight* is assigned or a *Pline* with a *Dashed* or *Phantom* linetype is used. Arrows at the ends of the cutting plane line indicate the line of sight for the section view.

10. If any aspect of the hatch lines needs to be edited at a later time, *Hatchedit* can be used to change those properties. If the hatch boundary is changed by *Stretch, Rotate, Scale, Move, Properties,* etc., the hatched area will conform to the new boundary.

BHATCH

Pull-down Menu	COMMAND (TYPE)	ALIAS (TYPE)	Short-cut	Screen (side) Menu	Tablet Menu
Draw *Hatch...*	*BHATCH* or *-BHATCH*	*BH* or *H*	...	*DRAW 2* *Bhatch*	*P,9*

Bhatch allows you to create hatch lines for a section view (or for other purposes) by simply PICKing inside a closed boundary. A closed boundary refers to an area completely enclosed by objects. *Bhatch* locates the closed boundary automatically by creating a <u>temporary *Pline*</u> that follows the outline of the hatch area, fills the area with hatch lines, and then deletes the boundary (default option) after hatching is completed. *Bhatch* ignores all objects or parts of objects that are not part of the boundary.

Any method of invoking *Bhatch* yields the *Boundary Hatch and Fill* dialog box (Fig. 26-3). Typically, the first step in this dialog box is the selection of a hatch pattern.

Figure 26-3

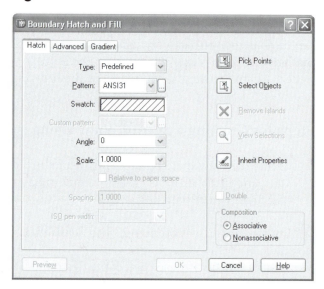

After you select the *Pattern* and other parameters for the hatch in the *Boundary Hatch and Fill* dialog box, you select *Pick Points* or *Select Objects* to return to the drawing and indicate the area to fill with the pattern. While in the drawing, you can right-click to produce a shortcut menu, providing access to other options without having first to return to the dialog box (Fig. 26-4).

Figure 26-4

Boundary Hatch Dialog Box—*Hatch* Tab

Type
This option allows you to specify the type of the hatch pattern: *Predefined, User-defined,* or *Custom.* Use *Predefined* for standard hatch pattern styles that AutoCAD provides (in the ACAD.PAT file).

Predefined
There are two ways to select from *Predefined* patterns: (1) select the *Swatch* tile to produce the *Hatch Pattern Palette* dialog box displaying hatch pattern names and image tiles (Fig. 26-5), or (2) select the *Pattern:* drop-down list to PICK the pattern name.

The *Hatch Pattern Palette* dialog box allows you to select a predefined pattern by its image tile or by its name. There are four tabs of image tiles: *ANSI, ISO, Other Predefined,* and *Custom.*

Figure 26-5

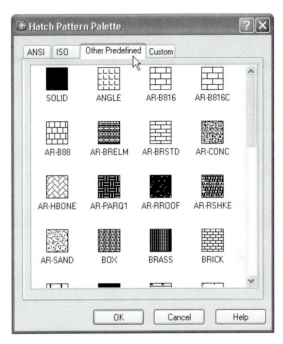

User-defined
To define a simple hatch pattern "on the fly," select the *User-defined* tile. This causes the *Pattern* and *Scale* options to be disabled and the *Angle, Spacing,* and *Double* options to be enabled. Creating a *User-defined* pattern is easy. Specify the *Angle* of the lines, the *Spacing* between lines, and optionally create *Double* (perpendicular) lines. All *User-defined* patterns have continuous lines.

Custom
Custom patterns are previously created user-defined patterns stored in other than the ACAD.PAT file. Custom patterns can contain continuous, dashed, and dotted line combinations. See the AutoCAD Customization Guide for information on creating and saving custom hatch patterns.

Pattern...
Selecting the *Pattern* drop-down list (see Figure 26-3) displays the name of each predefined pattern. Making a selection dictates the current pattern and causes the pattern to display in the *Swatch* window. The small button with ellipsis (...) just to the right of the *Pattern:* name drop-down list produces the *Hatch Pattern Palette.*

Swatch:
Click in the *Swatch* tile to produce the *Hatch Pattern Palette* dialog box (see Figure 26-5). You can select from *ANSI, ISO, Other Predefined,* and *Custom* (if available) hatch patterns.

Figure 26-6 displays each of the AutoCAD hatch patterns defined in the ACAD.PAT file. Note that the patterns are not shown to scale.

Figure 26-6

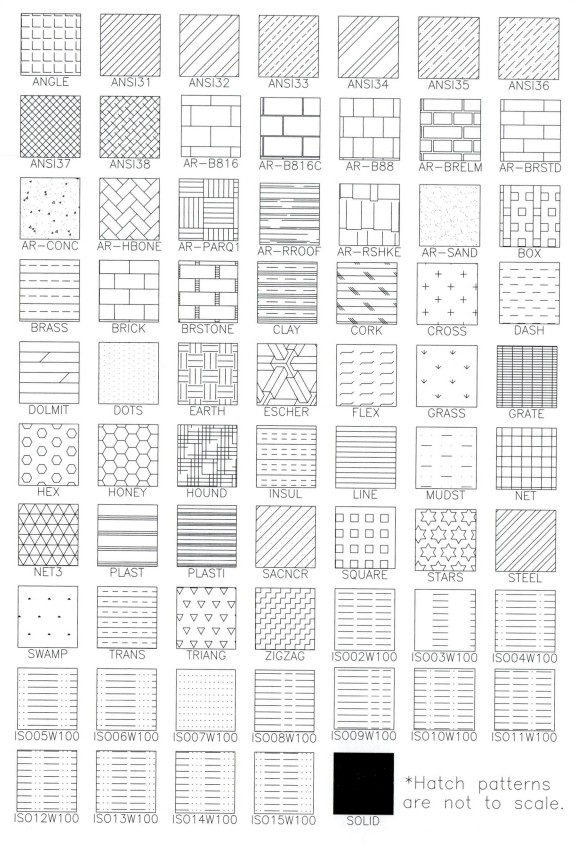

Hatch patterns are created using the current linetype. Therefore, the <u>*Continuous* linetype should be set</u> <u>current</u> when hatching to ensure the selected area is filled with the pattern as it appears in the image tile. After selecting a pattern, specify the desired *Scale* and *Angle* of the pattern or ISO pen width.

Scale

The value entered in this edit box is a scale factor that is applied to the existing selected pattern. Normally, this scale factor should be changed proportionally with changes in the drawing *Limits*. Like many other scale factors (*LTSCALE, DIMSCALE*), AutoCAD defaults are set to a value of 1, which is appropriate for the default *Limits* of 12 x 9. If you have calculated the drawing scale factor, enter that value in the *Scale* edit box (see Chapter 12 for information on the "Drawing Scale Factor"). The *Scale* value is stored in the *HPSCALE* system variable.

ISO hatch patterns are intended for use with metric drawings; therefore, the scale (spacing between hatch lines) is much greater than for inch drawings. Because *Limits* values for metric sheet sizes are greater than for inch-based drawings (25.4 times greater than comparable inch drawings), the ISO hatch pattern scales are automatically compensated. If you want to use an ISO pattern with inch-based drawings, calculate a hatch pattern scale based on the drawing scale factor and multiply by .039 (1/25.4).

Angle

The *Angle* specification determines the angle (slant) of the hatch pattern. The default angle of 0 represents whatever angle is displayed in the pattern's image tile. Any value entered deflects the existing pattern (as it appears in the swatch) by the specified value (in degrees). The value entered in this box is held in the *HPANG* system variable.

Relative to Paper Space

This option changes the hatch pattern scale relative to paper space units. Checking this box automatically changes the *HPSCALE* variable based on the proportion of the paper space units to model space units (viewport scale or *Zoom XP* factor). Using this option, you can easily display hatch patterns at a scale that is appropriate for your layout. You can use this option only when you have a layout with a viewport and invoke *Bhatch* from the layout.

Spacing

This option is enabled if *User-defined* pattern is specified. Enter a value for the distance between lines.

ISO Pen Width

You must select an ISO hatch pattern for this tile to be enabled. Selecting an *ISO Pen Width* from the drop-down list automatically sets the scale and enters the value in the *Scale* edit box. See "*Scale*."

Double

Only for a *User-defined* pattern, check this box to have a second set of lines drawn at 90 degrees to the original set.

Selecting the Hatch Area

Once the hatch pattern and options have been selected, you must indicate to AutoCAD what area(s) should be hatched. Either the *Pick Points* method, *Select Objects* method, or a combination of both can be used to accomplish this.

Pick Points

This tile should be selected if you want AutoCAD to automatically determine the boundaries for hatching. You only need to select a point <u>inside</u> the area you want to hatch. The point selected must be inside a <u>completely closed shape</u>. When the *Pick Points* tile is selected, AutoCAD gives the following prompts:

> Select internal point: **PICK**
> Selecting everything...
> Selecting everything visible...
> Analyzing the selected data...
> Analyzing the internal islands...
> Select internal point: **PICK** another area or **Enter**

When an internal point is PICKed, AutoCAD traces and highlights the boundary (Fig. 26-7). The interior area is then analyzed for islands to be included in the hatch boundary (if you select the *Flood* island detection method). Multiple boundaries can be designated by selecting multiple internal points. Type *U* to undo the last one, if necessary.

Figure 26-7

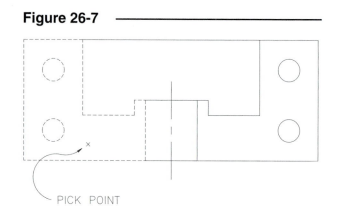

PICK POINT

The location of the point selected is usually not critical. However, the point must be PICKed inside the expected boundary. If there are any gaps in the area, a complete boundary cannot be formed and a boundary error message appears (Fig. 26-8).

Figure 26-8

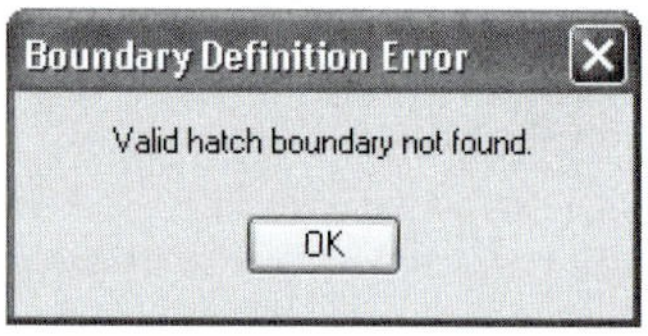

Boundary Definition Error ☒

Valid hatch boundary not found.

OK

Select Objects

Alternately, you can designate the boundary with the *Select Objects* method. Using the *Select Objects* method, you specify the boundary objects rather than let AutoCAD locate a boundary. With the *Select Objects* method, no temporary *Pline* boundary is created as with the *Pick Points* method. Therefore, the selected objects must form a closed shape with no gaps or overlaps. If gaps exist (Fig. 26-9) or if the objects extend past the desired hatch area (Fig. 26-10), AutoCAD cannot interpret the intended hatch area correctly, and problems will occur.

Figure 26-9

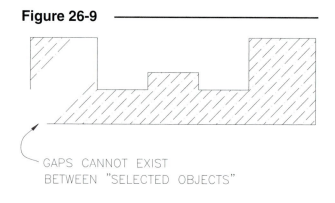

GAPS CANNOT EXIST
BETWEEN "SELECTED OBJECTS"

The *Select Objects* option can be used after the *Pick Points* method to select specific objects for *Bhatch* to consider before drawing the hatch pattern lines. For example, if you had created text objects or dimensions within the boundary found by the *Pick Points* method, you may then have to use the *Select Objects* option to select the text and dimensions (if *Ray Casting* is your choice of island detection method). Using this procedure, the hatch lines are automatically "trimmed" around the text and dimensions.

Figure 26-10

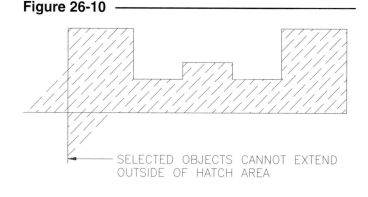

SELECTED OBJECTS CANNOT EXTEND
OUTSIDE OF HATCH AREA

Remove Islands
PICKing this tile allows you to select specific islands (internal objects) to remove from those AutoCAD has found within the outer boundary. If hatch lines have been drawn, be careful to *Zoom* in close enough to select the desired boundary and not the hatch pattern.

View Selections
Clicking the *View Selections* tile causes AutoCAD to highlight all selected boundaries. This can be used as a check to ensure the desired areas are selected.

Inherit Properties
This option allows you to select a hatch pattern from one existing in the drawing. This option operates similarly to *Matchprop* because you PICK the existing hatch pattern from the drawing and copy it to another area by selecting an internal point.

```
Select associative hatch object: PICK
Inherited Properties: Name <ANSI32>, Scale <1.0000>, Angle <0>
Select internal point:
```

Associative / Nonassociative
This checkbox toggles (on or off) the associative property for newly applied hatch patterns. Associative hatch patterns automatically update by conforming to the new boundary shape when the boundary is changed (see Figure 26-2). A non-associative hatch pattern is static even when the boundary changes.

Preview
You should always use the *Preview* option after specifying the hatch parameters and selecting boundaries, but before you apply the hatch. This option allows you to temporarily look at the hatch pattern in your drawing with the current settings applied and allows you to adjust the settings, if necessary, before using *OK*. After viewing the drawing, press Enter to redisplay the *Boundary Hatch* dialog box, allowing you to make adjustments.

Advanced **Tab**

The *Advanced* tab produces the options shown in Figure 26-11 and described below.

Island Detection Style
This section allows you to specify how the hatch pattern is drawn when the area inside the defined boundary contains text or closed areas (islands). Select the text from the list or PICK the icons displayed in the window. These options are applicable only when interior objects (islands) have been included in the selection set (considered for hatching). Otherwise, if only the outer shape is included in the selection set, the results are identical to the *Ignore* option. (See also "*Island Detection Method*.")

Figure 26-11

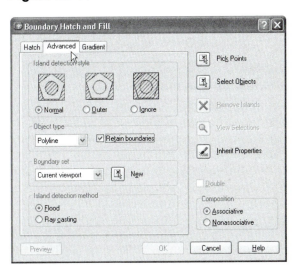

Normal

This should be used for most applications of *Bhatch*. Text or closed shapes within the outer border are considered in such cases. Hatching will begin at the outer boundary and move inward alternating between applying and not applying the pattern as interior shapes or text are encountered (Fig. 26-12).

Figure 26-12

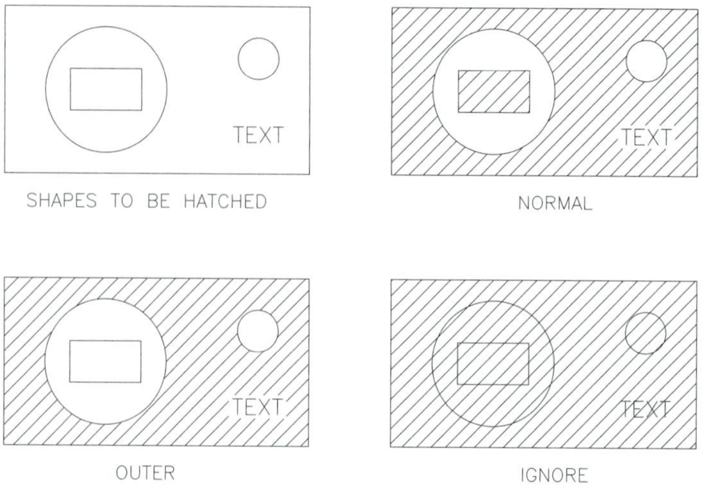

Outer

This option causes AutoCAD to hatch only the outer closed shape. Hatching is turned off for all interior closed shapes (Fig. 26-12).

Ignore

Ignore draws the hatch pattern from the outer boundary inward ignoring any interior shapes. The resulting hatch pattern is drawn through the interior shapes (Fig. 26-12).

Object Type

Bhatch creates *Polyline* or *Region* boundaries. This option is enabled only when the *Retain Boundaries* box is checked (see Figure 26-11, on the previous page).

Retain Boundaries

When AutoCAD uses the *Pick Points* method to locate a boundary for hatching, a temporary *Pline* or *Region* is created for hatching, then discarded after the hatching process. Checking this box forces AutoCAD to keep the boundary. When the box is checked, *Bhatch* creates two objects—the hatch pattern and the boundary object. (You can specify whether you want to create a *Pline* or a *Region* boundary in the *Object Type* drop-down list.) Using this option and erasing the hatch pattern accomplishes the same results as using the *Boundary* command.

After using *Bhatch*, the *Pline* or *Region* can be used for other purposes. To test this function, complete a *Bhatch* with the *Retain Boundaries* box checked; then use *Move* (make sure you select the boundary) to reveal the new boundary object (Fig. 26-13).

Figure 26-13

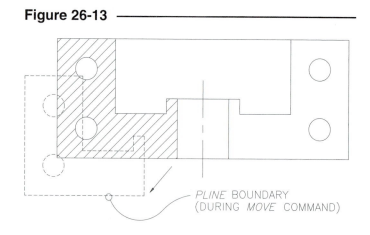

PLINE BOUNDARY
(DURING *MOVE* COMMAND)

Boundary Set

By default, AutoCAD examines all objects in the viewport when determining the boundaries by the *Pick Points* method. (*Current Viewport* is selected by default when you begin the *Bhatch* command.) For complex drawings, examining all objects can take some time. In that case, you may want to specify a smaller boundary set for AutoCAD to consider. Clicking the *New* tile clears the dialog boxes and permits you to select objects or select a window to define the new set.

Island Detection Method

This section specifies whether to include objects within the outermost boundary as boundary objects. These internal objects are known as islands.

Flood

Selecting the *Flood* method automatically <u>includes islands</u> as boundary objects and places the hatch lines around them, so the *Normal* or *Outer* island detection style can be applied.

Ray Casting

When you pick an internal point, a ray is cast from the point you specify to the nearest object. It then traces the boundary in a counterclockwise direction, thus <u>excluding islands</u> as boundary objects. If you want to include islands using this method, you must select them using *Select Objects*.

Gradient **Tab**

Gradient fills are new for AutoCAD 2004. The *Gradient* tab (Fig. 26-14) allows you to create a solid fill in a specified closed area (similar to the *Solid* pattern). However, rather than filling the area with one solid color, applying a gradient fill creates a gradual transition between one color and white or black (*One color*) or a transition between two colors (*Two color*).

Figure 26-14

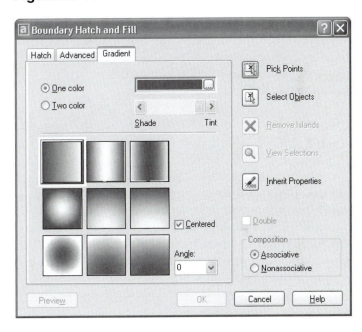

Gradient fill image tiles

Make your selection of possible gradient fill patterns from the nine image tiles (Figure 26-14, lower-left corner). The resulting gradient fill will be spread across the boundary(s) you select. For example, if you select the gradient fill displayed in the upper-left box (and apply *One color*), the pure color will appear at the extreme left edge of the selected boundary and no color will appear at the extreme right.

One Color

To create a gradient fill mix of one color with black or white, select the *One color* radio button. Select your desired color by using the button just to the right of the color swatch (see Figure 26-14, top-center) to produce the *Select Color* dialog box where you can select from the *Index Color* (ACI), *True Color*, or *Color Books* tabs (see "*Color*" in Chapter 11 for information on the *Select Color* dialog box). Use the *Shade* to *Tint* slider to adjust the amount of white or black to mix with your color selection (*Tint* means white and *Shade* means black). Positioning the slider in the center produces no gradient, only pure hue.

Two Color

To create a gradient fill mix of two colors, select the *Two color* radio button, then make your selection for each of the two colors by using the buttons just to the right of the color swatches to produce the *Select Color* dialog box. The resulting gradient fill is a transition between the two selected colors.

2004

Centered

Checking the *Centered* radio button ensures that the gradient fill is centered within (or among) the selected boundary(s). For example, Figure 26-15 displays two circles, each with a circular gradient fill applied. The left circle was created with *Centered* checked, the right with *Centered* unchecked.

Figure 26-15 ——————————

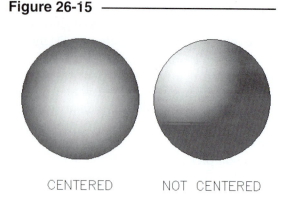

CENTERED   NOT CENTERED

Angle

Normally, the gradient fills are horizontally or vertically oriented (with the default setting of 0). You can produce gradient fills at 15-degree increments by selecting any setting in the *Angle* drop-down list. The nine pattern tiles display the fills at the selected angle. The *Angle* setting can be affected also by the *Centered* option. Figure 26-16 illustrates the left circle filled at 0 degrees and the right circle filled at 45 degrees.

Figure 26-16 ——————————

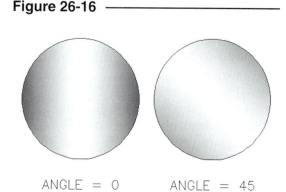

ANGLE = 0   ANGLE = 45

Keep in mind that the resulting gradient fills are determined by the boundary(s) selected. For example, in Figure 26-17 both circles were selected for hatching with one gradient fill (one use of the *BHATCH* command), whereas the objects in Figures 26-15 and 26-16 were each hatched individually.

Figure 26-17 ——————————

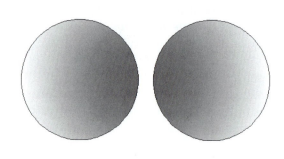

ONE HATCH OBJECT, TWO BOUNDARIES

HATCH

Pull-down Menu	COMMAND (TYPE)	ALIAS (TYPE)	Short-cut	Screen (side) Menu	Tablet Menu
...	*HATCH*	-H	...	...	...

Hatch must be typed at the keyboard since it is not available from the menus. *Hatch* operates in a Command line format with many of the options available in the *Bhatch* dialog box. However, *Hatch* creates a <u>non-associative</u> hatch pattern and <u>does not use the *Pick points*</u> method ("select internal point:") to create a boundary. Generally, *Bhatch* would be used instead of *Hatch* except for special cases.

You can use *Hatch* to find a boundary from existing objects. The objects must form a complete closed shape, just as with *Bhatch*. The shape can be composed of <u>one</u> closed object such as a *Pline, Polygon, Spline,* or *Circle* or composed of <u>several</u> objects forming a closed area such as *Lines* and *Arcs.* For example, the shape in Figure 26-18 can be hatched correctly with *Hatch,* whether it is composed of one object (*Pline, Region,* etc.) or several objects (*Lines, Arc,* etc.).

Figure 26-18

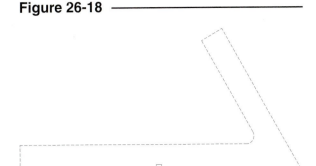

SELECT OBJECTS

If you select several objects to define the boundary, the objects must comprise <u>only the boundary shape</u> and not extend past the desired boundary. If objects extend past the desired boundary (Fig. 26-19) or if there are gaps, the resulting hatch will be <u>incorrect</u> as shown in Figures 26-9 and 26-10.

Figure 26-19

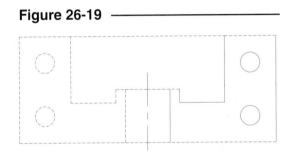

```
Command: hatch
Enter a pattern name or [?/Solid/User defined]
<ANSI31>: (pattern name or style)
Specify a scale for the pattern <1.0000>: (value) or
Enter
Specify an angle for the pattern <0>: (value) or Enter
Select objects to define hatch boundary or <direct hatch>,
Select objects: PICK
Select objects: Enter
Command:
```

Direct Hatch

With the *Direct Hatch* method of *Hatch*, you <u>specify points</u> to define the boundary, <u>not objects</u>. The significance of this method is that you can create a hatch pattern without using existing objects as the boundary. In addition, you can select whether you want to <u>retain or discard the boundary</u> after the pattern is applied (Fig. 26-20).

NOTE: To access the *Direct hatch* options, press **Enter** at the "Select objects:" prompt. See below.

Figure 26-20

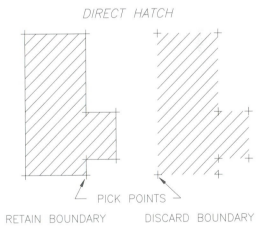

DIRECT HATCH

PICK POINTS

RETAIN BOUNDARY   DISCARD BOUNDARY

```
Command: hatch
Enter a pattern name or [?/Solid/User defined] <ANSI31>:
     (pattern name or style)
Specify a scale for the pattern <1.0000>: (value) or
     Enter
Specify an angle for the pattern <0>: (value) or Enter
Select objects to define hatch boundary or <direct hatch>,
Select objects: Enter
Retain polyline boundary? [Yes/No] <N>: (option)
Specify start point: PICK
Specify next point or [Arc/Close/Length/Undo]: PICK or (option)
Specify next point or [Arc/Close/Length/Undo]: PICK or (option)
Specify next point or [Arc/Close/Length/Undo]: c
Specify start point for new boundary or <apply hatch>: Enter
Command:
```

For special cases when you do not want a hatch area boundary to appear in the drawing (Fig. 26-21), you can use the *Direct Hatch* method to specify a hatch area and discard the boundary *Pline*.

Figure 26-21

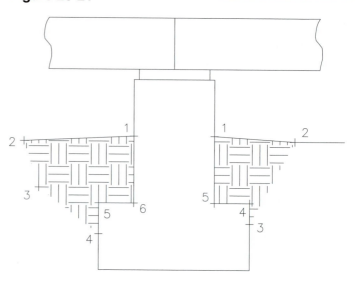

TOOLPALETTES

Pull-down Menu	COMMAND (TYPE)	ALIAS (TYPE)	Short-cut	Screen (side) Menu	Tablet Menu
Tools *Tool Palettes Window*	*TOOLPALETTES*	*TP*	*Ctrl +3*	...	...

In AutoCAD 2004 you can use Tool Palettes to drag and drop hatch patterns into your drawing. The process is simple. Open Tool Palettes by any method and locate the palette that contains the hatch pattern you want to use. Next, drag and drop the hatch from the palette directly into the closed area you want to fill with the pattern (Fig. 26-22).

If you right-click on any tool in the palette, its *Tool Properties* dialog box appears (Fig. 26-23). This dialog box is specific to the particular tool you right-clicked, and allows you to specify settings for the hatch such as *Angle* and *Scale*.

Figure 26-22

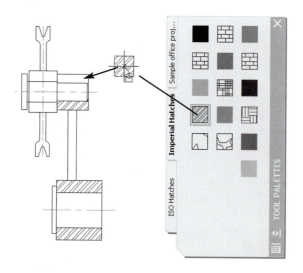

Figure 26-23

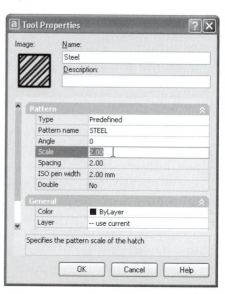

2004

Creating Tool Palettes

Figure 26-24

AutoCAD 2004 provides only three palettes: *ISO Hatches*, *Imperial Hatches*, and *Sample Office Project*. Therefore, you will most likely want to create new palettes containing the hatch patterns you use often. To do this, follow these steps:

1. Open Tool Palettes by any method. Right-click in the palette area to produce the shortcut menu (Fig. 26-24). Select *New Tool Palette*.
2. A new palette appears with an edit box (not shown). Enter the title you want to appear on the tab of the new palette, for example, "ANSI Hatches" (the other standard palette titles are *ISO Hatches*, *Imperial Hatches*, and *Sample Office Project*).
3. Open DesignCenter. Use the *Search* button. In the *Search* dialog box that appears (not shown), select *Hatch Pattern Files* in the *Look for* drop-down list and enter ACAD.PAT (or another .PAT file) in the *Search for the name* edit box. Select *Search Now*. When ACAD.PAT appears in the list at the bottom of the *Search* dialog box, double-click on ACAD.PAT. This action loads the hatch patterns into the DesignCenter Content Area.

4. From the DesignCenter Content Area, drag and drop the desired hatch patterns individually into your new palette (Fig. 26-25). If you want to specify other than the default *Scale* and *Angle*, right-click on any tool to produce its *Tool Properties* dialog box (see Figure 26-23).

Figure 26-25

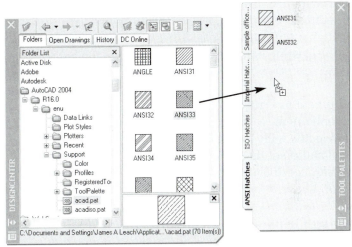

Alternately, you can create a new tool palette fully populated with all the hatch patterns from the ACAD.PAT file directly from DesignCenter. This is a quicker and simpler method; however, all the patterns from the file are included in your new palette. Follow these steps:

1. Open Tool Palettes by any method.
2. Open DesignCenter. Use the *Search* button. In the *Search* dialog box that appears (not shown), select *Hatch Pattern Files* in the *Look for* drop-down list and enter ACAD.PAT (or another .PAT file) in the *Search for the name* edit box. Select *Search Now*. When ACAD.PAT appears in the list at the bottom of the *Search* dialog box, double-click on ACAD.PAT.
3. In the DesignCenter <u>Tree View area</u>, right-click on the ACAD.PAT file name to produce a shortcut menu (Fig. 26-26). Select *Create Tool Palette of Hatch Patterns*. A new palette is automatically created fully populated with all hatch patterns from the file.

Figure 26-26

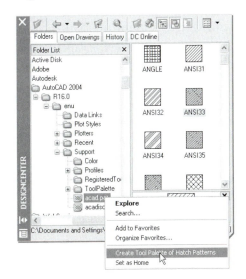

See Chapter 21, Blocks, DesignCenter, and Tool Palettes, for information on *Deleting*, *Renaming*, and other ways to customize tool palettes.

EDITING HATCH PATTERNS AND BOUNDARIES

HATCHEDIT

Pull-down Menu	COMMAND (TYPE)	ALIAS (TYPE)	Short-cut	Screen (side) Menu	Tablet Menu
Modify Object > Hatch...	*HATCHEDIT* or *-HATCHEDIT*	*HE*	...	*MODIFY1 Hatchedt*	*Y,16*

Hatchedit allows you to modify an <u>existing associative</u> hatch pattern in the drawing. This feature of AutoCAD makes the hatching process more flexible because you can hatch several areas to quickly create a "rough" drawing, then retroactively fine-tune the hatching parameters when the drawing nears completion with *Hatchedit*.

You can produce the *Hatchedit* dialog box (Fig. 26-27) by any method shown in the Command table as well as double-clicking on any hatched area in the drawing (assuming *DBLCLKEDIT* is set to *On* and *PICKSTYLE* is set to 1). The dialog box provides options for changing the *Pattern, Scale, Angle,* and *Type* properties of the existing hatch. You can also change the hatch to *Associative* or *Nonassociative*.

Apparent in Figure 26-27, the *Hatch Edit* dialog box is essentially the same as the *Boundary Hatch and Fill* dialog box with some of the options disabled. For an explanation of the options that are available with *Hatchedit*, see *Boundary Hatch* earlier in this chapter.

Figure 26-27 ——————————

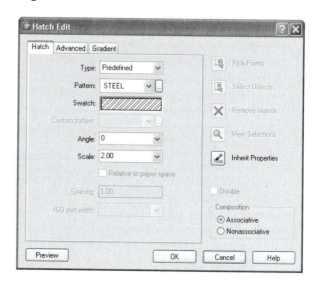

Using Grips with Hatch Patterns

Grips can be used effectively to edit hatch pattern boundaries of associative hatches. Do this by selecting the boundary with the pickbox or automatic window/crossing window (as you would normally activate an object's grips). Associative hatches <u>retain the association</u> with the boundary after the boundary has been changed using grips (see Figure 26-2, earlier in this chapter).

Beware, when selecting the hatch boundary, ensure you <u>edit the boundary and not the hatch object</u>. If you edit the hatch object only, the associative feature is lost. This can happen if you select the <u>hatch object</u> grip (Fig. 26-28).

Figure 26-28 ——————————

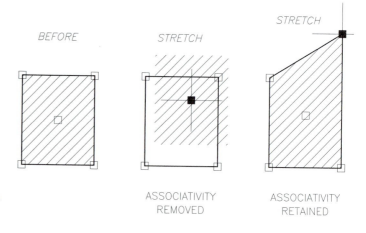

In this case, AutoCAD issues the following prompt:

Command: (select grip on hatch object)
STRETCH
Specify stretch point or [Base point/Copy/Undo/eXit]:
Hatch boundary associativity removed.
Command:

If you activate grips on <u>both</u> the hatch object and the boundary and make a boundary grip hot, the boundary and related hatch retain associativity (see Figure 26-28).

Object Selection Features of Hatch Patterns

When you select a hatch object for editing (PICK only one line of an entire hatch pattern), you actually select the entire hatch object. By default, when you PICK any element of the hatch pattern, the hatch object can be *Moved, Copied,* grip edited, etc. However, when you edit the hatch object <u>but not the boundary</u> associativity is removed (see "Using Grips with Hatch Patterns").

You can easily control whether just the hatch object or the <u>hatch object and boundary together</u> are selected when you PICK the hatch object. Do this by selecting *Associative Hatch* in the *Selection* tab of the *Options* dialog box (Fig. 26-29, lower left). Checking the *Associative Hatch* option sets the *PICKSTYLE* variable to a value of 2 (see Appendix A).

Figure 26-29

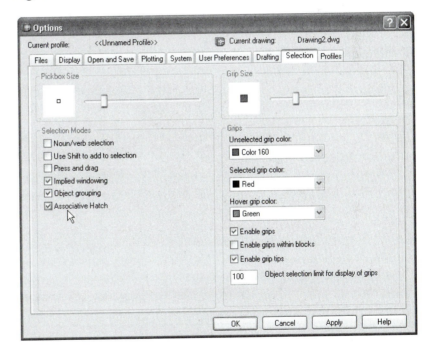

With *Associative Hatch* selected, PICKing the hatch <u>object</u> results in highlighting both the hatch object and boundary; therefore, you always edit the hatch object and the boundary together (Fig. 26-30). When *Associative Hatch* is not selected, PICKing the hatch object results in only the hatch object selection; therefore, you run the risk of editing the hatch object independent of the boundary and losing associativity.

Figure 26-30

OBJECT SELECTION WITH
ASSOCIATIVE HATCH
CHECKED

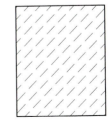

OBJECT SELECTION WITH
ASSOCIATIVE HATCH
NOT CHECKED

FILLMODE for Solid *Hatch* Fills

A *Solid* hatch pattern is available. This feature is welcomed by professionals who create hatch objects such as walls for architectural drawings and require the solid filled areas. There are many applications for this feature.

Solid-filled hatch areas have versatility because you can control the *Color* of the solid fill. By default, the color of the *Solid* hatch pattern (just like any other hatch pattern) assumes the current *Color* setting. In most cases, the current *Color* setting is *ByLayer*; therefore, the *Solid* hatch pattern assumes the current layer color. You can also assign the *Solid* hatches object-specific *Color*, in which case it is possible to have multiple *Solid* filled areas in different colors all on the same layer.

You can use the *Fill* command (or *FILLMODE* system variable) to control the visibility of the solid filled areas. *Fill* also controls the display of solid filled TrueType fonts (see Chapter 18). If you want to display solid filled *Hatch* objects as solid, set *Fill* to *On* (Fig. 26-31). Setting *Fill* to *Off* causes AutoCAD not to display the solid filled areas. *Regen* must be used after *Fill* to display the new visibility state. This solid fill control can be helpful for saving toner or ink during test prints or speeding up plots when much solid fill is used in a drawing.

Figure 26-31

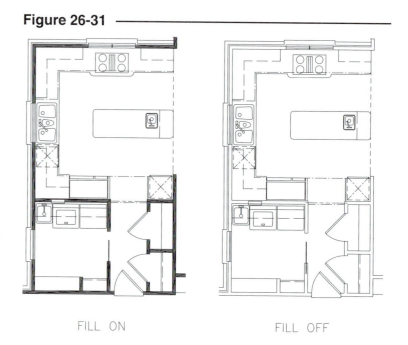

FILL ON   FILL OFF

DRAWORDER

Pull-down Menu	COMMAND (TYPE)	ALIAS (TYPE)	Short-cut	Screen (side) Menu	Tablet Menu
Tools Display Order >	DRAWORDER	DR	...	TOOLS 1 Drawordr	T,9

It is possible to use solid hatch patterns and raster images in combination with filled text, other hatch patterns and solid images, etc. With these solid areas and images, some control of which objects are in "front" and "back or "above" and "under" must be provided. This control is provided by the *Draworder* command.

For example, if a company logo were created using filled *Dtext*, a *Sand* hatch pattern, and a *Solid* hatched circle, the object that was created last would appear "in front" (Fig. 26-32, top logo). The *Draworder* command is used to control which objects appear in front and back or above and under. This capability is necessary for creating prints and plots in black and white and in color.

Figure 26-32

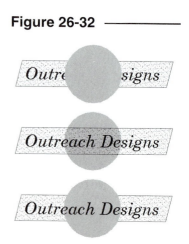

The *Draworder* command is simple to use. Simply select objects and indicate *Front*, *Back*, *Above*, or *Under*.

```
Command: draworder
Select objects: PICK
Select objects: Enter
Enter object ordering option [Above object/Under object/Front/Back] <Back>: (option)
Regenerating drawing.
Command:
```

In Figure 26-32, three (of four) possibilities are shown for the three objects—*Dtext*, *Solid* hatch and boundary, and *Sand* hatch and boundary.

DRAWING CUTTING PLANE LINES

Most section views (full, half, and offset sections) require a cutting plane line to indicate the plane on which the object is cut. The cutting plane line is drawn in a view <u>adjacent</u> to the section view because the line indicates the plane of the cut from its edge view. (In the section view, the cutting plane is perpendicular to the line of sight, therefore, not visible as a line.)

Standards provide two optional line types for cutting plane lines. In AutoCAD, the two linetypes are *Dashed* and *Phantom*, as shown in Figure 26-33. Arrows at the ends of the cutting plane line indicate the <u>line-of-sight</u> for the section view.

Figure 26-33

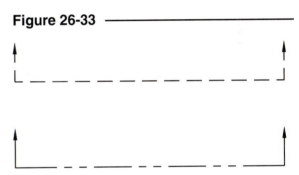

Cutting plane lines should be drawn or plotted in a heavy lineweight. Two possible methods can be used to accomplish this: use a *Pline* with *Width* or assign a *Lineweight* to the line or layer. If you prefer drawing the cutting plane line using a *Pline*, use a *Width* of .02 or .03 times the drawing scale factor. Assigning a *Lineweight* to the line or layer is a simpler method, but the *Lineweight* appears only in a plot preview. A *Lineweight* of .8mm or .031" is appropriate for cutting plane lines.

For the example section view, a cutting plane line is created in the top view to indicate the plane on which the object is cut and the line of sight for the section view (Fig. 26-34). (The cutting plane could be drawn in the side view, but the top would be much clearer.) First, a *New* layer named CUT is created and the *Dashed linetype* is assigned to the layer. The layer is *Set* as the *Current* layer. Next, construct a *Pline* with the desired *Width* to represent the cutting plane line.

Figure 26-34

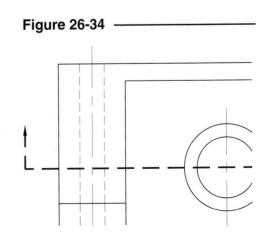

The resulting cutting plane line appears as shown in Figure 26-35 but without the arrow heads. The horizontal center line for the hole (top view) was *Erased* before creating the cutting plane line.

Figure 26-35

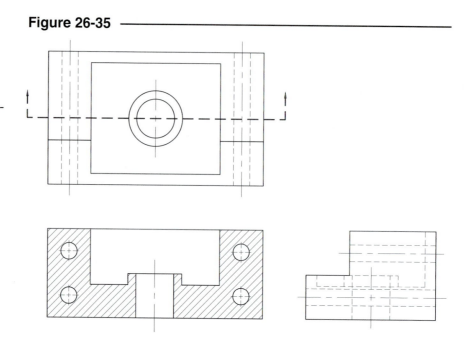

The last step is to add arrowheads to the ends of the cutting plane line. Arrowheads can be drawn by three methods: (1) use the *Solid* command to create a three-sided solid area; (2) use a tapered *Pline* with beginning width of 0 and ending width of .08, for example; or (3) use the *Leader* command to create the arrowhead (see Chapter 28, Dimensioning, for details on *Leader*). The arrowhead you create can be *Scaled*, *Rotated*, and *Copied* if needed to achieve the desired size, orientation, and locations.

CHAPTER EXERCISES

For the following exercises, create the section views as instructed. Use an appropriate template drawing for each unless instructed otherwise. Include a border and title block in the layout for each drawing.

1. *Open* the **SADDLE** drawing that you created in Chapter 24. Convert the front view to a full section. *Save* the drawing as **SADL-SEC**. *Plot* the drawing at **1=1** scale.

2. Make a multiview drawing of the Bearing shown in Figure 26-36. Convert the front view to a full section view. Add the necessary cutting plane line in the top view. *Save* the drawing as **BEAR-SEC**. Make a *Plot* at full size.

Figure 26-36

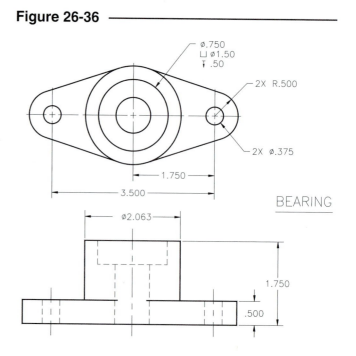

BEARING

3. Create a multiview drawing of the Clip shown in Figure 26-37. Include a side view as a full section view. You can use the **CLIP** drawing you created in Chapter 24 and convert the side view to a section view. Add the necessary cutting plane line in the front view. *Plot* the finished drawing at **1=1** scale and *SaveAs* **CLIP-SEC**.

Figure 26-37 ─────────────────

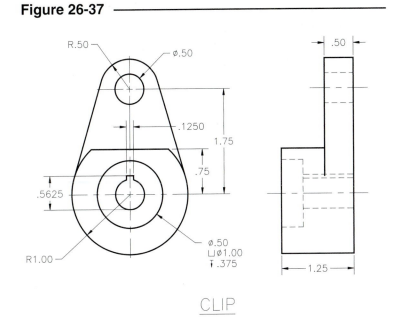

CLIP

4. Create a multiview drawing, including two full sections of the Stop Block as shown in Figure 26-38. Section B–B′ should replace the side view shown in the figure. *Save* the drawing as **SPBK-SEC**. *Plot* the drawing at **1=2** scale.

Figure 26-38 ──────────────────────────────────

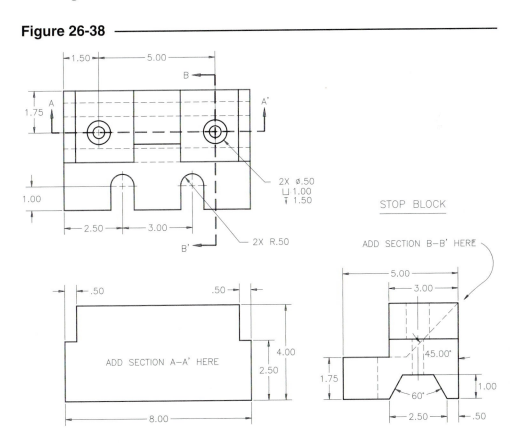

5. Make a multiview drawing, including a half section of the Pulley (Fig. 26-39). All vertical dimensions are diameters. Two views (including the half section) are sufficient to describe the part. Add the necessary cutting plane line. *Save* the drawing as **PUL-SEC** and make a *Plot* at **1:1** scale.

Figure 26-39

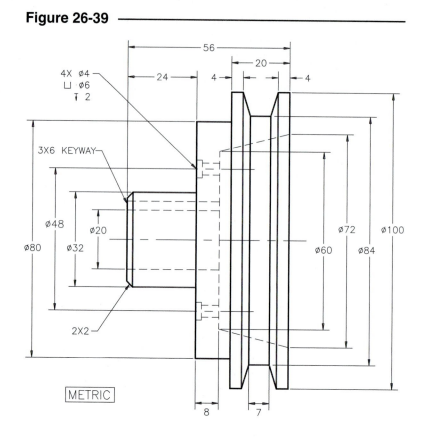

6. Draw the Grade Beam foundation detail in Figure 26-40. Do not include the dimensions in your drawing. Use the *Bhatch* command to hatch the concrete slab with **AR-CONC** hatch pattern. Use *Sketch* to draw the grade line and *Hatch* with the **EARTH** hatch pattern. *Save* the drawing as **GRADBEAM**.

Figure 26-40

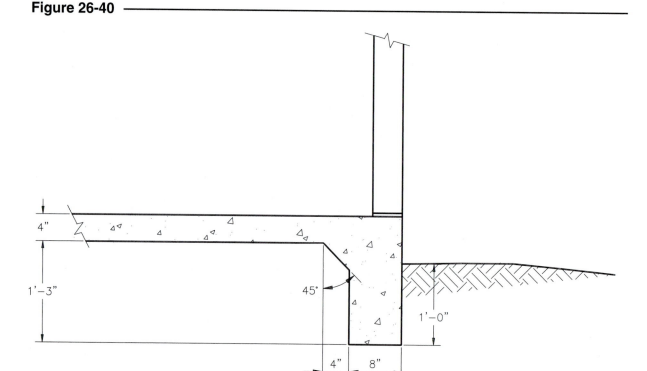

7. **Tool Palettes, Solid Fill**

   A. *Open* the **OFFICE-REV** drawing that you last worked on in Chapter 21. Next, open the **Tool Palettes**.

   B. Make a new *Layer* and name it **CARPET**. Accept the default color, linetype, etc. Make **CARPET** the *Current* layer.

   C. Draw a *Line* across the doorway threshold of one of the two offices. Next, drag and drop the *Entry Carpet* pattern from the standard *Sample Office Project* palette into the office. If problems occur, check to ensure that *Flood* is the *Island Detection Method* set in the *Advanced* tab of the *Boundary Hatch* dialog box. Also make sure there are no gaps in the office walls.

   D. When the solid *Entry Carpet* pattern appears in the office, notice how the pattern covers the outlines of the walls and furniture items. Use the *Draworder* command to move the new hatch pattern to the *Back*. Your drawing should look similar to that in Figure 26-41.

   Figure 26-41

   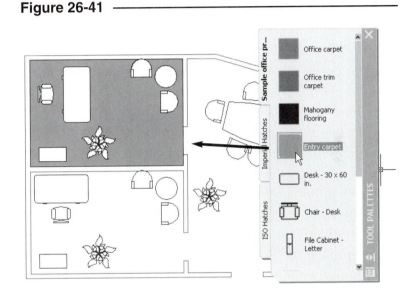

   E. Using the same technique, fill the other two office spaces with carpet from the tool palette or use another wood floor pattern such as *AR-HBONE* from the *Boundary Hatch* dialog box.

   F. Use *SaveAs* to save and rename the drawing to **OFFICE-REV-2**.

8. **Solid Fill, *Fill* Command**

   A. *Open* the **OFFICE-REV-2** drawing if not already open. Make a new *Layer* and assign the name **WALLS**. Accept the default color, linetype, etc. Make the **WALLS** layer *Current*.

   B. Use the *Boundary Hatch* dialog box to apply the *Solid* hatch pattern to the walls of the office.

   C. Use the *Fill* command and turn *Off* the solid fill. Use the *Regen* command to apply the new *Fill* setting. Does *Fill* affect only the walls, or does it affect the other carpet solid fill patterns?

   D. Turn *Fill On* and *Regen*. *Save* the drawing.

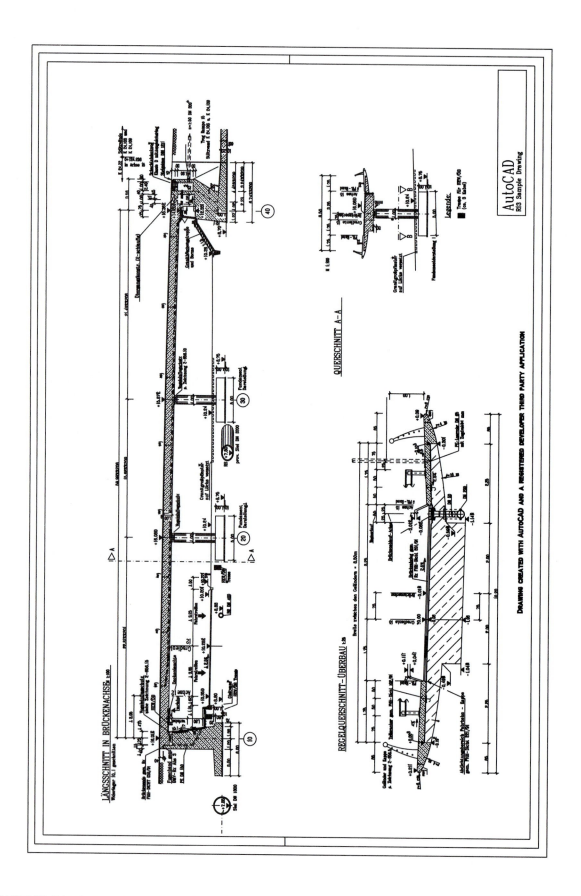

27

AUXILIARY VIEWS

Chapter Objectives

After completing this chapter you should:

1. be able to use the *Rotate* option of *Snap* to change the angle of the *SNAP*, *GRID*, and *ORTHO*;

2. be able to set the *Increment angle* and *Additional angles* for creating auxiliary views using *Polar Tracking*;

3. know how to use the *Offset* command to create parallel line copies;

4. be able to use *Xline* and *Ray* to create construction lines for auxiliary views.

CONCEPTS

AutoCAD provides no features explicitly for the creation of auxiliary views in a 2D drawing. No new commands are discussed in this chapter. However, four particular features that have been discussed earlier can assist you in construction of auxiliary views. Those features are the *SNAP* rotation, *Polar Tracking*, the *Offset* command, and the *Xline* and *Ray* commands.

An auxiliary view is a supplementary view among a series of multiviews. The auxiliary view is drawn in addition to the typical views that are mutually perpendicular (top, front, side). An auxiliary view is one that is normal (the line-of-sight is perpendicular) to an inclined surface of the object. Therefore, the auxiliary view is constructed by projecting in a 90 degree direction from the edge view of an inclined surface in order to show the true size and shape of the inclined surface. The edge view of the inclined surface could be at any angle (depending on the object), so lines are typically drawn parallel and perpendicular relative to that edge view. Hence, the *SNAP* rotation feature, *Polar Tracking*, the *Offset* command, and the *Xline* and *Ray* commands can provide assistance in this task.

Figure 27-1

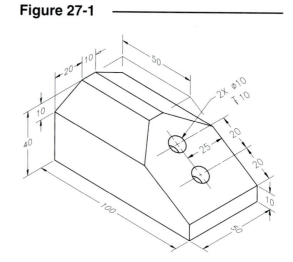

An example mechanical part used for the application of these AutoCAD features related to auxiliary view construction is shown in Figure 27-1. As you can see, there is an inclined surface that contains two drilled holes. To describe this object adequately, an auxiliary view should be created to show the true size and shape of the inclined surface.

This chapter explains the construction of a partial auxiliary view for the example object in Figure 27-1.

CONSTRUCTING AN AUXILIARY VIEW

Setting Up the Principal Views

Figure 27-2

To begin this drawing, the typical steps are followed for drawing setup (Chapter 12). Because the dimensions are in millimeters, *Limits* should be set accordingly. For example, to provide enough space to draw the views full size and to plot full size on an A sheet, *Limits* of 279 x 216 are specified.

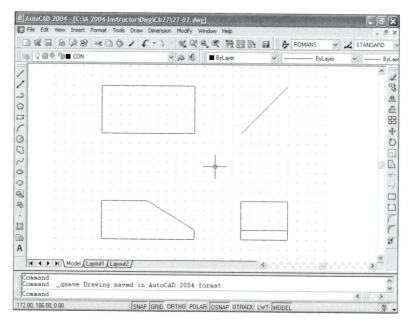

In preparation for the auxiliary view, the principal views are "blocked in," as shown in Figure 27-2. The purpose of this step is to ensure that the desired views fit and are optimally spaced within the allocated *Limits*. If there is too little or too much room, adjustments can be made to the *Limits*.

Notice that space has been allotted between the views for a partial auxiliary view to be projected from the front view.

Before additional construction on the principal views is undertaken, initial steps in the construction of the partial auxiliary view should be performed. The projection of the auxiliary view requires drawing lines perpendicular and parallel to the inclined surface. One or more of the three alternatives (explained next) can be used.

Using *Snap Rotate* and *ORTHO*

One possibility to construct an auxiliary view is to use the *Snap* command with the *Rotate* option. This action permits you to rotate the *SNAP* to any angle about a specified base point. The *GRID* automatically follows the *SNAP*. Turning *ORTHO ON* forces *Lines* to be drawn orthogonally with respect to the rotated *SNAP* and *GRID*.

Figure 27-3 displays the *SNAP* and *GRID* after rotation. The command syntax is given below.

In this figure, the cursor size is changed from the default 5% of screen size to 100% of screen size to help illustrate the orientation of the *SNAP, GRID,* and cursor when *SNAP* is *Rotated*. You can change the cursor size in the *Display* tab of the *Options* dialog box.

Figure 27-3

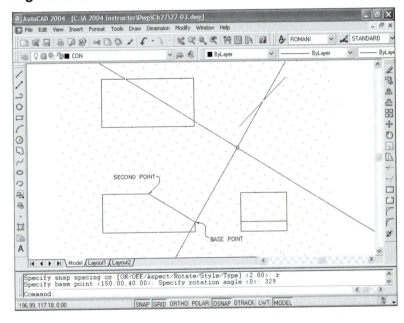

```
Command: snap
Specify snap spacing or [ON/OFF/Aspect/Rotate/Style/Type] <0.5000>: r
Specify base point <0.0000,0.0000>: PICK or (coordinates) (PICK starts a rubberband line.)
Specify rotation angle <0>: PICK or (value) (PICK to specify second point to define the angle. See
Figure 27-3.)
Command:
```

PICK (or specify coordinates for) the endpoint of the *Line* representing the inclined surface as the "base point." At the "rotation angle:" prompt, a value can be entered or another point (the other end of the inclined *Line*) can be PICKed. Use *OSNAP* when PICKing the *Endpoints*. If you want to enter a value but don't know what angle to rotate to, use *List* to display the angle of the inclined *Line*. The *GRID*, *SNAP*, and crosshairs should align with the inclined plane as shown in Figure 27-3.

(An option for simplifying construction of the auxiliary view is to create a new *UCS* [User Coordinate System] with the origin at the new base point. Use the *3Point* option and turn on *ORTHO* to select the three points. See Chapter 36, User Coordinate Systems, for more information on this procedure.)

After rotating the *SNAP* and *GRID*, the partial auxiliary view can be "blocked in," as displayed in Figure 27-4. Begin by projecting *Lines* up from and perpendicular to the inclined surface. (Make sure *ORTHO* is *ON*.) Next, two *Lines* representing the depth of the view should be constructed parallel to the inclined surface and perpendicular to the previous two projection lines. The depth dimension of the object in the auxiliary view is equal to the depth dimension in the top or right view. *Trim* as necessary.

Locate the centers of the holes in the auxiliary view and construct two *Circles*. It is generally preferred to construct circular shapes in the view in which they appear as circles, then project to the other views. That is particularly true for this type of auxiliary since the other views contain ellipses. The centers can be located by projection from the front view or by *Offsetting Lines* from the view outline.

Figure 27-4

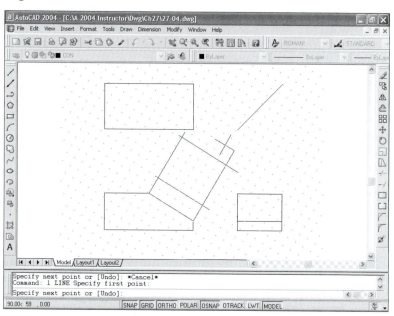

Next, project lines from the *Circles* and their centers back to the inclined surface. Use of a hidden line layer can be helpful here. While the *SNAP* and *GRID* are rotated, construct the *Lines* representing the bottom of the holes in the front view. (Alternately, *Offset* could be used to copy the inclined edge down to the hole bottoms; then *Trim* the unwanted portions of the *Lines*.)

Figure 27-5

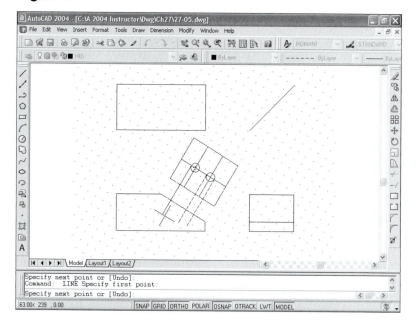

Rotating *SNAP* Back to the Original Position

Before details can be added to the other views, the *SNAP* and *GRID* should be rotated back to the original position. It is very important to rotate back using the same base point. Fortunately, AutoCAD remembers the original base point so you can accept the default for the prompt.

Next, enter a value of **0** when rotating back to the original position. (When using the *Snap Rotate* option, the value entered for the angle of rotation is absolute, not relative to the current position. For example, if the *Snap* was rotated to 45 degrees, rotate back to 0 degrees, not -45.)

Command: **snap**
Specify snap spacing or [ON/OFF/Aspect/Rotate/Style/Type] <2.00>: **r**
Specify base point <150.00,40.00>: **Enter** (AutoCAD remembers the previous base point.)
Specify rotation angle <329>: **0**
Command:

Construction of multiview drawings with auxiliaries typically involves repeated rotation of the *SNAP* and *GRID* to the angle of the inclined surface and back again as needed.

With the *SNAP* and *GRID* in the original position, details can be added to the other views as shown in Figure 27-6. Since the two circles appear as ellipses in the top and right side views, project lines from the circles' centers and limiting elements on the inclined surface. Locate centers for the two *Ellipses* to be drawn in the top and right side views.

Figure 27-6

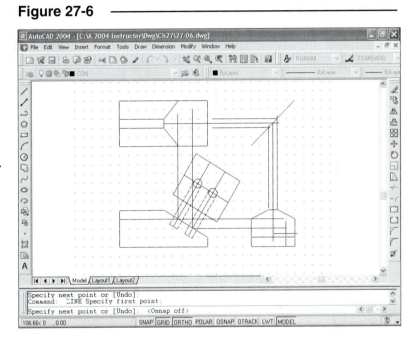

Use the *Ellipse* command to construct the ellipses in the top and right side views. Using the *Center* option of *Ellipse*, specify the center by PICKing with the *Intersection OSNAP* mode. *OSNAP* to the appropriate construction line *Intersection* for the first axis endpoint. For the second axis endpoint (as shown), use the actual circle diameter, since that dimension is not foreshortened.

Figure 27-7

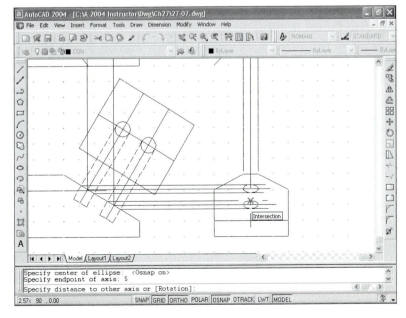

The remaining steps for completing the drawing involve finalizing the perimeter shape of the partial auxiliary view and *Copying* the *Ellipses* to the bottom of the hole positions. The *SNAP* and *GRID* should be rotated back to project the new edges found in the front view (Fig. 27-8).

At this point, the multiview drawing with auxiliary view is ready for centerlines, dimensioning, setting up a layout, and construction or insertion of a border and title block.

Figure 27-8

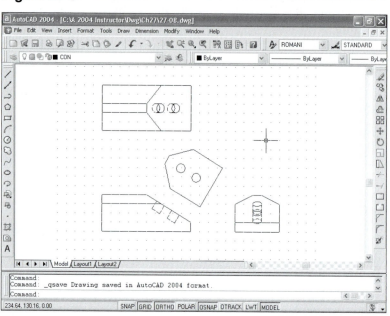

Using *Polar Tracking*

Polar Tracking can be used to create auxiliary views by facilitating construction of *Lines* at specific angles. To use *Polar Tracking* for auxiliary view construction, first use the *List* command to determine the angle of the inclined surface you want to project from. (Keep in mind that, by default, AutoCAD reports angles in whole numbers [no decimals or fractions], so use the *Units* command to increase the *Precision* of *Angular* units before using *List*.)

Once the desired angles are determined, specify the *Polar Angle Settings* in the *Drafting Settings* dialog box. Access the dialog box by right-clicking on the word *POLAR* at the Status Bar, typing *Dsettings*, or selecting *Drafting Settings* from the *Tools* pull-down menu. In the *Polar Tracking* tab of the dialog box, select the desired angles if appropriate from the *Increment angle* drop-down list. If the inclined plane is not at a regular angle offered from the drop-down list, specify the desired angle in the *Additional angles* edit box by selecting *New* and inputting the angles. Enter four angles in 90-degree increments (Fig. 27-9). Once the angles are set, ensure *POLAR* is on.

Figure 27-9

You may also want to set a *Polar Increment* in the *Snap and Grid* tab of the dialog box. This action makes the line lengths snap to regular intervals. (See Chapter 3 for more information on *Polar Snap* and *Polar Tracking*.)

Using *OSNAP*, begin constructing *Lines* perpendicular to the inclined plane. *Polar Tracking* should facilitate the line construction at the appropriate angles (Fig. 27-10).

Figure 27-10

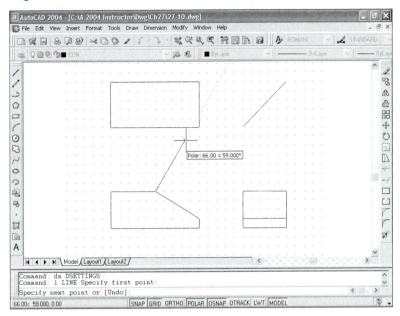

Perpendicular *Line* construction is also assisted by *Polar Tracking* (Fig. 27-11). Continue with this process, constructing necessary lines for the auxiliary view. The remainder of the auxiliary drawing, as described on previous pages (see Figures 27-5 through 27-8), could be constructed using *Polar Tracking*. The advantage of this method is that drawing horizontal and vertical *Lines* is also possible while *Polar Tracking* is on.

Object Snap Tracking can also be employed to construct objects aligned (at the specified polar angles) with object snap locations (*Endpoint*, *Midpoint*, *Intersection*, *Extension*, etc.). See Chapter 7 for more information on *Object Snap Tracking*.

Figure 27-11

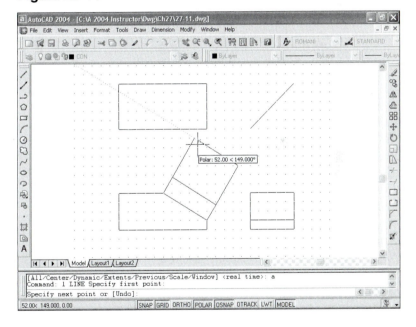

Using the *Offset* Command

Another possibility, and an alternative to the *SNAP* rotation, is to use *Offset* to make parallel *Lines*. This command can be particularly useful for construction of the "blocked in" partial auxiliary view because it is not necessary to rotate the *SNAP* and *GRID*.

OFFSET

	Pull-down Menu	COMMAND (TYPE)	ALIAS (TYPE)	Short-cut	Screen (side) Menu	Tablet Menu
	Modify *Offset*	*OFFSET*	*O*	...	*MODIFY1* *Offset*	*V,17*

Invoke the *Offset* command and specify a distance. The first distance is arbitrary. Specify an appropriate value between the front view inclined plane and the nearest edge of the auxiliary view (20 for the example). *Offset* the new *Line* at a distance of 50 (for the example) or PICK two points (equal to the depth of the view).

 Note that the *Offset* lines have lengths equal to the original and therefore require no additional editing (Fig. 27-12).

Figure 27-12 ——————

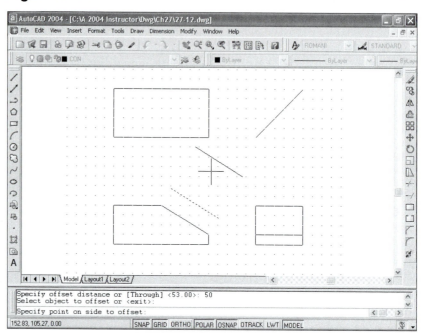

Next, two *Lines* would be drawn between *Endpoints* of the existing offset lines to complete the rectangle. *Offset* could be used again to construct additional lines to facilitate the construction of the two circles in the partial auxiliary view (Fig 27-13).

From this point forward, the construction process would be similar to the example given previously (Figs. 27-5 through 27-8). Even though *Offset* does not require that the *SNAP* be *Rotated*, the complete construction of the auxiliary view could be simplified by using the rotated *SNAP* and *GRID* in conjunction with *Offset*.

Figure 27-13 ——————

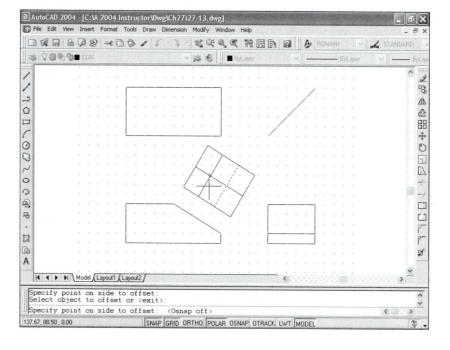

Using the *Xline* and *Ray* Commands

As a fourth alternative for construction of auxiliary views, the *Xline* and *Ray* commands could be used to create construction lines.

XLINE

Pull-down Menu	COMMAND (TYPE)	ALIAS (TYPE)	Short-cut	Screen (side) Menu	Tablet Menu
Draw Construction Line	*XLINE*	*XL*	...	*DRAW 1 Xline*	*L,10*

RAY

Pull-down Menu	COMMAND (TYPE)	ALIAS (TYPE)	Short-cut	Screen (side) Menu	Tablet Menu
Draw Ray	*RAY*	...	...	*DRAW 1 Ray*	*K,10*

The *Xline* command offers several options shown below:

Command: **xline**
Specify a point or [Hor/Ver/Ang/Bisect/Offset]:

The *Ang* option can be used to create a construction line at a specified angle. In this case, the angle specified would be that of the inclined plane or perpendicular to the inclined plane. The *Offset* option works well for drawing construction lines parallel to the inclined plane, especially in the case where the angle of the plane is not known.

Figure 27-14 illustrates the use of *Xline Offset* to create construction lines for the partial auxiliary view. The *Offset* option operates similarly to the *Offset* command described previously. Remember that an *Xline* extends to infinity but can be *Trimmed*, in which case it is converted to *Ray* (*Trim* once) or to a *Line* (*Trim* twice). See Chapter 15 for more information on the *Xline* command.

Figure 27-14

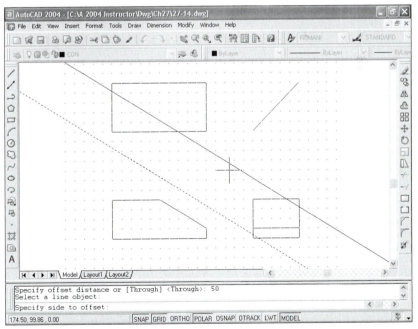

The *Ray* command also creates construction lines; however, the *Ray* has one anchored point and the other end extends to infinity.

```
Command: ray
Specify start point: PICK or (coordinates)
Specify through point: PICK or (coordinates)
```

Rays are helpful for auxiliary view construction when you want to create projection lines perpendicular to the inclined plane. In Figure 27-15, two *Rays* are constructed from the *Endpoints* of the inclined plane and *Perpendicular* to the existing *Xlines*. Using *Xlines* and *Rays* in conjunction is an <u>excellent method</u> for "blocking in" the view.

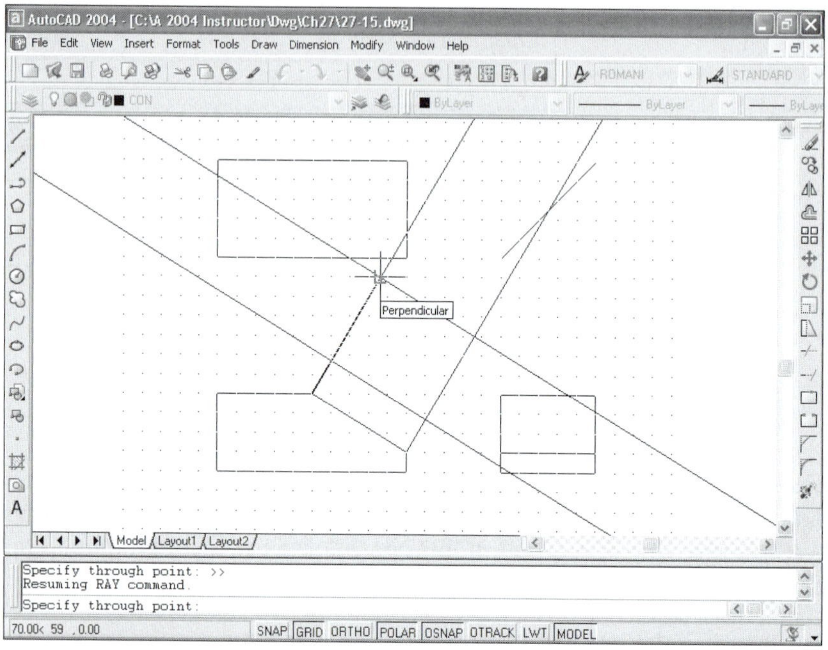

There are two strategies for creating drawings using *Xlines* and *Rays*. First, these construction lines can be created on a separate layer and set up as a framework for the object lines. The object lines would then be drawn on top of the construction lines using *Osnaps*, but would be drawn on the object layer. The construction layer would be *Frozen* for plotting. The other strategy is to create the construction lines on the object layer. Through a series of *Trims* and other modifications, the *Xlines* and *Rays* are transformed to the finished object lines.

Figure 27-15 ────────────

Now that you are aware of several methods for constructing auxiliary views, use any one method or a combination of methods for your drawings. No matter which methods are used, the final lines that are needed for the finished auxiliary view should be the same. It is up to you to use the methods that are the most appropriate for the particular application or are the easiest and quickest for you personally.

Constructing Full Auxiliary Views

The construction of a full auxiliary view begins with the partial view. After initial construction of the partial view, begin the construction of the full auxiliary by projecting the other edges and features of the object (other than the inclined plane) to the existing auxiliary view.

The procedure for constructing full auxiliary views in AutoCAD is essentially the same as that for partial auxiliary views. Use of the *Offset, Xline,* and *Ray* commands, *SNAP* and *GRID* rotation, and *Polar Tracking* should be used as illustrated for the partial auxiliary view example. Because a full auxiliary view is projected at the same angle as a partial, the same rotation angle and basepoint would be used for the *SNAP,* or the same *Increment angle* or *Additional angles* should be used for *Polar Tracking*.

CHAPTER EXERCISES

For the following exercises, create the multiview drawing, including the partial or full auxiliary view as indicated. Use the appropriate template drawing based on the given dimensions and indicated plot scale.

1. Make a multiview drawing with a partial auxiliary view of the example used in this chapter. Refer to Figure 27-1 for dimensions. *Save* the drawing as **CH27EX1**. *Plot* on an A size sheet at **1=1** scale.

2. Recreate the views given in Figure 27-16 and add a partial auxiliary view. *Save* the drawing as **CH27EX2** and *Plot* on an A size sheet at **2=1** scale.

Figure 27-16

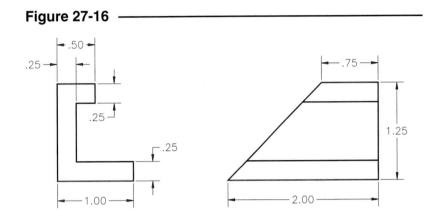

3. Recreate the views shown in Figure 27-17 and add a partial auxiliary view. *Save* the drawing as **CH27EX3**. Make a *Plot* on an A size sheet at **1=1** scale.

Figure 27-17

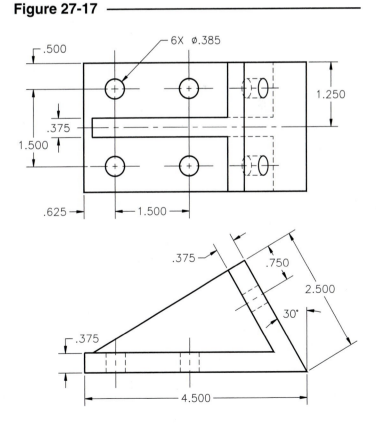

4. Make a multiview drawing of the given views in Figure 27-18. Add a full auxiliary view. *Save* the drawing as **CH27EX4**. *Plot* the drawing at an appropriate scale on an A size sheet.

Figure 27-18

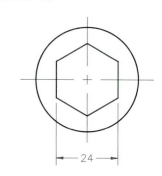

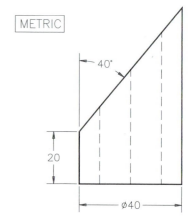

METRIC

5. Draw the front, top, and a partial auxiliary view of the holder. During construction, note that the back corner indicated with hidden lines in Figure 27-19 is not a "clean" intersection, but is actually composed of two separate vertical edges. Make a *Plot* full size. Save the drawing as **HOLDER**.

Figure 27-19

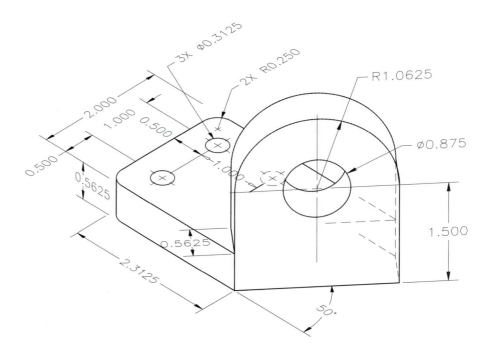

6. Draw three principal views and a full auxiliary view of the V-block shown in Figure 27-20. *Save* the drawing as **VBLOCK**. *Plot* to an accepted scale.

Figure 27-20

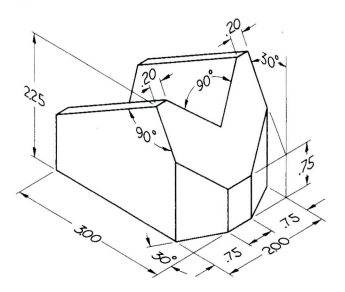

7. Draw two principal views and a partial auxiliary of the angle brace. *Save* as **ANGLBRAC**. Make a plot to an accepted scale on an A or B size sheet. Convert the fractions to the current ANSI standard, decimal inches.

Figure 27-21

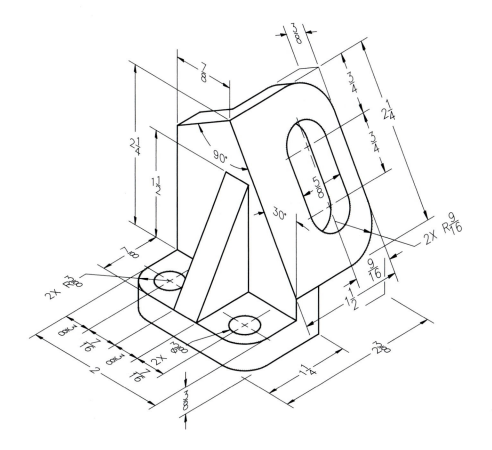

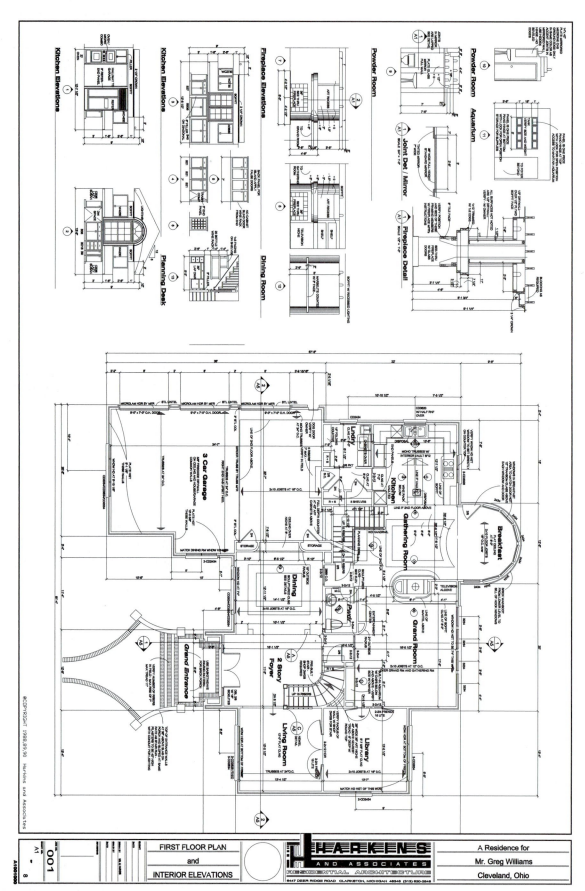

WILHOME.DWG Courtesy of Autodesk, Inc.

DIMENSIONING

Chapter Objectives

After completing this chapter you should:

1. be able to create linear dimensions with *Dimlinear*;

2. be able to append *Dimcontinue* and *Dimbaseline* dimensions to existing dimensions;

3. be able to create *Angular*, *Diameter*, and *Radius* dimensions;

4. know how to affix notes to drawings with *Leaders* and *Qleaders*;

5. know that *Dimordinate* can be used to specify Xdatum and Ydatum dimensions;

6. be able to use the new *Qdim* command to quickly create a variety of associative dimensions;

7. be able to create and apply geometric dimensioning and tolerancing symbols using the *Tolerance* command and dialog boxes;

8. know the possible methods for editing associative dimensions and dimensioning text.

CONCEPTS

As you know, drawings created with CAD systems should be constructed with the same dimensions and units as the real-world objects they represent. In this way, the features of the object that you apply dimensions to (lengths, diameters, angles, etc.) are automatically measured by AutoCAD and the correct values are displayed in the dimension text. So if the object is drawn accurately, the dimension values will be created correctly and automatically. Generally, dimensions should be created on a separate layer named DIMENSIONS, or similar, and dimensions should be created in model space (see Chapter 29 for information on dimensioning in paper space).

The main components of a dimension are:

1. Dimension line
2. Extension lines
3. Dimension text (usually a numeric value)
4. Arrowheads or tick marks

Figure 28-1

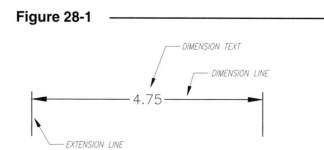

AutoCAD dimensioning is <u>semi-automatic</u>. When you invoke a command to create a linear dimension, AutoCAD only requires that you PICK an object or specify the extension line origins (where you want the extension lines to begin) and PICK the location of the dimension line (distance from the object). AutoCAD then measures the feature and draws the extension lines, dimension line, arrowheads, and dimension text.

Figure 28-2

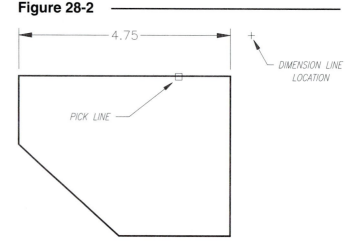

 For linear dimensioning commands, there are <u>two ways</u> to specify placement for a dimension in AutoCAD: you can PICK the <u>object</u> to be dimensioned or you can PICK the two <u>extension line origins</u>. The simplest method is to select the object because it requires only one PICK (Fig. 28-2):

> Command: *dimlinear*
> Specify first extension line origin or <select object>: **Enter**
> Select object to dimension: **PICK**

The other method is to PICK the extension line origins (Fig. 28-3). *Osnaps* should be used to PICK the object (endpoints in this case) so that the dimension is <u>associated</u> with the object.

Figure 28-3

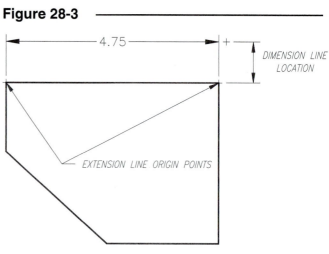

> Command: *dimlinear*
> Specify first extension line origin or <select object>: **PICK**
> Specify second extension line origin: **PICK**

Once the dimension is attached to the object, you specify how far you want the dimension to be placed from the object (called the "dimension line location").

Dimensioning in AutoCAD is <u>associative</u> (by default). Because the extension line origins are "associated" with the geometry, the dimension text automatically updates if the geometry is *Moved*, *Stretched*, *Rotated*, *Scaled*, or edited using grips.

Because dimensioning is semi-automatic, <u>dimensioning variables</u> are used to control the way dimensions are created. Dimensioning variables can be used to control features such as text or arrow size, direction of the leader arrow for radial or diametrical dimensions, format of the text, and many other possible options. Groups of variable settings can be named and saved as <u>Dimension Styles</u>. Dimensioning variables and dimension styles are discussed in Chapter 29.

Dimensioning commands can be invoked by the typical methods. The *Dimension* pull-down menu contains the dimension creation and editing commands (Fig. 28-4). A *Dimension* toolbar (Fig. 28-5) can also be activated by using the *Toolbars* list.

Figure 28-4

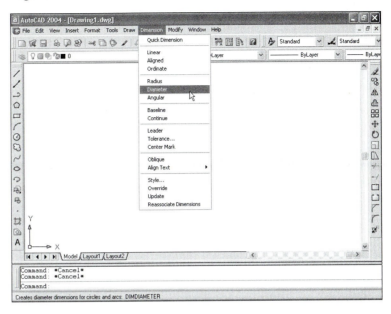

Figure 28-5

DIMENSION DRAWING COMMANDS

DIMLINEAR

Pull-down Menu	COMMAND (TYPE)	ALIAS (TYPE)	Short-cut	Screen (side) Menu	Tablet Menu
Dimension Linear	*DIMLINEAR*	*DIMLIN or DLI*	...	*DIMNSION Linear*	*W,5*

Dimlinear creates a <u>horizontal, vertical, or rotated</u> dimension. If the object selected is a horizontal line (or the extension line origins are horizontally oriented), the resulting dimension is a horizontal dimension. This situation is displayed in the previous illustrations (see Figures 28-2 and 28-3), or if the selected object or extension line origins are vertically oriented, the resulting dimension is vertical (Fig. 28-6).

Figure 28-6

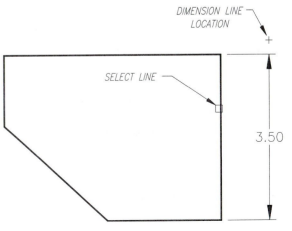

When you dimension an inclined object (or if the selected extension line origins are diagonally oriented), a vertical <u>or</u> horizontal dimension can be made, depending on where you drag the dimension line in relation to the object. If the dimension line location is more to the side, a vertical dimension is created (Fig. 28-7), or if you drag farther up or down, a horizontal dimension results (Fig. 28-8).

Figure 28-7

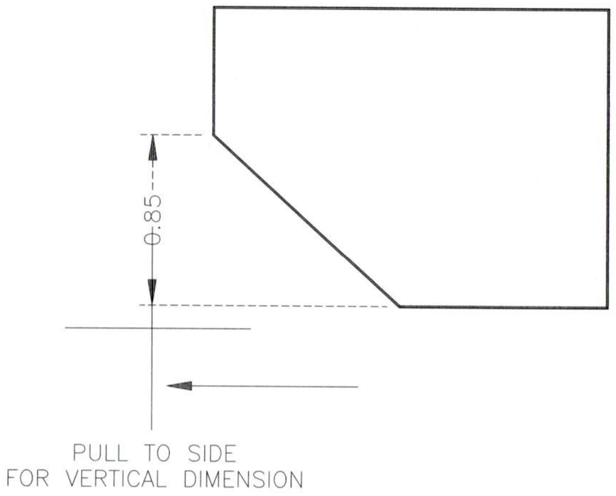

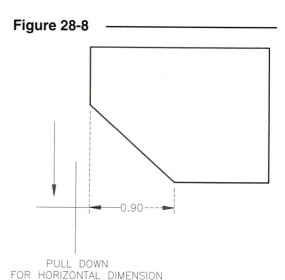

PULL TO SIDE
FOR VERTICAL DIMENSION

 If you select the extension line origins, it is very important to PICK the <u>object's endpoints</u> if the dimensions are to be truly associative (associated with the geometry). *OSNAP* should be used to find the object's *Endpoint, Intersection,* etc.

Command: **dimlinear**
Specify first extension line origin or <select object>: **PICK** (use *Osnaps*)
Specify second extension line origin: **PICK** (use *Osnaps*)
Specify dimension line location or [Mtext/Text/Angle/Horizontal/Vertical/Rotated]: **PICK** (where you want the dimension line to be placed)
Dimension text = *n.nnnn*
Command:

When you pick the location for the dimension line, AutoCAD automatically measures the object and inserts the correct numerical value. The other options are explained next.

Figure 28-8

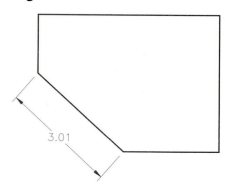

PULL DOWN
FOR HORIZONTAL DIMENSION

Rotated
If you want the dimension line to be drawn at an angle instead of vertical or horizontal, use this option. Selecting an inclined line, as in the previous two illustrations, would normally create a horizontal or vertical dimension. The *Rotated* option allows you to enter an <u>angular value</u> for the dimension line to be drawn. For example, selecting the diagonal line and specifying the appropriate angle would create the dimension shown in Figure 28-9. This object, however, could be more easily dimensioned with the *Dimaligned* command (see "*Dimaligned*").

Figure 28-9

A *Rotated* dimension should be used when the geometry has "steps" or any time the desired dimension line angle is different than the dimensioned feature (when you need extension lines of different lengths). Figure 28-10 illustrates the result of using a *Rotated* dimension to give the correct dimension line angle and extension line origins for the given object. In this case, the extension line origins were explicitly PICKed. The feature of *Rotated* that makes it unique is that you specify the <u>angle</u> that the dimension line will be drawn.

Figure 28-10

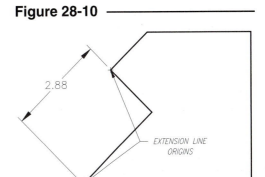

Text

Using the *Text* option allows you to enter any value or characters in place of the AutoCAD-measured text. The measured value is given as a reference at the command prompt.

> Command: **dimlinear**
> Specify first extension line origin or <select object>: **PICK**
> Specify second extension line origin: **PICK**
> Specify dimension line location or
> [Mtext/Text/Angle/Horizontal/Vertical/Rotated]: **T**
> Enter dimension text <*n.nnnn*>: (**text**) or (**value**)

Figure 28-11

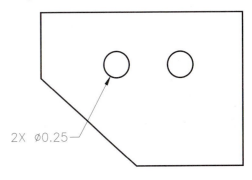

Entering a value or text at the prompt (above) causes AutoCAD to display that value or text instead of the AutoCAD-measured value. If you want to keep the AutoCAD-measured value but add a prefix or suffix, use the less-than, greater-than symbols (< >) to represent the actual (AutoCAD) value:

> Enter dimension text <0.25>: **2X <>**
> Specify dimension line location or [Mtext/Text/Angle/Horizontal/Vertical/Rotated]: **PICK**
> Dimension text = 0.25
> Command:

The "2X <>" response produces the dimension text shown in Figure 28-11.

NOTE: Changing the AutoCAD-measured value should be discouraged. If the geometry is drawn accurately, the dimensional value is correct. If you specify other dimension text, the text value is <u>not</u> updated in the event of *Stretching, Rotating,* or otherwise editing the associative dimension.

Mtext

This option allows you to change the existing or insert additional text to the AutoCAD-supplied numerical value. The text is entered by the *Multiline Text Editor.* The less-than and greater-than symbols (< >) represent the AutoCAD-supplied dimensional value. Place text or numbers inside the symbols if you want to override the correct measurement (not advised), or place text

Figure 28-12

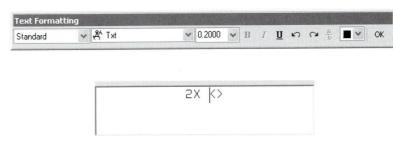

outside the symbols if you want to add annotation to the numerical value. For example, entering "2X " before the symbols (Fig. 28-12) creates a dimensional value as shown in Figure 28-11. Keep in mind—the power of using the *Multiline Text Editor* is the availability of all the text creation and editing features in the dialog box. For example, you can specify fonts, text height, bold, italic, underline, and stacked text or fractions.

Angle

This creates text drawn at the angle you specify. Use this for special cases when the text must be drawn to a specific angle other than horizontal. (It is also possible to make the text automatically align with the angle of the dimension line using the *Dimension Style Manager*. See Chapter 29.)

Horizontal

Use the *Horizontal* option when you want to force a horizontal dimension for an inclined line and the desired placement of the dimension line would otherwise cause a vertical dimension.

Vertical

This option forces a *Vertical* dimension for any case.

DIMALIGNED

Pull-down Menu	COMMAND (TYPE)	ALIAS (TYPE)	Short-cut	Screen (side) Menu	Tablet Menu
Dimension Aligned	DIMALIGNED	*DIMALI* or *DAL*	...	*DIMNSION Aligned*	W,4

An *Aligned* dimension is aligned with (at the same angle as) the selected object or the extension line origins. For example, when *Aligned* is used to dimension the angled object shown in Figure 28-13, the resulting dimension aligns with the *Line*. This holds true for either option—PICKing the object or the extension line origins. If a *Circle* is PICKed, the dimension line is aligned with the selected point on the *Circle* and its center. The command syntax for the *Dimaligned* command accepting the defaults is:

Figure 28-13 —————

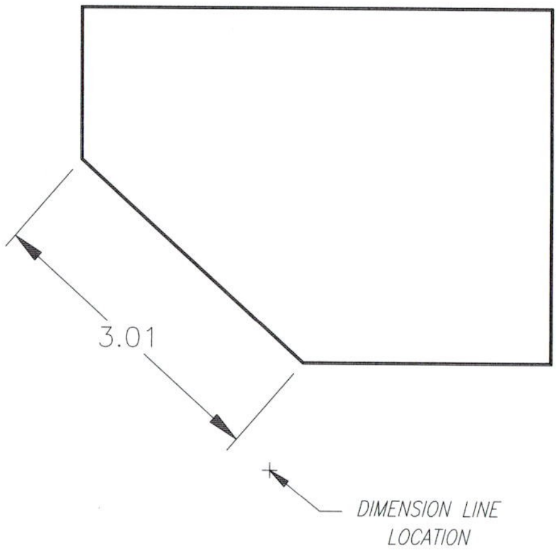

Command: **dimaligned**
Specify first extension line origin or <select object>: **PICK**
Specify second extension line origin: **PICK**
Specify dimension line location or [Mtext/Text/Angle]: **PICK**
Dimension text = *n.nnnn*
Command:

The three options (*Mtext/Text/Angle*) operate similar to those for *Dimlinear*.

Mtext

The *Mtext* option calls the *Multiline Text Editor*. You can alter the AutoCAD-supplied text value or modify other visible features of the dimension text such as fonts, text height, bold, italic, underline, stacked text or fractions, and text style (see "*Dimlinear*," *Mtext*).

Text

You can change the AutoCAD-supplied numerical value or add other annotation to the value in command line format (see "*Dimlinear*," *Text*).

Angle

Enter a value for the angle that the text will be drawn.

The typical application for *Dimaligned* is for dimensioning an angled but <u>straight</u> feature of an object, as shown in Figure 28-13. *Dimaligned* should not be used to dimension an object feature that contains "steps," as shown in Figure 28-10. *Dimaligned* always draws <u>extension lines of equal length</u>.

DIMBASELINE

	Pull-down Menu	COMMAND (TYPE)	ALIAS (TYPE)	Short-cut	Screen (side) Menu	Tablet Menu
	Dimension Baseline	DIMBASELINE	DIMBASE or DBA	...	DIMNSION Baseline	...

Dimbaseline allows you to create a dimension that uses an extension line origin from a previously created dimension. Successive *Dimbaseline* dimensions can be used to create the style of dimensioning shown in Figure 28-14.

A baseline dimension must be connected to an existing dimension. If *Dimbaseline* is invoked immediately after another dimensioning command, you are required only to specify the second extension line origin since AutoCAD knows to use the <u>previous</u> dimension's <u>first</u> extension line origin:

    Command: **dimbaseline**
    Specify a second extension line origin or [Undo/Select]
    <Select>: **PICK**
    Dimension text = *n.nnnn*

Figure 28-14

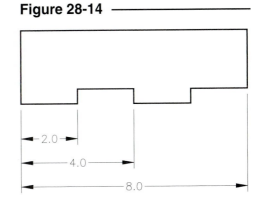

The <u>previous dimension's first extension line</u> is used also for the baseline dimension (Fig. 28-15). Therefore, you specify only the second extension line origin. Note that you are not required to specify the dimension line location. AutoCAD spaces the new dimension line automatically, based on the setting of the dimension line increment variable (Chapter 29).

Figure 28-15

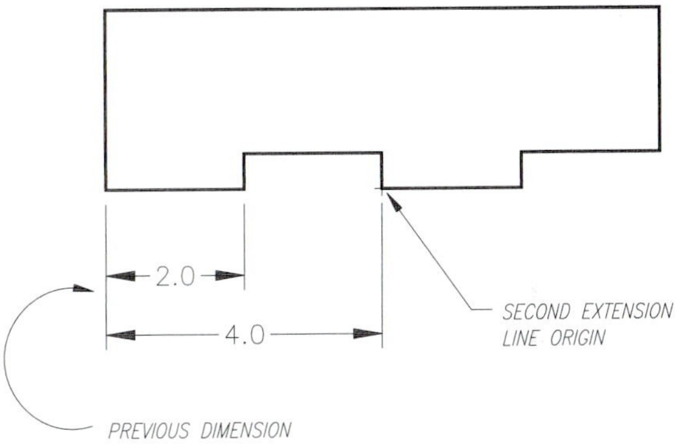

If you wish to create a *Dimbaseline* dimension using a dimension other than the one just created, use the "Select" option (Fig. 28-16):

    Command: **dimbaseline**
    Specify a second extension line origin or
    [Undo/Select] <Select>: **S** or **Enter**
    Select base dimension: **PICK**
    Specify a second extension line origin or
    [Undo/Select] <Select>: **Enter**
    Dimension text = *n.nnnn*

Figure 28-16

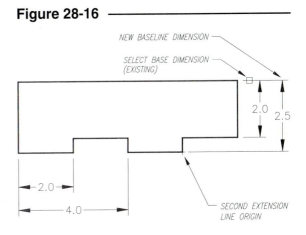

The extension line selected as the base dimension becomes the first extension line for the new *Dimbaseline* dimension.

The *Undo* option can be used to undo the last baseline dimension created in the current command sequence.

Dimbaseline can be used with rotated, aligned, angular, and ordinate dimensions.

DIMCONTINUE

Pull-down Menu	COMMAND (TYPE)	ALIAS (TYPE)	Short-cut	Screen (side) Menu	Tablet Menu
Dimension Continue	DIMCONTINUE	DIMCONT or DCO	...	DIMNSION Continue	...

Dimcontinue dimensions continue in a line from a previously created dimension. *Dimcontinue* dimension lines are attached to, and drawn the same distance from, the object as an existing dimension.

Dimcontinue is similar to *Dimbaseline* except that an existing dimension's <u>second</u> extension line is used to begin the new dimension. In other words, the new dimension is connected to the <u>second</u> extension line, rather than to the <u>first</u>, as with a *Dimbaseline* dimension (Fig. 28-17).

Figure 28-17

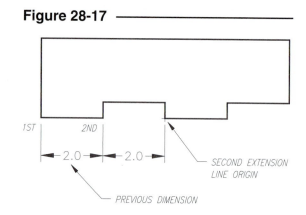

The command syntax is as follows:

Command: **dimcontinue**
Specify a second extension line origin or [Undo/Select]
<Select>: **PICK**
Dimension text = *n.nnnn*

Assuming a dimension was just drawn, *Dimcontinue* could be used to place the next dimension, as shown in Figure 28-17.

Figure 28-18

If you want to create a continued dimension and attach it to an extension line <u>other</u> than the previous dimension's second extension line, you can use the *Select* option to pick an extension line of any other dimension. Then, use *Dimcontinue* to create a continued dimension from the selected extension line (Fig. 28-18).

The *Undo* option can be used to undo the last continued dimension created in the current command sequence. *Dimcontinue* can be used with rotated, aligned, angular, and ordinate dimensions.

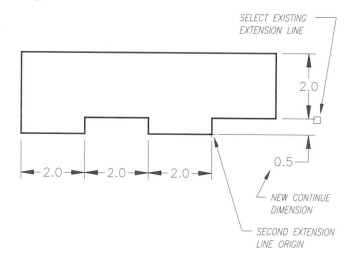

DIMDIAMETER

Pull-down Menu	COMMAND (TYPE)	ALIAS (TYPE)	Short-cut	Screen (side) Menu	Tablet Menu
Dimension Diameter	*DIMDIAMETER*	*DIMDIA* or *DDI*	...	*DIMNSION Diameter*	*X,4*

The *Dimdiameter* command creates a diametrical dimension by selecting any *Circle*. Diametrical dimensions should be used for full 360 degree *Circles* and can be used for *Arcs* of more than 180 degrees.

    Command: **dimdiameter**
    Select arc or circle: **PICK**
    Dimension text = *n.nnnn*
    Specify dimension line location or [Mtext/Text/Angle]: **PICK**
    Command:

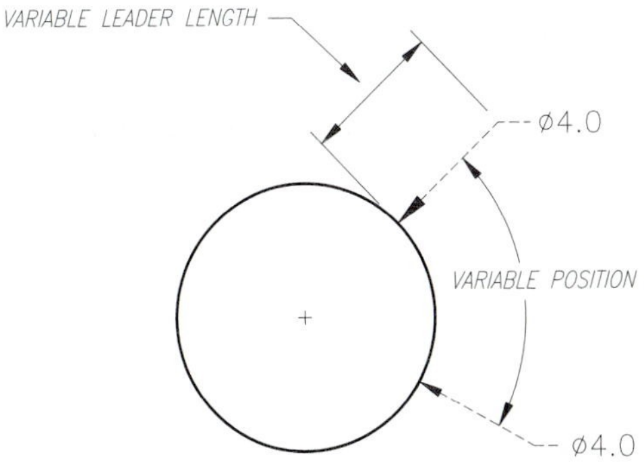

You can PICK the circle at any location. AutoCAD allows you to adjust the position of the dimension line to any angle or length (Fig. 28-19). Dimension lines for diametrical or radial dimensions should be drawn to a regular angle, such as 30, 45, or 60 degrees, never vertical or horizontal.

A typical diametrical dimension appears as the example in Figure 28-19. According to ANSI standards, a diameter dimension line and arrow should point inward (toward the center) for holes and small circles where the dimension line and text do not fit within the circle. Use the default variable settings for *Dimdiameter* dimensions such as this.

For dimensioning large circles, ANSI standards suggest an alternate method for diameter dimensions where sufficient room exists for text and arrows inside the circle (Fig. 28-20). To create this style of dimensioning, set the variables to *Text* and *Place text manually when dimensioning* in the *Fit* tab of the *Dimension Style Manager* (see Chapter 29).

Notice that the *Diameter* command creates center marks at the *Circle's* center. Center marks can also be drawn by the *Dimcenter* command (discussed later). AutoCAD uses the center and the point selected on the *Circle* to maintain its associativity.

Mtext/Text
The *Mtext* or *Text* options can be used to modify or add annotation to the default value. The *Mtext* option summons the *Multiline Text Editor* and the *Text* option uses Command line format. Both options operate similar to the other dimensioning commands (see *Dimlinear*, *Mtext*, and *Text*).

Notice that with diameter dimensions AutoCAD automatically creates the Ø (phi) symbol before the dimensional value. This is the latest ANSI standard for representing diameters. If you prefer to use a prefix before or suffix after the dimension value, it can be accomplished with the *Mtext* or *Text* option. Remember that the < > symbols represent the AutoCAD-measured value so the additional text should be inserted on <u>either side</u> of the symbols. Inserting a prefix by this method <u>does not override</u> the Ø (phi) symbol (Fig. 28-21).

Figure 28-19

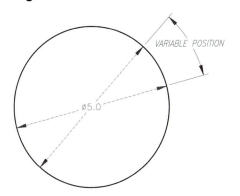

Figure 28-20

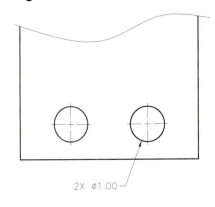

Figure 28-21

A prefix or suffix can alternately be added to the measured value by using the *Dimension Style Manager* and entering text or values in the *Prefix* or *Suffix* edit boxes. Using this method, however, <u>overrides</u> the Ø symbol (see Chapter 29).

Angle
With this option, you can specify an angle (other than the default) for the text to be drawn by entering a value.

DIMRADIUS

Pull-down Menu	COMMAND (TYPE)	ALIAS (TYPE)	Short-cut	Screen (side) Menu	Tablet Menu
Dimension Radius	*DIMRADIUS*	*DIMRAD* or *DRA*	...	*DIMNSION Radius*	X,5

Dimradius is used to create a dimension for an arc of anything less than half of a circle. ANSI standards dictate that a *Radius* dimension line should point outward (from the arc's center), unless there is insufficient room, in which case the line can be drawn on the outside pointing inward, as with a leader. The text can be located inside an arc (if sufficient room exists) or is forced outside of small *Arcs* on a leader.

```
Command: dimradius
Select arc or circle: PICK
Dimension text = n.nnnn
Specify dimension line location or [Mtext/Text/Angle]: PICK
Command:
```

Figure 28-22 —————————————

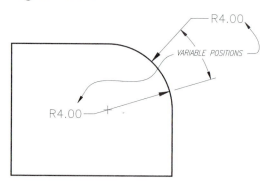

Assuming the defaults, a *Dimradius* dimension can appear on either side of an arc, as shown in Figure 28-22. Placement of the dimension line is variable. Dimension lines for arcs and circles should be positioned at a regular angle such as 30, 45, or 60 degrees, never vertical or horizontal.

When the radius dimension is dragged outside of the arc, a center mark is automatically created (Fig. 28-22). When the radius dimension is dragged inside the arc, no center mark is created. (Center marks can be created using the *Center* command discussed next.)

Figure 28-23 —————————————

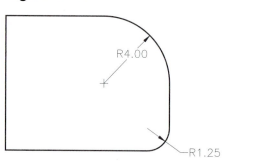

According to ANSI standards, the dimension line and arrow should point <u>outward</u> from the center for radial dimensions (Fig. 28-23). The text can be placed inside or outside of the *Arc*, depending on how much room exists. To create a *Dimradius* dimension to comply with this standard, dimension variables must be changed from the defaults. To achieve *Dimradius* dimensions as shown in Figure 28-23, set the variables to *Text* and *Place text manually when dimensioning* in the *Fit* tab of the *Dimension Style Manager* (see Chapter 29).

For very small radii, such as that shown in Figure 28-24, there is insufficient room for the text and arrow to fit inside the *Arc*. In this case, AutoCAD automatically forces the text outside with the leader pointing inward toward the center. <u>No changes</u> have to be made to the default settings for this to occur.

Figure 28-24 —————————————

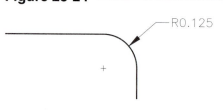

Mtext/Text
These options can be used to modify or add annotation to the default value. AutoCAD automatically inserts the letter "R" before the numerical text whenever a *Dimradius* dimension is created. This is the correct notation for radius dimensions. The *Mtext* option calls the *Multiline Text Editor*, and the *Text* option uses the Command line format for entering text. Remember that AutoCAD uses the < > symbols to represent the AutoCAD-supplied value. Entering text inside the < > symbols overrides the measured value. Entering text before the symbols adds a prefix <u>without overriding the "R" designation</u>. Alternately, text can be added by using the *Prefix* and *Suffix* options of the *Dimension Style Manager* series; however, a *Prefix* entered in the edit box <u>replaces</u> the letter "R." (See Chapter 29.)

Angle
With this option, you can specify an angle (other than the default) for the <u>text</u> to be drawn by entering a value.

DIMCENTER

Pull-down Menu	COMMAND (TYPE)	ALIAS (TYPE)	Short-cut	Screen (side) Menu	Tablet Menu
Dimension Center Mark	DIMCENTER	DCE	...	DIMNSION Center	X,2

The *Dimcenter* command draws a center mark on any selected *Arc* or *Circle*. As shown earlier, the *Dimdiameter* command and the *Dimradius* command sometimes create the center marks automatically.

The command requires you only to select the desired *Circle* or *Arc* to acquire the center marks:

```
Command: dimcenter
Select arc or circle: PICK
Command:
```

No matter if the center mark is created by the *Center* command or by the *Diameter* or *Radius* commands, the center mark can be either a small cross or complete center lines extending past the *Circle* or *Arc* (Fig. 28-25). The type of center mark drawn is controlled by the *Mark* or *Line* setting in the *Lines and Arrows* tab of the *Dimension Style Manager*. It is suggested that a new *Dimension Style* be created with these settings just for drawing center marks. (See Chapter 29.)

When you are dimensioning, short center marks should be used for *Arcs* of less than 180 degrees, and full center lines should be drawn for *Circles* and for *Arcs* of 180 degrees or more (Fig. 28-26).

NOTE: Since the center marks created with the *Dimcenter* command are <u>not</u> associative, they may be *Trimmed*, *Erased*, or otherwise edited, as shown in Figure 28-27. The center marks created with the *Dimradius* or *Dimdiameter* commands <u>are</u> associative and cannot be edited.

The lines comprising the center marks created with *Dimcenter* can be *Erased* or otherwise edited. In the case of a 180 degree *Arc*, two center mark lines can be shortened using *Break*, and one line can be *Erased* to achieve center lines as shown in Figure 28-27.

Figure 28-25

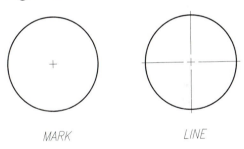

MARK   LINE

Figure 28-26

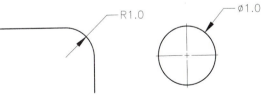

Figure 28-27

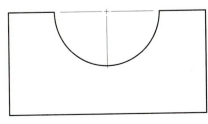

DIMANGULAR

	COMMAND (TYPE)	ALIAS (TYPE)	Short-cut	Screen (side) Menu	Tablet Menu
Pull-down Menu					
Dimension Angular	DIMANGULAR	*DIMANG* or *DAN*	...	*DIMNSION Angular*	X,3

The *Dimangular* command provides many possible methods of creating an angular dimension.

A typical angular dimension is created between two *Lines* that form an angle (of other than 90 degrees). The dimension line for an angular dimension is radiused with its center at the vertex of the angle (Figure 28-28). A *Dimangular* dimension automatically adds the degree symbol (°) to the dimension text. The dimension text format is controlled by the current settings for *Units* in the *Dimension Style* dialog box.

Figure 28-28 ———————————

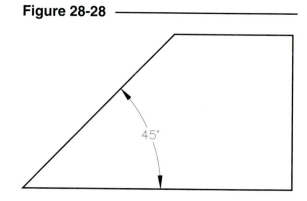

AutoCAD automates the process of creating this type of dimension by offering options within the command syntax. The default options create a dimension, as shown here:

```
Command: dimangular
Select arc, circle, line, or <specify vertex>: PICK
Select second line: PICK
Specify dimension arc line location or [Mtext/Text/Angle]: PICK
Dimension text = nn
Command:
```

Dimangular dimensioning offers some very useful and easy-to-use options for placing the desired dimension line and text location.

At the "Specify dimension arc line location or [Mtext/Text/Angle]:" prompt, you can move the cursor around the vertex to dynamically display possible placements available for the dimension. The dimension can be placed in any of four positions as well as any distance from the vertex (Fig. 28-29). Extension lines are automatically created as needed.

Figure 28-29 ———————————

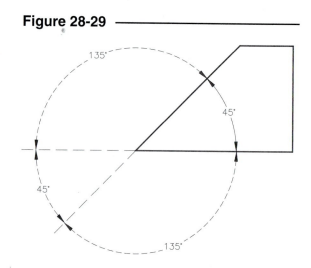

The *Dimangular* command offers other options, including dimensioning angles for *Arcs*, *Circles*, or allowing selection of any three points.

If you select an *Arc* in response to the "Select arc, circle, line, or <specify vertex>:" prompt, AutoCAD uses the *Arc's* center as the vertex and the *Arc's* endpoints to generate the extension lines. You can select either angle of the *Arc* to dimension (Fig. 28-30).

Figure 28-30

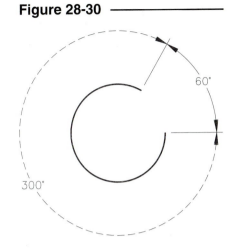

If you select a *Circle*, AutoCAD uses the PICK point as the first extension line origin. The second extension line origin does not have to be on the *Circle*, as shown in Figure 28-31.

Figure 28-31

If you press Enter in response to the "Select arc, circle, line, or <specify vertex>:" prompt, AutoCAD responds with the following:

Specify angle vertex: **PICK**
Specify first angle endpoint: **PICK**
Specify second angle endpoint: **PICK**

This option allows you to apply an *Angular* dimension to a variety of shapes.

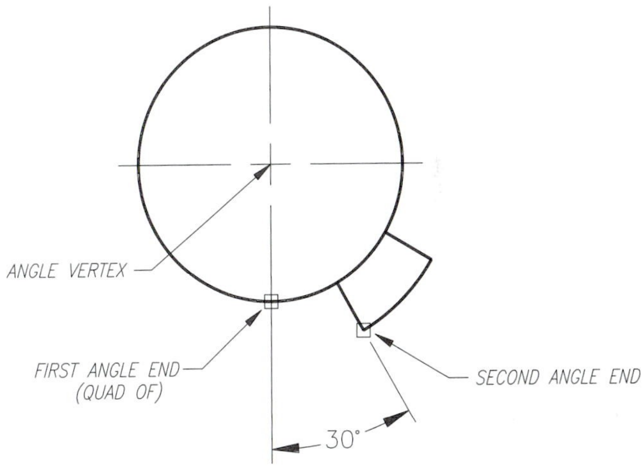

LEADER

Pull-down Menu	COMMAND (TYPE)	ALIAS (TYPE)	Short-cut	Screen (side) Menu	Tablet Menu
...	*LEADER*	*LEAD*	...	...	...

The *Leader* command (not *Qleader*) allows you to create an associative leader similar to that created with the *Diameter* command. The *Leader* command is intended to give dimensional notes such as the manufacturing or construction specifications shown here:

Command: **leader**
Specify leader start point: **PICK**
Specify next point: **PICK**
Specify next point or [Annotation/Format/Undo] <Annotation>: **Enter**
Enter first line of annotation text or <options>: **CASE HARDEN**
Enter next line of annotation text: **Enter**
Command:

Figure 28-32

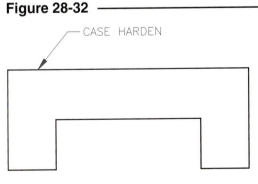

CASE HARDEN

At the "start point:" prompt, select the desired location for the arrow. You should use an *OSNAP* option (such as *Nearest*, in this case) to ensure the arrow touches the desired object.

A short horizontal line segment called the "hook line" is automatically added to the last line segment drawn if the leader line is 15 degrees or more from horizontal. Note that command syntax for the *Leader* in Figure 28-32 indicates only one line segment was PICKed. A *Leader* can have as many segments as you desire (Fig. 28-33).

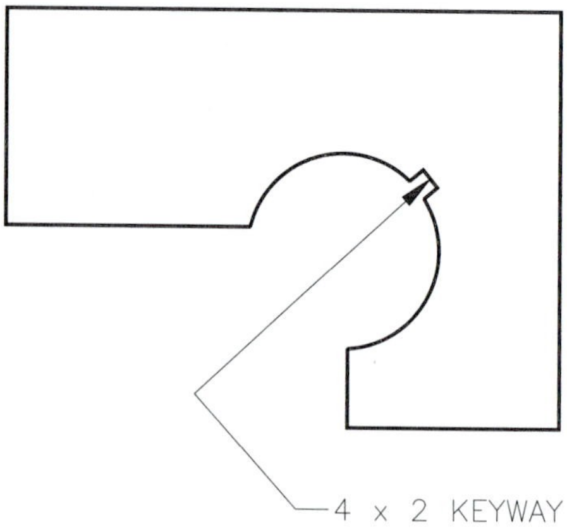

If you do not enter text at the "Annotation:" prompt, another series of options are available:

> Command: **leader**
> Specify leader start point: **PICK**
> Specify next point: **PICK**
> Specify next point or [Annotation/Format/Undo] <Annotation>: **Enter**
> Enter first line of annotation text or <options>: **Enter**
> Enter an annotation option [Tolerance/Copy/Block/None/Mtext] <Mtext>:

Figure 28-33

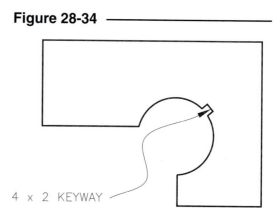

Format
This option produces another list of choices:

> Enter leader format option [Spline/STraight/Arrow/None] <Exit>:

Spline/STraight
You can draw either a *Spline* or straight version of the leader line with these options. The resulting *Spline* leader line has the characteristics of a normal *Spline*. An example of a *Splined* leader is shown in Figure 28-34.

Figure 28-34

Arrow/None
This option draws the leader line with or without an arrowhead at the start point.

Annotation
This option prompts for text to insert at the end of the *Leader* line.

Mtext
The *Multiline Text Editor* appears with this option. Text can be entered into paragraph form and you can use the *Mtext* options (see "*Mtext*," Chapter 18).

Tolerance
This option produces a feature control frame using the *Geometric Tolerances* dialog boxes (see "*Tolerance*").

Copy
You can copy existing *Text*, *Dtext*, or *Mtext* objects from the drawing to be placed at the end of the *Leader* line. The copied object is associated with the *Leader* line.

Block
An existing *Block* of your selection can be placed at the end of the *Leader* line. The same prompts as the *Insert* command are used.

None
Using this option draws the *Leader* line with no annotation.

Undo
This option undoes the last vertex point of the *Leader* line.

A *Leader* is affected by the current dimension style settings. You can control the *Leader's* arrowhead type, scale, color, etc., with the related dimensioning variables (see Chapter 29).

QLEADER

Pull-down Menu	COMMAND (TYPE)	ALIAS (TYPE)	Short-cut	Screen (side) Menu	Tablet Menu
Dimension Leader	QLEADER	LE	...	DIMNSION Leader	W,2

Qleader (quick leader) creates a leader with or without text similar in appearance to a leader created with *Leader*. However, *Qleader* allows you to specify <u>preset</u> parameters for the configuration of the text and the leader. These presets prevent you from having to specify the same parameters repeatedly when you create several leaders in a similar fashion for the current drawing. For example, you can specify what type of text object (*Mtext*, *Block*, *Tolerance*, etc.), how many line segments for the leader, what angle to draw the leader lines, what type of arrowheads, and other parameters for the leaders you create.

```
Command: qleader
Specify first leader point, or [Settings]: PICK (arrow location)
Specify next point: PICK (end of leader)
Specify next point: Enter
Specify text width <0.0000>: Enter
Enter first line of annotation text <Mtext>: Enter text or press Enter
Enter next line of annotation text: Enter text or press Enter
Command:
```

The command sequence above is used to create a one-segment leader with one string of text using the default *Qleader* settings. If you enter a value in response to the "Specify text width <0.0000>:" prompt, the value determines the width of the *Mtext* paragraph. As an alternative, you can press Enter to produce the *Mtext* dialog box.

With *Qleader*, you specify the preset configurations by pressing Enter or entering the letter "s" at the first prompt: "Specify first leader point, or [Settings]<Settings>:." This action produces the *Leader Settings* dialog box. Settings you make in this dialog box control the appearance of subsequent leaders you create in the drawing until a setting is changed. The settings are <u>saved in the current drawing</u> and are not registered in your system. There are three tabs in the dialog box. The *Annotation* tab is described next.

Annotation **Tab**

This tab allows you to control presets for the appearance of the leader text as further explained below (Fig. 28-35).

Annotation Type
This section specifies the type of text objects created (*Mtext, Copy an Object, Tolerance, Block Reference,* or *None*) and the related prompts when you use *Qleader.*

The *Mtext* option creates an *Mtext* object. Depending on your settings in the *Mtext Options* section, you can specify the width of each paragraph you create by responding to the "Specify text width <0.0000>:" prompt, use the *Mtext* dialog box, or press Enter to input several individual lines of text.

Figure 28-35

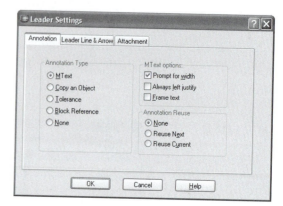

Specifying *Copy an Object* causes *Qleader* to prompt you to select an existing text object to copy. The text object is then automatically attached to the leader end.

Use the *Tolerance* button to cause *Qleader* to produce the *Geometric Tolerance* dialog box. Here you specify text and symbol content for feature control frames that are attached to the leader end (see *"Tolerance"* for details on geometric tolerancing and this dialog box).

Use the *Block Reference* option if you want to attach a *Block* to the leader end. *Qleader* prompts you to specify an existing *Block* reference defined in the current drawing by producing the "Enter block name or [?]:" prompt. Enter a question mark (?) to list existing *Blocks* in the current drawing.

To create leaders without text attached, select *None* as the *Annotation Type.*

Mtext Options
This area is enabled only when *Mtext* is the specified *Annotation Type.* You can select none, one or two options in the section.

When *Prompt for width* is checked, this setting produces the "Specify text width <0.0000>:" prompt at the command line. Enter a value for the paragraph width or press Enter to produce the *Mtext* dialog box. When this setting is not checked, the width prompt does not appear, but you can still produce the *Mtext* dialog box by pressing Enter at the "Enter first line of annotation text:" prompt.

When a leader is drawn from right to left so it is attached to the right side of the text paragraph (Fig. 28-36), AutoCAD normally right-justifies the text. Checking the *Always Left Justify* option always draws the text left justified.

Figure 28-36

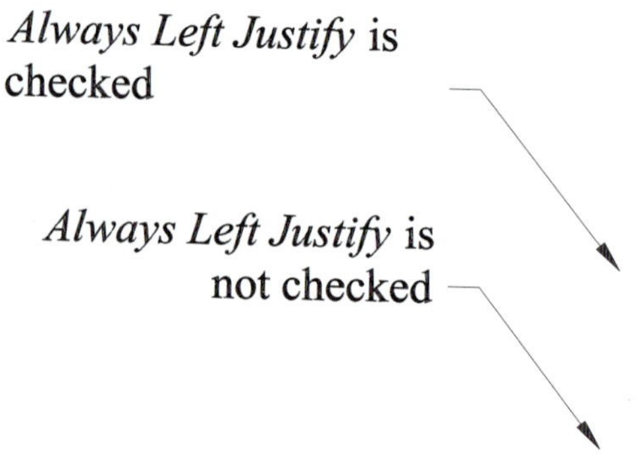

Always Left Justify is
checked

Always Left Justify is
not checked

The *Frame Text* option creates a frame around the text (Fig. 28-37).

Figure 28-37

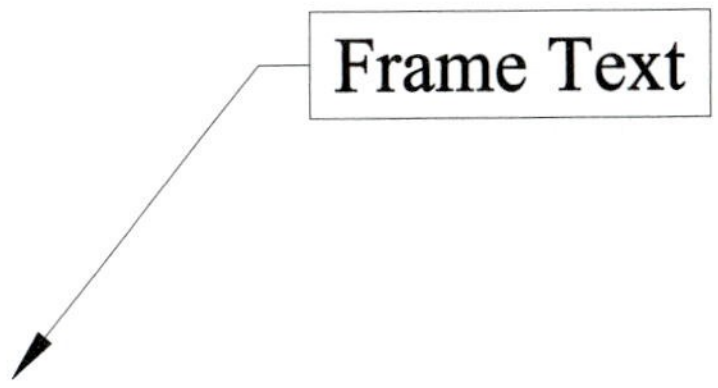

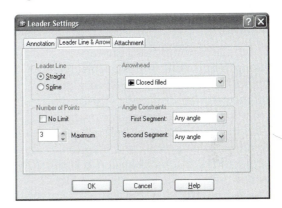

Frame Text

Annotation Reuse

When you want to add the same specification note or manufacturing note at several locations in a drawing, the *Annotation Reuse* option can save considerable time. To use the same annotation repeatedly, check *Reuse Next*. After setting this option, create one leader with the desired text. Subsequent leaders automatically appear with the same text. The *Reuse Current* button becomes active as a reminder.

Leader Line & Arrow **Tab**

Figure 28-38

This tab controls the appearance of the leader line and allows you to specify an arrowhead type to use (Fig. 28-38).

Leader Line

You can select from a *Straight* leader (*Line* segments) or a *Spline* (smooth curved) leader. Set the *Number of Points* to 2 (minimum setting) if you want one line segment (two end-points) or check *No Limit* if you are creating *Spline* leaders, since you need several points to control the shape of a curved leader.

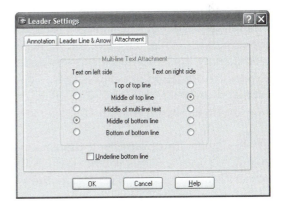

Arrowhead

In the *Arrowhead* section, use the drop-down list to select from many different possibilities for the leaders.

Angle Constraints

This section allows you to specify (force) an angle for the *Qleader* segments to be drawn. For example, selecting 45 causes the *Qleader* to always be drawn at 45 degrees.

Figure 28-39

Attachment Tab

The *Attachment* tab (Fig. 28-39) is used for setting the attachment point when *Mtext* is the selected annotation type. This tab is disabled unless *Mtext* is specified in the *Annotation* tab.

Figure 28-40

The *Multiline Text Attachment* options can be specified for cases when the text is located on the right or left side of the leader (text is located on the right when a leader is drawn from left to right and vice versa). The default position for text attachment (Fig. 28-40) is *Middle of top line* for text on the right side and *Middle of bottom line* for text on the left side.

Text on the right side
(Middle of top line)

Text on the left side
(Middle of bottom line)

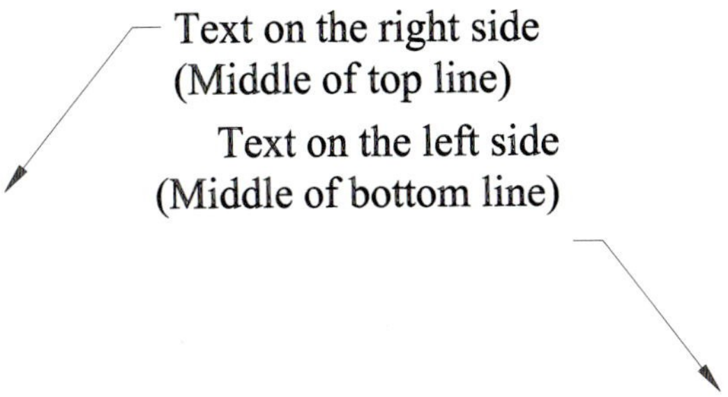

DIMORDINATE

Pull-down Menu	COMMAND (TYPE)	ALIAS (TYPE)	Short-cut	Screen (side) Menu	Tablet Menu
Dimension Ordinate	DIMORDINATE	DIMORD or DOR	...	DIMNSION Ordinate	W,3

Ordinate dimensioning is a specialized method of dimensioning used in the manufacturing of flat components such as those in the sheet metal industry. Because the thickness (depth) of the parts is uniform, only the width and height dimensions are specified as Xdatum and Ydatum dimensions.

Figure 28-41

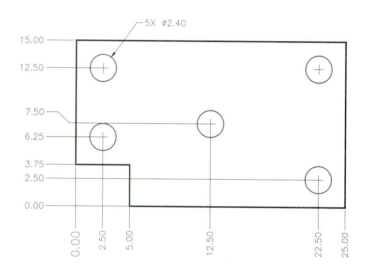

Dimordinate dimensions give an Xdatum or a Ydatum distance between object "features" and a reference point on the geometry treated as the origin, usually the lower-left corner of the part. This method of dimensioning is relatively simple to create and easy to understand. Each dimension is composed only of one leader line and the aligned numerical value.

To create ordinate dimensions in AutoCAD, the *UCS* command should be used first to establish a new 0,0 point. *UCS*, which stands for user coordinate system, allows you to establish a new coordinate system with the origin and the orientation of the axes anywhere in 3D space (see Chapters 34 and 36 for complete details). In this case, we need to change only the location of the origin and leave the orientation of the axes as is. Type *UCS* and use the *Origin* option to PICK a new origin as shown (Fig. 28-42).

Figure 28-42

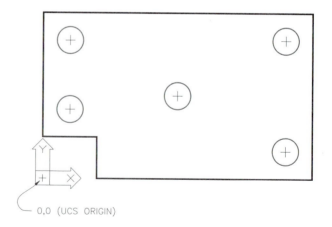

When you create a *Dimordinate* dimension, AutoCAD only requires you to (1) PICK the object "Feature" and then (2) specify the "Leader endpoint:". The dimension text is automatically aligned with the leader line.

It is not necessary in most cases to indicate whether you are creating an Xdatum or a Ydatum. Using the default option of *Dimordinate*, AutoCAD makes the determination based on the direction of the leader you specify (step 2). If the leader is <u>perpendicular</u> (or almost perpendicular) to the X axis, an Xdatum is created. If the leader is (almost) <u>perpendicular</u> to the Y axis, a Ydatum is created.

The command syntax for a *Dimordinate* dimension is this:

    Command: **dimordinate**
    Specify feature location: **PICK**
    Specify leader endpoint or
    [Xdatum/Ydatum/Mtext/Text/Angle]:
    PICK
    Dimension text = *n.nnnn*
    Command:

A *Dimordinate* dimension is created in Figure 28-43 by PICKing the object feature and the leader endpoint. That's all there is to it. The dimension is a Ydatum; yet AutoCAD automatically makes that determination, since the leader is perpendicular to the Y axis. It is a good practice to turn *ORTHO* or *POLAR ON* in order to ensure the leader lines are drawn horizontally or vertically.

An Xdatum *Dimordinate* dimension is created in the same manner. Just PICK the object feature and the other end of the leader line (Fig. 28-44). The leader is perpendicular to the X axis; therefore, an Xdatum dimension is created.

The leader line does not have to be purely horizontal or vertical. In some cases, where the dimension text is crowded, it is desirable to place the end of the leader so that sufficient room is provided for the dimension text. In other words, draw the leader line at an angle. (*ORTHO* must be turned *Off* to specify an offset leader line.) AutoCAD automatically creates an offset in the leader as shown in the 7.50 Ydatum dimension in Figure 28-45. As long as the leader is more perpendicular to the X axis, an Xdatum is drawn and vice versa.

Figure 28-43

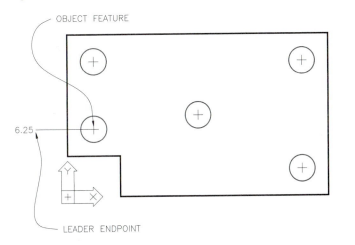

Figure 28-44

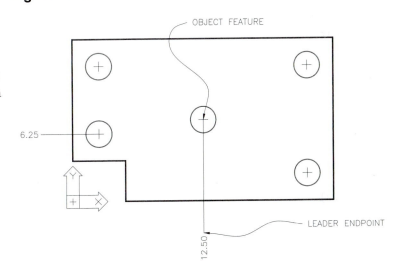

Figure 28-45

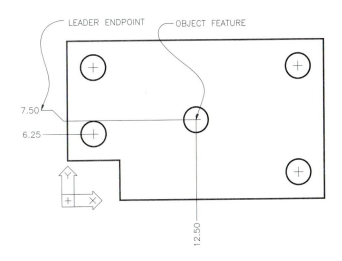

Xdatum/Ydatum

The *Xdatum* and *Ydatum* options are used to specify one of these dimensions explicitly. This is necessary in case the leader line that you specify is more perpendicular to the <u>other</u> axis that you want to measure along. The command syntax would be as follows:

Command: ***dimordinate***
Specify feature location: **PICK**
Specify leader endpoint or [Xdatum/Ydatum/Mtext/Text/Angle]: ***X***
Specify leader endpoint or [Xdatum/Ydatum/Mtext/Text/Angle]: **PICK**
Dimension text = *n.nnnn*
Command:

Mtext/Text

The *Mtext* and *Text* options operate similar to the same options of other dimensioning commands. The *Mtext* option summons the *Multiline Text Editor* and the *Text* option uses Command line format. Any text placed inside the < > characters overrides the AutoCAD-measured dimensional value. Any additional text entered on either side of the < > symbols is treated as a prefix or suffix to the measured value. All other options in the *Mtext* dialog box are usable for ordinate dimensions.

QDIM

Pull-down Menu	COMMAND (TYPE)	ALIAS (TYPE)	Short-cut	Screen (side) Menu	Tablet Menu
Dimension *QDIM*	*QDIM*	...	...	...	*W,1*

2004

Qdim (Quick Dimension) creates associative dimensions in AutoCAD 2004. *Qdim* simplifies the task of dimensioning by creating <u>multiple dimensions with one command</u>. *Qdim* can create *Continuous, Baseline, Radius, Diameter,* and *Ordinate* dimensions.

Command: ***qdim***
Associative dimension priority = Endpoint
Select geometry to dimension: **PICK**
Select geometry to dimension: **Enter**
Specify dimension line position, or [Continuous/Staggered/Baseline/Ordinate/Radius/Diameter/datumPoint/Edit/seTtings] <Continuous>: **PICK**

When *Qdim* prompts to "Select geometry to dimension," you can specify the geometry using any selection method such as a window, crossing window, or pickbox. For example, *Qdim* is used to dimension the geometry shown in Figure 28-46. Here a crossing window is used to select several features on one side of the shape. The only other step is to select where (how far from the geometry) you want the dimensions to appear. The resulting dimensions differ based on the *Qdim* options used and the current *Dimension Style* settings (see Chapter 29 for information on Dimension Styles).

Each *Qdim* option and the resulting dimensions are described next.

Figure 28-46

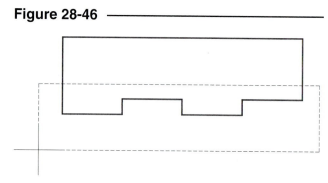

SeTtings

This option specifies the priority object snap *Qdim* uses to automatically draw extension line origins. *Endpoint* is the default; however, *Intersection* is useful if you want to dimension to lines (such as centerlines) that cross your objects.

Continuous

The *Continuous* option creates a string of dimensions in one row, similar to using *Dimlinear* followed by *Dimcontinue*. The resulting dimensions for our example are shown in Figure 28-47.

Baseline

Use the *Baseline* option of *Qdim* to create a stack of dimensions all generated from one baseline, similar to using *Dimlinear* followed by *Dimbaseline* (Fig. 28-48). The baseline end of the stack of dimensions is nearest the UCS (User Coordinate System) origin (0,0) by default. To change the baseline end of the stack of dimensions, use the *datumPoint* option of *Qdim* to specify a new origin point for the desired baseline end of the geometry (see the *datumPoint* option later in this section).

Staggered

The *Staggered* option creates a stack of dimensions alternating from one end or alternating outward from a central feature (Fig. 28-49). This type of dimension series is generally used only for parts that are symmetrical (denoted by a centerline), otherwise the dimension set would be incomplete. Note that the extension line sets (two extension lines for each dimension) are generated from the selected geometric features; therefore, dimensioning geometry with an odd number of features results in one feature not being dimensioned.

Ordinate

Using *Qdim* with the *Ordinate* option creates a string of ordinate dimensions, similar to using *Dimordinate* (Fig. 28-50). This style of dimensioning is used in manufacturing primarily for stamping or cutting parts of uniform thickness, such as in the sheet metal industry.

In ordinate dimensioning, each dimensional value specifies one "ordinate," or distance from a reference corner or datum point. By default, AutoCAD uses the current UCS or WCS origin as the datum point (note the values in Figure 28-50). You can use the *datumPoint* option to designate one corner of the part to use as the reference origin (see *datumPoint* next).

Figure 28-47

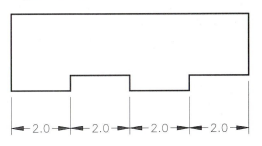

Figure 28-48

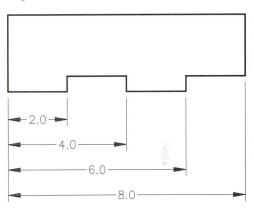

Figure 28-49

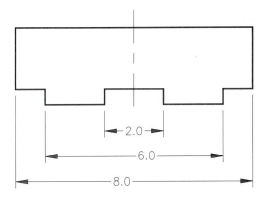

Figure 28-50

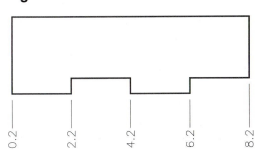

datumPoint

Typically, use the *datumPoint* option to specify a reference corner, or origin, of the geometry before creating ordinate dimensions. At the "Select new datum point:" prompt, use an *Endpoint* or other *Osnap* mode to select the desired datum point on the part. In Figure 28-51, the lower-left corner of the geometry was specified as the *datumPoint* before creating the ordinate dimensions.

Figure 28-51 ───────

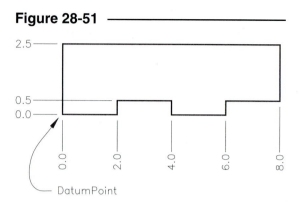

Radius

The *Continuous* option creates a series of radial dimensions similar to using *Dimradius* (Fig. 28-52). You can select multiple arcs to dimension with a window rather than picking the arcs individually because any linear geometry is automatically filtered out and not dimensioned. The resulting dimensions all have similar dimension line features, such as leader line length and other variables.

Figure 28-52 ───────

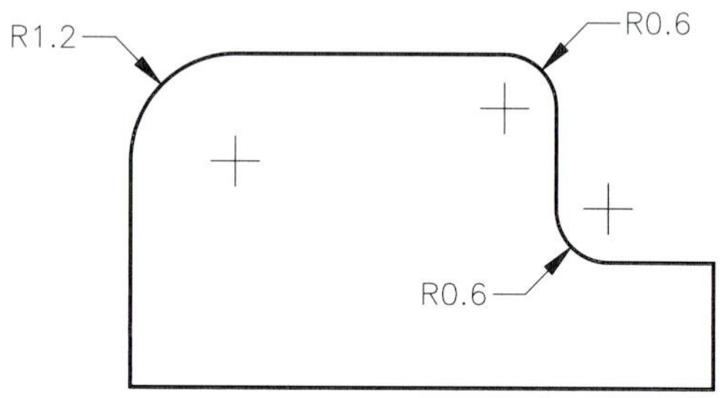

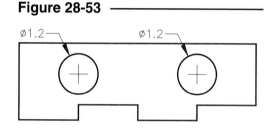

Diameter

Use the *Diameter* option to create a series of diametrical dimensions for circles or holes, similar to using *Dimdiameter*. Since any linear geometry is automatically filtered out when you select geometry to dimension, you can use a window rather than picking the circles individually. For example, selecting all geometry displayed in Figure 28-53 results in only two diametrical dimensions.

Figure 28-53 ───────

Edit

Once multiple dimensions are created, such as in Figure 28-47, you can use the *Edit* option to *Add* or *Remove* dimensions. Enter *E* at the *Qdim* prompt to select the options.

Figure 28-54 ───────

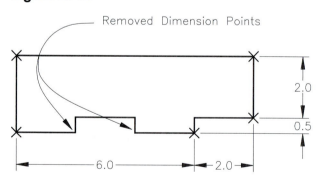

```
Command: qdim
Select geometry to dimension: PICK
Select geometry to dimension: Enter
Specify dimension line position, or
[Continuous/Staggered/Baseline/Ordinate/Radius/ Diameter/datumPoint/Edit] <Continuous>: e
Indicate dimension point to remove, or [Add/eXit] <eXit>:
```

When the *Edit* option is invoked, small markers appear at the extension line origins called dimension points (Fig. 28-54). You can *Add* or *Remove* dimensions by adding or removing dimension points. For example, using the *Remove* option, three dimensions are reduced to one by removing the indicated dimension points.

TOLERANCE

Pull-down Menu	COMMAND (TYPE)	ALIAS (TYPE)	Short-cut	Screen (side) Menu	Tablet Menu
Dimension Tolerance...	*TOLERANCE*	*TOL*	...	*DIMNSION Toleranc*	*X,1*

Geometric dimensioning and tolerancing (GDT) has become an essential component of detail drawings of manufactured parts. Standard symbols are used to define the geometric aspects of the parts and how the parts function in relation to other parts of an assembly. The *Tolerance* command in AutoCAD produces a series of dialog boxes for you to create the symbols, values, and feature control frames needed to dimension a drawing with GDT. Invoking the *Tolerance* command produces the *Symbol* dialog box and the *Geometric Tolerance* dialog box (see Figures 28-57 and 28-58 on the next page). These dialog boxes can also be accessed by the leader commands. Unlike general dimensioning in AutoCAD, geometric dimensioning is <u>not associative</u>; however, you can use *Ddedit* to edit the components of an existing feature control frame.

Some common examples of geometric dimensioning may be a control of the flatness of a surface, the parallelism of one surface relative to another, or the positioning of a group of holes to the outside surfaces of the part. Most of the conditions are controls that cannot be stated with the other AutoCAD dimensioning commands.

This book does not provide instruction on how to use geometric dimensioning; rather, it explains how to use AutoCAD to apply the symbology. The ASME Y14.5M - 1994 Dimensioning and Tolerancing standard is the authority on the topic in the United States. The symbol application presented in AutoCAD 2004 is actually based on the 1982 release of the Y14.5 standard. Some of the symbols in the 1994 version are modified from the 1982 standard.

The dimensioning symbols are placed in a feature control frame (FCF). This frame is composed of a minimum of two sections and a maximum of three different sections (Fig. 28-55).

Figure 28-55

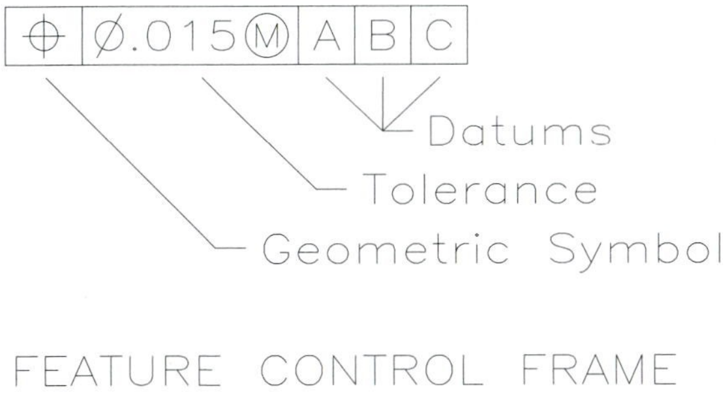

$\oplus$ | $\emptyset$.015 Ⓜ | A | B | C

Datums
Tolerance
Geometric Symbol

FEATURE CONTROL FRAME

The first section of the FCF houses one of 14 possible geometric characteristic symbols (Fig. 28-56).

The second section of the FCF contains the tolerance information. The third section includes any datum references.

Figure 28-56

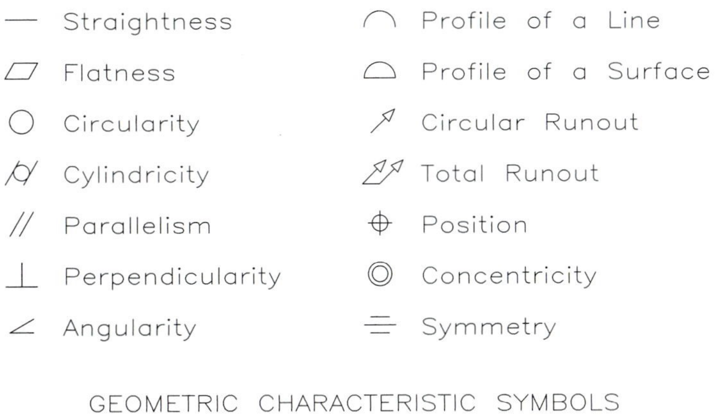

—	Straightness	⌒	Profile of a Line
▱	Flatness	⌒	Profile of a Surface
○	Circularity	⟋	Circular Runout
⌀	Cylindricity	⟋⟋	Total Runout
//	Parallelism	⊕	Position
⊥	Perpendicularity	◎	Concentricity
∠	Angularity	=	Symmetry

GEOMETRIC CHARACTERISTIC SYMBOLS

In AutoCAD, the *Tolerance* command allows you to specify the values and symbols needed for a feature control frame. The lines comprising the frame itself are automatically generated. Invoking *Tolerance* by any method produces the *Geometric Tolerance* dialog box (Fig. 28-57).

The *Geometric Tolerance* dialog box (Fig. 28-57) appears for you to specify the tolerance values and datum. The areas of the *Geometric Tolerance* dialog box are explained below.

Figure 28-57 ———————

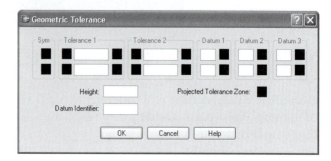

Sym
Click in this empty tile to produce the *Symbol* dialog box (Fig. 28-58). Here you select the geometric characteristic symbol to use for the new feature control frame. The selected symbol then appears in the *Sym* tile in the *Geometric Tolerance* dialog box.

Figure 28-58 ———————

Tolerance 1
This area is used to specify the first tolerance value in the feature control frame. This value specifies the amount of allowable deviation for the geometric feature. The three sections in this cluster are:

Dia (first box)
If a cylindrical tolerance zone is specified, a diameter symbol can be placed before the value by clicking in this area.

Value (second box)
Enter the tolerance value in this edit box.

MC (third box)
A material condition symbol can be placed after the value by choosing this tile. The *Material Condition* dialog box appears for your selection (Fig. 28-59):

Figure 28-59 ———————

M   Maximum material condition
L   Least material condition
S   Regardless of feature size

Once you select a material condition or cancel, the *Geometric Tolerance* dialog box reappears.

Tolerance 2
Using this section creates a second tolerance area in the feature control frame. GDT standards, however, require only one tolerance area. The options in this section are identical to *Tolerance 1*.

Datum 1
This section allows you to specify the primary datum reference for the current feature control frame. A datum is the exact (theoretical) geometric feature that the current feature references to establish its tolerance zone.

Datum (first box)
Enter the primary datum letter (A, B, C, etc.) for the FCF in this edit box.

MC (second box)
Choosing the *MC* tile allows you to specify a material condition modifier for the datum using the *Material Condition* dialog box (Fig. 28-59).

Datum 2
This section is used if you need to create a second datum reference in the feature control frame. The options are identical to those for *Datum 1*.

Datum 3
Use this section if you need to specify a third datum reference for the current feature control frame.

Height
The edit box allows entry of a value for a projected tolerance zone in the feature control frame. A projected tolerance zone specifies a permissible cylindrical tolerance that is projected above the surface of the part. The axis of the control feature must remain within the stated tolerance.

Projected Tolerance Zone
Click this section to insert a projected tolerance zone symbol (a circled "P") after the value.

Datum Identifier
This option creates a datum feature symbol. The edit box allows entry of the value (letter) indicating the datum. A hyphen should be included on each side of the datum letter, such as "-A-".

After using the dialog boxes and specifying the necessary values and symbols, AutoCAD prompts for you to PICK a location on the drawing for the feature control symbol or datum identifier:

    Enter tolerance location: **PICK**

The feature control symbol or datum identifier is drawn as specified.

Editing Feature Control Frames

The AutoCAD-generated feature control frames are not associative and can be *Erased*, *Moved*, or otherwise edited without consequence to the related geometry objects. Feature control frames created with *Tolerance* are treated as one object; therefore, editing an individual component of a feature control frame is not possible unless *Ddedit* is used. If *Dblclkedit* is *On*, you can double-click on the FCF.

Ddedit is normally used to edit text, but if an existing feature control frame is selected in response to the "Select an annotation object" prompt, the *Geometric Tolerance* dialog box appears. The edit boxes in this dialog box contain the values and symbols from the selected FCF. Making the appropriate changes updates the selected FCF. (See Chapter 18 for detailed information on *Ddedit*.)

Basic Dimensions

Basic dimensions are often required in drawings using GDT. Basic dimensions in AutoCAD are created by using the *Dimension Style Manager* (or by entering a negative value in the *DIMGAP* variable). The procedure is explained briefly in the following example (application 5) and discussed further in Chapter 29, Dimension Styles and Variables.

GDT-Related Dimension Variables

Some aspects of how AutoCAD draws the GDT symbols can be controlled using dimension variables and dimension styles (discussed fully in Chapter 29). These variables are:

DIMCLRE	Controls the color of the FCF
DIMCLRT	Controls the color of the tolerance text
DIMGAP	Controls the gap between the FCF and the text, and controls the existence of a basic dimension box
DIMTXT	Controls the size of the tolerance text
DIMTSTSTY	Controls the style of the tolerance text

Geometric Dimensioning and Tolerancing Example

Six different geometric dimensioning and tolerancing examples are shown on the SPACER drawing in Figure 28-60. Each example is indicated on the drawing and in the following text by a number, 1 through 6. The purpose of the examples is to explain how to use the geometric dimensioning features of AutoCAD. In each example, any method shown in the *Tolerance* command table can be used to invoke the command and dialog boxes.

Figure 28-60 ————————————————————————————

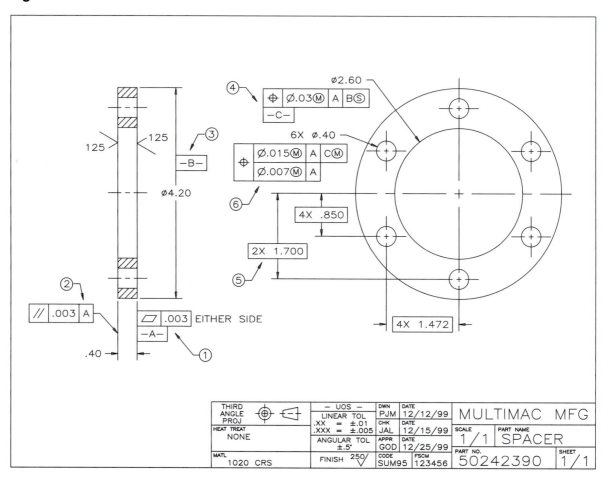

1. Flatness Application

The first specification applied is flatness. In addition to one of the surfaces being controlled for flatness, it is also identified as a datum surface. Both conditions are applied in the same application:

Command: **tolerance** (The *Geometric Tolerance* dialog box appears.)
PICK the **Sym** box. When the *Symbol* dialog box appears, select the *flatness* symbol. The *Geometric Tolerance* dialog box reappears.
PICK the **Tolerance 1** edit box and type ".003" as the tolerance.
PICK the **Datum Identifier** box and type "-A-".
PICK the **OK** button. (The *Geometric Tolerance* dialog box disappears.)
Enter tolerance location: **PICK** (PICK a point on the right extension line where you want the FCF to be attached, Fig. 28-62, on the next page.)
Command:

Figure 28-61 ——————

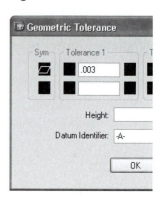

Because either side of the SPACER can be chosen for the flatness specification, the note "EITHER SIDE" is entered next to the FCF in Figure 28-60.

Figure 28-62

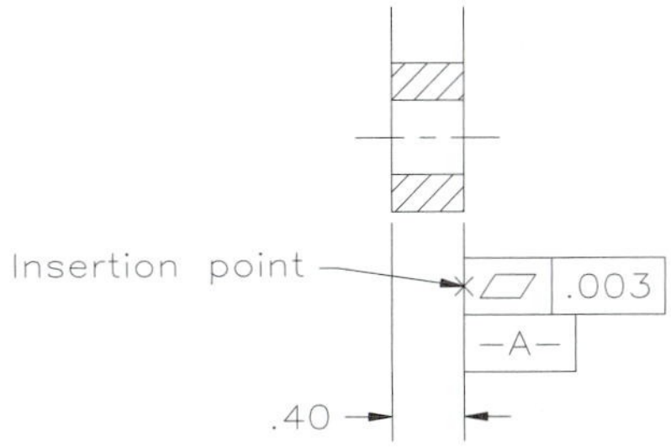

2.   Parallelism Application
The second application is a parallelism specification to the opposite side of the part. This control is applied using a *Leader*:

> Command: ***leader***
> Specify leader start point: **PICK** (PICK a point on the left extension line, Figure 28-60.)
> Specify next point: **PICK** (PICK a location slightly up and to the left.)
> Specify next point or [Annotation/Format/Undo] <Annotation>: **Enter**
> Enter first line of annotation text or <options>: **Enter**
> Enter an annotation option [Tolerance/Copy/Block/None/Mtext] <Mtext>: ***t*** (The *Geometric Tolerance* dialog box appears.) PICK the ***Sym*** box. When the *Symbol* dialog box appears, select the ***parallelism*** symbol. The *Geometric Tolerance* dialog box reappears.
> PICK the ***Tolerance 1*** edit box and type ".003" as the tolerance.
> PICK the ***Datum 1*** edit box and type "**A**".
> PICK the ***OK*** button. (The *Geometric Tolerance* dialog box clears and the parallelism FCF is placed to the left of the leader.)
> Command:

If the leader had projected to the right of the controlled surface, the FCF would be placed on the right of the leader.

3.   Datum Application
The third application is the B datum on the 4.20 diameter. This datum is used later in the position specification.

> Command: ***tolerance*** (The *Geometric Tolerance* dialog box appears.)
> PICK the ***Datum Identifier*** edit box and type "**-B-**". PICK the ***OK*** button.
> Enter tolerance location: **PICK** (PICK a point on the dimension line of the 4.20 diameter.)
> Command:

4.   Position Application
The fourth application is a position specification of the 2.60 diameter hole. It is identified as a C datum because it will be used in the position specification of the six mounting holes. This specification creates a relationship between inside and outside diameters and a perpendicularity requirement to datum A.

> Command: ***tolerance*** (The *Geometric Tolerance* dialog box appears.)
> PICK the ***Sym*** box so the *Symbol* dialog box appears.
> PICK the ***Position*** symbol. (The *Geometric Tolerance* dialog box reappears.)
> PICK the *Tolerance 1* ***Dia*** box. (This places a diameter symbol in front of the tolerance.)
> PICK the *Tolerance 1* ***Value*** box and type ".03" as a tolerance.
> PICK the *Tolerance 1* ***MC*** box. (The *Material Condition* dialog box appears.)
> PICK the circled ***M***.
> PICK the *Datum 1* ***Datum*** box and type "**A**".
> PICK the *Datum 2* ***Datum*** box and type "**B**".
> PICK the *Datum 2* ***MC*** box to produce the *Material Condition* dialog box.
> PICK the circled ***S***.
> PICK the ***Datum Identifier*** box and type "**-C-**".
> PICK the ***OK*** button to return to the drawing.
> Enter tolerance location: **PICK** (PICK a point that places the FCF below the 2.60 diameter [Fig. 28-60].)

If you need to move the FCF with *Move* or grips, select any part of the FCF or its contents since it is treated as one object.

5. **Basic Dimension Application**

    The fifth application concerns basic dimensions. Position uses basic (theoretically exact) location dimensions to locate holes because the tolerances are stated in the FCF. Before applying these dimensions, the *Dimension Style* must be changed. (See Chapter 29 for a complete discussion of Dimension Styles.)

    Command: ***ddim*** (The *Dimension Style Manager* appears.)
    PICK the *Modify* button.
    PICK the *Tolerances* tab.
    PICK the *Method* drop-down list and select *Basic*.
    PICK the *OK* button, then *Close* the *Dimension Style Manager*.
    Command: ***dimlinear*** (Use the *dimlinear* command to apply each of the three linear basic dimensions shown in Fig. 28-60.)

6. **Composite Position Application**

    The sixth application is the composite position specification. A composite specification consists of two separate lines but only one geometric symbol. Achieving this result requires entering data on the top and bottom lines in the *Geometric Tolerance* dialog box.

    Command: ***tolerance*** (The *Geometric Tolerance* dialog box appears.)
    PICK the *Sym* box so the *Symbol* dialog box appears.
    PICK the *position* symbol. (The *Geometric Tolerance* dialog box reappears.)
    PICK the *Tolerance 1 Dia* box in the *Geometric Tolerance* dialog box.
    PICK the *Tolerance 1 Value* box and enter ".015" as a tolerance.
    PICK the *Tolerance 1 MC* box. (The *Material Condition* dialog box appears.)
    PICK the circled *M*.
    PICK the *Datum 1 Datum* box and type "A".
    PICK the *Datum 2 Datum* box and type "C".
    PICK the *Datum 2 MC* box. (The *Material Condition* dialog box appears.)
    PICK the circled *M*.
    PICK the *Sym* box under the position symbol to produce the *Symbol* dialog box.
    PICK the *position* symbol.
    PICK the *Tolerance 1 Dia* box on the second line.
    PICK the *Tolerance 1 Value* box on the second line and type ".007" as a tolerance.
    PICK the *Tolerance 1 MC* box on the second line. (The *Material Condition* dialog box appears.)
    PICK the circled *M*.
    PICK the *Datum 1 Datum* box and type "A".
    PICK the *OK* button. (The *Geometric Tolerance* dialog box clears.)
    Enter tolerance location: **PICK** (PICK a point that places the FCF below the .40 diameter, Fig. 28-60.)
    Command:

Datum Targets

The COVER drawing (Fig. 28-63) uses datum targets to locate the part in 3D space. AutoCAD provides no commands to apply datum targets. The best way to apply these symbols is to use *Blocks* and attributes. The target circle diameter is 3.5 times the letter height. The dividing line is always drawn horizontally through the center of the circle.

Figure 28-63

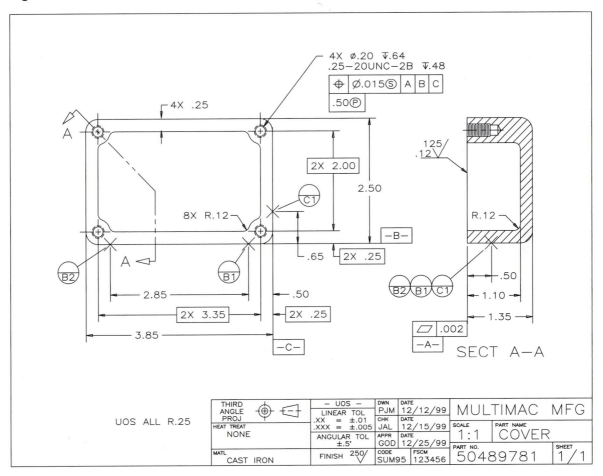

Projected Tolerance Zone Application

The COVER drawing also uses a projected tolerance with the position specification. A projected tolerance zone is used primarily with internally threaded holes and dowel holes. In this case, the concern is not the position and orientation of the threaded hole, but rather the position and orientation of the fastener inserted into the hole, specifically the shank of the fastener; therefore, the tolerance zone projects above the surface of the part and not within the part. This condition is especially important when the hole allowance is small.

Command: ***tolerance*** (The *Geometric Tolerance* dialog box appears.)
PICK the *Sym* box so the *Symbol* dialog box appears.
PICK the *Position* symbol. (The *Geometric Tolerance* dialog box reappears.)
PICK the *Height* box in the *Geometric Tolerance* dialog box and enter ".50".
PICK the *Projected Tolerance Zone* box. A circled *P* appears.
PICK the *OK* button in the *Geometric Tolerance* box.
Enter tolerance location: **PICK** (PICK a point that places the FCF below the .20 diameter, Fig. 28-63.)
Command:

This section has presented only the mechanics of GDT symbol application in AutoCAD. Geometric dimensioning and tolerancing can be a complicated and detailed subject to learn. Many different combination possibilities may appear in a feature control frame. However, for any one feature, there are very few possibilities. Knowing which of the possibilities is best comes from a fundamental knowledge of GDT. Knowledge of GDT increases your understanding of design, tooling, manufacturing, and inspection concepts and processes.

DIM

In AutoCAD releases previous to Release 13, dimensioning commands had to be entered at the "Dim:" prompt and could not be entered at the "Command:" prompt. The "Dim:" prompt puts AutoCAD in the dimensioning mode so earlier release dimensioning commands can be used. This convention can also be used in AutoCAD 2004. The *DIM* command must be typed in AutoCAD 2004 to produce the "Dim:" prompt. Enter *E* or *Exit* or press Escape to exit the dimensioning mode and return to the "Command:" prompt. The syntax is as follows:

    Command: **dim**
    Dim:

Release 12 and previous dimensioning command names do not have the "dim" prefix; therefore, using the "Dim:" prompt is required for them. Almost all of the Release 12 commands have the same name as the AutoCAD 2004 version but without the "dim" prefix. For example, *Dimdiameter* is *"Diameter"* in Release 12. In AutoCAD 2004, you can enter these older dimensioning command names (without the "dim" prefix) at the "Dim:" prompt; for example,

    Command: **dim**
    Dim: **diameter**

With the exception of a few newer dimensioning commands and options, the Release 12 dimensioning commands are the same as the AutoCAD 2004 versions without the "dim" prefix. However, several other older commands can be used in AutoCAD 2004 at the "Dim:" prompt.

UNDO or *U*
Erases the last dimension objects or dimension variable setting.

UPDATE
Performs the same action as *Dimstyle, Apply.*

STYLE
Changes the current <u>text</u> style.

HORIZONTAL
Performs the same action as *Dimlinear, Horizontal.*

VERTICAL
Performs the same action as *Dimlinear, Vertical.*

ROTATED
Performs the same action as *Dimlinear, Rotate.*

EDITING DIMENSIONS

Associative Dimensions

AutoCAD creates associative dimensions by default. Associative dimensions contain "definition points" that are associated to the related geometry. For example, when you use the *Dimlinear* command and then *Osnap* to two points on the drawing geometry in response to the "Specify first extension line origin" and "Specify second extension line origin," two definition points are created at the ends of the extension lines and are attached, or "associated," to the geometry. You may have noticed these small dots at the ends of the extension lines. The definition points are generally located where you PICK, such as at the extension line origins for most dimensioning commands, or the arrowhead tips for the leader commands and the arrowhead tips and centers for diameter and radius commands. All of the dimensioning commands are associative by default except *Qleader*, *Leader*, *Center Mark*, and *Tolerance*.

When you create the first associative dimension, AutoCAD automatically creates a new layer called DEF-POINTS. The layer is set to not plot and that property cannot be changed. AutoCAD manages this layer automatically, so do not attempt to alter this layer.

There are two advantages to associative dimensions. First, if the associated geometry is modified by typical editing methods, the associated definition points automatically change; therefore, the dimension line components (dimension lines and extension lines) automatically change and the numerical value automatically updates to reflect the geometry's new length, angle, radius, or diameter. Second, existing associative dimensions in a drawing automatically update when their dimension style is changed (see Chapter 29 for information on dimension styles).

Examples of typical editing methods that can affect associative dimensions are: *Chamfer*, *Extend*, *Fillet*, *Mirror*, *Move*, *Rotate*, *Scale*, *Stretch*, *Trim* (linear dimensions only), *Array* (if rotated in a *Polar* array), and grip editing options.

Since AutoCAD 2002, dimensions can have three possible associativity settings controlled by the *DIMAS-SOC* system variable.

Associative dimensions	(*DIMASSOC* = 2) Associative dimensions automatically adjust their locations, orientations, and measurement values when the geometric objects associated with them are modified. Associative dimensions in a drawing will update when the dimension style is modified. These fully associative dimensions are used in AutoCAD 2002 and 2004.
Nonassociative dimensions	(*DIMASSOC* = 1) Nonassociative dimensions do not change automatically when the geometric objects they measure are modified. These dimensions have definition points, but the definition points must be included (selected) if you want the dimensions to change when the geometry they measure is modified. The existing dimensions in a drawing will update when the dimension style is modified. In AutoCAD 2000 and previous releases, these dimensions were called associative.
Exploded dimensions	(*DIMASSOC* = 0) This type of dimension is actually a group of separate objects (lines, arrows, text) rather than a single dimension object. These dimensions do not have definition points and do not change when the geometry is modified. Exploded dimensions do not update when the dimension style is changed. These dimensions can also be created using *Explode* on existing associative or nonassociative dimensions.

2002

Characteristics of the three types of dimensions (associative, nonassociative, and exploded) are explained in the three illustrations here.

An associative dimension in AutoCAD 2002 and 2004 is fully associative; that is, if the geometry is changed (in this case by the *Scale* command) the related dimensions components (dimension lines, extension lines, and measured value) automatically update. The shape in Figure 28-64 is a *Pline*, so the *Scale* command affects the entire object. Note that only the single *Pline* object is selected to scale, not the dimensions.

Figure 28-65 displays a similar action—using *Scale* to change an object—but with nonassociative dimensions. Nonassociative dimensions do not automatically change when the related objects are modified unless the dimension definition points are included in the selection set. Note in this case the *Pline* and only two vertical dimensions (1.25 and 0.75) are selected and are therefore scaled, while the horizontal dimensions (1.50 and 1.00) are not selected and therefore are not scaled.

The characteristics of exploded dimensions are explained in Figure 28-66. Exploded dimensions are composed of several individual and unrelated components (arrows, text, and lines). When *Scale* is used, only the selected individual dimension components are affected, so the

Figure 28-64 ────────────────────────────────

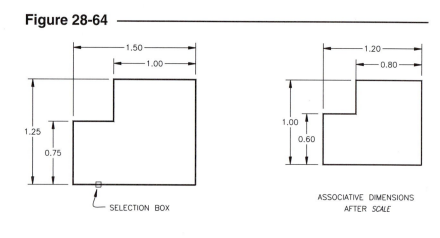

ASSOCIATIVE DIMENSIONS
AFTER *SCALE*

Figure 28-65 ────────────────────────────────

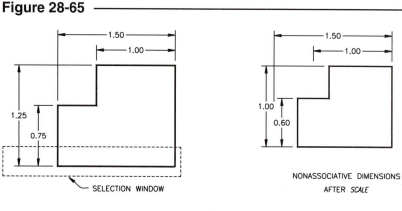

NONASSOCIATIVE DIMENSIONS
AFTER *SCALE*

Figure 28-66 ────────────────────────────────

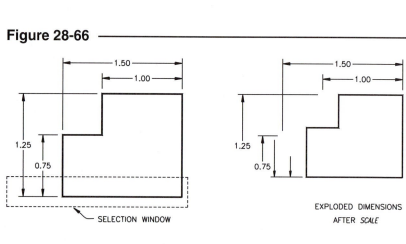

EXPLODED DIMENSIONS
AFTER *SCALE*

resulting dimensions are unusable—essentially, the dimensions fall apart. In addition, exploded dimensions do not update if their original dimension style is modified.

If you want to determine whether a dimension is associative or nonassociative, you can use the *Properties* palette or the *List* command. You can also use the *Quick Select* dialog box to filter the selection of associative or nonassociative dimensions. A dimension is considered associative even if only one end of the dimension is associated with a geometric object. Associativity is *not* maintained between a dimension and a block reference if the block is redefined. Associativity is not possible when dimensioning *Multiline* objects.

Remember that the associativity status for newly created dimensions (associative, nonassociative, or exploded) is controlled by the *DIMASSOC* system variable (2, 1, 0, respectively). You can set the *DIMASSOC* system variable at the Command line or use the *Options* dialog box (Fig. 28-67). In the *User Preferences* tab, a check in the *Make new dimensions associative* box (lower right) sets *DIMASSOC* to 2, while no check in this box sets *DIMASSOC* to 1. The *DIMASSOC* setting does not affect existing dimensions, only subsequently created ones. The *DIMASSOC* system variable is new with AutoCAD 2002 and replaces the *DIMASO* variable. The *DIMASSOC* variable is not saved with a dimension style. The *DIMASO* variable is obsolete since AutoCAD 2002 (see Chapter 29 for more information).

Figure 28-67

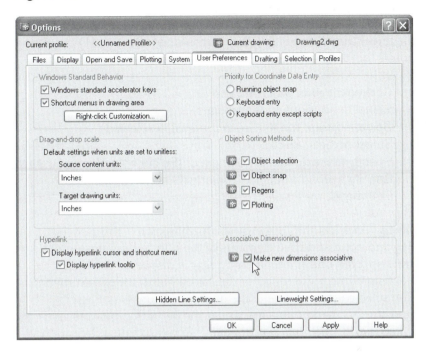

DIMREASSOCIATE

Pull-down Menu	COMMAND (TYPE)	ALIAS (TYPE)	Short-cut	Screen (side) Menu	Tablet Menu
Dimension Reassociate Dimension	*DIMREASSOCIATE*	...	...	...	...

AutoCAD 2002 and 2004 provide the *Dimreassociate* command. With *Dimreassociate*, you can convert nonassociative dimensions to associative dimensions. In addition, you can change the feature of the drawing geometry (line endpoint, etc.) that an existing definition point is associated with. This command is especially helpful in several situations, such as:

if you are updating a drawing completed in AutoCAD 2000 or a previous release and want to convert the dimensions to fully associative AutoCAD 2002 and 2004 dimensions;

if you created nonassociative dimensions in AutoCAD 2002 and 2004 but want to convert them to associative dimensions;

if you altered the position of an associative dimension's definition points accidentally with grips or other method and want to reattach them;

if you want to change the attachment point for an existing associative dimension.

2002

The *Dimreassociate* command prompts for the geometric features (line endpoints, etc.) that you want the dimension to be associated with. Depending on the type of dimension (linear, radial, angular, diametrical, etc.), you are prompted for the appropriate geometric feature as the attachment point. For example, if a linear dimension is selected to reassociate, AutoCAD prompts you to specify (select attachment points for) the extension line origins.

```
Command: dimreassociate
Select dimensions to reassociate ...
Select objects: PICK  (linear dimension selected)
Select objects: Enter
Specify first extension line origin or [Select object] <next>: PICK
Specify second extension line origin <next>: PICK
Command:
```

When you select the "dimensions to reassociate," the associative and nonassociative elements of the selected dimension(s) are displayed one at a time. A marker is displayed for each extension line origin (or other definition point) for each dimension selected (Fig. 28-68). If the definition point is not associated, an "X" marker appears, and if the definition point is associated, an "X" within a box appears. To reassociate a definition point, simply pick the new object feature you want the dimension to be attached to. If an extension line is already associated and you do not want to make a change, press Enter when its definition point marker is displayed. Figure 28-68 displays a partly associated linear dimension with its extension line origin markers during the *Dimreassociate* command.

Figure 28-68

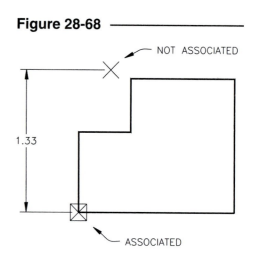

DIMDISASSOCIATE

Pull-down Menu	COMMAND (TYPE)	ALIAS (TYPE)	Short-cut	Screen (side) Menu	Tablet Menu
...	DIMDISASSOCIATE	...	...	...	...

The *Dimdisassociate* command automatically converts selected associative dimensions to nonassociative. The command prompts for you to select only the dimensions you want to disassociate, then AutoCAD makes the conversion and reports the number of dimensions disassociated.

```
Command: dimdisassociate
Select dimensions to disassociate ...
Select objects: PICK
Select objects: Enter
nn disassociated.
Command:
```

You can use any selection method including *Qselect* to "Select objects." *Dimdisassociate* filters the selection set to include only associative dimensions that are in the current space (model space or paper space layout) and not on locked layers.

DIMREGEN

Pull-down Menu	COMMAND (TYPE)	ALIAS (TYPE)	Short-cut	Screen (side) Menu	Tablet Menu
...	*DIMREGEN*	...	...	...	

Dimregen updates the locations of all associative dimensions in the current drawing. *Dimregen* does not alter the associativity features of dimensions; it affects only the <u>display of associative dimensions</u>. *Dimregen* may be needed to regenerate the display of associative dimensions in three cases:

after panning or zooming with a wheel mouse within a viewport in a paper space layout when dimensions created in paper space are associated with drawing objects in model space (see Chapter 29, "Dimensioning in Paper Space Layouts");

after opening an AutoCAD 2002 and 2004 drawing that has been modified with a previous version of AutoCAD and the dimensioned objects have been modified;

after opening a drawing containing *Xrefs* if the external reference geometry is dimensioned in the current drawing and the *Xrefs* have been modified.

Grip Editing Dimensions

Grips can be used effectively for editing dimensions. Any of the grip options (STRETCH, MOVE, ROTATE, SCALE, and MIRROR) is applicable. Depending on the type of dimension (linear, radial, angular, etc.), grips appear at several locations on the dimension when you activate the grips by selecting the dimension at the Command: prompt. Associative dimensions offer the most powerful editing possibilities, although nonassociative dimension components can also be edited with grips. There are many ways in which grips can be used to alter the measured value and configuration of dimensions.

Figure 28-69 shows the grips for each type of associative dimension. Linear dimensions (horizontal, vertical, and rotated) and aligned and angular dimensions have grips at each extension line origin, the dimension line position, and a grip at the text. A diameter dimension has grips defining two points on the diameter as well as one defining the leader length. The radius dimension has center and radius grips as well as a leader grip.

With dimension grips, a wide variety of editing options are possible. Any of the grips can be PICKed to make them **hot** grips. All grip options are valid methods for editing dimensions.

Figure 28-69

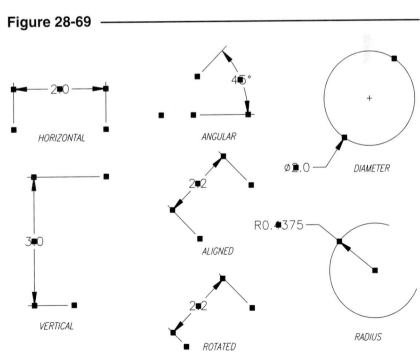

For example (Fig. 28-70), a horizontal dimension value can be increased by stretching an extension line origin grip in a horizontal direction. A vertical direction movement changes the length of the extension line. The dimension line placement is changed by stretching its grips. The dimension text can be stretched to any position by manipulating its grip.

An angular dimension can be increased by stretching the extension line origin grip. The numerical value automatically updates.

Figure 28-70

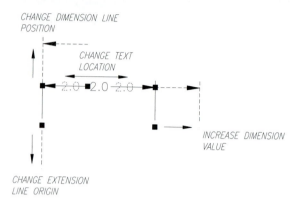

Stretching a rotated dimension's extension line origin allows changing the length of the dimension as well as the length of the extension line. An aligned dimension's extension line origin grip also allows you to change the aligned <u>angle</u> of the dimension.

Figure 28-71

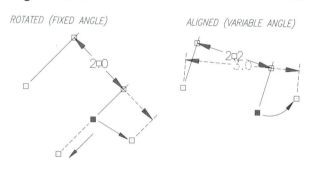

Rotating the <u>center</u> grip of a radius dimension (with the ROTATE option) allows you to reposition the location of the dimension around the *Arc*. Note that the text remains in its original horizontal orientation (Fig. 28-72).

Figure 28-72

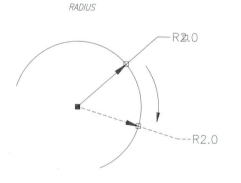

You can move the dimension <u>text</u> independent of the dimension line with grips for dimensions that have a *DIMTMOVE* variable setting of 2. Use an existing dimension, and change the *DIMTMOVE* setting to 2 (or change the *Fit* setting in the *Dimension Style Manager* to *Over the dimension line without a leader*). The dimension text can be moved to any location without losing its associativity. See "*Fit (DIMTMOVE)*," Chapter 29.

Many other possibilities exist for editing dimensions using grips. Experiment on your own to discover some of the possibilities that are not shown here.

Exploding Associative Dimensions

Associative dimensions are treated as one object. If *Erase* is used with an associative dimension, the entire dimension (dimension line, extension lines, arrows, and text) is selected and *Erased*.

Explode can be used to break an associative dimension into its component parts. The individual components can then be edited. For example, after *Exploding*, an extension line can be *Erased* or text can be *Moved*.

There are two main drawbacks to *Exploding* associative dimensions. First, the associative property is lost. Editing commands or *Grips* cannot be used to change the entire dimension but affect only the component objects. More important, Dimension Styles cannot be used with unassociative dimensions.

Dimension Editing Commands

Several commands are provided to facilitate easy editing of existing dimensions in a drawing. Most of these commands are intended to allow variations in the appearance of dimension <u>text</u>. These editing commands operate with associative and nonassociative dimensions, but not with exploded dimensions.

DIMTEDIT

Pull-down Menu	COMMAND (TYPE)	ALIAS (TYPE)	Short-cut	Screen (side) Menu	Tablet Menu
Dimension Align Text >	*DIMTEDIT*	*DIMTED*	...	*DIMNSION Dimtedit*	*Y,2*

Dimtedit (text edit) allows you to change the position or orientation of the text for a single associative dimension. To move the position of text, this command syntax is used:

Figure 28-73

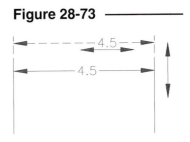

```
Command: Dimtedit
Select dimension: PICK
Specify new location for dimension text or
[Left/Right/Center/Home/Angle]: PICK
```

At the "Specify new location for dimension text" prompt, drag the text to the desired location. The selected text and dimension line can be changed to any position, while the text and dimension line retain their associativity.

Angle
The **Angle** option works with any *Horizontal, Vertical, Aligned, Rotated, Radius,* or *Diameter* dimensions. You are prompted for the new <u>text</u> angle (Fig. 28-74):

Figure 28-74

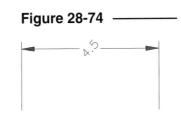

```
Command: dimtedit
Select dimension: PICK
Specify new location for dimension text or
[Left/Right/Center/Home/Angle]: a
Specify angle for dimension text: 45
Command:
```

Home
The text can be restored to its original (default) rotation angle with the **Home** option. The text retains its right/left position.

Figure 28-75

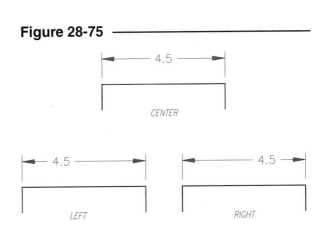

Right/Left
The **Right** and **Left** options automatically justify the dimension text at the extreme right and left ends, respectively, of the dimension line. The arrow and a short section of the dimension line, however, remain between the text and closest extension line (Fig. 28-75).

Center
The *Center* option brings the text to the center of the dimension line and sets the rotation angle to 0 (if previously assigned; Fig. 28-75).

DIMEDIT

Pull-down Menu	COMMAND (TYPE)	ALIAS (TYPE)	Short-cut	Screen (side) Menu	Tablet Menu
Dimension Oblique	*DIMEDIT*	*DIMED* or *DED*	...	*DIMNSION Dimedit*	*Y,1*

The *Dimedit* command allows you to change the angle of the extension lines to an obliquing angle and provides several ways to edit dimension text. Two of the text editing options (*Home, Rotate*) duplicate those of the *Dimtedit* command. Another feature (*New*) allows you to change the text value and annotation.

    Command: **dimedit**
    Enter type of dimension editing [Home/New/Rotate/Oblique] <Home>:

Home

This option moves the dimension text back to its original angular orientation angle without changing the left/right position. *Home* is a duplicate of the *Dimtedit* option of the same name.

New

You can change the text value and annotation of existing text. The same mechanism appears when you create the dimension and use the *Mtext* option—that is, the *Multiline Text Editor* appears. Remember that AutoCAD draws the dimension value in place of the < > characters. Add a prefix before these characters or a suffix after them. Entering text to replace the < > characters (erase them) causes your new text to appear instead of the AutoCAD-supplied value. Placing text or numbers <u>inside</u> the symbols overrides the correct measurement and creates static, nonassociative text.

The original AutoCAD-measured value <u>can be restored</u>, however, using the *New* option. Simply invoke *Dimedit* and *New* but do not enter any value in the *Multiline Text Editor*. After selecting the desired dimension, the AutoCAD-measured value is restored.

Rotate

AutoCAD prompts for an angle to rotate the text. Enter an absolute angle (relative to angle 0). This option is identical to the *Angle* option of *Dimtedit*.

Oblique

This option is unique to *Dimedit*. Entering an angle at the prompt affects the extension lines. Enter an absolute angle:

    Enter obliquing angle (press ENTER for none):

Normally, the extension lines are perpendicular to the dimension lines. In some cases, it is desirable to set an obliquing angle for the extension lines, such as when dimensions are crowded and hard to read (Fig. 28-76) or for dimensioning isometric drawings (see Chapter 29, "Dimensioning Isometric Drawings").

Figure 28-76 ——————

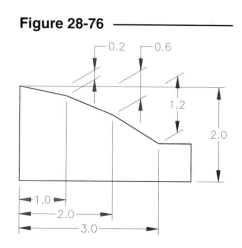

PROPERTIES

	COMMAND (TYPE)	ALIAS (TYPE)	Short-cut	Screen (side) Menu	Tablet Menu
Pull-down Menu					
Modify Properties...	*PROPERTIES*	*MO or PR*	**(Edit Mode)** *Properties*	*MODIFY1 Modify*	*Y,14*

The *Properties* command (discussed in Chapter 16) can be used effectively for a wide range of editing purposes. Double-clicking on any dimension produces the *Properties* palette. The palette gives a list of all properties of the dimension including dimension variable settings.

The dimension variable settings for the selected dimension are listed in the *Lines & Arrows*, *Text*, *Fit*, *Primary Units*, *Alternate Units*, and *Tolerances* sections. Each section is expandable, so you can change any property (dimension variable setting) in each section for the dimension (Fig. 28-77). Each entry in the left column represents a dimension variable; changing the entries in the right column changes the variable's setting.

Each of the six categories (*Lines & Arrows*, *Text*, *Fit*, *Primary Units*, *Alternate Units*, and *Tolerances*) corresponds to the tabs in the *Dimension Style Manager*. Typically, you create dimension styles in the *Dimension Style Manager* by selecting settings for the <u>dimensioning variables</u>, then draw the dimensions using one of the styles. Each style has different settings for dimension variables, so the resulting dimensions for each style appear differently.

Figure 28-77

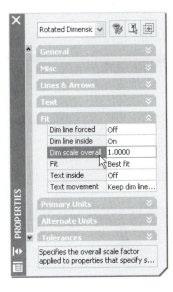

Using the *Properties* palette, you can change the dimension variable settings for any one or more dimensions retroactively. Making a change to a dimension variable with this method does not change the previously created dimension style, but it does create an <u>override for that particular dimension</u>. For more information on dimension variables, dimension styles, and dimension style overrides, see Chapter 29, Dimension Styles and Variables.

Customizing Dimensioning Text

As discussed earlier, you can specify dimensioning text other than what AutoCAD measures and supplies as the default text, using the *Text* option of the individual dimension creation commands. In addition, the text can be modified at a later time using the *Properties* command or the *New* option of *Dimedit*.

The less-than and greater-than symbols (< >) represent the AutoCAD-supplied dimensional value. Placing text or numbers <u>inside</u> the symbols overrides the correct measurement (not advised) and creates static text. Text placed inside the < > symbols is not associative and is <u>not</u> updated in the event of *Stretching*, *Rotating*, or otherwise editing the associative dimension. The original text can only be retrieved using the *New* option of *Dimedit*.

Text placed outside of the < > symbols, however, acts only as a prefix/suffix to the measured value. In the case of a diameter or radius dimension, AutoCAD automatically inserts a diameter symbol (Ø) or radius designator (R) before the dimension value. Inserting a prefix with the *Multiline Text Editor* <u>does not override</u> the AutoCAD symbols; however, entering a *Prefix* or *Suffix* using the *Dimension Style Manager* <u>overrides</u> the symbols. If you use the *Properties* palette to change the text using the *Text override* edit box, enter a text string and include the < and > symbols (such as "2X <>") to retain the AutoCAD-measured value.

This feature is important in correct ANSI standard dimensioning for entering the number of times a feature occurs. For example, one of two holes having the same diameter would be dimensioned "2X Ø1.00" (see Figure 28-21).

You can also enter special characters by using the Unicode values or "%%" symbols. The following codes can be entered <u>outside</u> the < > symbols for *Dimedit New, Mtext,* the *Text* option of dimensioning commands, or *Properties* palette:

Enter:	Enter:	Result	Description
\U+2205	%%c	Ø	diameter (metric)
\U+00b0	%%d	°	degrees
	%%o	‾‾‾	overscored text
	%%u	___	underscored text
\U+00b1	%%p	±	plus or minus
	%%*nnn*	varies	ASCII text character number
\U+*nnnn*		varies	Unicode text hexadecimal value

DIMENSIONING VARIABLES INTRODUCTION

Now that you know the commands used to create dimensions, you need to know how to control the appearance of dimensions. For example, after you create dimensions in a drawing, you may need to adjust the size of the dimension text and arrowheads, or change the text style, or control the number of decimal places appearing on the dimension text.

The appearance of dimensions is controlled by setting the dimensioning variables. You can ensure an entire group of dimensions appear the same by using one dimension style for the dimensions. A dimension style is the set of variables that all the dimension objects in the group reference.

Although you can change many settings that control the appearance of dimensions, there is one dimensioning variable that is the most universal, that is, the *Overall Scale (DIMSCALE)* of the dimensions. The *Overall Scale* controls all of the size-related features of dimension objects, such as the arrowhead size, the text size, the gap between extension lines and the object, and so on. Even though each of those features can be changed individually, changing the *Overall Scale* controls <u>all</u> of the size-related features of dimensions together.

Figure 28-78 ————

To change the *Overall Scale,* open the *Dimension Style Manager* (by selecting it from the *Dimension* pull-down menu, *Dimension* toolbar, or by typing *D*). Select the *Modify* button, access the *Fit* tab, and enter the desired value in the *Use overall scale of* edit box (Fig. 28-78). This action changes the overall size for all dimensions created using that particular dimension style.

For example, assume you created a drawing and applied the first few dimensions, but noticed that the dimension text was so small that it was barely readable. Instead of changing the dimension variable that controls the size of just the dimension text, change the *Overall Scale.*

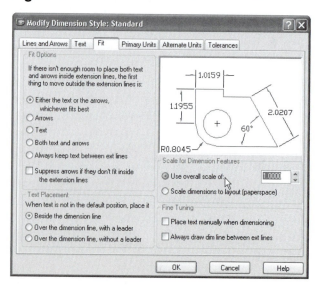

This practice ensures that the text, arrowheads, gaps, extension, and all size-related features of the dimension are changed together and are therefore correctly proportioned. Using the *Dimension Style Manager* to change this variable also ensures that all the dimensions created in that dimension style will have the same size.

Chapter 29 explains all the dimensioning variables, including *Overall Scale,* as well as dimension styles. While working on the exercises in this chapter, you will be able to create dimensions without having to change any dimension variables or dimension styles, although you may want to experiment with the *Overall Scale.* Knowledge and experience with <u>both</u> the dimensioning commands and dimensioning variables is essential for AutoCAD users in a real-world situation. Make sure you read Chapter 29 before attempting to dimension drawings other than those in these exercises.

CHAPTER EXERCISES

Only four exercises are offered in this chapter to give you a start with the dimensioning commands. Many other dimensioning exercises are given at the end of Chapter 29, Dimension Styles and Variables. Since most dimensioning practices in AutoCAD require the use of dimensioning variables and dimension styles, that information should be discussed before you can begin dimensioning effectively.

The units that AutoCAD uses for the dimensioning values are based on the *Unit format* and the *Precision* settings in the *Primary Units* tab of the *Dimension Style Manager.* You can dimension these drawings using the default settings, or you can change the *Units* and *Precision* settings for each drawing to match the dimensions shown in the figures. (Type *D,* select *Modify,* then the *Primary Units* tab.)

Other dimensioning features may appear different than those in the figures because of the variables set in the AutoCAD default STANDARD dimension style. For example, the default settings for diameter and radius dimensions may draw the text and dimension lines differently than you desire. After reading Chapter 29, those features in your exercises can be changed retroactively by changing the dimension style.

1. *Open* the **PLATES** drawing that you created in Chapter 9 Exercises. *Erase* the plate on the right. Use *Move* to move the remaining two plates apart, allowing 5 units between. Create a *New* layer called **DIM** and make it *Current.* Dimension the two plates, as shown in Figure 28-79. *Save* the drawing as **PLATES-D**.

Figure 28-79

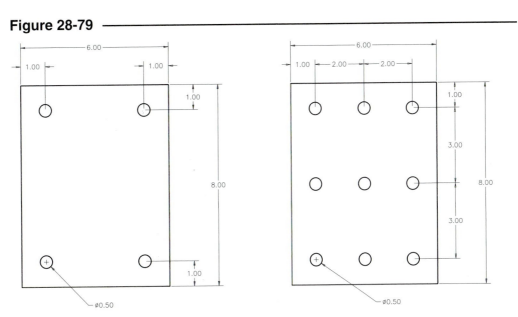

2. *Open* the **PEDIT1** drawing. Create a *New* layer called **DIM** and make it *Current*. Dimension the part as shown in Figure 28-80. *Save* the drawing as **PEDIT1-DIM**. Draw a *Pline* border of .02 *width* and *Insert* **TBLOCK** or **TBLOCKAT**. *Plot* on an A size sheet using *Scale to Fit*.

Figure 28-80

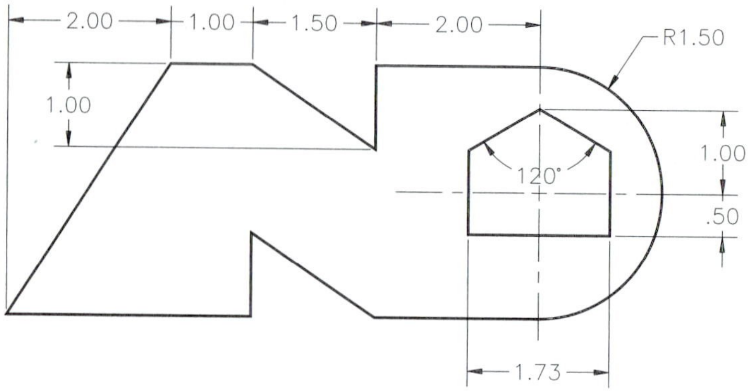

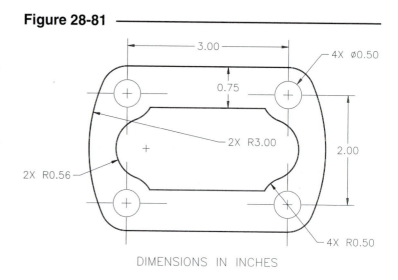

3. *Open* the **GASKETA** drawing that you created in Chapter 9 Exercises. Create a *New* layer called **DIM** and make it *Current*. Dimension the part as shown in Figure 28-81. *Save* the drawing as **GASKETD**.

Draw a *Pline* border with .02 *width* and *Insert* **TBLOCK** or **TBLOCKAT** with an **8/11** scale factor. *Plot* the drawing *Scaled to Fit* on an A size sheet.

Figure 28-81

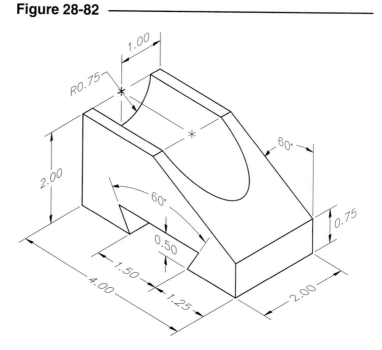

DIMENSIONS IN INCHES

4. *Open* the **BARGUIDE** multiview drawing that you created in Chapter 24 Exercises. Create the dimensions on the **DIM** layer. Keep in mind that you have more possibilities for placement of dimensions than are shown in Figure 28-82.

Figure 28-82

29

DIMENSION STYLES AND VARIABLES

Chapter Objectives

After completing this chapter you should:

1. be able to control dimension variable settings using the *Dimension Style Manager*;

2. be able to save and restore dimension styles with the *Dimension Style Manager*;

3. know how to create dimension style families and specify variables for each child;

4. know how to create and apply dimension style overrides;

5. be able to modify dimensions using *Update*, *Dimoverride*, *Properties*, and *Matchprop*;

6. know the guidelines for dimensioning;

7. be able to create associative dimensions in paper space;

8. know how to dimension isometric drawings.

CONCEPTS

Dimension Variables

Since a large part of AutoCAD's dimensioning capabilities are automatic, some method must be provided for you to control the way dimensions are drawn. A set of 70 <u>dimension variables</u> allows you to affect the way dimensions are drawn by controlling sizes, distances, appearance of extension and dimension lines, and dimension text formats.

An example of a dimension variable is *DIMSCALE* (if typed) or *Overall Scale* (if selected from a dialog box, Fig. 29-1). This variable controls the overall size of the dimension features (such as text, arrowheads, and gaps). Changing the value from 1.00 to 1.50, for example, makes all of the size-related features of the drawn dimension 1.5 times as large as the default size of 1.00 (Fig. 29-2). Other examples of features controlled by dimension variables are arrowhead type, orientation of the text, text style, units and precision, suppression of extension lines, fit of text and arrows (inside or outside of extension lines), and direction of leaders for radii and diameters (pointing in or out). The dimension variable changes that you make affect the <u>appearance</u> of the dimensions that you create.

Figure 29-1

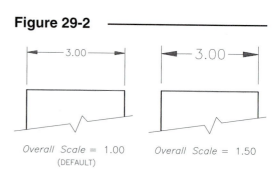

Figure 29-2

Overall Scale = 1.00 (DEFAULT)   Overall Scale = 1.50

There are two basic ways to control dimensioning variables:

1. Use the *Dimension Style Manager* (Fig. 29-3).

2. Type the dimension variable name in Command line format.

The dialog boxes employ "user-friendly" terminology and selection, while the Command line format uses the formal dimension variable names.

Changes to dimension variables are usually made <u>before</u> you create the affected dimensions. Dimension variable changes are <u>not always retroactive</u>, in contrast to *LTSCALE,* for example, which can be continually modified to adjust the spacing of existing non-continuous lines. Changes to dimensioning variables affect <u>existing</u> dimensions only when those changes are *Saved* to an existing *Dimension Style* that was in effect when previous dimensions were created. Generally, dimensioning variables should be set <u>before</u> creating the desired dimensions although it is possible to modify dimensions and *Dimension Styles* retroactively.

Dimension Styles

Associative and nonassociative dimensions are part of a <u>dimension style</u>. The default (and the only supplied) dimension style is named STANDARD (Fig. 29-3). Logically, this dimension style has all of the default dimension variable settings for creating a dimension with the typical size and appearance. Similar to layers, you can create, name, and specify settings for any number of dimension styles. Each dimension style contains the dimension variable settings that you select. <u>A dimension style is a group of dimension variable settings that has been saved under a name you assign</u>. When you set a dimension *Current*, AutoCAD remembers and resets that <u>particular combination of dimension variable settings</u>.

Figure 29-3

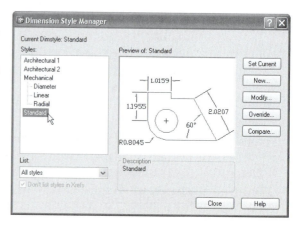

Imagine dimensioning a complex drawing <u>without</u> having dimension styles (Fig. 29-4). In order to create a dimension using limit dimensioning (as shown in the diameter dimension), for example, you would change the desired variable settings, then "draw" the dimension. To draw another dimension without extension lines (as shown in the interior slot), you would have to reset the previous variables and make changes to other variables in order to place the special dimensions as you prefer. This same process would be repeated each time you want to create a new type of dimension. If you needed to add another dimension with the limits, you would have to reset the same variables as before.

Figure 29-4

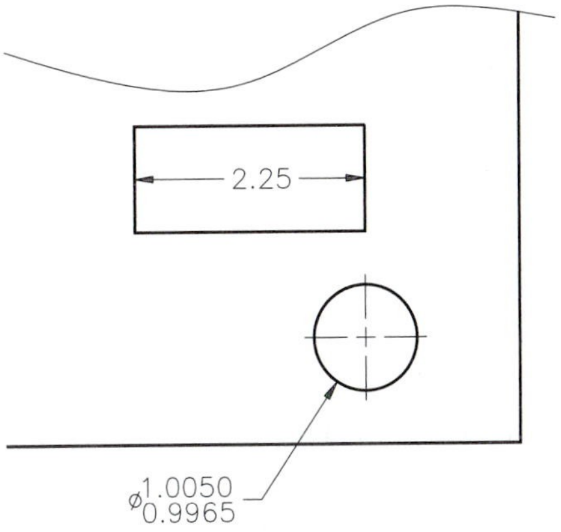

To simplify this process, you can create and save a dimension style for each particular "style" of dimension. Each time you want to draw a particular style of dimension, select the style name from the list and begin drawing. A dimension style could contain all the default settings plus only one or two dimension variable changes or a large number of variable changes.

To create a new dimension style, select the *New* button in the *Dimension Style Manager* (see Figure 29-3), then use the six tabs that appear in the dialog box to specify the appearance of the dimensions. Selecting options in these tabs (see Figure 29-14) actually sets values for related dimension variables that are saved in the drawing. When a dimension style is *Set Current*, the dimensions you create with that style appear with the dimension variable settings you specified. Creating dimensions with the current *Dimension Style* is similar to creating text with the current *Text Style*—that is, the objects created take on the current settings assigned to the style.

Another advantage of using dimension styles is that <u>existing</u> dimensions in the drawing can be globally modified by making a change and saving it to the dimension style(s). To do this, select *Modify* in the *Dimension Style Manager* or use the *Save* option of *–Dimstyle* after changing a dimension variable in Command line format. This action saves the changes for newly created dimensions as well as <u>automatically updating existing dimensions in the style</u>.

Dimension Style Families

In the previous chapter, you learned about the various <u>types</u> of dimensions, such as linear, angular, radial, diameter, ordinate, and leader. You may want the appearance of each of these types of dimensions to vary, for example, all radius dimensions to appear with the dimension line arrow pointing out and all diameter dimensions to appear with the dimension line arrow inside pointing in. It is logical to assume that a new dimension style would have to be created for each variation. However, a new dimension style name is not necessary for each type of dimension. AutoCAD provides <u>dimension style families</u> for the purpose of providing variations within each named dimension style.

The dimension style <u>names</u> that you create are assigned to a dimension style <u>family</u>. Each dimension style family can have six <u>children</u>. The children are <u>linear, angular, diameter, radial, ordinate, and leader</u>. Although the children take on the dimension variable settings assigned to the family, you can assign special settings for one or more children. For example, you can make a radius dimension appear slightly different from a linear dimension in that family. Do this by selecting the *New* button in the *Dimension Style Manager*, then selecting the type of dimension (*linear, radius, diameter*, etc.) you want to change from the *Create New Dimension Style* dialog box that appears (Fig. 29-5).

Figure 29-5

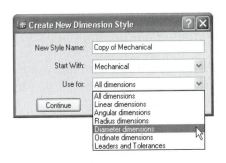

Select *Continue* to specify settings for the child. When a dimension is drawn, AutoCAD knows what type of dimension is created and applies the selected settings to that child.

Dimension styles that have children (special variable settings set for the *linear, radial, diameter*, etc. dimensions) appear in the list of styles in the *Dimension Style Manager* as sub-styles branching from the parent name. For example, notice the "Mechanical" style in Figure 29-3 has special settings for its *Diameter, Linear,* and *Radial* children.

For example, you may want to create the "Mechanical" dimension style to draw the arrows and dimension lines inside the arc but drag the dimension text outside of the arc for *Radial* dimensions (as shown in Fig. 29-6). To do this, create a new child for the "Mechanical" family as previously described, and set "Mechanical" as the current style. When a *Radius* dimension is drawn, the dimension line, text, and arrow should appear as shown in Figure 29-6.

Figure 29-6 ———————————

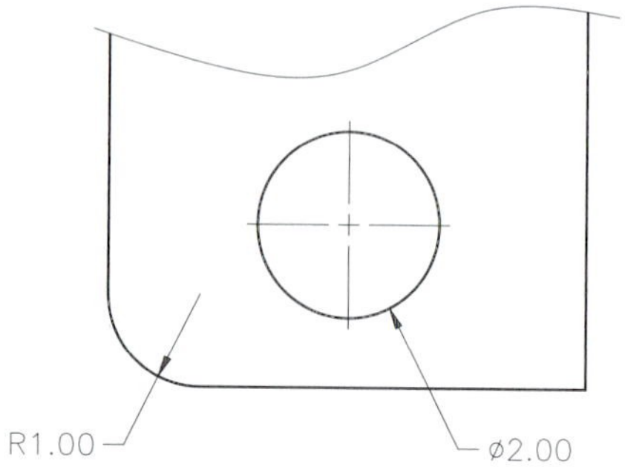

R1.00   ⌀2.00

In summary, a dimension style family is simply a set of dimension variables related by name. Variations within the family are allowed, in that each child (type of dimension) can possess its own subset of variables. Therefore, each child inherits all the variables of the family in addition to any others that may be assigned individually.

AutoCAD refers to these children as $0, $2, $3, $4, $6, and $7 as suffixes appended to the family name such as "Standard$0." AutoCAD automatically applies the appropriate suffix code to the dimension style name when a child is created. Although the codes do not appear in the *Dimension Style Manager*, you can display the code by using the *List* command and selecting an existing child dimension.

Child Type	Suffix Code
linear	$0
angular	$2
diameter	$3
radial	$4
ordinate	$6
leader	$7

Dimension Style Overrides

If you want to create one or two dimensions with a special appearance, but do not want to permanently save the new settings to the dimension style, you can create a dimension style override, then create the new dimensions. A dimension style override is a temporary dimension variable setting for the current dimension style. Existing dimensions (in the drawing) are not affected by an override. When another style is made current, the override settings are lost by default.

A dimension style override is created by selecting the style you want to change from the list in the *Dimension Style Manager*, then pressing *Set Current*, then select the *Override* button. Proceed to specify the variable settings from the six tabs that appear. When the override is made, a new branch under the family name appears in the *Dimension Style Manager* named "<style overrides>" (see Figure 29-3 under "Architectural 2"). Finally, create the new dimensions. To clear the overrides, simply make another style current.

You can also create an override by typing any dimension variable name at the Command line and making a change to the value. For example, type *DIMSCALE* to change the overall scale for new dimensions created in the style. Existing dimensions in the style are not affected. The override is applied only to the newly created dimensions in the current style and clears when another style is made current. Additionally, you can use the *Properties* palette to change any variable setting for an existing dimension. Doing so creates a dimension style override <u>for that particular dimension only</u>, not for the dimension style.

DIMENSION STYLES

DIMSTYLE
and DDIM

Pull-down Menu	COMMAND (TYPE)	ALIAS (TYPE)	Short-cut	Screen (side) Menu	Tablet Menu
Dimension Style...	*DIMSTYLE* or *DDIM*	*DIMSTY, DST,* or *D*	...	*DIMNSION Dimstyle*	*Y,3*

The *Dimstyle* or *Ddim* command produces the *Dimension Style Manager* (Fig. 29-7). This is the primary interface for creating new dimension styles and making existing dimension styles current. This dialog box also gives access to six tabs that allow you to change dimension variables. Features of the *Dimension Style Manager* that appear on the opening dialog box (shown in Figure 29-7) are described in this section. The tabs that allow you to change dimension variables are discussed in detail later in this chapter in the section "Dimension Variables."

The *Dimension Style Manager* makes the process of creating and using dimension styles easier and more visual than using Command line format. Using the *Dimension Style Manager*, you can *Set Current*, create *New* dimension styles, *Modify* existing dimension styles, and create an *Override*. You also have the ability to view a list of existing styles including children and overrides; see a preview of the selected style, child, or override; examine a description of a dimension style as it relates to any other style; and make an in-depth comparison between styles. You can also control the entries in the list to include or not include Xreferenced dimension styles. The options are explained below.

Figure 29-7

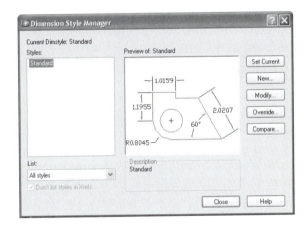

Styles
This list displays all existing dimension styles in the drawing, including *Xrefs*, depending on your selection in the *List* section below. The list includes the parent dimension styles, the children that are shown on a branch below the parent style, and any overrides also branching from family styles.

The buttons on the right side of the *Dimension Style Manager* affect the highlighted style in the list. For example, to modify an existing style, select the style name from the list and PICK the *Modify* button. If you want to create a *New* style based on an existing style, select the style name you want to use as the "template," then select *New*.

Select (highlight) any dimension style from the *Styles* list to display the appearance of dimensions for that style in the *Preview* tile. Selecting a style from the list also makes the *Description* area list the dimension variables settings compared to the current dimension style.

You can right-click to use a shortcut menu for the selected dimension style (Fig. 29-8). The shortcut menu allows you to *Set Current*, *Rename*, or *Delete* the selected dimension style. Only dimension styles that are unreferenced (no dimensions have been created using the style) can be deleted.

If you have created a style override (by pressing the *Override* button, then setting variables), you can save the overrides to the dimension style using the right-click shortcut menu. Normally, overrides are automatically discarded when another dimension style is made current. This shortcut menu is the only method available in the *Dimension Style Manager* to save overrides. (You can also use the *–Dimstyle* command to save overrides.)

Figure 29-8

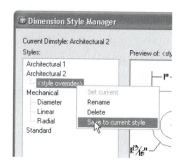

List
Select *All Styles* to display the list of all the previously created or imported dimension styles in the drawing. Selecting *Styles In Use* displays only styles that have been used to create dimensions in the drawing—styles that are saved in the drawing but are not referenced (no dimensions created in the style) do not appear in the list.

Don't List Styles in Xrefs
If the current drawing references another drawing (has an *Xref* attached), the *Xref's* dimension styles can be listed or not listed based on your selection here.

Preview
The *Preview* tile displays the appearance of the highlighted dimension style. When a dimension style name is selected, the *Preview* tile displays an example of all dimension types (linear, radial, etc.) and how the current variable settings affect the appearance of those dimensions. If a child is selected, the *Preview* tile displays only that dimension type and its related appearance. For example, Figure 29-9 displays the appearance of the selected child, *Mechanical: Radial*.

Figure 29-9

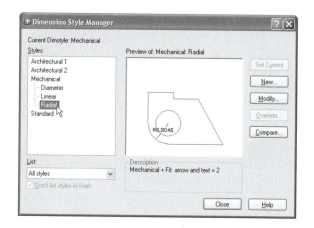

Description
The description area is a valuable tool for determining the variable settings for a dimension style. When a dimension style from the list is selected, the *Description* area lists the differences in variable settings from the current style. For example, assume the "Architectural 2" dimension style was created using "Standard" as the template. To display the dimension variable settings that have been changed (the differences between Standard and Architectural 2), make Standard the current style, then select Architectural 2 and examine the description area, as shown in Figure 29-10. You can use the same procedure to display the variable settings assigned to a child (the differences between the family style and the child), as shown in the *Description* area in Figure 29-9.

Figure 29-10

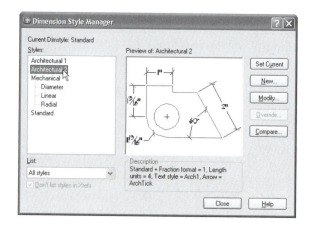

Set Current
Select the desired existing style from the list on the left, then select *Set Current*. The current style is listed at the top of the *Dimension Style Manager*. When you *Close* the dialog box, the current style is used for drawing new dimensions. If you want to set an *Override* to an existing style, you must first make it the current style.

New
Use the *New* button to create a new dimension style. The *New* button invokes the *Create New Dimension Style* dialog box (Fig. 29-11). The options are discussed next.

Figure 29-11

New Style Name
Enter the desired new name in the *New Style Name* edit box. Initially the name that appears is "Copy of (current style)."

Start With
After assigning a new name, select any existing dimension style to use as a "template" from the *Start With* drop-down list. Initially (until dimension variable changes are made for the new style), the new dimension style is actually a copy of the *Start With* style (has identical variable settings). By default, the current style appears in the *Start With* edit box.

Use For
If you want to create a dimension style family, select *All Dimensions*. In this way, all dimension types (*linear, radial, diameter,* etc.) take on the new variable settings. If you want to specify special dimension variable settings for a child, select the desired dimension type from the drop-down list (see Figure 29-5).

Continue
When the new name and other options have been selected, press the *Continue* button to invoke the *New Dimension Style* dialog box. Here you select from the six tabs to specify settings for any dimension variable to apply to the new style. See the "Dimension Variables" section for information on setting dimension variables.

NOTE: Generally you should not change the Standard dimension style. Rather than change the Standard style, create *New* dimension styles using Standard as the base style. In this way, it is easy to restore and compare to the default settings by making the Standard style current. If you make changes to the Standard style, it can be difficult to restore the original settings.

Modify
Use this button to change dimension variable settings for an existing dimension style. First, select the desired style name from the list, then press *Modify* to produce the *Modify Dimension Style* dialog box. Using *Modify* to change dimension variables automatically saves the changes to the style and updates existing dimensions in the style, as opposed to using *Override*, which does not change the style or update existing dimensions. See the "Dimension Variables"section for information on setting dimension variables.

Override
A dimension style override is a temporary variable setting that affects only new dimensions created with the override. An override does not affect existing dimensions previously created with the style. You can set an override only for the current style, and the overrides are automatically cleared when another dimension style is made current, unless you use the right-click shortcut menu and select *Save to current style* (see Figure 29-8).

To create a dimension style override, you must first select an existing style from the list and make it the current style. Next, select *Override* to produce the *Override Current Style* dialog box. See the "Dimension Variables" section for information on setting dimension variables to act as overrides.

Compare

This feature is very useful for examining variable settings for any dimension style. Press the *Compare* button to produce the *Compare Dimension Styles* dialog box (Fig. 29-12). This dialog box lists only dimension variable settings; changes to variable settings cannot be made from this interface.

Figure 29-12

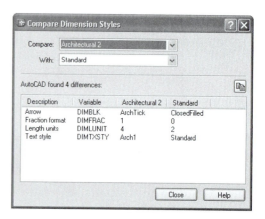

The *Compare Dimension Styles* dialog box lists the differences between the style in the *Compare* box and the style in the *With* box. For example, assume you created a new style named "Architectural 2" from the "Standard" style, but wanted to know what variable changes you had made to the new style. Select Architectural 2 in the *Compare* box and Standard in the *With* box. The variable changes made to Architectural 2 are displayed in the central area of the box. Also note that the settings for each variable are listed for each of the two styles.

A useful feature of this dialog box is that the formal variable names are listed in the *Variable* column. Many experienced users of AutoCAD prefer to use the formal variable names since they do not change from release to release as the dialog box descriptions do.

You can also use the *Compare Dimension Styles* dialog box to give the entire list of variables and related settings for one dimension style. Do this by selecting the desired dimension style name from the *Compare* drop-down list and select *<none>* from the *With* list (Fig. 29-13).

Figure 29-13

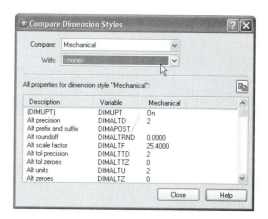

-DIMSTYLE

Pull-down Menu	COMMAND (TYPE)	ALIAS (TYPE)	Short-cut	Screen (side) Menu	Tablet Menu
Dimension Update	-DIMSTYLE	...	...	DIMNSION Dimstyle	Y,3

The *-Dimstyle* command is the Command line format equivalent to the *Dimension Style Manager*. Operations that are performed by the dialog box can also be accomplished with the *-Dimstyle* command. The *Apply* option offers one other important feature that is <u>not</u> available in the dialog box:

    Command: **-dimstyle**
    Current dimension style: Standard
    Enter a dimension style option [Save/Restore/STatus/Variables/Apply/?] <Restore>:

Save

Use this option to <u>create a new</u> dimension style. The new style assumes all of the current dimension variable settings comprised of those from the current style plus any other variables changed by Command line format (overrides). The new dimension style becomes the current style.

    Enter a dimension style option [Save/Restore/STatus/Variables/Apply/?] <Restore>: **s**
    Enter name for new dimension style or [?]:

Restore

This option prompts for an existing dimension style name to be restored. The restored style becomes the current style. You can use the "select dimension" option to PICK a dimension object that references the style you want to restore.

STatus

Status gives the <u>current settings</u> for all dimension variables. The displayed list comprises the settings of the current dimension style and any overrides. All dimension variables that are saved in dimension styles are listed:

    Enter a dimension style option [Save/Restore/STatus/Variables/Apply/?] <Restore>: **st**

```
DIMASO      Off         Create dimension objects
DIMSTYLE    Standard    Current dimension style (read-only)
DIMADEC     0           Angular decimal places
DIMALT      Off         Alternate units selected
DIMALTD     2           Alternate unit decimal places
etc.
```

Variables

You can use this option to <u>list</u> the dimension variable settings for <u>any existing dimension style</u>. You cannot modify the settings.

    Enter a dimension style option [Save/Restore/STatus/Variables/Apply/?] <Restore>: **v**
    Enter a dimension style name, [?] or <select dimension>:

Enter the name of any style or use "select dimension" to PICK a dimension object that references the desired style. A list of all variables and settings appears.

~stylename

This variation can be entered in response to the *Variables* option. Entering a dimension style name preceded by the tilde symbol (~) displays the <u>differences between the current style and the (~) named style</u>.

For example, assume that you wanted to display the <u>differences</u> between STANDARD and the current dimension style (Mechanical, for example). Use *Variables* with the ~ symbol. (The ~ symbol is used as a wildcard to mean "all but.") This option is very useful for keeping track of your dimension styles and their variable settings.

```
Command: -dimstyle
Current dimension style: Mechanical
Enter a dimension style option [Save/Restore/STatus/Variables/Apply/?] <Restore>: v
Enter a dimension style name, [?] or <select dimension>: ~standard
Differences between STANDARD and current settings:
```

```
            STANDARD    Current Setting
DIMSCALE    1.0000      0.7500
DIMUPT      Off         On
DIMTXSTY    Standard    Mech1
```

Apply

Use *Apply* to <u>update dimension objects</u> that you PICK with the current variable settings. The current variable settings can contain those of the current dimension style plus any overrides. The selected object loses its reference to its original style and is "adopted" by the applied dimension style family.

> Enter a dimension style option [Save/Restore/STatus/Variables/Apply/?] <Restore>: *a*
> Select objects:

This is a useful tool for <u>changing an existing dimension object from one style to another</u>. Simply make the desired dimension style current; then use *Apply* and select the dimension objects to change to that style.

Using the *Apply* option of *-Dimstyle* is identical to using the *Update* command. The *Update* command can be invoked by using the *Dimension Update* button, using *Update* from the *Dimension* pull-down menu, or typing *Dim*: (Enter), then *Update*.

DIMENSION VARIABLES

Now that you understand how to create and use dimension styles and dimension style families, let's explore the dimension variables. <u>Two methods</u> can be used to set dimension variables: the *Dimension Styles Manager* and Command line format. First, we will examine the dialog box (accessible through the *Dimension Styles Manager*) that allows you to set dimension variables. Dimension variables can alternately be set by typing the formal name of the variable (such as *DIMSCALE*) and changing the desired value (discussed after this section on dialog boxes). The dialog box offers a more "user-friendly" terminology, edit boxes, checkboxes, drop-down lists, and a preview image tile that automatically reflects the variables changes you make.

Changing Dimension Variables Using the Dialog Box Method

In this section, the dialog box that contains the six tabs for changing variables is explained (see Figure 29-14). The six tabs indicate different groups of dimension variables. The tab names are:

> *Lines and Arrows*
> *Text*
> *Fit*
> *Primary Units*
> *Alternate Units*
> *Tolerances*

Access to this dialog box from the *Dimension Style Manager* is accomplished by selecting *New, Modify,* or *Override*. Practically, there is only one dialog box that allows you to change dimension variables; however, you might say there are three dialog boxes since the title of the box changes based on your selection of *New, Modify,* or *Override*. Depending on your selection, you can invoke the *Create Dimension Style, Modify Dimension Style,* or *Override Current Style* dialog box. The options in the boxes are identical and only the titles are different; therefore, we will examine only one dialog box.

In the following pages, the heading for the paragraphs below include the dialog box option and the related formal dimension variable name in parentheses. The following figures typically display the AutoCAD default setting for a particular variable and an example of changing the setting to another value. These AutoCAD default settings (in the "Standard" dimension style) are for the ACAD.DWT template drawing and for *Start from Scratch, Imperial Default Settings*. If you use the *Metric Default Settings* or other template drawings, dimension variable settings may be different based on dimension styles that exist in the drawing other than Standard. See "Using *Setup Wizards* and Template Drawings" later in this chapter.

Lines and Arrows Tab

The options in this tab change the appearance of dimension lines, extension lines, and arrowheads (Fig. 29-14). The preview image automatically changes based on the settings you select in the *Dimension Lines*, *Extension Lines*, *Arrowheads*, and *Center Marks for Circles* sections.

Figure 29-14

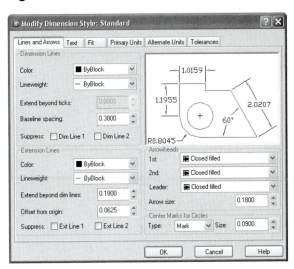

Dimension Lines Section (*Lines and Arrows* Tab)

Color (DIMCLRD)
The *Color* drop-down list allows you to choose the color for the dimension line (Fig. 29-15). Assigning a specific color to the dimension lines, extension lines, and dimension text gives you more control when color-dependent plot styles are used because you can print or plot these features with different line widths, colors, etc. This feature corresponds to the *DIMCLRD* variable (dim color dimension line).

Figure 29-15

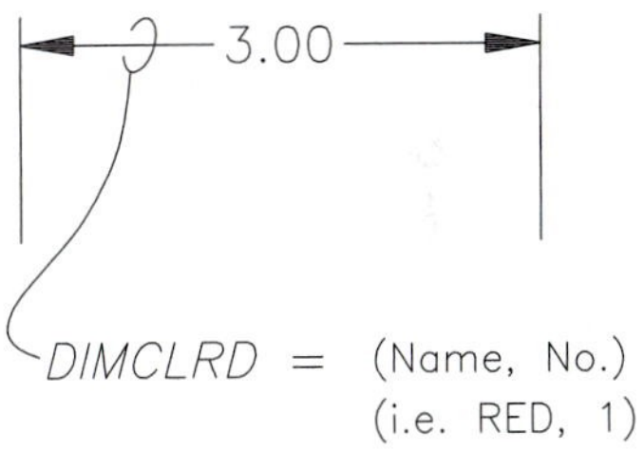

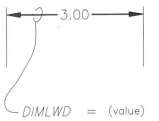

Lineweight (DIMLWD)
Use this option to assign lineweight to dimension lines (Fig. 29-16). Select any lineweight from the drop-down list or enter values in the *DIMLWD* variable (dim lineweight dimension line). Values entered in the variable can be -1 (*ByLayer*), -2 (*ByBlock*), 25 (*Default*), or any integer representing 100th of mm, such as 9 for .09mm.

Figure 29-16

Extend beyond ticks (DIMDLE)
The *Extend beyond ticks* edit box is disabled unless the *Oblique, Integral,* or *Architectural Tick* arrowhead type is selected (in the *Arrowheads* section). The *Extend* value controls the length of dimension line that extends past the dimension line (Fig. 29-17). The value is stored in the *DIMDLE* variable (dimension line extension). Generally, this value does not require changing since it is automatically multiplied by the *Overall Scale* (DIMSCALE).

Figure 29-17

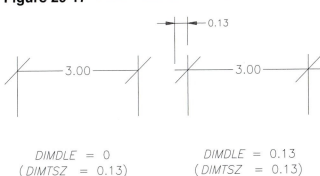

Baseline spacing **(DIMDLI)**
The *Baseline spacing* edit box reflects the value that AutoCAD uses in baseline dimensioning to "stack" the dimension line above or below the previous one (Fig. 29-18). This value is held in the *DIMDLI* variable (dimension line increment). This value rarely requires input since it is affected by *Overall Scale* (*DIMSCALE*).

Figure 29-18

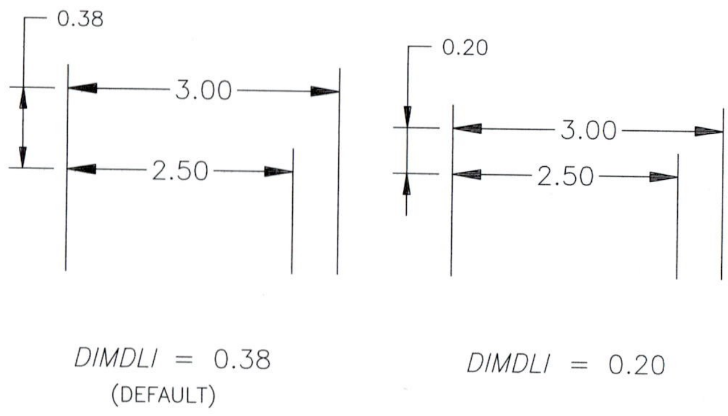

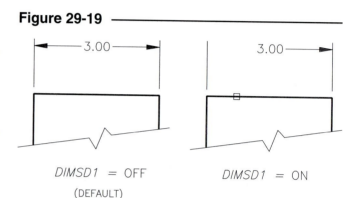

DIMDLI = 0.38
(DEFAULT)

DIMDLI = 0.20

Suppress Dim Line 1, Dim Line 2 **(DIMSD1, DIMSD2)**
This area allows you to suppress (not draw) the *1st* or *2nd* dimension line or both (Fig. 29-19). The first dimension line would be on the "First extension line origin" side or nearest the end of object PICKed in response to "Select object to dimension." These toggles change the *DIMSD1* and *DIMSD2* dimension variables (suppress dimension line 1, 2).

Figure 29-19

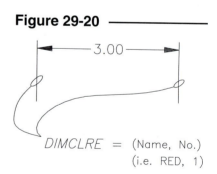

DIMSD1 = OFF
(DEFAULT)

DIMSD1 = ON

Extension Line **Section** (*Lines and Arrows* **Tab**)

Color **(DIMCLRE)**
The *Color* drop-down list allows you to choose the color for the extension lines (Fig. 29-20). You can also activate the standard *Select Color* dialog box. The color assignment for extension lines is stored in the *DIMCLRE* variable (dim color extension line).

Figure 29-20

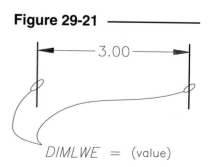

DIMCLRE = (Name, No.)
(i.e. RED, 1)

Lineweight **(DIMLWE)**
This option is similar to that for dimension lines, only the lineweight is assigned to extension lines (Fig. 29-21). Select any lineweight from the drop-down list or enter values in the *DIMLWE* (dim lineweight extension line) variable (*ByLayer* = -1, *ByBlock* = -2, *Default* = 25, or any integer representing 100th of mm).

Figure 29-21

DIMLWE = (value)

Extend beyond dim lines **(DIMEXE)**

The *Extend beyond dim lines* edit box reflects the value that AutoCAD uses to set the distance for the extension line to extend beyond the dimension line (Fig. 29-22). This value is held in the *DIMEXE* variable (extension line extension). Generally, this value does not require changing since it is automatically multiplied by the *Overall Scale* (*DIMSCALE*).

Figure 29-22

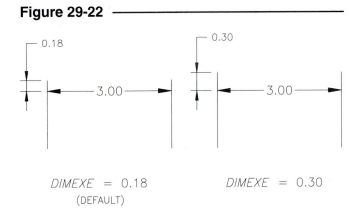

Offset from origin **(DIMEXO)**

The *Offset from origin* value specifies the distance between the origin points and the extension lines (Fig. 29-23). This offset distance allows you to PICK the object corners, yet the extension lines maintain the required gap from the object. This value rarely requires input since it is affected by *Overall Scale* (*DIMSCALE*). The value is stored in the *DIMEXO* variable (extension line offset).

Figure 29-23

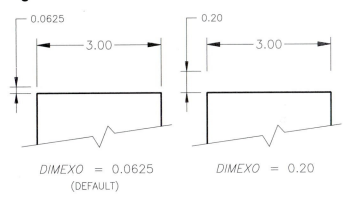

Suppress Ext Line 1, Ext Line 2
(DIMSE1, DIMSE2)

This area is similar to the *Dimension Line* area that controls the creation of dimension lines but is applied to extension lines. This area allows you to suppress the *1st* or *2nd* extension line or both (Fig. 29-24). These options correspond to the *DIMSE1* and *DIMSE2* dimension variables (suppress extension line 1, 2).

Figure 29-24

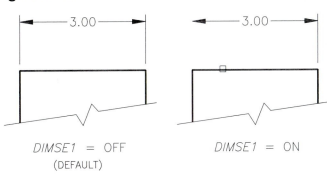

<u>*Arrowheads* Section (*Lines and Arrows* **Tab)**</u>

1st, 2nd **(DIMBLK, DIMSAH,**
DIMBLK1, DIMBLK2)

This area contains two drop-down lists of various arrowhead types, including dots and ticks. Each list corresponds to the *1st* or *2nd* arrowhead created in the drawn dimension (Fig. 29-25). The image tiles display each arrowhead type selected. Click in the first image tile to change both arrowheads, or click in each to change them individually. The variables affected are *DIMBLK*, *DIMSAH*, *DIMBLK1*, and *DIMBLK2*.

Figure 29-25

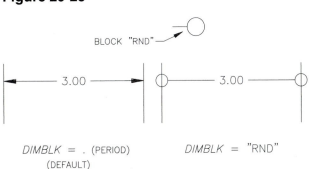

DIMBLK specifies the *Block* to use for arrowheads if both are the same (Fig. 29-25). When *DIMSAH* is on (separate arrow heads), separate arrow heads are allowed for each end defined by *DIMBLK1* and *DIMBLK2*. You can enter any existing *Block* name in the *DIMBLK* variables. See "Dimension Variables Table" for details.

Leader (DIMLDRBLK)

This drop-down list specifies the arrow type for leaders (Fig. 29-26). This setting does not affect the arrowhead types for dimension lines. Select any option from the drop-down list or enter a value in the *DIMLDRBLK* variable (dim leader block). (For a list of arrowhead entries, see "Dimension Variables Table.")

Figure 29-26

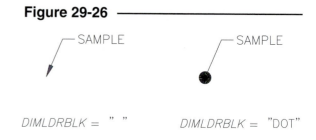

Arrow size (DIMASZ)

The size of the arrow can be specified in the *Arrow size* edit box. The *DIMASZ* variable (dim arrow size) holds the value (Fig. 29-27). Remember that this value is multiplied by *Overall Scale (DIMSCALE)*.

Figure 29-27

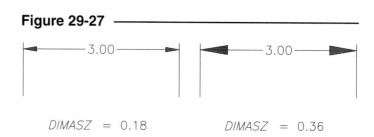

Center Marks for Circles **Section** (*Lines and Arrows* **Tab**)

Type (DIMCEN)

The list here determines how center marks are drawn when the dimension commands *Dimcenter, Dimdiameter,* or *Dimradius* are used. The image tile displays the *Mark, Line,* or *None* feature specified. This area actually controls the value of <u>one</u> dimension variable, *DIMCEN* by using a 0, positive, or negative value (Fig. 29-28). The *None* option enters a *DIMCEN* value of 0.

Figure 29-28

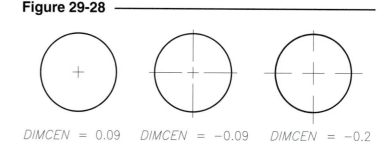

Size (DIMCEN)

The *Size* edit box controls the size of the short dashes and extensions past the arc or circle. The value is stored in the *DIMCEN* variable (Fig. 29-28). Only positive values can be entered in this edit box.

Text **Tab**

The options in this tab change the appearance, placement, and alignment of the dimension text (Fig. 29-29). The preview image automatically reflects changes you make in the *Text Appearance, Text Placement,* and *Text Alignment* sections.

Figure 29-29

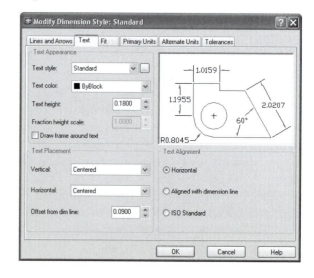

Text Appearance **Section (***Text* **Tab)**

Text style **(DIMTXSTY)**
This feature of AutoCAD allows you to have different text styles for different dimension styles. The text styles are chosen from a drop-down list of <u>existing</u> styles in the drawing. You can also create a text style "on the fly" by picking the button just to the right of the *Text style* drop-down list, which produces the standard *Text Style* dialog box. The text style used for dimensions remains constant (as defined by the dimension style) and does not change when other text styles in the drawing are made current (as defined by *Style, Dtext,* or *Mtext*). The *DIMTXSTY* variable (dim text style) holds the text style name for the dimension style.

Text color **(DIMCLRT)**
Select this drop-down list to select a color or to activate the standard *Select Color* dialog box. The color choice is assigned to the dimension text only (Fig. 29-30). This is useful for controlling the text appearance for printing or plotting when color-dependent plot styles are used. The setting is stored in the *DIMCLRT* variable (dim color text).

Figure 29-30

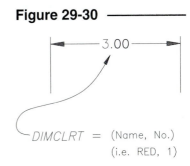

Text height **(DIMTXT)**
This value specifies the primary text height; however, this value is multiplied by the *Overall Scale* (*DIMSCALE*) to determine the actual drawn text height. Change *Text height* <u>only</u> if you want to increase or decrease the text height in relation to the other dimension components (Fig. 29-31). Normally, change the *Overall Scale* (*DIMSCALE*) to change all size-related features (arrows, gaps, text, etc.) proportionally. The text height value is stored in *DIMTXT*.

Figure 29-31

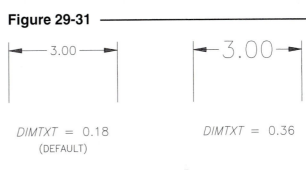

Fraction height scale **(DIMTFAC)**
When fractions or tolerances are used, the height of the fractional or tolerance text can be set to a proportion of the primary text height. For example, 1.0000 creates fractions and tolerances the same height as the primary text; .5000 represents fractions or tolerances at one-half the primary text height (Fig. 29-32). The value is stored in the *DIMTFAC* variable (dim tolerance factor).

Figure 29-32

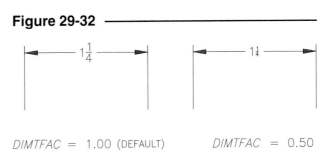

Draw frame around text **(DIMGAP)**
The "frame" or "gap" is actually an invisible box around the text that determines the offset from text to the dimension line. This option sets the *DIMGAP* variable to a negative value which makes the box visible (Fig. 29-33). This practice is standard for displaying a basic dimension. See *"Tolerance"* (geometric dimensioning and tolerancing) in Chapter 28. See also *Offset from dimension line* in the *"Text Placement"* section.

Figure 29-33

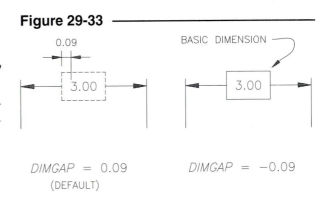

Text Placement **Section (*Text* Tab)**

Vertical (*DIMTAD*)
The *Vertical* option determines the vertical location of the text with respect to the dimension line. There are four possible settings that affect the *DIMTAD* variable (dim text above dimension line). *Centered*, the default option (*DIMTAD* = 0), centers the dimension text between the extension lines. The *Above* option places the dimension text above the dimension line except when the dimension line is not horizontal (*DIMTAD* = 1). *Outside* places the dimension text on the side of the dimension line farthest away from the extension line origin points—away from the dimensioned object (*DIMTAD* = 2). The *JIS* option places the dimension text to conform to Japanese Industrial Standards (*DIMTAD* = 3).

Horizontal (*DIMJUST*)
The *Horizontal* section determines the horizontal location of the text with respect to the dimension line (dimension justification). The default option (*DIMJUST* = 0) centers the text between the extension lines. The other four choices (*DIMJUST* = 1-3) place the text at either end of the dimension line in parallel and perpendicular positions (Fig. 29-34). You can use the *Horizontal* and *Vertical* settings together to achieve additional text positions.

Figure 29-34

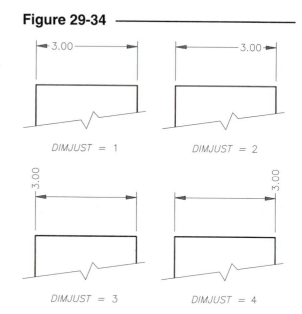

Offset from dim line (*DIMGAP*)
This value sets the distance between the dimension text and its dimension line. The offset is actually determined by an invisible box around the text (Fig. 29-35). Increasing or decreasing the value changes the size of the invisible box. The *Offset from dim line* value is stored in the *DIMGAP* variable. (Also see "*Draw frame around text.*")

Figure 29-35

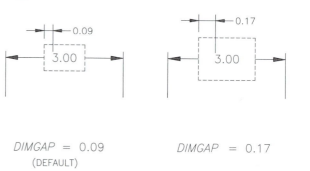

Text Alignment **Section, (*Text* Tab)**

The *Text Alignment* settings can be used in conjunction with the *Text Placement* settings to achieve a wide variety of dimension text placement options.

Horizontal (*DIMTIH*, *DIMTOH*)
This radio button turns on the *DIMTIH* (dim text inside horizontal) and *DIMTOH* (dim text outside horizontal) variables. The text remains horizontal even for vertical or angled dimension lines (see

Figure 29-36

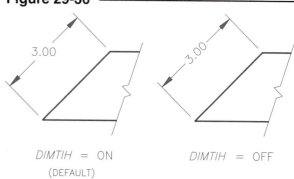

Figures 29-36 and 29-37). This is the correct setting for mechanical drawings (other than ordinate dimensions) according to the ASME Y14.5M-1994 standard, section 1.7.5, Reading Direction.

Aligned with dimension line **(DIMTIH, DIMTOH)**
Pressing this radio button turns off the *DIMTIH* and *DIMTOH* variables so the text aligns with the angle of the dimension line (see Figures 29-36 and 29-37).

ISO Standard **(DIMTIH, DIMTOH)**
This option forces the text inside the dimension line to align with the angle of the dimension line (*DIMTIH* = off), but text outside the dimension line is horizontal (*DIMTOH* = on).

Figure 29-37

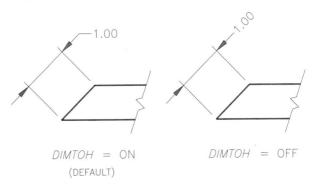

DIMTOH = ON
(DEFAULT)

DIMTOH = OFF

Fit Tab

The *Fit* tab allows you to determine the *Overall Scale* for dimensioning components, how the text, arrows, and dimension lines fit between extension lines, and how the text appears when it is moved (Fig. 29-38).

Scale for Dimension Features **Section (*Fit* Tab)**

Although this is not the first section in the dialog box, it is presented first because of its importance.

Figure 29-38

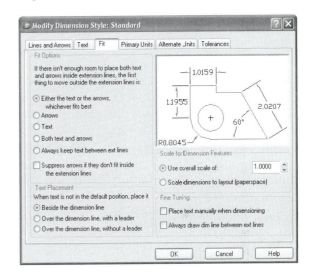

Overall Scale **(DIMSCALE)**
The *Overall Scale* value globally affects the scale of all size-related features of dimension objects, such as arrowheads, text height, extension line gaps (from the object), extensions (past dimension lines), etc. All other size-related (variable) values appearing in the dialog box series are multiplied by the *Overall Scale*. Notice how all dimensioning features (text, arrows, gaps, offsets) are all increased proportionally with the *Overall Scale* value (Fig. 29-39). Therefore, to keep all features proportional, change this one setting rather than each of the others individually.

Figure 29-39

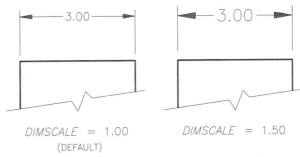

DIMSCALE = 1.00
(DEFAULT)

DIMSCALE = 1.50

Although this area is located on the right side of the box, it is probably the most important option in the entire series of tabs. Because the *Overall Scale* should be set as a family-wide variable, setting this value is typically the first step in creating a dimension style (Fig. 29-40).

Figure 29-40

Changes in this variable should be based on the *Limits* and plot scale. You can use the drawing scale factor to determine this value. (See "Drawing Scale Factor," Chapter 12.) The *Overall Scale* value is stored in the *DIMSCALE* variable.

Scale dimensions to layout (paper space) (DIMSCALE)

Checking this box forces dimension components to appear in the same size (*DIMSCALE*) for all viewports in a layout (Paper Space viewports created with *Vports*). Toggling this on sets *DIMSCALE* to 0. See Chapter 33 for detailed information on this subject.

Fit Options Section (Fit Tab)

This section determines which dimension components are <u>forced outside the extension lines only if there is insufficient room</u> for text, arrows, and dimension lines. In most cases where space permits all components to fit inside, the *Fit* settings have no effect on placement. *Linear, Aligned, Angular, Baseline, Continue, Radius,* and *Diameter* dimensions apply. See "*Radius* and *Diameter* Variable Settings" at the end of this section for information on recommended settings for *Radius* and *Diameter* dimensions.

Either the text or the arrows (DIMATFIT)

This is the default setting for the STANDARD dimension style. AutoCAD makes the determination of whether the text or the arrows are forced outside based on the size of arrows and the length of the text string (Fig. 29-41). This option often behaves similarly to *Arrows*, except that if the text cannot fit, the text is placed outside the extension lines and the arrows are placed inside. However, if the arrows cannot fit either, both text and arrows are placed outside the extension lines. The *DIMATFIT* (dimension arrows/text fit) setting is 3.

Figure 29-41

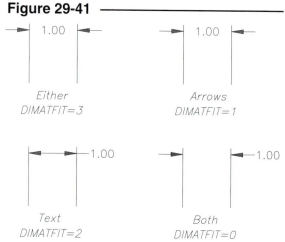

Arrows (DIMATFIT)

The *Arrows* option forces the arrows on the outside of the extension lines and keeps the text inside. If the text absolutely cannot fit, it is also placed outside the extension lines (see Figure 29-41). This option sets *DIMATFIT* to 1.

Text (DIMATFIT)

The *Text* option places the text on the outside and keeps the arrows on the inside unless the arrows cannot fit, in which case they are placed on the outside as well (see Figure 29-41). For this option, *DIMATFIT* = 2.

Both text and arrows (DIMATFIT)

The *Both text and arrows* option keeps the text and arrows together always. If space does not permit <u>both</u> features to fit between the extension lines, it places the text and arrows outside the extension lines (see Figure 29-41). You can set *DIMATFIT* to 0 to achieve this placement.

Always keep text between ext lines (DIMTIX)

If you want the text to be forced between the extension line no matter how much room there is, use this option (Fig. 29-42). Pressing this radio button turns *DIMTIX* on (text inside extensions).

Figure 29-42

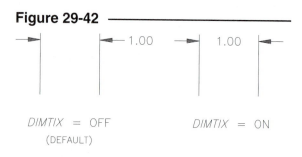

Suppress arrows if they don't fit **(DIMSOXD)**
When dimension components are forced outside the extension lines and there are many small dimensions aligned in a row (such as with *Continue* dimensions), the text, arrows, or dimension lines may overlap. In this case, you can prevent the arrows and the dimension lines from being drawn entirely with this option. This option suppresses the arrows and dimension lines <u>only</u> when they are forced outside (Fig. 29-43). The setting is stored as *DIMSOXD* = on (suppress dimension lines outside extensions).

Figure 29-43

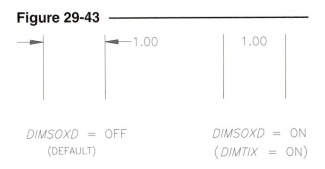

DIMSOXD = OFF
(DEFAULT)

DIMSOXD = ON
(DIMTIX = ON)

Text Placement **Section (*Fit* Tab)**

This section of the dialog box sets dimension text movement rules. When text is moved either by being automatically forced from between the dimension lines based on the *DIMATFIT* setting (the *Fit Options* above this section) or when you actually move the text with grips or by *Dimtedit*, these rules apply.

Beside the dimension line **(DIMTMOVE)**
This is the normal placement of the text—aligned with and beside the dimension line (Fig. 29-44). The text always moves when the dimension line is moved and vice versa. *DIMTMOVE* (dimension text move) = 0.

Figure 29-44

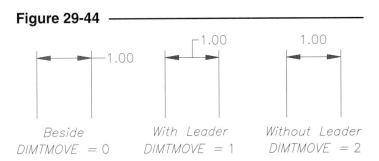

Beside
DIMTMOVE = 0

With Leader
DIMTMOVE = 1

Without Leader
DIMTMOVE = 2

Over the dimension line, with a leader **(DIMTMOVE)**
This option creates a leader between the text and the center of the dimension line whenever the text cannot fit between the extension lines or is moved using grips (see Figure 29-44). *DIMTMOVE* = 1.

Over the dimension line, without a leader **(DIMTMOVE)**
Use this setting to have the text appear above the dimension line, similar to *DIMTMOVE* = 1, but without a leader. This occurs only when there is insufficient room for the text between the extension lines or when you move the text with grips. *DIMTMOVE* = 2.

There is an important benefit to this setting (*DIMTMOVE* = 2). When there is sufficient room for the text and arrows between extension lines, this setting has no effect on the placement of the text. When there is insufficient room, the text moves above without a leader. In either case, if you prefer to move the text to another location with grips or using *Dimtedit*, the text moves as if were "detached" from the dimension line. The text can be moved independently to any location and the dimension retains its associatively. Figure 29-45 illustrates the use of grips to edit the dimension text.

Figure 29-45

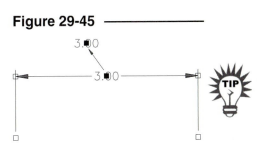

Without Leader (DIMTMOVE=2)

Fine Tuning **Section (***Fit* **Tab)**

Figure 29-46

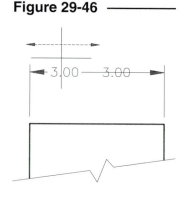

Place text manually when dimensioning **(DIMUPT)**
When you press this radio button, you can create dimensions and move the text independently in relation to the dimension and extension lines as you place the dimension line in response to the "specify dimension line location" prompt (Fig. 29-46). Using this option is similar to using *Dimtedit* after placing the dimension. *DIMUPT* (dimension user-positioned text) is on when this box is checked.

DIMUPT = ON

Always draw dim line between ext lines **(DIMTOFL)**
Occasionally, you may want the dimension line to be drawn inside the extension lines even when the text and arrows are forced outside. You can force a line inside with this option (Fig. 29-47). A check in this box turns *DIMTOFL* on (text outside, force line inside).

Figure 29-47

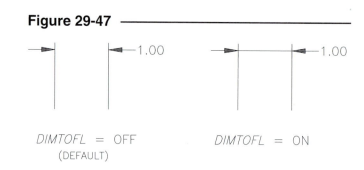

DIMTOFL = OFF
(DEFAULT)

DIMTOFL = ON

***Radius* and *Diameter* Variable Settings**
For creating mechanical drawing dimensions according to ANSI standards, the default settings in AutoCAD are correct for creating *Diameter* dimensions but not for *Radius* dimensions. The following variable settings are recommended for creating *Radius* and *Diameter* dimensions for mechanical applications.

For *Diameter* dimensions, the default settings produce ANSI-compliant dimensions that suit most applications—that is, text and arrows are on the outside of the circle or arc pointing inward toward the center. For situations where large circles are dimensioned, *Fit Options* (*DIMATFIT*) and *Place text manually* (*DIMUPT*) can be changed to force the dimension inside the circle (Fig. 29-48).

Figure 29-48

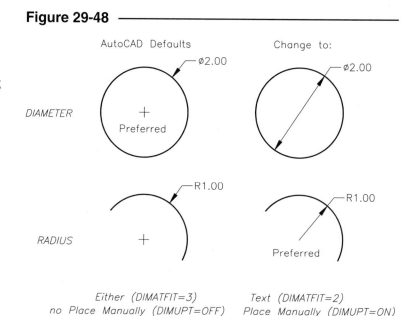

For *Radius* dimensions the default settings produce incorrect dimensioning practices. Normally (when space permits) you want the dimension line and arrow to be inside the arc, while the text can be inside or outside. To produce ANSI-compliant *Radius* dimensions, set *Fit Options* (*DIMAT-FIT*) and *Place text manually* (*DIMUPT*) as shown in Figure 29-48. Radius dimensions can be outside the arc in cases where there is insufficient room inside.

Primary Units Tab

The *Primary Units* tab controls the format of the AutoCAD-measured numerical value that appears with a dimension (Fig. 29-49). You can vary the numerical value in several ways such as specifying the units format, precision of decimal or fraction, prefix and/or suffix, zero suppression, and so on. These units are called primary units because you can also cause AutoCAD to draw additional or secondary units called *Alternate Units* (for inch <u>and</u> metric notation, for example).

Figure 29-49

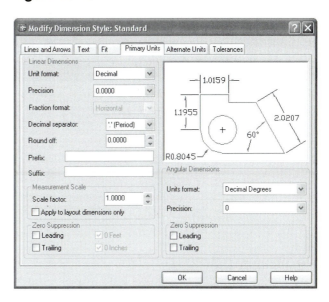

Linear Dimensions **Section (*Primary Units* Tab)**

This section controls the format of all dimension types except *Angular* dimensions.

Unit format **(*DIMLUNIT*)**
The *Units format* section drop-down list specifies the type of units used for dimensioning. These are the same unit types available with the *Units* dialog box (*Decimal, Scientific, Engineering, Architectural,* and *Fractional*) with the addition of *Windows Desktop*. The *Windows Desktop* option displays AutoCAD units based on the settings made for units display in Windows Control Panel (settings for decimal separator and number grouping symbols). Remember that your selection affects the <u>units drawn in dimension objects, not the global drawing units</u>. The choice for *Units format* is stored in the *DIMLUNIT* variable (dimension linear unit). This drop-down list is disabled for an *Angular* family member.

Precision **(*DIMDEC*)**
The *Precision* drop-down list in the *Dimension* section specifies the number of places for decimal dimensions or denominator for fractional dimensions. This setting does not alter the drawing units precision. This value is stored in the *DIMDEC* variable (dim decimal).

Fraction format **(*DIMFRAC*)**
Use this drop-down list to set the fractional format. The choices are displayed in Figure 29-50. This option is enabled only when *DIMLUNIT* (*Unit format*) is set to 4 (*Architectural*) or 5 (*Fractional*).

Figure 29-50

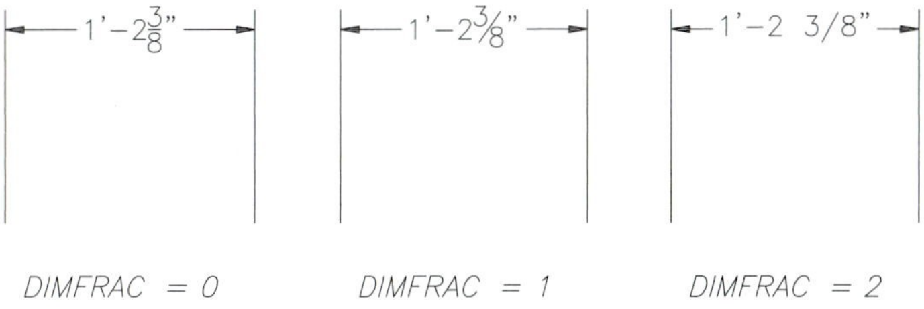

$\leftarrow 1'-2\frac{3}{8}" \rightarrow$	$\leftarrow 1'-2\frac{3}{8}" \rightarrow$	$\leftarrow 1'-2\ 3/8" \rightarrow$
DIMFRAC = 0	*DIMFRAC = 1*	*DIMFRAC = 2*

Decimal separator **(*DIMDSEP*)**
When you are creating dimensions whose unit format is decimal, you can specify a single-character decimal separator. Normally a decimal (period) is used; however, you can also use a comma (,) or a space. The character is stored in the *DIMDSEP* (dimension decimal separator) variable.

Round off **(DIMRND)**

Use this drop-down list to specify a precision for dimension values to be rounded. Normally, AutoCAD values are kept to 14 significant places but are rounded to the place dictated by the dimension *Precision* (*DIMDEC*). Use this feature to round up or down appropriately to the nearest specified decimal or fractional increment (Fig. 29-51).

Figure 29-51

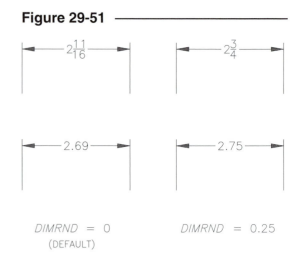

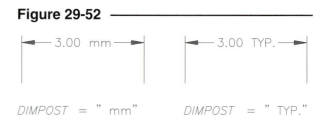

Prefix/Suffix **(DIMPOST)**

The *Prefix* and *Suffix* edit boxes hold any text that you want to add to the AutoCAD-supplied dimensional value. A text string entered in the *Prefix* edit box appears before the AutoCAD-measured numerical value and a text string entered in the *Suffix* edit box appears after the AutoCAD-measured numerical value. For example, entering the string " mm" or a " TYP." in the *Suffix* edit box would produce text as shown in Figure 29-52. (In such a case, don't forget the space between the numerical value and the suffix.) The string is stored in the *DIMPOST* variable.

Figure 29-52

If you use the *Prefix* box to enter letters or values, any AutoCAD-supplied symbols (for radius and diameter dimensions) are overridden (not drawn). For example, if you want to specify that a specific hole appears twice, you should indicate by designating a "2X" before the diameter dimension. However, doing so by this method overrides the phi (Ø) symbol that AutoCAD inserts before the value. Instead, use the *Mtext/Dtext* options within the dimensioning command or use *Dimedit* or *Ddedit* to add a prefix to an existing dimension having an AutoCAD-supplied symbol. Remember that AutoCAD uses the < > brackets in the *Mtext Editor* to represent the AutoCAD-measured value, so place a prefix in front of the brackets. Do not overwrite the brackets unless you want to lose the AutoCAD-measured value.

Measurement Scale Section (*Primary Units* Tab)

Scale factor **(DIMLFAC)**

Any value placed in the *Scale factor* edit box is a <u>multiplier</u> for the AutoCAD-measured numerical value. The default is 1. Entering a 2 would cause AutoCAD to draw a value two times the actual measured value (Fig. 29-53). This feature might be used when a drawing is created in some scale other than the actual size, such as an enlarged detail view in the same drawing as the full view. Since AutoCAD 2002, changing this setting is <u>unnecessary</u> when you create associative dimensions in paper space attached to objects in model space (see "Dimensioning in Paper Space").

Figure 29-53

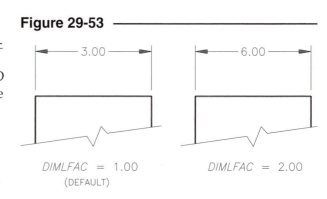

However, if you are using versions of AutoCAD previous to 2002, or if you are using AutoCAD 2002 or later nonassociative dimensions in paper space, change this setting to create dimensions that display other than the actual measured value. This setting is stored in the *DIMLFAC* variable (dimension length factor).

Apply to layout dimension only (DIMLFAC)

Use this checkbox to apply the *Scale factor* to dimensions placed in a layout for AutoCAD 2002 or later nonassociative dimensions or for drawings in versions earlier than AutoCAD 2002. For example, assume you have a detail view displayed at 2:1 in a viewport and want to place nonassociative dimensions in paper space, set the *Scale factor* to .5 and check *Apply to layout dimension only* so the measured values adjust for the viewport scale. A check in this box sets the *DIMLFAC* variable to the negative of the *Scale factor* value. Associative dimensions in AutoCAD 2002 and 2004 drawn in paper space automatically adjust for the viewport scale, so changing this setting is unnecessary (see "Dimensioning in Paper Space").

Zero Suppression Section (*Primary Units* Tab)

Figure 29-54

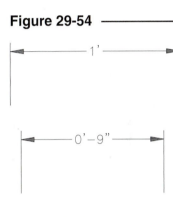

The *Zero Suppression* section controls how zeros are drawn in a dimension when they occur. A check in one of these boxes means that zeros are <u>not drawn for that case</u>. This sets the value for *DIMZIN* (dimension zero indicator).

Leading/Trailing/0 Feet/0 Inches (DIMZIN)

Leading and *Trailing* are enabled for *Scientific, Decimal, Engineering* and *Fractional* units. The *0 Feet* and *0 Inches* checkboxes are enabled for *Architectural* and *Engineering* units.

For example, assume primary *Units format* was set to *Architectural* and *Zero Suppression* was checked for *0 Inches* only. Therefore, when a measurement displays feet and no inches, the 0 inch value is suppressed (Fig. 29-54, top dimension). On the other hand, since *0 Feet* is not checked, a measurement of less than 1 foot would report 0 feet (Fig. 29-54, bottom dimension).

Angular Dimensions Section (*Primary Units* Tab)

Units format (DIMAUNIT)

The *Units format* drop-down list sets the unit type for angular dimensions, including *Decimal degrees, Degrees Minutes Seconds, Gradians,* and *Radians*. This list is enabled only for parent dimension style and *Angular* family members. The variable used for the angular units is *DIMAUNIT* (dimension angular units).

Precision (DIMADEC)

This option sets the number of places of precision (decimal places) for angular dimension text. The selection is stored in the *DIMADEC* variable (dimension angular decimals). This option is enabled only for the parent dimension style and *Angular* family member.

Zero Suppression (DIMAZIN)

Use these two checkboxes to set the desired display for angular dimensions when there are zeros appearing before or after the decimal. A check in one of these boxes means that zeros are <u>not drawn for that case</u>.

Alternate Units Tab

This tab allows you to display alternate, or secondary, dimensioning values along with the primary AutoCAD-measured value when a dimension is created (Fig. 29-55). Typically, alternate units consist of millimeter values given in addition to the decimal inch values. This practice is often called "dual dimensioning." The options in this tab control the display and format of alternate units. Notice that most of these options are equivalent to those found in the *Primary Units* tab.

Figure 29-55 ────────────

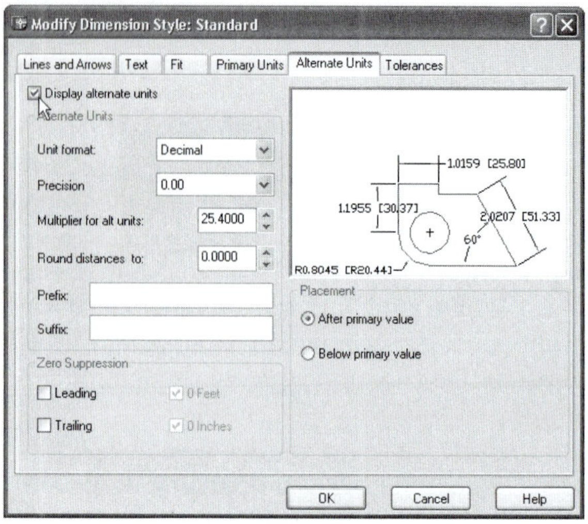

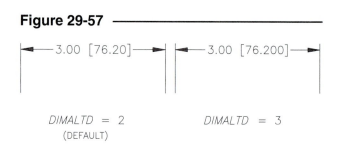

Display alternate units (DIMALT)
By default, AutoCAD displays only one value for a dimension—the primary unit. If you want to have AutoCAD measure and create an additional value for each dimension, check this box (Fig. 29-56). All options in this tab are disabled unless *Display alternate units* is checked. The presence of alternate units is controlled by setting the *DIMALT* variable on (dimension alternate).

Figure 29-56 ────────────

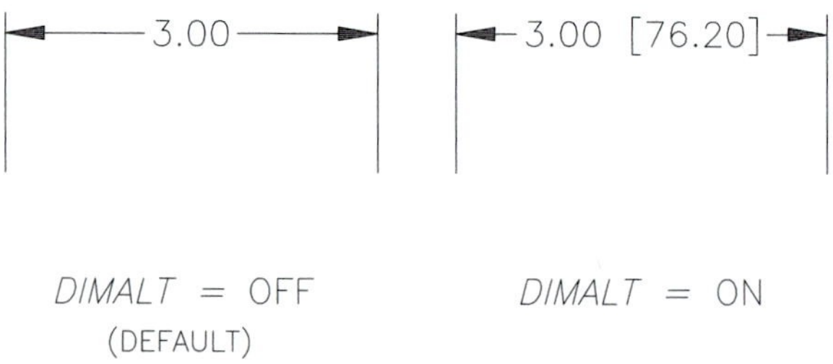

|←──── 3.00 ────→| |←─ 3.00 [76.20] →|

DIMALT = OFF   DIMALT = ON
(DEFAULT)

Alternate Unit Section (*Alternate Units* Tab)

Unit format (DIMALTU)
This drop-down list sets the format for alternate units. The setting is stored in the *DIMALTU* variable (dimension alternate units). This is the alternate units equivalent to *Units format* for *Primary Units*.

Precision (DIMALTD)
Use this drop-down list to set the decimal precision for alternate units. Note that the alternate units precision is controlled independently of the primary units precision (Fig. 29-57). The value is stored in the *DIMALTD* variable (dimension alternate decimals).

Figure 29-57 ────────────

|←─ 3.00 [76.20] →| |←─ 3.00 [76.200] →|

DIMALTD = 2   DIMALTD = 3
(DEFAULT)

Multiplier for alt units (DIMALTF)
This is the alternate units equivalent for the primary units scale factor (*DIMLFAC*). In other words, the AutoCAD-measured primary units value is multiplied by this factor to determine the displayed value for alternate units (Fig. 29-58). You can also enter the desired multiplier in the *DIMALTF* (dimension alternate factor) variable.

Figure 29-58 ────────────

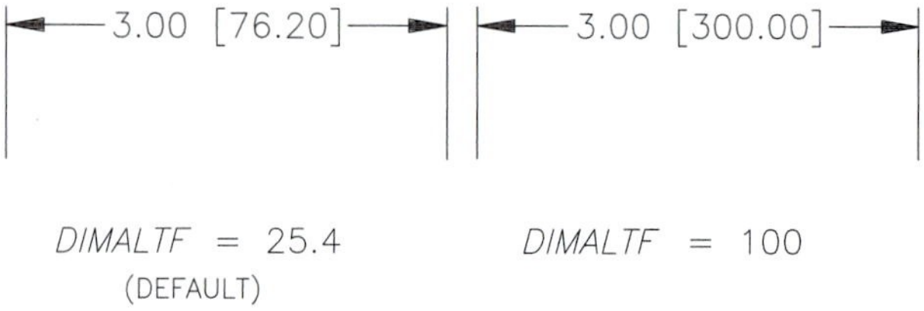

|←─ 3.00 [76.20] →| |←─ 3.00 [300.00] →|

DIMALTF = 25.4   DIMALTF = 100
(DEFAULT)

Round distances to (DIMALTRND)

If you do not want the alternate units value to be displayed as the actual measurement, but to be rounded to nearest regular increment, enter the desired increment in this edit box. For example, you may want alternate units to be displayed with two places of precision (to the right of the decimal) but to round to the nearest millimeter. Alternately, set the increment using the *DIMALTRND* variable (alternate rounding). This is the alternate units equivalent to the *Round off* option in the *Primary Units* tab (see Figure 29-51).

Prefix/Suffix (DIMAPOST)

This section allows you to include a prefix and/or suffix with the alternate units measured value. For example, you may want to display " mm" after the alternate units value (Fig. 29-59). The prefix and/or suffix is stored in the *DIMAPOST* variable. (See also *Prefix/Suffix* in the *Primary Units* tab for more information.)

Figure 29-59

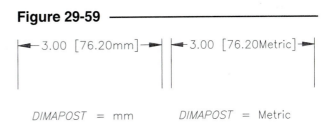

Zero Suppression Section (Alternate Units Tab)

The *Zero Suppression* section controls how zeros are drawn in alternate units values when they occur. A check in one of these boxes means that zeros are <u>not drawn for that case</u>. Your selection sets the value for *DIMALTZ*. See the description for *Zero Suppression, Primary Units* Tab for more information and illustration.

Placement Section (Alternate Units Tab)

This section only has two options; both are related to the *DIMPOST* variable. Normally, the alternate units values are placed *After primary value*. You can instead toggle *Below primary value* to yield a display as shown in Figure 29-60, right dimension. Selecting *Below primary value* sets the *DIMPOST* variable to "\x."

Figure 29-60

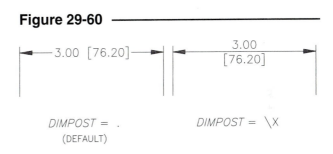

Tolerances Tab

The *Tolerances* tab allows you to create several formats of tolerance dimensions such as limits, two forms of plus/minus dimensions, and basic dimensions (Fig. 29-61). Most options in this tab are disabled until you select a *Method*.

Figure 29-61

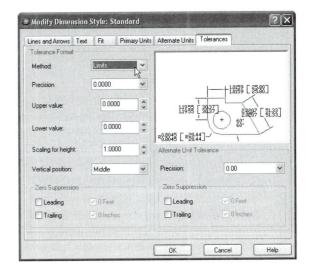

Tolerance Format Section (*Tolerances* Tab)

Method (DIMTOL, DIMLIM, DIMGAP)

The *Method* option displays a drop-down list with five types: *None, Symmetrical, Deviation, Limits,* and *Basic*. The four possibilities (other than *None*) are illustrated in Figure 29-62. The *Symmetrical* and *Deviation* methods create plus/minus dimensions and turn on the *DIMTOL* variable (dimension tolerance). The *Limits* method creates limit dimensions and turns the *DIMLIM* variable on (dimension limits). The *Basic* method creates a basic dimension by drawing a box around the dimensional value, which is accomplished by changing the *DIMGAP* to a negative value.

Figure 29-62

3.00±0.0050
Method: = Symmetrical
DIMTOL = ON
DIMTP = 0.0050
3.00+0.0050 −0.0030
Method: = Deviation
DIMTOL = ON
DIMTP = 0.0050
DIMTM = 0.0030
3.0050
2.9970
Method: = Limits
DIMLIM = ON
DIMTP = 0.0050
DIMTM = 0.0030
3.00
Method: = Basic
DIMGAP = −.09

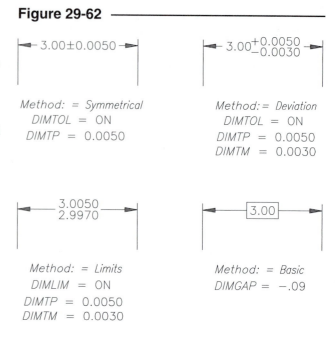

Precision (DIMTDEC)

Use the *Precision* drop-down list to set the precision (number of places to the right of the decimal place) for values when drawing *Symmetrical, Deviation,* and *Limits* dimensions. The precision is stored in the *DIMTDEC* variable (dimension tolerance decimals).

Upper value/Lower value (DIMTP, DIMTM)

An *Upper Value* and *Lower Value* can be entered in the edit boxes for the *Deviation* and *Limits* method types. In these cases, the *Upper Value* (*DIMTP*—dimension tolerance plus) is added to the measured dimension and the *Lower Value* (*DIMTP*—dimension tolerance minus) is subtracted (Fig. 29-62). An *Upper Value* only is needed for *Symmetrical* and is applied as both the plus and minus value.

Scaling for height (DIMTFAC)

The height of the tolerance text is controlled with the *Scaling for height* edit box value. The entered value is a proportion of the primary dimension value height. For example, a value of .50 would draw the tolerance text at 50% of the primary text height (Fig. 29-63). The setting affects the *Symmetrical, Deviation,* and *Limits* tolerance methods. The value is stored in the *DIMTFAC* (dimension tolerance factor) variable. Note that this is the same variable that controls the *Fraction height scale* for architectural and fractional values.

Figure 29-63

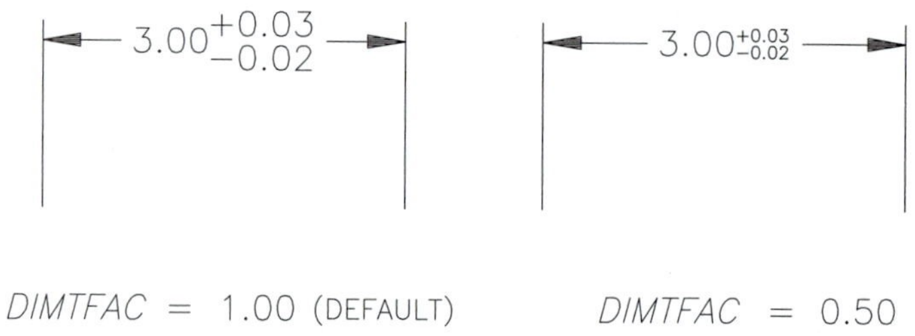

3.00+0.03 −0.02
DIMTFAC = 1.00 (DEFAULT)
3.00+0.03 −0.02
DIMTFAC = 0.50

Vertical position (DIMTOLJ)

For *Deviation, Symmetrical,* and *Limits* tolerance methods, you can control the placement of the tolerance values in relation to the primary units values (Fig. 29-64). The choices are *Top, Middle,* and *Bottom* and set the *DIMTOLJ* variable (dimension tolerance justification) to 2, 1, and 0, respectively. The *Top* option aligns the primary text and the top tolerance value, and the *Bottom* option aligns the primary text and the bottom tolerance value.

Figure 29-64

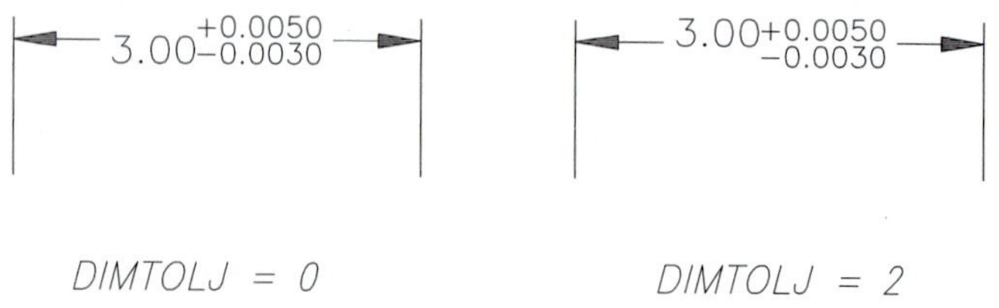

3.00+0.0050 −0.0030
DIMTOLJ = 0
3.00+0.0050 −0.0030
DIMTOLJ = 2

Zero Suppression (DIMTZIN) (*Tolerance Format* Section)

The *Zero Suppression* section controls how zeros are drawn in *Symmetrical, Deviation,* and *Limits* tolerance dimensions when they occur. A check in one of these boxes means that zeros are not drawn for that case. The *Zero Suppression* section here operates identically to the *Zero Suppression* section of the *Primary Units* tab, except that it is applied to tolerance dimensions (see *Zero Suppression, Primary Units* Tab for more information). The affected variable is *DIMTZIN* (dimension tolerance zero indicator).

Alternate Unit Tolerance Section (*Tolerances* Tab)

When you have specified that alternate units are to be drawn (in the *Alternate Units* tab) <u>and</u> you have turned on some form of tolerance *Method* (*Symmetrical, Limits,* etc.), the alternate units will automatically display as a tolerance along with the primary units. In other words, when alternate units are on, both primary and alternate units are drawn the same—with or without tolerances.

Precision (DIMALTTD)

This drop-down list sets the decimal precision for the alternate units tolerances (when alternate units and tolerances are on). The *DIMALTTD* variable (dimension alternate tolerance decimal) holds the setting.

Zero Suppression (DIMALTTZ)

When alternate units and tolerances are on, use these check boxes to determine how leading and trailing zeros are treated. The setting is stored in the *DIMALTTZ* variable (dimension alternate tolerance zeros). See *Zero Suppression* in the *Primary Units* Tab section for more information on zero suppression.

Changing Dimension Variables Using the Command Line Format

DIM...
(VARIABLE NAME)

Pull-down Menu	COMMAND (TYPE)	ALIAS (TYPE)	Short-cut	Screen (side) Menu	Tablet Menu
Tools *Inquiry >* *Set Variable*	*(VARIABLE NAME)*	...	...	TOOLS 1 *Setvar*	U,10

As an alternative to setting dimension variables through the *Dimension Style Manager,* you can type the dimensioning variable name at the Command: prompt. There is a noticeable difference in the two methods—the dialog boxes use <u>different nomenclature</u> than the formal dimensioning variable names used in Command line format; that is, the dialog boxes use descriptive terms that, if selected, make the appropriate change to the dimensioning variable. The formal dimensioning variable names accessed by Command line format, however, all begin with the letters *DIM* and are accessible only by typing.

Another important but subtle difference in the two methods is the act of saving dimension variable settings to a dimension style. Remember that all drawn dimensions are part of a dimension style, whether it is STANDARD or some user-created style. When you change dimension variables by the Command line format, the changes become <u>overrides</u> until you use the *Save* option of the -*Dimstyle* command or the *Dimension Style Manager.* When a variable change is made, it becomes an <u>override that is applied to the current dimension style</u> and affects only the newly drawn dimensions. Variable changes must be *Saved* to become a permanent part of the style and to retroactively affect all dimensions created with that style.

In order to access and change a variable's setting by name, simply type the variable name at the Command: prompt. For dimension variables that require distances, you can enter the distance (in any format accepted by the current *Units* settings) or you can designate by (PICKing) two points.

For example, to change the value of the *DIMSCALE* to .5, this command syntax is used:

```
Command: dimscale
Enter new value for dimscale <1.0000>: .5
```

Associative Dimensions

Since AutoCAD 2002, dimensions have full associativity. The *DIMASSOC* variable controls the associative feature. The *DIMASSOC* variable setting cannot be saved in a dimension style; therefore, one dimension style can contain associative and nonassociative dimensions, but not exploded dimensions (exploded dimensions can be created from a dimension style, but cannot be updated since they do not reference the style). The *Dimension Style Manager* does not provide access to the *DIMASSOC* variable. *DIMASSOC* must be changed by Command line format or by using the *Options* dialog box (see "Associative Dimensions" in Chapter 28).

Dimension Variables Table

This table is a summary of the 70 dimensioning variables in AutoCAD 2004. Each variable has a brief description, the type of variable it is, and its <u>default setting</u> (taken from the *AutoCAD Command Reference*). Remember that the dimension variables and the current settings can be listed using the *STatus* and *Variables* options of the *-Dimstyle* command.

Variable	Characteristics	Description
DIMADEC	Type: Integer Saved in: Drawing Initial value: 0	Controls the number of places of precision displayed for angular dimension text. When *DIMADEC* is set to a value from 0 to 8, angular dimensions display precisions that differ from their linear dimension values: 0-8 Angular dimension is drawn using the number of decimal places corresponding to the *DIMADEC* setting.
DIMALT	Type: Switch Saved in: Drawing Initial value: Off	When turned on, enables alternate units dimensioning. See also *DIMALTD, DIMALTF, DIMALTZ (DIMALTTZ, DIMALTTD)*, and *DIMAPOST*.
DIMALTD	Type: Integer Saved in: Drawing Initial value: 2	Controls alternate units decimal places. If *DIMALT* is enabled, *DIMALTD* governs the number of decimal places displayed in the alternate measurement.
DIMALTF	Type: Real Saved in: Drawing Initial value: 25.4000	Controls alternate units scale factor. If *DIMALT* is enabled, *DIMALTF* multiplies linear dimensions by a factor to produce a value in an alternate system of measurement.
DIMALTRND	Type: Real Saved in: Drawing Initial value: 0.00	Rounds off the alternate dimension units.
DIMALTTD	Type: Integer Saved in: Drawing Initial value: 2	Sets the number of decimal places for the tolerance values of an alternate units dimension. *DIMALTTD* sets this value when entered on the command line or set in the *Dimension Style Manager*.

Variable	Characteristics	Description
DIMALTTZ	Type: Integer Saved in: Drawing Initial value: 0	Toggles suppression of zeros in tolerance values. 0 Suppresses zero feet and precisely zero inches 1 Includes zero feet and precisely zero inches 2 Includes zero feet and suppresses zero inches 3 Includes zero inches and suppresses zero feet To the preceding values, add: 4 Suppresses leading zeros 8 Suppresses trailing zeros
DIMALTU	Type: Integer Saved in: Drawing Initial value: 2	Sets the units format for alternate units of all dimension style family members except angular 1 Scientific 2 Decimal 3 Engineering 4 Architectural (stacked) 5 Fractional (stacked) 6 Architectural 7 Fractional 8 Windows® Desktop (decimal format using Control Panel settings for decimal separator and number grouping symbols)
DIMALTZ	Type: Integer Saved in: Drawing Initial value: 0	Controls the suppression of zeros for alternate unit dimension values. *DIMALTZ* values 0–3 affect feet-and-inch dimensions only. 0 Suppresses zero feet and precisely zero inches 1 Includes zero feet and precisely zero inches 2 Includes zero feet and suppresses zero inches 3 Includes zero inches and suppresses zero feet 4 Suppresses leading zeros in decimal dimensions (for example, 0.5000 becomes .5000) 8 Suppresses trailing zeros in decimal dimensions (for example, 12.5000 becomes 12.5) 12 Suppresses both leading and trailing zeros (for example, 0.5000 becomes .5)
DIMAPOST	Type: String Saved in: Drawing Initial value: ""	Specifies a text prefix or suffix (or both) to the alternate dimension measurement for all types of dimensions except angular. For instance, if the current *Units* mode is Architectural, *DIMALT* is enabled, *DIMALTF* is 25.4, *DIMALTD* is 2, and *DIMAPOST* is set to "mm," a distance of 10 units would be edited as 10"[254.00mm]. To disable an established prefix or suffix (or both), set it to a period.
DIMASO	Type: Switch Saved in: Drawing Initial value: On Obsolete in AutoCAD 2002 and 2004. (see *DIMASSOC*)	Controls the associativity of dimension objects. Off Creates no association between the various elements of the dimension. The lines, arcs, arrowheads, and text of a dimension are drawn as separate objects. On Creates an association between the elements of the dimension. The elements are formed into a single object. If the definition point on the object moves, the dimension value is updated. *DIMASO* is not stored in a dimension style.

Variable	Characteristics	Description
DIMASSOC	Type: Integer Saved in: Drawing Initial value: 2	Controls the associativity of dimension objects. 0 Creates exploded dimensions. There is no association between the various elements of the dimension. The lines, arcs, arrowheads, and text of a dimension are drawn as separate objects. 1 Creates nonassociative dimension objects. The elements of the dimension are formed into a single object. If the definition point on the object moves, the dimension value is updated. 2 Creates associative dimension objects. The elements of the dimension are formed into a single object, and one or more definition points of the dimension are coupled with association points on geometric objects. If the association point on the geometric object moves, the dimension location, orientation, and value is updated. DIMASSOC is not stored in a dimension style. Drawings saved in a format previous to AutoCAD 2002 retain the setting of the DIMASSOC system variable. When the drawing is reopened in AutoCAD 2002 or later, the dimension associativity setting is restored. If a legacy drawing is opened in AutoCAD 2002, the DIMASSOC system variable takes on the value of the legacy drawing's DIMASO system variable.
DIMASZ	Type: Real Saved in: Drawing Initial value: 0.1800	Controls the size of dimension line and leader line arrowheads. Also controls the size of hook lines. Multiples of the arrowhead size determine whether dimension lines and text are to fit between the extension lines. Also used to scale arrowhead blocks if set to DIMBLK. DIMASZ has no effect when DIMTSZ is other than zero.
DIMATFIT	Type: Integer Saved in: Drawing Initial value: 3	Determines how dimension text and arrows are a arranged when space is not sufficient to place both within the extension lines. 0 Places both text and arrows outside extension lines 1 Moves arrows first, then text 2 Moves text first, then arrows 3 Moves either text or arrows, whichever fits best AutoCAD adds a leader to moved dimension text when DIMTMOVE is set to 1.
DIMAUNIT	Type: Integer Saved in: Drawing Initial value: 0	Sets the angle format for angular dimension: 0 Decimal degrees 1 Degrees/minutes/seconds 2 Gradians 3 Radians

Variable	Characteristics	Description

DIMAZIN	Type: Integer
	Saved in: Drawing
	Initial value: 0
		Suppresses zeros for angular dimensions.
		0 Displays all leading and trailing zeros
		1 Suppresses leading zeros in decimal dimensions
		     (for example, 0.5000 becomes .5000)
		2 Suppresses trailing zeros in decimal dimensions
		     (for example, 12.5000 becomes 12.5)
		3 Suppresses leading and trailing zeros
		     (for example, 0.5000 becomes .5)

DIMBLK	Type: String Saved in: Drawing Initial value: ""	Sets the arrowhead block displayed at the ends of dimension lines or leader lines. To turn off arrowheads, enter a single period (.). Arrowhead block entries and *Override* Dimension Style dialog boxes are shown below. You can also enter the names of user-defined arrowhead blocks. "" closed filled "_DOT" dot "_DOTSMALL" dot small "_DOTBLANK" dot blank "_ORIGIN" origin indicator "_ORIGIN2" origin indicator 2 "_OPEN" open "_OPEN90" right angle "_OPEN30" open 30 "_CLOSED" closed "_SMALL" dot small blank "_NONE" none "_OBLIQUE" oblique "_BOXFILLED" box filled "_BOXBLANK" box "_CLOSEDBLANK" closed blank "_DATUMFILLED" datum triangle filled "_DATUMBLANK" datum triangle "_INTEGRAL" integral "_ARCHTICK" architectural tick
DIMBLK1	Type: String Saved in: Drawing Initial value: ""	If *DIMSAH* is on, *DIMBLK1* specifies user-defined arrowhead blocks for the first end of the dimension line. This variable contains the name of a previously defined block. To disable an established block name, set it to a single period (.). For a list of arrowheads, see *DIMBLK*.
DIMBLK2	Type: String Saved in: Drawing Initial value: ""	If *DIMSAH* is on, *DIMBLK2* specifies user-defined arrowhead blocks for the second end of the dimension line. This variable contains the name of a previously defined block. To disable an established block name, set it to a single period (.). For a list of arrowheads, see *DIMBLK*.

Variable	Characteristics	Description
DIMCEN	Type: Real Saved in: Drawing Initial value: 0.0900	Controls drawing of circle or arc center marks and centerlines by the *DIMCENTER*, *DIMDIAMETER*, and *DIMRADIUS* dimensioning commands. 0 No center marks or lines are drawn. <0 Centerlines are drawn. >0 Center marks are drawn. The absolute value specifies the size of the mark portion of the centerline. *DIMRADIUS* and *DIMDIAMETER* draw the center mark or line only if the dimension line is placed outside the circle or arc.
DIMCLRD	Type: Integer Saved in: Drawing Initial value: 0	Assigns colors to dimension lines, arrowheads, and dimension leader lines. Also controls the color of leader lines created with the *LEADER* command. The color can be any valid color number or special color label *BYBLOCK* or *BYLAYER*. Using the *SETVAR* command, supply the color number. Integer equivalents for *BYBLOCK* and *BYLAYER* are 0 and 256, respectively. From the Command: prompt, set the color values by entering *DIMCLRD* and then a standard color name or *BYBLOCK* or *BYLAYER*.
DIMCLRE	Type: Integer Saved in: Drawing Initial value: 0	Assigns colors to dimension extension lines. The color can be any valid color number or the special color label *BYBLOCK* or *BYLAYER*. See *DIMCLRD*.
DIMCLRT	Type: Integer Saved in: Drawing Initial value: 0	Assigns colors to dimension text. The color can be any valid color number or the special color label *BYBLOCK* or *BYLAYER*. See *DIMCLRD*.
DIMDEC	Type: Integer Saved in: Drawing Initial value: 4	Sets the number of decimal places for the tolerance values of a primary units dimension. The precision is based on the units or angle format you have selected.
DIMDLE	Type: Real Saved in: Drawing Initial value: 0.0000	Extends the dimension line beyond the extension line when oblique strokes are drawn instead of arrowheads.
DIMDLI	Type: Real Saved in: Drawing Initial value: 0.3800	Controls the dimension line spacing for baseline dimensions. Each baseline dimension is offset by this amount, if necessary, to avoid drawing over the previous dimension. Changes made with *DIMDLI* are not applied to existing dimensions.
DIMDSEP	Type: Single character Saved in: Drawing Initial value: Decimal point	Specifies a single-character decimal separator to use when creating dimensions whose unit format is decimal. When prompted, enter a single character at the Command line. If dimension units is set to *Decimal*, the *DIMDSEP* character is used instead of the default decimal point. If *DIMDSEP* is set to NULL (default value, reset by entering a period), AutoCAD uses the decimal point as the dimension separator.

Variable	Characteristics	Description
DIMEXE	Type: Real Saved in: Drawing Initial value: 0.1800	Determines how far to extend the extension line beyond the dimension line.
DIMEXO	Type: Real Saved in: Drawing Initial value: 0.0625	Determines how far extension lines are offset from origin points. If you point directly at the corners of an object to be dimensioned, the extension lines stop just short of the object.
DIMFIT	Type: Integer Saved in: Drawing Initial value: 3	<u>Obsolete</u>. Has no effect in AutoCAD 2000 and later releases except to preserve the integrity of pre-AutoCAD 2000 scripts and AutoLISP routines. *DIMFIT* is replaced by *DIMATFIT* and *DIMTMOVE*.
DIMFRAC	Type: Integer Saved in: Drawing Initial value: 0	Sets the fraction format when *DIMLUNIT* is set to 4 (Architectural) or 5 (Fractional). 0 Horizontal 1 Diagonal 2 Not stacked (for example, 1/2)
DIMGAP	Type: Real Saved in: Drawing Initial value: 0.0900	Sets the distance around the dimension text when you break the dimension line to accommodate dimension text. Also sets the gap between annotation and a hook line created with the *LEADER* command. A negative *DIMGAP* value creates basic dimensioning—dimension text with a box around its full extents. AutoCAD also used *DIMGAP* as the minimum length for pieces of the dimension line. When calculating the default position for the dimension text, it positions the text inside the extension lines only if doing so breaks the dimension lines into two segments at least as long as *DIMGAP*. Text placed above or below the dimension line is moved inside if there is room for the arrowheads, dimension text, and a margin between them at least as large as *DIMGAP*: 2 * (*DIMASZ* + *DIMGAP*). *DIMGAP* also sets the gap between a tolerance symbol and its feature control frame.
DIMJUST	Type: Integer Saved in: Drawing Initial value: 0	Controls horizontal dimension text position: 0 Center justifies the text between the extension lines. 1 Positions the text next to the first extension line. 2 Positions the text next to the second extension line. 3 Positions the text above and aligned with the first extension line. 4 Positions the text above and aligned with the second extension line.
DIMLDRBLK	Type: String Saved in: Drawing Initial value: ""	Specifies the arrow type for leaders. To turn off arrowhead display, enter a single period (.). For a list of arrowhead entries, see *DIMBLK*.

2002

Variable	Characteristics	Description
DIMLFAC	Type: Real Saved in: Drawing Initial value: 1.0000	Sets a global scale factor for linear dimensioning measurements. All linear distances measured by dimensioning (including radii, diameters, and coordinates) are multiplied by the *DIMLFAC* setting before being converted to dimension text. *DIMLFAC* has no effect on angular dimensions, and it is not applied to the values held in *DIMTM*, *DIMTP*, or *DIMRND*. Since AutoCAD 2002, this variable is not needed if you create <u>associative</u> dimensions in paper space attached to objects in model space. However, if you are using AutoCAD 2002-style <u>nonassociative</u> dimensions in paper space, change this setting to a negative value equal to the reciprocal of the viewport scale so the measured values adjust for the viewport scale. The *Viewport* option of *DIMLFAC* is nonexistent in AutoCAD 2002 and 2004.
DIMLIM	Type: Switch Saved in: Drawing Initial value: Off	When turned on, generates dimension limits as the default text. Setting *DIMLIM* on forces *DIMTOL* to be off.
DIMLUNIT	Type: Integer Saved in: Drawing Initial value: 2	Sets units for all dimension types except Angular. 1 Scientific 2 Decimal 3 Engineering 4 Architectural 5 Fractional 6 Windows desktop
DIMLWD	Type: Enum Saved in: Drawing Initial value: -2	Assigns lineweight to dimension lines. Values are standard lineweights. -3 *BYLAYER* -2 *BYBLOCK* 1-211 integer representing 100th of mm
DIMLWE	Type: Enum Saved in: Drawing Initial value: -2	Assigns lineweight to extension lines. Values are standard lineweights. -3 *BYLAYER* -2 *BYBLOCK* 1-211 integer representing 100th of mm
DIMPOST	Type: String Saved in: Drawing Initial value: ""	Specifies a text prefix or suffix (or both) to the dimension measurement. For example, to establish a suffix for millimeters, set *DIMPOST* to "mm"; a distance of 19.2 units would be displayed as "19.2mm." If tolerances are enabled, the suffix is applied to the tolerances as well as to the main dimension. Use <> to indicate placement of the text in relation to the dimension value. For example, enter <>mm to display a 5.0 millimeter radial dimension as "5.0mm." If you entered "mm <>", the dimension would be displayed as "mm 5.0." Use the <> mechanism for angular dimensions.

Variable	Characteristics	Description
DIMRND	Type: Real Saved in: Drawing Initial value: 0.0000	Rounds all dimensioning distances to the specified value. For instance, if *DIMRND* is set to 0.25, all distances round to the nearest 0.25 unit. If you set *DIMRND* to 1.0, all distances round to the nearest integer. Note that the number of digits edited after the decimal point depends on the precision set by *DIMDEC*. *DIMRND* does not apply to angular dimensions.
DIMSAH	Type: Switch Saved in: Drawing Initial value: Off	Controls use of user-defined arrowhead blocks at the ends of the dimension line: Off Normal arrowheads or user-defined arrowhead blocks set by *DIMBLK* are used. On User-defined arrowhead blocks are used. *DIMBLK1* and *DIMBLK2* specify different user-defined arrowhead blocks for each end of the dimension line.
DIMSCALE	Type: Real Saved in: Drawing Initial value: 1.0000	Sets the overall scale factor applied to dimensioning variables that specify sizes, distances, or offsets. It is not applied to tolerances or to measured lengths, coordinates, or angles. Also affects the scale of leader objects created with the *LEADER* command. 0.0 AutoCAD computes a reasonable default value based on the scaling between the current Model Space viewport and Paper Space. If you are in Paper Space, or in Model Space and not using the Paper Space feature, the scale factor is 1.0. >0 AutoCAD computes a scale factor that leads text sizes, arrowhead sizes, and other scaled distances to plot at their face values.
DIMSD1	Type: Switch Saved in: Drawing Initial value: Off	When turned on, suppresses drawing of the first dimension line and arrowhead.
DIMSD2	Type: Switch Saved in: Drawing Initial value: Off	When turned on, suppresses drawing of the second dimension line and arrowhead.
DIMSE1	Type: Switch Saved in: Drawing Initial value: Off	When turned on, suppresses drawing of the first extension line.
DIMSE2	Type: Switch Saved in: Drawing Initial value: Off	When turned on, suppresses drawing of the second extension line.
DIMSHO	Type: Switch Saved in: Drawing Initial value: On	Obsolete. Has no effect in AutoCAD 2000 and later releases except to preserve the integrity of pre-AutoCAD 2000 scripts and AutoLISP routines.
DIMSOXD	Type: Switch Saved in: Drawing Initial value: Off	When turned on, suppresses drawing of dimension lines outside the extension lines. If the dimension lines would be outside the extension lines and *DIMTIX* is on, setting *DIMSOXD* to on suppresses the dimension line. If *DIMTIX* is off, *DIMSOXD* has no effect.

2002

Variable	Characteristics	Description
DIMSTYLE	(Read-only) Type: String Saved in: Drawing	*DIMSTYLE* is both a command and a system variable. The *DIMSTYLE* system variable shows the current dimension style. To display the *DIMSTYLE* system variable, use the *Setvar* command. The *DIMSTYLE* system variable is read-only; you cannot change its value on the command line. To change the current dimension style, use the *DIMSTYLE* command.
DIMTAD	Type: Integer Saved in: Drawing Initial value: 0	Controls vertical position of text in relation to the dimension line: 0 Centers the dimension text between the extension lines. 1 Places the dimension text above the dimension line except when the dimension line is not horizontal and text inside the extension lines is forced horizontal (*DIMTIH* = 1). The distance from the dimension line to the baseline of the lowest line of text is the current *DIMGAP* value. 2 Places the dimension text on the side of the dimension line farthest away from the defining points. 3 Places the dimension text to conform to a JIS representation.
DIMTDEC	Type: Integer Saved in: Drawing Initial value: 4	Sets the number of decimal places for the tolerance values for a primary units dimension.
DIMTFAC	Type: Real Saved in: Drawing Initial value: 1.0000	Sets a scale factor used to calculate the height of text for dimension fractions and tolerances. AutoCAD multiplies *DIMTXT* by *DIMTFAC* to set the fractional or tolerance text height.
DIMTIH	Type: Switch Saved in: Drawing Initial value: On	Controls the position of dimension text inside the extension lines for all dimension types except ordinate dimensions: Off Aligns text with the dimension line. On Draws text horizontally.
DIMTIX	Type: Switch Saved in: Drawing Initial value: Off	Draw text between extension lines: Off The result varies with the type of dimension. For linear and angular dimensions, AutoCAD places text inside the extension lines if there is sufficient room. For radius and diameter dimensions, setting *DIMTIX* off forces the text outside the circle or arc. On Draws dimension text between the extension lines even if AutoCAD would ordinarily place it outside those lines.
DIMTM	Type: Real Saved in: Drawing Initial value: 0.0000	When *DIMTOL* or *DIMLIM* is on, sets the minimum (or lower) tolerance limit for dimension text. AutoCAD accepts signed values for *DIMTM*. If *DIMTOL* is on and *DIMTP* and *DIMTM* are set to the same value, AutoCAD draws a ± symbol followed by the tolerance value.

continued on the next page...

Variable	Characteristics	Description
DIMTM (continued...)		If *DIMTM* and *DIMTP* values differ, the upper tolerance is drawn above the lower and a plus sign is added to the *DIMTP* value if it is positive. For *DIMTM*, AutoCAD uses the negative of the value you enter (adding a minus sign if you specify a positive number and a plus sign if you specify a negative number). No sign is added to a value of zero.
DIMTMOVE	Type: Integer Saved in: Drawing Initial value: 0	Sets dimension text movement rules. 0 Moves the dimension line with dimension text. 1 Adds a leader when dimension text is moved. 2 Allows text to be moved freely without a lead.
DIMTOFL	Type: Switch Saved in: Drawing Initial value: Off	When turned on, draws a dimension line between the extension lines, even when the text is placed outside the extension lines. For radius and diameter dimensions (while *DIMTIX* is off), draws a dimension line and arrowheads inside the circle or arc and places the text and leader outside.
DIMTOH	Type: Switch Saved in: Drawing Initial value: On	When turned on, controls the position of dimension text outside the extension lines: 0 Aligns text with the dimension line. 1 Draw text horizontally.
DIMTOL	Type: Switch Saved in: Drawing Initial value: Off	When turned on, appends dimension tolerances to dimension text. Setting *DIMTOL* on forces *DIMLIM* off.
DIMTOLJ	Type: Integer Saved in: Drawing Initial value: 1	Sets the vertical justification for tolerance values relative to the nominal dimension text: 0 Bottom 1 Middle 2 Top
DIMTP	Type: Real Saved in: Drawing Initial value: 0.0000	When *DIMTOL* or *DIMLIM* is on, sets the maximum (or upper) tolerance limit for dimension text. AutoCAD accepts signed values for *DIMTP*. If *DIMTOL* is on and *DIMTP* and *DIMTM* are set to the same value, AutoCAD draws a ± symbol followed by the tolerance value. If *DIMTM* and *DIMTP* values differ, the upper tolerance is drawn above the lower and a plus sign is added to the *DIMTP* value if it is positive.
DIMTSZ	Type: Real Saved in: Drawing Initial value: 0.0000	Specifies the size of oblique strokes drawn instead of arrowheads for linear, radius, and diameter dimension: 0 Draws arrows. >0 Draws oblique strokes instead of arrows. Size of oblique strokes is determined by this value multiplied by the *DIMSCALE* value.
DIMTVP	Type: Real Saved in: Drawing Initial value: 0.0000	Adjusts the vertical position of dimension text above or below the dimension line. AutoCAD uses the *DIMTVP* value when *DIMTAD* is off. The magnitude of the vertical

continued on the next page...

Variable	Characteristics	Description
DIMTVP (continued...)		offset of text is the product of the text height and DIMTVP. Setting DIMTVP to 1.0 is equivalent to setting DIMTAD to on. AutoCAD splits the dimension line to accommodate the text only if the absolute value of DIMTVP is less than 0.7.
DIMTXSTY	Type: String Saved in: Drawing Initial value: "STANDARD"	Specifies the text style of the dimension.
DIMTXT	Type: Real Saved in: Drawing Initial value: 0.1800	Specifies the height of dimension text, unless the current text style has a fixed height.
DIMTZIN	Type: Integer Saved in: Drawing Initial value: 0	Controls the suppression of zeros in tolerance values. DIMTZIN stores this value when you enter it on the Command line or set it under *Primary Units* in the *Dimension Style Manager*. DIMTZIN values 0–3 affect feet-and-inch dimensions only. 0 Suppresses zero feet and precisely zero inches 1 Includes zero feet and precisely zero inches 2 Includes zero feet and suppresses zero inches 3 Includes zero inches and suppresses zero feet 4 Suppresses leading zeros in decimal dimensions (for example, 0.5000 becomes .5000) 8 Suppresses trailing zeros in decimal dimensions (for example, 12.5000 becomes 12.5) 12 Suppresses both leading and trailing zeros (for example, 0.5000 becomes 0.5)
DIMUNIT		Obsolete. Has no effect since AutoCAD 2000 except to preserve the integrity of pre-AutoCAD 2000 scripts and AutoLISP routines. DIMUNIT is replaced by DIMLUNIT and DIMFRAC.
DIMUPT	Type: Switch Saved in: Drawing Initial value: Off	Controls cursor functionality for User Positioned Text: Off Cursor controls only the dimension line location. On Cursor controls the text position as well as the dimension line location.
DIMZIN	Type: Integer Saved in: Drawing Initial value: 0	Controls the suppression of zeros in the primary unit value. DIMZIN stores this value when you enter it on the Command line or set it under *Primary Units* in the *Dimension Style Manager*. DIMZIN values 0–3 affect feet-and-inch dimensions only. 0 Suppresses zero feet and precisely zero inches 1 Includes zero feet and precisely zero inches 2 Includes zero feet and suppresses zero inches 3 Includes zero inches and suppresses zero feet 4 Suppresses leading zeros in decimal dimensions (for example, 0.5000 becomes .5000)

continued on the next page...

Variable	Characteristics	Description
DIMZIN (continued...)		8 Suppresses trailing zeros in decimal dimensions (for example, 12.5000 becomes 12.5) 12 Suppresses both leading and trailing zeros (for example, 0.5000 becomes 0.5) *DIMZIN* also affects real-to-string conversions performed by the AutoLISP rtos and angtos functions.

Using *Setup Wizards* and Template Drawings

ACAD.DWT Template Drawing

The previous dimension variable table gives default settings for the STANDARD dimension style in the ACAD.DWT template drawing. This drawing is basically the same that is traditionally used when you begin a drawing session in previous releases of AutoCAD. In AutoCAD 2004, the ACAD.DWT drawing is the template or basis used when you begin AutoCAD with the *Startup* dialog box toggled off, choose *Start from Scratch, Imperial Default Settings* in the *Create New Drawing* dialog box, select the ACAD.DWT as a template drawing, or choose all the inch defaults in the *Setup Wizards*. In other cases, one or more dimension variables are automatically changed. In some template drawings, dimension styles other than STANDARD are available.

ANSI, ISO-25, DIN, and JIS Dimension Styles

The ANSI (American National Standards Institute) template drawings contain the STANDARD dimension style, which has the same variable settings as in the ACAD.DWT and as listed in the previous Dimension Variables table. If you select an ISO, DIN, or JIS template drawing, one or more dimension styles other than STANDARD may exist in the drawing. The following table lists the <u>differences</u> in the dimension variable settings listed by dimension style. All dimension variables not listed in the table have the same setting as the STANDARD style.

ISO-25 is the default ISO (International Standards Organization) template dimension style and the default style when you select *Metric* in *Start From Scratch*. DIN is the default dimension style for DIN (Deutsches Institut für Normung eV) templates. JIS is the default dimension style for JIS (Japanese Industrial Standard) templates.

Variable	Description	ANSI	ISO-25	DIN	JIS
DIMALTD	Alternate unit decimal places	2	4	2	2
DIMALTF	Alternate unit scale factor	25.4000	0.0394	0.0394	0.0394
DIMALTTD	Alternate tolerance decimal places	2	3	2	2
DIMALTU	Alternate units	2	2	8	2
DIMASZ	Arrow size	0.1800	2.5000	2.5000	2.5000
DIMCEN	Center mark size	0.0900	2.500	2.500	0.0000
DIMDEC	Decimal places for dimensions	4	2	4	4
DIMDLI	Dimension line spacing	0.3800	3.7500	3.7500	7.0000
DIMDSEP	Decimal separator	.	,	.	.
DIMEXE	Extension above dimension line	0.1800	1.2500	1.2500	1.0000
DIMEXO	Extension line origin offset	0.0625	0.625	0.625	1.0000
DIMGAP	Gap from dimension line to text	0.0900	0.6250	0.6250	0.0000
DIMLUNIT	Linear unit format	2	2	6	2
DIMSOXD	Suppress outside dimension lines	Off	Off	Off	On
DIMTAD	Place text above the dimension line	0	1	1	1

continued on the next page...

(continued...)

Variable	Description	ANSI	ISO-25	DIN	JIS
DIMTDEC	Tolerance decimal places	4	2	4	4
DIMTIH	Text inside extensions is horizontal	On	Off	On	Off
DIMTIX	Place text inside extensions	Off	Off	Off	On
DIMTOFL	Force line inside extension lines	Off	On	On	On
DIMTOH	Text outside horizontal	On	Off	Off	Off
DIMTOLJ	Tolerance vertical justification	1	0	1	1
DIMTXT	Text height	0.1800	2.5000	2.5000	2.5000
DIMTZIN	Tolerance zero suppression	0	8	0	0
DIMZIN	Zero suppression	0	8	8	8

As you notice, most of the differences in variable settings are related to size. Sizes of individual dimension components are changed to show the arrows, gaps, offsets, increments, etc. appropriately for the English or metric system used. In this way, the *Overall Scale* (*DIMSCALE*) for each dimension style is set to 1 by default and can be easily changed to adjust for different plot scales and plot sheets. See "Table of AutoCAD-Supplied Template Drawing Settings," Chapter 12, for a more information on the template drawings.

Setup Wizards

When you use the *Advanced Setup Wizard*, no dimension styles are changed from the default settings; however, dimension style <u>overrides may be created</u>. For example, if you use the *Advanced Setup Wizard* to begin a drawing based on the ACAD.DWT template, then select *Architectural* units, an override to the STANDARD dimension style is created with *DIMLUNIT* set to 4 (*Architectural*). Other changes may occur depending on which template is used and which units are selected in the *Advanced Setup Wizard*. The *MEASUREINIT* system variable defines which template is used for the *Setup Wizards* (ACAD.DWT or ACADISO.DWT) and, therefore, which dimension style is used (STANDARD or ISO-25). No changes to dimensioning variables are made using the *Quick Setup Wizard*.

MODIFYING EXISTING DIMENSIONS

Even with the best planning, a full understanding of dimension variables, and the correct use of *Dimension Styles*, it is probable that changes will have to be made to existing dimensions in the drawing due to design changes, new plot scales, or new industry/company standards. There are several ways that changes can be made to existing dimensions while retaining associativity and membership to a dimension style. The possible methods are discussed in this section.

Modifying a Dimension Style

Existing dimensions in a drawing can be modified by making one or more variable changes to the dimension style family or child, then *Saving* those changes to the dimension style. This process can be accomplished using either the *Dimension Style Manager* or Command line format. When you use the *Modify* option in the *Dimension Style Manager* or the *Save* option of the *-Dimstyle* command, the existing dimensions in the drawing <u>automatically update</u> to display the new variable settings.

Creating Dimension Style Overrides and Using *Update*

You can modify existing dimensions in a drawing without making permanent changes to the dimension style by creating a dimension style override, then using *Update* to apply the new setting to an existing dimension. This method is preferred if you wish <u>to modify one or two dimensions</u> without modifying all dimensions referencing (created in) the style.

To do this, use either the Command line format or *Dimension Style Manager* to set the new variable. In the *Dimension Style Manager*, select *Override* and make the desired variable settings in the *Override Current Style* dialog box. This creates an override to the current style. In Command line format, simply enter the formal dimension variable name and make the change to create an override to the current style, then use *Update* to apply the current style settings plus the override settings to existing dimensions that you PICK. The overrides remain in effect for the current style unless the variables are reset to the original values or until the overrides are cleared. You can <u>clear overrides</u> for a dimension style by making another dimension style current in the *Dimension Style Manager*.

UPDATE

Pull-down Menu	COMMAND (TYPE)	ALIAS (TYPE)	Short-cut	Screen (side) Menu	Tablet Menu
Dimension *Update*	DIM *UPDATE*	DIM *UP*	...	*DIMNSION* *Update*	*Y,3*

Update can be used to update existing dimensions in the drawing to the current settings. The current settings are determined by the current dimension style and any dimension variable overrides that are in effect (see previous explanation, "Creating Dimension Style Overrides and Using *Update*"). This is an excellent method of modifying one or more existing dimensions without making permanent changes to the dimension style that the selected dimensions reference. *Update* has the same effect as using *-Dimstyle, Apply*.

Update is actually a Release 12 command. In Release 12, dimensioning commands could only be entered at the Dim: prompt. For example, to create a linear dimension, you had to type *Dim* and press Enter, then type *Linear*. In Release 13, all dimensioning commands were upgraded to top-level commands with the *Dim-* prefix added, so you can type *Dimlinear* at the command prompt, for example. *Update*, although very useful, was never upgraded. In AutoCAD 2004, the command is given prominence by making available an *Update* button and an *Update* option in the *Dimension* pull-down menu. However, if you prefer to type, you must first type *Dim*, press Enter, and then enter *Update*. (See *DIM* in Chapter 28.) The command syntax is as follows:

```
Command: Dim
Dim: Update
Select objects: PICK (select a dimension object to update)
Select objects: Enter
Dim: press Esc or type Exit
Command:
```

For example, if you wanted to change the *DIMSCALE* of several existing dimensions to a value of 2, change the variable by typing *DIMSCALE*. Next use *Update* and select the desired dimension. That dimension is updated to the new setting. The command syntax is the following:

```
Command: Dimscale
New value for DIMSCALE <1.0000>: 2
Command: Dim
Dim: Update
Select objects: PICK (select a dimension object to update)
Select objects: PICK (select a dimension object to update)
Select objects: Enter
Dim: Exit
Command:
```

Beware, *Update* creates an override to the current dimension style. You should reset the variable to its original value unless you want to keep the override for creating other new dimensions. You can <u>clear overrides</u> for a dimension style by making another dimension style current in the *Dimension Style Manager*.

DIMOVERRIDE

	COMMAND (TYPE)	ALIAS (TYPE)	Short-cut	Screen (side) Menu	Tablet Menu
Pull-down Menu					
Dimension Override	DIMOVERRIDE	*DIMOVER or DOV*	...	...	*Y,4*

Dimoverrride grants you a great deal of control to <u>edit existing dimensions</u>. The abilities enabled by *Dimoverride* are similar to the effect of using the *Properties* palette.

Dimoverride enables you to <u>make variable changes</u> to dimension objects that exist in your drawing <u>without creating an override to the dimension style</u> that the dimension references (was created under). For example, using *Dimoverride,* you can make a variable change and select existing dimension objects to apply the change. The existing dimension does not lose its reference to the parent dimension style nor is the dimension style changed in any way. In effect, you can <u>override the dimension styles for selected dimension objects</u>. There are two steps: set the desired variable and select dimension objects to alter.

```
Command: dimoverride
Enter dimension variable name to override or [Clear overrides]: (variable name)
Enter new value for dimension variable <current value>: (value)
Enter dimension variable name to override: Enter
Select objects: PICK
Select objects: PICK or Enter
Command:
```

The *Dimoverride* feature differs from creating dimension style overrides in two ways: (1) *Dimoverride* applies the changes to the selected dimension objects only, so the overrides are not appended to the parent dimension styles, only the objects; and (2) *Dimoverride* can be used once to change dimension objects referencing multiple dimension styles, whereas to make such changes to dimensions by creating dimension style overrides requires changing all the dimension styles, one at a time. The effect of using *Dimoverride* is essentially the same as using the *Properties* palette.

Dimoverride is useful as a "backdoor" approach to dimensioning. Once dimensions have been created, you may want to make a few modifications, but you do not want the changes to affect dimension styles (resulting in an update to all existing dimensions that reference the dimension styles). *Dimoverride* offers that capability. You can even make one variable change to affect all dimensions globally without having to change multiple dimension styles. For example, you may be required to make a test plot of the drawing in a different scale than originally intended, necessitating a new global *Dimscale*. Use *Dimoverride* to make the change for the plot:

```
Command: dimoverride
Enter dimension variable name to override or [Clear overrides]: dimscale
Enter new value for dimension variable <1.0000>: .5
Enter dimension variable name to override: Enter
Select objects: (window entire drawing) Other corner: 128 found
Select objects: Enter
Command:
```

This action results in having all the existing dimensions reflect the new *DIMSCALE*. No other dimension variables or any dimension styles are affected. Only the selected objects contain the overrides. *Dimoverride* does not append changes (overrides) to the original dimension styles, so no action must be taken to clear the overrides from the styles. After making the plot, *Dimoverride* can be used with the *Clear* option to change the dimensions (by object selection) back to their original appearance. The *Clear* option is used to clear overrides from <u>dimension objects, not from dimension styles</u>.

Clear

The *Clear* option removes the overrides from the <u>selected dimension objects</u>. It does not remove overrides from the current dimension style:

    Command: **dimoverride**
    Enter dimension variable name to override or [Clear overrides]: **c**
    Select objects: **PICK**
    Select objects: **Enter**
    Command:

The dimension then displays the variable settings as specified by the dimension style it references without any overrides (as if the dimension were originally created without the overrides). Using *Clear* does not remove any overrides that are appended to the dimension style so that if another dimension is drawn, the dimension style overrides apply.

MATCHPROP

Pull-down Menu	COMMAND (TYPE)	ALIAS (TYPE)	Short-cut	Screen (side) Menu	Tablet Menu
Modify Match Properties	*MATCHPROP*	*MA*	...	*MODIFY1 Matchprp*	*Y,14 and Y,15*

Matchprop can be used to "convert" an existing dimension to the style (including overrides) of another dimension in the drawing. For example, if you have two linear dimensions, one has *Oblique* arrows and *Romans* text font (*Dimension Style* = "Oblique") and one is a typical linear dimension (*Dimension Style* = "Standard") as in Fig. 29-65, "before." You can convert the typical dimension to the "Oblique" style by using *Matchprop*, selecting the "Oblique" dimension as the "source object" (to match), then selecting the typical dimension as the "destination object" (to convert). The typical dimension then references the "Oblique" dimension style and changes appearance accordingly (Fig. 29-65, "after"). Note that *Matchprop* does not alter the dimension text value.

Figure 29-65

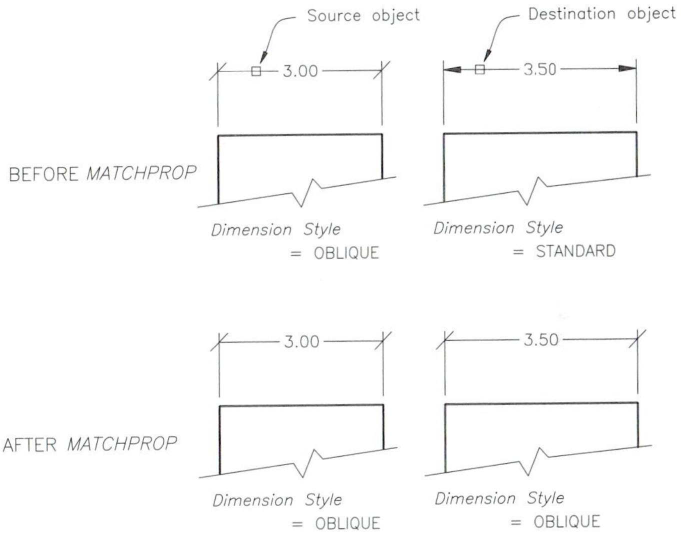

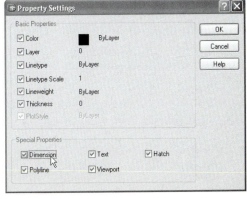

Using the *Settings* option of the *Matchprop* command, you can display the *Property Settings* dialog box (Fig. 29-66). The *Dimension* box under *Special Properties* must be checked to "convert" existing dimensions as illustrated above.

Using *Matchprop* is a fast and easy method for modifying dimensions from one style to another. However, this method is applicable only if you have existing dimensions in the drawing with the desired appearance that you want others to match.

Figure 29-66

PROPERTIES

Pull-down Menu	COMMAND (TYPE)	ALIAS (TYPE)	Short-cut	Screen (side) Menu	Tablet Menu
Modify Properties...	*PROPERTIES*	PR or CH	*(Edit Mode)* Properties or Crtl+1	MODIFY1 *Property*	Y,12 to Y,13

Remember that *Properties* can be used to edit existing dimensions. Using *Properties* and selecting one dimension displays the *Properties* palette with all of the selected dimension's properties and dimension variables (see Figure 28-77). Here you can modify any aspect of one or more dimensions, including text, and access is given to the *Lines and Arrows, Text, Fit, Primary Units, Alternate Units,* and *Tolerances* categories. Any changes to dimension variables through *Properties* result in <u>overrides to the dimension object only</u> but do not affect the dimension style. *Properties* has essentially the same result as *Dimoverride* (see "*Dimoverride*"). *Properties* can also be used to modify the dimension text value. *Properties* of dimensions are also discussed in Chapter 28.

DIMTEDIT, DIMEDIT

Dimension text of existing dimensions can be modified using either *Dimtedit* or *Dimedit*. These commands allow you to change the text position with respect to the dimension line and the angle of the text. *Dimtedit* can be used with the *New* option to restate the original AutoCAD-measured text value if needed. See Chapter 28 for a full explanation of these commands.

Grips

Grips can be used to effectively alter dimension length, position, text location, and more. Keep in mind the possibilities of moving the dimension text with grips for dimensions with *DIMTMOVE* set to 2. See Chapter 28 for a full discussion of the possibilities of Grip editing dimensions.

GUIDELINES FOR DIMENSIONING IN AutoCAD

Listed in this section are some guidelines to use for dimensioning a drawing using dimensioning variables and dimension styles. Although there are other strategies for dimensioning, two strategies are offered here as a framework so you can develop an organized approach to dimensioning.

In almost every case, dimensioning is one of the last steps in creating a drawing, since the geometry must exist in order to dimension it. You may need to review the steps for drawing setup, including the concept of drawing scale factor (Chapter 12).

Strategy 1. Dimensioning a Single Drawing

This method assumes that the fundamental steps have been taken to set up the drawing and create the geometry. Assume this has been accomplished:

> Drawing setup is completed: *Units, Limits, Snap, Grid, Ltscale, Layers,* border, and titleblock.
> The drawing geometry (objects comprising the subject of the drawing) has been created.

Now you are ready to dimension the drawing subject (of the multiview drawing, pictorial drawing, floor plan, or whatever type of drawing).

1. Create a *Layer* (named DIM, or similar) for dimensioning if one has not already been created. Set *Continuous* linetype and appropriate color. Make it the *current* layer.

2. Set the *Overall Scale (DIMSCALE)* based on drawing *Limits* and expected plotting size.

    For plotted dimension text of 3/16":

    Multiply *Overall Scale (DIMSCALE)* times the drawing scale factor. The default *Overall Scale* is set to 1, which creates dimensioning text of approximately 3/16" (default *Text Height:* or *DIMTXT* =.18) when plotted full size. All other size-related dimensioning variables' defaults are set appropriately.

    For plotted dimension text of 1/8":

    Multiply *Overall Scale* times the drawing scale factor, <u>times .7</u>. Since the *Overall Scale* times the scale factor produces dimensioning text of .18, then .18 x .7 = .126 or approximately 1/8". (See "Optional Method for Fixed Dimension Text Height.") (*Overall Scale* [*DIMSCALE*] is one variable that usually remains constant throughout the drawing and therefore should generally have the same value for every dimension style created.)

3. Make the other dimension variable changes you expect you will need to create <u>most dimensions in the drawing</u> with the appearance you desire. When you have the basic dimension variables set as you want, *Save* this new dimension style family as TEMPLATE or COMPANY_STD (or other descriptive name) style. If you need to make special settings for types of dimensions (*Linear, Diameter,* etc.), create "children" at this stage and save the changes to the style. This style is the fundamental style to use for creating most dimensions and should be used as a template for creating other dimension styles. If you need to reset or list the original (default) settings, the STANDARD style can be restored.

4. Create all the relatively simple dimensions first. These are dimensions that are easy and fast and require <u>no other dimension variable changes</u>. Begin with linear dimensions; then progress to the other types of dimensions.

5. Create the special dimensions next. These are dimensions that require variable changes. Create appropriate dimension styles by changing the necessary variables, then save each set of variables (relating to a particular style of dimension) to an appropriate dimension style name. Specify dimension variables for the classification of dimension (children) in each dimension style when appropriate. The dimension styles can be created "on the fly" or as a group before dimensioning. Use TEMPLATE or COMPANY_STD as your base dimension style when appropriate.

    For the few dimensions that require variable settings unique to that style, you can create dimension variable overrides. Change the desired variable(s), but do not save to the style so the other dimensions in the style are not affected.

6. When all of the dimensions are in place, make the final adjustments. Several methods can be used:

    A. If modifications need to be made <u>familywide</u>, change the appropriate variables and save the changes to the dimension style. This action automatically updates existing dimensions that reference that style.

    B. To modify the appearance of selected dimensions, you can create dimension style overrides, then use *Update* or *-Dimstyle, Apply* to update the selected dimensions to the new settings. Keep in mind that the dimension style overrides are still in effect and are applied to new dimensions. Clear the overrides by setting the original style current.

    C. Alternately, use *Properties* to change variable settings for selected dimensions. These changes are made as overrides to the selected object only and do not affect the dimension style. This action has the same result as using *Dimoverride* but for only one dimension.

D. To change one dimension to adopt the appearance of another, use *Matchprop*. Select the dimension to match first, then the dimension(s) to convert. *-Dimstyle, Apply* can be used for this same purpose.

E. Use *Dimoverride* to make changes to selected dimensions or to all dimensions <u>globally</u> by windowing the entire drawing. *Dimoverride* has the advantage of allowing you to assign the variables to change and selecting the objects to change all in one command. What is more, the changes are applied as overrides only to the selected dimensions but <u>do not alter the original dimension styles</u> in any way.

F. If you want to change only the dimension text value, use *Properties* or *Dimedit, New*. *Dimedit, New* can also be used to reapply the AutoCAD-measured value if text was previously changed. The location of the text can be changed with *Dimtedit* or Grips.

G. Grips can be used effectively to change the location of the dimension text or to move the dimension line closer or farther from the object. Other adjustments are possible, such as rotating a *Radius* dimension text around the arc.

Strategy 2. Creating Dimension Styles as Part of Template Drawings

1. Begin a *New* drawing or *Open* an existing *Template*. Assign a descriptive name.

2. Create a DIM *Layer* for dimensioning with *continuous* linetype and appropriate color and lineweight (if one has not already been created).

3. Set the *Overall Scale* accounting for the drawing *Limits* and expected plotting size. Use the guidelines given in Strategy 1, step 2. Make any other dimension variable changes needed for general dimensioning or required for industry or company standards.

4. Next, *Save* a dimension style named TEMPLATE or COMPANY_STD. This should be used as a template when you create most new dimension styles. The *Overall Scale* is already set appropriately for new dimension styles in the drawing.

5. Create the appropriate dimension styles for expected drawing geometry. Use TEMPLATE or COMPANY_STD dimension style as a base style when appropriate.

6. *Save* and *Exit* the newly created template drawing.

7. Use this template in the future for creating new drawings. Restore the desired dimension styles to create the appropriate dimensions.

Using a template drawing with prepared dimension styles is a preferred alternative to repeatedly creating the same dimension styles for each new drawing.

Optional Method for Fixed Dimension Text Height in Template Drawings

To summarize Strategy 1, step 2., the default *Overall Scale* (DIMSCALE =1) times the default *Text Height* (DIMTXT =.18) produces dimensioning text of approximately 3/16″ when plotted to 1=1. To create 1/8″ text, multiply *Overall Scale* times .7 (.18 x .7 = .126). As an alternative to this method, try the following.

For 1/8" dimensions, for example, multiply the initial values of the size-related variables by .7; namely

Text Height	*(DIMTXT)*	.18 x .7 = .126
Arrow Size	*(DIMASZ)*	.18 x .7 = .126
Extension Line Extension	*(DIMEXE)*	.18 x .7 = .126
Dimension Line Spacing	*(DIMDLI)*	.38 x .7 = .266
Text Gap	*(DIMGAP)*	.09 x .7 = .063

Save these settings in your template drawing(s). When you are ready to dimension, simply multiply *Overall Scale* (1) times the drawing scale factor.

Although this method may seem complex, the drawing setup for individual drawings is simplified. For example, assume your template drawing contained preset *Limits* to the paper size, say 11 x 8.5. In addition, the previously mentioned dimension variables were set to produce 1/8" (or whatever) dimensions when plotted full size. Other variables such as *LTSCALE* could be appropriately set. Then, when you wish to plot a drawing to 1=1, everything is preset. If you wish to plot to 1=2, simply multiply all the size-related variables (*Limits*, *Overall Scale*, *Ltscale*, etc.) times 2!

DIMENSIONING IN PAPER SPACE LAYOUTS

Generally, dimensions are created in model space and are attached to model space geometry. In this way, you can make one or more layouts, each with one or more viewports that display the model space geometry, and the dimensions are visible by default in each of the layouts and viewports. Assuming the dimensions are created on a dimensioning layer or layers, you can control the display of the dimensions in each viewport using viewport-specific layer visibility controls in the *Layer Manager*. This strategy is used for almost all AutoCAD drawings previous to AutoCAD 2002, and will most likely be continued for most drawings in the future except for certain cases.

Since AutoCAD 2002, it is possible to create dimensions in a paper space layout <u>associated</u> with (attached to) model space geometry inside a viewport. To explain, a typical drawing includes geometry in model space (such as mechanical part views, a floor plan, or a diagram) and at least one paper space layout used to display and print the model geometry. Since AutoCAD 2002, you can create the dimensions in a layout (in paper space) and *Osnap* to objects in model space (inside a viewport). These new dimensions are fully associative and display the actual measurement value of the drawing objects <u>in model space units</u>.

Figure 29-67

In previous versions of AutoCAD, it was possible to create dimensions in paper space; however, those dimensions displayed the measurement value in paper space units by default. For example, consider the AutoCAD 2000 drawing of a plate shown with two viewports in Figure 29-67. All but one of the dimensions are in model space. The left viewport displays the entire plate including the model space dimensions at 1:1 scale, while the right viewport displays the plate at 2:1 scale. Notice that the dimension features (text, arrows, gaps, etc.) in the right viewport are shown twice as large since the view is scaled 2:1. Below the right viewport a dimension has been

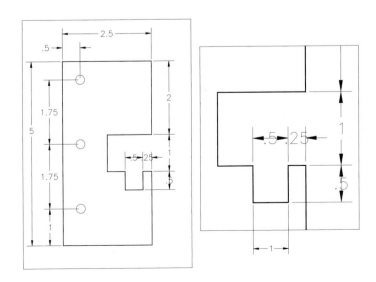

created in paper space ("1"). For that dimension, the dimension features (text, arrows, gaps, etc.) appear in the correct size, but the measured value is in paper space units—which is incorrect since the detail view is 2:1 (the measured value should be "2"). To compensate for this inconsistency, the *DIMLFAC* (dimension length factor) system variable had to be set in AutoCAD 2000 and previous versions to convert paper space units to model space units. Using *DIMLFAC* for this purpose became obsolete in AutoCAD 2002.

In contrast, a similar drawing is shown in Figure 29-68 created in AutoCAD 2002, but this drawing has some dimensions in model space (left viewport) and some dimensions in paper space (right viewport). Note that the dimensions for the small cutout have been omitted in model space (left viewport). All of the dimensions for the detail (right viewport) are created in paper space. Some of the dimensions for the detail are actually outside of the viewport border, but the two short horizontal dimensions (.5 and .25) appear to be in model space even though they were actually created in paper space. Note that in AutoCAD 2004, all dimensions associated to model space geometry, whether created in model space or paper space, display the correct value of the model feature they are measuring.

Figure 29-68

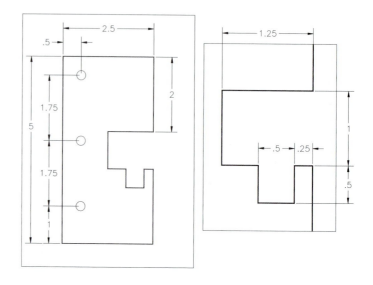

Figure 29-69 displays the completed drawing after adding some annotation and freezing the viewport border layer. Note that you cannot detect which dimensions are in model space and which are in paper space.

Figure 29-69

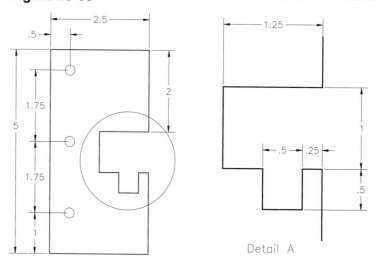

When to Dimension in Paper Space

Although it is still recommended to create dimensions in model space for most situations, there are some cases where creating dimensions in paper space could be used. Generally, whenever you want specific dimensions to appear for only one of several viewports, it may be useful to create those dimensions in paper space. Here are two specific examples.

If you have created a 3D solid model, then generated 2D views (a multiview layout) from the model with each view in a separate paper space viewport, consider creating the dimensions for each view in paper space.

If you want one or more detail views (small, enlarged sections of a larger drawing) each in a separate viewport or layout and at different scales than the full drawing display, consider creating dimensions for the detail views in paper space.

For the situations listed above, the advantages of creating dimensions in paper space attached to objects in model space (inside a viewport) are these:

Setting *DIMSCALE* or *Overall Dimension Scale* (the size of the dimension text, arrowheads, etc.) for paper space dimensions is simplified since you are concerned only with paper space units and not with the viewport scale (the scale of the geometry that appears in the viewports). This is especially important when you have several detail views of a drawing and each detail view is at a different scale than the full drawing view. Paper space dimensions appear in one size, whereas model space dimensions appear in different sizes when the drawing is scaled differently in each viewport. This problem occurs in Figure 29-67 but is remedied in Figure 29-68.

Creating dimensions in paper space ensures that those dimensions appear only for that viewport but not for other displays of the model geometry appearing in other viewports and layouts. Therefore, you do not have to use different layers for different sets of dimensions and the *Layer Manager* to set viewport-specific visibility (which dimension layers you want to appear in which viewports). This problem and the remedy are also shown in Figures 29-67 and 29-68, respectively.

Dimensioning a 3D model requires that dimensions for each side or view of the model appear on different planes (created by using construction planes called User Coordinate Systems or UCSs in AutoCAD). For a multiview-type setup of a 3D model, dimensioning in paper space prevents having to use different UCSs as well as different layers for the dimensions that should appear in each view (on each plane of the 3D model). Also, you do not have to control viewport-specific layer visibility to ensure that dimensions for the top view do not appear in the front view, and so on.

Although it appears that these advantages might outweigh the practice of dimensioning in model space, most drawings require that the dimensions appear in multiple viewports or layouts; therefore, dimensioning in model space is the more common practice. Dimensioning in paper space eliminates the possibility of showing the same dimensions in multiple viewports or layouts. In addition, remember that creating <u>associative</u> dimensions in paper space is available only since AutoCAD 2002, so the practice of dimensioning in paper space is not recommended if you collaborate with clients using previous releases. See Chapters 33 and 42 for more information and applications of dimensioning in paper space.

Procedure for Dimensioning in Paper Space

Dimensioning in paper space differs from dimensioning in model space only in a few respects such as determining the *DIMSCALE* (*Overall Scale*) and the need to use *Dimregen* when zooming or panning in the viewport; therefore, the procedure for dimensioning in paper space is similar to that for dimensioning in model space. Here is a typical procedure to use when you need paper space dimensions.

1. Complete the model space geometry.

2. (Optional, depending on the situation and desired result.) Create a layer named DIM or DIMENSIONS and set it *Current*. Use the *Dimension Style Manager* to create a dimension style for the <u>model space</u> dimensions. Set the *Overall Scale* and other variables and save the style. Create the dimensions you need in model space. These dimensions are those that you want to appear in multiple viewports.

3. Set up the desired layouts and viewports and set the desired viewport scale for each viewport (so the model geometry appears in the desired scale in each viewport).

4. Create a layer named DIM-PS or DIMENSIONS-PAPERSPACE and set the layer *Current*.

5. Use the *Dimension Style Manager* to create a dimension style for the <u>paper space</u> dimensions by copying the style for model space dimensions and rename the style DIM-PS or similar. Set the *Overall Scale* so the dimensions print in paper space units. (Normally, an *Overall Scale* of 1 is used for paper space dimensions since the drawing is printed from the layout at 1:1.) The other dimension variables normally would not change (from the model space settings).

6. Activate the desired layout and ensure that paper space is active (the cursor appears in paper space, not in the viewport). With *Osnap* on, create the dimensions in paper space but *Osnap* the extension line origins, etc. to the objects in model space. The location of the dimension text can be placed outside or inside the viewport borders but the actual dimension objects are in paper space. Figure 29-70 displays two paper space dimensions, one has text inside the viewport border.

Figure 29-70 ————————————————

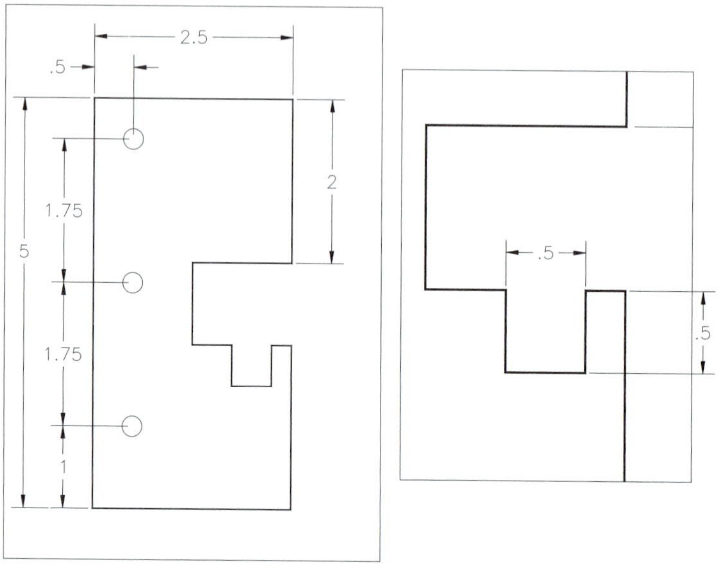

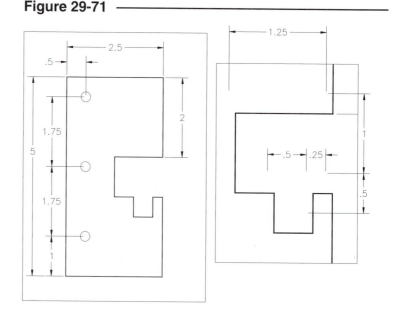

7. Complete the paper space dimensions. If you need to *Zoom* or *Pan* inside the viewport or change the viewport scale (using *Zoom XP* or the *Viewport Scale* drop-down edit box), you should use *Dimregen* to force AutoCAD to redisplay the location of the paper space dimension objects to align correctly with the new location of the model space geometry. Figure 29-71 illustrates the drawing after using *Pan* but <u>before</u> using *Dimregen*.

Figure 29-71 ————————————————

8. Complete any additional notation or objects in paper space. In the *Layer Manager*, *Freeze* or disable plotting for the viewport border layer (see Figure 29-69).

DIMENSIONING ISOMETRIC DRAWINGS

Dimensioning isometric drawings in AutoCAD is accomplished using *Dimaligned* dimensions and then adjusting the angle of the extension lines with the *Oblique* option of *Dimedit*. The technique follows two basic steps:

1. Use *Dimaligned* or the *Vertical* option of *Dimlinear* to place a dimension along one edge of the isometric face. Isometric dimensions should be drawn on the isometric axes lines (vertical or at a 30° rotation from horizontal; Fig. 29-72).

Figure 29-72

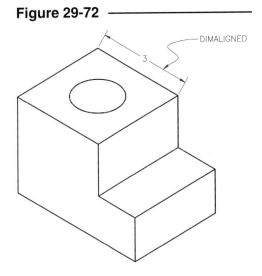

2. Use the *Oblique* dimensioning option of *Dimedit* and select the dimension just created. When prompted to "Enter obliquing angle," enter the desired value (**30** in this case) or PICK two points designating the desired angle. The extension lines should change to the designated angle. In AutoCAD, the possible obliquing angles for isometric dimensions are **30**, **150**, **210**, or **330**.

Figure 29-73

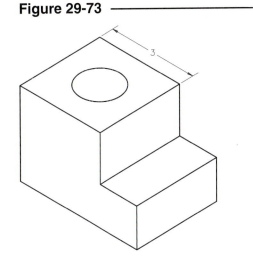

Place isometric dimensions so that they align with the face of the particular feature. Dimensioning the object in the previous illustrations would continue as follows.

Create a *Dimlinear, vertical* dimension along a vertical edge.

Figure 29-74

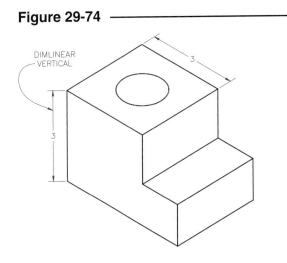

Use *Dimedit*, *Oblique* to force the dimension to an isometric axis orientation. Enter a value of **150** or PICK two points in response to the "Enter obliquing angle:" prompt.

Figure 29-75

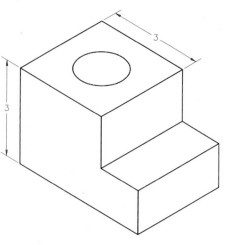

For isometric dimensioning, the extension line origin points <u>must</u> be aligned with the isometric axes. If not, the dimension is not properly oriented. This is important when dimensioning an isometric ellipse (such as this case) or an inclined or oblique edge. Construct centerlines for the ellipse on the isometric axes. Next, construct an *Aligned* dimension and *Osnap* the extension line origins to the centerlines. Finally, use *Dimedit Oblique* to reorient the angle of the extension lines.

Figure 29-76

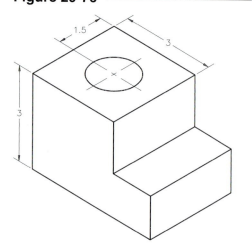

Using the same technique, other appropriate *Dimlinear vertical* or *Aligned* dimensions are placed and reoriented with *Oblique*. Use a *Leader* to dimension a diameter of an *Isocircle*, since *Dimdiameter* cannot be used for an ellipse.

Figure 29-77

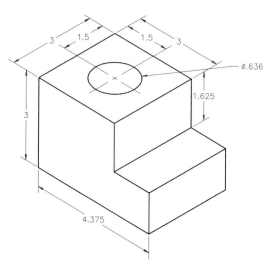

The dimensioning text can be treated two ways. (1) For quick and simple isometric dimensioning, <u>unidirectional</u> dimensioning (all values read from the bottom) is preferred since there is no automatic method for drawing the numerical values in an isometric plane (Fig. 28-80). (2) As a better alternative, text *Styles* could be created with the correct obliquing angles (30 and -30 degrees). The text must also be rotated to the correct angle using the *Rotate* option of the dimension commands or by using *Dimedit Rotate* (Fig. 29-78). Optionally, *Dimension Styles* could be created with the correct variables set for text style and rotation angle for dimensioning on each isoplane.

Figure 29-78 ─────────

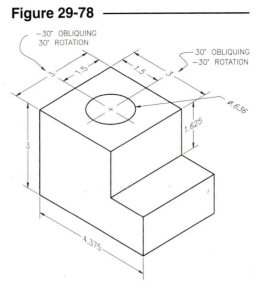

CHAPTER EXERCISES

For each of the following exercises, use the existing drawings, as instructed. Create dimensions on the DIM (or other appropriate) layer. Follow the "Guidelines for Dimensioning in AutoCAD" given in the chapter, including setting an appropriate *Overall Scale* based on the drawing scale factor. Use dimension variables and create and use dimension styles when needed.

1. **Dimension One View**

    Open the **HAMMER** drawing you created in Chapter 13 Exercises and use the *Saveas* command to rename it **HAMMER-DIM**. Add all necessary dimensions as shown in Chapter 13 Exercises, Figure 13-30. Create a *New* dimension style and set the variables to generate dimensions as they appear in the figure (set *Precision* to **0.00**, *Text Style* to *Romans* font, and *Overall Scale* to **1** or **1.3**). Create the dimensions in model space. Your completed drawing should look like that in Figure 29-79.

Figure 29-79 ─────────

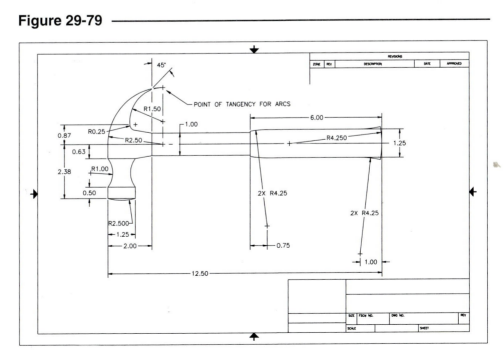

2. **Dimensioning a Multiview**

Figure 29-80

Open the **SADDLE** drawing that you created in Chapter 24 Exercises. Set the appropriate dimensional *Units* and *Precision*. Add the dimensions as shown. Because the illustration in Figure 29-80 is in isometric, placement of the dimensions can be improved for your multiview. Use optimum placement for the dimensions. *Save* the drawing as **SADDL-DM** and make a *Plot* to scale.

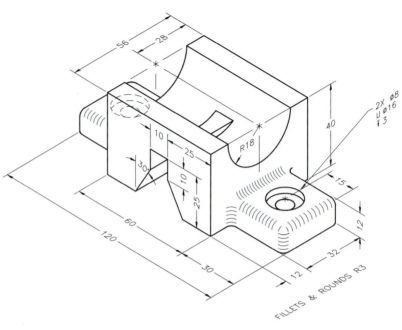

3. **Architectural Dimensioning**

Open the **OFFICE** drawing that you completed in Chapter 21. Dimension the floor plan as shown in Figure 21-44. Add *Text* to name the rooms. *Save* the drawing as **OFF-DIM** and make a *Plot* to an accepted scale and sheet size based on your plotter capabilities.

4. **Dimensioning an Auxiliary**

Figure 29-81

Open the **ANGLBRAC** drawing that you created in Chapter 27. Dimension as shown in Figure 29-81, but <u>convert the dimensions to *Decimal*</u> with **Precision** of **.000.** Use the "Guidelines for Dimensioning." Dimension the slot width as a *Limit* dimension—**.6248/.6255**. *Save* the drawing as **ANGL-DIM** and *Plot* to an accepted scale.

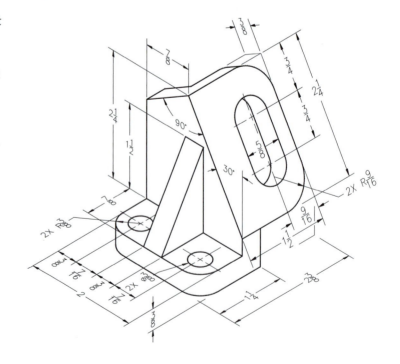

5. **Isometric Dimensioning**

Figure 29-82

Dimension the support bracket that you created as an isometric drawing named **SBRACKET** in the Chapter 25 Exercises. All of the dimensions shown in Figure 29-82 should appear on your drawing in the optimum placement. *Save* the drawing as **SBRCK-DM** and *Plot* to **1=1** on and A size sheet.

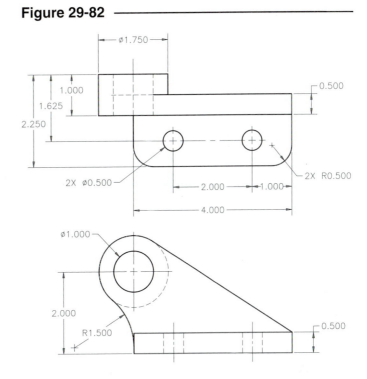

6. **Dimensioning a Multiview**

Figure 29-83

Open the **ADJMOUNT** drawing that you completed in Chapter 24. Add the dimensions shown in Figure 29-83 but <u>convert the dimensions to *Decimal*</u> with **Precision** of .000 and check **Trailing** under **Zero Suppression.** Set appropriate dimensional *Units* and *Precision*. Calculate and set an appropriate *Overall Scale*. Use the "Guidelines for Dimensioning" given in this chapter. Save the drawing as **ADJM-DIM** and make a *Plot* to an accepted scale.

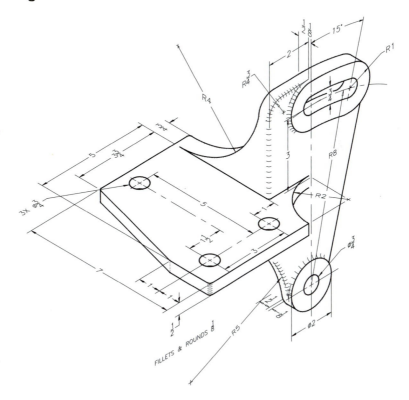

7. **Geometric Dimensioning and Tolerancing**

This exercise involves Flatness, Profile of a Surface, and Position applications. *Open* the dimensioned **ANGL-DIM** drawing you worked on in Exercise 3 and add the following geometric dimensions:

A. **Flatness** specification of **.002** to the bottom surface.

B. **Profile of a Surface** specification of **.005** to the angled surface relative to the bottom surface and the right side surface of the 1/4 dimension. The 30° angle must be *Basic*.

C. **Position** specification of **.01** for the two mounting holes relative to the bottom surface and two other surfaces that are perpendicular to the bottom. Remember, the location dimensions must be *Basic* dimensions.

Use *SaveAs* to save and name the drawing **ANGL-TOL**.

8. **Geometric Dimensioning and Tolerancing**

This exercise involves a cylindricity and runout application. Draw the idler shown in Figure 29-84. Apply a **Cylindricity** specification of **.002** to the small diameter and a **Circular Runout** specification of **.007** to the large diameter relative to the small diameter as shown. The P730 neck has a **.07** radius and **30°** angle. *Save* the drawing as **IDLERDIM**.

Figure 29-84 ─────────────────────────────────────

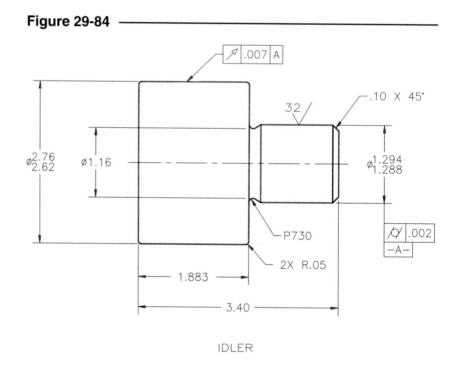

IDLER

9. **Dimension in Paper Space**

A. *Open* the **2BR-APT** drawing you worked with in Chapter 20 Exercises. Make a *New* dimension style and set the variables to generate dimensions as they appear in Chapter 15 Exercises, Figure 15-45. Create the dimensions in model space on a layer called **DIM**.

B. Create a *New Layout*. Use *Pagesetup* for the layout and select an appropriate *Plot Device* that can use a "**B**" size sheet, such as an HP 7475 plotter. Set the layout to a "**B**" size sheet. Next, use DesignCenter to locate and insert the **ANSI-B title block** from the Template folder. Make a new layer named **VPORTS** and create one viewport as shown in Figure 29-85. Set the viewport scale to display the apartment floor plan at **1/4"=1'** scale.

Figure 29-85

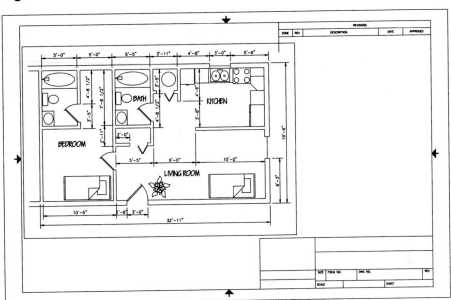

C. Create a smaller viewport on the right to display only one of the bathrooms at 1/2"=1' scale. Use the *Layer Manager* to turn off the display of the **DIM** layer for the new viewport. Create a new layer named **DIM-PS**. Create a *New* dimension style by copying the dimension style you created for model space, but change the *Overall Scale* to **1**. Create the dimensions for the plumbing wall (distance between centers of the fixtures) in paper space as shown in Figure 29-86. Label each viewport giving the scale. *Freeze* the **VPORTS** layer to achieve a drawing like that shown in Figure 29-86. Save the drawing as **APT-DIM**.

Figure 29-86

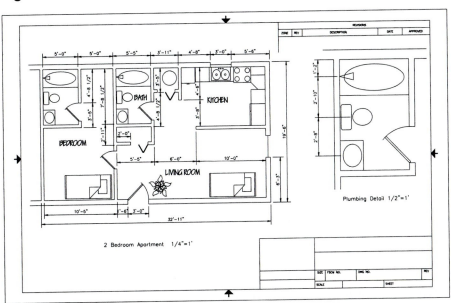

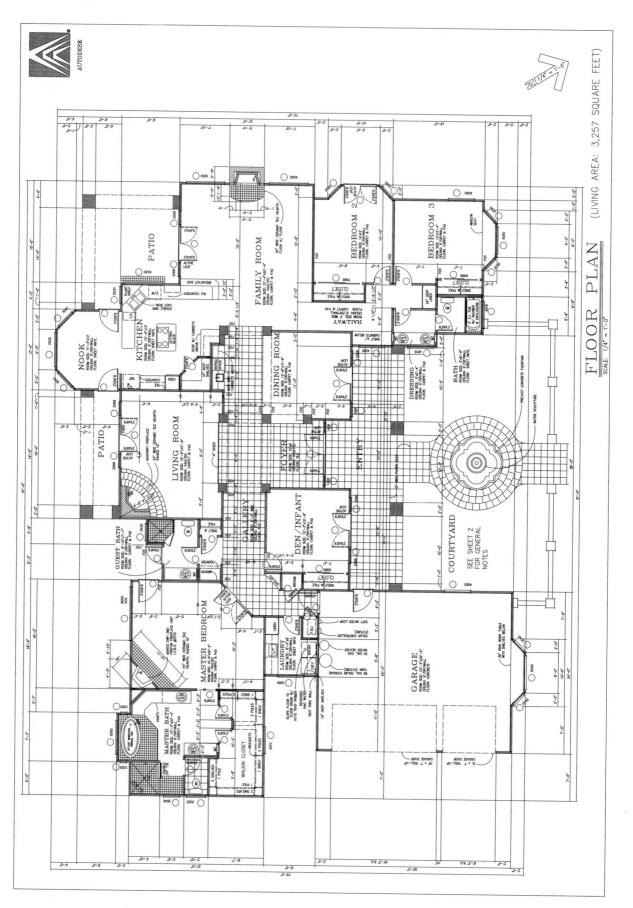

FLOOR PLAN (LIVING AREA: 3,257 SQUARE FEET)

SCALE: 1/4" = 1'-0"

HOUSEPLN.DWG Courtesy of Autodesk, Inc.

XREFERENCES

Chapter Objectives

After completing this chapter you should:

1. know the differences and similarities between an *Xrefed* drawing and an *Inserted* drawing;

2. be able to *Attach* an externally referenced drawing to the current drawing;

3. understand that an *Overlay* drawing cannot be nested;

4. be able to *Reload*, *Unload*, or *Detach* an externally referenced drawing;

5. be able to *Bind* an entire *Xrefed* drawing and *Xbind* individual named objects;

6. be able to *Xclip* an externally referenced drawing so only a portion of the *Xref* appears;

7. be able to control demand loading with *INDEXCTL* and *XLOADCTL*;

8. be able to use *Xopen* to open an *Xref* drawing for editing in a separate window;

9. know how to edit *Blocks* and *Xrefs* "in place" using *Refedit* and how to control in-place editing with *XEDIT*.

CONCEPTS

When you *Insert* a drawing as a *Block*, the inserted drawing becomes part of the current drawing—it is considered an "internal" reference. An *Inserted Block* (internal reference) is static—it does not change unless you *Explode*, edit, and redefine it. In contrast, an *Xref* (external reference) is a reference to another drawing. An *Xrefed* drawing is viewable from within the current drawing, but it is not *Inserted*. The *Xref* geometry and associated "named objects" do not become a part of the current drawing by default, although the *Xrefed* layer visibility can be controlled. When the original *Xref* drawing is changed, the changes are visible from within the current drawing.

INSERT	XREF
Any drawing can be *Inserted* as a *Block*.	Any drawing can be *Xrefed*.
The drawing "comes in" as a *Block*.	The drawing is visible as an *Xref*.
The *Block* drawing is a permanent part of the current drawing.	The *Xref* is not permanent, only "attached" or "overlayed."
The current drawing file size increases approximately equal to the *Block* drawing size.	The current drawing increases only by a small amount (enough to store information about loading the *Xref*).
The *Inserted Block* drawing is static. It does not change.	Each time the current drawing is opened, it loads the most current version of the *Xref* drawing.
If the original drawing that was used as the *Block* is changed, the *Inserted Block* does not change because it is not linked to the original drawing.	If the original *Xref* drawing is changed, the changes are automatically reflected when the current drawing is *Opened*.
Objects of the *Inserted* drawing cannot be changed in the current drawing unless *Refedit* or *Explode* is used.	The *Xref* can be changed in the current drawing if you use *Refedit* or use *Xopen* to edit the *Xref* in a separate window.
The current drawing can contain multiple *Block*s.	The current drawing can contain multiple *Xrefs*.
A *Block* drawing cannot be converted to an *Xref*.	An *Xref* can be converted to a *Block*. The *Bind* option makes it a *Block*—a permanent part of the current drawing.
A *Block* is *Inserted* on the current layer.	An *Xref* drawing's layers are also *Xrefed* and visibility of its layers can be controlled independently.
Any associated "named objects" of a *Block* can be used in the current drawing.	Any "named objects" of an *Xref* can be used in the current drawing only if *Xbind* is used.
Blocks can be nested.	An *Attached Xref* can be nested.
Any *Block* that is referenced by another *Block* is "nested."	An *Overlay Xref* cannot be nested.
A "circular" reference is not allowed with *Blocks* (X references Y and Y references X) because nested *Blocks* are also referenced.	Circular references are automatically detected (when the drawing you are *Xrefing* contains a nested *Xref* of the current drawing). Any nested *Xrefs* will load, but circular *Xrefs* are terminated.

INSERT	*XREF*
When a *Block* is inserted, you can control the properties of the block's layers.	When a drawing is *Attached* or *Overlayed*, you can control the properties of the *Xref's* layers.
You can *Xclip* the *Block* so that only portions of the entire *Block* drawing are visible.	You can *Xclip* the *Xref* so that only portions of the entire *Xref* drawing are visible.

As you can see, an *Xrefed* drawing has similarities and differences to an *Inserted* drawing. The following figures illustrate both the differences and similarities between an *Xrefed* drawing and an *Inserted* one.

The relationship between the current (parent) drawing and a drawing that has been *Inserted* and one that has been *Xreferenced* is illustrated here (Fig. 30-1). The *Inserted* drawing becomes a permanent part of the current drawing. No link exists between the parent drawing and the original *Inserted* drawing. In contrast, the *Xrefed* drawing is <u>not</u> a permanent part of the parent drawing but is a dynamic link between the two drawings. In this way, when the original *Xrefed* drawing is edited, the changes are reflected in the parent drawing (if a *Reload* is invoked or when the parent drawing is *Opened* next). The file size of the parent drawing for the *Inserted* case is the sum of both the drawings whereas the file size of the parent drawing for the *Xref* case is only the original size plus the link information.

Figure 30-1

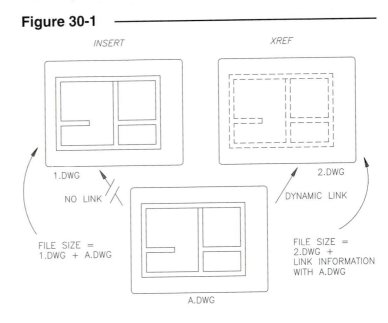

Some of the <u>similarities</u> between *Xreferenced* drawings and *Inserted* drawings are shown in the following figures.

Any number of drawings can be *Xrefed* or *Inserted* into the current drawing (Fig. 30-2). For example, component parts can be *Xrefed* to compile an assembly drawing.

Figure 30-2

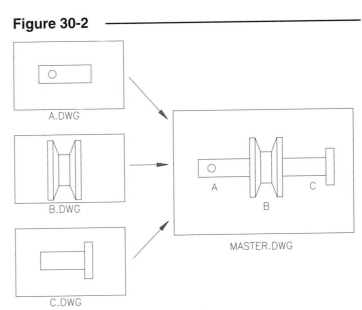

The current drawing can contain nested *Xrefs*. This feature is also similar to *Blocks*. For example, the OFFICE drawing can *Xref* a TABLE drawing and the TABLE drawing can *Xref* a CHAIR drawing. Therefore, the CHAIR drawing is considered a "nested" *Xref* in the OFFICE drawing (Fig. 30-3).

Figure 30-3

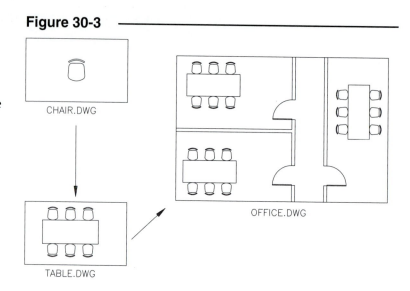

Xref drawings have many applications. *Xref*s are particularly useful in a networked office or laboratory environment. For example, several people may be working on one project, each person constructing individual components of a project. The components may be mechanical parts of an assembly; or electrical, plumbing, and HVAC layouts for a construction project; or several areas of a plant layout. In any case, each person can *Xref* another person's drawing as an external reference without fear of the original drawing being edited.

As an example of the usefulness of *Xref*s, a project coordinator could *Open* a new drawing, *Xref* all components of an assembly, analyze the relationships among components, and plot the compilation of the components (see Figure 30-2). The master drawing may not even contain any objects other than *Xref*s, yet each time it is *Opened*, it would contain the most up-to-date component drawings.

In another application, an entire team can access (*Xref, Attach*) the same master layout, such as a floor plan, assembly drawing, or topographic map. Figure 30-4 represents a mechanical design team working on an automobile assembly and all accessing the master body drawing. Each team member "sees" the master drawing, but cannot change it. If any changes are made to the original master drawing, all team members see the updates whenever they *Reload* the *Xref* or *Open* their drawing.

Figure 30-4

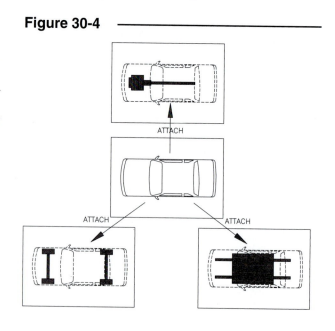

You also have the capability to clip portions of the *Xref* drawing that you do not want to see. Once you attach an *Xref*, you can use the *Xclip* command to draw a rectangular or polygonal (any shaped) boundary to determine what area of the *Xref* to view—all areas outside the boundary become invisible. For example, assume a site plane is *Xrefed* in order to calculate drainage around a house (Fig. 30-5).

Figure 30-5

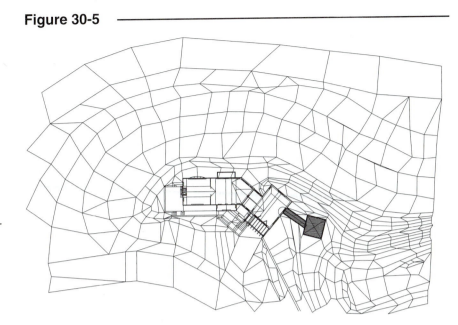

Xclip can be used to display only a portion of the Xref drawing needed for the calculations (Fig. 30-6).

Figure 30-6

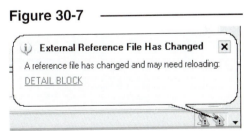

Internal features known as "demand loading," "spatial indexing," and "layer indexing" allow AutoCAD to <u>load only the portion of the drawing that appears in the clip boundary</u> into the current drawing. Demand loading and spatial and layer indexing can save a considerable amount of time and memory when Xreferencing large drawings such as maps or large floor plans, because only the small portion of the drawing that is needed (in the clip boundary) is actually loaded.

The named objects (*Layers, Text Styles, Blocks, Views*, etc.) that may be part of an *Xref* drawing become <u>dependent</u> objects when they are *Xrefed* to the parent drawing. These dependent objects cannot be renamed, changed, or used in the parent drawing. They can, however, be converted individually (permanently attached) to the parent drawing with the *Xbind* command.

Even though an *Xrefed* drawing's named objects (layers, linetypes, text styles, etc.) are attached, you cannot normally draw on or edit any of its layers. You do, however, have complete control over the <u>visibility</u> of the *Xref* drawing's layers. The *State* of the layers (*On, Off, Freeze*, and *Thaw*) of the *Xref* drawing can be controlled like any layers in the parent drawing.

If a drawing used as an *Xref* needs editing, there are three possibilities if you are using AutoCAD 2004. First, you can close the parent drawing, then *Open* the original external drawing, make the edits and *Save* the drawing. When the parent drawing is opened again, the new *Xref* is loaded. Second, <u>while in the parent drawing</u>, you can use the *Xopen* command to open the original (used as an *Xref*) to make the changes. In this case, when the *Xref* drawing is changed and saved, AutoCAD instructs you that a change has been made

Figure 30-7

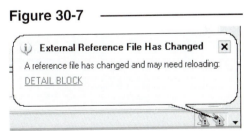

so you can reload the *Xref* in the parent drawing (Fig. 30-7). The parent drawing then reflects the new changes in the *Xref*. Third, you can use the *Refedit* command (in AutoCAD 2000 or newer) to edit the *Xref* from <u>within the parent drawing</u>, but only if the *Xedit* variable is set to 1 in the *Xrefed* drawing. Any changes made are "sent back" to the original *Xref* drawing.

An *Xref* Example

Assume that you are an interior designer in an architectural firm. The project drawings are stored on a local network so all team members have access to all drawings related to a particular project. Your job is to design the interior layout comprised of chairs, tables, desks, and file cabinets for an office complex. You use a prototype drawing named INTERIOR.DWG for all of your office interiors. It is a blank drawing that contains only block definitions (not yet *Inserted*) of each chair, desk, etc., as shown in Figure 30-8. The *Block* definitions are CHAIR, CONCHAIR, DESK, TABLE, FILECAB, and CONTABLE.

Figure 30-8

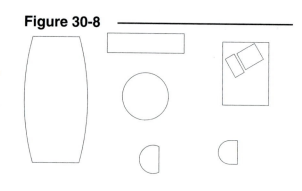

An architect on the team has almost completed the floorplan for the project. You use the *Attach* option of the *Xref* command to "see" the architect's floor plan drawing, OFFICEX.DWG. The visibility of the individual dependent layers can be controlled with the *Layer* command or *Layer Control* drop-down list. For example, layers showing the HVAC, electrical layout, and other details are *Frozen* to yield only the floor plan layer needed for the interior layout (Fig. 30-9).

Figure 30-9

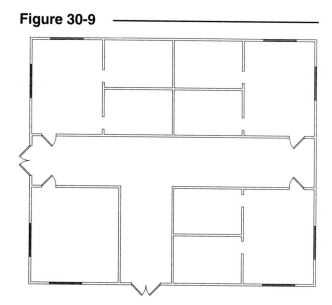

Next, you draw objects in the current drawing or, in this case, *Insert* the office furniture *Block*s that are defined in the current drawing (INTERIOR.DWG). The resulting drawing would display a complete layout of the office floor plan plus the *Inserted* furniture—as if it were one drawing (Fig. 30-10). You *Save* your drawing and go home for the evening.

Figure 30-10

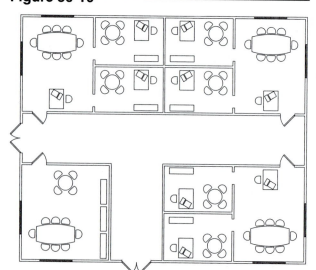

That evening, the architect makes a change to the floor plan. He works on the original OFFICEX drawing that you *Xrefed*. The individual office entry doors in the halls are moved to be more centrally located. The new layout appears in Figure 30-11.

Figure 30-11

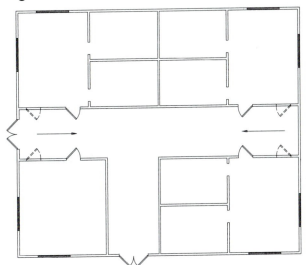

The next day, you *Open* the INTERIOR drawing. The automatic loading of the OFFICEX drawing displays the latest version of the office floor plan with the design changes. The appropriate changes to the interior drawing must be made, such as relocating the furniture by the hall doors (Fig. 30-12). When all parts of the office complex are completed, the INTERIOR (parent) drawing could be plotted to show the office floor plan and the interior layout. The resulting plot would include objects in the parent drawing and the visible *Xref* drawing layers.

Figure 30-12

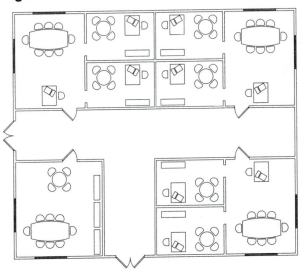

AutoCAD's ability to externally reference drawings makes team projects much more flexible and efficient. If the *Xref* capability did not exist, the OFFICEX drawing would have to be *Inserted* as a *Block*, and design changes made later to the OFFICEX drawing would not be apparent. The original OFFICEX *Block* would have to be *Erased*, *Purged*, and the new drawing *Inserted*.

The *Xclip* feature gives you the capability to isolate only a portion of an *Xrefed* drawing. For example, assume you wanted to discuss a particular room with the architect, so you open a *New* drawing and *Xref* the INTERIOR drawing. The OFFICEX drawing also is *Xrefed* because it is nested (previously *Xrefed* into the INTERIOR drawing) so the new drawing appears the same as the original INTERIOR drawing (see Figure 30-12). To isolate only the room in discussion, *Xclip* is used to draw a rectangular boundary around the desired area (Fig. 30-13).

Figure 30-13

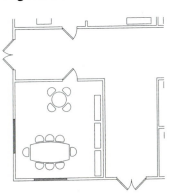

Often in the termination of a project, the final set of drawings is sent to the client. Because the *Xref* capability makes it feasible to work on a design as a set of individual component drawings, it is necessary to send the complete set of drawings to the customer or to combine the set into one drawing to submit. There are several strategies you can use to transfer the set of drawings.

If you want to keep the set of drawings as *Xrefs* to the parent drawing, the *Xrefs* can be saved with a *Relative Path*. In this way the parent drawing can always locate and load the *Xrefs* as long as they were loaded onto another machine in the same directory structure relative to the location of the parent. A similar method is to set the *PROJECTNAME* variable to locate any *Xrefs*. Another strategy is to combine all the drawings into one drawing using *Bind*. *Bind* brings an *Xrefed* drawing into the current drawing as if it were *Inserted*, so it becomes a permanent part of the parent drawing (as a *Block*).

CREATING AND USING *XREFS*

The *Xref* command and related commands can be invoked through the *Reference* toolbar (Fig. 30-14). If you prefer the pull-down menus, the *Insert* and *Modify* pull-downs contain *Xref* and related commands.

Figure 30-14

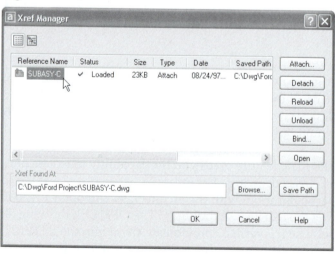

XREF

Pull-down Menu	COMMAND (TYPE)	ALIAS (TYPE)	Short-cut	Screen (side) Menu	Tablet Menu
Insert *Xref Manager...*	XREF or -XREF	XR or -XR	...	*INSERT* *Xref*	T,4

The *Xref* command invoked by typing or icon button produces the *Xref Manager* (Fig. 30-15). This one dialog box provides you with a means to create and manage *Xrefs*. The central area of the dialog box (discussed later) lists drawings that are currently *Attached* or *Overlayed* (see *List View* and *Tree View*). The *Xref* options are activated by the buttons on the right side of the dialog box.

If you prefer to type, you can use the -*Xref* command (notice the hyphen prefix). Using -*Xref* displays the following prompt:

Figure 30-15

Command: **-xref**
Enter an option [?/Bind/Detach/Path/Unload/Reload/Overlay/Attach] <Attach>:

Whether you use *Xref* or -*Xref*, the *Attach, Overlay, Detach, Reload, Unload, Bind*, and *Path* options are the same.

Attach

This (the default) option attaches one drawing to another by making the *Xref* drawing visible in the "parent" drawing. Even though the *Xref* drawing is visible, it cannot be edited from within the parent drawing unless *Refedit* is used. *Attach* creates a link between two drawings (Fig. 30-16).

Figure 30-16

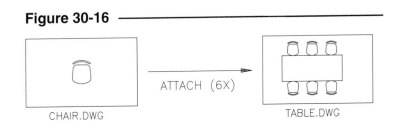

CHAIR.DWG   ATTACH (6X)   TABLE.DWG

Selecting the *Attach* button produces the standard *Select File* dialog box where you locate and select the .DWG file to attach (not shown). When you have selected the desired file to *Xref*, select *OK* to dismiss the *Select File* dialog box and produce the *External Reference* dialog box (see Figure 30-17).

When the *External Reference* dialog box appears (Fig. 30-17), specify the *Reference Type—Attachment* or *Overlay*. Ensure the *Attachment* button is pressed. (See "*Overlay*" next.) (Typing the *-Xref* command and using the *Attach* option produces the *Select File* dialog box directly—the *Xref Manager* does not appear.) Once the desired file is specified, the *External Reference* dialog box appears again for further action.

Figure 30-17

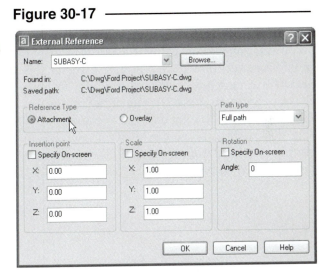

Notice the similarity of the edit boxes (Fig. 30-17) to the *Insert* command. The *Insertion Point, Scale,* and *Rotation* options operate identically to the *Insert* command. Similar to the action of *Insert*, the base point of the *Xref* drawing is its 0,0 point unless a different base point has been previously defined with the *Base* command.

You can also specify the *Path Type* to save with the *Xref* by selecting *Full Path, Relative Path* or *No Path* from the drop-down list. If you expect no changes to occur for the directory system used for the drawings and related *Xrefs*, *Full Path* is a suitable choice. However, if the drawing and related *Xrefs* might be sent to a client or another office or if the location on your system for the parent drawing and related *Xrefs* is expected to change, a better selection might be *Relative Path* or *No Path* (see "Managing Xref Drawings").

Attachment causes the specified drawing to appear in the current drawing similar to the way a drawing appears when it is *Inserted* as a *Block*. Each time the current drawing is *Opened*, the *Attached* drawing is loaded as an *Xref*. More than one reference drawing can be *Attached* to the current drawing. A single reference drawing can be *Attached* to any number of insertion points in the current drawing. The dependent objects in the reference drawing (*Layers, Blocks, Text styles*, etc.) are assigned new names, "external drawing|old name" (see "Dependent Objects and Names").

Overlay

An *Xref Overlay* is similar to an *Attached Xref* with one main difference—an *Overlay* cannot be nested. In other words, if you *Attach* a drawing that (itself) has an *Overlay*, the *Overlay* does not appear in your drawing. On the other hand, if the first drawing is *Attached*, it appears when the parent drawing is *Attached* to another drawing (Fig. 30-18).

Figure 30-18

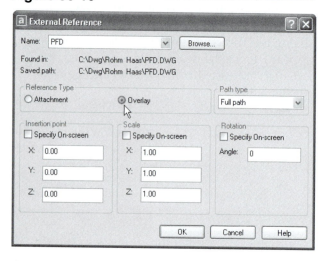

To produce an *Overlay* using the *Xref Manager*, you must first select the *Attach* button to produce the *Select File* dialog box. After selecting the desired file to overlay, press the *Overlay* button in the *External Reference* dialog box that appears (Fig. 30-19). If you type the *-Xref* command, simply use the *Overlay* option and the standard *Select File* dialog box appears.

Figure 30-19

The *Overlay* option prevents "circular" *Xrefs* from appearing by preventing unwanted nested *Xrefs*. This is helpful in a networking environment where many drawings *Xref* other drawings. For example, assume drawing B has drawing A as an *Overlay*. As you work on drawing C, you *Xref* and view only drawing B without drawing B's overlays—namely drawing A. This occurs because drawing A is an *Overlay* to B, not *Attached*. If all drawings are *Overlays*, no nesting occurs (Fig. 30-20).

Figure 30-20

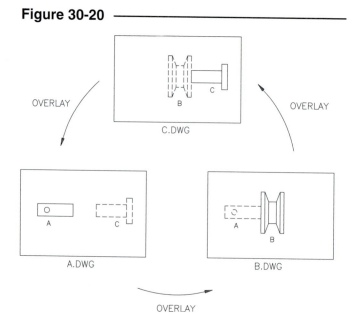

The overlay concept is particularly useful in a concurrent engineering project, where all team members must access each other's work simultaneously. In the case of a design team working on an automobile assembly, each team member can *Attach* the master body drawing and also *Overlay* the individual subassembly drawings (Fig. 30-21). *Overlay* enables each member to *Xref* all drawings without bringing in each drawing's nested Xreferences.

Figure 30-21

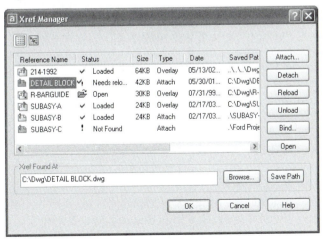

There are times when you cannot foresee the potential use of a drawing and therefore do not *Overlay* rather than *Attach*. For example, in a large design office it is quite possible to *Xref* a drawing on the network that includes your current drawing as an *Attached Xref*, thus causing a circular reference. If this occurs, an *AutoCAD Alert* dialog box appears giving notification of the circular reference and presents the option to continue or not (Fig. 30-22). If you respond with *Yes*, AutoCAD reads in the *Xref* and any nested *Xrefs* to the point where it detects the circularity, then terminates. Answering *No* cancels the *Attach*.

Figure 30-22

AutoCAD Alert

Circular references detected.

Continue?

[Yes]   [No]

Detach

Detach breaks the link between the parent drawing and the *Xref*. Using this option causes the *Xrefed* drawing to "drop" from view immediately. When the parent drawing is *Opened*, the previously *Attached* drawing is no longer loaded as an *Xref*. Nested *Xrefs* cannot be *Detached* or selected to *Bind* since they are actually part of another drawing. You can *Detach* the drawing that contains the nested *Xref*.

Xref found at

You can specify the drive and directory of the external reference using the *Xref found at* section (or the *Path* option of the *–Xref* command). This is necessary if an *Xrefed* drawing has been relocated to another drive or directory on the computer drive or network drives. For example, if the SUBASY-C drawing (that your parent drawing referenced) was relocated to another drive or directory, it could not be loaded when the parent drawing was opened and would be listed as "not found" in the *Xref Manager* (Fig. 30-23).

Use the *Path* option to specify the new location. Enter the path in the edit box, or use the *Browse...* tile to invoke the *Select new path* dialog box (essentially the same interface as the *Select File* dialog box). When you use the *Path* option of the typed version of the command, *-Xref*, the path must also be typed.

Figure 30-23

Xref Manager

Reference Name	Status		Size	Type	Date	Saved Pat
214-1992	✓	Loaded	64KB	Overlay	05/13/02...	..\..\..\Dwg
DETAIL BLOCK	✓!	Needs relo...	42KB	Attach	05/30/01...	C:\Dwg\DE
R-BARGUIDE		Open	30KB	Overlay	07/31/99...	C:\Dwg\R-
SUBASY-A	✓	Loaded	24KB	Overlay	02/17/03...	C:\Dwg\SL
SUBASY-B	✓	Loaded	24KB	Attach	02/17/03...	.\SUBASY-
SUBASY-C	!	Not Found		Attach		.\Ford Proje

Attach...
Detach
Reload
Unload
Bind...
Open

Xref Found At
C:\Dwg\DETAIL BLOCK.dwg   [Browse...]   [Save Path]

[OK]   [Cancel]   [Help]

When you archive (back up) drawings, remember also to back up the drives and directories where the *Xref* drawings are located (or use the *Bind* option when the project is finished). One possible file management technique is to specify a directory especially for *Xrefs* (see "Managing Xref Drawings"). In a networking environment, access rights to *Xref* directories must be granted.

Reload

This option forces a *Reload* of the external drawing at any time while the current drawing is in the Drawing Editor. This ensures that the current drawing contains the <u>most recent version</u> of the external drawing. In AutoCAD 2004 you are notified if a drawing has been changed and requires *Reloading* (see Figure 30-24, R-BARGUIDE drawing). (See also "*XREFNOTIFY*.") In a networking environment, you may not be able to *Attach*, *Overlay*, *Bind*, or *Reload* a drawing based on whether the referenced drawing is currently being edited (open) by another person or by the setting of the *XLOADCTL* system variable in the referenced drawing (see "*XLOADCTL*").

Unload

The *Unload* option can be used to increase computing speed in a drawing. In cases when you need to view (load) drawings only on an as-need basis, *Unload* them. This causes the current (parent) drawing to open more quickly, speeds regenerations, and uses less memory. The unloaded *Xref* is not visible, but the link information remains in the current drawing. To view the *Xref* again, use *Reload*. This method is preferred over using *Detach* when you expect to use the *Xref* again.

Bind

The *Bind* option <u>converts</u> the *Xref* drawing to a *Block* in the current drawing, then terminates the external reference partnership. The original *Xrefed* drawing is not affected. The names of the dependent objects in the external drawing are changed to avoid possible conflicts in the case that the parent and *Xref* drawing have dependent objects with the same name (see "Dependent Objects and Names"). If you want to only bind selected named objects, use the *Xbind* command.

When the *Bind* button is selected (in the *Xref Manager*), the *Bind Xrefs* dialog box appears for you to select the *Bind Type* (see Figure 30-28). This option determines how the <u>names of dependent objects</u> (such as layers and blocks) appear in the drawing after they become a permanent part of the parent drawing (see "Dependent Objects and Names").

Open

This option is a new feature for AutoCAD 2004 (see also *Xopen* command in "Editing *Xrefs* and *Blocks*"). Using this button, you can open an externally referenced drawing in a separate window (of the same AutoCAD session), make any needed changes, then *Save* the drawing. When the opened drawing has been changed and saved, AutoCAD notifies you (in the system tray) that the drawing has been changed and needs reloading (see Figure 30-7). (See also "*XREFNOTIFY*.") In addition, a message appears in the Status column of the *Xref Manager* reminding you to *Reload* (see Figure 30-24, R-BARGUIDE). This method is generally easier than using *Refedit* for editing an *Xref*. *Opening* a drawing through the *Xref Manager* (*Xopen*) <u>overrides</u> the setting of the *XLOAD-CTL* and *XEDIT* variables (see "Demand Loading" and "Editing *Xrefs* and *Blocks*").

Figure 30-24

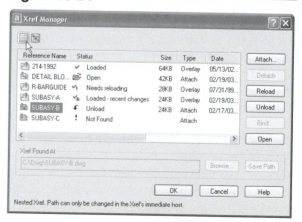

List View and Tree View

The central area of the *Xref Manager* is used to list and give the status of drawings that have been *Xrefed*. Use the two buttons in the upper-left corner of the dialog box (Fig. 30-24) to display the information in a list that can be ordered alphabetically by column (*List View*) or in a hierarchical tree structure (*Tree View*).

List View

This option (default) displays the *Xrefs* alphabetically by name. You can sort the list by column in forward or reverse order (*Reference Name, Size, Type, Date,* etc.) by selecting the column heading and clicking once or twice. You can resize the column width by dragging the column separator left or right. The *Reference Name* can also be edited (up to 256 characters, no spaces). The *Type* can be *Attach* or *Overlay*. The *Status* column displays the state of each reference. The following states are possible:

Status	Description
Loaded	*Xref* was found when drawing was opened or reloaded
Unload	*Xref* will be unloaded when dialog box is closed
Unloaded	*Xref* is currently unloaded
Needs Reloading	*Xref* has changed, use *Reload* to force reload when dialog box is closed
Reload	*Xref* will be reloaded when dialog box is closed
Loaded, recent changes	*Xref* was loaded but has been changed recently
Not found	*Xref* was not found in search paths when drawing was opened or reloaded
Unresolved	*Xref* was found but could not be read by AutoCAD
Unreferenced	*Xref* was attached but is erased
Orphaned	Nested *Xref* is attached to an *Xref* that is not loaded
Open	*Xref* will be opened for editing in another window when dialog is closed

Tree View

The *Tree View* shows the *Xrefs* in a hierarchical tree structure. This option is particularly useful if you have <u>nested</u> *Xrefs* (Fig. 30-25). Note that the icons also indicate the type and status of each referenced drawing (*Overlay, Attach, Reload, Unload, Not found,* etc.).

Figure 30-25

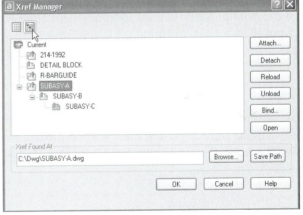

Dependent Objects and Names

The named objects that have been created as part of a drawing become dependent objects when that drawing is *Xrefed*. The named objects in an *Xref* drawing can be any of the following:

> *Blocks*
> *Layers*
> *Linetypes*
> *Dimension Styles*
> *Text Styles*

These dependent objects <u>cannot</u> be renamed or changed in the parent drawing. Dependent text or dimension styles cannot be used in the parent drawing. You cannot draw on dependent layers from the parent drawing; you can only control the visibility of the dependent layers. You <u>can</u>, however, *Bind* the entire *Xref* and the dependent named objects, or you can bind individual dependent objects with the *Xbind* command. *Xbind* converts an individual dependent object into a permanent part of the parent drawing. (See "*Xbind*.")

When an *Xref* contains named objects, the names of the dependent objects are changed (in the parent drawing). A new naming scheme is necessary because conflicts would occur if the *Xref* drawing and the parent drawing both contained *Layers* or *Blocks*, etc., with the same names, as would happen if both drawings used the same template drawing. When a drawing is *Attached*, its original object names are prefixed by the drawing name and separated by a pipe (|) symbol.

For example, if the DB_SAMP drawing is *Xrefed* and it contains a layer named PHONES, it is listed (in the parent drawing) as DB_SAMP|PHONES (Fig. 30-26).

Figure 30-26

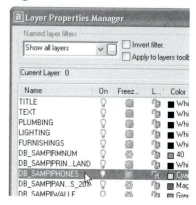

Because the dependent objects cannot be used in the parent drawing (you cannot insert a dependent block nor can you draw on a dependent layer), these objects cannot be renamed or deleted either. If you attempt to delete or rename a dependent layer a warning appears (Fig. 30-27). (Keep in mind that dependent blocks, layers, etc. can be used if *Bind* or *Xbind* is used to make them a permanent part of the parent drawing. Additionally, DesignCenter can be used to drag-and-drop named objects into the parent drawing.)

Figure 30-27

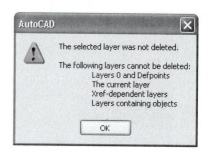

Likewise, if the *Xref* drawing contains a block named DESK2, it is renamed in the parent drawing to DB_SAMP|DESK2. Using the *-Block* command to reveal a listing of blocks in the current drawing yields the following display. Note the tally below of "User Blocks," "External References," and "Dependent Blocks":

```
Command: -block
Enter block name or [?]: ?
Enter block(s) to list <*>:

Defined blocks.
  "CENOS"                  Xref: resolved
  "DB_SAMP"                Xref: resolved
  "DB_SAMP|CHAIR7"         Xdep: "DB_SAMP"
  "DB_SAMP|COMPUTER"       Xdep: "DB_SAMP"
  "DB_SAMP|DESK2"          Xdep: "DB_SAMP"
  "DB_SAMP|DESK3"          Xdep: "DB_SAMP"
  "DB_SAMP|DOOR"           Xdep: "DB_SAMP"
  "DB_SAMP|DR-36"          Xdep: "DB_SAMP"
  "DB_SAMP|DR-69P"         Xdep: "DB_SAMP"
  "DB_SAMP|DR-72P"         Xdep: "DB_SAMP"
  "DB_SAMP|KEYBOARD"       Xdep: "DB_SAMP"
  "DB_SAMP|RMNUM"          Xdep: "DB_SAMP"
  "DB_SAMP|SOFA2"          Xdep: "DB_SAMP"
```

"DOD-part306"	Xref: resolved
"SUBASY-A"	Xref: resolved
"SUBASY-B"	Xref: resolved
"SUBASY-C"	Xref: resolved

User Blocks	External References	Dependent Blocks	Unnamed Blocks
0	6	11	0

Command:

The same naming scheme operates with all dependent objects—*Dimension Styles*, *Text Styles*, *Views*, etc.

Dependent Object Names with *Bind* and *Xbind*

If you want to *Bind* the *Xref* drawing, the link is severed and the *Xref* drawing becomes a *Block* in the parent drawing. Or, if you want to bind only <u>individual named objects</u>, the *Xbind* command can be used to bring in specific *Blocks* or *Layer* names. The names are converted from the previous dependent naming scheme (using the *Xref* drawing name as a prefix with a | [pipe] separator) to another name depending on the command and option you use.

If you use *Xbind* or use *Bind* with the default *Bind* option, the names of dependent objects keep the *Xref* drawing name, but the | (pipe) symbol is changed to a $#$ separator, where # represents a random number. For example, assume you use the *Bind* button (in the *Xref Manager*) with the DB_SAMP drawing from the previous example. The *Bind Xrefs* dialog box appears (Fig. 30-28). The default (*Bind*) option is the traditional naming scheme. That is, the *Xref* drawing name prefix stays with the object name.

Figure 30-28

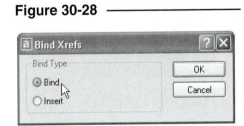

Using the *Bind* option, the resulting layer names would appear in the *Layer Properties Manager* listing (Fig. 30-29). Notice the dependent layer previously named DB_SAMP|CHAIRS is changed to DB_SAMP0CHAIRS. These layers <u>can be renamed</u> at this point because they are now a permanent part of the current drawing. Using the *Xbind* command to "import" selected named objects results in the same layering scheme illustrated here.

Figure 30-29

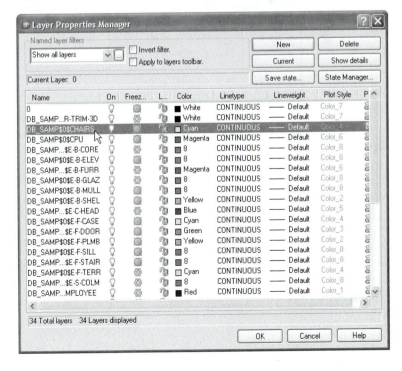

On the other hand, if you select the *Insert* option in the *Bind Type* dialog box, the *Xref* drawing name prefix is dropped from the object names when they are "imported." Therefore, the named objects assume their <u>original name</u>, assigned when they were created in the original drawing. Using the previous *Xref* example (DB_SAMP), the layer names would appear in the *Layer Properties Manager* listing as A-DOOR-TRIM-3D, CHAIRS, CPU, E-B-CORE, etc. (Fig. 30-30).

Figure 30-30

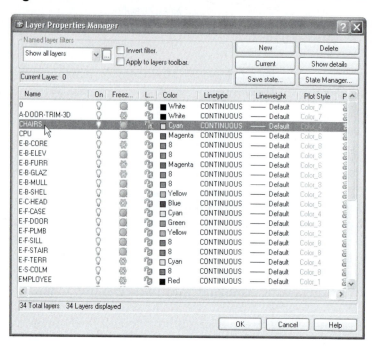

This method (*Bind, Insert*) is preferred if you want to *Bind* an *Xrefed* drawing that was created from the same template drawing as the current (parent) drawing. In this way, the same layers, dimension styles, text styles, etc. are not duplicated when you *Bind* the *Xref*. However, this method should <u>not be used</u> when <u>two different</u> *Blocks* exist with the same name or two layers exist with different linetype and color settings that should be maintained. When a dependent *Block* or layer having the same name as that in the parent drawing is "imported" with *Bind, Insert*, the *Block* definition and the layer properties of the parent drawing take precedence. Therefore, it is possible to accidentally redefine a *Block* or lose linetype and color properties of a layer with this scheme. There is no *Insert* option for *Xbind*.

XATTACH

	COMMAND (TYPE)	ALIAS (TYPE)	Short-cut	Screen (side) Menu	Tablet Menu
Pull-down Menu					
Insert *External Reference* *Manager...* *Attach...*	*XATTACH*	*XA*	...	*INSERT* *Xref* *Attach...*	*T,4*

Xattach is a separate command from *Xref* but accomplishes the same action as *Xref, Attach*. If you use *Xref*, the *Xref Manager* appears, then you must select the *Attach...* tile to invoke the *Select File* and *External Reference* dialog boxes. However, you can use *Xattach* to invoke the *Select File* and *External Reference* dialog boxes directly (Fig. 30-31). Therefore, use the *Xattach* command when you want to <u>*Attach or Overlay*</u> an *Xref* but do not want to view or manage the list of current Xrefs.

Figure 30-31

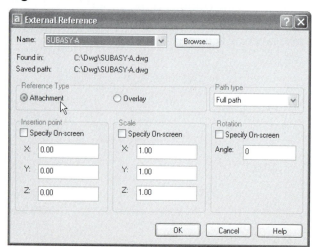

XBIND

	COMMAND (TYPE)	ALIAS (TYPE)	Short-cut	Screen (side) Menu	Tablet Menu
Pull-down Menu					
Modify Object> External Reference>	XBIND or -XBIND	XB or -XB	...	MODIFY1 Xbind	X,19

The *Xbind* command is a separate command; it is not an option of the *Xref* command as are most other Xreference controls. *Xbind* is similar to *Bind*, except that *Xbind* binds an <u>individual</u> dependent object, whereas *Bind* converts the <u>entire</u> *Xref* drawing to a *Block*. Any of the listed dependent named objects (see "Dependent Objects and Names") that exist in a drawing can be converted to a permanent and usable part of the current drawing with *Xbind*. In effect, *Xbind* makes a copy of the named object since the original named object is not removed from the dependent drawing.

The *Xbind* command produces the *Xbind* dialog box (Fig. 30-32). All *Attached* and *Overlayed Xrefs* are listed. Expand any *Xref* to reveal the dependent object types (*Block*, *Dimstyle*, etc.) that can be imported. Expand each dependent object type to list the named objects of the type for that particular drawing. Select the desired object, then pick *Add ->* to add the object to the current (parent) drawing. The following object types are valid.

Figure 30-32

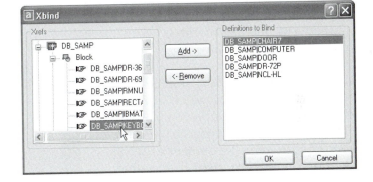

Block
Select this option to *Xbind* a *Block* from the *Xrefed* drawing. The *Block* can then be inserted into the current drawing and has no link to the original *Xref* drawing.

Dimstyle
This option allows you to make a dependent Dimension Style a permanent part of the current drawing. Dimensions can be created in the current drawing that reference the new Dimension Style.

Layer
Any dependent layer can be brought into the current drawing with this option. The geometry that resides on that layer does <u>not</u> become part of the current drawing, only the layer name and its properties. *Xbinding* a layer also *Xbinds* the layer's color and linetype.

Linetype
A linetype from an *Xrefed* drawing can be bound to the current drawing using this option. This action has the same effect as loading the linetype.

Textstyle
If a text style exists in an *Xrefed* drawing that you want to use in the current drawing, use this option to bring it into the current drawing. New text can be created or existing text can be modified to reference the new text style.

If you use (type) the *-Xbind* command, the Command line version of the command is invoked and individual dependent objects must be typed:

```
Command: -xbind
Enter symbol type to bind [Block/Dimstyle/LAyer/LType/Style]:
```

You must know and type the exact name of the dependent object name, including the dependent drawing name prefix and pipe character; therefore, this method is cumbersome compared to using the *Xbind* dialog box.

ADCENTER

Pull-down Menu	COMMAND (TYPE)	ALIAS (TYPE)	Short-cut	Screen (side) Menu	Tablet Menu
Tools *DesignCenter*	ADCENTER	ADC	Ctrl+2	...	...

As an alternative to using *Xref*, *Xattach*, and *Xbind*, DesignCenter can be used to drag-and-drop drawings or named objects into the parent drawing. DesignCenter is accessed by any method shown in the table above.

Using DesignCenter to *Attach* or *Overlay Xrefs*
Normally when you drag-and drop a .DWG file from the DesignCenter Content Area (window on the right) directly into the current drawing, it is *Inserted* as a *Block*. You can, however, use DesignCenter to attach *Xrefs*.

To *Attach* or *Overlay* an *Xref*, use the Tree View window (left window) to locate the folder that contains the drawing file you want to *Xref*, so all drawings contained in the folder appear in the Content Area (right window). Next, right-click on the desired drawing in the Content Area. Alternately right-drag (drag-and-drop, but holding down the right button) the desired drawing. This causes a shortcut menu to appear (Fig. 30-33). Select *Attach as Xref* from the menu. This causes the *External Reference* dialog box to appear (see Figure 30-31). Note that you can specify *Attach* or *Overlay* in the *External Reference* dialog box. Proceed to specify the desired options in the dialog box and bring in the new *Xref*. The result is identical to using *Xattach* or the *Xref Manager* to *Attach* or *Overlay* the *Xref*.

Figure 30-33

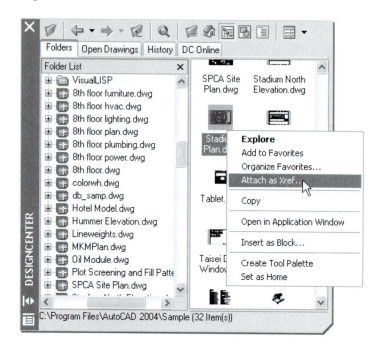

Using DesignCenter as an Alternative to *Xbind*
DesignCenter can also be used to drag-and-drop named objects from any drawing into the current drawing. *Xbind* allows you to bring named objects into the current drawing but only from *Xrefed* drawings. With DesignCenter, you can drag-and-drop from any drawing. Named objects that can be imported with DesignCenter are *Blocks*, *Dimstyles*, *Layers*, *Layouts*, *Linetypes*, *Textstyles* and *Xrefs*.

To bring named objects into the current drawing, use the Tree View window (Fig. 30-34, left window) to locate the drawing file that contains the named objects you want to import, then expand the drawing (plus sign) so all named objects contained in the drawing appear in the Content Area. (right window). Next, drag-and-drop the desired named object from the Content Area. into the drawing area. This causes the selected named object to become part of the current drawing.

See Chapter 21 for more detailed information on DesignCenter.

Figure 30-34

	Pull-down Menu	COMMAND (TYPE)	ALIAS (TYPE)	Short-cut	Screen (side) Menu	Tablet Menu
VISRETAIN	*Tools* *Options...* *Open and Save* *Retain changes to Xref layers*	*VISRETAIN*	...	...	*TOOLS2* *Option* *Open and Save* *Retain changes to Xref layers*	...

This variable controls the dependent (*Xref* drawing) layer visibility, color, linetype, lineweight, and plot style settings in the parent drawing. Remember that you cannot draw or edit on a dependent layer without *Refedit*; however, you can change the layer's settings (*Freeze/Thaw, On/Off, Color, Linetype*, etc.) with the *-Layer* command, *Layer Properties Manager*, or *Layer Control* drop-down box. The setting of *VISRETAIN* determines if the settings that have been made are retained when the *Xref* is reloaded.

When *VISRETAIN* is set to 1 in the parent drawing, all the layer settings (*Freeze/Thaw, On/Off, Color, Linetype*, etc.) are retained for the Xreferenced dependent layers. When you *Save* the drawing, the settings for dependent layers are saved with the current (parent) drawing, regardless of whether the settings have changed in the *Xref* drawing itself. In other words, if *VISRETAIN=1* in the parent drawing, the layer settings for dependent layers are saved. If *VISRETAIN=0*, then dependent layer properties settings are determined by the settings in the *Xref* drawing.

VISRETAIN also determines whether or not changes to the path of nested *Xrefs* are saved. You can change the path of an *Xref* or of a nested *Xref* (location AutoCAD searches for *Xrefs*) with the *Path* option of *-Xref* or the *Path type* drop-down list in the *Xref Manager*.

You can set *VISRETAIN* in the *Open and Save* tab of the *Options* dialog box. To set *VISRETAIN* to 1, select the checkbox next to *Retain changes to Xref layers* (Fig. 30-35, right center). No check appearing in the box means that *VISRETAIN* is set to 0.

Figure 30-35 ─────────────

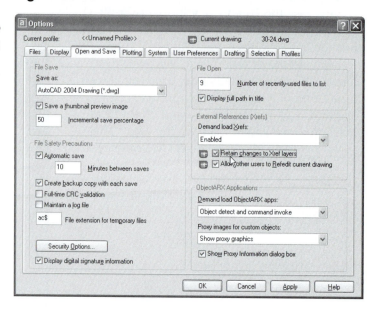

XCLIP

Pull-down Menu	COMMAND (TYPE)	ALIAS (TYPE)	Short-cut	Screen (side) Menu	Tablet Menu
Modify *Clip>* *Xref*	*XCLIP*	*XC*	...	*MODIFY1* *Xclip*	*X,18*

When working with *Xrefs* you may want to display only a specific part of the attached drawing. For example, you may *Xref* an entire manufacturing plant layout (floor plan) in order to redesign an office space, or you may need to *Xref* a large map of a city just to design a new interchange at one location. In these cases, loading and displaying the entire *Xref* would occupy most of the system resources just for display, even though only a small area is needed.

Using the *Xclip* command, you can visually clip an *Xref drawing or a Block* to display only the portion inside the area defined by the clip boundary (Fig. 30-36). You can define the boundary as a rectangle or as a many-sided polygon of virtually any shape. All geometry inside the boundary is displayed normally. The portions of the *Xrefed* drawing (or *Block*) that fall outside the boundary are not only restricted from the display, but a feature called "demand loading" can be used to prevent the unwanted portions of the *Xref* from loading so system resources are not used unnecessarily (see "Demand Loading," "*INDEXCTL*" and "*XLOADCTL*").

Figure 30-36 ─────────────

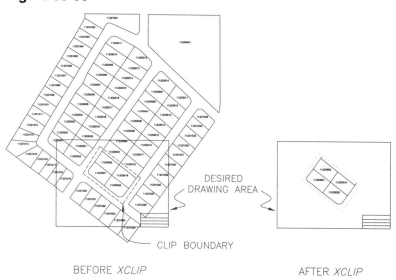

Clipping does not edit or change the *Xref* in any way, it only prevents portions from being displayed:

```
Command: xclip
Select objects: PICK (the Xref Object)
Select objects: Enter
Enter clipping option [ON/OFF/Clipdepth/Delete/generate Polyline/New boundary] <New>: n
Specify clipping boundary: [Select polyline/Polygonal/Rectangular] <Rectangular>: r
Specify first corner: PICK
Specify opposite corner: PICK
Command:
```

The options of *Xclip* are as follows.

New Boundary

Rectangular
You define a rectangular boundary by PICKing diagonal corners. The clipping boundary is on the XY plane of the current UCS.

Polygonal
Select points as the vertices of a polygon. The polygon can have any shape and number of sides.

Select Polyline
Select an existing polyline (*Pline*) to use for the boundary. The existing polyline must consist of straight segments but can be open.

ON/OFF

ON displays only the *Xref* geometry inside the clipping boundary. *OFF* displays the entire *Xref*, ignoring the clipping boundary.

Clipdepth

Use this option to set front and back clipping planes on the *Xref* object. This option is useful for 3D geometry when you want to clip portions of the object behind a back clipping plane and in front of a front clipping plane. You must specify a clip distance from and parallel to the clipping boundary.

Delete

You can delete an existing boundary for the selected *Xref* or *Block* with this option. *Erase* cannot be used to delete a clipping boundary.

Generate Polyline

Automatically draws a *Pline* along an existing clipping boundary. The new *Pline* assumes the current layer, linetype, and color settings. This option is helpful if you want to change the clipping boundary using *Pedit* with the new *Pline*, then use the *Polyline* option to establish the new boundary.

XCLIPFRAME

Pull-down Menu	COMMAND (TYPE)	ALIAS (TYPE)	Short-cut	Screen (side) Menu	Tablet Menu
Modify *Object>* *External Reference>* *Frame*	XCLIPFRAME	...	...	...	...

XCLIPFRAME is a system variable that can be used to control the display of the clipping boundary created by the *Xclip* command. For example, in some cases it may be desirable to display the boundary, while in other cases no boundary may be preferred (Fig. 30-37). The options are 0 (turns clipping boundary off) and 1 (turns clipping boundary on).

Figure 30-37

XCLIPFRAME = 1

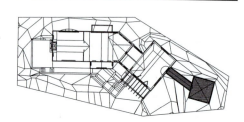

XCLIPFRAME = 0

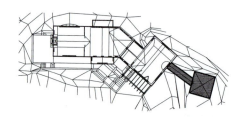

Demand Loading

When *Xref* clipping is used to display only a selected area of an *Xrefed* drawing or block, it logically follows that the area outside the clipping boundary is of little or no interest in the current drawing. In order to conserve system resources and save time loading and regenerating these unwanted drawing areas, AutoCAD provides a feature called "demand loading."

AutoCAD 2000 introduced the *Partialopen* and *Partialload* commands (only drawings saved in AutoCAD 2000 or later format are enabled for this feature). These commands allow you to open only part of a drawing based on *Layers, Views,* or by selecting a particular area of a drawing to load by specifying a boundary (see *"Partialopen"* and *"Partialload"* in Chapter 2, Working with Files, for more information). These features are similar to the *Xclip* feature for *Xrefs* and can also take advantage of demand loading.

Demand loading allows AutoCAD to load only the information (about the *Xref* or partially opened drawing) that is needed to display the geometry in the clipping boundary. To take maximum advantage of demand loading, the demand loading feature must be enabled in the parent or partially opened drawing and the *Xref* drawings must be saved with layer and spatial indexes enabled. When a clipping boundary is formed, AutoCAD uses the <u>layer index</u> and <u>spatial index</u> in the *Xrefed* or partially opened drawing to categorize which objects are fully within, partially within, and not within the clipping boundary and which layers are visible and not visible.

| | Spatial Index | Updated in the *Xref* or partially opened drawing when a clipping boundary is formed to display a portion of the drawing's objects |

| | Layer Index | Updated in the *Xref* or partially opened drawing when layer visibility is changed (*Freeze, Thaw, On, Off*) |

Using the layer index and spatial index, AutoCAD determines which portions of the *Xrefed* or partially opened drawing file to load. This process provides performance advantages; however, certain demand loading controls must be used. The controls for demand loading are the *INDEXCTL* and *XLOADCTL* system variables.

INDEXCTL enables spatial indexing and layer indexing and should be set in the *Xref* drawings and in drawings intended to be used with *Partialopen* and *Partialload*. *XLOADCTL* is intended to be set in the parent drawing for an *Xref*, but has no usefulness for the *Partialopen* and *Partialload* features.

INDEXCTL

Pull-down Menu	COMMAND (TYPE)	ALIAS (TYPE)	Short-cut	Screen (side) Menu	Tablet Menu
File *Saveas...* *Tools >* *Options...* *Index type*	*INDEXCTL*	...	...	*FILE* *Saveas* *Options...* *Index type*	...

When a drawing is used as an *Xref*, and particularly a clipped *Xref*, demand loading can increase performance in the parent drawing. Similarly, when *Partialopen* or *Partialload* is used to display a drawing, demand loading can also increase performance. In order for this to occur, spatial indexing and/or layer indexing must be enabled in the *Xref* or partially opened drawing(s). *INDEXCTL* controls how a drawing is loaded <u>when it is used as an *Xref*</u>. In other words, <u>*INDEXCTL* is set in the *Xref* drawing</u> (the drawing to be referenced), <u>not in the parent drawing</u> (the current drawing when using the *Xref* command).

Considering this in another way, assume you *Open* a drawing (parent drawing) and *Xref* two other drawings. Next, you create a clipping boundary around each *Xref*, so only a portion of each *Xref* is visible. Then you *Save* the parent drawing. In order to take advantage of demand loading and realize a performance increase when you *Open* the parent drawing again, layer and spatial indexing must first be enabled in <u>each of the two *Xrefed* drawings</u> for demand loading to occur. *Open* each (original) *Xref* drawing, and set *INDEXCTL* to 1, 2, or 3, then *Save*.

The *INDEXCTL* variable can be set by Command line format or can be set by using *SaveAs*, selecting *Tools*, then selecting the *Options...* button. The *Saveas Options* dialog box appears (Fig. 30-38) with four options for *Index Type*.

Figure 30-38

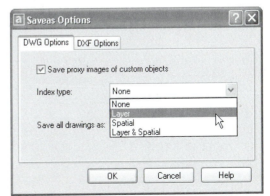

The *INDEXCTL* system variable is set to 0 by default, in which no spatial or layer indexes are created. The settings for *INDEXCTL* are as follows.

Setting	*Export Options* Setting	Effect
0	*None*	no indexes created
1	*Layer*	layer indexes created
2	*Spatial*	spatial indexes created
3	*Layer & Spatial*	both layer and spatial indexes created

To take full advantage of demand loading, set *INDEXCTL* to 3 in the drawing you anticipate using as an *Xrefed* or partially opened drawing.

NOTE: <u>Only drawings in Release 14 or newer format</u> can be *Saved* with an *INDEXCTL* setting. Drawings created and saved in previous releases are loaded in their entirety, so no benefit is obtained from demand loading.

XLOADCTL

Pull-down Menu	COMMAND (TYPE)	ALIAS (TYPE)	Short-cut	Screen (side) Menu	Tablet Menu
Tools *Options...* *Open and Save* *Demand* *Load Xrefs:*	*XLOADCTL*	...	...	*TOOLS2* *Options...* *Open and Save* *Demand* *Load Xrefs...*	...

The *XLOADCTL* variable set in the parent drawing (1) turns on or off demand loading for the current drawing and (2) controls what access other users have to *Xrefed* drawings. *XLOADCTL* is intended to be set in the parent drawing for an *Xref*, but has no usefulness for the *Partialopen* and *Partialload* features.

Demand loading is dynamic. If the clipping boundary is changed, the spatial index in the *Xrefed* drawing is automatically updated and if an *Xref* layer is frozen or thawed, the layer index is updated. This unique feature allows information from the *Xrefed* drawing to be loaded "on demand." However, there is one drawback to demand loading for *Xrefs*—that is, when demand loading is enabled, AutoCAD places a lock on all *Xrefed* drawings so that it can read and modify the information it needs (indexes) on demand. Therefore, other users can *Open* the *Xrefed* drawings as "read only" but cannot *Save* changes to them. Alternately, you can use the *Copy* option of demand loading, which creates a temporary copy of the *Xrefed* drawing(s) for use by the parent drawing and leaves the original *Xref* drawing available for use by others to edit and *Save*.

NOTE: *XLOADCTL* has no effect on <u>your</u> ability to use *Xopen* (or the *Open* button in the *Xref Manager*). In other words, if *XLOADCTL* is set to 1 (others cannot edit), you can still open and edit a currently *Xrefed* drawing using *Xopen*; however, other users can only open the original *Xref* drawing as "read-only."

The options of *XLOADCTL* are given in the following table.

Setting	*Options* Setting	Effect
0	*Disabled*	demand loading is disabled *Xref* drawing is loaded in its entirety others can access, edit, and save the *Xref* file
1	*Enabled*	demand loading is enabled locks original *Xref* drawing for use by parent drawing only
2	*Enabled with copy*	demand loading is enabled creates a temporary copy of the *Xref* for use by the parent drawing others can access, edit, and save original *Xref* file

The default setting for *XLOADCTL* is 2. If you are in a networking environment, a setting of 1 ensures others cannot alter the *Xref*, while a setting of 2 allows other users involved in a particular project (related to the *Xrefed* drawing) to continue working on the original *Xref* files.

XLOADCTL can also be set in the *Open and Save* tab of the *Options* dialog box (Fig. 30-39, upper right). The three options (*Disabled*, *Enabled*, and *Enabled with copy*) are explained in the previous table.

Figure 30-39

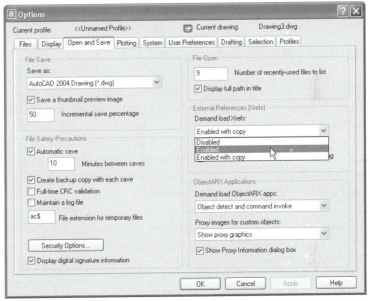

XLOADPATH	Pull-down Menu	COMMAND (TYPE)	ALIAS (TYPE)	Short-cut	Screen (side) Menu	Tablet Menu
	Tools *Options...* *Files* *Temporary External* *Reference File Location*	*XLOADPATH*	...	...	*TOOLS2* *Options...* *Files* *Temporary* *External Refs..*	...

This system variable is used when *XLOADCTL* is set to 2 (*Enabled with copy*). Normally, when AutoCAD creates the copy of *Xref* drawings the temporary copy is placed in the directory specified as *Temporary External Reference File Location* in the *Files* tab of the *Options* dialog box. You can use the *XLOADPATH* variable directly (in Command line format) or the *Options* dialog box to specify another location for temporary files. In the *Files* tab of the *Options* dialog box, expand *Temporary External Reference File Location*, highlight the previously specified location, then select the *Browse...* button to specify a new location for temporary files.

As a summary for setting demand loading and related controls, Figure 30-40 illustrates the appropriate drawing files (parent drawing and related *Xref* drawings) for setting the *XLOADCTL*, *XLOADPATH*, and *INDEXCTL* variables.

Figure 30-40

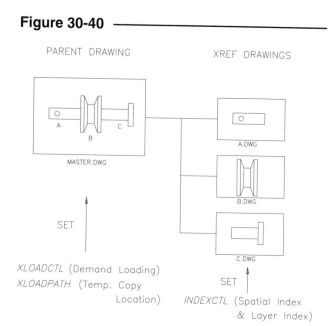

EDITING FOR *XREFS* AND *BLOCKS*

The strategy for editing external drawings has changed over the last several versions of AutoCAD. When external references were first introduced, the innovation was that any drawing could be viewed (externally referenced) from within another drawing, but the external drawing could not be changed from within the parent drawing. In this way, the drawings could be viewed by many users to facilitate working collaborative design environments, yet the lack of editing capabilities in the parent drawing prevented the externally referenced drawings from being edited by unauthorized users. This was AutoCAD's built-in protection for *Xrefs*.

In AutoCAD 2000, a new concept was introduced that allowed *Xrefs* and *Blocks* to be edited "in place," or from within the parent drawing. First, use the *Refedit* command to select individual *Xrefs* or *Blocks* to be edited, then use *Refset* to add to or remove from the working set of objects, make the changes, and finally use *Refclose* to close the editing session and save the changes. The changes are "sent back" to the original external drawing. The *XEDIT* system variable (set in the original *Xref* drawing) controls whether or not *Refedit* can be used to make edits from the parent drawing, so security for the *Xrefs* can be maintained when desired. *Refedit* is still usable in AutoCAD 2004, but is best used only for editing *Blocks*.

In AutoCAD 2004, a simpler and more direct method of editing *Xrefs* is available—the *Xopen* command (also see "*Xref*" and the *Open* button in the *Xref Manager*). Using *Xopen* <u>from within the parent drawing</u>, you can open any external reference drawing in a separate window, make changes, *Save*, and view the changes in the parent drawing immediately upon reloading the *Xref*. The *Xopen* command <u>overrides</u> the *XEDIT* variable, which in previous versions disabled editing capabilities from within the parent drawing. This method is clean and simple; however, there are no AutoCAD commands or system variables that can prevent unauthorized users from changing the original externally referenced drawings. Controls must be put in place by the CAD managers using network security measures to prevent unauthorized editing. (See Chapter 46, CAD Management.)

Figure 30-41

In AutoCAD 2004, you can select any *Xref* and right-click to produce a shortcut menu (Fig. 30-41) with options for *Edit Xref in-place* (*Refedit* command), *Open Xref* (*Xopen* command), *Clip Xref* (*Xclip* command), and for producing the *Xref Manager*.

XOPEN

Pull-down Menu	COMMAND (TYPE)	ALIAS (TYPE)	Short-cut	Screen (side) Menu	Tablet Menu
Insert Xref Manager... Open	XOPEN	...	...	...	...

The *Xopen* command allows a drawing currently being viewed as an *Xref* from within the parent drawing to be opened and edited in a separate window. In previous versions of AutoCAD (2000 and 2002), only the *Refedit* command could be used to edit an *Xref* drawing from within the parent drawing.

If you use *Xopen* in Command line format, you are prompted to select the desired *Xref* you want to edit. After selection, you are presented with the new drawing window and the original externally referenced drawing.

```
Command: xopen
Select Xref: PICK
Command:
```

Alternately, use the *Open* button in the *Xref Manager* (Fig. 30-42). A new "open folder" icon appears next to the selected *Xref*. You can tag multiple drawings for opening. When the *Xref Manager* is closed by picking *OK*, a new drawing window opens for each tagged drawing. Use the *Window* pull-down menu or the Ctrl+Tab key sequence to view the new drawing window(s).

Selecting the shortcut menu option (*Open Xref*) immediately opens the new window with the selected *Xref* drawing. You have full editing capabilities when you invoke *Xopen* by any method to open an *Xref* drawing.

When the changes are made and *Saved*, you are normally notified in the parent drawing that the *Xref* has changed. A notification icon or bubble appears at the system tray (depending on the setting of *XREFNOTIFY* system variable) indicating the name of the drawing that has changed and that it requires reloading (Fig. 30-43). To reload the drawing, use the *Reload* button in the *Xref Manager* (see Figure 30-42). Also see "*XREFNOTIFY.*"

Figure 30-42

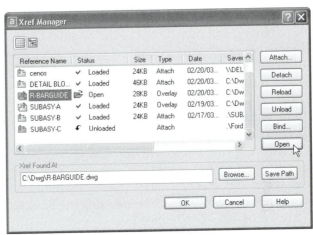

Figure 30-43

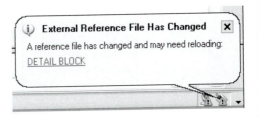

NOTE: Opening a drawing using *Xopen* (by any method) overrides the setting of both the *XLOADCTL* and *XEDIT* variables. When *XEDIT* is set to 0 (in the original external drawing), the drawing currently being viewed as an *Xref* cannot be edited using *Refedit* in the parent drawing. When *XLOADCTL* is set to 1 (in the parent drawing), the drawing currently being viewed as an *Xref* cannot be edited by another user (technically, by another AutoCAD session, even if it is the same user as the person viewing the *Xref*). In other words, when viewing an *Xref* (in which its *XEDIT* variable is set to 0 and the parent's *XLOADCTL* variable is set to 1), you can use the *Xopen* command to edit the original drawing, but you cannot use the normal *Open* command or the *Refedit* command to edit the drawing.

In-Place *Xref* and *Block* Editing

In AutoCAD 2000 and newer versions you can edit *Xrefs* and *Blocks* "in place," or from within the parent drawing using the *Refedit* command and a series of other commands and variables described next. Although these features can be used in AutoCAD 2004, the *Xopen* command is a simpler method for editing *Xrefs*; however, using *Refedit* to change *Blocks* from within the parent drawing is still a viable method since *Blocks* cannot be edited "in place" using *Xopen*. The commands and variables for editing Blocks or Xrefs "in place" are:

Refedit	(command)	Allows you to select individual objects in an *Xref* or *Block* and edit that geometry from within the current "parent" drawing
Refset	(command)	Allows you to add or remove objects to and from a working set while editing *Blocks* and *Xrefs* from the current drawing
Refclose	(command)	Closes the current reference editing session and saves or discards changes made to the *Xref* or *Block* geometry back to the original *Xref* drawing file or *Block* definition
XEDIT	(variable)	Controls whether a drawing file can be edited in place when referenced by another drawing
XFADECTL	(variable)	Controls the percentage of fading that occurs to objects not in the working set to edit

With the "in-place *Xref* and *Block* editing" capability, anyone who can *Xref* a drawing can also change that drawing and save it back to the original file <u>unless *XEDIT* is set to 0 in the original *Xref* drawing</u>. This situation presents a potential danger since the default setting for *XEDIT* is 1, which allows in-place editing. Therefore, it is suggested that the *XEDIT* system variable be changed to 0 for all template drawings in multi-user project environments.

Refedit is used to select *Xrefs*, *Blocks*, or <u>nested</u> *Xrefs* and *Blocks* to edit. When *Refedit* is used to change an *Xref* or nested *Xref* from within the current drawing and *Refclose* is used to save the changes, the original *Xref* drawing is updated with the new changes, which has the same effect as opening the original *Xref* drawing and saving changes. When *Blocks* defined in the current drawing or nested *Blocks* are edited and saved with this capability, the *Block* definition is saved in the original drawing containing the *Block*, which has the same effect as redefining the *Block* in its original drawing by the traditional method (see "Redefining *Blocks*," Chapter 21).

REFEDIT

Pull-down Menu	COMMAND (TYPE)	ALIAS (TYPE)	Short-cut	Screen (side) MenuMenu	Tablet
Modify *In place Xref and Block Edit>* *Edit Reference*	*REFEDIT*	...	...	*Modify2* *Refedit*	...

The capability known as "in-place editing" allows you to change geometry in an *Xrefed* drawing or nested *Block* from within the current "parent" drawing. Previous to AutoCAD 2000, *Xref* drawings could be changed only by opening the original *Xref* drawing and *Blocks* could be redefined only if *Exploded*. With in-place editing, changes to an *Xref* or nested *Block* can be made from within the parent drawing, then saved back to the original file.

<u>The *Refedit* command simply allows you to select which *Xref* or *Block* you want to edit and to pick the specific geometry you want to change.</u> Only one *Xref* or *Block* at a time (and its nested objects) can be selected for editing. The changes you make to the selected *Xref* or *Block* are accomplished using typical

editing commands. Use the *Refclose* command to save the changes back to the original drawing file. *Xrefs* and nested *Blocks* can be edited in place only if their *XEDIT* system variable (in the original *Xrefed* drawing) is set to 1 to allow external editing.

Invoking *Refedit* by any method produces the "Select reference" prompt. Select any *Block* or *Xref* from the drawing area using the pickbox or window.

> Command: **refedit**
> Select reference: **PICK** (*Xref* or *Block*)
> (*Reference Edit* dialog box appears)

After selecting an *Xref* or *Block*, the *Reference Edit* dialog box appears (Fig. 30-44). The name of the selected reference appears under *Reference name:* along with its preview image on the right.

You can choose to *Automatically select all nested objects*, or you can choose that AutoCAD *Prompt to select nested objects*. If you want to edit only a few *Lines* or *Circles* in your *Xref* or *Block*, you should choose to be prompted. Select *OK* to return to the drawing editor.

Figure 30-44 ——————————————

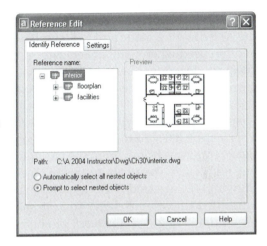

The "Select nested objects" prompt requests that you select geometry <u>contained in</u> the selected reference object such as a *Line*, *Circle*, or *Block* that you want to edit. This becomes your working set. After selection, the working set becomes visually distinct because all other objects not in the working set become faded. For example, Figure 30-45 displays geometry comprising the table and chairs as the "nested objects," or working set, and the remainder of the drawing is faded. The amount of fading is controlled by the *XFADECTL* system variable (see "*XFADECTL*" later in this section).

Figure 30-45 ——————————————

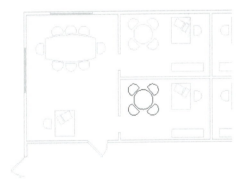

The working set can then be edited and changed in any way using normal editing commands and grips. To save changes made to the working set, use *Refclose* with the *Save* option. The changes are saved back to the original *Xref* drawing file or to the *Block* definition in the drawing that contains the *Block*. If a warning is issued when you select "nested objects" (Fig. 30-46), editing is not allowed because the *XEDIT* system variable is set to 0 in the original *Xrefed* drawing or original drawing containing the nested *Block*.

Figure 30-46 ——————————————

Refedit allows you to edit nested *Xrefs* and nested *Blocks*. When you select a reference object in response to the first *Refedit* prompt, "Select reference," the *Reference Edit* dialog appears and a hierarchical nesting structure is displayed for the reference objects (Fig. 30-47). Select the desired nesting level from the list to cause AutoCAD to highlight the selected reference in the drawing.

Settings **Tab**
The *Settings* tab (Fig. 30-48) provides the following options.

Create Unique Layer, Style, and Block Names
When this box is checked (default position), the layer and dependent object names of selected *Xrefs* are prefixed with the *Xref* drawing name and x, similar to the naming convention used when you *Bind* an *Xref*. If the box is not checked, dependent layer and object names appear in the current drawing as they would in the original *Xref* drawing.

Display Attribute Definitions for Editing
This feature lets you to edit <u>attribute definitions (tags)</u> in *Block* references from within the current drawing. When this box is checked, the attribute tags in *Blocks* (in the current drawing) or in nested *Blocks* are displayed instead of the attribute values themselves. You can select the attribute tags for editing along with the *Block* geometry. When changes are saved back to the *Block* reference, only <u>new</u> insertions of the *Block* are affected. Attributes of the previous *Block* insertions are not changed. See Chapter 22, Block Attributes, for more details on this subject.

Figure 30-47

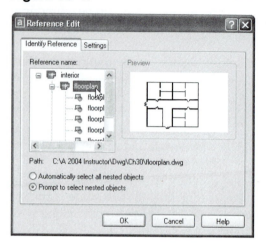

Figure 30-48

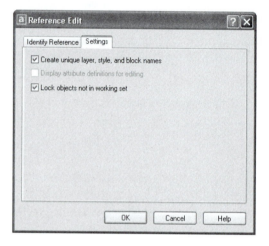

Lock Objects Not in Working Set
This option automatically locks all objects not in the working set including objects in the parent drawing (see *Refset*). Any selected locked objects behave as if they are on a locked layer. This feature prevents you from accidentally selecting and editing objects in the parent drawing.

If you select objects in an *Xref* as the working set, but they are on a locked layer, you must *Unlock* the layer to edit the geometry. When changes are saved back to the *Xref*, the layer state of the original *Xref* drawing is not changed—it remains locked.

The *Refedit* feature cannot be used with individual *Blocks* that have been inserted using *Minsert*.

Using *-Refedit* at the Command Line

Typing *-Refedit* produces the Command line version shown below.

```
Command: -refedit
Select reference: PICK
Select nesting level [Ok/Next] <Next>: O
Select nested objects: PICK
Select nested objects: Enter
Display attribute definitions [Yes/No] <No>: Enter
Use REFCLOSE or the Refedit toolbar to end reference editing session.
Command:
```

The Command line prompt to "Select nesting level [Ok/Next] <Next>:" serves the same function as the *Next* button or selecting from the hierarchical list in the *Reference Edit* dialog box. Note the last prompt, "Display attribute definitions [Yes/No] <No>:", provides the same level of access for editing attribute tags as provided in the dialog box version of the command.

REFSET

Pull-down Menu	COMMAND (TYPE)	ALIAS (TYPE)	Short-cut	Screen (side) Menu	Tablet Menu
Modify *In place Xref and Block Edit>* *Add to Workset* or *Remove from Workset*	*REFSET*	...	...	...	...

You can *Add* or *Remove* objects from the working set using *Refset* while editing in place. You initially define a working set of objects to edit when you select objects at the "Select nested objects:" prompt during the *Refedit* command. The working set is visually distinct while the other objects in the drawing become faded. You can add or remove to or from this working set using *Refset*.

```
Command: refset
Transfer objects between the reference set and drawing...
Enter an option [Add/Remove] <Add>: PICK
1 Added to working set
Command:
```

Add

 Use this option or button from the *Refedit* toolbar to select objects to add to the working set. The selected objects change from faded to distinct and can then be edited.

Remove

 This option or button from the *Refedit* toolbar allows you to select objects to remove from the working set. Objects removed from the working set become faded and cannot be edited.

REFCLOSE

Pull-down Menu	COMMAND (TYPE)	ALIAS (TYPE)	Short-cut	Screen (side) Menu	Tablet Menu
Modify *In-place Xref and Block Edit>* *Save Reference Edits* or *Discard Reference Edits*	*REFCLOSE*	...	...	...	...

Use the *Refclose* command to *Save* back or *Discard* changes made to objects during an in-place editing session. In other words, use *Refedit* to select *Xrefs* and *Blocks* and objects within those references to edit, modify the geometry, then use *Refclose* to close the editing session and save or discard the modifications to the working set.

```
Command: refclose
Enter option [Save/Discard reference changes] <Save>: Enter
Regenerating model.
1 block instances updated
(Block name) redefined.
Command:
```

You can also *Save* or *Discard* edits by using the options on the *Modify* pull-down menu or the *Refedit* toolbar. The *Refedit* toolbar appears after a reference is selected for editing and disappears after changes to the reference are saved or discarded.

Save

Saving changes saves the modifications made to the working set back to the original reference (*Xref* or *Block*) drawing file. The change is made without actually opening the reference drawing or recreating the block. If you decide to *Save* changes, you must confirm the dialog box that appears to complete the desired action (Fig. 30-49).

Figure 30-49

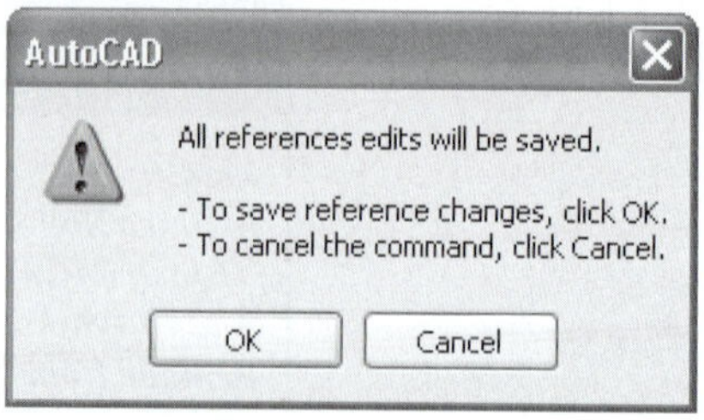

Using the *Save* option of *Refclose* does not save changes made to other objects (not in the in-place editing working set)—you must use the full *Save* command to save those changes. If you use *Refedit* and then *Refclose Save* to make edits to *Blocks* contained only in the current drawing, you must also *Save* the drawing.

Discard Reference Changes

Use this option if you want to close the in-place editing session and discard all changes made to the working set. A dialog box appears similar to that shown in Figure 30-49. Any changes you have made to other objects in the current drawing (not in the working set) are not discarded.

Immediately after using *Refclose*, you can use the *Undo* command to return to the editing session.

XEDIT

Pull-down Menu	COMMAND (TYPE)	ALIAS (TYPE)	Short-cut	Screen (side) Menu	Tablet Menu
Tools *Options...* *Open and Save* *Allow others to Refedit current drawing*	XEDIT	...	...	*Tools2* *Options...* *Open and Save* *Allow others to Refedit current drawing*	...

The value of this system variable determines whether or not a drawing can be edited in place when it is used as an *Xref*. <u>XEDIT should be set in the original Xref drawing</u>, not in the "parent" drawing. The possible settings are:

      XEDIT = 0   The drawing cannot be edited while *Xrefed* by another drawing.
      XEDIT = 1   The drawing can be edited in place when used as an *Xref* in another drawing.

You can change this system variable two ways: enter *XEDIT* at the Command line and change the value, or use the *Open and Save* tab of the *Options* dialog box (see Figure 30-35). If you use the *Options* dialog box, placing a check in *Allow others to Refedit current drawing* sets the *XEDIT* variable to 1. The variable is saved in the drawing.

The default setting of *XEDIT* in AutoCAD template drawings is 1, so in-place <u>editing can occur</u> without changing this variable. Keep in mind that in this case anyone who can *Xref* a drawing can also make an accidental or unauthorized change to that drawing and save it back to the original file. This situation presents a potential danger in multi-user project environments. Therefore, it is suggested that the *XEDIT* system variable be changed to 0 for all template drawings in multi-user project environments.

XFADECTL

	Pull-down Menu	COMMAND (TYPE)	ALIAS (TYPE)	Short-cut	Screen (side) Menu	Tablet Menu
	...	XFADECTL	...	...	...	...

This system variable controls the percentage of fading during an in-place editing session. When *Refedit* is used and reference geometry is selected for editing, it becomes the working set and remains distinct (not faded). All other objects in the current drawing become faded (see Figure 30-45). The default value is 50. The *XFADECTL* range is from 0 (0%, no fading) to 90 (90% fading). Fading ends when *Refclose* is used to close the in-place editing session.

You can set this variable by entering *XFADECTL* at the Command line or by accessing the *Display* tab of the *Options* dialog box (Fig. 30-50, lower right). Use the *Reference Edit fading intensity* slider to control the *XFADECTL* variable. *XFADECTL* is saved in the system registry, so once the variable is set, you do not have to reset it whenever new *Xrefs* are edited.

NOTE: *Shademode* must be set to *2D wireframe* for fading to occur (see Chapter 35, 3D Display and Viewing).

Figure 30-50 ———————————

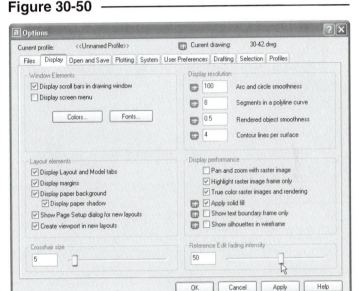

XREFNOTIFY

	Pull-down Menu	COMMAND (TYPE)	ALIAS (TYPE)	Short-cut	Screen (side) Menu	Tablet Menu
	...	XFADECTL	...	...	...	...

XREFNOTIFY is a system variable that controls the display of the notification icon and balloon message that appear at the lower-right corner of the Status bar when *Xrefs* have been updated or are not found. The settings are described below. This variable is saved in the system registry.

XREFNOTIFY = 0	*Xref* notification is disabled.
XREFNOTIFY = 1	*Xref* icon is enabled. This setting turns on the *Xref* icon in the Status bar when *Xrefs* are attached to the current drawing. If *Xrefs* are not found, the icon is displayed with a yellow alert symbol (!).
XREFNOTIFY = 2	*Xref* notification and balloon messages are enabled. This setting turns on the *Xref* icon, as in a setting of 1, but also displays balloon messages in the same area when *Xrefs* are modified.

MANAGING *XREF* DRAWINGS

Because of the external reference feature, the contents of a drawing can be stored in multiple drawing files and directories. In other words, if a drawing has *Xrefs*, the drawing is actually composed of several drawings, each possibly located in a different directory. This means that special procedures to handle drawings linked in external reference partnerships should be considered when drawings are to be backed up or sent to clients. Several considerations are listed here:

1. Create a project name for the drawing with the *PROJECTNAME* variable. The project name is stored in the drawing file. Notify the client so the client can create the same project name on his or her system (with the *Files* tab of the *Options* dialog box) to store the drawings and *Xrefs* (see "*PROJECT-NAME*").

2. When *Attaching Xrefs* to a drawing intended to be used by clients, use the *No path* or *Relative path* options (in the *External Reference* dialog box). If the *No path* option is used, you can copy the *Xref* drawings on another system into the same directory (folder) as the parent drawing. If the *Relative path* option is used, back up and copy the same directory structure (from the parent drawing down only) on the new system as on the original system.

3. Modify the current drawing's path to the external reference drawing so both are stored in the same directory; then archive them together.

4. Archive the directory structure of the external reference drawing along with the drawing which references it.

5. Make the external reference drawing a permanent part of the current drawing with the *Bind* option of the *Xref* command prior to archiving. This option is preferred when a finished drawing set is sent to a client.

6. Use the *Reference Manager* that comes with AutoCAD 2004 to manage the paths of drawings and the related *Xrefs*. For example, if you copied the parent drawing and related *Xrefs* onto another system with a different directory structure than that saved in the *Full path* on the original system when the *Xrefs* were *Attached*, launch the *Reference Manager* on the new system. Using *Reference Manager*, you can change the path for a group of *Xrefs* globally. See "Reference Manager."

PROJECTNAME

Pull-down Menu	COMMAND (TYPE)	ALIAS (TYPE)	Short-cut	Screen (side) Menu	Tablet Menu
Tools *Options...* *Files* *Project Files...* *Search Path*	PROJECTNAME	...	...	*TOOLS2* *Options* *Files* *Project Files..* *Search Path*	...

Project names make it easier for AutoCAD users to exchange drawings that contain *Xrefs*. The *PRO-JECTNAME* variable allows you to specify a project name that AutoCAD uses to point to an alternate search path when trying to locate *Xrefs*.

If you load a drawing containing several *Xrefs* given to you by a colleague from another company, AutoCAD first attempts to locate the *Xrefs* by searching the hard-coded paths. (The path of an *Xref* is "hard coded" when the drawing is originally *Attached* or *Overlayed* or if the *Path type* option is used in the *Xref Manager*). The *Xrefs* are not found initially unless you have an identical directory structure on your system or network, and the *Xrefs* have been copied into the directories accordingly. If an *Xref* cannot be found because the drawing is not located in the hard-coded path, AutoCAD then uses the drawing's *PROJECTNAME* to search for the *Xref* in the location specified by your system registry.

There are two components to this scheme: (1) the *PROJECTNAME* variable is saved in the parent drawing file, and (2) the directory location for that project name is saved in your system registry. This scheme makes it possible for collaborating designers to exchange drawings containing *Xrefs* when the designers have different directory structures. The drawing that is exchanged contains the project name, but the location (path) for that project name can be different for each system (saved in the system registries). Each designer's system contains the same project name but has a different drive and/or directory mapping. In other words, if an *Xref* cannot be found in the hard-coded path, AutoCAD then searches for the *Xref* by using the *PROJECTNAME* variable in the drawing file, but the <u>location</u> for the project name is saved in your system registry.

The value for the drawing's *PROJECTNAME* variable (*Project1*, for example) can only be set using the *PROJECTNAME* variable in command line format. The *PROJECTNAME* variable is saved in the <u>parent drawing file</u>. However, you must use the *Files* tab of the *Options* dialog box (but <u>not</u> the *PROJECTNAME* variable in command line format) to specify the directory mapping (location), which is then stored in <u>your system registry</u>.

System Registry Project Name

To assign a search path to be saved in your system registry, you <u>must</u> use the *Files* tab of the *Options* dialog box (Fig. 30-51). You <u>cannot</u> use the *PROJECTNAME* variable or any other method of Command line entry to add, remove, or modify the directory location for a project name in the system registry. Using the dialog box, follow these steps.

Figure 30-51

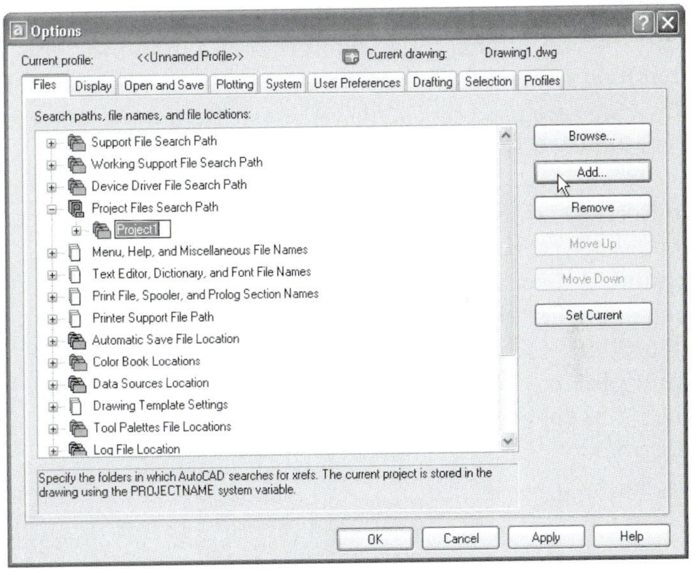

1. Double-click *Project Files Search Path*, then select *Add*. A project folder named *Project1* appears (or *Projectx*, where x is the next available number).
2. Enter a new name or press Enter to accept *Projectx*.
3. Double-click to expand *Project1* (or other name you entered).
4. Select *Browse*. Use the *Browse for Folder* dialog box that appears (not shown) to locate the desired folder on your system where you want to store *Xref* drawings for the project.
5. Select *OK* in both dialog boxes to save the project name location to your system registry.

The action above assigns a project name to the <u>system registry</u>.

Saving the *PROJECTNAME* in the Drawing File

Use the *PROJECTNAME* variable at the command line to save the drawing's component of the project name. Do this <u>after</u> registering the project name in your system using the *Options* dialog box. The Command prompt appears as follows:

    Command: **projectname**
    Enter new value for PROJECTNAME, or . for none <" ">: **project1**
    Command:

The action on the previous page assigns a project name to the <u>drawing file</u>.

If you try this step <u>before</u> using the *Options* dialog box to register the project name on your system, the following prompt appears:

> Command: **projectname**
> Enter new value for PROJECTNAME, or . for none <" ">: **project1**
> "project1" not found in the registry. Use the Options dialog Files tab to create the project name and set the project search paths.
> Command:

Reference Manager

Reference Manager, a separate application that comes with AutoCAD 2004, allows you to manage all references to a drawing contained in external files. The paths to all referenced files are saved in each AutoCAD drawing. However, Reference Manager allows you to list referenced files in selected drawings and to modify the saved reference paths without opening the drawing files in AutoCAD because Reference Manager is a stand-alone application that you run outside of AutoCAD. Use the Start Menu on your computer to launch Reference Manager from the Autodesk group.

References to a drawing contained in external files can be text fonts, attached images, plot configurations, and externally referenced drawings (*Xrefs*). The Reference Manager helps you manage the paths (directory locations) for all references to a drawing. This is especially helpful when drawings or the files that they reference (such as *Xrefs*) are moved to different folders or to different disk drives, such as when drawings are used by clients. Also use Reference Manager to fix groups of unresolved references. See Chapter 46, CAD Management, for more information.

In Reference Manager, use the *Edit Selected Paths* option to specify new paths for *Xref* drawings (Fig. 30-52). You can select multiple *Xrefs* and change the paths for all of them together.

Figure 30-52

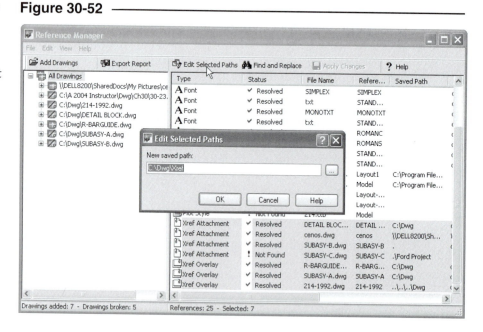

2004

XREFCTL

	COMMAND (TYPE)	ALIAS (TYPE)	Short-cut	Screen (side) Menu	Tablet Menu
Pull-down Menu					
...	XREFCTL	...	...	...	...

AutoCAD has a tracking mechanism for the *Xref* activity in drawings. An external ASCII log file can be kept for all drawings that contain information on external references. The *XREFCTL* variable controls the creation of the *Xref* log files.

If *XREFCTL* is set to 1, a log file registers each *Attach*, *Overlay*, *Bind*, *Detach*, *Reload*, and *Unload* of each external reference for every drawing. These log files have the same name as the related drawing with a file extension ".XLG." The log files are placed in the same directory as the related drawing. If a drawing is *Opened* again and the *Xref* command is used, the *Xref* activity (*Attach*, *Overlay*, *Bind*, etc.) is appended to the existing log file. The .XLG files have no <u>direct</u> connection to the drawing, so they can be deleted if desired without consequence to the related drawing file. If *XREFCTL* is set to 0 (default setting), no log is created or appended. The *XREFCTL* variable is saved in the system registry, so once it is set, it affects all drawings.

An example .XLG file is shown here. This file lists the activity from an earlier example in this chapter when the OFFICEX drawing was *Xrefed* to the INTERIOR drawing.

```
================================

Drawing: D:\Dwg\Interior.dwg
Date/Time: 08/24/03 08:55:16
Operation: Attach Xref


================================

Attach Xref OFFICEX: D:\Dwg\Officex.dwg

   Searching in ACAD search path

   Update block symbol table:
    Appending symbol: OFFICEX|DUPOUT
    Appending symbol: OFFICEX|L1-TBLCK
    Appending symbol: OFFICEX|DESK
    Appending symbol: OFFICEX|CHAIR
    Appending symbol: OFFICEX|CONFCHAIR
    Appending symbol: OFFICEX|TABLE
    Appending symbol: OFFICEX|CONFTABLE
    Appending symbol: OFFICEX|FILECAB
    Appending symbol: OFFICEX|DOOR
    Appending symbol: OFFICEX|WINDOW
   Block update complete.

   Update layer symbol table:
    Appending symbol: OFFICEX|FLOORPLAN
    Appending symbol: OFFICEX|FACILITIES
etc.
```

CHAPTER EXERCISES

1. *Xref Attach*

   Use the **SLOTPLATE2** drawing that you modified last (with *Stretch*) in Chapter 9 Exercises. The slot plate is to be manufactured by a stamping process. Your job is to nest as many pieces as possible within the largest stock sheet size of 30" x 20" that the press will handle.

   A. *Open* the **SLOTPLATE2** drawing and use the *Base* command to specify a basepoint at the plate's lower-left corner. *Save* and *Close* the drawing.

   B. Begin a *New* drawing using the C-D-SHEET prototype and assign the name **SLOTNEST**. Set *Limits* to **34 x 22**. Draw the boundary of the stock sheet (30" x 20"). *Xref Attach* the **SLOTPLATE2** into the sheet drawing multiple times as shown in Figure 30-53 to determine the optimum nesting pattern for the slot plate and to minimize wasted material. The slot plate can be rotated to any angle but cannot be scaled. Can you fit 12 pieces within the sheet stock? *Save* and *Close* the drawing.

   Figure 30-53 ————————————

   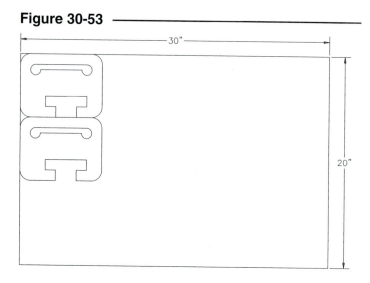

   C. In the final test of the piece, it was determined that a symmetrical orientation of the "T" slot in the bottom would allow for a simplified assembly. *Open* the **SLOTPLATE2** drawing and use *Stretch* to center the "T" slot about the vertical axis as shown in Figure 30-54. *Save* the new change and *Close* the drawing.

   D. *Open* the **SLOTNEST** drawing. Is the design change reflected in the nested pieces?

   Figure 30-54 ————————————

   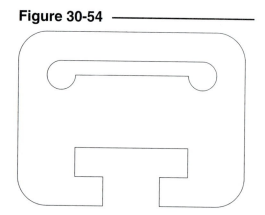

2. *Attach, Detach, Xbind, VISRETAIN*

   As the interior design consultant for an architectural firm, you are required to place furniture in an efficiency apartment. You can use some of the furniture drawings (*Blocks*) that you specified for a previous job, the office drawing, and *Insert* them into the apartment along with some new *Blocks* that you create.

   A. Use the **C-D-SHEET** as a *template* and set *Architectural Units* and set *Limits* of **48' x 36'**. (Scale factor is 24 and plot scale is 1/2"=1' or 1=24). Assign the name **INTERIOR**. *Xref* the **OFFICE** drawing that you created in Chapter 21 Exercises. Examine the layers (using the *Layer Control* drop-down list) and list (?) the *Blocks*. *Freeze* the **TEXT** layer. Use *Xbind* to bring the **CHAIR**,

DESK, and **TABLE** *Blocks* and the **FURNITURE** *Layer* into the INTERIOR drawing. *Detach* the OFFICE drawing. Then *Rename* the new *Blocks* and the new *Layer* to the original names (without the **OFFICE0** prefix). In the INTERIOR drawing, create a new *Block* called **BED**. Use measurements for a queen size (60" x 80"). *Save* the INTERIOR drawing.

B. *Xref* the **EFF-APT2** drawing (from Chapter 18 Exercises) into the INTERIOR.DWG. Change the *Layer* visibility to turn *Off* the **EFF-APT2|TEXT** layer. *Insert* the furniture *Blocks* into the apartment on *Layer* **FURNITURE**. Lay out the apartment as you choose; however, your design should include at least one insertion of each of the furniture *Blocks* as shown in Figure 30-55. When you are finished with the design, *Save* and *Close* the INTERIOR drawing.

Figure 30-55

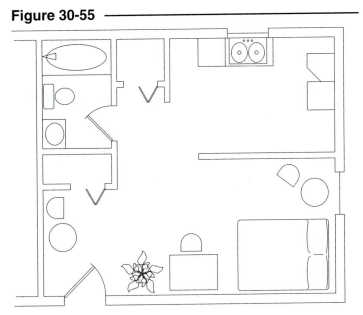

C. Assuming you are now the architect, a design change is requested by the client. In order to meet the fire code, the entry doorway must be moved farther from the end of the hall within the row of several apartments. *Open* the **EFF-APT2** drawing and use *Stretch* to relocate the entrance **8'** to the right as shown in Figure 30-56. *Save* and *Close* the EFFAPT2 drawing.

Figure 30-56

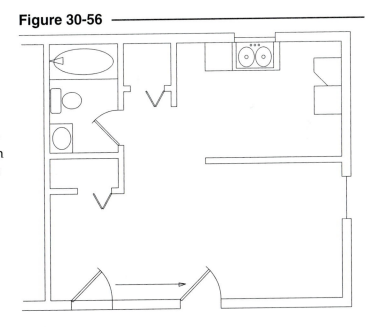

D. Next (as the interior design consultant), *Open* the **INTERIOR** drawing. Notice that the design change made by the architect is reflected in your INTERIOR drawing. Does the text appear in the drawing? *Freeze* the **TEXT** layer again, but this time change the *VISRETAIN* variable to **1**. Make the necessary alterations to the interior layout to accommodate the new entry location. *Save* the INTERIOR drawing.

3. *Xref Attach*

As project manager for a mechanical engineering team, you are responsible for coordinating an assembly composed of 5 parts. The parts are being designed by several people on the team. You are to create a new drawing and *Xref* each part, check for correct construction and assembly of the parts, and make the final plot.

A. The first step is to create two new parts (as <u>separate</u> drawings) named **SLEEVE** and **SHAFT**, as shown in Figure 30-57. Use appropriate drawing setup and layering techniques. In each case, draw the two views on <u>separate layers</u> (so that appearance of each view can be controlled by layer visibility). Use the *Base* command to specify an insertion point at the right end of the rectangular views at the centerline.

Figure 30-57

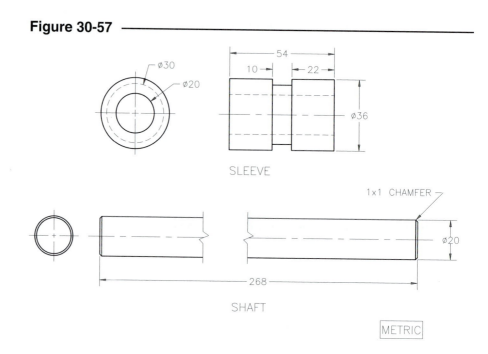

B. You will eventually *Xref* the drawings **SHAFT**, **SLEEVE**, **PUL-SEC** (from Chapter 26 Exercises), **SADDLE** (from Chapter 24 Exercises), and **BOLT** (from Chapter 9 Exercises) to achieve the assembly as shown in Figure 30-58. Before creating the assembly, *Open* each of the drawings and use *Properties* to move the desired view (with all related geometry, no dimensions) to a new layer. You may also want to set a *Base* point for each drawing.

Figure 30-58

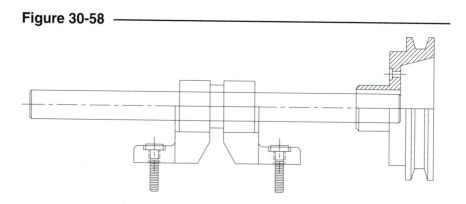

C. Finally, begin a *New* drawing (or use a *Template*) and assign the name **ASSY**. Set appropriate *Limits*. Make a *Layer* called **Xref** and set it *Current*. *Xref* the drawings to complete the assembly. The **BOLT** drawing must be scaled to acquire a diameter of **6mm**. Use the *Layer Control* drop-down list to achieve the desired visibility for the *Xrefs*. Set *VISRETAIN* to **1**. Make a *Plot* to scale. *Save* and *Close* the **ASSY** drawing.

4. *Xref Overlay*

A. Assume you are now the project coordinator/checker for the assembly. (If it is possible in your lab or office, exchange the SHAFT drawings [by diskette or by network] with another person so that you can

Figure 30-59 ────────────────────

perform this step for each other.) Begin a *New* drawing and make a *Layer* called **CHECK**. Next, *Xref Overlay* the **SHAFT** drawing (of your partner). On the CHECK layer, insert some *text* by any method and a dimension similar to that shown in Figure 30-59. *Save* and *Close* the CHECK drawing. (When completed, exchange the CHECK and SHAFT drawings back.)

B. *Open* the **SHAFT** drawing. In order to see the notes from the checker, *Xref Overlay* the **CHECK** drawing. The instructions for the change in design should appear. (Notice that the *Overlay* option prevents a circular *Xref* since SHAFT references CHECK and CHECK references SHAFT.) Make the changes by adding the 3 x 3 NECK to the geometry (Fig. 30-60). *Save* and *Close* the **SHAFT** drawing (with the *Overlay*).

Figure 30-60 ────────────────────

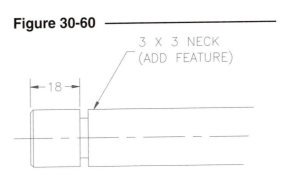

C. Now *Open* the **ASSY** drawing. The most recent *Xrefs* are automatically loaded, so the design change should appear (Fig. 30-61). Why don't the checker notes (from CHECK drawing) appear? Finally, *Save* the drawing and make another plot.

Figure 30-61 ────────────────────

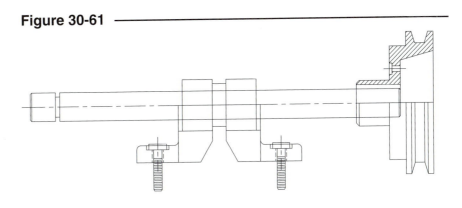

5.   *Layouts, Vports*

This exercise serves as a review of the basic procedures and commands for using an *Xref* and creating layouts and viewports to display and print an existing drawing.

A.   *Open* the **ASSY** drawing that you created in previous exercises (make sure all of the *Xrefed* drawings are also located in your working directory). Use *Xref* with the *Bind* option to bind all of the *Xrefed* drawings. Use *SaveAs* to save and name the drawing as **ASSY-PS**.

B.   Activate a *Layout* tab. Use *Pagesetup* to select the paper size and to set the plotting options for the device you use to plot with. Make a *New Layer* named **TITLE** and set it as the *Current* layer. Draw a border and draw, *Insert*, or *Xref* a title block (in paper space).

C.   Make a *New Layer* called **VIEWPORT** or **VPORT** and set it as *Current*. Next, use *Vports* to create one viewport. Make the viewport as large as possible while staying within the confines of the title block and border. Double-click inside the viewport to activate model space in the viewport, and use *Zoom XP* or the *Viewport scale* drop-down list (in the *Viewports* toolbar) to scale the model geometry times paper space (remember that the geometry is in millimeter units). Enter the scale in the title block (scale = XP factor or viewport scale).

D.   Turn *Off* the **VIEWPORT** layer and *Save* the drawing (as ASSY-PS). *Plot* the drawing <u>from the layout</u> at **1=1**. Your drawing should appear similar to Figure 30-62.

Figure 30-62 ───────────────

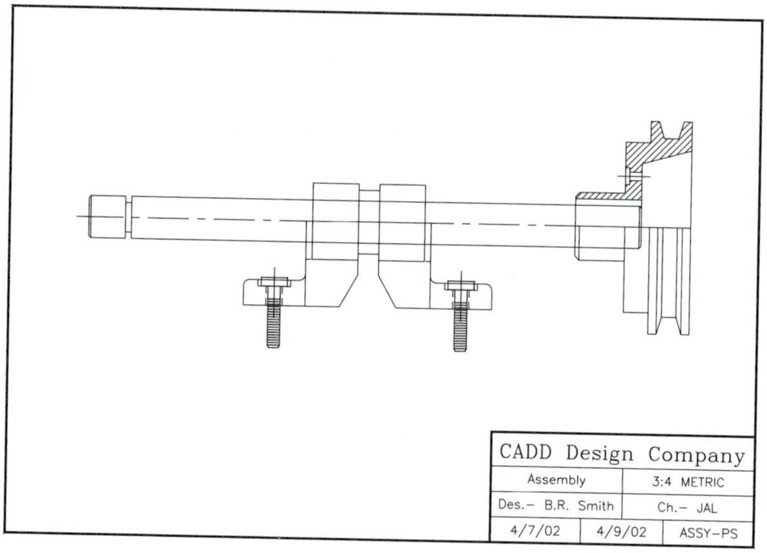

CADD Design Company		
Assembly	3:4 METRIC	
Des.− B.R. Smith	Ch.− JAL	
4/7/02	4/9/02	ASSY−PS

6.   *DesignCenter, XCLIP, XLOADCTL, INDEXCTL*

A.   *Open* the **DB_SAMP** drawing from the sample drawings provided with AutoCAD (if AutoCAD was installed with the "Full" installation and accepting all the other defaults, DB_SAMP.DWG is located in the C:\Program Files\AutoCAD 2004\Sample directory). Use *SaveAs* to change the name to **SAMP-OFFICE** and change the location of the drawing (using *SaveAs*) to your working directory.

B.   Begin a *New* drawing. Use *Save* and assign the name **SAMP-CLIP**. Next, use *DesignCenter* to *Attach* as an *Xref* the **SAMP-OFFICE** drawing. Use an *Insertion point* of **0,0**, accept the defaults for scale factors and rotation angle, then *Zoom All*. The entire SAMP-OFFICE drawing should appear.

C. Create a clipping boundary using the **Xclip** command. Make a **Rectangular** boundary around room 6150 (lower right corner office). **Zoom** in to display only the displayed office, 6150. Now, use the **Layer** command or *Layer Control* drop-down list to **Freeze** the **SAMP-OFFICE|EMPLOYEE** layer. Your drawing should look like Figure 30-63.

Figure 30-63

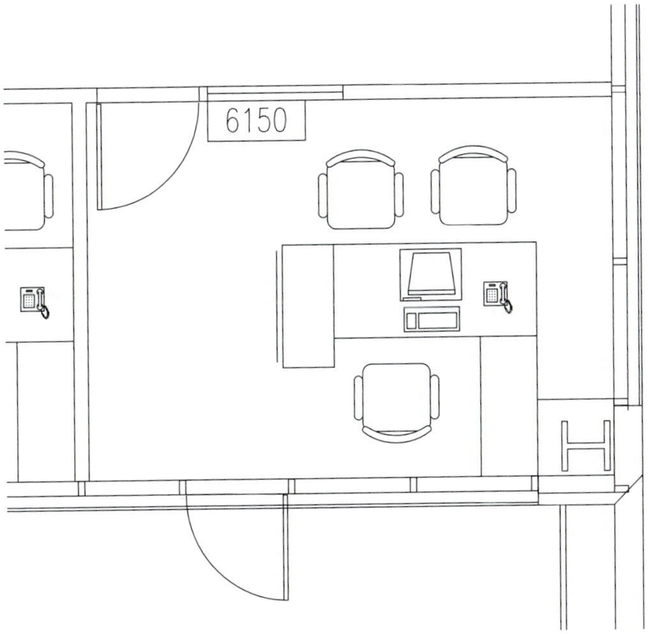

D. Use **XLOADCTL** or use the **Open and Save** tab of the **Options** dialog box to set the variable to **0** or **Disabled**. **Save** and **Close** the drawing (SAMP-CLIP).

E. Now **Open SAMP-CLIP** again and use your watch or count seconds to time how long it takes to load the drawing and Xref drawing. At this point, demand loading is disabled and layer and spatial indexing are not being utilized so the drawing and Xref loading time should be at its maximum for your system.

F. Next, **Close** the **SAMP-CLIP** drawing and **Open** the **SAMP-OFFICE** drawing. Use the **INDEX-CTL** variable or the **SaveAs** command with the **Options...** button to set indexing on for both layer and spatial indexes (a setting of 3). **Save** the SAMP-OFFICE drawing.

G. **Close** the **SAMP-OFFICE** drawing and **Open** the **SAMP-CLIP** drawing again. Use **XLOADCTL** or use the **Open and Save** tab of the **Options** dialog box to set the variable to **1** or **Enabled**. This action turns on demand loading, and you should now take advantage of the spatial and layer indexing. **Save** the drawing.

H. **Close**, then **Open** the **SAMP-CLIP** drawing again and time the loading process. You should experience a slight increase in the loading time speed now that demand loading is enabled. When using large (file size) Xrefs or multiple Xrefs, this feature can offer a big performance increase.

7. *Refedit*

A. **Open** the **SAMP-CLIP** drawing. Assuming you are planning to move into office 6150, your employer gives you the responsibility to arrange the furniture in your new office to your liking.

B. Use *Refedit* and select any wall at the "Select reference:" prompt. The *Reference Edit* dialog box should appear with the SAMP-OFFICE drawing listed as the *Reference name*. Select **OK** to dismiss the dialog box. At the "Select nested reference:" prompt, select two chairs. You should see all but the two chairs fade to a lighter color shade. Move the chairs to new locations in the office.

C   Use *Refset* to add other items, such as the desk, computer, and phone, to the working set as needed. Use *Move* and *Rotate* to position the other furniture items to new locations. Figure 30-64 shows one possible office arrangement. When you are finished, use *Refclose* to save your changes back to the SAMP-OFFICE drawing.

Figure 30-64

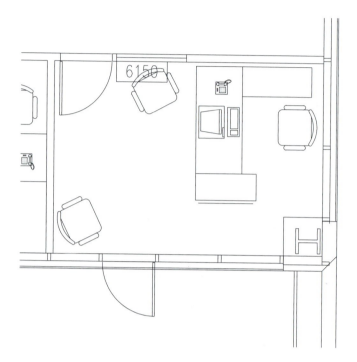

D. *Close* the **SAMP-CLIP** drawing. *Open* the **SAMP-OFFICE** drawing to verify that the changes you made to office 6150 were saved.

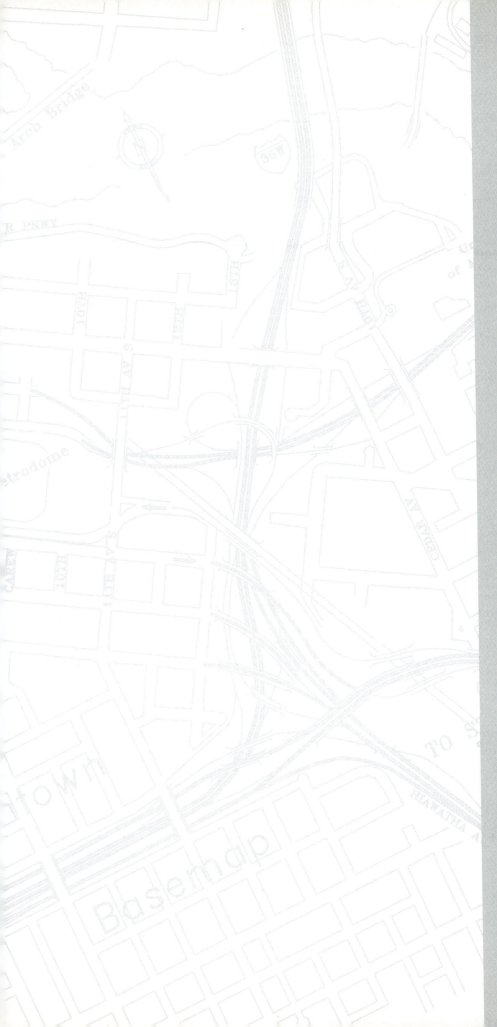

31

OBJECT LINKING AND EMBEDDING (OLE)

Chapter Objectives

After completing this chapter you should:

1. be able to copy and paste objects from one open AutoCAD drawing to another open AutoCAD drawing;

2. be able to use AutoCAD's Object Linking and Embedding (OLE) feature to copy information from a document (in a software program such as AutoCAD, Microsoft Word, or Microsoft Excel) and temporarily store it in the Clipboard;

3. be able to paste information from the Clipboard into a drawing;

4. be able to embed information from another software program into an AutoCAD drawing;

5. be able to link information from another software program into an AutoCAD drawing;

6. be able to manage OLE information in a drawing.

CONCEPTS

The term "graphics communication tool" is often used to describe AutoCAD's many advanced graphic capabilities. For many years people have been communicating design and conceptual information using AutoCAD's 2D and 3D graphic features. The Microsoft® Windows operating systems have now enabled AutoCAD users to extend their graphic communication abilities within, and beyond, the Drawing Editor.

Microsoft Windows introduced the capability to "copy" and "paste" objects. As an AutoCAD user, there are two basic ways in which you can copy and paste objects. First, since AutoCAD 2000, it is possible to copy and paste AutoCAD objects between multiple open drawings. Secondly, you can copy and paste between AutoCAD and other Windows software applications.

Cut and Paste Between AutoCAD Drawings

AutoCAD allows you to open multiple drawings at the same time in one AutoCAD session. One of the greatest advantages for doing this is the ability to copy objects from one drawing and paste them into another drawing. This feature can save hours of redrawing the same or similar geometry and increase your ability to collaborate with others on design ideas.

When objects are copied from one AutoCAD drawing, the pasted objects in the receiving drawing are <u>still normal AutoCAD objects</u>. Once the objects have been pasted, there are no special functions or commands that you need to manipulate and edit them. There are also commands in AutoCAD specifically intended for you to handle the objects during the cut and paste process, such as *Copy with Base Point* and *Copy to Original Coordinates*. For example, Figure 31-1 displays the AutoCAD Drawing Editor with two drawings open. Objects from the drawing on the left are copied with the *Copyclip* command and placed into the drawing on the right with the *Pasteclip* command.

Figure 31-1

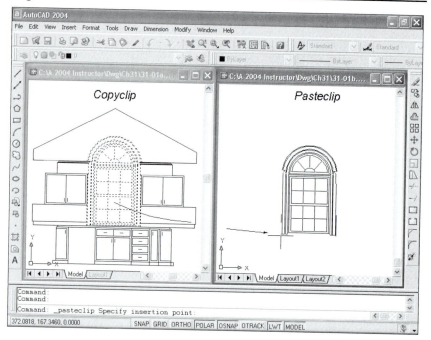

In AutoCAD Release 14, all pasted objects behaved as *Blocks* and had to be *Exploded* to allow editing. In some ways, these objects behaved as if they were copied from one software application into a different one, mostly due to the fact that you had to have two separate AutoCAD sessions running to do this! However, since AutoCAD 2000, (1) you can have multiple drawings open in one AutoCAD session, and (2) copying and pasting between drawings is transparent—that is, even though the basic Windows Object Linking and Embedding (OLE) mechanism is being utilized, there are no special functions you need to use.

These commands can be used to cut (erase), copy, and paste objects between AutoCAD drawings:

Menu Command	Formal Command	Result
Copy	*Copyclip*	copies selected objects to the Clipboard
Cut	*Cutclip*	copies object to the Clipboard and cuts (erases) them from the original drawing
Copy with Base Point	*Copybase*	copies objects to the Clipboard and asks for a base point
Copy Link	*Copylink*	copies all objects to the Clipboard
Paste	*Pasteclip*	pastes Clipboard objects into the drawing
Paste as Block	*Pasteblock*	pastes Clipboard objects into the drawing as a *Block*
Paste to Original Coordinates	*Pasteorig*	pastes Clipboard objects into the new drawing at the same location as the original drawing

Cut and Paste Between AutoCAD and Other Windows Applications (OLE)

You can also easily communicate 2D and 3D AutoCAD designs with other Windows applications and benefit from the Windows Object Linking and Embedding (OLE) feature. Unlike using only one software application (AutoCAD) to copy and paste, this concept combines objects created in two different software applications but displayed in only one. Microsoft developed the method called object linking and embedding, or OLE, to handle dissimilar objects. The standard OLE features of linking and embedding operate as smoothly and easily from within AutoCAD as they do in other Windows software such as Microsoft® Word, Excel, and others.

AutoCAD's support of the OLE features allows you to include data and information created in an application, such as Word, within your AutoCAD drawing. You can also take AutoCAD drawing information and use it within Word documents, Excel spreadsheets, and so on. This feature works best if all the applications using or sharing the common information support the OLE feature. But it is not absolutely necessary for a software application to support the OLE feature for that application to display AutoCAD drawing information by way of the copy and paste functions available within Windows.

AutoCAD's OLE features utilize the standard Windows *Copy, Cut,* and *Paste* commands common to all Windows programs. OLE, however, extends these Windows commands so information copied from one application (such as a Word document) and pasted into another application (like an AutoCAD drawing) can be *linked* back to the original source document, also known as the server. The source document is a separate document; it is only displayed within the client document (the AutoCAD drawing in this example).

Object Linking and Embedding enables you to create a bill of materials within, for example, Word, which is designed specifically for word processing, then paste it into an AutoCAD drawing. First, create and format the bill of materials within Word (Fig. 31-2). Next, *Copy* the formatted text onto the Clipboard and then *Paste* it into the AutoCAD drawing. In AutoCAD (the client), it will have the same appearance that it has in the Word (the source) document.

Figure 31-2 —————————

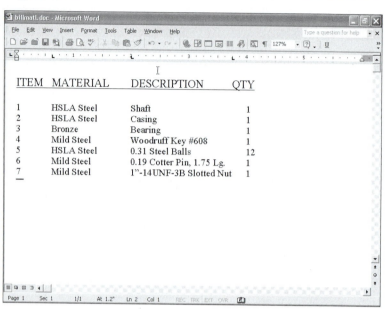

If the bill of materials is pasted into the AutoCAD drawing as a <u>linked</u> object, then the text document that appears in AutoCAD is simply a picture of the original source document (Fig. 31-3). If the source document is changed within Word (its original software application), then you can cause that source to update automatically or "manually" in AutoCAD, depending on the update options you set within the OLE *Links* dialog box. While the drawing file is open, you can also double-click on the linked object's picture in AutoCAD to cause the source application to be opened (in this case, Word) with the document loaded and ready for editing. Upon saving and closing the bill of materials docu-

Figure 31-3 —————————

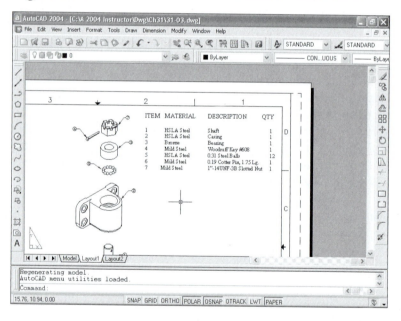

ment, the "picture" of the bill of materials in the AutoCAD drawing is updated automatically. You can also select *Update* from the *File* pull-down menu of the source application program to update the linked OLE object dynamically while editing it.

There are two strategies for pasting information from other applications into your AutoCAD drawing:

1. <u>Embed the object when pasting</u>
    Embedding is similar to using the AutoCAD *Insert* command. When embedded, the object becomes a permanent part of the drawing. If the source application that created the object supports OLE, you can double-click on the embedded object and the source application is started so you can edit the contents of the object for the insertion of the (embedded) objects in AutoCAD only. The original source document remains unchanged.

2. <u>Link the object when pasting</u>
    Linking is similar to the AutoCAD *Xref* command. When the object is linked, it is not part of the drawing; it is only displayed in the drawing. When you double-click on the linked object, the source application is started with the original document loaded and ready for editing. Any changes made to the original source document are reflected in the drawing automatically or upon a manual update, depending on your update setting in the OLE *Links* dialog box. Similar to an *Xref*, when the original document is edited, the updated version appears in the client application.

These AutoCAD commands are used to copy, cut, link, or embed objects between AutoCAD and other applications:

Menu Command	Formal Command	OLE Result
Copy	*Copyclip*	Embedded or linked
Cut	*Cutclip*	Embedded
Paste	*Pasteclip*	Embedded
Paste Special	*Pastespec*	Embedded or linked
Paste as Hyperlink	*Pasteashyperlink*	Hyperlink

The main difference between linked and embedded is how the information is stored—with the client (application that "displays" the document) or server (source application that created the document).

The Windows Clipboard

Windows manages the OLE features. The Windows Clipboard is a key mechanism used throughout the process of cutting and pasting within AutoCAD and linking and embedding other information in an AutoCAD drawing. The Clipboard is a Windows feature which stores information temporarily in memory so that it can be utilized (or pasted) into another application. This enables you to create the information only once and then share it with other documents where it is needed.

For example, assume the bill of materials mentioned earlier was created, spell checked, and formatted in Word, then selected and copied to the common memory area called the Clipboard. Then the information was pasted (similar to the AutoCAD *Insert* command) into an AutoCAD drawing with the same content and appearance it had in the original document. The same information could then be pasted (inserted) into a spreadsheet with the identical content and appearance it has in the AutoCAD drawing and in the original Word document that created it.

The Clipboard retains the most recent information that was cut or copied from an application currently running in the Windows environment. When another piece of information (e.g., graphical objects or text) is cut or copied, it replaces the last group of information stored within the Clipboard's memory area.

The Clipboard's information can be saved to a file (normally with a .CLP extension) if you activate the Windows Clipboard utility program and manually save the information.

The options encountered when performing a paste vary depending upon the AutoCAD paste command being used (*Pasteclip, Pastespec*) and the type of object information being pasted from the Clipboard. The following table lists some common object types and options that are available when pasting objects into an AutoCAD drawing.

Object Type:	Paste	Objects	As:			
	AutoCAD Vectors	AutoCAD Block	AutoCAD Text	Bitmap	Picture	Image
AutoCAD Vector Objects (*Lines, Arcs, Circles, Blocks, Plines*, etc.)	✔	✔				
Other Vector Objects (primarily Windows metafile)		✔				
AutoCAD Text Objects (*Dtext, Mtext, Text*)		✔	✔		✔	✔
Formatted Text Objects (such as Word text or Excel spreadsheet)		✔	✔	✔	✔	✔

To summarize, first create the document (to be copied) in its source application (Word, for example). While the document is open, use *Copy* or similar OLE command in that application to copy the document contents (called an OLE object) to the Clipboard. Next, open the client application (AutoCAD, for example) and use *Paste, Paste Special,* or a similar OLE command to paste the OLE object. The OLE object appears in the client application just as it did in its original format (in the source application). Later, you can double-click on the OLE object to open the original (source) application and modify the object. If the object is <u>linked,</u> the original source document is updated; if the object is <u>embedded,</u> only the OLE object is changed, but the source document (file) is not changed.

OLE RELATED COMMANDS

AutoCAD OLE related commands covered in this chapter are:

Cutclip, Copyclip, Copylink, Copybase, Pasteclip, Pasteblock, Pasteorig, Pastespec, Insertobj, Pasteashyperlink, Olelinks, and *Olescale*

Most OLE related commands are available from the *Edit* pull-down menu (Fig. 31-4). The *Insertobj* command is accessible from the *Insert* pull-down menu.

Figure 31-4

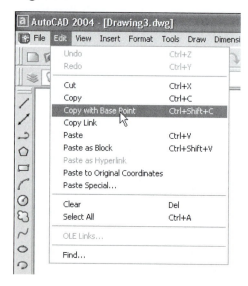

In addition, right-clicking in the drawing area when no commands are in use produces the default shortcut menu displaying the OLE commands (Fig. 31-5).

Figure 31-5

	Pull-down Menu	COMMAND (TYPE)	ALIAS (TYPE)	Short-cut	Screen (side) Menu	Tablet Menu
COPYCLIP	*Edit* *Copy*	*COPYCLIP*	...	*Ctrl+C or* (Default Menu) *Copy*	*EDIT* *Copyclip*	*T,14*

You can use the *Copyclip* command to select the objects from the current drawing to be placed on the Clipboard. *Copyclip* prompts you to "Select objects:". You can use any of the normal AutoCAD selection options (like *Window, Crossing, Fence, Group,* etc.). You can also pre-select the objects (known as Noun-Verb selection) you want placed on the Clipboard, and then issue the *Copyclip* command to avoid the "Select Objects:" prompt:

```
Command: copyclip
Select objects: PICK (Select objects to be copied from the drawing and placed on the Clipboard)
Select objects: Enter
Command:
```

AutoCAD objects that are selected with *Copyclip* remain in the drawing. Copies of the objects are temporarily stored on the Clipboard.

The *Copyclip* command can be used when you plan to paste the objects back into AutoCAD (using *Pasteclip, Pastespec,* or *Pasteblock*) or when you intend to link or embed them within other applications. When the objects are pasted into another application, you may have little control over where they are placed, depending on the host application.

When you paste the objects into the same or other AutoCAD drawing, you can drag the objects into the desired location. The *Copyclip* command does not prompt for a base point; therefore, pasting the objects involves using a base point at the lower-left corner of the set of objects (Fig. 31-6).

Figure 31-6

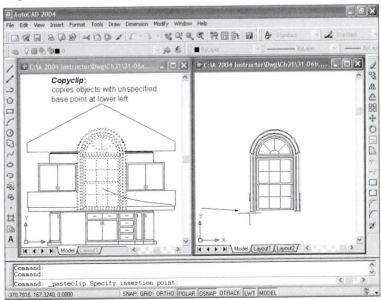

COPYBASE

	Pull-down Menu	COMMAND (TYPE)	ALIAS (TYPE)	Short-cut	Screen (side) Menu	Tablet Menu
	Edit Copy with Base Point	COPYBASE	...	(Default Menu) Copy with Base Point	EDIT CopyBase	...

Copybase, similar to *Copyclip*, copies objects onto the Clipboard; however, *Copybase* allows you to specify a base point for the object(s). When you later paste the set of *Copybase* objects into a drawing using *Pasteclip*, *Pastespec*, or *PasteBlock*, you can locate the point of insertion using the previously specified base point (Fig. 31-7).

In contrast, *Copyclip* does not allow you to specify a base point but automatically determines a base point, usually in the lower left corner of the object set (see Figure 31-6). Therefore, *Copybase* is generally preferred to *Copyclip* for typical operations within AutoCAD because you have more direct control of the insertion point.

Figure 31-7

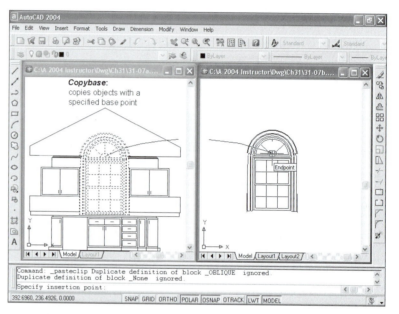

If you intend to use *Pasteorig* to paste the object set, it does not matter if you use *Copyclip* or *Copybase* since the object set is pasted to the original coordinates regardless of the base point. (See "*Pasteorig*," later in this chapter.)

COPYLINK

	Pull-down Menu	COMMAND (TYPE)	ALIAS (TYPE)	Short-cut	Screen (side) Menu	Tablet Menu
	Edit Copy Link	COPYLINK	...	...	EDIT Copylink	...

The *Copylink* command is used to <u>copy the entire contents of the current drawing</u>, including objects on *Frozen* or *Locked* layers. You are not prompted to select individual objects. Whatever is contained in the drawing is copied to the Clipboard. This information can then be pasted into AutoCAD as normal AutoCAD objects or into another application as linked or embedded objects.

    Command: **copylink**
    Command:

No change is made to the objects in the current drawing. Copies of the objects are temporarily stored in the Clipboard.

CUTCLIP

Pull-down Menu	COMMAND (TYPE)	ALIAS (TYPE)	Short-cut	Screen (side) Menu	Tablet Menu
Edit Cut	CUTCLIP	...	Ctrl+X	EDIT Cutclip	T,13

The *Cutclip* command places selected objects onto the Clipboard. However, unlike *Copyclip*, it <u>deletes</u> (*Erases*) the selected objects from the drawing. If desired, you can use the AutoCAD *Oops* command to "unerase" the objects deleted from the drawing by *Cutclip*. If you use *Oops*, you must use it before you delete any other objects from the drawing since *Oops* brings back only the last set of *Erased* objects.

```
Command: cutclip
Select objects: PICK (Select objects to be removed from the drawing and placed on the Clipboard)
Select objects: Enter
Command:
```

This command is typically used only when you do not need the selected items in the current drawing any longer but wish to use them in another drawing or another application. In this case, *Cutclip* allows you to clear (*Erase*) them out of the current drawing and then paste them into another application. For example, you may want to open a second AutoCAD drawing, then cut objects from the first drawing and paste them into the second drawing. This action helps to reduce the current drawing file size and also eliminates the need to draw the items again for another drawing. Since the objects temporarily remain on the Clipboard, the deleted items can be pasted into another drawing or document where they are needed.

PASTECLIP

Pull-down Menu	COMMAND (TYPE)	ALIAS (TYPE)	Short-cut	Screen (side) Menu	Tablet Menu
Edit Paste	PASTECLIP	...	Ctrl+V	EDIT Pasteclp	U,13

You can copy information <u>from AutoCAD or another application</u> and paste it into your drawing with AutoCAD's *Pasteclip* command. If you copy AutoCAD objects and paste them into the same or another AutoCAD drawing, the objects are inserted as the original AutoCAD objects having the same properties. If you *Pasteclip* text from a word processing application or data from a spreadsheet application, the pasted object is embedded within the AutoCAD drawing.

Pasting AutoCAD Objects
First, you must use *Copyclip*, *Copybase*, *Copylink*, or *Cutclip* to copy AutoCAD objects to the Clipboard. Next, use *Pasteclip* to insert the objects into the same or another open AutoCAD drawing. AutoCAD prompts for the base point.

```
Command: pasteclip
Specify insertion point: PICK
Command:
```

The objects are inserted into the new location or new drawing. AutoCAD objects are pasted as the original objects and retain their original properties such as layer, color, linetype, text style, dimension style, and so on. Therefore, layers on which the pasted objects reside and other named objects associated with the pasted objects (linetypes, dimension styles, etc.) are added to the new drawing if they did not previously exist in the host drawing.

Pasting Non-AutoCAD Objects

If you have used Copy or similar OLE command from a word processing or spreadsheet program, for example, those objects are copied to the Clipboard. In AutoCAD, use *Pasteclip* to bring the objects into AutoCAD as <u>embedded</u> objects. No insertion base point is requested.

Command: ***pasteclip***

When *Pasteclip* is used with non-AutoCAD objects, the *OLE Properties* dialog box appears (Fig. 31-8). This dialog box gives you control of the *Size, Scale*, and other properties of the non-AutoCAD OLE objects.

You can change the OLE object's size by entering values in <u>either</u> the *Size* edit boxes or the *Scale* edit boxes. Changing one set automatically changes the other.

The two *Height* edit boxes can be "tied" together so that changing one *Height* value automatically changes the other *Height* value. The *Width* edit boxes operate similarly. If the *Lock Aspect Ratio* box is checked, the *Height* and *Width* values remain proportional when either is changed. Remove the check if you want to vary the *Height* independently of the *Width*, and vice versa. Press the *Reset* button to return the edit boxes to the original values.

Figure 31-8

If the pasted OLE object is composed of text, the *Text Size* edit boxes are enabled. This section allows you to change only the text independently of other objects in the set. You can select from fonts and point sizes that are included in the pasted objects.

In the *OLE Plot Quality* drop-down list, select the plot quality option that matches the pasted OLE object file type. (See "*OLEQUALITY*.")

The checkbox at the bottom left of the dialog box allows you to disable the dialog box when *Pasteclip* is used. In this case, the OLE objects are inserted into AutoCAD at their original size. You can then use the *Olescale* command to produce this dialog box to modify the object's size, scale, and other properties at a later time (see "*Olescale*," later in this chapter).

Once the non-AutoCAD object is inserted, it becomes an embedded object. If the embedded object's source application supports OLE, the object is displayed in the original application's format. The embedded object can be edited using the original application that created it by double-clicking on the object.

Embedded objects are <u>copies</u> of the original information. They have no connection, or link, to the original source document in which they were created. Therefore, editing the original objects using the source application does not update the embedded (copied) object. Alternately, if you use the *Pastespec* command instead of *Pasteclip*, you have the choice of embedding or linking the objects.

Keep in mind that using *Pasteclip* to paste an AutoCAD object into an AutoCAD 2000 or later drawing <u>does not create an embedded object</u>. In releases previous to AutoCAD 2000, pasted AutoCAD objects became *Block* insertions. However, since AutoCAD 2000 pasted AutoCAD objects are inserted as exact copies of the original object set, complete with all associated properties and named objects, unless *Pasteblock* is used.

PASTEORIG

Pull-down Menu	COMMAND (TYPE)	ALIAS (TYPE)	Short-cut	Screen (side) Menu	Tablet Menu
Edit Paste to Original Coordinates	PASTEORIG	...	(Default Menu) Paste to Original Coordinates	EDIT PasteOri	...

The *Pasteorig* command is designed specifically to be used for <u>copying AutoCAD objects into other AutoCAD drawings</u>. *Pasteorig* inserts the object(s) into the new drawing at the same coordinate position as the original drawing. Therefore, you are not prompted for an insertion point.

    Command: **pasteorig**
    Command:

To illustrate, open two AutoCAD drawings and use *Cutclip, Copyclip, Copylink,* or *Copybase* to copy objects from one drawing onto the Clipboard. Next, make the other AutoCAD drawing current by clicking in it and then use *Pasteorig.* The object(s) are pasted to the same coordinate location as in the original drawing (Fig. 31-9). This feature makes it simple to create or edit a second drawing based on geometry from a previously created drawing.

Note that there is no added control by using *Copybase* to specify a base point when you capture the original objects since *Pasteorig* automatically locates the new objects at their original coordinates regardless of a specified base point. *Pasteorig* does not operate for copying objects back to the original drawing. *Pasteorig* does not function for copying non-AutoCAD objects from other applications into AutoCAD.

Figure 31-9

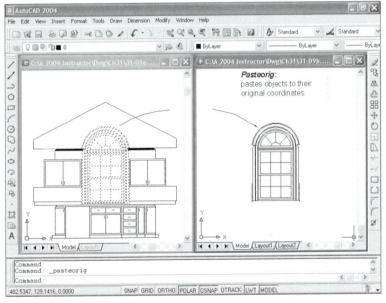

PASTEBLOCK

Pull-down Menu	COMMAND (TYPE)	ALIAS (TYPE)	Short-cut	Screen (side) Menu	Tablet Menu
Edit Paste as Block	PASTEBLOCK	...	(Default Menu) Paste as Block	Edit PasteBlk	...

Pasteblock is intended to be used for <u>copying AutoCAD objects into the same or other AutoCAD drawings</u>. *Pasteblock* operates essentially the same as *Pasteclip* except that *Pasteblock* inserts AutoCAD objects into AutoCAD as *Blocks.* This can be useful if you want to manipulate all objects from one paste operation as one object—a *Block.*

In releases of AutoCAD previous to AutoCAD 2000, all AutoCAD objects that were pasted back into AutoCAD became *Block* references. In order to edit the pasted geometry, the *Blocks* had to be *Exploded*. Since AutoCAD 2000, *Pasteblock* provides the same functionality that *Pasteclip* served in previous releases. Use *Explode* to edit the individual entities within the *Block* (see Chapter 21, Blocks and DesignCenter).

PASTESPEC

Pull-down Menu	COMMAND (TYPE)	ALIAS (TYPE)	Short-cut	Screen (side) Menu	Tablet Menu
Edit Paste special...	PASTESPEC	...	...	EDIT Pastespe	...

Pastespec is intended to be used primarily to paste objects <u>from other applications</u> into AutoCAD.

The *Pastespec* command gives you the choice of embedding or linking objects that are pasted into the drawing. *Pastespec* produces the *Paste Special* dialog box (Fig. 31-10). Your selection of *Paste* or *Paste Link* determines if the objects from the Clipboard are <u>embedded</u> or <u>linked</u>, respectively.

Figure 31-10

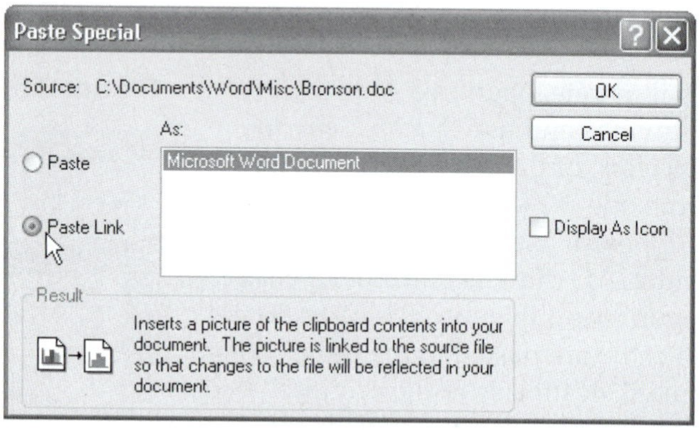

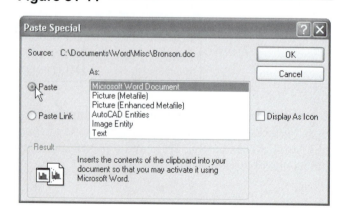

If the information copied from an application that supports OLE is <u>linked</u> within your AutoCAD drawing, a "picture" of the original document is displayed that links (or references) the original source document that was used to create it.

To use *Pastespec*, first you must use *Copy* or similar OLE command from another application (Word, Excel, AutoCAD, etc.) to bring objects onto the Clipboard. Next, use the *Pastespec* command in AutoCAD:

Command: **pastespec** (The *Paste Special* dialog box appears)
Command:

Displayed near the top of the *Paste Special* dialog box is the *Source:* application which created the object that is currently contained on the Clipboard (see Figure 31-10). The two radio buttons on the left allow you to embed (*Paste*) or link (*Paste Link*) the objects.

The list of (Paste) *As:* options varies depending on the type of information contained on the Clipboard and the current *Paste/Paste Link* radio button you have selected (Fig. 31-11). If the current information on the Clipboard is from an AutoCAD drawing, the *Paste Link* option is <u>not</u> available (grayed out). All of the options in this dialog box are described in detail next.

Figure 31-11

Paste

When the *Paste* radio button is selected, the object becomes <u>embedded</u> within the drawing. You can select the type of object to be inserted in the drawing by making a selection from the (Paste) *As:* list (see Fig. 31-11). The list of object types varies depending on the source of the objects on the Clipboard.

Paste Link

Selecting the *Paste Link* button causes AutoCAD to display a "picture" of the object in the drawing. The information displayed is not actually copied into the drawing; it is <u>linked</u> back to the source document from which it was originally copied. This option is available only when the information contained on the Clipboard was created with an application that supports OLE. When *Paste Link* is selected, only one choice appears in the (Paste) *As:* list—the source application (see Fig. 31-10).

Result

The *Result* box at the bottom of the dialog box describes the current object on the Clipboard. It also indicates how the pasted objects are managed in the drawing based on your selections within the dialog box (*Paste/Paste Link* and *Paste As:* list). This is a very helpful feature that dynamically explains your options as they are selected.

Display As Icon

You can display the source application's icon instead of displaying the actual information pasted from the Clipboard by choosing *Display As Icon* (Fig. 31-12). Since only the application's icon is visible in the drawing, you must double-click on it to view the linked or embedded information. This icon is a <u>shortcut</u> that points to the location of the information pasted from the Clipboard. This can be useful in a situation where the linked information is relevant to the project depicted in your drawing, but it is not necessary to display that information directly in the drawing.

Figure 31-12 —————————————————————————

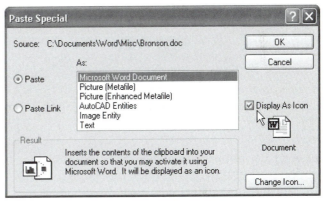

For example, you may be displaying a *Block* of a pump in the drawing. Specific text information such as the pump's specifications, manufacturer's model number, cost, and ordering information could be included within a Word document which is pasted into your drawing and displayed as an icon. The information is important for reference, but it is not necessary to display the information directly in your drawing.

If the information represented by the icon is <u>embedded</u>, double-clicking on the icon activates the application used to create it and displays the contents of the information that was pasted into the drawing in its original format. Changes made to the information as a result of this action are changed for the embedded information only and <u>are not reflected</u> in the source (original) document.

If the *Icon* information is linked, double-clicking on the icon activates the application used to create it and in turn displays <u>the original file</u> that contains the information. The information referenced by the pasted icon becomes highlighted in the source document. Once the source document is saved and closed, the updated version is linked to the icon displayed within your drawing. In this case, the source document itself has actually been changed.

Change Icon

Whenever you have selected the *Display As Icon* checkbox, a preview of the application's default icon is displayed directly beneath the checkbox area. A new button also appears labeled *Change Icon.* This allows you to change the icon if you so desire.

OLESCALE

Pull-down Menu	COMMAND (TYPE)	ALIAS (TYPE)	Short-cut	Screen (side) Menu	Tablet Menu
...	OLESCALE	...	...	...	...

The *Olescale* command produces the *OLE Properties* dialog box (see Figure 31-8), which allows you to alter the *Size, Scale,* and other properties of OLE objects in AutoCAD. If you used *Pasteclip* or *Pastespec* to insert an OLE (non-AutoCAD) object into a drawing, you can modify its properties with this feature. This dialog box is the same dialog box that appears by default when you use *Pastespec* (see "*Pastespec*" earlier in this chapter).

You can also produce this dialog box by highlighting the OLE object, right-clicking it, and selecting *Properties* from the shortcut menu that appears.

INSERTOBJ

Pull-down Menu	COMMAND (TYPE)	ALIAS (TYPE)	Short-cut	Screen (side) Menu	Tablet Menu
Insert OLE Object...	INSERTOBJ	IO	...	INSERT Insertob	T,1

Insertobj also allows you to paste specific information <u>from another application</u> into your drawing as <u>either a linked or embedded</u> object. It functions in much the same way as the *Pastespec* command. The key difference between *Insertobj* and *Pastespec* is that *Insertobj* allows you to paste a <u>file</u> without having the application or document currently open and a particular portion of it selected. Because you are not copying objects from the Clipboard, you do not have to use *Copy* or similar OLE command before using *Insertobj. Insertobj* produces the *Insert Object* dialog box (Fig. 31-13).

Figure 31-13

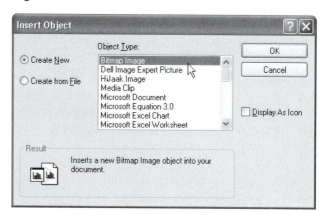

```
Command: insertobj (produces the Insert Object dialog box)
Command:
```

Create New

The options in this command afford you some flexibility. If the file doesn't exist, you can create a new file by selecting *Create New,* then the application you want to use to create it. The applications available to you (registered on your system) are listed in the *Object Type* listing. The selected application is opened when you select *OK* so that you can create the new document. When you are finished creating the new document and you close the application, the new document (object) is <u>embedded</u> into your drawing.

Create from File

When the *Create from File* radio button is selected (Fig. 31-14), you can select the *Browse* button to look for files available in your available drives and folders. After selecting a specific file, you can paste it into your drawing as an <u>embedded</u> or <u>linked</u> object. If the object is to be linked, the display of the linked object may be limited to an icon based on the file type and the OLE capabilities of the source application.

Figure 31-14

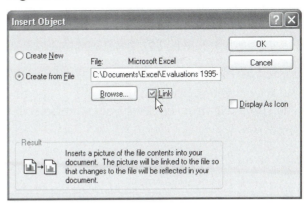

PASTEASHYPERLINK						
Pull-down Menu	**COMMAND (TYPE)**	**ALIAS (TYPE)**	**Short-cut**	**Screen (side) Menu**	**Tablet Menu**
Edit *Paste as Hyperlink*	PASTEASHYPERLINK	...	...	...	...

The *Hyperlink* command (not *Pasteashyperlink*) produces the *Insert Hyperlink* dialog box. That dialog box is used to create a link between an AutoCAD object and a Web page or another document. A hyperlink has a similar function to an embedded OLE object, in that opening the hyperlink opens the related document (usually your Web browser) and displays the Web page or other document. A hyperlink, however, is attached to an AutoCAD <u>object</u>, so passing the cursor over the object displays the hyperlink symbol and right-clicking on the object allows you to open the hyperlink. (See Chapter 19, Internet Tools, for more information on Hyperlinks.)

The *Pasteashyperlink* command or *Paste as Hyperlink* option from the *Edit* pull-down menu is another method of creating a hyperlink as an alternative to the *Insert Hyperlink* dialog box; however, you can use <u>*Pasteashyperlink* to link only to documents, not to Web addresses</u>. The resulting hyperlink is attached to an AutoCAD object. The main difference is that with the *Pasteashyperlink* command, you first open the document you want to link to, use *Copy*, then open AutoCAD and use *Paste as Hyperlink*. AutoCAD simply asks you to select an object to attach the link to.

    Command: **pasteashyperlink**
    Select objects: **PICK**
    Select objects: **Enter**
    Command:

The resulting hyperlink (to the remote document) is an <u>embedded</u> object in the drawing attached to a specific AutoCAD object. The procedure is as follows:

1. open the source program, then open the desired document you want to link to;
2. highlight some text, such as the title or first line of the document;
3. select *Copy* from the program's *Edit* pull-down menu while leaving the document open;
4. toggle to AutoCAD and select *Paste as Hyperlink* from the *Edit* menu;
5. select the AutoCAD object you want to attach the document to.

Two conditions must exist for *Pasteashyperlink* to operate: 1) the document that you *Copy* from must be a previously saved document (not a Web page), and 2) the document that you copy from must be open at the time that you *Paste*.

2002

The text that you select in step 2 appears at the cursor when you pass the cursor over the AutoCAD object. Opening the link produces the source program and displays the entire document that contains the text that you selected to *Copy*. Therefore, the selected text has significance only in that it appears in AutoCAD at the cursor, so you should select text that indicates the name or purpose of the linked document.

Pasteashyperlink is essentially another way to hyperlink a document to an object, but with this method you select the link from within the linked document rather than browsing for the document (file name) when establishing a *Hyperlink* from AutoCAD in the *Insert Hyperlink* dialog box.

MANAGING OLE OBJECTS

OLELINKS

Pull-down Menu	COMMAND (TYPE)	ALIAS (TYPE)	Short-cut	Screen (side) Menu	Tablet Menu
Edit *OLE Links...*	*OLELINKS*	...	...	*EDIT* *OLElinks*	...

When you have *Linked* an object within an AutoCAD drawing, you can control if it is automatically or manually updated whenever changes are made to the original (source) document. The *Olelinks* command activates the *Links* dialog box with options for the maintenance of object links. If no objects are currently *Linked* within the drawing, the *OLE Links...* option in the *Edit* menu is grayed out and typing *Olelinks* invokes no response from AutoCAD.

Command: **olelinks** (activates the *Links* dialog box if links exist in the current drawing)
Command:

The following are features included in the *Links* dialog box (Fig. 31-15).

Figure 31-15 ————————————————————

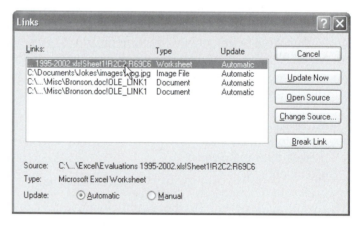

Links, *Type*, and *Update* Columns
The list box in the main section of the dialog provides a listing of all linked objects in the drawing. The *Links* column displays the source information (file name and path). The *Type* is the source application type. The *Update* column lists the current status of the update option.

Source
Found in the lower-left corner of the dialog, this description specifies the drive, path, application, and additional information such as object type or selection area, depending on the source application.

Type
Also listed in the lower-left corner of the dialog is the *Type* of link for the highlighted item. This is the application-specific name (source application) of the currently selected linked object. The application-specific name is sometimes followed by an OLE link number assigned to the object by AutoCAD.

Update

Two *Update* radio buttons are located at the bottom of the dialog. The *Automatic* option causes updating of the linked object whenever its source document is modified while the drawing is open. Setting the *Update* function to *Manual* prevents the link from updating when revisions are made to the source. When *Manual* is selected, use *Update Now* to update the link.

Update Now

This button forces the links selected from the list to be updated. When you close the dialog box, listed data reflects the newly updated status of the source object.

Open Source

This option opens the currently selected link with the source application so it can be modified. The linked information is highlighted when the source application is opened.

Break Link

Use this option to sever the link of the object to its source file and automatically convert it to a "Static OLE object" or Windows metafile object. It can no longer be updated if the source is changed. The object can no longer be edited with the source application or with AutoCAD commands.

For example, if a spreadsheet displaying a bill of materials for the drawing is created in Excel and then linked inside the drawing, it can be modified at any time with Excel simply by double-clicking on the spreadsheet picture displayed on the drawing. If, however, the link is broken, the spreadsheet picture is converted to a metafile picture object and can no longer be edited using Excel. Static OLE objects cannot be edited or removed with AutoCAD commands. To remove a static OLE object, right-click on the object and select *Clear* from the cursor menu that appears.

MOUSE-ACTIVATED OLE EDITING FEATURES

You can access two additional editing features for embedded and linked objects by locating the mouse pointer on an OLE object and selecting it.

Resizing and Repositioning an OLE Object

By clicking once on an OLE object you can activate the object for editing its size or location. After selecting the linked or embedded object once, you will see "handles" around the boundary of the object (Fig. 31-16).

You can reposition an OLE object in the drawing by clicking once and moving it to a new location. When you left-click on the object once, simply place the cursor anywhere inside the boundary of the object so a four-arrow cursor appears (see Figure 31-16). Press the left button to move (reposition) the OLE object in the drawing, much like you would use *Real Time Pan* to reposition an entire drawing.

Figure 31-16

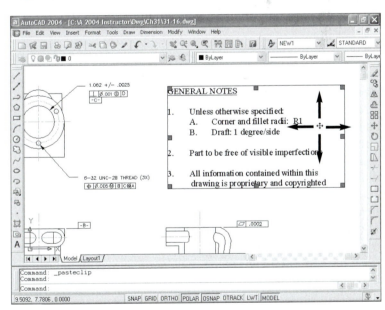

Selecting a midpoint handle allows you to stretch the object in one direction at a time, vertically or horizontally. A two-arrow cursor appears when the pointer moves near a midpoint or corner handle. Selecting a corner handle scales the object proportionally, allowing you to maintain its aspect (width-to-height) ratio. The latter is generally the preferred method since stretching the object in only one direction changes the proportions of the text or image, often making it appear distorted or unreadable.

You can restore a resized or distorted OLE object to its original size and proportion if needed. Do this by right-clicking on the OLE object, then choose *Properties* to produce the *OLE Properties* dialog box. In the dialog box, select *Reset* to reset the proportions of the OLE object back to its original size and height-to-width ratio.

Right-Click Shortcut Menu

Cut, Copy, Clear, Undo

When you place the cursor inside the boundary of an OLE object and right-click on the mouse, the OLE shortcut menu appears (Fig. 31-17). You can select the standard OLE *Cut* and *Copy* commands through this menu to cut or copy the OLE object to the Clipboard. You can also *Clear* the OLE object, which deletes (*Erases*) it from the drawing. *Undo* undoes the last action taken on the object.

Figure 31-17

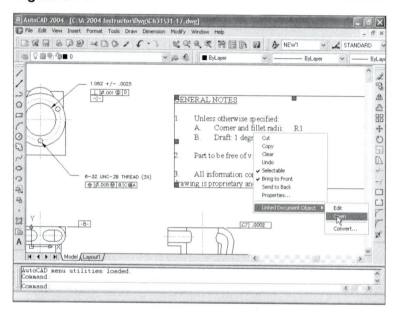

Selectable

An OLE pasted object is *Selectable* by default. When the object is *Selectable* (a check appears in the cursor menu before the word *Selectable*), you can PICK it (place the cursor on the object and left-click) so its size/position handles appear. An <u>OLE object cannot be included in an AutoCAD selection set</u> nor can it be edited by typical AutoCAD commands, even when it is *Selectable*; however, it <u>can</u> be *Erased*. You can remove the check to make the OLE object unselectable. When an OLE object is not selectable, you cannot resize or reposition it by left-clicking it. You can right-click on it to produce the OLE cursor menu to make it *Selectable* again.

Bring to Front, Send to Back

The drawing order in which the OLE object is displayed relative to other objects in the drawing can be controlled through the cursor menu by using *Bring to Front* or *Send to Back*. These two options accomplish the same action as using the *Draworder* command options (see Chapters 26 and 32 for details on the *Draworder* command).

(Object Type) Object>

The option description (*Object Type*) that appears at the bottom of the cursor menu differs based on the source application of the selected OLE object. Selecting this option may produce a cascading menu so you can *Edit, Open,* or *Convert...* the OLE object. Selecting *Edit* or *Open* starts the source application so the document can be edited in the application in which it was created. The *Convert...* option produces the *Convert* dialog box (Fig. 31-18) which operates only for embedded objects. The options available in this dialog differ based on the selected OLE object type. Figure 31-18 displays options for converting a Microsoft Excel document object.

Figure 31-18

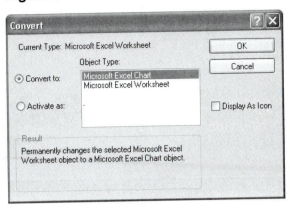

OLE-RELATED SYSTEM VARIABLES

Three system variables affect the appearance of OLE objects in the Drawing Editor and in plots: *OLEHIDE, OLEQUALITY,* and *OLESTARTUP.*

OLEHIDE

OLEHIDE controls the visibility of OLE objects in AutoCAD. Your choice for this variable setting affects the AutoCAD Drawing Editor screen as well as printing and plotting. The possible settings are:

0   All OLE objects are visible (initial setting)
1   OLE objects are visible in paper space only
2   OLE objects are visible in model space only
3   No OLE objects are visible

The *OLEHIDE* system variable setting is saved in the system registry.

OLEQUALITY

OLEQUALITY controls the default quality level of embedded OLE objects for plotting and printing. You can set the quality level for plotting and printing at the time of creating the OLE object by making one of the five selections in the *OLE plot quality* drop-down list in the *OLE Properties* dialog box (see Figure 31-8). Changing the *OLEQUALITY* variable changes the default option in the drop-down list. Changing the setting in the drop-down list does not reset the *OLEQUALITY* variable. The five options are:

0   Line art quality, such as an embedded spreadsheet
1   Text quality, such as an embedded Word® document (initial setting)
2   Graphics quality, such as an embedded pie chart
3   Photograph quality
4   High quality photograph

The *OLEQUALITY* setting is saved in the system registry.

OLESTARTUP

OLESTARTUP determines whether the source application of an embedded OLE object loads when plotting. Loading the OLE source application may improve the plot quality, depending on the source application and type of OLE object. The possible settings are:

0   Does not load the OLE source application (initial setting)
1   Loads the OLE source application when plotting

The *OLESTARTUP* system variable setting is saved in the drawing file.

CHAPTER EXERCISES

1. **AutoCAD Objects:** *Copylink, Copyclip, Copybase, Pasteclip, Pasteorig*

   In this exercise you will gain experience copying and pasting between two open AutoCAD drawings. You will use existing drawings to copy most of the needed geometry to develop a half section view.

   A. *Open* the **SADDL-DM** drawing you created in the Chapter 29 exercises. Activate the **Model** tab, or if you created a titleblock in model space, *Freeze* its layer so you see only the geometry (including center lines and hidden lines) and dimensions. Also, *Freeze* layer **DIM**.

   B. With SADDL-DM open, begin a *New* drawing based on the *Start from Scratch*, *Imperial* settings. Use the *Window* pull-down menu to *Tile Vertically*. *Zoom Extents* in the SADDL-DM window.

   C. Use the *Copylink* command to automatically copy all objects from SADDL-DM. Use *Pasteorig* to bring the objects into the new drawing. This should create a display similar to that shown in Figure 31-19. Notice that all layers come into the new drawing when *Copylink* is used, including layers that are frozen. Since you do not want all these layers, *Exit* the new drawing without saving changes and begin a *New* drawing again using the *Imperial* default settings.

   Figure 31-19 ————————————

   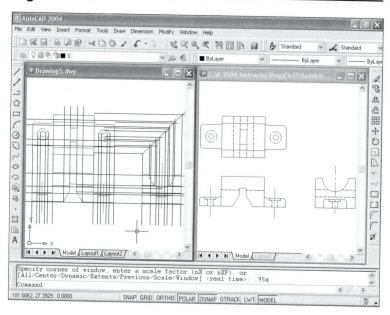

D. Now use *Copyclip* and select only the visible geometry from the SADDL-DM drawing. Activate the new drawing and use *Pasteorig. Zoom Extents*. The two drawings should appear identical, except note that the new drawing's linetypes appear to be continuous. Change the *Ltscale* to **15**. *Save* the new drawing as **SADDLE-HALF SECTION**.

E. Next, *Close* SADDL-DM and <u>do not save changes</u>. *Open* the **SADL-SEC** drawing from Chapter 26 exercises, and *Tile Vertically*. In order to copy the cutting plane line from SADL-SEC, use *Copybase* and specify a base point at the *Center* of one of the two holes in the top view. Since the two drawings (SADDL-DM and SADL-SEC) may have different coordinate locations for the geometry, do not use *Pasteorig*, but use *Pasteclip* to place the cutting plane line in the SADDLE-HALF SECTION drawing at the *Center* of the appropriate hole. Your display should appear similar to that shown in Figure 31-20.

Figure 31-20

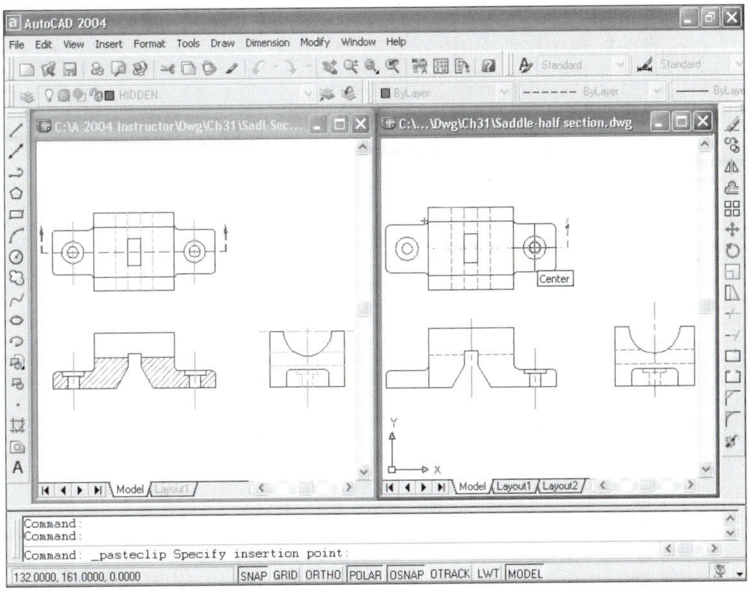

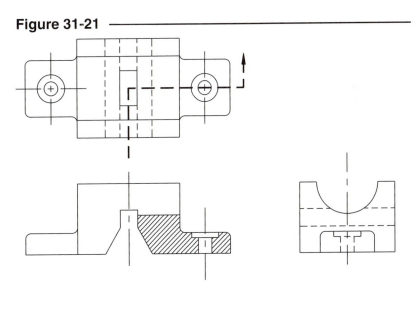

F. *Close* the SADL-SEC drawing and <u>do not save changes</u>. Maximize the new drawing and continue to generate a half section view. Your results should appear similar to that shown in Figure 31-21.

Figure 31-21

G. This technique (using copy and paste between AutoCAD drawings) is useful especially in cases such as this when you need geometry from <u>two or more drawings</u>. What is a better method to use when you need to create a drawing similar to only one other drawing?

2.   *Copy, Pasteclip*

For this exercise, you will write a short paragraph of General Notes to be used in an AutoCAD drawing. Since the notes are text objects, it is easier to use a word processor to write and format the notes, *Copy* them to the Clipboard, and then use *Pasteclip* in AutoCAD to insert the notes into your drawing.

A.   Begin a *New* drawing (*Start from Scratch, Imperial* default settings). Also, while AutoCAD is running, start **Word** and compose a short paragraph of General Notes similar to the note shown in Figure 31-22 (the word "copy-righted" is intentionally mis-spelled). After typing the notes, highlight the entire block of text and the select *Copy* from the *Edit* pull-down menu. This action copies the Word document onto the Clipboard.

Figure 31-22 ───────

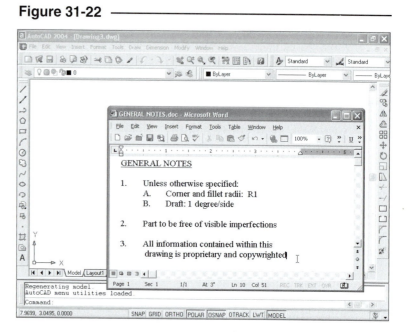

B.   Now return to the new drawing you started in step A (you can use the **Ctrl+Tab** key sequence to switch to AutoCAD). In AutoCAD, select *Pasteclip*. The information from the Clipboard (the General Notes copied previously from Word) is pasted into the top left portion of the drawing area. You can see the "handles" (blue boxes) that allow you to adjust the size of the pasted object. Select one of the <u>corners</u> of the General Notes border and drag it in or out to resize the note. Notice that the text within the pasted image changes proportionally as you resize the image from one of the corner resize buttons. Now select a resize button at one of the <u>midpoints</u> of the image. What happens to the text within the image?

C.   The notes from the Word document are still stored within the current memory of the Clipboard. Select the *Mtext* command. When the *Multiline Text Editor* appears, do not enter in any text, but type a Ctrl+V (hold down the Control key and press the "v" key). Select the *OK* button in the *Multiline Text Editor*. What happened? Save this drawing as **PASTETXT**.

D.   *Close* the PASTE TEXT drawing. *Open* the **SADDLE-HALF SECTION** drawing that you created in Exercise 1. Use *Pasteclip* to paste the note contained on the Clipboard. Use the resize buttons to resize the note and move it to the open space in the upper-right corner. *Save* the drawing.

3.   **Double-clicking on an Embedded Object**

Assume that you printed the **SADDLE-HALF SECTION** drawing and noticed the misspelled word. Follow this procedure to change the spelling of the word "copywrighted" to "copy-righted."

A.   *Open* the **SADDLE-HALF SECTION** drawing if not already opened. Point to the pasted (embedded) note object and then double-click on it using the mouse. What happens?

B. Now correct the misspelled word in **Word**. Then select *Exit* from the *File* pull-down menu in Word. Next, view the notes in AutoCAD. What happened to the notes? *Save* the drawing, but do not *Exit*.

4. *Copylink, Paste*

If you want to copy the entire contents of the display to the Clipboard and then paste the objects into another application, you can accomplish this using *Copylink*.

A. Select *Copy Link* from the *Edit* pull-down menu. The contents of the current display are copied onto the Clipboard.

B. Next, open **Word**. Then select *Paste* from the *Edit* pull-down menu in Word. This action pastes the contents of the Clipboard (AutoCAD objects in this case) into the Word document. You can also choose *Paste Special* instead of *Paste* from the Word *Edit* pull-down to allow you to choose whether to embed or link the contents of the Clipboard into this document.

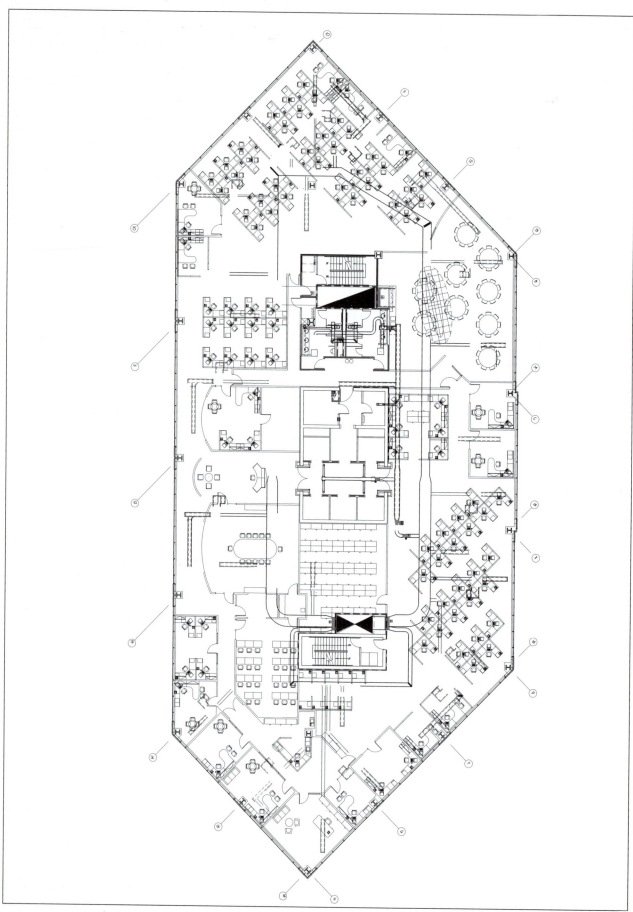

8TH FLOOR.DWG Courtesy, Autodesk, Inc.

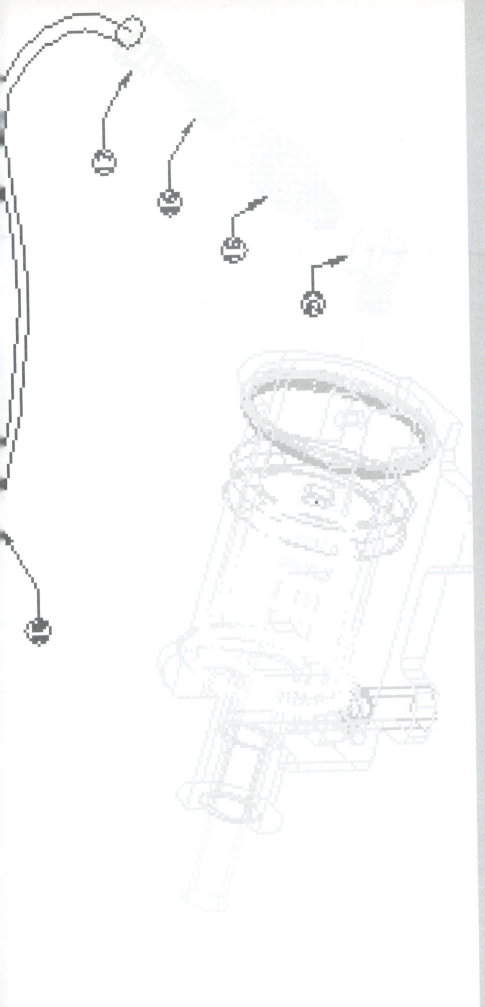

32

RASTER IMAGES AND VECTOR FILES

Chapter Objectives

After completing this chapter you should:

1. know the difference between raster files and vector files;

2. be able to *Imageattach* raster images within your drawing;

3. be able to use *Imageadjust* and *Imagequality* to control the appearance of an attached image;

4. be able to use *Imageclip* to create a clip boundary for an attached image;

5. know that certain images can be controlled for *Transparency*;

6. be able to *Import* vector file formats into your drawing and convert the contents into AutoCAD objects;

7. know the individual commands for importing and exporting specific file types (*Bmpout*, *3dsin*, etc.);

8. be able to *Export* a variety of vector file formats from your drawing with this one command.

CONCEPTS

The previous chapter discussed how to utilize textual and graphic data from other applications within your AutoCAD drawings using AutoCAD's Object Linking and Embedding (OLE) feature. This chapter discusses how to utilize graphic information in the form of raster file and vector file formats within your drawings.

This chapter discusses two different but related topics: (1) attaching raster files and (2) importing and exporting vector file formats. When a raster image is attached to your drawing, the image that appears is actually linked to the original raster file similar to the way text or spreadsheet information in an AutoCAD drawing is linked to the original file when using OLE. When other types of graphic file formats are imported into your drawings, such as vector files, AutoCAD converts the contents of the file into AutoCAD objects and permanently adds them to the drawing.

Raster Files versus Vector Files

Raster images are "pictures" composed of many dots of various shades of gray to produce black and white images or many dots of various colors to produce color images. For example, when you look closely at a color or black and white picture in a newspaper, you can actually see the small dots that make up the image. The quality, or resolution, of the image depends upon the process used to create the image. Some software can produce images with many dots per inch (DPI) yielding a high quality (resolution) image that looks very distinct and clear. Images that have lower resolution have fewer dots per inch and produce a lower quality image. The number of dots used to create the image determines the resolution, or quality, of the picture. Figure 32-1 displays an enlarged view of a raster image to reveal the dots composing the lines and arcs.

Figure 32-1 ─────────────

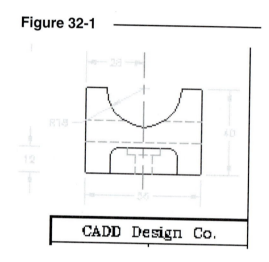

A few of the applications that create raster images frequently attached within AutoCAD drawings are computer graphics or desktop publishing programs (used for creating company logos, for example), document management software, and mapping and geographic information systems (which produce aerial photographs or satellite photographs).

In contrast, CAD programs such as AutoCAD, which are the primary graphics creation and storage media for the engineering, architecture, design, and construction industries, generate vector data. A vector file defines a line, for example, as a vector between two endpoints. Therefore, lines in vector drawings do not enlarge when zoomed in like the dots in a raster image do (Fig. 32-2). The thickness of the displayed line on a computer monitor (for a vector drawing) is normally one pixel, the size of which is determined by the display device. Therefore, vector drawings are used when great detail is needed. Raster images are useful for insertion into a CAD (vector) drawing and can be helpful in the development of a CAD drawing but cannot be used for drawings that require detailed information because of the static configuration of the dots.

Figure 32-2 ─────────────

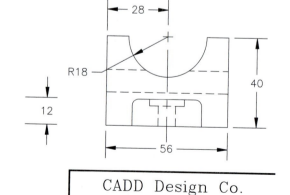

AutoCAD stores its vector data in the .DWG format but can also create other forms of vector files such as .SAT or .WMF discussed in this chapter. (See Chapter 19, Internet Tools, for more information on vector and raster files.)

Attaching Raster Images and Importing Vector Files

Two techniques for utilizing raster file and vector file formats within AutoCAD drawings discussed in this chapter are:

1. attaching a raster image, which links the image to the original file, and
2. importing a vector file, which converts the file contents into AutoCAD objects.

Because AutoCAD can store graphic information only in a vector format, it treats an attached raster image differently because it is <u>not</u> vector information (made up of lines, arcs, and circles). An attached raster image in AutoCAD is actually a picture of the original raster file that was selected when attaching it to your drawing. Attached raster images are displayed within a <u>frame</u> in your drawing and are *linked* to (or reference) the original file. On the other hand, if an image is imported into AutoCAD, it must be converted to AutoCAD objects for AutoCAD to store the geometry as vector data.

Exporting Raster and Vector Files

Because AutoCAD's primary function is to create and manipulate vector data (AutoCAD objects), it specializes in handling vector file formats, especially the .DWG format. Although AutoCAD can create raster files such as .BMP, .TIF, .JPG, and .PNG, these formats have little usefulness for storing drawing information. Only images generated by AutoCAD's rendering capabilities are well suited for saving in a raster format (see Chapter 41, Rendering). Therefore, AutoCAD's file exportation capabilities are centered around vector formats such as .DWG, .DWF, .DXF, .SAT, .STL, .WMF, and .EPS. These file formats are also discussed in this chapter.

Related Commands

This chapter discusses the following AutoCAD commands related to using raster images and vector file formats within your drawing:

Image	Inserts various types of raster images into your drawing
Imageattach	Attaches, or links, a new raster image object and definition within your drawing
Imageframe	Controls whether the image frame is displayed on screen or hidden from view
Imageadjust	Allows you to control the brightness, contrast, and fade values of a raster image
Imageclip	Enables you to create new clipping boundaries for raster image objects
Imagequality	Allows you to control the display quality of raster images
Import	Imports a variety of vector file formats into AutoCAD
3dsin, 3dsout	Imports and exports 3D Studio files
Acisin, Acisout	Imports and exports ACIS solid models
Dxfin, Dxfout	Imports and exports AutoCAD drawings in an exchange format
Psout	Exports Encapsulated PostScript files
Bmpout	Exports bitmap (.BMP) raster files
Jpgout	Exports JPEG raster files
Tifout	Exports Tagged Image File Format (.TIF) files
Pngout	Exports Portable Network Graphic (.PNG) files
Export	Exports a variety of vector file formats from AutoCAD

A few other commands, which are covered in more detail in previous or later chapters, are also discussed in this chapter in the context of attaching, importing, and manipulating raster images and vector files.

ATTACHING RASTER IMAGES

The procedure for attaching a raster image in AutoCAD is very similar to using the *Xref* command (see Chapter 30, Xreferences). A raster image can be *Attached, Detached, Reloaded,* and the *Path* modified, just as you might do with an *Xref.* When you attach an image, it is displayed within a frame in the drawing. You can view the original image, but the image (and all the dots that define it) does not become a permanent part of the drawing. Like an *Xref,* you see only the picture of the original file within its frame. Since raster images normally have large file sizes, this method of "attaching" can save a tremendous amount of space within your drawing.

A few example applications of attaching a raster image to an AutoCAD drawing are listed here:

A. You want to display a digital (raster) picture of the true, or intended, object. For example, you can take a digital photograph of a house and attach it within your construction drawing of an addition to the house.
B. You want to use images of company or client logos within a titleblock.
C. You need to compare and verify site plan information, such as overlaying an aerial photograph to verify the contours shown on your AutoCAD drawing.
D. You need to compare or reconstruct an old "paper" drawing by scanning it (creating a raster image) and attaching it in an AutoCAD drawing.

You can edit the appearance of the attached image but not the individual parts (or dots) of the image within AutoCAD. You can adjust the quality, color, contrast, brightness, and transparency of an attached image with AutoCAD's various image editing commands and system variables. Attached raster images can also be edited with typical AutoCAD editing commands such as *Copy, Move, Rotate,* and *Scale.* Attached images can be edited with Grips as well. The special image attaching and editing commands are discussed next.

Activate the *Reference* toolbar to access buttons for attaching and manipulating raster images. The *Insert* and *Modify* pull-down menus also contain commands for attaching and manipulating raster images.

IMAGE

Pull-down Menu	COMMAND (TYPE)	ALIAS (TYPE)	Short-cut	Screen (side) Menu	Tablet Menu
Insert *Image Manager...*	*IMAGE* or *-IMAGE*	*IM* or *-IM*	...	*INSERT* *Image*	*T,3*

Use the *Image* command to control the attachment of raster images within your drawing. There is no practical limit to the number or size of images you can attach within your drawing. You can have more than one image displayed within any viewport.

Although you can attach a variety of file types (BMP, TIF, JPG, GIF, TGA, etc.), AutoCAD determines how to handle the image (for display and editing) based upon the image file contents not the image file type. The following raster file types are accepted by AutoCAD's *Image* and *Imageattach* commands:

Image File Type	File Extensions	Description
BMP	*.BMP, *.RLE, *.DIB	Windows device-independent bitmap format
CALS-1	*.RST, *.GP4, *.MIL, *.CAL, *.CG4	Mil R-Raster-1
FLIC	*.FLC, *.FLI	Autodesk Animator FLIC
GEOSPOT	*.BIL	GeoSPOT with HDR and PAL files
GIF	*.GIF	Compuserve Graphic Interchange Format
IG4	*.IG4	Image Systems Group 4
IGS	*.IGS	Image Systems Grayscale
JPEG, JFIF	*.JPG	JPEG raster
PCX	*.PCX	PC Paintbrush Exchange
PICT	*.PCT	Macintosh PICT1, PICT2
PNG	*.PNG	Portable Network Graphics
RLC	*.RLC	Run-Length Compressed
TARGA	*.TGA	Truevision
TIFF	*.TIF	Tagged Image File Format

The *Image* command invokes the *Image Manager* (Fig. 32-3). In this dialog box you can *Attach, Detach, Reload, Unload,* locate a file using *Browse,* and save a new path using *Save Path.* The operation of and options within this dialog box are similar to those in the *External Reference* dialog box (see Chapter 30).

Figure 32-3

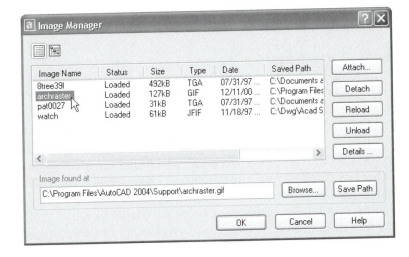

Tree View, List View

The two buttons located in the top-left corner of this dialog box allow you to control the appearance of the list of currently attached raster images in the center of the dialog box. The list can be displayed in a *List View* or *Tree View.*

List View is the default setting for the list when you open the *Image Manager.* When *List View* is active, column headings appear at the top of the list for *Image Name, Status, Size, Type, Date,* and *Saved Path* for each image currently attached to the drawing.

When *Tree View* is selected, AutoCAD displays the list of attached images in a tree structure. This format allows you to view the list of images currently attached to the drawing in a hierarchical structure.

Attach

When you first activate the *Image Manager* the only available button is *Attach*. Use this button to attach an image to the drawing in the same way you attach an *Xref*. Selecting the *Attach* button invokes the *Select Image File* dialog box (Fig. 32-4). Alternately, this dialog box can be opened directly using the *Imageattach* command.

From the *Select Image File* dialog box you can select from a variety of raster image file types (listed in the previous table). The features and functionality of the *Select Image File* dialog box are the same as those discussed in Chapter 30 for attaching *Xref* files. However, the *Select Image File* dialog box has the additional option of *Hide Preview*/*Show Preview* that allows you to turn on and off the *Preview* display area. If the preview is turned off, the *Preview* display box does not display the thumbnail picture of the file currently selected from the file list. Use this feature to increase the speed at which you can select files from the dialog box when you know the image files by name. Turn the *Preview* display on again to view the thumbnail of the selected image file.

Figure 32-4

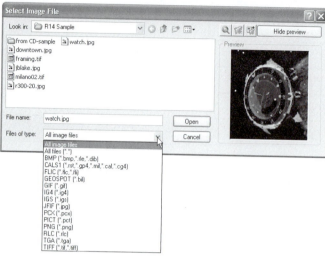

When an image file has been selected for attachment (select *Open* in the *Select Image File* dialog box), the *Image* dialog box appears for you to specify the attachment parameters (Fig. 32-5). Specifying the *Insertion Point* and *Rotation* angle is similar to inserting a *Block* or attaching an *Xref*. Use *Scale* factor to match the image geometry scale to the scale of the geometry in the drawing.

The *Scale* factor is based on the ratio of image units to the current AutoCAD units. You can enter a value in the edit box if you know the scale of the geometry contained within the image file, otherwise use the default setting, *Specify on screen*. Select the *Details* button to display image size and resolution information (Fig. 32-5, bottom).

Figure 32-5

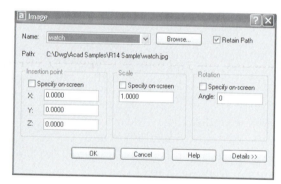

If you want to attach a second image in a drawing, selecting the *Attach* button may produce the *Image* dialog box directly. You must then select *Browse* to produce the *Select Image File* dialog box to select from folders and files. When the same image is attached more than once in a drawing, only one reference is displayed in the list in the *Select Image File* dialog box, but a different insertion point, scale, clipping boundary, etc. can be specified for each instance of the image in your drawing.

Detach

You can detach image files appearing in the list by highlighting the desired file(s), then selecting *Detach*. This action deletes all references to the image within the drawing (breaks the link to the image file), but does not delete the file from the storage device (local or network drive). This is essentially the same as detaching an *Xref* from a drawing.

Reload

You can *Unload* an image to remove its display while you work on the drawing, but the attachment information is not deleted from the drawing. You can then *Reload* the image picture by selecting the *Reload* button. *Reload* causes AutoCAD to reread the file referenced in the *Saved Path* and redisplay it within the image frame. This is also useful if the referenced image file has been changed and you want the changes to be displayed during the current drawing session.

Unload

This feature unloads the raster image picture from the display without deleting the attachment information from your drawing. Only the image's frame remains visible in the drawing. Unloading an image can dramatically decrease the amount of time AutoCAD needs to redisplay, or regenerate, the drawing since the image is not resident in current memory. Use this feature when you do not need to see the image in order to work on the drawing.

Unloading a drawing is not the same as turning off the display of the image. You can turn off the display of an image within the drawing using *Properties* (see *"Properties,"* this chapter).

Details...

You can quickly access specific information about an image by selecting an image from the list and selecting the *Details* button (in the *Image Manager*). This activates the *Image File Details* dialog box that lists details of the currently selected image as shown in Figure 32-6.

Figure 32-6

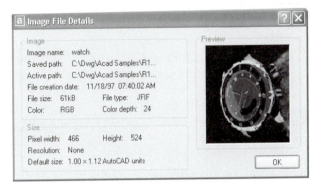

Image Found At

A text box at the bottom of the *Image Manager* displays the current location of the highlighted image file in the list. You can choose another image file to attach to the selected *Image name* (and keep the same name from the list) by selecting the *Browse* button to the right of this text box.

You may want to do this, for example, if you have an older version of an aerial photograph currently attached and you want to temporarily display a newer photo of the same aerial view. You can *Browse*, find and select the newer raster image, then exit the *Image Manager* to verify that it is a more appropriate image to display in your drawing.

If you prefer the newer file, you can then return to the *Image Manager* and save the current path for the attached image by selecting the attached image from the list, then selecting *Save Path*. In this case, the *Image name* originally stored by AutoCAD does not change, but the path and file name change.

Browse

Selecting this button takes you to the *Select Image File* dialog box, enabling you to search folders and locate raster files to attach.

Save Path

Select this function when you have used *Browse* to specify a new image file to display in your drawing and you want to make it a permanent change. If you do not use *Save Path* prior to exiting the drawing, any changes made to the original path or file name specified for the named image are lost.

IMAGEATTACH

	Pull-down Menu	COMMAND (TYPE)	ALIAS (TYPE)	Short-cut	Screen (side) Menu	Tablet Menu
	Insert *Image Manager...* *Attach...*	*IMAGEATTACH*	*IAT*	...	*INSERT* *Image* *Attach...*	...

Imageattach allows you to attach a raster image directly without having to use the *Image* command first. When you use *Imageattach,* the *Select Image File* dialog box appears (see Figure 32-4). The capabilities and operation of this dialog box were discussed previously regarding the *Image* command (see *Image*).

IMAGEFRAME

Pull-down Menu	COMMAND (TYPE)	ALIAS (TYPE)	Short-cut	Screen (side) Menu	Tablet Menu
Modify Object> Image> Frame	IMAGEFRAME	...	...	MODIFY1 *Imagefrm*	...

The *Imageframe* command allows you to control whether the image frames are displayed on the screen or hidden from view. When you attach an image within your drawing it is displayed inside a "frame" that defines the image's outermost boundary. The *Imageframe* options are *Off* and *On*. Setting *Imageframe* to *Off* hides the frames so they are no longer visible in the AutoCAD Drawing Editor, but the images remain visible:

```
Command: imageframe
Enter image frame setting [ON/OFF] <ON>: (option)
Command:
```

The *Imageframe* setting is global; that is, it affects all images in the drawing. All image frames are *Off* or *On*.

Images can be modified by using the general editing commands (*Move, Copy, Scale, Rotate, Grips*, etc.) or by using special image editing commands (discussed next). The general editing commands require you to select the image's frame; therefore, turning the image frame *Off* makes the image unselectable for editing with general editing commands. You must select the image's frame whenever AutoCAD prompts you to "Select objects:" to modify. The image becomes selectable again once you turn the display of its frame *On*.

The special image adjusting commands (*Imageadjust, Imageclip,* and *Imagequality*) allow you to select images whether the image frames are *On* or *Off*.

IMAGEADJUST

Pull-down Menu	COMMAND (TYPE)	ALIAS (TYPE)	Short-cut	Screen (side) Menu	Tablet Menu
Modify Object> Image> Adjust	IMAGEADJUST	IAD	...	MODIFY1 *Imageadj*	X,20

You can adjust an image's *Brightness, Contrast,* and *Fade* values by using *Imageadjust.* Use the *Imageadjust* command or double-click on an image's frame (if *Dblclkedit* is *On* and *Imageframe* is *On*) to produce the *Imageadjust* dialog box (Fig. 32-7). You can change the following settings.

Figure 32-7

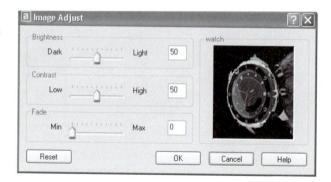

Brightness
This slider and edit box control the brightness of the image. The edit box values range from 0 to 100. The higher the value, the brighter the image becomes. *Brightness* indirectly affects the range of *Contrast* for the image.

Contrast

Use this slider or edit box to control the contrast of the image. *Contrast* is the ratio of light to dark (for black and white images) or pure hue to no color (for color images). The higher the value, the more each pixel is forced to its primary or secondary color. *Contrast* indirectly affects *Fade*. The WATCH.JPG image is displayed in Figure 32-8 with *Contrast* set to 90 and *Brightness* set to 70.

Figure 32-8

Fade

The *Fade* effect of the image controls how much the image blends with the background color. The higher the value, the more the image is blended with the current background color. A value of 100 allows you to blend the image completely into the background. If you change the screen background color, the image will fade to the new color. For plotting, white is used for the background fade color.

Reset

Select this button to reset the values for *Brightness, Contrast,* and *Fade* back to the default settings of 50, 50, and 0, respectively.

IMAGECLIP

Pull-down Menu	COMMAND (TYPE)	ALIAS (TYPE)	Short-cut	Screen (side) Menu	Tablet Menu
Modify *Clip>* *Image*	*IMAGECLIP*	*ICL*	...	*MODIFY1* *Imageclp*	*X,22*

You can clip an image with *Imageclip* in the same manner that you clip *Xrefs* with *Xclip*. When specifying the area to clip, you can use a rectangle or polygon to define the clip boundary. Everything outside the clip boundary is hidden from view. Figure 32-9 displays the WATCH.JPG image with a polygonal clip boundary.

Specifying the clip boundary rectangle or polygon is similar to using AutoCAD's *Rectangle* command or the *Window Polygon* option when selecting objects. *Rectangle* is the default boundary.

Figure 32-9

```
Command: imageclip
Select image to clip: PICK (select an image frame)
Enter image clipping option [ON/OFF/Delete/New boundary]
    <New>: Enter
Enter clipping type [Polygonal/Rectangular] <Rectangular>: Enter
Specify first corner point: PICK
Specify opposite corner point: PICK
Command:
```

Once a clipping boundary has been specified for an image you can turn the boundary *Off*. This action makes the entire image visible again (Fig. 32-10). You can also turn the boundary back *On*, which displays the image with its original clip boundary (see Fig. 32-9). Alternately, you can *Delete* the clip boundary so the entire image is visible.

Figure 32-10

If you use *Imageclip* and select an image that already has a boundary, the following prompt appears:

```
Command: imageclip
Select image to clip: PICK
Enter image clipping option [ON/OFF/Delete/New boundary]
    <New>: Enter
Delete old boundary? [No/Yes] <Yes>: Enter
```

IMAGEQUALITY

Pull-down Menu	COMMAND (TYPE)	ALIAS (TYPE)	Short-cut	Screen (side) Menu	Tablet Menu
Modify *Object>* *Image>* *Quality*	IMAGEQUALITY	...	...	*MODIFY1* *Imagequa*	...

This command allows you to adjust the quality setting that affects display performance. High-quality images take longer to display. When you change the *Imagequality* setting, the display updates immediately without causing a *Regen*. The images attached to your drawing are always plotted using a high-quality display, regardless of the *Imagequality* setting. The setting you choose for *Imagequality* affects all images in the drawing globally.

```
Command: imagequality
Enter image quality setting [High/Draft] <High>: (option)
Command:
```

TRANSPARENCY

Pull-down Menu	COMMAND (TYPE)	ALIAS (TYPE)	Short-cut	Screen (side) Menu	Tablet Menu
Modify *Object>* *Image>* *Transparency*	*TRANSPARENCY*	...	...	*MODIFY1* *Transpar*	...

Transparency allows you to control whether the <u>background</u> pixels in an image are transparent or opaque. Many image file formats allow images with transparent pixels. When setting image transparency to *On*, AutoCAD recognizes transparent pixels so that graphics on the screen (AutoCAD objects or another image) show through those pixels.

Transparency can be set for both bitonal and non-bitonal (Alpha RGB or gray-scale) images. When you attach an image to your drawing, the image's transparency setting is *Off* by default:

Figure 32-11

> Command: **transparency**
> Select image(s): **PICK**
> Select image(s): **Enter**
> Enter transparency mode [ON/OFF] <OFF>: **(option)**
> Command:

This unique feature allows you to put one image on top of another, then set *Transparency* for the top image *On*. (Use *Draworder* to specify which image is "on top.") The transparency feature makes the "top" image appear in the same scene as the "background" image (Fig. 32-11, right image).

PROPERTIES

Pull-down Menu	COMMAND (TYPE)	ALIAS (TYPE)	Short-cut	Screen (side) Menu	Tablet Menu
Modify *Properties*	*PROPERTIES*	*MO* or *PR*	(Edit Menu) *Properties...*	*MODIFY1* *Property*	*Y,12* to *Y,13*

Using *Properties* and selecting an image gives you the power to change many of the image's properties as you might with other image-related commands. The *Imageframe* must be *On* for you to select the image. The adjustable features of the image are listed in the *Image Adjust* section of the *Properties* palette. You can change *Brightness*, *Contrast*, and *Fade*, similar to using the *Imageadjust* command. Select the ellipsis (...) just to the right of each of these edit boxes to produce the *Imageadjust* dialog box.

In the *Misc* section, you can set *Show clipped* and set *Transparency*—identical to the *On/Off* options of the *Imageclip* and *Transparency* commands. *Show image* allows you to turn on and off the display of the image.

Figure 32-12

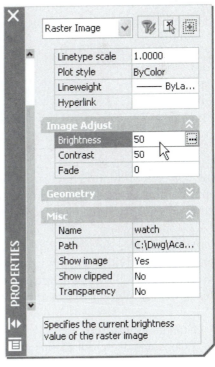

DRAWORDER

Pull-down Menu	COMMAND (TYPE)	ALIAS (TYPE)	Short-cut	Screen (side) Menu	Tablet Menu
Tools Display Order>	DRAWORDER	DR	...	*TOOLS1 Drawordr*	T,9

The *Draworder* command is very useful when utilizing raster images in your drawing. Often you will attach the raster image so that it appears on top of another image or other AutoCAD objects. The *Draworder* command allows you to select an attached image and then specify the drawing order placement of the raster image as *Above, Under, Front,* or *Back*:

Command: **draworder**
Select objects: **PICK**
Select objects: **Enter**
Enter object ordering option [Above object/Under object/Front/Back] <Back>: **(option)**
Command:

For additional information about *Draworder,* see Chapter 26, Section Views.

SAVEIMG

Pull-down Menu	COMMAND (TYPE)	ALIAS (TYPE)	Short-cut	Screen (side) Menu	Tablet Menu
Tools Display Image> Save...	SAVEIMG	...	...	*TOOLS1 Saveimg*	...

The *Saveimg* (save image) command allows you to save the current drawing or rendered image as either a .BMP, .TGA, or .TIF (raster) format. The saved image can later be viewed by using the *Replay* command or can be attached by using *Imageattach*. See Chapter 41, Rendering, for more information on this command.

BMPOUT

Pull-down Menu	COMMAND (TYPE)	ALIAS (TYPE)	Short-cut	Screen (side) Menu	Tablet Menu
...	BMPOUT	...	...	...	...

The *Bmpout* command enables you to create a .BMP (bitmap) file of the drawing or of selected objects from the drawing. The resulting file can be read by many viewers such as Microsoft Internet Explorer or viewed as a raster image (using *Imageattach*) in an AutoCAD drawing.

Bmpout produces the *Create Raster File* dialog box (not shown), which is similar to the *Save Drawing As* dialog box but without options for saving as other file types. You can specify a location and a name for the new file. A .BMP file extension is automatically appended to the file name. When the dialog box clears, the following prompt appears at the Command line:

Command: **bmpout**
Select objects or <all objects and viewports>: **PICK** or **Enter**

You can select one or more objects or press Enter to write the entire drawing contents to the file. No other options exist for the command. See Chapter 41, Rendering, for other information on saving .BMP images generated by AutoCAD.

JPGOUT

Pull-down Menu	COMMAND (TYPE)	ALIAS (TYPE)	Short-cut	Screen (side) Menu	Tablet Menu
...	*JPGOUT*	...	...	...	...

The *Jpgout* command allows you to generate a raster file in JPEG format. *Jpgout* produces the *Create Raster File* dialog box, identical to that appearing for the *Bmpout* command. When the dialog box clears, the prompt appears at the Command line, like the *Bmpout* command, allowing you to select specific objects or the entire drawing. (See *"Bmpout."*)

PNGOUT

Pull-down Menu	COMMAND (TYPE)	ALIAS (TYPE)	Short-cut	Screen (side) Menu	Tablet Menu
...	*PNGOUT*	...	...	...	...

The *Pngout* command allows you to generate a .PNG (Portable Network Graphics) raster file of selected objects or of the entire drawing. The process is the same as that for using the *Bmpout* command or *Jpgout* command. (See *"Bmpout."*)

TIFOUT

Pull-down Menu	COMMAND (TYPE)	ALIAS (TYPE)	Short-cut	Screen (side) Menu	Tablet Menu
...	*TIFOUT*	...	...	...	...

The *Tifout* command operates identically to the *Bmpout*, *Pngout*, and *Jpgout* commands but creates a raster file in Tagged Image File Format. See *"Bmpout"* for more information on this process.

IMPORTING AND EXPORTING VECTOR FILES

Importing vector file formats into your drawing converts the graphic information, or objects, within the selected file into AutoCAD objects and inserts them onto your drawing. The converted information becomes a permanent part of your drawing. Depending on the file being imported, this can dramatically increase the file size of your AutoCAD drawing.

Since importing vector files converts the contents into AutoCAD objects, you can edit them using typical AutoCAD editing methods. An imported object is initially defined in your drawing as a *Block* reference. The first step in editing imported vector files with AutoCAD is to *Explode* the block. You will find that AutoCAD often converts the information into *Lines, Arcs, Circles, Plines, Solids* (2D), or solids (3D).

IMPORT

Pull-down Menu	COMMAND (TYPE)	ALIAS (TYPE)	Short-cut	Screen (side) Menu	Tablet Menu
...	IMPORT	IMP	...	...	T,2

The *Import* command is used for importing vector information into your drawing and produces the *Import File* dialog box (Fig. 32-13). The *Import File* dialog box has the same layout and basic functionality as the *Select Image File* dialog box. The available file types you can import to your drawing are given in the following list.

.WMF   Windows Metafile
.SAT   ACIS Technology
.3DS   3D Studio

NOTE: The *Preview* section of the *Import File* dialog box in the first release of AutoCAD 2004 does not operate for .WMF files.

Figure 32-13

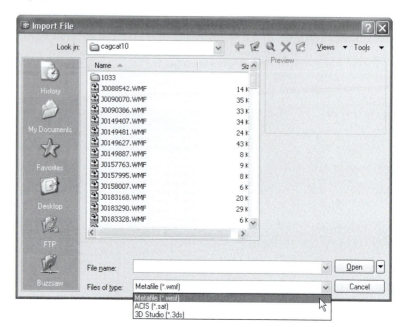

There is an additional feature in the *Import File* dialog box that is not in the *Select Image File* dialog box. Selecting *Options* from the *Tools* menu produces the *WMF In Options* dialog box (Fig. 32-14) but is active only when the *Metafile (WMF)* type is specified in the *Files of type* list box. The *Wide Lines* option lets you control whether the imported .WMF file maintains relative line widths or imports line widths of 0. The *Wire Frame (No Fills)* option controls whether the metafile is converted into a wireframe (lines only) or keeps solid-filled objects.

Figure 32-14

TIP When the metafile is imported, it is converted to a *Block* reference. Use *Explode* to break the *Block* into component objects such as *Lines*, *Arcs*, and *Plines*.

3DSIN

Pull-down Menu	COMMAND (TYPE)	ALIAS (TYPE)	Short-cut	Screen (side) Menu	Tablet Menu
Insert *3D Studio...*	3DSIN	...	...	*INSERT* *3DSin*	...

3D Studio Max® is a 3D modeling, rendering, and animation product that is produced by Kinetix®, a division of Autodesk. 3D Studio Max creates models, complete with materials, lights, cameras, etc., that can be stored in .3DS file format.

The *3dsin* command in AutoCAD enables you to read in geometry and rendering data from 3D Studio (.3DS) files. *3dsin* produces the *3D Studio File Import* dialog box (Fig. 32-15). AutoCAD is capable of importing 3D Studio information such as meshes, materials, mapping, lights, and cameras. 3D Studio procedural materials and smoothing groups <u>are not</u> imported.

Figure 32-15

Once you have selected a .3DS file to import and select the *Open* button, the *3D Studio File Import Options* dialog box is displayed, enabling you to specify the objects to import, the layers to place them on, and how to handle objects constructed of multiple materials (Fig. 32-16).

Figure 32-16

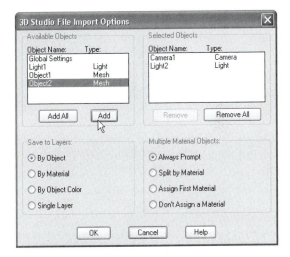

3DSOUT

Pull-down Menu	COMMAND (TYPE)	ALIAS (TYPE)	Short-cut	Screen (side) Menu	Tablet Menu
File *Export...* **.3ds*	*3DSOUT*	...	...	*FILE* *EXPORT* **.3ds*	...

You can convert your AutoCAD geometry and rendering data into 3D Studio data by using the *3dsout* command. *3dsout* first opens the *3D Studio Output File* dialog box (similar to the *3D Studio File Import* dialog box, Fig. 32-15). Once you give the output file name and select *Save*, the *3D Studio File Export Options* dialog box appears, giving you the opportunity to specify the division method, output, mode, smoothing angle, and welding threshold for the export process (Fig. 32-17).

Figure 32-17

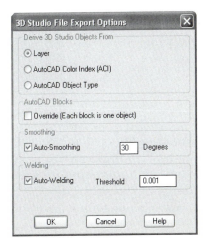

ACISIN

Pull-down Menu	COMMAND (TYPE)	ALIAS (TYPE)	Short-cut	Screen (side) Menu	Tablet Menu
Insert ACIS File...	*ACISIN*	...	...	*INSERT ACISin*	...

AutoCAD uses the ACIS® solid modeling engine to construct solids. This modeler is used also by other CAD programs. The *Acisin* and *Acisout* commands allow you to exchange solid model geometry with other programs that use the ACIS modeling engine. ACIS solid model geometry can be stored as a .SAT (ASCII) file format.

In AutoCAD, use the *Acisin* command to produce the *Select ACIS File* dialog box for importing ACIS (.SAT) files into AutoCAD. The *Select ACIS File* dialog box is the typical file navigation dialog box except the *Files of type* section is set to *ACIS (*.sat)*.

ACISOUT

Pull-down Menu	COMMAND (TYPE)	ALIAS (TYPE)	Short-cut	Screen (side) Menu	Tablet Menu
*File Export... *.sat*	*ACISOUT*	...	...	*FILE Export *.sat*	...

You can use the *Acisout* command to export AutoCAD solid or *Region* data from your drawing into an ACIS format. The ACIS file format allows AutoCAD to store solid model objects as an ASCII (.SAT) file type. The *Create ACIS File* dialog box that appears is the same as the *Select ACIS File* dialog box. *Acisout* ignores all selected objects that are not solids or *Regions*.

STLOUT

Pull-down Menu	COMMAND (TYPE)	ALIAS (TYPE)	Short-cut	Screen (side) Menu	Tablet Menu
*File Export... *.stl*	*STLOUT*	...	...	*FILE Export *.stl*	...

You can create a file to be used with Stereo Lithography Apparatus for creating prototypes from AutoCAD solid models. The *Stlout* command creates an ASCII format STL file defining the 3D geometry. For more information, see Chapter 39, Advanced Solids Features.

DXFIN

Pull-down Menu	COMMAND (TYPE)	ALIAS (TYPE)	Short-cut	Screen (side) Menu	Tablet Menu
...	*DXFIN*	...	...	...	...

Autodesk initiated an interchangeable CAD drawing (vector file) standard in the early 1980s. This standard, called the Drawing Interchange File Format (.DXF), was intended to be utilized by AutoCAD and other CAD programs for exchanging vector drawing data in a simple ASCII format. Because the ASCII format can be used, DXF files can be very large. The DXF format has become one of the most widely used formats (along with IGES format) for exchanging drawing data.

Importing vector data in .DXF format from other CAD applications is accomplished by using the *Dxfin* command. This command produces the *Select File* dialog box (not shown), similar to the other file import dialog boxes. <u>The *Dxfin* command must be used in a new drawing to function properly</u>.

DXFOUT

Pull-down Menu	COMMAND (TYPE)	ALIAS (TYPE)	Short-cut	Screen (side) Menu	Tablet Menu
Save As...	*DXFOUT*	...	...	...	...

You can use the *Dxfout* command or the *SaveAs* command to export your drawing to a Drawing Interchange Format (*.DXF) file. *Dxfout* produces the *Save Drawing As* dialog box (not shown), the same dialog box that appears with the *SaveAs* command. To save the drawing as a .DXF, use the *Files of type* drop-down list and select one of the .DXF file type options. You can choose to export the drawing as an AutoCAD 2004, previous AutoCAD, or LT format for the .DXF.

When the *Save as type* box indicates the desired .DXF format you want to use, you can select *Options* from the *Tools* menu to produce the *Saveas Options* dialog box. Activate the *DXF Options* tab (Figure 32-18).

Here you can specify whether you want the interchange file to be created in ASCII format or binary format as well as the degree of accuracy used (in decimal places) for defining the geometry. Drawing Interchange Files, whether created as ASCII or binary format, are saved as a .DXF file type.

Figure 32-18

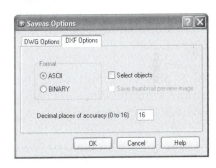

Because an ASCII file is composed of recognizable alphanumeric characters, the ASCII format .DXF files can be quite large, so the binary format is sometimes used. The advantage to the ASCII file is that it is readable in a text editor or word processor (see Figure 1-4), whereas the binary format is not.

DXBIN

Pull-down Menu	COMMAND (TYPE)	ALIAS (TYPE)	Short-cut	Screen (side) Menu	Tablet Menu
Insert Drawing Exchange Binary...	*DXBIN*	...	...	*INSERT DXBin*	...

The *Dxbin* command allows you to import Drawing Exchange Binary (.DXB) files into AutoCAD. These .DXB files are formatted with a specially coded binary format and are produced by programs such as AutoShade. *Dxbin* produces the *Select DXB File* dialog box (not shown).

PSOUT

Pull-down Menu	COMMAND (TYPE)	ALIAS (TYPE)	Short-cut	Screen (side) Menu	Tablet Menu
...	*PSOUT*	...	...	...	...

You can export parts or all of your drawing as a PostScript (.EPS) file using *Psout*. PostScript files generated by *Psout* can contain a PostScript rendering of an AutoCAD model, for example. AutoCAD normally exports *Arcs, Circles, Plines,* and filled *Regions* <u>as PostScript primitives instead of vectors</u>. One exception is when the AutoCAD objects cannot be represented in PostScript, such as extruded objects, in which case they are output as vectors. In this case, the information is exported as wireframe images just as AutoCAD displays them in your drawing. *HIDE* and *SHADE* have no effect on the output created when you use *Psout*.

Psout produces the *Create PostScript File* dialog box (not shown). Selecting *Options* from the *Tools* menu allows you to set PostScript specific output options via the *Postscript Out Options* dialog box.

In the *Postscript Out Options* dialog box (Fig. 32-19) you can specify what part of the drawing should be written to the .EPS file by selecting *Display, Extents, Limits, View,* or *Window* in the *What to plot* section. EPS files can contain a preview which speeds and simplifies the selection and display of a PostScript image in other software programs. The preview can be saved in EPSI or TIFF format, along with the size in pixels for the preview image. The *Prolog Section Name* specifies a name for the prolog section to be read from the ACAD.PSF file. Because a PostScript file can contain all information necessary for printing to a specific size paper, that information can also be specified in this dialog box. For more information, see Using PostScript Files in the AutoCAD *User's Guide*.

Figure 32-19 ——————————

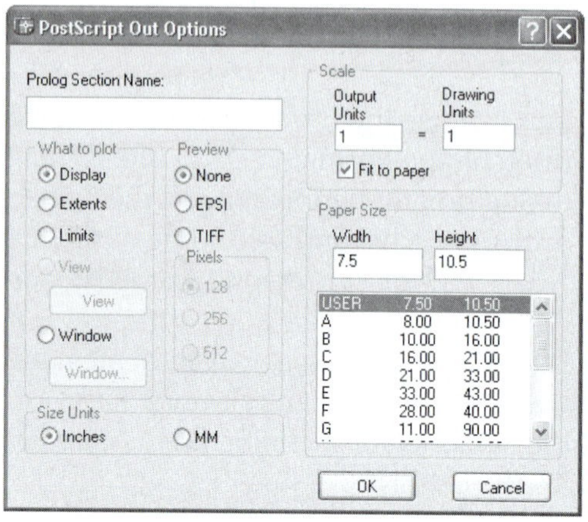

As an alternative to *Psout*, if you configure a PostScript driver in the *Add-a-Plotter* wizard, you can output your drawings in PostScript format. To configure the PostScript driver, in the *Add-a-Plotter* wizard, select *Adobe* from the *Manufacturer* list, and select a *PostScript* level from the *Model* list.

EXPORT

Pull-down Menu	COMMAND (TYPE)	ALIAS (TYPE)	Short-cut	Screen (side) Menu	Tablet Menu
File Export...	EXPORT	EXP	...	FILE Export	...

AutoCAD can *Export* your entire drawing or portions of it to many vector file formats other than .DWG. This can be accomplished by two methods: (1) use the *Export* command and select the desired file type to export, or (2) use the specific export command as discussed earlier, such as *Dxfout, 3dsout,* or *Psout.* (.BMP raster files and .DXX data extract files can also be created using *Export*.)

Use the *Export* command to produce the *Export Data* dialog box (Fig. 32-20). Here you can choose the file format you want AutoCAD to create during the export process.

Figure 32-20 ——————————

The available file formats that AutoCAD can export are listed in the following table along with the equivalent specific AutoCAD commands.

File Extensions	AutoCAD Equivalent Command	Description
*.3DS	*3DSOUT*	3D Studio format
*.BMP	*BMPOUT*	Windows device-independent bitmap format
*.DWG	*WBLOCK*	AutoCAD drawing file format
*.DXX	*ATTEXT*	Attribute extract DXF format
*.EPS	*PSOUT*	Encapsulated PostScript format
*.SAT	*ACISOUT*	ACIS solid object format
*.STL	*STLOUT*	Solid object stereolithography format
*.WMF	*WMFOUT*	Windows Metafile format

CHAPTER EXERCISES

1. *Image Attach*

   You can use the *Image* command to enhance the appearance of your drawing by attaching a raster image such as a company logo, for example.

   A. *Open* the **SADDL-DM** drawing from Chapter 29 Exercises. *Zoom* in to the titleblock area in the lower-right corner of the drawing. You will attach a raster image in the titleblock as a company logo.

   B. Use the *Image* command, and in the *Image Manager* choose the **Attach...** button. Locate a raster image file called **ADESK2.TGA** in the **Textures** folder. (To find where your *Textures* folder is located, use *Options*, *Files* tab, then expand *Texture Maps Search Path*.) Select this file and *Open* it. In the *Image* dialog box choose **OK**. PICK the insertion point as shown in Figure 32-21 and scale the image by dragging it so that the image frame fits within the first text area of the title block.

   Figure 32-21

   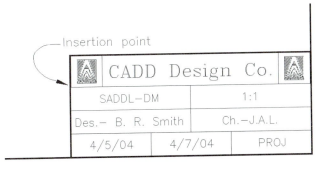

   C. Select the attached image and *Copy* it to the other side of this title area. Compare your results to Figure 32-21.

   D. Use *Imageadjust* and select one of the image's frames. Set *Brightness* to **75** and *Contrast* to **100**. The logo should now appear as a pure black and white image. Do the same for the other image. *Zoom All* and save this drawing as **SADDL-DM2**.

2. ***Image Attach, Imageclip***

It is often useful to display only a portion of an attached image in your drawing. You can accomplish this with *Imageclip*.

A. Begin a *New* drawing and select *Start from Scratch, Imperial* default settings. Use the *Image* command and select *Attach....* Locate the **WORLDMAP**.**TIF** file from the **AutoCAD 2004/Sample/VBA** folder, then select *Open*. In the *Image* dialog box, ensure *Retain path* is checked. Ensure *Specify on screen* is checked in all boxes and attach the watch image at an *Insertion point* of **2,2**, a *Scale factor* of **5**, and *Rotation angle* of **0**.

B. Now use the *Image* command again to attach the watch a second time to your drawing at the point **8,2**. Use the same *Scale factor* and *Rotation angle* as the first attachment, but this time specify these parameters in the *Image* dialog box.

C. Issue the *Image* command again and notice that although you attached the WORLDMAP.TIF image twice it is referenced only once in the list of attached images. Select *Cancel* in the *Image Manager*.

D. Now use the *Imageclip* command to create the two clipping boundaries as shown in Figure 32-22.

E. After you have created the clip boundaries, use the *Imageclip* command again to turn the boundaries *OFF* and then back *ON* again. You will find that the clip boundary is saved even though you turn it *OFF* to display the entire image. Save this drawing as **CLIPPING**.

Figure 32-22 ——————————————————

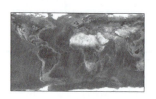

First attachment of WORLDMAP.TIF with a rectangular clip boundary turned On.

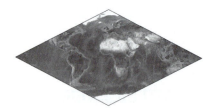

Second attachment of WORLDMAP.TIF with a polygonal clip boundary turned on.

3. ***Draworder***

Raster images are also useful for reference purposes during drawing construction. You can utilize aerial photographs, for example, in order to construct your drawing in relation to existing structures. When using images as reference information, you often need to reorder the display of objects on your drawing.

A. Begin a *New* drawing and select *Start from Scratch, Imperial* default settings. Use the *Image* command and select *Attach*. Locate and open the **BIGLAKE**.**TGA** file from the **Textures** folder. (To find where your *Textures* folder is located, use *Options, Files* tab, then expand *Texture Maps Search Path.*) Attach the image to your drawing at **2,2** with a *Scale* factor of **5** and *Rotation* angle of **0**

B. Use *Imageadjust*, select the image, and set the *Brightness* to **75**.

C. Using the *Rectangle* command, draw a closed object as shown in Figure 32-23. Use the *Hatch* command to hatch the object with a *Solid* pattern as shown.

Figure 32-23

Draw an enlcosed boundary
using the Rectangle command.

Hatch the enlcosed boundary
using a solid hatch pattern.

D. Notice that the solid hatch pattern is displayed on top of the lake image. Now use *Draworder*, select the solid hatch pattern, and send it to the *Back*. It is no longer visible on screen since it is now in back of the lake image (you may have to *Regen*). You can reverse the display of these two objects by using *Draworder* again, and selecting the lake image (by selecting its frame), and sending it to the *Back* of the display order. Save this drawing as **DWGORDER**.

4. *Image Found At*

You might encounter situations when you already have an image attached to your drawing and you need to redisplay a more current or totally different version of the image on your drawing.

A. *Open* the drawing called **CLIPPING**. Use the *Image* command and select **WORLDMAP** from the list of attached images. Choose the *Browse* button in the *Image Found At* section, and *Open* the **8PEOP08L.TGA** file from the **Textures** folder. Then select *Open* to close the dialog box.

B. The WORLDMAP image name, attached at two different locations, now displays the contents of the 8PEOP08L file. You did not use the *Save Path* option in the *Image Manager*, so this is not a permanent change of the image picture. *Close* the drawing but <u>do not save </u>the changes. Use *Open*, then select the same CLIPPING drawing to *Open*. Notice that the WORLDMAP.TIF file is still the image that is displayed for the WORLDMAP attachment name.

World Boundary Map
Digital Chart of the World
Coordinate System: Latitude/Longitude
Drawing Unit: Decimal Degrees
Constructed from digital maps provided by
American Digital Cartography, Inc.
115 W. Washington Street
Appleton, WI 54911-4775
Phone: (920) 733-6678
Copyright 1994, American Digital Cartography, Inc.

MAP2GLOBE.DWG Courtesy, Autodesk, Inc.

33

ADVANCED LAYOUTS AND PLOTTING

Chapter Objectives

After completing this chapter you should:

1. be able to use the *Layer Properties Manager* or the *Vplayer* command to control layer visibility in viewports;

2. be able to set *PSLTSCALE* for multiple viewports and layouts;

3. know how to dimension in paper space;

4. know the "Reverse Method" for determining the viewport drawing scale factor and how to use it to specify size-related settings;

5. know several applications for creating multiple viewports and layouts and considerations for doing so;

6. know the difference between color-dependent and named Plot Style Tables;

7. be able to create a new Plot Style Table using *Add-A-Plot Style Table Wizard*;

8. be able to edit a Plot Style Table using the Plot Style Table Editor;

9. be able to assign named plot styles to layers and objects;

10. know how to apply plot stamps to your prints and plots.

CONCEPTS

This chapter is concerned with advanced concepts of using layouts and plotting those layouts—specifically creating multiple viewports per layout, creating multiple layouts, attaching Plot Style Tables to layouts, and assigning plot styles to layers and objects. To isolate these complex ideas as an attempt to simplify their explanation, the chapter is divided into two sections: Advanced Layouts and Advanced Plotting.

Topics discussed in this chapter are arranged as follows:

Advanced Layouts

> Layer Visibility for Viewports (*Layer, Current VP Freeze, New VP Freeze, Vplayer*)
> Linetype Scale in Viewports (*Psltscale*)
> Dimensioning in Paper Space
> The "Reverse Method" for Calculating Drawing Scale Factor
> Applications for Multiple Viewports
> Considerations for Using Multiple Viewports and Layouts

Advanced Plotting

> Plot Style Tables and Plot Styles
> Plot Stamping

ADVANCED LAYOUTS

You are already familiar with creating and setting up drawings (Chapters 6 and 12), creating layouts and viewports (Chapter 13), and plotting layouts (Chapter 14). The ideas and examples used in those chapters are fundamental to your ability to set up and create drawings in AutoCAD; however, they were based on the premise of using only one large viewport per layout. The first half of this chapter deals with concepts related to setting up multiple viewports and multiple layouts.

To this point, you have learned the following. Objects that represent the subject of the drawing (model geometry) are normally drawn in model space. Dimensioning and related annotation (text) are traditionally performed in model space because they are directly associated to the model geometry. A layout simulates the sheet of paper you will print or plot on, so it is used to prepare the desired display of the model geometry for plotting. A viewport is created in a layout with the *Vports* command in order to "see" the model geometry. The display of the geometry in the viewport should be scaled to achieve an appropriate scale for the drawing. Since a layout represents the plotted sheet, you normally plot the layout at 1:1 scale. Therefore, the geometry displayed in the viewports is scaled to the desired plot scale by using the *Viewports* tool bar, the *Properties* palette for the viewport, or a *Zoom XP* factor. With the layout, you can use the *Page Setup* and *Plot* dialog boxes to specify and save the print or plot parameters such as plot device, paper size, and orientation, etc.

Now that you are experienced with those fundamentals, consider the following possibilities. There are many applications for creating multiple viewports per layout and/or creating multiple layouts in one drawing, each displaying different drawing geometry or the same geometry at different scales. For example, you may want to create one plot to show several views of one drawing, such as an overall view of an assembly and several "detail" views of individual components. Or you may want to show a floor plan and several views of individual rooms. You may have the need to show several separate drawings on one plot, which can be accomplished by *Xrefing* several drawings, creating several viewports, and controlling each viewport to display only the layers associated with one *Xref.* In another application, you may need to set up multiple layouts for creating multiple sets of plots, each layout displaying a different combination of layers. Using this scheme, for example, you can create and save a layout to plot the floor plan and the plumbing details for the plumbing contractor, a layout to plot the floor plan and the electrical layout for the electrical contractor, a layout for the floor plan and the HVAC layout, and so on.

When using multiple viewports and multiple layouts, the setup becomes more complex but offers more alternatives. Remember, you must scale the display of the geometry in each viewport and you must control the layer visibility in each viewport (which layers are visible in which viewports). When you have multiple viewports, you cannot effectively use the "drawing scale factor" to determine the viewport scale since that idea is based on preparing the drawing for only one large viewport. In addition, you cannot effectively control which layers are visible in which viewports using the global *Freeze/Thaw* or *On/Off* controls.

This chapter explains how you can control the viewport-specific layer visibility using the *Current VP Freeze* button and *New VP Freeze* button in the *Layer Properties Manager* or *Layer Control* drop-down list. In addition, another strategy for calculating the drawing scale factor for viewports is discussed. The "reverse method" is used to help determine appropriate values for *Ltscale*, *Dimscale*, text height, etc. based on the viewport scale.

This chapter also presents a totally different strategy you may want to consider when using multiple viewports and layouts, that is, creating dimensions in paper space, creating text in paper space, and setting *Psltscale* to paper space. This strategy prevents you from having to calculate and set the text height, *Dimscale*, and *Ltscale* for each viewport and having to control the layer visibility specific to each viewport.

Historically, the use of paper space, viewports, and layouts has changed from early AutoCAD releases having only a few capabilities (and therefore few applications) to current releases offering advanced capabilities and applications. In contrast to pre-AutoCAD 2000 releases, the common use of paper space, viewports, and layouts represents a major paradigm shift in the way drawings are prepared and plotted. Although this technology continues to advance and become more "user-friendly," these concepts are not simple to understand nor easy to carry out. Because there are currently different applications, strategies, and personal preferences for creating and setting up multiple viewports and layouts, this chapter presents the alternatives and discusses the advantages and disadvantages of each. In this way, you can make the appropriate decisions based on your specific applications, preferences, and level of technology.

Layer Visibility for Viewports

AutoCAD provides you with the capability to create multiple layouts. In each layout, you can create one or several viewports using *Vports* or *–Vports.* Even though you can have multiple viewports in a drawing, each viewport displays the same model space geometry, right? Well, yes and no. That is, there is only one model space, and by default the same model geometry is displayed in every viewport. You can, however, control what parts of the model space geometry are displayed in each viewport by (1) using *Zoom, Pan,* and other viewing commands to display different areas of model space and (2) using layer visibility to control what layers appear in which viewports. This section discusses layer visibility control for viewports.

LAYER

	Pull-down Menu	COMMAND (TYPE)	ALIAS (TYPE)	Short-cut	Screen (side) Menu	Tablet Menu
	Format Layer...	*LAYER* or *-LAYER*	*LA* or *-LA*	...	*FORMAT Layer*	*U,5*

The *Layer* command produces the *Layer Properties Manager* dialog box. This dialog box can be used to control which layers are visible in which viewports. Two small icon buttons in the *Layer Properties Manager* dialog box are used for viewport-specific layer visibility, *Current VP Freeze* and *New VP Freeze*.

Figure 33-1 displays the *Layer Properties Manager* showing the *Current VP Freeze* and *New VP Freeze* columns (last two columns). <u>These icons do not appear when the *Model* tab is current.</u> If a layout tab is selected and a viewport is active (the cursor appears in a viewport), use these icons to freeze or thaw layers <u>in the current viewport</u> or to freeze or thaw layers <u>in new viewports</u> (the selected layer will be frozen or thawed for any new viewports that are created).

Figure 33-1 ───────────

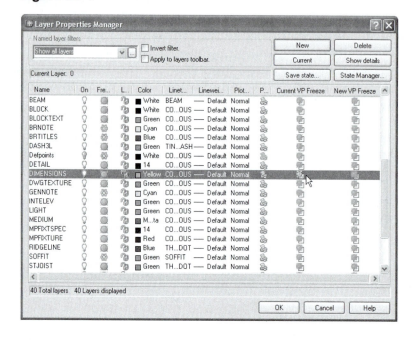

Current VP Freeze
For example, assume you have two viewports created in a layout and each viewport displays the same model space geometry (Fig. 33-2). Notice that in the left viewport all layers are visible compared to the right viewport, in which the dimension layer is not visible. To freeze the "Dimensions" layer in the right viewport only, first make the right viewport current, then use *Layer* to produce the *Layer Properties Manager* and select the *Current VP Freeze* icon to change this layer to a frozen state <u>for that viewport</u> (snowflake; see Figure 33-1). If the left viewport were made active and then the *Layer Properties Manager* invoked, these icons for all layers would appear thawed (sunshine) for the current (left) viewport.

Figure 33-2 ───────────

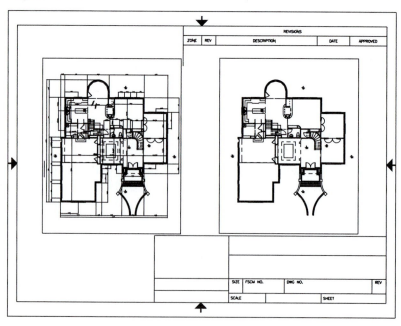

When a viewport is current, the *Current VP Freeze* icons that appear give the visibility state of layers <u>specific to the viewport that is current</u> when the list is displayed. In other words, the icons (*Current VP Freeze*) may display different information depending on which viewport is current when the list is viewed. Changing the *Freeze/Thaw* icon affects the layer's state globally (for all viewports in the drawing), while changing the *Current VP Freeze* icon affects the layer's state only <u>for the current viewport</u>.

A *Layout* tab itself is considered a paper space viewport when dealing with layer visibility. That is, if you are in paper space (*Pspace* is active, not in a viewport) and you PICK the *Current VP Freeze* icon, the highlighted layer becomes frozen for the entire layout (the current viewport), but not in the individual viewports or in the *Model* tab.

Since AutoCAD 2002, a *Freeze or Thaw in Current Viewport* icon appears in the *Layer Control* drop-down list (Fig. 33-3). This icon has the identical function as the *Current VP Freeze* icon in the *Layer Properties Manager*. This feature was not available in AutoCAD 2000. However, the *New VP Freeze* icon that appears in the *Layer Properties Manager* does not exist in the *Layer Control* drop-down list.

Figure 33-3

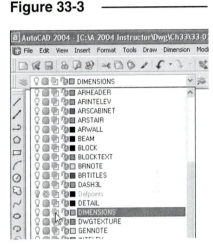

New VP Freeze

This icon prevents the selected layer from appearing in any viewports that are created from that time on. Any layers selected with the *New VP Freeze* attribute do not appear in subsequently created viewports.

This feature is helpful when you need several viewports, each to display a separate drawing. For example, assume you want to *Xref* several drawings, but only want one drawing to appear in each viewport. Assume you created one viewport and *Xrefed* a drawing into model space so it (all its layers) appeared in the viewport (Fig. 33-4). If you were to create a second viewport, the same drawing would normally appear in the second viewport just as all model space geometry normally appears in all viewports. If you were to *Xref* a second drawing into model space, it would also normally appear in <u>both</u> viewports. (See Chapter 30, Xreferences, for details on *Xref*.)

Figure 33-4

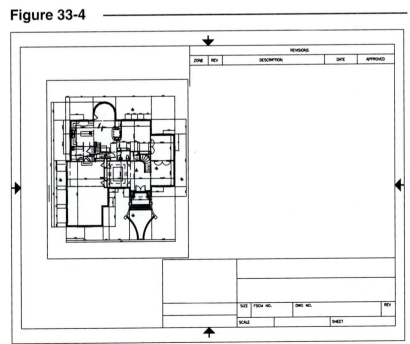

To avoid this problem, after you create one viewport and *Xref* the first drawing but before you create the second viewport, use *New VP Freeze* for all layers of the *Xref* drawing (Fig. 33-5, all "Xref1|*" layers). (Unlike the *Current VP Freeze* icon, the *New VP Freeze* icon displays the same information for a specific layer no matter which viewport is current when the list is displayed.)

Figure 33-5

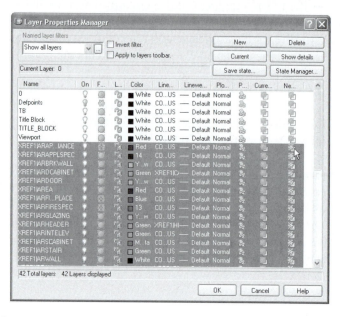

Next, create the second viewport. The original *Xref* drawing does not appear in the second viewport (right) since all of its layers are frozen for that viewport and all other new viewports (Fig. 33-6).

Figure 33-6

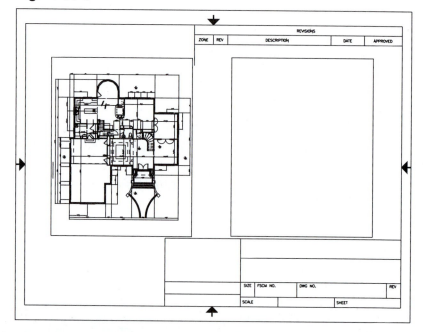

VPLAYER

Pull-down Menu	COMMAND (TYPE)	ALIAS (TYPE)	Short-cut	Screen (side) Menu	Tablet Menu
...	*VPLAYER*	...	...	...	...

As an alternative to using the *Layer Properties Manager* or the *Layer Control* drop-down list, the *Vplayer* (Viewport Layer) command provides options for control of the layers you want to see in each viewport (viewport-specific layer visibility control). With *Vplayer* you can specify the state (*Freeze/Thaw* and *Off/On*) of the drawing layers in any existing or new viewport. Although the *Layer* command allows you to *Freeze/Thaw* layers <u>globally</u> and for <u>viewports</u>, *Vplayer* allows you to *Freeze/Thaw* only the <u>viewport-specific</u> layer setting.

A *Layout* tab must be set current for the *Vplayer* command to operate, and a layer's global (*Layer*) settings must be *On* and *Thawed* to be affected by the *Vplayer* settings. You can use *Vplayer* while you are in paper space or in any viewport. The simplest method, however, is to make the desired viewport current, then use *Vplayer*.

For example, you may have two viewports set up in a layout, such as in Figure 33-7. Assume you want to *Freeze* the DIM (dimensioning) layer for the right viewport, so you make that the current viewport (your cursor is in the right viewport). In order to control the visibility of the DIM layer for the current viewport, the command syntax shown below would be used.

Figure 33-7

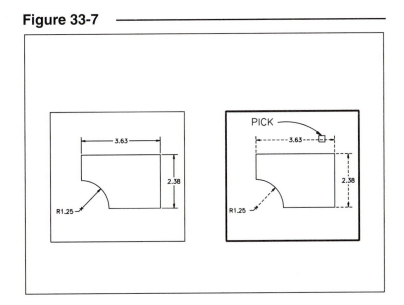

```
Command: vplayer
Enter an option [?/Freeze/Thaw/Reset/Newfrz/Vpvisdflt]: f
Enter layer name(s) to freeze or <select objects>: Enter
Select objects: (Select an object on the desired layer. See Figure 33-7.)
Select objects: Enter
Enter an option [All/Select/Current] <Current>: Enter
Enter an option [?/Freeze/Thaw/Reset/Newfrz/Vpvisdflt]: Enter
Command:
```

This action results in *Freezing* the DIM layer (containing the highlighted objects) in the right viewport.

You <u>do not</u> have to make the desired viewport current before using *Vplayer*—your cursor can be in paper space or in another viewport. Invoke *Vplayer* at any time. Assume that you are in the left viewport, but want to *Freeze* the DIM layer for the right viewport. In the command sequence on the next page, notice that AutoCAD allows you to switch to paper space to PICK the desired <u>viewport object</u> (viewport borders can be PICKed only from paper space). This sequence results in *Freezing* the DIM layer for the right viewport (Fig. 33-8).

Figure 33-8

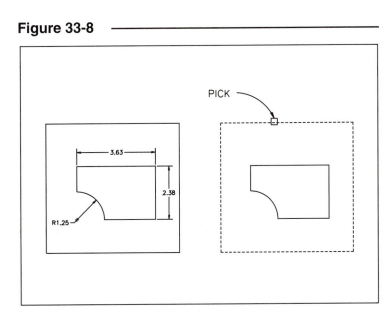

```
Command: vplayer
Enter an option [?/Freeze/Thaw/Reset/Newfrz/Vpvisdflt]: f
Enter layer name(s) to freeze or <select objects>: dim
Enter an option [All/Select/Current] <Current>: s
Switching to Paper space.
Select objects: PICK (Select the desired viewport object. See Figure 33-8.)
Select objects: Enter
Switching to Model space.
Enter an option [?/Freeze/Thaw/Reset/Newfrz/Vpvisdflt]: Enter
Command:
```

Note that the *Select* option switches to paper space if necessary so that the viewport objects (borders) can be selected. The *Freeze* or *Thaw* does not take place until the command is completed and the drawing is automatically regenerated.

The *Vplayer* **Current** option automatically selects the current viewport. The **All** option automatically selects all viewports. The other options are explained next.

?

This option lists the frozen layers in the current viewport or a selected viewport. AutoCAD automatically switches to paper space if you are in model space to allow you to select the viewport object.

Freeze

This option controls the visibility of specified layers in the current or selected viewports. You are prompted for layer(s) to *Freeze*, then to select which viewports for the *Freeze* to affect.

Thaw

This option turns the visibility control back to the global settings of *On/Off/Freeze/Thaw* in the *Layer* command. The procedure for specification of layers and viewports is identical to the *Freeze* option procedure.

Reset

Reset returns the layer visibility status to the default if one was specified with *Vpvisdflt*, which is described next. You are prompted for "Layer(s) to Reset" and viewports to select by "All/Select/<Current>:".

Newfrz

This option creates new layers. The new layers are frozen in all viewports. This is a shortcut when you want to create a new layer that is visible only in the current viewport. Use *Newfrz* to create the new layer, then use *Vplayer Thaw* to make it visible only in that (or other selected) viewport(s). You are prompted for "New viewport frozen layer name(s):". You may enter one name or several separated by commas.

Vpvisdflt

This option allows you to set up layer visibility for new viewports. This is handy when you want to create new viewports but do not want any of the existing layers to be visible in the new viewports. This option is particularly helpful for *Xrefs*. For example, suppose you created a viewport and *Xrefed* a drawing "into" the viewport. Before creating new viewports and *Xrefing* other drawings "into" them, use *Vpvisdflt* to set the default layer visibility to *Off* for the existing *Xrefed* layers in new viewports. (See the example for *Layer, New VP Freeze*.) You are prompted for a list of layers and whether they should be *Thawed* or *Frozen*.

Linetype Scale in Viewports

Layouts give you the capability of creating several viewports that look into model space. These viewports can contain several views of one or more drawings at different scales. Because each viewport can display linetype scaling differently based on viewport scale, the *PSLTSCALE* variable is needed to control how multiple linetype scales appear in one layout.

PSLTSCALE

Pull-down Menu	COMMAND (TYPE)	ALIAS (TYPE)	Short-cut	Screen (side) Menu	Tablet Menu
Format Linetype... Details >> Use paper space units	PSLTSCALE	...	...	*FORMAT Linetype... Details >> Use p.s. units*	...

LTSCALE controls linetype scaling <u>globally</u> for the drawing, whereas the *PSLTSCALE* variable controls the linetype scaling for non-continuous lines for viewing in <u>paper space</u>. *LTSCALE* is a variable that accepts any value other than 0, whereas *PSLTSCALE* is a variable that accepts either a 0 (*Off*) or 1 (*On*). *LTSCALE* can still be used to change linetype scaling globally regardless of the *PSLTSCALE* setting.

If you want the linetype scale to always appear the same with respect to the drawing geometry, set *PSLTSCALE* to 0. This ensures that non-continuous (dashed) lines have a constant scale (based on the *LTSCALE*) with respect to the geometry regardless of how those lines are viewed. In other words, assume you created a hidden line between two points such that the line contained three short dashes. When *PSLTSCALE* is 0, you would always see three dashes whether they were viewed from the *Model* tab or whether the line appeared in two different viewports at two different scales. Even when you *Zoomed*, the line would have three dashes. With *PSLTSCALE* set to 0, the scale of non-continuous lines is controlled only by *LTSCALE*. A *PSLTSCALE* setting of 0 is suggested for drawings using only one viewport or when multiple viewports or layouts display the geometry at one scale.

If *PSLTSCALE* is set to 0, linetype spacing is controlled exclusively by the *LTSCALE* variable and is relative to the drawing units in model space (or in paper space when lines are created there). In other words, if *PSLTSCALE* is 0, the dashed lines would always have the same number of dashes for one *LTSCALE* setting. However, when the same lines are viewed in different viewports in different scales, the linetypes appear different lengths as in Figure 33-9.

Figure 33-9

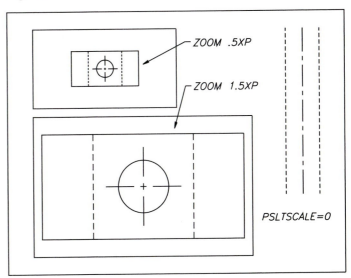

ZOOM .5XP

ZOOM 1.5XP

PSLTSCALE=0

A *PSLTSCALE* setting of 1 (the default setting) is recommended when you are viewing the same geometry in multiple viewports at different scales. A *PSLTSCALE* setting of 1 makes all non-continuous lines appear to have the same linetype scaling, regardless of viewport scale (or *Zoom XP*) settings. Technically, *PSLTSCALE* set to 1 displays all non-continuous lines relative to paper space units so that all lines in paper space and in different viewports appear the same. Therefore, the model geometry in viewports can have different viewport scales (or *Zoom XP* factors), yet the linetypes display with the same scale (Fig. 33-10).

Figure 33-10

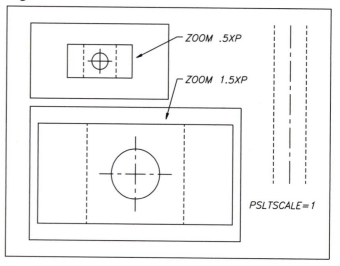

Set *PSLTSCALE* at the Command prompt by entering "Psltscale" and typing 1 or 0. *PSLTSCALE* can also be set using the *Linetype* command which produces the *Linetype Manager* (Fig. 33-11, lower left). You must use the *Show Details* button for this feature to appear. A check appearing in the *Use paper space units for scaling* checkbox sets *PSLTSCALE* to 1. The *Global scale factor* represents the *LTSCALE* value and the *Current object scale* represents the *CELTSCALE* value (see Chapter 12 for more information). If *PSLTSCALE* is changed, a *Regenall* must be used to display the effects of the new setting.

Figure 33-11

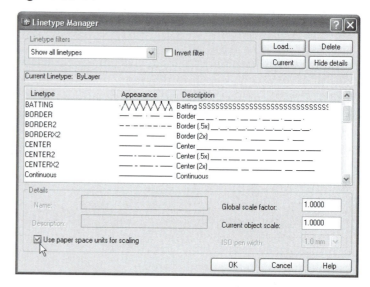

As an example, consider the crankshaft drawing (Fig. 33-12) that appears in its entirety in the top viewport at 1:1 and as a detail in the lower viewport at 2:1. Notice that the hidden line dashes are almost imperceptible in the top viewport, whereas the hidden lines in the lower viewport are much clearer. In this case, *PSLTSCALE* = 0, so the linetype scale is proportional to the model space geometry, no matter at what scale it is viewed.

Figure 33-12

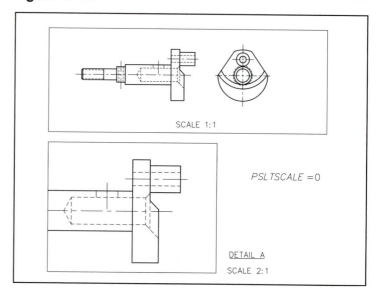

In contrast, when *PSLTSCALE* is set to 1 (Fig. 33-13), hidden line dashes in both viewports appear the same—they are automatically adjusted for the viewport scale. Another way to grasp this is that the linetype scale is proportional to paper space units, not model space units. In most cases where a layout contains multiple viewports, a *PSLTSCALE* setting of 1 is more practical.

Figure 33-13

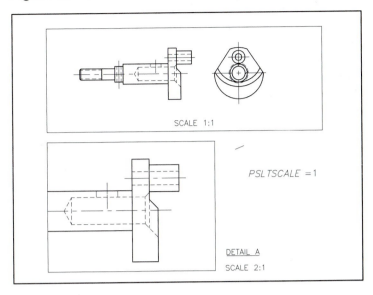

Dimensioning in Paper Space

When a drawing is presented in one viewport, or is presented in more than one viewport but at the same scale, dimensioning in <u>model space</u> is recommended. In these cases dimensioning in model space is the simplest and most versatile option. However, in cases where a drawing is displayed in multiple viewports at different scales, dimensioning is more complex. Even though the geometry appears in different scales, dimensions should always appear at the same size (dimension text is typically .125″ or 6mm).

There are two basic alternatives for creating dimensions displayed in multiple viewports at different scales: 1) put different sets of dimensions in <u>model space</u> on different layers and with different *Overall scales* (dimension styles), then control the viewport-specific visibility so the correct dimensions appear in the appropriate viewports, and 2) create the dimensions in <u>paper space</u> all in one size (for AutoCAD 2002 and 2004 only). (See Chapter 29 for basic concepts on dimensioning in paper space.)

As an example for dimensioning in paper space, assume you created a mechanical drawing of a plate to be stamped (Fig. 33-14). You chose to show an overall view of the plate and a detail view giving the dimensions specific to the small notch. Therefore, a layout with two viewports is created, with the viewport scale for the full view set to 1:1 (left viewport) and viewport scale for the detail view set to 2:1.

Figure 33-14

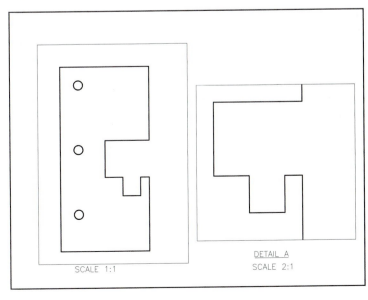

Assume you then activate the *Model* tab and begin to dimension the object as you would normally. Checking the layout to see how the first few dimensions look, you notice that the dimensions appear in both viewports, and those in the detail view are also scaled 2:1 with the geometry (Fig. 33-15).

Figure 33-15

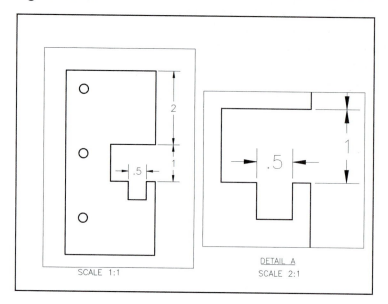

As one solution to the problem, you erase all the model space dimensions. Then, in the layout, you begin dimensioning <u>in paper space</u> with the dimensions associated with (attached to) the model space geometry (Fig. 33-16). You can use the same dimension style as before with an *Overall scale* of 1, since the layout will be printed 1:1. All dimensions have the same size, since they are all in the same space. Note that it <u>appears</u> that some dimensions are actually inside the viewport while others fall outside the viewport (when the viewport layers are frozen, this is not noticeable). In addition, dimensions attached to the geometry in the detail view reflect the actual model space measurement even though the geometry is scaled differently. Dimensions for the notch and for the overall part are placed in the appropriate viewport.

Figure 33-16

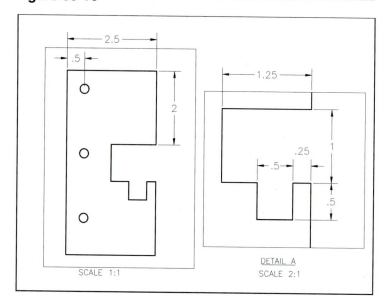

As the last step, the viewport borders layer is *Frozen* and detail area is indicated in the full view. Dimensioning in paper space allows you to dimension a drawing shown in different viewports at different scales.

Figure 33-17

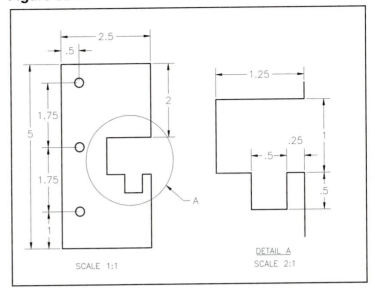

Although dimensioning in paper space appears to be a simple solution to the complexities that arise when using multiple viewports, it is not always as simple as the previous example. Keep in mind that you may need more than two viewports or more than one layout. Some advantages and disadvantages of dimensioning in paper space are listed below.

Advantages to creating dimensions in a paper space layout include:
1. you can use an *Overall scale* of 1, so calculation of the drawing scale factor (for dimensions, at least) is simpler;
2. you can attach dimensions to objects in multiple viewports at different scales and all dimensions appear at one size, so you do not necessarily need more than one dimension style for one layout; and
3. layer control is simplified, since all paper space dimensions appear in the layout and do not appear in other layouts.

Disadvantages to dimensioning in paper space include:
1. each viewport and each layout requires its own set of dimensions, so you may have to draw all dimensions for every viewport and for every layout, even when you show the same geometry;
2. you cannot *Copy* associative dimensions effectively from one viewport to another since copies are not associative;
3. you cannot copy associative dimensions effectively from one layout to another layout unless you copy the entire layout with the *Layout* command;
4. paper space dimensions are not useable if the drawing is used as an *Xref*, since only the model geometry of an *Xref* is displayed; and
5. truly associative dimensioning in paper space is available only in AutoCAD 2002 and 2004, so you cannot exchange drawings effectively using AutoCAD 2000 or older.

For more ideas and a more complete discussion of this topic, see "Applications for Multiple Viewports."

The ability to create paper space dimensions displaying model space measurements is available only since AutoCAD 2002. In previous releases of AutoCAD, dimensioning in paper space resulted in values measured in paper space units. Therefore, for example, the notch shown in the detail view in Figure 33-17 measuring .5 x .5 would be measured as 1 x 1 by a paper space dimension since the geometry is shown at 2:1 scale. In AutoCAD 2000 and previous releases, this condition could be corrected by changing the *DIMLFAC* variable to serve as a multiplier for the paper space units.

The "Reverse Method" for Calculating Drawing Scale Factor

Chapter 12 in this text gives the steps for drawing setup, including an explanation for calculating the "drawing scale factor" (DSF). Briefly, the DSF is determined when you set the *Limits*. Set the *Limits* to the sheet size you intend to print or plot on times some factor that will provide adequate space for you to draw the subject of the drawing full size. This factor, or multiplier, is the drawing scale factor. The reciprocal of the DSF is the plot scale. For example, if you plan to print on an 11 x 8.5 sheet and need enough space to draw an object 36" long, set the *Limits* to 11 x 8.5 times 4, or 44 x 34. So, the drawing scale factor is 4. The plot scale is then 1:4.

The drawing scale factor is important because there are many size-related variables and features (many that are set to a value of 1) that should change based on the size of the drawing and the intended plot scale. For example, the drawing scale factor gives you a guideline for the initial settings for *Ltscale*, dimension *Overall scale*, hatch pattern scale, text height, and so on.

This principle is useful for drawings that are to be set up in the model tab or in a layout with one viewport that occupies the sheet size (printable area of the layout). However, when you plan to have multiple viewports, this scheme does not work effectively. As a matter of fact, setting *Limits* is not a necessary step to begin drawing—setting *Limits* only prepares the drawing area with enough space to "see" those first several lines. Nevertheless, the method described in Chapter 12 does give you an understanding of the relationships and some guidelines to use for the initial settings.

Regardless of whether you plan to use only one viewport or multiple viewports, you need some method to help you determine how to set sizes of such things as text, dimension *Overall scale* (assuming you intend to create text and dimensions in model space), hatch pattern scale, etc. A different method for calculating drawing scale factor, called the "reverse method," is given here. The reverse method works well when:

· you plan to display the drawing in multiple viewports at different scales;
· you plan on using one viewport to fill the layout, but have not yet decided on a plot device or sheet size; or
· you are ready to begin drawing, but do not yet know some information such as what plot devices or sheet sizes you will use, what sizes you will create the viewports, or what scales you will display the geometry.

Steps for Calculating the Drawing Scale Factor by the "Reverse Method"

This method works by finding the viewport scale by trial-and-error using the *Viewport Scale* drop-down list. The viewport drawing scale factor is the reciprocal of the viewport scale.

1. Determine the approximate space you will need to create the subject of the drawing full size in the actual units. Set the *Limits* accordingly, but only to provide enough space. *Limits* do not have to be in any proportion or size. *Zoom All*.

   (As an alternative, you do <u>not</u> have to set the *Limits*. You can instead, *Zoom* out, estimating about how much area you will need to "see" the entire drawing.)

2. Begin creating the model geometry. Draw at least an outline of the area you'll need, "box in" the views, or draw the parts of the geometry that occupy the majority of the drawing area. You can continue to draw the geometry, until completion if desired, or at least until you are ready to dimension, hatch, add text, or for some reason require a drawing scale factor.

3. Activate a *Layout* tab. Use *Pagesetup* to assign a plot device and a sheet size. Activate the *Viewports* toolbar (including the *Viewport Scale* drop-down list. Use *Vports* to create the viewport or viewports that you need.

4. Activate the viewport for which you want to determine the drawing scale factor. *Zoom Extents*, or use real-time *Zoom*, to display the approximate area of the drawing you want in the viewport. Use one of the layer tools to set the desired viewport-specific layer visibility to ensure all objects that you intend to show appear in the viewport.

5. Use the *Viewport Scale* drop-down list to set (by trial and error) an appropriate scale for the viewport. You may need to *Pan* occasionally to "see" the desired area of the drawing. (If you are planning to create dimensions in model space or paper space, ensure you allow enough room either inside or outside the viewport.) Once the viewport scale is set, the drawing scale factor for that viewport is the reciprocal of the scale appearing in the *Viewport Scale* box. (Optional—once you have settled on the desired scale for the viewport, activate paper space, highlight the viewport border, and *Lock* the viewport using the *Properties* palette.)

6. To determine the <u>drawing scale factor for the viewport</u>, convert the viewport scale (calculate the reciprocal). Proportional scales (such as 1:2, 1:4, 1:8) can be converted directly by using the second number as the DSF (2, 4, 8, respectively). For feet and inch scales (for example, 1/4"=1'), multiply the denominator (bottom number of the fraction) by 12 (such as 4 x 12 = 48).

   As an alternate method, activate paper space if not already active. Highlight the viewport border and use the *List* command to give the "Scale relative to Paper space: *n.nnnn* xp" (where *n.nnnn* represents a value). This value is the viewport scale as a decimal value. To find the drawing scale factor for the viewport, calculate 1/*n.nnnn*. This is the drawing scale factor for that viewport.

7. Use the viewport drawing scale factor to set the size-related scales such as dimension *Overall scale* (if you are dimensioning in model space), *Ltscale* (if this is the main viewport), hatch pattern scale, text size (for model space text), and so on. For dimensions and text you create for the viewport in model space, use layers named DIM-1, TEXT-1, or similar, so visibility can be controlled for this and other viewports.

8. Repeat steps 4 through 7 for each viewport that will display the geometry at a different scale.

This method can be used for any case when you plot from a layout, including when you have only one viewport that occupies the entire layout.

Applications for Multiple Viewports

The capability of AutoCAD to create multiple viewports in one layout and multiple layouts in one drawing offers many possibilities for displaying a drawing. Several features of the same geometry can be shown using multiple viewports in concert with layer control for the viewports. In addition, one drawing can be printed and plotted in multiple ways using multiple layouts. It is even possible to print or plot several drawings in one layout using *Xrefs*.

A few examples of typical applications that make use of multiple viewports and/or layouts are:

· architectural drawings that display several aspects of one drawing, such as an overall view of a floor plan and enlarged views of particular rooms or areas
· mechanical drawings that display one view of the entire part or assembly and additional detail views of particular aspects of the part or individual components of the assembly

- civil engineering or construction drawings that show an overall plan and one or more detailed views of a specific aspect of the site or structure
- 3D mechanical drawings that show several views of one part or assembly, each view showing a different side or "view" of the 3D model, as in a multiview drawing (see Chapter 42)
- multiple architectural, engineering, construction, or design drawings displayed on one sheet, accomplished by *Xrefing* the drawings and displaying one drawing per viewport

Three examples of using multiple viewports or layouts are given next: 1) using two layouts to plot the same geometry at different scales, 2) using one layout to show the same geometry at different scales, and 3) using one layout to display *Xref* drawings in multiple viewports at different scales.

Using Two Layouts to Plot the Same Geometry at Different Scales

For the first problem, consider the drawing shown in Figure 33-18. This mechanical part was originally set up to plot only on a "C" size engineering sheet (22" x 17"). *Limits* were set to 22 x 17 so the geometry could be drawn full size, then the layout was set up with an ANSI C title block. Since the *Limits* size equaled the sheet size, viewport scale was set to 1:1 and the layout was plotted at 1:1. *Dimscale, Ltscale,* and other size-related variables were set to 1. Dimensions were created in model space. The resulting plot is a full size drawing of the part. The dimension text and other text as well as hidden line dashes are plotted at the ANSI standard .125" (1/8").

Figure 33-18 ───────────────────────────────────

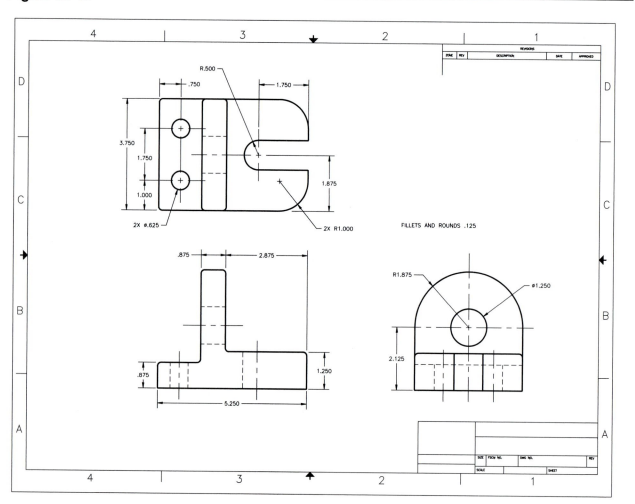

Assume the client calls and requests a fax of the drawing. You are now faced with printing the drawing on an ANSI "A" size 11 x 8.5 sheet in order to send the fax. A new layout is created with an ANSI A title-block and the geometry is scaled to fit the viewport at .375, or 3/8"=1" scale. However, making a print at this scale generates dimension text, other text, and hidden line dashes at about .047" (3/64")—much too small to be read easily. Here are several possible solutions.

Solution 1

Display the same model space dimensions in the new layout at a new *Overall scale*. In this solution, you have to make several changes such as setting the dimension *Overall scale* and *Ltscale* to 2.666 (the reciprocal of .375) in order to make the text readable and hidden lines recognizable. In addition, the placement of several dimensions must be adjusted to accommodate the (relatively) increased text size (Fig. 33-19). One problem with this solution now is that the drawing is prepared and saved only for A size prints since the *Ltscale* and dimension scale have changed.

Figure 33-19

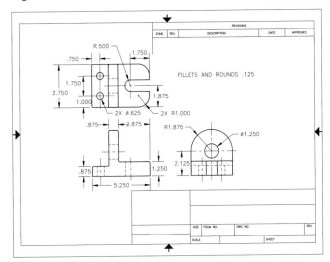

Solution 2

Create a duplicate file. Before making any changes (keeping the settings for the ANSI C layout) use *SaveAs*. Create the ANSI A layout and related settings in the new, separate drawing. Now you have a drawing prepared for A size prints and one for C size plots. The resulting problem is there are two drawings to update and maintain when needed.

Solution 3

Create a second layout. Create a <u>separate</u> set of the <u>model space</u> dimensions, and place them on a separate layer. Assign the new dimensions to a new dimension style and change the *Overall scale* appropriately for the new layout. Use viewport-specific layer visibility to display the correct dimension set in the appropriate viewports (layouts). Set *PSLTSCALE* to 1. The result is one drawing with all settings saved, ready to plot or print using either layout. The disadvantage is the amount of time to create, adjust, and manipulate visibility for the new dimensions.

Solution 4

Erase all model space dimensions. Set up the two layouts displaying the viewport geometry at different scales, and then create the dimensions in paper space. Two sets of dimensions are required, one for each layout. Set *PSLTSCALE* to 1. Again you have one drawing with all settings saved and ready to print or plot either layout. Once again, much time is needed to create the two sets of dimensions and the dimensions are unusable for new viewports, layouts, or *Xrefs*.

Solution 5

Assume that the dimensions were originally created in paper space. The existing ANSI C layout can be copied with the *Layout* command. This action gives you the needed second set of dimensions. However, these dimensions are <u>not associative</u>. The titleblock and viewport must be erased, the layout must be set up to the new device and paper size, and a new titleblock and viewport must be created. Then the existing (new) dimensions must be reassociated—almost as much trouble as creating all new dimensions from scratch. Once again, this solution has the same shortcomings as Solution 4.

Using Two Viewports to Display the Same Geometry at Different Scales

For the second application, consider the drawing of a plate (Fig. 33-20) showing a layout with two viewports, each viewport displaying some of the same geometry but at different scales. (This is the same example used to discuss creating dimensions in paper space.) Here, dimensions are <u>already created in model space</u> on a layer named DIM, or similar name. As you can see, the right viewport is intended as a detail view to show an enlarged view of the notched area of the plate at 2:1 scale. As a result, the dimensions are also shown at 2:1 scale.

Figure 33-20

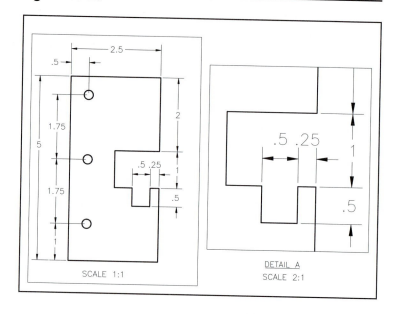

Solution 1

As shown in Figures 33-15, 33-16, and 33-17, *Erase* all of the model space dimensions and create new dimensions in paper space. See the example in "Dimensioning in Paper Space."

Solution 2

Erase <u>only</u> the model space dimensions for the notch (appearing in the detail view), and then re-create them in <u>paper space</u> (Fig. 33-21). Note that the dimensions for the detail have been erased (see left viewport), and then new dimensions for the detail view were added, but <u>placed in paper space</u> (right viewport). Even though these dimensions appear to be in the viewport, they are actually in paper space—"on top of" the viewport. Create these dimensions on a separate layer, for example, DIM-2. Note that part of one model space dimension (large arrowhead) appears in the detail view. To correct for this, the DIM layer is frozen for the right viewport only (*Current VP Freeze*). The DIM-2 layer does not have to be frozen for the left viewport, since these dimensions appear only in the layout. One shortcoming is that you cannot display any view of the drawing with all dimensions.

Figure 33-21

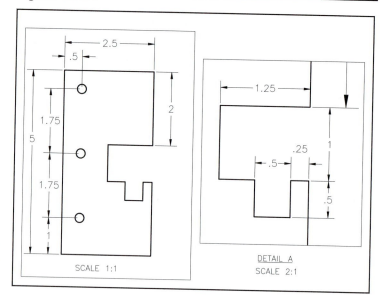

Solution 3
Similar to Solution 2, *Erase* only the model space dimensions for the notch. Re-create these dimensions in <u>model space</u>, but on a new layer and in a new dimension style with the *Overall scale* set to .5, since these dimensions will be displayed a 2:1. You can make the new dimension style by copying the existing one, then changing the *Overall scale*. Use viewport-specific layer visibility to display the appropriate dimensions in the appropriate viewports. In this case you can display all dimensions if needed, but at two different sizes. You must use *Matchprop* or a similar method to make all the dimensions the same scale if you need to show all in one view.

Solution 4
Use either Solution 2 or 3, but instead of erasing the dimensions for the notch, create a new layer named DIM-3 and move these dimensions to the new layer. This gives you the possibility of showing all dimensions in one view if needed. This is the only solution that provides a full set of model space dimensions.

Using One Layout to Plot *Xref* Drawings in Multiple Viewports at Different Scales

In this problem, one layout is created with three viewports to display and plot three *Xrefed* drawings at different scales. Although this sounds complex, it is a fairly straightforward procedure. The drawing typically contains no geometry (other than the *Xrefs*) and little or no text other than simple notes, title-block and border.

Begin a *New* drawing, immediately activate a *Layout* tab and use *Pagesetup*. Set up the layout for the desired plot device and sheet size. In most cases, a large sheet size is required since multiple drawings will be displayed and plotted. *Insert* the desired title-block and border. Create the first viewport with the *Vports* command on layer VIEWPORTS. *Save* the drawing.

Figure 33-22

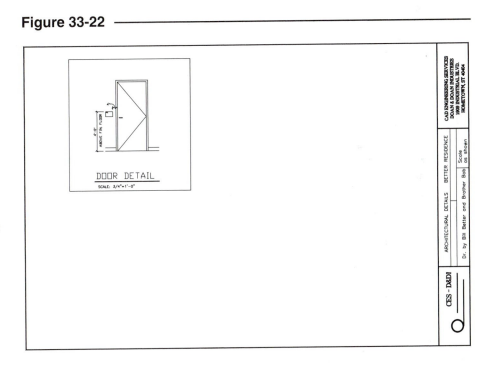

Before *Xrefing* the first drawing, it is important to set a common layer current, such as layer 0. (If layer VIEWPORTS is current when the drawings are *Xrefed* and that layer is later *Frozen*, the *Xrefs* will not appear.) Also, double-click inside the viewport since you want to *Xref* the drawing into model space, not paper space. Use *Xref* and *Attach* the desired drawing. In this example, the DOOR drawing is referenced (Fig. 33-22). Use the *Viewports* toolbar or other method to set the viewport scale.

A second viewport is created on layer VIEW-PORTS. As soon as the new viewport is created, the DOOR drawing appears in the new viewport. The layer is changed to layer 0, and *Xref* is used to *Attach* the second drawing—WALL, in this case (Fig. 33-23). Note that without controlling for viewport-specific layer visibility, both drawings appear in both viewports. With the second viewport still active (your cursor is inside the viewport), use the *Layer Control* drop-down list or the *Layer Properties Manager* to

Figure 33-23 ───────────

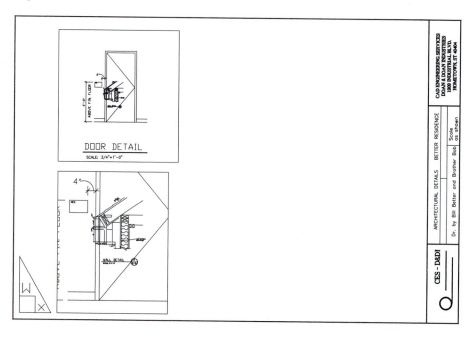

freeze the DOOR|* layers (*Current VP Freeze*). Next, click in the DOOR viewport and freeze the WALL|* layers. <u>Additionally</u>, to prevent the *Xref* drawings from appearing in any <u>new</u> viewports you create, select *New VP Freeze* for all the DOOR|* and WALL|* layers. Use the *Viewports* toolbar or other method to set the viewport scale in the new viewport.

As in the previous step, create a new viewport on the appropriate layer, reset the current layer, then *Xref* the third drawing (SECTION) into model space. In this case, the previous two drawings do not appear in the new viewport; however, the third drawing appears on all three viewports (not shown). Use *Current VP Freeze* to freeze the SECTION drawing in the first two viewports. Set the viewport scale for the last viewport. *Freeze* the VIEWPORTS layer (globally) to complete the drawing (Fig. 33-24).

Remember to set the *VIS-RETAIN* variable to 1; otherwise, none of the viewports' layer visibility settings will be saved with the drawing.

Since, in this case, it was known for each of the drawings at what scales they were to be plotted, the scale for each could be set independently using the viewport scale. If dimensions were created for these drawings in model space and the recommended methods for setting DSF were used, the dimensions could be displayed

Figure 33-24 ───────────

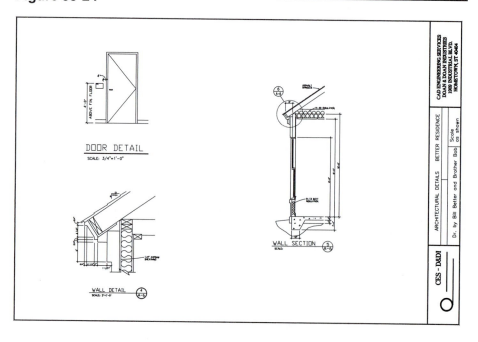

effectively in this parent drawing and the dimension layers could be controlled for visibility. If the dimensions for the *Xrefed* drawings were created in paper space, they would be unusable in the parent drawing since you cannot display any layouts of an externally referenced drawing.

Considerations for Using Multiple Viewports and Layouts

It is difficult to provide guidelines you can use to set up drawings such as these because there are so many possible combinations of viewports and layouts <u>and</u> there are so many possible strategies to use for any one setup based on your preferences and the particular application. However, here are some general considerations:

1. If you need to show several aspects of one drawing on one sheet (print or plot), use multiple viewports in one layout.

2. If you need to show the same geometry but print or plot it in different scales or on different devices, create and save multiple layouts.

3. If you need to show several drawings on one sheet (print or plot), create multiple viewports on a layout, *Xref* the drawings, and control layer visibility to display one drawing per viewport.

4. Linetype scale can be shown either relative to model space geometry in all viewports and layouts (*PSLTSCALE* = 0), or can be scaled to appear the same for all viewports relative to paper space (*PSLTSCALE* = 1). *LTSCALE* and *PSLTSCALE*, however, are easily changed to suit a specific layout as needed.

5. Model space dimensions, text, and hatching can appear in multiple viewports.

6. Paper space dimensions and text appear only for the specific viewport or layout.

7. Model space dimensions, text, and hatching can appear in different viewports at the same scale without alteration.

8. Model space dimensions, text, and hatching intended to appear in multiple viewports of one layout in different scales must be duplicated at different scales, placed on separate layers, and controlled for viewport visibility.

9. Model space dimensions, text, and hatching intended to appear in multiple layouts in different scales can be reset (*Overall* scale) for each print, but may require adjustment and cannot be saved in one drawing file (unless duplicated on other layers).

10. Paper space dimensions and text can be easily created for each viewport or for each layout, but can be used effectively only for that particular viewport or layout.

PLOT STYLE TABLES AND PLOT STYLES

A plot style is an object property. A plot style controls how an object looks on a plotted drawing. By modifying an object's plot style, you can <u>override</u> that object's drawing color, linetype, and lineweight for the plot. You can also specify additional features such as line end styles, join styles, fill styles, dithering, gray scale, pen assignment, and screening. You can use plot styles if you need to plot the same drawing with different "looks."

When a drawing is plotted in AutoCAD 2000 or later release, three main components are referenced to develop the finished plot:

1. layout settings you set up in the *Plot* dialog box and *Page Setup* dialog box (saved in the .DWG file);
2. plotter configuration information you specify for the plot device (saved in the device's .PC3 file); and
3. settings that are assigned in a Plot Style Table you attach (saved in .CTB and .STB files).

The third component, Plot Style Tables, are optional for plotting. This section discusses Plot Style Tables and plot styles.

Before explaining plot styles, it would help to consider how the appearance of objects is controlled in plotted drawings in pre-AutoCAD 2000 releases. In these releases, the <u>color</u> of an object determined how objects appeared in plotted drawings. For example, if you used a plotter with colored pens, it was normally configured so the object's color would plot with a pen of the same color. Alternately, you could configure the print or plot device so colors in the drawing would correspond to certain lineweights on the finished page. Normally, the objects' color is set to *BYLAYER*, so you can consider that the <u>layer's color</u> rather than the <u>object's color</u> determined the plotted appearance. The practice of associating colors with certain plotter pen numbers is known as making "pen assignments."

This strategy is still valid in AutoCAD 2004 and will most likely be used for the majority of plots. As a matter of fact, several <u>color-dependent Plot Style Tables</u> are supplied with AutoCAD that utilize this same concept; that is, an object's color determines how the object appears in the plot. There is one limitation to this system, however. With a color-dependent Plot Style Table, all objects in the same color generally have the same appearance on the plotted sheet.

AutoCAD 2004 also offers <u>named Plot Style Tables</u>, in which the contained plot styles are not automatically assigned by color but can have names that you designate and can be assigned to <u>any object</u>. With named Plot Style Tables, objects that appear in one color in the drawing can be plotted with different appearances.

Plot styles, even the new color-dependent Plot Style Tables, offer new advantages not available in releases of AutoCAD previous to 2000. With plot styles you can assign line end styles, joint styles, fill patterns, gray scales, screen pattern percentages, and pen assignments. Therefore, a color-dependent Plot Style Table is simply a list of colors, and for each color in the drawing you can assign a specific lineweight, linetype, line end style, joint style, fill pattern, gray scale, screen percentage, color, and pen assignment for the plot.

Unlike some other object properties, plot style is <u>optional</u>. Objects will print or plot based on color even if no plot style is assigned to it or if no Plot Style Table is attached. If you assign a plot style to an object, and then detach or delete the Plot Style Table that defines that plot style, the plot style will have no effect on the object.

What is the Difference between a Plot Style and a Plot Style Table?

It is important to make the distinction between a plot style and a Plot Style Table because of how they are assigned or what they are attached to. This book uses lowercase letters for the terms "plot style" and upper- and lowercase for "Plot Style Table" to help make that distinction.

plot style	A property assigned to objects that determines the way the object appears in a plot. A plot style contains several appearance characteristics such as color, linetype, lineweight, line end style, joint style, fill pattern, gray scale, screen percentage, and pen assignment.
Plot Style Table	A Plot Style Table is a collection of many plot styles. A color-dependent Plot Style Table contains 255 plot styles, each plot style is a color. A Plot Style Table is "attached" to the *Model* tab or a *Layout* tab.

The Plot Style Table is attached to a <u>layout</u>. You can attach the same or a different Plot Style Table to each layout (the *Model* tab and all *Layout* tabs). You attach a Plot Style Table to a layout using the *Page Setup* dialog box. Since different layouts can have different Plot Style Tables attached, two similar layouts could be plotted with two different "looks." Alternately, one layout could be plotted with one Plot Style Table and later plotted with another Plot Style Table to achieve a different "look."

For example, Figure 33-25 displays the AutoCAD-supplied ACAD.CTB color-dependent Plot Style Table. The colors listed in the table (Color 1, Color 2, Color 3, etc.) are plot styles. Each color in the drawing takes on the appearance defined by its plot style, so everything drawn in Color 1 can have a particular plotted color, grayscale, pen number, screening, linetype, and so on, whereas everything drawn in Color 2 can take on a different appearance.

Figure 33-25

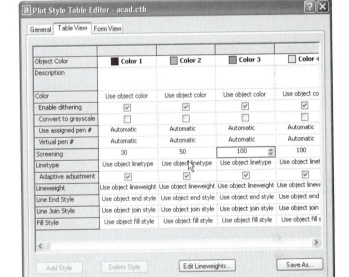

Color Plot Style Table Example

As an example, consider a construction drawing set up in one *Layout*, but the project has two phases. Using two Plot Style Tables, the layout could be plotted for the demolition phase showing one appearance, then later plotted during the construction phase with a different appearance.

The floor plan in Figure 33-26 illustrates the demolition phase. Walls to be demolished are most apparent, shown in 100% screen (dark), while unchanged walls are in 50% screen (medium), and new construction is barely noticeable in 30% screen (light). Note the Plot Style Table in Figure 33-25 is set up to plot Color 3 (demolition) in 100%, Color 2 (unchanged) in 50%, and Color 1 (new construction) in 30% screen.

Figure 33-26

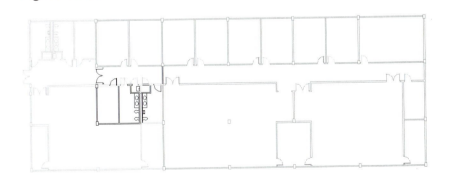

Figure 33-27 shows the same layout but for the construction phase. In this case, a second Plot Style Table is attached to the layout that applies the opposite screen percentages to Color 1 and Color 3, thereby emphasizing the walls needed for new construction (100% screen, dark) and fading the demolished walls (30% screen, light).

Figure 33-27

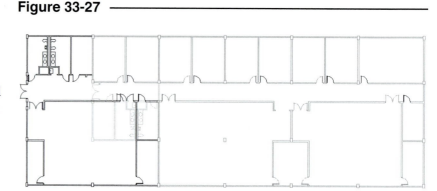

When you use plot styles you have the option to display the plot styles in the layout so you don't have to plot the drawing to see the results. You can choose to display plot styles by selecting *Display Plot Styles* on the *Plot Device* tab in the *Page Setup* dialog box. If you choose not to display plot styles in the drawing, you can view them using the *Full Preview* option in the *Plot* dialog box or by using the *Preview* command.

Color-Dependent and Named Plot Style Tables

There are two plot style modes: Color-Dependent and Named. Each drawing you open in AutoCAD is in either one mode or the other. You can change the mode for new drawings using the *Plotting* tab in the *Options* dialog box to change the plot style mode setting (see Figure 33-28 and following explanation).

Color-Dependent Plot Style Tables

By default, AutoCAD continues to use object color to control output effects by creating color-dependent Plot Style Tables. For each color-dependent Plot Style Table there are 255 plot styles; each plot style is a color. You assign plot characteristics (lineweight, screening, dithering, end and joint types, fill patterns, etc.) to each plot style (see Figure 33-25). When your drawing plots, the appearance of each line is based on its color in the drawing and the characteristics assigned to that color. In a color-dependent Plot Style Table, you cannot add, delete, or rename color-dependent plot styles. Color-dependent Plot Style Tables are stored in files with the extension .CTB. AutoCAD supplies several color-dependent Plot Style Tables (see "*Stylesmanager*").

Named Plot Style Tables

A named Plot Style Table is not based on a list of colors, but is based on a list of plot style names that you define. You can assign any plot style to any object regardless of that object's color. Each named Plot Style Table can contain many named plot styles. For each plot style name you define, you can assign specific line characteristics (lineweight, screening, dithering, end and joint types, fill patterns, etc.). A named Plot Style Table is similar to that shown in Figure 33-25, except the individual plot style names can be assigned instead of using color numbers (see Figure 33-34, top row). Normally, you would then assign the named plot styles to layers in your drawing (although plot styles can be assigned to any object). Named Plot Style Tables are stored in files with the extension .STB. AutoCAD supplies several named Plot Style Tables.

Since AutoCAD can use only one of the two types of Plot Style Tables at a time (named or color-dependent), use the *Plotting* tab of the *Options* dialog box to make the desired choice (Fig. 33-28, next page). The options, located on the right side of the tab, are explained briefly here.

Default Plot Style Behavior for New Drawings

Use this section to control how plot styles affect drawings. This setting is saved with the drawing. Making a change here <u>affects new drawings but not the current drawing</u>. If you want to change the default plot style behavior for a drawing, select an option <u>before</u> opening or creating a drawing. Changing the default plot style affects only <u>new</u> drawings and Release 14 and earlier drawings when first opened in AutoCAD 2004.

Figure 33-28

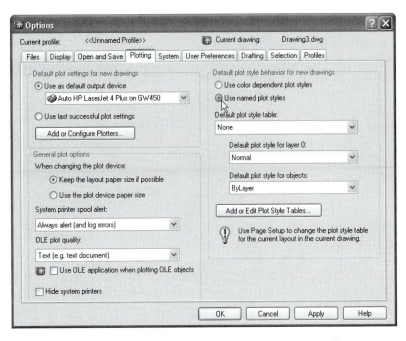

Use Color Dependent Plot Styles

By default, AutoCAD uses color-dependent plot styles for new drawings and pre-AutoCAD 2000 drawings. You can also create and edit color-dependent plot styles by setting the *PSTYLEPOLICY* system variable to 1. If a drawing is saved with *Use Color Dependent Plot Styles* as the default, you can change it to *Use Named Plot Styles* later. However, once a drawing is saved with *Use Named Plot Styles* as the default, you can change it to use color dependent plot styles with *Convertpstyles*.

Use Named Plot Styles

This check causes AutoCAD to use named plot styles as the default in both new drawings and pre-AutoCAD 2000 drawings. This allows you to use the AutoCAD-supplied named Plot Style Tables or create new named Plot Style Tables. You can also set the *PSTYLEPOLICY* system variable to 0 to enable named Plot Style Tables. Once a drawing is saved with *Use Named Plot Styles* as the default, you can change it to use color dependent plot styles using *Convertpstyles*.

Default Plot Style Table

From this drop-down list, select the default Plot Style Table to attach to new drawings. If you are using color-dependent plot styles (*Use Color Dependent Plot Styles* is checked), this option lists all color-dependent Plot Style Tables (.CTB files) found in the search path as well as the value of *None*. If you are using named plot styles (*Use Named Plot Styles* is checked), this option lists all named Plot Styles Tables (.STB files).

Default Plot Style for Layer 0

Use Named Plot Styles must be checked to enable this drop-down list. Your choice sets the default plot style for Layer 0 for new drawings or pre-AutoCAD 2000 drawings. The list displays the default value *Normal* and any plot styles defined in the currently loaded Plot Style Table. You can also control the default plot style for Layer 0 using the *DEFLPLSTYLE* system variable (saved in the drawing).

Default Plot Style for Objects

Use Named Plot Styles must be checked to enable this drop-down list. Here you set the default plot style that is assigned when you create new objects. The list displays a *BYLAYER*, *BYBLOCK*, and *Normal* style, and lists any plot styles defined in the currently loaded Plot Style Table. Alternately, you can set the *DEFPLSTYLE* system variable.

Add or Edit Plot Style Tables

Selecting this button invokes the *Plot Style Table Manager*, which is a separate system window. See *"Stylesmanager"* later in this chapter.

Attaching a Plot Style Table to a Layout

To attach a Plot Style Table to a *Layout* or *Model* tab, first select the desired *Model* tab or *Layout* tab. Invoke the *Page Setup* dialog box (Fig. 33-29). In the *Plot Device* tab, use the *Plot Style Table (pen assignments)* drop-down list to select a the desired table (.CBT or .STB file). Select the *OK* button to save the new attachment.

Use the *Edit...* button to invoke the *Plot Style Table Editor*. (See the section titled "Editing Plot Styles Assigned to Objects.") Selecting the *New...* button invokes the *Add-A-Plot Style Table Wizard*. (See "Creating a Plot Style Table.") You can choose to *Display Plot Styles* if you want the layout to reflect the appearance of the plot styles in the layout (so the layout looks like a *Full Preview*).

Figure 33-29

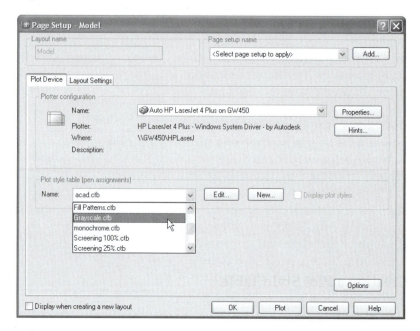

If you want to change the name of a named Plot Style Table from Style 1, Style 2, etc. to something more descriptive (such as walls, electrical, plumbing), it should be done before opening a drawing that references the table. If the names are changed after the Plot Style Tables have been attached to a drawing, the plot style names in the drawing will not match the names of the styles in the attached Plot Style Table.

Attaching to a Viewport
In AutoCAD 2000 you could attach a Plot Style Table to a viewport. This feature is <u>no longer available in AutoCAD 2004</u>. In AutoCAD 2000, you can attach a Plot Style Table to a viewport by highlighting a viewport border, then using the *Properties* window; however, any Plot Style Table attached to the layout would override the Plot Style Table attached to a viewport.

STYLESMANAGER	Pull-down Menu	COMMAND (TYPE)	ALIAS (TYPE)	Short-cut	Screen (side) Menu	Tablet Menu
	File Plot Style Manager...	*STYLESMANAGER*	...	...	...	...

The *Stylesmanager* command produces the *Plot Style Manager*. You create Plot Style Tables and edit Plot Style Tables with the *Plot Style Manager*. In AutoCAD 2004, selecting the *Plot Styles Manager* produces the Windows Explorer displaying the *Add-A-Plot Style Table Wizard* and the list of available Plot Style Tables (.CTB and .STB files). This is a separate application window that runs independently of AutoCAD.

Color-dependent Plot Style Tables are stored as .CTB files, and named Plot Style Tables are stored as .STB files. To find the location of the .CTB and .STB files on your system, use the drop-down list in the *Address* section at the top of the window, or use *Options*, *Files* tab, and expand the *Printer Support File Path*, then *Plot Style Table Search Path* sections.

All Plot Style Tables on your system are listed in the window, including both color-dependent and named Plot Style Tables. The AutoCAD-supplied Plot Style Tables are displayed in Figure 33-30.

Figure 33-30

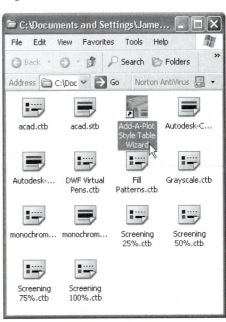

To create a new Plot Style Table using the *Plot Style Manager*, select the *Add-A-Plot Style Table Wizard* (see Figure 33-30). After creating new Plot Style Tables, they automatically appear in the window. See "Creating a Plot Style Table."

You can also use the *Plot Style Manager* to edit existing Plot Style Tables. Using the manager, select a Plot Style Table to edit by double-clicking its icon. Editing a table involves assigning the appearance characteristics (color, pen assignment, screening, linetype, lineweight, line end style, line join style, and fill style) for each plot style. See "Editing Plot Styles."

Creating a Plot Style Table

Many AutoCAD-supplied Plot Style Tables are available and have "generic" plot styles. These tables are intended for your use and can be edited for your needs using the *Plot Style Table Editor*. You can use the existing Plot Style Tables or use the *Add-A-Plot Style Table Wizard* to create other tables.

To create a new Plot Style Table, access the *Add-A-Plot Style Table Wizard* from the *Plot Style Manager* (see Figure 33-30) or select the *Wizards* from the *Tools* pull-down menu.

In the *Add-A-Plot Style Table Wizard*, you can *Start from scratch*, *Use an existing plot style table*, *Use My R14 Plotter Configuration*, or *Use a PCP or PC2 file*. Selecting the first choice, *Start from Scratch*, you can create a color-dependent Plot Style Table or a named Plot Style Table (Fig. 33-31).

Figure 33-31

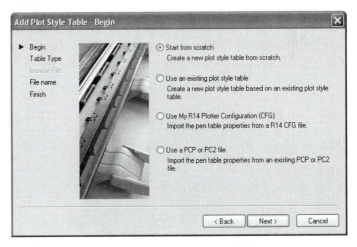

The *Add-A-Plot Style Table Wizard* guides you through the steps to easily create the new table. There are only a few steps involved with creating a new Plot Style Table. Once the table is created, your specific preferences on how objects appear in plots are achieved by editing the table and creating the individual plot styles using the *Plot Styles Table Editor*. The wizard contains explanatory text on each page, so no additional explanation is required here. If you feel that you do need help, look in the *AutoCAD User's Guide*. Once the table is created, edit it using the *Plot Style Table Editor* as explained in the next section.

Plot Style Table Conversion from Previous Releases

In releases of AutoCAD previous to 2000, the components needed for plotting were saved in the ACAD.CFG and the .PCP or .PC2 files. If you want, you can use the *Add-A-Plot Style Table Wizard* to convert your older Release 14 pen assignments to Plot Style Tables.

These new Plot Style Tables will then be available in the list in addition to the supplied Plot Style Tables. See Figure 33-31, last two options (*Use My R14 Plotter Configuration (CFG)* and *Use a PCP or PC2 File*). (See also *"Pcinwizard,"* Chapter 14.)

Editing Plot Styles

To edit plot styles, use the *Edit...* button in the *Page Setup* dialog box or double-click on a plot style file name (.CTB or .STB) from the *Plot Style Manager* or from Windows Explorer. This action invokes the *Plot Style Table Editor* (Fig. 33-32).

If the selected Plot Style Table is color-dependent, the assignment to objects is automatic; that is, each of the 255 plot styles in the table is already assigned to an object color. However, you can set the *Fill, Line Join Style, Line End Style, Lineweight, Line Type, Screening,* and so on for each color as shown. Therefore, each drawing color takes on the appearance of the selected options. You cannot change the plot style names (Color 1, Color 2, etc.) for a color-dependent Plot Style Table.

Figure 33-32

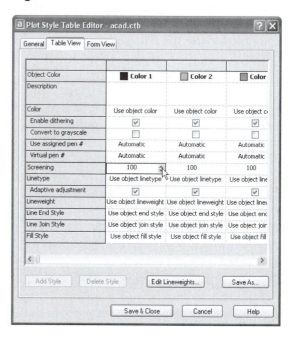

You can make the settings in either the *Table View* (see Figure 33-32) or *Form View* (Fig. 33-33) tabs. These two tabs have identical options. Setting an option in one tab also makes the matching setting in the other tab.

Figure 33-33

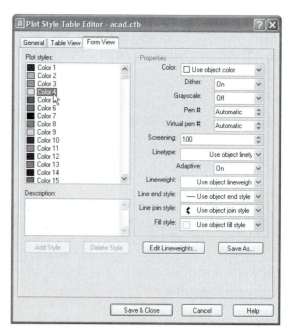

If the selected Plot Style Table is a named Plot Style Table, the names are not automatically assigned as with the color-dependent tables. You assign the names for each plot style based on some descriptive feature or application. For example, you may want to name plot styles for an application, such as Plumbing, HVAC, Electrical, or possibly Phase 1 and Phase 2. You could also name the plot styles according to a feature you assign; for example, you may use the names Wide, Medium, and Thin to describe lineweights that are defined in the styles.

Figure 33-34 displays a named Plot Style Table in the *Plot Style Table Editor*. All of the same features (color, dither, grayscale, pen number, etc.) are available as in the color-dependent Plot Style Tables. The only additional feature is that <u>you assign the name</u> of each plot style, instead of the styles being automatically named for a color (Color 1, Color 2, etc.) as in a color-dependent table. By default, the plot styles are named Style 1, Style 2, etc. You can rename the styles by double-clicking on the name (top row).

Figure 33-34

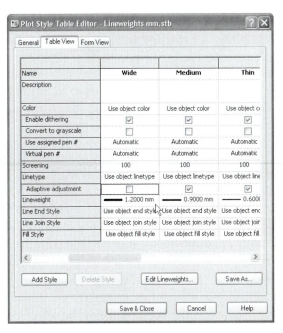

Create new plot styles by pressing the *Add Style* button. This action creates a new style named Style#, where # is the next number in the sequence.

You can make the settings in either the *Table View* (Fig. 33-34) or *Form View* (see Figure 33-33) tabs. These two tabs have identical options. Setting an option in one tab also makes the matching setting in the other tab.

Named Plot Style Tables have some advantages over color-dependent tables. With named tables, you can assign specific characteristics to objects regardless of the objects' screen color, whereas with color-dependent tables all objects with the same screen color must appear the same in the plot (have the same plot style). Named plot styles are usually easier to assign to objects in complex drawings (when the drawing contains many colors and layers). See the discussion in the next section.

Assigning Named Plot Styles to Objects

Keep in mind that each drawing can have only one of the two types of Plot Style Tables attached—color-dependent or named Plot Style Tables. The desired table type should be set in the *Options* dialog box before creating a drawing or before opening a pre-AutoCAD 2000 drawing (see "Color-Dependent Plot Style Tables" and "Named Plot Style Tables").

Figure 33-35

Color-Dependent Plot Style Tables
The process of assigning plot styles to objects is necessary only for named Plot Style Tables. This is based on the fact that <u>color-dependent plot styles are already assigned to colors</u>, not objects. In other words, each color-dependent plot style (named Color 1, Color 2, etc.) is automatically assigned to its matching color. Figure 33-35 explains this idea. Notice that in the figure displaying the *Layer Properties Manager*, the *Plot Styles* column is disabled since assignment is automatic. You will also notice that drawings that have color-dependent Plot Style Tables attached display a <u>disabled</u> *Plot Style* drop-down list on the *Object Properties* toolbar (Figure 33-36 shows an enabled drop-down list for named Plot Style Tables).

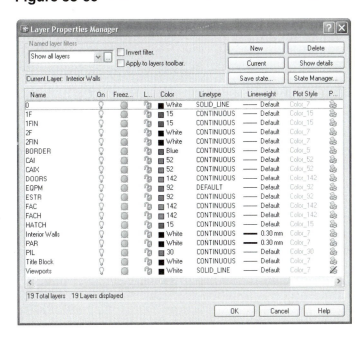

Named Plot Style Tables
<u>Plot styles in a named Plot Style Table can be attached to any object</u>. Individual plot styles within a named table are often assigned to layers. Assigning plot styles to individual objects offers much more flexibility, but can be a tedious job and create a fairly complex layout. If plot styles are assigned both to layers and objects, the plot style assigned to an object overrides the plot style assigned to the layer.

If you want to assign plot styles to individual objects (rather than to layers), three methods are available: the *Plot Style* drop-down list on the *Object Properties* toolbar, the *Properties* palette, or the *Plotstyle* command.

Figure 33-36

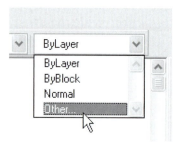

Figure 33-36 displays the *Plot Style* drop-down list on the *Object Properties* toolbar. Using this method, you would select the desired plot style from the list. The selected style becomes the <u>current</u> object-specific plot style—that is, any object created after the setting is made will have the selected plot style. This method is similar to specifying an object-specific color, linetype, or lineweight (see Chapter 11). The *Plot Style* drop-down list on the *Object Properties* toolbar cannot be used to assign color-dependent plot styles.

You can also use the *Plotstyle* command to assign object-specific plot styles (see "*Plotstyle*" next). Using *Plotstyle* has the same result as using the *Plot Style* drop-down list. The *Plotstyle* command can only be used to assign named plot styles.

Figure 33-37 displays the *Properties* palette. To use this method, select the desired object(s), then invoke the *Properties* palette. Select the desired plot style from the drop-down list. This method differs from the two previously described methods in that using the *Properties* palette is <u>retroactive for the selected objects only</u>, whereas the *Plotstyle* command and the *Object Properties* toolbar *Plot Style* drop-down list set plot styles for <u>newly created objects</u>.

Figure 33-37

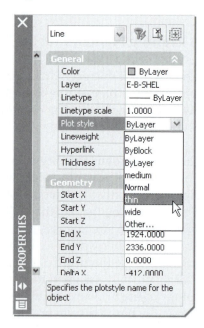

Making an assignment using any one of these three methods is similar to assigning colors, linetypes, and lineweights to individual objects rather than assigning the properties by the object's layer (*BYLAYER*). With this method, you lose the simplicity of organizing objects by layer and color; however, that is one of the underlying reasons for using named plot styles—so objects of different colors can be plotted in different ways.

Alternately, named plot styles can be assigned to layers. This is generally a simpler method, particularly for complex drawings with many objects. Figure 33-38 displays *the Layer Properties Manager* in which three plot styles, named Wide, Medium, and Thin, have been assigned to various layers. In this example, each of the three plot styles has a matching lineweight designated in the style.

Figure 33-38

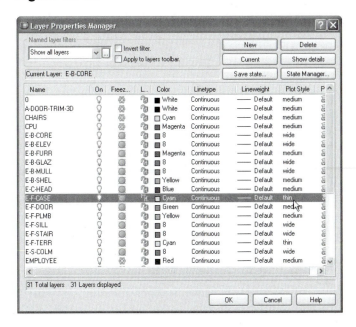

To assign plot styles from a named Plot Style Table to layers, click on the plot style name (usually *Normal*) in the *Plot Style* column for a specific layer. The *Select Plot Style* dialog box appears (Fig. 33-39) providing the list of plot styles to select from. All objects on the selected layer are then plotted with the parameters (screening, pen number, lineweight, line end joints, etc.) as previously designated in the plot style.

Figure 33-39

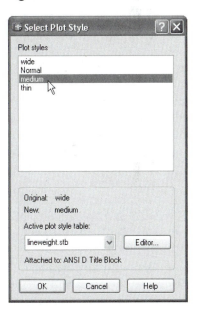

Note that the information given at the bottom of the dialog box indicates the *Active Plot Style Table* and the layout or viewport that it is *Attached to*. You can select a different Plot Style Table from this list to attach to the layout or viewport; however, doing so changes the setting previously made in the *Page Setup* dialog box (for the layout) or in the *Properties* palette (for the viewport). In other words, you can assign plot styles to objects and layers, but you cannot attach Plot Style Tables to individual objects or layers, only to layouts and viewports.

PLOTSTYLE	Pull-down Menu	COMMAND (TYPE)	ALIAS (TYPE)	Short-cut	Screen (side) Menu	Tablet Menu
	Format *Plot Style...*	*PLOTSTYLE* or *-PLOTSTYLE*	...	...	...	...

As previously described briefly, the *Plotstyle* command is used to set the current named plot style for newly created objects. It has the same function as selecting a current plot style from the *Plot Style* drop-down list on the *Object Properties* toolbar. That is, the selected named plot style becomes the plot style assigned to all newly created objects.

The *Plotstyle* command produces the *Current Plot Style* dialog box (not shown). This dialog box is essentially the same as the *Select Plot Style* dialog box that appears when assigning plot styles to layers (see Figure 33-39), except with *Plotstyle* the styles are assigned to objects. The *Plotstyle* command operates only for drawings that were created in named Plot Style Table mode. See the previous discussion, "Assigning Named Plot Styles to Objects."

To use the Command line version, type *–Plotstyle*. The command prompts are shown here.

        Command: **-plotstyle**
        Current plot style is "ByLayer"
        Enter an option [?/Current] :

Enter the name of the desired plot style. The question (?) option lists all plot styles in the currently attached table. The current drawing must be in named Plot Style Table mode and you must have a named Plot Style Table attached to the layout or viewport for this command to operate.

Named Plot Style Table Example

The following section is an example of using one <u>named</u> Plot Style Table to plot a drawing two times with different lineweights. Compare this example with the example given early in the chapter.

In the earlier example, two similar <u>color-dependent</u> Plot Style Tables were used to assign different screen percentages to different colors in the drawing. The two plots were created by attaching a different Plot Style Table to the layout before creating each plot. Using several color-dependent Plot Style Tables for one layout is a reasonable practice since the individual plot style names (color numbers) cannot change between Plot Style Tables. Therefore, all color-dependent Plot Style Tables have the same plot style names, although the specifications for each style (screen percentage in this case) can be different.

In the following example, only one named Plot Style Table is attached to a layout. To achieve different plots, the lineweight value is changed for individual plot styles to achieve the desired effects. In addition, objects to be plotted in different plot styles (lineweights) are drawn in the same color; therefore, color-dependent Plot Style Tables could not be used.

A drawing of an office building is completed and a layout is created to display and plot only the HVAC plan (Fig. 33-40). A plot displaying only the building core and the HVAC information is sent to the HVAC contractors for developing bids.

After a contractor is retained, it is discovered that one large duct unit must be specially manufactured. In order to make the unit more apparent for the manufacturer, a special plot must be made by applying a heavier lineweight to the unit.

Figure 33-40

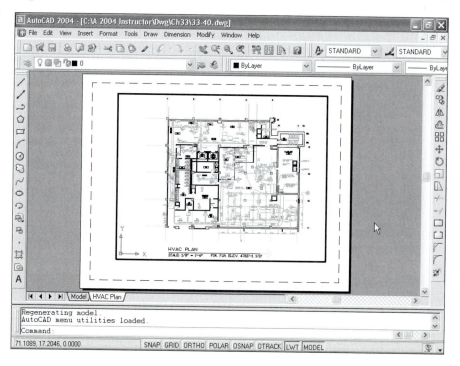

Since all of the HVAC objects are drawn in one color (see Figure 33-40, light colored objects), a color-dependent Plot Style Table cannot be used. Instead, a named Plot Style Table is created called "LW mm" (Fig. 33-41). Several plot styles are created with different lineweight settings; each plot style is named according to weight (1.20mm, .90mm, .60mm, etc.). The Plot Style Table is then attached to the layout using the *Page Setup* dialog box.

Figure 33-41

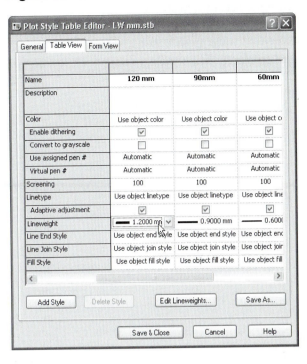

Next, objects representing the duct unit are selected and the *Properties* palette is invoked (Fig. 33-42). In the *Plot Style* row for all 20 objects [note the "All (20)" displayed at the top of the box], the desired plot style (named *90mm*) is selected and assigned to the objects.

Figure 33-42

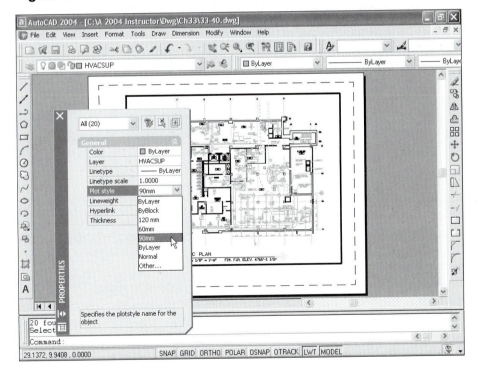

The finished plot reveals the special duct unit plotted in the wide lines (Fig. 33-43). Note that these objects are <u>plotted differently even though they share the layer and color of the other HVAC components.</u>

Figure 33-43

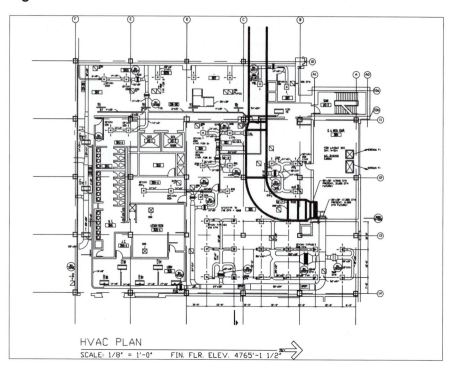

HVAC PLAN
SCALE: 1/8" = 1'-0"   FIN. FLR. ELEV. 4765'-1 1/2"

If another plot must be made to reveal other duct units in the layout, the process of applying the lineweight-based plot styles to particular objects within one layer can be used again. If you wanted to prepare a plot with uniform lineweights for all objects, it would be simple to select all objects and then use the *Properties* palette to assign the same plot style to all objects.

CONVERTPSTYLES

Pull-down Menu	COMMAND (TYPE)	ALIAS (TYPE)	Short-cut	Screen (side) Menu	Tablet Menu
...	CONVERTPSTYLES	...	...	...	...

Convertpstyles allows you to change <u>drawings</u> from using one type of Plot Style Table (named or color-dependent) to the other type of Plot Style Table. This feature was introduced in AutoCAD 2002. *Convertpstyles* is a Command line-driven utility. Type *Convertpstyles* in the drawing you want to convert.

Converting Drawings from Named Plot Style Tables to Color-Dependent Plot Style Tables

If a named Plot Style Table drawing is current when you use *Convertpstyles*, a small dialog box advises you that the named plot styles attached to objects, model space, and layouts will be detached.

```
Command: convertpstyles
Drawing converted from Named plot style mode to Color Dependent mode.
Command:
```

After a drawing is converted to use color-dependent Plot Style Tables, you can assign a color-dependent Plot Style Table as you would normally. Plot styles are assigned by color.

Converting Drawings from Color-Dependent Plot Style Tables to Named Plot Style Tables
Before you use *Convertpstyles* in a color-dependent drawing, you should use *Convertctb* to convert the attached color-dependent tables to named tables. You can convert the color-dependent Plot Style Tables assigned to the drawing to named Plot Style Tables using *Convertctb* (see *Convertctb* next). After converting the tables, use *Convertpstyles* to reattach the newly converted named Plot Style Tables.

Assuming *Convertctb* has been used first in the drawing, type *Convertpstyles*. AutoCAD displays the standard file navigation dialog box for you to select the named Plot Style Table file to attach to the converted drawing. You must select a named Plot Style Table that was converted using *Convertctb* or created from a PC2 or PCP file. Normally you should select the named Plot Style Table that was converted from the color-dependent Plot Style Table that was assigned to the same drawing. The selected named Plot Style Table is assigned to model space and to all layouts. Drawing layers are each assigned a named plot style (from the converted Plot Style Table) that has the same plot properties that their color-dependent plot style had. After going through this conversion, you can change the named Plot Style Table assignment or assign other named Plot Style Tables to model space or layouts.

	Pull-down Menu	COMMAND (TYPE)	ALIAS (TYPE)	Short-cut	Screen (side) Menu	Tablet Menu
CONVERTCTB	...	*CONVERTCTB*	...	...	...	...

Convertctb is intended to be used in a drawing that you want to convert from using color-dependent Plot Style Tables to use named Plot Style Tables. First use *Convertctb* to convert the attached color-dependent <u>tables</u> to named tables, then use *Convertpstyles* in the drawing to convert the <u>drawing</u> (see *Convertpstyles*).

When you type *Convertctb* at the Command prompt in the drawing, the standard file navigation dialog box appears for you to select the desired color-dependent Plot Style Table, then reappears for you to specify a <u>name</u> (you should use the same name as the .CTB) and location for the new (now named) Plot Style Table. You should use *Convertctb* to convert all attached tables. The original color-dependent tables are not changed.

Convertctb creates one named plot style for each color that has unique plot properties, one named plot style for each group of colors that are assigned the same plot properties, and a default named plot style called NORMAL.

PLOT STAMPING

Plot stamping provides options for you to have an informational text "stamp" plotted on each drawing. For example, the drawing name, date, and time of plot can be added to each plot for tracking and verification purposes.

Plot stamping was available in AutoCAD Release 14's Batch Plot utility. This feature was not available in AutoCAD 2000, but appears again in AutoCAD 2002 and AutoCAD 2004.

PLOTSTAMP

Pull-down Menu	COMMAND (TYPE)	ALIAS (TYPE)	Short-cut	Screen (side) Menu	Tablet Menu
File *Plot* *Plot Device* *Plot Stamp* *Settings...*	PLOTSTAMP	...	...	...	...

To enable plot stamping, you must use the *Plot* dialog box *Plot Device* tab, then check *On* under *Plot Stamp* (not shown). To specify what text you want to appear on the plot, select *Settings...* under *Plot Stamp* in the *Plot* dialog box or enter *Plotstamp* at the Command line. This action produces the *Plot Stamp* dialog box (Fig. 33-44).

Saving plot stamp parameters involves changing a plot stamp parameter (.PSS) file. Two .PSS files are supplied by AutoCAD, INCHES.PSS and MM.PSS. You can change either of these files or create new files. A simple method for creating your own .PSS files is to use either of the AutoCAD-supplied files as a template, make the desired changes in the *Plot Stamp* dialog box, then use the *Save As* button to save the changes to a new file name. (See "*Plot stamp parameter file.*")

Figure 33-44

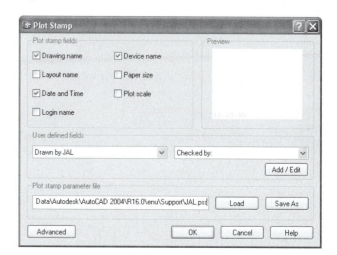

In AutoCAD 2004, .PSS files are stored (by default for stand-alone installations) under your profile name in the Documents and Settings folder. To find the location of the .PSS files on your system, see the edit box in the *Plot stamp parameter file* section (lower-left corner), or use *Options, Files* tab, and expand the *Working Support File Search Path* section.

Plot stamp fields
In this area, check any items you want to appear in the plotted drawing. Each piece of text is added to the plot stamp separated by a comma. Content for *Drawing name, Layout name, Device name, Paper size,* and *Plot scale* is automatically obtained from the AutoCAD drawing. *Plot scale* may not be an appropriate item to select when you plot a layout since this text is actually the plot scale that appears in the *Plot* dialog box (normally set to 1:1), not the viewport scale (the actual plot scale of the geometry). Content for *Date and Time* and *Login name* are obtained from the computer's system data.

Preview
This image tile indicates only the location and orientation of the plot stamp, not the text content. You cannot see a preview of the actual plot stamp using the *Full Preview* button or the *Preview* command; you must make a plot.

User defined fields
You can specify additional text to include in the plot stamp by selecting the *Add/Edit* button. The *User Defined Fields* dialog box appears (not shown) for you to enter new text or edit existing user text fields. Once specified, this personalized text appears in the *User defined fields* drop-down lists (see Figure 33-44). Two user defined fields can be added to a plot stamp.

Plot stamp parameter file

You can store the current plot stamp information in a file with a .PSS extension. For example, if you wanted to create a standard plot stamp for your company or school, first specify the plot stamp fields and advanced options you want, then save them to a .PSS file. Other users can then access the saved file and stamp their plots based on those standard settings.

To create a plot stamp parameter file, use either the AutoCAD-provided MM.PSS or INCHES.PSS file as a template, then save your changes to a new name using the *Save As* button. The *Load* button displays the standard file selection dialog box in which you can specify the location of the parameter file you want to use. The MM.PSS or INCHES.PSS files differ only in millimeter or inch units used for the default text height and offset.

Advanced

This button produces the *Advanced Options* dialog box (Fig. 33-45) where you can specify the *Location, Orientation*, and *Text properties* of the plot stamp. The *Location* and *Orientation* are reflected in the *Preview* tile of the *Plot Stamp* dialog box. The *X offset* and *Y offset* are the distances from the edge of the printable area or paper border you want the stamp to appear.

In the *Log file location* area you can indicate if you want to *Create a log file* and specify the name and location of the file. A log file is an ASCII text file that contains a record for each plot created, and each record lists the same items you specified for the plot stamp such as the date and time, drawing name, device name, and so on.

Figure 33-45

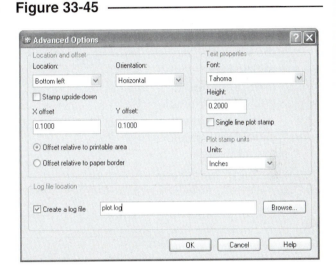

CHAPTER EXERCISES

1. **Multiple Viewports and Viewport-Specific Layer Visibility**

   A. *Open* the **Single Cavity Mold** drawing (download from the www.mhhe.com/leach Web site). Use *SaveAs* to rename the drawing **Ch33 SCM** and locate it in your working directory. Activate each of the layout tabs and the model tab to familiarize yourself with the drawing.

   B. Activate the **Plan of Ejector** tab. Right-click and select *Move or Copy* from the shortcut menu. In the *Move or Copy* dialog box, select **Plan of Ejector** and *Create a copy*, then *OK*. Now activate the new layout, right-click and *Rename* the tab **Ejector Details**.

   C. With the new layout activated, use *Pagesetup* to select an available printer as the *Plot Device*. Select *Letter* as the *Paper Size*. Select *OK* to save the settings for the layout. Next, use the *Layer Control* drop-down list and select *Freeze or thaw in current viewport* to freeze layer **Bourder**. Make a test print of the layout. Note in the print that the dimensions are barely readable at that size, and the linetype scale is too large.

D. Using **grips**, select the existing viewport and **STRETCH** it to modify the viewport size to occupy only the right half of the printable area as shown in Figure 33-46. Next, make layer **LAYOUT-VIEWPORTS** the *Current* layer. Use *Vports* to make *Two: Horizontal* viewports and set the *Viewport Spacing* to **.25**. *Fit* the viewports into the left half of the printable area. Your drawing should look similar to that in Figure 33-46.

Figure 33-46

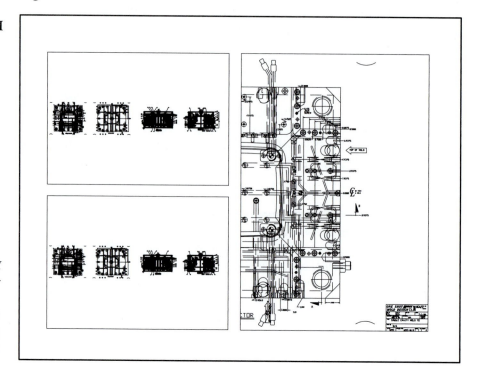

E. Use *Zoom* and *Pan* to display the upper-right corner of the mold in the small top viewport and the lower-right corner of the mold in the bottom viewport. Activate the *Viewports* toolbar and set the viewport scales as follows:

Figure 33-47

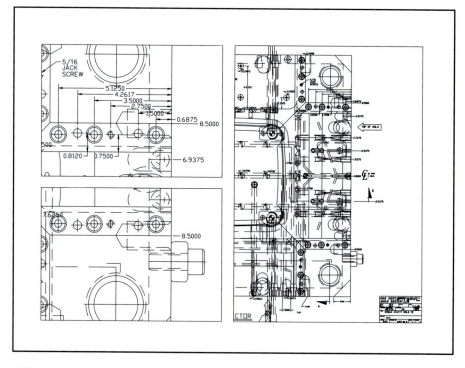

top left viewport	**1:2**
bottom left viewport	**1:2**
right viewport	**3:16** (.1875)

The resulting display should look like Figure 33-47.

F. Activate the right viewport and use the *Layer Properties Manager* and select *Current VP Freeze* to freeze these layers for the right viewport only:

A-H20
AIR
B-H20
DIMF
SLID1H20
SLID2H20

Figure 33-48

G. Activate paper space. On layer **NOTES**, add two circles and leaders to indicate Detail A and B. Make the text size **.12** and the text font *Romans*. Add the labels and scales for the two small viewports as shown in Figure 33-48. Set *LTSCALE* to **.5**. Finally, ensure *PSLTSCALE* is set to **1** so the linetype dashes appear equal in all viewports. *Save* the drawing and make a print of the layout.

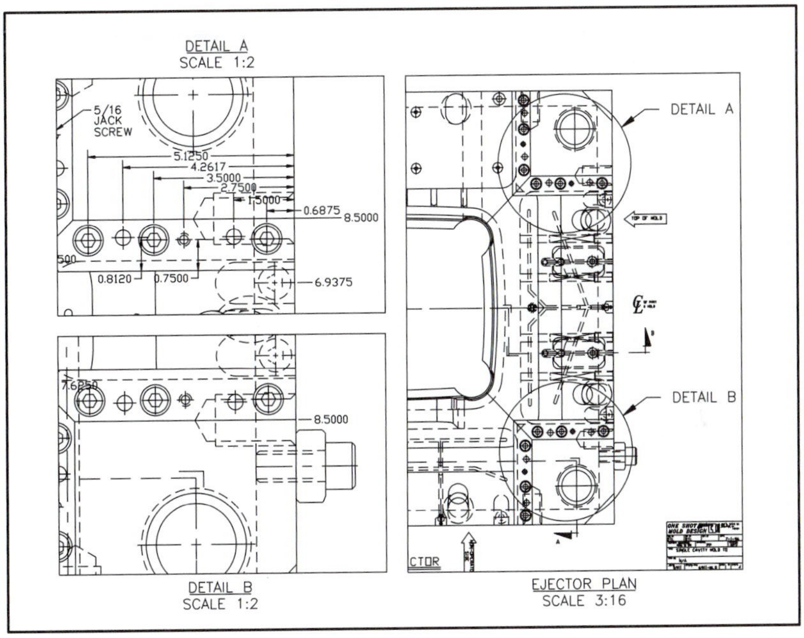

2. **Multiple Viewports and Dimensioning in Paper Space**

In this exercise, you will create a layout with two viewports to show an overall view of a drawing and a detail in a different scale. You will dimension the drawing both in model space and in paper space.

A. *Open* the **2BR-APT** you worked on last in Chapter 20 Exercises. Create a text *Style* using *Roman Simplex* font. Create a new dimension style named **ARCH-1** using the new font and with other settings to create architectural style dimensions as shown in Chapter 15, Figure 15-45. The drawing scale factor is **48**, so set the appropriate *Overall scale*. In model space, create dimensions for the apartment on the **DIM** layer. Make sure you add dimensions also to the new bedroom in your drawing. If you need assistance placing the dimensions, refer to Figure 33-49, on the next page. Use *SaveAs* and name the drawing **2BR-APT-DIM**.

B. Create a new layout using the *Template* option. Select the *ANSI B template* with an *ANSI B title block*. Next, use *Pagesetup* to specify a plot device and paper size for an *ANSI B* size sheet.

C. In the layout, create one viewport on the **VPORTS** layer that occupies about 2/3 of the layout area. *Zoom* and *Pan* as needed to show the entire floor plan. Set the viewport scale to **1/4"=1'**. If the entire floor plan does not appear in the viewport, make the necessary adjustments to the viewport border using **grips**.

D. Next, make a second smaller viewport on the right side of the layout. *Zoom* and *Pan* to display only the bathroom at the far left of the floor plan. Set the viewport scale to 1/2"=1'.

E. Make a new layer named **DIM-PS** and set the properties the same as layer DIM. Make a *New* dimension style called **ARCH-2** using ARCH-1 as a "template." Change the *Overall scale* to **1**. In paper space and on layer DIM-PS, create dimensions to indicate the distances between the wall and the three fixtures as shown in Figure 33-49. Label the viewports as shown, *Freeze* layer **VPORTS**, and *Save* the drawing. Make a *Plot*.

Figure 33-49

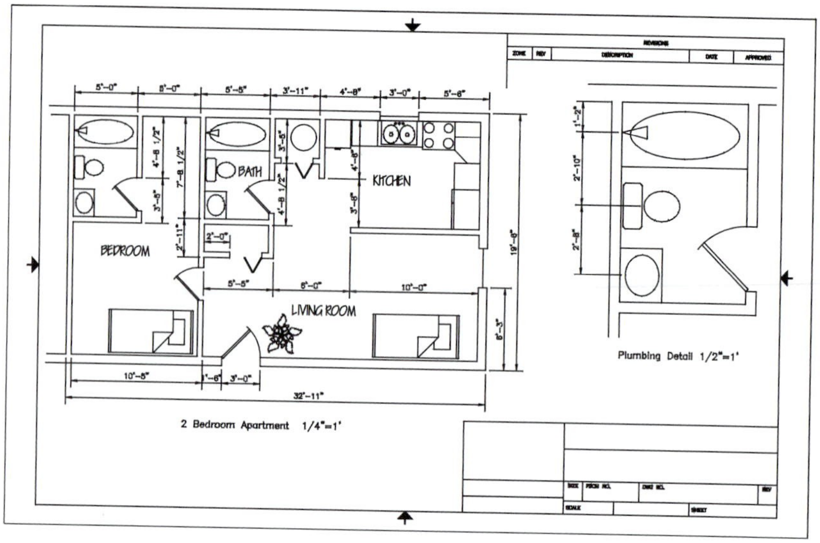

3. **Multiple Viewports, Dimensioning in Model Space, and Viewport-Specific Layer Control**

In this exercise, you will create a layout with three viewports to show an overall view of a drawing and two details in a different scale. You will dimension the drawing in model space and control the visibility for each viewport.

A. *Open* the **EFF-APT2** drawing that you worked on in Chapter 30 Exercises (with the door *Stretched* to the center of the front wall). *Erase* the plant. Create a text *Style* using **Roman Simplex** font. In model space, create dimensions for the interior of each room on the **DIM** layer, similar to those dimensions you created in Exercise 2. (You may consider also opening the **2BR-APT-DIM** drawing from the last exercise and using copy and paste to copy the needed dimensions to the new drawing. In this case, however, the dimensions will have to be reassociated.) *Zoom* in, and on another *New Layer* named **DIM2** give the dimensions for the wash basin in the bath and the counter for the kitchen sink.

B. To set up the layout, you have two choices: (1) use a *Layout Template,* or (2) create and set up the layout from scratch. In either case, use the template, plot device, and paper size of your choice.

C. Make a *New Layer* or use the existing layer named **VIEWPORT** and set it as the *Current* layer. Use *Vports* to make one viewport on the right side of the layout, occupying approximately 2/3 of the page. Use *Vports* again to make two smaller viewports in the remaining space on the left.

D. Activate the large viewport (in *Mspace*) and use *Zoom XP* or other viewport scale method to achieve the largest possible display of the apartment to an accepted scale. In the top-left viewport, use display controls to produce a scaled detail of the bath. Produce a scaled detail of the kitchen sink in the lower-left viewport, similar to that shown in Figure 33-50.

Figure 33-50

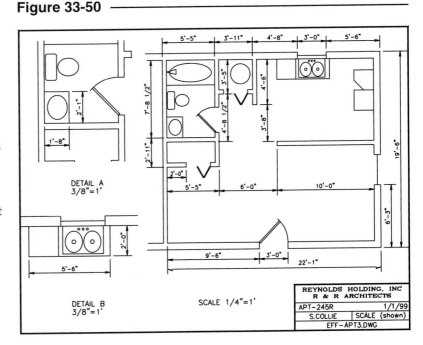

E. Use the *Layer Properties Manager* to *Current VP Freeze* layer **DIM2** in the large viewport and *Current VP Freeze* layer **DIM** in the two small viewports. Use *Properties* and select the dimensions that appear in the two detailed views and set a new *Dim scale overall* (in the *Fit* category) so the dimensions appear at the appropriate size for the viewport scale.

F. Return to paper space and *Freeze* layer **VIEWPORT**. Use *Text* (in paper space) to label the detail views and give the scale for each. *Save* the drawing as **EFF-APT3** and *Plot* from paper space at **1:1**. The final plot should look similar to Figure 33-50.

4. **Multiple Viewports, *Xref*, and Viewport-Specific Layer Visibility**

In this exercise, you will create three paper space viewports and display three *Xrefed* drawings, one in each viewport. Viewport-specific layer visibility control will be exercised to produce the desired display.

A. Begin a *New* drawing using a *Template* to correspond to the sheet size you intend to plot on. Activate a *Layout* tab. Use *Pagesetup* to select the paper size and to set the plotting options for the device you use to plot with. Create *New Layers* named **TITLE**, **XREF**, and **VIEWPORT**. On the **VIEWPORT** layer, use *Vports* to create a viewport occupying about 1/2 of the page on the right side.

B. Make the **XREF** layer *Current*. Double-click inside the viewport to activate model space in the viewport. In the viewport, *Xref* the **GASKETD** drawing from Chapter 28 Exercises. Use *Zoom* with an *XP* value or use the *Viewport scale* drop-down list (in the *Viewports* toolbar) to scale the gasket appropriately to paper space. Use the *Layer Properties Manager* to *Freeze* (globally) the **GASKETD|DIM** layer, then use *New VP Freeze* for all **GASKETD*** layers.

C. Option 1. Change to *PS* (double-click in paper space) and create two more viewports on *Layer* **VIEWPORT** on the left side, each equal in size to about half of the viewport on the right. Set *Layer* **XREF** *Current* and *Xref* the **GASKETB** and the **GASKETC** drawings. Change to *MS* and use the *Layer Properties Manager* or *Vplayer* to constrain only one gasket to appear in each viewport.

Option 2.   Change to *PS* and create one more viewport on *Layer* **VIEWPORT** equal to one half the size of the viewport on the right. Set *Layer* **XREF** *Current* and *Xref* only the **GASKETB** drawing. Change to *MS* and use the *Layer Properties Manager* dialog box or *Vplayer* to set the **GASKETB** visibility for the existing and new viewports. Repeat these steps for **GASKETC**.

D. Use *Zoom XP* or the *Viewport Scale* drop-down list to produce a scaled view of each new gasket. Change to *PS* and *Freeze* layer **VIEWPORT**. Use *Text* to label the gaskets. *Plot* the drawing to scale. Your plot should look similar to Figure 33-51. *Save* the drawing as **GASKETS**.

Figure 33-51 ——————————

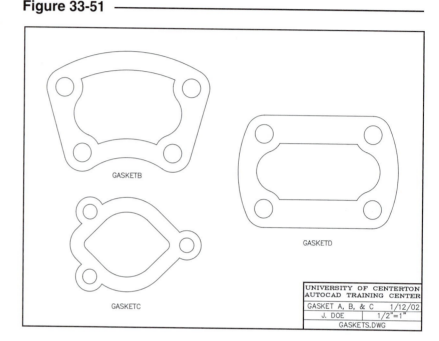

5. **Create a Color-Dependent Plot Style Table**

Start AutoCAD and begin a *New* drawing by any method. Open the *Plot Style Manager*. Use the *Add-A-Plot Style Table Wizard* to create a new Plot Style Table. In the *Begin* step, select *Start from Scratch*. In the next step, select *Color-Dependent Plot Style Table*, then proceed to assign the name **SCREEN PERCENTS**. When you are finished, check to make sure the new Plot Style Table appears in the *Plot Style Manager*.

6. **Edit a Color-Dependent Plot Style Table**

Open the *Plot Style Manager*. Double-click on the **SCREEN PERCENTS** Plot Style Table to cause the *Plot Style Table Editor* to appear. Use the *Table View* or *Form View* to assign *Screening* values of **100**, **75**, **50**, and **25** to plot styles **Color 1**, **Color 2**, **Color 3**, and **Color 4**, respectively. Select *Save & Close* to save the new settings. Reopen the new Plot Style Table to ensure your settings were saved.

7. **Create and Edit a Named Plot Style Table**

Open the *Plot Style Manager* if not yet open. Use the *Add-A-Plot Style Table Wizard* to create a new Plot Style Table. In the *Begin* step, select *Start from Scratch*. In the next step, select *Named Plot Style Table*, then proceed to assign the name **LINEWEIGHTS.** In the *Finish* step, select the *Plot Style Table Editor* button.

In the *Plot Style Table Editor*, create five plot styles named **1.20mm**, **.90mm**, **.60mm**, **.30mm**, and **.15mm**. Assign the respective settings from the *Lineweight* drop-down list. Select the *Save & Close* button. Finally, select the *Finish* button to close the wizard. In the *Plot Style Table Manager*, double-click on the new **LINEWEIGHTS.STB** file and check your settings.

8. **Set the Default Plot Style Behavior**

In AutoCAD, invoke the *Options* dialog box and select the *Plotting* tab. In the *Default Plot Style Behavior* section (upper right corner), select *Use Color Dependent Plot Style Tables*. Under *Default Plot Style Table*, select *None*. Select the *OK* button to close the *Options* dialog box and save your changes. Now, new drawings that you create will use color-dependent Plot Style Tables by default, but can later be changed to use named Plot Style Tables.

9. **Assigning Color-Dependent Plot Styles**

A. ***Open*** the **HAMMER-DIM** drawing you created in Chapter 29 Exercises. Make a plot from the *ANSI B Title Block* layout as it exists from the previous exercise (before assigning plot styles). Note that the drawing plots with all lines in the same lineweight, making it difficult to discern between object lines and dimension lines (Fig. 33-52).

Figure 33-52

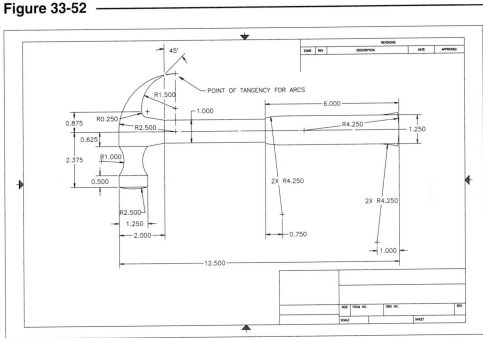

B. From the layout tab, invoke the *Page Setup* dialog box and attach the **SCREEN PERCENTS** Plot Style Table you created and edited in Exercises 5 and 6. Select the *Edit* button to view the plot styles settings (Color 1 =100, Color 2=75, etc.). Use *SaveAs* and assign the name **HAMMER-PSTYLES**.

C. Use the *Layer Properties Manager* to see which plot styles are assigned to which layers. Since the plot styles are assigned by color, you will have to change the layer colors to achieve the desired screen percentages. Make the **GEOMETRY** layer color 1 (red) and the **DIMENSIONS** layer color 3 (green).

D. Use the *Page Setup* dialog box again and check *Display Plot Styles* in the *Plot Device* tab. Check the layout to ensure the plot styles are assigned correctly. Your display should look similar to Figure 33-53. Finally, make a *Plot*. The plot should display the geometry at a 100 percent screen and the dimensions in a lighter screening. Save the drawing.

Figure 33-53

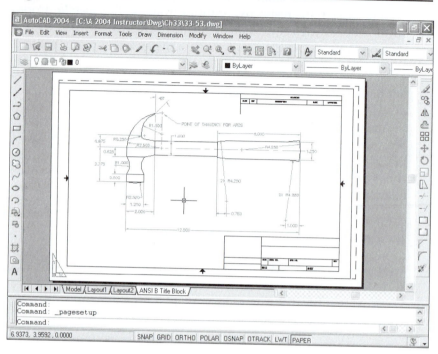

E. Experiment by changing the **DIMENSIONS** layer to color 3 or 4. Make plots to see the results.

10. **Assign Named Plot Styles**

A. *Open* the **DB_SAMP** drawing from the AutoCAD 2004/Samples folder. This drawing was created to use a named Plot Style Table. Assuming you are the general contractor, you are required to create a plot that displays the entire plan but highlights the building's concrete core.

B. From the layout tab, invoke the *Page Setup* dialog box and attach the **LINEWEIGHTS.STB** Plot Style Table you created and edited in Exercise 7. Select the *Edit* button to view the plot styles settings and style names. Also check *Display Plot Styles*. Use *SaveAs* and assign the name **VP_SAMP-PSTYLES** and save the file to your working directory.

Figure 33-54

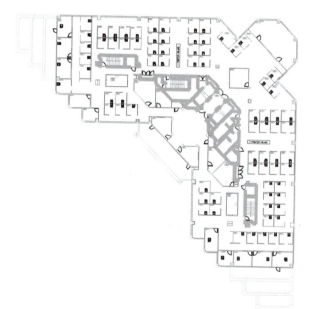

C. Open the *Layer Properties Manager*. *Freeze* all the *Cyan* layers except **E-F-TERR**. Also *Freeze* layers **FILE_CABINETS** and **FURNITURE**.

D. Assign the **.15mm** plot style to all layers. Then assign the **.90mm** plot style to layer **E_B_CORE**. Your display should look like Figure 33-54 (this figure does not display the viewport). *Save* the drawing. Make a *Plot* and check your results.

11. **Assign Named Plot Styles**

A. *Open* the **SADDL-DM** drawing from Chapter 29 Exercises. The model geometry and dimensions are drawn in the *Model* tab, and the titleblock and border are inserted into a layout tab.

B. Use Windows Explorer to search for the **ACAD.CTB**. Make a *Copy* of the file and *Rename* it **LINEWEIGHTS.CTB**. *Exit* Explorer.

C. In AutoCAD, select the layout tab and use the *Page Setup* dialog box to attach the new Plot Style Table that you created in the previous step. Also select *Display Plot Styles*. There should be no apparent change in the layout. Use *SaveAs* and assign the name **SADDL-DIM-PLOTSTYLES**.

D. Use any method to edit the new Plot Style Table so that the color of the geometry object lines have a heavy (thick) lineweight assigned and the dimension, hidden line, and center line colors have a light (thin) lineweight assigned. If the layout displays the geometry as you intend, make a *Plot* of the drawing. The resulting plot should look similar to Figure 33-55. *Save* the drawing.

Figure 33-55

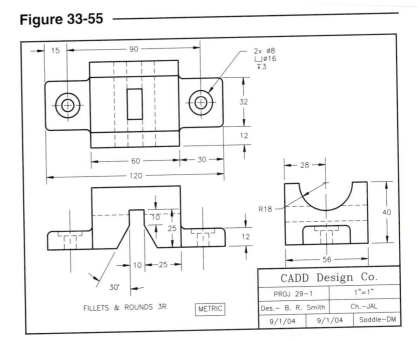

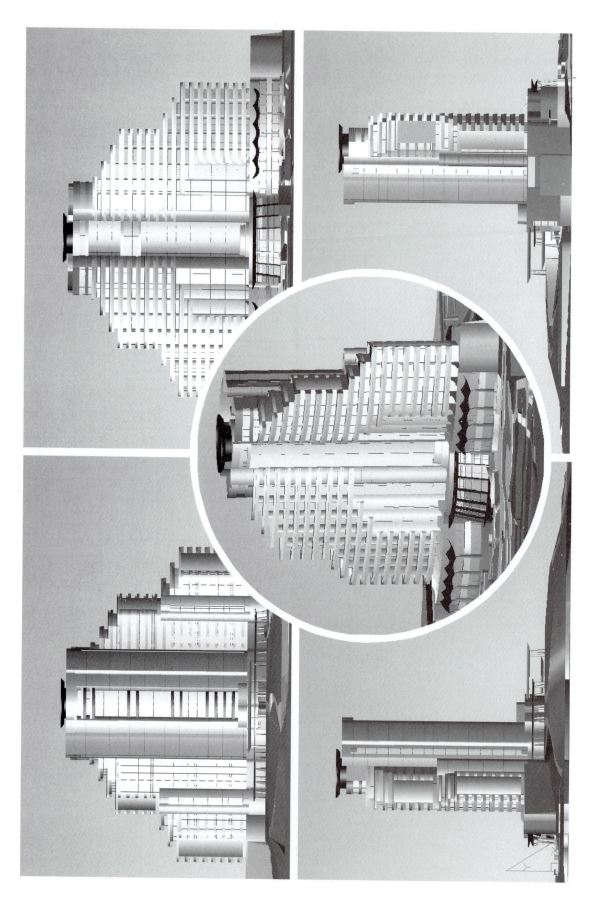

HOTEL MODEL.DWG, Courtesy of Autodesk, Inc.

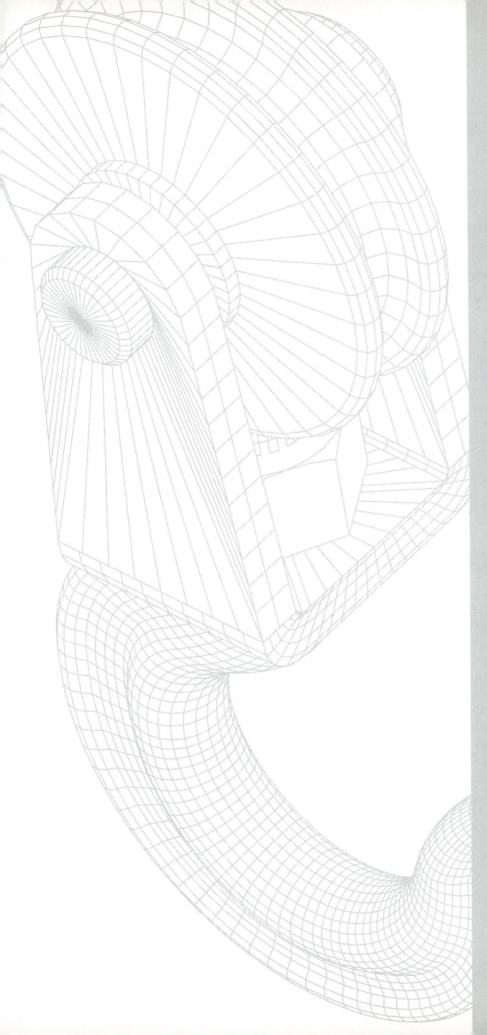

3D MODELING BASICS

Chapter Objectives

After completing this chapter you should:

1. know the characteristics of wire-frame, surface, and solid models;

2. know the six formats for 3D coordinate entry;

3. understand the orientation of the World Coordinate System (WCS);

4. be able to use the right-hand rule for orientation of the X, Y, and Z axes and for determining positive and negative rotation about an axis;

5. be able to control the appearance and positioning of the Coordinate System icon with the *Ucsicon* command.

CONCEPTS

Three basic types of 3D (three-dimensional) models created by CAD systems are used to represent actual objects. They are:

1. Wireframe models
2. Surface models
3. Solid models

These three types of 3D models range from a simple description to a very complete description of an actual object. The different types of models require different construction techniques, although many concepts of 3D modeling are the same for creating any type of model on any type of CAD system.

Wireframe Models

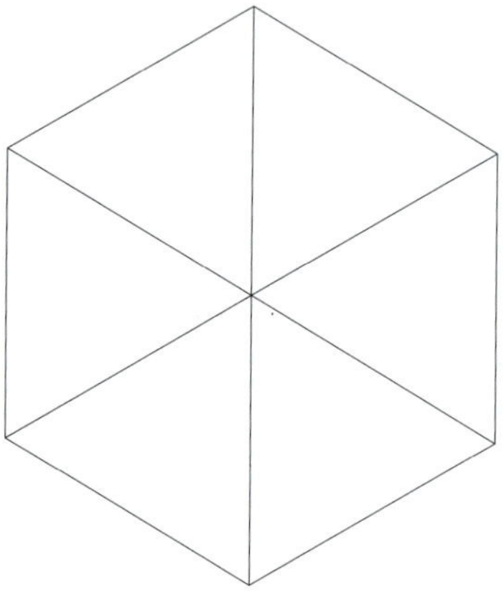

"Wireframe" is a good descriptor of this type of modeling. A wireframe model of a cube is like a model constructed of 12 coat-hanger wires. Each wire represents an <u>edge</u> of the actual object. The <u>surfaces</u> of the object are <u>not</u> defined; only the boundaries of surfaces are represented by edges. No wires exist where edges do not exist. The model is see-through since it has no surfaces to obscure the back edges. A wireframe model has complete dimensional information but contains no volume. Examples of wireframe models are shown in Figures 34-1 and 34-2.

Wireframe models are relatively easy and quick to construct; however, they are not very useful for visualization purposes because of their "transparency." For example, does Figure 34-1 display the cube as if you are looking toward a top-front edge or looking toward a bottom-back edge? Wireframe models tend to have an optical illusion effect, allowing you to visualize the object from two opposite directions unless another visual clue such as perspective is given.

With AutoCAD a wireframe model is constructed by creating 2D objects in 3D space. The *Line, Circle, Arc,* and other 2D *Draw* and *Modify* commands are used to create the "wires," but 3D coordinates must be specified. The cube in Figure 34-1 was created with 12 *Line* segments. AutoCAD provides all the necessary tools to easily construct, edit, and view wireframe models.

Figure 34-1 ——————

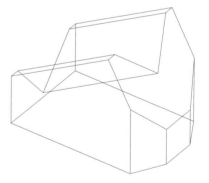

Figure 34-2 ——————

A wireframe model offers many advantages over a 2D engineering drawing. Wireframe models are useful in industry for providing computerized replicas of actual objects. A wireframe model is dimensionally complete and accurate for all three dimensions. Visualization of a wireframe is generally better than a 2D drawing because the model can be viewed from any position or a perspective can be easily attained. The 3D database can be used to test and analyze the object <u>three dimensionally</u>. The sheet metal industry, for example, uses wireframe models to calculate flat patterns complete with bending allowances. A wireframe model can also be used as a foundation for construction of a surface model. Because a wireframe describes edges but not surfaces, wireframe modeling is appropriate to describe objects with planar or single-curved surfaces, but not compound curved surfaces.

Surface Models

Surface models provide a better description of an object than a wireframe, principally because the surfaces as well as the edges are defined. A surface model of a cube is like a cardboard box—all the surfaces and edges are defined, but there is nothing inside. Therefore, a surface model has volume but no mass. A surface model provides an excellent visual representation of an actual 3D object because the front surfaces obscure the back surfaces and edges from view. Figure 34-3 shows a surface model of a cube, and Figure 34-4 displays a surface model of a somewhat more complex shape. Notice that a surface model leaves no question as to which side of the object you are viewing.

Surface models require a relatively tedious construction process. Each surface must be constructed individually. Each surface must be created in, or moved to, the correct orientation with respect to the other surfaces of the object. In AutoCAD, a surface is constructed by defining its edges. Often, wireframe models are used as a framework to build and attach surfaces. The complexity of the construction process of a surface is related to the number and shapes of its edges. AutoCAD is not a complete surface modeler. The tools provided allow construction of simple planar and single curved surfaces, but there are few capabilities for construction of double-curved or other complex surfaces. No NURBS (Non-Uniform Rational B-Splines) surfacing capabilities exist in AutoCAD, although these capabilities are available in other Autodesk products such as Mechanical Desktop® and Inventor®.

Most CAD systems, including AutoCAD, can display surface and solid models in wireframe, hidden, and shaded representation. Figure 34-4 displays an object in "hidden" representation. Figure 34-5 shows the same object in "wireframe" representation. Surface and solid modeling systems can generally display objects in any of the three modes (wireframe, shaded, and hidden) at any time. Although hidden and shaded modes enhance your ability to visualize the object, wireframe representation is often used during the construction process so all edges can be seen and selected, if needed. (See Chapter 35, 3D Display and Viewing, for more information.)

Figure 34-3

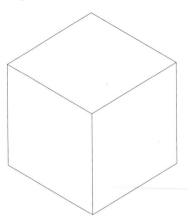

Figure 34-4

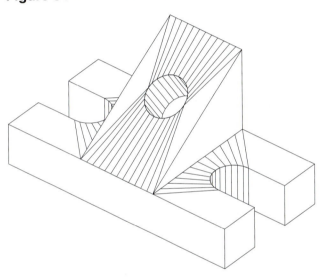

Figure 34-5

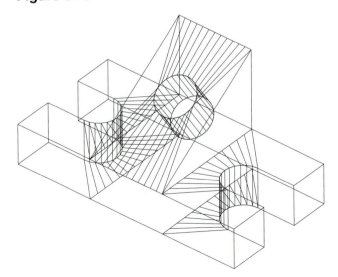

Solid Models

Solid modeling is the most complete and descriptive type of 3D modeling. A solid model is a complete computerized replica of the actual object. A solid model contains the complete surface and edge definition, as well as description of the interior features of the object. If a solid model is cut in half (sectioned), the interior features become visible. Since a solid model is "solid," it can be assigned material characteristics and is considered to have mass. Because solid models have volume and mass, most solid modeling systems include capabilities to automatically calculate volumetric and mass properties.

Solid model construction techniques are generally much simpler (and much more fun) than those of surface models. AutoCAD's solid modeler, called ACIS, is a <u>hybrid</u> modeler. That is, ACIS is a combination CSG (Constructive Solid Geometry) and B-Rep (Boundary Representation) modeler. CSG is characterized by its simple and straightforward construction techniques of combining primitive shapes (boxes, cylinders, wedges, etc.) utilizing Boolean operations (*Union*, *Subtract*, and *Intersect*, etc.). Boundary Representation modeling defines a model in terms of its edges and surfaces (boundaries) and determines the solid model based on which side of the surfaces the model lies. The user interface and construction techniques used in ACIS (primitive shapes combined by Boolean operations) are CSG-based, whereas the B-Rep capabilities are invoked automatically to display models in mesh representation and are transparent to the user. Figure 34-6 displays a solid model constructed of simple primitive shapes combined by Boolean operations.

Figure 34-6

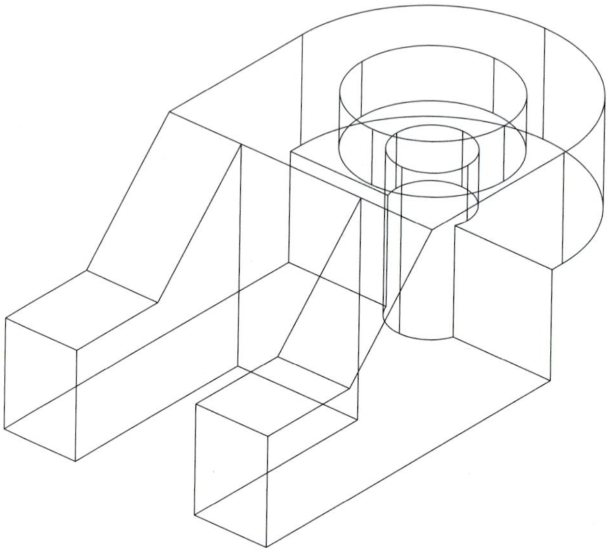

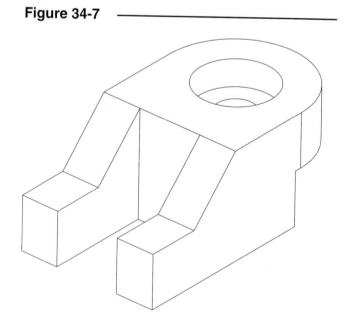

The CSG modeling techniques offer you the advantage of complete and relatively simple editing. CSG construction typically begins by specifying dimensions for simple primitive shapes such as boxes or cylinders, then combines the primitives using Boolean operations to create a "composite" solid. Other primitives and/or composite solids can be combined by the same process. Several repetitions of this process can be continued until the desired solid model is finally achieved. CSG construction techniques are discussed in detail in Chapter 38.

Solid models, like surface models, are capable of wireframe, shaded, or hidden display. Generally, wireframe display is used during construction (see Figure 34-6), and hidden or shaded representation is used to display the finished model (Fig. 34-7).

Figure 34-7

3D COORDINATE ENTRY

When creating a model in three-dimensional drawing space, the concept of the X and Y coordinate system, which is used for two-dimensional drawing, must be expanded to include the third dimension, Z, which is measured from the origin in a direction perpendicular to the plane defined by X and Y. Remember that two-dimensional CAD systems use X and Y coordinate values to define and store the location of drawing elements such as *Lines* and *Circles*. Likewise, a three-dimensional CAD system keeps a database of X, Y, and Z coordinate values to define locations and sizes of two- and three-dimensional elements. For example, a *Line* is a two-dimensional object, yet the location of its endpoints in three-dimensional space must be speci-fied and stored in the database using X, Y, and Z coordinates (Fig. 34-8). The X, Y, and Z coordinates are always defined in that order, delineated by commas. The AutoCAD Coordinate Display (*Coords*) dis-plays X, Y, and Z values.

Figure 34-8

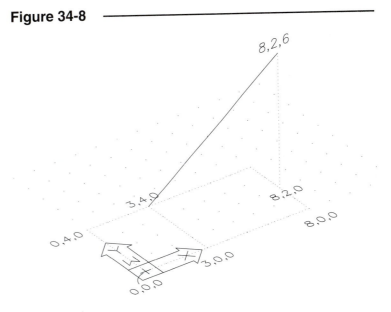

The icon that appears in the lower-left corner of the AutoCAD Drawing Editor is the Coordinate System icon (sometimes called the UCS icon) (Fig. 34-9). This UCS icon displays the directions for the X, Y, and the Z axes and can be made to locate itself at the origin, 0,0. The X coordinate values increase going to the right along the X axis, and the Y values increase going upward along the Y axis. This is the UCS default icon that appears when AutoCAD is in the *2D wireframe* option of *Shademode*.

Figure 34-9

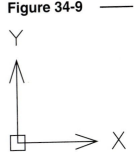

A second form of the UCS icon is available that also indicates directions for all three axes, X, Y, and Z (Fig. 34-10). In addition to the 3D arrows and "X," "Y," and "Z" labels, this icon has color to assist in indicating orientation. The X axis is red, the Y axis is green, and the Z axis is cyan (light blue). The setting for the *Shademode* command determines the display of this or the other icons. To display this icon, change the *Shademode* setting from *2D wireframe* to *3D wire-frame* or to one of the other "shaded" settings (see "*Shademode*" in Chapter 35).

Figure 34-10

A third Coordinate System icon is available in AutoCAD (Fig. 34-11). This icon is available using the *Ucsicon* command (see "UCS Icon Control" at the end of this chapter). This icon is the traditional icon that has been used for releases of AutoCAD previous to AutoCAD 2000. Although this icon does not give the direction of the Z axis, it illustrates the orientation of the XY plane at a glance. Sadly, this icon is available only during the *2D Wireframe* option of *Shademode*.

Figure 34-11

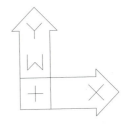

2002

You can use any one of the three icons when you construct 3D models. It is the opinion of this author that, although the Z axis is not shown, the "2D" icon gives the user immediate orientation of the XY plane and, therefore, orientation in 3D space. For the exact reason that the newer, 3-pole icons display all three axes, it is more confusing to process which pole is which and, therefore, takes longer to distinguish 3D orientation—the labels must be read or the colors memorized. Displaying only two of the three axes as well as the "planar" effect of the "2D" icon makes it the better choice for immediately conveying the orientation of the XY plane. However, the "2D" icon is not available in AutoCAD 2004 for shaded display modes.

NOTE: Most of the illustrations in this book depicting 3D objects display the "2D" icon for these reasons.
1.   This textbook does not have the advantage of displaying color (needed for the 3-pole icon).
2.   The 2-pole, "planar" feature helps you distinguish the orientation of the XY plane easier.
3.   The "W" on the icon communicates when the World Coordinate System is current.

3D Coordinate Entry Formats

Because construction in three dimensions requires the definition of X, Y, <u>and</u> Z values, the methods of coordinate entry used for 2D construction must be expanded to include the Z value. The five methods of command entry used for 2D construction are valid for 3D coordinates with the addition of a Z value specification. Relative polar coordinate specification (@dist<angle) is expanded to form two other coordinate entry methods available explicitly for 3D coordinate entry. The six methods of coordinate entry for 3D construction follow:

1.	**Interactive coordinates**	PICK	Use the cursor to select points on the screen. *OSNAP*, Object Snap Tracking, or point filters must be used to select a point in 3D space; otherwise, points selected are <u>on the XY plane</u>.
2.	**Absolute coordinates**	X,Y,Z	Enter explicit X, Y, and Z values relative to point 0,0.
3.	**Relative rectangular coordinates**	@X,Y,Z	Enter explicit X, Y, and Z values relative to the last point.
4.	**Cylindrical coordinates (relative)**	@dist<angle,Z	Enter a distance value, an angle in the XY plane value, and a Z value, all relative to the last point.
5.	**Spherical coordinates (relative)**	@dist<angle<angle	Enter a distance value, an angle <u>in</u> the XY plane value, and an angle <u>from</u> the XY plane value, all relative to the last point.
6.	**Direct distance entry**	dist,direction	Enter (type) a value, and move the cursor in the desired direction. To draw in 3D effectively, <u>ORTHO</u> must be <u>On</u>.

Cylindrical and spherical coordinates can be given <u>without</u> the @ symbol, in which case the location specified is relative to point 0,0,0 (the origin). This method is useful if you are creating geometry centered around the origin. Otherwise, the @ symbol is used to establish points in space relative to the last point.

Examples of each of the six 3D coordinate entry methods are illustrated in the following section. In the illustrations, the orientation of the observer has been changed from the default plan view in order to enable the visibility of the three dimensions.

Interactive Coordinate Specification

Figure 34-12 illustrates using the interactive method to PICK a location in 3D space. *OSNAP* must be used in order to PICK in 3D space. Any point PICKed with the input device without *OSNAP* will result in a location on the XY plane. In this example, the *Endpoint OSNAP* mode is used to establish the "Specify next point:" of a second *Line* by snapping to the end of an existing vertical *Line* at 8,2,6:

Figure 34-12

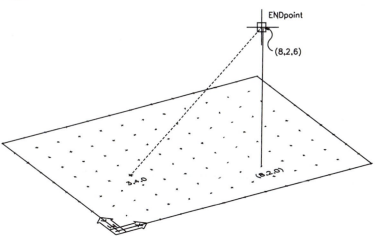

```
Command: line
Specify first point: 3,4,0
Specify next point or [Undo]: endpoint of PICK
```

Absolute Coordinates

Figure 34-13 illustrates the absolute coordinate entry to draw the *Line*. The endpoints of the *Line* are given as explicit X,Y,Z coordinates:

Figure 34-13

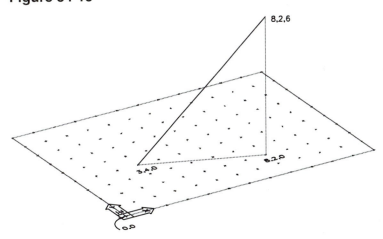

```
Command: line
Specify first point: 3,4,0
Specify next point or [Undo]: 8,2,6
```

Relative Rectangular Coordinates

Relative rectangular coordinate entry is displayed in Figure 34-14. The "Specify first point:" of the *Line* is given in absolute coordinates, and the "Specify next point:" end of the *Line* is given as X,Y,Z values relative to the last point:

Figure 34-14

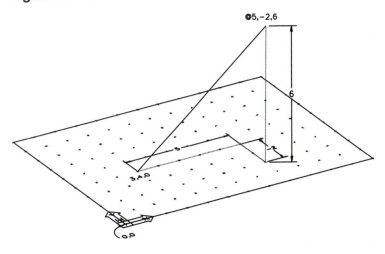

```
Command: line
Specify first point: 3,4,0
Specify next point or [Undo]: @5,-2,6
```

Cylindrical Coordinates (Relative)

Cylindrical and spherical coordinates are an extension of polar coordinates with a provision for the third dimension. Relative cylindrical coordinates give the distance <u>in</u> the XY plane, angle <u>in</u> the XY plane, and Z dimension and can be relative to the last point by prefixing the @ symbol. The *Line* in Figure 34-15 is drawn with absolute and relative cylindrical coordinates. (The *Line* established is approximately the same *Line* as in the previous figures.)

    Command: *line*
    Specify first point: 3,4,0
    Specify next point or [Undo]:
    @5<-22,6

Spherical Coordinates (Relative)

Spherical coordinates are also an extension of polar coordinates with a provision for specifying the third dimension in angular format. Spherical coordinates specify a distance, an angle <u>in</u> the XY plane, and an angle <u>from</u> the XY plane and can be relative to the last point by prefixing the @ symbol. The distance specified is a <u>3D distance</u>, not a distance in the XY plane. Figure 34-16 illustrates the creation of approximately the same line as in the previous figures using absolute and relative spherical coordinates.

    Command: *line*
    Specify first point: 3,4,0
    Specify next point or [Undo]:
    @8<-22<48

Direct Distance Entry

Direct distance entry operates as it does in 2D, except <u>ORTHO must be On</u> in order to draw effectively by this method in 3D space. When *ORTHO* is *On*, objects drawn in 3D space are limited to a <u>plane parallel with the current XY plane.</u> For example, to draw a square in 3D space with sides of 3 units (Fig. 34-17), begin by using *Line* and specifying the "Specify first point:" at the

Figure 34-15

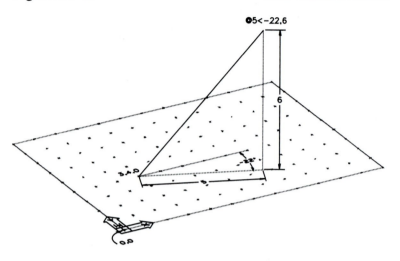

Figure 34-16

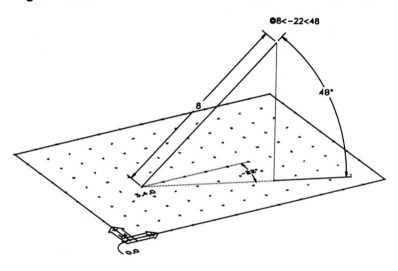

Figure 34-17

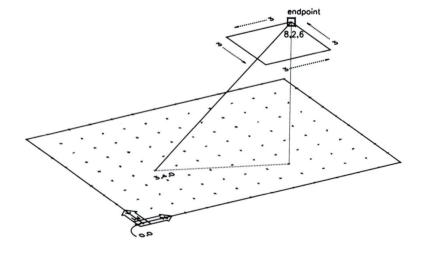

Endpoint of the existing diagonal *Line* shown (point 8,2,6). To draw the first 3 unit segment, turn on *ORTHO*, move the cursor in the desired direction, and enter "3." The *ORTHO* feature "locks" the *Line* to a plane parallel to the current XY plane. Next, move the cursor in the desired (90 degree) direction and enter "3," and so on. The command syntax follows:

```
Command: line
Specify first point: endpoint of PICK
Specify next point or [Undo]: <Ortho on> 3
Specify next point or [Undo]: 3
Specify next point or [Undo]: 3
Specify next point or [Undo]: 3
Specify next point or [Undo]: Enter
Command:
```

Point Filters

Point filters are used to filter X and/or Y and/or Z coordinate values from a location PICKed with the pointing device. Point filtering makes it possible to build an X,Y,Z coordinate specification from a combination of point(s) selected on the screen and point(s) entered at the keyboard. A .XY (read "point XY") filter would extract, or filter, the X and Y coordinate value from the location PICKed and then prompt you to enter a Z value. Valid point filters are listed below.

.X	Filters (finds) the X component of the location PICKed with the pointing device.
.Y	Filters the Y component of the location PICKed.
.Z	Filters the Z component of the location PICKed.
.XY	Filters the X and Y components of the location PICKed.
.XZ	Filters the X and Z components of the location PICKed.
.YZ	Filters the Y and Z components of the location PICKed.

The .XY filter is the most commonly used point filter for 3D construction and editing. Because 3D construction often begins on the XY plane of the current coordinate system, elements in Z space are easily constructed by selecting existing points on the XY plane using .XY filters and then entering the Z component of the desired 3D coordinate specification by keyboard. For example, in order to draw a line in Z space two units above an existing line on the XY plane, the .XY filter can be used in combination with *Endpoint* *OSNAP* to supply the XY component for the new line. See Figure 34-18 for an illustration of the following command sequence:

Figure 34-18

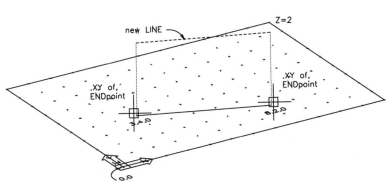

```
Command: line
Specify first point: .XY of
endpoint of PICK (Select one Endpoint of the existing line on the XY plane.)
(need Z) 2
Specify next point or [Undo]: .XY of
endpoint of PICK (Select the other Endpoint of the existing line.)
(need Z) 2
```

Typing **.XY** and pressing **Enter** cause AutoCAD to respond with "of" similar to the way "of" appears after typing an *OSNAP* mode. When a location is specified using an .XY filter, AutoCAD responds with "(need Z)" and, likewise, when other filters are used, AutoCAD prompts for the missing component(s).

Object Snap Tracking

Object Snap Tracking functions in 3D similar to the way that *OSNAP* and Direct Distance entry operate—that is, when a point is "acquired" by one of the *OSNAP* modes, tracking occurs in the plane of the acquired point. To be more specific, when you acquire a point in 3D space (not on the XY plane), tracking vectors appear originating from the acquired object <u>in a plane parallel to the current XY plane</u>.

Figure 34-19

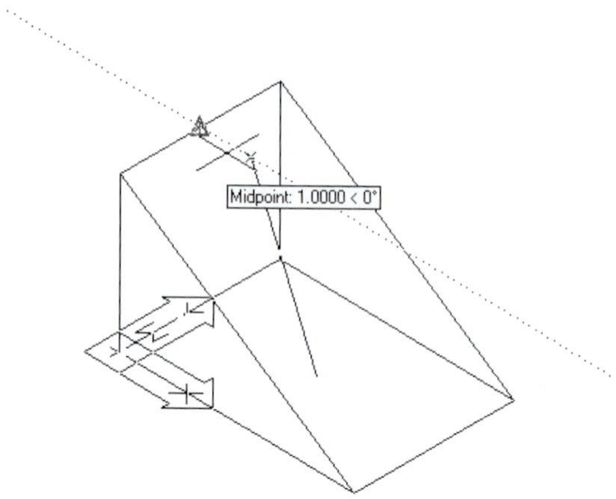

For example, assume a wedge was created with its base on the XY plane. You can track on a plane in 3D space parallel to the current XY plane that passes through the acquired point (Fig. 34-19). In other words, begin the "first point" of a line at a point on the XY plane (at the center of the base, in this case). For the "next point," acquire a point in 3D space with Object Snap Tracking. Note that tracking vectors are generated from that acquired point into space on an imaginary plane that is parallel with the current XY plane (and parallel with the X axis in this case).

COORDINATE SYSTEMS

In AutoCAD two kinds of coordinate systems can exist, the <u>World Coordinate System</u> (WCS) and one or more <u>User Coordinate Systems</u> (UCS). The World Coordinate System <u>always</u> exists in any drawing and cannot be deleted. The user can also create and save multiple User Coordinate Systems to make construction of a particular 3D geometry easier. <u>Only one coordinate system can be active</u> at any one time in any one view or viewport, either the WCS or one of the user-created UCSs. You can have more than one UCS active at any one time if you have several viewports active (model space or layout viewports).

The World Coordinate System (WCS) and WCS Icon

The World Coordinate System (WCS) is the default coordinate system in AutoCAD for defining the position of drawing objects in 2D or 3D space. The WCS is always available and cannot be erased or removed but is deactivated temporarily when utilizing another coordinate system created by the user (UCS). The icon that appears (by default) at the lower-left corner of the Drawing Editor (Fig. 34-9) indicates the orientation of the WCS. The icon, whose appearance (*ON, OFF*) is controlled by the *Ucsicon* command, appears for the WCS and for any UCS.

Remember that three possible UCS icons are available and determined by both your choice in the *Ucsicon* command *Properties* dialog box and your current *Shademode* setting. The <u>"3D" (3-pole) colored icon</u> (see Figure 34-10) that appears with all *Shademode* options except *2D Wireframe* <u>does not indicate</u> if the World Coordinate System is current. Both of the other icons (see Figures 34-9 and 34-11) <u>do indicate if the WCS is active</u>, but these two icons are displayed only during the *2D wireframe* option of *Shademode*. The "3D" version of the icon (see Figure 34-9) displays a small square on the XY plane if the WCS is current, but this square disappears when another UCS is current. The "2D" version of the icon (see Figure 34-11) displays the letter "W" when the WCS is current. This feature is another reason why the older "2D" icon is better for determining 3D orientation. For this reason and other reasons (see discussion after Figure 34-11), the traditional "2D" icon is displayed in most of the figures in this text.

The orientation of the WCS with respect to Earth may be different among CAD systems. In AutoCAD the WCS has an architectural orientation such that the XY plane is a horizontal plane with respect to Earth, making Z the height dimension. A 2D drawing (X and Y coordinates only) is thought of as being viewed from above, sometimes called a <u>plan</u> view. Therefore, in a 3D AutoCAD drawing, X is the width dimension, Y is the depth dimension, and Z is height. This default orientation is like viewing a floor <u>plan</u>—from above (Fig. 34-20).

Some CAD systems that have a mechanical engineering orientation align their World Coordinate Systems such that the XY plane is a vertical plane intended for drawing a front view. In other words, some mechanical engineering CAD systems define X as the width dimension, Y as height, and Z as depth.

Figure 34-20

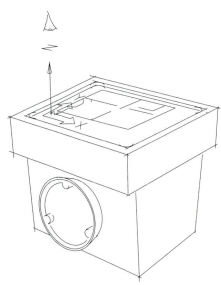

User Coordinate Systems (UCS) and Icons

There are no User Coordinate Systems that exist as part of the AutoCAD default template drawings (ACAD.DWT and ACADISO.DWT) as they come "out of the box." UCSs are created to suit the 3D model when and where they are needed.

Creating geometry is relatively simple when having to deal only with X and Y coordinates, such as in creating a 2D drawing, or when creating simple 3D geometry with uniform Z dimensions. However, 3D models containing complex shapes on planes not parallel with the XY plane are good candidates for UCSs (Fig. 34-21).

A User Coordinate System is thought of as a construction plane created to simplify creation of geometry on a specific plane or surface of the object. The user creates the UCS, aligning its XY plane with a surface of the object, such as along an inclined plane, with the UCS origin typically at a corner or center of the surface.

The user can then create geometry aligned with that plane by defining only X and Y coordinate values of the current UCS (Fig. 34-22). The *SNAP*, *Polar Tracking*, and *GRID* automatically align with the current coordinate system, providing *SNAP* points and enhancing visualization of the construction plane. Practically speaking, it is easier in some cases to specify only X and Y coordinates with respect to a specific plane on the object rather than calculating X, Y, and Z values with respect to the World Coordinate System.

Figure 34-21

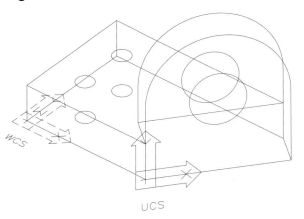

Figure 34-22

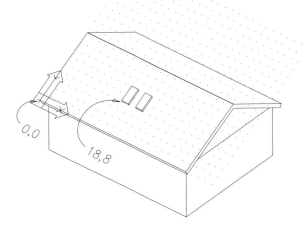

User Coordinate Systems can be created by any of several options of the *UCS* command. Once a UCS has been created, it becomes the current coordinate system. <u>Only one coordinate system can be active in one view or viewport</u>; therefore, it is suggested that UCSs be saved (by using the *Save* option of *UCS*) for possible future geometry creation or editing.

When a UCS is created, the icon at the lower-left corner of the screen can be made to automatically align itself with the UCS along with *SNAP* and *GRID*. The letter "W," however, appears only on the "2D" icon and only when the WCS is active (when the WCS is the current coordinate system). When creating 3D geometry, it is recommended that the *ORigin* option of the *Ucsicon* command be used to place the icon always at the origin of the current UCS, rather than in the lower-left corner of the screen. Since the origin of the UCS is typically specified as a corner or center of the construction plane, aligning the Coordinate System icon with the current origin aids your visualization of the UCS orientation. The Coordinate Display (*Coords*) in the Status Bar always displays the X, Y and Z values of the <u>current</u> coordinate system, whether it is WCS or UCS.

THE RIGHT-HAND RULE

AutoCAD complies with the right-hand rule for defining the orientation of the X, Y, and Z axes. The right-hand rule states that if your right hand is held partially open, the thumb, first, and middle fingers define positive X, Y, and Z directions, respectively, and positive rotation about any axis is like screwing in a light bulb.

More precisely, if the thumb and first two fingers are held out to be mutually perpendicular, the thumb points in the positive X direction, the first finger points in the positive Y direction, and the middle finger points in the positive Z direction (Fig. 34-23).

In this position, looking toward your hand from the tip of your middle finger is like the default AutoCAD viewing orientation—positive X is to the right, positive Y is up, and positive Z is toward you.

Figure 34-23 ——————————————

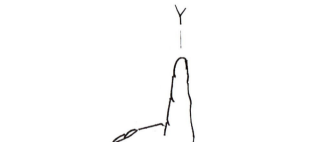

Positive rotation about any axis is <u>counterclockwise looking down the axis toward the origin</u>. For example, when you view your right hand, positive rotation about the X axis is as if you look down your thumb toward the hand (origin) and twist your hand counterclockwise (Fig. 34-24).

Figure 34-24 ——————————————

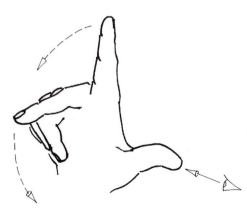

Figure 34-25 shows the 2D UCS icon in a +90 degree rotation about the X axis (like the previous figure). This orientation is typical for setting up a <u>front</u> view UCS for drawing on a plane parallel with the front surface of an object.

Figure 34-25 ————

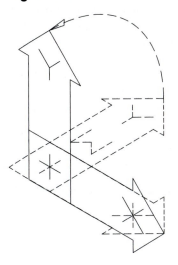

Positive rotation about the Y axis would be as if looking down your first finger toward the hand (origin) and twisting counterclockwise (Fig. 34-26).

Figure 34-26 ————

Figure 34-27 illustrates the 2D UCS icon with a +90 degree rotation about the Y axis (like the previous figure). To avoid confusion in relating this figure to Figure 34-26, consider that the icon is oriented on the XY plane as a horizontal plane in its original (highlighted) position, whereas Figure 34-26 shows the hand in an upright position. (Compare Figures 34-20 and 34-10.)

Figure 34-27 ————

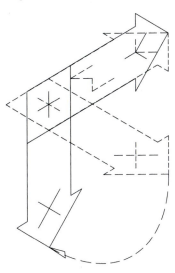

Positive rotation about the Z axis would be as if looking
toward your hand from the end of your middle finger and
twisting counterclockwise (Fig. 34-28).

Figure 34-28 —————————

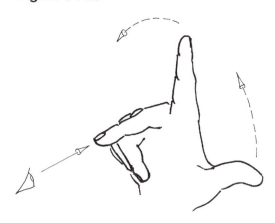

Figure 34-29 shows the 2D UCS icon with a +90 degree rotation
about the Z axis. Again notice the orientation of the icon is horizontal, whereas the hand (Fig. 34-28) is upright.

Figure 34-29 —————————

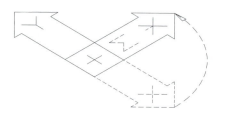

UCS ICON CONTROL

UCSICON

Pull-down Menu	COMMAND (TYPE)	ALIAS (TYPE)	Short-cut	Screen (side) Menu	Tablet Menu
View *Display >* *UCS Icon >*	*UCSICON*	...	...	VIEW 2 UCSicon	L,2

The *Ucsicon* command controls the appearance and positioning of the Coordinate System icon. In order
to aid your visualization of the current UCS or the WCS, it is <u>highly</u> recommended that the Coordinate
System icon be turned *ON* and positioned at the
ORigin.

Figure 34-30 —————————

```
Command: ucsicon
Enter an option
[ON/OFF/All/Noorigin/ORigin/Properties] <ON>: on
Command:
```

This option causes the Coordinate System icon to
appear (Fig. 34-30).

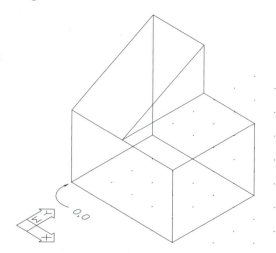

The *Origin* setting causes the icon to move to the origin of the current coordinate system (Fig. 34-31).

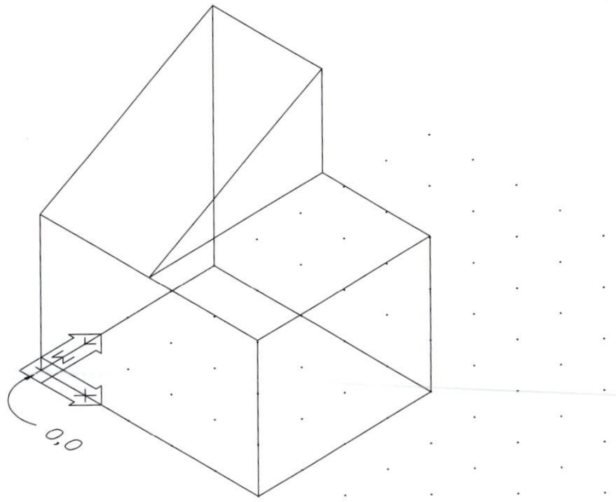

The icon does not always appear at the origin after using some viewing commands like *3Dorbit*. This is because *3Dorbit* may force the geometry against the border of the graphics screen (or viewport) area, which prevents the icon from aligning with the origin. In order to cause the icon to appear at the origin, try using *Zoom* with a **.9X** magnification factor. This action usually brings the geometry slightly in from the border and allows the icon to align itself with the origin.

Options of the *Ucsicon* command are:

Figure 34-31

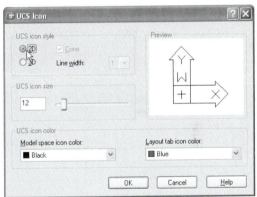

ON   Turns the Coordinate System icon on.

OFF   Turns the Coordinate System icon off.

All   Causes the *Ucsicon* settings to be effective for all viewports.

Noorigin   Causes the icon to appear always in the lower-left corner of the screen, not at the origin.

ORigin   Forces the placement and orientation of the icon to align with the origin of the current coordinate system.

Properties

The *Properties* option produces the *UCS Icon* dialog box (Fig. 34-32). Here you can select the *UCS icon style* in the top section. Select from *2D* or *3D*. If *3D* is selected, you also can use the *Cone* option to produce a 3D effect at the X and Y pole ends and increase the *Line width* to make the icon more prominent.

This text displays most figures using the *2D UCS icon style* because 1) it displays whether the WCS or other UCS is current, which is not communicated by the 3-pole axis, 2) this textbook does not have the advantage of displaying color—needed for the 3-pole icon, and 3) the 2-pole, "planar" feature helps you distinguish the orientation of the XY plane easier.

Figure 34-32

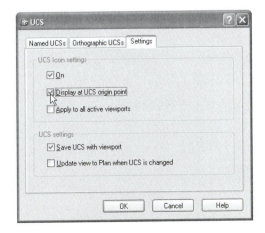

Other options in the *UCS Icon* dialog box include an adjustment for the *UCS icon size* (in pixels) and the *UCS icon color* for model space and paper space. (The default option for *Model space icon color* is either *Black* or *White*—the opposite of whatever you set for the *Model* tab background color in the *Options* dialog box.)

Figure 34-33

The *UCS Manager* can be invoked (type *Ucsman*) to control the appearance of the UCS icon (Fig. 34-33). Note the same options are available in the *UCS Manager* to turn the icon on or off, display it at the origin or not, and to apply the settings to all viewports. You can also control the appearance of the *Ucsicon* by selecting *Display* from the view pull-down menu.

Now that you know the basics of 3D modeling, you will develop your skills in viewing and displaying 3D models (Chapter 35, 3D Display and Viewing). It is imperative that you are able to view objects from different viewpoints in 3D space before you learn to construct them.

CHAPTER EXERCISES

1. What are the three types of 3D models?

2. What characterizes each of the three types of 3D models?

3. What kind of modeling techniques does AutoCAD's solid modeling system use?

4. What are the five formats for 3D coordinate specification?

5. Examine the 3D geometry shown in Figure 34-34. Specify the designated coordinates in the specified formats below.

   A. Give the coordinate of corner D in absolute format.

   B. Give the coordinate of corner F in absolute format.

   C. Give the coordinate of corner H in absolute format.

   D. Give the coordinate of corner J in absolute format.

   E. What are the coordinates of the line that define edge F-I?

   F. What are the coordinates of the line that define edge J-I?

   G. Assume point E is the "last point." Give the coordinates of point I in relative rectangular format.

   H. Point I is now the last point. Give the coordinates of point J in relative rectangular format.

   I. Point J is now the last point. Give the coordinates of point A in relative rectangular format.

   J. What are the coordinates of corner G in cylindrical format (from the origin)?

   K. If E is the last point, what are the coordinates of corner C in relative cylindrical format?

Figure 34-34 ——————————————

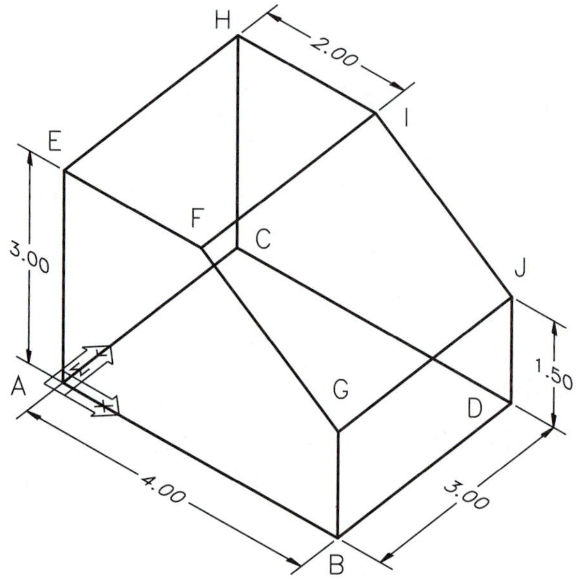

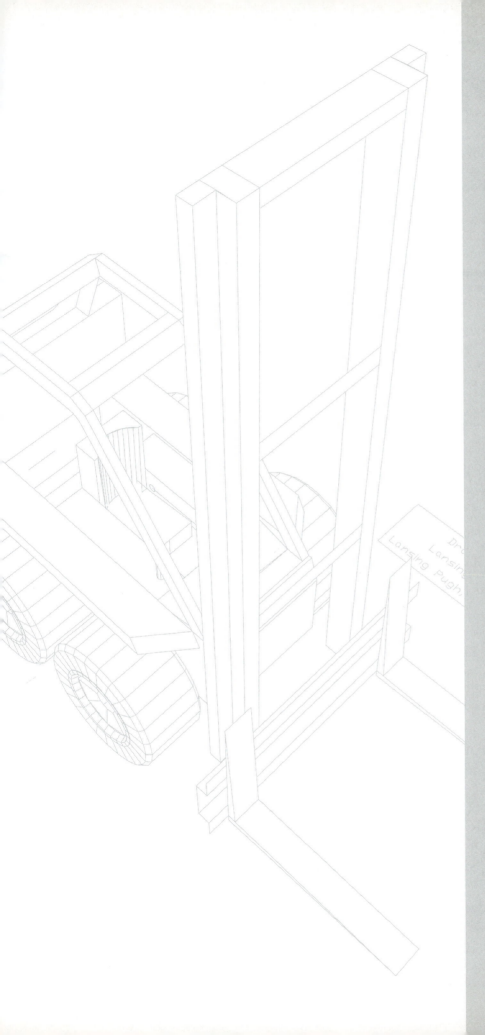

35

3D DISPLAY AND VIEWING

Chapter Objectives

After completing this chapter you should:

1. be able to use *Shademode* to generate a wireframe, hidden, and four shaded displays of an object;

2. be able to use *HIsettings* to specify the display of hidden lines;

3. be able to generate a front, top, side, and several isometric viewpoints of a 3D model with the 3D views;

4. be able to use the *3Dorbit* command to dynamically change the viewpoint of the 3D model;

5. be able to generate perspective views, adjust distance, zoom, pan, and put an object in continuous motion with the 3D orbit commands;

6. be able to use the *Vpoint* command with the *Vector*, *Rotate*, and *Tripod* options to view a 3D model from various viewpoints;

7. be able to create 3D configurations (top, front, side, and isometric views) of viewports in model space with the *Vports* command.

3D VIEWING AND DISPLAY

Display Commands

AutoCAD provides you with many alternatives for displaying surface and solid models. The default display is a "wireframe" representation in which all edges are visible. During the construction process, use a wireframe display or a "hidden" display in which the hidden edges are shown in a different line-type or color. Use a more realistic, "shaded" display for checking the model and for the final presentation. These commands provide the display alternatives:

Shademode   The *Shademode* command allows you to display surface or solid objects in several shaded or unshaded representations, including wireframe, hidden, and four shaded alternatives. The shading options fill the surfaces with the object color, calculate light reflection, and apply gradient shading.

Hlsettings   New for AutoCAD 2004, this command allows you to specify a different color or linetype to use for "hidden" lines. In this way you can "see" the edges that would normally be obscured from view, thus making the construction process easier.

Viewing Commands

These commands allow you to change the direction from which you view a 3D model or otherwise affect the view of the 3D model:

View   The *View* command has several options. You can select from preset orthographic views, including isometric viewpoints. *View* allows you to save and restore 3D viewpoints. *View* also includes an option for saving the current UCS with the view.

3Dorbit   *3Dorbit* controls interactive viewing of objects in 3D. You can dynamically view the object from any point, even while the object remains shaded.

3Dcorbit   Like *3Dorbit*, the *3Dcorbit* command enables you to set the 3D objects into continuous motion while shaded, hidden, or in any other *Shademode*.

3Dpan   Like real-time *Pan*, *3Dpan* enables you to drag the 3D objects about the view interactively while in any *Shademode*.

3Dzoom   Like real-time *Zoom*, *3Dzoom* keeps the 3D objects in their current *Shademode*.

3Dswivel   This command simulates the effect of turning the camera.

3Dclip   *3Dclip* allows you to establish front and/or back clipping planes in the *Adjust Clipping Planes* window.

3Ddistance   Use this command to make objects appear closer or farther away. If the objects are in perspective mode, you can control the amount of perspective.

Vpoint   *Vpoint* allows you to change your viewpoint of a 3D model. The object remains stationary while the viewpoint of the observer changes. Three options are provided: *Vector*, *Tripod*, and *Rotate*.

Plan   The *Plan* command automatically gives the observer a plan (top) view of the object. The plan view can be with respect to the WCS (World Coordinate System) or an existing UCS (User Coordinate System).

Zoom   Although *Zoom* operates in 3D just as in 2D, the *Zoom Previous* option restores the previous 3D view.

Vports   You can use *Vports* effectively during construction of 3D objects to divide the screen into several sections, each displaying a different standard 3D view.

3D DISPLAY COMMANDS

Commands that are used for changing the appearance of a surface or solid model in AutoCAD are *Shademode, Hide, Hlsettings, Shade,* and *Render.* By default, surface and solid models are shown in wireframe representation in order to speed computing time and enhance visibility during construction. As you know, wireframe representation can somewhat hinder your visualization of a model because it presents the model as transparent. To display surfaces and solid models as opaque and to remove or change the display of the normally obscured edges, *Shademode, Hide, Hlsettings, Shade,* and *Render* can be used.

In AutoCAD 2004 you can print a surface or a solid model as a shaded image. Select from *As Displayed, Wireframe, Hidden,* or *Rendered* options in the *Plot* dialog box. See "Printing Shaded 3D Models" near the end of this chapter.

Wireframe models are not affected by display commands since they do not contain surfaces. Wireframe models can be displayed only in wireframe representation.

SHADEMODE

Pull-down Menu	COMMAND (TYPE)	ALIAS (TYPE)	Short-cut	Screen (side) Menu	Tablet Menu
View *Shade>*	*SHADEMODE*	*SHA*	...	...	N,2

Shademode applies a shaded appearance to surface and solid models and allows you to select the mode, or type, of shading to apply. Shaded surfaces on an object increase the realistic appearance and can greatly enhance your visibility and understanding of the model, especially for complex geometry.

When *Shademode* is used, the model <u>retains</u> the specified shade mode until you change it. For example, if you use *Shademode* to generate a *Hidden* display, you can edit the geometry, add geometry, select a new viewpoint, or use *3Dorbit* and the display mode is retained. In releases previous to AutoCAD 2000, the *Shade* command was used instead of *Shademode*; however, once the model was shaded, the display was temporary and remained only until the next drawing regeneration.

Shademode operates in the current viewport or full-screen display, but is allowed only in model space. If you create a new layout, the layout displays the model in the *Shademode* that was current when the layout was created.

Shademode allows you to specify one of seven options and then displays the objects in the current viewport using the option you choose.

```
Command: shademode
Current mode: Hidden
Enter option [2D wireframe/3D wireframe/Hidden/Flat/Gouraud/fLat+edges/gOuraud+edges] <Hidden>:
(option)
Regenerating model.
Command:
```

The options are explained and illustrated next.

2D Wireframe

This option displays only lines and curves. The 3D geometry is defined only by the edges representing the surface boundaries. The UCS icon is displayed as a "wireframe" icon. Also, <u>linetypes and lineweights are visible</u>. If the *Compass* is on (*COMPASS* system variable is set to 1), it does not appear in the *2D Wireframe* view. Raster and OLE objects are visible in the display. Generally used for 2D geometry, this option can be used for 3D geometry when lineweights and linetypes are important (Fig. 35-1).

Figure 35-1

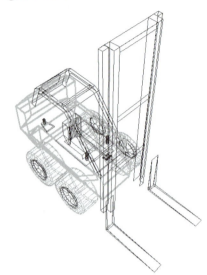

3D Wireframe

Generally used to display 3D geometry when a wireframe representation is needed, this mode displays the objects using lines and curves to represent the surface edges (Fig. 35-2). Since all edges are displayed, this option is useful for 3D construction and editing. The shaded 3D UCS icon appears in this mode. Linetypes and lineweights are not displayed and raster and OLE objects are not visible. The *Compass* is visible if turned on. The lines are displayed in the material colors if you have applied *Materials*, otherwise they appear in the defined line or layer color. (See Chapter 41, Rendering, for information on using *Material*.)

Figure 35-2

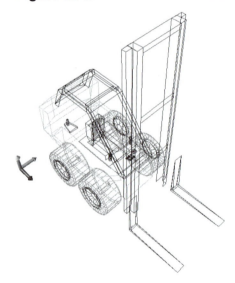

Hidden

Use this mode to display 3D objects similar to the *3D Wireframe* representation but with hidden lines suppressed. In other words, edges that would normally be obscured from view by opaque surfaces become hidden. This option is similar to using the *Hide* command (Fig. 35-3). This option enhances visualization but is not recommended for construction and editing unless *Hlsettings* is used.

Figure 35-3

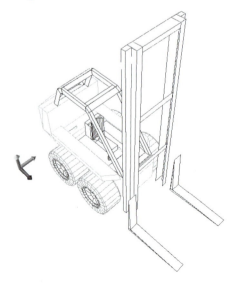

The *Hlsettings* command is intended to be used in conjunction with this option of *Shademode*. First, use *Hlsettings* to specify a different color or linetype to use for "hidden" lines. Then use the *Hidden Shademode* option to calculate which edges become "hidden" (see "*Hlsettings*").

Flat Shaded

 This and the other three color shaded options (described next) fill the surfaces with the defined color (by object, by layer, or *Material*). The surfaces are shaded as if a light source were placed at the camera (observer) position. When 3D objects contain curved geometry, this option displays curves as being faceted. For example, a cylinder is displayed as having many small flat surfaces (notice the forklift wheels in Figure 35-4). Therefore, curved objects appear flatter and less smooth than *Gouraud Shaded*. If you have applied *Materials* to objects, they are displayed accordingly with the *Flat Shaded* option.

Figure 35-4

Gouraud Shaded

 With this option the surfaces are also filled and shaded like *Flat Shaded*, except here the curved surfaces are smoothed to display true curved geometry (notice the forklift wheels in Figure 35-5). This feature gives the objects a smooth, realistic appearance. Objects are shaded in the layer or object color unless *Materials* have applied to the objects.

Figure 35-5

Flat Shaded, Edges On

 Select this option to achieve a combination of the *Flat Shaded* and *3D Wireframe* options. The resulting display has the effect of flat-shading with the edges highlighted (Fig. 35-6).

Figure 35-6

Gouraud Shaded, Edges On

This choice combines the *Gouraud Shaded* and *3D Wireframe* options (Fig. 35-7). The objects are *Gouraud Shaded* with the surface edges showing through. This option is good for cases where visualization is important but minor editing may be necessary and is made easier since edges are visible.

Figure 35-7

 The four shaded options of *Shademode* display lights, materials, textures, and transparency applied with AutoCAD's rendering capabilities; however, some limitations occur such as the inability to display shadows, reflection, refraction, and other features. See Chapter 41, Rendering, for more information.

HIDE

Pull-down Menu	COMMAND (TYPE)	ALIAS (TYPE)	Short-cut	Screen (side) Menu	Tablet Menu
View *Hide*	*HIDE*	*HI*	...	VIEW 2 *Hide*	M,2

The *Hide* command generates a display of a 3D model similar to the *Hidden* option of *Shademode*. However, *Hide* operates differently based on your setting for *Shademode*:

 2D Wireframe   When *Shademode* has this setting, *Hide* operates as a separate command. It removes obscured (hidden) lines from the display; however, *Hide* in this mode is not persistent—using *Regen* or *3Dorbit* cancels the display generated by *Hide*. Also, *Hide* in this mode is affected by the setting of the *DISPSILH* variable; that is, if *DISPSILH* is set to 1, no mesh lines appear on curved surfaces (see "Display Variables" in Chapter 39).

3D Wireframe   If *Shademode* has this setting, *Hide* performs a normal *Shademode Hidden* display.

The application of the *DISPSILH* variable also affects printed drawings (of a 3D solid). If *DISPSILH* is set to 1, mesh (facet) lines do not appear on the print when displayed in hidden mode. To generate a print of a 3D solid in hidden display in AutoCAD 2004, use the *Hidden* option of *Shade Plot* in the *Plot* dialog box or in the viewport's *Properties* palette or shortcut menu. In previous versions of AutoCAD, use the *Hide Objects* option in the *Plot* dialog box or the *Hideplot* option for the viewport. See *DISPSILH*, Chapter 39.

HLSETTINGS

Pull-down Menu	COMMAND (TYPE)	ALIAS (TYPE)	Short-cut	Screen (side) Menu	Tablet Menu	Menu
...		*HLSETTINGS*	...	...	...	...

AutoCAD 2004 includes this useful new utility command for controlling the appearance of hidden lines. Hidden lines are those edges of 3D solids and surfaces that would normally be obscured from a particular view. Hidden lines are calculated only when you use the *Hide* command and the *Hidden* options of *Shademode* and *Shade Plot*. (When you use the *2D Wireframe* or *3D Wireframe* options of *Shademode*,

2004

normally obscured edges appear in the display along with all other lines.) Although *Shademode* options such as *Hidden* or *Gouraud Shaded* are persistent and display in real-time, the *Hlsettings* are persistent but do not display in real-time during a *3Dorbit* operation. Not available through any menus or toolbars, you must type *Hlsettings* to produce the *Hidden Line Settings* dialog box (Fig. 35-8).

Figure 35-8

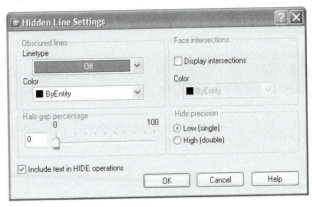

Although this new feature appears straightforward, there are no controls for linetype spacing or lineweights. Also, because of the subtle differences in the display of hidden lines for *Hide, Hidden Shademode*, and *Shadeplot*, attaining the exact display or plot you want may require experimentation. The dialog box options are explained next.

Obscured Lines
This section specifies the linetype and color you want to display for obscured lines. An obscured line is a hidden edge that is made visible by changing its *Color* or *Linetype*. Instead of using this dialog box, you can change the *Linetype* and *Color* settings using the *OBSCUREDLTYPE* and *OBSCUREDCOLOR* system variables, respectively.

Linetype
When this drop-down list is set to *Off*, hidden edges are not shown (when the hidden display options are used). For example, the solid model in Figure 35-9 is displayed with the *Off* setting for *Linetype*, then using the *Hidden* option of *Shademode*.

When any other linetype option such as *Solid, Dashed, Dotted*, etc. is selected, the hidden edges are displayed in the selected linetype. For example, the solid model in Figure 35-10 is displayed with the *Dashed* linetype setting, then the *Hidden* option of *Shademode* performed.

A *Linetype* setting is similar to an object-specific linetype setting in that the linetype overrides, and does not affect, the layer's linetype setting. Also, you cannot control the lineweight for these hidden lines individually—they can be controlled only for the entire object.

Figure 35-9

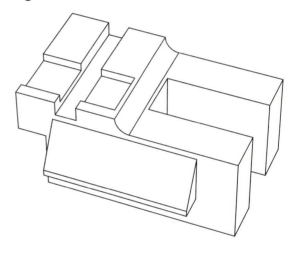

Figure 35-10

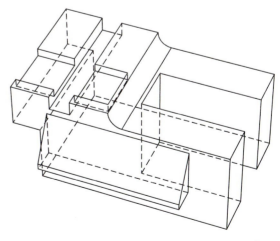

No options exist for controlling the scale of these lines globally. Unlike regular linetypes, *Ltscale* does not affect the scale or spacing for dashes or dots. Also, hidden linetypes are not affected by zooming—the dashes or dots maintain the same size and spacing relative to screen size when zooming.

NOTE: *Halo Gap Percentage* must be set to *0* for hidden linetypes to appear in a hidden display.

Color

You can change the color of the obscured edges using this drop-down list. For example, it may be useful to set the *Linetype* setting to *Solid*, then set the *Color* to gray or another lower intensity color. In this way, after performing a *Hidden* display, the visible object lines would appear bright while the hidden lines appear less intense.

Selecting a specific *Color* (red, blue, green) is similar to using an object-specific color setting in that it overrides the layer's color setting. Selecting *Bylayer* is useful only if the object itself has a different object-specific color. You cannot control the lineweight for hidden lines based on color—you can only control lineweight for the entire object.

Halo Gap Percentage

The "halo gap" affects only lines that are partially hidden and partially visible. The "gap" begins at the point where a line becomes visible and extends past ("comes out" from behind) another visible edge of an object. The *Halo Gap Percentage* shortens these lines by creating a gap at the point where it becomes visible to give a "halo" effect. For example, a *Halo Gap Percentage* of 10 is applied to the object shown in Figure 35-11.

Figure 35-11

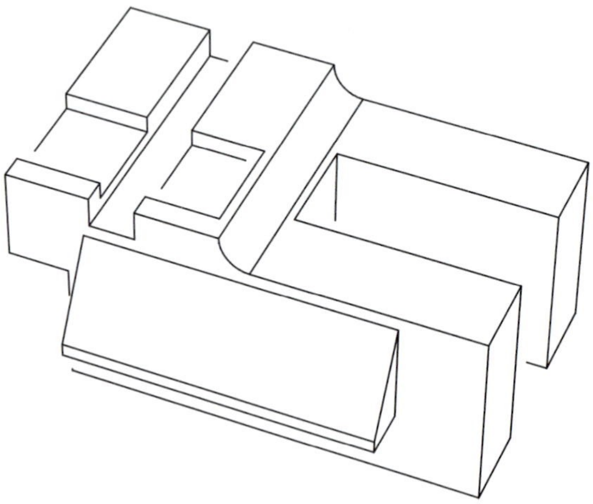

You can specify the *Halo Gap Percentage* either by using the slider or entering a value in the edit box. Although AutoCAD *Help* indicates the distance is specified as a percentage of one inch on the screen, this is not exact and appears to be affected by variables such as your screen resolution or display drivers. *Halo Gap Percentage* is relative to screen size and is not affected by zooming. You can also change this setting by using the *HALOGAP* system variable.

Include Text in HIDE Operations

Your setting for this checkbox specifies whether text objects (*Dtext* and *Mtext* objects) are included when a *Hidden* display is calculated. This option is also affected by your selection for *Linetype*. In other words, if this box is checked, and if text appears behind another object (it would be normally obscured), and the *Linetype* setting is *Off*, the text is hidden from view. For these same parameters but with a *Linetype* setting of *Dashed*, the text is drawn in a dashed linetype. You can also set this option by using the *HIDE-TEXT* system variable.

Face Intersections

In AutoCAD, unlike reality, two or more separate solid objects can occupy the same space (overlap), usually as an intermediate step to joining the objects into one composite solid (using the *Union*, *Subtract*, or *Intersection* command). When AutoCAD displays these situations, normally no lines of intersection appear between the objects. Using the *Face Intersections* options, you can display the edges that would appear at the intersection of the objects.

Display Intersections

Checking this box forces AutoCAD to calculate and display lines of intersections between solid objects that overlap. If no *Color* setting is made, the lines of intersection appear in the same color as the objects. You can also set this option by using the *INTERSECTIONDISPLAY* system variable.

Color

If *Display Intersections* is checked, you can specify a specific color to apply to the lines of intersection. You can also change this setting by using the *INTERSECTIONCOLOR* system variable.

Hide Precision
This setting controls the accuracy of hidden or shaded displays. The *Low (single)* option uses less memory. Setting the hide precision to *High (double)* produces more precise hides, but also uses more memory and can affect performance, especially for displaying complex solids. You can also change this setting by using the *HIDEPRECISION* system variable.

NOTE: The options set in the *Hlsettings* dialog box (technically, the related system variables) are saved in the current drawing file.

SHADE, SHADEDIF, SHADEDGE

The command *Shade* and the system variables *SHADEDIF* and *SHADEDGE* are older shading techniques that still appear in AutoCAD, most likely for legacy scripts and customized menus. Although these older shading methods are generally more difficult to use, they do not provide any additional benefits over using *Shademode*. It is recommended that you use the newer method of shading, *Shademode*, and do not use these older shading controls. *Shade* and *SHADEDGE* are explained briefly here.

Shade is the pre-AutoCAD 2000 version of *Shademode*. Issuing the *Shade* command causes the surface or solid models in the view to be shaded, similar to the effects of the *Shademode* command. Similar to the *Shademode* command, *Shade* is persistent (not affected by a *Regen*).

The *SHADEDGE* variable setting controls what type of shading results when the *Shade* command is used. The four choices are similar to four options of the *Shademode* command.

    0   equivalent to the *Gouraud* option of *Shademode* (see Figure 35-5)
    1   similar to the *Gouraud+edges Shademode* option (see Figure 35-7)
    2   the same as the *Hidden* option of *Shademode* (see Figure 35-3)
    3   (default setting) equivalent to the *Flat+edges* option of *Shademode* (see Figure 35-6)

RENDER

The *Render* command is used to create a realistic rendering of the 3D model (solid or surface) model. AutoCAD's rendering capabilities provide you with an amazing amount of control for generating photo-realistic renderings, including backgrounds, shadows, reflections, texture mapping, and landscaping effects. See Chapter 41, Rendering, for the full explanation of all rendering features included in AutoCAD.

3D VIEWING COMMANDS

When using the viewing commands *View, 3Dorbit, Vpoint, Plan*, etc., it is important to imagine that the <u>observer moves about the object</u> rather than imagining that the object rotates. The object and the coordinate system (WCS and icon) always remain <u>stationary</u> and always keep the same orientation with respect to Earth. Since the observer moves and not the geometry, the objects' coordinate values retain their integrity, whereas if the geometry rotated within the coordinate system, all coordinate values of the objects would change as the object rotated. The viewing commands change only the viewpoint of the <u>observer</u>.

Although UCSs have not been discussed in detail to this point, the viewing command you use can affect the UCS if you generate an orthographic viewpoint such as the top, front, side, etc. view. The *View* command (discussed next) can automatically change the UCS to align with the view if the *UCSORTHO* system variable is set to 1 (default setting) and you use the *Top, Front, Right* side, etc. option. All other viewing commands such as the *3Dorbit* commands, *Vpoint*, and *Plan* do not affect the UCS. It is recommended that you set *UCSORTHO* to **0** while you practice with the 3D viewing commands. The *UCSORTHO* system variable is discussed in detail in Chapter 36.

VIEW

	COMMAND (TYPE)	ALIAS (TYPE)	Short-cut	Screen (side) Menu	Tablet Menu
Pull-down Menu					
View *Named Views...*	*VIEW or* *-VIEW*	*V or* *-V*	...	*VIEW 1* *Ddview*	*M,5*

The *View* command provides several functions. You can assign a name, save, and restore viewed areas of a drawing, and you can control whether you want to save the current UCS with the view. (See *View* in Chapter 10 for the *Save* and *Restore* functions of *View*. See Chapter 36 for the UCS controls included in the *View* command.) This chapter explains how to use *View* to generate preset viewing directions for 3D objects.

The view command can be invoked by several methods. You can also use *View* in command line version by typing –*View*. Invoking *View* (not –*View*) by any method produces the *View* dialog box. The *View* dialog box (not shown) gives access to *Orthographic and Isometric Views* described on the following pages.

However, if you only want to produce a 3D view, you can bypass the dialog box and generate the desired viewpoint using the icon buttons from the Standard toolbar (Fig. 35-12) or using the *3D Views* option from the *Views* pull-down menu (Fig. 35-13).

Figure 35-12

Keep in mind that if you use the *Top, Bottom, Front, Back, Right,* or *Left* option, and the *UCSORTHO* system variable is set to 1 (default setting), the UCS automatically changes to align with the view.

Figure 35-13

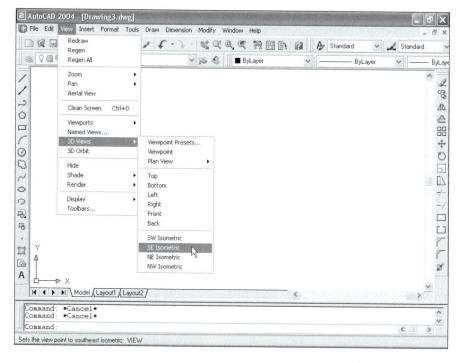

Each of the 3D View options is described next. The FORKLIFT drawing is displayed for several viewpoints (the hidden shademode was used in these figures for clarity). It is important to remember that for each view imagine you are viewing the object from the indicated position in space. The object does not rotate.

Top

The object is viewed from the top (Fig. 35-14). Selecting this option shows the XY plane from above (the default orientation when you begin a *New* drawing). Notice the position of the WCS icon. This orientation should be used periodically during construction of a 3D model to check for proper alignment of parts.

Figure 35-14

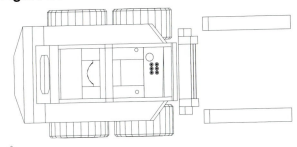

Bottom

This option displays the object as if you are looking up at it from the bottom. The Coordinate System icon appears backward when viewed from the bottom.

Left

This is looking at the object from the left side.

Figure 35-15

Right

Imagine looking at the object from the right side (Fig. 35-15). This view is used often as one of the primary views for mechanical part drawing. If *UCSORTHO* is set to 0, the "2D" Coordinate System icon does not appear in this view, but a "broken pencil" is displayed instead (meaning that it is not a good idea to draw on the XY plane from this view).

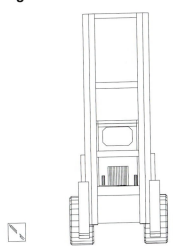

Front

Selecting this option displays the object from the front view. This is a common view that can be used often during the construction process. The front view is usually the "profile" view for mechanical parts (Fig. 35-16).

Figure 35-16

Back

Back displays the object as if the observer is behind the object. Remember that the object does not rotate; the observer moves.

SW Isometric

Isometric views are used more than the "orthographic" views (front, top, right, etc.) for constructing a 3D model. Isometric viewpoints give more information about the model because you can see three dimensions instead of only two dimensions (as in the previously discussed views).

SE Isometric

The southeast isometric is generally the first choice for displaying 3D geometry. If the object is constructed with its base on the XY plane so that X is length, Y is depth, and Z equals height, this orientation shows the front, top, and right sides of the object. Try to use this viewpoint as your principal mode of viewing during construction. Note the orientation of the WCS icon (Fig. 35-17).

NE Isometric

The northeast isometric shows the right side, top, and back (if the object is oriented in the manner described earlier).

NW Isometric

This viewpoint allows the observer to look at the left side, top, and back of the 3D object (Fig. 35-18).

NOTE: 3D View displays the object (or orients the observer) with respect to the World Coordinate System by default. For example, the *Top* option always shows the plan view of the WCS XY plane. Even if another coordinate system (UCS) is active, AutoCAD temporarily switches back to the WCS to attain the selected viewpoint.

You can also generate a 3D view (as previously illustrated) by invoking the *View* command instead of using the quicker *3D View* pull-down menu options or icons from the Standard toolbar. Invoking the *View* command produces the *View* dialog box. The *Orthographic & Isometric Views* tab gives access to the 3D views (Fig. 35-19).

To generate a 3D view using the dialog box, several steps are required. First, select the desired view from the list. Next, make the desired view current by selecting the *Set Current* button. Alternately, you can double-click on the desired view to make it current. Finally, select the *OK* button to produce the view.

NOTE: It is strongly suggested that you ensure the *Relative to:* option is set to *World*. This causes the 3D view that is generated to be relative to the World Coordinate System. Using another option can make the resulting view difficult to predict (*UCSBASE* system variable). See Chapter 36 for more information on this option. Also see Chapter 36 for information concerning the *Restore orthographic UCS with View* option (*UCSORTHO* system variable).

Figure 35-17

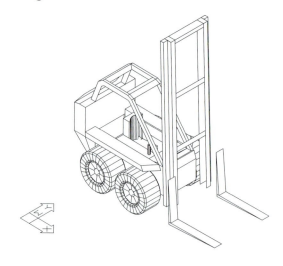

Figure 35-18

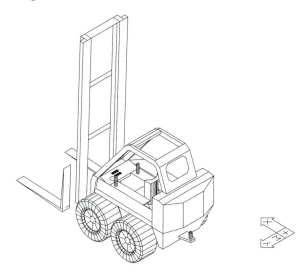

Figure 35-19

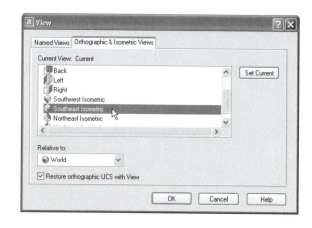

If you prefer to use the *View* command in Command line format, type *–View* (use a hyphen prefix). This action produces the following prompts.

    Command: **-view**
    Enter an option [?/Orthographic/Delete/Restore/Save/Ucs/Window]:

All you can see, all of the options available in the dialog box as well as the 3D views are available in Command line format. To generate a 3D view, use the *Orthographic* option.

    Command: **-view**
    Enter an option [?/Orthographic/Delete/Restore/Save/Ucs/Window]: **o**
    Enter an option [Top/Bottom/Front/BAck/Left/Right]<Top>: (**option**)
    Regenerating model.
    Command:

3D Orbit Commands

AutoCAD 2000 introduced a set of new <u>interactive</u> 3D viewing commands, all centered around the *3Dorbit* command. With these commands you can manipulate the view of 3D models by dragging your cursor. When one of the new 3D view commands is activated by typing or selecting from the 3D Orbit toolbar (Fig. 35-20) all commands in the 3D Orbit group are available through a right-click shortcut menu (Fig. 35-21). The new 3D commands all issue the same simple command prompt (see "*3Dorbit*") because they are interactive, not Command line driven. *3Dorbit* and the related viewing commands are listed and briefly explained at the beginning of this chapter and described in detail on this and the following pages.

Figure 35-20

Figure 35-21

The new *3Dorbit* and related commands essentially replace the function of the older *Dview* (Dynamic View) command and its options. Although the older *Dview* command is still available in AutoCAD, the new 3D commands are more interactive and user-friendly. Because these new commands more directly serve the same function as the older technology, a full description of *Dview* is not discussed in this text.

3DORBIT

Pull-down Menu	COMMAND (TYPE)	ALIAS (TYPE)	Short-cut	Screen (side) Menu	Tablet Menu
View 3D Orbit	3DORBIT	3do	...	VIEW 3dorbit	R,5

3Dorbit controls viewing of 3D objects interactively. By holding down the left button on your pointing device and dragging your cursor, you can control the view of 3D objects by several methods depending on where you place your cursor with respect to the "arc ball."

You can control the view of the entire 3D model or of selected 3D objects. If you want to see only a few objects during the interactive manipulation, select those objects <u>before you invoke the command</u>, otherwise the entire 3D model will be seen. Only a simple command prompt is given since the options are interactive. The same command prompt is issued for all the new 3D viewing commands.

Command: **3dorbit**
Press ESC or ENTER to exit, or right-click to display shortcut-menu.
Regenerating model.
Command:

The shortcut menu (see Figure 35-21) gives access to the commands described on the following pages.

When you invoke *3Dorbit*, the current viewport displays the selected objects and the 3D Orbit "arc ball" (Fig. 35-22). The arc ball is a circle with each quadrant denoted by a smaller circle. Moving the cursor over different areas of the arc ball changes the direction in which the view rotates. The center of the arc ball is the target. (Remember that you are changing the <u>view</u>. When moving the cursor about the arc ball, it appears that the 3D objects rotate; however, in theory the <u>target</u> remains stationary while the observer, or <u>camera</u>, moves around the objects.) Also notice when *3Dorbit* is invoked, the *Ucsicon* (if on) appears as a shaded 3D icon with the X axis as a red arrow, Y axis as green, and Z axis as cyan.

Figure 35-22

There are four possible methods of rotating your view depending on where you place the cursor and the direction you move it. The cursor icon changes to indicate one of these four methods. The four positions are as follows.

Inside the Arc Ball

 This icon is displayed when you move the cursor inside the arc ball. Imagine a sphere around the model. Click and drag the pointing device (hold down the left button) to rotate the sphere (and 3D model) around the target point (center of the sphere) in any direction.

Outside the Arc Ball

 Move the cursor outside the arc ball, then and click and drag to "roll" the display. This action rotates the view around the center of the arc ball. Imagine rotating the view about a Z axis protruding perpendicular from the screen.

Left and Right Quadrants

 This icon is displayed when you move the cursor over either of the small circles on the left or right quadrants of the arc ball, then click and drag to the left or right. This allows you to rotate the view about a vertical axis that passes through the center of the arc ball or about an imaginary Y axis on the screen.

Top and Bottom Quadrants

 To rotate the view about a horizontal axis passing through the center of the arc ball, place the cursor over either of the small circles on the top or bottom quadrants of the arc ball, then click and drag up or down. This is similar to rotation about an imaginary X screen axis.

Remember that the center of the arc ball is the target and remains stationary. The arc ball always appears in the center of your screen or current viewport. To center your objects in the arc ball, use the *Pan* (*3Dpan*) option or *Orbit uses AutoTarget* option from the shortcut menu.

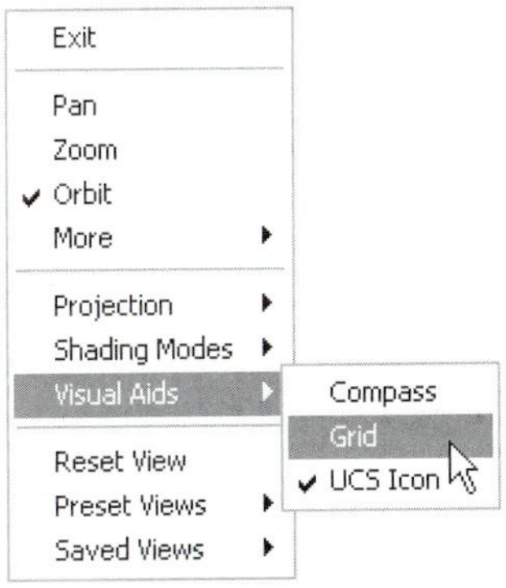

3Dorbit Shortcut Menu Options

During the *3Dorbit* or related command, you can use the shortcut menu to access other 3D viewing commands and options (Fig. 35-23). Using commands from the shortcut menu does not exit the *3Dorbit* session (unless, of course, you select *Exit*).

More, Orbit Maintains Z

Use this option to keep the Z axis in its current orientation when dragging horizontally inside the arc ball. For example, if the Z axis is vertical before using *3Dorbit*, this option keeps the Z axis in a vertical orientation while using *3Dorbit*. This setting is recommended.

More, Orbit Uses AutoTarget

This option forces the center of the orbit (target) in the center of the objects, rather than on the center of the viewport. This setting is particularly helpful when the objects are not located in the center of the viewport.

Visual Aids, Compass

Selecting the *Compass* toggle from the shortcut menu produces a sphere within the arc ball composed of three circles in 3D space. The three circles represent rotation about each of the three axes, X, Y, and Z (Fig. 35-24).

Visual Aids, Grid

Select *Visual Aids*, then *Grid* from the shortcut menu to produce a grid on the XY plane of the current UCS (see Figures 35-26 and 35-27). The number of major grid lines is controlled by the *Grid Spacing* value specified in the *Grid* command or in the *GRIDUNIT* system variable. The number of grid lines between the major lines defaults to 10 but changes as you zoom out, and the grid appears smaller. The *Grid* command cannot be activated during a *3Dorbit* session.

Figure 35-23

Figure 35-24

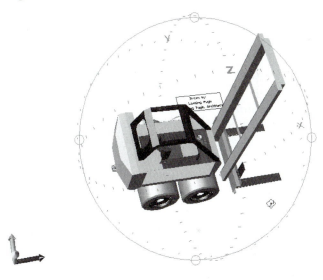

Visual Aids, UCS Icon

Use this option to toggle the UCS icon on and off. The UCS icon can enhance your orientation of the view because it changes as the view changes, even during perspective projection. When *3Dorbit* is invoked, the UCS icon (if on) appears as a shaded 3D icon with the X axis as a red arrow, Y axis as green, and Z axis as cyan. Alternately, use the *Ucsicon* command to control the icon visibility.

Reset View

You can use *Reset View* from the shortcut menu to undo all changes to the view during the *3Dorbit* session and reset to the previous view (when *3Dorbit* was started) and still remain in the *3Dorbit* session.

Exit

If you have achieved the view you want, you must press Enter, Escape, or select *Exit* from the shortcut menu to save the view and use other (non-*3Dorbit*) commands.

Other 3D viewing commands in the shortcut menu are discussed on the following pages.

3DPAN

	Pull-down Menu	COMMAND (TYPE)	ALIAS (TYPE)	Short-cut	Screen (side) Menu	Tablet Menu
	...	3DPAN	...	...	...	...

The cursor changes to a hand cursor when this command is active. Similar to the real-time *Pan* command, you can click and drag the view of the objects to any location in your drawing area. This tool is particularly useful with *3Dorbit* since the center of the arc ball is the target unless you use *Orbit uses AutoTarget*. Therefore, change the target of your 3D model by dragging it to the center of the arc ball. Keep in mind the arc ball temporarily disappears during *Pan*. Use *3Dpan* with *Perspective* toggled on to get a true 3D effect.

3DZOOM

	Pull-down Menu	COMMAND (TYPE)	ALIAS (TYPE)	Short-cut	Screen (side) Menu	Tablet Menu
	...	3DZOOM	...	...	...	...

3Dzoom operates similarly to the real-time version of the *Zoom* command. The cursor changes to a magnifying glass. Move the cursor up to zoom in (enlarge image) and down to zoom out (reduce image). If you were previously using *3Dorbit*, the arc ball disappears until you select *Orbit* from the shortcut menu.

Zoom Window
From the shortcut menu (Fig. 35-25) you can also select *More, Zoom Window* to zoom into a smaller area by defining a window, similar to the normal *Zoom* command with the *Window* option. With the *3Dorbit Zoom Window*, you must click and drag to define two diagonal corners of the window rather than click at each corner.

Zoom Extents
Select *More, Zoom Extents* from the shortcut menu to perform a typical *Zoom Extents*, similar to the normal *Zoom* command with the *Extents* option. The view is centered and sized so all selected objects are displayed in the 3D view.

Figure 35-25

3DCORBIT

	Pull-down Menu	COMMAND (TYPE)	ALIAS (TYPE)	Short-cut	Screen (side) Menu	Tablet Menu
	...	3DCORBIT	...	...	...	...

The *3Dcorbit* command allows you to set the selected 3D objects into continuous motion. The cursor changes to a sphere with two arrows orbiting it. To set the 3D objects in motion, click in the drawing area and hold down the left button, drag the cursor in any direction, then release the button while the cursor is in motion. The objects continue to spin in the direction of your cursor motion. The speed of the cursor movement determines the speed of the spin. Change the direction and speed of the object movement by clicking and dragging the cursor again. You can stop the motion at any position and discontinue *3Dcorbit* by pressing Escape or Enter. Alternately, right-click and select *Exit* or other option.

3DDISTANCE

	COMMAND (TYPE)	ALIAS (TYPE)	Short-cut	Screen (side) Menu	Tablet Menu
Pull-down Menu					
...	3DDISTANCE	...	...	...	...

The *3Ddistance* command adjusts the amount of perspective applied to the 3D view. In other 3D views, such as those generated using *Vpoint* or the 3D views obtained through the menus, you see a parallel projection, not a perspective projection. Perspective projection takes into account the distance the observer is from the object, whereas a parallel projection does not.

A parallel projection displays 3D objects unrealistically because parallel edges of an object remain parallel in the display (Fig. 35-26). A good example of this theory is to activate the grid (from the shortcut menu select *Visual Aids*, then *Grid*), then use *3Dorbit* to generate a view displaying the parallel grid lines.

Figure 35-26

A perspective projection simulates the way we actually see 3D objects because it takes into account the distance from our eyes (camera) to the objects (target). Therefore, since objects that are farther away appear smaller, parallel edges on a 3D object converge toward a point as they increase in distance. Figure 35-27 shows the same objects as the previous figure but in perspective projection.

The *3Ddistance* command adjusts the distance between the camera and the target. In other words, it increases or decreases the amount of perspective. Before using *3Ddistance*, you can achieve a reasonable amount of perspective by using the shortcut menu to select *Projection*, then *Perspective*. This action sets the selected 3D objects in perspective mode. Then use *3Ddistance* to increase or decrease the distance, or amount of perspective.

Figure 35-27

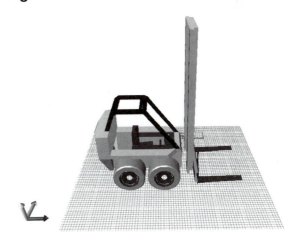

When you invoke *3Ddistance* by typing or selecting from the *3D Orbit* toolbar or shortcut menu, the icon changes to a double-sided arrow shown in perspective. Hold the left mouse button down and move the cursor upward to achieve more perspective (the view becomes closer), or move the cursor down to achieve less perspective (the view becomes farther away).

A very important fact to keep in mind as you use *3Ddistance* is that <u>you should not be concerned with the size of the view</u>, only the distance the view is from the objects(s) and, therefore, the amount of perspective. <u>Use *3Dzoom* to adjust the size of the view after using *3Ddistance*</u>. It is easy to confuse the action of *3Ddistance* with that of *3Dzoom*—the two are related but achieve different results.

3DCLIP

Pull-down Menu	COMMAND (TYPE)	ALIAS (TYPE)	Short-cut	Screen (side) Menu	Tablet Menu
...	3DCLIP	...	...	...	...

Clipping planes are invisible planes that can pass through the view. Anything in front of the front clipping plane or in back of the back clipping plane becomes invisible when the planes are toggled on. The clipping planes are always parallel with the screen and perpendicular to the viewing direction (the screen Z axis).

Using *3Dclip* produces the *Adjust Clip Planes* window (Fig. 35-28). The window displays the current view from above so the clip planes can be seen on edge (appear as horizontal lines). The white (or black) line on the bottom of the window represents the front clipping plane and the green line on the top (if visible) represents the back clipping plane. As you move the clip planes through the 3D objects, the results are displayed in the drawing area. To see the results of adjusting clipping planes, the clipping planes must be toggled on. The controls for the *Adjust Clip Planes* window are available if you select the icon buttons or right-click in the window to produce the shortcut menu.

Figure 35-28

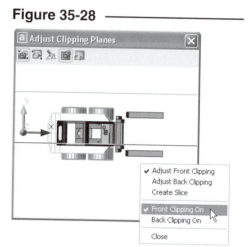

Front Clipping On

You must toggle the front clipping plane visibility on (depress the icon button or check the shortcut menu) if you want to see the results of adjusting the front clipping plane in the drawing area. Toggle this option on before using *Adjust Front Clipping*. Once the desired clipping is set, you can toggle this on or off for the display in the drawing area.

Back Clipping On

Toggle this option on before using *Adjust Back Clipping*. Like *Front Clipping On*, you must toggle the back clipping plane on to see the results of adjusting the back clipping plane in the drawing area. Once the desired clipping is set, you can toggle this on or off for the display in the drawing area.

Adjust Front Clipping

Move the horizontal black line near the bottom of the window upward to adjust the front clipping plane into the 3D objects. Any objects or parts of objects in front of the front clipping plane disappear. Notice how most of the fork and lift sections of the forklift are removed by the front clipping plane in Figure 35-29.

Figure 35-29

Adjust Back Clipping

Move the horizontal green line near the top of the window to adjust the back clipping plane. Generally, move the plane downward to pass through the 3D objects. Any objects or parts of objects in back of the back clipping plane disappear as shown in Figure 35-30.

Figure 35-30 ────────

Slice

Make this selection if you want to move both front and back clipping planes together. First, adjust either clipping plane to determine the thickness of the slice, then toggle on *Slice*. You can achieve a view to display only a thin section of your 3D model as shown in Figure 35-31.

Figure 35-31 ────────

When the *Adjust Clip Planes* window is dismissed, the clipping planes remain active (control the display) <u>during and after</u> the *3Dorbit* session. At any time, control the visibility of clipping planes using the *3Dorbit* shortcut menu by selecting *More*, then the *Front Clipping On* toggle or *Back Clipping On* toggle.

3DSWIVEL

	Pull-down Menu	COMMAND (TYPE)	ALIAS (TYPE)	Short-cut	Screen (side) Menu	Tablet Menu
	...	3DSWIVEL	...	...	...	...

The 3D view is determined by the location of the observer, known in AutoCAD as the camera, and the target, which is generally in the center of the 3D objects. The *3Dswivel* command changes the target of the view. Using *3Dswivel* simulates the effect of turning a camera on a tripod to change the viewing area. For example, if you moved the cursor (arched arrow) to tilt the camera upward, the objects in the viewfinder would move down and vice versa. Likewise, if you turned the camera to the right, objects in the viewfinder would move to the left.

Action of the *3Dswivel* command is similar to that of *3Dpan* because it changes the target for the view; however, the cursor controls of *3Dpan* and *3Dswivel* affect movement of the view in opposite directions.

CAMERA

Pull-down Menu	COMMAND (TYPE)	ALIAS (TYPE)	Short-cut	Screen (side) Menu	Tablet Menu
...	*CAMERA*	...	...	...	...

Although the *Camera* command is not one of the *3Dorbit* group and does not appear in the *3D Orbit* toolbar or shortcut menu, it is related and should be discussed in this context. *Camera* is not as interactive as the *3Dorbit* group and requires Command line input or point specification.

Camera allows you to set different <u>camera and target</u> locations by specifying 3D coordinate values or by picking points in 3D space. You must be familiar with your geometry locations to enter coordinate values. If you PICK points to define locations in 3D space, <u>use *Osnap* modes</u>.

```
Command: camera
Specify new camera position <current>: (value) or PICK
Specify new camera target <current>: (value) or PICK
Command:
```

To set the camera and target locations for a 3D orbit view, use the *Camera* command before starting *3Dorbit*.

A good application for the *Camera* command is an architectural 3D model. In this case, the model may represent a large object such as a house or building—an object that you might want to see from inside. In contrast, most mechanical applications are relatively small and only viewed from outside. For an architectural model, it would be appropriate to set the camera location inside a room and set different target locations at each end of the room as shown in Figures 35-32 and 35-33.

Figure 35-32 ───────────────

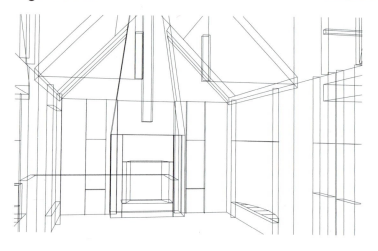

Figure 35-33 ───────────────

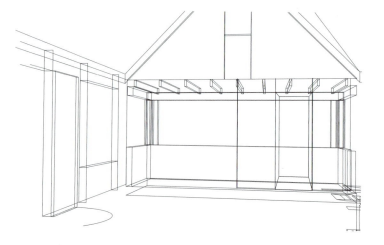

VPOINT

	COMMAND (TYPE)	ALIAS (TYPE)	Short-cut	Screen (side) Menu	Tablet Menu
Pull-down Menu					
View *3D Views >* *Viewpoint*	*VPOINT*	*-VP*	...	VIEW 1 *Vpoint*	N,4

Vpoint, like *3Dorbit*, allows you to achieve a specific view (or viewpoint) of a 3D object; however, *Vpoint* is an older command. Nevertheless, *Vpoint* can be used to specify an exact viewing angle or direction, whereas *3Dorbit* is strictly interactive. *Vpoint* offers fewer options than *3Dorbit* and displays the objects in parallel projection only.

Rotate

The name of this option is somewhat misleading because the object is not rotated. The *Rotate* option prompts for two angles in the WCS (by default) which specify a vector indicating the direction of viewing. The two angles are (1) the angle in the XY plane and (2) the angle from the XY plane. The observer is positioned along the vector looking toward the origin.

```
Command: vpoint
Current view direction: VIEWDIR=0.0000,0.0000,1.0000
Specify a view point or [Rotate] <display compass and tripod>: r
Enter angle in XY plane from X axis <270>: 315
Enter angle from XY plane <90>: 35.3
Command:
```

The first angle is the angle <u>in</u> the XY plane at which the observer is positioned looking toward the origin. This angle is just like specifying an angle in 2D. The second angle is the angle that the observer is positioned <u>up or down from</u> the XY plane. The two angles are given with respect to the WCS (Fig. 35-34).

Figure 35-34 ————————

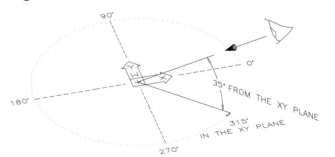

Angles of **315** and **35** specified in response to the *Rotate* option display an almost <u>perfect isometric</u> viewing angle. An isometric drawing often displays some of the top, front, and right sides of the object. For some regularly proportioned objects, perfect isometric viewing angles can cause visualization difficulties, while a slightly different angle can display the object more clearly. Figure 35-35 and Figure 35-36 display a cube from an almost perfect isometric viewing angle (**315** and **35**) and from a slightly different viewing angle (**310** and **40**), respectively.

Figure 35-35 ———— **Figure 35-36** ————

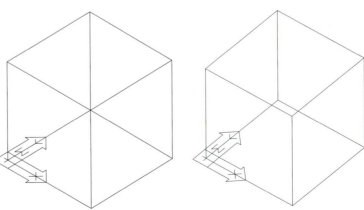

The *Vpoint Rotate* option can be used to display 3D objects from a front, top, or side view. *Rotate* angles for common views are:

Top	270, 90
Front	270, 0
Right side	0,0
Southeast isometric	315, 35.27
Southwest isometric	225, 35.27

View Direction

Another option of the *Vpoint* command is to enter X,Y,Z coordinate values. The coordinate values indicate the position in 3D space at which the observer is located. The coordinate values do not specify an absolute position but rather specify a <u>vector</u> passing through the coordinate position and the origin. In other words, the observer is located at <u>any point along the vector looking toward the origin</u>.

Because the *Vpoint* command generates a parallel projection and since parallel projection does not consider a distance (which is considered only in perspective projection), the magnitude of the coordinate <u>values</u> is of no importance, only the relationship among the values. Values of 1,-1,1 would generate the same display as 2,-2,2.

The command syntax for specifying a perfect isometric viewing angle is as follows:

```
Command: vpoint
Current view direction:  VIEWDIR=0.0000,0.0000,1.0000
Specify a view point or [Rotate] <display compass and tripod>: 1,-1,1
Command:
```

Coordinates of 1,-1,1 generate a display from a perfect isometric viewing angle.

Figure 35-37 illustrates positioning the observer in space, using coordinates of 1,-1,1. Using *Vpoint* coordinates of 1,-1,1 generates an isometric display similar to *Rotate* angles of 315, 35, or *SE Isometric*.

Other typical views of an object can be easily achieved by entering coordinates at the *Vpoint* command. Consider the following coordinate values and the resulting displays:

Figure 35-37 ————————————

Coordinates	Display
0,0,1	Top
0,-1,0	Front
1,0,0	Right side
1,-1,1	Southeast isometric
-1,-1,1	Southwest isometric

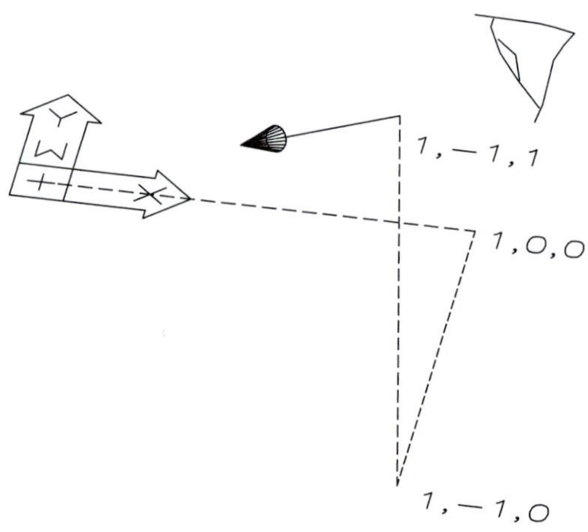

Tripod

This option displays a three-pole axes system which rotates dynamically on the screen as the cursor is moved within a "globe." The axes represent the X, Y, and Z axes of the WCS. When you PICK the desired viewing position, the geometry is displayed in that orientation. Because the viewing direction is specified by PICKing a point, it is difficult to specify an exact viewpoint.

When the *Tripod* method is invoked, the current drawing temporarily disappears and a three-pole axes system appears at the center of the screen. The axes are dynamically rotated by moving the cursor (a small cross) in a small "globe" at the upper-right of the screen (Fig. 35-38).

Figure 35-38

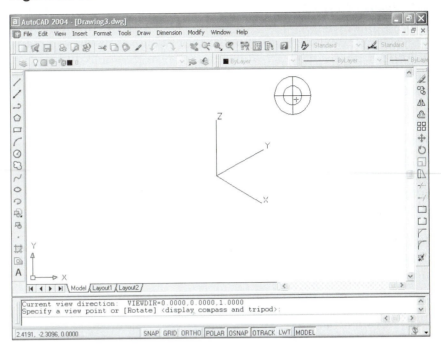

The three-pole axes indicate the orientation of the X, Y, and Z axes for the new *Vpoint*. The center of the "globe" represents the North Pole, so moving the cursor to that location generates a plan, or top, view. The small circle of the globe represents the Equator, so moving the cursor to any location on the Equator generates an elevation view (front, side, back, etc.). The outside circle represents the South Pole, so locating the cursor there shows a bottom view. When you PICK, the axes disappear and the current drawing is displayed from the new viewpoint. The command format for using the *Tripod* method is as follows:

```
Command: vpoint
Current view direction:  VIEWDIR=0.0000,0.0000,1.0000
Specify a view point or [Rotate] <display compass and tripod>: Enter  (tripod appears)
PICK (Select desired viewing direction.)
Regenerating model
Command:
```

Using the *Tripod* method of *Vpoint* is quick and easy. Because no exact *Vpoint* can be given, it is difficult to achieve the exact *Vpoint* twice. Therefore, if you are working with a complex drawing that requires a specific viewpoint of a 3D model, use another option of *Vpoint* or save the desired *Vpoint* as a named *View*.

DDVPOINT

Pull-down Menu	COMMAND (TYPE)	ALIAS (TYPE)	Short-cut	Screen (side) Menu	Tablet Menu
View *3D Views >* *Viewpoint Presets...*	*DDVPOINT*	*VP*	...	VIEW 1 *Ddvpoint*	N,5

The *Ddvpoint* command produces the *Viewpoint Presets* dialog box (Fig. 35-39 on the next page). This tool is an interface for attaining viewpoints that could otherwise be attained with the *3D Viewpoint* or the *Vpoint* command.

Ddvpoint serves the same function as the *Rotate* option of *Vpoint*. You can specify angles *From: X Axis* and *XY Plane*. Angular values can be entered in the edit boxes, or you can PICK anywhere in the image tiles to specify the angles. PICKing in the enclosed boxes results in a regular angle (e.g., 45, 90, or 10, 30) while PICKing near the sundial-like "hands" results in irregular angles (Fig. 35-39). The *Set to Plan View* tile produces a plan view.

The *Relative to UCS* radio button calculates the viewing angles with respect to the current UCS rather than the WCS. Normally, the viewing angles should be absolute to WCS, but certain situations may require this alternative. Viewing angles relative to the current UCS can produce some surprising viewpoints if you are not completely secure with the model and observer orientation in 3D space.

Figure 35-39

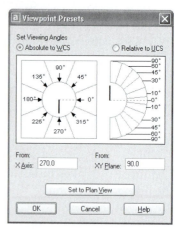

 NOTE: When you use this tool, ensure that the *Absolute to WCS* radio button is checked unless you are sure the specified angles should be applied relative to the current UCS.

PLAN

Pull-down Menu	COMMAND (TYPE)	ALIAS (TYPE)	Short-cut	Screen (side) Menu	Tablet Menu
View *3D Views >* *Plan View >*	*PLAN*	…	…	VIEW 1 *Plan*	N,3

 This command is useful for quickly displaying a plan view of any UCS:

    Command: **plan**
    Enter an option [Current ucs/Ucs/World] <Current>: (**option**) or **Enter**
    Regenerating model.
    Command:

Responding by pressing Enter causes AutoCAD to display the plan (top) view of the current UCS. Typing *W* causes the display to show the plan view of the World Coordinate System. The *World* option does <u>not</u> cause the WCS to become the active coordinate system but only displays its plan view. Invoking the *UCS* option displays the following prompt:

    Enter name of UCS or [?]:

The *?* displays a list of existing UCSs. Entering the name of an existing UCS displays a plan view of that UCS.

ZOOM

The *Zoom* command can be used effectively with 3D models just as with 2D drawings. However, the *Previous* option of *Zoom* is particularly applicable to 3D work.

 Previous

This option of *Zoom* restores the previous display. When you are using 3D Views, *Vpoint*, *3Dorbit*, and *Plan* for viewing a 3D model, *Zoom Previous* will display the previous viewpoint.

VPORTS

	Pull-down Menu	COMMAND (TYPE)	ALIAS (TYPE)	Short-cut	Screen (side) Menu	Tablet Menu
	View *Viewports >*	*VPORTS* or *-VPORTS*	...	...	*VIEW 1* *Vports*	**M,3** and **M,4**

The *Vports* command creates viewports. Either model space viewports or paper space (layout) viewports are created, depending on which space is current when you activate *Vports*. See Chapters 10 and 12 for basic information about *Vports* and using viewports for 2D drawings.

This section discusses the use of viewports to enhance the construction of 3D geometry. Options of *Vports* are available with AutoCAD specifically for this application.

When viewports are created in the *Model* tab (model space), the screen is divided into sections like tiles, unlike paper space viewports. Tiled viewports (viewports in model space) <u>affect only the screen display</u>. The viewport configuration <u>cannot be plotted</u>. If the *Plot* command is used, only the current viewport is plotted.

Figure 35-40 ————————

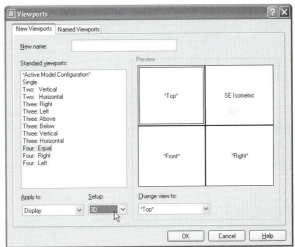

To use viewports for construction of 3D geometry, use the *Vports* command to produce the *Viewports* dialog box (Fig. 35-40). Make sure you invoke *Vports* in the *Model* tab (model space). To generate a typical orthographic view arrangement in your drawing, first select three or four viewports from the list in the *Standard viewports* list, then select the *3D* option in the *Setup:* drop-down list (see Figure 35-40, bottom center). This action produces a typical top, front, southeast isometric, and side view, depending on whether you selected three or four viewports.

Assuming the options shown in Figure 35-40 were used with the FORKLIFT drawing, the configuration shown in Figure 35-41 would be generated automatically. Note, however, that the objects are not sized proportionally in the viewports. Since a *Zoom Extents* is automatically performed, each view is sized in relation to its viewport, not in relation to the other views.

Figure 35-41 ————————

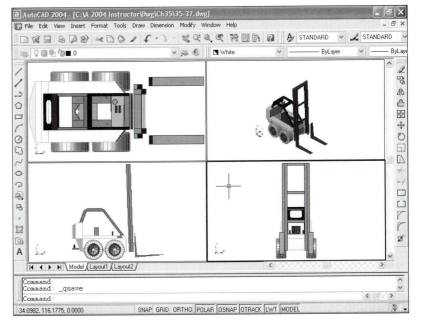

Next, if you desire all views to display the object proportionally, use *Zoom* with a constant scale factor (such as 1, not 1x) in each viewport. This action would generate a display similar to that shown in Figure 35-42.

This *3D Setup* option is available only from the *Viewports* dialog box. If you use one of the other options from the *View* pull-down menu (*3 Viewports, 4 Viewports*, etc.) or type the command line version (*-Vports*), you cannot automatically generate a typical orthographic view setup.

NOTE: The setting of *UCSORTHO* has an effect on the resulting UCSs when the *3D Setup* option of the *Viewport* dialog box is used. If *UCSORTHO* is set to 0, no new UCSs

Figure 35-42

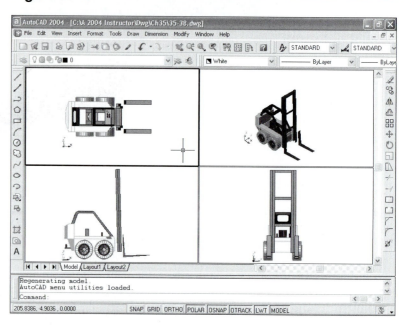

are created when *Vports* is used. However, if *UCSORTHO* is set to 1, a new UCS is generated for each new orthographic view created so that the XY plane of each new UCS is parallel with the view. In other words, each orthographic view (not the isometric) now shows a UCS with the XY plane parallel to the screen. See Chapter 36 for more information on the *UCSORTHO* system variable.

Printing Shaded 3D Models

In AutoCAD 2004 you can print 3D models in shaded mode. If you have a laserjet, inkjet, or electrostatic type printer configured as the plot device (not a pen plotter), you can select from *As Displayed, Wireframe, Hidden*, or *Rendered* options from the *Plot* dialog box or from the *Properties* palette for the viewport.

Figure 35-43

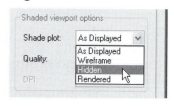

Printing a Shaded Image From the *Model* Tab
If you want to print a shaded 3D drawing from the *Model* tab, use the *Shaded Viewport Options* section of the *Plot* dialog box, *Plot Settings* tab. Use the *Shade Plot* drop-down list to produce the desired *Wireframe, Hidden*, or *Rendered* display in the print (Fig. 35-43). Alternately, you can use the *Shademode* or *Render* commands to produce the desired display on screen, then select *As Displayed* from this list to print the drawing as it appears on screen. You can also select from several resolution qualities for the print (see the "Shaded Viewport Options" section in Chapter 14).

Figure 35-44

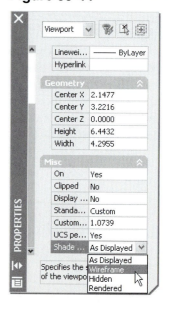

Printing a Shaded Image From a *Layout* Tab
If you are printing from a layout and want to specify a shaded image for one or more viewports, use the *Shadeplot* option of the viewport's *Properties* palette. Here you can select from the same options (Fig. 35-44) as in the *Plot* dialog box.

Alternately, you can type the *–Vports* command to assign the *Shadeplot* option for one or more viewports. You must be in a *Layout* for the *Shadeplot* options to appear in the Command prompt.

> Command: *–VPORTS*
> Specify corner of viewport or [ON/OFF/Fit/Shadeplot/Lock/Object/Polygonal/Restore/2/3/4] <Fit>: *s*
> Shade plot? [As displayed/Wireframe/Hidden/Rendered] <As displayed>:

In either case (printing a viewport in a *Layout* tab or from the *Model* tab), select from the *As Displayed*, *Wireframe*, *Hidden*, or *Rendered* options to produce the desired print. To make a print of a shaded image (shaded with *Shademode*), apply the desired *Shademode* to the screen display, then select the *As Displayed* option in the *Plot* dialog box. Alternately, you can use the *Rendered* option in the *Plot* dialog box as long as no rendering options have been specified for the drawing. The *Rendered* option prints the drawing according to the settings made in the *Render* dialog box, including any specified lights, materials, background, and view direction (see Chapter 41 for more information on these features). If no rendering settings have been specified, the *Rendered* option produces a rendered image similar to first generating a *Gouraud Shaded* screen display, then making a plot with the *As Displayed* option.

Printing a Hidden Display for Solid Models

If you intend to make a print of a solid model in a "hidden" display (with *Hlsettings* applied), the simplest way to control the resulting print (whether printing from a *Layout* tab or the *Model* tab) is to set *Shade Plot* option to *As Displayed*, generate the desired display on screen, then print. Special considerations should be made because of the effect of the *DISPSILH* and *FACETRES* system variables.

1. If you want to print a "hidden" display with no mesh (facet) lines (for curved surfaces) appearing in the print, first set the *Shade Plot* option to *As Displayed*. Next, set *DISPSILH* to 1, set *Shademode* to 2D *Wireframe*, then use *Hide*. Make the print. (The *Hidden* option of *Shademode* results in the same print, but is confusing because the screen display shows facet lines.)

2. If you want to print a "hidden" display with mesh lines (for curved surfaces), set the *Shade Plot* option to *As Displayed*. Then set *DISPSILH* to 0 and use the *Hidden* option of *Shademode*. Control the density of the mesh lines by changing the value for the *FACETRES* system variable. Always generate a *Plot Preview* before making the print since the *FACETRES* may appear differently in the print than on screen.

See Chapter 39, Advanced Solids Features, for more information on *DISPSILH* and *FACETRES*.

TIP

2004

CHAPTER EXERCISES

For these exercises you need to download the drawings from the McGraw-Hill/Leach Web site. Go to **www.mhhe.com/leach**, then find the page for **AutoCAD 2004 Instructor** and locate the download drawings. Four of the drawings are used in these exercises: TRUCK MODEL, OPERA, WATCH, and R300-20. These drawings are AutoCAD 2002 sample drawings and are provided courtesy of Autodesk.

1. In this exercise you will use a sample drawing and practice with the 3D display commands *Shademode* and *Hide*.

    A. *Open* the **TRUCK MODEL** drawing. Select the *Model* tab if not already current.

    B. Use the *Hide* command to generate a hidden display. Your display should look like that in Figure 35-45.

    Figure 35-45

    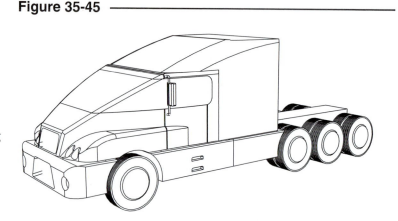

    C. Now use *Shademode* and change the display to *Hidden*. Is there any difference between this *Shademode* setting and the results of the *Hide* command?

    D. Type *DISPSILH* and change the setting to **1**. Set *Shademode* to 2D Wireframe. Use the *Hide* command. Do mesh lines appear? Now use the *Hidden* option of *Shademode*. Do mesh lines appear? Change *DISPSILH* back to **0** and try these options again.

    E. Change the *Shademode* setting to *Flat*. Now use the *Flat+Edges* option.

    F. Finally, change the setting to *Gouraud*. Now make the edges visible using the *Gouraud+Edges* option.

    G. *Exit* the Truck Model drawing and <u>do not save the changes</u>.

2. In this exercise, you use an AutoCAD sample drawing to practice with the *View* command. *Open* the **OPERA** drawing. Select the *Model* tab. *Freeze* layers **E-AGUAS, E-PREDI, E-TERRA, E-VIDRO,** and **O-PAREDE**. Use *Shademode* to generate a *Gouraud* display.

    A. Use any method to generate a *Top* view. Examine the view and notice the orientation of the coordinate system icon.

    B. Generate a *SE Isometric* viewpoint to orient yourself. Examine the icon and find north (Y axis). Consider how you (the observer) are positioned as if viewing <u>from</u> the southeast.

    C. Produce a *Front* view. This is a view looking north. Notice the icon. The Y axis should be pointing away from you.

    D. Generate a *Right* view. Produce a *Top* view again.

E. Next, view the OPERA from a *SE Isometric* viewpoint. Your view should appear like that in Figure 35-46. Notice the orientation of the WCS icon on your screen.

Figure 35-46

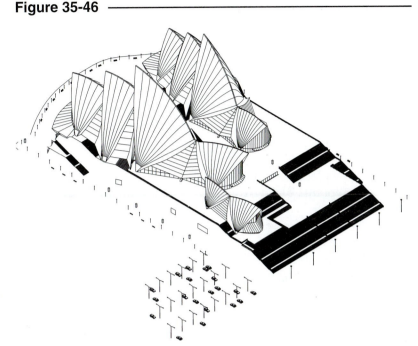

F. Finally, generate a *NW Isometric* and *NE Isometric*. In each case, examine the WCS icon to orient your viewing direction.

G. Experiment and practice more with *3D Viewpoint* if you desire. <u>Do not *Save*</u> the drawing.

3. In this exercise you will practice using the **3D views**, *Ucsicon*, *Vpoint*, and *Zoom*. **Open** the **TRUCK MODEL** again that you used in Exercise 1.

   A. First, type "*UCS*," then use the *World* option. This action makes the World Coordinate System the current coordinate system. Use the *Ucsicon* command and ensure the icon is at the *Origin*. Notice the orientation of the truck with respect to the WCS. The X axis points outward from the right side of the truck, Y positive is to the rear of the truck, and Z positive is up.

   B. Generate a *Top* view. Note the position of the X, Y, and Z axes. You should be looking down on the XY plane and Z points up toward you.

   C. Generate a *SE Isometric* view. Now use *Zoom* with the *Previous* option two times to give you the original view.

   D. Generate a *Front* view. Notice that AutoCAD performs a *Zoom Extents* so all of the drawing appears on the screen. *Freeze* layer **BASE**. Now generate a *Front* view again.

   E. Use *Vpoint* with the *Tripod* option to view the truck from the southeast direction and slightly from above. HINT: Pick the lower right (southeast) quadrant inside the small circle.

   F. Use *Vpoint* with the *Tripod* option again to view the truck from the southwest and slightly above. HINT: Pick the lower left (southwest) quadrant inside the small circle.

   G. Close the drawing but <u>do not save</u> the changes.

4. Now you will get some practice with the **3Dorbit** commands. **Open** the **WATCH** drawing. Activate the *Model* tab.

    A. First generate a **SE Isometric** view to get a good look at the watch. Notice that the components are disassembled, similar to an exploded view assembly drawing.

    B. Now invoke **3Dorbit**. Note that the *Grid* is on. Right-click to produce the shortcut menu, then under *Visual Aids* toggle off the *Grid*. In the shortcut menu, select *Shading Modes*, then *Gouraud Shaded*.

    C. Now place your cursor inside the "arc ball" and drag your pointing device to change your viewpoint. Next, try "rolling" the display by using your cursor outside the arc ball.

    D. Next, try placing your cursor on the small circle at the right or left quadrant of the arc ball. Hold down the left mouse button and drag left and right to spin the watch about an imaginary vertical axis. Try the same technique from the top and bottom quadrant to spin the watch about an imaginary horizontal axis.

    E. Finally generate your choice of views that shows the watch so all components are visible. Change the *Projection* to *Perspective*. Under the *More* option in the shortcut menu, select *Continuous Orbit*. Hold down the left mouse button, drag the mouse, and release the button to put the watch in a continuous spin.

    F. *Close* the drawing and do not save changes.

5. Use the R300-20 drawing to practice with the **3Dorbit** commands. **Open** the **R300-20** drawing. Activate the *Model* tab.

    A. Carefully <u>select only the assembled model</u> (not the exploded components), then activate **3Dorbit**. Only the selected components should appear in the arc ball. Right-click to produce the shortcut menu and turn off the *Grid*. Do not exit *3Dorbit*.

    B. Use **Pan** to move the assembly to the center of the arc ball. Next, use **Zoom** to enlarge the display. Use **Pan** again if necessary. Then turn on the *Gouraud Shaded* mode. Do not exit *3Dorbit*.

    C. Now, select *Orbit* from the shortcut menu and change your viewpoint. When you rotate the arc ball, does the assembly stay inside the arc ball? If not, select *More* and *Orbit uses AutoTarget* from the shortcut menu.

    D. Adjust the orbit view to place the assembly in the arc ball. Turn on *Perspective*. Now use *Adjust Distance* to increase the amount of perspective. You will have to *Zoom* then to reduce the size of the assembly. Finally, select *Reset* view to return to the original view.

    E. *Close* the drawing, <u>do not save</u> changes, then *Open* the same drawing again. This time select all the components of the exploded assembly (not the assembled model).

    F. Start **3Dorbit**, right-click, and toggle off the *Grid*. Turn on *Gouraud Shaded*.

G. Select ***Adjust Clipping*** planes from the shortcut menu. The *Adjust Clipping Planes* window should appear. Select the ***Adjust Front Clipping Plane*** button (you can right-click inside the window to select from a shortcut menu). Notice how adjusting the clipping plane inside the window changes the display of the model in the drawing area to dynamically display the cutting plane as you pass it through the model. Notice also that when you close the window and ***Exit*** *3Dorbit*, the clipping plane remains on in the drawing area and only the portion of the model you selected for *3Dorbit* remains visible.

H. ***Close*** the drawing and do not save the changes.

6. In this exercise you will create four model space viewports with the *3D* option. ***Open*** the **WATCH** drawing again. Activate the ***Model*** tab. Check the setting of the **UCSORTHO** variable by typing it at the command line and examining the setting. A setting of **0** means no new UCSs will be created when new views are generated.

A. Invoke the ***Vports*** command (make sure you are in model space). In the *New Viewports* tab of the *Viewports* dialog box, select ***Four: equal***. At the bottom of the dialog box in the *Setup* drop-down list, select *3D*. Select the ***OK*** button.

B. Your display should show a top, front, right side, and southeast isometric view. Notice that the watch is not sized proportionally with respect to the other views. Activate the front viewport (bottom left). Type ***Zoom*** at the command line and enter a value of **1.5**. Do the same for the other two orthographic views. Your display should look like that shown in Figure 35-47.

C. Activate the isometric viewport. Enter ***Shademode*** at the command line and change the setting to ***Gouraud***.

D. This is a good arrangement to use when you are constructing 3D models because you can see several views of the object on your screen at once. Remember this idea when you begin working with the exercises in the next several chapters. ***Close*** the drawing but <u>do not save</u> the changes.

Figure 35-47

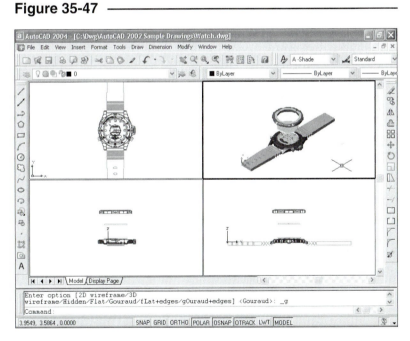

7. In this exercise, you will create a simple solid model that you can use for practicing creation of User Coordinate Systems in Chapter 36. You will also get an introduction to some of the solid modeling construction techniques discussed in Chapter 38.

A. Begin a ***New*** drawing and name it ***SOLID1***. Turn ***On*** the ***Ucsicon*** and force it to appear at the ***ORigin***.

B. Type *Box* to create a solid box. When the prompts appear, use **0,0,0** as the "Corner of box." Next, use the *Length* option and give dimensions for the Length, Width, and Height as **5, 4,** and **3**. A rectangle should appear. Change your viewpoint to *SE Isometric* to view the box. *Zoom* with a magnification factor of *.6X*.

Figure 35-48

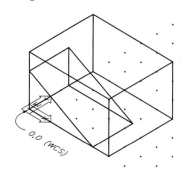

C. Type the *Wedge* command. When the prompts appear, use **0,0,0** as the "Corner of wedge." Again use the *Length* option and give the dimensions of **4, 2.5,** and **2** for *Length, Width,* and *Height*. Your solid model should appear as that in Figure 35-48.

D. Use *Rotate* and select only the wedge. Use **0,0** as the "Base point" and enter **-90** as the "Rotation angle." The model should appear as Figure 35-49.

Figure 35-49

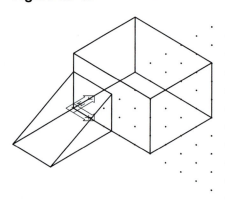

E. Now use *Move* and select the wedge for moving. Use **0,0** as the "Base point" and enter **0,4,3** as the "Second point of displacement."

F. Finally, type *Union*. When prompted to "Select objects:", PICK both the wedge and the box. The finished solid model should look like Figure 35-50. *Save* the drawing.

Figure 35-50

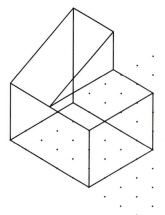

8. In this exercise you will use the SOLID1 drawing from the previous exercise and practice with the *Hlsettings* command in conjunction with *Shademode*.

A. First, use **Hlsettings** to specify the *Dashed* linetype to use for "hidden" lines. Then use the **Hidden Shademode** option to calculate which edges become "hidden." Your display should appear like that in Figure 35-51.

Figure 35-51 ───────────────────

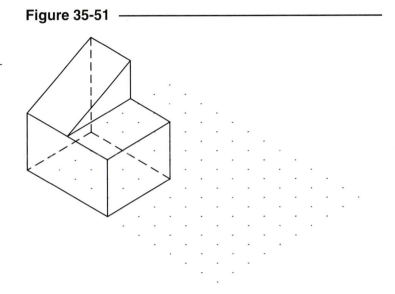

B. Generate a **SW Isometric** view. Do the "hidden" lines change correctly?

C. Invoke **Hlsettings**. Change the **Linetype** to **Solid** and change the **Color** to any color other than the current object color. Then use the **Hidden** option of **Shademode** to display the colored hidden edges.

D. Use **Hlsettings** again. Change the **Linetype** to **Off**. Set the **Halo Gap Percentage** to **15**. Next, use the **Hidden** option of **Shademode**. The resulting display should look similar to that in Figure 35-52. Note the "gap" in the edge line extending from behind the inclined plane.

Figure 35-52 ───────────────────

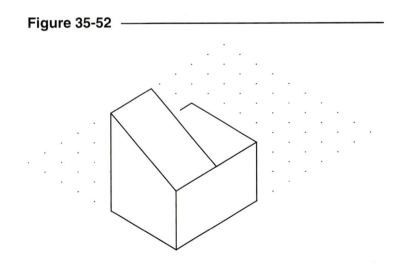

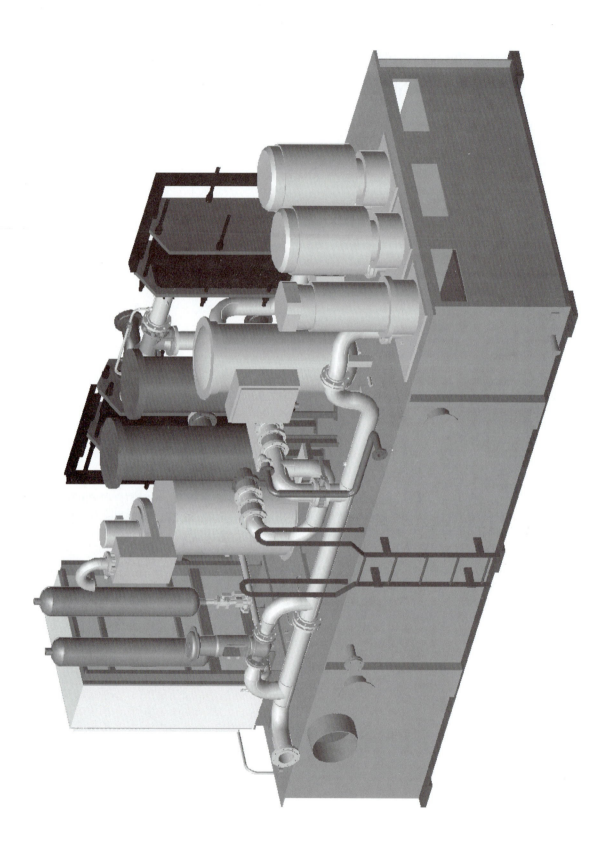

OIL MODULE.DWG Courtesy, Autodesk, Inc.

USER COORDINATE SYSTEMS

Chapter Objectives

After completing this chapter you should:

1. know how to create, *Save*, *Restore*, and *Delete* User Coordinate Systems using the *UCS* command;

2. know how to use the *UCS Manager* to create, *Save*, *Restore*, and *Delete* User Coordinate Systems;

3. be able to create UCSs by the *Origin*, *Zaxis*, *3point*, *Object*, *Face*, *View*, *X*, *Y*, and *Z* methods;

4. be able to create *Orthographic* UCSs using the *UCS* command and UCS Manager;

5. know the function of the *UCSVIEW*, *UCSORTHO*, *UCSFOLLOW*, and *UCSVP* system variables.

CONCEPTS

No User Coordinate Systems exist as part of any of the AutoCAD default template drawings as they come "out of the box." UCSs are created to simplify construction of the 3D model when and where they are needed.

When you create a multiview or other 2D drawing, creating geometry is relatively simple since you only have to deal with X and Y coordinates. However, when you create 3D models, you usually have to consider the Z coordinates, which makes the construction process more complex. Constructing some geometries in 3D can be very difficult, especially when the objects have shapes on planes not parallel with, or perpendicular to, the XY plane or not aligned with the WCS (World Coordinate System).

UCSs are created when needed to simplify the construction process of a 3D object. For example, imagine specifying the coordinates for the centers of the cylindrical shapes in Figure 36-1, using only world coordinates. Instead, if a UCS were created on the face of the object containing the cylinders (the inclined plane), the construction process would be much simpler. To draw on the new UCS, only the X and Y coordinates (of the UCS) would be needed to specify the centers, since anything drawn on that plane has a Z value of 0. The Coordinate Display (*Coords*) in the Status Bar always displays the X, Y, and Z values of the <u>current</u> coordinate system, whether it is the WCS or a UCS.

Figure 36-1

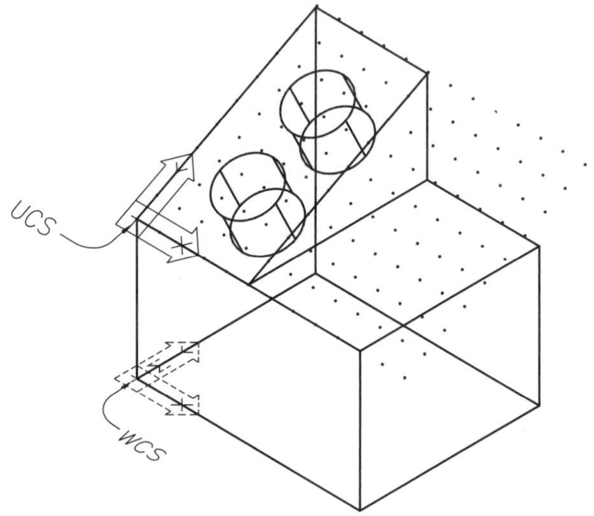

Generally, you would create a UCS aligning its XY plane with a surface of the object, such as along an inclined plane, with the UCS origin typically at a corner or center of the surface. Any coordinates that you specify (in any format) are assumed to be user coordinates (coordinates that lie in or align with the current UCS). Even when you PICK points interactively, they fall on the XY plane of the current UCS. The *SNAP*, *Polar Snap*, and *GRID* automatically align with the current coordinate system XY plane, enhancing your usefulness of the construction plane.

UCS COMMANDS

User Coordinate Systems are created by using any of several options of the *Ucs* command or UCS Manager (*Ucsman* command). You can create multiple UCSs in one drawing. The *Save* option of the *Ucs* command allows you to assign a name and save the UCSs that you create so you can *Restore* them at a later time.

In releases of AutoCAD previous to AutoCAD 2000, only one UCS could be active in a drawing at one time. If you chose to use viewports (*Vports* command), only one UCS was active and visible in all viewports at a time. This process was relatively simple to understand and manage, but had limitations.

Since AutoCAD 2000, several features are available that make UCSs much more flexible, but also more difficult to manage. These features include saving UCSs with *Views*, automatically changing UCSs to align with views when they are created or restored, having multiple UCSs visible when using *Model* tab viewports, and locking UCSs to specific viewports. This chapter will explain the features and assist you in using UCSs wisely.

When creating 3D models, it is recommended that you use the *Ucsicon* command with the *On* option to make the icon visible and the *Origin* option to force the icon to the origin of the current coordinate system rather than its default placement in the lower-left corner of the screen.

NOTE: The "2D" icon is used for most of the figures in this text because 1) this textbook does not display color needed for the 3-pole icon, 2) the 2-pole, "planar" feature helps you distinguish the orientation of the XY plane easier, and 3) the "W" on the "2D" icon communicates when the World Coordinate System is current.

UCS

Pull-down Menu	COMMAND (TYPE)	ALIAS (TYPE)	Short-cut	Screen (side) Menu	Tablet Menu
Tools *Move UCS ,* *New UCS> or* *Orthographic UCS>*	*UCS*	...	...	TOOLS 2 UCS	W,7

The *UCS* command allows you to create, save, restore, and delete User Coordinate Systems. There are many options available for creating UCSs. The *UCS* command always operates only in Command line mode.

Figure 36-2

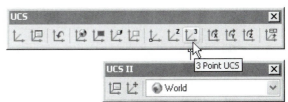

```
Command: ucs
Current ucs name:  *WORLD*
Enter an option
[New/Move/orthoGraphic/Prev/Restore/Save/Del/Apply/?/World]
<World>:
```

Although the *UCS* command operates only in Command line mode, two *UCS* toolbars are available (Fig. 36-2). Choosing one of these icons activates the particular option of the *UCS* command.

You can also use the *Tools* pull-down menu to select an option of the *UCS* command. There are two main options on the *Tools* pull-down menu: *Orthographic UCS* and *New UCS*. Suboptions are on cascading menus (Fig. 36-3).

Figure 36-3

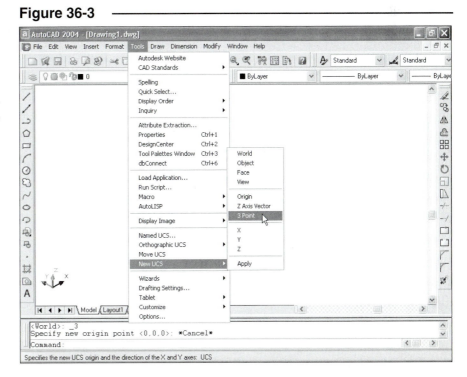

When you create UCSs, AutoCAD prompts for points. These points can be entered as coordinate values at the keyboard or you can PICK points on existing 3D objects. *OSNAP* modes should be used to PICK points in 3D space (not on the current XY plane). In addition, understanding the right-hand rule is imperative when creating some UCSs (see Chapter 34, 3D Modeling Basics).

The *UCS* options are discussed next.

Move, Origin

The *Move* option is essentially the same as the *Origin* option. This option defines a new UCS by allowing you to specify a new X, Y, and Z location. The orientation of the new UCS (direction for the X, Y, and Z axes) remains the same as its previous position, only the origin location changes (Fig. 36-4).

Figure 36-4

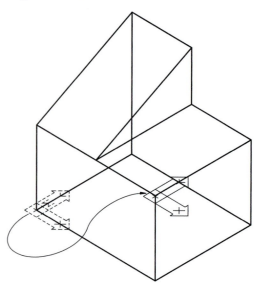

```
Command: ucs
Current ucs name:  *WORLD*
Enter an option [New/Move/orthoGraphic/Prev/Restore/Save/Del/Apply/?/World] <World>: m
Specify new origin point or [Zdepth]<0,0,0>: PICK or (value) or Z
Command:
```

Coordinates can be specified in any format or can be PICKed (use *OSNAPs* in 3D space).

New

The *New* option does not actually create a new UCS. It merely gives access to the other options for creating new UCSs. This option is necessary since there are too many *UCS* options to fit on one Command line, so *New* displays a second level of options.

```
Command: ucs
Current ucs name:  *WORLD*
Enter an option [New/Move/orthoGraphic/Prev/Restore/Save/Del/Apply/?/World] <World>: n
Specify origin of new UCS or [ZAxis/3point/OBject/Face/View/X/Y/Z] <0,0,0>:
```

The options displayed on the Command line above are explained next.

ZAxis

You can define a new UCS by specifying an <u>origin</u> and a direction for the <u>Z axis</u>. Only the two points are needed. The X <u>or</u> Y axis generally remains parallel with the current UCS XY plane, depending on how the Z axis is tilted. AutoCAD prompts:

Specify new origin point <0,0,0>: **PICK** or (**coordinates**)
Specify point on positive portion of Z-axis <*n.nnnn, n.nnnn, n.nnnn*>: **PICK** or (**coordinates**)

Figure 36-5

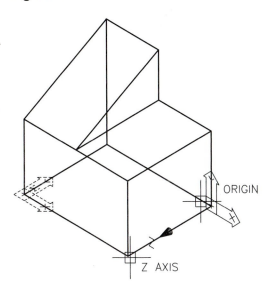

3point

This new UCS is defined by (1) the <u>origin</u>, (2) a point on the <u>X axis</u> (positive direction), and (3) a point on the <u>Y axis</u> (positive direction) or <u>XY plane</u> (positive Y). This is the most universal of all the UCS options (works for most cases). It helps if you have geometry established that can be PICKed with *OSNAP* to establish the points. The prompts are:

Specify origin of new UCS or
[ZAxis/3point/OBject/Face/View/X/Y/Z] <0,0,0>: **3**
Specify new origin point <0,0,0>: **PICK** or (**coordinates**)
Specify point on positive portion of X-axis <*n.nnnn, n.nnnn, n.nnnn*>: **PICK** or (**coordinates**)
Specify point on positive-Y portion of the UCS XY plane <*n.nnnn, n.nnnn, n.nnnn*>: **PICK** or (**coordinates**)
Command:

Figure 36-6

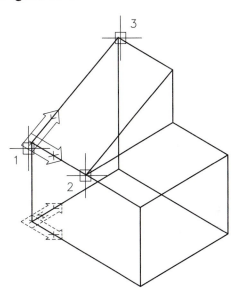

Object

This option creates a new UCS aligned with the selected object. You need to PICK only one point to designate the object. The origin of the new UCS is nearest the endpoint of the edge (or center of circular shape) you select. The orientation of the new UCS is based on the type of object selected and the XY plane <u>that was current when the object was created</u>.

For example, assume you created a solid model using only the World Coordinate System. Then, using the *Object* option, you could quickly create new UCSs aligned with edges of the model; however, the XY plane of the new coordinate systems would remain parallel with the WCS XY plane (that was current when those edges were created). This option is extremely fast, but takes some experience and may require some trial-and-error experimentation to achieve the UCS orientation you intend.

AutoCAD prompts:

Figure 36-7

> Select object to align UCS:

The following list gives the orientation of the UCS using the *Object* option for each type of object.

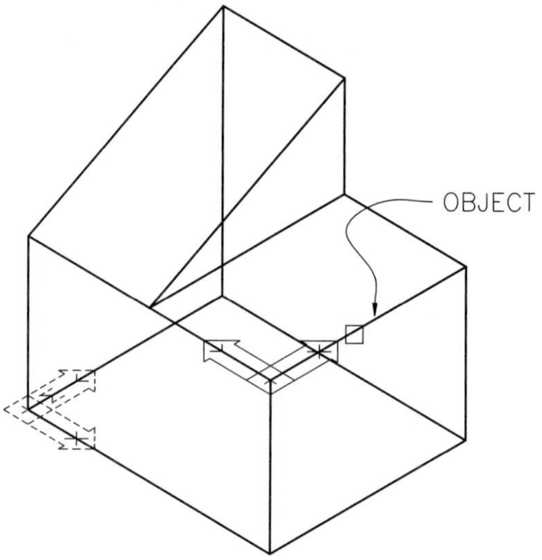

Object	Orientation of New UCS
Line	The end nearest the point PICKed becomes the new UCS origin. The new X axis aligns with the *Line*. The XY plane keeps the same orientation as the previous UCS.
Circle	The center becomes the new UCS origin with the X axis passing through the PICK point.
Arc	The center becomes the new UCS origin. The X axis passes through the endpoint of the *Arc* that is closest to the pick point.
Point	The new UCS origin is at the *Point*. The X axis is derived by an arbitrary but consistent "arbitrary axis algorithm." (See the *AutoCAD Customization Guide*.)
Pline (2D)	The *Pline* start point is the new UCS origin with the X axis extending from the start point to the next vertex.
2D Solid	The first point of the *Solid* determines the new UCS origin. The new X axis lies along the line between the first two points.
Dimension	The new UCS origin is the middle point of the dimension text. The direction of the new X axis is parallel to the X axis of the UCS that was current when the dimension was drawn.
3Dface	The new UCS origin is the first point of the *3Dface*, the X axis aligns with the first two points, and the Y positive side is on that of the first and fourth points. (See Chapter 40, Surface Modeling, "*3Dface*.")
Text, Insertion, or *Attribute*	The new UCS origin is the insertion point of the object, while the new X axis is defined by the rotation of the object around its extrusion direction. Thus, the object you PICK to establish a new UCS will have a rotation angle of 0 in the new UCS.

Face

The *Face* option operates with <u>solids only</u>. A *Face* is a planar or curved surface on a solid object, as opposed to an edge. When selecting a *Face*, you can place the cursor directly on the desired face when you PICK, or select an edge of the desired face.

    Specify origin of new UCS or [ZAxis/3point/OBject/Face/View/X/Y/Z] <0,0,0>: **f**
    Select face of solid object: **PICK**
    (Select desired face of 3D solid to attach new UCS. Edges or faces can be selected.)
    Enter an option [Next/Xflip/Yflip] <accept>:

Next
Since you cannot actually PICK a surface, or if you PICK an edge, AutoCAD highlights a surface near where you picked or adjacent to a selected edge (Fig. 36-8). Since there are always two surfaces joined by one edge, the *Next* option causes AutoCAD to highlight the other of the two possible faces. Press Enter to accept the highlighted face.

Xflip
Use this option to flip the UCS icon 180 degrees about the X axis. This action causes the Y axis to point in the opposite direction (see Figure 36-8).

Yflip
This option causes the UCS icon to flip 180 degrees about the Y axis. The X axis then points in the opposite direction.

Figure 36-8

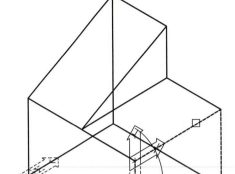

FACE, XFLIP

View

 This *UCS* option creates a UCS parallel with the screen (perpendicular to the viewing angle). The UCS <u>origin</u> remains unchanged. This option is handy if you wish to use the current viewpoint and include a border, title, or other annotation. There are no options
or prompts.

Figure 36-9

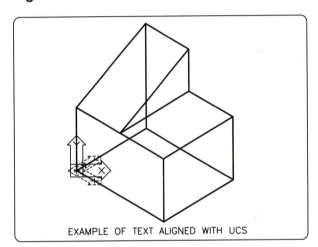

EXAMPLE OF TEXT ALIGNED WITH UCS

X   Y   Z

   Each of these options rotates the UCS about the indicated axis according to the right-hand rule. The Command prompt is:

Specify rotation angle about *n* axis <90>:

The angle can be entered by PICKing two points or by entering a value. This option can be repeated or combined with other options to achieve the desired location of the UCS. It is imperative that the right-hand rule is followed when rotating the UCS about an axis (see Chapter 34).

Figure 36-10

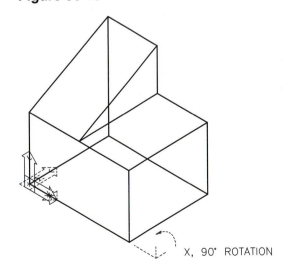

X, 90° ROTATION

Notice that in the prompt shown above for the *X*, *Y*, or *Z*, axis rotate options, the default rotation angle is 90 degrees ("<90>" as shown above). If you press Enter at the prompt, the axis rotates by the displayed amount (90 degrees, in this case). You can set the *UCSAXISANG* system variable to another value which will be displayed in brackets when you next use the *X*, *Y*, or *Z*, axis rotate options. Valid values are: 5, 10, 15, 18, 22.5, 30, 45, 90, 180. See *"UCSAXISANG"* in the System Variables section of this chapter.

Orthographic

The *Orthographic* option is used to create new UCSs in an orthographic orientation so that the XY plane of the new UCS is <u>parallel</u> to the *Top, Front, Right, Bottom, Back,* or *Left* faces. For example, if a *Right* UCS were created, its XY plane would be visible when viewing from the right side view of the object; or a *Plan* view of the *Right* UCS would yield a normal right side view.

When an *Orthographic* option is used, the UCS does not move to the designated side of the object (*Right, Top, Front,* etc.). Instead, the UCS only rotates to that orientation. The UCS generally rotates in relation to the WCS (when *UCSBASE* system variable is set to the default, *World*). For example, if the current coordinate system were the WCS, selecting the *Right* option would rotate the UCS to the correct orientation <u>keeping the same origin</u>, but would not move to the rightmost face of the object (Fig. 36-11).

Figure 36-11

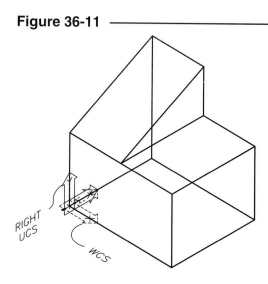

ORTHOGRAPHIC, RIGHT

Previous

Use this option to restore the previous UCS. AutoCAD remembers the ten previous UCSs used. *Previous* can be used repeatedly to step back through the UCSs.

World

Using this option makes the WCS (World Coordinate System) the current coordinate system.

Save

Invoking this option prompts for a name you want to assign for saving the current UCS. Up to 256 characters can be used in the name. Entering a name causes AutoCAD to save the current UCS. The *?* option and the *Dducs* command list all previously saved UCSs.

Restore

Any *Saved* UCS can be restored with this option. The *Restored* UCS becomes the <u>current</u> UCS. The *?* option and the UCS Manager list the previously saved UCSs.

Delete

You can remove a *Saved* UCS with this option. Entering the name of an existing UCS causes AutoCAD to delete it.

Apply
The *Apply* option allows you to apply the UCS in the current viewport to all or selected viewports.

Since AutoCAD 2000, each view or viewport can have a different UCS. With one view on the screen, only one UCS can be active. If, however, you are using viewports in model space (as described in Chapter 35), you can have a different UCS in each viewport. The *Apply* option allows you to use the UCS in the current viewport and make it the active UCS for any other, or all, viewports.

```
Command: ucs
Current ucs name:  *WORLD*
Enter an option [New/Move/orthoGraphic/Prev/Restore/Save/Del/Apply/?/World] <World>: a
Pick viewport to apply current UCS or [All]<current>: PICK (PICK inside the viewport you want to
change to the current viewport's UCS.)
Command:
```

First, make whichever viewport current that contains the UCS you want to apply. Next, use the *UCS* command and the *Apply* option. Then select the viewport you want to accept the new UCS (from the previous current viewport).

?
This option lists the named UCSs (UCSs that you *Saved*). Also listed are the origin coordinates and directions for the axes.

NOTE: It is important to remember that changing a UCS or creating a new UCS does not change the display (unless the *UCSFOLLOW* system variable is set to 1). Only one coordinate system can be active in a view. If you are using viewports, each viewport can have its own UCS if desired.

DDUCSP

	COMMAND (TYPE)	ALIAS (TYPE)	Short-cut	Screen (side) Menu	Tablet Menu
Pull-down Menu					
Tools *Orthographic* *UCS...*	*DDUCSP*	...	...	TOOLS 2 *Ucsp*	W,9

The *Dducsp* command invokes the UCS Manager. It can be used to create new UCSs or to restore existing UCSs. The UCS Manager can also be used to set specific UCS-related system variables, described later.

The *Dducsp* command activates specifically the *Orthographic UCSs* tab of the UCS Manager (Fig. 36-12). Here you can create new UCSs, similar to the *Orthographic* option of the *Ucs* command. *Orthographic UCSs* are not automatically saved as named UCSs but are created as you select them; that is, you can create, save, and restore additional but separate UCSs with the names Front, Top, Right, etc.

Figure 36-12

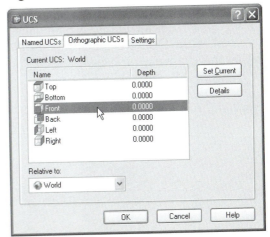

To create one of these *Orthographic UCSs*, you can <u>double-click</u> the desired UCS name, or highlight the desired UCS name (one click) and select the *Current* button. Press *OK* to return to the drawing to see the new UCS. Keep in mind that although the new UCSs are normal to the front, top, right, etc. orthographic views, they are <u>not necessarily attached to the front, top, or right faces of the objects</u> (see Figure 36-11 and previous explanation under *UCS, Orthographic*). The new UCSs typically have the same origin as the WCS, unless a new *Depth* is specified or a named UCS is selected in the *Relative to:* drop-down list.

Depth

Figure 36-13

The *Orthographic UCS* tab has one feature not available with the *Ucs* command. Notice in the *Orthographic UCS* tab that each UCS can have a specified depth. *Depth* is a Z dimension or distance perpendicular to the XY plane to locate the new UCS origin. Select the *Depth* <u>value</u> of any UCS to produce the *Orthographic UCS Depth* dialog box (Fig. 36-13).

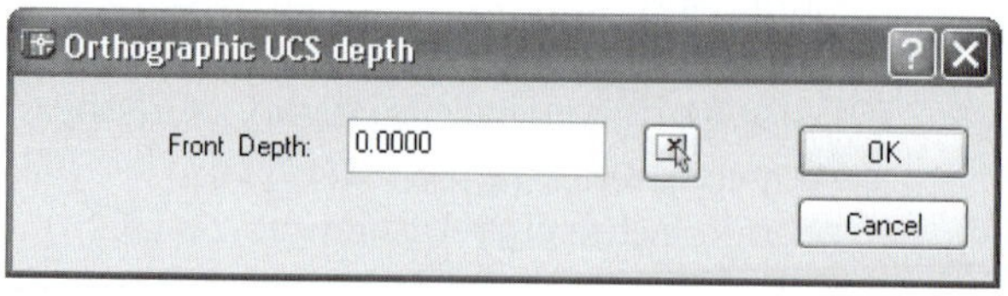

Details

Select the *Details* button to see the coordinates for the origin of the highlighted UCS. The direction vector coordinates for the X, Y, and Z axes are also listed. This has the same function as the *?* option of the *UCS* command.

Relative to:

The *Relative to:* drop-down list allows you to select any named UCS as the base UCS. When the *Front*, *Top*, *Right*, or other UCS option is selected, it is created in a orthographic orientation with respect to the origin and orientation of the selected named UCS. Typically, you want to <u>ensure *World* is selected as the base UCS</u> unless you have some other specific application. The setting in the *Relative to:* list is stored in the *UCSBASE* system variable.

UCSMAN

Pull-down Menu	COMMAND (TYPE)	ALIAS (TYPE)	Short-cut	Screen (side) Menu	Tablet Menu
Tools *Named UCS...*	*UCSMAN*	*UC*	...	*TOOLS2* *Ucsman*	*W,8*

The *Ucsman* command invokes the UCS Manager. The UCS Manager has three tabs: *Named UCSs*, *Orthographic UCSs* and *Settings*. The UCS Manager can be accessed by several methods shown in the table above. However, the *Orthographic UCSs* tab of the UCS Manager is also accessed by the *Dducsp* command (see *Dducsp* command earlier this chapter).

Figure 36-14

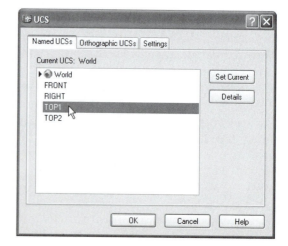

Named UCSs **Tab**

Ucsman produces the *Named UCS* tab of the UCS Manager (Fig. 36-14). The list contains all named UCSs. Use this tab to make named UCSs *Current,* just as you would use the *Restore* option of the *Ucs* command. (You must first create named UCSs using the *Save* option of the *Ucs* command.) In this tab you can double-click the desired UCS name or highlight (single-click) the desired name and then select *Set Current.* Either way, you must then select *OK* to restore the desired UCS. The *Details* button lists the origin coordinates and X, Y, and Z direction vectors of the highlighted UCS.

Settings **Tab**

The *Settings* tab can be accessed directly only by invoking the UCS Manager. However, you can use the *Dducsp* or *Dducs* commands or buttons explained earlier to produce the other tabs, then change tabs. The *Settings* tab (Fig. 36-15) allows you to control the UCS icon and two system variables.

UCS Icon settings

Settings in this section are identical to controls of the *Ucsicon* command (see the *Ucsicon* command in Chapter 34, 3D Modeling Basics). The *On* checkbox toggles the UCS icon on or off. *Display at UCS origin point* moves the icon to the coordinate system origin (otherwise it is located in the lower left corner of the screen) and has the identical functions as the *Origin* and *Noorigin* options of the *Ucsicon* command. *Apply to all viewports* is the same as the *All* option of the *Ucsicon* command; that is, the settings made in this section apply to the icons in all active viewports when multiple viewports are used.

Figure 36-15 ───────────

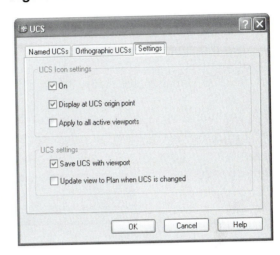

UCS Settings

These two settings control two UCS-related system variables that affect only the currently active viewport when multiple viewports are used. *Save UCS with viewport* controls the *UCSVP* system variable and affects the UCS of the current viewport. When you are using multiple viewports, you have the option of locking the orientation of a UCS with a specific viewport so that when a new UCS is created or an existing UCS is restored, the locked UCS remains unchanged (that is, the UCS of the viewport that is active when the box is checked remains unchanged). *Update view to Plan when UCS is changed* controls the *UCS-FOLLOW* system variable which affects the View of the active viewport. The viewport that is active when this box is checked will automatically change to a plan view of a new UCS whenever a new UCS is created or an existing UCS restored. When both boxes are checked, *Update view to Plan when UCS is changed* overrides *Save UCS with viewport.* (See "UCS System Variables" next in this chapter.)

UCS SYSTEM VARIABLES

In releases previous to AutoCAD 2000, UCS control was relatively simple and straight-forward. Because of the added UCS flexibility since AutoCAD 2000, it is easy for first-time users to become overwhelmed. This section is provided to help avoid confusion and to offer guidelines for UCS use. Not all UCS system variables are explained here, but only those that are typically accessed by a user.

UCSBASE

Pull-down Menu	COMMAND (TYPE)	ALIAS (TYPE)	Short-cut	Screen (side) Menu	Tablet Menu
Tools *Orthographic UCSs* *Presets...* *Relative to*	*UCSBASE*	...	...	*TOOLS2* *Ucsp* *Relative to*	...

UCSBASE stores the name of the UCS that defines the origin and orientation when *Orthographic* UCSs are created using the *UCS, Orthographic* option or the *Orthographic UCSs* tab of the UCS Manager. Normally the World Coordinate System is used as the base UCS, so when a new orthographic UCS is created it is simply rotated about the WCS origin to be normal to the selected orthographic view (top, front, right, etc.). *UCSBASE* can also be set in the *Relative to:* drop-down list of the *Orthographic UCSs* tab of the UCS Manager (see Figure 36-12).

For beginning users, it is suggested that this variable be left at its initial value, *World*. This variable is saved in the current drawing.

UCSVIEW

Pull-down Menu	COMMAND (TYPE)	ALIAS (TYPE)	Short-cut	Screen (side) Menu	Tablet Menu
View *Named Views* *Named Views* *New* *Save UCS with view*	*UCSVIEW*	...	...	*VIEW1* *Ddview* *Named Views* *New* *Save UCS with view*	...

UCSVIEW determines whether the current UCS is saved with a <u>named</u> view. For example, suppose you create a new 3D viewpoint, such as a southeast isometric, to enhance your visualization while you construct one area of a model. You also create a UCS on the desired construction plane. With *UCSVIEW* set to 1, you then *Save* that view and assign a name such as RIGHT. Whenever the RIGHT view is later restored, the UCS is also restored, no matter what UCS is current before restoring the view. Without this control, the current UCS would remain active when the RIGHT view is restored. *UCSVIEW* has two possible settings.

0   The current UCS is not saved with a named view.
1   The current UCS is saved whenever a named view is saved (initial setting).

The *UCSVIEW* variable can be accessed in the *New View* dialog box (Fig. 36-16). This dialog box is produced when you create a *New* view from the *Named Views* tab of the *View* dialog box (*View* command). *UCSVIEW* is saved in the current drawing.

UCSVIEW is easily remembered because it saves the *UCS* with the *VIEW*. Remember that <u>the view must be named to save its UCS.</u>

Figure 36-16

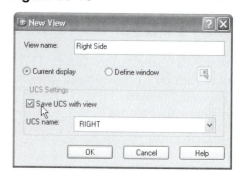

UCSORTHO

	Pull-down Menu	COMMAND (TYPE)	ALIAS (TYPE)	Short-cut	Screen (side) Menu	Tablet Menu
	View Named Views... Restore orthographic UCS with View	*UCSORTHO*	...		*VIEW1 Ddview Restore orthographic UCS with View*	...

This variable determines whether an orthographic UCS is automatically created when an orthographic view is created or restored. For example, if *UCSORTHO* is set to 1 and you select or create a front view, a UCS is automatically set up to be parallel with the view. In this way, the XY plane of the UCS is always parallel with the screen when an orthographic view is restored. The settings are as follows.

0   Specifies that the UCS setting remains unchanged when an orthographic view is restored.
1   Specifies that the related orthographic UCS is automatically created or restored when an orthographic view is created or restored (initial setting).

In order for *UCSORTHO* to operate as expected, you must select a preset orthographic view from the *View* pull-down menu, *Orthographic* option of the *View* command, or an orthographic view icon button. *UCSORTHO* does <u>not</u> operate with views created using *Vpoint*, *Plan*, the *3Dorbit* commands, or for any isometric preset views. *UCSORTHO* operates for views as well as viewports. *UCSORTHO* is saved in the current drawing.

A good exercise you can use to understand *UCSORTHO* is to use *Vports* to create four tiled viewports in model space and select the *3D Setup*. With *UCSORTHO* set to 1, the arrangement shown in Figure 36-17 would result. Here, each orthographic view has its own orthographically aligned UCS.

Figure 36-17

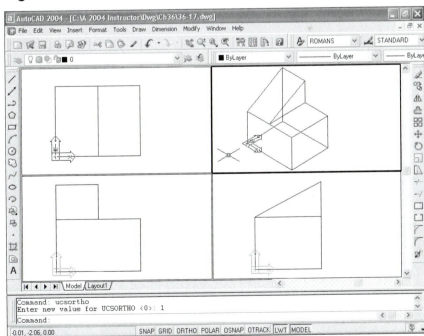

If *UCSORTHO* is set to 0 and the *Vports* command is used to create a *3D Setup*, the arrangement shown in Figure 36-18 would be created. Notice that the WCS is the only coordinate system in all viewports.

Figure 36-18

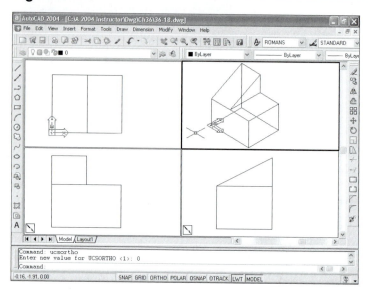

The *UCSORTHO* variable can be set at the Command line as well as in the *View* dialog box (Fig. 36-19, lower left).

 A point to help remember *UCSORTHO* is <u>when the view changes to an orthographic view, the UCS changes</u>.

Figure 36-19

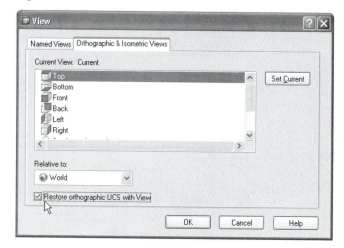

UCSFOLLOW

Pull-down Menu	COMMAND (TYPE)	ALIAS (TYPE)	Short-cut	Screen (side) Menu	Tablet Menu
Tools *Named UCS...* *Settings* *Update view to* *Plan when UCS* *is changed*	UCSFOLLOW	...	...	*TOOLS2* *UCS* *Follow:*	...

The *UCSFOLLOW* system variable, if set to 1 (*On*), causes the <u>plan view</u> of a UCS to be displayed automatically <u>when a UCS is made current</u>. *UCSFOLLOW* can be set separately for each viewport.

For example, consider the case shown in Figure 36-20. Two tiled viewports (*Vports*) are being used, and the *UCSFOLLOW* variable is turned *On* for the left viewport only. Notice that a UCS created on the inclined surface is current. The left viewport shows the plan view of this UCS automatically, while the right viewport shows the model in the same orientation no matter what coordinate system is current. If the WCS were made active, the right viewport would keep the same viewpoint, while the left would automatically show a plan view of the new UCS when the change was made.

NOTE: When *UCSFOLLOW* is set to 1, the display does not change until a new UCS is created or restored.

Figure 36-20

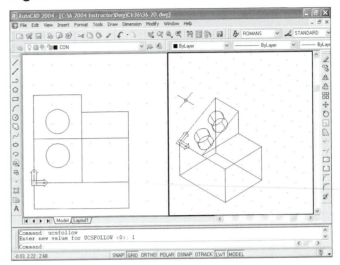

The *UCSFOLLOW* settings are as follows.

0   Changing the UCS does not affect the view (initial setting).
1   Any UCS change causes a change to plan view of the new UCS in the viewport current when *UCSFOLLOW* is set.

The setting of *UCSFOLLOW* affects only model space. *UCSFOLLOW* is saved in the current drawing and is viewport specific.

A point to help remember *UCSFOLLOW* is when the UCS changes, the view follows.

UCSVP

Pull-down Menu	COMMAND (TYPE)	ALIAS (TYPE)	Short-cut	Screen (side) Menu	Tablet Menu

Tools
Named Views...
Settings
UCS Settings
Save UCS
   with viewport
	UCSVP	...	...	*TOOLS2*
				Ucsman
				Settings
				UCS Settings
				Save UCS
				   with viewport
					...

UCSVP operates with model space and paper space viewports and is viewport specific (can be set for each viewport). It determines whether the UCS in the active viewport remains locked or changes when another UCS is made current in another viewport. *UCSVP* is similar to *UCSVIEW* where the UCS is saved with the *View*, only with *UCSVP* the UCS is saved with the viewport. The possible *UCSVP* settings are shown below.

0   The UCS setting is not locked to the viewport so the UCS reflected becomes that of any other viewport that is made active.
1   The viewport and UCS are locked. Therefore, the UCS is saved in that viewport and does not change when another viewport displaying another UCS becomes active (initial setting).

For example, suppose *UCSVP*, *UCSORTHO*, and *UCSVIEW* are all set to 0 (off) so there is no relationship established between views, viewports, and UCSs. In that case, only one UCS in the drawing could be active and would therefore appear in all viewports, similar to the arrangement shown in Figure 36-21. If a new UCS were created, it would appear in all viewports.

Figure 36-21

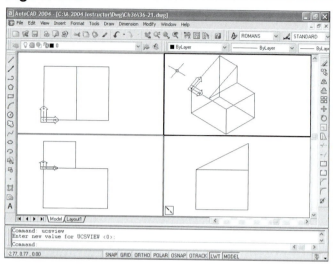

Assume *UCSVP* is then set to 1 in the upper-right (isometric) viewport. Suppose then another viewport were made active and the WCS restored. Now the WCS appears in all viewports <u>except</u> the isometric viewport where the UCS is locked (*UCSVP*=1) as shown in Figure 36-22.

Figure 36-22

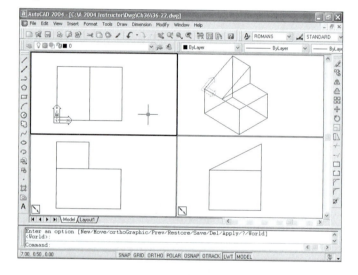

Remember that *UCSVP* saves a UCS to a <u>view-port</u>, whereas *UCSVIEW* saves a UCS to a named *View*.

UCSAXISANG

Pull-down Menu	COMMAND (TYPE)	ALIAS (TYPE)	Short-cut	Screen (side) Menu	Tablet Menu
...	*UCSAXISANG*	...	...	...	...

This variable stores the default angle when rotating the UCS around one of its axes using the *X*, *Y*, or *Z* options of the *UCS* command. When one of these options is used, the following prompt is displayed.

   Specify rotation angle about *n* axis <90>:

The initial value is 90, so when you use one of these options to rotate the UCS, you can press Enter to rotate 90 degrees instead of entering a value. If you want another angle to appear as the default (in the < > brackets), enter the value in the *UCSAXISANG* variable. Valid values are: 5, 10, 15, 18, 22.5, 30, 45, 90, 180. The variable setting is saved in the system registry.

3D Basics and Wireframe, Surface, and Solid Modeling

Now that you understand the basics of 3D modeling, you are ready to progress to constructing and editing wireframe, surface, and solid models. Almost all the topics discussed in Chapters 34, 35, and 36 are applicable to <u>all types</u> of 3D modeling.

In the progression from an elementary to a complete description of an actual object, the three types of models range from wireframe to surface to solid. The next several chapters, however, present the modeling types in a different order: wireframe models, solid models, and surface models. This is a logical order from simple to complex with respect to command functions and construction and editing complexity. That is to say, wireframe construction and editing is the simplest, and surface construction and editing is the most tedious and complex.

CHAPTER EXERCISES

1. Using the model in Figure 36-23, assume a UCS was created by rotating about the X axis 90 degrees from the existing orientation (World Coordinate System).

    A. What are the absolute coordinates of corner F?

    B. What are the absolute coordinates of corner H?

    C. What are the absolute coordinates of corner J?

Figure 36-23

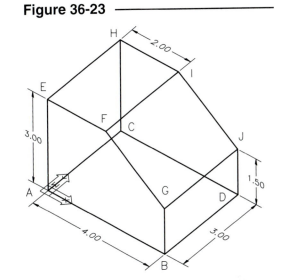

2. Using the model in Figure 36-23, assume a UCS was created by rotating about the X axis 90 degrees from the existing orientation (WCS) and then rotating 90 degrees about the (new) Y.

    A. What are the absolute coordinates of corner F?

    B. What are the absolute coordinates of corner H?

    C. What are the absolute coordinates of corner J?

3. *Open* the **SOLID1** drawing that you created in Chapter 35, Exercise 7. View the object from a *SE Isometric* view. Make sure that the *UCSicon* is *On* and set to the *ORigin*.

    A. Create a *UCS* with a vertical XY plane on the front surface of the model as shown in Figure 36-24. *Save* the UCS as **FRONT**.

Figure 36-24

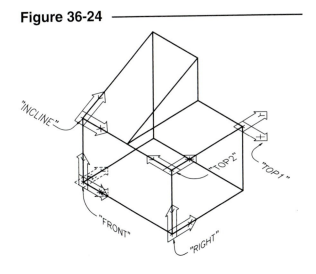

B. Change the coordinate system back to the *World*. Now, create a *UCS* with the XY plane on the top horizontal surface and with the orientation as shown in Figure 36-24 as TOP1. *Save* the UCS as **TOP1**.

C. Change the coordinate system back to the *World*. Next, create the *UCS* shown in the figure as TOP2. *Save* the UCS under the name **TOP2**.

D. Activate the *WCS* again. Create and *Save* the **RIGHT** UCS.

E. Use *SaveAs* and change the drawing name to **SOLIDUCS**.

4. In this exercise, you will create two viewport configurations, each with a different setting for *UCSORTHO*. In the first configuration, only one UCS exists, and in the second configuration, a new UCS is created for each viewport.

   A. *Open* the **SOLID1** drawing again. Use *SaveAs* and change the name to **VPUCS1**. Set the *UCSORTHO* variable to **0** so no new UCSs are created when you set up viewports. Ensure you are in model space and use the *Vports* command to create *Four Equal* viewports. Also in the *Viewports* dialog box, select the **3D Setup**. The resulting drawing should look like that in Figure 36-18. Notice that only one UCS appears in all viewports. Next, use any viewport and create a new UCS in the same orientation as the "Front" UCS shown in Figure 36-24. What happens in the other viewports? *Close* **VPUCS1** and *Save* the changes.

   B. *Open* the **SOLID1** drawing again. Use *SaveAs* and change the name to **VPUCS2**. This time, set the *UCSORTHO* variable to **1** so when you set up viewports new orthographically oriented UCSs are created. Set up the same viewport configuration as in the previous step. Your new viewport configuration and the resulting new orthographic UCSs should look like that in Figure 36-17. In the isometric view create a new UCS on the inclined plane (like the "Incline" UCS in Figure 36-24). What happens to the UCSs in the other viewports? *Close* **VPUCS2** and *Save* the changes.

5. Using the same drawing, make the *WCS* the active coordinate system. In this exercise, you will create a model and screen display like that in Figure 36-20.

   A. Set the *UCSORTHO* variable to **0** so no new UCSs are created when you set up viewports. Create *2 Vertical* tiled viewports. Use *Zoom* in each viewport if needed to display the model at an appropriate size. Make sure the *UCSicon* appears at the origin. Activate the <u>left</u> viewport. Set *UCSFOLLOW On* in that viewport only. (The change in display will not occur until the next change in UCS setting.)

   B. Use the right viewport to create the **INCLINE UCS,** as shown in Figure 36-24. *Regenall* to display the new viewpoint.

   C. In either viewport create two cylinders as follows: Type *cylinder*; specify **1.25,1** as the *center*, **.5** as the *radius*, and **-1** (negative 1) as the *height*. The new cylinder should appear in both viewports.

   D. Use *Copy* to create the second cylinder. Specify the existing center as the "Basepoint, of displacement." To specify the basepoint, you can enter coordinates of **1.25,1** or **PICK** the center with *Osnap*. Make the copy **1.5** units above the original (in a Y direction). Again, you can PICK interactively or use relative coordinates to specify the second point of displacement.

   E. Use *SaveAs* and assign the name **SOLID2**.

WIREFRAME MODELING

Chapter Objectives

After completing this chapter you should:

1. understand how common 2D *Draw* and *Modify* commands are used to create 3D wireframe models;

2. gain experience in specification of 3D coordinates;

3. gain experience with 3D viewing commands;

4. be able to apply point filters to 3D model construction;

5. gain fundamental experience with creating, *Saving*, and *Restoring* User Coordinate Systems;

6. be able to create, *Save*, and *Restore* tiled viewport configurations;

7. be able to manipulate the Coordinate System icon.

CONCEPTS

Wireframe models are created in AutoCAD by using common draw commands that create 2D objects. When draw commands are used with X ,Y, and Z coordinate specification, geometry is created in 3D space. A true wireframe model is simply a combination of 2D elements in 3D space.

A *Line*, for example, can be drawn in 3D space by specifying 3D coordinates in response to the "Specify first point:" and "Specify next point or [Undo]:" prompts (Fig. 37-1):

Figure 37-1

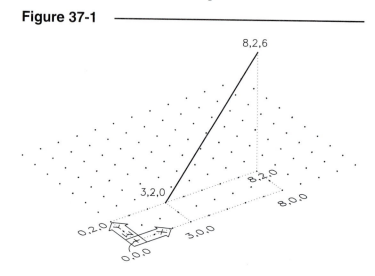

```
Command: line
Specify first point: 3,2,0
Specify next point or [Undo]: 8,2,6
Specify next point or [Undo]: Enter
Command:
```

The same procedure of entering 3D coordinate values can be applied to other draw commands such as *Arc, Circle,* etc. These 2D objects are combined in 3D space to create wireframe models. No draw commands are used <u>specifically</u> for creating 3D wireframe geometry (with the exception of *3Dpoly*). The use of 3D viewing and display, User Coordinate Systems (UCS), and 3D coordinate entry are the principal features of AutoCAD, along with the usual 2D object creation commands, and they allow you to create wireframe models. Specification of 3D coordinate values (Chapter 34), 3D display and viewing (Chapter 35), and the use of UCSs (Chapter 36) aid in creating wireframe models as well as surface and solid models. The *3Dpoly* command is discussed at the end of this chapter.

WIREFRAME MODELING TUTORIAL

Because you are experienced in 2D geometry creation, it is efficient to apply your experience, combined with the concepts already given in the previous three chapters, to create a wireframe model in order to learn fundamental modeling techniques. Using the tutorial in this chapter, you create the wireframe model of the stop plate shown in Figure 37-2 in a step-by-step fashion. The Wireframe Modeling Tutorial employs the following fundamental modeling techniques:

Figure 37-2

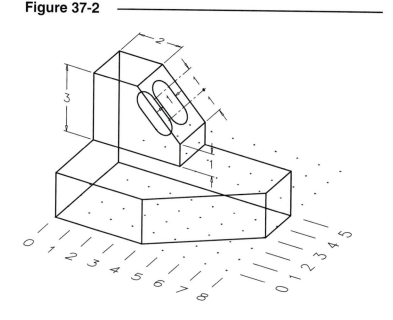

3D viewing commands
3D coordinate entry
Point filters
Polar Snap and *Polar Tracking*
Vports
User Coordinate Systems
Ucsicon manipulation

It is suggested that you follow along and work on your computer as you read through the pages of this tutorial. The Wireframe Modeling Tutorial has 22 steps. Additional hints, such as explicit coordinate values and menus that can be used to access the commands, are given on pages immediately following the tutorial (titled "Hints"). Refer to "Hints" only if you need assistance.

The figures in this tutorial display the "2D" UCS icon to more clearly depict the orientation of the XY plane. If you want to activate this icon, use the *Properties* option of *Ucsicon*, then select *2D*.

1. Set up your drawing.

   A. Start AutoCAD. Use **Start from Scratch, Imperial** default settings. Use **Save** and name the file **WIREFRAME**.
   B. Examine the figure on the previous page showing the completed wireframe model. Notice that all of the dimensions are to the nearest unit. For this exercise, assume generic units.
   C. Keep the default **Decimal Units**. Set **Units Precision .00** (two places to the right of the decimal).
   D. Keep the default *Limits* (assuming they are already set to 0,0 and 12,9 or close to those values). It appears from Figure 37-2 that the drawing space needed is a minimum of 8 units by 5 units on the XY plane.
   E. Turn on **Grid Snap** or **Polar Snap** and **Polar Tracking**. Set the increments to 1.
   F. Turn on the *GRID*.
   G. Make a layer named **MODEL**. Set it as the **Current** layer. Set a **Color**, if desired.

2. Draw the base of the part on the XY plane.

   Figure 37-3 ————————

   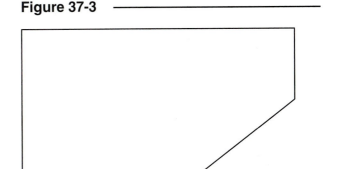

   A. Use **Line**. Begin at the origin (0,0,0) and draw the shape of the base of the stop plate. (It is a good practice to begin construction of any 3D model at 0,0,0.) Use **Grid Snap** or **Polar Snap** as you draw. See Figure 37-3.
   B. Change your **Vpoint** with the **Rotate** option. Specify angles of **305, 30**. Next, **Zoom .9X**.
   C. Turn the **Ucsicon ON** and force it to the **ORigin**. See Figure 37-4 to see if you have the correct *Vpoint*.

3. Begin drawing the lines bounding the top "surface" of the base by using absolute coordinates and .XY filters or *Polar Tracking*.

   Figure 37-4 ————————

   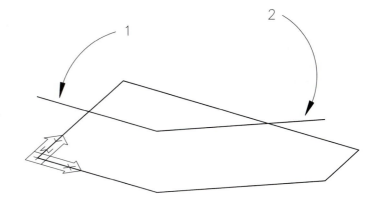

   A. Draw the first two *Lines*, as shown in Figure 37-4, by entering <u>absolute</u> coordinates. You cannot PICK points interactively (without *OSNAP*) since it results in selections only <u>on the XY plane</u> of the current coordinate system. Do not exit the *Line* command when you finish.

B. Draw *Lines* 3 and 4 using *.XY* point filters or *Polar Snap* and *Polar Tracking* (Fig. 37-5). Do not exit the *Line* command.

C. Complete the shape defining the top surface (*Line* 5) by using the *Close* option. The completed shape should look like Figure 37-5.

Figure 37-5

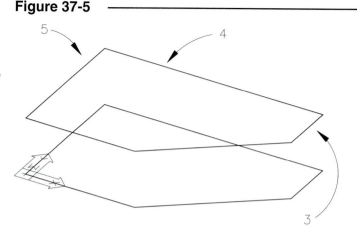

4. There is an easier way to construct the top "surface"; that is, use *Copy* to create the top "surface" from the bottom "surface."

A. *Erase* the 5 lines defining the top shape.

B. Use *Copy* and select the 5 lines defining the bottom shape. You can PICK **0,0,0** as the "base point." The "second point of displacement" can be specified by entering absolute coordinates or relative coordinates or by using an .XY filter. Compare your results with Figure 37-5.

5. Check your work and *Save* the drawing.

A. Use *Plan* to see if all of the *Lines* defining the top shape appear to be aligned with those on the bottom. The correct results at this point should look like Figure 37-3 since the top shape is directly above the bottom.

B. Use *Zoom Previous* to display the previous view (same as *Vpoint, Rotate*, 305,30.)

C. If you have errors, correct them. You can use *Undo* or *Erase* the incorrect *Lines* and redraw them. See "Hints" for further assistance. You may have to use *OSNAP* or enter absolute coordinates to begin a new *Line* at an existing *Endpoint*.

D. *Save* your drawing.

6. Draw the vertical *Lines* between the base and top "surfaces" using the following methods.

Figure 37-6

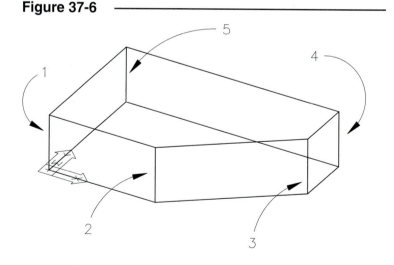

A. Enter absolute coordinates (X, Y, and Z values) to draw the first two vertical *Lines* shown in Figure 37-6 (1, 2).

B. Use another method (interactive coordinate entry) to draw the remaining three *Lines* (3, 4, 5). You have learned that if you PICK a point, the selected point is on the XY plane of the current coordinate system. *OSNAP*, however, allows you to PICK points in 3D space. Begin each of the three *Lines* ("Specify first point:") by using **Grid Snap** and picking the appropriate point on the XY plane. The second point of each *Line* ("Specify next point or [Undo]:") can be specified by using *OSNAP* (use *Endpoint*).

C. Once again, there is an easier method. *Erase* four lines and <u>leave one</u>. You can use *Copy* with the *Multiple* option to copy the one vertical *Line* around to the other positions. Make sure you select the *Multiple* option. In response to the "base point or displacement:" prompt, select the base of the remaining *Line* (on the XY plane) with *Grid Snap* on. In response to "second point of displacement:" make the multiple copies.

D. Use a *Plan* view to check your work. *Zoom Previous*. Correct your work, if necessary.

7. Draw two *Lines* defining a vertical plane. Often in the process of beginning a 3D model (when geometry has not yet been created in some positions), it may be necessary to use absolute coordinates. This is one of those cases.

A. Use absolute coordinates to draw the two *Lines* shown in Figure 37-7. Refer to Figure 37-2 for the dimensions.

B. *Zoom Extents* and then *Zoom .9X* so that the coordinate system icon appears at the origin. Your screen should look like Figure 37-7.

Figure 37-7

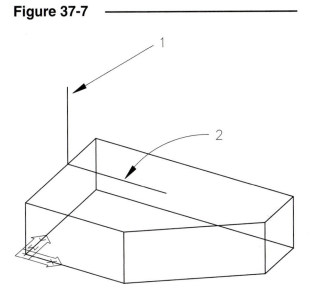

8. Rather than drawing the next objects by defining absolute coordinates, you can establish a vertical construction plane and then PICK points with *Grid Snap* and *GRID* or *Polar Snap* and *Polar Tracking* on the construction plane.

A. This can be done by creating a new UCS (User Coordinate System). Invoke the *UCS* command by typing *UCS* or by selecting it from the menus. Use the *3point* option and select the line *Endpoint*s in the order indicated (Fig. 37-8) to define the origin, point on X axis, and point on Y axis.

Figure 37-8

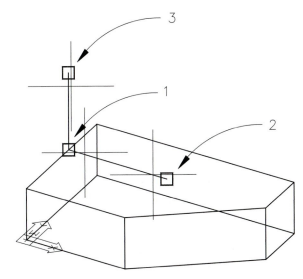

B. The Coordinate System icon should appear in the position shown in Figure 37-9. Notice that the *GRID* and *SNAP* have followed the new UCS. Move the cursor and note the orientation of the cursor. You can now draw on this vertical construction plane simply by PICKing points on the XY plane of the new UCS!

If your UCS does not appear like that in the figure, *UNDO* and try again.

Figure 37-9

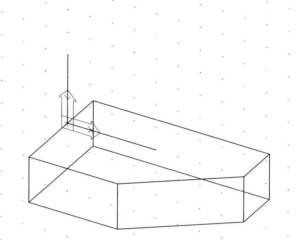

C. Draw the *Lines* to complete the geometry on the XY plane of the UCS. Refer to Figure 37-2 for the dimensions. The stop plate should look like that in Figure 37-10.

9. Check your work; save the drawing and the newly created UCS.

A. View your drawing from a plan view of the UCS. Invoke *Plan* and accept the default option "*<current>*." If your work appears to be correct, use *Zoom Previous*.

B. Save your new UCS. Invoke the *UCS* command as before. Use the *Save* option and assign the name **FRONT**.

Figure 37-10

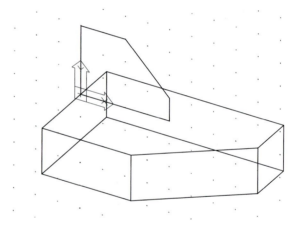

10. Create a second vertical surface behind the first with *Copy* (Fig. 37-11). You can give coordinates relative to the (new) current UCS.

A. Invoke *Copy*. Select the four lines (highlighted) comprising the vertical plane (on the current UCS). Do <u>not</u> select the bottom line defining the plane. Specify **0,0** as the "base point:" and give either absolute or relative coordinates (with respect to the current UCS) as the "second point of displacement." Note that the direction to *Copy* is <u>negative</u> Z.

Figure 37-11

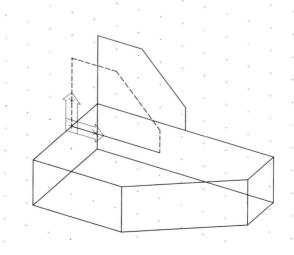

B. Alternately, using *Copy* with *OSNAP* would be valid here since you can PICK (with *OSNAP*) the back corner.

11. Draw the horizontal lines between the two vertical "surfaces" using one of several possible methods.

A. Use the **Line** command to connect the two vertical surfaces with four horizontal edges as shown in Figure 37-12. With *SNAP* on, you can **PICK** points on the XY plane of the UCS. Use *OSNAP* to **PICK** the **Endpoints** of the lines on the back vertical plane.

B. As another possibility, you could draw only one **Line** by the previous method and **Copy** it with the **Multiple** option to the other three locations. In this case, **PICK** the "base point" on the XY plane (with *SNAP ON*). Specifying the "second point(s) of displacement" is easily done (also at *SNAP* points).

Figure 37-12

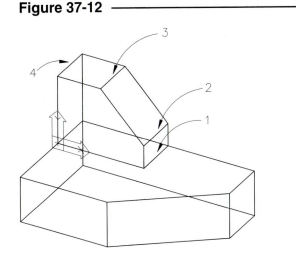

12. A wireframe model uses 2D objects to define the bounding edges of "surfaces." There should be no objects that do not fulfill this purpose. If you examine Figure 37-13, you notice two lines that should be removed (highlighted). These lines do not define edges since they each exist between coplanar "surfaces." Remove these lines with *Trim*.

Trimming 2D objects in 3D space requires special attention. You have two choices for using *Trim* in this case: *Projection = None* or *View*. You can use the *View* option of *Projection* (to trim any objects that appear to intersect from the current view) or you must select the vertical *Line* as a cutting edge (because with the *None* option of *Projection*, you cannot select a cutting edge that is perpendicular to the XY plane of the current coordinate system). In other words, when *Projection=None*, you can select only the vertical *Line* as the cutting edge, but when *Projection=View*, either the vertical or horizontal *Line* can be used as the cutting edge (Fig. 37-14).

Figure 37-13

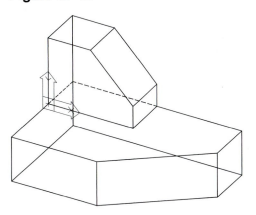

A. Use **Trim** and select either *Line* (highlighted, Fig. 37-14) as the *cutting edge*. Change the **Projection** to **View** and *Trim* the indicated *Lines*. Use *Trim* again to trim the other indicated horizontal *Line* (previous figure).

B. Change back to the WCS. Invoke **UCS**; make the **World** Coordinate System current.

Figure 37-14

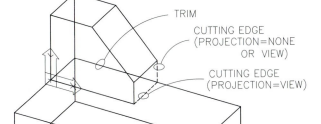

TRIM

CUTTING EDGE
(PROJECTION=NONE
OR VIEW)

CUTTING EDGE
(PROJECTION=VIEW)

13. Check your work and *Save* your drawing.

Figure 37-15

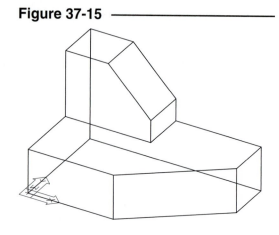

   A. Make sure your wireframe model looks like that in Figure 37-15.
   B. Use *Save* to secure your drawing.

14. It may be wise to examine your model from other viewpoints. As an additional check of your work and as a review of 3D viewing, follow these steps.

   A. Invoke *UCS* and *Restore* the UCS you saved called **FRONT**. Now use *Plan* with the *Current UCS* option. This view represents a front view and should allow you to visualize the alignment of the two vertical planes constructed in the last several steps.
   B. Next, use *UCS* and restore the **World** Coordinate System. Then use *Plan* again with the *Current UCS* option. Check your model from this viewing angle.
   C. Finally, use *Vpoint Rotate* and enter angles of **305** and **30**. This should return to your original viewpoint.

15. Instead of constantly changing *Vpoints*, it would be convenient to have several *Vpoints* of your model visible on the screen <u>at one time</u>. This can be done with the *Vports* (Viewports) command. In order to enhance visualization of your model for the following constructions, divide your screen into four sections or viewports with *Vports* and then use the **3D Setup** option.

   A. First, type **UCSORTHO** at the Command line and change the setting to **0**. This ensures no new UCSs will be created when new orthographic viewports are created. Next, use the *Vports* command to produce the *Viewports* dialog box. In the *New Viewports* tab, select *Four Equal*. Also select *3D* in the *Setup* drop-down list. When you select *OK*, your screen should display the three typical orthographic views (top, front, and right side) and an isometric view.

   B. Use the *Ucsicon* command with the *All* option to make the icon appear at the origin for all viewports. Then make the upper right viewport active and change the *Vpoint* angles to **305** and **30** with the *Rotate* option as before (step 2.B.). Also, use *Zoom .9X* in each of the three orthographic viewports to achieve a display as shown in Figure 37-16.

Figure 37-16

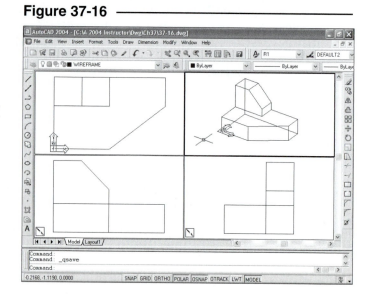

Note that by using viewports you can achieve multiple views of the model; however, since *UCSORTHO* is set to 0, only one UCS appears in all four viewports. The Coordinate System icon appears in all viewports, indicating the current coordinate system orientation. The *Ucsicon ORigin* and *All* settings force the icons to appear at the origin in all viewports.

16. In the next construction, add the milled slot in the inclined surface of the stop plate. Because the slot is represented by geometry parallel with the inclined surface, a UCS should be created on the surface to facilitate the construction.

A. Make the upper-right viewport current. Use the **UCS 3point** option. Select the three points indicated in Figure 37-17 for (1) the "origin," (2) "point on the positive X axis," and (3) "point on the positive Y axis portion on the XY plane." Make sure you *OSNAP* with the **Endpoint** option.

Figure 37-17

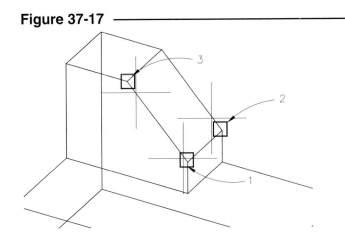

B. The icon should move to the new UCS origin. Its orientation should appear as that in Figure 37-18. Notice that the new coordinate system most likely appears in all viewports (unless the *UCSVP* variable has been changed to 0). (If all your viewports do not display the new UCS, proceed anyway.)

C. Save this new UCS by using the **UCS** command with the **Save** option. Assign the name **AUX** (short for "auxiliary view").

(Remember to check the "Hints" if you have trouble.)

Figure 37-18

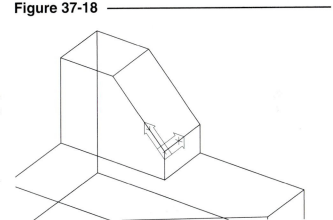

17. To enhance your visualization of the inclined surface, change the UCS and viewpoint for the lower-right viewport.

A. Ensure the upper-right viewport is current. To make the new UCS appear in all viewports, use the **UCS** command with the **Apply** option, then answer **All** to the "Pick viewport to apply current UCS or [All]<current>:" prompt. All viewports should display the new "AUX" UCS.

B. Now make the lower-right viewport current. Use the *Plan* command with the *Current* option to create an auxiliary view looking directly at the inclined surface. Your screen display should look like that in Figure 37-19.

Figure 37-19

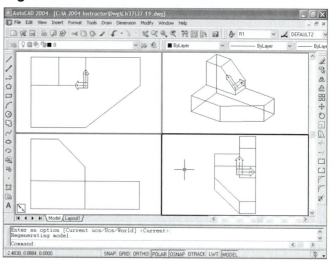

C. It would also be helpful to *Zoom* in with a Window to only the inclined surface. Do so. Your results (for that viewport) should look like Figure 37-20.

Figure 37-20

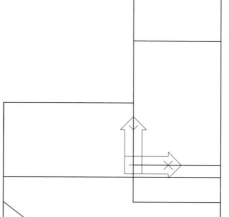

18. The next several steps guide you through the construction of the slot (Fig. 37-21). Refer to Figure 37-2 for the dimensions of the slot.

   A. Turn on the *Grid Snap* and *GRID* for the viewport.
   B. Draw two *Circles* with a .5 radius. The center of the lower circle is at coordinate **1,1** (of the UCS). The other circle center is one unit above.
   C. Draw two vertical *Lines* connecting the circles. Use *Tangent* or *Quadrant* OSNAP.
   D. Use *Trim* to remove the inner halves of the *Circles* (highlighted). Notice that changes made in one viewport are reflected in the other viewports.

Figure 37-21

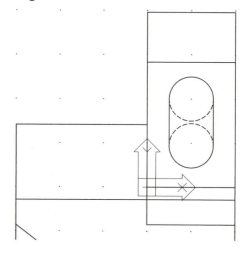

19. The slot has a depth of 1 unit. Create the bottom of the slot by *Copying* the shape created on the top "surface."

   A. With the current UCS, the direction of the *Copy* is in a <u>negative</u> Z direction. Use *Copy* and select the two *Arcs* and two *Lines* comprising the slot. Use the *Center* of an *Arc* or the origin of the UCS as the "base point." The "second point of displacement" can be specified using absolute or relative coordinates.

   B. Check to ensure that the copy has been made correctly. Compare your work to that in Figure 37-22. No other lines are needed to define the slot. The "surface" on the inside of the slot has a smooth transition from curved to straight, so no additional edges (represented by lines) exist.

Figure 37-22

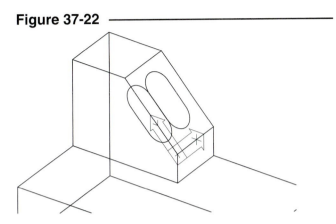

20. The geometry defining the stop plate is complete at this point. No other edges are needed to define the "surfaces." However, you may be called on to make a design change to this part in the future. Before you change your display back to a *Single* viewport and save your work, it is helpful to *Save* the viewport configuration so that it can be recalled in a future session. Unless you save the current viewport configuration, you have to respecify the *Vpoints* for each viewport.

   A. Use the *-Vports* command and the *Save* option (use the hyphen prefix). You can specify any name for the viewport configuration. A descriptive name for this configuration is **FTIA** (short for <u>F</u>ront, <u>T</u>op, <u>I</u>sometric, and <u>A</u>uxiliary). Using this convention tells you how many viewports (by the number of letters in the name) and the viewpoints of each viewport when searching through a list of named viewports.

   B. When changing back to a single viewport, the <u>current</u> viewport becomes the single viewport. Make the upper-right viewport (isometric-type viewpoint) current. Invoke the *Vports* command and the *Single* option. The resulting screen should display only the isometric-type viewpoint (like Fig. 37-23). If needed, the viewport configuration can be recalled in the future by using the *Restore* option of *Vports*.

Figure 37-23

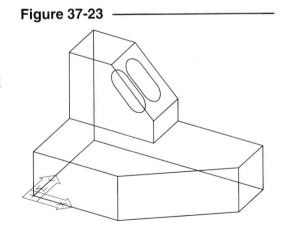

   C. Before saving your drawing, you should make the WCS current by using the *UCS* command. Note that changing the UCS does not generally change the display (unless *UCSFOLLOW* is set to 1), but creating a new orthographic view can create or change the UCS (if *UCSORTHO* is set to 1).

21. *Save* your drawing. Use this drawing for practicing with display commands covered in the previous chapters, 3D Display and Viewing, and User Coordinate Systems. The stop plate is especially good for practicing with *3Dorbit*.

22. When you are finished experimenting with display and view commands, *Exit* AutoCAD and do not save your changes.

Hints

H1.   A. Start AutoCAD by the method you normally use at your computer workstation. Select the *File* pull-down menu, *New....* In the dialog box, select *Start from Scratch*, then *Imperial* default settings. Next, use *Save* and assign the name **WIREFRAME**.
      B. (No help needed.)
      C. *Format* pull-down menu, *Units....*
      D. (No help needed.)
      E. Press **F9**. Make sure the word **SNAP** appears on the Status Bar. To set *Polar Snap*, right-click on the word *SNAP* on the Status Bar, then select *Settings* from the shortcut menu. In the *Snap and Grid* tab, set *Polar Distance* to **1.000** and **PICK** the *Polar Snap* radio button in *Snap type & style*.
      F. Press **F7**. The *GRID* should be visible.
      G. *Format* pull-down menu, *Layer...*, select *New*, type **MODEL** in the edit box, then *Current* and *OK*.

H2.   A. *Draw* pull-down menu, *Line*. Make sure *Grid Snap* or *Polar Snap* is *ON*. The *Coords* display or tracking tip helps you keep track of the line lengths as you draw.
      B. Type *Vpoint*. Type *r*. Enter **305** for the first angle and **30** for the second. Type **Z** (for *Zoom*). Enter **.9X**
      C. *View* pull-down, *Display>*, then *UCS icon*, then check *On*. Do this again, but check *Origin*.

H3.   A.

	Line 1	Specify first point:	**0,0,2**
		Specify next point or [Undo]:	**4,0,2**
	Line 2	Specify next point or [Undo]:	**8,3,2**
		Specify next point or [Undo]:	do not press **Enter** yet

      B. If you happened to exit the *Line* command, you can use *Line* again and attach to the last end-point by entering an "@" in response to the "Specify first point:" prompt. If that does not work, use *OSNAP Endpoint* to locate the end of the last *Line*.

<u>Using point filters:</u> At the "Specify next point or [Undo]:" prompt, type **.XY** and press **Enter**. AutoCAD responds with "of." **PICK** a point on the XY plane <u>directly below</u> the desired location for the *Line* end (Fig. 37-24). AutoCAD responds with "(need Z)." Enter a value of **2**. The resulting *Line* end coordinate has the X and Y values of the point

Figure 37-24

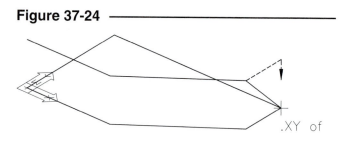

you PICKed and the Z value of 2. In other words, since you can PICK points only on the XY plane, using the .XY point filter locates only the X and Y coordinates of the selected point and allows you to specify the Z coordinate separately.

Using *Polar Snap* and *Polar Tracking*: Make sure the words **SNAP** and **POLAR** appear recessed on the Status Bar and **Polar Snap** is **On** (not *Grid Snap*). Move the cursor in the desired direction and watch the tracking tip as you draw (Fig. 37-25).

Figure 37-25

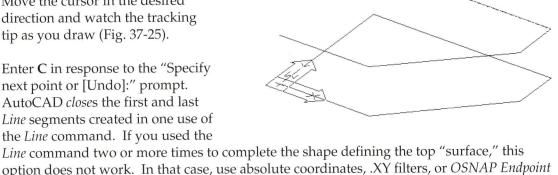

C. Enter **C** in response to the "Specify next point or [Undo]:" prompt. AutoCAD *close*s the first and last *Line* segments created in one use of the *Line* command. If you used the *Line* command two or more times to complete the shape defining the top "surface," this option does not work. In that case, use absolute coordinates, .XY filters, or *OSNAP Endpoint* to connect to the final endpoint.

H4. A. Type **E** for *Erase* or use the **Modify** pull-down, **Erase**. *Erase* the 5 *Lines*.

B. Type **CP** for *Copy* or use the **Modify** pull-down, **Copy**. Select the 5 *Lines* in response to "Select objects:." PICK or enter **0,0,0** as the "base point or displacement." Enter either **0,0,2** (absolute coordinates) or **@0,0,2** (relative coordinates) in response to "second point of displacement:".

H5. A. Type **Plan** and then the **W** option or select **View** pull-down, **3D Views >**, **Plan View >**, **World UCS**.

B. Type **Z** for *Zoom* and type **P** for *Previous*.

C. If you have mistakes, correct them by first *Erasing* the incorrect *Lines*. Begin redrawing at the last correct *Line Endpoint*. You have to use absolute coordinates, .XY filters, or *OSNAP Endpoint* to attach the new *Line* to the old.

D. Type **Save** or select **File** pull-down, **Save**.

H6. A. Line 1   Specify first point: **0,0,0**
             Specify next point or [Undo]:   **0,0,2**
       Line 2   Specify first point: **4,0,0**
             Specify next point or [Undo]:   **4,0,2**

B. Line 3   Specify first point:
             PICK point a. (with *SNAP* on*)*
             Specify next point or [Undo]:
             Endpoint of, **PICK** point b.
       Line 4   Specify first point:
             PICK point c.
             Specify next point or [Undo]:
             Endpoint of, **PICK** point d.
       Line 5   Specify first point:
             PICK point e.
             Specify next point or [Undo]:
             Endpoint of, **PICK** point f.

Figure 37-26

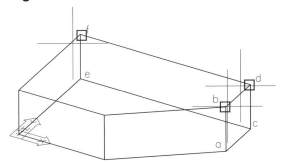

C. Type *E* for *Erase*. *Erase* the last four vertical lines, leaving the first. Type *CP* for *Copy*. Select the remaining vertical line. At the next prompt, type *M* for *Multiple*. When asked for the "base point:," select the bottom of the line (on the XY plane—make sure *SNAP* is *On*). At the "second point of displacement:" prompt, pick the other locations for the new lines.

D. *View* pull-down, **3D Views >, Plan View >, World UCS**. Examine your work. Type *Z* for *Zoom*. Type *P* for *Previous*.

H7. A. Line 1 Specify first point: **0,3,2**
            Specify next point or [Undo]:   **0,3,5**
         Line 2 Specify first point: **0,3,2**
            Specify next point or [Undo]:   **4,3,2**

B. Type *Z*. Type *E* for *Extents*. Press **Enter** to invoke the last command (*Zoom*). Type **.9X**.

H8. A. Type **UCS**. Type *N* for *New*, then **3** for *3point*. When prompted for the first point ("origin"), invoke the cursor (*OSNAP*) menu and select **Endpoint**. **PICK** point 1 (Fig. 37-8). When prompted for the second point ("point on positive portion of the X-axis"), again use **Endpoint** and **PICK** point 2. Repeat for the third point ("point on positive Y portion of the UCS XY plane").

B. If you have trouble establishing this UCS, you can type the letter *U* to undo one step. Repeat until the icon is returned to the origin of the WCS orientation. If that does not work for you, type **UCS** and use the **World** option. Repeat the steps given in the previous instructions. Make sure you use *OSNAP* correctly to **PICK** the indicated **Endpoint**s.

C. When constructing the *Lines* representing the vertical plane, ensure that *Grid Snap* and *GRID* (F7) are *ON*. Notice how you can PICK points only on the XY plane of the UCS with the cursor. This makes drawing objects on the plane very easy. Use the **Line** command to draw the remaining edges defining the vertical "surface."

H9. A. *View* pull-down menu, **3D Views >, Plan View >, Current UCS**. Check your work. Then type *Z*, then *P*. Correct your work if necessary.

B. Type **UCS**. Type *S* for the *Save* option. Enter the name **FRONT**.

H10. A. Type *CP* or select **Copy** from the **Modify** pull-down menu. Select the indicated four *Lines* (see Figure 37-11). For the base point, enter or **PICK 0,0**. For the second point of displacement, enter **0,0,-2**.

B. Alternately, perform the same sequence except, when prompted for the "second point of displacement:", select **Endpoint** from the *OSNAP* menu and **PICK** the farthest corner of the top plane.

H11. A. Type *L* for *Line*. At the "Specify first point:" prompt **PICK** point (a) on the UCS XY plane (Fig. 37-27). You should be able to *SNAP* to the point. At the "Specify next point or [Undo]:" prompt, select the **Endpoint** *OSNAP* option and **PICK** point (b). Repeat the steps for the other 3 *Lines*.

Figure 37-27 ────────────

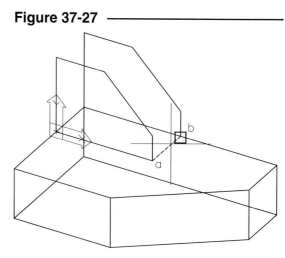

B. Alternately, draw only the first *Line* (line 1, Fig. 37-12). Type *CP* for *Copy*. Select the first *Line*. Type *M* for *Multiple*. At the "base point:" prompt, **PICK** point (a) (Fig. 37-27). Since this point is on the XY plane, **PICK** (with *Grid Snap ON*). For the second points of displacement, **PICK** the other corners of the vertical surface on the XY plane (with *SNAP ON*).

H12.   A. *Modify* pull-down, **Trim**. Select either the vertical or horizontal *Line* (as indicated in Fig. 37-14) as the cutting edge. Change **Projection** to **View**. Complete the trim operation. With *Projection* now set to *View*, **Trim** the other *Line* indicated (see Figure 35-13).

      B. Type **UCS**, then **W** for *World*. The WCS icon should appear at the origin (see Figure 37-15).

H13.   A. If your wireframe model is not correct, make the necessary changes before the next step.

      B. *File* pull-down, **Save**.

H14.   A. *Tools* pull-down, **Named UCS...**. In the *Named UCS* tab, select **FRONT**. **PICK** the *Current* tile, then **OK**. Make sure you PICK <u>*Current*</u>. Then, from the *View* pull-down, select *3D Views >*, *Plan View >*, *Current UCS*. Examine your figure from this viewpoint.

      B. Invoke the *UCS* dialog box again and set **World** as **Current**. Next, use the *View* pull-down menu again to see a *Plan View* of the *World* coordinate system.

      C. Type **Vpoint**, then **R** for the *Rotate* option. Enter **305** for the "angle in the XY plane from the X axis." Enter **30** for the "angle from the XY plane."

H15.   A. No help needed for creating viewports. To experiment with creating and erasing *Lines*, invoke the **Line** command and **PICK** points in the current viewport (wherever the cursor appears). Now, use **Line** again and **PICK** only the "Specify first point:." Move your cursor to another viewport and make it current (**PICK**). **PICK** the "Specify next point or [Undo]:" in the new viewport. Notice that you can change the current viewport <u>within</u> a command to facilitate using the same draw command (or edit command) in multiple viewports. **Erase** the experimental *Lines*.

      B. Type **Ucsicon**, then **A** for the *All* option, then **OR** for *Origin*. Next, click in the upper right viewport to make it active. Type **Vpoint**, then **R** for the *Rotate* option. Enter angles of **305**, then **30**. Make one of the other three viewports current, then type **Z** for *Zoom*, then enter a value of **.9X**. *Zoom* **.9X** also in the other two orthographic viewports.

H16.   A. **PICK** the lower-right viewport. Type **UCS**, then **N**, then **3**. When prompted for the "origin," select *Endpoint* from the *OSNAP* menu and **PICK** point 1 (see Fig. 37-18). Use *Endpoint OSNAP* to select the other two points.

      B. No help needed.

      C. Select the *Tools* pull-down menu, **Named UCS...**. When the *UCS* dialog appears, **PICK** the current UCS "**Unnamed**." In the edit box, double-click and type over the **Unnamed** with the name **AUX**. **PICK OK**.

H17.   A. (No help needed.)

      B. **PICK** the lower-right viewport. Select *View* pull-down, *3D Views >*, *Plan View >*, *Current UCS*.

      C. Type **Z**. Make a Window around the entire inclined surface.

H18.   A. Check to see if *Grid Snap* and *GRID* are *ON*. If not, right-click on the word **SNAP** on the Status Bar and select **Grid Snap**. Make sure the word *SNAP* appears recessed. Turn on *GRID* (press **F7**).

B. Select the *Draw* pull-down menu, then *Circle >* with the *Center, Radius* option. For the center, **PICK** point **1,1** (watch the *Coords* display). Enter a value of **.5** for the radius. Repeat this for the second *Circle*, but the center is at **1,2**.

C. Turn *SNAP* off. Type *L* for *Line*. At the "Specify first point:" prompt, select the *Quadrant OSNAP* mode and **PICK** point (a) (Fig. 37-28). At the "Specify next point or [Undo]:" prompt, use *Quadrant* again and **PICK** point (b).

Figure 37-28

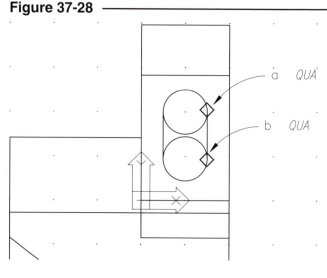

D. *Modify* pull-down, *Trim*. Select the two *Lines* as *Cutting edges* and *Trim* the inner halves of the circles, as shown in Figure 37-21.

H19. A. Type *CP*. Select the two arcs and two lines comprising the slot. At the "base point or displacement:" prompt, enter **0,0,0**. At the "second point of displacement:" prompt, enter **0,0,-1**.
B. No help needed.

H20. A. Type *-Vports*, then *S* for *Save*. At the "name for new viewport configuration prompt," enter **FTIA**.
B. **PICK** the upper right viewport. Type *Vports*. Choose the *Single* option. Activate *Vports* again. The name FTIA should appear in the *Named Viewports* tab. Select *FTIA*, then *OK*. Finally, type *U* to undo the last action.
C. Select the *Tools* pull-down menu and then *Named UCS...*. **PICK** *World* from the list and make it *Current*.

H21. *File* pull-down, *Save*.

H22. *File* pull-down, *Exit* AutoCAD.

COMMANDS

The draw and edit commands that operate for 2D drawings are used to create 3D elements by entering X,Y,Z values, or by using UCSs. *Pline* cannot be used with 3D coordinate entry. Instead, the *3Dpoly* command is used especially for creating 3D polylines.

3DPOLY

	COMMAND (TYPE)	ALIAS (TYPE)	Short-cut	Screen (side) Menu	Tablet Menu
Pull-down Menu					
Draw *3D Polyline*	*3DPOLY*	*3P*	...	DRAW 1 *3dpoly*	O,10

A line created by *3Dpoly* is a *Pline* with 3D coordinates. The *Pline* command allows only 2D coordinate entry, whereas the *3Dpoly* command allows you to create <u>straight</u> polyline segments in 3D space by specifying 3D coordinates. A *3Dpoly* line has the "single object" characteristic of normal *Plines*; that is, several *3Dpoly* line segments created with one *3Dpoly* command are treated as one object by AutoCAD.

Figure 37-29

```
Command: 3dpoly
Specify start point of polyline: PICK or (coordinates)
Specify endpoint of line or [Undo]: PICK or (coordinates)
Specify endpoint of line or [Undo]: PICK or (coordinates)
Specify endpoint of line or [Close/Undo]: Enter
Command:
```

If you PICK points or specify coordinate values, AutoCAD simply connects the points with straight polyline segments in 3D space. For example, *3Dpoly* could be used to draw line segments in a spiral fashion by specifying coordinates or PICKing points (Fig. 37-29).

Close
This option closes the last point to the first point entered in the command sequence.

Undo
This option deletes the last segment and allows you to specify another point for the new segment.

A *3Dpoly* does <u>not</u> have the other features or options of a *Pline* such as *Arc* segments and line *Width*.

Special options of *Pedit* can be used with *3Dpoly* lines to *Close, Edit Vertex, Spine curve,* or *Decurve* the *3Dpoly*. You can also use grips to edit the *3Dpoly*.

Figure 37-30

Although some applications may require a helix, *3Dpoly* and *Pedit* <u>cannot</u> be used in conjunction to do this. The *Spine curve* option of the *Pedit* command converts the *3Dpoly* to a polyline version of a B-Spline, not a true NURBS spline. A *Spline*-fit *Polyline* does not pass through the vertices that were specified when the *Pline* or *3DPoly* was created. The *Spline* command, however, <u>can</u> be used to specify points in 3D to create a true helix (Fig. 37-30).

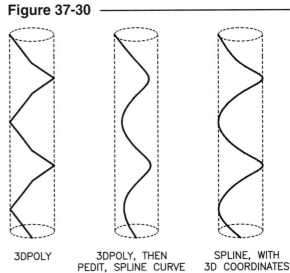

3DPOLY   3DPOLY, THEN PEDIT, SPLINE CURVE   SPLINE, WITH 3D COORDINATES

Lines and *Arcs* <u>cannot</u> be converted to *3Dpoly* lines using *Pedit*. You can, however, convert *Lines* and *Arcs* to *Plines* if they lie on the XY plane of the current UCS. Spline-fit *Plines* and *3Dpoly* lines can be converted to NURBS splines with the *Object* option of the *Spline* command, but the conversion does not realign the new *Spline* to pass through the original vertices.

CHAPTER EXERCISES

1–4.   **Introduction to Wireframes**

Create a wireframe model of each of the objects in Figures 37-31 through 37-34. Use each dimension marker in the figure to equal one unit in AutoCAD. All holes are 1 unit in diameter. Begin with the lower-left corner of the model located at 0,0,0. Assign the names **WFEX1**, **WFEX2**, **WFEX3**, and **WFEX4**.

Figure 37-31 ────────────────

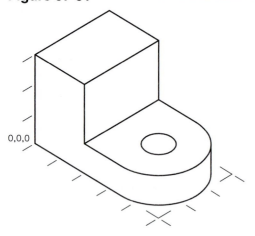

Figure 37-32 ────────────────

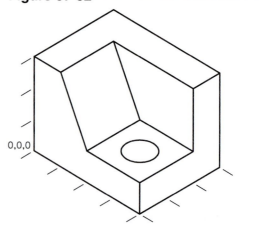

Figure 37-33 ────────────────

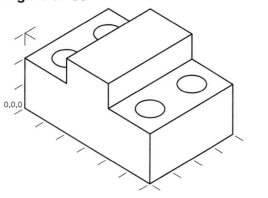

Figure 37-34 ────────────────

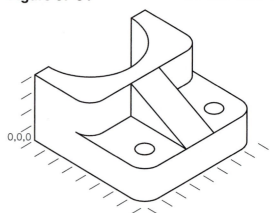

5. Create a wireframe model of the corner brace shown in Figure 37-35. *Save* the drawing as **CBRAC-WF**.

Figure 37-35

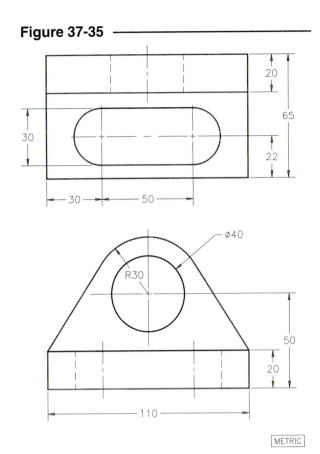

METRIC

6. Make a wireframe model of the V-block (Figure 37-36). *Save* as **VBLCK-WF**.

Figure 37-36

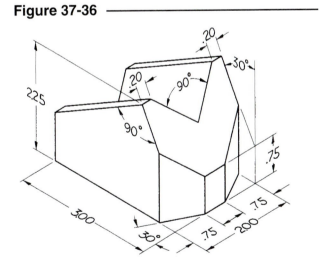

7. Make a wireframe model of the bar guide (Figure 37-37). *Save* as **BGUID-WF**.

Figure 37-37

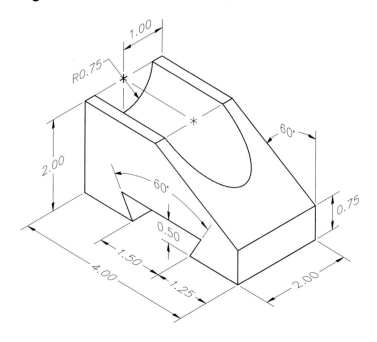

8. Make a wireframe model of the angle brace (Figure 37-38). *Save* as **ANGLB-WF**.

Figure 37-38

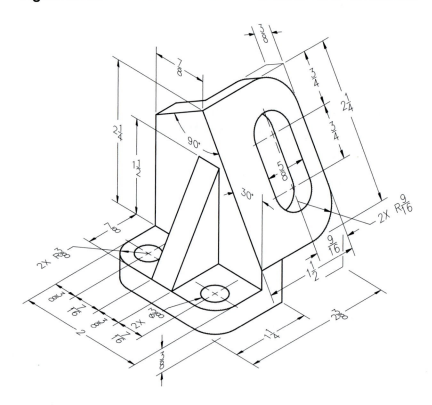

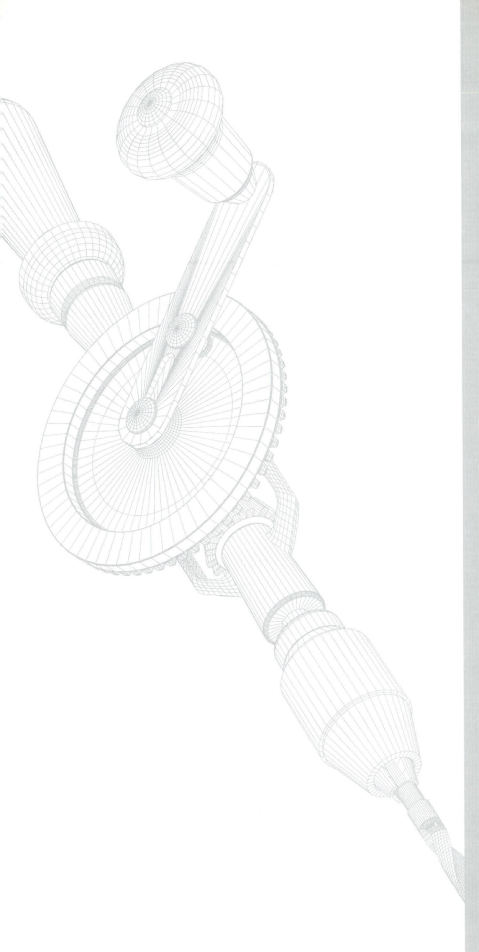

38

SOLID MODELING CONSTRUCTION

Chapter Objectives

After completing this chapter you should:

1. be able to create solid model primitives using the following commands: *Box*, *Wedge*, *Cone*, *Cylinder*, *Torus*, and *Sphere*;

2. be able to create swept solids from existing 2D shapes using the *Extrude* and *Revolve* commands;

3. be able to move solids in 3D space using *Move*, *Align*, and *Rotate3D*;

4. be able to create new solids in specific relation to other solids using *Mirror3D* and *3DArray*;

5. be able to combine multiple primitives into one composite solid using *Union*, *Subtract*, and *Intersect*;

6. know how to create beveled edges and rounded corners using *Chamfer* and *Fillet*;

7. be able to change composite solids with the vast array of editing tools in *Solidedit*.

CONCEPTS

The ACIS solid modeler is included in AutoCAD. With the ACIS modeler, you can create complex 3D parts and assemblies using Boolean operations to combine simple shapes, called "primitives." This modeling technique is often referred to as CSG or Constructive Solid Geometry modeling. ACIS also provides a method for analyzing and sectioning the geometry of the models (Chapter 39).

The techniques used with ACIS for construction of many solid models follow four general steps:

1. Construct simple 3D <u>primitive</u> solids, or create 2D shapes and convert them to 3D solids by extruding or revolving.

2. Create the primitives <u>in location</u> relative to the associated primitives or <u>move</u> the primitives into the desired location relative to the associated primitives.

3. Use <u>Boolean operations</u> (such as *Union*, *Subtract*, or *Intersect*) to combine the primitives to form a <u>composite solid</u>.

4. Make necessary design changes to features of a composite solid using the variety of editing tools in *Solidedit*.

A solid model is an informationally complete representation of the shape of a physical object. Solid modeling differs from wireframe or surface modeling in two fundamental ways: (1) the information is more complete in a solid model and (2) the method of construction of the model itself is relatively easy. Solid model construction is accomplished by creating regular geometric 3D shapes such as boxes, cones, wedges, cylinders, etc. (primitive solids) and combining them using union, subtract, and intersect (Boolean operations) to form composite solids.

CONSTRUCTIVE SOLID GEOMETRY TECHNIQUES

AutoCAD uses Constructive Solid Geometry (CSG) techniques for construction of solid models. CSG is characterized by solid primitives combined by Boolean operations to form composite solids. The CSG technique is a relatively fast and intuitive way of modeling that imitates the manufacturing process.

Primitives

Solid primitives are the basic building blocks that make up more complex solid models. The ACIS primitives commands are:

BOX	Creates a solid box or cube
CONE	Creates a solid cone with a circular or elliptical base
CYLINDER	Creates a solid cylinder with a circular or elliptical base
EXTRUDE	Creates a solid by extruding (adding a Z dimension to) a closed 2D object (*Pline, Circle, Region*)
REVOLVE	Creates a solid by revolving a shape about an axis
SPHERE	Creates a solid sphere
TORUS	Creates a solid torus
WEDGE	Creates a solid wedge

Primitives can be created by entering the command name or by selecting from the menus or icons.

Boolean Operations

Primitives are combined to create complex solids by using Boolean operations. The ACIS Boolean operators are listed below. An illustration and detailed description are given for each of the commands.

UNION   Unions (joins) selected solids.

SUBTRACT   Subtracts one set of solids from another.

INTERSECT   Creates a solid of intersection (common volume) from the selected solids.

Primitives are created at the desired location or are moved into the desired location before using a Boolean operator. In other words, two or more primitives can occupy the same space (or share some common space), yet are separate solids. When a Boolean operation is performed, the solids are combined or altered in some way to create one solid. AutoCAD takes care of deleting or adding the necessary geometry and displays the new composite solid complete with the correct configuration and lines of intersection.

Figure 38-1

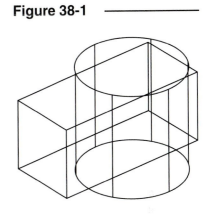

Consider the two solids shown in Figure 38-1. When solids are created, they can occupy the same physical space. A Boolean operation is used to combine the solids into one composite solid and it interprets the resulting utilization of space.

UNION
Union creates a union of the two solids into one composite solid (Fig. 38-2). The lines of intersection between the two shapes are calculated and displayed by AutoCAD. (*Hidden* shademode has been used for this figure.)

Figure 38-2

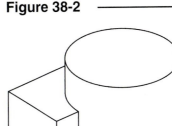

SUBTRACT
Subtract removes one or more solids from another solid. ACIS calculates the resulting composite solid. The term "difference" is sometimes used rather than "subtract." In Figure 38-3, the cylinder has been subtracted from the box.

Figure 38-3

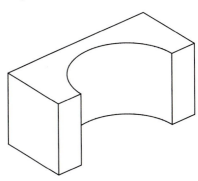

INTERSECT

Intersect calculates the intersection between two or more solids. When *Intersect* is used with *Regions* (2D surfaces), it determines the shared <u>area</u>. Used with solids, as in Figure 38-4, *Intersect* creates a solid composed of the shared <u>volume</u> of the cylinder and the box. In other words, the result of *Intersect* is a solid that has only the volume which is part of both (or all) of the selected solids.

Figure 38-4

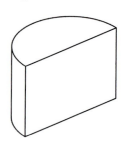

SOLID PRIMITIVES COMMANDS

This section explains the commands that allow you to create primitives used for construction of composite solid models. The commands allow you to specify the dimensions and the orientation of the solids. Once primitives are created, they are combined with other solids using Boolean operations to form composite solids.

NOTE: If you use a pointing device to PICK points, use *OSNAP* when possible to PICK points in 3D space. If you do not use *OSNAP*, the selected points are located on the current XY construction plane, so the true points may not be obvious. It is recommended that you use *OSNAP* or enter values.

The solid modeling commands are located in groups. The commands for creating solid primitives are located in a group in the *Draw* menus under *Solids* (Fig. 38-5). Commands for editing solids and the Boolean commands are found in the *Modify* pull-down menu. You can bring the *Solids* toolbar and the *Solids Editing* toolbar (which contains the Boolean commands) to the screen by selecting *Toolbars…* from the *View* pull-down menu (Fig. 38-6).

Figure 38-5

Figure 38-6

NOTE: When you begin to construct a solid model, <u>create the primitives at location 0,0,0</u> (or some other <u>known point</u>) rather than PICKing points anywhere in space. It is very helpful to know where the primitives are located in 3D space so they can be moved or rotated in the correct orientation with respect to other primitives when they are assembled to composite solids.

BOX

Pull-down Menu	COMMAND (TYPE)	ALIAS (TYPE)	Short-cut	Screen (side) Menu	Tablet Menu
Draw *Solids >* *Box*	BOX	…	…	DRAW 2 SOLIDS Box	J,7

Box creates a solid box primitive to your dimensional specifications. You can specify dimensions of the box by PICKing or by entering values. The box can be defined by (1) giving the corners of the base, then height, (2) by locating the center and height, or (3) by giving each of the three dimensions. The base of the box is oriented parallel to the current XY plane:

> Command: **box**
> Specify corner of box or [CEnter] <0,0,0>: **PICK** or (**coordinates**) or (**CE**) or **Enter**

Options for the *Box* command are listed as follows.

Corner of box
Pressing **Enter** begins the corner of a box at 0,0,0 of the current coordinate system. In this case, the box can be moved into its desired location later. As an alternative, a coordinate position can be entered or **PICK**ed as the starting corner of the box. AutoCAD responds with:

> Specify corner or [Cube/Length]:

The other corner can be PICKed or specified by coordinates. The *Cube* option requires only one dimension to define the cube.

Length
The *Length* option prompts you for the three dimensions of the box in the order of X, Y, and Z. AutoCAD prompts for *Length*, *Width*, and *Height*. These terms are vague since they do not define a specific 3D orientation (in fact, most dictionaries define "length" as the longer of two dimensions). Therefore, substitute X, Y, and Z for AutoCAD's prompts of *Length*, *Width*, and *Height*.

Figure 38-7 shows a box created at 0,0,0 with X, Y, and Z dimensions of 5, 4, and 3.

Figure 38-7

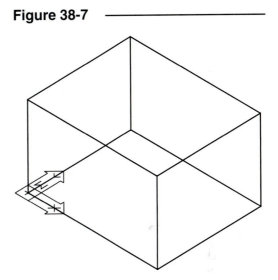

Center
With this option, you first locate the center of the box, then specify the three dimensions of the box. AutoCAD prompts:

> Specify center of box <0,0,0>: **PICK** or (**coordinates**)

Specify the center location. AutoCAD responds with:

> Specify corner or [Cube/Length]:

The resulting solid box is centered about the specified point (0,0,0 for the example; Fig. 38-8).

Figure 38-8

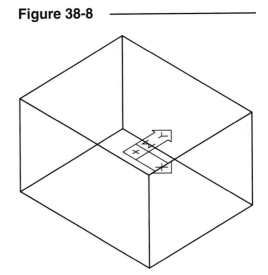

Figure 38-9 illustrates a *Box* created with the *Center* option using *OSNAP* to snap to the *Center* of the top of the cylinder. Note that the center of the box is the <u>volumetric</u> center, not the center of the base.

Figure 38-9

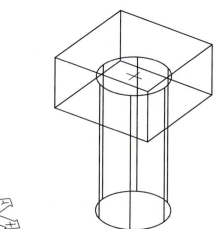

CONE

Pull-down Menu	COMMAND (TYPE)	ALIAS (TYPE)	Short-cut	Screen (side) Menu	Tablet Menu
Draw Solids > Cone	CONE	...	...	DRAW 2 SOLIDS Cone	M,7

Cone creates a right circular or elliptical solid cone ("right" means the axis forms a right angle with the base). You can specify the center location, radius (or diameter), and height. By default, the orientation of the cylinder is determined by the current UCS so that the base lies on the XY plane and height is perpendicular (in a Z direction). Alternately, the orientation can be defined by using the *Apex* option.

Center Point
Using the defaults (center at 0,0,0, PICK a radius, enter a value for height), the cone is generated in the orientation shown in Figure 38-10. The cone here may differ in detail (number of contour lines), depending on the current setting of the *ISOLINES* variable. The default prompts are shown as follows:

Figure 38-10

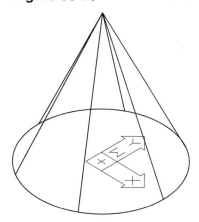

```
Command: cone
Current wire frame density:  ISOLINES=4
Specify center point for base of cone or [Elliptical] <0,0,0>: PICK or
(coordinates)
Specify radius for base of cone or [Diameter]: PICK or (coordinates)
Specify height of cone or [Apex]: PICK or (value)
Command:
```

Invoking the *Apex* option (after *Center point* of the base and *Radius* or *Diameter* have been specified) displays the following prompt:

```
Specify apex point: PICK or (coordinates)
```

Locating a point for the apex defines the height and orientation of the cone (Fig. 38-11). The axis of the cone is aligned with the line between the specified center point and the *Apex* point, and the height is equal to the distance between the two points.

Figure 38-11

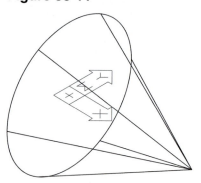

The solid model in Figure 38-12 was created with a *Cone* and a *Cylinder*. The cylinder was created first, then the cone was created using the *Apex* option of *Cone*. The orientation of the *Cone* was generated by PICKing the *Center* of one end of the cylinder for the "Center point" of the base of the cone and the *Center* of the cylinder's other end for the *Apex*. *Union* created the composite model.

Figure 38-12

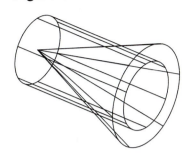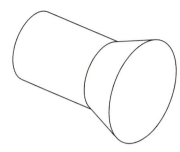

Elliptical

This option draws a cone with an elliptical base (Fig. 38-13). You specify two axis endpoints to define the elliptical base. An elliptical cone can be created using the *Center, Apex,* or *Height* options:

Figure 38-13

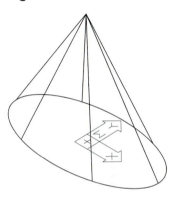

Specify center point for base of cone or [Elliptical] <0,0,0>: *e*
Specify axis endpoint of ellipse for base of cone or [Center]: *c*
Specify center point of ellipse for base of cone <0,0,0>: **PICK** or (**coordinates**)
Specify axis endpoint of ellipse for base of cone: **PICK** or (**coordinates**)
Specify length of other axis for base of cone: **PICK** or (**coordinates**)
Specify height of cone or [Apex]:

CYLINDER

Pull-down Menu	COMMAND (TYPE)	ALIAS (TYPE)	Short-cut	Screen (side) Menu	Tablet Menu
Draw *Solids >* *Cylinder*	*CYLINDER*	...	...	DRAW 2 SOLIDS *Cylinder*	L,7

Cylinder creates a cylinder with an elliptical or circular base with a center location, diameter, and height you specify. Default orientation of the cylinder is determined by the current UCS, such that the circular plane is coplanar with the XY plane and height is in a Z direction. However, the orientation can be defined otherwise by the *Center of other end* option.

Center Point

The default options create a cylinder in the orientation shown in Figure 38-14:

Figure 38-14

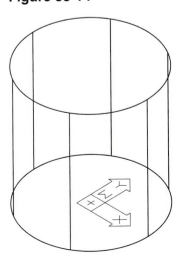

Command: *cylinder*
Current wire frame density: ISOLINES=4
Specify center point for base of cylinder or [Elliptical] <0,0,0>: **PICK** or (**coordinates**)
Specify radius for base of cylinder or [Diameter]: **PICK** or (**coordinates**)
Specify height of cylinder or [Center of other end]: **PICK** or (**coordinates**)

Elliptical

This option draws a cylinder with an elliptical base (Fig. 38-15). Specify two axis endpoints to define the elliptical base. An elliptical cylinder can be created using the *Center, Center of other end,* or *Height* options.

Figure 38-15

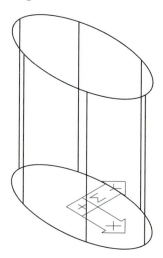

> Specify center point for base of cylinder or [Elliptical] <0,0,0>: *e*
> Specify axis endpoint of ellipse for base of cylinder or [Center]: *c*
> Specify center point of ellipse for base of cylinder <0,0,0>: **PICK** or (**coordinates**)
> Specify axis endpoint of ellipse for base of cylinder: **PICK** or (**coordinates**)
> Specify length of other axis for base of cylinder: **PICK** or (**coordinates**)
> Specify height of cylinder or [Center of other end]: **PICK** or (**value**)

The **Center of other end** option of *Cylinder* is similar to *Cone Apex* option in that the height and orientation are defined by the *Center point* and the *Center of other end*. In Figure 38-16, a hole is created in the *Box* by using *Center of other end* option and *OSNAP*ing to the diagonal lines' *Midpoints*, then *Subtract*ing the *Cylinder* from the *Box*.

Figure 38-16

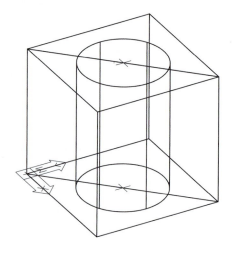

WEDGE

Pull-down Menu	COMMAND (TYPE)	ALIAS (TYPE)	Short-cut	Screen (side) Menu	Tablet Menu
Draw *Solids >* *Wedge*	WEDGE	WE	...	DRAW 2 SOLIDS *Wedge*	N,7

Corner of Wedge

Wedge creates a wedge solid primitive. The base of the wedge is always parallel with the current UCS XY plane, and the slope of the wedge is always along the X axis.

Figure 38-17

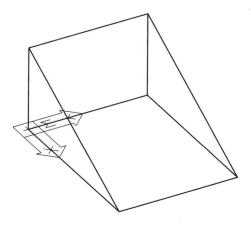

> Command: **wedge**
> Specify first corner of wedge or [CEnter] <0,0,0>: **PICK** or (**coordinates**)
> Specify corner or [Cube/Length]: **PICK** or (**coordinates**)
> Specify height: **PICK** or (**value**)

Accepting all the defaults, a *Wedge* can be created as shown in Figure 38-17, with the slope along the X axis.

Invoking the **Length** option prompts you for the *Length, Width,* and *Height.* More precisely, AutoCAD means width (X dimension), depth (Y dimension), and height (Z dimension).

Center
The point you specify as the center is actually in the center of an imaginary box, half of which is occupied by the wedge. Therefore, the center point is actually at the <u>center of the sloping side of the wedge</u> (Fig. 38-18).

Figure 38-18 ———

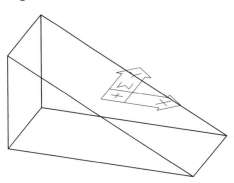

SPHERE

Pull-down Menu	COMMAND (TYPE)	ALIAS (TYPE)	Short-cut	Screen (side) Menu	Tablet Menu
Draw *Solids >* *Sphere*	*SPHERE*	...	...	*DRAW 2* *SOLIDS* *Sphere*	K,7

Sphere allows you to create a solid sphere by defining its center point and radius or diameter:

Figure 38-19 ———

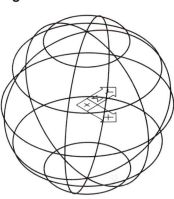

```
Command: sphere
Current wire frame density: ISOLINES=4
Specify center of sphere <0,0,0>: PICK or (coordinates)
Specify radius of sphere or [Diameter]: PICK or (coordinates)
Command: PICK or (value)
```

Creating a *Sphere* with the default options would yield a sphere similar to that in Figure 38-19.

TORUS

Pull-down Menu	COMMAND (TYPE)	ALIAS (TYPE)	Short-cut	Screen (side) Menu	Tablet Menu
Draw *Solids >* *Torus*	*TORUS*	*TOR*	...	*DRAW 2* *SOLIDS* *Tours*	O,7

Torus creates a torus (donut shaped) solid primitive using the dimensions you specify. Two dimensions are needed: (1) the radius or diameter of the tube and (2) the radius or diameter from the axis of the torus to the center of the tube. AutoCAD prompts:

```
Command: torus
Current wire frame density: ISOLINES=4
Specify center of torus <0,0,0>: PICK or (coordinates)
Specify radius of torus or [Diameter]: PICK or (value)
Specify radius of tube or [Diameter]: PICK or (value)
Command:
```

Figure 38-20 shows a *Torus* created using the default orientation with the axis of the tube aligned with the Z axis of the UCS and the center of the torus at 0,0,0. The *Torus* has a torus radius of 4 and a tube radius of 1. (*Hidden* shademode was used for this display.)

Figure 38-20 ─────────

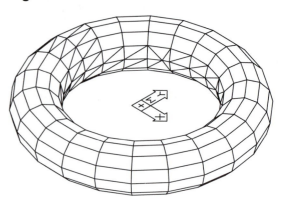

A self-intersecting torus is allowed with the *Torus* command. A self-intersecting torus is created by specifying a torus radius less than the tube radius. Figure 38-21 illustrates a torus with a torus radius of 3 and a tube radius of 4.

Figure 38-21 ─────────

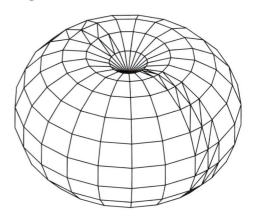

EXTRUDE

Pull-down Menu	COMMAND (TYPE)	ALIAS (TYPE)	Short-cut	Screen (side) Menu	Tablet Menu
Draw *Solids >* *Extrude*	*EXTRUDE*	*EXT*	...	DRAW 2 SOLIDS *Extrude*	P,7

Extrude (like *Revolve*) is a "sweeping operation." Sweeping operations use existing 2D objects to create a solid.

Extrude means to add a Z (height) dimension to an otherwise 2D shape. This command extrudes an existing closed 2D shape such as a *Circle, Polygon, Rectangle, Ellipse, Pline, Spline,* or *Region.* Only <u>closed 2D shapes</u> can be extruded. The closed 2D shape cannot be self-intersecting (crossing over itself).

Two methods determine the direction of extruding: <u>perpendicular</u> to the shape and along a <u>*Path*</u>. With the default method, the selected 2D shape is extruded <u>perpendicular to the plane</u> of the shape <u>regardless</u> of the current UCS orientation. Using the *Path* method, you can extrude the existing closed 2D shape along any existing path determined by a *Line, Arc, Spline,* or *Pline.*

The versatility of this command lies in the fact that <u>any closed shape</u> that can be created by (or converted to) a *Pline, Spline, Region,* etc., no matter how complex, can be transformed into a solid and can be extruded perpendicularly or along a *Path. Extrude* can simplify the creation of many solids that may otherwise take much more time and effort using typical primitives and Boolean operations. Create the closed 2D shape first, then invoke *Extrude:*

```
Command: extrude
Current wire frame density:  ISOLINES=4
Select objects: PICK
Select objects: Enter
Specify height of extrusion or [Path]: PICK or (value)
Specify angle of taper for extrusion <0>: ENTER or (value)
Command:
```

Figure 38-22 shows a *Pline* before and after using *Extrude*.

Figure 38-22

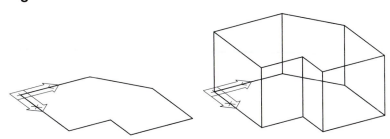

A taper angle can be specified for the extrusion. The resulting solid has sides extruded inward at the specified angle (Fig. 38-23). This is helpful for developing parts for molds that require a slight draft angle to facilitate easy removal of the part from the mold. Entering a negative taper angle results in the sides of the extrusion sloping outward.

Figure 38-23

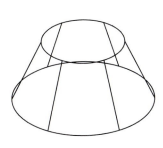
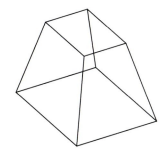

Many complex shapes based on closed 2D *Splines* or *Plines* can be transformed to solids using *Extrude* (Fig. 38-24).

Figure 38-24

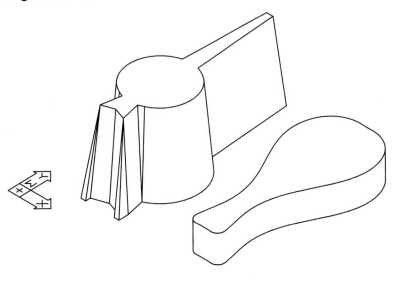

The *Path* option allows you to sweep the 2D shape along an existing line or curve called a "path." The *Path* can be composed of a *Line, Arc, Ellipse, Pline,* or *Spline* (or can be different shapes converted to a *Pline*). This path must lie in a plane. Figure 38-25 shows a closed *Pline* extruded along a *Pline* path and a curved *Spline* path:

Figure 38-25

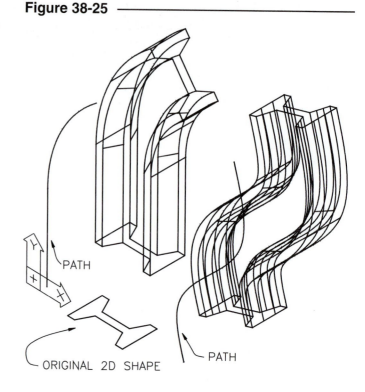

```
Command: extrude
Current wire frame density: ISOLINES=4
Select objects: PICK
Select objects: Enter
Specify height of extrusion or [Path]: path
Select extrusion path: PICK
```

The path cannot lie in the same plane as the 2D shape to be extruded, since the plane of the 2D shape is always extruded perpendicular along the path. If one of the endpoints of the path is not located on the plane of the 2D shape, AutoCAD will automatically move the path to the center of the profile temporarily. Notice how the original 2D shape was extruded perpendicularly to the path (Fig. 38-26).

Figure 38-26

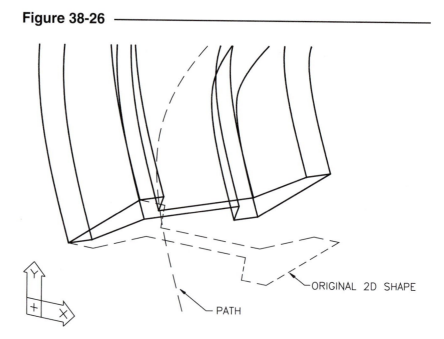

REVOLVE

Pull-down Menu	COMMAND (TYPE)	ALIAS (TYPE)	Short-cut	Screen (side) Menu	Tablet Menu
Draw Solids > Revolve	*REVOLVE*	*REV*	...	*DRAW 2 SOLIDS Revolve*	Q,7

Revolve creates a swept solid. *Revolve* creates a solid by revolving a 2D shape about a selected axis. The 2D shape to revolve can be a *Pline, Polygon, Circle, Ellipse, Spline,* or a *Region* object. Only one object at a time can be revolved. *Splines* or *Plines* selected for revolving <u>must be closed</u>. The command syntax for *Revolve* (accepting the defaults) is:

```
Command: revolve
Current wire frame density:  ISOLINES=4
Select objects: PICK
Select objects: Enter
Specify start point for axis of revolution or define axis by [Object/X (axis)/Y (axis)]: PICK
Specify endpoint of axis: PICK
Specify angle of revolution <360>: Enter or (value)
```

Figure 38-27 illustrates a possibility for an existing *Pline* shape and the resulting *revolved* shape generated through a full circle.

Because *Revolve* acts on an existing object, the 2D shape intended for revolution should be created in or moved to the desired orientation. There are multiple options for selecting an axis of revolution.

Figure 38-27 ────────────────

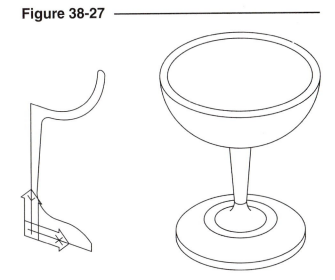

Object
A *Line* or single segment *Pline* can be selected for an axis. The positive axis direction is from the closest endpoint PICKed to the other end.

X or Y
Uses the positive *X* or *Y* axis of the current UCS as the positive axis direction.

Start point of axis
Defines two points in the drawing to use as an axis (length is irrelevant). Select any two points in 3D space. The two points do <u>not</u> have to be coplanar with the 2D shape.

If the *Object* or *Start point* options are used, the selected object or the two indicated points do <u>not</u> have to be coplanar with the 2D shape. The axis of revolution used is <u>always on the plane of the 2D shape</u> to revolve and is aligned with the direction of the object or selected points.

Figure 38-28 demonstrates a possible use of the *Object* option of *Revolve*. In this case, a *Pline* square is revolved about a *Line* object. Note that the *Line* used for the axis is <u>not</u> on the same plane as the *Pline*. *Revolve* uses an axis <u>on the plane</u> of the revolved shape aligned with the direction of the selected axis. In this case, the endpoints of the *Line* are 0,-1,-1 and 0,1,1 so the shape is actually revolved about the Y axis.

Figure 38-28

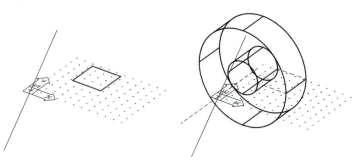

After defining the axis of revolution, *Revolve* requests the number of degrees for the object to be revolved. Any angle can be entered.

Figure 38-29

Another possibility for revolving a 2D shape is shown in Figure 38-29. The shape is generated through 270 degrees.

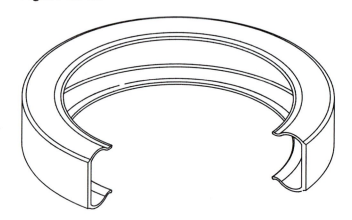

COMMANDS FOR MOVING SOLIDS

When you create the desired primitives, Boolean operations are used to construct the composite solids. However, the primitives must be in the correct position and orientation with respect to each other before Boolean operations can be performed. You can either create the primitives in the desired position during construction (by using UCSs) or move the primitives into position after their creation. Several methods that allow you to move solid primitives are explained in this section.

 When constructing 3D geometry, it is critical that the objects are located in space at some <u>known</u> position. Do <u>not</u> create primitives at any convenient place in the drawing; know the position. The location is important when you begin the process of moving primitives to assemble 3D composite solids. Of course, *OSNAPs* can be used, but sometimes it is necessary to use coordinate values in absolute, rectangular, or polar format. A good practice is to <u>create primitives at the final location</u> if possible (use UCSs when needed) or <u>create the primitives at 0,0,0</u>, then <u>move</u> them preceding the Boolean operations.

MOVE

Pull-down Menu	COMMAND (TYPE)	ALIAS (TYPE)	Short-cut	Screen (side) Menu	Tablet Menu
Modify *Move*	*MOVE*	*M*	...	*MODIFY2* *Move*	*V,19*

The *Move* command that you use for moving 2D objects in 2D drawings can also be used to move 3D primitives. Generally, *Move* is used to change the position of an object in one plane (translation), which is typical of 2D drawings. *Move* can also be used to move an ACIS primitive in 3D space <u>if</u> *OSNAPs* or 3D coordinates are used.

Move operates in 3D just as you used it in 2D. Previously, you used *Move* only for repositioning objects in the XY plane, so it was only necessary to PICK or use X and Y coordinates. Using *Move* in 3D space requires entering X, Y, and Z values or using *OSNAPs*.

For example, to create the composite solid used in the figures in Chapter 36, a *Wedge* primitive was *Moved* into position on top of the *Box*. The *Wedge* was created at 0,0,0, then rotated. Figure 38-30 illustrates the movement using absolute coordinates described in the syntax below:

```
Command: move
Select objects: PICK
Select objects: Enter
Specify base point or displacement: 0,0
Specify second point of displacement or <use
first point as displacement>: 0,4,3
Command:
```

Alternately, you can use *OSNAPs* to select geometry in 3D space. Figure 38-31 illustrates the same *Move* operation using *Endpoint OSNAPs* instead of entering coordinates.

Figure 38-30

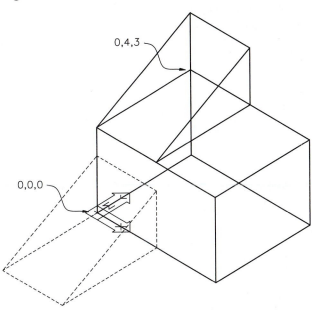

Figure 38-31

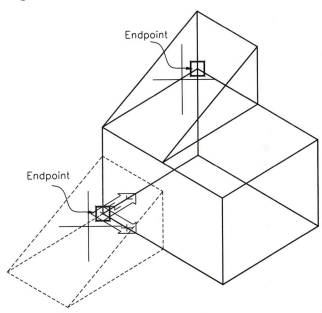

ALIGN

Pull-down Menu	COMMAND (TYPE)	ALIAS (TYPE)	Short-cut	Screen (side) Menu	Tablet Menu
Modify *3D Operation >* *Align*	*ALIGN*	*AL*	...	*MODIFY2* *Align*	*X,14*

Align is an application that is loaded automatically when typing or selecting this command from the menus. *Align* is discussed in Chapter 16 but only in terms of 2D alignment.

Align is, however, a very powerful 3D command because it automatically performs 3D <u>translation and rotation</u> if needed. *Align* is more intuitive and in many situations is easier to use than *Move*. All you have to do is select the points on two 3D objects that you want to align (connect).

Align provides a means of aligning one shape (an object, a group of objects, a block, a region, or a 3D solid) with another shape. The alignment is accomplished by connecting source points (on the shape to be moved) to destination points (on the stationary shape). You can use *OSNAP* modes to select the source and destination points, assuring accurate alignment. Either a 2D or 3D alignment can be accomplished with this command. The command syntax for 3D alignment is as follows:

```
Command: align
Select objects: PICK
Select objects: Enter
Specify first source point: PICK (use OSNAP)
Specify first destination point: PICK (use OSNAP)
Specify second source point: PICK (use OSNAP)
Specify second destination point: PICK (use OSNAP)
Specify third source point or <continue>: PICK (use OSNAP)
Specify third destination point: PICK (use OSNAP)
Command:
```

Figure 38-32

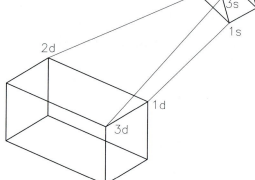

After the source and destination points have been designated, lines connecting those points temporarily remain until *Align* performs the action (Fig. 38-32).

Align performs a translation (like *Move*) and two rotations (like *Rotate*), each in separate planes to align the points as designated. The motion automatically performed by *Align* is actually done in three steps.

Figure 38-33

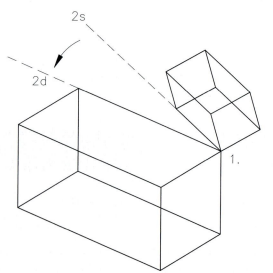

Initially, the first source point is connected to the first destination point (translation). These two points always physically <u>touch</u>.

Next, the vector defined by the first and second source points is aligned with the vector defined by the first and second destination points. The length of the segments between the first and second points on each object is of no consequence because AutoCAD only considers the <u>vector direction</u>. This second motion is a rotation along one axis.

Finally, the third set of points are aligned similarly. This third motion is a rotation along the other axis, completing the alignment.

Figure 38-34

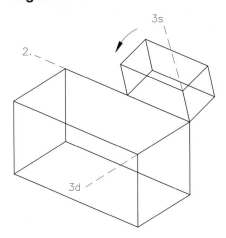

In some cases, such as when cylindrical objects are aligned, only two sets of points have to be specified. For example, if aligning a shaft with a hole (Fig. 38-35), the first set of points (source and destination) specify the attachment of the base of the shaft with the bottom of the hole (use *Center OSNAPs*). The second set of points specify the alignment of the axes of the two cylindrical shapes. A third set of points is <u>not</u> required because the radial alignment between the two objects is not important. When only two sets of points are specified, AutoCAD asks if you want to "continue" based on only two sets of alignment points. The command syntax is as follows:

Figure 38-35

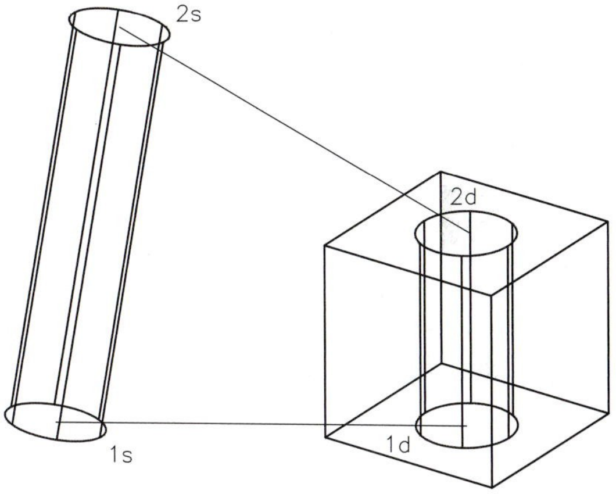

Command: *align*
Select objects: **PICK**
Select objects: **Enter**
Specify first source point: **PICK** (use *OSNAP*)
Specify first destination point: **PICK** (use *OSNAP*)
Specify second source point: **PICK** (use *OSNAP*)
Specify second destination point: **PICK** (use *OSNAP*)
Specify third source point or <continue>: *c*
Scale objects based on alignment points? [Yes/No] <N>: *y* or *n*
Command:

You can also <u>scale the source object</u> to fit between the two selected destination points. If you choose to "Scale objects based on alignment points," the source object is scaled to the two destination points (all dimensions scaled proportionally) and the destination object retains its size.

ROTATE3D

	COMMAND (TYPE)	ALIAS (TYPE)	Short-cut	Screen (side) Menu	Tablet Menu
Pull-down Menu					
Modify *3D Operation >* *Rotate 3D*	*ROTATE3D*	...	...	*MODIFY2* *Rotate3D*	*W,22*

Rotate3D is very useful for any type of 3D modeling, particularly with CSG, where primitives must be moved, rotated, or otherwise aligned with other primitives before Boolean operations can be performed.

Rotate3D allows you to rotate a 3D object about any axis in 3D space. Many alternatives are available for defining the desired rotational axis. Following is the command sequence for rotating a 3D object using the default (*2points*) option:

```
Command: rotate3d
Current positive angle: ANGDIR=counterclockwise ANGBASE=0
Select objects: PICK
Select objects: Enter
Specify first point on axis or define axis by [Object/Last/View/Xaxis/Yaxis/Zaxis/2points]: PICK or
(option) (Select the first point to define the rotational axis.)
Specify second point on axis: PICK
Specify rotation angle or [Reference]: (value) or r
Command:
```

The options are explained next.

2points

The power of *Rotate3D* (over *Rotate*) is that any points or objects in <u>3D space</u> can be used to define the axis for rotation. When using the default (*2points* option), remember you can use *OSNAP* to select points on existing 3D objects. Figure 38-36 illustrates the *2points* option used to select two points with *OSNAP* on the solid object selected for rotating.

Figure 38-36

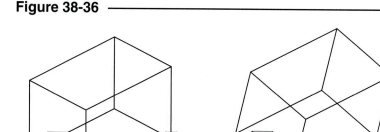

A. B.

Object

This option allows you to rotate about a selected 2D object. You can select a *Line, Circle, Arc,* or 2D *Pline* segment. The rotational axis is aligned with the selected *Line* or *Pline* segment. Positive rotation is determined by the right-hand rule and the "arbitrary axis algorithm." When selecting *Arc* or *Circle* objects, the rotational axis is perpendicular to the plane of the *Arc* or *Circle* passing through the center. You <u>cannot</u> select the edge of a solid object with this option.

Last

This option allows you to rotate about the axis used for the last rotation.

Figure 38-37

View

The *View* option allows you to pick a point on the screen and rotates the selected object(s) about an axis perpendicular to the screen and passing through the selected point (Fig. 38-37).

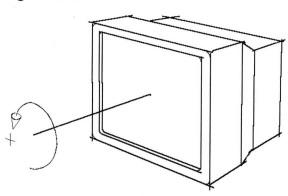

Xaxis

With this option, you can rotate the selected objects about the X axis of the current UCS or any axis parallel to the X axis of the current UCS. You are prompted to pick a point on the X axis. The point selected defines an axis for rotation parallel to the current X axis passing through the selected point. You can use *OSNAP* to select points on existing 3D objects (Fig. 38-38). The current X axis can be used if the point you select is on the X axis.

Figure 38-38

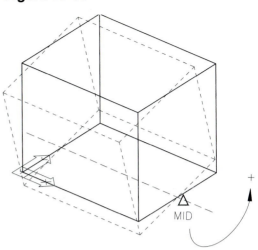

Yaxis

This option allows you to use the Y axis of the current UCS or any axis parallel to the Y axis of the current UCS as the axis of rotation. The point you select defines a rotational axis parallel to the current Y axis passing through the point. *OSNAP* can be used to snap to existing geometry (Fig. 38-39).

Figure 38-39

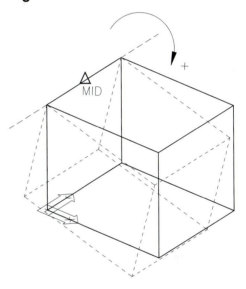

Zaxis

With this option, you can use the Z axis of the current UCS or any axis parallel to the Z axis of the current UCS as the axis of rotation. The point you select defines a rotational axis parallel to the current Z axis passing through the point. Figure 38-40 indicates the use of the *Midpoint OSNAP* to establish a vertical (parallel to Z) rotational axis.

Figure 38-40

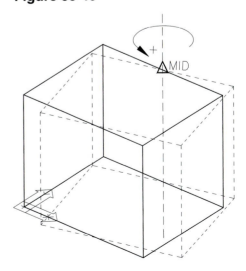

Reference
After you have specified the axis for rotation, you must specify the rotation angle. You are presented with the following prompt:

> Specify rotation angle or [Reference]: **r**
> Specify the reference angle <0>: **PICK** or (**value**) (PICK two points; *OSNAPs* can be used. You can also enter a value.)
> Specify the new angle: **PICK** or (**value**)
> Command:

The angle you specify for the reference (relative) is used instead of angle 0 (absolute) for the starting position. You can enter either a value or PICK two points to specify the angle. You then specify a new angle. The *Reference* angle you select is rotated to the absolute angle position you specify as the "new angle."

Figure 38-41 illustrates how *Endpoint OSNAPs* are used to select a *Reference* angle. The "new angle" is specified as **90**. AutoCAD rotates the reference angle to the 90 degree position.

Figure 38-41

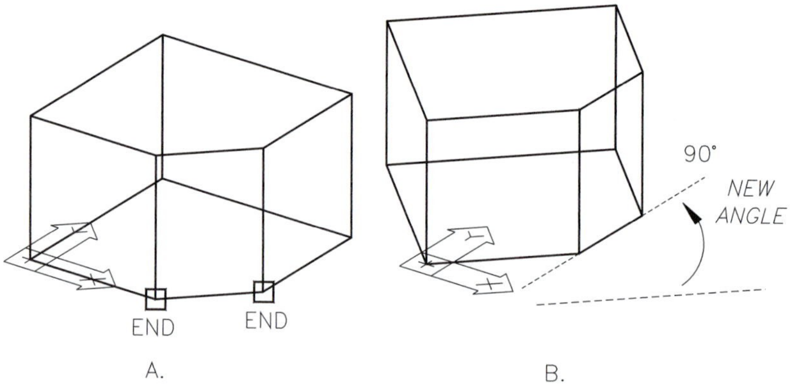

A.

B.

90°
NEW ANGLE

END   END

MIRROR3D

Pull-down Menu	COMMAND (TYPE)	ALIAS (TYPE)	Short-cut	Screen (side) Menu	Tablet Menu
Modify 3D Operation > Mirror 3D	*MIRROR3D*	...	...	*MODIFY2 Mirror3D*	*W,21*

Mirror3D operates similar to the 2D version of the command *Mirror* in that mirrored replicas of selected objects are created. With *Mirror* (2D) the selected objects are mirrored about an axis. The axis is defined by a vector lying in the XY plane. With *Mirror3D*, selected objects are mirrored about a <u>plane</u>. *Mirror3D* provides multiple options for specifying the plane to mirror about:

> Command: **mirror3d**
> Select objects: **PICK**
> Select objects: **Enter**
> Specify first point of mirror plane (3 points) or [Object/Last/Zaxis/View/XY/YZ/ZX/3points] <3points>: **PICK** or (**option**)

The options are listed and explained next. A phantom icon is shown in the following figures <u>only</u> to aid your visualization of the mirroring plane.

3points

The *3points* option mirrors selected objects about the plane you specify by selecting three points to define the plane. You can PICK points (with or without *OSNAP*) or give coordinates. *Midpoint OSNAP* is used to define the 3 points in Figure 38-42 A to achieve the result in B.

Figure 38-42

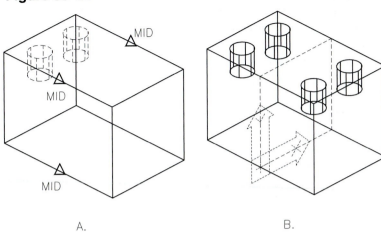

A.

B.

Object

Using this option establishes a mirroring plane with the plane of a 2D object. Selecting an *Arc* or *Circle* automatically mirrors selected objects using the plane in which the *Arc* or *Circle* lies. The plane defined by a *Pline* segment is the XY plane of the *Pline* when the *Pline* segment was created. Using a *Line* object or edge of an ACIS solid is not allowed because neither defines a plane. Figure 38-43 shows a box mirrored about the plane defined by the *Circle* object. Using *Subtract* produces the result shown in B.

Figure 38-43

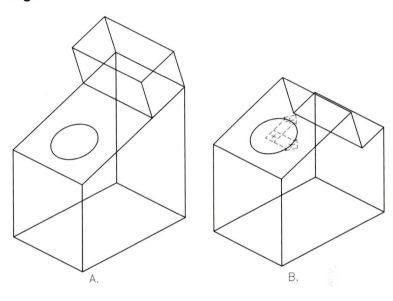

A.

B.

Last

Selecting this option uses the plane that was last used for mirroring.

Zaxis

With this option, the mirror plane is the <u>XY</u> plane perpendicular to a Z vector you specify. The first point you specify on the Z axis establishes the location of the XY plane origin (a point through which the plane passes). The second point establishes the Z axis and the orientation of the XY plane (perpendicular to the Z axis). Figure 38-44 illustrates this concept. Note that this option requires only two PICK points.

Figure 38-44

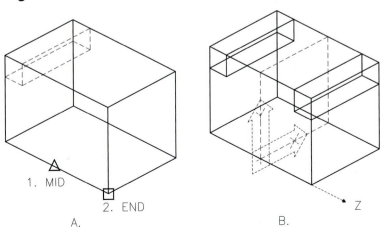

A.

B.

View

The *View* option of *Rotate3D* uses a mirroring plane <u>parallel</u> with the screen and perpendicular to your line of sight based on your current viewpoint. You are required to select a point on the plane. Accepting the default (0,0,0) establishes the mirroring plane passing through the current origin. Any other point can be selected. You must change your viewpoint to "see" the mirrored objects (Fig. 38-45).

Figure 38-45

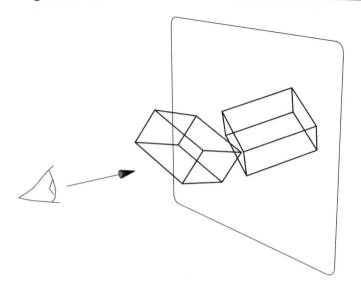

XY

This option situates a mirroring plane parallel with the current XY plane. You can specify a point through which the mirroring plane passes. Figure 38-46 represents a plane established by selecting the *Center* of an existing solid.

Figure 38-46

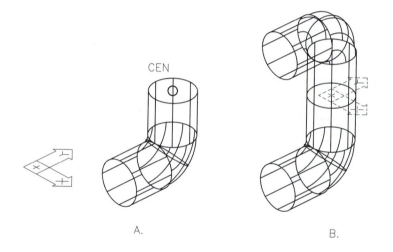

CEN

A.

B.

YZ

Using the *YZ* option constructs a plane to mirror about that is parallel with the current YZ plane. Any point can be selected through which the plane will pass (Fig. 38-47).

Figure 38-47

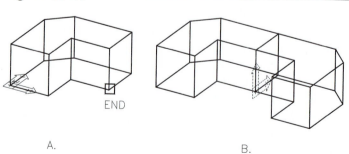

END

A.

B.

ZX

This option uses a plane parallel with
the current ZX plane for mirroring.
Figure 38-48 shows a point selected on the
Midpoint of an existing edge to mirror two
holes.

Figure 38-48

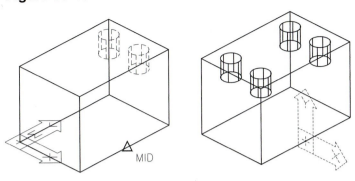

A.   B.

3DARRAY

Pull-down Menu	COMMAND (TYPE)	ALIAS (TYPE)	Short-cut	Screen (side) Menu	Tablet Menu
Modify *3D Operation >* *3D Array*	*3DARRAY*	*3A*	...	*MODIFY2* *3Darray*	*W,20*

Rectangular

With this option of *3Darray*, you create a 3D array specifying three dimensions—the number of and
distance between rows (along the Y axis), the number/distance of columns (along the X axis), and the
number/distance of levels (along the Z axis). Technically, the result is an array in a <u>prism</u> configuration <u>rather than a rectangle</u>.

```
Command: 3darray
Select objects: PICK
Select objects: Enter
Enter the type of array [Rectangular/Polar] <R>: r
Enter the number of rows (—-) <1>: (value)
Enter the number of columns (III) <1>: (value)
Enter the number of levels (...) <1>: (value)
Specify the distance between rows (—-): PICK or (value)
Specify the distance between columns (III): PICK or (value)
Specify the distance between levels (...): PICK or (value)
Command:
```

The selection set can be one or more objects. The entire set is treated as one object for arraying. All
values entered must be positive.

Figures 38-49 and 38-50 illustrate creating a *Rectangular 3Darray* of a cylinder with 3 rows, 4 columns, and 2 levels.

Figure 38-49

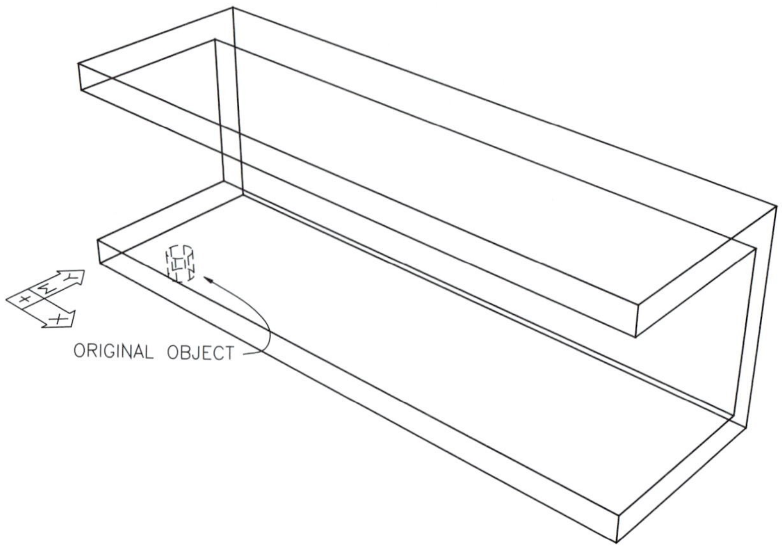

ORIGINAL OBJECT

The cylinders are *Subtracted* from the extrusion to form the finished part.

Figure 38-50

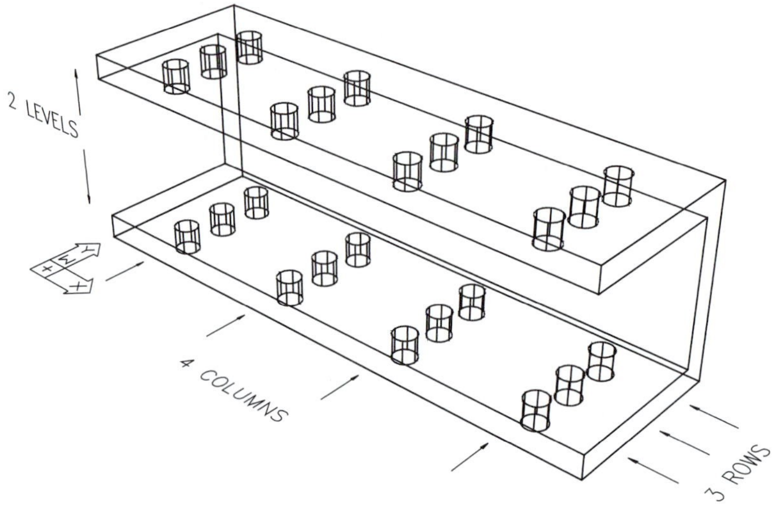

2 LEVELS

4 COLUMNS

3 ROWS

Polar

Similar to a *Polar Array* (2D), this option creates an array of selected objects in a circular fashion. The only difference in the 3D version is that an array is created about an axis of rotation (3D) rather than a point (2D). Specification of an axis of rotation requires two points in 3D space:

```
Command: 3darray
Select objects: PICK
Select objects: Enter
Enter the type of array [Rectangular/Polar] <R>: p
Enter the number of items in the array: (value)
Specify the angle to fill (+=ccw, -=cw) <360>: PICK or (value)
Rotate arrayed objects? [Yes/No] <Y>: y or n
Specify center point of array: PICK or (coordinates)
Specify second point on axis of rotation: PICK or (coordinates)
Command:
```

In Figures 38-51 and 38-52, a *3Darray* is created to form a series of holes from a cylinder. The axis of rotation is the center axis of the large cylinder specified by PICKing the *Center* of the top and bottom circles.

Figure 38-51

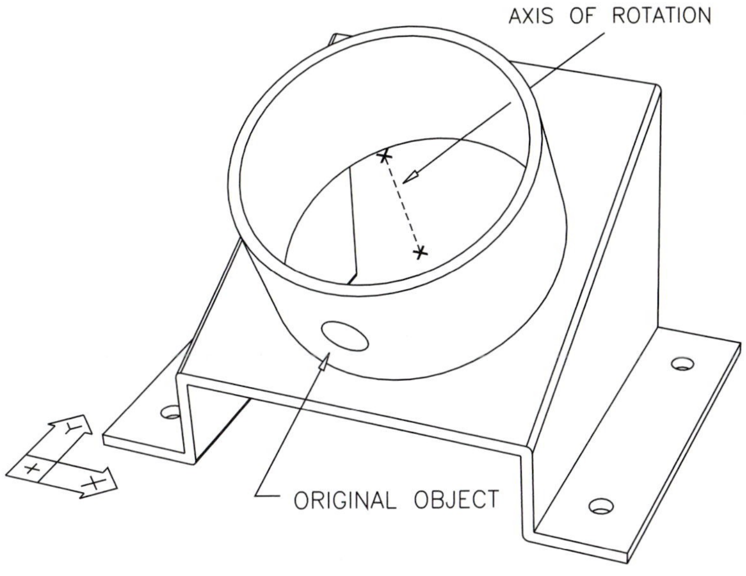

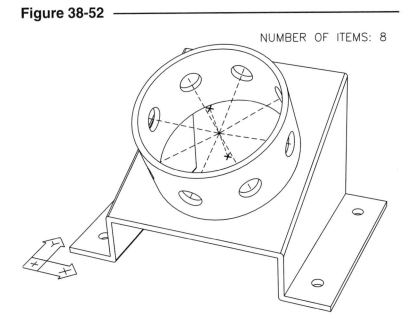

After the eight items are arrayed, the small cylinders are subtracted from the large cylinder to create the holes.

Figure 38-52

BOOLEAN OPERATION COMMANDS

Once the individual 3D primitives have been created and moved into place, you are ready to put together the parts. The primitives can be "assembled" or combined by Boolean operations to create composite solids. The Boolean operations found in AutoCAD are listed in this section: *Union*, *Subtract*, and *Intersect*.

UNION

Pull-down Menu	COMMAND (TYPE)	ALIAS (TYPE)	Short-cut	Screen (side) Menu	Tablet Menu
Modify *Solids Editing >* *Union*	UNION	UNI	...	MODIFY2 *Union*	X,15

Union joins selected primitives or composite solids to form one composite solid. Usually, the selected solids occupy portions of the same space, yet are separate solids. *Union* creates one solid composed of the total encompassing volume of the selected solids. (You can union solids even if the solids do not overlap.) All lines of intersections (surface boundaries) are calculated and displayed by AutoCAD. Multiple solid objects can be unioned with one *Union* command:

    Command: **union**
    Select objects: **PICK** (Select two or more solids.)
    Select objects: **Enter** (Indicate completion of the selection process.)
    Command:

Two solid boxes are combined into one composite solid with *Union* (Fig. 38-53). The original two solids (A) share the same physical space. The resulting union (B) consists of the total contained volume. The new lines of intersection are automatically calculated and displayed. *Hidden* shademode was used to enhance visualization in B.

Figure 38-53

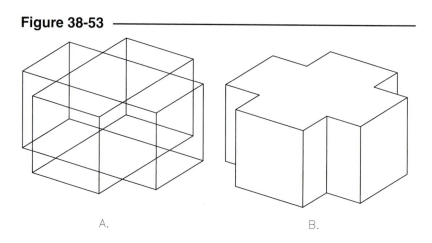

A.

B.

Figure 38-54

Because the volume occupied by any one of the primitives is included in the resulting composite solid, any redundant volumes are immaterial. The two primitives in Figure 38-54 A yield the same enclosed volume as the composite solid B. (*Hidden* shademode has been used for this figure.)

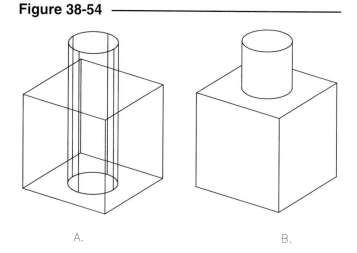

A.

B.

Multiple objects can be selected in response to the *Union* "Select objects:" prompt. It is not necessary, nor is it efficient, to use several successive Boolean operations if one or two can accomplish the same result.

Two primitives that have coincident faces (touching sides) can be joined with *Union.* Several "blocks" can be put together to form a composite solid.

Figure 38-55 illustrates how several primitives having coincident faces (A) can be combined into a composite solid (B). Only one *Union* is required to yield the composite solid.

Figure 38-55

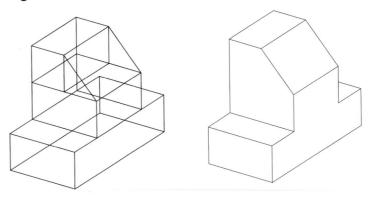

A.   B.

SUBTRACT

Pull-down Menu	COMMAND (TYPE)	ALIAS (TYPE)	Short-cut	Screen (side) Menu	Tablet Menu
Modify Solids Editing > Subtract	SUBTRACT	SU	...	MODIFY2 Subtract	X,16

Subtract takes the difference of one set of solids from another. *Subtract* operates with *Regions* as well as solids. When using solids, *Subtract* subtracts the <u>volume</u> of one set of solids from another set of solids. Either set can contain only one or several solids. *Subtract* requires that you first select the set of solids that will remain (the "source objects"), then select the set you want to subtract from the first:

```
Command: subtract
Select solids and regions to subtract from...
Select objects: PICK  (select what you want to keep)
Select objects: Enter
Select solids and regions to subtract...
Select objects: PICK  (select what you want to remove)
Select objects: Enter
Command:
```

The entire volume of the solid or set of solids that is subtracted is completely removed, leaving the remaining volume of the source set.

To create a box with a hole, a cylinder is located in the same 3D space as the box (see Figure 38-56). *Subtract* is used to subtract the entire volume of the cylinder from the box. Note that the cylinder can have any height, as long as it is at least equal in height to the box.

Because you can select more than one object for the objects "to subtract from" and the objects "to subtract," many possible construction techniques are possible.

Figure 38-56

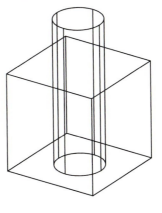 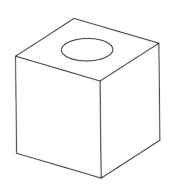

A.   B.

If you select multiple solids in response to the select objects "to subtract from" prompt, they are <u>automatically</u> unioned. This is known as an <u>*n*-way Boolean</u> operation. Using *Subtract* in this manner is very efficient and fast.

Figure 38-57 illustrates an *n*-way Boolean. The two boxes (A) are selected in response to the objects "to subtract from" prompt. The cylinder is selected as the objects "to subtract...". *Subtract* joins the source objects (identical to a *Union*) and subtracts the cylinder. The resulting composite solid is shown in B. (*Hidden* shademode has been used for this figure.)

Figure 38-57 ————————————————

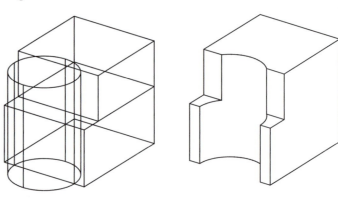

A. B.

INTERSECT

	COMMAND (TYPE)	ALIAS (TYPE)	Short-cut	Screen (side) Menu	Tablet Menu
Pull-down Menu					
Modify Solids Editing > Intersect	INTERSECT	IN	...	MODIFY2 Intrsect	X,17

Intersect creates composite solids by calculating the intersection of two or more solids. The intersection is the common volume <u>shared</u> by the selected objects. Only the 3D space that is <u>part of all</u> of the selected objects is included in the resulting composite solid. *Intersect* requires only that you select the solids from which the intersection is to be calculated.

    Command: **intersect**
    Select objects: **PICK** (Select all desired solids.)
    Select objects: **Enter** (Indicates completion of the selection process.)
    Command:

An example of *Intersect* is shown in Figure 38-58. The cylinder and the box share common 3D space (A). The result of the *Intersect* is a composite solid that represents that common space (B). (*Hidden* shademode has been used for this figure.)

Figure 38-58 ————————————————

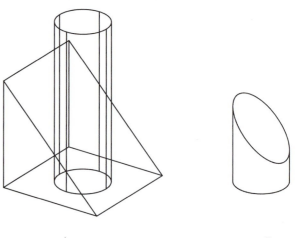

A. B.

Intersect can be very effective when used in conjunction with *Extrude*. A technique known as <u>reverse drafting</u> can be used to create composite solids that may otherwise require several primitives and several Boolean operations. Consider the composite solid shown in Figure 38-55 A. Using *Union*, the composite shape requires four primitives.

A more efficient technique than unioning several box primitives is to create two *Pline* shapes on vertical planes (Fig. 38-59). Each *Pline* shape represents the outline of the desired shape from its respective view: in this case, the front and side views. The *Pline* shapes are intended to be extruded to occupy the same space. It is apparent from this illustration why this technique is called reverse drafting.

Figure 38-59 ─────────

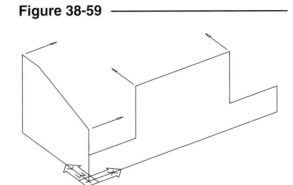

The two *Pline* "views" are extruded with *Extrude* to comprise the total volume of the desired solid (Fig. 38-60 A). Finally, *Intersect* is used to calculate the common volume and create the composite solid (B).

Figure 38-60 ─────────

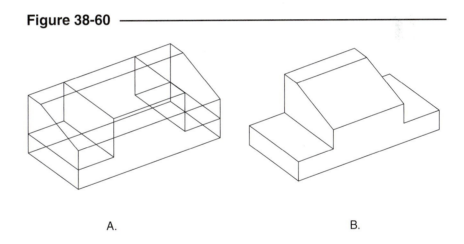

A.   B.

CHAMFER

Pull-down Menu	COMMAND (TYPE)	ALIAS (TYPE)	Short-cut	Screen (side) Menu	Tablet Menu
Modify *Chamfer*	*CHAMFER*	*CHA*	...	*MODIFY2* *Chamfer*	*W,18*

Chamfering is a machining operation that bevels a sharp corner. *Chamfer* chamfers selected edges of an AutoCAD solid as well as 2D objects. Technically, *Chamfer* (used with a solid) is a Boolean operation because it creates a wedge primitive and then adds to or subtracts from the selected solid.

When you select a solid, *Chamfer* recognizes the object as a solid and <u>switches to the solid version of prompts and options</u>. Therefore, all of the 2D options are not available for use with a solid, <u>only the "distances" method</u>. When using *Chamfer*, you must both select the "base surface" and indicate which edge(s) on that surface you wish to chamfer:

```
Command: chamfer
(TRIM mode) Current chamfer Dist1 = 0.5000, Dist2 = 0.5000
Select first line or [Polyline/Distance/Angle/Trim/Method/mUltiple]: PICK (Select solid at desired edge)
Base surface selection...
Enter surface selection option [Next/OK (current)] <OK>: N or Enter
Specify base surface chamfer distance <0.5000>: (value)
Specify other surface chamfer distance <0.5000>: (value)
Select an edge or [Loop]: PICK (Select edges to be chamfered)
Select an edge or [Loop]: Enter
Command:
```

When AutoCAD prompts to select the "base surface," only an edge can be selected since the solids are displayed in wireframe. When you select an edge, AutoCAD highlights one of the two surfaces connected to the selected edge. Therefore, you must use the "Next/<OK>:" option to indicate which of the two surfaces you want to chamfer (Fig. 38-61 A). The two distances are applied to the object, as shown in Figure 38-61 B.

Figure 38-61

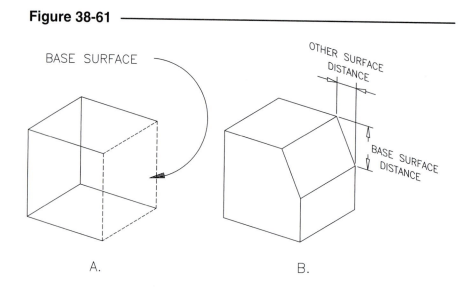

BASE SURFACE

OTHER SURFACE DISTANCE

BASE SURFACE DISTANCE

A.

B.

You can chamfer multiple edges of the selected "base surface" simply by PICKing them at the "Select an edge or [Loop]:" prompt (Fig. 38-62). If the base surface is adjacent to cylindrical edges, the bevel follows the curved shape.

Figure 38-62

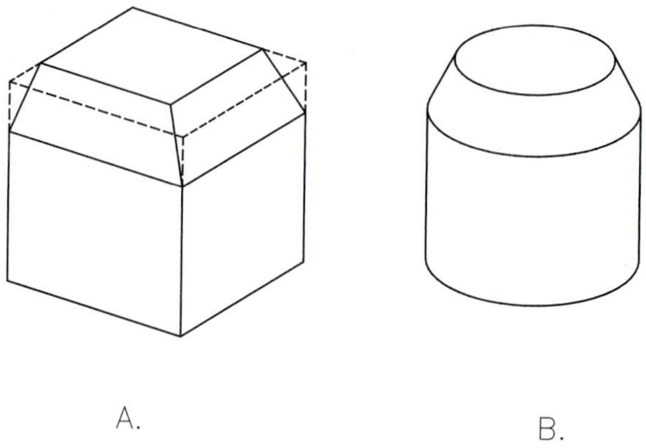

A.

B.

Loop

The *Loop* option chamfers the entire perimeter of the base surface. Simply PICK any edge on the base surface.

> Select an edge or [Loop]: *l*
> Select an edge loop or [Edge]: **PICK** (Select edges to form loop)
> Select an edge or [Loop]: **Enter**
> Command:

Figure 38-63

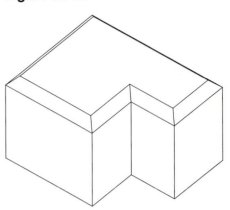

Edge

The *Edge* option switches back to the "Select edge" method.

Multiple

The *Multiple* option simply repeats the *Chamfer* command.

FILLET

Pull-down Menu	COMMAND (TYPE)	ALIAS (TYPE)	Short-cut	Screen (side) Menu	Tablet Menu
Modify *Fillet*	*FILLET*	*F*	...	*MODIFY 2* *Fillet*	*W,19*

Fillet creates fillets (concave corners) or rounds (convex corners) on selected solids, just as with 2D objects. Technically, *Fillet* creates a rounded primitive and automatically performs the Boolean needed to add or subtract it from the selected solids.

When using *Fillet* with a solid, the command <u>switches</u> to a special group of prompts and options for 3D filleting, and the <u>2D options become invalid</u>. After selecting the solid, you must specify the desired radius and then select the edges to fillet. When selecting edges to fillet, the edges must be PICKed individually. Figure 38-64 depicts concave and convex fillets created with *Fillet*. The selected edges are highlighted.

Figure 38-64

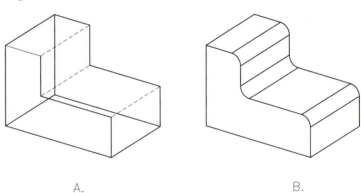

A.   B.

> Command: *fillet*
> Current settings: Mode = TRIM, Radius = 0.5000
> Select first object or [Polyline/Radius/Trim/mUltiple]: **PICK** (Select desired edge to fillet)
> Enter fillet radius <0.5000>: (**value**)
> Select an edge or [Chain/Radius]: **Enter** or **PICK** (Select additional edges to fillet)
> Command:

Curved surfaces can be treated with *Fillet*, as shown in Figure 38-65. If you want to fillet intersecting concave or convex edges, *Fillet* handles your request, provided you specify all edges in <u>one</u> use of the command. Figure 38-65 shows the selected edges (highlighted) and the resulting solid. Make sure you select <u>all</u> edges together (in one *Fillet* command).

Figure 38-65

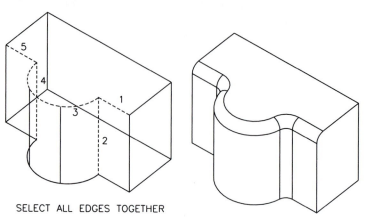

SELECT ALL EDGES TOGETHER

Chain
The *Chain* option allows you to fillet a series of connecting edges. Select the edges to form the chain (Fig. 38-66). If the chain is obvious (only one direct path), you can PICK only the ending edges, and AutoCAD will find the most direct path (series of connected edges):

Figure 38-66

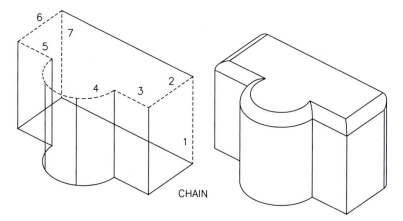

CHAIN

> Select an edge or [Chain/Radius]: **c**
> Select an edge chain or
> [Edge/Radius]: **PICK**

Edge
This option cycles back to the "<Select edge>:" prompt.

Radius
This method returns to the "Enter radius:" prompt.

Multiple
The *Multiple* option causes the *Fillet* command to repeat.

DESIGN EFFICIENCY

Now that you know the complete sequence for creating composite solid models, you can work toward improving design efficiency. The typical construction sequence is (1) create primitives, (2) ensure the primitives are in place by using UCSs or any of several move and rotate options, and (3) combine the primitives into a composite solid using Boolean operations. The typical step-by-step, "building-block" strategy, however, may not lead to the most efficient design. In order to minimize computation and construction time, you should <u>minimize the number of Boolean operations</u> and, if possible, the <u>number of primitives</u> you use.

For any composite solid, usually several strategies could be used to construct the geometry. You should plan your designs ahead of time, striving to minimize primitives and Boolean operations.

For example, consider the procedure shown in Figure 38-55. As discussed, it is more efficient to accomplish all unions with one *Union*, rather than each union as a separate step. Even better, create a closed *Pline* shape of the profile; then use *Extrude*. Figure 38-57 is another example of design efficiency based on using an *n*-way Boolean. Multiple solids can be unioned automatically by selecting them at the select objects "to subtract from" prompt of *Subtract*. Also consider the strategy of reverse drafting, as shown in Figure 38-59. Using *Extrude* in concert with *Intersect* can minimize design complexity and time.

In order to create efficient designs and minimize Boolean operations and primitives, keep these strategies in mind:

- Execute as many subtractions, unions, or intersections as possible within one *Subtract*, *Union*, or *Intersect* command.

- Use *n*-way Booleans with *Subtract*. Combine solids (union) automatically by selecting <u>multiple</u> objects "to subtract from," and then select "objects to subtract."

- Make use of *Plines* or regions for complex profile geometry; then *Extrude* the profile shape. This is almost always more efficient for complex curved profile creation than using multiple primitives and Boolean operations.

- Make use of reverse drafting by extruding the "view" profiles (*Plines* or *Regions*) with *Extrude*, then finding the common volume with *Intersect*.

SOLIDS EDITING

In AutoCAD Releases 13 and 14, editing of 3D solids was not feasible. Once Boolean operators were used to combine primitives, the original 3D primitives could not be edited. For example, if you wanted to change the diameter of a hole, you could only create a new *Cylinder* and *Union* it to "plug" the hole, then make another *Cylinder* with the desired new diameter and *Subtract* it.

AutoCAD 2000 introduced the *Solidedit* command which has options for extruding, moving, rotating, offsetting, tapering, copying, coloring, separating, shelling, cleaning, checking, and deleting faces and edges of 3D solids. The unique feature of this command is that <u>you can change individual primitives and partial internal or external geometry of composite solids</u>. This single command contains all the options which are available on the *Solids Editing* menu from the *Modify* pull-down (Fig. 38-67) and on the *Solids Editing* toolbar (see Figure 38-6).

Figure 38-67

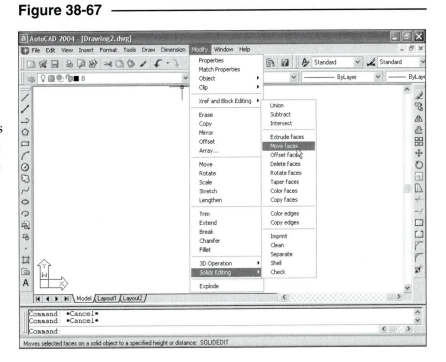

SOLIDEDIT

Pull-down Menu	COMMAND (TYPE)	ALIAS (TYPE)	Short-cut	Screen (side) Menu	Tablet Menu
Modify Solids Editing >	*SOLIDEDIT*	...	...	...	...

The *Solidedit* command has several "levels" of options. The first level prompts you to specify the type of geometry you want to edit—*Face, Edge,* or *Body.* The second level of options depends on your first level response as shown below.

```
Command: solidedit
Solids editing automatic checking: SOLIDCHECK=1
Enter a solids editing option [Face/Edge/Body/Undo/eXit] <eXit>: f
Enter a face editing option
[Extrude/Move/Rotate/Offset/Taper/Delete/Copy/coLor/Undo/eXit] <eXit>:
```

Solidedit operates on three types of geometry: *Edges, Faces* and *Bodies.* An *Edge* is defined as the common edge between two surfaces which has the appearance of a line, arc, circle, or spline (Fig. 38-68 A). *Faces* are planar or curved surfaces of a 3D object (Fig. 38-68 B). A *Body* is defined as an existing 3D solid or a non-solid shape created with a *Solidedit* option.

Figure 38-68

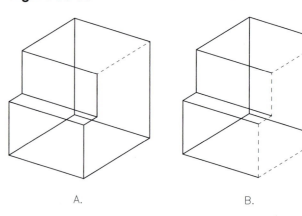

A.   B.

 Object selection is an integral part of using *Solidedit*. For example, once you have specified the type of geometry and the particular editing option, the *Solidedit* command prompts to select *Edges, Faces,* or a *Body.*

```
Select faces or [Undo/Remove]: Select a face or enter an option
Select faces or [Undo/Remove/ALL]: Select a face or enter an option
```

When you are prompted to select *Faces* or *Edges,* you will most likely go through a series of adding and removing geometry until the desired set of lines is highlighted. In addition to the pickbox, you can use the following selection options.

```
Crossing/Fence/CPolygon/Undo/Remove/ALL
```

To use a selection method other than the pickbox, you must type the desired choice at the "Select faces or [Undo/Remove]:" prompt since no automatic window or crossing window is available. As an alternative to *Remove,* you can hold down the Shift key and select objects.

 Ensure that you highlight exactly the intended geometry before you proceed with editing. If an incorrect set is selected, you can get unexpected results or an error message can appear such as that below.

```
Modeling Operation Error:
  No solution for an edge.
```

For example, many of the *Face* options allow selection of one or multiple faces. In Figure 38-69 both selection sets are valid faces or face combinations, but each will yield different results. Figure 38-69 A indicates selection of one face and B indicates selection of several faces defining an entire primitive. Add and remove faces or edges until you achieve the desired geometry.

Figure 38-69

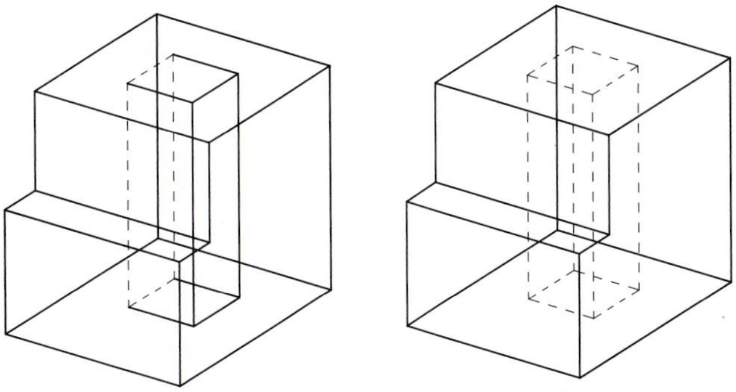

A.   B.

Face Options

The *Face* options edit existing surfaces and create new surfaces. A *Face* is a planar or curved surface. *Faces* that are part of existing 3D solids can be altered to change the configuration of the 3D composite solid. Individual *Faces* can be edited and multiple *Faces* comprising a primitive can be edited. New surfaces can be created from existing *Faces*, but entirely new independent solids cannot be created using these editing tools.

Extrude

The *Extrude* option of *Solidedit* allows you to extrude any *Face* of a 3D object in a similar manner to using the *Extrude* command to create a 3D solid from a *Pline*. This capability is extremely helpful if you need to make a surface on a 3D solid taller, shorter, or longer. You can select one or more faces to extrude at one time.

```
Command: solidedit
Solids editing automatic checking: SOLIDCHECK=1
Enter a solids editing option [Face/Edge/Body/Undo/eXit] <eXit>: f
Enter a face editing option
[Extrude/Move/Rotate/Offset/Taper/Delete/Copy/coLor/Undo/eXit] <eXit>: e
Select faces or [Undo/Remove]: PICK
Select faces or [Undo/Remove/ALL]: PICK or remove
Select faces or [Undo/Remove/ALL]: Enter
Specify height of extrusion or [Path]: (value)
Specify angle of taper for extrusion <0>: Enter or (value)
Solid validation started.
Solid validation completed.
```

For example, Figure 38-70 illustrates the extrusion of a face to make the 3D object taller, where A indicates the selected (highlighted) face and B shows the result. This extrusion has a 0 degree taper angle.

Figure 38-70

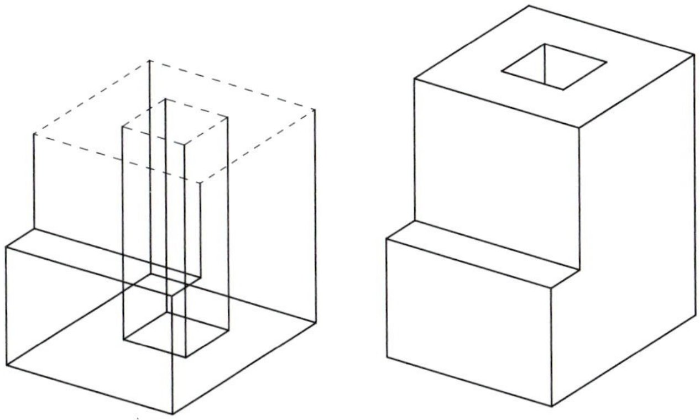

A.   B.

Specifying a positive angle tapers the face inward (Fig. 38-71 A) and specifying a negative angle tapers the face outward (B) *Height* is always perpendicular to the selected *Face*, not necessarily vertical.

Figure 38-71

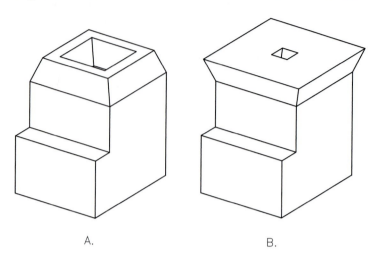

A.   B.

An internal face can be selected for extrusion. In Figure 38-72, a single face is selected (A) and the *Height* value specified is greater than the distance to the outer face of the solid, creating an open side on the object (B). *Extrude* is the only option that allows an internal face to "pass through" an external face. With all other options, the outermost "bounding box" of the solid cannot be extended or opened by editing an internal primitive or face.

Figure 38-72

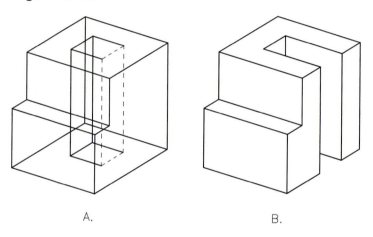

A.   B.

Faces can be extruded along a *Path*. The *Path* object can be a *Line*, *Circle*, *Arc*, *Ellipse*, *Pline*, or *Spline*. For example, Figure 38-73 illustrates extrusion of a face along a *Line* path (A) to yield the result shown on the right (B).

NOTE: Entering a <u>positive</u> value at the "Specify height of extrusion or [Path]:" prompt <u>increases</u> the volume of the solid and entering a <u>negative</u> value <u>decreases</u> the volume of the solid. For example, entering a positive value for extruding the face in Figure 38-72 A would result in a smaller hole.

Figure 38-73

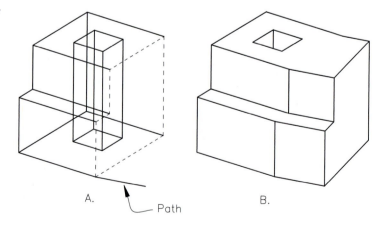

A.   Path   B.

Move

With the *Move* option, you can move a single face of a 3D solid or move an entire primitive within a 3D solid. This capability is particularly helpful during geometry construction or when there is a design change and it is required to alter the location of a hole or other feature within the confines of the composite 3D model.

```
[Extrude/Move/Rotate/Offset/Taper/Delete/Copy/coLor/Undo/eXit] <eXit>: m
Select faces or [Undo/Remove]: PICK
Select faces or [Undo/Remove/ALL]: PICK  or remove
Select faces or [Undo/Remove/ALL]: Enter
Specify a base point or displacement: PICK
Specify a second point of displacement: PICK
```

For example, Figure 38-74 illustrates using the *Move* option to relocate a hole primitive (four faces) within a composite solid. In this case you must ensure all faces, and only the faces, comprising the primitive are selected.

With the *Move* option the internal features <u>cannot</u> be moved to extend into or past the "bounding box" of the composite solid, as they can with *Extrude* (see Figure 38-72). If this condition is attempted, the following message appears.

> Modeling Operation Error:
>   Improper edge/edge intersection.

Another possibility for *Move* is to move only one or selected faces to alter the configuration of an internal feature of a composite solid. An example is shown in Figure 38-75, where only one face is selected to move. Compare the results in Figures 38-74 B and 38-75 B.

Keep in mind that *Move* can be used to move a singular external face, achieving the same result as *Extrude* (see previous Figure 38-70).

Figure 38-74

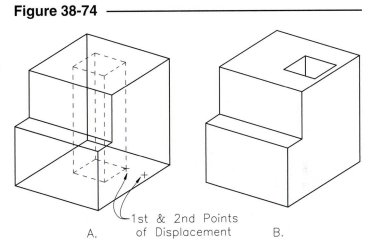

A.   1st & 2nd Points of Displacement   B.

Figure 38-75

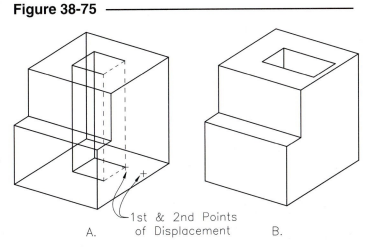

A.   1st & 2nd Points of Displacement   B.

Offset

The 2D *Offset* command makes a parallel copy of a *Line*, *Arc*, *Circle*, *Pline*, etc. The *Offset* option of *Solidedit* makes a 3D offset. A simple application is making a hole larger or smaller.

```
[Extrude/Move/Rotate/Offset/Taper/Delete/Copy/coLor/Undo/eXit] <eXit>: o
Select faces or [Undo/Remove]: PICK
Select faces or [Undo/Remove/ALL]: PICK  or remove
Select faces or [Undo/Remove/ALL]: Enter
Specify the offset distance: PICK or (value)
```

You can offset through a point or specify an offset distance. Specify a positive value to increase the volume of the solid or a negative value to decrease the volume of the solid.

Figure 38-76

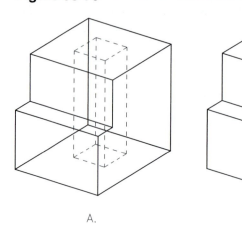

 Figure 38-76 demonstrates how an internal feature, such as a hole, can be *Offset* to effectively change the volume of the hole. Since positive values increase the size of the solid, a negative value was used here. In other words, holes inside a solid become smaller when the solid is *Offset* larger.

A.   B.

Delete

 This option of *Solidedit* deletes faces from composite solids. Although you cannot delete one planar face from a solid, you can delete one curved face, such as a cylinder, that comprises an entire primitive. You can also delete multiple faces that comprise a primitive or entire feature.

```
[Extrude/Move/Rotate/Offset/Taper/Delete/Copy/coLor/Undo/eXit] <eXit>: d
Select faces or [Undo/Remove]: PICK
Select faces or [Undo/Remove/ALL]: PICK  or remove
Select faces or [Undo/Remove/ALL]: Enter
Solid validation started.
Solid validation completed.
```

As an example, Figure 38-77 A displays two selected faces to be deleted from a composite solid. The result is displayed in B. In this case, both faces must be highlighted for the deletion to operate.

Figure 38-77

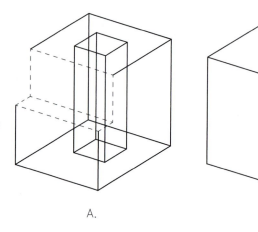

For the same composite solid, if all four internal faces comprising the hole are selected, as shown in previous Figure 38-76 A, the hole could be deleted.

A.   B.

Rotate

 Rotate is helpful during geometry construction and for design changes when components within a composite solid must be rotated in some way.

```
[Extrude/Move/Rotate/Offset/Taper/Delete/Copy/coLor/Undo/eXit] <eXit>: r
Select faces or [Undo/Remove]: PICK
Select faces or [Undo/Remove/ALL]: PICK  or remove
Select faces or [Undo/Remove/ALL]: Enter
Specify an axis point or [Axis by object/View/Xaxis/Yaxis/Zaxis] <2points>: PICK
Specify the second point on the rotation axis: PICK
Specify a rotation angle or [Reference]: PICK or (value)
```

Several methods of rotation are possible based on the selected axis. These options are essentially the same as those available with the *Rotate3D* command (see *"Rotate3D"* discussed previously).

For example, Figure 38-78 illustrates the rotation of a primitive within a composite solid. Here, the primitive is rotated about two points defining one edge of the hole.

NOTE: Positive and negative rotation complies with the right-hand rule. Also keep in mind that when PICKing two points to define the axis of rotation, positive direction on the axis is determined from the first to the second point picked.

One face of a composite solid can be selected for rotation within a composite solid (Fig. 38-79 A). The selected face is rotated about two points defining one edge of the face.

Figure 38-78

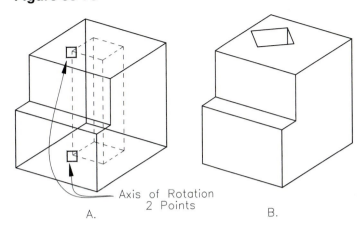

Axis of Rotation
2 Points

A.

B.

Figure 38-79

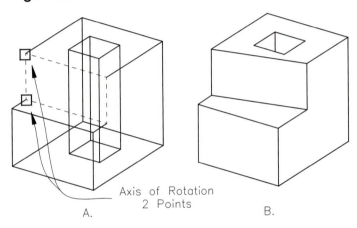

Axis of Rotation
2 Points

A.

B.

Taper

Taper angles a face. You can use *Taper* to change the angle of planar or curved faces.

[Extrude/Move/Rotate/Offset/Taper/Delete/Copy/coLor/Undo/eXit] <eXit>: **t**
Select faces or [Undo/Remove]: **PICK**
Select faces or [Undo/Remove/ALL]: **PICK** or remove
Select faces or [Undo/Remove/ALL]: **Enter**
Specify the base point: **PICK**
Specify another point along the axis of tapering: **PICK**
Specify the taper angle: **Enter** or (**value**)

The rotation of the taper angle is determined by the order of selection of the base point and second point. Although AutoCAD prompts for the "axis of tapering," the two points actually determine a reference line from which the taper angle is applied. The taper starts at the first base point and tapers away from the second base point. The rotational axis passes through the first base point and is <u>perpendicular</u> to the line between the base points. <u>The line between the base points does not specify the rotational axis, as implied.</u>

Figure 38-80

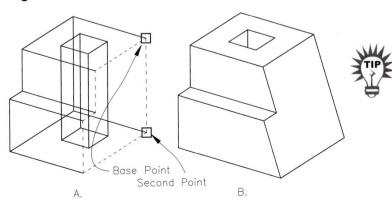

Base Point
Second Point

A.

B.

To explain this idea, Figure 38-80 A (previous page) illustrates the selection of a planar face, the base point, and the second point. An angle of –20 degrees (negative value) is entered to achieve the results shown in B. Positive angles taper the face inward (toward the solid) and negative values taper the face outward (away from the solid).

For some cases, either *Taper* or *Rotate* could be used to achieve the same results. For example, both Figure 38-80 B and previous Figure 38-79 B could be attained using *Taper* or *Rotate* (although different points must be selected depending on the option used).

Curved faces can also be selected for applying a *Taper*. Figure 38-81 A illustrates a cylindrical hole primitive. The entire cylinder primitive is selected as one *Face* object. Note the selection of the base point and second point. *Taper* converts the cylindrical hole into a conical hole (B).

Figure 38-81

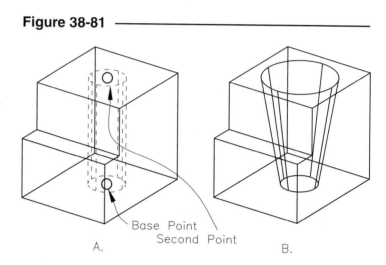

Base Point
Second Point

A.   B.

Copy

As you would expect, the *Copy* option copies *Faces*. However, the resulting objects are *Regions* or *Bodies*.

```
[Extrude/Move/Rotate/Offset/Taper/Delete/Copy/coLor/Undo/eXit] <eXit>: c
Select faces or [Undo/Remove]: PICK
Select faces or [Undo/Remove/ALL]: PICK  or remove
Select faces or [Undo/Remove/ALL]: Enter
Specify the base point or displacement: PICK  or (value)
Specify a second point of displacement: PICK
```

Although any *Face* can be selected (planar or curved), Figure 38-82 shows how *Copy* can be used to create a *Region* from a planar face. Typically, the *Copy* option would be used to create a new *Region* from existing 3D solid geometry. The *Extrude* command could then be used on the *Region* to form a new 3D solid from the original face.

If a curved *Face* is selected to *Copy*, the resulting object is a *Body*. This type of body is a curved surface, not a solid.

Figure 38-82

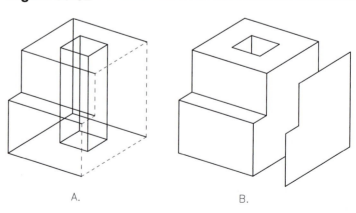

A.   B.

Using the same composite model as in previous figures, assume you were to *Copy* only the five vertical faces indicated in Figure 38-83 A. After selecting the second point of displacement (B), notice only those five vertical surfaces result. Not included are the faces defining the hole or any horizontal faces. Therefore, you can select *All* faces comprising a composite solid to *Copy,* and the resulting model is a surface model composed of *Regions*.

Figure 38-83

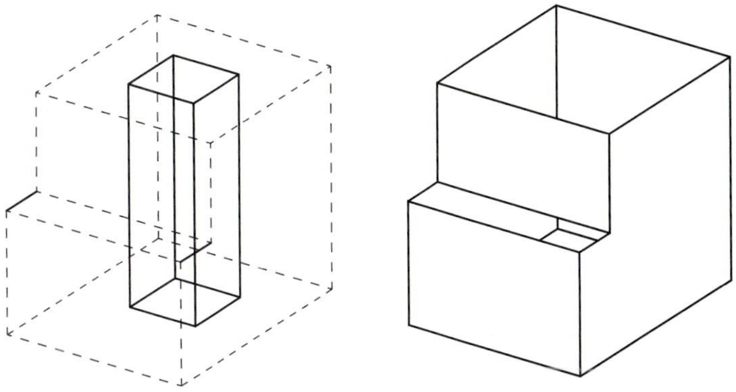

A.

B.

Color

The *Color Face* option of *Solidedit* simply changes the color of selected faces.

```
[Extrude/Move/Rotate/Offset/Taper/Delete/Copy/coLor/Undo/eXit] <eXit>: 1
Select faces or [Undo/Remove]: PICK
Select faces or [Undo/Remove/ALL]: PICK  or remove
Select faces or [Undo/Remove/ALL]: Enter
Enter new color <BYLAYER>: (color)
```

The resulting display of the 3D solid depends on the *Shademode* setting. Only the object edges appear in the designated color when *2D wireframe, 3D wireframe,* or *Hidden* settings are used. For the *Flat* and *Gouraud* options of *Shademode,* all surfaces are displayed in shaded intensities of the designated color. For *Flat+edges* or *Gouraud+edges,* the surfaces are shaded in the designated colors and edges are displayed in a lighter intensity of the colors (unless the *Color Edge* option is used).

Edge Options

Edges are lines or curves that define the boundary between *Faces*. Copying an *Edge* would result in a 2D object. The selection process for *Edges* is critical, but not as involved as selecting *Faces*. You can use multiple selection methods (the same options as with *Edges*), and selection may involve a process of adding and removing objects. As with *Face* selection, make sure you have the exact desired set highlighted before continuing with the procedure. The command syntax is as follows.

```
Command: solidedit
Solids editing automatic checking: SOLIDCHECK=1
Enter a solids editing option [Face/Edge/Body/Undo/eXit] <eXit>: edge
Enter an edge editing option [Copy/coLor/Undo/eXit] <eXit>:
```

Copy

The *Copy* option of *Edge* creates wireframe elements, not surfaces or solids. Copied edges become 2D objects such as a *Line, Arc, Circle, Ellipse,* or *Spline*. These elements could be used to create other 3D models such as wireframes, surfaces, or solids.

```
Enter an edge editing option [Copy/coLor/Undo/eXit] <eXit>: c
Select edges or [Undo/Remove]: PICK
Select edges or [Undo/Remove]: PICK  or remove
Select edges or [Undo/Remove]: Enter
Specify a base point or displacement: PICK
Specify a second point of displacement: PICK
```

One or multiple *Edges* can be selected. After indicating the second point of displacement, the selected edges are extracted from the solid and copied to the new location (Fig. 38-84).

Keep in mind that the *Copied Edges* are wireframe elements. These new elements can be used for construction of other 3D geometry, such as with the *Body, Imprint* option (see "*Body* Options" later in this section).

Figure 38-84

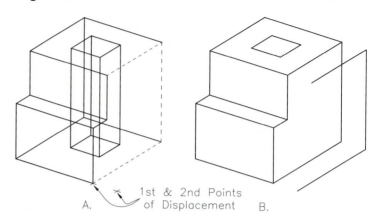

1st & 2nd Points
of Displacement
A.   B.

Color

The *Color* option for *Edges* operates similarly to the *Color* option for *Faces* except that only the *Edges* are affected. This option is useful in cases where you need the edges to be more or less visible to bring attention to certain features or components of a solid, or in cases when you want the surfaces of a solid to be shaded in one color and the edges to appear in another.

```
Enter an edge editing option [Copy/coLor/Undo/eXit] <eXit>: l
Select edges or [Undo/Remove]: PICK
Select edges or [Undo/Remove]: PICK or remove
Select edges or [Undo/Remove]: Enter
Enter new color <BYLAYER>: (color)
```

The selected edges appear in the new color. The *Shademode* setting also affects the appearance of the surfaces and edges. When *Shademode* is set to any option that causes the edges to appear (*2D wireframe, 3D wireframe, Hidden, Flat+edges,* or *Gouraud+edges*), the edges appear in the selected color. In the *Flat* and *Gouraud* modes, only the surfaces appear without edges. You can use the *Flat+edges* and *Gouraud+edges* modes to display the edges in one color and the surfaces in another.

Body Options

A *Body* is typically a 3D solid. Any primitive or composite solid can be selected as a *Body*. However, some *Solidedit* options can create a *Body* that is a non-solid. For example, selecting a curved *Face* and using *Copy* creates a curved surface that AutoCAD lists as a *Body*. Several options are offered here to alter the configuration of a *Body* or to create or edit 2D geometry used to interact with a *Body*.

Imprint

Imprint is used to attach a 2D object on an existing face of a solid (*Body*). The new *Imprint* can then be used with *Extrusion* to create a new 3D solid.

Objects that can be used to make the *Imprint* can be an *Arc, Circle, Line, Pline, Ellipse, Spline, Region,* or *Solid*. The selected object to *Imprint* must touch or intersect the solid in some way, such as when a *Circle* lies partially on a *Face*, or when two solid volumes overlap. When *Imprint* is used, one or more 2D components becomes attached, or imprinted, to an existing 3D solid face. The resulting *Imprint* has little usefulness of itself, but is an intermediate step to creating a new 3D solid shape on the existing solid.

```
[Imprint/seParate solids/Shell/cLean/Check/Undo/eXit] <eXit>: i
Select a 3D solid: PICK
Select an object to imprint: PICK
Delete the source object <N>: y
Select an object to imprint: Enter
```

For example, Figure 38-85 A displays a 3D solid with a *Circle* that is on the same plane as the vertical right face of the solid. The *Imprint* option is used, the solid is selected, and the *Circle* is selected as the "object to imprint." Answering "Y" to "Delete the source object," the resulting *Imprint* is shown in B.

Figure 38-85

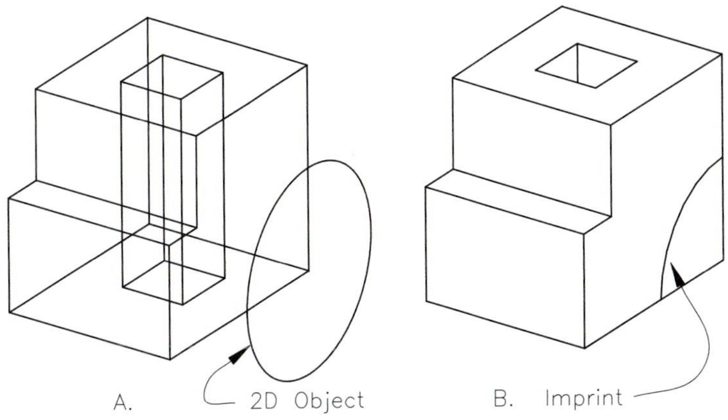

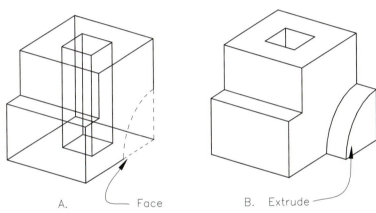

A.   2D Object   B.   Imprint

The *Imprint* creates a separate *Face* on the object—in this case, a total of two coplanar *Faces* are on the same vertical surface of the object. The new *Face* can be treated independently for use with other *Solidedit* options. For example, *Extrude* could be used to create a new solid feature on the composite solid (Fig. 38-86).

Figure 38-86

A.   Face   B.   Extrude

Clean

 The *Clean* option deletes any 2D geometry on the solid. For example, you may use *Imprint* to "attach" 2D geometry to a *Body* (solid) for the intention of using *Extrude*. *Imprinted* objects become permanently attached to the *Body* and cannot be removed by *Erase*. If you need to remove an *Imprint*, use the *Clean* option to do so. One use of *Clean* removes all *Imprints*.

```
[Imprint/sePatate solids/Shell/cLean/Check/Undo/eXit] <eXit>: l
Select a 3D solid: PICK
```

Separate

 It is possible in AutoCAD to *Union* two solids that do not occupy the same physical space and have two distinct volumes. In that case, the two shapes appear to be separate, but are treated by AutoCAD as one object (if you select one, both become highlighted).

Occasionally you may intentionally or inadvertently use *Union* to combine two or more separate (not physically touching) objects. *Shell* and *Offset*, when used with solids containing holes, can also create two or more volumes from one solid. *Separate* can then be used to disconnect these into discrete independent solids. *Separate* <u>cannot</u> be used to disconnect or "break down" *Unioned*, *Subtracted*, or *Intersected* solids that form one volume.

```
[Imprint/sePatate solids/Shell/cLean/Check/Undo/eXit] <eXit>: p
Select a 3D solid: PICK
```

Shell

Shell converts a solid into a thin-walled "shell." You first select a solid to shell, then specify faces to remove, and enter an offset distance. An example would be to convert a solid cube into a hollow box—the thickness of the walls equal the "shell offset distance." If no faces are removed from the selection set, the box would have no openings. Faces that are removed become open sides of the box.

```
[Imprint/seParate solids/Shell/cLean/Check/Undo/eXit] <eXit>: s
Select a 3D solid: PICK
Remove faces or [Undo/Add/ALL]: PICK
Remove faces or [Undo/Add/ALL]: PICK
Remove faces or [Undo/Add/ALL]: Enter
Enter the shell offset distance: (value)
Solid validation started.
Solid validation completed.
```

A positive value entered at the "shell offset distance" prompt creates the wall thickness outside of the original solid, whereas a negative value creates the wall thickness inside the existing boundary.

An example is shown in Figure 38-87. The left object (A) indicates the solid with all the selected faces highlighted and the faces removed from the selection set not highlighted. A negative offset distance is used to yield the shape shown on the right (B).

Don't expect to achieve the desired results the first time. As with most *Solidedit* options, the process of adding and removing faces is an inexact procedure since any edge you select can highlight either one of two faces.

Figure 38-87

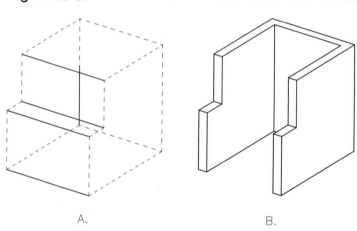

A.   B.

Check

Use this option to validate the 3D solid object as a valid ACIS solid, independent of the *SOLIDCHECK* system variable setting.

```
[Imprint/seParate solids/Shell/cLean/Check/Undo/eXit] <eXit>: c
Select a 3D solid: PICK
This object is a valid ACIS solid.
```

The *SOLIDCHECK* variable turns the automatic solid validation on and off for the current AutoCAD session. By default, *SOLIDCHECK* is set to 1, or on, to validate the solid. The *Solidedit* command displays the current status of the variable as shown in the command syntax below.

```
Command: solidedit
Solids editing automatic checking:  SOLIDCHECK=0
Enter a solids editing option [Face/Edge/Body/Undo/eXit] <eXit>
```

SLICE

Pull-down Menu	COMMAND (TYPE)	ALIAS (TYPE)	Short-cut	Screen (side) Menu	Tablet Menu
Draw *Solids >* *Slice*	*SLICE*	*SL*	...	DRAW 2 SOLIDS *Slice*	...

The *Slice* command is a separate command (not part of the *Solidedit* options). *Slice* is also discussed in Chapter 39, but only for creating sections from solids. However, *Slice* is a useful command for solids construction and editing for special cases and is therefore discussed in this chapter for those purposes.

A solid that contains oblique surfaces is a likely candidate for using *Slice*. For example, the simplest method of construction for the solid shown in Figure 38-88 is to create the rectilinear shapes (using *Box*), then use *Slice* to create the oblique surfaces.

Figure 38-88

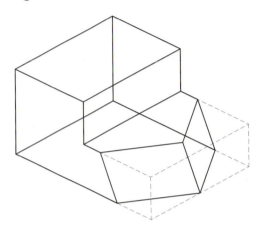

With the *Slice* command you can cut a solid into two "halves" by defining a plane in any orientation. You can keep both sides of the resulting solid(s) or PICK one side to keep. (If in doubt as to which side to keep or how to select it, keep both sides, then use *Erase* to remove the unwanted side.)

```
Command: slice
Select objects: PICK
Select objects: Enter
Specify first point on slicing plane by [Object/Zaxis/View/XY/YZ/ZX/3points] <3points>: (option)
(option prompts)
Specify a point on desired side of the plane or [keep Both sides]: PICK or Enter
Command:
```

One oblique surface for the solid (Fig. 38-89) is created by first drawing a 2D *Line* on the solid, then defining a *3point* slicing plane using *OSNAPs* (*Endpoint, Endpoint* of *Line*, then *Midpoint* of solid edge). This figure shows the resulting solids after using *Slice* and keeping both sides. Duplicating this procedure for the other oblique surface creates the solid shown previously in Figure 38-88.

Slice can be used for many applications, such as an alternative to using *Wedge* (Fig. 38-17), *Chamfer* (Fig. 38-61), or *Solidedit, Taper* (Fig. 38-80). Additionally, if you wanted to cut a section of a solid away, such as creating a notch or a half section (cutting away one quarter of an object), consider making multiple slicing planes with *Slice*, keeping all sides, *Erase* the unwanted section, then using *Union* to join the sections back into one solid.

Figure 38-89

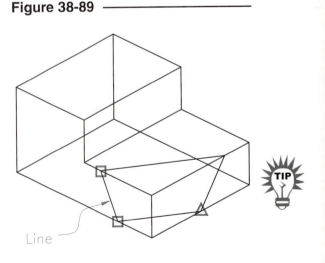

Line

CHAPTER EXERCISES

1. What are the typical three steps for creating composite solids?

2. Consider the two solids in Figure 38-90. They are two extruded hexagons that overlap (occupy the same 3D space).

    Figure 38-90 ————————

    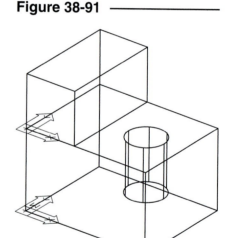

    A. Sketch the resulting composite solid if you performed a *Union* on the two solids.

    B. Sketch the resulting composite solid if you performed an *Intersect* on the two solids.

    C. Sketch the resulting composite solid if you performed a *Subtract* on the two solids.

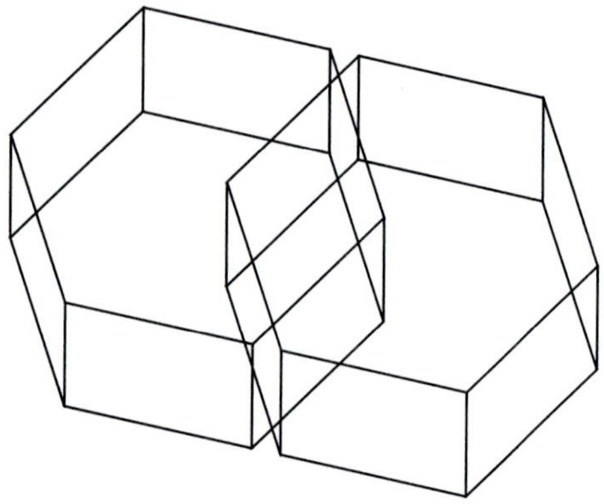

For the following exercises, use a *Template* drawing or begin a *New* drawing. Turn *On* the *Ucsicon* and set it to the *Origin*. Set the *Vpoint* with the *Rotate* option to angles of **310, 30**.

3. Begin a drawing and assign the name **CH38EX3**.

    Figure 38-91 ————————

    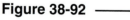

    A. Create a *box* with the lower-left corner at **0,0,0**. The *Lengths* are **5, 4**, and **3**.

    B. Create a second *box* at a new UCS as shown in Figure 38-91 (use the *ORigin* option). The *box* dimensions are **2 x 4 x 2**.

    C. Create a *Cylinder*. Use the same UCS as in the previous step. The *cylinder Center* is at **3.5,2** (of the *UCS*), the *Diameter* is **1.5**, and the *Height* is **-2**.

    D. *Save* the drawing.

    E. Perform a *Union* to combine the two boxes. Next, use *Subtract* to subtract the cylinder to create a hole. The resulting composite solid should look like that in Figure 38-92. *Save* the drawing.

    Figure 38-92 ————————

    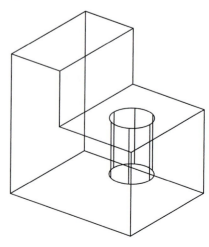

4. Begin a drawing and assign the name **CH38EX4**.

    A. Create a *Wedge* at point **0,0,0** with the *Lengths* of **5**, **4**, **3**.

    B. Create a *3point UCS* option with an orientation indicated in Figure 38-93. Create a *Cone* with the *Center* at **2,3** (of the *UCS*) and a *Diameter* of **2** and a *Height* of **-4**.

Figure 38-93

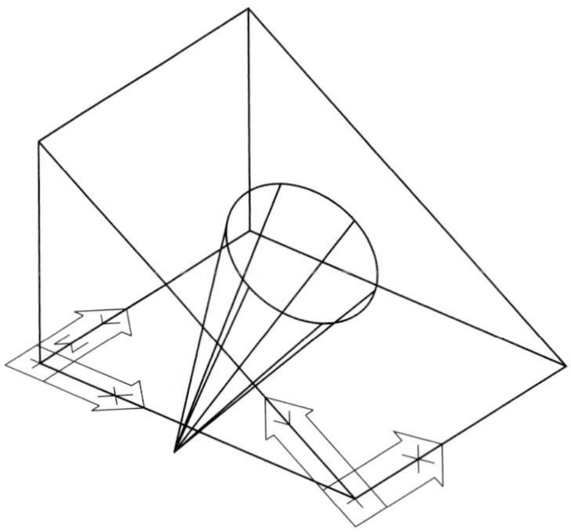

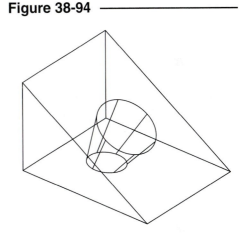

    C. *Subtract* the cone from the wedge. The resulting composite solid should resemble Figure 38-94. *Save* the drawing.

Figure 38-94

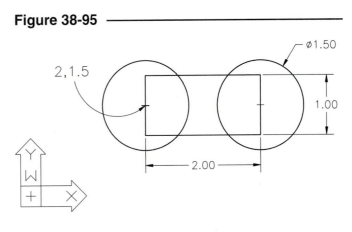

5. Begin a drawing and assign the name **CH38EX5**. Display a *Plan* view.

    A. Create 2 *Circles* as shown in Figure 38-95, with dimensions and locations as specified. Use *Pline* to construct the rectangular shape. Combine the 3 shapes into a *Region* by using the *Region* and *Union* commands <u>or</u> converting the <u>outside</u> shape into a *Pline* using *Trim* and *Pedit*.

    B. Change the display to an isometric-type *Vpoint*. *Extrude* the *Region* or *Pline* with a *Height* of **3** (no *taper angle*).

Figure 38-95

C. Create a *Box* with the lower-left corner at **0,0**. The *Lengths* of the box are **6, 3, 3**.

D. *Subtract* the extruded shape from the box. Your composite solid should look like that in Figure 38-96. *Save* the drawing.

Figure 38-96 ——————

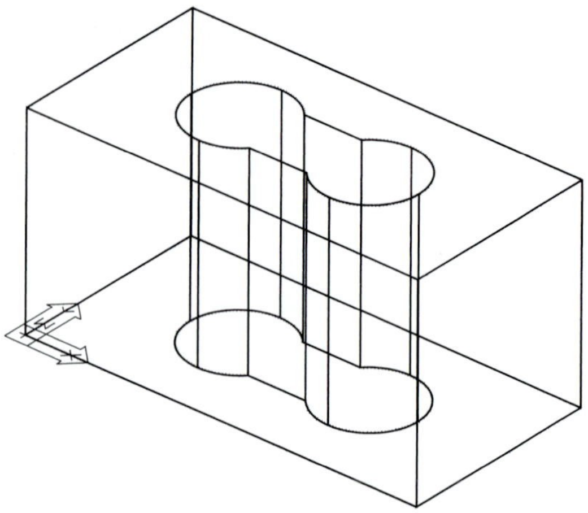

6. Begin a drawing and assign the name **CH38EX6**. Display a *Plan* view.

Figure 38-97 ——————

A. Create a closed *Pline* shape symmetrical about the X axis with the locational and dimensional specifications given in Figure 38-97.

B. Change to an isometric-type *Vpoint*. Use *Revolve* to generate a complete circular shape from the closed *Pline*. Revolve about the **Y** axis.

C. Create a *Torus* with the *Center* at **0,0**. The *Radius of torus* is **3** and the *Radius of tube* is **.5**. The two shapes should intersect.

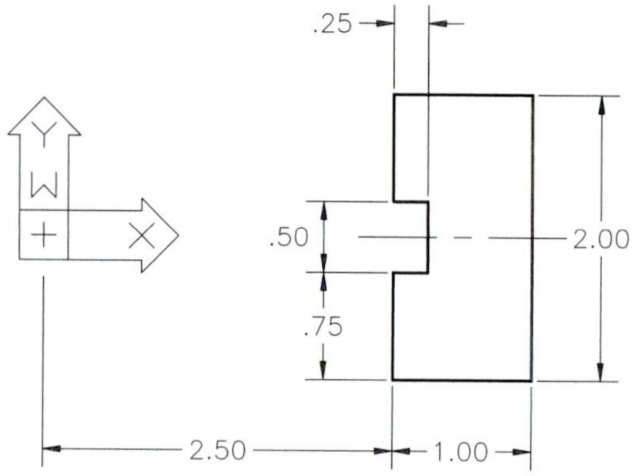

D. Use *Hidden* shademode to generate a display like Figure 38-98.

Figure 38-98 ——————

E. Create a *Cylinder* with the *Center* at **0,0,0**, a *Radius* of **3**, and a *Height* of **8**.

F. Use *Rotate3D* to rotate the revolved *Pline* shape **90** degrees about the X axis (the *Pline* shape that was previously converted to a solid—not the torus). Next, move the shape up (positive Z) **6** units with *Move*.

G. Move the torus up **4** units with *Move*.

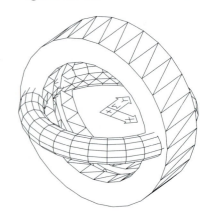

H. The solid primitives should appear as those in Figure 38-99. (*Hidden* shademode has been used for the figure.)

Figure 38-99

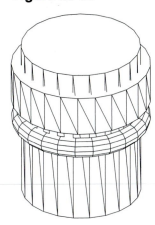

I. Use *Subtract* to subtract both revolved shapes from the cylinder. Use *Shademode* with the *Hidden* option. The solid should resemble that in Figure 38-100. *Save* the drawing.

Figure 38-100

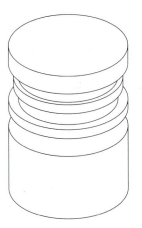

7. Begin a *New* drawing or use a *Template.* Assign the name **FAUCET**.

A. Draw 3 closed *Pline* shapes, as shown in Figure 38-101. Assume symmetry about the longitudinal axis. Use the WCS and create 2 new *UCS*s for the geometry. Use *3point Arcs* for the "front" profile.

Figure 38-101

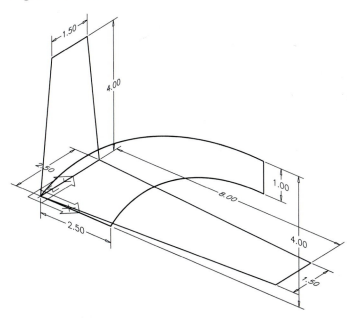

B. *Extrude* each of the 3 profiles into the same space. Make sure you specify the correct positive or negative *Height* value.

Figure 38-102 ───────────

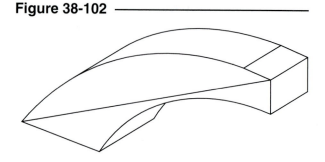

C. Finally, use *Intersect* to create the composite solid of the Faucet. *Save* the drawing.

D. (Optional) Create a nozzle extending down from the small end. Then create a channel for the water to flow (through the inside) and subtract it from the faucet.

8. Construct a solid model of the bar guide in Figure 38-103. Strive for the most efficient design. It is possible to construct this object with one *Extrude* and one *Subtract*. Save the model as **BGUID-SL**.

Figure 38-103 ─────────────────

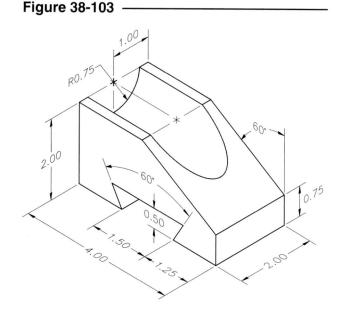

9. Make a solid model of the V-block shown in Figure 38-104. Several strategies could be used for construction of this object. Strive for the most efficient design. Plan your approach by sketching a few possibilities. Save the model as **VBLOK-SL**.

Figure 38-104 ───────────

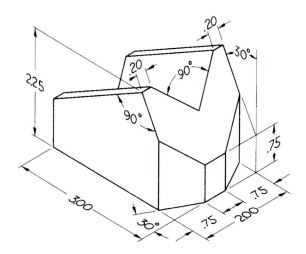

10. Construct a composite solid model of the support bracket using efficient techniques. Save the model as **SUPBK-SL**.

Figure 38-105

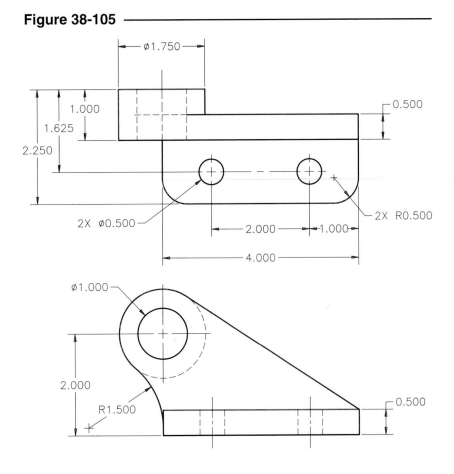

11. Construct the swivel shown in Figure 38-106. The center arm requires *Extruding* the 1.00 x 0.50 rectangular shape along an arc path through 45 degrees. Save the drawing as **SWIVEL**.

Figure 38-106

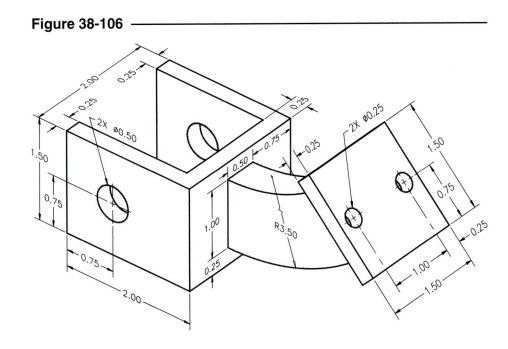

12. Construct a solid model of the angle brace shown in Figure 38-107. Use efficient design techniques. Save the drawing as **AGLBR-SL**.

Figure 38-107

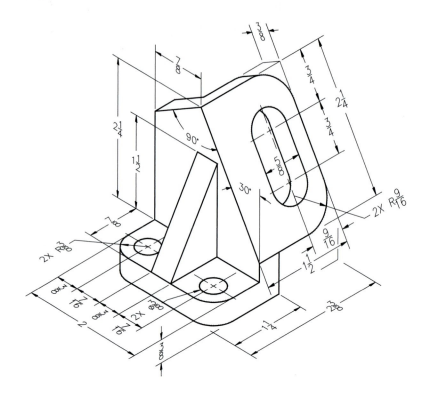

13. Construct a solid model of the saddle shown in Figure 38-108. An efficient design can be utilized by creating *Pline* profiles of the top "view" and the front "view," as shown in Figure 38-109. Use *Extrude* and *Intersect* to produce a composite solid. Additional Boolean operations are required to complete the part. The finished model should look similar to Figure 38-110. Save the drawing as **SADL-SL**.

Figure 38-108

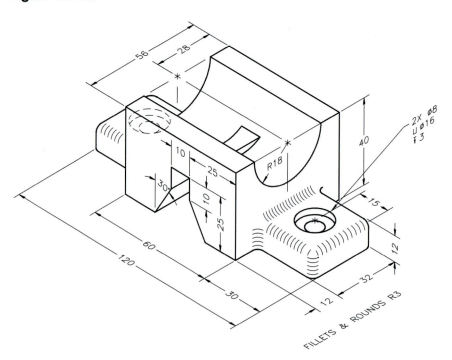

Figure 38-109

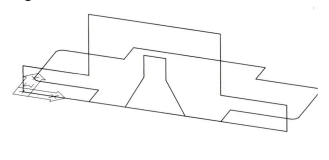

Figure 38-110

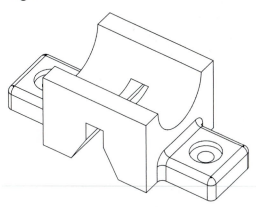

14. Construct a solid model of a bicycle handle bar. Create a center line (Fig. 38-111) as a *Path* to extrude a *Circle* through. Three mutually perpendicular coordinate systems are required: the **WORLD**, the **SIDE**, and the **FRONT**. The center line path consists of three separate *Plines*. First, create the 370 length *Pline* with 60 radii arcs on each end on the WCS. Then create the drop portion of the bars using the SIDE *UCS*. Create three *Circles* using the FRONT *UCS*, and extrude one along each *Path*.

Figure 38-111

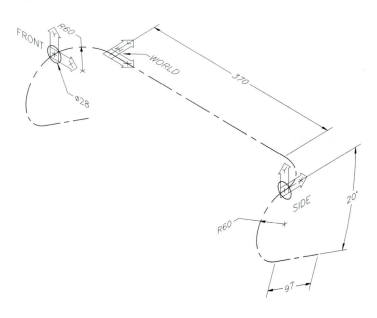

Plot the bar using *Shadeplot, Hidden* as shown in Figure 38-112. *Save* the drawing as **DROPBAR**.

Figure 38-112

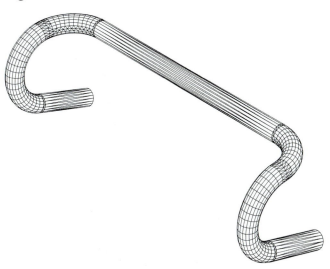

15. Create a solid model of the pulley. All vertical dimensions are diameters. Orientation of primitives is critical in the construction of this model. Try creating the circular shapes on the XY plane (circular axis aligns with Z axis of the WCS). After the construction, use *Rotate3D* to align the circular axis of the composite solid with the Y axis of the WCS. *Save* the drawing as **PULLY-SL**.

Figure 38-113

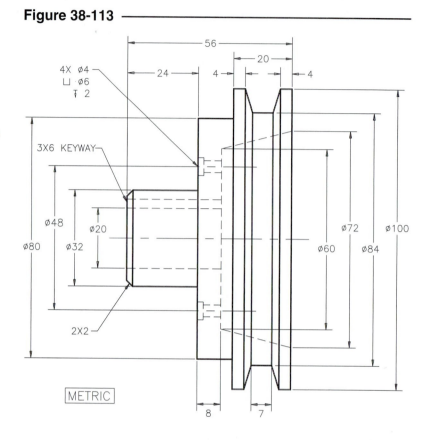

16. Create a composite solid model of the adjustable mount (Fig. 38-114). Use *Move* and other methods to move and align the primitives. Use of efficient design techniques is extremely important with a model of this complexity. Assign the name **ADJMT-SL**.

Figure 38-114

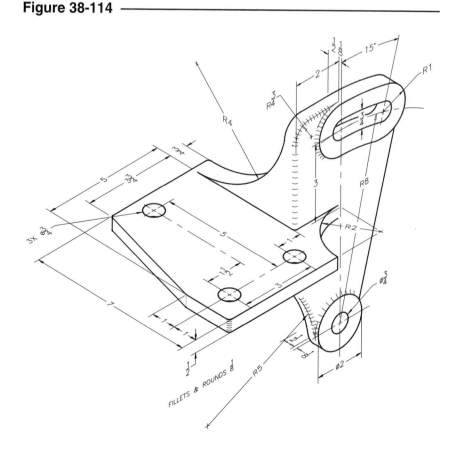

39

ADVANCED SOLIDS FEATURES

Chapter Objectives

After completing this chapter you should be able to:

1. use the *ISOLINES*, *DISPSILH*, *FACETRES*, and *FACETRATIO* variables to control the display of tessellation lines, silhouette lines, and mesh density for solid models;

2. calculate mass properties of a solid model using *Massprop*;

3. determine if *Interference* exists between two or more solids and create a solid equal in volume to the interference;

4. create a 2D section view for a solid model using *Section*;

5. use *Slice* to cut a solid model at any desired cutting plane and retain one or both halves;

6. use *STLOUT* to create a file suitable for use with rapid prototyping apparatus.

CONCEPTS

Several topics related to solid modeling capabilities for AutoCAD ACIS models are discussed in this chapter. The topics are categorized in the following sections:

Solid Modeling Display Variables
Analyzing Solid Models
Creating Sections from Solids
Converting Solids

SOLID MODELING DISPLAY VARIABLES

AutoCAD solid models are displayed in wireframe representation by default. Wireframe representation requires less computation time and less complex file structure, so your drawing time can be spent more efficiently. When you use *Hide, Shade, Shademode,* or *Render,* the solid models are automatically meshed before they are displayed with hidden lines removed or as a shaded image. This meshed version of the model is apparent when you use *Hide* on cylindrical or curved surfaces.

Four variables control the display of solids for wireframe, hidden, and meshed representation. The *ISOLINES* variable controls the number of tessellation lines that are used to visually define curved surfaces for wireframes. The *DISPSILH* variable can be toggled on or off to display silhouette lines for wireframe displays. *FACETRES* is the variable that controls the density of the mesh apparent with *Hide.* *FACETRATIO* creates a two-dimensional mesh for cylinders and cones. The *ISOLINES, DISPSILH,* and *FACETRES* display variables are saved in the drawing file. *FACETRATIO,* however, is not saved.

ISOLINES

Pull-down Menu	COMMAND (TYPE)	ALIAS (TYPE)	Short-cut	Screen (side) Menu	Tablet Menu
Tools *Options...* *Display* *Coutour lines* *per surface*	ISOLINES	...	...	*TOOLS2* *Options* *Display* *Contour lines* *per surface*	...

This variable sets the <u>number of tessellation lines</u> that appear on a curved surface when shown in <u>wireframe</u> representation. The default setting for *ISOLINES* is 4 (Fig. 39-1). A solid of extrusion shows fewer tessellation lines (the current *ISOLINES* setting less 4) to speed regeneration time.

A higher setting gives better visualization of the curved surfaces but takes more computing time (Fig. 39-2). After changing the *ISOLINES* setting, *Regen* the drawing to see the new display.

Figure 39-1 ——————— **Figure 39-2** ———————

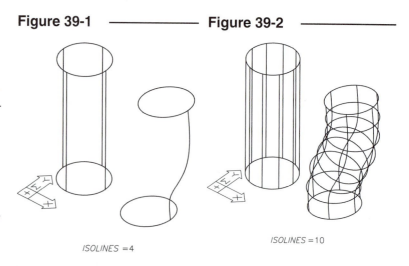

ISOLINES = 4

ISOLINES = 10

NOTE: In an isometric view attained by *3D Viewpoint,* the 4 lines (like Fig. 39-1) appear to overlap—they align when viewed from any perfect isometric angle.

DISPSILH

Pull-down Menu	COMMAND (TYPE)	ALIAS (TYPE)	Short-cut	Screen (side) Menu	Tablet Menu
Tools *Options...* *Display* *Show silhouettes in wireframe*	DISPSILH	...	...	*TOOLS2* *Options* *Display* *Show silhouettes in wireframe*	...

This variable can be turned on to display the limiting element contour lines, or silhouette, of curved shapes for a wireframe display (Fig. 39-3). The default setting is 0 (off). Since the silhouette lines are <u>viewpoint dependent</u>, some computing time is taken to generate the display. You should <u>not</u> leave DISPSILH on during construction if you are working with a complex model with many curved surfaces.

DISPSILH has a special function when used with the *Hide* command (when *Shademode* is set to *2D Wireframe*) and when creating prints and plots with hidden lines removed (*Shade Plot, Hidden*). When DISPSILH is set to 0 and a hidden display is generated, or when making a print or plot with hidden lines calculated, the solids appear opaque and display mesh (facet) lines (Fig. 39-4). If DISPSILH is set to 1, then a *2D Wireframe Hide* is performed or the drawing is printed with a hidden display, the solids appear opaque but do <u>not</u> display the mesh lines (Fig 39-5). See "Printing a Hidden Display for Solid Models" in Chapter 35 for more information.

NOTE: The *Hidden* option of *Shademode* may or may not show mesh lines (regardless of the DISP-SILH setting) depending on variables in your operating system, display driver, or graphics display settings.

Figure 39-3

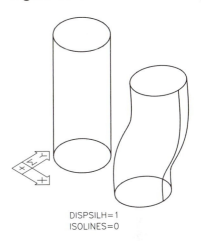

DISPSILH=1
ISOLINES=0

Figure 39-4

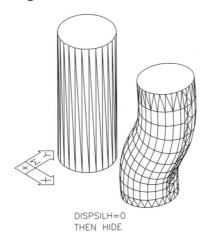

DISPSILH=0
THEN HIDE

Figure 39-5

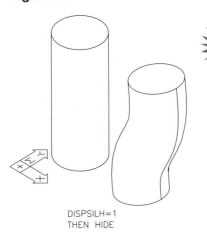

DISPSILH=1
THEN HIDE

FACETRES

Pull-down Menu	COMMAND (TYPE)	ALIAS (TYPE)	Short-cut	Screen (side) Menu	Tablet Menu
Tools *Options...* *Display* *Rendered object smoothness*	FACETRES	...	...	*TOOLS2* *Options* *Display* *Rendered object smoothness*	...

FACETRES controls the mesh density of curved surfaces on solid objects. The mesh is most apparent when an object with curved surfaces is displayed using *Hide, Shade, Shade Plot,* and *Shademode* (*Flat, Flat+edges,* and *Gouraud+edges* options). The default setting is .5, as shown in Figure 39-6 on the next page.

Decreasing the value produces a coarser mesh (Fig. 39-7), while increasing the value produces a finer mesh. The higher the value, the more computation time involved to generate the display or plot. The density of the mesh is actually a factor of both the *FACETRES* setting and the *VIEWRES* setting. Increasing *VIEWRES* also makes the mesh more dense. *FACETRES* can be set to any value between .01 and 10.

Figure 39-6 ——————— **Figure 39-7** ———————

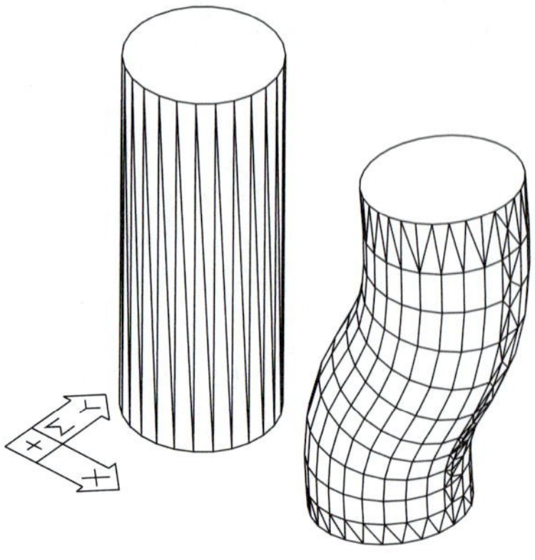

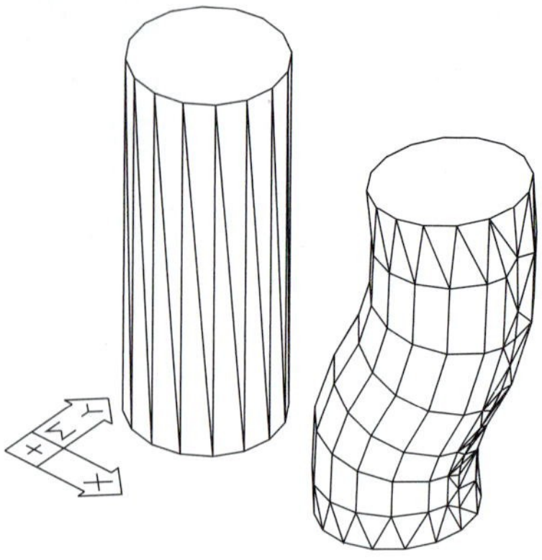

FACETRES =0.5   FACETRES =0.2

FACETRATIO

Pull-down Menu	COMMAND (TYPE)	ALIAS (TYPE)	Short-cut	Screen (side) Menu	Tablet Menu
...	*FACETRATIO*	...	...	...	...

FACETRATIO is a variable that affects only cylindrical and conical solids. When *FACETRATIO* is on (set to 1), an *n* by *m* mesh (two-dimensional mesh) is created. When *FACETRATIO* is off (set to 0), a 1 by *n* mesh (one-dimensional mesh) is created. Compare the two cylinders in Figures 39-7 and 39-8. The cylinder in Figure 39-7 displays a mesh generated in only one dimension—around the circumference—whereas the cylinder in Figure 39-8 displays a cylinder with a 2-dimensional mesh.

Figure 39-8 ———————

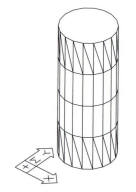

FACETRATIO = 1

The *Options* dialog box can be used to change the settings for the *ISOLINES, DISPSILH,* and *FACETRES* variables (Fig. 39-9). The right side of the *Display* tab has two edit boxes and one checkbox that allow changing these variables as follows.

Rendered object smoothness   *FACETRES*
Contour lines per surface   *ISOLINES*
Show silhouettes in wireframe   *DISPSILH*

Figure 39-9 ———————

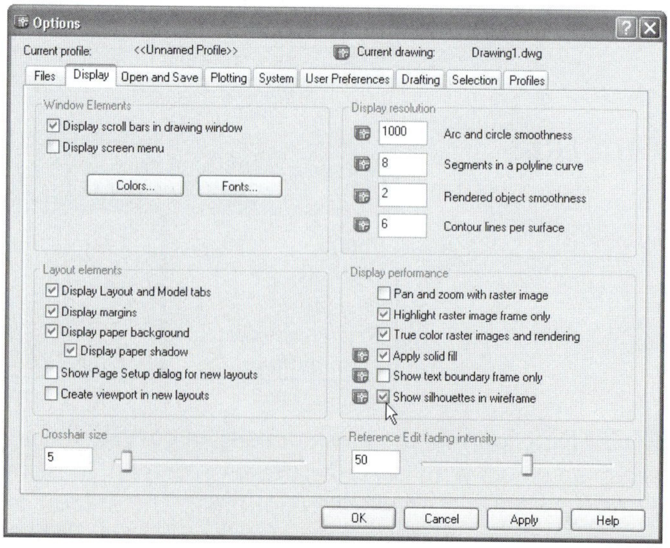

ANALYZING SOLID MODELS

Two commands in AutoCAD allow you to inquire about and analyze the solid geometry. *Massprop* calculates a variety of properties for the selected ACIS solid model. AutoCAD does the calculation and lists the information in screen or text window format. The data can be saved to a file for future exportation to a report document or analysis package. The *Interfere* command finds the interference of two or more solids and highlights the overlapping features so you can make necessary alterations.

MASSPROP

Pull-down Menu	COMMAND (TYPE)	ALIAS (TYPE)	Short-cut	Screen (side) Menu	Tablet Menu
Tools *Inquiry >* *Mass Properties*	*MASSPROP*	...	...	TOOLS 1 *Massprop*	U,7

Since solid models define a complete description of the geometry, they are ideal for mass properties analysis. The *Massprop* command automatically computes a variety of mass properties.

Mass properties are useful for a variety of applications. The data generated by the *Massprop* command can be saved to an .MPR file for future exportation in order to develop bills of material, stress analysis, kinematics studies, and dynamics analysis.

Applying the *Massprop* command to a solid model produces a text screen displaying the following list of calculations (Fig. 39-10):

Figure 39-10

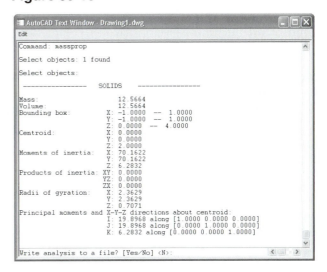

Mass	Mass is the quantity of matter that a solid contains. Mass is determined by density of the material and volume of the solid. Mass is not dependent on gravity and, therefore, different from but proportional to weight. Mass is also considered a measure of a solid's resistance to linear acceleration (overcoming inertia).
Volume	This value specifies the amount of space occupied by the solid.
Bounding Box	These lengths specify the extreme width, depth, and height of the selected solid.
Centroid	The centroid is the geometrical center of the solid. Assuming the solid is composed of material that is homogeneous (uniform density), the centroid is also considered the center of mass and center of gravity. Therefore, the solid can be balanced when supported only at this point.
Moments of Inertia	Moments convey how the mass is distributed around the X, Y, and Z axes of the current coordinate system. These values are a measure of a solid's resistance to <u>angular</u> acceleration (mass is a measure of a solid's resistance to <u>linear</u> acceleration). Moments of inertia are helpful for stress computations.

Products of Inertia	These values specify the solid's resistance to <u>angular</u> acceleration with respect to two axes at a time (XY, YZ, or ZX). Products of inertia are also useful for stress analysis.
Radii of Gyration	If the object were a concentrated solid mass without holes or other features, the radii of gyration represent these theoretical dimensions (radius about each axis) such that the same moments of inertia would be computed.
Principal Moments and X, Y, Z Directions	In structural mechanics, it is sometimes important to determine the orientation of the axes about which the moments of inertia are at a maximum. When the moments of inertia about centroidal axes reach a maximum, the products of inertia become zero. These particular axes are called the principal axes, and the corresponding moments of inertia with respect to these axes are the principal moments (about the centroid).

Notice the "Mass:" is reported as having the same value as "Volume." This is because AutoCAD solids cannot have material characteristics assigned to them, so AutoCAD assumes a density value of 1. To calculate the mass of a selected solid in AutoCAD, use a reference guide (such as a machinist's handbook) to find the material density and multiply the value times the reported volume (mass = volume x density).

INTERFERE

Pull-down Menu	COMMAND (TYPE)	ALIAS (TYPE)	Short-cut	Screen (side) Menu	Tablet Menu
Draw *Solids >* *Interference*	*INTERFERE*	*INF*	...	*DRAW 2* *SOLIDS* *Interfer*	...

In AutoCAD, unlike real life, it is possible to create two solids that occupy the same physical space. *Interfere* checks solids to determine whether or not they interfere (occupy the same space). If there is interference, *Interfere* reports the overlap and allows you to create a new solid from the interfering volume, if you desire. Normally, you specify two sets of solids for AutoCAD to check against each other:

```
Command: interfere
Select first set of solids:
Select objects: PICK
Select objects: Enter
1 solid selected.
Select second set of solids:
Select objects: PICK
Select objects: Enter
1 solid selected.
Comparing 1 solid against 1 solid.
Interfering solids (first  set): 1
            (second set): 1
Interfering pairs:           1
Create interference solids? [Yes/No] <N>: y
Command:
```

If you answer "yes" to the last prompt, a new solid is created equal to the exact size and volume of the interference. The original solids are not changed in any way. If no interference is found, AutoCAD reports "Solids do not interfere."

For example, consider the two solids shown in Figure 39-11. (The parts are displayed in wireframe representation.) The two shapes fit together as an assembly. The locating pin on the part on the right should fit in the hole in the left part.

Figure 39-11

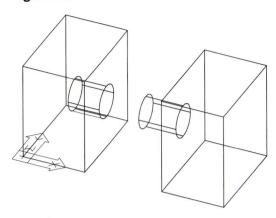

Sliding the parts together until the two vertical faces meet produces the assembly shown in Figure 39-12. There appears to be some inconsistency in the assembly of the hole and the pin. Either the pin extends beyond the hole (interference) or the hole is deeper than necessary (no interference). Using *Interfere*, you can find an overlap and create a solid is created by answering "yes" to "Create interference solids?" Use *Move* with the *Last* selection option to view and analyze the solid of interference.

Figure 39-12

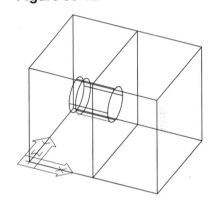

You can compare <u>more than two</u> solids against each other with *Interfere*. This is accomplished by selecting <u>all desired solids at the first prompt</u> and none at the second:

> Select the first set of solids: **PICK**
> Select the second set of solids: **Enter**

AutoCAD then compares all solids in the first set against each other. If more than one interference is found, AutoCAD highlights intersecting solids, one pair at a time.

CREATING SECTIONS FROM SOLIDS

Two AutoCAD commands are intended to create sections from solid models. *Section* is a drafting feature that creates a 2D "section view." The cross-section is determined by specifying a cutting plane. A cross-section view is automatically created based on the solid geometry that intersects the cutting plane. The original solid is not affected by the action of *Section*. *Slice* actually cuts the solid at the specified cutting plane. *Slice* therefore creates two solids from the original one and offers the possibility to retain both halves or only one. Many options are available for placement of the cutting plane.

SECTION

Pull-down Menu	COMMAND (TYPE)	ALIAS (TYPE)	Short-cut	Screen (side) Menu	Tablet Menu
Draw *Solids >* *Section*	*SECTION*	*SEC*	...	*DRAW 2* *SOLIDS* *Section*	...

Section creates a 2D cross-section of a solid or set of solids. The cross-section created by *Solsect* is considered a traditional 2D section view. The cross-section is defined by a cutting plane, and the resulting section is determined by any solid material that passes through the cutting plane.

The cutting plane can be specified by a variety of methods. The options for establishing the cutting plane are listed in the command prompt:

```
Command: section
Select objects: PICK
Select objects: Enter
Specify first point on Section plane by [Object/Zaxis/View/XY/YZ/ZX/3points] <3points>:
```

For example, assume a cross-section is desired for the geometry shown in Figure 39-13. To create the section, you must define the cutting plane.

Figure 39-13

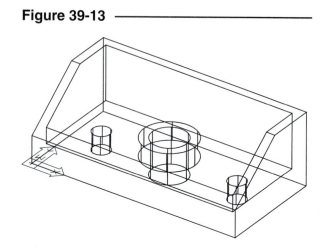

For this case, the *ZX* option is used and the requested point is defined to establish the position of the plane, as shown in Figure 39-14. The cross-section will be created on this plane.

Once the cutting plane is established, a cross-section is automatically created by *Section*. The resulting geometry is a *Region* created on the <u>current</u> layer.

Figure 39-14

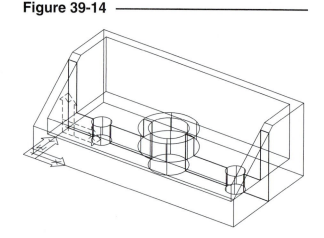

The resulting *Region* can be moved if needed as shown in Figure 39-15. If you want to use the *Region* to create a 2D section complete with hatch lines, you may need to create a *UCS* on the plane of the shapes before using *Bhatch*. Other *Lines* may be needed to make a complete section view.

Figure 39-15

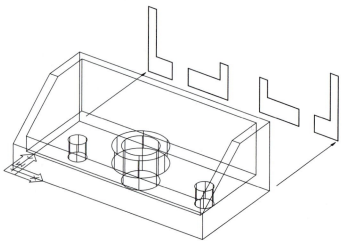

SLICE

	COMMAND (TYPE)	ALIAS (TYPE)	Short-cut	Screen (side) Menu	Tablet Menu
Pull-down Menu					
Draw *Solids >* *Slice*	*SLICE*	*SL*	...	DRAW 2 SOLIDS *Slice*	...

Slice creates a true solid section. *Slice* cuts an ACIS solid or set of solids on a specified cutting plane. The original solid is converted to two solids. You have the option to keep both halves or only the half that you specify. (Also see *Slice* in Chapter 38.) Examine the following command syntax:

```
Command: slice
Select objects: PICK
Select objects: Enter
Specify first point on slicing plane by [Object/Zaxis/View/XY/YZ/ZX/3points] <3points>: (option)
(option prompts)
Specify a point on desired side of the plane or [keep Both sides]: PICK or Enter
Command:
```

Entering *B* at the "keep Both sides" prompt retains the solids on both sides of the cutting plane. Otherwise, you can pick a point on either side of the plane to specify which half to keep.

Figure 39-16

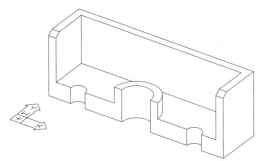

For example, using the solid model shown previously in Figure 39-13, you can use *Slice* to create a new sectioned solid shown here. The ZX method is used to define the cutting plane midway through the solid. Next, the new solid to retain was specified by PICKing a point on that geometry. The resulting sectioned solid is shown in Figure 39-16. Note that the solid on the near side of the cutting plane was not retained.

CONVERTING SOLIDS

AutoCAD provides several utilities for converting solids to and from other file formats. Older AutoCAD solid models created with AME (Advanced Modeling Extension) can be converted to an ACIS model with some success. Other file utilities allow you to import and export .SAT files and to create files for use with stereolithography apparatus.

AMECONVERT

	COMMAND (TYPE)	ALIAS (TYPE)	Short-cut	Screen (side) Menu	Tablet Menu
Pull-down Menu					
...	*AMECONVERT*	...	...	...	...

Ameconvert is a conversion utility to convert older solid models (created with AME Release 2 or 2.1) to ACIS solid models. The command is simple to use; however, the conversion may change the model slightly. If the selected solids to convert are not AME Release 2 or 2.1, AutoCAD ignores the request.

The newer AutoCAD solid modeler, ACIS, creates solids of higher accuracy than the older AME modeler. Because of this new accuracy, objects that are converted may change in appearance and form. For example, two shapes that were originally considered sufficiently close and combined by Boolean operations in the old modeler may be interpreted as slightly offset in the ACIS modeler. This can occur for converted features such as fillets, chamfers, and through holes. Occasionally, a solid created in the older AME model is converted into two or more solids in the ACIS modeler.

Solids that are converted to ACIS solids can be edited using AutoCAD 2004 solids commands. Converted solids that do not convert as expected can usually be recreated to produce an equivalent, but more accurate, solid.

STLOUT

Pull-down Menu	COMMAND (TYPE)	ALIAS (TYPE)	Short-cut	Screen (side) Menu	Tablet Menu
File *Export…* **.stl*	*STLOUT*	…	…	*FILE* *Export* **.stl*	…

The *Stlout* command converts an ACIS solid model to a file suitable for use with rapid prototyping apparatus. Use *Stlout* if you want to use an ACIS solid to create a prototype part using stereo lithography or sintering technology. This technology reads the CAD data and creates a part by using a laser to solidify microthin layers of plastic or wax polymer or powdered metal. A complex 3D prototype can be created from a CAD drawing in a matter of hours.

Stlout writes an ASCII or binary .STL file from an ACIS solid model. The model must reside entirely within the positive X,Y,Z octant of the WCS (all part geometry coordinates must be positive):

```
Command: stlout
Select a single solid for STL output:
elect objects: PICK
Create a binary STL file ? [Yes/No] <Y>: Enter N to create an ASCII rather than binary file.
Command:
```

AutoCAD displays the *Create .STL File* dialog box for you to designate a name and path for the file to create.

When you design parts with AutoCAD to be generated by stereolithography apparatus, it is a good idea to create the profile view (that which contains the most complex geometry) parallel to the XY plane. In this way, the aliasing (stair-step effect) on angled or curved surfaces, caused by incremental passes of the laser, is minimized.

ACISIN

Pull-down Menu	COMMAND (TYPE)	ALIAS (TYPE)	Short-cut	Screen (side) Menu	Tablet Menu
Insert *ACIS File…*	*ACISIN*	…	…	*INSERT* *ACISin*	…

The *Acisin* command imports an ACIS solid model stored in a .SAT (ASCII) file format. This utility can be used to import .SAT files describing a solid model created by AutoCAD or other CAD systems that create ACIS solid models. The *Select ACIS File* dialog box appears, allowing you to select the desired file. AutoCAD reads the file and builds the model in the current drawing.

ACISOUT

	Pull-down Menu	COMMAND (TYPE)	ALIAS (TYPE)	Short-cut	Screen (side) Menu	Tablet Menu
	File *Export...* **.sat*	ACISOUT	...	...	FILE Export **.sat*	...

This utility is used to export an ACIS solid model to a .SAT (ASCII) file format that can later be read by AutoCAD or other CAD systems utilizing the ACIS modeler. The *Create ACIS File* dialog box is used to define the desired name and path for the file to be created.

CHAPTER EXERCISES

1. *Open* the **SADL-SL** drawing that you created in Chapter 38 Exercises. Calculate *Mass Properties* for the saddle. Write the report out to a file named **SADL-SL.MPR**. Use a text editor or the DOS TYPE command to examine the file.

2. *Open* the **SADL-SL** drawing again. Use *Slice* to cut the model in half longitudinally. Use an appropriate method to establish the "slicing plane" in order to achieve the resulting model, as shown in Figure 39-17. Use *SaveAs* and assign the name **SADL-CUT**.

Figure 39-17 ──────────

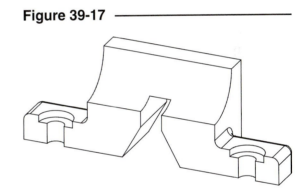

3. *Open* the **PULLY-SL** drawing that you created in Chapter 38 Exercises.

Make a *New Layer* named **SECTION** and set it *Current*. Then use the *Section* command to create a full section "view" of the pulley. Establish a vertical cutting plane through the center of the model. Remove the section view object (*Region*) with the *Move* command, translating **100** units in the **X** direction. The model and the new section view should appear as in Figure 39-18 (*Hide* was performed on the pulley to enhance visualization). Complete the view by establishing a *UCS* at the section view, then adding the *Bhatch*, as shown in Figure 39-18. Finally, create the necessary *Lines* to complete the view. *SaveAs* **PULLY-SC**.

Figure 39-18 ──────────

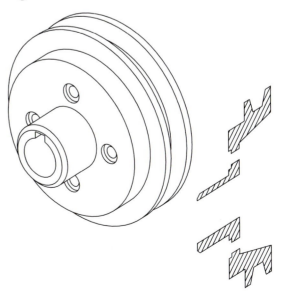

4. *Open* the **SADL-SL** drawing that you created in the Chapter 38 Exercises. Change the *ISOLINES* setting to display **10** tessellation lines. Use the *Shademode* command with the *Hidden* option to create a meshed hidden display. Change the *FACETRES* setting to display a coarser mesh and *Regen*. Make a plot of the model with the coarse mesh and with hidden lines (use *Shade Plot* in the *Plot* dialog box).

    Next, change the *FACETRES* setting to display a fine mesh. Use *Regen* to reveal the change. Make a plot of the model with *Shade Plot* to display the fine mesh. Then set the *DISPSILH* variable to **1** and make another plot with lines hidden. What is the difference in the last two plots? *Save* the **SADL-SL** file with the new settings.

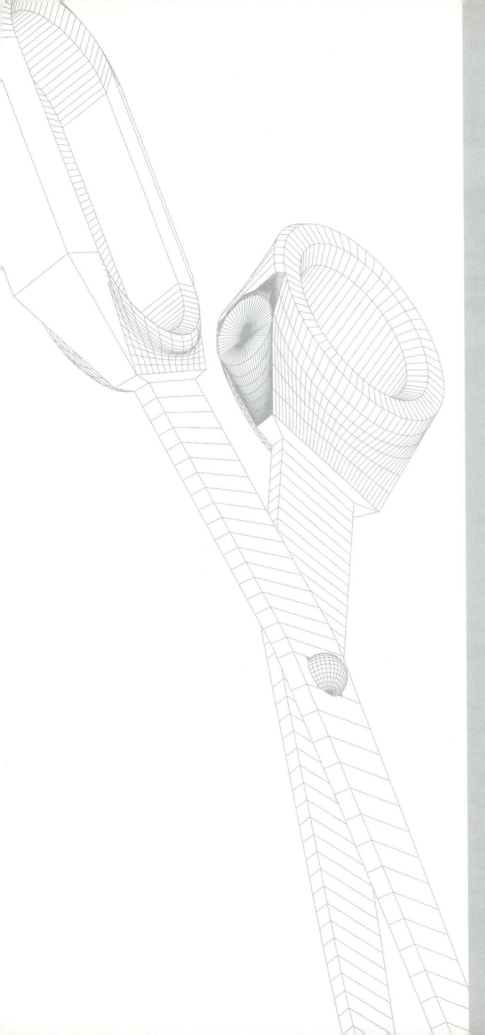

SURFACE MODELING

Chapter Objectives

After completing this chapter you should:

1. be able to create planar surfaces bounded by straight edges using *3Dface*;

2. be able to edit *3Dface*s with grips and create invisible edges between *3Dface*s using *Edge*;

3. be able to create meshed surfaces bounded by straight sides using *3Dmesh*;

4. be able to create geometrically defined meshed surfaces using *Rulesurf*, *Tabsurf*, *Revsurf*, and *Edgesurf*;

5. be able to edit polygon meshes (*3Dmesh*, *Rulesurf*, *Tabsurf*, *Revsurf*, and *Edgesurf*) using *Pedit*, *Explode*, and Grips;

6. be able to create *Regions* for use with surface models;

7. be able to use surface modeling primitives (*3D Objects*) to aid in construction of complex surface models;

8. be able to use *Thickness* and *Elevation* to create surfaces from 2D draw commands.

CONCEPTS

A surface model is more sophisticated than a wireframe model because it contains the description of surfaces as well as the description of edges. Surface models can contain descriptions of complex curves, whereas surfaces on a wireframe are assumed to be planar or single curved. Surface models are superior to wireframe models in that they provide better visualization cues. The surface information describes how the surfaces appear and gives the model a "solid" look, since surfaces that are nearer to the observer naturally obscure the surfaces and edges that are behind.

Generally, surface models do not describe physical objects as completely and as accurately as solid models. A surface model can be compared to a cardboard box—it has surfaces but is hollow inside. If a surface model is sectioned (cut in half), there is only air inside, whereas a solid model can be sectioned to reveal its interior features. Therefore, surface models have volume, whereas solid models have volume and mass. Curved surfaces are defined by discrete meshes composed of straight edges; therefore, small "gaps" may exist where a curved surface is adjacent to a planar surface.

Surface modeling is similar to wireframe and solid modeling in that 3D coordinate values must be used to create and edit the geometry. However, the construction and editing process of surface modeling is complex and somewhat tedious compared to wireframe and solid modeling. Each surface must be defined by describing the edges that bound the surface. Each surface must be constructed individually in location or moved into location with respect to other surfaces that comprise the 3D model.

Surface modeling is the type of 3D modeling that is best suited for efficiently defining complex curved shapes such as automobile bodies, aircraft fuselages, and ship hulls. Thus, surface modeling is a necessity for 3D modeling in many industries.

The surface modeling capabilities of AutoCAD provide solutions for only a small portion of the possible applications. AutoCAD does not utilize NURBS (Non-Uniform Rational B-Spline) surfacing techniques, which is the current preferred surface modeling technology. NURBS technology is utilized, however, in Autodesk's Mechanical Desktop and Inventor product lines.

The categories of surfaces and commands that AutoCAD provides to create surface models are as follows:

Surfaces with straight edges (usually planar)

3Dface	A surface defined by 3 or 4 straight edges

Meshed surfaces (polygon meshes)

3Dmesh	A planar, curved, or complex surface defined by a mesh
Pface	A surface with any number of vertices and faces

Geometrically defined meshed surfaces (polygon meshes)

Edgesurf	A surface defined by "patching" together 4 straight or curved edges
Rulesurf	A surface created between 2 straight or curved edges
Revsurf	A surface revolved from any 2D shape about an axis
Tabsurf	A surface created by sweeping a 2D shape in the direction specified by a vector

Surface model primitives

3D Objects...	A menu of complete 3D primitive surface models (box, cone, wedge, sphere, etc.) is available. These simple 3D shapes can be used as a basis for construction of more complex shapes.
Thickness, Solid	AutoCAD's early methods for creation of planar "extrusions" of 2D shapes are discussed briefly.

Regions can be used to simplify construction of complex planar surfaces for use with surface models. The region modeler utilizes Boolean operations, which makes creation of surfaces with holes and/or complex outlines relatively easy.

A visualization enhancement for surfaces is provided by the *Shademode* command. Normally, during construction, surfaces are displayed by default (like solids) in <u>wireframe</u> representation. Other options of *Shademode* cause a regeneration of the display showing surfaces as <u>opaque</u>, therefore obscuring other surfaces or objects behind them.

The surface modeling commands can be accessed easily from the *Draw* pull-down menu (Fig. 40-1). A *Surfaces* toolbar can also be invoked (Fig. 40-2) by selecting *Toolbars...* from the *View* pull-down menu.

The surfacing commands are discussed first in this chapter. Then, application of the surfacing commands to create a complete 3D model is discussed.

Figure 40-1

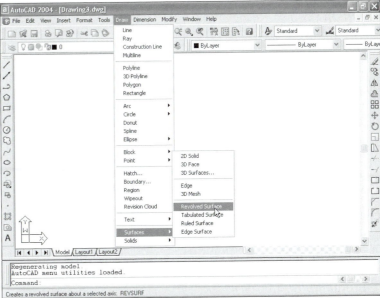

Figure 40-2

SURFACES WITH STRAIGHT EDGES

3DFACE

Pull-down Menu	COMMAND (TYPE)	ALIAS (TYPE)	Short-cut	Screen (side) Menu	Tablet Menu
Draw *Surfaces >* *3D Face*	*3DFACE*	*3F*	...	*DRAW 2* *SURFACES* *3Dface*	*M,8*

3Dface creates a surface bounded by three or four straight edges. The three- or four-sided surface can be connected to other *3Dfaces* within the same *3Dface* command sequence, similar to the way several *Line* segments created in one command sequence are connected. Beware, there is <u>no *Undo* option</u> for *3Dface*. The command sequence is:

```
Command: 3dface
Specify first point or [Invisible]: PICK or (coordinates)
Specify second point or [Invisible]: PICK or (coordinates)
Specify third point or [Invisible] <exit>: PICK or (coordinates)
Specify fourth point or [Invisible] <create three-sided face>: PICK or (coordinates)
Specify third point or [Invisible] <exit>: PICK or (coordinates)
Specify fourth point or [Invisible] <create three-sided face>: PICK or (coordinates)
Specify third point or [Invisible] <exit>: Enter
Command:
```

After completing four points, AutoCAD connects the next point ("third point:") to the previous fourth point. The next fourth point is connected to the previous third point, and so on. Figure 40-3 shows the sequence for attaching several *3Dface* segments in one command sequence.

Figure 40-3

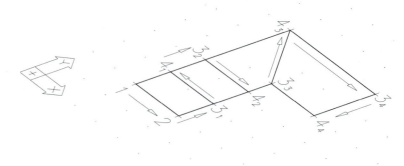

In the example, each edge of the *3Dface* is <u>coplanar</u>, as is the entire sequence of *3Dfaces* (the Z value of every point on each edge is 0). The edges must be <u>straight</u> but not necessarily coplanar. Entering specific coordinate values, using point filters, or using *OSNAP* in 3D space allows you to create geometry with *3Dface* that is not on a plane. The following command sequence creates a *3Dface* as a complex curve by entering a different Z value for the fourth point (Fig. 40-4):

Figure 40-4

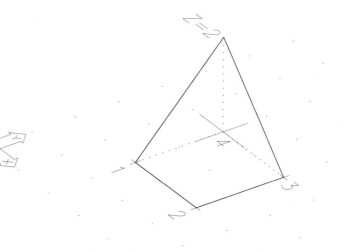

```
Command: 3dface
Specify first point or [Invisible]: PICK
Specify second point or [Invisible]:
PICK
Specify third point or [Invisible]
<exit>: PICK
Specify fourth point or [Invisible]
<create three-sided face>: .XY
of PICK
(need Z): (value)
Specify third point or [Invisible]
<exit>: Enter
Command:
```

Although it is possible to use *3Dface* to create complex curved surfaces, the surface curve is not visible as it would be if a mesh were used. A *3Dmesh* provides superior visibility and flexibility and is recommended for such a surface.

Figure 40-5 displays a possibility for creating nonplanar geometry in one command sequence. .XY filters can be used to PICK points, or absolute X,Y,Z values can be entered.

Figure 40-5

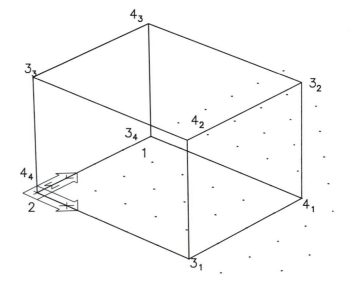

The lines between *3Dface* segments in Figure 40-3 are visible even though the segments are coplanar. AutoCAD provides two methods for making invisible edges. The first method requires that you enter the letter *"I"* immediately before the <u>first</u> of the two points that define an invisible edge. For the shape shown in Figure 40-6, enter the letter *"I"* <u>before</u> points 3_1, 3_2, and 3_3. The second, and much easier, method for defining invisible edges of *3Dfaces* is to use the *Edge* command.

EDGE

Pull-down Menu	COMMAND (TYPE)	ALIAS (TYPE)	Short-cut	Screen (side) Menu	Tablet Menu
Draw Surfaces > Edge	*EDGE*	...	...	DRAW 2 SURFACES *Edge:*	...

The first method for creating invisible edges described above takes careful planning. The same action shown in Figure 40-6 can be accomplished much more easily and <u>retroactively</u> by this second method. AutoCAD provides the *Edge* command to create invisible edges after construction of the *3Dface*. The Command prompt is:

Command: **edge**
Specify edge of 3dface to toggle visibility or [Display]: **PICK**

Selecting edges converts visible edges to invisible edges. <u>Any</u> edges on the *3Dface* can be made invisible by this method. The *Midpoint OSNAP* option is automatically invoked to simplify selection of the edges (Fig. 40-6). The *Display* option causes selected invisible edges to become visible again.

Figure 40-6

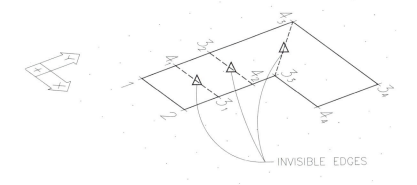
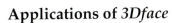

INVISIBLE EDGES

Applications of *3Dface*

3Dface is used to construct 3D surface models by creating individual (planar) faces or connected faces. For relatively simple models, the faces are placed in space as they are constructed. A construction strategy to use for more complex models is to first construct a wireframe model and then attach surfaces (*3Dfaces* or other surfaces) to the wireframe. The wireframe geometry can be constructed on a separate layer; then the layer can be turned off after the surfacing is complete.

Figure 40-7

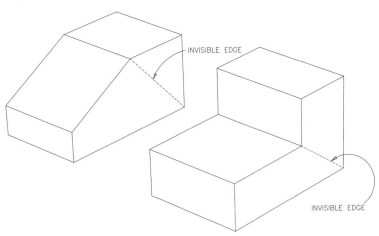

INVISIBLE EDGE

INVISIBLE EDGE

Figures 40-7 and 40-8 show some applications of *3Dface* for construction of surface models. Remember, *3Dface* can be used <u>only</u> for construction of surfaces with straight edges.

The edges made invisible by *Edge* are displayed in these figures in a hidden linetype. *Hidden* shademode has also been used to enhance visualization. You may want to use *Shademode* periodically to enhance the visibility of surfaces.

Figure 40-8

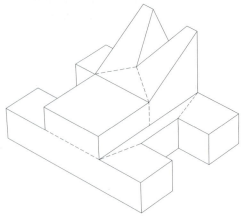

Editing *3Dfaces*

Grips can be used to edit *3Dfaces*. Each *3Dface* has three or four grips (one for each corner). Activating grips (PICKing the *3Dface* at the open Command prompt), makes the individual *3Dface* surface become highlighted and display its grips. If you then make one of the three or four grips **hot**, all of the normal Grip editing options are available, that is, STRETCH, MOVE, ROTATE, SCALE, and MIRROR. The single activated *3Dface* surface can then be edited. MOVE, for example, could be used to move the single activated surface to another location.

Figure 40-9

The STRETCH option, however, is generally the most useful Grip editing option, since all of the other options change the entire surface. The STRETCH option allows you to change one (or more) corner(s) of the *3Dface* rather than the entire surface, as with the other options. Grips in AutoCAD 2004 always appear <u>parallel to the current UCS XY plane</u> and operate in their respective planes. For example, in Figure 40-9 the grips are aligned parallel to the current coordinate system. Therefore, STRETCHing the hot grip (shown solid) allows you to change the shape of the *3Dface*, but restricts the movement parallel to the current XY plane. Changing to a different UCS orientation allows you to STRETCH in another plane.

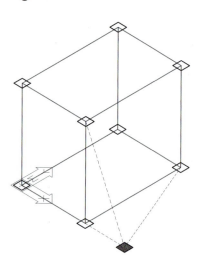

CREATING MESHED SURFACES

Meshes are versatile methods for generating a surface. *Ai_mesh* can be used to create a simple planar surface bounded by <u>four</u> straight edges (Fig. 40-10), or *3Dmesh* can be used to create a complex surface defining an irregular shape bounded by <u>four</u> sides (Fig. 40-11). The surface created with *Ai_mesh* or *3Dmesh* differs from a *3Dface* because the surface is defined by a "mesh." A mesh is a series of vertices (sometimes called "nodes") arranged in rows and columns connected by lines. *Pedit* or grips can be used to edit the resulting mesh (see "Editing Polygon Meshes," this chapter).

Figure 40-10

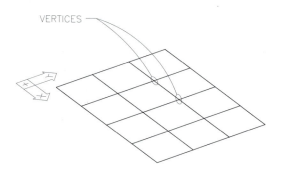

VERTICES

Figure 40-11

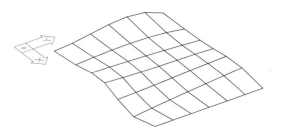

AI_MESH

	Pull-down Menu	COMMAND (TYPE)	ALIAS (TYPE)	Short-cut	Screen (side) Menu	Tablet Menu
	Draw Surfaces > 3D Surfaces... Mesh	*AI_MESH*	...	...	*DRAW 2 SURFACES 3Dobjec: Mesh*	...

Although no icon button is provided for this command, you can invoke *Ai_mesh* through the *3D Objects* dialog box (see "Surface Modeling Primitives") or by typing. *Ai_mesh* can be used to create a simple planar surface bounded by <u>four</u> straight edges (Fig. 40-10). Using this command causes:

Command: **ai_mesh**
Specify first corner point of mesh: **PICK** or
(**coordinates**)
Specify second corner point of mesh: **PICK** or
(**coordinates**)
Specify third corner point of mesh: **PICK** or
(**coordinates**)
Specify fourth corner point of mesh: **PICK** or
(**coordinates**)
Enter mesh size in the M direction: (**value**)
Enter mesh size in the N direction: (**value**)
Command:

Figure 40-12 ───────

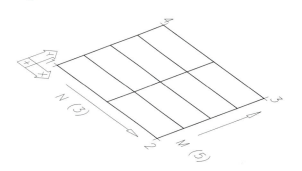

First, specify the four corners of the mesh. The number of vertices along each side is defined by the *M* size and the *N* size. The *M* direction is <u>perpendicular</u> to the first side specified, and the *N* direction is <u>perpendicular</u> to the second side specified (Fig. 40-12).

The values specified for the *M* size and *N* size <u>include</u> the vertices along the edges of the surface. The *M* and *N* sizes are the numbers of rows and columns of vertices (density of the mesh).

Ai_mesh can quickly generate a complex curved surface with straight edges by specifying nonplanar coordinates for the corners. The following command sequence is used to generate the surface shown in Figure 40-13:

Figure 40-13 ───────

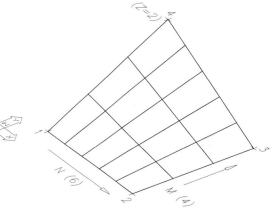

Command: **ai_mesh**
Specify first corner point of mesh: **1,1,0**
Specify second corner point of mesh: **5,1,0**
Specify third corner point of mesh: **5,5,0**
Specify fourth corner point of mesh: **1,5,2**
Enter mesh size in the M direction: **4**
Enter mesh size in the N direction: **6**
Command:

Ai_mesh visually defines the curvature of a surface better than *3Dface* because of the mesh lines. Figure 40-14 illustrates an application for a nonplanar *Ai_mesh*.

Figure 40-14 ───────

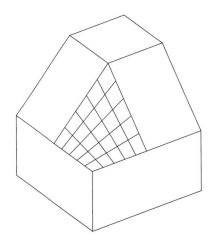

3DMESH

Pull-down Menu	COMMAND (TYPE)	ALIAS (TYPE)	Short-cut	Screen (side) Menu	Tablet Menu
Draw Surfaces > 3D Mesh	*3DMESH*	...	...	*DRAW 2 SURFACES 3Dmesh:*	...

3Dmesh is a versatile tool for generating a complex surface. It is used to create a complex surface defining an irregular shape bounded by <u>four</u> sides (see Figure 40-11). *Pedit* or grips can be used to edit the *3Dmesh* retroactively.

3Dmesh prompts for coordinate values for the placement of <u>each</u> vertex, allowing you to create a variety of shapes. A numbering scheme is used to label each vertex. The first vertex in the <u>first</u> row in the M direction is labeled *0,0*, the second in the first row is labeled *0,1*, and so on. The first vertex in the <u>second</u> row is labeled *1,0*, the second in that row is *1,1*, and so on.

Since each vertex can be given explicit coordinates, a surface can be created to define any shape. An example is shown in Figure 40-15. This command is useful for developing geographical topographies. The command syntax for this method of defining a *3Dmesh* is as follows:

Figure 40-15

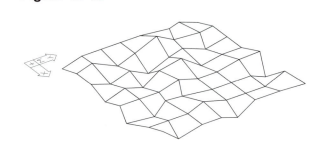

```
Command: 3dmesh
Enter size of mesh in M direction: (value)
Enter size of mesh in N direction: (value)
Specify location for vertex (0, 0): PICK or
(coordinates)
Specify location for vertex (0, 1): PICK or (coordinates)
Specify location for vertex (0, 2): PICK or (coordinates)
(continues sequence for all vertices in the row)
Specify location for vertex (1, 0): PICK or (coordinates)
Specify location for vertex (1, 1): PICK or (coordinates)
Specify location for vertex (1, 2): PICK or (coordinates)
(continues sequence for all vertices in the row)
Specify location for vertex (2, 0): PICK or (coordinates)
Specify location for vertex (2, 1): PICK or (coordinates)
Specify location for vertex (2, 2): PICK or (coordinates)
(continues sequence for all vertices in the row)
(sequence continues for all rows)
Command:
```

Obviously, using *3Dmesh* is very tedious. For complex shapes such as topographical maps, an AutoLISP program can be written to read coordinate data from an external file to generate the mesh.

PFACE

Pull-down Menu	COMMAND (TYPE)	ALIAS (TYPE)	Short-cut	Screen (side) Menu	Tablet Menu
...	*PFACE*	...	...	...	...

The *Pface* (polyface mesh) command can be used to create a polygon mesh of virtually any topology. *Pface* is the most versatile of any of the 3D mesh commands. It allows creation of any shape, any number of vertices, and even objects that are not physically connected, yet are considered the same object.

The *Pface* command, however, is designed primarily for use by applications that run with AutoCAD and for use by other software developers. *Pedit* cannot be used to edit *Pface*s.

The procedure for creation of geometry with *Pface* is somewhat like the previous procedure for *3Dmesh*. There are two basic steps to *Pface*. AutoCAD first prompts for the coordinate location for each vertex. Any number of vertices can be defined.

```
Command: pface
Specify location for vertex 1: PICK or (coordinates)
Specify location for vertex 2 or <define faces>: PICK or (coordinates)
(This sequences continues until pressing Enter.)
```

Secondly, AutoCAD prompts for the vertices to be assigned to each face. A face can be attached to any set of vertices.

```
Face 1, vertex 1:
Enter a vertex number or [Color/Layer]: (number) or (option)
Face 1, vertex 2:
Enter a vertex number or [Color/Layer]: (number) or (option)
```

Specifying a *Pface* mesh of any size can be tedious. The geometrically defined mesh commands (*Rulesurf, Tabsurf, Revsurf,* and *Edgesurf*) are more convenient to use for most applications.

CREATING GEOMETRICALLY DEFINED MESHES

Geometrically defined meshes are meshed surfaces that are created by "attaching" a surface to existing geometry. In other words, geometrically defined meshed surfaces require existing geometry to define them. *Edgesurf, Rulesurf, Tabsurf,* and *Revsurf* are the commands that create these meshes. The geometry used may differ for the application but always consists of either two or four objects (*Lines, Circles, Arcs,* or *Plines*).

Controlling the Mesh Density

Geometrically defined mesh commands (*Edgesurf, Rulesurf, Revsurf,* and *Tabsurf*) do not prompt the user for the number of vertices in the *M* and *N* direction. Instead, the number of vertices is determined by the settings of the *SURFTAB1* and *SURFTAB2* variables. The number of vertices includes the endpoints of the edge:

```
Command: surftab1
Enter new value for SURFTAB1 <6>: (value)
Command:
```

SURFTAB1 and *SURFTAB2* must be set before using *Edgesurf, Rulesurf, Revsurf,* and *Tabsurf*. These variables are not retroactive for previously created surface meshes.

The individual surfaces (1 by 1 meshes between vertices) created by geometrically defined meshes are composed of a series of straight edges connecting the vertices. These edges do not curve but can only change direction between vertices to approximate curved edges. The higher the settings of *SURFTAB1* and *SURFTAB2*, the more closely these straight edges match the defining curved edges.

Note that the defining edges of geometrically defined mesh <u>do not become part of the surface</u>; they remain separate objects. Therefore, you can construct surface models by utilizing existing wireframe model objects as the defining edges. If the wireframe model is on a separate layer, that layer can be *Frozen* after constructing the surfaces to reveal only the surfaces.

EDGESURF

Pull-down Menu	COMMAND (TYPE)	ALIAS (TYPE)	Short-cut	Screen (side) Menu	Tablet Menu
Draw Surfaces > Edge Surface	EDGESURF	…	…	DRAW 2 SURFACES *Edgsurf:*	R,8

An *Edgesurf* is a meshed surface generated between <u>four existing</u> edges. The four edges can be of any shape as long as they have connecting endpoints (no gaps, no overlaps). The four edges can be *Lines*, *Arcs*, or *Plines*. The four edges can be selected in any order. The surface that is generated between the edges is sometimes called a "Coon's surface patch." *Edgesurf* interpolates the edges and generates a smooth transitional mesh, or patch, between the four shapes.

The Command syntax is as follows:

```
Command: edgesurf
Current wire frame density:  SURFTAB1=6  SURFTAB2=6
Select object 1 for surface edge: PICK
Select object 2 for surface edge: PICK
Select object 3 for surface edge: PICK
Select object 4 for surface edge: PICK
Command:
```

Figure 40-16 ———————————

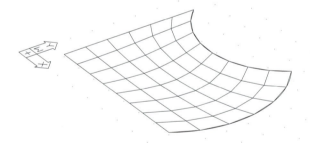

Figure 40-16 displays an *Edgesurf* generated from four planar edges. Settings for *SURFTAB1* and *SURFTAB2* are 6 and 8.

The relation of *SURFTAB1* and *SURFTAB2* for any *Edgesurf* are given here:

SURFTAB1	first edge picked
SURFTAB2	second edge picked

Figure 40-17 ———————————————

Edges used to define an *Edgesurf* may be <u>nonplanar</u>, resulting in a smooth, nonplanar surface, as displayed in Figure 40-17. Remember, the edges may be any *Pline* shape in 3D space. Use *Shademode* to enhance your visibility of nonplanar surfaces.

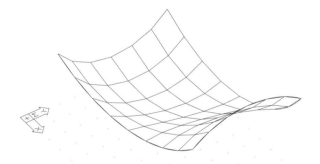

RULESURF

	COMMAND (TYPE)	ALIAS (TYPE)	Short-cut	Screen (side) Menu	Tablet Menu
Pull-down Menu					
Draw Surfaces > Ruled Surface	*RULESURF*	...	...	*DRAW 2 SURFACES Rulsurf:*	Q,8

A *Rulesurf* is a polygon meshed surface created between <u>two edges</u>. The edges can be *Lines*, *Arcs*, or *Plines*. The command syntax only asks for the two "defining curves." The two defining edges must <u>both</u> be open or closed.

Figure 40-18 displays two planar *Rulesurfs*.

Figure 40-18

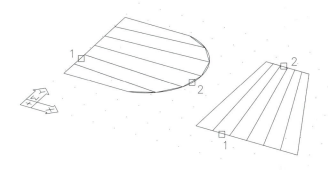

Figure 40-19 displays possibilities for creating *Rulesurfs* between two edges that are nonplanar.

The Command syntax for *Rulesurf* is as follows:

    Command: **rulesurf**
    Current wire frame density: SURFTAB1=6
    Select first defining curve: **PICK**
    Select second defining curve: **PICK**
    Command:

Figure 40-19

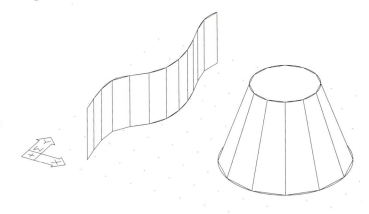

A cylinder can be created with *Rulesurf* by utilizing two *Circles* lying on different planes as the defining curves. A cone can be created by creating a *Rulesurf* between a *Circle* and a *Point* (Fig. 40-20), or a complex shape can be created using two identical closed *Pline* shapes (Fig. 40-20).

Figure 40-20

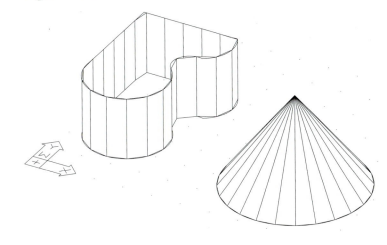

For open defining edges (not *Circles* or closed *Plines*), the *Rulesurf* is generated connecting the two <u>selected</u> ends of the defining edges. Figure 40-21 illustrates the two possibilities for generating a *Rulesurf* between the same two straight nonplanar *Lines* based on the endpoints selected.

Since the curve is stretched between only two edges, the number of vertices along the *defining curve* is determined only by the value previously specified for *SURFTAB1*.

Figure 40-21

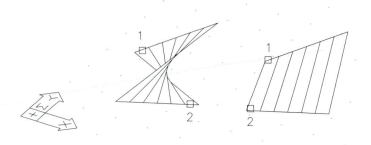

SURFTAB1	defining curve
SURFTAB2	not used

TABSURF

Pull-down Menu	COMMAND (TYPE)	ALIAS (TYPE)	Short-cut	Screen (side) Menu	Tablet Menu
Draw Surfaces > Tabulated Surface	TABSURF	...	...	DRAW 2 SURFACES Tabsurf:	P,8

A *Tabsurf* is generated by an existing *path curve* and a *direction vector*. The *curve path* (generatrix) is extruded in the direction of, and equal in length to, the *direction vector* (directrix) to create a swept surface. The path curve can be a *Line, Arc, Circle, 2D Polyline,* or *3D Polyline*. The direction vector can be a *Line* or an open 2D *Pline*. If a curved *Pline* is used, the surface is generated in a direction connecting the two end vertices of the *Pline*.

A *Circle* can be swept with *Tabsurf* to generate a cylinder (Fig. 40-22). The *Tabsurf* has only *N* direction; therefore, the current setting of *SURFTAB1* controls the number of vertices along the *path curve*.

The relation of *SURFTAB1* and *SURFTAB2* for *Tabsurf* are:

SURFTAB1	Path curve
SURFTAB2	not used

TIP The *Tabsurf* is generated in the direction <u>away from</u> the end of the direction vector that is selected. This is evident in both Figures 40-22 and 40-23.

The Command sequence is:

```
Command: tabsurf
Select object for path curve: PICK
Select object for direction vector: PICK
```

Figure 40-22

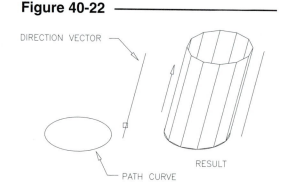

DIRECTION VECTOR

RESULT

PATH CURVE

Figure 40-23

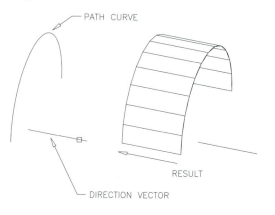

PATH CURVE

RESULT

DIRECTION VECTOR

REVSURF

	COMMAND (TYPE)	ALIAS (TYPE)	Short-cut	Screen (side) Menu	Tablet Menu
Pull-down Menu					
Draw Surfaces > Revolved Surface	*REVSURF*	...	...	DRAW 2 SURFACES *Revsurf:*	O,8

This command creates a surface of revolution by revolving an existing path curve (*object to revolve*) around an *axis*. The path curve can be a *Line, Arc, Circle*, 2D or 3D *Pline*. The *axis* can be a *Line* or open *Pline*. If a curved *Pline* is used as the *axis*, only the endpoint vertices are considered for the axis of revolution.

```
Command: revsurf
Current wire frame density: SURFTAB1=6  SURFTAB2=6
Select object to revolve: PICK
Select object that defines the axis of revolution: PICK
Specify start angle <0>: Enter or (value)
Specify included angle (+=ccw, -=cw) <360>: Enter or (value)
Command:
```

The structure of the command allows many variations. The *object to revolve* can be <u>open or closed</u>. The *axis* can be in any plane. The *start angle* and *included angle* allow for complete (closed) or partial (open) revolutions of the surface in any orientation. Figures 40-24 and 40-25 show possible *Revsurfs* created with path curves generated through 360 degrees.

The wine glass was created with an open path curve generated through 360 degrees (Fig. 40-24).

The number of vertices in the direction of revolution is controlled by the setting of *SURFTAB2*. The *SURFTAB1* setting controls the number of vertices along the length of the revolved object.

The relation of *SURFTAB1* and *SURFTAB2* for *Revsurf* is given here:

SURFTAB1   *Axis of revolution*
SURFTAB2   *Object to revolve*

A torus can be created by using a closed path curve generated through 360 degrees (Fig. 40-25). In this case, the surface begins at the path curve, is revolved all the way around the axis, and closes on itself.

When generating a *Revsurf* through <u>less than</u> 360 degrees, you must specify the included angle. Entering a positive angle specifies a counterclockwise direction of revolution, and a negative angle specification causes a clockwise revolution. As an alternative, you can pick different points on the *axis of revolution*. The end of the line PICKed represents the end nearest the origin for positive rotation using the right-hand rule.

Figure 40-24

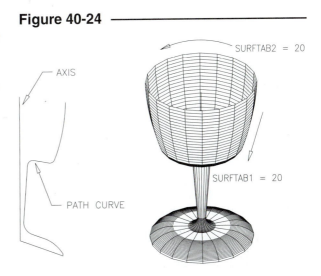

Figure 40-25

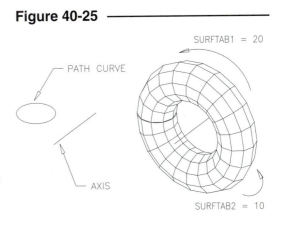

Figure 40-26 displays a *Revsurf* generated through 90 degrees. The end of the *axis of revolution* PICKed specifies the direction of positive rotation.

Figure 40-26

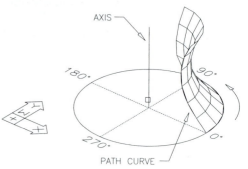

Figure 40-27 displays a *Revsurf* generated in a negative revolution direction, accomplished by specifying a negative angle or by PICKing a point specifying an inverted origin (of the right-hand rule) for rotation.

Figure 40-27

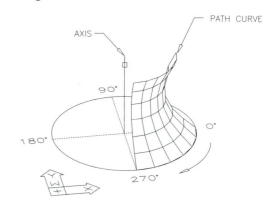

EDITING POLYGON MESHES

There are several ways that existing AutoCAD polygon meshed surfaces (*3Dmesh, Rulesurf, Tabsurf, Revsurf,* and *Edgesurf*) can be changed. The use of *Pedit*, Grips, and *Explode* each provide special capabilities.

PEDIT with Polygon Meshes

One of the options of *Pedit* is *Edit vertex*, which allows editing of individual vertices of a *Pline*. Vertex editing can also be accomplished with *3Dmeshes, Revsurfs, Edgesurfs, Tabsurfs,* and *Rulesurfs*. *Pedit* does <u>not</u> allow editing of *3Dfaces*. See Chapter 16 for information on using *Pedit* with 2D objects.

    Command: *pedit*
    Select polyline: **PICK** (select the surface)
    Enter an option [Edit vertex/Smooth surface/Desmooth/Mclose/Nclose/Undo]:

Each *Pedit* option for editing Polygon Meshes is explained and illustrated here.

Edit Vertex
Invoking this option causes the display of another set of options at the Command prompt, which allows you to locate the vertex that you wish to edit:

    Current vertex (0,0).
    Enter an option [Next/Previous/Left/Right/Up/Down/Move/REgen/eXit] <N>:

Pressing Enter locates the marker (X) at the next vertex. Selecting *Left/Right* or *Up/Down* controls the direction of the movement in the *N* and *M* directions (as specified by *SURFTAB1* and *SURFTAB2* when the surfaces were created; Fig. 40-28).

Figure 40-28

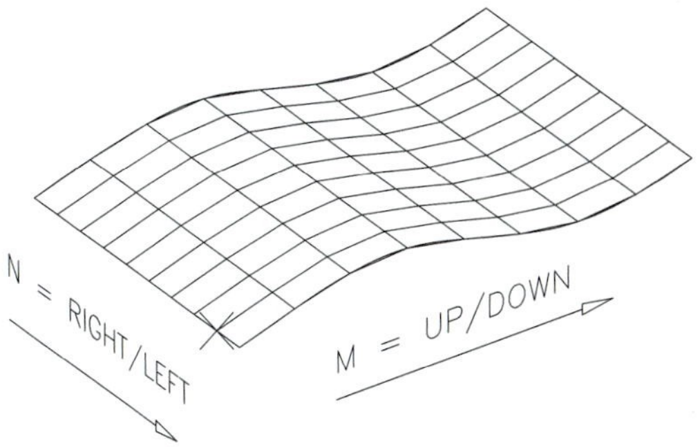

Once the vertex has been located, use the *Move* option to move the vertex to any location in 3D space (Fig. 40-29).

Figure 40-29

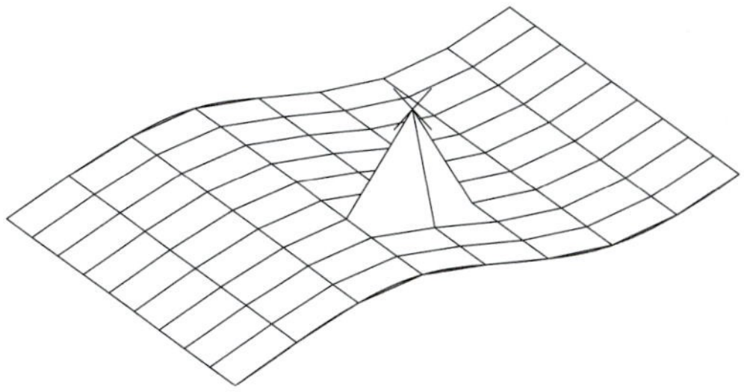

Smooth surface
Invoking this option causes the surface to be smoothed using the current setting of *SURFTYPE* (described later) (Fig. 40-30).

Figure 40-30

Desmooth
The *Desmooth* option reverses the effect of *Smooth*.

Mclose/Nclose
The *Mclose* and *Nclose* options cause the surface to close in either the *M* or *N* directions. The last and first set of vertices automatically connect. Figure 40-31 displays a half-cylindrical surface closed with *Nclose*.

Figure 40-31

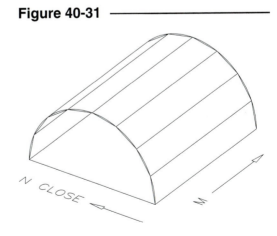

Undo
Undo reverses the last operation with *Pedit*.

eXit
This option is used to keep the editing changes, exit the *Pedit* command, and return to the Command prompt.

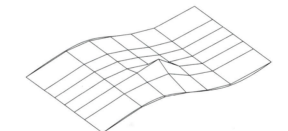

Variables Affecting Polygon Meshes

SURFTYPE

Pull-down Menu	COMMAND (TYPE)	ALIAS (TYPE)	Short-cut	Screen (side) Menu	Tablet Menu
...	SURFTYPE	...	...	...	...

This variable affects the degree of smoothing of a surface when using the *Smooth* option of *Pedit*. The Command prompt syntax is shown here:

Command: **surftype**
Enter new value for Surftype<6>: (**value**)

The values allowed are **5**, **6**, and **8**. The options are illustrated in the following figures.

Original polygon mesh (Fig. 40-32)

Figure 40-32

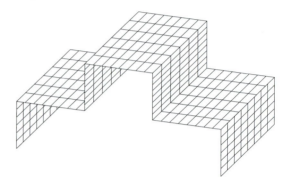

SURFTYPE=5
Quadratic B-spline surface (Fig. 40-33)

Figure 40-33

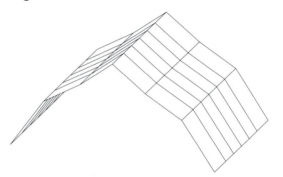

SURFTYPE=6
Cubic B-spline surface (Fig. 40-34)

Figure 40-34

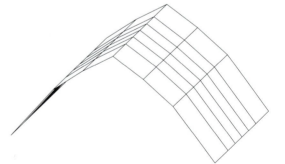

SURFTYPE=8
Bezier surface (Fig. 40-35)

Figure 40-35

NOTE: Changing *SURFTYPE* by Command line format sets the smoothing only for <u>new</u> polygon meshes. If you want to change an <u>existing</u> polygon mesh, use *Properties*, then select *Quadratic, Cubic*, or *Bezier* from the *Fit/Smooth* drop-down list.

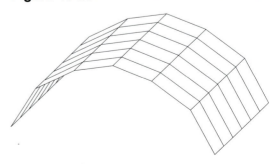

SPLFRAME

	COMMAND (TYPE)	ALIAS (TYPE)	Short-cut	Screen (side) Menu	Tablet Menu
Pull-down Menu					
...	*SPLFRAME*	...	...	...	...

The *SPLFRAME* (spline frame) variable causes the original polygon mesh to be displayed along with the smoothed version of the surface. The values allowed are **0** for off and **1** for on. *SPLFRAME* can also be used with 2D polylines when using the *Spline* option.

SPLFRAME is set to **1** in Figure 40-36 to display the original polygon mesh.

Figure 40-36 —————————

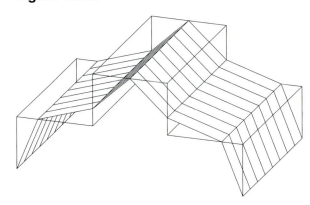

Editing Polygon Meshes with Grips

Grips can be used to edit the individual vertices of a polygon mesh (*3Dmesh, Edgesurf, Rulesurf, Tabsurf,* and *Revsurf*). Normally, the STRETCH option of Grip editing would be used to stretch (move) the location of a single vertex grip. The other Grip editing options (MOVE, SCALE, ROTATE, MIRROR) affect the entire polygon mesh. Selecting the mesh at the open Command prompt activates all of the grips (one for each vertex). Selecting one of the individual grips again makes that one **hot** and able to be relocated with the STRETCH option.

For example, a *3Dmesh* could be generated to define a surface with a rough texture, such as a chipped stone (Fig. 40-37).

Grip editing STRETCH option could be used to relocate individual vertices of the mesh. The result is a surface with a rough or bumpy surface (Fig. 40-38).

Although the concept is simple, some care should be taken to assure that the individual vertices are moved in the desired plane.

Keep in mind that grips in AutoCAD 2004 always appear <u>parallel to the current UCS XY plane</u> and operate in their respective planes. Therefore, you can control which plane the grips are STRETCHed by setting up the appropriate UCS (see previous Figure 40-9).

Figure 40-37 —————————

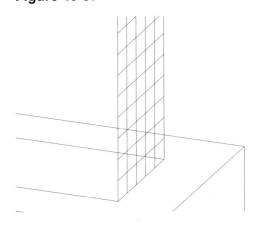

Figure 40-38 —————————

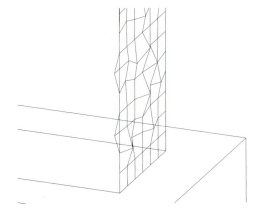

Editing Polygon Meshes with *Explode*

The *Explode* command can be used to allow editing 3Dmeshes, geometrically defined meshes (*Edgesurf, Revsurf,* etc.), 3D primitives (*3D Objects...*), and *Regions*.

When used with other AutoCAD objects, *Explode* will "break down" the object group into a lower level of objects; for example, a *Pline* can be broken into its individual segments with *Explode*. Likewise, *Explode* will "break down" a 3D polygon mesh, for example, into individual 1 by 1 meshes (in effect, *3Dfaces*). *Explode* will affect any polygon meshes (*3Dmesh, Edgesurf, Revsurf, Tabsurf, Rulesurf*). *Explode* converts a polygon mesh into its individual 1 by 1 meshes (*3Dfaces*) (Fig. 40-39).

Figure 40-39

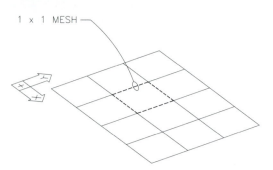

CONSTRUCTION TECHNIQUES FOR POLYGON MESHES AND *3DFACES*

Surfaces are created individually and combined to form a complete surface model. The surfaces can be created and moved into place in order to "assemble" the complete model, or surfaces can be created in place to form the finished model. Because polygon meshes require existing geometry to define the surfaces, the construction technique generally used is:

1. Construct a wireframe model (or partial wireframe) defining the edges of the surfaces on a separate layer.
2. Attach polygon mesh and other surfaces to the wireframe using another layer.

The creation of UCSs greatly assists in constructing 3D elements in 3D space during construction of the wireframe and surfaces. The sequence in the creation of a complete surface model is given here as a sample surface modeling strategy.

Surface Model Example

Consider the object in Figure 40-40. To create a surface model of this object, first create a partial wireframe.

Figure 40-40

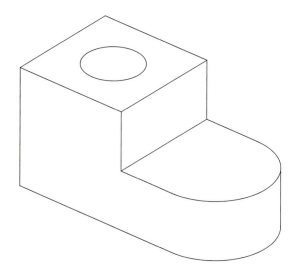

1. Create part of a wireframe model using *Line* and *Circle* elements. Create the wireframe on a separate layer. Begin the geometry at the origin. Use *Vpoint* to enable visualization of three dimensions (Fig. 40-41).

Figure 40-41

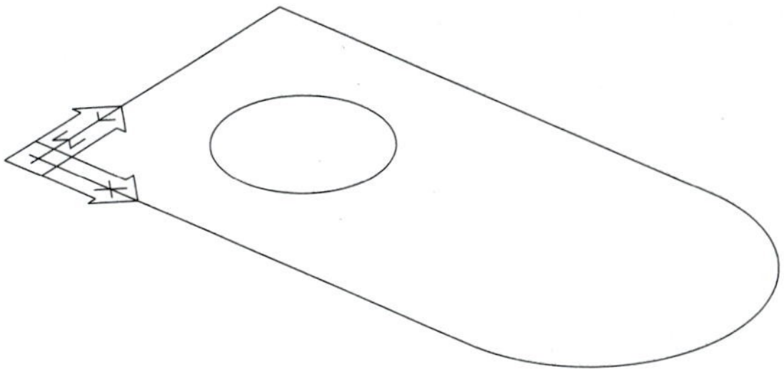

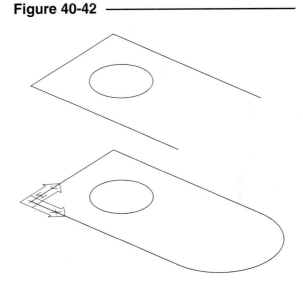

2. *Copy* the usable objects to the top plane by specifying a Z dimension at the "second point of displacement" prompt (Fig. 40-42).

Figure 40-42

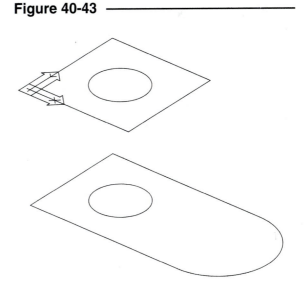

3. Create a UCS on this new plane using the *Origin* option. *OSNAP* the *Origin* to the existing geometry. Draw the connecting *Line* and *Trim* the extensions (Fig. 40-43).

Figure 40-43

4. *Copy* the *Arc* and two attached *Lines* specifying a Z dimension for the "second point of displacement." *Copy* the *Line* drawn last (on the top plane) specifying a negative Z dimension. *Trim* the extending *Line* ends. A new *UCS* is not necessary on this plane (Fig. 40-44).

This is not a complete wireframe, but it includes all the edges that are necessary to begin attaching surfaces.

Figure 40-44

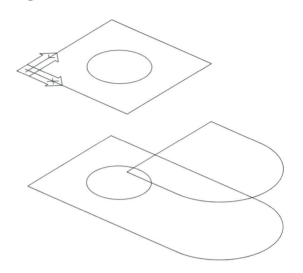

5. On a new layer, a 2-segment *3Dface* is created by PICKing the *Endpoints* of the wireframe objects in the order indicated. Remember that *3Dface* segments connect at the third and fourth points (Fig. 40-45). (*Hidden Shademode* is used in this figure to provide visibility of the *3Dface*.)

Figure 40-45

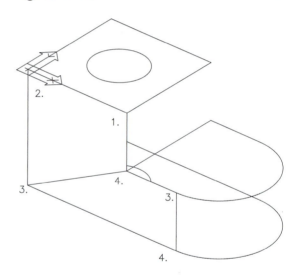

6. The *3Dface* is *Copied* to the opposite side of the model. A vertical *3Dface* is created between the top and the intermediate plane (Fig. 40-46). Another vertical *3Dface* is created on the back (not visible). (*Hidden Shademode* is used again in this figure. Notice how the *Circle* is treated.)

Figure 40-46

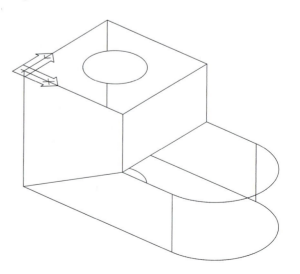

7. *Rulesurf* is used to create the rounded surface shown between the two *Arcs* (Fig. 40-47). The *SURFTAB1* and *SURFTAB2* variables were previously set to 12.

Figure 40-47

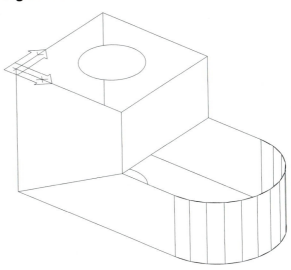

8. Next, a *Rulesurf* is used to create the cylindrical hole surface between the *Circles* on the top and bottom planes. The wireframe layer was *Frozen* before *Hidden Shademode* was used for this figure (Fig. 40-48).

Figure 40-48

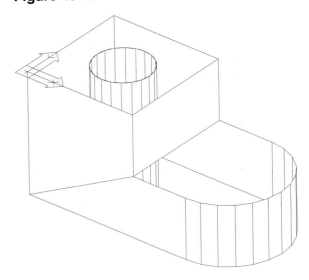

Figure 40-49

9. The surface connecting the *Arc* and three *Lines* is created by *Edgesurf* (Fig. 40-49).

10. The only remaining surfaces to create are the top and bottom. <u>There is no</u> polygon mesh, *3Dface*, or *3Dmesh* that can be generated to create a simple surface configuration such as that on top—a surface with a hole! It must be accomplished by creating several adjoining surfaces.

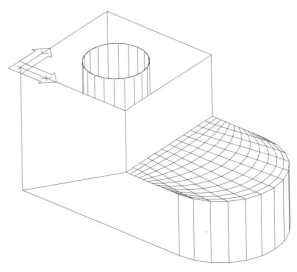

Several methods are possible for construction of a surface with a hole. (The top surface has been isolated here for simplicity.) Probably the simplest method is shown here (Fig. 40-50). First, the *Circle* must be <u>replaced</u> by two 180 degree *Arcs*. Next a *Rulesurf* is created between one *Arc* and one edge as shown.

Figure 40-50

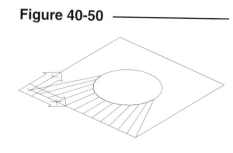

11. The top plane is completed by creating another *Rulesurf* and two triangular *3Dfaces* adjoining the *Rulesurfs* (Fig. 40-51). (Wireframe elements are <u>not</u> required for the *3Dface*.)

Figure 40-51

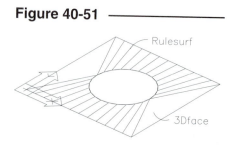

12. The surface model is shown here (Fig. 40-52). The bottom surface (not visible here) was created by *Copying* the top plane surfaces and the intermediate plane *Edgesurf*. The layer containing the wireframe elements has been *Frozen* and *Hidden Shademode* has been used to enhance visibility.

Figure 40-52

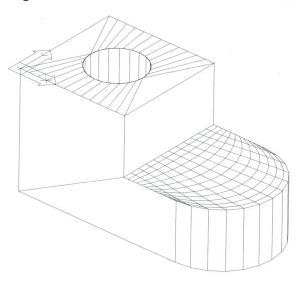

13. The mesh lines shown in Figure 40-52 are not visible when a *Render* is performed. This figure illustrates the completed model after using *Render*. If you use *Shademode*, the mesh lines are displayed with the *2D Wireframe, 3D Wireframe, Hidden, Flat+Edges*, and *Gouraud+Edges* options, but mesh lines are not displayed with the *Flat* and *Gouraud* options.

Figure 40-53

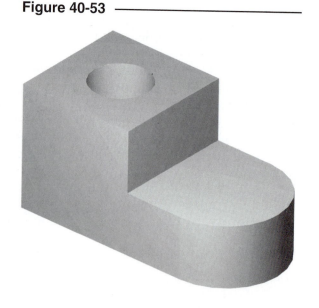

Limitations of Polygon Meshes and *3Dface*

From the previous example of the creation of a relatively simple 3D surface model, you can imagine how construction of some shapes with *3Dface* or polygon meshes can be somewhat involved. The particular characteristics of a surface that cause some difficulty in creation are the surfaces that embody many edges or that have holes, slots, or other "islands." Examine the surfaces shown in Figure 40-54. Although some of these shapes are fairly common, the construction techniques using *3Dfaces* and polygon meshes could be very involved.

3Dface and polygon meshes would <u>not</u> be an efficient construction technique to use in the creation of shapes such as these. Another AutoCAD feature, Region Modeling, can be used to easily create such shapes. Region models can be combined with *3Dfaces*, polygon meshes, and other surfacing techniques to create surface models.

Figure 40-54

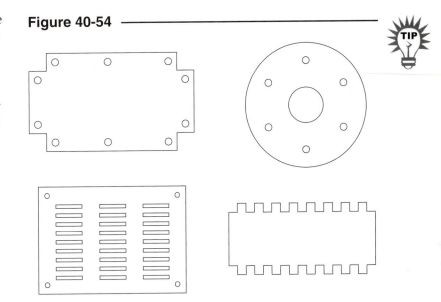

REGION MODELING APPLICATIONS

Region modeling is sometimes defined in the AutoCAD documentation as "2D solid modeling," <u>not</u> "surface modeling." This is because region modeling applies solid modeling techniques, namely, Boolean operations, to 2D closed objects. Practically, however, *Regions* can be considered surfaces.

A region is created by applying the *Region* command to <u>closed</u> 2D objects such as *Plines, Circles, Ellipses,* and *Polygons* (see Chapters 15 and 16). *Regions* are always <u>planar</u>. <u>Composite</u> regions can be constructed by combining multiple *Regions* using Boolean operations. AutoCAD allows you to utilize the Boolean operations *Union, Subtract,* and *Intersect* with *Regions* to create planar surfaces with holes and complex shapes. These surfaces can then be combined with other surfaces (regions, *3Dfaces*, or polygon meshes) on other planes to create complex surface models.

Region modeling solves the problem of creating relatively simple surfaces that are not easily accomplished with the polygon mesh commands such as *3Dmesh, Rulesurf, Revsurf, Tabsurf,* and *Edgesurf.* Region modeling makes creation of surfaces with islands (holes) especially easy.

For example, consider the creation of a rectangular surface with a circular hole. This problem is addressed in the previous Surface Model Example (see Figure 40-51). Using the polygon mesh commands, this relatively simple surface requires four *Lines,* two *Arcs* as a wireframe structure, then two *Rulesurfs* and two *3Dfaces* to complete the surface.

With the Region Modeler, the same surface can be created by using *Subtract* with a circular *Region* (create a *Circle*, then use *Region*) and a rectangular *Region* (draw a closed rectangle with *Line*, *Pline*, or *Rectangle,* then use *Region*; Fig. 40-55).

Figure 40-55

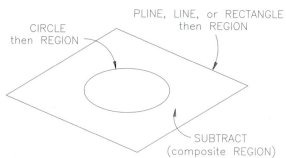

Consider the surfaces shown in Figure 40-54. Imagine the amount of work involved in creating the surfaces using commands such as *3Dface*, *3Dmesh, Rulesurf, Revsurf, Tabsurf,* and *Edgesurf.* Next, consider the work involved to create these shapes as regions. Surfaces such as these are very common for mechanical applications for surface models.

SURFACE MODELING PRIMITIVES

3D

Pull-down Menu	COMMAND (TYPE)	ALIAS (TYPE)	Short-cut	Screen (side) Menu	Tablet Menu
Draw *Surfaces >* *3D Surfaces...*	*3D or* *AI_BOX,* *AI_WEDGE, etc.*	...	...	*DRAW 2* *SURFACES* *3Dobjec:*	...

This command allows you to create 3D surface <u>primitives</u>. A primitive is a simple 3D geometric shape such as a *box, cone, dish, dome, mesh, pyramid, sphere, torus,* or *wedge* (the mesh is discussed earlier). These primitives are pre-created shapes composed of *3Dfaces* or polygon meshes. The primitives in the *3D Objects* image tiles represent surface models (Fig. 40-56).

Surface primitives can be created by selecting the image tiles in the *3D Objects* dialog box, by selecting from the *SURFACES* screen menu, or by typing the object name prefaced by "AI_"; for example "*AI_BOX*" summons the prompts for constructing a box.

Figure 40-56

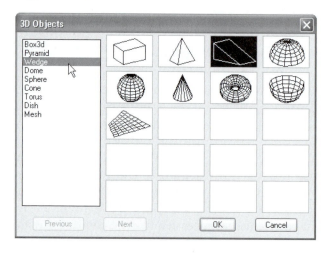

AutoCAD prompts you for the dimensions of the primitive. Depending on the type of primitive chosen, the prompts differ. The prompts for a *box* are given here:

```
Command: ai_box
Specify corner point of box: PICK or (coordinates)
Specify length of box: PICK or (value)
Specify width of box or [Cube]: PICK or (value)
Specify height of box: PICK or (value)
Specify rotation angle of box about the Z axis or [Reference]: PICK or (value)
Command:
```

These primitives are composed of AutoCAD surfaces. The 3D primitives (*3D Surfaces*) are provided to simplify the process of creating complex surface models. Generally, the primitives have to be edited and combined with other surfaces to build the complex model you need for a particular application. In this case, editing commands (*Pedit, Explode*, Grip editing, etc.) discussed in this chapter can be used along with other surfacing capabilities (*3Dmesh, 3Dface, Revsurf, Edgesurf, Region*, etc.). The editing methods differ depending on the type of surface. *Pedit, Explode*, and Grips can be used with any polygon mesh (*torus, sphere, dome, dish, cone*). *Pedit* <u>cannot</u> be used with *3Dfaces* (*box, wedge, pyramid*); however, *Explode* and Grip editing can be used. See "Editing Polygon Meshes" and "Editing *3Dfaces*" in this chapter.

For example, the *Box* can be *Exploded* into its individual *3Dfaces*. Figure 40-57 exhibits the box in its original form. Figure 40-58 illustrates the *box* after it has been *Exploded* and the top *3Dface* removed to reveal the inside. (*Hidden Shademode* has been performed in these figures to enhance visualization.)

Figure 40-57 —————————————— **Figure 40-58** ——————————————

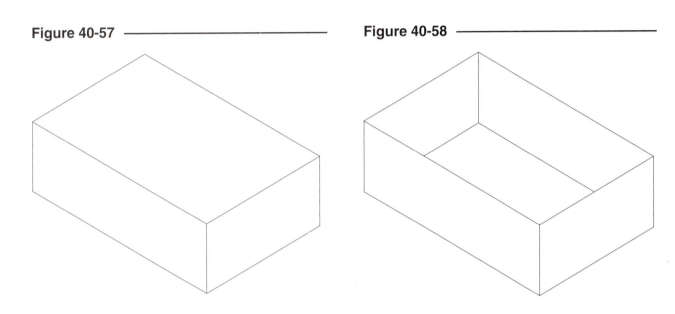

USING *THICKNESS* AND *ELEVATION*

THICKNESS

Pull-down Menu	COMMAND (TYPE)	ALIAS (TYPE)	Short-cut	Screen (side) Menu	Tablet Menu
Format Thickness	THICKNESS	TH	...	FORMAT Thicknes	V,3

ELEVATION

Pull-down Menu	COMMAND (TYPE)	ALIAS (TYPE)	Short-cut	Screen (side) Menu	Tablet Menu
...	ELEVATION	...	...	...	...

Thickness and *Elevation* were introduced with Version 2.1 (1985) as AutoCAD's first capability for creating 3D objects. *Thickness* and *Elevation* are still useful for creating simple surfaces. Most 2D objects (*Line, Circle, Arc, Point, Pline, Rectangle, Polygon*, etc.) can be assigned *Thickness* or *Elevation* property. Only newer (since Release 13) objects such as NURBS curves (*Spline, Ellipse*), *Xline, Ray,* and *Mlines* cannot possess *Thickness* but <u>can</u> possess *Elevation*. *Thickness* and *Elevation* are usually assigned before using a 2D command; however, they can also be changed later using *Change* or *Properties*.

Thickness is a value representing a <u>Z dimension</u> for a 2D object. The new object is created using the typical draw command (*Line, Circle,* or *Arc,* etc.), but the resulting object has a uniform Z dimension as assigned. For example, a two unit high cylinder can be created by setting *Thickness* to 2 and then using the *Circle* command (Fig. 40-59).

Figure 40-59

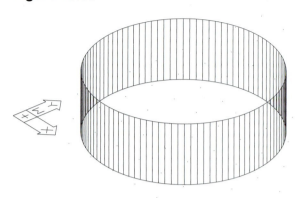

Thickness is <u>always perpendicular</u> to the XY plane of the object. Until later versions of AutoCAD, this posed a major limitation with the use of *Thickness*. This limited cylinders, for example, to having a vertical orientation <u>only</u>; no cylinders could be created with a horizontal orientation. The term "2 ½-D" was used to describe this capability of CAD packages during this phase of development.

Today, with the ability to create User Coordinate Systems (UCSs), *Thickness* can be used to create surfaces in any orientation. Therefore, the effectiveness of *Thickness* and *Elevation* commands is increased. Figure 40-60 displays a cylinder oriented horizontally, achieved by creating a UCS having a vertical XY plane and a horizontal Z dimension. UCSs can be created to achieve any orientation for using *Thickness*.

Figure 40-60

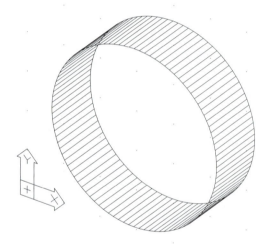

With a *Thickness* set to a value greater than **0**, commands normally used to create 2D objects can be used to create 3D surfaces having a uniform Z dimension perpendicular to the current XY plane. Figure 40-61 shows a *Line, Arc,* and *Point* possessing *Thickness* property. Note that a *Point* object with *Thickness* creates a line perpendicular to the current XY plane.

Figure 40-61

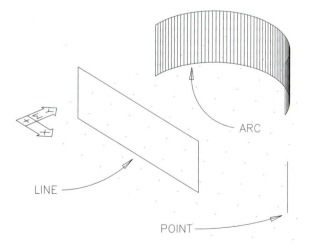

ARC

LINE

POINT

The ability to create *Plines* with *Thickness* offers many possibilities for developing complex surface models (Fig. 40-62). (Keep in mind that 2D shapes created with *Line* and *Arc* objects can be converted to *Plines* with *Pedit*.)

Figure 40-62

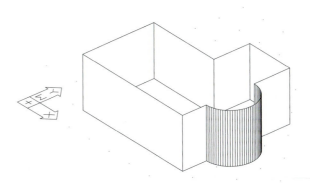

Using the *Shademode* command with surfaces created with *Thickness* enhances your visibility. Figures 40-62 and 40-63 show examples of *Hidden Shademode*. Notice that *Circles* created with *Thickness* have a "top." Other closed shapes created with *Line* or *Pline* do <u>not</u> have a "top."

Assigning a value for *Thickness* causes all subsequently created objects to have the assigned *Thickness* property. *Thickness* must be set back to **0** to create normal 2D objects.

Figure 40-63

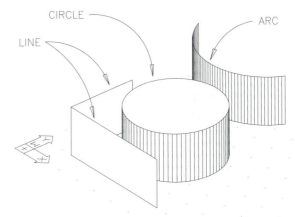

Elevation controls the Z dimension of the "base plane" for subsequently created objects. Normally, unless an explicit Z value is given, 2D objects (with or without *Thickness*) are created on the XY plane or at an *Elevation* of **0**. Changing the value of the *Elevation* changes the "base plane," or the elevation, of the construction plane for newly created objects.

Figure 40-64 displays the same objects as in the previous figure except the *Circle* has an *Elevation* of 2 (level with the top edges of the other objects). It may be necessary to use *Shademode* and *3Dorbit* to "see" the relationship of objects with different *Elevations*.

Like *Thickness, Elevation* should be set back to 0 to create geometry normally on the XY plane.

Figure 40-64

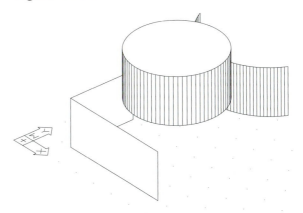

Changing *Thickness* and *Elevation*

Because *Thickness* and *Elevation* are properties that 2D objects may possess (like *Color* and *Linetype*), they can be changed or assigned <u>retroactively</u>. *Elevation* can only be changed retroactively using the *Change* command with the *Properties* option (see Chapter 16). *Thickness* can be changed using the *Properties* palette.

Figure 40-65

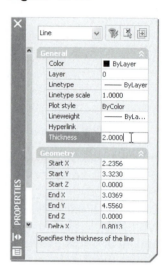

PROPERTIES

Using the *Properties* command (see Chapter 16) produces a dialog box providing the ability to change the *Thickness* of <u>existing</u> objects (Fig. 40-65).

As an alternative to setting *Thickness* before creating surfaces, the 2D geometry can first be created and then changed to 3D surfaces by using this method to retroactively set *Thickness*.

Figure 40-66 displays 2D *Pline* shapes. The *Properties* palette can be used to change the *Thickness* of the shape to create a 3D surface model.

Figure 40-66

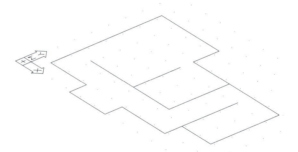

Figure 40-67 shows the same *Pline* shapes as in the previous figure after the *Thickness* property was changed to generate the 3D surface model.

If you want to change an object's *Elevation*, the *Change* command can be used. The Command syntax is as follows:

Figure 40-67

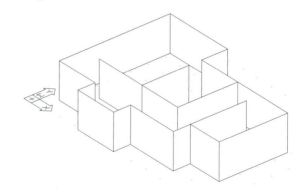

```
Command: change
Select objects: PICK
Select objects: Enter
Specify change point or [Properties]: p
Enter property to change
[Color/Elev/LAyer/LType/ltScale/LWeight/Thickness]: e
Specify new elevation <0.0000>: (value)
Enter property to change [Color/Elev/LAyer/LType/ltScale/LWeight/Thickness]: Enter
Command:
```

CHAPTER EXERCISES

1–3. Create a surface model of each of the objects in
Figures 40-68 through 40-71. These are the same
objects you worked with in Chapter 37,
Wireframe Modeling. You can use each of the
Chapter 37 Exercises wireframes (**WFEX1**,
WFEX2, and **WFEX3**) as a "skeleton" on which to
attach surfaces. Make sure you create a *New
Layer* for the surfaces. Use each dimension
marker in the figure to equal one unit in
AutoCAD. Locate the lower-left corner of the
model at coordinate **0,0,0**. Assign the names
SURFEX1, **SURFEX2**, and **SURFEX3**. Make a
plot of each model with hidden lines removed.

Figure 40-68 ————————————

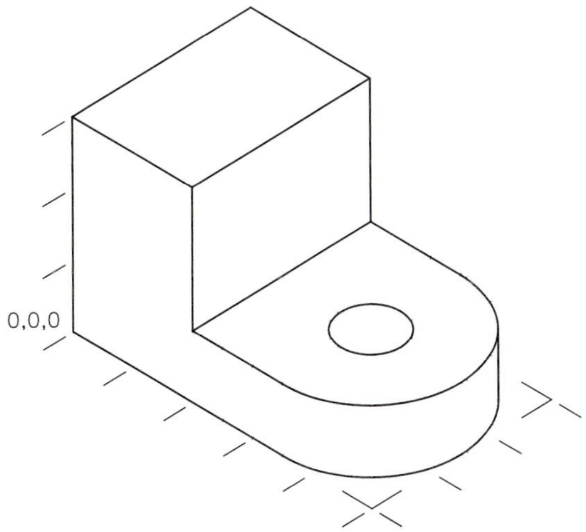

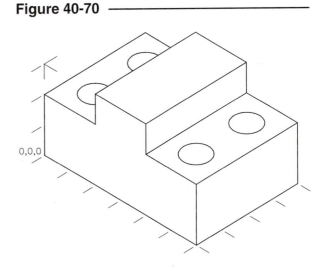

Figure 40-69 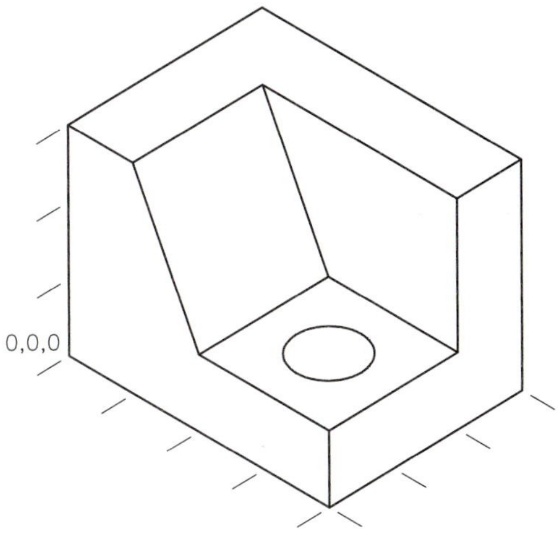 ————————————   **Figure 40-70** ————————————

Figure 40-71 ————————————

4. *Open* the **WFEX4** drawing that you created in
Chapter 37 Exercises. *SaveAs* **SURFEX4**. For
this exercise, create a surface model using a
combination of regions and polygon meshes.
Use the existing wireframe geometry.
Convert the *Lines, Arcs,* and *Circles* representing planar surfaces to closed *Plines*, then to
Regions. Use polygon meshes for the curved
surfaces. *Save* the drawing. Generate a
descriptive viewpoint with *Vpoint* or
3Dorbit. Activate a *Layout* and insert a titleblock and border. Create one *Vport* to
display the model. *Plot* the drawing with
hidden lines removed.

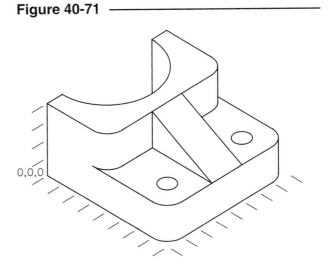

5. Create a surface model of the bar guide shown in Figure 40-72. You may want to use the **BGUID-WF** drawing that you created in Chapter 37 Exercises as a wireframe foundation. Make a *New Layer* for the surface model. Generate a descriptive viewpoint with *Vpoint* or *3Dorbit*. Activate a *Layout* and insert a titleblock and border. Create one *Vport* to display the model. *Plot* the model with hidden lines removed. *Save* the surface model as **BGD-SURF**.

Figure 40-72

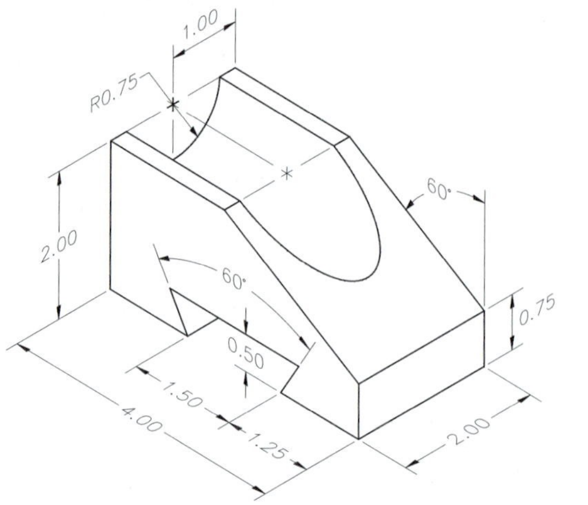

6. Create a surface model of the saddle in Figure 40-73. There is no previously created wireframe model, so you must begin a *New* drawing or use a *Template*. The fillets and rounds should be created using *Rulesurf* or *Tabsurf*. *Revsurf* must be used to create the complex fillets and rounds at the corners. Generate a descriptive viewpoint with *Vpoint* or *3Dorbit*. Activate a *Layout* and insert a titleblock and border. Create one *Vport* to display the model. Plot the drawing with hidden lines removed. *Save* the drawing as **SDL-SURF**.

Figure 40-73

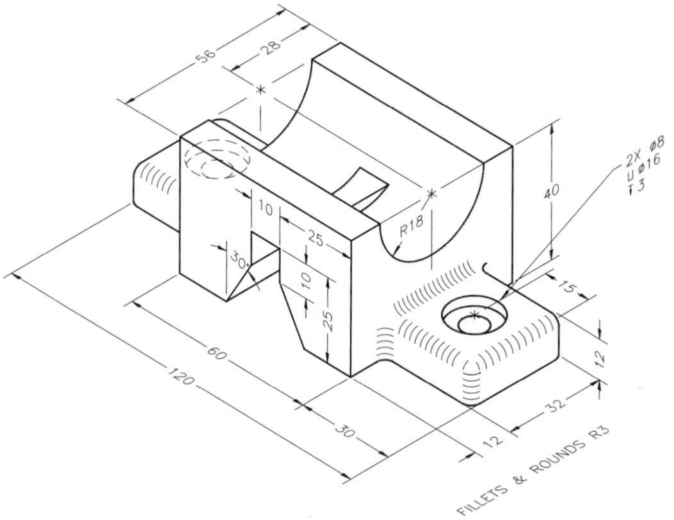

7. Begin a *New* drawing (or use a *Template*) for the pulley shown in Figure 40-74. This is the same object that you worked on in Chapter 38 Exercises. As a reminder, the vertical dimensions are diameters. Use *Revsurf* to create the curved (cylindrical) surfaces. Use *Regions* to create the vertical circular faces that would appear on the front and back (the surfaces containing the 4 counterbored holes or the keyway). Generate a descriptive viewpoint with *Vpoint* or *3Dorbit*. Activate a *Layout* and insert a titleblock and border. Create one *Vport* to display the model. *Save* the drawing as **PUL-SURF**. *Plot* the model with the *Hideplot* option.

Figure 40-74

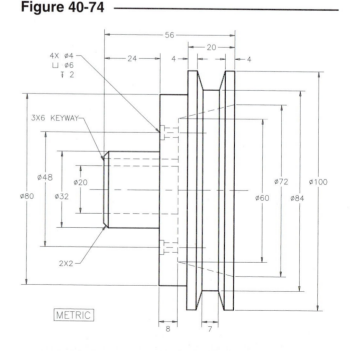

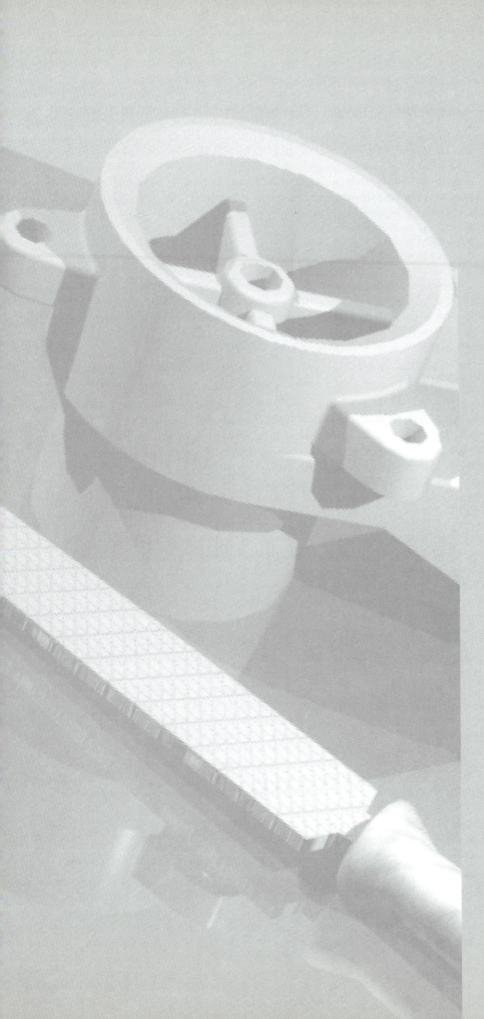

RENDERING

Chapter Objectives

After completing this chapter you should:

1. know the typical steps for rendering a surface or solid model using *Render*;

2. be able to create a variety of backgrounds for a rendering with *Fog* and *Background*;

3. be able to select and insert landscape images such as trees and shrubs;

4. understand and be able to create and set parameters for the four types of lighting: *Ambient Light*, *Point Light*, *Distant Light*, and *Spotlight*;

5. be able to select *Materials* from the *Materials Library*, *Import* them into the drawing, and *Attach* them to objects, colors, or layers;

6. be able to create *New* and *Modify* existing material properties of *Color*, *Ambient*, *Reflection*, *Roughness*, and other attributes;

7. be able to map raster images to objects and adjust the image orientation with *Setuv*;

8. be able to specify parameters and create a rendering using the *Render* command;

9. be able to use *Saveimg* to save and *Replay* to replay rendered images.

CONCEPTS

AutoCAD provides the ability to enhance the visualization of a <u>surface</u> or a <u>solid</u> model with the *Hide, Shade, Shademode,* and *Render* commands. Because the default representation for surface and solid models is wireframe representation, it is difficult for the viewer to sense the three-dimensional properties of the model in this mode. *Hide, Shade, Shademode,* and *Render* change the appearance of a model from wireframe to "solid." *Hide* and *Hidden Shademode* calculate which surfaces obscure others and displays only the visible ones, and *Shade* and *Shademode* add the object color to the model. *Render* adds another level of realism to surface and solid models by giving you control of backgrounds, lights, colors, materials, and other effects.

Render involves the use of *Light* sources and the application of *Materials* for the 3D model. You control the type, intensity, color, shadows, and positioning of lights in 3D space. Render also provides a variety of materials for attachment to objects, colors, or layers and allows you to adjust the properties of the material. With Render you can add and adjust *Background, Fog, Landscape* images, and images mapped onto surfaces of objects.

Render is intended to be used as a visualization tool for 3D models. Visualization of a design before it has been constructed or manufactured can be of significant value. Presentation of a model with light sources and materials can reveal aspects of a design that may otherwise be possible only after construction or manufacturing. For example, a rendered architectural setting may give the designer and client previews of color, lighting, or spatial relationships. Or a rendered model may allow the engineer to preview the functional relationship of mechanical parts in a manner otherwise impossible with wireframe representation. Thus, this "realistic" presentation, made possible by *Render,* can shorten the design feedback cycle and improve communication among designers, engineers, architects, contractors, and clients.

With AutoCAD Render, you can render to a viewport, to a separate rendering window, or to a file. The *Render* command uses the *Lights* and *Materials* you assign and calculates the display based on the parameters that you specify. If no *Lights* or *Materials* are created for the drawing, a default light and material are used to render the current viewport. If you have created several *Lights, Materials,* and *Views,* you can use the *Scene* command to specify a particular named *View* and a combination of *Lights.*

Since AutoCAD Release 14, Render has been enhanced to provide "photo-realistic" rendering capabilities. The ability to calculate shadows and reflection of color and light from one surface to another supplies this "photo-realistic" quality.

The rendered screen image created with the *Render* command can be saved as a TIFF, TGA, and/or a BMP file and replayed at a later time. The *Render* and *Rendering Preferences* dialog boxes provide control for format and quality of the rendered image.

In AutoCAD 2004 you can print a rendered image. Do this by using the *Shade Plot* options in the *Plot* dialog box or by using the *Shadeplot* options for a viewport. See "Saving, Printing, and Replaying Renderings."

In AutoCAD 2002, *Shademode* was enhanced to display any lights, materials, textures, and transparency applied in the drawing. Lights, materials, textures, and transparency can be applied to a shaded image by first using the rendering features in AutoCAD, then using the *Options* dialog box to configure your graphics display device to recognize these rendered features for *Shademode.* Rendered objects in *Shademode* cannot display shadows, 3D textures, bump maps, opacity maps, refraction, reflection, background, and fog. See "Setting Rendering Options for *Shademode*" near the end of the chapter.

Rendering commands can be accessed by typing or selecting from the *View* screen pull-down menu (Fig. 41-1) and the *VIEW2* screen menu.

Figure 41-1

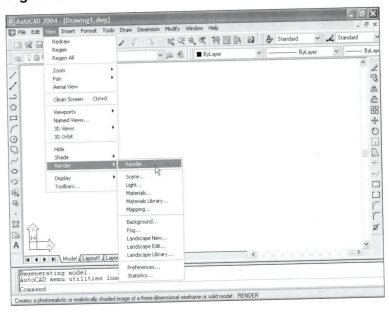

A *Render* toolbar (Fig. 41-2) is available by accessing the *Toolbar* dialog box. Optionally, several rendering commands can be selected from the digitizing menu.

Figure 41-2

Creating a Default Rendering

Creating a realistic rendering often takes more time that creating the 3D model. Although many parameters, adjustments, and steps can be used to create a realistic rendering (discussed later), you can use the *Render* command to create a rendering without specifying any background, lights, or materials.

To create a rendering, you must first create a 3D model, then use *Vpoint* or *3Dorbit* (by any method) to generate a representative view of the function or feature of the model you want to display. For example, Figure 41-3 displays the support bracket from the desired viewpoint in wireframe representation.

Figure 41-3

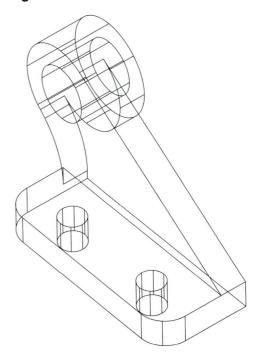

Next use the *Render* command which produces the *Render* dialog box (Fig. 41-4). Using all the defaults, press the *Render* button to generate a default rendering. AutoCAD creates the rendering using the current background color, the current object color, and a virtual "over-the-shoulder" *Distant* light source (Fig. 41-5).

Figure 41-4 ────────────────

Figure 41-5 ────────────────

Typical Steps for Using Render

Although a rendering of a 3D surface or solid model can be created easily using default settings, the following steps are normally taken to create a rendering with the desired effect:

1. You must first create or open an existing 3D model. Both surface and solid models can be rendered.

2. Use *Vpoint* or *3Dorbit* to specify the desired viewing position. Perspective generations (created with *3Ddistance* or *Dview Distance*) are allowed. If you want to create several renderings from different viewpoints, use the *Save* option of *View* to save each display to a named view so it can be *Restored* at a later time.

3. Set the *Background* for the object to be rendered. You can use a solid color or gradient color as the background. Optionally, specify an image (.BMP, .TIF, or other file) as the background to merge with the objects to be rendered. You can select from supplied *Landscape* images if needed and even adjust a *Fog* appearance.

4. Use the *Lights* dialog box series to create one or more lights. Parameters such as light type, intensity, color, and location (position in space) can be specified. A light source representing the Sun can be specified by geographic location and time of day. You can specify if the *Lights* cast shadows.

5. Use the *Materials* dialog box series to select the desired material(s) for the object(s). The material properties such as specification of color, reflective qualities, and other attributes can be adjusted. You can *Preview* the material as you adjust the parameters.

6. If you have created several *Views* and *Lights,* use the *Scenes* dialog box to select the specific combination of *Lights* and the desired *View* for the rendered scene.

7. Specify the desired format and quality for the rendering by the *Render* or *Rendering Preferences* dialog box. Several options are available for the rendering.

8. Use the *Render* command to create the rendered image on the screen.

9. Because rendering normally involves a process of continual adjustment to *Light* and *Material* parameters, any or all of steps 2 through 8 can be repeated to make the desired adjustments. Don't expect to achieve exactly what you want the first time.

10. When the final adjustments have been made, use the *Render* or *Rendering Preferences* dialog box to specify parameters such as *Photo Raytrace, Shadows,* and other rendering quality settings to create a realistic image.

11. Use the *Render* command again to create the rendered display or file. Usually, you should render to the screen until you have just the image you want; then save to a file if desired.

12. If you rendered to the screen, you can then save the rendered image to a .BMP, .TIF, or .TGA file using the *Save Image* dialog box, or save the image as a .BMP file from the rendering window. Alternately, you can write the rendering directly to a file without generating the render to the display. You <u>cannot</u> use the *Tifout, Pngout,* or *Jpgout* commands to save a rendered image.

Color Systems

There are several methods for choosing and assigning colors in the Render for such applications as background colors, material colors, and light colors. Because these systems appear throughout the many Render dialog boxes, these concepts are presented here.

Color for a specific application (background, material, or light) can be assigned by one or more of the systems described next. These systems appear in several dialog boxes. In each case, current color adjustments are displayed in a square color box or *Preview* image tile appearing in the dialog box.

RGB

The *RGB* system determines the color by the amount of *Red, Green,* and *Blue* components (Fig. 41-6). Values for each can be entered into the edit boxes or changed using the slider bars. View the "color box" to see the changes in real time or press *Preview* to see the image tile. Pure red, green, or blue hues can be achieved by moving <u>only</u> that color to the right, or entering 1.0 in the edit box for that value and 0 for the other two.

Figure 41-6

Moving all of the *Red, Green,* and *Blue* color bars to the left produces black, and moving all to the right produces white (although gray may appear in the color box).

HLS

The slider bars adjust the *Hue, Lightness,* and *Saturation* visible in the small boxes (Fig. 41-7). The mix of *HLS* appears in the "color box."

Figure 41-7

Hue	Controls the color (red, blue, yellow, etc.).
Lightness	Controls the value (white to black). White is all the way to the right (although the color box may appear gray), or a value of 1.0, and black is to the left, or 0.
Saturation	Controls the purity of the color (mix of *Hue* and *Lightness*). Pure color is to the right, or a value of 1.0, and pure gray scale is to the left, or 0.

A pure hue is achieved by a *Lightness* of .5 and *Saturation* of 1.

The *Select Color* option produces the *Select Color* dialog box. The three tabs are described below. Also see Chapter 11, Layers and Object Properties, for more information on these options.

True Color

You can pick the desired color from the spectrum area and use the slider to change the white-black value or enter values in the edit boxes. Select from either the *HLS* (Hue, Lightness, Saturation) model (Fig. 41-8) or *RGB* (Red, Green, Blue) model.

Color Books

Use this section (not shown) to select from industry-standard colors from the Pantone or RAL color systems.

Index Color

Selecting this option invokes the 256-color ACI (AutoCAD Color Index) palette (Fig. 41-9). Select the desired color from the palette or enter the name (for colors 1-16) or number in the edit box.

Photorealism, Raytracing, and Shadows

AutoCAD offers enhancements that allow Render to create images that are very photo-realistic: raytracing and the ability to calculate shadows. These features increase the scene's realism but can increase rendering time.

Photo Real

This option of *Render* is called a "scanline renderer" because it "paints" one horizontal row of pixels at a time. It can display transparent materials and bitmapped images. Bitmapped images are images from an external file that can be planar (as for a background) or mapped to geometry such as a sphere, box, or other shape. *Photo Real* can also generate volumetric and mapped shadows (see explanation next).

Photo Raytrace

Raytracing is the process of calculating or following the paths of light rays as they are generated from a source and cast shadows or bounce off objects and illuminate other objects. Raytracing is needed to calculate reflections of light and color.

Volumetric Shadows

This type of shadow is found by calculating the volume of space cast by the shadow of an object, then generating a shadow based on this volume. These shadows have a sharp edge between the shadow and the illuminated area. Volumetric shadows are sensitive to transparent and translucent objects and calculate a shadow based on the color (light to dark) of the object.

Shadow Maps

Shadow mapping allows you to adjust the "fuzziness" or softness of the edge of a shadow (as opposed to a hard or sharp edge). You can also adjust the size or accuracy of the shadow map that is generated—the greater the size (in pixels), the more accurate. Shadow maps do not display the color cast by translucent objects.

Raytraced Shadows

Raytraced shadows are generated by tracing the path of light rays generated by a light source. These shadows have hard edges but can transmit color when passing through translucent objects.

Figure 41-8

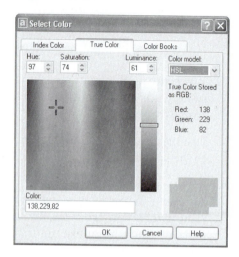

Figure 41-9

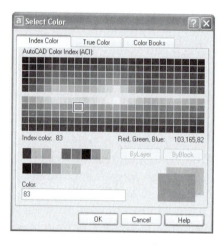

Rendering Time
Although Render is relatively fast for simple rendering, shadows always increase rendering time. Shadow maps are particularly time consuming. Consider turning *Shadows* on only after setting background effects and adjusting lights.

Now that you know the concepts and general procedure for selecting colors and creating a rendered image, background, lighting, and material concepts and commands are presented next. Examples of the dialog boxes used to achieve the desired effects are included in the following information.

CREATING A BACKGROUND

Once the 3D model is created, you can use the *Background* command to specify the background you want to appear in the rendering. There are five choices for the background:

1. the current *AutoCAD Background* (in the Drawing Editor),
2. a *Solid* color background,
3. a *Gradient* color blended from two or three color bands,
4. an *Image* in the form of a .TIF, .GIF, .JPG, .BMP, .TGA, .PNG, or .PCX file,
5. a previous rendering to *Merge* with the current rendering.

When you specify a background and generate a rendering, the object(s) in the rendering is automatically displayed "in front of" the background color or scene.

BACKGROUND						
	Pull-down Menu	**COMMAND (TYPE)**	**ALIAS (TYPE)**	**Short-cut**	**Screen (side) Menu**	**Tablet Menu**
	View Render > Background...	BACKGROUND	...	...	*VIEW 2 Backgrnd*	Q,2

Use the *Background* command to set up and adjust the background for the rendering. The background consists of colors or an image that appear behind the rendered geometry. Consider the background as a 2D plane behind the geometry always oriented normal to the viewing angle.

Background produces the *Background* dialog box (Fig. 41-10). At the top of the dialog box are four options: *Solid, Gradient, Image,* and *Merge.* The *AutoCAD Background* option is located on the middle left. The *HLS, RGB,* or *Select Color* dialog box (in the central area) can be used to set background colors.

Figure 41-10

AutoCAD Background
Select this option to use the current color in the Drawing Editor as the render background. This option is accessible only if the *Solid* radio button is pressed.

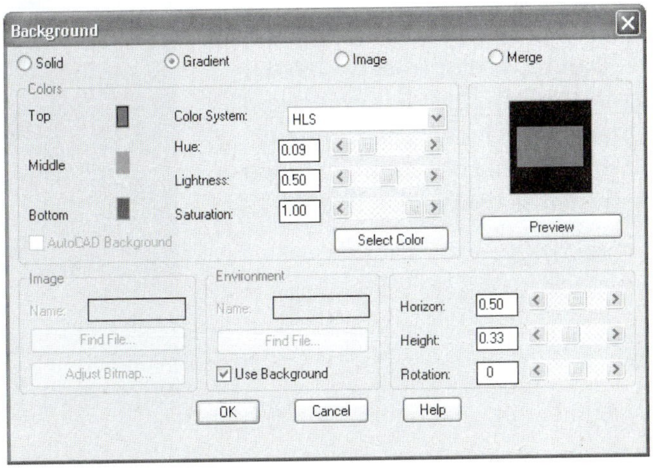

Solid

Select *Solid* and remove the check from *AutoCAD Background* to create a specific color background. This action enables the *RGB* and *HLS* color systems. The selected color appears in the *Top* color box in "real time" as you make adjustments. Press *Preview* to see a large image of the solid color.

Gradient

This background type produces two or three colors that blend across the background (Fig. 41-11). Select the *Top, Middle,* and *Bottom* color boxes to set and view each color as you change it with one of the three available color systems.

When *Gradient* is selected, the *Horizon, Height,* and *Rotation* adjustments are enabled.

Horizon

This setting determines where the *Middle* color is located in the mix. A value of 50 puts the *Middle* color in the middle, a lower value places the *Middle* color near the bottom (when *Rotation* is 0), and a higher value moves the *Middle* color near the top (when *Rotation* is 0).

Height

Use this adjustment to specify the width of the band of *Middle* color. The greater the value, the wider the band; the smaller the value, the narrower the band. To use only two gradient colors (*Top* and *Bottom*), set *Height* to 0.

Rotation

You can change the orientation of the bands of color from horizontal to any other angle. For example, enter 45 to achieve a background with bands of color running diagonally across the display (Fig. 41-12).

Figure 41-11

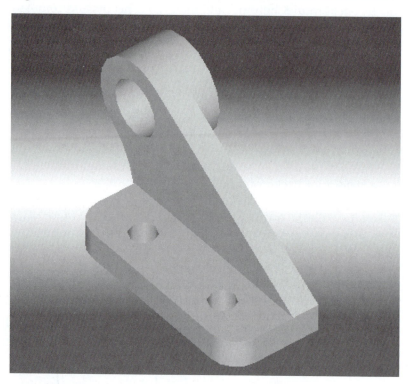

Figure 41-12

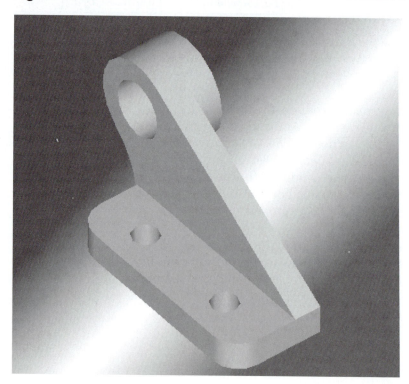

Image

Selecting the *Image* button enables the *Image* section of the *Background* dialog box (Fig. 41-13) and disables the other sections. With this option you can insert an image to use as the background for your rendering. Any .TIF, .GIF, .JPG, .BMP, .TGA, .PNG, or .PCX file can be used as the image.

Figure 41-13

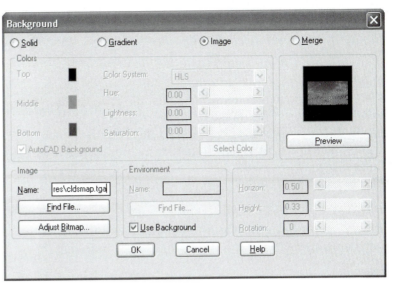

Several TGA files are supplied with AutoCAD and are located in the Textures folder. To find the location of this folder, use *Options*, *Files* tab, and expand *Texture Maps Search Path*. You may also find .BMP and other image files in some of the Windows or Winnt folders or folders of other applications.

Figure 41-14

Environment

You can create additional reflection and refraction effects on the rendered objects by defining the *Environment*. This option causes the rendered objects to reflect or mirror the background image. For *Environment* to have an effect, *Materials* selected for the rendered objects must be reflective and the *Rendering Type* must be *Photo Real* or *Photo Raytrace*. You can use the selected *Solid, Gradient, Image,* or *Merge* background (check *Use Background*), or you can select a specific image such as a .TIF, .GIF, .JPG, .BMP, .TGA, .PNG, or .PCX file (remove the check from *Use Background*, then enter a *Name* or pick *Find File...*).

Adjust Bitmap...

Select *Adjust Bitmap...* to produce the *Adjust Background Bitmap Placement* dialog box (Fig. 41-15). When an image is used for the background, this dialog box provides numerous tools for you to adjust the placement and proportions of the image.

Fit to Screen

Select this option to automatically adjust the aspect ratio (width to height) of the image to fit the proportion of the screen. If *Use Image Aspect Ratio* is <u>also</u> checked, the image's larger dimension is fit to match the screen; therefore, the other dimension may not fill the screen. *Fit to Screen* disables all other options.

Use Image Aspect Ratio

This option prevents the width-to-height ratio of the image from being altered. You can still change *Offset* and *Scale*.

Figure 41-15

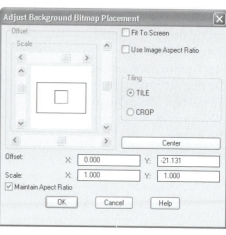

Offset and Scale

These values can be changed only when *Fit to Screen* is not checked. Use the slider bars or the edit boxes to change values. The red rectangle in the image tile represents the screen (render window) and the two-colored (black and magenta) rectangle represents the selected background image. Use *Offset* to move the image up, down, left, or right. Use *Scale* to make the selected image larger or smaller (this feature is misleading in that a larger value creates a smaller image and vice versa). If an image is *Offset* or *Scaled* so it is smaller than the entire screen area, the area outside the background image is controlled by *Tile* or *Crop*.

Center

Press *Center* to center the selected image file on the screen and reset the *Offset* values to 0.

Maintain Aspect Ratio

A check in this box locks the *X* and *Y* values proportionally. With this box checked, the image width-to-height ratio is maintained even if *Scale* values are altered. If you want to stretch the image, remove the check and change the *X* and *Y* *Scale* values independently. If *X* and *Y* values are changed (so the image proportions are distorted), you can check *Maintain Aspect Ratio* again to lock the new proportions while using *Scale*.

Figure 41-16

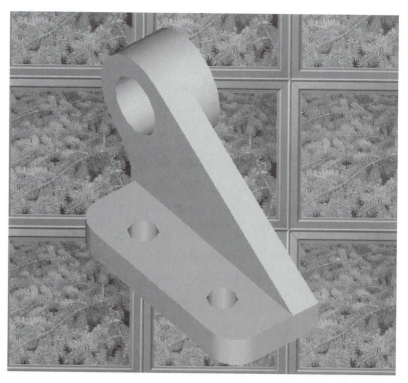

Tile and Crop

When the selected image is smaller (or *Scaled* to be smaller) than the screen or render window, these options have an effect. *Tile* causes the image to be repeated so it fills the screen (Fig. 41-16). *Crop* causes the rendered screen to display only one instance of the background image and all areas outside the background are black (cropped).

Merge

Merge is another method of using an external image file as a background. To do this, use the *Replay* command to display a .BMP, .TGA, or .TIF image. The display is stored in the "frame buffer." Then select *Merge* and render the model geometry. You must have 24-bit color enabled on your video display for this option to be available. However, a better method is to use the *Image* option, which allows a wider variety of file types as a background and automatically merges the image with the model geometry.

FOG

Pull-down Menu	COMMAND (TYPE)	ALIAS (TYPE)	Short-cut	Screen (side) Menu	Tablet Menu
View *Render >* *Fog...*	*FOG*	...	...	VIEW 2 Fog	P,2

Using the *Fog* command displays the *Fog/Depth Cue* dialog box (Fig. 41-17). *Fog* adds a visual effect that simulates distance information in the rendering. You can specify any color for this effect. The amount of fog (color) applied to the geometry is based on the distance different parts of the geometry are from the camera (viewpoint). For most applications, the farther the geometry, the more fog applied. The terms "fog" and "depth cueing" are use to describe two extremes of this one effect—that is, fog is white, and depth cueing is black.

Figure 41-17

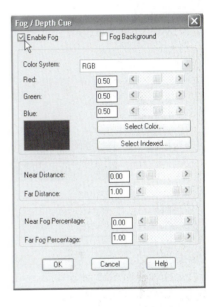

Enable Fog
Check this option to apply fog to the rendering. *Enable Fog* must be checked to enable all other options in the dialog box. Fog is applied only to the model geometry for *Photo Real* and *Photo Raytrace* rendering options.

Fog Background
Check this box to apply fog to the background image as well as the model geometry. Fog is applied to the background for any rendering option (*Render*, *Photo Real*, or *Photo Raytrace*).

Color System and Color Controls
Use these settings to adjust the color you want to use for fog. Select from *RGB*, *HLS*, or *Select Color* dialog box (see "Color Systems" earlier this chapter).

Near Distance and Far Distance
Use these sliders or edit boxes to set the values to define where the fog begins and ends. The values are percentages of the distance from the camera to the background or back clipping plane.

Near Fog Percentage and
Far Fog Percentage

Figure 41-18

These adjustments set the density of the
fog for the near and far distances
(ranging from 0 to 100 percent). For
example, to make the model geometry
appear less visible in the distance, set
the *Near Fog Percentage* low and the *Far
Fog Percentage* high (Fig. 41-18).

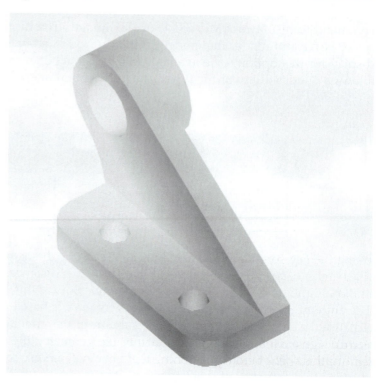

LSNEW

Pull-down Menu	COMMAND (TYPE)	ALIAS (TYPE)	Short-cut	Screen (side) Menu	Tablet Menu
View *Render >* *Landscape New...*	LSNEW	...	...	VIEW 2 *Lsnew*	...

Lsnew produces the *Landscape New* dialog box (Fig. 41-19). This feature
allows you to add photo-realistic images, such as trees, bushes, people,
and signs to your renderings. Related commands *Lsedit* and *Lslib*
allow you to modify landscape objects and maintain landscape object
libraries, respectively.

Figure 41-19

A landscape object has a bitmap image mapped onto it. You can
manipulate the placement of the landscape object in the drawing
directly (with grips, for example) or you can use the *View Aligned*
option in the *Landscape New* dialog box to align the object. Each landscape object has grips at the top, bottom, and at each corner. Use the
base grip to move the entire object, top grips to adjust height, and
bottom corner grips to scale and rotate it. Landscape objects in the
drawing (unrendered) appear as planar objects such as a triangle, rectangle, or two intersecting triangles, depending on the alignment and
single or double face specifications.

Library

The central area lists the AutoCAD-supplied landscape objects (in the RENDER.LLI file). Select one you
want to insert into the drawing.

Height

Use this edit box or slider to specify the height of the object in the current drawing units. *Height* is always oriented as the positive Z direction in the current USC.

Position

Use this button to temporarily exit the dialog box and PICK a location in the drawing to place the landscape object. The default position is 0,0,0 of the current UCS.

Geometry

This section specifies the quality of the object and alignment of the object in the drawing.

Single Face or Crossing Faces

A *Single Face* object is faster to render but less realistic. *Crossing Face* objects result in better images, especially for *Photo Raytrace* renderings.

View Aligned

With this box checked, the object is positioned to be normal to the current view. The position of the landscape object changes to be normally aligned if the view changes. Use this option for trees and bushes. If you want planar objects such as road signs to retain the orientation you specify, remove this check before positioning them.

Landscape objects are planar objects that appear in the drawing depending on these settings (Fig. 41-20). A *Single Face, View Aligned* object appears as a triangle. You cannot rotate it with grips. A *Single Face* positioned object appears as a rectangle and can be rotated with grips. A *Crossing Face* object always appears as intersecting triangles. These objects cannot be rotated when *View Aligned* but can be when positioned (fixed).

The landscape objects appear as normal trees, bushes, people, and signs in the rendered image (Fig. 41-21).

Figure 41-20

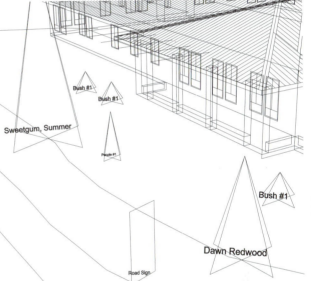

Figure 41-21

LSEDIT

	Pull-down Menu	COMMAND (TYPE)	ALIAS (TYPE)	Short-cut	Screen (side) Menu	Tablet Menu
	View Render > Landscape Edit...	*LSEDIT*	...	...	VIEW 2 Lsedit	...

You can position landscape objects when you create them with *Lsnew* or when you manipulate a land-scape object directly in the drawing (with grips, for example). Alternately, you can use *Lsedit* to reposi-tion a selected object in the same manner as when you created the object. The *Lsedit* command produces the *Landscape Edit* dialog box, which is identical to the *Landscape New* dialog box (see Fig. 41-19), except you can only change options for the selected object—not create new objects.

LSLIB

	Pull-down Menu	COMMAND (TYPE)	ALIAS (TYPE)	Short-cut	Screen (side) Menu	Tablet Menu
	View Render > Landscape Library...	*LSLIB*	...	...	VIEW 2 Lslib	...

The *Lslib* command produces the *Landscape Library* dialog box (Fig. 41-22). Use this tool to manage libraries of landscape objects. Only a few objects are supplied with AutoCAD. These objects can be modified and other objects can be created or purchased, then managed using this dialog box.

Figure 41-22 ——————

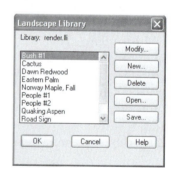

Library:
The central area displays objects in the current library. The current *Library* file is listed above.

Modify
Select an object, then select this option to produce the *Landscape Library Edit* dialog box (see Fig. 41-23). All landscape objects are defined by an image file and an opacity map (to specify see-through areas of the image).

New
If you have an image to import into the library, use this option to produce the *New Landscape Library* dialog box. This dialog is identical to the *Landscape Library Edit* dialog except the edit boxes are blank. Enter or *Find* an image file and opacity map file for your new object.

Delete
Select an item to delete from the list, then select this button. The object is deleted from the library, but the referenced image and opacity map files are not deleted.

Open
Use this option to locate and open another library (.LLI file). The contents of the library appear in the list.

Save

When changes are made to the current library that you want to keep, click this button to save the .LLI file under the current name or create a new name.

Landscape Library Edit **Dialog Box**

This dialog box (Fig. 41-23) appears when you use the *Modify* option from the *Landscape Library* dialog box (see Figure 41-22). The information about the selected object is displayed in the edit boxes. In the *Default Geometry* section, you can modify the *Single* or *Crossing Faces* parameter and change the *View Aligned* assignment for the selected object. If you want the named object to reference another image file or opacity map file, enter or *Find* the desired file name. Change the name of the object in the *Name* edit box if desired.

Figure 41-23

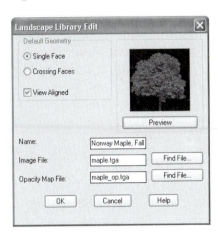

CREATING LIGHTS FOR THE MODEL

AutoCAD recognizes four types of light: *Ambient Light, Point Light, Distant Light,* and *Spotlight*. *Point Lights, Distant Lights,* and *Spotlights* can be positioned anywhere in space and adjusted for color and intensity. *Distant Lights* and *Spotlights* can be set to shine in a specific direction, whereas *Point Lights* shine in all directions. *Ambient Light* has no location and affects all surfaces equally. *Ambient Light* can be controlled only for color and intensity. *Point Lights, Distant Lights,* and *Spotlights* can cast shadows. *Ambient Light* cannot cast shadows.

Ambient Light

Ambient Light is the overall lighting of surfaces. Each surface is illuminated with an equal amount of ambient light, regardless of the surface's position. Thus, you cannot distinguish the adjoining faces of a box or curvature of a sphere when this is the only type of light in the model. Figure 41-24 illustrates the rendering of a box and sphere with only ambient light.

Figure 41-24

Adjusting ambient light would be like using a dimmer switch to darken or lighten a room except that the light has no source or direction. Ambient light is the most unrealistic type of light, but it is needed to display "background" surfaces that may not be illuminated by *Point* or *Distant* lights. You do not create or position an *Ambient Light* in the model; it is ever-present unless its intensity has been set to zero.

Distant Light

Figure 41-25

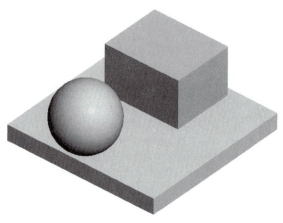

A *Distant Light* shines in one direction and has parallel rays. It is used to simulate sunlight. Because *Distant Light* represents the sun, you specify only a direction vector, not a light location. The vector determines the direction of the parallel beams of light which extend to infinity. Similar to sunlight, the light intensity does not fall off (grow dimmer) for surfaces farther away from the "source." Illumination of a surface is determined only by the angle between the surface and the light beams—surfaces that are perpendicular to the beams are more brightly illuminated.

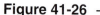

Figure 41-25 illustrates a box and sphere illuminated by <u>only</u> one *Distant Light* shining from above and to the right. Notice how each surface is illuminated differently because of the angle of the surface to the direction vector, yet each individual flat surface is illuminated evenly across the entire surface.

Point Light

Figure 41-26

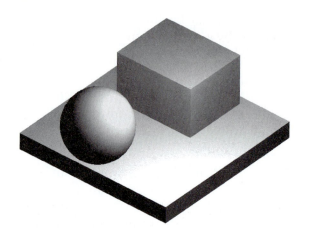

A *Point Light* is inserted at the location you specify, but no direction vector is needed since this light shines in all directions. Point light would be typical of the light emitted from a light bulb. Point light falls off (grows dimmer) as the distance between the light source and surfaces increases, so surfaces close to the light are brighter than those at a distance. The box in Figure 41-26 is illuminated by <u>only</u> one *Point Light* above and slightly to the right side. Notice how the light falls off the farther the surfaces are from the light source.

Spotlight

Figure 41-27

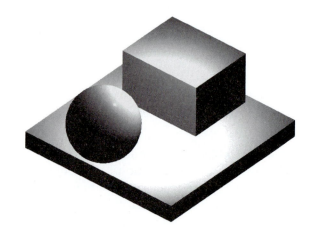

Spotlights have characteristics much like a spotlight used for stage lighting. You can specify the light's *Location* in space as well as the point that it shines toward (*Target*). Like point lights, the light falls off as the illuminated surfaces become farther from the light *Location*. Similar to stage spotlights, the beams of light from the source (*Location*) to the *Target* form a "cone" of light. Figure 41-27 shows the box and sphere illuminated by a *Spotlight* shining from directly above the objects. Notice the circular illuminated area formed by the cone of light. The angle of a *Spotlight's* cone of light can be adjusted for full illumination and for light fall-off.

For *Point Lights*, *Distant Lights*, and *Spotlights*, the brightness of a rendered surface depends on the following parameters:

1. The *Intensity* of the light can be specified. The higher the *Intensity*, the brighter the rendered surfaces appear.

2. The *Color* of the light can be specified. The lighter the *Color*, the brighter the rendered surfaces appear.

3. The angle between the light source and the surface affects the brightness of the surface. A surface that shines the brightest is one in which the light is shining perpendicular to the surface. The more angled the surface is from perpendicular, the darker it appears.

4. The distance between the light and the surface affects brightness (*Point Lights* and *Spotlights* only).

5. When surfaces have a high *Reflective* attribute, the brightness of the reflected light is determined by the angle formed by the light source, object surface, and observer's viewpoint.

6. A surface's brightness can also be affected by a shadow or light reflected from another surface (for *Raytraced* displays).

For dramatic effects, only one light might be used; however, for most renderings several lights or a combination of types of lights is used. Typically, a mix of *Ambient Light* and one or two other lights is used. Figure 41-28 illustrates the box and sphere illuminated by a combination of *Point*, *Distant*, and *Ambient* lighting.

Now that you understand the types of lights and the effects of each type on a rendered object, examine the commands that let you insert and control the lights.

Figure 41-28

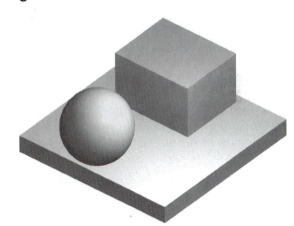

LIGHT	Pull-down Menu	COMMAND (TYPE)	ALIAS (TYPE)	Short-cut	Screen (side) Menu	Tablet Menu
	View *Render >* *Light...*	LIGHT	...	...	VIEW 2 Light	O,1

When you invoke the *Light* command by any method, the *Lights* dialog box appears, as shown in Figure 41-29.

New...
To create a new light, select the type of light (*Distant Light*, *Point Light*, or *Spotlight*) from the drop-down list and PICK the *New* tile. The appropriate dialog box for the type of light appears and allows you to set the light's parameters. The light types and related dialog boxes are discussed later.

Figure 41-29

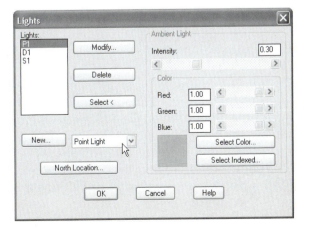

Lights:

The *Lights:* section in the upper-left corner displays any existing (previously created) lights. Using the related buttons beside the list, you can *Modify…, Delete,* or *Select* (from the drawing) an existing light. Modifying a light offers the <u>same dialog box</u> and options used to create a new light.

Ambient Light

The *Ambient Light* section (upper-right) is where the percentage of ambient light is set by using the slider bar or by entering a value in the edit box. The setting is <u>global</u> and affects all surfaces equally. Generally a percentage of .30 is a good starting value.

You can specify the *Color* of light that is emitted by the ambient source by using the slider bars in the *Lights* dialog box (*RGB* color system), or by using the *Select Color* dialog box.

Light Colors

All lights can be assigned a color using either the *RGB* color system or the *Select Color* dialog box (see "Color Systems" earlier this chapter). The selected light color is the color that the light emits. Although each <u>object</u> in the model can have a color (determined by the object's assigned material), the light color casts a hue on the object's surfaces. Normally, white or bright colors are used for lights.

Creating New and Modifying Existing Lights

To insert new lights in your drawing, select the type of light (*Point Light, Distant Light,* or *Spotlight*) from the drop-down list in the *Lights* dialog box (see Figure 41-29); then pick the *New…* button. A different dialog box titled *New (type) Light* appears depending on your light choice.

If you want to modify an existing light, select the light name from the *Lights:* list (in the *Lights* dialog box) and PICK the *Modify…* button. The dialog box that appears is titled *Modify (type) Light* but is otherwise identical to the dialog box used to create the new light.

New Point Light and *Modify Point Light* Dialog Boxes

Light Name

When creating a new light, assign a name for the light in the *Light Name:* edit box as a first step (Fig. 41-30). In this example, "P1" is assigned as the name for a new point light. Existing light names can be renamed by changing the name in the edit box and selecting *OK*.

Intensity

Intensity is a value that specifies the brightness of the <u>light</u>; the higher the value, the brighter the light. The intensity can be set with the slider bar or by entering a value in the edit box. The resulting brightness of a <u>surface</u> illuminated by a *Point* light is based on several variables: intensity, color, attenuation, distance to the surface, and angle between the surface and the light beams.

Figure 41-30 ————————————

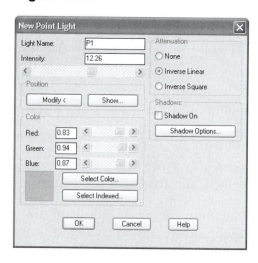

Position

When a light is created, AutoCAD automatically assigns a default location for the light based on the existing geometry. Selecting the **Show** tile displays the default *Location* of the light (Fig. 41-31). The *Location* is the coordinate position of the light in 3D space (*Target* is used only for *Spotlights*). You can change the position by selecting the **Modify** tile, then entering the desired coordinates or PICKing a new location (use *OSNAP* to PICK in 3D space).

Figure 41-31 ————————————

Color

The color of a *Point* light (as any other light type) can be assigned using the *RGB* color system or *Select Color* dialog box (see previous explanation in "Color Systems").

Attenuation

Point Lights possess a fall-off quality similar to actual lighting. That is, the farther the light beams travel from the source (*Location*), the dimmer the light becomes. *Point Light Attenuation* has three options.

None

This option forces the light emitted from a point light to keep full intensity no matter how far the rendered surfaces are from the source. This is similar to a *Distant Light*.

Inverse Linear

This option sets an inverse linear relation between the distance and the illumination. The brightness of a rendered surface is inversely proportional to the distance from the source. Put another way, there is a 1 to 1 relationship between distance and darkness. Generally, *Inverse Linear* attenuation is the easiest to work with and appears realistic for most AutoCAD renderings.

Inverse Square

This is the greatest degree of light fall-off. The amount of light reaching a surface is inversely proportional to the square of the distance. In other words, if the rendered surfaces are 2, 3, and 4 units from the light source, the illumination for each will be $1/2^2$, $1/3^2$, and $1/4^2$ (1/4, 1/9, and 1/16) as strong. This option simulates the physical law explaining light fall-off.

Shadows

Check the *Shadows On* checkbox if you want Render to calculate shadows cast by this light. (*Photo Real* or *Photo Raytrace* and the [global] *Shadows* checkbox in the *Render* dialog box must also be selected for shadows to be displayed in the model.) Pressing the *Shadow Options* produces a dialog box with the following options (Fig. 41-32).

Figure 41-32

Shadow Volumes/Ray Traced Shadows

When this box is checked, *Photo Real* renderer produces volumetric shadows and *Photo Raytrace* produces raytraced shadows for this light. The other options are disabled.

Shadow Map Size

This setting controls the size in pixels of one side of the shadow map. The possible range is 64 to 4096. The larger the map size, the more accurate the shadows but the longer it takes to calculate them.

Shadow Softness

This value determines the "fuzziness" of the edges of shadows for shadow-mapped shadows. The value can range from 1 to 10 and represents the number of pixels along the edge of the shadow that are blended from the illuminated to the non-illuminated area. A value of 2 to 4 usually gives the best results.

Shadow Bounding Objects

You can reduce rendering time by selecting a set of objects that determine a bounding box. The bounding box is used to clip the shadow maps.

New Distant Light and Modify Distant Light Dialog Boxes

Distant Light simulates sunlight. With distant lights, parallel beams of light illuminate the solid or surface objects. Distance lights have no location or target, only a direction vector. Distance light has no attenuation, so the beams of light maintain the same intensity regardless of the location of the surfaces. The direction vector is specified by using either the Azimuth and Altitude or Light Source Vector sections of the Distant Light dialog box (Fig. 41-33) or the Sun Angle Calculator. The model in the Drawing Editor is oriented in space such that the XY plane of the WCS represents horizontal (Earth's surface), and north is a positive Y direction.

Figure 41-33

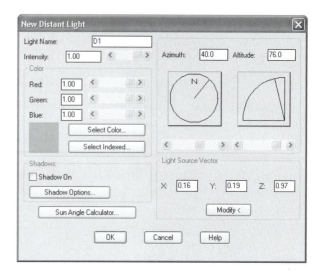

Light Name, Intensity, Color, and Shadows

The Name, Intensity, Color, and Shadows sections in the dialog boxes operate identically to the same sections in the other dialog boxes (see "Point Light").

Azimuth

Azimuth is the bearing in degrees from north in a clockwise direction; for example, due east is 90 degrees. PICK in the image tile, use the slider bars, or enter a value in the edit box to make your selection.

Altitude

Altitude is the slope of the direction vector specified in degrees from horizontal; for example, 90 degrees specifies a light source from directly above (high noon) and 20 to 30 degrees simulates sunlight in the early morning or late afternoon. PICK in the image tile, use the slider bars, or enter a value in the edit box to make your selection.

Light Source Vector

An alternative to the Azimuth and Altitude method, the Light Source Vector edit boxes specify the location of a "source" related to 0,0,0 of the WCS. You can enter values in the edit boxes.

Sun Angle Calculator

Since a Distant Light is used to simulate sunlight, use the Sun Angle Calculator to determine the location of the sun with respect to the model by specifying a date, time, and time zone (Fig. 41-34). Making these adjustments automatically computes the Azimuth and Altitude of the Point Light. The location of the model on Earth can be specified if you know the Longitude and Latitude, or you can select a Geographic Location.

Figure 41-34

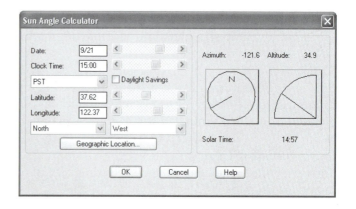

Geographic Location

This dialog allows you to select a city from the list or PICK a location on the map to determine the location on Earth of the model (Fig. 41-35). The drop-down list allows you to select from the seven continents. Selecting from the map or city list automatically enters the related values in the *Longitude* and *Latitude* edit boxes.

Figure 41-35

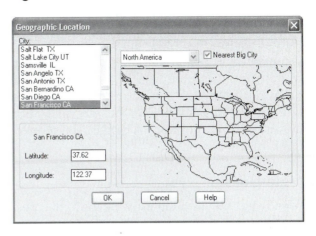

New Spotlight and *Modify Spotlight* **Dialog Boxes**

A *Spotlight* has characteristics similar to a spotlight used for stage lighting. Because you can specify a *Location* and a *Target* for the *Spotlight,* any surface or area of a surface or solid model can be illuminated from any direction. In addition, the angle of the "cone" of light (formed as the light beams travel from the source) can be adjusted. Spotlights also possess attenuation, color, and intensity. All options are available from the *New Spotlight* (Fig. 41-36) and *Modify Spotlight* dialog boxes.

Light Name, Intensity, Color, Attenuation, **and** *Shadows*

These options operate identically to the same options appearing in the *Distant* and *Point Light* dialog boxes. See *Distant Light* dialog boxes and *Point Light* dialog boxes.

Figure 41-36

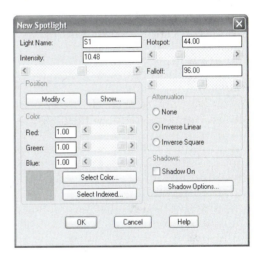

Hotspot **and** *Falloff*

These two sections determine characteristics of the cone of light emitted from a *Spotlight*. The *Hotspot* is the area of the model that receives full illumination (Fig. 41-37). The size of this area is determined by the angular value entered in the edit box or appearing there as a result of moving the slider bars. The light begins to fall off just outside the *Hotspot* until approaching no illumination at the *Falloff* cone. The area outside the *Falloff* cone receives no illumination. The *Falloff* angle must be equal to or greater than the *Hotspot* angle.

Figure 41-37

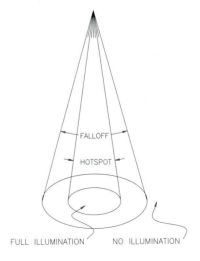

Position

You can indicate the *Location* and *Target* for the *Spotlight* in this cluster. A default *Location* and *Target* are calculated by AutoCAD when you create the light and those coordinates can be viewed by selecting the *Show...* tile (see Fig. 41-31). To specify another *Location* and *Target*, select *Modify <*. Enter coordinate values or PICK points (use *OSNAP* when PICKing in 3D space).

You can use the *Move* command to reposition the lights; however, it is recommended that you use the *Location* and *Target* options in the light dialog boxes. Changing a *Distance Light* location with *Move* also moves the *Target* position. A light can be deleted by *Erasing* its icon or by using the *Delete* option in the *Lights* dialog box.

Light Icons

For any light, an icon representing the light type and position appears in the drawing at the light *Location* (Figs. 41-38 and 41-39). These icons are actually *Block* insertions that appear in the drawing but not in renderings. The light icons are inserted on layer ASHADE, which is automatically created by AutoCAD when you place the first light. This layer can be *Frozen* to prevent the icons from appearing in drawings or plots.

Figure 41-38

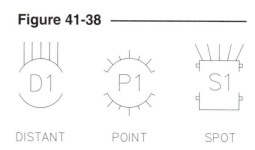

DISTANT   POINT   SPOT

Figure 41-39

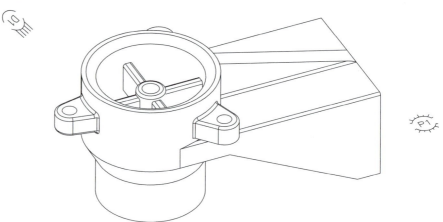

USING MATERIALS IN THE MODEL

When the background has been set and the lights have been inserted and adjusted for position, color, attenuation, and intensity, you are ready to select *Materials* and attach them to the model. Because the default template drawing has no materials loaded, one or more materials must be selected from the *Material Library*. Once the desired materials have been imported into the drawing, you must attach them to components of the model. Different materials can be attached to different components of the model. A material can be attached to an object, layer, or ACI (AutoCAD Color Index number). If a material is attached to a layer or ACI, every object on the layer or having the ACI assumes the material. Materials can be created and modified in a variety of ways to attain a particular color and reflective qualities.

MATLIB	Pull-down Menu	COMMAND (TYPE)	ALIAS (TYPE)	Short-cut	Screen (side) Menu	Tablet Menu
	View *Render >* *Materials Library...*	*MATLIB*	...	...	*VIEW 2* *Matlib*	*Q,1*

The *Matlib* command summons the dialog box that provides access to the AutoCAD-supplied materials (Fig. 41-40). The names in the *Current Library* list (right side) represent the choices from the RENDER.MLI file. A material is brought into the drawing by highlighting it and selecting the *Import* button. The selected material(s) then appear in the *Current Drawing* list (left side) representing all of the materials currently in the drawing and ready for attachment to the model components. Importing materials increases file size, so import only those that are needed. This dialog box can also be summoned from the *Materials* dialog box (see next).

Figure 41-40

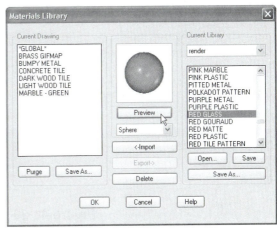

Preview
You can highlight <u>one material at a time</u> (from either list) and select the *Preview* button to see how the material appears attached to the sphere or cube. Select either the *Sphere* or *Cube* for preview from the drop-down list.

Import
Import brings material descriptions into the drawing from the RENDER.MLI or other .MLI files. After materials are imported, select *OK*; then use the *Rmat* command (*Materials* dialog box) to attach materials to objects, layers, or ACI numbers. If you try to import a name that exists in the drawing, the *Reconcile Imported Material Names* dialog box appears, asking if you want to overwrite the existing material in the drawing by the same name.

Export
Use this button to add selected materials from the *Current Drawing* list to the *Current Library* list. This is useful when you have created a new material or modified an existing material in the drawing and want to save it in the material library for use in other drawings.

Delete
This option deletes selected materials from the *Current Drawing* list or the *Current Library* list. Deleting materials from the *Current Library* list does not delete them from the RENDER.MLI file.

Purge (*Current Drawing* list)
Use this button to purge all unattached materials from the material list. Unattached materials are those that have been imported but have not been attached to objects, layers, or ACI numbers using the *Rmat* command (*Materials* dialog box).

SaveAs (*Current Drawing* list)
The materials listed on this side can be saved to a .MLI file for use in other drawings. Assign the desired file name to save in the *Library File* dialog box that appears.

Open (*Current Library* list)
Material library (.MLI) files other than the default RENDER.MLI can be opened for material selection and used in the current drawing. Other .MLI files could be those you have created or those supplied with other software.

Save (*Current Library* list)
The materials listed in the library list can be saved to a .MLI file for use in other drawings. If you have created new or modified existing materials, they can be saved in the RENDER.MLI or other .MLI file. Assign the desired file name to save in the subsequent *Library File* dialog box.

RMAT

Pull-down Menu	COMMAND (TYPE)	ALIAS (TYPE)	Short-cut	Screen (side) Menu	Tablet Menu
View *Render >* *Materials...*	*RMAT*	...	...	VIEW 2 *Rmat*	*P,1*

After materials have been imported from the materials library (see *Matlib*), this dialog box is used to attach selected materials to AutoCAD objects, layers, or ACI numbers. You can also gain access to the tools for creating new and modifying existing materials from this dialog box (Fig. 41-41).

Figure 41-41

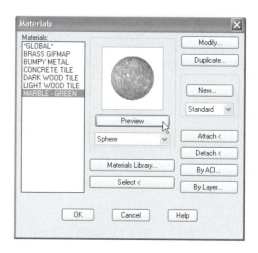

Materials
This list represents the materials that are available for attachment. The list includes the materials imported previously using the *Materials Library* dialog box and a material named *GLOBAL*, which is the default material used for objects when you *Render* without attaching materials.

Preview
Use this button to see how the selected material (one at a time) appears on the sphere or cube. Select either a *Sphere* or *Cube* to use in the preview image tile from the drop-down list.

Materials Library...
This button invokes the *Materials Library* dialog box for importation of other materials (see *Matlib*).

Select <
This option temporarily removes the dialog box to allow selection of an object. When an object is selected, the *Materials Library* dialog box reappears with the selected object's material name and attachment method used displayed at the bottom of the dialog box.

Attach <
Use this button to attach the highlighted material from the *Materials:* list to an object. The dialog box is temporarily removed to allow selection of an object. When the *Render* command is used, the object is rendered with the attached material.

Detach<

The *Detach* option allows you to unattach an assigned material by selecting an object.

By ACI...

You can attach a material to an *ACI* (AutoCAD Color Index number) using this button. With this method of attachment, every object in the drawing that has the selected color number displays the attached material when the *Render* command is used. The *Attach by AutoCAD Color Index* dialog box appears (Fig. 41-42).

To attach a material, you must select <u>both</u> the desired material from the list on the left and the desired color or number from the list on the right. Then PICK the *Attach* button and the material name appears next to the color in the *Select ACI:* list (see Fig. 41-42). If you want to *Detach* a material from a color number, select the choice from the *Select ACI:* list only; then PICK *Detach.*

Figure 41-42

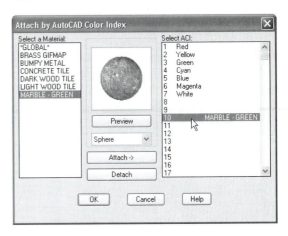

By Layer...

Using this option produces the *Attach by Layer* dialog box (not shown), which looks and operates identically to the *Attach by AutoCAD Color Index* dialog box. To attach a material, select <u>both</u> the desired material and the desired layer; then PICK *Attach.* Attached materials appear in the list with the layer name.

Modify...

Select a material from the list and choose the *Modify...* button to produce the *Modify Standard Material* dialog box. See "Creating New and Modifying Existing Materials."

Duplicate...

If you want to create a new material but use an existing material as a starting point, choose this method. The highlighted name from the *Materials List:* is used as a template. The *New Standard Material* dialog box appears. See "Creating New and Modifying Existing Materials."

New...

This button produces the *New Standard Material* dialog box and enables you to create a new material "from scratch." You can select from four template materials (in the drop-down list), *Standard, Granite, Marble,* and *Wood.* See "Creating New and Modifying Existing Materials," next.

CREATING NEW AND MODIFYING EXISTING MATERIALS

Render allows you to control seven attributes that define a material: *Color/Pattern, Ambient, Reflection, Roughness, Transparency, Refraction,* and *Bump Map.* For many attributes, the *Value* (intensity) and *Color* can be adjusted. Other attributes such as *Transparency, Refraction,* and *Bump Map* have special adjustments. These adjustments are made in the *Modify Standard Material* and *New Standard Material* dialog boxes (see Figure 41-50). The attributes are first explained here.

Color/Pattern

Color/Pattern is the global color of the material. The color (hue) and the value (intensity) can be adjusted. Technically, *Color* represents the light that is reflected in a diffuse manner. Diffuse light affects all surfaces equally, not accounting for highlights (highlights are controlled by the *Reflection* attribute). Setting a low *Color* value makes the surfaces appear dark and dull (Fig. 41-43, left sphere). Increasing the *Color* value makes the surface reflect more light; therefore, the assigned color appears brighter (right sphere). Texture maps can be used in combination with this attribute. When a texture map is applied to the object it appears to be "painted" on the object's surface (see "Image Mapping").

Figure 41-43

Ambient

Ambient light is background light that illuminates all surfaces. Because most of the surfaces are illuminated by *Point Lights, Distant Lights,* and *Spotlights,* adjusting the *Ambient* attribute controls the color and intensity of material "in the shadows." A low *Ambient* setting makes surfaces in the "shadows" appear dark and adds contrast (Fig. 41-44, left sphere), while a high *Ambient* setting increases the <u>overall lighting</u> and makes the "shadows" brighter (right sphere).

Figure 41-44

Reflection

This attribute controls the light that is reflected from *Point Lights, Distant Lights,* and *Spotlights. Reflection* accounts for the highlights on an object (technically called specular reflection). Controlling the value of *Reflection* increases or decreases the amount of reflected light. Use a low *Reflection* value to make the surfaces appear flat or matte, like blotting paper or soft material (Fig. 41-45, left sphere). Use a high *Reflection* value to increase light contrast on surfaces from *Point Lights, Distant Lights,* and *Spotlights* (right sphere). Reflection maps can be used in combination with this attribute.

Figure 41-45

Roughness

Roughness is a function of *Reflection*—if *Reflection* value is 0, *Roughness* has no effect. *Roughness* does not affect the amount of highlight, only the <u>size</u> of the highlight. A smaller *Roughness* value produces a small shiny spot on a sphere. Use a small *Roughness* value to produce a hard, shiny material (Fig. 41-46, left sphere). If a surface has a high *Roughness* value, its reflection is spread in a less perfect, larger cone. The larger the *Roughness* value, the larger the cone of reflection and the larger the highlight (right sphere).

Figure 41-46

Transparency

With this attribute you can make an object appear translucent or transparent. The *Value* adjusts the amount of transparency: from opaque (0.0), to translucent (0.1-0.9) , to transparent (1.0). Do not use this attribute if you want materials to appear opaque (solid), as in Figure 41-47, left sphere. Use a medium-high *Value* to make an object appear translucent (right sphere). Opacity maps can be used in combination with this attribute (see "Image Mapping"). *Photo Real* must be selected in the *Render* dialog box for *Transparency* to be calculated.

Figure 41-47

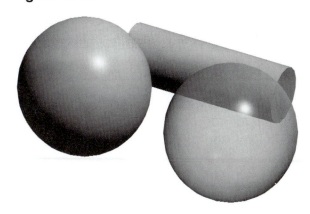

Refraction

This attribute is a function of *Transparency,* so this option can be used only when a positive *Transparency* value is set. Refraction is the bending of light waves as they pass through translucent or transparent material. Note the appearance of the cylinder through the right sphere in Figure 41-48 compared to the previous figure. *Refraction* calculates the effect of the light rays from the curvature of translucent objects. This feature requires the *Rendering Type* to be set to *Photo Raytrace* in the *Render* dialog box.

Figure 41-48

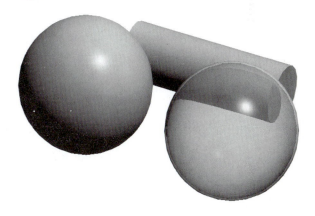

Bump Map

This attribute allows you to select a bump map to be applied to the object's surface. Bump maps create an embossed or bas-relief effect. For example, notice how the Autodesk "calipers" logo appears to be embossed on the surface of the right sphere (Fig. 41-49).

Figure 41-49

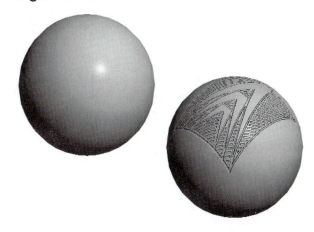

New Standard Material and *Modify Standard Material* Dialog Boxes

To make adjustments to a material's attributes, select either *Modify...*, *Duplicate...*, or *New...* from the *Materials* dialog box (Fig. 41-41). In any case, the resulting dialog box is the same; however, the title of the dialog box differs. For example, if you selected *Modify...*, the *Modify Standard Material* title is used (Fig. 41-50); otherwise, the *New Standard Material* title is used. Options in the dialog box are described here.

Figure 41-50

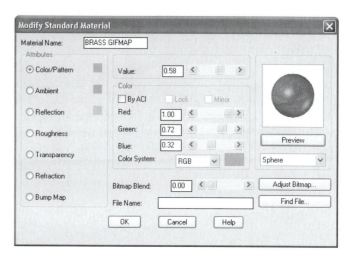

Material Name:

If you are modifying an existing material, the name appears here. You can rename an existing material, but it does not create a duplicate material. (Choose the *Duplicate...* button in the *Materials* dialog box to keep the original material and create a new one.)

Attributes

Only one of these radio buttons can be depressed. The selected *Color/Pattern, Ambient, Reflection, Roughness, Transparency, Refraction,* or *Bump Map* button determines which adjustments can be made in the *Value, Color,* and *Bitmap Blend* sections. See previous discussion of attributes.

Value

The magnitude or intensity of the selected attribute is adjusted by using the slider bar or entering a value in the edit box.

Color

This cluster enables you to set the color for each material attribute three ways: *RGB* system, *HLS* system, and the *ACI* pallet. These methods are identical to those used for setting light colors explained previously (see "Color Systems").

If *By ACI* is checked in the *Color* cluster, the material color is set to match the object's drawing color (assigned *BYLAYER* or as an object-specific setting). In this case, the color is fixed to the object's drawing color and cannot be changed. Removing the "X" in the checkbox enables the *RGB* and *HLS* systems.

Lock

This is a useful checkbox to use if you want the *Ambient* and *Reflection* attribute colors to remain the same as the main *Color*. In this way, you can make adjustments to the attribute *Values* while ensuring that all reflection colors are not changed. Remove the check from *Lock* if you want *Ambient* and *Reflection* colors to be different than the object's global color.

Mirror

This feature (enabled only for the *Reflection* attribute) allows you to create mirrored reflections. The *Photo Raytrace* renderer must be used for this effect. If an image file is specified for the reflection map, the resulting reflection includes the background and the reflection map.

Bitmap Blend

See "Image Mapping" for a discussion of mapping concepts and this dialog box section.

Dialog Boxes for *New Granite, Marble,* and *Wood Material*

When you create a new material, you can select from four AutoCAD-supplied materials to use as a template: *Standard, Granite, Marble,* or *Wood.* Select from the drop-down list in the *Materials* dialog box before selecting the *New* button (Fig. 41-51, right side).

Selecting the *Standard* material template, then *New...* produces the *New Standard Material* dialog box which is identical to the *Modify Standard Material* dialog box except for the title (see Fig. 41-50). This dialog box and all related options (except the *Bitmap Blend* section) were previously discussed.

Figure 41-51

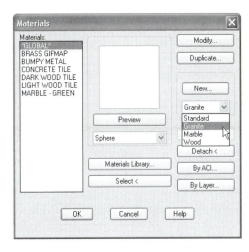

New Granite Material Dialog Box

If you want to create a new granite material, select *Granite* from the drop-down list (Fig. 41-51), then select *New.* The *New Granite Material* dialog box appears (Fig. 41-52). A template that simulates granite material appears.

Material Name, Value, Color, Preview, Bump Map, and *Bitmap Blend*
These options have identical functions to those options in the *New Standard Material* dialog box. See the previous section for information related to these options. See "Image Mapping" (next) for information on *Bitmap Blend.*

Figure 41-52

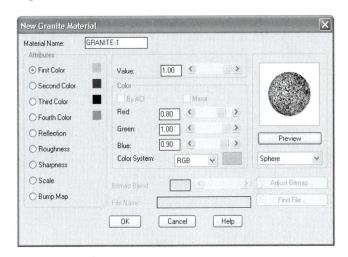

First, Second, Third, and *Fourth Color*
Granite can have four colors present. Select the desired button to change the color. The *Value* specifies the amount of each color in the granite pattern. To use fewer than four colors, enter a *Value* of zero for one or more colors. The color systems are the same as for *Standard* material.

Reflection and *Roughness*
Reflection adjusts the material's highlight (specular) *Color* and *Value.* In other words, *Reflection* is the amount of light reflected or the intensity of the highlight.

Roughness is a function of *Reflection.* *Roughness* adjusts the <u>size</u> of the highlight. A small *Roughness* value makes the material (granite) appear hard and shiny (like polished granite), whereas a high *Roughness* value makes the material appear rough and abrasive.

Sharpness
This setting defines the sharpness of the edges of color on the stone. A *Sharpness* of 0 makes the colors blurred, or blended together. A *Sharpness* of 1.0 makes the colors distinct with sharp edges between different colors.

Scale
Use this option to scale the pattern relative to the objects you attach it to. Larger values create a bigger pattern while smaller values create a tighter pattern.

New Marble Material Dialog Box

Select *Marble* from the drop-down list in the *Materials* dialog box (see Figure 41-51), then select *New* to produce the *New Marble Material* dialog box (Fig. 41-53). The template material is marble.

Material Name, Value, Color, Bump Map, and *Bitmap Blend*
These options have identical functions to the same options in the *New Standard Material* dialog box. See the previous sections for information related to these options. See "Image Mapping" (next) for information on *Bitmap Blend*.

Figure 41-53

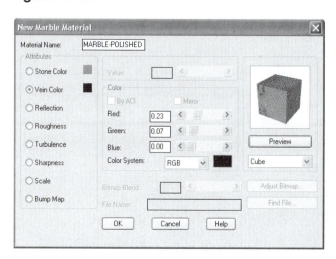

Reflection, Roughness, Scale, and *Sharpness*
These options are the same as discussed in the previous section, *New Granite Material* dialog box.

Preview
It may be helpful when creating marble to view both the *Sphere* and the *Cube* in the *Preview* window. The *Cube* gives a clear indication of the "vein" characteristics in the stone.

Stone Color and *Vein Color*
Marble has two color characteristics: the color of the body of the marble and the color of the long "veins" or "swirls" that run through the stone color. Usually a white or black color is typical for the veins in marble.

Turbulence
This feature adjusts the amount of "swirl" and color in the veins of the stone. Higher values produce more vein color and more swirling, but higher values increase rendering time. Values between 1 and 10 are recommended.

New Wood Material Dialog Box

Selecting *Wood* from the drop-down list in the *Materials* dialog box allows you to create a new wood pattern from a supplied template. The *New Wood Material* dialog box has the following settings (Fig. 41-54).

Figure 41-54

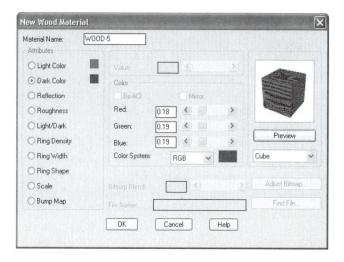

Material Name, Value, Color, Preview, Bump Map, Reflection, Roughness, Scale, **and** *Bitmap Blend*
These options have functions identical those options in the *Create New Standard Material* and other material dialog boxes. The *Bitmap Blend* section is discussed later in "Image Mapping."

Light Color **and** *Dark Color*
To visually describe wood, use two colors to define the main body of wood and the grain. *Light Color* is generally used as the color for the body of the wood (soft, wide grain), and *Dark Color* is used for the thin grain in the wood. Generally shades of brown are used.

Light/Dark
This adjustment controls the ratio of light color to dark color, that is, the size of the light grain in proportion to the dark grain. A low *Light/Dark Ratio* (adjusted by the *Value* slider) creates more dark grain color, while a high value produces more light color grain.

Ring Density
This adjustment varies the density of the grain, or viewing from the end of a "log," the number of rings. Larger values create more rings and a denser grain.

Ring Width
You can change the variation in the widths of grain by adjusting the value of *Ring Width*. A value of 0 produces uniform width for all grain (or rings, if viewed from the end). A high value creates grain (rings) of different widths.

Ring Shape
Use *Ring Shape* to control circularity of the rings of grain (when viewed from the end of a "log"). A value of 0 gives perfectly circular rings, whereas a high value gives distorted (not round) rings. Use a high value to create more "swirls" in the wood grain.

IMAGE MAPPING

Image mapping is the process of projecting a 2D image onto the surface of a 3D object. The image used for projecting can be any image in .BMP, .PNG, .JPG, .TGA, .TIF, .GIF, and .PCX file formats (Fig. 41-55, for example, is the CAMOFLAB.TGA file in the Textures directory.)

Figure 41-55 ——————

Another type of mapping accomplished with the *Setuv* command allows you to adjust the <u>orientation</u> of an image or of an AutoCAD material. *Setuv* can be used to change the orientation of <u>materials</u> as well as mapped <u>images</u> on objects. See "*Setuv*" at the end of this section.

<u>Image mapping is accomplished by using the *Rmat* command</u> in the *Create Material* or *Modify Material* dialog boxes (including those for creating new or modifying existing *Standard, Granite, Marble,* or *Wood* material templates) to attach any of the four types of maps. See Figure 41-50, *Modify Standard Material* dialog box.

You must render using the *Photo Real* or *Photo Raytrace* renderer for mapped images to be visible in the resulting rendering.

You can use four kinds of image maps in AutoCAD: texture maps, reflection maps, opacity maps, and bump maps.

Texture Maps

A texture map is the projection of an image (such as a tile pattern) onto an object (such as a table or floor). Texture maps define colors for the surface of the object, as if the bitmap image were "painted" onto the object. For example, you could apply an image of a brick pattern to a horizontal flat surface to create the appearance of a brick floor. Texture maps conform to the attached object's surface geometry and therefore are affected by light and shadow. These capabilities can result in some very realistic renderings.

Figure 41-56 displays the CAMOFLB.TGA image mapped onto the right sphere. You can attach a texture map using the *Modify Standard Materials* dialog box. This example was accomplished by using the *Modify Standard Materials* dialog box with the *Color/Pattern* attribute on and assigning the image file in the *Bitmap Blend* section (see Fig. 41-50).

Figure 41-56

Reflection Maps

A reflection map (also referred to as an "environment map") simulates the effect of a scene reflected on the surface of an object. For example, an image used as a background (maybe a sky scene) can be reflected on an object with a highly reflective surface (such as chrome-plated metal). For reflection maps to render well, the object's material should have high reflection and low roughness, and the image file used for the reflection bitmap should have a high resolution.

Figure 41-57 displays the pattern from the right sphere "mapped" on the left sphere. This was accomplished by selecting the *Mirror* checkbox and setting a high *Reflective* and low *Roughness* value to the material for the left sphere in the *Modify Standard Materials* dialog box. Although the pattern is not actually attached (mapped) to the left sphere, it is considered an environmental map since it affects the appearance of the left sphere.

Figure 41-57

Opacity Maps

Opacity maps create the illusion of opaque and transparent areas on the attached object. (Since most objects are naturally opaque, think of an opacity map as a "transparency" map.) AutoCAD uses the color (lightness or darkness) of the mapped image to determine opacity or transparency. Pure white areas in an opacity map are translated as completely opaque, while pure black areas are treated as transparent. For example, if an image of a white square with a black circle were mapped onto a surface, the white area would be seen as solid (opaque) and the black area would appear as a hole in the surface. If an opacity map is a color image (not black and white), only the equivalent gray-scale values of the colors are used for the opacity calculation. Opacity maps do not necessarily change the color of the opaque areas of the mapped object.

Figure 41-58

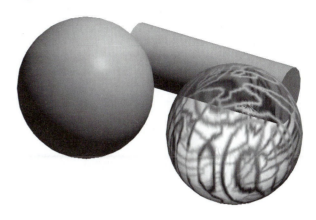

Figure 41-58 displays the previous image attached as an opacity map to the right sphere. This effect was created by setting a high *Transparency* attribute value in the *Modify Standard Materials* dialog box and using the *Bitmap Blend* section to map the CAMOFLAB.TGA image (with the *Transparency* attribute on).

Bump Maps

Bump maps are similar to opacity maps in that the value (gray-scale) of the image is translated to change the surface appearance of attached object. However, in this case the brightness values of a bump map image are translated into apparent changes in the height of the surface of an object. For example, consider mapping an image of white text with a black background onto a surface. The resulting rendering would give the text (white image) the appearance of being raised (or embossed) against a flat (black) background. The surface has not actually changed, although the appearance of the surface (due to the map) has changed. Bump mapping increases rendering time significantly.

Figure 41-59

The bump map in Figure 41-59 (right sphere) was created by selecting the *Bump Map* attribute, then using the *Bitmap Blend* section to specify and adjust the image. Use a small *Bitmap Blend* value (from 0.1 to 0.02).

Combination Maps

You can apply maps in combination. For example, you can apply a wood grain bitmap as both a bump map and a texture map on a paneled wall to give the wall both the "feel" and the color of wood. In addition, an opacity map (a white surface with black dots) could be used to create the illusion of worm holes in the wood.

All maps have a blend value that specifies how much they affect the rendering. For example, a texture map with a blend value less than the maximum (1.0) allows some of the material's surface colors to show through. Lower blend values reduce the effect of the mapped image.

Adjust Material Bitmap Placement Dialog Box

Invoke this dialog box (Fig. 41-60) by selecting the *Adjust Bitmap...* button in the *New Standard Material* or *Modify Standard Material* dialog box. This box is essentially the same as the *Adjust Background Bitmap Placement* dialog box used to place a background image with the *Background* command (see Figure 41-15).

Figure 41-60

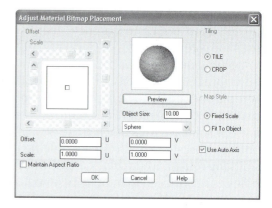

Offset and *Scale*
The image in the upper-left corner indicates the extents of the object (red box) and the bitmap image (black and magenta rectangle). Initially the two rectangles are the same size, indicating a mapping scale of 1:1. *Offset* values range from 1 to -1 in either dimension (U or V).

Maintain Aspect Ratio
This checkbox keeps the U and V dimensions together. Remove the check to scale the image more or less in one dimension.

Tile and *Crop*
If the size of the image you use is smaller than the objects, you can choose to create either a tiled or a cropped map. A tiled map repeats the pattern indefinitely, fitting together like "tiles," until the entire object is covered. This effect is used in Microsoft Windows when a small image is tiled repeatedly to produce a pattern on your desktop. Tiling is adjustable to obtain different tiling values along the mapping axes, U and V. With cropped projection, you fix one image in a specific location on an object. The rest of the object is rendered with the original colors of the material. This effect is sometimes known as a "decal" map because fixed-size image is attached to the object's surface.

Fixed Scale
If you are using tiled bitmaps (repeated patterns), such as bricks, stonework, tile, and wallpaper, this option specifies a fixed scale for the images. The *Scale U* and *Scale V* values control the scale at which the material is tiled onto drawing objects during rendering.

Fit to Object
Use this option when applying a single image to be mapped, such as a billboard or a wall painting. The bitmap placement offset and scale values must be adjusted manually.

Use Auto Axis
If this option is not selected, fixed scale mapping is applied only to XY-oriented surfaces. When checked, images are mapped on XY-, YZ-, and XZ-oriented surfaces. If some images appear backward (such as those containing text) on some faces, make a duplicate application of the image without using *Auto axis*, assign this new image explicitly to the desired surface, and then correct the projection orientation.

Setting Mapping Coordinates

If you want to change the orientation of an AutoCAD <u>material</u> or of a mapped <u>image</u> with respect to the object, you can change the mapping coordinates with the *Setuv* command. Normally, when a material such as wood or an image such as a checkerboard is attached to an object, AutoCAD automatically sets the orientation of the material or image on the objects. As you know from the previous discussion, you can adjust the position of an image only with the *Adjust Material Bitmap Placement* dialog box. In contrast, *Setuv* allows you to change the entire orientation of both materials and mapped images.

For example, assume you used the *Rmat* command to assign wood material to a 3D box to simulate a block of wood. Using the default AutoCAD material orientation of the material with respect to the box, the object would render with the results shown in Figure 41-61.

Figure 41-61

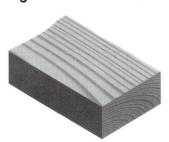

Let's assume that you wanted the block of wood to appear as if it were cut diagonally across the grain. You could use the *Setuv* command to change the orientation of the material to achieve a rendering such as that in Figure 41-62.

Figure 41-62

Mapping coordinates are also referred to as *UV* coordinates, hence the command *Setuv*. The letters *UV* are used because the coordinates are independent of the *XY* coordinates used to describe the AutoCAD geometry; the coordinates are related to the mapped objects, not to any traditional coordinate system.

SETUV

Pull-down Menu	COMMAND (TYPE)	ALIAS (TYPE)	Short-cut	Screen (side) Menu	Tablet Menu
View *Render >* *Mapping...*	*SETUV*	...	...	*VIEW 2* *Mapping*	*R,1*

Setuv allows you to change the orientation of an attached <u>material</u> or mapped <u>image</u> with respect to the object. *Setuv* sets the mapping coordinates (U, V, and W). See previous discussion, "Setting Mapping Coordinates."

When you use *Setuv*, AutoCAD first prompts you to select objects. When you have selected the AutoCAD objects to which you want to assign mapping coordinates, AutoCAD displays the *Mapping* dialog box (Fig. 41-63). You can use the *Preview* box in the *Mapping* and *Adjust Projection* dialog boxes to experiment to find the best result before you do a full rendering.

Figure 41-63

Projection
Select one of the buttons to specify *Planar, Cylindrical, Spherical,* or *Solid* projection. Normally, select the projection type that matches the overall shape of the object for which you want to specify coordinates.

Adjust Coordinates
Selecting this button displays a dialog box for adjusting projection coordinates relative to the object. The dialog box that is displayed is based on your choice of *Projection: Planar, Cylindrical, Spherical,* or *Solid*.

Acquire From

Use this option to return to the drawing so you can select one object to "copy" coordinates from. The object should already have mapping coordinates assigned to it—these become the current mapping settings. You can accept the acquired mapping coordinates as is or use *Adjust Coordinates* to modify them.

Copy To

This option allows you to copy the current mapping coordinates to any number of objects. When you finish the selection, then choose *OK*, the current mapping coordinate settings are applied to the objects you selected.

Adjust Planar Coordinates Dialog Box

When you select *Planar Projection,* then *Adjust Coordinates,* this dialog box appears (Fig. 41-64). Use this dialog box to specify the plane used for projecting the bitmap onto the object. *Planar* projection maps the material or image onto the object with a one-to-one correspondence, as if you were projecting the image from a slide projector onto the surface. This does not distort the texture; it just scales the image to fit the object.

Figure 41-64

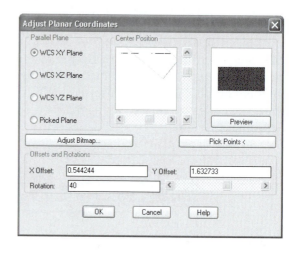

Parallel Plane

This section is used to select one of the three perpendicular reference planes of the WCS or a plane you pick using *Pick Points.* If you want to select a plane, use *Pick Points,* then select *Picked Plane.*

Center Position

Shows a diagram of normal to the object's current parallel plane (red image). The current projection square for the material or image is in light blue (the top of the projection square is indicated by a small blue tick mark, and the square's left edge is shown in green). You can move the material/image projection square by moving the sliders.

Adjust Bitmap

Choose this button to adjust the material/image offset and scaling. This dialog box operates similar to those for *Adjust Material Bitmap Placement* (see Fig. 41-60) and *Adjust Background Bitmap Placement.*

Pick Points

Selecting this button temporarily displays the Drawing Editor so you can specify a projection plane by picking points on an object. You must select three points: the lower-left, lower-right, and upper-left corners of the plane.

Offset and *Rotation*

You can change the *X* and *Y* offsets of the map by entering new values in the boxes. You can change the rotation by entering a new value or by moving the slider.

Adjust Cylindrical Coordinates Dialog Box

Selecting the *Cylindrical* projection type, then *Adjust Coordinates* displays the *Adjust Cylindrical Coordinates* dialog box (Fig. 41-65). Here you can define the axis of the cylindrical coordinate system and the wrap line for the bitmap projection. *Cylindrical* projection maps an image or material onto an object with a cylindrical projection; the horizontal edges are wrapped together but not the top and bottom edges. The height of the image is scaled along the cylinder's axis.

Figure 41-65

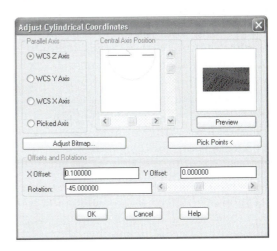

Parallel Axis
Use this option to select one of the three perpendicular axes of the WCS or an axis you pick using the *Pick Points* button.

Central Axis Position
This section shows a view of the selected objects' meshes (red image) on a plane perpendicular to the current axis. It also shows the projection axis with a blue circle (the center of the circle is the projection axis). A green radius shows the wrap line.

By default, the axis is placed at the center of the object's extents in both directions, with the base of the cylinder at the minimum vertical extent and its top at the maximum vertical extent. The wrap line is placed toward minimum Z—or minimum Y when the Z axis is parallel to the cylindrical coordinate system axis.

Pick Points
Use this option to select a projection cylinder by picking points in the drawing. You must select three points: the center-bottom of the mapping cylinder, the center top of the cylinder, and the direction toward the seam (wrap line).

Offset and *Rotation, Adjust Bitmap*
These options operate similarly to the other dialog boxes (see "*Adjust Planar Coordinates* Dialog Box," previous).

Adjust Spherical Coordinates Dialog Box

This dialog box is used to define the polar axis of the spherical coordinate system and the wrap line for the bitmap projection. *Spherical* projection wraps the texture both horizontally and vertically. The top edge of the texture map is compressed to a point at the "north pole" of the sphere as is the bottom edge at the "south pole."

The *Adjust Spherical Coordinates* dialog box has the same options and operates similarly to the *Adjust Cylindrical Coordinates* dialog box (see Fig. 41-65).

Adjust UVW Coordinates Dialog Box

Figure 41-66

Use this dialog box (Fig. 41-66) to adjust the three coordinates to shift a solid, 3D material: marble, granite, or wood. Because solid materials are three-dimensional, they have three mapping coordinates, U, V, and W, and can be applied from any angle. You do not always need to specify the mapping coordinates for these materials, but you can. For example, you might want to change the material's orientation for a particular rendering or skew a pattern along one dimension.

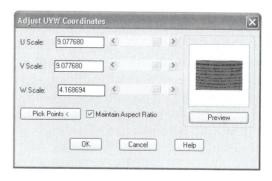

U Scale, V Scale, and W Scale
The material/image scale changes the scale of the material relative to the object mapping coordinates in any dimension (U, V, or W). Adjust the scale by using the three sliders or by entering a value in the edit boxes. The scale is the number of times the image fits onto the object in the U, V, or W direction. Select *Preview* to see the results of the new scale.

Maintain Aspect Ratio
The *Maintain Aspect Ratio* check box locks the U, V, and W scaling dimensions together. When *Maintain Aspect Ratio* is turned on, moving one slider moves the other, and a value you enter in one box changes the other. Remove the check to change the scale in any one, two, or three independent directions to distort the image or material for particular effects.

USING *SCENES*

When the *Background* has been set (if desired), *Lights* have been inserted, and *Materials* have been attached and adjusted, you can use the *Render* command to calculate the rendered view and display it in the viewport or rendering window, or write the image to a file (whichever is configured). If no *Scene* has been specified, AutoCAD uses the current view and all lights in the rendering. However, you can choose to use *Scene* to specify a particular view or lighting combination. *Scenes* are helpful if you want to create several renderings of a model using different viewpoints and different light combinations.

SCENE

Pull-down Menu	COMMAND (TYPE)	ALIAS (TYPE)	Short-cut	Screen (side) Menu	Tablet Menu
View *Render >* *Scene...*	SCENE	...	...	VIEW 2 *Scene*	N,1

If you want to render objects from one of several named *Views* or use different light combinations for a viewpoint, you can use the *Scene* command to select the *View* and *Light* combinations for each render. You can save particular *Views* and lighting configurations to a *Scene* with a name that you assign. This action allows you to create multiple renderings without having to recreate the configuration each time. *Scenes* do not have icons that are inserted into the drawing like *Light* icons.

Invoking the *Scene* command by any method displays the *Scenes* dialog box (Fig. 41-67, next page). Selecting the **NONE** entry renders the current viewpoint and uses all lights inserted in the drawing. If no lights are inserted, AutoCAD uses one from over the shoulder like a default rendering.

New...
Selecting the *New...* option allows you to create a new *Scene* for the drawing and produces the *New Scene* dialog box (see Fig. 41-68).

Modify...
This option produces the *Modify Scene* dialog box where you can modify existing *Scenes*. This dialog box is the same as the *New Scene* dialog.

Delete
Delete removes the highlighted scene from the drawing.

Figure 41-67

Creating New and Modifying Existing Scenes

The *New Scene* and the *Modify Scene* dialog boxes allow you to select from a list of existing named *Views* and inserted *Lights*. First, assign the desired name for the scene by entering it in the edit box in the upper-right (Fig. 41-68). The name is limited to eight characters, no spaces.

The *Views* column lists all previously created views. The **CURRENT** view can be selected if no *Views* have been previously assigned. You can only have <u>one *View*</u> in a *Scene*. Selecting an existing view makes it the current view for the drawing as well as that used for the subsequent *Render*.

You can also select *Lights* from the list to be included for the rendering of the scene. Any number of lights can be selected for a scene. Select multiple lights by holding down Ctrl while selecting. Highlighted lights are included in the scene. There is no limit to the number of *Scenes* that can be created.

Figure 41-68

RENDERING AND SPECIFYING PREFERENCES

Finally, after setting a *Background,* inserting *Lights,* attaching and modifying *Materials,* you are ready to set the rendering preferences and create the rendering. Using the *Render* command is the easiest step in creating a rendering because Render does the work.

Use the *Render* command to set the preferences and create the rendering. If you want to set preferences without creating a rendering, you can use *Rpref* to produce the *Rendering Preferences* dialog box (see next).

RENDER

Pull-down Menu	COMMAND (TYPE)	ALIAS (TYPE)	Short-cut	Screen (side) Menu	Tablet Menu
View *Render >* *Render...*	*RENDER*	*RR*	...	VIEW 2 *Render*	*M,1*

Using this command causes Render to calculate and display the rendered scene to the configured device. The rendering is generated according to the settings in the *Render* or *Rendering Preferences* dialog box.

Render uses the *Lights* and *View* selected in the *Scene* dialog box. If no *Scene* or selection set is specified, the current view is rendered with all *Lights*. If there are no *Lights* in the drawing, Render assumes an "over-the-shoulder" *Distant Light* source with an intensity of 1.

After issuing the *Render* command, the *Render* dialog box appears (Fig. 41-69) unless the *Skip Render Dialog* option is checked in the *Rendering Preferences* dialog box. This dialog box offers the same options as the *Rendering Preferences* dialog box.

After setting the preferences or accepting the defaults, select the *Render* button to generate the rendering. Rendering can take considerable time, depending on the complexity of the scene, your computer system capabilities, and the settings in the *Render* or *Rendering Preferences* dialog box. Once you have selected the *Render* button to initiate the rendering process, you can stop (cancel) the rendering by pressing Esc.

It is a good habit to determine or check the settings in this dialog box immediately before using the *Render* command. Since rendering takes time, taking a minute to make sure the desired options are set beforehand can save wasted rendering time.

Figure 41-69

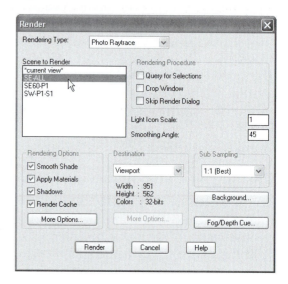

Render **Dialog Box**

Rendering Type
This drop-down box lists *Render, Photo Real,* and *Photo Raytrace.* Other rendering devices can be installed.

Render
This is the default option. *Render* creates the fastest rendering. Use this option to save time when you first begin adjusting *Background, Lights,* and *Materials* for the model.

Photo Real
This of *Render* can display transparent materials and bitmapped images. Mapped images are images from an external file that can be planar (as for a background) or mapped to geometry such as a sphere, box, or other shape. *Photo Real* can also generate volumetric and mapped shadows.

Photo Raytrace
Photo Raytrace produces the highest quality (most realistic) renderings. Raytracing is the process of calculating or following the paths of light rays as they are generated from a source and cast shadows or bounce off objects and illuminate other objects. Raytracing is needed to calculate reflections of light and color.

Scene to Render
This section lists all scenes previously created, including the current view. Select any scene from the list to render.

Rendering Procedure

Query for Selections
When you render, you are prompted to specify a selection set to render if this button is selected. This is an excellent way to save time when you need to render only one object or set of objects in the view as a test, such as when you are adjusting *Material* for one object.

Crop Window
When this option is checked and you then select *Render,* AutoCAD prompts you to select an area (window) on the screen to render. This is also a good way to save time when working on only one section of the model.

Skip Render Dialog
Normally, the *Render* dialog box appears when you use the *Render* command. Checking this box skips this procedure. In this case, you can use the *Rendering Preferences* dialog box to set your preferences.

Rendering Options

Smooth Shading
If *Smooth Shading* is checked, Render automatically smoothes out the multifaceted appearance of polygon meshes. With *Smooth Shading* off, the mesh edges appear in the final rendering. If an edge defines a corner of greater than 45 degrees, it is not affected by *Smooth Shading*.

Apply Materials
This option applies your materials when the rendering is calculated. If *Apply Materials* is not checked, Render uses the color, ambient, and reflection settings for the default *GLOBAL* material for the rendering.

Shadows
This is the global setting for calculating shadows. Each *Light* also has a *Shadows* checkbox. For a *Light* to calculate shadows, this button as well as the individual *Light's* button must be checked. *Photo Real* or *Photo Raytracing* must be used to calculate shadows.

More Options...
Selecting this tile produces a dialog box with advanced options. The dialog box that appears depends on whether you have *Render, Photo Real* or *Photo Raytrace* selected. See "*Render Options* Dialog Box" at the end of this section.

Destination

Viewport
If this option is checked, Render calculates the rendering and displays it in the Drawing Editor. If you are using tiled or paper space viewports, the rendering is in the current viewport.

Render Window
Selecting this option creates the render in a separate window, not in the Drawing Editor.

File
If this option is checked, the rendering is calculated and written to the file but is not displayed. You can specify the type of file to render to by selecting *More Options...* in this cluster, which produces the *File Output Configuration* dialog box. See"Saving a Rendered Image to a File" in "Saving, Printing, and Replaying Renderings."

If you specified that Render create a file, the rendering is calculated and written to the specified file or device but is not displayed on the screen (see "Saving, Printing, and Replaying Renderings"). Alternately, if the *Viewport* or *Rendering Window* is the selected *Destination* and you decide to save the rendered screen display after creating the rendering, you can use *Saveimg* (see "*Saveimg*") to save the image as a .BMP, .TGA, or .TIF, or you can use *Save* from the Render window to save the image as a .BMP file.

Lights Icon Scale

Enter a value to scale the *Light* icons that are inserted into the drawing (see Figure 41-38). Any value larger than 0 is valid.

Smoothing Angle

This value determines the angle at which smoothing is applied. For any angles on the object less than the value specified, AutoCAD renders as if no edge exists (see "*Smooth Shading*").

Sub Sampling

Use this option to reduce rendering time and image quality without dropping the effects (such as shadows). This feature speeds up rendering by rendering only a sample of the pixels. Select an option from the drop-down list ranging from 1:1 (slowest, best quality) to 8:1 (fastest, lowest quality).

Background...

This option invokes the *Background* dialog box (see the *Background* command).

Fog/Depth Cue...

Selecting this button produces the *Fog/Depth Cue* dialog box (see the *Fog* command).

Render Options Dialog Box

The *Rendering Options* dialog box (Fig. 41-70) appears when you select *More Options...* in the *Rendering Options* section of the *Rendering* dialog box, and <u>Render</u> is the selected *Rendering Type*.

Figure 41-70

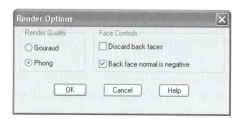

Render Quality

Gouraud

This method calculates light intensity at each vertex and interpolates the intensities between vertices. As a result, a sphere, for example, may have a diamond-shaped highlight.

Phong

This creates the most realistic type of rendering with higher-quality highlights. Phong calculates light intensity at each pixel.

Face Controls

Discard Back Faces

This setting prevents Render from reading the back faces of surfaces when rendering, which can save considerable time calculating the rendering (see "*Back Face Normal Is Negative*").

Back Face Normal Is Negative

Checking this setting reverses which faces Render considers as back faces. Normally, a positive normal vector points outward from the interior of the object toward the viewer. When constructing surfaces such as a *3Dface*, specifying vertices in a counterclockwise direction creates a positive normal vector toward the observer. Turning off this setting reverses the direction AutoCAD considers back faces. If surfaces in a surface model rendering are black and *Discard Back Faces* is checked, reverse this setting or turn off *Discard Back Faces*.

Photo Real and *Photo Raytrace Render Options* Dialog Boxes

These dialog boxes appear when either <u>*Photo Real*</u> or <u>*Photo Raytrace*</u> is selected and you select *More Options…* in the *Rendering Options* section of the *Render* or *Rendering Preferences* dialog box. The *Photo Real* options are the same as the *Photo Raytrace* options, excluding *Adaptive Sampling* and *Ray Tree Depth*. Only the *Photo Raytrace Rendering Options* dialog box is shown here (Fig. 41-71).

Figure 41-71

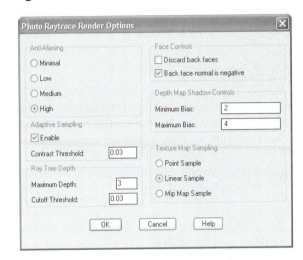

Anti-Aliasing
This section offers four levels of anti-aliasing resolution. Aliasing is the jagged edges on a diagonal edge caused by rows and columns of pixels. The range (*Minimal* to *High*) controls how many multiple samples are rendered per pixel. The trade-off in quality is the increased rendering time.

Adaptive Sampling
Anti-aliasing can be accelerated with *Adaptive Sampling* enabled. This feature prevents anti-aliasing from samples with low color contrast. The *Contrast Threshold* value determines the level of contrast (from 0 to 1.0) before anti-aliasing takes place. Higher values increase rendering time at the expense of image quality.

Ray Tree Depth
This option controls the depth of the tree to track reflected and refracted rays. Greater values for *Maximum Depth* give better images (10 is the recommended maximum value). The *Cutoff Threshold* value (percentage) determines how strong the rays must be for each "bounce" before they are terminated. Higher values create a greater tree depth.

Depth Map Shadow Controls
Use this option to prevent "detached" shadows (from the object). Set *Minimum Bias* from 2 to 20 and *Maximum Bias* from 4 to 10.

Texture Map Sampling
When a texture map is projected onto an object smaller than itself, use this section to control how the image is sampled. *Point Sample* causes render to use the closest pixel, *Linear Sample* is a linear average of pixels, and *Mip Map Sample* takes a triangular sample.

Face Controls
See "*Render Options* Dialog Box."

RPREF

Pull-down Menu	COMMAND (TYPE)	ALIAS (TYPE)	Short-cut	Screen (side) Menu	Tablet Menu
View Render > Preferences...	*RPREF*	*RPR*	...	VIEW 2 *Rpref*	R,2

The *Rpref* command invokes the *Rendering Preferences* dialog box (not shown). This box is identical to the *Render* dialog box except you cannot generate a rendering from this box (there is no *Render* button). Use this box when you want to make changes to the rendering preferences, then press *OK* to save the changes. (The *Render* dialog box provides no *OK* button to save changes; *Render* and *Cancel* are the only two methods to leave the *Render* dialog box.)

Because the options in the *Rendering Preferences* dialog are identical to those in the *Render* dialog box (other than the *Render* button), see the previous sections for information on these options.

Rendering Example

A solid model of an intake housing and machinist's file are used here as an example for creating a *Background*, inserting *Lights*, creating *Materials*, using a *Bump Map*, setting rendering preferences, and creating a rendering (refer to "Typical Steps for Using Render").

1. After the geometry is completed, *3Dorbit* or *Vpoint* is used to provide an appropriate viewing orientation for the rendering. If desired, the viewpoint can be saved to a named *View* to be used later for defining *Scenes*. Figure 41-72 displays the solid models ready to begin the rendering process. Notice that the drawing includes 3 objects: the intake housing, a machinist's file, and a surface (only the back edge of the *Box* appears) that is used to simulate a table top, which will provide a surface for shadows and reflection for the intake housing and file.

Figure 41-72 ───

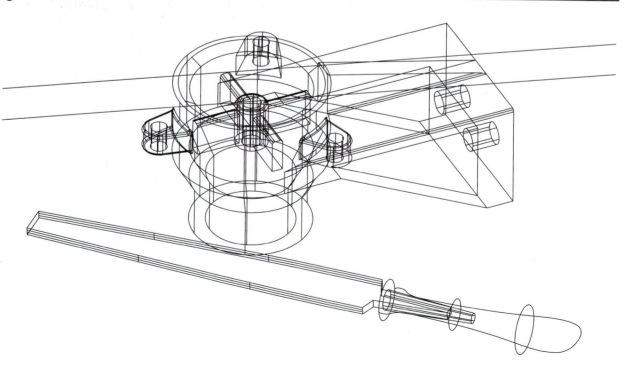

2. To begin the rendering setup, a *Background* is created (Fig. 41-73). Using the *Gradient* option, three colors are selected. A 90-degree *Rotation* is specified for the bands of color to give the illusion of vertically oriented background objects and to provide directional contrast to the horizontal orientation of the objects to be rendered. A quick, "rough" rendering is generated for checking the background using the *Render* option of *Rendering Type* (not *Photo Real* or *Photo Raytrace*). The rendering is generated with no lights or materials specified (using the "over-the-shoulder" light and GLOBAL material). The *Query for Selections* option is checked in the *Render* dialog box to allow selection of only the intake housing to speed rendering time. All other options in the *Render* dialog box are set to the minimum, fastest, or default values.

Figure 41-73 ——————————

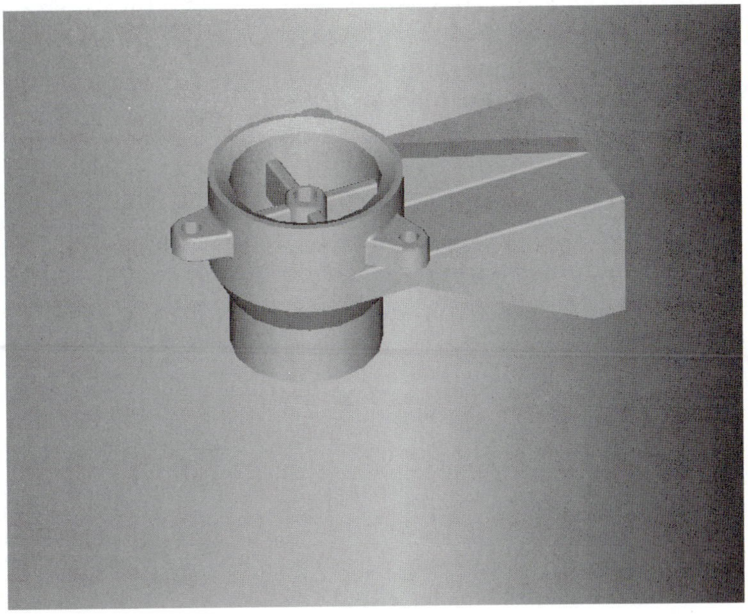

3. Next, the *Lights* dialog box series is used to insert a light in the drawing (Fig. 41-74). For this example, the first light is a *Point Light* positioned in the upper right. Setting the *Point Light* color to a cool color (very light shade of purple or blue) can be used to reinforce the appearance of metallic surfaces. *Shadows On* is selected for the light when created, but the (global) *Shadows* option (in the *Render* dialog box) is off for this preliminary work. The light position illuminates the front, top, and right sides of the intake housing. Notice how the point light illumination falls off along the top surface. Since only one light is used, the top and right side surfaces appear very bright, while the back of the cylinder has no light. Also notice the "floating" effect of the model due to the lack of shadows.

Figure 41-74 ——————————

A material is selected and modified to represent a smooth, reflective metal for the intake housing (see Fig. 41-74). In the *Modify Standard Material* dialog box, the *Reflection* attribute is set with a high value and *Roughness* is given a low value. A hard, dark, and reflective material is selected for the table top. Only the intake housing and table top are selected for the rendering (use *Query for Selection*). *Photo Real* is set as the *Rendering Type* to enhance the light reflective characteristics of the materials. Generating a *Render* (Fig. 41-74) displays the results of the light position and intensity as well as material characteristics. You can make several adjustments to *Light* and *Material* settings, then generate "rough" renderings until the desired effect is obtained.

4. Next, materials are assigned to the file and handle (each created as a separate model). The handle has a supplied wood material and the metal portion of the file has a standard metallic material attached (Fig. 41-75). The *Photo Real* renderer is used, but *Shadows* are toggled off to speed rendering time when checking and adjusting material characteristics for the file and handle.

Figure 41-75

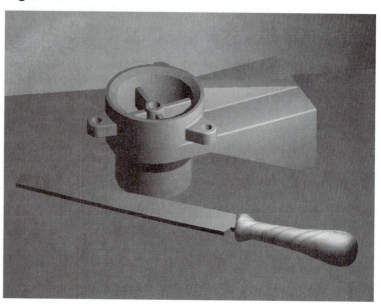

5. A rendering with only one point light can create drastic contrast from one side of the scene to the other—some portions of the model are well illuminated, while others are dark. Therefore, a second *Light* is inserted into the drawing to provide more even and realistic illumination over the surfaces of the model. This second light is a *Distant* light positioned on the left side and having a direction set to illuminate the left side of the models (Fig. 41-76). The new light compensates for the previously dark left side of the models; however, with two or more new lights, a scene can have a bright, washed-out effect. *Intensity* for the lights should be adjusted to

Figure 41-76

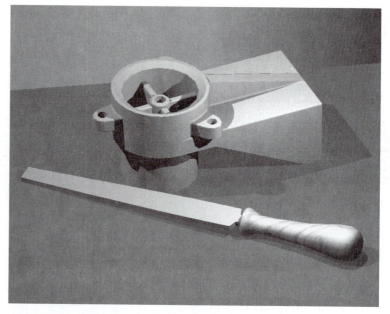

provide the desired illumination. Make several *Intensity, Location,* and *Color* adjustments, if necessary, to achieve the desired effects. A useful technique is to use a slightly different color for the two lights; for example, with the *Distant Light,* use a warm tone (light yellow).

The *Shadows* checkbox is selected in the *Render* dialog box to generate shadows (both lights were created with *Shadow On* in the light dialog boxes). Shadows can make a dramatic difference in the realism of the rendered image. Note that the objects no longer "float," but now appear to be sitting on the table. Also note both sets of shadows, one from each light source.

6. For the final adjustments, an image file (of a grid pattern) is used as a *Bump Map* to provide the serrated edges on the surface of the file model. The *Setuv* command is then used to position (rotate) the bitmapped image at the desired angle with respect to the body of the file model.

For this final rendering (Fig. 41-77), *Photo Raytrace* is the desired *Rendering Type* so reflections can be generated. The table surface material *Reflection* attribute is modified by toggling *Mirror* in the *Modify Standard Material* dialog box. With these options, rendering time increases considerably.

Also note that a perspective view is used in this rendering. This view can be accomplished by using *3Ddistance* to create the perspective view initially, but turning the perspective *Off* for preliminary renderings, then restoring perspective for the final rendering. Alternately, create two similar *Views*—one with and one without perspective.

The *Shadows* options can be adjusted for the final rendering in the individual *Light's* dialog boxes. For example, you may want to generate shadow maps, which are necessary for "soft" shadows. Figure 41-77 displays *Shadow Volumes/Raytraced Shadows.*

Other rendering options can be set at this time. *Anti-aliasing* can be set to *Medium* or *High* to minimize the "stair-step" effect of the table edge (this problem occurs with model edges that are "almost" horizontal or vertical).

Don't expect to achieve excellent results when you create your first few renderings. Many options are available; therefore, a surprising amount of time can be spent making adjustments and generating "rough" renderings. Please consider using all time-saving techniques except for "final" renderings.

Figure 41-77

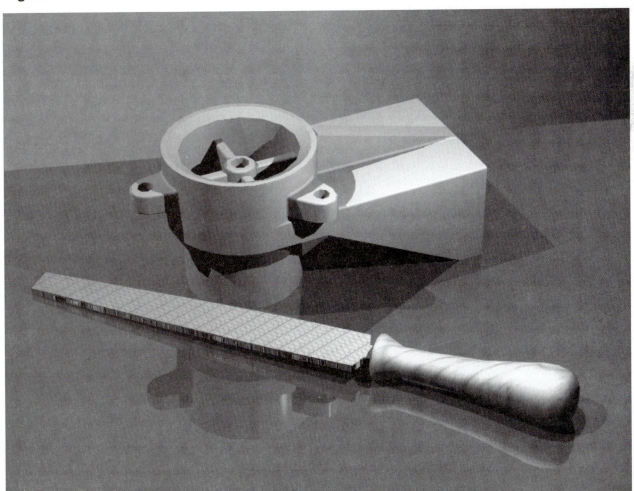

SAVING, PRINTING, AND REPLAYING RENDERINGS

Renderings can be saved to files in a number of ways. If you have used the *Render* command to generate a rendered display in the Drawing Editor, then want to save the image to a .BMP, .TGA, or .TIF file, use *Saveimg*. If you created the rendering in the Render window, then you can save the image to a .BMP file only. If you want to render directly to a file (without rendering to the display first), the *File* option of the *Destination* cluster of the *Render* or *Rendering Preferences* dialog box can be checked. To replay an image from a file, use *Replay*. If you want to print a rendering, first select *Render Window* as the *Destination*, then select *Print*. Alternately, use *Shade Plot* with the *Rendered* or *As Displayed* option.

SAVEIMG

Pull-down Menu	COMMAND (TYPE)	ALIAS (TYPE)	Short-cut	Screen (side) Menu	Tablet Menu
Tools Display Image > Save...	SAVEIMG	...	...	TOOLS 1 Saveimg	...

The *Saveimg* command allows you to save the rendered image in the Drawing Editor (you selected *Viewport* as the rendering *Destination*) to a .BMP, .TIF, or .TGA file format. The image can be displayed at a later time with the *Replay* command. No matter how long it took for *Render* to calculate and display the original rendering, when it is saved to a .BMP, .TIF, or .TGA file, it can be *Replayed* in a matter of seconds.

If you want to create a rendering and save it to a file directly (without first rendering to the screen), you can select *File* as the *Destination* in the *Rendering Preferences* dialog box (see "Saving a Rendered Image to a File," next).

Figure 41-78

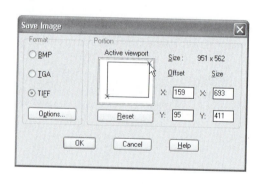

The *Saveimg* command invoked by any method produces the *Save Image* dialog box where you can select the file format and the portion of the image you want to save (Fig. 41-78).

Format
BMP
   A BMP file is a Windows bitmap raster file format.
TGA
   This is the Truevision v2.0 format (.TGA file extension). It is a 32-bit format (16 million colors).
TIFF
   Also a 32-bit format, this option is a Tagged Image File Format (.TIF file extension).

Options
Depending on the type of file (TIF and TGA only), file compression is available.

Offset
This option sets the lower-left corner for the image to save. The X and Y values are in screen pixels. You can also PICK in the *Active viewport* image tile to set *Size* and *Offset*.

Size
This option sets the upper-right corner for the image in pixels.

Reset
This option resets the *Offset* and *Size* values to the full-screen defaults.

Saving a Rendered Image to a File

If your choice for *Destination* is *File* (in the *Render* or *Rendering Preferences* dialog box) you can render directly to the selected raster file format without rendering to the screen. In this case, when *Render* is invoked, the screen display keeps its previous drawing or image and the calculated rendering is written directly to the designated file.

When *File* is selected as the *Destination* in either the *Render* dialog box (see *Render*, Fig. 41-69) or in the *Rendering Preferences* dialog box, you can choose the file format for the rendering to be written by selecting *More Options...* button. The *File Output Configuration* dialog box appears (Fig. 41-79).

Figure 41-79

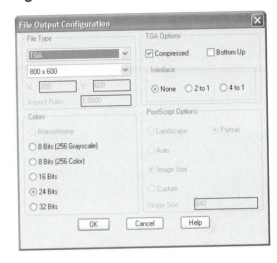

File Type

This option specifies the type of output file and rendering resolution. Supported option types are *BMP, TGA, PCX, TIFF* (.TIF), and *PostScript* (.EPS). Screen resolution is the number of columns and rows of pixels created for the image.

Aspect Ratio

This area sets the aspect ratio (ratio of horizontal/vertical).

Colors

There are several options for the number of colors based on the file type.

Options

Compression turns on file compression if supported by the selected file type. *Bottom Up* starts the scan lines from the bottom left instead of the top left.

Interlace

None turns off interlacing. Checking *2 to 1* or *4 to 1* sets those interlacing modes on.

PostScript Options

These options are available only if *PostScript* is selected as a file type.

Rendering and Printing from a Rendering Window

Figure 41-80

If you have selected *Render Window* as the *Destination* in either the *Render* dialog box or in the *Rendering Preferences* dialog box, your renderings will appear in a separate window (Fig. 41-80). You can toggle between the rendering window and AutoCAD using the Windows Task Bar or the Alt+Tab key sequence.

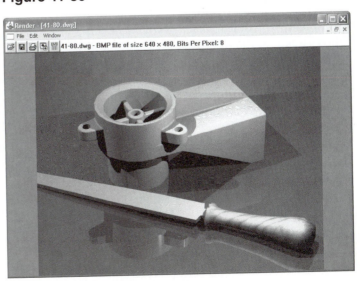

Several options are here that are not available by other means. For example, several options for printing a rendered image are available by selecting either the print icon or *Print* from the *File* pull-down menu. If you have configured AutoCAD to use the configured Windows printer, the *Print* dialog box appears (Fig. 41-81).

Printing is allowed across several pages by using the *Tile Pages* slider bars. You can also drag the position squares at the corners of the image to change or reproportion the size or reposition the image on the selected sheet(s).

Figure 41-81

Selecting the options icon or *Option* from the *File* pull-down menu produces the *Windows Render Options* dialog box (Fig. 41-82). Here you can change the resolution (*Size in Pixels*) and *Color Depth* of the bitmap image.

Figure 41-82

Printing a Rendered Image with *Shade Plot*

In AutoCAD 2004 you can print rendered 3D models using the *Shade Plot* option in the *Plot* dialog box or using the *Shadeplot* option for a viewport. See also "Printing Shaded 3D Models" in Chapter 35 and "Shaded Viewport Options" in Chapter 14.

To make a print of an image already rendered on the screen, select the *As Displayed* option of *Shade Plot*. You can also make a rendered print (with or without previously generating a rendered display to the screen) using the *Rendered* option of *Shade Plot*. The *Rendered* option prints the drawing according to the settings made in the *Rendering Preferences* dialog box, including any specified lights, materials, background, and views.

Printing a Rendered Image From the *Model* Tab
To print a shaded 3D drawing from the *Model* tab, use the *Shaded Viewport Options* section of the *Plot* dialog box, *Plot Settings* tab. Use the *Shade Plot* drop-down list and select *Rendered* (Fig. 41-83). Alternately, you can first use the *Render* command to produce the desired display on screen, then select *As Displayed* from this list to print the drawing as it appears on screen. You can select from several resolution qualities (*Draft, Preview, Normal, Presentation, Maximum, Custom*) for the print using this method only (not from the *Properties* palette or using the *-Vports* command).

Figure 41-83

2004

Printing a Rendered Image From a *Layout* Tab

If you are printing from a layout and want to specify a rendered image for one or more viewports, use the *Shadeplot* option of the viewport's *Properties* palette (Fig. 41-84). Here you can select from the same options as in the *Plot* dialog box. If you prefer to type, use the *Shadeplot* option of *–Vports*.

Figure 41-84

REPLAY	Pull-down Menu	COMMAND (TYPE)	ALIAS (TYPE)	Short-cut	Screen (side) Menu	Tablet Menu
	Tools *Display Image >* *View...*	*REPLAY*	...	...	*TOOLS 1* *Replay*	*V,8*

The *Replay* command opens the *Replay* dialog box (not shown) where you specify the file you want to replay. You can replay any .BMP, .TGA, or .TIF files. Enter the desired file format to replay in the *Files of Type* or *Pattern* edit box to make the existing files appear in the files list. After you select the file to replay, the *Image Specifications* dialog box appears.

Figure 41-85

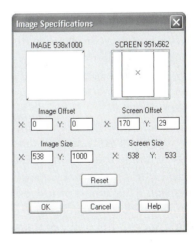

In the *Image Specifications* dialog box (Fig. 41-85), you can specify that the entire image or a portion of the image be displayed. You can also offset the image in the screen area.

The *Image* area of the dialog box displays the full image size (listed in the *Image Size* edit boxes). You can display the full image or PICK the lower-left and upper-right corners in this box to define a portion of the image to display. Optionally, you can enter the lower-left corner X,Y values (in pixels) in the *Image Offset* edit box and the image size (in pixels) in the *Image Size* edit box.

The *Screen* image box allows you to PICK the location on the screen for the image to be replayed. Just PICK a location in this image tile to specify the new location for the <u>center</u> of the image. Alternately, the lower-left corner position can be specified by entering X,Y values (in pixels) in the *Screen Offset* edit boxes. When AutoCAD displays the image, the size is limited to the size (in pixels) of the window in which it is displayed.

MISCELLANEOUS RENDERING TIPS AND FEATURES

STATS

Pull-down Menu	COMMAND (TYPE)	ALIAS (TYPE)	Short-cut	Screen (side) Menu	Tablet Menu
View *Render >* *Statistics...*	*STATS*	...	...	VIEW 2 *Stats*	...

The *Stats* command reports information about the last rendering that you created. The *Statistics* dialog box appears (Fig. 41-86) giving information such as time spent rendering and the status of all options used (as set in the *Render* or *Rendering Preferences* dialog box).

The data collected by *Stats* can be saved to a file that you designate in ASCII format. Any file extension can be assigned. In many cases it is a good idea to save this information along with the rendering file (.BMP, .TGA, .TIF, etc.) to keep a record of the settings used to create the image.

Figure 41-86 ————————

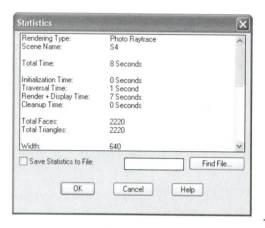

Setting Rendering Options for *Shademode*

Since AutoCAD 2002, *Shademode* has been enhanced to display any lights, materials, textures, and transparency applied in the drawing. Lights, materials, textures, and transparency are applied using the <u>rendering</u> features in AutoCAD.

Two basic steps are used to display rendered features using *Shademode*. First, use the rendering commands to apply lights, materials, and textures, and, second, use *Options* to configure your graphics display device to recognize these rendered features for *Shademode*. In other words, there is nothing new you must do in the *Shademode* command to enable the new features—use *Options* and the rendering commands instead.

Figure 41-87 ————————

To configure your display device, select the *System* tab in the *Options* dialog box. Next, select *Properties* under *Current 3D Graphics Display*. This action produces the *3D Graphics System Configuration* dialog box. In the upper-right corner of this dialog box (Fig. 41-87), select *Render options* to enable selection of any or all of the other options below it. Once enabled, rendering features applied in the drawing (lights, materials, and textures) are displayed whenever you use *Flat Shaded, Gouraud Shaded, Flat Shaded Edges On,* or *Gouraud Shaded Edges On*.

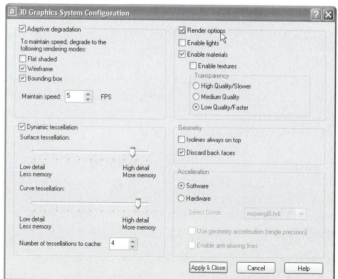

The rendering options in the *3D Graphics System Configuration* dialog box are described here.

Render Options
Checking this option enables the options listed below it.

Enable Lights
For the *Shademode* options, this checkbox applies lights to objects that were defined with the *Light* command. If this option is not selected or if *Lights* have not been used for the drawing, then the "default" lighting for 3D views is used.

Enable Materials
For the *Shademode* options, this checkbox displays any materials attached to objects using the *Rmat* command. If the objects have no attached material, the default "Global" material is used. If this option is not selected or if *Rmat* has not been used in the drawing, then no materials are displayed.

Enable Textures
This option shows textures (for shaded objects) that are attached using the *Rmat* and *Setuv* commands. *Enable Materials* must also be selected for textures to be visible.

Transparency
These settings adjust the transparency quality for shaded objects. You can improve image quality at the expense of redraw time. *Enable Materials* must also be selected for transparency to be turned on.

Rendered objects in *Shademode* are subject to the following limitations. Shadows, 3D textures, bump maps, opacity maps, refraction, and reflection are not shown. Lights have no attenuation and have a default light intensity. 2D textures (bitmaps and bitmap blending) are displayed for the color and pattern only. Other rendering features such as background and fog are not supported in 3D views.

VIEWRES and FACETRES

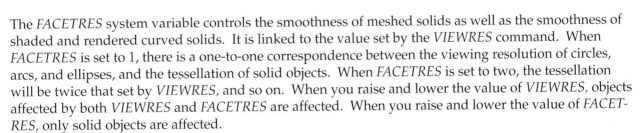

The value you set with the *VIEWRES* command controls the accuracy of the display of circles, arcs, and ellipses. To increase performance while you draw, set a low value for *VIEWRES*. However, to make sure you get a good-quality rendering, raise the value before rendering drawings that contain arcs or circles. (See "Solid Modeling Display Variables" in Chapter 39.)

The *FACETRES* system variable controls the smoothness of meshed solids as well as the smoothness of shaded and rendered curved solids. It is linked to the value set by the *VIEWRES* command. When *FACETRES* is set to 1, there is a one-to-one correspondence between the viewing resolution of circles, arcs, and ellipses, and the tessellation of solid objects. When *FACETRES* is set to two, the tessellation will be twice that set by *VIEWRES*, and so on. When you raise and lower the value of *VIEWRES*, objects affected by both *VIEWRES* and *FACETRES* are affected. When you raise and lower the value of *FACET-RES*, only solid objects are affected.

Drawing Outward-Facing Surfaces for Surface Models

When you prepare surface models for rendering, consider these factors. AutoCAD uses the "normal" on each face to determine which is a front face and which is a back face. A normal is a vector that is perpendicular to each polygon face on your model and points outward from the surface. If you draw the face (such as a *3Dface*) by PICKing counterclockwise, the normals point outward; if you draw the face clockwise, the normals point inward. You should draw all faces consistently. Mixing methods produces unexpected rendering results. AutoCAD calculates all the normals in the drawing during rendering.

If all surface model faces are drawn consistently, all normals should point outward from (or inward toward) the model. If problems still result in rendering, you can select *Back Face Normal Is Negative* in the *Rendering Preferences* dialog box (see *Rpref*).

If all faces are created consistently, you can save time in the rendering process by turning on *Discard Back Faces* in the *Render* or *Rendering Preferences* dialog box. This action discards the faces with normals pointing away from your viewpoint because they wouldn't be visible from that viewpoint. The time saved is proportional to the number of faces discarded and the total number of faces.

In cases when you are rendering transparent objects or when you can see two sides of an object (such as looking inside an open container), you may want to leave back faces in the rendering (do not check *Discard Back Faces*).

CHAPTER EXERCISES

1. *Render, Lights, Materials, Saveimg, Replay*

   Open the **PULLY-SL** drawing that you created in Chapter 38 Exercises. Refer to the "Typical Steps for Using Render" in this chapter to create a rendering. As an aid, follow these brief suggestions.

   Use *3Dorbit* to show the best view of the pulley. Create a *Distant Light* located to the left side and slightly above the pulley. Create a *Point Light* located to the right side and above. Define and *Attach* a *Material* representing steel. Modify the material to have high *Reflective* and low *Roughness* values. Using the *Render* option of *Rendering Type*, perform several renders and adjust the *Light* positions and intensity until you find the desired results. Experiment also with the *Material* properties. When you have the best settings, perform a *Phong Render* (use *More Options…*) to achieve results similar to those shown in Figure 41-88. Finally, use *Saveimg* to save the rendering as **PULLY-REND.TIF**. Use *Replay* to ensure the image is saved. Use *SaveAs* and rename the drawing **PULLY-REND**.

   Figure 41-88 ——————————

2. *Lights, Materials, Saveimg, Replay, Phong Render*

   Open the **SADL-SL** model that you created in Chapter 38 Exercises. For this rendering, use *3Dorbit* to generate a useful viewpoint and a small amount of perspective (*Distance*). Set up two lights—a cool-colored *Distant Light* with an *Azimuth* of **-130** and an *Altitude* of **50,** and a warm-colored *Point Light* at **300,-200,300**. *Attach* the *Blue Metallic Material* to represent metal. *Modify* the material to have a high *Reflective* value (**.80**) and a low *Roughness* value (**.25**).

Create several *Gouraud Renderings* to adjust the light intensities. The *Point Light* should provide slightly more light (at the near end of the model), and the *Distant Light* should provide slightly less (on the far side). The varying light intensities with the color effects give the proper impression of distance. Also make necessary adjustments to the *Material* parameters. When you have a good combination, compare your image to Figure 41-89 and create a *Phong Render*. Save the rendering to a **.TIF** file with the file name **SADL-REND**. Use *SaveAs* and rename the drawing **SADL-REND**.

Figure 41-89

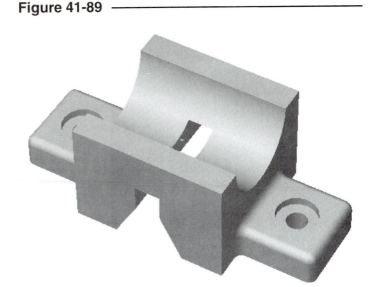

3. *Background, Lights, Materials, Saveimg, Replay, Shadows, Photo Real*

Open the **PULLY-REND** drawing again. Save the existing viewpoint (from the previous rendering, exercise 1) to a named *View*. Create another view to show the back of the part. Try to show several of the through-holes in the view. Save the new viewpoint to a named *View*. Use *Background* and create a *Gradient* background with three colors. Use the *Rotation* option to bring the bands of color into a vertical orientation. Create a few quick *Renders* to check and adjust the background.

Create one more *Light* with a *Location* of your choosing. Turn *Shadows On* in each of the *Light's* dialog boxes, but leave *Shadows* off in the *Render* dialog box while making preliminary adjustments. In order to create the new rendering, create a *Scene* by selecting the new *View*, the new *Light*, and one of the *Lights* previously created for the first rendering. Experiment to see which of the original two lights gives the best results. Adjust the light *Intensities* and *Colors* if necessary. When you have your best settings, turn *Shadows* on and create a *Photo Real* rendering. Use *Saving* and name the rendering **PULLY-R-BK.TIF**. Save the drawing as **PULLY-R-BK**.

4. *Spotlight, Shadows, New Standard Material, Photo Raytrace*

Open **SADL-REND** again. Change *VIEWRES* to **1000** and *FACETRES* to **1.0**. Make a solid *Box* using the *Center* option (center located at **0,-70,0**) with *Length* **600**, *Width* **300**, and *Height* **-20**. Then use *Rotate* with the *Last* selection option and rotate **20** degrees. This *Box* should serve as a base or "table top" for the saddle to rest on and provide a background for the *Spotlight* you will soon create. Create a *New Standard Material* with high *Reflection* (**.60**) and low *Roughness* (**.25**) and select a dark *Color* with a low *Value* (**.20**). *Attach* the new material to the "table." Create a *Background* for the scene using the *Gradient* option, or use the *Image* option and find a suitable image in the Textures or Windows folders. Create a few "rough" *Photo Real* renderings with *Shadows* off to test and adjust the background.

Next, insert a *Spotlight* with a *Target* of **60,0,0** and a *Location* of **200,-200,300.** Assign a warm color to the *Spotlight.* Turn *Shadows On* for the light and set the *Shadow Map Size* to **1024.** Adjust the *Intensity* of the *Point Light* to **0** so that the *Spotlight* provides the primary illumination from this viewing direction. Perform a few *Photo Real* renderings with *Shadows* off to test the angles of the *Hotspot* and *Falloff.* Make adjustments to those values so the saddle is fully illuminated but the light falls off just beyond the saddle. Perform several *Photo Real* renderings to test and adjust the lighting parameters. Any of several parameters can be adjusted—*Intensities, Colors,* and light *Locations.*

Turn on *Shadows* for all lights and globally for the scene. *Modify* the material you created for the table by selecting the *Reflective* attribute and toggling on *Mirror.* Perform a *Photo Raytrace* rendering. Try to achieve a reflective effect on the table top similar to that shown in Figure 41-90. Use *SaveAs* and rename the drawing **SADDL-REND2.**

Optional (time permitting): Create several renderings and use *Saving* to assign the names **SADL-REND2, SADL-REND3,** and so on. Use *Replay* to compare and determine the best renderings.

Figure 41-90

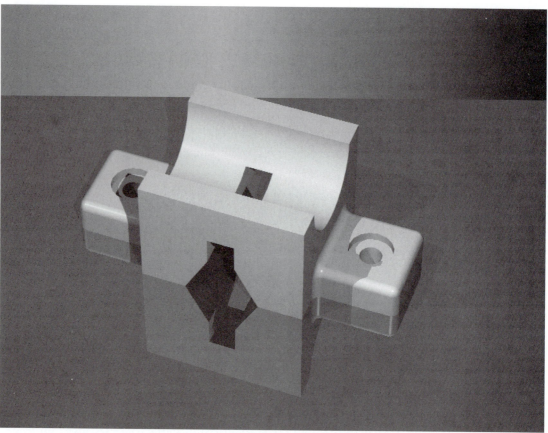

5. **Image Mapping,** *Setuv*

Create a machinist's file with handle similar to that used previously as a rendering example earlier in the chapter. Use the dimensions shown in Figure 41-91, top, on the next page, to create half of the file using *Pline* and *Spline,* then *Mirror* the shape along the longitudinal axis. Create a *Region* from the resulting shape, then use *Extrude* with a height of **3/16** to create a solid model of the file.

Create the handle solid model. First, draw the shape shown in Figure 41-91, bottom, with *Pline* and *Spline*, then use *Region* to create one closed shape. Make the solid by using *Revolve*. Use *Move* to attach the handle to the file. Set *VIEWRES* to **1000** and *FACETRES* to **1.0**. Save the drawing as **FILE**.

Figure 41-91

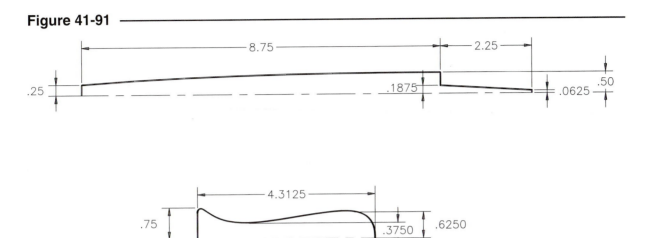

Next, attach appropriate wood and metallic materials to the two solids. Create one *Point Light* and one *Distant Light*, then adjust the intensities and colors. Make adjustments to the materials and lights as needed by creating several rough *Renders*.

Modify the file material by using the *Bump Map* attribute. Locate the **GRID.TGA** image file in the **Textures** directory and set *Bitmap Blend* to **.02**. Use *Setuv* to change the *Rotation* angle to **30** and the *Scale* to **2.0**. Set the *Rendering Type* to *Photo Raytrace* and create a rendering. Make adjustments as necessary. Try to achieve a realistic file such as that shown in Figure 41-92. *Save* the drawing and use *Saveimg* to save the rendering as **FILE.TIF**.

Figure 41-92

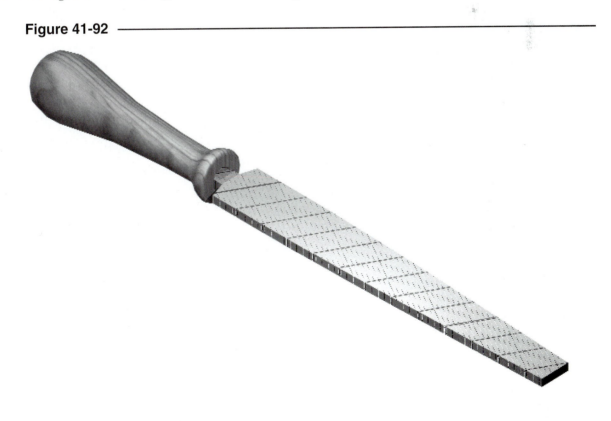

6. *Lsnew, Lsedit*

   Open the **CAMPUS.DWG.** (Download the CAMPUS.DWG from the **www.mhhe.com/leach** Web site.) Rename the file to **CAMPUS2** when you use *SaveAs* to save it to your working directory. Activate the *Model* tab. *Zoom* in to any area (see Fig. 41-21 for an example), then insert several landscape objects using *Lsnew.* Use *Lsedit* and **grips** to adjust the position and location of the images. Generate several preliminary *Renders* while you create new and adjust existing *Lights.* Use *Saveimg* to save your best rendering as **CAMPUS2.TIF.**

7. *Setuv*

   Create a scene similar to that shown in Figure 41-93. Use *Spheres* for the balls; create the glass by *Revolving* a profile; create the rack by *Extruding* two shapes and *Subtracting* the inner shape; and create the pool cue by *Revolve, Cone,* or *Extrude* with a taper angle. Create appropriate *Lights* and *Attach* appropriate materials. *Save* the drawing and assign the name **POOL.**

   Here is a challenge: Use AutoCAD to create several TIFF images (stripes and numbers) to attach to the balls. Use *Modify Material* with the *Color/Pattern* attribute to locate and assign the images. Use *Setuv* to adjust the orientation of the mapped images.

   Figure 41-93

8. *Open* the **ADJMT-SL** or other solid or surface model that you would like to render. Achieving skill with Render requires time and experimentation. Experiment with different *Backgrounds, Lights, Materials,* and *Scenes,* and *Images.* Have fun.

42

CREATING 2D DRAWINGS FROM 3D MODELS

Chapter Objectives

After completing this chapter you should:

1. be able to use *Mvsetup* to create viewports showing the standard engineering "views" of an existing 3D model;

2. be able to use *Solview* to create viewports and layers for use with *Soldraw*;

3. be able to use *Soldraw* to project 3D solids onto 2D planes, complete with hidden and visible lines;

4. be able to use layers created with *Soldraw* to complete a dimensioned 2D multiview drawing;

5. be able to create a 2D or 3D profile, complete with hidden lines, from any view of a solid model using *Solprof*;

6. be able to use *Solprof* in conjunction with *Mvsetup* to create a 2D drawing from a solid model.

CONCEPTS

This chapter discusses the use of AutoCAD features that assist in the creation of a 2D drawing from a 3D model. Several features in AutoCAD provide a means for creating a 2D drawing from a 3D model: *Mvsetup, Solview, Soldraw,* and *Solprof. Mvsetup* can be used with any type of 3D model—wireframe, surface, or solid. *Solview, Soldraw,* and *Solprof* are used specifically with AutoCAD solid models for converting 3D geometry to 2D drawings.

Mvsetup is an automated routine that can be invoked only by typing at the command prompt. It operates with <u>any type</u> of 3D model. This routine automatically sets up paper space viewports. It also has a series of options that allow alternate arrangements for the viewports. If you use the *Standard Engineering* option, *Mvsetup* creates the viewports and automatically places the 3D model correctly with a top, front, right side, and isometric viewpoints, similar to the *3D Setup* option of *Vports.* Although the viewports are set up and the 3D geometry is arranged for you, other operations are required to create a 2D drawing complete with hidden lines and dimensions with correct visibility within viewports.

The *Solview* and *Soldraw* commands used together are far more automated and powerful than any other alternative for creating 2D drawings from AutoCAD solid models. *Solview* is similar to *Mvsetup* in that it automatically sets up the views in paper space viewports. It differs from *Mvsetup,* however, because you can select the desired views and their placement, including section and auxiliary views. *Solview* also creates layers for use with *Soldraw.* The *Soldraw* command is used as the next step to project the geometry (of AutoCAD solid models only) onto a 2D plane and automatically uses the newly created layering scheme for creating the 2D geometry. These new layers are complete with hidden lines and correct viewport-specific visibility settings. Dimensions can be drawn as an additional "manual" step on the layers provided by *Solview* or can be drawn in paper space.

Solprof (solid profile) creates a profile of an AutoCAD solid model. The profiled geometry can be used as a wireframe or projected onto a 2D plane. *Solprof* creates new layers for the profile geometry, complete with correct hidden lines. *Solprof* can be used in conjunction with *Mvsetup* or *Vports, 3D Setup* to create standard engineering views. *Solprof* can then be used as a second operation to <u>profile</u> the 3D model (convert the 3D model in each view into 2D line drawings).

In summary, using *Solview* and *Soldraw* is the superior method for creating 2D drawings from AutoCAD ACIS solid models. It is the most universal method and accomplishes the most for you automatically, including setting up layers for dimensioning. *Mvsetup* can be used with wireframe or surface models to semi-automatically set up standard engineering views, but it cannot project the model onto a 2D plane with hidden and visible lines.

Mechanical Desktop and Inventor

Mechanical Desktop® and Inventor® are separate software products sold by Autodesk intended exclusively for solid modeling. Mechanical Desktop includes AutoCAD, whereas Inventor is a stand-alone product. Although the two products are similar, Inventor is based on newer technology.

Mechanical Desktop and Inventor are parametric-based solid modelers, meaning that parametric relationships rather than dimensional values can be specified for geometric features. In this way, when you change a dimension, the related features also change. The model construction method is features-based rather than Boolean-based. With features modeling, typical manufacturing operation terms are used to create geometry. For example, you may use the *Hole* command instead of creating a *Cylinder* and *Subtracting* it. In addition, Mechanical Desktop and Inventor automatically create 2D drawings with dimensions from the 3D model. The 2D and 3D geometries are bidirectionally linked so that if you make a (dimensional) change to one, the other is automatically updated.

USING *MVSETUP* FOR STANDARD ENGINEERING DRAWINGS

Mvsetup is a program that assists you in setting up paper space viewports. The *Standard Engineering* option can be used to set up a 3D model in viewports, each viewport having a different standard view. This application of *Mvsetup* accomplishes similar results as the *3D Setup* option of *Vports*. See Chapters 13, 33, and 35 if you need more information about layouts, paper space, and *Vports*.

MVSETUP

Pull-down Menu	COMMAND (TYPE)	ALIAS (TYPE)	Short-cut	Screen (side) Menu	Tablet Menu
...	*MVSETUP*	...		...	...

This section discusses the use of *Mvsetup* using the *Standard Engineering* option—the option for creating engineering drawings from 3D models. This option of *Mvsetup* operates with <u>any type</u> of 3D model. It automatically creates four viewports in a layout and displays the model from four *Vpoints* (front, top, side, and isometric views).

Mvsetup is a similar but older method of creating paper space viewports than the *Vports* command with the *3D* option; however, *Mvsetup* provides other options that allow you to insert a titleblock and scale and align the geometry. The sequence is given here.

Typical Steps for Using the *Standard Engineering* Option of *Mvsetup*

1. Create the 3D part geometry in model space.

2. Create a layer for viewports and a layer for the titleblock (named VPORTS and TITLE, for example). Set the viewports layer current.

3. Activate a *Layout* tab and use the *Page Setup* dialog box to set the *Plot device* and *Paper size*.

4. Invoke *Mvsetup* and use *Options* to set the *Mvsetup* preferences.

5. Use *Title block* to *Insert* or *Xref* one of many AutoCAD-supplied or user-supplied borders and titleblocks.

6. Use the *Create* option to make the paper space viewports. Select the *Standard Engineering* option from the list.

7. Use *Scale viewports* to set the viewport scale factor (*Zoom XP* factor) for the model geometry displayed in the viewports.

8. The model geometry that appears in one viewport can be aligned with the model in adjacent viewports if necessary using the *Align* option.

9. From this point, other AutoCAD commands must be used to create dimensions for the views or to convert some lines to "invisible" lines.

Mvsetup **Example**

To illustrate these steps, a 3D wireframe model (WFEX3 drawing from Chapter 37 Exercises) is used as an example.

1. Create the 3D part geometry in model space. The wireframe is shown here as it exists in model space viewed from an isometric-type *Vpoint*.

2. Create a VPORTS layer for viewports and a TITLE layer for the titleblock. Set the VPORTS layer current.

3. Activate the *Layout1* tab. Erase any viewports. Use the *Page Setup* dialog box. Set the *Plot device* and the *Paper size* you intend to use.

4. Invoke *Mvsetup* and use *Options* to set the *Mvsetup* preferences. Specifically, the *Layer* option is used to specify the layer for the titleblock.

Figure 42-1

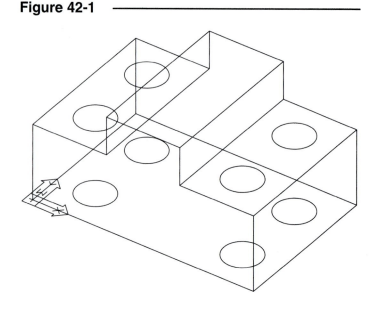

```
Command: mvsetup
Enter an option [Align/Create/Scale viewports/Options/Title block/Undo]: o
Enter an option [Layer/LImits/Units/Xref] <exit>: l
Enter layer name for title block or [. (for current layer)] <title>: title
Enter an option [Layer/LImits/Units/Xref] <exit>: Enter
Enter an option [Align/Create/Scale viewports/Options/Title block/Undo]:
```

5. The *Title* block option is used to insert a titleblock and border in paper space.

```
Enter an option [Align/Create/Scale viewports/Options/Title block/Undo]: t
Enter title block option [Delete objects/Origin/Undo/Insert] <Insert>: i
Available title blocks:...
  0:  None
  1:  ISO A4 Size(mm)
  2:  ISO A3 Size(mm)
  3:  ISO A2 Size(mm)
```

(Thirteen titleblock options are displayed here.)

```
Enter number of title block to load or [Add/Delete/Redisplay]: 8
Create a drawing named ansi_b.dwg? <Y>: n
```

The previously selected options produce a titleblock and border appearing in paper space, as shown in Figure 42-2. The 3D model is not visible because viewports have not yet been created.

Figure 42-2

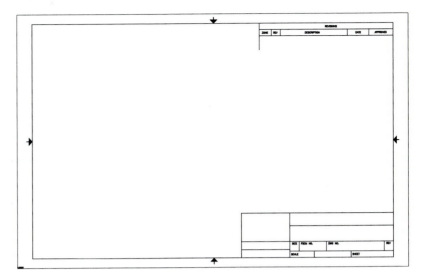

6. Use the *Create* option of *Mvsetup* to create the desired viewport configuration:

> Enter an option [Align/Create/Scale viewports/Options/Title block/Undo]: **c**
> Enter option [Delete objects/Create viewports/Undo] <Create>: **Enter**
> Available layout options: . . .
> 0:   None
> 1:   Single
> 2:   Std. Engineering
> 3:   Array of Viewports
> Enter layout number to load or [Redisplay]: **2**
> Bounding area for viewport(s). Default/<First point >: **PICK**
> Specify opposite corner: **PICK**
> Specify distance between viewports in X direction <0.0>: **Enter** or **(value)**
> Specify distance between viewports in Y direction <0.2>: **Enter** or **(value)**

Select the *"2: Std. Engineering"* option. PICK two corners to define the bounding area for the four new viewports. The action produces four new viewports and automatically defines a *Vpoint* for each viewport. The resulting drawing displays the standard engineering front, top, and right side views of the 3D model in addition to an isometric-type view, as shown in Figure 42-3.

Figure 42-3

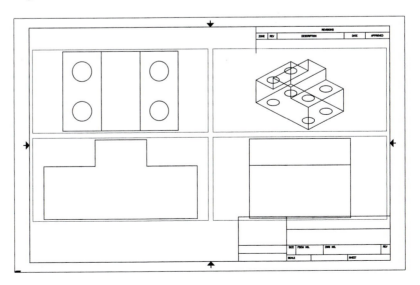

7.   Notice in Figure 42-3 that the geometry in each viewport is displayed at its maximum size, as if a *Zoom Extents* were used. The *Scale viewports* option is used to set a scale for the geometry in each viewport:

> Enter an option [Align/Create/Scale viewports/Options/Title block/Undo]: ***s***
> Select the viewports to scale...
> Select objects: **PICK** (Select the viewport border objects)
> Select objects: **Enter**
> Set zoom scale factors for viewports. Interactively/<Uniform>: ***u***
> Set the ratio of paper space units to model space units...
> Enter the number of paper space units <1.0>: **Enter** or (**value**)
> Enter the number of model space units <1.0>: **Enter** or (**value**)
> Enter an option [Align/Create/Scale viewports/Options/Title block/Undo]:

Make sure you select the <u>viewport objects</u> (borders) at the "Select viewports to scale:" prompt. The *Uniform* option ensures that the 3D model will be scaled to the same proportion in each viewport. The ratio of paper space units to model space units is like a *Zoom XP* factor; for example, a ratio of 1 paper space unit to 2 model space units is equivalent to a *Zoom 1/2XP*.

The above action produces the scaling of the 3D model space units relative to paper space units, as shown in Figure 42-4.

Figure 42-4 ————————————————————————

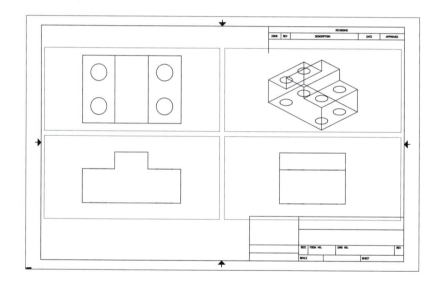

8.   The *Align* option of *Mvsetup* can be used to orthogonally align the views if they are not automatically aligned by the *Scale viewports* option. The *Align* option allows you to *OSNAP* to an object in one view to accomplish a *Horizontal* or *Vertical* alignment with an object in another viewport. The *Horizontal* option can be used to align the model appearing in the front and side views and the *Vertical* option is used to select model geometry for alignment of the top and front views.

An additional step is required for this example in order to display the isometric view as desired within the viewport. *Zoom Extents, Zoom XP,* and *Pan* can be used to locate and size the model appropriately.

Finally, the VPORTS layer created for inserting the viewports can be turned *Off* or *Frozen* to display the views without the viewport borders, as shown in Figure 42-5.

Figure 42-5

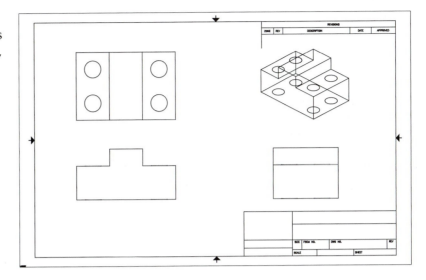

9. This is as far as *Mvsetup* goes. As you notice, the result is not a complete conventional multiview drawing because there are no dimensions nor are there hidden lines representing the invisible edges of the object. Since the views are actually different *Vpoints* of a 3D model, hidden lines are not possible for solid models using *Mvsetup,* but hidden lines can be created for wireframe models.

Creating Hidden Lines for a 3D Wireframe Shown with *Mvsetup*

If a complete 2D drawing is to be constructed from the 3D wireframe model, some lines from the model must be converted to "invisible" lines, and/or other lines may have to be added. It is not possible to convert edges of a solid or surface model to a hidden linetype, but edges of a wireframe model can be converted. Because a wireframe model is composed of simple objects such as *Line, Arc,* and *Circle,* these objects can easily be changed to a hidden linetype or to a layer with a hidden linetype assigned. Probably the simplest way to create the "invisible" lines is to use *Properties,* select the desired objects, and change them to a layer having a hidden linetype assigned.

Better yet, it is desirable to create a separate hidden layer for each view (viewport). In this way, you have control over each view's hidden-layer visibility. For example, you may want to turn off just the isometric viewport's hidden lines or change that layer's linetype to continuous. Creating separate layers for each viewport's hidden linetypes may take more effort but affords you the most flexibility for displaying the drawing. Refer to Chapter 33, Advanced Layouts and Plotting, for more information on viewport-specific layer visibility.

Creating Dimensions for a 3D Model Shown with *Mvsetup*

If you use *Mvsetup* or any other method of creating 2D "views" of a 3D model in paper space viewports, dimensions can be created for each view. Two methods are possible.

1. Create dimensions in paper space. This method is by far the simplest method; however, it is available only since AutoCAD 2002. In AutoCAD 2002 and 2004, dimensions can be drawn in paper space that are associated with (attached to) objects inside a viewport. Therefore, create the dimensions in paper space for each viewport to indicate measurements for each "view." In this way, separate layers, viewport-specific layer visibility, and multiple UCSs are <u>not</u> required to create and

display each "view's" dimensions in the correct viewports and in the correct orientation. See Chapters 29 and 33 for information on creating dimensions in paper space and the advantages and disadvantages of doing so.

2. Create dimensions for each view on a separate layer and make each dimensioning layer visible only in the appropriate view (viewport). This method is required if you are using AutoCAD 2000 or earlier.

   A. Create three new layers such as DIM-TOP, DIM-FRONT, and DIM-SIDE.

   B. If not existing, create and save three UCSs named TOP, FRONT, and SIDE, each with its XY plane parallel to the matching view.

   C. Starting with the front viewpoint, make the FRONT UCS active and make layer DIM-FRONT the current layer. Create the dimensions for the front view on this UCS and layer.

   D. Make layer DIM-FRONT frozen for all <u>other</u> viewports so the dimensions for the front view appear only in the front viewport.

   E. Moving to the TOP viewport, set its matching UCS and layer current. Create the dimensions for the view. Make layer DIM-TOP frozen for all other viewports.

   F. Do the same for the remaining side view. Finally, turn off the viewport layer and insert the title block.

USING *SOLVIEW* AND *SOLDRAW* TO CREATE MULTIVIEW DRAWINGS

The *Solview* command creates new layers and new views in paper space viewports for the existing 3D model space geometry. The *Soldraw* command creates new 2D objects on the new layers. *Soldraw* projects the model geometry onto a 2D plane with the appropriate continuous or hidden linetype. *Soldraw* can operate only with viewports that have been created with *Solview* (not with those created with *Mvsetup*). *Solview* and *Soldraw* operate for solid models only. The typical steps for using *Solview* and *Soldraw* to create a 2D drawing from a 3D model are described here.

Typical Steps for Using *Solview* and *Soldraw*

1. Create the part geometry in model space. Set up a UCS parallel to the desired profile (front) plane of the object. Also ensure that the HIDDEN linetype is loaded in the drawing.

2. Activate a *Layout* tab. Erase any viewports. Use the *Page Setup* dialog box to set the *Plot device* and the *Paper size* you intend to use.

3. Type *Solview* at the command prompt, select the *Setup View* icon button, or select *Solids > Setup > View* from the *Draw* pull-down menu.

4. Use the *UCS* option to create the profile (front) view. You can select the location and scale for the view.

5. Use the *Ortho* option to create the other principal orthographic views. Usually a top and/or side view is needed.

6. If a section or auxiliary view is desired, use the *Section* or *Auxiliary* option to create the view in the desired location.

7. Invoke *Soldraw* by any method. *Soldraw* is used to project the "views" to a 2D plane, thus creating new 2D geometry on the appropriate layers (created by *Solview*).

8. *Freeze* the VPORTS layer. Create a new layer named TITLE and draw, *Insert*, or *Xref* a titleblock and border in paper space.

9. You can then create dimensions for the 2D drawing on the "*-DIM" layers prepared by *Solview*. Use the normal commands for creating and editing dimensions.

SOLVIEW

Pull-down Menu	COMMAND (TYPE)	ALIAS (TYPE)	Short-cut	Screen (side) Menu	Tablet Menu
Draw *Solids >* *Setup >* *View*	*SOLVIEW*	...	...	DRAW 2 SOLIDS *Solview*	...

Solview creates paper space viewports and automatically specifies the correct viewing angle (*Vpoint*) for each *viewport*. You select which views you want and the location for each. *Solview* also prepares new layers for subsequent use with the *Soldraw* command:

    Command: **solview**
    Enter an option [Ucs/Ortho/Auxiliary/Section]:

Ucs

The *Ucs* option creates a <u>view</u> normal (perpendicular) to the XY plane of a User Coordinate System. The *Ucs* option is generally the best way to create the first viewport from which other viewports can be created. All other *Solview* options require an existing viewport. You can select and readjust the view's location. You can also specify the size of the viewport.

Ortho

This option creates a principal orthographic view (top, side, bottom) from an existing viewport. New *Ortho* viewports are created by selecting an <u>edge</u> of an existing viewport object to project from. You can select and readjust the view's location and set the size of the viewport.

Auxiliary

This option is used to create an auxiliary view from an existing view. You are switched to model space to PICK two points to define the inclined plane used for the auxiliary projection. *OSNAPs* should be used for selection.

Section

This option creates a section view. *Solview* uses *Slice* to create the section at the cutting plane you define. The 3D geometry behind the cutting plane remains visible. Similar to the *Auxiliary* option, you PICK two points to define a cutting plane.

Solview creates the layers that *Soldraw* uses to place the visible lines, hidden lines, and section hatching for <u>each view</u>. *Solview* also creates layers for dimensioning that are set for visibility per viewport. Because of the possible complexity of the drawing, *Solview* places the visible lines, hidden lines, dimensions, and section hatching for each view on <u>separate layers</u>. This provides you with complete visibility control. The following naming convention is used for the layering scheme.

<u>Layer Name</u>	<u>Object Type</u>
view name–VIS	Visible lines
view name–HID	Hidden lines
view name–DIM	Dimensions
view name–HAT	Hatch patterns

(The *view name* is the original name that you specified for the view when it was created.)

Solview also creates a layer called VPORTS for the viewport objects. This layer should be reserved exclusively for the use of *Solview* and *Soldraw*. Do not draw or alter information on this layer.

SOLDRAW

Pull-down Menu	COMMAND (TYPE)	ALIAS (TYPE)	Short-cut	Screen (side) Menu	Tablet Menu
Draw *Solids >* *Setup >* *Drawing*	*SOLDRAW*	...	...	DRAW 2 SOLIDS *Soldraw*	...

Soldraw is generally used immediately after *Solview*. The *Soldraw* command uses the 3D model in each viewport created by the *Solview* command and projects the profiles onto a 2D plane. (*Soldraw* actually uses the *Solprof* command.) New 2D objects are created as profiles and sections in the viewports. The new 2D objects are drawn in the appropriate *Continuous* or *Hidden* linetypes. If sectional views are included, hatch patterns are drawn using the current values of the *HPNAME, HPSCALE,* and *HPANG* variables. You can use *Soldraw* only with previously created *Solview* viewports.

Soldraw is very easy to use because all you have to do is select the viewports that you want to be affected by *Soldraw*. There are no options for *Soldraw*:

> Command: **soldraw**
> Select viewports to draw...
> Select objects: **PICK** viewport objects
> Command:

Solview and *Soldraw* Example

As the steps (given previously) for using *Solview* and *Soldraw* are explained here, an example AutoCAD ACIS solid model is used to illustrate the process (Fig. 42-6). The AutoCAD ACIS solid model is the angle brace shown in previous chapter exercises.

Figure 42-6

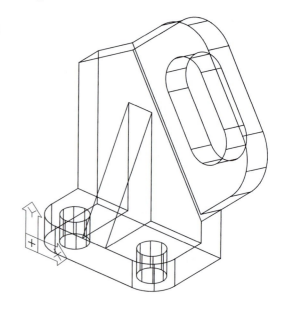

1. The first step to use *Solview* and *Soldraw* is to create a 3D model. When the geometry is complete, create a UCS parallel to the desired front view of your model, as shown in Figure 42-6. Your first view selected using *Solview* should be the profile (front) view.

2. Next, activate a *Layout* tab. Erase any viewports. Use the *Page Setup* dialog box to set the *Plot device* and the *Paper size* you intend to use. If desired, *Insert* or *Xref* a titleblock.

3. Invoke *Solview* at the Command prompt or select the *Setup View* icon button.

4. The *Current UCS* option is used to create a paper space viewport displaying the profile (front) view (see Fig. 42-7). The *UCS* option creates a view normal (perpendicular) to the current UCS XY plane.

> Command: *solview*
> Enter an option [Ucs/Ortho/Auxiliary/Section]: *u*
> Enter an option [Named/World/?/Current] <Current>: **Enter** (to use the current UCS)
> Enter view scale <1.0000>: **Enter** or (**value**)
> Specify view center: **PICK**
> Specify view center <specify viewport>: **PICK** (to readjust if necessary)
> Specify view center <specify viewport>: **Enter**
> Specify first corner of viewport: **PICK**
> Specify opposite corner of viewport: **PICK**
> Enter view name: **front**
> Enter an option [Ucs/Ortho/Auxiliary/Section]:

The "Enter view scale<1>:" value is the same value as a *Zoom XP* factor. That is, enter the proportion of model geometry units to paper space units.

PICK the desired "view center" to specify the desired location of the view. The model geometry appears after PICKing the view center. The "view center" can be readjusted if necessary.

At the "Specify first corner:" and "Specify opposite corner:" prompts, select the desired paper space viewport corners as indicated in Figure 42-7. Make sure you allow sufficient room for dimensions or other drawing objects you may want to include later. (It is OK for the viewports to overlap.) Finally, you must name the view. In this case, "FRONT" is used as the name of the new viewport.

Figure 42-7

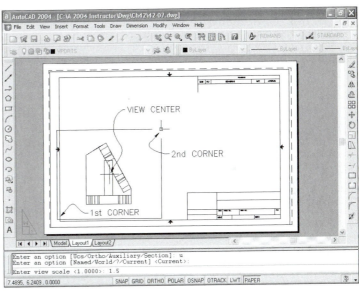

5. After the front view is created, you can create other views with the *Ortho* option. This option creates a new viewport orthographically aligned with the <u>edge</u> of the viewport object that you PICK:

> Enter an option [Ucs/Ortho/Auxiliary/Section]: *o*
> Specify side of viewport to project: **PICK** (the desired edge of viewport)
> Specify view center: **PICK**
> Specify view center <specify viewport>: **PICK** (to adjust if necessary)
> Specify view center <specify viewport>: **Enter**
> Specify first corner of viewport: **PICK**
> Specify opposite corner of viewport: **PICK**
> Enter view name: **top**
> Enter an option [Ucs/Ortho/Auxiliary/Section]:

As shown in Figure 42-8, PICK the <u>edge of the viewport object</u> representing the viewing direction for the new *Ortho* viewport. In other words, PICK the top edge of the front viewport to produce a top view. Notice the *Midpoint OSNAP* option is automatically invoked when you PICK the edge of the viewport. Next, PICK the view center for the top view (as before when the front viewport was established).

Figure 42-8 ————————

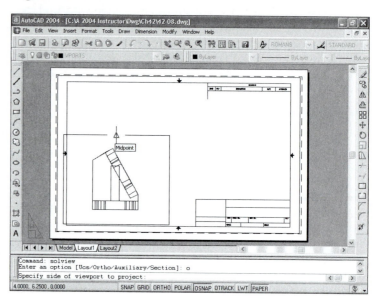

The new view then appears (Fig. 42-9). Specify the size for the viewport. Assign a name for this viewport such as "TOP."

You can continue creating viewports with the *Ortho* option until the necessary principal views are established. For example, a right side view could be created by using the *Ortho* option, then PICKing the right edge of the front viewport object.

Figure 42-9 ————————

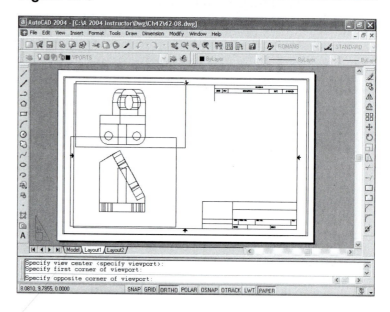

6. In this example, an auxiliary view is required to provide the necessary visual and dimensional information for the drawing. *Solview* is used with the *Auxiliary* option to create the viewport.

For this procedure you must select two points to define the auxiliary surface. You are automatically switched to model space for PICKing the surface's first and second points (*OSNAPs* should be used). For the example, the auxiliary surface is established as the upper right inclined plane as shown in Figure 42-10.

Figure 42-10

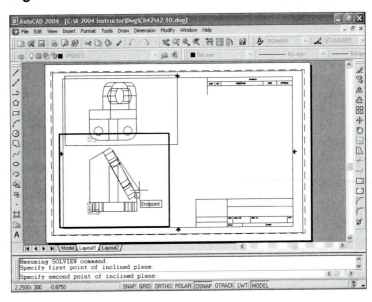

Next, PICK a point to define which side of the auxiliary surface to view the object from:

    Enter an option [Ucs/Ortho/Auxiliary/Section]: **a**
    Specify first point of inclined plane: **PICK** (use *Osnaps*)
    Specify second point of inclined plane: **PICK** (use *Osnaps*)
    Specify side to view from: **PICK** (on desired side of inclined plane)
    Specify view center: **PICK**
    Specify view center <specify viewport>: **PICK** (to adjust if necessary)
    Specify view center <specify viewport>: **Enter**
    Specify first corner of viewport: **PICK**
    Specify opposite corner of viewport: **PICK**
    Enter view name: **aux**
    Enter an option [Ucs/Ortho/Auxiliary/Section]:

The resulting viewport with auxiliary view is shown in Figure 42-11.

The model geometry visible in the viewports is the existing 3D model. The model has not been changed in any way by *Solview*. However, as well as the viewports that are new, *Solview* creates new layers complete with viewport-specific layer visibility.

Figure 42-11

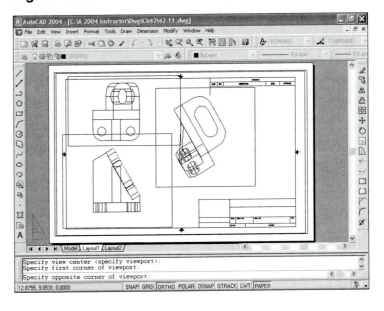

For our example, listing the layers reveals the work that was done by *Solview*. The following list shows <u>all</u> and <u>only</u> layers created by the previous options of *Solview*:

Layer Name	State	Color	Linetype
FRONT-DIM	On	7(white)	CONTINUOUS
FRONT-HID	On	7(white)	HIDDEN
FRONT-VIS	On	7(white)	CONTINUOUS
AUX-DIM	On	7(white)	CONTINUOUS
AUX-HID	On	7(white)	HIDDEN
AUX-VIS	On	7(white)	CONTINUOUS
TOP-DIM	On	7(white)	CONTINUOUS
TOP-HID	On	7(white)	HIDDEN
TOP-VIS	On	7(white)	CONTINUOUS
VPORTS	On	7(white)	CONTINUOUS

Current layer: VPORTS

Solview uses the name that you specify for the views as prefixes for the layer names. The layer visibility has automatically been set so that layers are visible only in the associated viewports (TOP-* layers are only visible in the top view, for example).

In addition, *Solview* creates a new UCS for each viewport with its XY plane parallel with the view (the *UCSVP* system variable is automatically set to 1 so each new UCS is saved with its viewport). In this way, when dimensions are created for each view, they are correctly aligned with the plane of the view.

7. Invoke the *Soldraw* command. Now that the views have been established by the *Solview* command, the model geometry is ready for projection onto 2D planes.

    ```
    Command: soldraw
    Select viewports to draw:
    Select objects: PICK
    Select objects: Enter
    ```

The viewport objects are PICKed in response to the "Select viewports to draw:" prompt. Normally you would <u>select all viewports</u>. PICK the viewport <u>objects</u> (borders) with the pickbox or Crossing Window, as shown in Figure 42-12. Because much computation is involved, *Soldraw* may take some time, depending on the complexity of the model and the speed of your computer system. When *Solview* has finished, the results may not be apparent—hidden lines may not display correctly if the HIDDEN linetype was not previously loaded and the *LTSCALE* variable is not set appropriately. You can complete these actions retroactively if needed to make the hidden lines appear appropriately in the views.

Figure 42-12

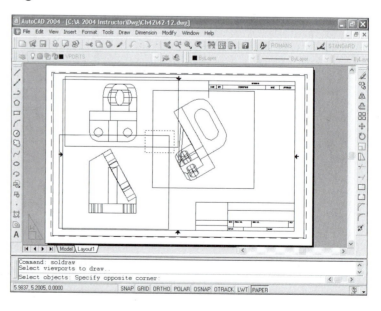

8. Use the *Layer Control* drop-down list or the *Layer Properties Manager* dialog box to control the visibility of the layers. *Freeze* the VPORTS layer to prevent the viewport borders from displaying, as shown in Figure 42-13.

9. If not done previously, create a new layer (called TITLE or BORDER, for example) and draw, *Insert,* or *Xref* a titleblock and border in paper space.

10. If you want to create dimensions for the new 2D views, use the normal dimensioning commands.

Figure 42-13

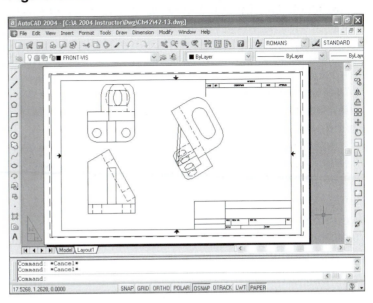

Dimensioning is accomplished in one of two ways:

A. Create dimensions in paper space. If you are using AutoCAD 2002 or 2004, you can use this method to "attach" paper space dimensions directly to objects in model space (in the viewport). See Chapters 29 and 33 for more discussion on dimensioning in paper space.

B. Create dimensions inside each viewport. Since layers have already been created complete with the correct viewport-specific layer visibility and appropriate UCSs have been created with the correct orientation for each viewport, simply click in each viewport and create the needed dimensions for each view on the appropriate layer.

USING *SOLPROF* WITH AutoCAD SOLID MODELS

SOLPROF

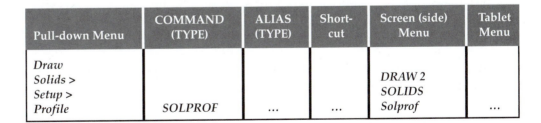

Pull-down Menu	COMMAND (TYPE)	ALIAS (TYPE)	Short-cut	Screen (side) Menu	Tablet Menu
Draw *Solids >* *Setup >* *Profile*	*SOLPROF*	...	...	DRAW 2 SOLIDS *Solprof*	...

Solprof is a command that creates a profile of AutoCAD solids. *Solprof* generates a separate profile containing only the edges of straight and curved surfaces of a solid as seen from a particular viewpoint (parallel projection only—perspectives cannot be profiled). The "profile" can be projected to a plane or can be generated as a wireframe model. The profile is generated on two new <u>layers that are automatically created</u>, one with hidden lines and one with visible lines. *Solprof* determines which profile edges should be projected as hidden or as visible lines and draws them on the appropriate layer. *Solprof* can be used in conjunction with *Mvsetup* to create standard engineering 2D drawings.

Solprof must operate <u>within a paper space viewport</u>. Activate a *Layout* tab and use the *Page Setup* dialog box to set the *Plot device* and the *Paper size* you intend to use. If desired, *Insert* or *Xref* a title block. Use *Vports* or *Mvsetup* to create the desired viewports. Then use the *Model* command or double-click inside the viewport to activate model space (inside the viewport). Then use *Solprof* and select the solids that you want profiled. Remember to <u>select the solids, not the viewport objects</u> as with *Soldraw*.

> Command: **solprof**
> Select objects: **PICK** (Select the <u>solid or solids</u> that you want profiled.)
> Select objects: **Enter**
> Display hidden profile lines on separate layer? [Yes/No] <Y>: **Enter** or **N**
> Project profile lines onto a plane? [Yes/No] <Y>: **Enter** or **N**
> Delete tangential edges? [Yes/No] <Y>: **Enter** or **N**

Accepting the *Yes* default for "Display hidden profiles on a separate layer?" prompt causes *Solprof* to create two *Blocks,* one with hidden lines and one with the linetype of the existing solid. Two new layers are created for the *Block* insertions:

> PV-*nx*   Visible profile layer
> PH-*nx*   Hidden profile layer

(Where PV designates the Profile Visible objects, PH designates the Profile Hidden objects, and *nx* represents a number and letter AutoCAD assigns to the viewports. This letter and number combination is called the "viewport handle." You can *List* the viewport object to display its handle.)

The *Hidden* linetype must be loaded into the drawing before *Solprof* can use it, or you can assign the HIDDEN linetype to layer PH-* after using *Solprof*. Answering *No* to the "Display hidden profiles on a separate layer?" prompt causes *Solprof* to create one *Block* drawn as visible lines with the linetype of the existing solid.

The next prompt, "Project profile lines onto a plane? <Y>:," allows you to create a 3D wireframe or a 2D profile. Answering *Yes* causes AutoCAD to project the profile onto a plane parallel to the screen, similar to the *UCS View* option. In other words, the solid as it appears from the current viewpoint is transformed to a 2D projection. This option is particularly useful for creating a 2D "view" from a solid model *Vpoint*. Answering *No* creates a 3D wireframe of the model.

The last prompt allows you to delete "tangential edges." These edges are lines between two tangent faces. Answering *No* to this prompt creates a line between a planar surface and a tangent curved surface. You should answer *Yes* to delete tangential edges for most applications.

When *Solprof* has been used, the new objects may not be visible because *Solprof* does not change the layer visibility. To see the new *Solprof* layers, use the *Layer Properties Manager* dialog box or the *Layer Control* drop-down list to *Freeze* the original model layer(s). The new *Solprof* layers can then be viewed and edited if you choose.

Solprof **Example**

As an example, consider the solid model of the angle brace shown in Figure 42-14. In this case, the model has been completed and is displayed in model space. Also assume that the *Hidden* linetype has been loaded into the drawing.

Assuming the model is displayed in the *Model* tab with the desired viewpoint, activate a *Layout* tab and use the *Page Setup* dialog box to set the *Plot device* and the *Paper size* you intend to use. If desired, *Insert* or *Xref* a titleblock. Use *Vports* to create the desired viewport(s). Then use the *Model* command or double-click inside the viewport to activate model space (inside the viewport).

Invoke *Solprof* while model space is active (the cursor is <u>in</u> the viewport) and select the solid. All of the defaults are accepted. The resulting new layers are created as shown:

PH-1DD   *Hidden* linetype
PV-1DD   *Continuous* linetype

Layer visibility is <u>automatically</u> set to display the profile geometry only in the specific viewport (*Frozen* for new viewports).

Using the *Layer* command, the model layer must be *Frozen* to reveal the two profile layers. The resulting profile geometry is displayed in its paper space viewport in Figure 42-15.

This procedure can also be used to generate a 2D multiview drawing from an AutoCAD ACIS solid model. The process involves creating the viewports in a *Layout* using the *3D Setup* option of *Vports* or individually creating the viewports for the desired views—top, front, side, etc. Alternately, *Mvsetup* could be used to create the views. Next, use *Solprof* to create the views, complete with hidden and visible lines, for each viewport (see the following example).

Figure 42-14

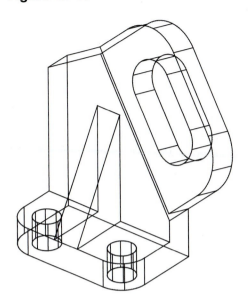

Figure 42-15

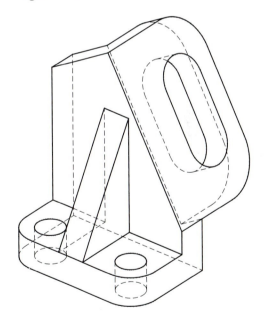

Keep in mind the power of using *Solprof* to automatically create a wireframe model from a solid model (answering *No* to, "Project profile lines onto a plane?"). In many cases when a wireframe is desired, it may be easier to create a solid model using the relatively simple CSG techniques, then use *Solprof* to automatically create the wireframe model from the solid. This method can be used when the model consists of multiple curved surfaces and complex lines of intersection (edges) between surfaces.

Using *Solprof* with *Mvsetup*

The procedure for creating 2D drawings from AutoCAD ACIS solid models using *Solprof* can be automated by using *Mvsetup* or the *3D Setup* option of *Vports* in conjunction with *Solprof*. *Solprof* is then used to create the hidden and visible lines for each "view."

For example, consider this case to illustrate *Mvsetup*. Given the adjustable slide solid model shown in Figure 42-16, assume the same procedure was followed to create the layout as that illustrated in Figures 42-1 through 42-5 earlier this chapter (for a wireframe model).

Figure 42-16

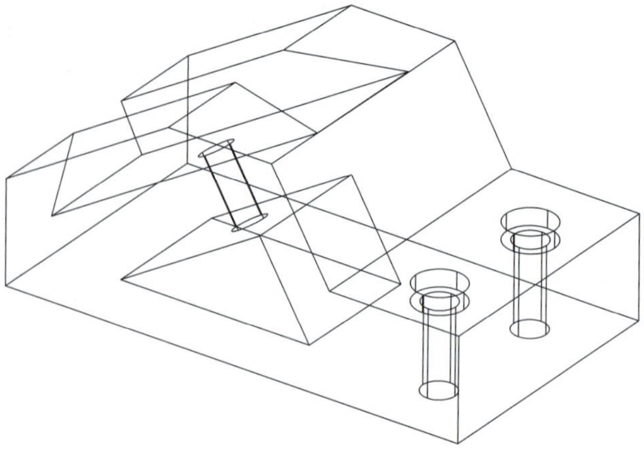

In this case, either of two methods could be used to develop the "views." *Mvsetup* could be used with the "standard engineering" option to create the top, front, side, and isometric views. Alternately, *Vports* with the *3D Setup* option could be used to fulfill this purpose. In either case, the resulting setup should be that shown in Figure 42-17. (The viewports layer is *Frozen* in this figure.)

Figure 42-17

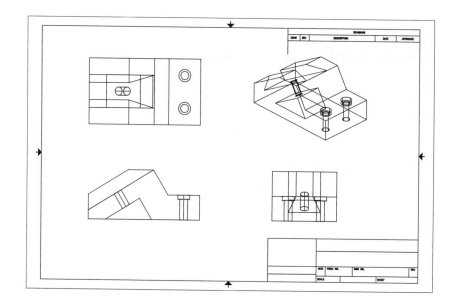

Solprof can be used next. The *Mspace* command is used to activate any viewport. Invoke *Solprof* and select the geometry in the current viewport to profile:

```
Command: solprof
Select objects: PICK (Select the solid or solids that you want profiled.)
Select objects: Enter
Display hidden profiles lines on separate layer? <Y>: Enter
Project profile lines onto a plane? <Y>: Enter
Delete tangential edges? <Y>: Enter
```

This procedure is followed for each viewport. *Solprof* creates the profiles for each viewport. Remember that *Solprof* creates two layers for <u>each profile</u> (or for each viewport), one for hidden lines and one for visible lines. For this example, the following layers exist after the profile geometry is created. (Only PH-* and PV-* are created by *Solprof*.)

Layer Name	State	Color	Linetype
0	On	7 (white)	CONTINUOUS
MODEL	On	7 (white)	CONTINUOUS
PH-A61	On	7 (white)	HIDDEN
PH-A62	On	7 (white)	HIDDEN
PH-A63	On	7 (white)	HIDDEN
PH-A64	On	7 (white)	HIDDEN
PV-A61	On	7 (white)	CONTINUOUS
PV-A62	On	7 (white)	CONTINUOUS
PV-A63	On	7 (white)	CONTINUOUS
PV-A64	On	7 (white)	CONTINUOUS
TITLE	On	7 (white)	CONTINUOUS
VPORTS	Frozen	7 (white)	CONTINUOUS

Solprof automatically determines the viewport-specific layer visibility. Each of the profile layers (PV-* and PH-*) is *Frozen* for the *Current* and *New* viewports. In other words, only the front view profile is visible in the front viewport, only the top view profile is visible in the top viewport, and so on.

Freezing the MODEL layer reveals the profile geometry (PV-* and PH-* layers). Figure 42-18 displays the drawing after *Solprof* has been used in each viewport and the MODEL layer has been *Frozen*. The drawing is now ready for adding centerlines, dimensions, and annotation.

Figure 42-18 ─────────

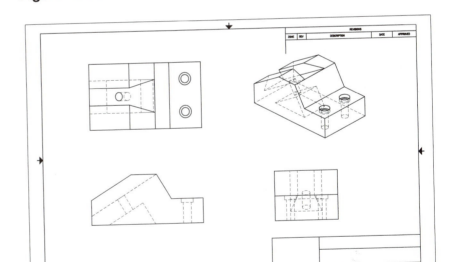

If you want to dimension a drawing, such as that in Figure 42-18, it is possible to do so by using the normal dimensioning commands. If you use *Mvsetup* to create 2D "views" of a 3D model in a layout, dimensions can be created for each view. Two methods are possible: 1) create dimensions in paper space, and 2) create dimensions for each view on a separate layer and make each dimensioning layer visible only in the appropriate view (viewport). See "Creating Dimensions for a 3D Model Shown with *Mvsetup*" earlier in this chapter. See Chapters 29 and 33 for information on creating dimensions in paper space and the advantages and disadvantages of doing so.

CHAPTER EXERCISES

1. **Using *Mvsetup* with a Solid Model**

Open the **BGUID-SL** drawing that you created in Chapter 38 Exercises. Use *SaveAs* to assign a new name, **BGUD-MVS**. Create two new *Layers* named **VPORTS** and **TITLE**. Make **VPORTS** the *Current* layer. Activate a *Layout* tab. Use the *Page Setup* dialog box to set the *Plot device* and the *Paper size* for plotting on a **B-size** sheet. Use *Mvsetup* with the following *Options*: *Layer* and *Title*. Insert the *B-size* sheet titleblock. Then use the *Create* option for *Standard Engineering* setup. *Scale* the geometry in each viewport *Uniformly* to a factor of **1**. The drawing should look like Figure 42-19. *Save* the drawing.

Figure 42-19

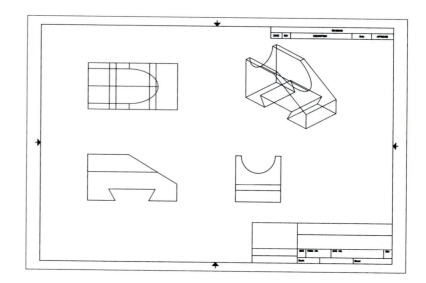

2. **Using *Mvsetup* with a Wireframe Model**

A. *Open* the **BGUID-WF** drawing that you created in Chapter 36 Exercises. Use *SaveAs* to assign a new name, **BGUD-MV2**. Make a *Layer* for the viewports and one for the titleblock. Set the viewports layer *Current*. Use *Mvsetup* with the same procedure as in the previous exercise.

B. Convert the necessary lines to "invisible" lines by creating four new *Layers* for hidden lines, one for each viewport. Assign the *Hidden* linetype to each new layer. Use the *Current VP Freeze* and *New VP Freeze* options to assign the correct viewport-specific layer visibility. Using *Properties*, change the appropriate lines to the matching layers.

Figure 42-20

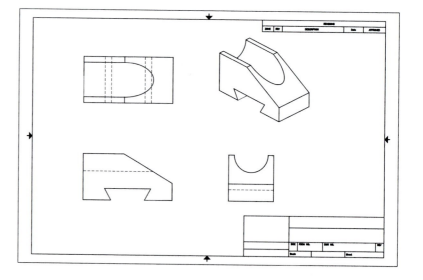

C. *Freeze* the invisible edges of the wireframe model in the isometric viewport. *Save* the drawing. The completed drawing should look like Figure 42-20.

3. **Creating Dimensions in Paper Space**

Open the **BGUD-MV2** drawing if not already open. Use *Saveas* to save the drawing as **BGUD-MV3**. Create dimensions for the views <u>in paper space</u>. Create the dimensions on the DIM layer and draw the dimensions as shown in Figure 42-21. *Save* the drawing and *Plot* to scale.

Figure 42-21

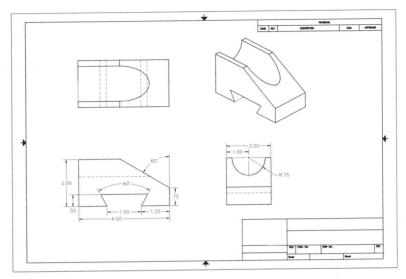

4. **Creating a 2D Drawing Using** *Solview* **and** *Soldraw*

A. *Open* the **SADL-SL** drawing from the Chapter 38 Exercises. Set up a **UCS** parallel with the front profile of the model (to create the front view as shown in Fig. 42-22). Activate a *Layout* tab. Use the *Page Setup* dialog box to set the *Plot device* and the *Paper size* for plotting on an **A**-size sheet. Change the *Printable area* units to *mm*. Next, make sure the **Hidden Linetype** is loaded and **LTSCALE** is set to **13**. Use *SaveAs* to give the name **SADDLE-SV**.

Figure 42-22

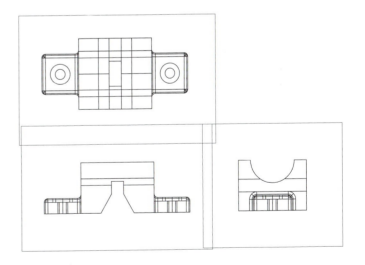

B. Use *Solview* with the **UCS** option to create the **FRONT** view. A *Scale* factor of **1** can be used. Use the *Ortho* option for the **TOP** and **RIGHT** views. If the views do not align orthogonally, use *Zoom Extents,* then *Zoom 1XP* in each viewport, or use *Mvsetup* with the *Align* option. Your results should appear like Figure 42-22. *Save* the drawing (as **SADDLE-SV**).

C. Next, use *Soldraw* to create the visible and hidden line geometry. When finished, *Freeze* the layer that the solid model is on. Insert or draw a title block and border. If the model does not fit within the border, use *Move* to move the viewport objects accordingly (make sure you *Move* two viewports together to ensure alignment). The drawing should appear like Figure 42-23. Use *SaveAs* and change the name to **SADDLE-SD**.

D. Next, dimension the drawing and add centerlines <u>in model space</u> (inside the viewports). Remember to use the layers that *Solview* created for the dimensions (one for each viewport). *Save* the drawing when you are finished (as **SADDLE-SD**). Make a *Plot* to scale.

Figure 42-23 ——————————————

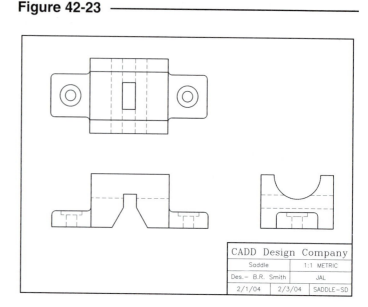

5. Create a multiview drawing from the solid model you created of the angle brace (**AGLBR-SL** drawing from Chapter 38). Use *Solview* and *Soldraw* to create the multiviews. Use the *Auxiliary* option to create an auxiliary view. Create the dimensions on the **DIM** layers <u>in model space</u> (inside the viewports). *Save* the drawing as **AGLBR-SV**.

6. Use the solid model of the **SWIVEL** from Chapter 38 Exercises. Set up a layout for a B-size sheet and plan to print or plot the drawing at full size (1 = 1). Use *Solview* and *Soldraw* to create a front, top, and auxiliary view in a manner similar to that used in the previous exercises (4 and 5). Dimension the drawing (use Figure 38-104 as a guide). Use *SaveAs* to assign the new name **SWIVEL-SD**.

MISCELLANEOUS COMMANDS AND FEATURES

Chapter Objectives

After completing this chapter you should:

1. be able to manage named objects and files using wildcards;

2. be able to control the display with *REGENAUTO*, *BLIPMODE*, *DRAGMODE*, and *Fill*;

3. be able to reinitialize I/O ports and the ACAD.PGP file using *Reinit* and be able to check the drawing for errors using *Audit*;

4. know how to use *Multiple* to automatically repeat commands;

5. be able to control dialog boxes with the *FILEDIA* and *MAXSORT* system variables;

6. be able to create and view slides using *Mslide* and *Vslide*;

7. know how to create and run a *Script* for viewing slide shows;

8. know how to use *Cal* to invoke the geometry calculator for computing arithmetic expressions, locating points, and creating and editing geometry.

CONCEPTS

This chapter discusses several unrelated AutoCAD commands and features that are helpful for intermediate-level users. The following commands, variables, and features are covered in this chapter:

Miscellaneous Features
  Wildcards
  REGENAUTO
  BLIPMODE
  DRAGMODE
  Fill
  Multiple
  FILEDIA
  MAXSORT
  Reinit
  Audit

Using Slides and Scripts
  Mslide
  Vslide
  Creating and Using Scripts
  Creating a Slide Show
  Script

The Geometry Calculator
  Cal
  Calculator Functions and Examples

The *Options* Dialog Box
  All tabs of the *Options* dialog box are explained and referenced to other chapters.

MISCELLANEOUS FEATURES

AutoCAD keeps objects in a drawing file other than those we know as graphical entities (*Line, Circle, Arc*, etc.). These are called "named objects." The named objects are:

Blocks
Dimension Styles
Layers
Layouts
Linetypes
Text Styles
User Coordinate Systems
Views
Viewport configurations

Wildcard characters, the *Rename* command, and the *Purge* command can be used to access or alter named objects.

Using Wildcards

Whenever AutoCAD prompts for a list of names, such as file names, system variables, *Block* names, *Layer* names, or other named objects, any of the wildcards in the following wildcard list can be used to access those names. These wildcards help you specify a select group of named objects from a long list without having to repeatedly type or enter the complete spelling for <u>each</u> name in the list.

For example, the asterisk (*) is a common wildcard that is used to represent any alphanumeric string (group of letters or numbers). You may choose to specify a list of layers all beginning with the letters "DIM" and ending with any string. Using the *-Layer* command and entering "DIM*" in response to the *?* option yields a list of only the layer names beginning with DIM.

Commands That Accept Wildcards

Several AutoCAD commands prompt you for a name or list of names or display a list of names to match your specification. Any command that has a ? option provides a list. Wildcards can be used with any of these commands and with many dialog boxes:

Attedit
-Block
-Dimstyle
-Insert
-Layer
-Linetype
-Lweight
-Rename
Setvar
-Style
UCS
-View
Vplayer
-Xref

Valid Wildcards That Can Be Used in AutoCAD

The following list defines valid wildcard characters that can be used in AutoCAD:

Character	Definition
# (pound)	Matches any numeric digit
@ (at)	Matches any alpha character
. (period)	Matches any nonalphanumeric character
* (asterisk)	Matches any string, including the null string. It can be used anywhere in the search pattern—at the beginning, middle, or end of the string
? (question mark)	Matches any single character
~ (tilde)	Matches anything but the pattern
[...]	Matches any one of the characters enclosed
[~...]	Matches any character not enclosed
- (hyphen)	Used with brackets to specify a range for one character
' (reverse quote)	Escapes special characters (reads next character literally)

Below are listed one or more examples for each application of wildcard patterns:

Pattern	Will match or include . . .	But not . . .
ABC	Only ABC	
~ABC	Anything but ABC	
A?C	Any 3-character sequence beginning with A and ending with C	AC, ABCD, AXXC, or XABC
AB?	ABA, AB3, ABZ, etc.	AB, ABCE, or XAB
?BC	ABC, 3BC, XBC, etc.	AB, ABCD, BC, or XXBC
A*C	AAC, AC, ABC, AX3C, etc.	XA or ABCD
A*	Anything starting with A	XAAA
*AB	Anything ending with AB	ABX
AB	AB anywhere in string	AXB
~*AB*	All strings without AB	AB, ABX, XAB, or XABX
'*AB	*AB	AB, XAB, or *ABC
[AB]C	AC or BC	ABC or XAC
[~AB]C	XC or YC	AC, BC, or XXC
[A-J]C	AC, BC, JC, etc.	ABC, AJC, or MC
[~A-J]C	Any character not in the range A–J, followed by C	AC, BC, or JC

Wildcard Examples

For example, assume that you have a drawing of a two-story residential floor plan with the following *Layers* and related settings:

Layer name	State		Color	Linetype	Lineweight
"0"	on	-P	7 (white)	"CONTINUOUS"	Default
"1-ELEC-DIM"	on	-P	7 (white)	"CONTINUOUS"	Default
"1-ELEC-LAY"	on	-P	7 (white)	"CONTINUOUS"	Default
"1-ELEC-TXT"	on	-P	7 (white)	"CONTINUOUS"	Default
"1-FLPN-DIM"	on	-P	7 (white)	"CONTINUOUS"	Default
"1-FLPN-LAY"	on	-P	7 (white)	"CONTINUOUS"	Default
"1-FLPN-TXT"	on	-P	7 (white)	"CONTINUOUS"	Default
"1-HVAC-DIM"	on	-P	7 (white)	"CONTINUOUS"	Default
"1-HVAC-LAY"	on	-P	7 (white)	"CONTINUOUS"	Default
"1-HVAC-TXT"	on	-P	7 (white)	"CONTINUOUS"	Default
"2-ELEC-DIM"	on	-P	7 (white)	"CONTINUOUS"	Default
"2-ELEC-LAY"	on	-P	7 (white)	"CONTINUOUS"	Default
"2-ELEC-TXT"	on	-P	7 (white)	"CONTINUOUS"	Default
"2-FLPN-DIM"	on	-P	7 (white)	"CONTINUOUS"	Default
"2-FLPN-LAY"	on	-P	7 (white)	"CONTINUOUS"	Default
"2-FLPN-TXT"	on	-P	7 (white)	"CONTINUOUS"	Default
"2-HVAC-DIM"	on	-P	7 (white)	"CONTINUOUS"	Default
"2-HVAC-LAY"	on	-P	7 (white)	"CONTINUOUS"	Default
"2-HVAC-TXT"	on	-P	7 (white)	"CONTINUOUS"	Default

If you wanted to turn *Off* all the layers related to the second floor (names beginning with "2"), you can use the asterisk (*) wildcard as follows:

```
Command: -layer
Current layer: "0"
Enter an option
?/Make/Set/New/ON/OFF/Color/Ltype/LWeight/Plot/PStyle/Freeze/Thaw/LOck/Unlock]: off
Enter name list of layer(s) to turn off: 2*
Enter an option
[?/Make/Set/New/ON/OFF/Color/Ltype/LWeight/Plot/PStyle/Freeze/Thaw/LOck/Unlock]: ?
```

Layer name	State		Color	Linetype	Lineweight
"0"	on	-P	7 (white)	"CONTINUOUS"	Default
"1-ELEC-DIM"	on	-P	7 (white)	"CONTINUOUS"	Default
"1-ELEC-LAY"	on	-P	7 (white)	"CONTINUOUS"	Default
"1-ELEC-TXT"	on	-P	7 (white)	"CONTINUOUS"	Default
"1-FLPN-DIM"	on	-P	7 (white)	"CONTINUOUS"	Default
"1-FLPN-LAY"	on	-P	7 (white)	"CONTINUOUS"	Default
"1-FLPN-TXT"	on	-P	7 (white)	"CONTINUOUS"	Default
"1-HVAC-DIM"	on	-P	7 (white)	"CONTINUOUS"	Default
"1-HVAC-LAY"	on	-P	7 (white)	"CONTINUOUS"	Default
"1-HVAC-TXT"	on	-P	7 (white)	"CONTINUOUS"	Default
"2-ELEC-DIM"	off	-P	7 (white)	"CONTINUOUS"	Default
"2-ELEC-LAY"	off	-P	7 (white)	"CONTINUOUS"	Default
"2-ELEC-TXT"	off	-P	7 (white)	"CONTINUOUS"	Default
"2-FLPN-DIM"	off	-P	7 (white)	"CONTINUOUS"	Default
"2-FLPN-LAY"	off	-P	7 (white)	"CONTINUOUS"	Default
"2-FLPN-TXT"	off	-P	7 (white)	"CONTINUOUS"	Default
"2-HVAC-DIM"	off	-P	7 (white)	"CONTINUOUS"	Default
"2-HVAC-LAY"	off	-P	7 (white)	"CONTINUOUS"	Default
"2-HVAC-TXT"	off	-P	7 (white)	"CONTINUOUS"	Default

You may want to *Freeze* all layers (both floors) related to the electrical layout (having ELEC in the layer name). (Assume all the layers are *On* again.) You could use the question mark (?) to represent any floor number and an asterisk (*) for any string after ELEC as follows.

```
Command: -layer
Current layer: "0"
Enter an option
?/Make/Set/New/ON/OFF/Color/Ltype/LWeight/Plot/PStyle/Freeze/Thaw/LOck/Unlock]: f
Enter name list of layer(s) to freeze: ?-elec*
Enter an option
[?/Make/Set/New/ON/OFF/Color/Ltype/LWeight/Plot/PStyle/Freeze/Thaw/LOck/Unlock]: ?
```

Layer name	State		Color	Linetype	Lineweight
"0"	on	-P	7 (white)	"CONTINUOUS"	Default
"1-ELEC-DIM"	Frozen	-P	7 (white)	"CONTINUOUS"	Default
"1-ELEC-LAY"	Frozen	-P	7 (white)	"CONTINUOUS"	Default
"1-ELEC-TXT"	Frozen	-P	7 (white)	"CONTINUOUS"	Default
"1-FLPN-DIM"	on	-P	7 (white)	"CONTINUOUS"	Default
"1-FLPN-LAY"	on	-P	7 (white)	"CONTINUOUS"	Default
"1-FLPN-TXT"	on	-P	7 (white)	"CONTINUOUS"	Default
"1-HVAC-DIM"	on	-P	7 (white)	"CONTINUOUS"	Default

"1-HVAC-LAY"	on	-P	7 (white)	"CONTINUOUS"	Default
"1-HVAC-TXT"	on	-P	7 (white)	"CONTINUOUS"	Default
"2-ELEC-DIM"	Frozen	-P	7 (white)	"CONTINUOUS"	Default
"2-ELEC-LAY"	Frozen	-P	7 (white)	"CONTINUOUS"	Default
"2-ELEC-TXT"	Frozen	-P	7 (white)	"CONTINUOUS"	Default
"2-FLPN-DIM"	on	-P	7 (white)	"CONTINUOUS"	Default
"2-FLPN-LAY"	on	-P	7 (white)	"CONTINUOUS"	Default
"2-FLPN-TXT"	on	-P	7 (white)	"CONTINUOUS"	Default
"2-HVAC-DIM"	on	-P	7 (white)	"CONTINUOUS"	Default
"2-HVAC-LAY"	on	-P	7 (white)	"CONTINUOUS"	Default
"2-HVAC-TXT"	on	-P	7 (white)	"CONTINUOUS"	Default

 You may want to use only the layout layers (names ending with LAY) and *Freeze* all the other layers. (Assume all layers are *On* and *Thawed*.) The tilde (~) character can be used to match anything but the pattern given. The tilde (~) character is translated as "anything except."

Command: **-layer**
Current layer: "0"
Enter an option
?/Make/Set/New/ON/OFF/Color/Ltype/LWeight/Plot/PStyle/Freeze/Thaw/LOck/Unlock]: **f**
Enter name list of layer(s) to freeze: **~*lay**
Enter an option
[?/Make/Set/New/ON/OFF/Color/Ltype/LWeight/Plot/PStyle/Freeze/Thaw/LOck/Unlock]: **?**

Layer name	State		Color	Linetype	Lineweight
------------------	----------		-------------	------------	------------
"0"	on	-P	7 (white)	"CONTINUOUS"	Default
"1-ELEC-DIM"	Frozen	-P	7 (white)	"CONTINUOUS"	Default
"1-ELEC-LAY"	on	-P	7 (white)	"CONTINUOUS"	Default
"1-ELEC-TXT"	Frozen	-P	7 (white)	"CONTINUOUS"	Default
"1-FLPN-DIM"	Frozen	-P	7 (white)	"CONTINUOUS"	Default
"1-FLPN-LAY"	on	-P	7 (white)	"CONTINUOUS"	Default
"1-FLPN-TXT"	Frozen	-P	7 (white)	"CONTINUOUS"	Default
"1-HVAC-DIM"	Frozen	-P	7 (white)	"CONTINUOUS"	Default
"1-HVAC-LAY"	on	-P	7 (white)	"CONTINUOUS"	Default
"1-HVAC-TXT"	Frozen	-P	7 (white)	"CONTINUOUS"	Default
"2-ELEC-DIM"	Frozen	-P	7 (white)	"CONTINUOUS"	Default
"2-ELEC-LAY"	on	-P	7 (white)	"CONTINUOUS"	Default
"2-ELEC-TXT"	Frozen	-P	7 (white)	"CONTINUOUS"	Default
"2-FLPN-DIM"	Frozen	-P	7 (white)	"CONTINUOUS"	Default
"2-FLPN-LAY"	on	-P	7 (white)	"CONTINUOUS"	Default
"2-FLPN-TXT"	Frozen	-P	7 (white)	"CONTINUOUS"	Default
"2-HVAC-DIM"	Frozen	-P	7 (white)	"CONTINUOUS"	Default
"2-HVAC-LAY"	on	-P	7 (white)	"CONTINUOUS"	Default
"2-HVAC-TXT"	Frozen	-P	7 (white)	"CONTINUOUS"	Default

Remember that wildcards can be used with many AutoCAD commands. For example, you may want to load a certain set of *Linetypes*, perhaps all the HIDDEN variations. The following sequence could be used:

Command: **-linetype**
 ?/Create/Load/Set: **l**
Linetype(s) to load: **hid***
Linetype HIDDEN loaded.
Linetype HIDDEN2 loaded.
Linetype HIDDENX2 loaded.

Or you may want to load all of the linetypes except the "X2" variations. This syntax could be used:

```
Command: -linetype
?/Create/Load/Set: l
Linetype(s) to load: ~*x2

Linetype BORDER loaded.
Linetype BORDER2 loaded.
Linetype CENTER loaded.
Linetype CENTER2 loaded.
Linetype DASHDOT loaded.
Linetype DASHDOT2 loaded.
Linetype DASHED loaded.
Linetype DASHED2 loaded.
etc.
```

RENAME

The *Rename* and *-Rename* commands allow you to rename <u>any named object</u> that is part of the current drawing. See Chapter 21, Blocks, DesignCenter, and Tool Palettes.

PURGE

Purge allows you to selectively delete any named object that is not referenced in the drawing. In other words, if the drawing has any named objects defined but not appearing in the drawing, they can be deleted with *Purge*. See Chapter 21, Blocks, DesignCenter, and Tool Palettes.

MAXSORT

The *MAXSORT* system variable controls the maximum number of named objects or files that are alphabetically sorted when a list is produced in a dialog box or Command line listing. The default setting is 1000, meaning that a maximum of 1000 items are sorted. If the list contains more than the number specified in *MAXSORT*, no sorting is done.

If *MAXSORT* is set to 0, no alphabetical sorting is done and the list is displayed in the order that the <u>items were created</u>. It may be helpful in some cases to set *MAXSORT* to 0 if you want items to be displayed in the order they were created. For example, if *MAXSORT* is set to 0, the display of layers may appear, as shown in Figure 43-1. Change the variable by typing "maxsort" at the Command: prompt or by using the *Setvar* command.

Keep in mind that display of <u>external files</u> in AutoCAD for Windows may appear in alphabetical order, even though *MAXSORT* is set to 0. The Windows sorting overrides the *MAXSORT* setting for external files only but should not affect named objects (internal). The setting for *MAXSORT* is saved in the system registry.

Figure 43-1

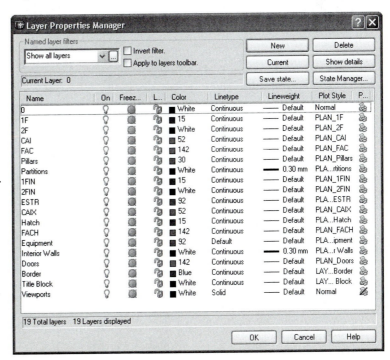

REGENAUTO

Pull-down Menu	COMMAND (TYPE)	ALIAS (TYPE)	Short-cut	Screen (side) Menu	Tablet Menu
...	*REGENAUTO*	...	...	...	...

When changes are made to a drawing that affect the appearance of the objects on the screen, the drawing is generally regenerated automatically. A regeneration can be somewhat time consuming if you are working on an extremely complex drawing or are using a relatively slow computer. In some cases, you may be performing several operations that cause regenerations between intermediate steps that you feel are unnecessary.

Regenauto allows you to control whether automatic regenerations are performed. *Regenauto* is *On* by default, but for special situations you may want to turn the automatic regenerations *Off* temporarily. The setting of *Regenauto* is stored in the *REGENMODE* variable (1=*On* and 0=*Off*). If *Regenauto* is *Off* and a regeneration is needed, AutoCAD prompts:

    About to regen—proceed? <Y>

If you prompt with a *No* response, the regeneration will be aborted.

BLIPMODE

Pull-down Menu	COMMAND (TYPE)	ALIAS (TYPE)	Short-cut	Screen (side) Menu	Tablet Menu
...	*BLIPMODE*	...	...	...	...

"Blips" are small markers that can be made to appear whenever you PICK a point. Most users prefer that the blips do not appear; therefore, Autodesk decided to set *Blipmode* to *Off* in recent releases of AutoCAD. The *BLIPMODE* variable is saved with the drawing file.

AutoCAD Release 13 and earlier versions have a *Blipmode* setting of *On* by default. In the case that you work with drawings created in Release 13 or earlier, it is likely that the blips may appear when you PICK points. If you prefer, you can turn *Blips Off*. The command format is this:

    Command: **blipmode**
    ON/OFF <ON>: (option)

The setting is saved with the drawing file in the *BLIPMODE* system variable.

DRAGMODE

Pull-down Menu	COMMAND (TYPE)	ALIAS (TYPE)	Short-cut	Screen (side) Menu	Tablet Menu
...	*DRAGMODE*	...	...	...	...

AutoCAD dynamically displays some objects such as *Circles, Blocks, Polygons*, etc., as you draw or insert them. This dynamic feature is called "dragging." If you are using a slow computer or for certain operations such as complex *Block* insertions, you may wish to turn dragging off.

The *DRAGMODE* system variable provides three positions for the dynamic dragging feature. The command format is as follows:

    Command: **dragmode**
    ON/OFF/Auto <Auto>: (option)

By default, *DRAGMODE* is set to *Auto*. In this position, dragging is automatically displayed whenever possible, based on the ability of the command in use to support it. Commands that support this feature issue a request for dragging when the command is used. This request is automatically filled when *DRAGMODE* is in the *Auto* position.

When *DRAGMODE* is *On*, you must enter *"drag"* at the command prompt when the request is issued if you want to enable dragging. If *DRAGMODE* is *Off*, dragging is not displayed and all requests are automatically ignored. The *DRAGMODE* variable setting is stored with the current drawing file.

FILL

Pull-down Menu	COMMAND (TYPE)	ALIAS (TYPE)	Short-cut	Screen (side) Menu	Tablet Menu
...	*FILL*	...	...	...	...

The *Fill* command controls the display of objects that are filled with solid color. Commands that create solid filled objects are *Donut*, *Solid*, and *Pline* (with *width*), *Bhatch*, and text commands using TrueType fonts. *Fill* can be toggled *On* or *Off* to display or plot the objects with or without solid color. The command format produces this prompt:

```
Command: fill
ON/OFF <On>: (option)
Command:
```

Figure 43-2 displays several objects with *Fill On* and *Off*. The default position for the *Fill* command is *On*. If *Fill* is changed, the drawing must be regenerated to display the effects of the change. The setting for *Fill* is stored in the *FILLMODE* system variable.

Figure 43-2

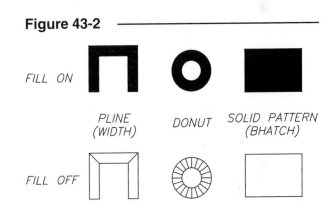

Since solid filling a drawing with many wide *Plines*, *Donuts*, and *Solid* hatch patterns may be time consuming to regenerate, it may be useful to turn *Fill Off* temporarily. *Fill Off* would be <u>especially</u> helpful for speeding up test plots.

See also Chapters 18 and 26 for using *Fill* with solid-filled TrueType fonts and solid hatch patterns.

REINIT

Pull-down Menu	COMMAND (TYPE)	ALIAS (TYPE)	Short-cut	Screen (side) Menu	Tablet Menu
...	*REINIT*	...	...	...	...

Many software programs require you to exit and restart the program in order for reinitialization to be performed. AutoCAD provides the *Reinit* command to prevent having to exit AutoCAD to accomplish this task.

The *Reinit* command reinitializes the input/output ports of the computer (to the digitizer) and reinitializes the AutoCAD Program Parameters (ACAD.PGP) file. Reinitialization may be necessary if you physically switch the port cable from the digitizer or if it loses power temporarily. If you edit the ACAD.PGP file, it must be reinitialized before you can use the new changes in AutoCAD (see Chapter 44 for information on the ACAD.PGP file).

Invoking the *Reinit* command causes the *Re-initialization* dialog box to appear (Fig. 43-3). You may check one of the boxes. PICKing the *OK* tile causes an immediate reinitialization.

Figure 43-3

If you have configured the digitizer as the "Current System Pointing Device," AutoCAD uses the Windows mouse driver; therefore, the "Digitizer" checkbox is disabled. If your pointing device is disabled and you need to use this dialog box, use TAB to move to the desired options, the space bar to check the boxes, and Enter to execute an *OK*.

MULTIPLE

Pull-down Menu	COMMAND (TYPE)	ALIAS (TYPE)	Short-cut	Screen (side) Menu	Tablet Menu
...	MULTIPLE	...	...	...	...

Multiple is not a command, but rather a <u>command modifier</u> (or adjective) since it is used in conjunction with another command. Entering *"multiple"* at the Command prompt <u>as a prefix</u> to almost any command causes the command to automatically repeat until you stop the sequence with Escape. For example, if you wanted to use the *-Insert* command repetitively, you enter the following:

```
Command: multiple
Enter command name to repeat: -insert
Enter block name or [?]:
```

The *Insert* command would repeat until you stop it with Escape. This is helpful for inserting multiple *Blocks*.

The multiple modifier repeats only the command, not the command options. For example, if you wanted to repeatedly use *Circle* with the *Tangent, Tangent, Radius* option, the *TTR* would have to be entered each time the *Circle* command was automatically repeated.

AUDIT

Pull-down Menu	COMMAND (TYPE)	ALIAS (TYPE)	Short-cut	Screen (side) Menu	Tablet Menu
File Drawing Utilites > Audit	AUDIT	...	...	FILE Audit	Y,24

The *Audit* command is AutoCAD's diagnostic utility for checking drawing files and correcting errors. *Audit* will examine the current drawing. You can decide whether or not AutoCAD should fix any errors if they are found. The command may yield a display something like this:

```
Command: audit
Fix any errors detected? <N> y
 15    Blocks audited
Pass 1 100    objects audited
Pass 1 200    objects audited
Pass 1 282    objects audited
Pass 2 100    objects audited
Pass 2 200    objects audited
Pass 2 282    objects audited
Pass 3 100    objects audited
Pass 3 200    objects audited
```

Pass 3 282   objects audited
Total errors found 0 fixed 0
Command:

If errors are detected and you requested them to be fixed, the last line of the report indicates the number of errors found and the number fixed. If *Audit* cannot fix the errors, try the *Recover* command (see Chapter 2).

You can create an ASCII report file by changing the *AUDITCTL* system variable to 1. In this case, when the *Audit* command is used, AutoCAD automatically writes the report out to disk in the current directory using the current drawing file name and an .ADT file extension. The default setting for *AUDITCTL* is 0 (off).

FILEDIA

This system variable enables and disables FILE-related DIAlog boxes. File-related dialog boxes are those that are used for reading or writing files. For example, when you request to *Open* a file, the *Select File* dialog box appears (Fig. 43-4).

Figure 43-4

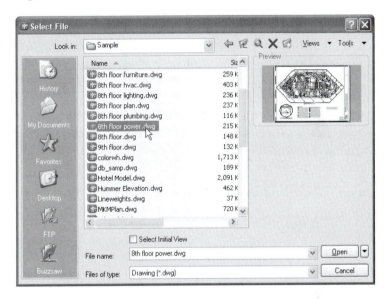

The settings for *FILEDIA* are as follows:

FILEDIA = 1   File dialog boxes appear when a file command is invoked (default setting).
FILEDIA = 0   File dialog boxes are disabled and command prompts are used.

The *FILEDIA* variable can be set to 0 to present the Command line prompt shown below instead of the *Select File* dialog box:

Command: **open**
Enter name of drawing to open:

If file dialog boxes are disabled, they can be invoked at the Command line by entering a tilde (~) symbol at the Command prompt. For example, when *FILEDIA* is set to 0, invoking the *Open* command by any method produces the command prompt; however, typing "~" (tilde) at the "Enter name of drawing to open:" prompt produces the *Select File* dialog box.

Disabling the dialog boxes can be helpful for running a script file. Since dialog boxes require user input with a pointing device, some scripts require the Command line interface to be used instead of dialog boxes.

The *FILEDIA* variable setting is saved in the system registry.

ATTDIA

The *ATTDIA* variable controls the display of the *Enter Attributes* dialog box. The default setting is 0 (off). Ensure the setting is off if you are using a script to enter attributes. The variable is saved in the drawing file. (See Chapter 22, Block Attributes, for more information.)

USING SLIDES AND SCRIPTS

A slide is a "snapshot" of an AutoCAD drawing that can be made and saved to a file using the *Mslide* command. The resulting slide file contains only one image with no other drawing information. Slide files can later be viewed in AutoCAD using the *Vslide* command. A prepared series of slides can be presented at a later time, much like a slide show. This section also explains how to create a self-running slide show from your slide files by creating and running a *Script*.

MSLIDE

Pull-down Menu	COMMAND (TYPE)	ALIAS (TYPE)	Short-cut	Screen (side) Menu	Tablet Menu
...	*MSLIDE*	...	...	...	...

Mslide is short for "make slide." A slide is a "screen capture" of the current AutoCAD display. The image that is saved is composed of the objects that appear in the drawing editor when the *Mslide* command is issued (excluding the menus, tool bars, grid, cursor, and Command line). If model or paper space viewports are active, the current viewport's image is captured. When a slide is made, the resulting image is saved as a file in the directory of your choice with an extension of .SLD.

Mslide is simple to use. Assume that you are working on a drawing—for example, Figure 43-5—and want to take a "snapshot" of the image using *Mslide*. Invoke the command by any method:

Command: **mslide** (The *Create Slide File* dialog box appears. Assign a name and directory.)
Command:

Figure 43-5

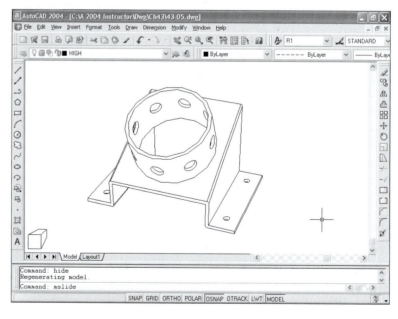

When the *Create Slide File* dialog box appears (Fig. 43-6, next page), assign a name for the slide. (If the *FILEDIA* variable is set to 0, the command line format is activated instead of the dialog box.) Only the file name is required because AutoCAD automatically assigns the .SLD file extension. The drawing image is saved without the grid, cursor, menus, toolbars, etc. Only the drawing image is saved in the .SLD; no other drawing information is saved. Therefore, an .SLD file is much smaller (file size in bytes) than the corresponding .DWG file. Use the *Vslide* command to view the slide again.

AutoCAD .SLD files are proprietary, meaning the file format was developed by AutoCAD to be used in AutoCAD. The .SLD file format is not widely used like a .TIF, .GIF, .JPG, etc. but can be viewed or converted by some software. Slide files are also a convenient format for saving an image created with the AutoCAD *Shade* and *Shademode* commands (see Chapter 35, 3D Display and Viewing).

Figure 43-6

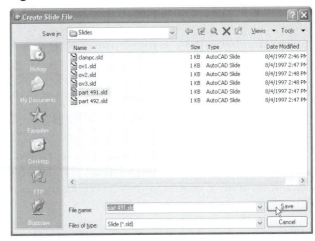

VSLIDE	Pull-down Menu	COMMAND (TYPE)	ALIAS (TYPE)	Short-cut	Screen (side) Menu	Tablet Menu
	...	*VSLIDE*	...	...	...	...

Vslide is short for "view slide." Use this command to view previously made AutoCAD slide (.SLD) files. Assuming previously created slide files are accessible, invoke *Vslide* by any method:

Command: **vslide** (The *Select Slide File* dialog box appears. Select the desired directory and slide name.)
Command:

When the *Select Slide File* dialog appears (Fig. 43-7), enter or select the desired directory and slide name. The selected slide then appears in the Drawing Editor (if viewports are active, it appears in the current viewport). Using the slide from the previous example (see *Mslide*), the slide image is "projected" in the Drawing Editor (Fig. 43-8).

Figure 43-7

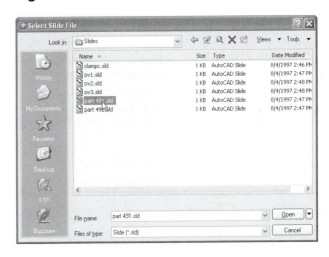

A slide file may be viewed during any AutoCAD session. If a drawing is in progress, the slide file is projected on top of the current drawing, much like a photographic slide is displayed on a projection screen or wall. Because a slide file contains only a description of the display, the slide file image cannot be edited. You cannot use *Pan* or *Zoom* to modify the display of the slide. In fact, any AutoCAD display command issued (*Redraw, Pan, Zoom*, etc.) will cause the slide to disappear and act upon the current drawing, not the slide. After a slide has been viewed, use *Redraw* to refresh the current drawing before using draw or edit commands. Because .SLD files are generally much smaller and contain less information than .DWG files, viewing a slide is much faster than loading a drawing.

Figure 43-8

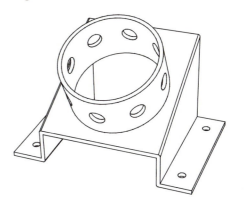

Slide files are helpful for keeping a visual record of drawings, since the .SLD file is much smaller and faster loading than the corresponding .DWG file. The most common use for a series of slides is giving a presentation or a demonstration. Using prepared slides to present a complex drawing is much faster and more reliable than using *Pan* and *Zoom*. A script file (*.SCR) file can be created to "run" a series of slide files—a scripted slide show (see "Creating and Using Scripts").

Creating and Using Scripts

Scripts are text files that contain AutoCAD commands. Similar to a batch routine written in DOS, a script file contains a user-specified listing of the desired AutoCAD commands. To run a script, use the *Script* command. AutoCAD reads and operates each command and option in the sequence it is listed in the script file. A script file contains essentially the same text that you would enter at the Command line during a drawing session. When a script is executed, each line of the script is echoed at the Command line.

Script files can be used for a number of purposes. If you want to give a presentation, you can create a self-running display of a series of slides by creating a script file. For this purpose, the script file repeats the *Vslide* command and the slide names. Since a script can contain any AutoCAD commands, you could make scripts to execute a series of frequently used commands—like a macro (this topic is discussed in Chapter 44, Basic Customization). You could even recreate an entire drawing by listing every command used to create the drawing in the script file.

There are three basic steps to creating and running a script:

1. In AutoCAD, determine the commands, options, and responses that you want to include in the script file. This can involve a "practice run" to determine each step. You may have to write down each command, option, and response. The *Logfileon* and *Logfileoff* commands can help with this process (see Chapter 44).

    If you are creating a slide show, the slides must be created first. A practice run of the show using *Vslide* can be helpful to ensure you know the correct slide sequence.

2. Use a text editor such as Windows Notepad or other program that can save an ASCII text file. Include all the desired commands. Save the file as an .SCR file extension.

3   In AutoCAD, use the *Script* command to locate, load, and run the script.

Creating a Slide Show

Assume that you have created the slides and know the sequence that you want to display them. You are ready to create the script file. Use any text editor or word processor that can create text files in ASCII format (without internal word processing codes). The script file should be saved using any descriptive file name but should have a .SCR file extension.

Begin the script file (in your text editor) by entering the *Vslide* command. When the *Vslide* command is issued in AutoCAD, the "Slide file:" prompt appears. Therefore, your next line of text (in the script file) should be the name of the desired slide file (SLIDE1, for example) without the .SLD extension. Upper- or lowercase letters can be used. For example, the script file at this point might look like this:

```
VSLIDE
SLIDE1
```

Each line or space in a script file is equivalent to one AutoCAD command, option, or response by the user. In other words, place a space or begin a new line for each time you would press Enter in AutoCAD. The file would be translated as:

Command: ***VSLIDE*** Enter
Slide file: **SLIDE1** Enter

You probably want to allow the slide to stay visible for a specific amount of time—for example, 5 seconds—then move on to the next slide. You may also want the slide show to be self-repeating automatically for use during presentations or an open house. A few script-controlling commands are given here that will enable you to more effectively present the slide show:

DELAY   With the *Delay* command, you can enter the number of milliseconds you wish to have the current slide displayed. For example, an approximate 5 second display of the slide would be accomplished by entering "*Delay 5000.*" When entering the delay time, consider that the slide will remain displayed while the next *Vslide* command is issued, as well as during the time that it takes your system to retrieve and display the next slide.

NOTE: The *Delay* time may vary based on the speed of your computer. A delay time of 1000 may be less than 1 second on some computers.

RSCRIPT   Typically entered as the last line of a script file, this command loops to the first line of the script and repeats the script indefinitely (until you interrupt with Backspace or Escape). Make sure you include an extra space or line after the *Rscript* line (to act as an Enter).

RESUME   This command cannot be used in the script file, but must be typed at the keyboard during a script execution. You can interrupt a script, then type *Resume* to pick up with the script again. See "*Script*" next for details on using *Resume*.

An example script file is given here for displaying two slides (SLIDE1 and SLIDE2), then the drawing in the background, then repeating the show:

```
VSLIDE
SLIDE1
DELAY 2000
VSLIDE
SLIDE2
DELAY 500
REDRAW
DELAY 1000
RSCRIPT
```

In the script above, SLIDE1 displays for 2 seconds, SLIDE2 displays for .5 seconds, the *Redraw* causes the drawing (in the background) to appear and display for 1 second, and then the entire script repeats.

Remember, for the script to operate correctly, no extra spaces can exist (at the end of lines, for example) and slide file names can have no spaces.

The drive and directory location of the slide files is significant. The slide files should be located in the path designated by the *Support File Search Path* section of the *Files* tab in the *Options* dialog box (Fig. 43-9). Otherwise, the path should be given in the script file before each slide name as shown below:

Figure 43-9

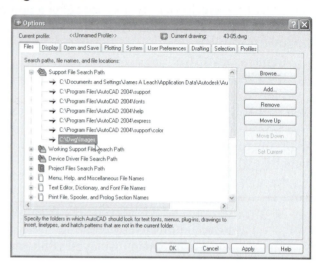

```
VSLIDE
C:\DWGS\IMAGES\SLIDE1
DELAY 2000
VSLIDE
C:\DWGS\IMAGES\SLIDE2
DELAY 500
REDRAW
DELAY 1000
RSCRIPT
```

Once the script is created and saved as an .SCR file, return to AutoCAD and test it by using the *Script* command.

SCRIPT

Pull-down Menu	COMMAND (TYPE)	ALIAS (TYPE)	Short-cut	Screen (side) Menu	Tablet Menu
Tools *Run Script...*	*SCRIPT*	*SCR*	...	*TOOLS 1* *Script*	*V,9*

The *Script* command allows you to locate, load, and run a script file. Invoking the *Script* command produces the *Select Script File* dialog box (Fig. 43-10):

Command: **script** (The *Select Script File* dialog box appears. Select the desired directory and script name.)
Command:

Figure 43-10

In the dialog box, the desired directory and file can be selected. Notice the default file extension of .SCR. (If the *FILEDIA* variable is set to 0, the dialog box does not appear and the "Script file:" prompt appears at the Command line instead.)

When the file is selected, the script executes. There is nothing more to do at this point other than watch the script (or begin your verbal presentation). If you are giving a verbal presentation, here are some other ideas:

1. Text slides can be created as an introduction, conclusion, and topic or section heading. Just use the *Mtext* or *Dtext* command to create the desired text screen in AutoCAD; then create a slide with *Mslide*.

2. To temporarily interrupt the script and view a particular slide, press the Backspace or Escape key. This enables you to discuss one slide indefinitely or stop to answer questions. Typing *Resume* causes the script to pick up again from the stopping point.

3. Remember that any AutoCAD drawing can be in the background while the slide show is projected over the drawing. The Backspace and *Resume* feature allows you to cut from the slides and return to the drawing for editing, *Pans*, or *Zooms*, then return to the slide show. You may want to have a drawing loaded in the background (an entire view of the drawing, for example) and show slides "on top" (close-ups, etc.).

NOTE: The *FILEDIA* variable does not have to be set to 0 before running a slide show script. Even though the *Vslide* command normally invokes a dialog box, the Command line format is automatically activated when using *Vslide* from a script. With scripts using other commands that activate dialog boxes, it may be helpful to set *FILEDIA* to 0.

THE GEOMETRY CALCULATOR

The geometry calculator can be used as a standard 10-key calculator and can also be used to calculate the location of points in a drawing. It can be used transparently (within a command) to calculate and pass values to the current command. The real power of the geometry calculator is the ability to be used with existing geometry utilizing *OSNAPs*.

The geometric calculator provides the following capabilities:

1. It can be used as a standard calculator to return values for arithmetic expressions.

2. The calculator can be used transparently to compute numbers, distances, angles, and vector directions as input for a command.

3. The calculator can be used transparently to pass coordinate data to the current command. This feature simplifies creating new geometry and editing existing geometry by enabling you to locate points that are not easily accessible by other means.

4. *OSNAPs* can be used in the expressions to further enhance your ability to use existing geometry (coordinates) in calculations.

5. Built-in functions are available as shortcuts for using *OSNAPs*, creating geometry, and making mathematical calculations and conversions.

6. The calculator can be used as a programmable calculator because of its ability to store variables and interact with AutoLISP defined functions and expressions.

CAL

Pull-down Menu	COMMAND (TYPE)	ALIAS (TYPE)	Short-cut	Screen (side) Menu	Tablet Menu
...	*CAL*	...	...	...	...

The calculator prompts you for the entry of an "expression." The calculator uses the Command line interface as shown in this example (divide 12 by 5):

```
Command: cal
Initializing...>> Expression: 12/5
2.4
Command:
```

The expression can contain arithmetic expressions, vector points, coordinates, and variables. When you enter the expression, the geometry calculator recognizes the following standard operation priorities:

1. Expressions in parentheses are considered first, beginning with the innermost.
2. Exponents are calculated first, any multiplication or division is calculated next, and addition or subtraction is calculated last.
3. Operators of equal precedence (for example, several multiplications) are taken from left to right.

These expressions can be applied to existing geometry using standard *OSNAP* functions or used in any AutoCAD command where points, vectors, and numbers are expected. The calculator supports many of the standard numeric functions found on typical scientific calculators.

Numeric Operators

Operator	Operation
()	group expressions
^	exponentiation
*,/	multiplication and division
+,-	addition and subtraction

A few examples of simple mathematical expressions using numeric operators are given here:

(5+6)*2	=	22
2^2*(14/2)	=	28
((6/2)-1)*PI	=	6.28319

Format for Feet and Inches
The default unit of measure for distances is one inch; however, feet and inches can be entered as follows:

feet'-inches" or *feet'inches"*

Expressions using feet and inch notation convert to real numbers as in the following examples:

3'	evaluates to	3*12=36.0
3.75"	evaluates to	3*12+ 9= 45.0
4'-5"	evaluates to	4*12+ 5= 53.0
6.5'3"	evaluates to	6*12+ 6+ 3=81.0

Format for Angles

The default unit of measure for angles is decimal degrees and is entered like this:

*deg***d***min'sec"*

Enter a number followed by an **r** to designate radians or **g** to designate grads. Examples are as follows:

45d15'22"
0d15'22"
1.55r
21g

Points and Vectors

Both points and vectors can be used in expressions. A point or vector is a set of real numbers enclosed in brackets [ ]. You can omit from your input string coordinates with a value of zero and the comma(s) immediately preceding the right bracket. Some examples of this syntax are shown below:

[2,5,8]	=	2,5,8
[1,2]	=	1,2,0
[, ,4]	=	0,0,4
[ ]	=	0,0,0

A point defines a location in space. A vector defines a direction in space. You can add a vector to a point to obtain another point. (In the examples later in this chapter, the notation p1,p2 is used to designate points and v1,v2 is used to designate vectors.)

Point Formats

The calculator also supports the standard input format for rectangular, polar, relative, and other input:

Coordinate entry method	Format
Relative rectangular	[@X,Y,Z}
Relative polar	[@dist<angle]
Relative cylindrical	[@dist<angle,Z]
Relative spherical	[@dist<angle<angle]
WCS (overrides UCS)	use the * (asterisk) prefix inside brackets: [*X,Y,Z]

Standard Numeric Functions

The geometric calculator supports these standard numeric functions:

Function	Description
sin(*angle*)	Sine of the angle.
cos(*angle*)	Cosine of the angle.
tang(*angle*)	Tangent of the angle.
asin(*real*)	Arcsine of the number. The number must be between -1 and 1.
acos(*real*)	Arccosine of the number. The number must be between -1 and 1.
atan(*real*)	Arctangent of the number.
ln(*real*)	Natural log of the number.

Function	Description
log(*real*)	Base-10 log of the number.
exp(*real*)	Natural exponent of the number.
exp10(*real*)	Base-10 exponent of the number.
sqr(*real*)	Square of a number.
sqrt(*real*)	Square root of a number. The number must be non-negative.
abs(*real*)	Absolute value of the number.
round(*real*)	Number rounded to the nearest integer.
trunc(*real*)	Integer potion of the number.
r2d(*angle*)	Angles in radians converted to degrees. For example, r2d(pi) converts the constant pi to 180 degrees.
d2r(*angle*)	Angles in degrees converted to radians. For example, d2r (180) converts 180 degrees to radians and returns the value of the constant pi.
pi	The constant pi.

OSNAP Modes

You can include *OSNAP* modes as part of an arithmetic expression. When *Cal* encounters an *OSNAP* mode, the aperature appears and you are prompted to "Select entity for (*OPTION*) snap:". The selected *OSNAP* point coordinate values are passed to the expression for evaluation. To use *OSNAP* modes, enter the three-character name in the expression:

CAL *OSNAP* Mode	AutoCAD *OSNAP* Mode
END	*Endpoint*
EXT	*Extension*
INS	*Insert*
INT	*Intersection*
MID	*Midpoint*
CEN	*Center*
NEA	*Nearest*
NOD	*Node*
QUA	*Quadrant*
PAR	*Parallel*
PER	*Perpendicular*
TAN	*Tangent*

Built-in Functions

The Geometry Calculator has a number of "built-in" functions that provide a variety of utilities to be used for conversions and inquiries of existing geometry.

Function	Description
abs(v)	Calculates the length of vector v, a non-negative real number.
abs([1,2,4])	Calculates the length of vector [1,2,4].
ang(p)	Angle between the X axis and vector p.
ang(p1, p2)	Angle between the X axis and line (p1, p2).
ang(apex, p1, p2)	Angle between the lines (apex, p1) and (apex, p2) projected onto the XY plane.
ang(apex, p1, p2, p)	Angle between the lines (apex, p1) and (apex, p2).
cvunit(1,inch,cm)	Converts the value 1 from inches to centimeters.
cur	Sets the value of the variable *LASTPOINT*.
cur(p)	Pick a point using the graphics cursor with a rubberband line from point p.
dist(p1, p2)	Distance between points p1 and p2.
dpl(p, p1, p2)	Distance between point p and the line (p1, p2).
dpp(p, p1, p2, p3)	Distance between point p and the plane defined by p1, p2 & p3.
ill(p1, p2, p3, p4)	Intersection of lines (p1, p2) and (p3, p4).
ilp(p1,p2,p3,p4,p5)	Intersection of line (p1, p2) and plane defined by p3, p4 & p5.
nor	Unit vector normal to a circle, arc, or arc polyline segment.
nor(v)	Unit vector in the XY plane and normal to the vector v.
nor(p1, p2)	Unit vector in the XY plane and normal to the line (p1, p2).
nor(p1, p2, p3)	Unit vector normal to plane defined by p1, p2, & p3.
pld(pl, p2, units)	Point on the line (p1, p2) that is drawing units away from the point p1.
plt(p1, p2, t)	Point on line (p1, p2) that is t segments from p1. A segment = distance from p1 to p2.
rad	Radius of the selected object.
rot(p, origin, ang)	Rotates point p through angle ang using a line parallel to the Z axis passing through origin as the axis of rotation.
rot(p1, p2, ang)	Rotates point p through angle ang using line (p1, p2) as the axis of rotation.
u2w(p1)	Converts point p1 expressed in WCS to the current UCS.
vec(p1, p2)	Vector from point p1 to point p2.
vec1(p1, p2)	Unit vector from point p1 to point p2.
w2u(p1)	Converts point p1 expressed in WCS to the current UCS.

Shortcut Functions

In addition to computing numbers and points, you can use the calculator to create and edit geometry in your drawing. This is one of the most powerful features of *CAL*. The following table lists the geometric functions supported by the calculator. Using these shortcuts prevents you from having to enter the three-letter *OSNAP* mode and parentheses in the expression.

Function	Shortcut for	Description
dee	dist(end, end)	Distance between two endpoints.
ille	ill(end, end, end, end)	Intersection of two lines defined by four endpoints.
mee	(end+end)/2	Midpoint between two endpoints.
nee	nor(end, end)	Unit vector in the XY plane and normal to two endpoints.
vee	vec(end, end)	Vector from two endpoints.
vee1	vec1(end, end)	Unit vector from two endpoints.

Point Filter Functions

You can use these fuctions to filter the X,Y, and Z components of a point or vector coordinate.

Function	Description
xyof(*p1*)	X and Y components of a point. The Z component is set to 0.0.
xzof(*p1*)	X and Z components of a point. The Y component is set to 0.0.
yzof(*p1*)	Y and Z components of a point. The X component is set to 0.0.
xof(*p1*)	X component of a point. The Y and Z components are set to 0.0.
yof(*p1*)	Y component of a point. The X and Z components are set to 0.0.
zof(*p1*)	Z component of a point. The X and Y components are set to 0.0.
rxof(*p1*)	X component of a point.
ryof(*p1*)	Y component of a point.
rzof(*p1*)	Z component of a point.

Transparent Use of *CAL*

The real power of the calculator is shown when you use the command <u>transparently</u>. In other words, you can use *CAL* during another command if you prefix it with an apostrophe (') symbol. In this way, *CAL* passes on the solution for the expression as data input for the command.

For example, using *CAL* to draw the first point of a *Line* as the midpoint between two endpoints could be accomplished as shown here:

```
Command: Line
Specify first point: 'cal
>>Expression: mee
```

Geometry Calculator Examples

Figure 43-11

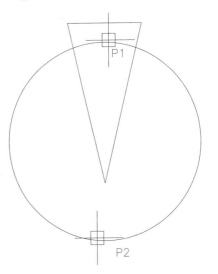

Point, Vector, and *OSNAP*s Example

The following example places the beginning of a *Line* [1,-.5] units over from the selected *Quadrant* of a *Circle*, then adds two more calculated vectors to create a wedge (Fig. 43-11). Don't forget to enter an apostrophe (') to use *Cal* transparently if you are typing the command.

```
Command: line
Specify first point: 'cal
>>Expression: qua+[-1,.5]
>>Select entity for QUA snap: PICK
(PICK the top quadrant on the circle, P1.)
```

To continue with the line from the previous sequence, a vector is added to a current point to obtain the next point on the line.

```
Specify next point or [Undo]: 'cal
>> Expression: cur+[,1.5]
>> Enter a point: qua
of PICK (PICK the bottom quadrant of the circle, P2.)
```

To finish the wedge, the next point of the vector is placed in the same fashion as the first point.

```
Specify next point or [Undo]: 'cal
>> Expression: qua+[1,.5]
>> Select entity for QUA snap: PICK  (PICK the top quadrant of the circle, P1 again.)
Specify next point or [Undo] C  (To close the wedge shape.)
```

Numeric Functions and Built-in Functions Example

The following example displays the use of numeric functions and built-in functions in finding the circumference and area of a *Circle* (no figure).

To find the circumference of a *Circle*, first create the variable (R1) which collects the circle radius. Entering any variable name (R1 in this case) followed by an equal (=) sign stores the value or point to an AutoLISP variable:

```
Command: cal
>> Expression: R1=rad
>> Select circle, arc or polyline segment for RAD function: PICK  (PICK anywhere on a Circle.)
2.58261
```

The value is stored to variable R1. Then enter the radius in a numeric function to determine the circumference:

```
Command: cal
>> Expression: 2*pi*R1
16.227
```

Now use the radius variable (R1) in a numeric function format to determine the area of a circle:

```
Command: cal
>> Expression: pi*sqr(R1)
20.9541
```

OSNAP **Examples**

In the following example, the calculator returns the point (Fig. 43-12, point A) halfway between the center of the circle and the endpoint of the object:

Figure 43-12

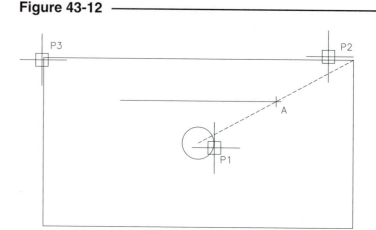

Command: *line*
Specify first point: *'cal*
>> Expression: **(cen+end)/2**
>> Select entity for CEN snap: **PICK**
(PICK any point on the circle, P1.)
>> Select entity for END snap: **PICK**
(PICK the upper-right corner of the rectangle, P2.)

AutoCAD finds the point halfway between the selected points (A) and passes the coordinate to the *Line* command as the "Specify first point:". Do the same to determine the "Specify next point or [Undo]" of the *Line*:

Specify next point or [Undo]: *'cal*
>> Expression: **(cen+end)/2**
>> Select entity for CEN snap: **PICK** (PICK the circle, P1 again.)
>> Select entity for END snap: **PICK** (PICK the upper-left corner of the rectangle, P3.)
Specify next point or [Undo]: **Enter**

The *Line* is drawn with its endpoints halfway between the *OSNAP*s.

This example uses *END* and *CEN* to calculate the centroid (Fig. 43-13, point B) defined by 3 points (2 endpoints and the center of a circle):

Figure 43-13

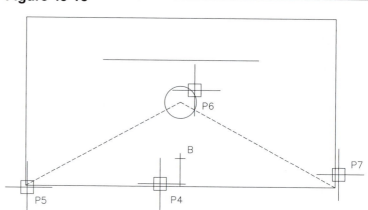

Command: *line*
Specify first point: *mid*
of **PICK** (PICK the bottom horizontal line of the rectangle, P4.)
Specify next point or [Undo]: *'cal*
>> Expression: **(end+cen+end)/3**
>> Select entity for END snap: **PICK** (PICK the lower-left corner of the rectangle, P5.)
>> Select entity for CEN snap: **PICK** (PICK the circle, P6.)
>> Select entity for END snap: **PICK** (PICK the lower-right corner of the rectangle, P7.)
Specify next point or [Undo]: **Enter**

For the next drawing example, *Cal* prompts for an object, then returns a point that is 1 unit to the left of the selected object's midpoint. In the second sequence, the point is placed 1 unit to the right of the selected object's midpoint.

Figure 43-14

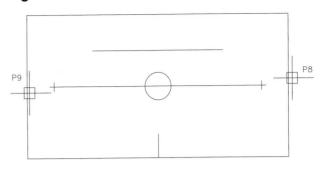

```
Command: Line
Specify first point: 'cal
>> Expression: mid+[-1,]
>> Select entity for MID snap: PICK  (PICK the
right vertical line of the rectangle, P8.)


Specify next point or [Undo]: 'cal
>> Expression: mid+[1,]
>> Select entity for MID snap: PICK  (PICK the left vertical line of the rectangle, P9.)
Specify next point or [Undo]: Enter
Command:
```

Shortcut Function Examples

The calculator shortcut functions are used to add text and circles to the previous drawing. In the first example, middle justified text is placed at the intersection of two lines defined by endpoints (*ille*) (Fig. 43-15).

Figure 43-15

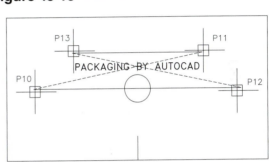

```
Command: dtext
Justify/Style/<Start point>: j
Align/Fit/Center/Middle/Right/TL/TC/
TR/ML/MC/MR/BL/BC/BR: m
Middle point: 'cal
>> Expression: ille
>> Select one endpoint for ILLE:First line: PICK  (PICK the left end of the long horizontal line, P10.)
>> Select another endpoint for ILLE:First line: PICK  (PICK the right end of the short
horizontal line, P11.)
>> Select one endpoint for ILLE:Second line: PICK  (PICK the right end of the long
horizontal line, P12.)
>> Select another endpoint for ILLE:Second line: PICK  (PICK the left end of the short horizontal line,
P13.)
Height <0.2000>: Enter
Rotation angle <0>: Enter
Text: PACKAGING BY AUTOCAD
Text: Enter
```

Use another built-in function to return the midpoint between two endpoints (*mee*) for the center of a circle (Fig. 43-16):

Figure 43-16

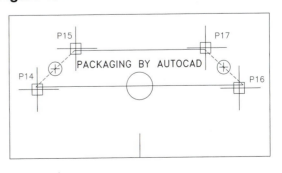

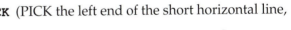

```
Command: Circle
3P/2P/TTR/<Center point>: 'CAL
>> Expression: mee
>> Select one endpoint for MEE: PICK  (PICK the
left end of the long horizontal line, P14.)
>> Select another endpoint for MEE: PICK  (PICK
the left end of the short horizontal line, P15.)
Diameter/<Radius> <0.5000>: .25
```

Command: CIRCLE 3P/2P/TTR/<Center point>: `CAL
>> Expression: **mee**
>> Select one endpoint for MEE: **PICK** (PICK the right end of the long horizontal line, P16.)
>> Select another endpoint for MEE: **PICK** (PICK the right end of the short horizontal line, P17.)
Diameter/<Radius> <0.2500>: **Enter**
Command:

The Geometry Calculator has tremendous power, but it can be beneficial only if you know how and when to use it. Review the function tables (particularly the Built-in and Shortcut Functions) and practice with the expressions. Many of the functions can save drawing time and allow you to construct and edit more efficiently than by other methods.

THE *OPTIONS* DIALOG BOX

OPTIONS

Pull-down Menu	COMMAND (TYPE)	ALIAS (TYPE)	Shortcut Menu	Screen (side) Menu	Tablet
Tool Options...	*OPTIONS*	*OP*	(Default Menu) *Options...*	*TOOLS2 Options*	*Y, 10*

The *Options* command invokes the *Options* dialog box (see Figure 43-17). The *Options* dialog box contains controls that allow you to specify how AutoCAD operates. Many of the options manage AutoCAD system variable settings that can also be changed by accessing the system variables directly at the Command line. Other options control the graphical user interface features that can only be changed through this dialog box.

Whether the *Options* dialog box settings change system variables or not, each setting is stored in one of two places:

1. the current drawing file
2. the system registry

Settings that are stored in the current drawing file are denoted in the dialog box with the blue AutoCAD icon. These settings can change when a new drawing is made current. Settings stored in the system registry (of the computer you are using) are generally user preferences and remain with that specific computer and therefore do not change when a new drawing is made current. These settings may be different on computers in a lab or office.

The following material explains the controls of the *Options* dialog box or refers you to other chapters of this text where the topics are discussed in more detail related to a specific feature of AutoCAD.

Files Tab

This tab allows you to specify the search paths, file names, and file locations of AutoCAD-related files. The default settings are automatically created upon installation and are required for AutoCAD to operate correctly. For most cases these settings do not need changing; however, a few options may need customizing to suit your needs. Click on the plus symbol (+) to expand each heading and display suboptions. Help is given at the bottom of the dialog box when you select a folder.

Figure 43-17

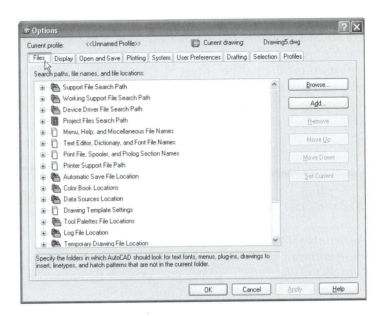

Support File Search Path
This section specifies where AutoCAD looks for text fonts, help files, menus, hatch patterns, plug-ins, linetypes, and drawings to insert when they are not found in the current drawing. For example, if you have a symbol library on a network drive, you can select the *Add* button to add the new folder. Do not *Remove* the default settings.

Working Support File Search Path
This set of directories is specific to your system and is the same as *Support File Search Path* for stand-alone configurations. For example, if you are running AutoCAD from a network, you can specify local directories to include in the search path.

Device Driver File Search Path
This option specifies where AutoCAD looks for device drivers for the video display, pointing devices, and printers and plotters. Do not change the defaults, but you can add other directories if additional driver directories are needed.

Project Files Search Path
The folder AutoCAD searches for *Xrefs* is listed here. See Chapter 30 for information on *XLOADPATH* and *PROJECTNAME.*

Menu, Help, and Miscellaneous File Names
This section specifies the location of the menu and help files AutoCAD must access. You may want to change the initial URL found when the *Browser* command is used by setting the *Default Internet Location* (see Chapter 19, Internet Tools). The *Configuration File* path specifies the location of the configuration file used to store hardware device driver information. The *License Server* path allows network administrators to list client license servers available to the network license manager program.

Text Editor, Dictionary, and Font File Names
This section specifies the name and path for the text editor (*MTEXTED*), the main and custom dictionaries (*DCTMAIN, DCTCUST*), and the alternate font file and font mapping file (*FONTALT, FONTMAP*). See Chapter 18, Creating and Editing Text.

Print File, Spooler, and Prolog Section Names

The *Print File Name* allows you to specify a name for .PLT files other than the current drawing name. PLT files are created when you check *Plot to File* in the *Plot* dialog box. The name can also be specified in the *Plot* dialog box when the plot is made. The *Print Spool Executable* tells AutoCAD what application to use for print spooling. *PostScript Prolog Section Name* assigns a name for a customized prolog section in the ACAD.PSF file. The prolog section is used with the *Psout* command. (The *PSPROLOG* system variable is obsolete in AutoCAD 2004).

Printer Support File Path

This section specifies search path settings for printer support files such as *Print Spooler File Location* (where AutoCAD writes print spool plots), *Printer Configuration Search Path* (the path for printer configuration files with a .PC3 extension), *Printer Description File Search Path* (the path for printer description files with a .PMP file extension), and *Plot Style Table Search Path* (the path for plot style table files with an .STB or .CTB file extension).

Automatic Save File Location

Here you can specify the path for the temporary automatic save file that is created if you select *Automatic Save* on the *Open and Save* tab. This value is also controlled by the *SAVEFILEPATH* system variable. See *"SAVETIME"* in Chapter 2, Working with Files.

Color Book Locations

AutoCAD supplies Pantone color books and RAL color books that can be used when specifying colors in the *Select Color* dialog box. You can define other color book files in multiple folders for each path specified. This option is saved with the user profile.

Data Sources Location

This section specifies the path for database source files that can be attached to a drawing using *DBCONNECT*. Change the path if you want to connect to database files you supply that are in a different location.

Drawing Template Settings

The *Drawing Template File Location* specifies the path for AutoCAD template files (.DWT files). Since only one folder can be specified, add your own templates to this folder instead of changing this setting (see Chapter 12, Advanced Drawing Setup). The *Drawing Template File Name for QNEW* specifies what template AutoCAD uses when the *Qnew* command is used (see Chapters 2 and 6 for more information on *Qnew*).

Tool Palettes File Locations

This setting specifies the path for tool palette support files. An .ATC file (AutoCAD Tool Catalog) is automatically created and saved in this location when you customize your own tool palettes.

Log File Location

If you create a log file (use *Logfileon* or select *Maintain a Log File* in the *Open and Save* tab), the file will be found in this location. This value is also controlled by the *LOGFILEPATH* system variable. See *"Logfileon"* in Chapter 44.

Temporary Drawing File Location

Here you can specify the location AutoCAD uses to store temporary files. Generally, a temporary folder created specially for temporary files such as C:\TEMP is suggested. The *TEMPPREFIX* (read-only) system variable also stores the current location of temporary drawing files. See "AutoCAD Backup Files" in Chapter 2.

Temporary External Reference File Location
Use this section to specify the location of your *Xref* files. See "*XLOADPATH*" in Chapter 30.

Texture Maps Search Path
The standard AutoCAD installation specifies the directories AutoCAD searches for rendering texture maps. Additional textures can be added to these directories. See Chapter 41, Rendering.

i-drop Associated File Location
Use this section to specify where files are placed when you drag and drop content from the Internet. The default setting is null; therefore, content is placed in the current directory (of the current drawing).

Display Tab

Figure 43-18

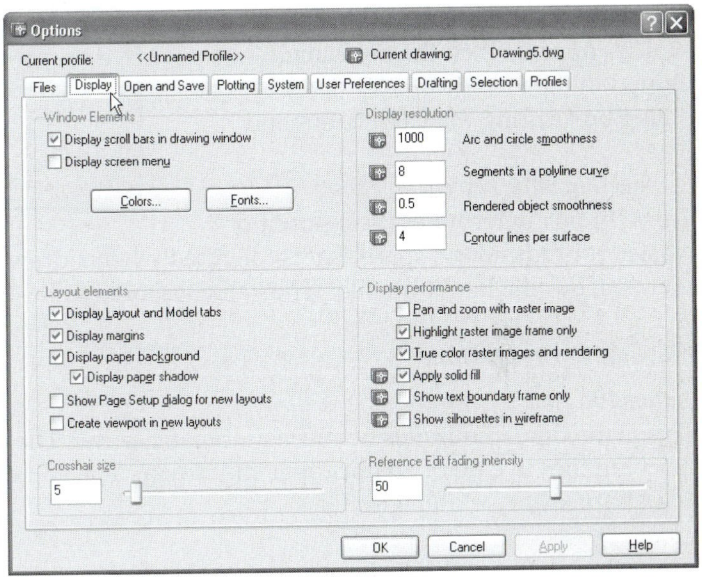

Windows Elements
See Chapter 1, Getting Started, for information on these options.

Layout Elements
See Chapter 13, Layouts and Viewports, for more information.

Crosshair Size
Use this edit box or slider bar to change the size of the crosshair cursor. The value represents a percentage of the screen size, so entering 100 causes the crosshairs to extend completely across the drawing area.

Display Resolution
The following options control the related system variables.

Arc and circle smoothness	*VIEWRES*	See Chapter 10
Segments in a polyline curve	*SPLINESEGS*	See Chapter 16
Rendered object smoothness	*FACETRES*	See Chapter 39
Contour lines per surface	*ISOLINES*	See Chapter 39

Display Performance
These options control the listed system variables.

Pan and zoom with raster image	*RTDISPLAY*	Controls the display of raster images during real time *Zoom* or *Pan*
Highlight raster image frame only	*IMAGEHLT*	See Chapter 32
True color raster images and rendering		Sets the AutoCAD Render window or images to true color if available on your system
Apply solid fill	*FILLMODE*	See Chapters 15, 26, and 43
Show text boundary frame only	*QTEXT*	See Chapter 18
Show silhouettes in wireframe	*DISPSILH*	See Chapter 39

Reference Edit Fading Intensity
This slider controls the value (0-90) of the fading effect of *Xrefs* controlled by *XFADECTL*. See Chapter 30, Xreferences.

Open and Save Tab

Figure 43-19

File Save

Save As:
Use this drop-down list to select the type of drawing format that is used when *Save* is invoked. You can also set this option in the *Save Drawing As* dialog box.

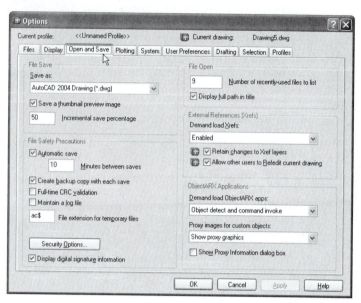

Save a Thumbnail Preview Image
When this box is checked, AutoCAD creates a "thumbnail sketch" of the drawing when it is saved. The thumbnail sketch appears when the file is highlighted in the *Preview* area of the *Select File* dialog box during the *Open* command.

Incremental Save Percent
This edit box determines the amount of wasted, or redundant, space that is tolerated when a drawing is saved. The higher the value, the faster the save, but the larger the file. The value is stored in the *ISAVEPERCENT* system variable.

File Safety Precautions

Automatic Save
This is the time in minutes between automatic saves. The automatic save feature creates temporary drawing files in case of an improper shutdown. The value is stored in the *SAVETIME* system variable. See Chapter 2, Working with Files.

Create Backup Copy with Each Save
When this box is checked AutoCAD creates a .BAK file from the previous .DWG file when a drawing is saved (*ISAVEBAK*). Speed up drawing saves by removing this check.

Full-Time CRC Validation
CRC (cyclic redundancy check) is an error-checking mechanism. If your drawings are being corrupted and you suspect a hardware problem or AutoCAD error, turn on this option. Checking this box causes a cyclic redundancy check to be performed each time an object is read into the drawing.

Maintain a Log File
This box determines if the command history is stored in a log file. See *"Logfileon"* and *"Logfileoff,"* Chapter 44.

File Extension for Temporary Files
Specify a file extension for files created when AutoCAD needs temporary file space.

Security Options
Selecting this button produces the *Security Options* dialog box where you can specify a password for the current drawing and attach a digital signature (ID certificate) to the current drawing. Passwords can also be specified with the *Securityoptions* command and in the *Save File As* dialog box. If *Display Digital Signature Information* is checked, when the drawing is opened, the *Digital Signature Contents* box is displayed. See Chapter 2 for information on passwords and Chapter 46 for information on digital signatures.

File Open

Number of Recently Used Files to List
This value controls the files listed at the bottom of the *Files* pull-down menu.

Display Full Path in Title
This check controls whether only the current drawing name or the name and full path (directory location)
is listed at the top of the AutoCAD window.

External References (Xrefs)

Demand Load Xrefs:
The options are *Disabled, Enabled,* and *Enabled with copy.* The selected option determines if demand
loading is enabled and if others can access a currently *Xrefed* drawing. The value is stored in the *XLOAD-
CTL* variable. See Chapter 30, Xreferences.

Retain Changes to Xref Layers
Use this check to control the *VISRETAIN* system variable. If checked, layer visibility settings of *Xrefed*
drawings made in the current drawing are saved with the current drawing (not the *Xref*). See Chapter 30,
Xreferences.

Allow Other Users to Refedit Current Drawing
Check this box if you want to allow others to be able to use in-place editing (*Refedit*) with the current
drawing. This setting controls the *XEDIT* system variable. See Chapter 30, Xreferences.

ObjectARX Applications

Demand Load ObjectARX Apps:
This drop-down box specifies if and when AutoCAD demand loads a third-party (add-on) application if
a drawing contains custom objects created in that application. This setting controls the *DEMANDLOAD*
system variable.

Proxy Images for Custom Objects
When third-party applications create new (not AutoCAD-native) objects, this setting determines how
these proxy objects are displayed. Alternately, you can access the *PROXYSHOW* system variable.

Show Proxy Information Dialog Box
Check this box if you want a notice to appear if you open a drawing containing proxy objects and the application that created the objects is not available. This setting is stored in the *PROXYNOTICE* system variable.

Plotting **Tab**

See Chapters 13, 14, and 33 for information regarding the options in this tab.

System **Tab**

Current 3D Graphics Display
This drop-down list (see Figure 43-20, next page) displays the 3D graphics display systems available on
your computer. The default is the Heidi 3D graphics display system (*GSHEIDI10*).

Properties...
This button provides access to the *3D Graphics System Configuration* dialog box. The dialog box is specific
for the current 3D graphics display system. See Chapters 35, 39, and 41.

Current Pointing Device
A list of the available pointing device drivers is accessible in this drop-down box. The options are as follows.

> *Current System Pointing Device:* AutoCAD uses the Windows system pointing device.
> *Wintab Compatible Digitizer:* This option is selected when a digitizer is used as the pointing device.

Accept Input From
If you are using a digitizer, you specify whether AutoCAD accepts input from both a mouse and a digitizer or ignores mouse input when a digitizer is set.

Figure 43-20

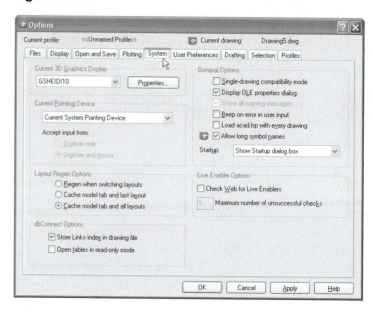

Layout Regen Options

These options specify how regenerations occur for the *Model* and *Layout* tabs. When you make changes to the drawing, the entire drawing is either regenerated when you switch tabs or the changes are saved to memory and displayed only for specific tabs that are activated. These options are also controlled by the *LAYOUTREGENCTL* system variable.

Regen When Switching Layouts
This setting regenerates the entire drawing each time you switch tabs.

Cache Model Tab and Last Layout
This option saves the changes to memory, displays the changes only for the two tabs, and suppresses a drawing regeneration. A full regeneration occurs when you switch to any other tab.

Cache Model Tab and All Layouts
Use this choice to regenerate the drawing only the first time you switch to each tab. For the remainder of the drawing session, the changes are displayed but saved to memory; regenerations are suppressed.

dbConnect Options

Store Links Index in Drawing File
When *Dbconnect* is used, this check stores the database index within the AutoCAD drawing file. This option can enhance performance during SQL queries. Speed up drawing opening and file size (for drawings with database information) by removing the check.

Open Tables in Read-Only Mode
When you connect to databases in AutoCAD, the database tables can appear within the AutoCAD drawing file. This choice specifies whether to open database tables in read-only mode.

General Options

Single-Drawing Compatibility Mode
Because AutoCAD allows multiple drawings to be open simultaneously, you can specify whether a Single-Drawing Interface (SDI) or a Multi-Drawing Interface (MDI) is enabled. You can also control this setting using the *SDI* system variable.

2002

Display OLE Properties Dialog
When you insert OLE objects into AutoCAD drawings, the *OLE Properties* dialog box appears by default. Remove this check to disable the dialog box display. See Chapter 31.

Show All Warning Messages
This check globally displays all dialog boxes that include a *Don't Display This Warning Again* option. If checked, all dialog boxes with warning options are displayed regardless of the settings in each dialog box.

Beep on Error in User Input
Check this if you want AutoCAD to sound an alarm beep when it detects an invalid entry.

Load Acad.lsp with Every Drawing
If checked, AutoCAD loads the ACAD.LSP file when every drawing is opened. If this option is cleared, only the ACADDOC.LSP file is loaded into all drawing files. Clear this option if you do not want to run certain LISP routines (listed in the ACAD.LSP file) in specific drawing files. This setting is saved in the *ACADLSPASDOC* system variable.

Allow Long Symbol Names
AutoCAD named objects can include up to 255 characters. This check determines if long names are enabled for named objects such as layers, dimension styles, blocks, linetypes, text styles, layouts, UCS names, views, and viewport configurations. The value is stored in the *EXTNAMES* system variable.

Startup
Use this list to control if the *Startup* dialog box is displayed when you start AutoCAD. See Chapter 6, Basic Drawing Setup, and Chapter 12, Advanced Drawing Setup, for more information.

Live Enabler Options

If you use a drawing that contains non-AutoCAD-native objects created by third-party ObjectARX programs, you can use Object Enablers to allow you to display and use the custom objects in AutoCAD drawings even when the ObjectARX application that created them is unavailable. This section controls how AutoCAD checks for Object Enablers that are available on the Autodesk Web site. These settings are stored in the *PROXYWEBSEARCH* system.

Check Web for Live Enablers
Select this option to continually check for Object Enablers on the Autodesk Web site.

Maximum Number of Unsuccessful Checks
Specifies the number of times AutoCAD will continue to check for Object Enablers after unsuccessful attempts.

Figure 43-21

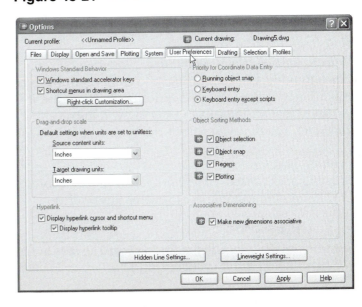

User Preferences Tab

Windows Standard Behavior

Windows Standard Accelerator Keys
The initial setting in AutoCAD follows the Windows standard in interpreting keyboard accelerators. For example, Ctrl+C invokes *Copyclip*. When this option is cleared, AutoCAD uses the older Release 12 and previous standard keyboard accelerators; for example, Ctrl+C issues a *Cancel*.

2002

Shortcut Menus in Drawing Area
This option controls whether right-clicking in the drawing area displays a shortcut menu or issues Enter.

Right-Click Customization
This button displays the *Right-Click Customization* dialog box (Fig. 43-22). Here you can set preferences for shortcut menus that appear in the drawing area. Three basic types of shortcut menus are available: *Default mode, Edit mode,* and *Command mode* menus (see Chapter 1). The options here are self-explanatory. The settings are stored in the *SHORTCUTMENU* system variable.

Figure 43-22

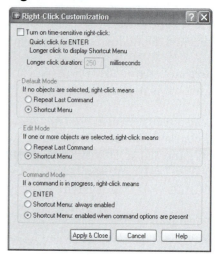

Drag and Drop Scale

Source Content Units
When an object is inserted into the current drawing using drag-and-drop, automatic scaling occurs. This variable sets which units are used for the source object being inserted into current drawing when they are set to *Unitless* or are unspecified (in the *Block* dialog box). The value is stored in the *INSUNITSDEFSOURCE* system variable. See Chapter 21, Blocks, DesignCenter, and Tool Palettes, for more information.

Target Drawing Units
You can also specify which units to use for automatic scaling when *Blocks* or drawings are inserted into the current drawing when no insert units are specified (with the *Drawing Units* dialog box or *INSUNITS* system variable). The value is stored in the *INSUNITSDEFTARGET* system variable.

Hyperlink

Display Hyperlink Cursor and Shortcut Menu
If you drag the cursor across a hyperlinked object, the hyperlink cursor appears alongside the crosshairs when this option is checked. When shortcut menus are enabled, a cascading hyperlink shortcut menu appears when you right-click. If this option is cleared, the hyperlink cursor is never displayed and the hyperlink option on shortcut menus is not available.

Display Hyperlink Tooltip
If this option is selected, a hyperlink tool tip is displayed when the pointing device moves over an object that contains a hyperlink. The tool tip contains the information you express in the *Description* section of the *Insert Hyperlink* dialog box.

Priority for Coordinate Data Entry
When Running Osnap is on and you enter coordinates at the keyboard, this setting determines which entry (keyboard or *Osnap*) has priority. For example, if *Endpoint Osnap* is on and you enter values at the keyboard that are close to an *Endpoint,* will the resulting point be at the specified coordinates or at the *Endpoint?* You can also set the priority using the *OSNAPCOORD* system variable.

Running Object Snap
Running object snaps have priority over specific coordinates at all times. You can also set the *OSNAPCOORD* system variable to 0 to enable this priority.

Keyboard Entry
Coordinates that you enter at the keyboard always override running object snaps. You can also set the *OSNAPCOORD* system variable to 1 to prioritize keyboard entry.

Keyboard Entry Except Scripts
AutoCAD uses the specific coordinates that you enter at the keyboard rather than running object snaps, except in scripts. You can also set the *OSNAPCOORD* system variable to 2.

Object Sorting Methods
When AutoCAD uses objects for particular processes, it can sort objects or not sort objects depending on your selections. When objects are sorted, they are ordered from the least recently created to the most recently created. If two overlapping objects exist, AutoCAD sorts the newest object first. If sorting is cleared, object selection is determined by a random sort order. You can also set this option using the *SORTENTS* system variable. You can decide to sort or not for the following processes.

Object Selection, Object Snap, Regens, and *Plotting*

Associative Dimensioning
Check this box to *Make New Dimensions Associative*. Otherwise, dimensions are created as AutoCAD 2000 and earlier "nonassociative" dimensions. You can also set this option by using the *DIMASSOC* system variable. See Chapters 28 and 29.

Hidden Line Settings
This button produces the *Hidden Line Settings* dialog box where you can specify how surface and solid objects are treated for *Hide* and for the *Hidden* options of *Shademode* and *Shade Plot*. See Chapter 35, 3D Display and Viewing.

Lineweight Settings
Use this button to specify options in the *Lineweight Settings* dialog box. See Chapter 11, Layers and Object Properties.

Drafting **Tab**

See Chapter 7, Object Snap and Object Snap Tracking, for information on this tab.

Selection **Tab**

See Chapter 20, Advanced Selection Sets, for information on this tab.

Profiles **Tab**

When you customize environment settings, such as color schemes and toolbar visibility and location, you can then save your settings as a profile using the *Profiles* tab in the *Options* dialog box. Profiles are saved on the workstation, not in drawing files. If several people use the same workstation and also use the same login name, you can restore your particular environment by making your profile current. You could also create and save profiles to use with different projects.

Figure 43-23

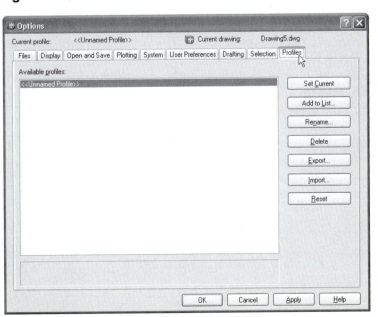

The profile information is stored in the system registry. Profiles can also be saved to a text file (an .ARG file), that can be imported to other workstations. Once you save a profile with the *Export* option, you can export or import the .ARG file to and from different computers. For example, you may want to transport your profile to different computers, or you may want to set up all stations in a lab or office with the same profile.

Any changes that are made to the environment are automatically updated to the current profile. By default, AutoCAD stores your current options in a profile named <<UNNAMED PROFILE>>. It is suggested, however, that you save your own profile under a different name.

If changes have been made to the environment and you want to restore the system defaults, highlight the <<UNNAMED PROFILE>>, then select *Reset* to make the <<UNNAMED PROFILE>> the system default profile. Next, make the <<UNNAMED PROFILE>> the *Current* profile.

To create your own profile, use the *Add to List* button to display the *Add Profile* dialog box. Enter the new name to first create the new profile. Then set up your desired environment and the settings are automatically added to your profile.

If you want to save those settings in an .ARG file, you must export the profile. For example, exporting a profile named "Myprofile" creates a MYPROFILE.ARG file. If changes are made to the environment and you then export the profile under the current profile name, AutoCAD updates the matching .ARG file with the new settings. If changes are unintentionally made to the environment, you can import the profile again into AutoCAD to restore the profile settings.

The individual buttons on the *Profiles* tab are explained next.

Available Profiles
This list displays a list of the available (previously created) profiles.

Set Current
This button makes the highlighted profile the current profile. To make a profile the current profile, select the desired profile from the list and choose *Set Current*.

Add to List
Use this button to create a new profile. First, select this button to display the *Add Profile* dialog box (Fig. 43-24). Enter the desired name in the *Profile Name* edit box, add a description, and *Apply & Close*. This action saves the current profile under a different name. Next, highlight the new profile from the list and set the new profile *Current*. Then set up your desired environment, and changes are automatically saved to the new current profile.

Figure 43-24

Rename
You can rename any profile. Selecting this button displays the *Change Profile* dialog box for changing the name and description of the selected profile. Use *Rename* when you want to rename a profile but keep its current settings.

Delete
You can delete any profile unless it is the current profile. First make any other profile current, then highlight the name to delete, then select the *Delete* button.

Export

This option allows you to export a profile, creating a file with an .ARG extension. Highlight the desired profile from the list and select *Export*. The *Export Profile* dialog box appears where you can enter a name for the .ARG file. Normally, you would want to assign the same file name as the name of the profile. The new profile can be used on other computers by *Importing* the profile. You can import the file on the same computer or a different computer.

Import

This option allows you to import a profile (an .ARG file) that has been saved with *Export*. You cannot import the current profile; therefore, first make any other profile *Current*. Select *Import* to produce the *Import Profile* dialog box (not shown). Locate and highlight the desired .ARG file and select *Open*. When the second (smaller) *Import Profile* dialog box appears, enter the name of the profile you want to create on your system (normally enter the same name as the assigned file name you are importing), then *Apply & Close*. To make the imported profile current, highlight that name in the *Available Profiles* list, and select *Current*.

Reset

Using this button *Resets* the values in the selected profile to the system default settings. Therefore, the profile that is highlighted in the *Available Profiles* list becomes reset to the defaults. Normally, you would want to *Reset* only the <<UNNAMED PROFILE>> to the system defaults.

Figure 43-25

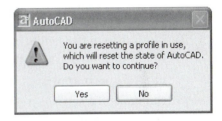

First, highlight the desired profile to reset from the *Available Profiles* list. Then select *Reset*. Finally, you must make the reset profile *Current* to see the changes. If the highlighted name is also the current profile, a warning message appears (Fig. 43-25). Selecting *Yes* resets the current profile to the system defaults.

CHAPTER EXERCISES

1. **Wildcards**

   In this exercise, you will use wildcards to manage layers for a complex drawing. *Open* the AutoCAD **WILHOME** drawing that is located in the \AutoCAD 2004\Sample directory. Use the *Layer Properties Manager* dialog box to view the layers. Examine the layer names; then *Cancel*.

   A. Type the *-Layer* command. *Freeze* all of the layers whose names begin with the letters **AR**. (Hint: Enter **AR*** at the "...layer(s) to freeze:" prompt.) Do you notice the change in the drawing? Use the *Layer Control* drop-down box or list the layers using the *?* option to view the *State* of the layers.

   B. *Thaw* all layers (use the asterisk *). Now *Freeze* all of the layers <u>except</u> the architectural layers (those beginning with **AR**). (Hint: Enter **~AR*** at the "...layer(s) to freeze:"prompt.) Do only the architectural features appear in the drawing now? Examine the list of layers and their *State*. *Thaw* all layers again.

   C. Next, *Freeze* all of the layers except the wall layers (layers that have a "WALL" string). (Hint: Enter **~*WALL*** at the ...layer(s) to freeze:" prompt.) Do only the grids appear in the drawing now? Examine the list of layers and their *State*. *(Layer 0 cannot be Frozen* because it is the Current layer.) *Thaw* all layers again and do <u>not</u> *Save* changes.

2. *Multiple* modifier

    Open the **OFF-ATT2** drawing and prepare to *Insert* several more furniture *Blocks* into the office. Type *Multiple* at the Command: prompt, then Enter, then *-Insert* and Enter. Proceed to *Insert* the *Blocks*. Notice that you do not have to repeatedly enter the *-Insert* command. Now try the *Multiple* modifier with the *Line* command. Remember this for use with other repetitive commands. (Do not *Save* the drawing.)

3. *Audit*

    With the OFF-ATT2 drawing open, use the *Audit* command to report any errors. If errors exist, use *Audit* again to fix the errors.

4. *BLIPMODE, DRAGMODE*

    A. Begin a *New* drawing using your **ASHEET** prototype. Turn *BLIPMODE On*. Experiment with this setting by drawing a few *Lines* and *Circles*. Use *Erase*. Notice that you have to *Redraw* often when *BLIPMODE* is *On*.

    B. Now change the setting of *DRAGMODE* to *Off*. Draw a few *Circles*, a *Polygon*, an *Ellipse*, and then **dimension** your figures. Do you prefer *DRAGMODE On* or *Off*?

    C. Since the *BLIPMODE* and *DRAGMODE* system variables are saved with the drawing file, you can make the settings in the template drawings so they will preset to your preference when you begin a *New* drawing (if you prefer other than the default settings). If you prefer *BLIP-MODE On* or *DRAGMODE Off*, *Close* the **ASHEET** and do not save. Use *Open* to open each of the template drawings (**ASHEET, BSHEET**, and **C-D-SHEET**), change the desired setting, and *Save*.

5. *Fill, Regen*

    Open the **SPCA Site Plan** sample drawing (from the C:\Program Files\AutoCAD 2004\Sample directory). Then type *Regenall* and check the amount of time it takes for the drawing to regenerate (use your stopwatch or count the seconds). Now turn *Fill Off*. Also turn *Qtext On*. Make another *Regenall* as before to check the time. Is there any improvement?

    Imagine the time required for plotting this drawing using a pen plotter. *Fill* and *Qtext* can be set to save time for test plots. If you have a pen plotter, make a plot of the drawing with *Fill* and *Qtext* set in each position. How many minutes are saved with *Qtext On* and *Fill Off*? (Do <u>not</u> *Save* the changes to **SPCA Site Plan.**)

6. *Mslide, Vslide*

    A. Open your choice of the following sample drawings: **DB_SAMP, 8TH FLOOR, MKMPLAN**, or **WILHOME**. (Assuming the default installation was used for AutoCAD, the drawings are located in the C:\Program Files\AutoCAD 2004\Sample directory.) Make eight to ten slides of the drawing, showing several detailed areas (using *Zoom* and *Pan*, etc.) and at least one slide showing the entire drawing. Experiment with *Freezing* layers and include several slides with some layers frozen. Assign descriptive names for the slides so you can determine the subject of each slide by its name. Strive to describe as much as possible about the drawing in the limited number of slides.

B.   When the slides of the drawing are complete, prepare at least two text slides to use as informative material such as an introduction, conclusion, or topic headings (make the text in one or more separate drawings, then use *Mslide*).

C   Finally, use *Vslide* to view your slides and determine the optimum sequence to describe the drawing. Write down the order of the slides that you choose.

7. *Script*

Use **Windows Notepad** or other text editor to create a script file to run your slide show. Include appropriate *delays* to allow viewing time or time for you to verbally describe the slides. Make the script self-repeating. Name the script **XXXSHOW.SCR** (where XXX represents your initials). In AutoCAD, use *Script* to run the show to "debug" it and optimize the delays.

8. **Geometry Calculator**

Figure 43-26 ————————————————

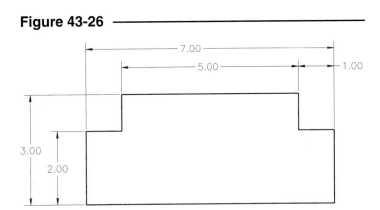

A.   This exercise utilizes several *Cal* functions. Begin by drawing the shape shown in Figure 43-26 using a *Pline*. Do not draw the dimensions. *Save* the drawing as **CAL-EX.**

B.   Draw the *Circle* in the center of the *Pline* shape with a **.5** unit *radius* (Fig. 43-27). The center of the *Circle* is located halfway between the *Midpoints* of the top and bottom horizontal line segments. Use the *Calculator* transparently to place the *Center* of the *Circle*:

Figure 43-27 ————————————————

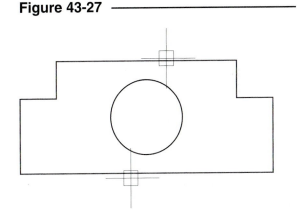

Hint: Command: `circle`
         Specify center point for circle or
         [3P/2P/Ttr (tan tan radius)]:
         > Expression: **(mid+mid)/2**

C.   Draw the small *Circle* (Fig. 43-28) with a **.1** *radius*. Its *Center* is located **.25** units to the right and **.25** units up from the lower-left corner of the *Pline* shape. Use *Cal* transparently to place the *Circle*:

Figure 43-28 ————————————————

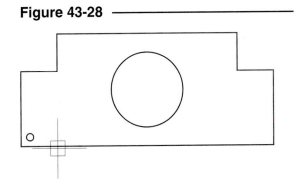

Hint: Command: `Circle`
         Specify center point for circle or
         [3P/2P/Ttr (tan tan radius)]:
         >> Expression: **end+[.25,.25]**

D. Create the *Circle* in the upper-right corner (Fig. 43-29, P1). It has a *Radius* of **1.5** times the existing small *Circle*. Store the existing small circle radius to a variable named **R1**. Then create the new *Circle* with the *Center* located **.5** units to the left and down from the upper-right corner:

Figure 43-29

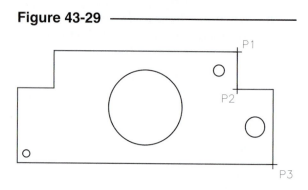

Hint 1: >> Expression: `R1=rad`
Hint 2: >> Expression: `end+[-.5,-.5]`

E. The last *Circle* (on the far right) is located midway between two endpoints (P2 and P3). It has a radius of **1.8** times the size of the last circle that you created. Use a shortcut function and variable (if necessary) to create the *Circle*. *Save* the drawing.

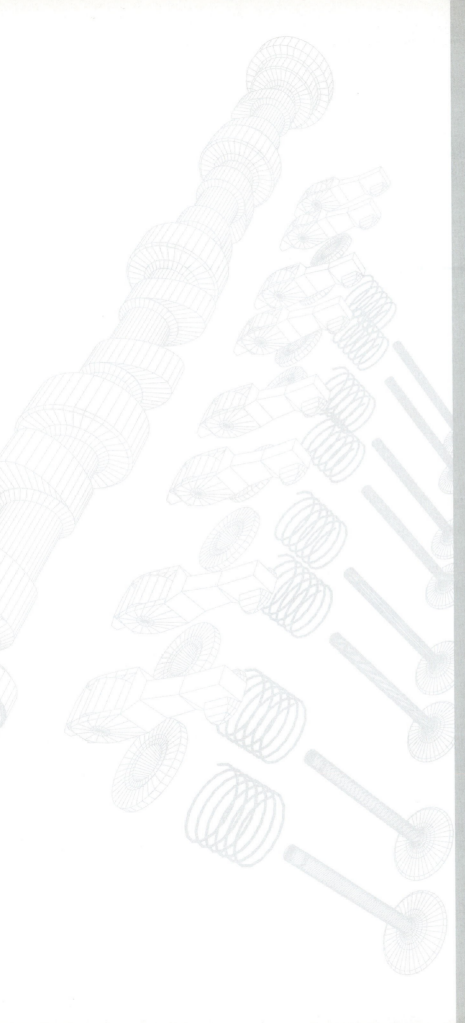

BASIC CUSTOMIZATION

Chapter Objectives

After completing this chapter you should:

1. be able to assign which pull-down menus appear on the menu bar;

2. be able to customize the existing toolbars and create new toolbars;

3. be able to create new and modify existing tool button icons and assign them to perform special tasks;

4. be able to assign your own shortcut keystrokes for commands;

5. be able to create your own simple and complex linetypes;

6. be able to customize the ACAD.PGP file to create your own command aliases;

7. be able to create script files for accomplishing repetitive tasks automatically.

CONCEPTS

This chapter is offered as an introduction to customizing AutoCAD. Autodesk has written AutoCAD with an "open" architecture that allows you to customize the way the program operates to suit your particular needs. You have the ability to change any of the menus, toolbars, tool buttons, shortcut keys, and tool palettes as well as create new linetypes.

The topics that are best for beginning your customization experience are introduced in this chapter.

>   Customizing which pull-down menus are visible on the menu bar
>   Customizing toolbars, icon buttons, and shortcut keys
>   Creating your own simple and complex linetypes
>   Customizing the ACAD.PGP file and creating command aliases
>   Creating script files to use for automated activities

In addition to the basic customization capabilities covered in this chapter, one other widely used customization feature is the ability to customize the menus. You can modify existing and create new pull-down menus, screen menus, and digitizing tablet menus. See Chapter 45, Menu Customization, available at www.mhhe.com/leach.

PARTIAL MENU LOADING AND CUSTOMIZING THE MENU BAR

AutoCAD has the ability to load "partial menus." These are menu files that can "overlay" an existing menu file. A menu file can contain all the information for a complete set of pull-down, screen, toolbar, and digitizing menus. This section describes the *Menuload* command used for partial loading and for customizing the existing menu bar. See Chapter 45, Menu Customization, for more information on menu loading.

MENULOAD

Pull-down Menu	COMMAND (TYPE)	ALIAS (TYPE)	Short-cut	Screen (side) Menu	Tablet Menu
Tools *Customize Menus...*	*MENULOAD*	...	...	*TOOLS 2* *Menuload*	*Y,9*

The *Menuload* command invokes the *Menu Customization* dialog box (Fig. 44-1). The *Menu Bar* tab allows you to customize the visibility of particular pull-down menus. You can also use the *Menu Groups* tab to load (or unload) one or more menu groups as an "overlay" to the existing menu.

In the *Menu Bar* tab, you can remove menu bars (individual pull-down menus) that currently appear on screen. For example, you may not be dimensioning and therefore want to remove the *Dimensioning* menu bar. To do this, select the desired *Menu Bar* from the list on the right, then select *Remove*. The selected bar disappears from the top of the AutoCAD screen.

Figure 44-1

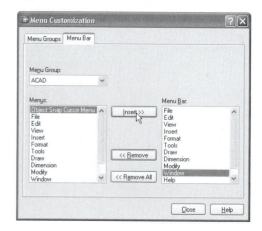

You can also add menus from the list on the left to the existing group of menus bars. Do this by making a selection from the *Menus*: list, then press *Insert*. For example, you may want to *Insert* the *Object Snap Cursor Menu* into the menu bar (see Figure 44-1). The highlighted menu in the *Menus* list is inserted just before the highlighted menu in the *Menu Bar* list.

You can use the *Menu Groups* tab to load partial menus to overlay the existing menus (Fig. 44-2). A partial menu can contain a complete set of pull-down, screen, toolbar, and digitizing menus, so using this tab installs all of those components. The menu loading feature is mainly used for your own special (customized) menus or third party (add-on products) menus. See Chapter 45, Menu Customization, for examples.

Figure 44-2

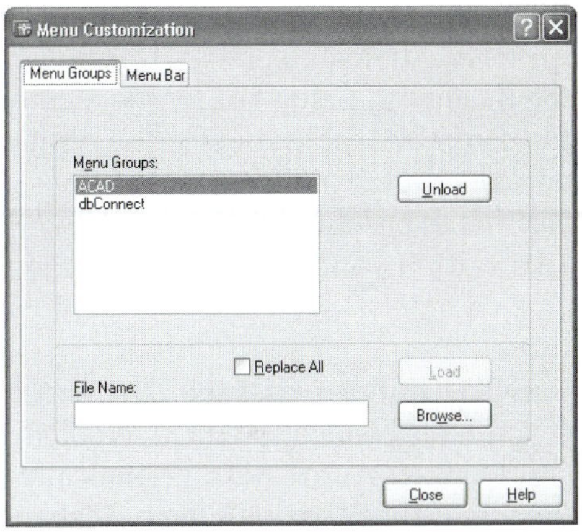

You can use this tab to unload partial menus, as well. For example, assume you installed the *Express* menu. This menu can be unloaded if you do not intend to use those features for the drawing session.

Partial menu loading and unloading is effective only for the current drawing session (until you *Exit* AutoCAD). Starting AutoCAD again loads all the original configured menus.

CUSTOMIZING TOOLBARS, TOOL BUTTONS, AND SHORTCUT KEYS

In AutoCAD, you have the ability to customize toolbars. By doing so, you can create toolbars that have your most frequently used commands or even create buttons (tools) that invoke "new" commands, that is, choices that activate a command, command option, system variable, or series of commands not previously offered in the existing toolbars or menus. Using the series of dialog boxes that provide for tool and toolbar customization, you can accomplish these actions:

    Modify existing toolbar groups
    Create new toolbar groups
    Modify existing tool buttons
    Create new tool buttons
    Customize shortcut keys

First, you should gain experience with modifying existing and creating new toolbars, then try creating new icon buttons.

Modifying Existing and Creating New Toolbars

You can modify existing toolbars or create your own toolbar by placing the buttons (also called "tools") you want into a group. For example, you could make a new toolbar that contains your most frequently used solid modeling commands <u>and</u> the Boolean tools, thus allowing you to turn off the *Solids* and *Solids Editing* toolbars. The *Toolbar* command provides access to a series of dialog boxes that allow you to specify which tools to include in an existing or new toolbar. This section explains the options of the *Toolbars* and related dialog boxes and suggests steps for modifying and creating toolbars.

TOOLBAR

	COMMAND (TYPE)	ALIAS (TYPE)	Short-cut	Screen (side) Menu	Tablet Menu
Pull-down Menu					
View Toolbars...	*TOOLBAR*	*TO*	...	*VIEW 2 Toolbar*	*R,3*

Using the *Toolbar* command produces the *Customize* dialog box (Fig. 44-3). You can access the *Customize* dialog box by the methods shown in the command table above or by right-clicking on any tool (button). There are five tabs: *Commands, Toolbars, Properties, Keyboard* and *Tool Palettes.* The tabs, buttons, and options are described here.

Figure 44-3

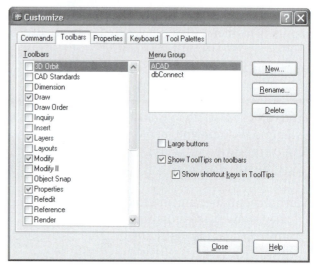

Close

The *Close* button closes the *Customize* dialog box. If you have created or edited a toolbar, button, or shortcut key, AutoCAD automatically compiles the menu structure with the new changes when you *Close*.

Toolbars **Tab**

Toolbars **list**

The *Toolbars* list (see Figure 44-3) gives the existing toolbar names. If you want to activate an existing toolbar group, check the name of the toolbar to cause it to appear on the screen. Clear the checkbox to remove the toolbar from the screen.

Menu Group

If you have more than the standard menu loaded (ACAD.MNU), you can select from the list to display the toolbar groups contained in the file.

New

New opens the *New Toolbar* dialog box and allows you to create a new toolbar (Fig. 44-4). Enter the name (title) of the new toolbar you want to create. Then select *OK* to have the new toolbar (containing no buttons) appear in the drawing area (usually above the *Customize* dialog box). The new toolbar name is added to the *Toolbars* list in the *Toolbars* tab.

Figure 44-4

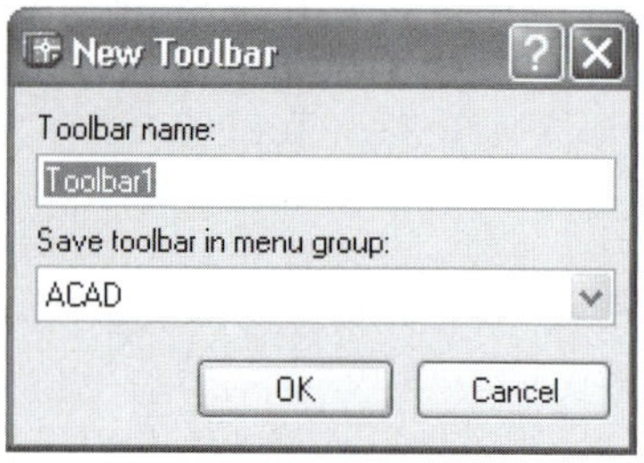

Rename

Highlight any toolbar from the *Toolbars* list, then select *Rename* to produce the *Rename* dialog box (not shown) where you can change the name of the selected toolbar (the name appears in the toolbar title bar).

Delete

The *Delete* option deletes the toolbar name currently highlighted in the list. You are prompted for a confirmation for the delete, at which point you can choose *Yes* or *Cancel*.

Large Buttons

Check this option to produce 24 x 24 pixel-sized buttons on the screen. The normal size is 16 x 16 pixels.

Show ToolTips on Toolbars
This checkbox causes a ToolTip (the small text string in a yellow box) to display whenever you rest the pointer on a tool button.

Show Shortcut Keys in ToolTips
If the command has an associated shortcut key (for example, *Ctrl+O* for *Open*), check this box to force the shortcut key to appear with the ToolTip.

Commands Tab

Figure 44-5

Use the *Commands* tab (Fig. 44-5) to customize the contents of any toolbar. Here you choose the tools (buttons) to be added or removed from any toolbar that is displayed in the Drawing Editor.

To add a tool to an open toolbar, simply drag the tool from the *Commands* list and drop it into a toolbar. Remove an existing tool from a toolbar by dragging it from the toolbar and releasing it into the graphics area of the Drawing Editor. In this case, you must answer *Yes* in the confirmation dialog box that appears.

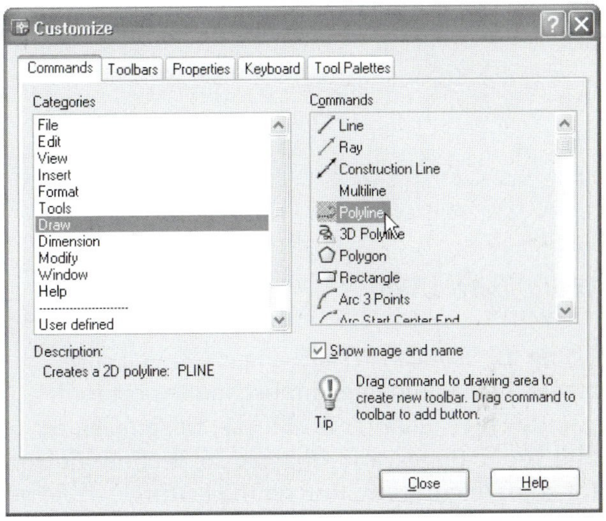

The toolbar must be visible in order to add or remove a tool. If the toolbar is not visible on the screen, use the *Toolbars* tab to display the toolbar. (To create a new toolbar, use *New* in the *Toolbars* tab.) To locate the tools you want to work with, use the *Categories* list to select the tool group you need. After selecting the category, find the desired tool in the *Commands* list and "drag and drop" into a toolbar.

When you are finished editing a toolbar or button, select *Close* to close the *Customize* dialog box and compile the menu structure with the new changes. See "Customizing Toolbars" and "Customizing Buttons" for more information.

Properties Tab

Figure 44-6

Use the *Properties* tab (or *Button Properties* tab) to customize the buttons (Fig. 44-6). This tab allows you to edit the image that appears on the button as well as change the command or set of commands that are activated when you select the button. You can change existing buttons and create new buttons. For a full explanation of this tab and the features within, refer to "Customizing Buttons" later in this chapter.

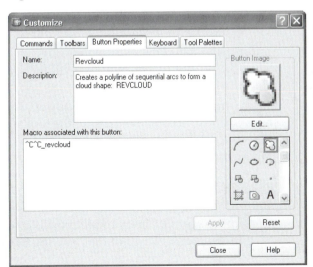

Keyboard Tab

Some commands in AutoCAD can be accessed by shortcut key strokes—key strokes that use the Alt, Ctrl, or Ctrl and Shift keys and another character or number. With this tab (Fig. 44-7), you can change the existing shortcut key stroke assignments and create new shortcut key assignments. See "Customizing Shortcut Keys" later in this chapter.

Figure 44-7

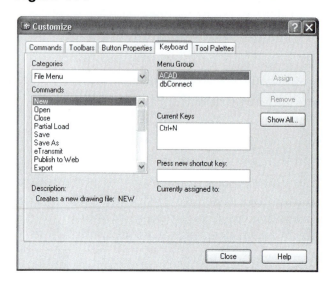

Tool Palettes Tab

Using the *Tool Palettes* tab (Fig. 44-8) you can create, delete, rename, import and export tool palettes. See Chapter 21, Blocks, DesignCenter, and Tool Palettes, and Chapter 26, Section Views, for complete information on the features.

Figure 44-8

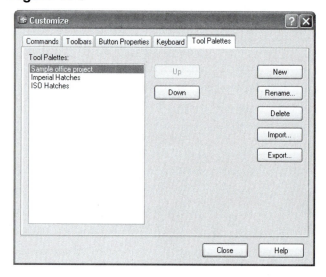

Customizing Toolbars

Steps for Modifying an Existing Toolbar

1. Type *Toolbar*, right-click on any tool and select *Customize*, or use any other method (listed in the *Toolbar* command table in this book) to invoke the *Customize* dialog box.

2. Ensure the toolbar you want to modify is visible. If that toolbar is not visible on the screen, select the checkbox next to its name to make the toolbar appear.

3. Choose the *Commands* tab in the *Customize* dialog box.

4. When the *Commands* tab is active, you can:

    A. Remove any tool from a visible toolbar by dragging it out of the toolbar group and dropping it into the graphics area of the Drawing Editor.
    B. Move any tool from one group to another by dragging and dropping.
    C. Add a tool from the *Commands* tab to a visible toolbar group. See Step 5.

5. To add a tool to a visible group (without removing it from another group), select the desired tool's group from the *Categories* list. This displays the available tool icons in the toolbar group ("category"). Drag the desired tool(s) from the dialog box and drop into the desired visible toolbar group. In Figure 44-9, the Boolean tools (*Union, Subtract, Intersect*) are being added to the *Solids* toolbar group.

6. *Close* the *Customize* dialog box. AutoCAD automatically compiles the ACAD.MNR and ACAD.MNS menu files. This enables the modi-fied toolbars to be used immediately.

Figure 44-9

Steps for Creating a New Toolbar Group

1. Invoke the *Customize* dialog box by any method. Activate the *Toolbars* tab.

2. Choose *New* to open the *New Toolbar* dialog box (see Figure 44-4). In the *Toolbar Name* field, enter the new name, for example, "My Toolbar." PICK the *OK* button. When the dialog box closes, your new toolbar name (My Toolbar) is highlighted in the toolbar listing. The new empty toolbar appears above the *Customize* dialog box.

3. From the *Commands* tab, drag and drop any tool icons onto the new toolbar (Fig. 44-10). Use the *Categories* list to highlight menu groups and display any tool icon that you need.

Figure 44-10

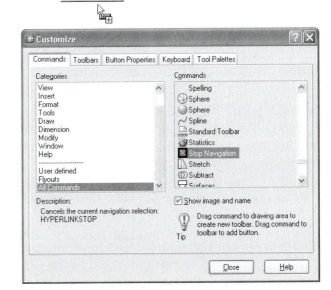

4. *Close* the *Customize* dialog box. AutoCAD automatically compiles the ACAD.MNR and ACAD.MNS menu files.

Note: If you are dragging a tool from a cate-gory in the *Customize* dialog box and release the tool in the graphics area prior to dropping it into an existing toolbar, a new toolbar is created with the name "Toolbar*n*" where *n* is a number. To close the toolbar, pick the toolbar's close "X" in the upper-right corner. Delete it from the *Toolbars* dialog box listing by scrolling down to "Toolbar*n*," then select the *Delete* button. *Close* the *Customize* dialog box.

Customizing Buttons

Not only can you modify existing toolbars and create new toolbar groups in AutoCAD, you can customize individual buttons. Existing buttons can be modified and new buttons can be created. You can modify or create the image that appears on the icon and assign the command(s) that are activated when you PICK the button. The *Button Properties* tab provides the interface for "drawing" the button image and assigning the associated command.

The *Button Properties* tab (Fig. 44-11) is invoked by opening the *Customize* dialog box, then clicking (once) on any visible tool button. The areas of the *Button Properties* tab are described as follows.

Figure 44-11

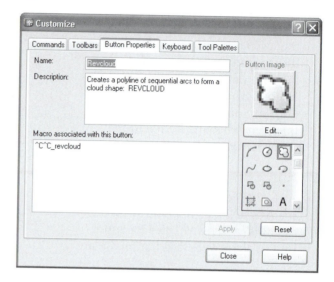

Name
This edit box defines the name of the button which also serves as the "tooltip" string that appears when the pointer rests on the button.

Description
This field contains the "help string" that appears at the bottom of the screen over the Status Bar when the pointer rests on the button.

Macro associated with this button
Enter any command, command and option, series of commands, system variable changes, etc. in this edit box. Similar to menu customization, all commands, options, or system variables must contain the exact spellings as you would enter at the Command prompt. Spaces in the string are treated as returns (pressing Enter). Two cancels (^C^C) are generally placed before every command in a macro.

Button Image
Use this area and slider bar below to select any of the AutoCAD-supplied icons for editing.

Edit
Selecting this button produces the *Button Editor* (Fig. 44-12). The *Button Editor* is like a miniature "paint" program that you can use to modify existing or create new button images. From the top of the editor, select a pencil for sketching, a line drawing instrument, a circle and ellipse drawing instrument, or an eraser. The color pallet appears on the right. Activate the *Grid* to "see" the pixels. You can *Clear* the entire button face, *Open* an existing button (.BMP file), or *Undo* the last drawing sequence. The drawing procedure is quite simple and fun. Use *Save* to save your work to a .BMP file before you *Close* the editor, or you can *Close* without saving. Use *SaveAs* to rename the button image to a .BMP file.

Figure 44-12

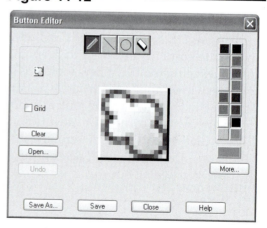

Apply

After you use the *Button Editor* to make changes to the button image, macro, name, or description, select the *Apply* option to force the new changes to be applied to the button.

Reset

This option is like an *Undo* for the *Button Properties* tab. Any changes you make to a particular button, including changes to the button image, macro, name, or description, are reset to the original state when you select *Reset*.

Steps for Changing Existing Buttons

1. Invoke the *Customize* dialog box by any method. Then click (once) on any existing tool button visible in the Drawing Editor. The *Button Properties* tab appears.

2. If it is necessary to change the appearance of the icon, select *Edit...* to produce the *Button Editor*. Make the necessary changes, *Save* the changes, and select *Apply* to have the new icon appear on the button in the Drawing Editor.

3. Make any desired changes in the *Name* and *Description* edit boxes.

4. Enter the new action desired for the button in the *Macro* area. You can alter the existing action or delete the existing action and create a new series of commands or variable changes. Select *Apply*.

5. *Close* the *Customize* dialog box. AutoCAD automatically compiles the ACAD.MNR and ACAD.MNS menu files.

Steps for Creating New Buttons

1. If you want the new button(s) to reside in a new toolbar group, create the new toolbar group first (see "Steps for Creating a New Toolbar Group"). If you want the new button to reside in an existing group, make sure the toolbar is visible in the Drawing Editor.

2. Invoke the *Customize* dialog box using any method.

3. In the *Commands* tab, select *User defined* from the *Categories* drop-down list. Drag and drop the words "User defined button" (an empty button) onto the desired toolbar (in the Drawing Editor).

4. Click (once) on the new empty button (in the Drawing Editor). The *Button Properties* tab appears.

5. Select *Edit...* to produce the *Button Editor*. Design the new button, *Save* the changes, and *Close* the *Button Editor*.

6. Make the desired entries in the *Name* and *Description* edit boxes.

7. Enter the new action desired for the button in the *Macro* area. Select *Apply* to have the new icon appear on the button in the Drawing Editor.

8. *Close* the *Customize* dialog box. AutoCAD automatically compiles the ACAD.MNR and ACAD.MNS menu files.

An Example Custom Toolbar Group and Buttons

Figure 44-13

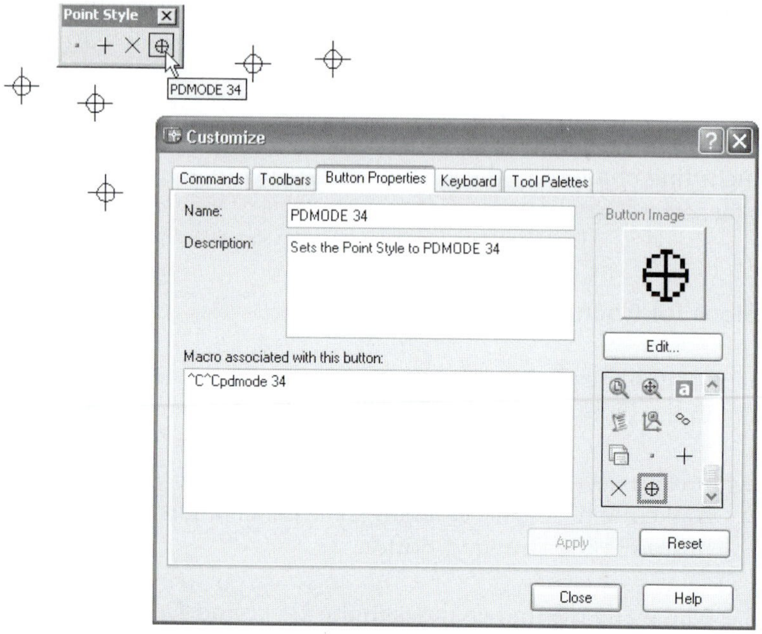

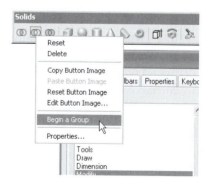

Assume you work for a civil engineering firm and utilize *Point* objects often in your work. You decide to create a new toolbar group to change the display of the *Point* objects with <u>one</u> PICK rather than accessing the *Point Style* dialog box from the *Format* pull-down menu, PICKing the desired style, and PICKing the *OK* button (four PICKs required). To create a custom toolbar, you must create new button icons and assign the buttons to set the *PDMODE* system variable to the appropriate value.

Figure 44-13 illustrates the custom toolbar group under construction. The new toolbar group (in the Drawing Editor, above) is titled *Point Style*. Only four buttons have been created at this time, with one under construction for changing the *Point* style to a circle with a cross (*PDMODE* 34). The first button creates a *Point*, the other two buttons change *PDMODE* to values of 2 and 3.

Note the contents of the *Button Properties* tab (see Figure 44-13). The *Button Editor (Edit...)* was used previously to create the new icon (circle and two crossing lines). The *Name* edit box contains the *Description* "tooltip" text (note the pointer above the dialog box resting on the button). The edit box contains the "help string" that appears at the Status Bar (not shown during construction). The *Macro* area contains two *Cancels* (^C^C), the *PDMODE* system variable call, a space to force a Return (like pressing Enter), and the desired value (34).

When this toolbar group is complete, selecting any one of the buttons causes the *Point* objects in the drawing to immediately display the new point style.

Right-Click Menu

Figure 44-14

Whenever the *Customize* dialog box is open, you can right-click on any tool to produce a shortcut menu (Fig. 44-14). Options in this shortcut menu are as follows.

Reset
If you used the *Button Properties* tab and/or the *Button Editor* to make changes to a button, use *Reset* to undo changes you made including changes to the button image, macro, name, and description. This is the same as using *Reset* in the *Button Properties* tab.

Delete
This option deletes the highlighted button. A confirmation dialog box appears.

Copy Button Image/Paste Button Image
These two options are used to copy the image on one button and paste it to another button. This action does <u>not</u> copy the macro, name, or description to the "pasted" button, only the image.

Reset Button Image
This option is similar to *Reset*; however, only the button <u>image</u> is reset, not the macro, name, or description.

Edit Button Image
Selecting this option produces the *Button Editor* with the highlighted button's image ready for editing.

Begin a Group
This is a very useful feature that is not found elsewhere. Selecting this option produces a separator (small vertical line) between sections in a toolbar (see Figure 44-9, the group separator between the solid commands and the Boolean commands). Move the cursor to the desired button, right-click, and select *Begin a Group* to produce a separator <u>to the left of the highlighted button</u>.

Properties
Selecting this option produces the *Button Properties* tab with the highlighted button's properties ready for editing.

Customizing Shortcut Keys

Using the *Keyboard* tab in the *Customize* dialog box (Fig. 44-15) you can change the shortcut key sequence that is assigned to a specific command or create new shortcut key sequences for commands. In AutoCAD, shortcut keys (Function keys, an Alt + a letter, or a Ctrl + a letter) can be pressed to access a particular command as an alternative to typing the command or selecting a button or menu option.

Almost all of the possible F keys, Alt+letter, and Ctrl+letter sequences are already assigned to commands in AutoCAD. Therefore, you can usually only reassign these shortcuts from one existing command to another. You can, however, use <u>Ctrl+Shift+another letter</u> to create new shortcuts. Options in the *Keyboard* tab are explained here.

Figure 44-15

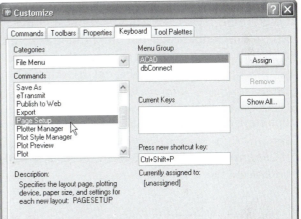

Categories
This list gives all the standard pull-down menus and toolbars. When you select a menu or toolbar from the list, the associated menu or toolbar items are displayed in the *Commands* list.

Commands
This list displays all the commands for a specific *Category*. Select the command you want to customize. If the selected command has a shortcut key already assigned to it, the key combination is displayed under *Current Keys*.

Menu Group
Specifies the current menu (.MNU) file. The AutoCAD standard menu group is called ACAD. If other custom menu groups are installed, they are listed in the order that they were installed.

Current Keys
If a shortcut exists for the highlighted command, this area displays the key combination previously defined.

Press New Shortcut Key

This is the area where you specify a key sequence to be used as the keyboard shortcut for the selected command. To specify a shortcut key sequence, click inside the edit box, then press the key sequence you want to use. You can simultaneously press Ctrl and a letter, or you can simultaneously press Ctrl+Shift and a letter. If the keyboard shortcut you specify is already assigned to another AutoCAD command, the "Currently assigned to" message is displayed. If you use an invalid key combination, AutoCAD does not show the combination in the edit box, so try using another key combination.

Assign

After you specify a new shortcut key sequence, use *Assign* to save the new shortcut key to the selected command.

Remove

This option removes the shortcut key assignment from the selected command.

Show All

Use this option to produce the *Shortcut Keys* dialog box that displays the shortcut key assignments for all menu groups (Fig. 44-16). The *Shortcut Keys* dialog box lists the command, menu group, and category associated with each shortcut key.

NOTE: You can reassign AutoCAD-programmed shortcut keys to perform different actions, but you cannot *Remove* them. You cannot reassign shortcut keys that are internally assigned to Windows, for example, F10, Ctrl+F4, Ctrl+F6, or Ctrl+Alt+Del.

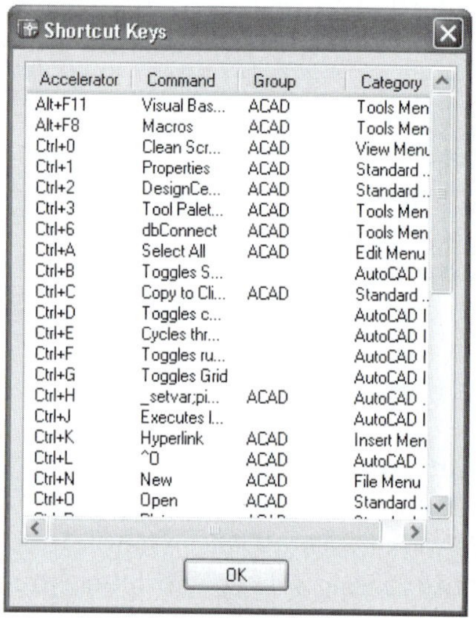

Figure 44-16

Steps for Creating a New Shortcut Key Sequence

1. Use any method to produce the *Customize* dialog box. Select the *Keyboard* tab.

2. Search the *Categories* list to find the menu or toolbar that contains the command you want to use for the new shortcut. Select the desired menu or toolbar.

3. Find the desired command in the *Commands* list.

4. Click inside the *Press New Shortcut Key* edit box. On the keyboard, simultaneously press the desired new key sequence you want to assign to the command. Use Ctrl+a letter or Ctrl+Shift+a letter.

5. If nothing appears in the *Press New Shortcut Key* edit box or the "Currently assigned to" message is displayed, try using another key combination.

6. Press *Assign* to save the new shortcut key sequence. Press *Close* in the *Customize* dialog box to compile the menu with the new shortcut key.

CREATING CUSTOM LINETYPES

AutoCAD has a wide variety of linetype definitions from which you can choose (Chapter 11, Layers and Object Properties). These definitions are stored in an ASCII file called ACAD.LIN. The ACAD.LIN file describes both simple and complex linetypes. Simple linetypes can be composed of line segments, dots, and spaces only. Complex linetypes include text or graphical shapes within the definition of the linetype pattern. However, you are not limited to these linetype definitions. New linetypes can be easily created that contain the line spacing, text, and shapes that you need.

Linetypes are created by using a text editor or word processor to describe the components of the linetype. The Windows Notepad is a sufficient text editor that creates ASCII files (only alphanumeric characters without word processing codes) and can be used for this purpose. New linetype definitions are stored in ASCII files that you name and assign with an .LIN file extension. The file names should indicate the nature of the linetype definitions. For example, ELECT.LIN can include linetype definitions for electrical layouts, or CIVIL.LIN can contain linetype definitions for the civil engineering discipline. This scheme promotes better file maintenance than appending custom linetypes to the ACAD.LIN file.

Creating Simple Linetypes

A linetype definition is created by typing the necessary characters using a text editor and saving as an .LIN file. Examine the basic components of a simple linetype definition below. Each element of the linetype definition is separated by a comma (with no spaces). A linetype definition includes the following items:

1. The <u>name</u> of the linetype with a prefix of *

2. A <u>description</u> of the linetype (either text or symbols)

3. The <u>alignment</u> field

4. Numerical <u>distances</u> for dashes, dots, and spaces

5. The <u>shape</u> or <u>text</u> creation elements (contained in square brackets [ ])

A Simple Linetype Example

The following code is an example of creating a line that looks like line definition "EXAMPLE1" in the figure:

```
*example1,a sample linetype ____ . . _____ . __
A,1.0,-0.25,0,-0.1,0,-0.25,1.5,-0.25,0,-0.25,0.5,-0.5
```

The previous definition can be analyzed as follows:

1. The name of the linetype is "EXAMPLE1."

2. A description of the linetype is a "sample" of the linetype that you create with the keyboard characters: ____ . . _____ . __ (Note that this "graphical" description is created using the underscore and period characters.)

3. A   This is the alignment field.

4. 1.0   A positive value designates a line segment. This segment has a 1 unit length.

5. -0.25   A negative value denotes a gap or blank space. This gap has a length of 0.25.

6. 0   A value of 0 creates a dot.

7. repeat of the previous elements.

Figure 44-17 shows the "EXAMPLE1" linetype defined above.

Figure 44-17 ───────────────

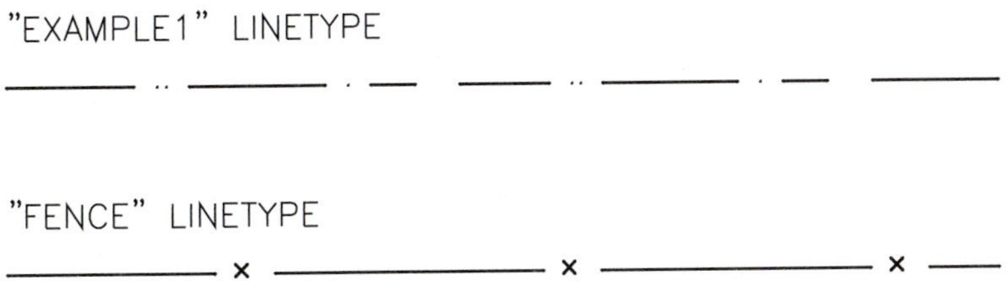

"EXAMPLE1" LINETYPE

──────── ·· ──────── · ── ──── ·· ───── · ── ────

"FENCE" LINETYPE

───────── × ─────────── × ─────────── × ──

Creating Complex Linetypes

In a complex linetype, two additional fields are possible in the definition and are placed in brackets [ ]. These fields are the Shape field and the String field. Each field has the following form:

Shapes: [shape description, shape file, scale, rotation, Xoffset, Yoffset]
Strings: ["text string," text style name, scale, rotation, Xoffset, Yoffset]

For linetypes containing shapes, the shape description and the compiled file name (.SHX) are provided in the first two places of the field. In comparison, a linetype containing a text string has the text string in quotes (" ") and the defined text style name in its first two places. If the text style does not exist, the current text style is used. The last four places of string and shape fields are the same. They are described as follows:

Scale	S=*value*	This is the scale factor by which the height of the text or shape definition is multiplied.
Rotation	R=*value* or A=*value*	A rotation angle for the text or shape is given here. The "A" designator is for an absolute rotational value, regardless of the the angle of the current line segment. The "R" designator is to specify a rotational value relative to the angle of the current line segment.
Xoffset	X= *value*	You can specify an X offset value from the end of the last element with this value.
Yoffset	Y=value	Specify a Y offset value from the end of the last element with this value.

Complex Linetype Examples

```
*PL,Property Line ----PL----PL----
A,0.5,-0.5,[PLSYM,SYMBOLS.SHX,S=1.5,R=0,X=-0.1,Y=-0.1],-0.5,0.5,1

*Fence,Fence line ----x----x----x----
A,1.5,-0.25,["x",STANDARD,S=0.2,R=0.0,X=0.05,Y=-0.1],-0.4,1.5
```

See Figure 44-17 for a sample of the "Fence" linetype defined above. For more examples of simple and complex linetypes, use the Windows Notepad to open the ACAD.LIN file. It is suggested that you make new files when you create your first linetypes rather than appending to the ACAD.LIN file.

CUSTOMIZING THE ACAD.PGP FILE

The ACAD.PGP (AutoCAD Program Parameter) file is an ASCII text file that provides two main features. It provides a link between AutoCAD and other programs by specifying what external programs can be accessed from within AutoCAD, and it defines command aliases—the one- or two-letter shortcuts for commands. In short, the ACAD.PGP file defines what commands can be typed at the AutoCAD Command prompt that are not native to AutoCAD.

Because the ACAD.PGP file is an ASCII file, you can edit the file to add your own command aliases and to define which external commands you want to use from within AutoCAD. Use any text editor, such as Windows Notepad or Wordpad, to view and modify the file. A portion of the AutoCAD ACAD.PGP file follows:

```
B,          *BLOCK
-B,         *-BLOCK
BH,         *BHATCH
BO,         *BOUNDARY
-BO,        *-BOUNDARY
BR,         *BREAK
C,          *CIRCLE
CH,         *PROPERTIES
-CH,        *CHANGE
CHA,        *CHAMFER
CHK,        *CHECKSTANDARDS
COL,        *COLOR
COLOUR,     *COLOR
CO,         *COPY
```

This is just a small section of the ACAD.PGP file that defines the command aliases. The actual file is many pages long and contains primarily command aliases for just about every AutoCAD command. (See Appendixes B and C for the command aliases sorted by command and sorted by alias.) Autodesk's intention is for you to modify the file and to define the external commands and command aliases for personal productivity, although most of that work has been done for you.

All lines of this file that begin with a semicolon (;) are remarks only. These lines (more than a page) give useful information about modifying the file, the format for specifying external commands, and the format for specifying command aliases. Only lines without the semicolon (;) are read by the AutoCAD program. Defining external commands and command aliases are explained next.

Specifying External Commands

The External Commands section of the ACAD.PGP file determines what commands or other programs can be accessed from <u>within AutoCAD</u> by typing the command at the AutoCAD Command prompt. By examining this section of the file, you can see that the following DOS commands are already usable from within AutoCAD:

```
CATALOG,    DIR /W,       8,File specification: ,
DEL,        DEL,          8,File to delete: ,
DIR,        DIR,          8,File specification: ,
EDIT,       START EDIT,   9,File to edit: ,
SH,         ,             1,*OS Command: ,
SHELL,      ,             1,*OS Command: ,
START,      START,        1,*Application to start: ,
TYPE,       TYPE,         8,File to list: ,
```

The following Windows commands are also usable from within AutoCAD:

```
EXPLORER,   START EXPLORER,  1,,
NOTEPAD,    START NOTEPAD,   1,*File to edit: ,
PBRUSH,     START PBRUSH,    1,,
```

For example, type DIR from the Command prompt to list the contents of the current directory. If you prefer using Windows Explorer for file management, just type "Explorer" at the AutoCAD Command prompt to open it.

In AutoCAD 2004, you can use the *Tools* pull-down menu to edit the ACAD.PGP file in Notepad (Fig. 44-18). Select *Tools, Customize, Edit Custom Files,* then *Program Parameters (acad.pgp)*. This new feature makes customizing ACAD.PGP possible while AutoCAD is running.

NOTE: If you edit ACAD.PGP while AutoCAD is running, use the *Reinit* command after editing to reinitialize ACAD.PGP to make the new changes usable (see Chapter 43 for information on *Reinit*).

If you use another text editor—Norton Editor, for example—you can change the ACAD.PGP file to allow "NE" to be typed at the command prompt. Do it by changing this:

```
EDIT,     START EDIT,    1,File to edit: ,
```

to this:

```
NE,       START NE,      1,File to edit: ,
```

Ensure you follow the format guidelines specified in the first page of the file:

```
<Command name>,[<DOS request>],<Bit flag>,[*]<Prompt>,
```

You should also use spaces (press the space bar) instead of tabs in the ACAD.PGP file.

Figure 44-18

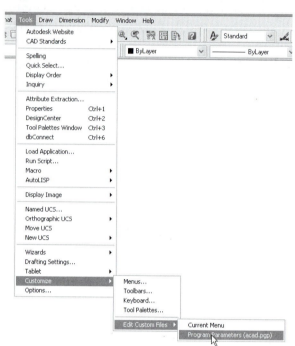

Defining Command Aliases

Autodesk intends for you to define your own command aliases. After questioning users, Autodesk decided to define a command alias for almost all commands in AutoCAD. However, you can add more aliases, delete the defined aliases that you don't use, or alter the aliases that AutoCAD provides. The command alias section begins with these entries:

```
3A,       *3DARRAY
3DO,      *3DORBIT
3F,       *3DFACE
3P,       *3DPOLY
A,        *ARC
ADC,      *ADCENTER
AA,       *AREA
```
and so on...

Because most commands already have an alias defined, it is probably more productive to learn the existing aliases rather than create new ones. There may be cases where it would be helpful to redefine existing aliases, however. For example, you may use *Trim* often but not *Mtext* (*TR* is the current alias for the *Trim* command, and *T* <u>and</u> *MT* for *Mtext*). In this case you could delete the *T* alias for *Mtext* (and just use *MT* for *Mtext*) and change *Trim* to *T*. To do this, follow the format specified at the beginning of the aliases section (give the alias and a comma, space over, and then give the AutoCAD command prefaced by an asterisk).

First, find and delete this:

```
    T,        *MTEXT
```

and change this:

```
    TR,       *TRIM
```

to this:

```
    T,        *TRIM
```

You may want to modify some of the other defined aliases. For example, you could change the alias C to invoke *Copy* rather than *Circle*. An alternative to *Circle* could be CI.

You can make comments anywhere in the ACAD.PGP by typing a semicolon (;) at the beginning of the line. This is helpful for making note of the changes that you have made.

One last suggestion: make a copy of the ACAD.PGP on diskette or in another directory before you begin experimenting, just in case.

CREATING SCRIPT FILES

A script file is an ASCII text file that contains a series of AutoCAD commands to be performed in succession. A script file is an external file that you create with a text editor and assign an .SCR extension. The script file is composed of commands in the sequence that you would normally use just as if you entered the commands at the AutoCAD command prompt. The *Script* command is used to tell AutoCAD to run the specified script file. When AutoCAD "runs" a script file, each command in the file is executed in the order that it is listed.

Script files have many uses, such as presenting a slide show, creating drawing geometry, or invoking a series of AutoCAD commands you use often—like a "macro." Script files are commonly used for repetitive tasks such as setting up a drawing. In fact, script "macros" may be incorporated directly into a menu file, thereby eliminating the need for an external .SCR file. However, in this chapter, only independent script files are addressed.

Chapter 43 discusses the *Script* command and the steps for creating scripts to present slide shows. A script for viewing slides uses the *Vslide* and *Delay* commands repetitively. If you are not familiar with the procedure for creating and running a script for this fundamental application, please review "Using Slides and Scripts" in Chapter 43.

Scripts for Drawing Setups

As you become familiar with AutoCAD, you realize several steps are repeated each time you set up a new drawing. For example, several commands are used and variables are set to prepare the drawing for geometry construction and for eventual plotting. Drawing setup is a good candidate for script application. Script files can be created to perform those repetitive drawing setup steps for you. Although an interactive AutoLISP routine would provide a more efficient and powerful method for this task, creating a script is relatively simple and does not require much knowledge or effort beyond that of the AutoCAD commands that would typically be used.

Keep in mind that template drawings that have many of the initial settings can be used as an alternative to creating a script for drawing setup. However, one advantage of a script (in general) is that it can be executed in any drawing and at any time during the drawing process.

For example, assume that you often set up drawings for plotting model space at a 1/4" = 1'-0" scale on a 24" x 36" sheet of paper. This involves setting *Limits, Grid, Snap, LTSCALE*, and *DIMSCALE*. Instead of executing each of these operations each time "manually," create a script to automate the process. Here are the steps:

1. Open your text editor from within AutoCAD or before starting AutoCAD. Creating scripts while AutoCAD is running is the most efficient method for writing and "debugging" scripts. In AutoCAD, use Notepad (or other text editor specified in the ACAD.PGP) to open your text editor.

2. Create the ASCII text file with the commands that you would normally use for setting up the drawing. Remember (from Chapter 43) that a space or a new line in the script is read as a Return (pressing the Enter key) while the script executes in AutoCAD. The script file may be written like this:

   ```
   LIMITS 0,0 1728,1152
   GRID 48
   SNAP 12
   LTSCALE 48
   DIMSCALE 48
   -DIMSTYLE SAVE STANDARD Y
   ```

3. Save the file with an .SCR extension. A good name for this file is SETUP48.SCR. Consider locating the file in a directory in the AutoCAD environment (as specified in the *Support Files Search Path* section of the *Files* tab in the *Options* dialog box) or in a location with other scripts.

4. Return to AutoCAD and use the *Script* command to test the script. Don't expect the script to operate the first time if this is your first experience writing scripts. It is common to have "bugs" in the script. For example, an extra space in the file can cause the script to get out of the intended sequence. An extra space in the script is hard to locate (since it is not visible). A good text editor allows the cursor to move only to places occupied by a character (including a space) and should be used for script writing.

Examining the commands in SETUP48.SCR, first, set the *Limits* using inch values. The "-DIMSTYLE SAVE STANDARD Y" line saves the *DIMSCALE* setting to the STANDARD dimension style. (If a dimension style other than STANDARD exists, the new *DIMSCALE* setting should be incorporated also into these dimension styles.) Keep in mind that scripts operate as if the commands were entered at the Command prompt, not as you might operate using dialog boxes.

Many other possibilities exist for drawing setup scripts. For example, you could set *-Units* (command) and the *DIMUNITS* variable. You may want to create a script for each plot scale and sheet size. It depends on your particular applications, hence the term "customization."

Possible applications for scripts are numerous. In addition to drawing setups, scripts can be used for geometry creation. It is possible to write scripts to create particular geometry "in real time" as the script executes. For example, repetitively drawn shapes could be created by scripts as an alternative to inserting *Blocks*. This idea is not generally used, but it may have advantages for certain applications. There are endless possibilities for script applications. Generally, anytime you find yourself performing the same tasks or series of commands repetitively, remember that a script can automate that process.

Special Script Commands

Creating scripts is a very straightforward procedure because the script file contains the AutoCAD commands that you would normally use at the command prompt for the particular operation. A few commands are intended for use specifically with scripts and other commands can be used as an aid to create or run script files. Those commands are listed next. Many of these commands are discussed in detail in "Using Slides and Scripts," Chapter 43.

SCRIPT	Use this command in AutoCAD to instruct AutoCAD to load and execute a script. The *Select Script File* dialog box appears, prompting you for the desired script file.
DELAY	Include this command in a script file to force a delay (specified in milliseconds) until the next command in the script is executed. *Delay* is commonly used in script files that display slide shows.
RSCRIPT	*Rscript* is short for "Repeat script." This command can be entered as the last line of a script file to make the script repeat until interrupted. *Rscript* is helpful for self-running slide shows.
RESUME	If you want to interrupt a script with the Break or backspace key, type *Resume* to resume the script from the point at which it was interrupted. This command has no purpose in the script file itself but is intended to be typed at the keyboard.
LOGFILEON/ LOGFILEOFF	These commands are used to turn on and off the log file creation utility in AutoCAD. When *Logfileon* is used, the conversation that appears at the command prompt (command history) is written to an external text file until *Logfileoff* is used (explained next in this chapter).

LOGFILEON/
LOGFILEOFF

Pull-down Menu	COMMAND (TYPE)	ALIAS (TYPE)	Short-cut	Screen (side) Menu	Tablet Menu
Tools *Options...* *Open and Save* *Maintain a log file*	*LOGFILEON* or *LOGFILEOFF*	...	...	*TOOLS 2* *Options...* *Open and Save* *Maintain a log file*	...

Logfileon and *Logfileoff* are used to turn on and off the creation of the AutoCAD log file. The log file is a text file that contains the command history—the text that appears at the Command line. When *Logfileon* is invoked, nothing appears to happen; however, any commands, prompts, or other text that appear at the command prompt from that time on are copied and written to the drawing's log file. Use *Logfileoff* to close the log file and discontinue recording the command history.

If the log file is on, AutoCAD continues to write to the log file, and each AutoCAD session in the log is separated by a dashed line. If you turn *Logfileoff* and *Logfileon* again, the new session is appended to the existing file. When a log file is created, it takes the current drawing's name by default. For example, if you are working on a drawing named PART35 and turn *Logfileon*, the log file is saved with a name similar to PART35_1_1_3281 (where _1_1_3281 are automatically generated numbers). You can use the *Options* dialog box, *Files* tab, *Log File Location* section to locate the current location of the log files or to set a different location for log files (see Figure 44-19).

The log file is an ASCII text file that can be edited. The file has no direct link to AutoCAD and can be deleted without consequence to AutoCAD operation. The log file cannot be accessed externally until *Logfileoff* is used or AutoCAD is exited.

A sample log file is shown here:

```
[ AutoCAD - Mon Jan 12 10:25:16 2004 ]──────────────────────────────────  -------------------

Command: line
Specify first point:
Specify next point or [Undo]:
Specify next point or [Undo]:
Specify next point or [Close/Undo]:
Specify next point or [Close/Undo]:
Command: circle
Specify center point for circle or [3P/2P/Ttr (tan tan radius)]:
Specify radius of circle or [Diameter]:
Command: copy
Select objects: 1 found
Select objects: 1 found, 2 total
Select objects: 1 found, 3 total
Select objects:
Specify base point or displacement, or [Multiple]: Specify second point of displacement or <use first point
as displacement>:
Command: trim
Current settings: Projection=UCS Edge=None
Select cutting edges ...
Select objects: 1 found
Select objects:
Select object to trim or [Project/Edge/Undo]:
Select object to trim or [Project/Edge/Undo]:
Select object to trim or [Project/Edge/Undo]:
Command: _qsave
```

The log file is especially helpful when you are writing scripts. When you create a script file, the commands and options listed in the file must be typed in the exact sequence that would occur if you were actually using AutoCAD. In order to write the script correctly, you have three choices: (1) remember every command and prompt exactly as it would occur in AutoCAD; (2) go through a "dry run" in AutoCAD and write down every step; (3) turn *Logfileon*; then execute a "dry run" to have all the steps and resulting text recorded.

 In other words, turn *Logfileon* to create a rough script file. Then turn *Logfileoff*, open the log file, and edit out the AutoCAD prompts such as "Command:" and "Select objects:," etc. Finally, save the file to the desired name with an .SCR extension.

If you want to change the location of the log file that is created when you use *Logfileon*, use the *Files* tab of the *Options* dialog box (Fig. 44-19). In the dialog box, expand the *Log File Location* tree, select *Browse* to find the desired directory location, select *Apply*, then *OK*. This directory location is stored in, and can also be changed using, the *LOGFILEPATH* system variable.

Although more cumbersome for most applications than using the *Logfileon* and *Logfileoff* commands, you can also turn on and off the log file by toggling the *Maintain a log file* check box in the *Open and Save* tab of the *Options* dialog box.

Figure 44-19

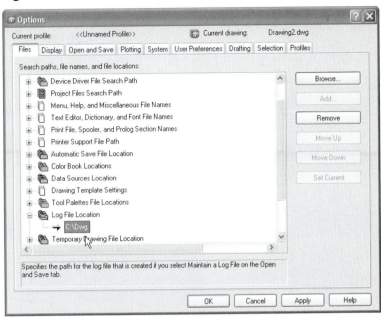

MENU CUSTOMIZATION

If you enjoy customizing AutoCAD using the techniques shown in this chapter, you may be ready to try customizing the menus in AutoCAD. Chapter 45, Menu Customization, explains how you can customize the pull-down menus, image tiles, screen menu, and digitizing tablet menu for your particular needs. Customizing the menus can help make you very efficient at performing specific complex tasks that may otherwise require you to use multiple commands. You can download Chapter 45 at no charge at www.mhhe.com/leach.

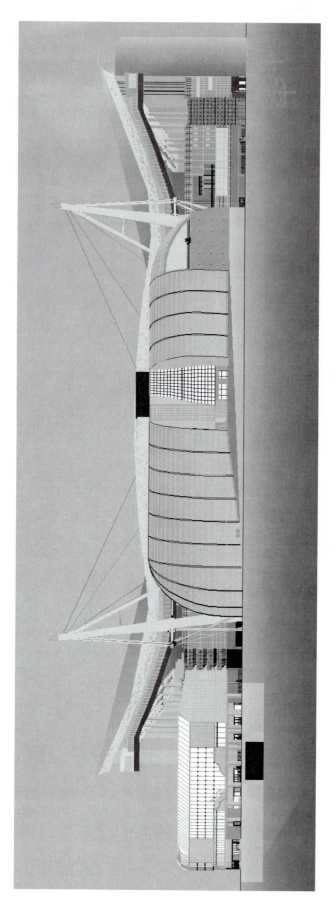

STADIUM NORTH ELEVATION.DWG, Courtesy of Autodesk, Inc.

APPENDIX A

System Variables

APPENDIX A

System Variables

Variable	Characteristic	Description
ACADLSPASDOC	Type: Integer Saved in: Registry Initial value: 0	Controls whether AutoCAD loads the acad.lsp file into every drawing or just the first drawing opened in an AutoCAD session. 0 Loads acad.lsp into just the first drawing opened in an AutoCAD session 1 Loads acad.lsp into every drawing opened
ACADPREFIX	(Read-only) Type: String Not saved	Stores the directory path, if any, specified by the ACAD environment variable, with path separators appended if necessary.
ACADVER	(Read-only) Type: String Not saved	Stores the AutoCAD version number. This variable differs from the DXF file $ACADVER header variable, which contains the drawing database level number.
ACISOUTVER	Type: Integer Not saved Initial value: 40	Controls the ACIS version of SAT files created using the *Acisout* command. Currently, *Acisout* supports only values of 15, 16, 17, 18, 20, 21, 30, 40, 50, 60, and 70.
ADCSTATE	(Read-only) Type: Integer Not saved Initial value: varies	Determines whether DesignCenter is active or not. For developers who need to determine status through AutoLISP. 0 DesignCenter is not active 1 DesignCenter is active
AFLAGS	Type: Integer Not saved Initial value: 0	Sets attribute flags for *Attdef* bit-code. It is the sum of the following: 0 No attribute mode selected 1 Invisible 2 Constant 4 Verify 8 Preset
ANGBASE	Type: Real Saved in: Drawing Initial value: 0.0000	Sets the base angle to 0 with respect to the current UCS.
ANGDIR	Type: Integer Saved in: Drawing Initial value: 0	Sets the positive angle direction from angle 0 with respect to the current UCS. 0 Counter-clockwise 1 Clockwise
APBOX	Type: Integer Saved in: Registry Initial value: 0	Turns the AutoSnap aperture box on or off. The aperture box is displayed in the center of the cursor when you snap to an object. This option is available only when Marker, Magnet, or Snaptip is selected. 0 Aperture box is not displayed 1 Aperture box is displayed

2004

Variable	Characteristic	Description
APERTURE	Type: Integer Saved in: Registry Initial value: 10	Sets *Aperture* command object snap target height, in pixels.
AREA	(Read-only) Type: Real Not saved	Stores the last area computed by *Area*, *List*, or *Dblist*.
ATTDIA	Type: Integer Saved in: Drawing Initial value: 0	Controls whether *Insert* uses a dialog box for attribute value entry. 0 Issues prompts on the command line 1 Uses a dialog box
ATTMODE	Type: Integer Saved in: Drawing Initial value: 1	Controls display of attributes. 0 Off makes all attributes invisible 1 Normal retains current visibility of each attribute: visible attributes are displayed; invisible attributes are not 2 On makes all attributes visible
ATTREQ	Type: Integer Saved in: Drawing Initial value: 1	Determines whether *Insert* uses default attribute settings during insertion of blocks. 0 Assumes the defaults for the values of all attributes 1 Turns on prompts or dialog box for attribute values, as specified at *Attdia*
AUDITCTL	Type: Integer Saved in: Registry Initial value: 0	Controls whether *Audit* creates an ADT file (audit report). 0 Prevents writing of ADT files 1 Writes ADT files
AUNITS	Type: Integer Saved in: Drawing Initial value: 0	Sets units for angles. 0 Decimal degrees 1 Degrees/minutes/seconds 2 Gradians 3 Radians 4 Surveyor's units
AUPREC	Type: Integer Saved in: Drawing Initial value: 0	Sets number of decimal places for angular units displayed on the Status line or when listing an object.
AUTOSNAP	Type: Integer Saved in: Registry Initial value: 63	Controls AutoSnap marker, SnapTip, and magnet. Also turns on Polar and Object Snap Tracking, and controls the display of Polar and Object Snap Tracking ToolTips. The system variable value is the sum of the following bit values: 0 Turns off the AutoSnap marker, SnapTips, and magnet. Also turns off Polar Tracking, Object Snap Tracking, and ToolTips for Polar and Object Snap Tracking 1 Turns on the AutoSnap marker 2 Turns on the AutoSnap SnapTips 4 Turns on the AutoSnap magnet 8 Turns on Polar Tracking

Variable	Characteristic	Description
AUTOSNAP *(continued)*		16 Turns on Object Snap Tracking 32 Turns on ToolTips for Polar Tracking and Object Snap Tracking
BACKZ	(Read-only) Type: Real Saved in: Drawing	Stores the back clipping plane offset from the target plane for the current viewport, in drawing units. Meaningful only if the back clipping bit in *VIEWMODE* is on. The distance of the back clipping plane from the camera point can be found by subtracting *BACKZ* from the camera-to-target distance.
BINDTYPE	Type: Integer Not Saved Initial value: 0	Controls how xref names are handled when binding xrefs or editing xrefs in place. 0 Traditional binding behavior ("xref1|one" becomes "xref0one") 1 Insert-like behavior ("xref1|one" becomes "one")
BLIPMODE	Type: Integer Saved in: Drawing Initial value: 0	Controls whether marker blips are visible. 0 Turns off marker blips 1 Turns on marker blips
CDATE	(Read-only) Type: Real Not saved	Sets calendar date and time.
CECOLOR	Type: String Saved in: Drawing Initial value: "BYLAYER"	Sets the color of new objects. Valid values include *Bylayer, Byblock*, and an integer from 1 to 255. Valid values for True Colors are a string of integers each from 1 to 255 separated by commas and preceded by RGB. The True Color setting is entered as follows: RGB:000,000,000.
CELTSCALE	Type: Real Saved in: Drawing Initial value: 1.0000	Sets the current object linetype scaling factor. This sets the linetype scaling for new objects relative to the *LTSCALE* setting. A line created with *CELTSCALE*=2 in a drawing with *LTSCALE* set to 0.5 would appear the same as a line created with *CELTSCALE*=1 in a drawing with *LTSCALE*=1.
CELTYPE	Type: String Saved in: Drawing Initial value: "BYLAYER"	Sets the linetype of new objects.
CELWEIGHT	Type: Integer Saved In: Drawing Initial value: "BYLAYER"	Sets the lineweight of new objects. -1 Sets the lineweight to "BYLAYER" -2 Sets the lineweight to "BYBLOCK" -3 Sets the lineweight to "DEFAULT" Other valid values that can be entered in millimeters include: 0, 5, 9, 13, 15, 18, 20, 25, 30, 35, 40, 50, 53, 60, 70, 80, 90, 100, 106, 120, 140, 158, 200, and 211. All values must be entered in millimeters. (Multiply a value by 2.54 to convert values from inches to millimeters.)

Variable	Characteristic	Description
CHAMFERA	Type: Real Saved in: Drawing Initial value: 0.0000	Sets the first chamfer distance.
CHAMFERB	Type: Real Saved in: Drawing Initial value: 0.0000	Sets the second chamfer distance.
CHAMFERC	Type: Real Saved in: Drawing Initial value: 0.0000	Sets the first chamfer length.
CHAMFERD	Type: Real Saved in: Drawing Initial value: 0.0000	Sets the first chamfer angle.
CHAMMODE	Type: Integer Not saved Initial value: 0	Sets the input method by which AutoCAD creates chamfers. 0 Requires two chamfer distances 1 Requires one chamfer distance and an angle
CIRCLERAD	Type: Real Not saved Initial value: 0.0000	Sets the default circle radius. A zero sets no default.
CLAYER	Type: String Saved in: Drawing Initial value: 0	Sets the current layer.
CMDACTIVE	(Read-only) Type: Integer Not saved	Stores the bit code that indicates whether an ordinary command, transparent command, script, or dialog box is active. It is the sum of the following: 1 Ordinary command is active 2 Ordinary command and a transparent command are active 4 Script is active 8 Dialog box is active 16 DDE is active 32 AutoLISP is active (only visible to an ObjectARX-defined command) 64 ObjectARX command is active
CMDECHO	Type: Integer Not saved Initial value: 1	Controls whether AutoCAD echoes prompts and input during the AutoLISP (command) function. 0 Turns off echoing 1 turns on echoing
CMDNAMES	(Read-only) Type: String Not saved	Displays the name of the currently active command and transparent command. For example, LINE'ZOOM indicates that the Zoom command is being used transparently during the Line command. This variable is designed for use with programming interfaces such as AutoLISP, DIESEL, and ActiveX Automation.

Variable	Characteristic	Description
CMLJUST	Type: Integer Saved in: Drawing Initial value: 0	Specifies multiline justification. 0 Top 1 Middle 2 Bottom
CMLSCALE	Type: Real Saved in: Drawing Initial value: 1.0000 (English) 20.0000 (Metric)	Controls the overall width of a multiline. A scale factor of 2.0 produces a multiline that is twice as wide as the style definition. A zero scale factor collapses the multiline into a single line. A negative scale factor flips the order of the offset lines (that is, the smallest or most negative is placed on top when the multiline is drawn from left to right).
CMLSTYLE	Type: String Saved in: Drawing Initial value: "STANDARD"	Sets the multiline style that AutoCAD uses to draw the multiline.
COMPASS	Type: Integer Initial value: 0	Controls whether the 3D compass is on or off in the current viewport. 0 Turns the 3D compass off 1 Turns the 3D compass on
COORDS	Type: Integer Saved in: Drawing Initial value: 1	Controls when coordinates are updated on the Status line. 0 Coordinate display is updated as you specify points with the pointing device 1 Display of absolute coordinates is continuously updated 2 Display of absolute coordinates is continuously updated, and distance and angle from last point are displayed when a distance or angle is requested.
CPLOTSTYLENAME	Type: String Saved in: Drawing	Controls the current plot style for new objects. If the current drawing is color-dependent (PSTYLEPOLICY is set to 1), CPLOTSTYLE is read-only and has a value of "BYCOLOR." If the (PSTYLEPOLICY is set to 0), CPLOTSTYLE can be set to the following values ("BYLAYER" is the default): "BYLAYER" "BYBLOCK" "NORMAL" "USER DEFINED"
CPROFILE	Type: (Read-only) Saved in: Registry Initial value: <<Unnamed Profile>>	Displays the name of the current profile.
CTAB	(Read-only) Type: String Saved in: Drawing	Returns the name of the current (model or layout) tab in the drawing. Provides a means for the user to determine which tab is active.

Variable	Characteristic	Description
CURSORSIZE	Type: Integer Saved in: Registry Initial value: 5	Determines the size of the cursor as a percentage of the screen size. Valid settings range from 1 to 100 percent. When set to 100, the cursor are full-screen and the ends of the cursor are never visible. When less than 100, the ends of the cursor may be visible when the cursor is moved to one edge of the screen.
CVPORT	Type: Integer Saved in: Drawing Initial value: 2	Sets the identification number of the current viewport. You can change this value, thereby changing the current viewport, if the following conditions are met: • The identification number you specify is that of an active viewport. • A command in progress has not locked cursor movement to that viewport. • Tablet mode is off.
DATE	(Read-only) Type: Real Not saved	Stores the current date and time represented as a Julian date and fraction in a real number: <Julian date>.<Fraction> For example, on January 29, 1993, at 2:29:35 in the afternoon, the *DATE* variable would contain 2446460.603877364. Your computer clock provides the date and time. The time is represented as a fraction of a day. To compute differences in time, subtract the times returned by *DATE*. To extract the seconds since midnight from the value returned by *DATE*, use AutoLISP expressions: **(setq s (getvar "DATE"))** **(setq seconds (* 86400.0 − s (fix s)))** The *DATE* system variable returns a true Julian date only if the system clock is set to UTC/Zulu (Greenwich Mean Time). *TDCREATE* and *TDUPDATE* have the same format as *DATE*, but their values represent the creation time and last update time of the current drawing.
DBCSTATE	(Read-only) Type: Integer Saved in: Drawing Initial value: 0	Stores the state of the *dbConnect Manager*, active or not active. 0 The *dbConnect Manager* is not displayed 1 The *dbConnect Manager* is displayed
DBMOD	(Read-only) Type: Integer Not saved	Indicates the drawing modification status using bit code. It is the sum of the following: 1 Object database modified 4 Database variable modified 8 Window modified 16 View modified
DCTCUST	Type: String Saved in: Registry Initial value: ""	Displays the path and file name of the current custom spelling dictionary.

2004

Variable	Characteristic	Description
DCTMAIN	Type: String Saved in: Registry Initial value: *varies by country*	Displays the file name of the current main spelling dictionary. The full path is not shown, because this file is expected to reside in the *support* directory. You can specify a default main spelling dictionary using *Setvar.* When prompted for a new value for *DCTMAIN*, you can enter one of the keywords below. *Keyword Language Name* enu American English ena Australian English ens British English (ise) enz British English (ize) ca Catalan cs Czech da Danish nl Dutch (primary) nls Dutch (secondary) fi Finnish fr French (unaccented capitals) fra French (accented capitals) de German (Scharfes s) ded German (Dopple s) it Italian no Norwegian (Bokmal) non Norwegian (Nynorsk) pt Portuguese (Iberian) ptb Portuguese (Brazilian) ru Russian (infrequent io) rui Russian (frequent io) es Spanish (unaccented capitals) esa Spanish (accented capitals) sv Swedish
DEFLPLSTYLE	Type: String Saved in: Registry	Specifies the default plot style for new layers. If the current drawing is color-dependent (*PSTYLEPOLICY* is set to 1), *DEFLPLSTYLE* is read-only and has a value of "BYCOLOR." If the current drawing is named plot style mode (*PSTYLEPOLICY* is set to 0), *DEFLPLSTYLE* is writable and has a default value of "NORMAL."
DEFPLSTYLE	Type: String Saved in: Registry	Specifies the default plot style for new objects. If the current drawing is color-dependent (*PSTYLE-POLICY* is set to 1), *DEFPLSTYLE* is read-only and has a value of "BYCOLOR." If the current drawing is named plot style mode (*PSTYLEPOLICY* is set to 0), *DEFPLSTYLE* is writable and has a default value of "BYLAYER."
DELOBJ	Type: Integer Saved in: Drawing Initial value: 1	Controls whether objects used to create other objects are retained or deleted from the drawing database. 0 Objects are retained 1 Objects are deleted

Variable	Characteristic	Description
DEMANDLOAD	Type: Integer Saved in: Registry Initial value: 3	Specifies if and when AutoCAD demand loads a third-party application if a drawing contains custom objects created in that application. 0 Turns off demand loading. 1 Demand loads the source application when you open a drawing that contains custom objects. This setting does not demand load the application when you invoke one of the application's commands. 2 Demand loads the source application when you invoke one of the application's commands. This setting does not demand load the application when you open a drawing that contains custom objects. 3 Demand loads the source application when you open a drawing that contains custom objects or when you invoke one of the application's commands.
DIASTAT	(Read-only) Type: Integer Not saved	Stores the exit method of the most recently used dialog box. 0 Cancel 1 OK
DIM...	See "Dimension Variables Table" in Chapter 29	
DISPSILH	Type: Integer Saved in: Drawing Initial value: 0	Controls display of silhouette curves of solid objects in wireframe mode. Also controls whether mesh is drawn or suppressed when a solid object is hidden. 0 Off 1 On
DISTANCE	(Read-only) Type: Real Not saved	Stores the distance computed by *Dist*.
DONUTID	Type: Real Not saved Initial value: 0.5000	Sets the default for the inside diameter of a donut.
DONUTOD	Type: Real Not saved Initial value: 1.0000	Sets the default for the outside diameter of a donut. It must be nonzero. If *DONUTID* is larger than *DONUTOD*, the two values are swapped by the next command.
DRAGMODE	Type: Integer Saved in: Drawing Initial value: 2	Controls display of objects being dragged. 0 Does not display an outline of the object as you drag it 1 Displays the outline of the object as you drag it only if you enter **drag** on the command line after selecting the object to drag 2 Auto; always displays an outline of the object as you drag it

Variable	Characteristic	Description
DRAGP1	Type: Integer Saved in: Registry Initial value: 10	Sets regen-drag input sampling rate.
DRAGP2	Type: Integer Saved in: Registry Initial value: 25	Sets fast-drag input sampling rate.
DWGCHECK	Type: Integer Saved in: Registry Initial Value: 0	Checks drawings for potential problems when opening them. 0 If a drawing that you try to open is found to have a potential problem, you are warned before the drawing is opened. 1 If a drawing that you try to open is found to have a potential problem, or if it was saved by an application other than AutoCAD or AutoCAD LT, you are warned before the drawing is opened. 2 If a drawing that you try to open is found to have a potential problem you are notified on the command line after the drawing is opened. 3 If a drawing that you try to open is found to have a potential problem or if it was saved by an application other than AutoCAD or AutoCAD LT, you are notified on the command line after the drawing is opened.
DWGCODEPAGE	(Read-only) Type: String Saved in: Drawing	Stores the same value as *SYSCODEPAGE* (for compatibility).
DWGNAME	(Read-only) Type: String Not saved	Stores the drawing name as entered by the user. If the drawing has not been named yet, *DWGNAME* defaults to "Drawing.dwg." If the user specified a drive/directory prefix, it is stored in the *DWGPREFIX* variable.
DWGPREFIX	(Read-only) Type: String Not saved	Stores the drive/directory prefix for the drawing.
DWGTITLED	(Read-only) Type: Integer Not saved	Indicates whether the current drawing has been named. 0 Drawing has not been named 1 Drawing has been named
EDGEMODE	Type: Integer Saved in: Registry Initial value: 0	Controls how *Trim* and *Extend* determine cutting and boundary edges. 0 Uses the selected edge without an extension 1 Extends or trims the selected object to an imaginary extension of the cutting or boundary edge Line, arc, elliptical arc, ray, and polyline are objects eligible for natural extension. The natural extension of a line or ray is an unbounded line (xline); an arc is a circle; and an elliptical arc is an ellipse. A polyline is broken down into its line and arc components, which are extended to their natural boundaries.

Variable	Characteristic	Description
ELEVATION	Type: Real Saved in: Drawing Initial value: 0.0000	Stores the current elevation relative to the current UCS for the current space.
ERRNO	(Read-only) Type: Integer Not saved Initial value: 0	Displays the number of the appropriate error code when an AutoLISP function call causes an error that AutoCAD detects. AutoLISP applications can inspect the current value of *ERRNO* with (getvar "errno"). The *ERRNO* system variable is not always cleared to zero. Unless it is inspected immediately after an AutoLISP function has reported an error, the error that its value indicates may be misleading. This variable is always cleared when starting or opening a drawing. See the *AutoLISP Developer's Guide* for more information.
EXPERT	Type: Integer Not saved Initial value: 0	Controls whether certain prompts are issued. 0 Issues all prompts normally. 1 Suppresses "About to regen, proceed?" and "Really want to turn the current layer off?" 2 Suppresses the preceding prompts and "Block already defined. Redefine it?" (*Block*) and "A drawing with this name already exists. Overwrite it?" (*Save* or *Wblock*). 3 Suppresses the preceding prompts and those issued by *Linetype* if you try to load a linetype that's already loaded or create a new linetype in a file that already defines it. 4 Suppresses the preceding prompts and those issued by *Ucs Save* and *Vports Save* if the name you supply already exists. 5 Suppresses the preceding prompts and those issued by the *Dimstyle Save* option and *Dimoverride* if the dimension style name you supply already exists (the entries are redefined). When a prompt is suppressed by *EXPERT*, the operation in question is performed as though you entered **y** at the prompt. The setting of *EXPERT* can affect scripts, menu macros, AutoLISP, and the command functions.
EXPLMODE	Type: Integer Not saved Initial value: 1	Controls whether *Explode* supports non-uniformly scaled (NUS) blocks. 0 Does not explode NUS blocks 1 Explodes NUS blocks
EXTMAX	(Read-only) Type: 3D Point Saved in: Drawing	Stores the upper-right point of the drawing extents. Expands outward as new objects are drawn, shrinks only with *Zoom All* or *Zoom Extents*. Reported in World coordinates for the current space.

2004

Variable	Characteristic	Description
EXTMIN	(Read-only) Type: 3D Point Saved in: Drawing	Stores the lower-left point of the drawing extents. Expands outward as new objects are drawn, shrinks only with *Zoom All* or *Zoom Extents*. Reported in World coordinates for the current space.
EXTNAMES	Type: Integer Saved in: Drawing Initial value: 1	Sets the parameters for non-graphical object names (such as linetypes and layers) stored in symbol tables. 0 Uses Release 14 parameters, which limit names to 31 characters in length. Names may include the letters A-Z, the numerals 0-9, and the special characters dollar sign ($), underscore (_), and hyphen (-). 1 Uses AutoCAD 2000 (and later releases) parameters. Names can be up to 255 characters in length, and may include the letters A-Z, the numerals 0-9, spaces, and any special characters not used by Microsoft Windows and AutoCAD for other purposes.
FACETRATIO	Type: Integer Not saved Initial Value: 0	Controls the aspect ratio of faceting for cylindrical and conic ACIS solids. A setting of 1 increases the density of the mesh to improve the quality of rendered and shaded models. 0 Creates an N by 1 mesh for cylindrical and conic ACIS solids 1 Creates an N by M mesh for cylindrical and conic ACIS solids
FACETRES	Type: Real Saved in: Drawing Initial value: 0.5	Further adjusts the smoothness of shaded and rendered objects and objects with hidden lines removed. Valid values are from 0.01 to 10.0.
FILEDIA	Type: Integer Saved in: Registry Initial value: 1	Suppresses display of the file dialog boxes. 0 Dialog boxes are not displayed. You can still request a file dialog box to appear by entering a tilde (~) in response to the command's prompt. The same is true for AutoLISP and ADS functions. 1 File dialog boxes are displayed. However, if a script or AutoLISP/ObjectARX program is active, AutoCAD displays an ordinary prompt.
FILLETRAD	Type: Real Saved in: Drawing Initial value: 0.0000	Stores the current fillet radius.
FILLMODE	Type: Integer Saved in: Drawing Initial value: 1	Specifies whether multilines, traces, solids, solid-fill hatches, and wide polylines are filled in. 0 Objects are not filled in 1 Objects are filled

Variable	Characteristic	Description
FONTALT	Type: String Saved in: Registry Initial value: "simple.shx"	Specifies the alternate font to be used when the specified font file cannot be located. If an alternate font is not specified, AutoCAD displays the Alternate Font dialog box for the following cases: 1. A Release 13 drawing is opened; *FONTALT* is not set or not found; and a TrueType, SHX, or PostScript font is not found for a defined text style. 2. A Release 14 drawing is opened, *FONTALT* is not set or not found, and a SHX or PostScript font is not found for a defined text style. For missing TrueType fonts in Release 14 drawings, AutoCAD automatically substitutes the closest TrueType font available. 3. The *Browse* button is pressed in the *Options* dialog box when you specify an alternate font. AutoCAD validates the alternate font specified for FONTALT. If the font name or font file name is not found, the message "Font not found" is displayed. Enter either a TrueType font name (for example, Times New Roman Bold) or a TrueType file name (for example *timebd.ttf*). When a TrueType file name is entered for *FONTALT*, AutoCAD returns the font name in place of the file name) if the font is registered with the operating system.
FONTMAP	Type: String Saved in: Registry Initial value: "acad.fmp"	Specifies the font mapping file to be used. A font mapping file contains one font mapping per line; the original font used in the drawing and the font to be substituted for it are separated by a semicolon (;). For example, to substitute the Times TrueType font for the Romans font, the line in the mapping file would read as follows: **romanc.shx;times.ttf** If *FONTMAP* does not point to a font mapping file, if the FMP file is not found, or if the font file name specified in the FMP file is not found, AutoCAD uses the font defined in the style. If the font in the style is not found, AutoCAD substitutes the font according to substitution rules.
FRONTZ	(Read-only) Type: Real Saved in: Drawing	Stores the front clipping plane offset from the target plane for the current viewport, in drawing units. Meaningful only if the front clipping bit in *VIEWMODE* is on and the front-clip-not-at-eye bit is also on. The distance of the front clipping plane from the camera point is found by subtracting *FRONTZ* from the camera-to-target distance.
FULLOPEN	(Read-only) Type: Integer Not saved	Indicates whether the current drawing is partially open. 0 Indicates a partially open drawing 1 Indicates a fully open drawing

Variable	Characteristic	Description
GFANG	Type: Integer Not saved Initial value: 0	Specifies the angle of a gradient fill. Valid values are 0 through 360 degrees.
GFCLR1	Type: String Not saved Initial value: "RGB 000, 000, 255"	Specifies the color for a one-color gradient fill or the first color for a two-color gradient fill. Valid values are "RGB 000, 000, 000" through "RGB 255, 255, 255."
GFCLR2	Type: String Not saved Initial value: "RGB 255, 255, 153"	Specifies the second color for a two-color gradient fill. Valid values are "RGB 000, 000, 000" through "RGB 255, 255, 255."
CFCLRLUM	Type: Real Not saved Initial value: 1.0000	Makes the color a tint (mixed with white) or a shade (mixed with black) in a one-color gradient fill. Valid values are 0.0 (darkest) to 1.0 (lightest).
GFCLRSTATE	Type: Integer Not saved Initial value: 1	Specifies whether a gradient fill uses one color or two colors. 0 Two-color gradient fill 1 One-color gradient fill
GFNAME	Type: Integer Not saved Initial value: 1	Specifies the pattern of a gradient fill. Valid values are 1 through 9. 1 Linear 2 Cylindrical 3 Inverted cylindrical 4 Spherical 5 Inverted spherical 6 Hemispherical 7 Inverted hemispherical 8 Curved 9 Inverted curved
GFSHIFT	Type: Integer Not saved Initial value: 0	Specifies whether the pattern in a gradient fill is centered or is shifted up and to the left. 0 Centered 1 Shifted up and to the left
GRIDMODE	Type: Integer Saved in: Drawing Initial value: 0	Specifies whether the grid is turned on or off. 0 Turns the grid off 1 Turns the grid on
GRIDUNIT	Type: 2D point Saved in: Drawing Initial value: 0.5000, 0.5000	Specifies the grid spacing (*X* and *Y*) for the current viewport.
GRIPBLOCK	Type: Integer Saved in: Registry Initial value: 0	Controls the assignment of grips in blocks. 0 Assigns a grip only to the insertion point of the block 1 Assigns grips to objects within the block
GRIPCOLOR	Type: Integer Saved in: Registry Initial value: 5	Controls the color of nonselected grips (drawn as box outlines). The valid range is 1-255.

2004

Variable	Characteristic	Description
GRIPHOT	Type: Integer Saved in: Registry Initial value: 1	Controls the color of selected grips (drawn as filled boxes). The valid range is 1-255.
GRIPHOVER	Type: Integer Saved in: Registry Initial value: 3	Controls the fill color of a grip when the cursor pauses over the grip. The valid range is 1 to 255.
GRIPOBJLIMIT	Type: Integer Saved in: Registry Initial value: 100	Suppresses the display of grips when the initial selection set includes more than the specified number of objects. The valid range is 1 to 32,767. When set to 1, grips are suppressed when more than one object is selected.
GRIPS	Type: Integer Saved in: Registry Initial value: 1	Controls use of selection set grips for the Stretch, Move, Rotate, Scale, and Mirror grip modes. 0 Turns off grips 1 Turns on grips
GRIPSIZE	Type: Integer Saved in: Registry Initial value: 3	Sets the size of the box drawn to display the grip, in pixels. The valid range is 1-255.
GRIPTIPS	Type: Integer Saved in: Registry Initial value: 1	Controls the display of grip tips when the cursor hovers over grips on custom objects that support grip tips. 0 Turns off the display of grip tips 1 Turns on the display of grip tips
HALOGAP	Type: Integer Saved in: Drawing Initial value: 0	Specifies the distance to shorten a haloed line. The distance is specified as a percentage of one inch and is independent of Zoom level. A haloed line is shortened at the point where it will be hidden and is visible only when the Hide or Hidden Shademode is used.
HANDLES	(Read-only) Type: Integer Saved in: Drawing Initial value: On	Reports whether object handles can be accessed by applications. Because handles can no longer be turned off, has no effect except to preserve the integrity of scripts.
HIDEPRECISION	Type: Integer Not saved Initial Value: 0	Controls the accuracy of hides and shades. Hides can be calculated in double precision or single precision. Setting HIDEPRECISION to 1 produces more accurate hides because it uses double precision, but it also uses more memory and can affect performance, especially when hiding solids. 0 Single precision; uses less memory 1 Double precision; uses more memory
HIDETEXT	Type: Switch Saved in: Drawing Initial Value: 0	Specifies whether text objects created by the Text, Dtext, or Mtext commands are processed during a Hide. 0 Disables Hide processing of text objects. Text is not hidden and does not hide other objects unless the text has a thickness assigned to it.

2004

2004

2002

2002

Variable	Characteristic	Description
HIDETEXT (*continued*)		1 Enables Hide processing of text objects. Text is hidden and text hides other objects. When dealing with text objects, legacy behavior is achieved by setting *HIDETEXT* to 0.
HIGHLIGHT	Type: Integer Not saved Initial value: 1	Controls object highlighting; does not affect objects selected with grips. 0 Turns off object selection highlighting 1 Turns on object selection highlighting
HPANG	Type: Real Not saved Initial value: 0	Specifies the hatch pattern angle.
HPASSOC	Type: Integer Saved in: Registry Initial value: 1	Controls whether hatch patterns and gradient fills are associative. 0 Hatch patterns and gradient fills are not associated with their boundaries 1 Hatch patterns and gradient fills are associated with their boundaries and are updated when the boundaries change
HPBOUND	Type: Integer Saved in: Registry Initial value: 1	Controls the object type created by *Bhatch* and *Boundary.* 0 Creates a region 1 Creates a polyline
HPDOUBLE	Type: Integer Not saved Initial value: 0	Specifies hatch pattern doubling for user-defined patterns. 0 Turns off hatch pattern doubling 1 Turns on hatch pattern doubling
HPNAME	Type: String Not saved Initial value: "ANSI31"	Sets a default hatch pattern name of up to 34 characters, no spaces allowed. Returns "" if there is no default. Enter a period (.) to set no default.
HPSCALE	Type: Real Not saved Initial value: 1.0000	Specifies the hatch pattern scale factor; must be nonzero.
HPSPACE	Type: Real Not saved Initial value: 1.0000	Specifies the hatch pattern line spacing for user-defined simple patterns; must be nonzero.
HYPERLINKBASE	Type: String Saved in: Drawing Initial value: " "	Specifies the path used for all relative hyperlinks in the drawing. If no value is specifed, the drawing path is used for all relative hyperlinks.
IMAGEHLT	Type: Integer Saved in: Registry Initial value: 0	Controls whether the entire raster image or only the raster image frame is hightlighted. 0 Highlights only the raster image frame 1 Highlights the entire raster image

Variable	Characteristic	Description
INDEXCTL	Type: Integer Saved in: Drawing Initial value: 0	Controls whether layer and spatial indexes are created and saved in drawing files. 0 No indexes are created 1 Layer index is created 2 Spatial index is created 3 Layer and spatial indexes are created
INETLOCATION	Type: Real Saved in: Registry Initial value: "http://www.autodesk.com"	Stores the Internet location used by *Browser*.
INSBASE	Type: 3D point Saved in: Drawing Initial value: 0.0000, 0.0000,0.0000	Stores insertion basepoint set by *Base*, expressed as a UCS coordinate for the current space.
INSNAME	Type: String Not saved Initial value: ""	Sets a default block name for *Insert* or *-Insert*. The name must conform to symbol naming conventions. Returns "" if no default is set. Enter a period (.) to set no default.
INSUNITS	Type: Integer Saved in: Drawing Initial value: 0	When you drag a block or image from Design-Center, specifies a drawing units value as follows: 0 Unspecified - Unitless 1 Inches 2 Feet 3 Miles 4 Millimeters 5 Centimeters 6 Meters 7 Kilometers 8 Microinches 9 Miles 10 Yards 11 Angstroms 12 Nanometers 13 Microns 14 Decimeters 15 Decameters 16 Hectometers 17 Gigameters 18 Astronomical Units 19 Light Years 20 Parsecs
INSUNITSDEFSOURCE	Type: Integer Saved in: Registry Initial value: 1	Sets source content units value. Valid range is 0 to 20. (See *INSUNITS*.)
INSUNITSDEFTARGET	Type: Integer Saved in: Registry Initial value: 1	Sets target drawing units value. Valid range is 0 to 20. (See *INSUNITS*.)

Variable	Characteristic	Description
INTERSECTIONCOLOR	Type: Integer Saved in: Drawing Initial value: 257	Specifies the color of intersection polylines. An intersection polyline is what displays as the face-to-face intersection of 3D solids when the *Hide* command is used or the *Shademode* command is set to *Hidden*. Value 0 designates entity color *Bbyblock*, value 256 designates entity color *Bylayer*, and value 257 designates entity color *Byentity*. Values 1-255 designate an AutoCAD color index (ACI). The *INTERSECTIONCOLOR* setting is visible only if the *INTERSECTIONDISPLAY* is turned on by setting it to a value of 1.
INTERSECTIONDISPLAY	Type: Switch Saved in: Drawing Initial value: Off	Specifies the display of intersection polylines. An intersection polyline is what displays as the face-to-face intersection of 3D solids when the *Hide* command is used or the *Shademode* command is set to *Hidden*. 0 or Off turns off the display of intersection polylines 1 or On turns on the display of intersection polylines
ISAVEBAK	Type: Integer Saved in: Registry Initial value: 1	Improves the speed of incremental saves, especially for large drawings. *ISAVEBAK* controls the creation of a backup file (BAK). In Windows, copying the file data to create a BAK file for large drawings takes a major portion of the incremental save time. 0 No BAK file is created (even for a full save) 1 A BAK file is created **WARNING:** In some cases (such as a power failure in the middle of a save), it's possible that drawing data can be lost.
ISAVEPERCENT	Type: Integer Saved in: Registry Initial value: 50	Determines the amount of wasted space tolerated in a drawing file. The value of *ISAVEPERCENT* is an integer between 0 and 100. The default value of 50 means that the estimate of wasted space within the file does not exceed 50% of the total file size. Wasted space is eliminated by periodic full saves. When the estimate exceeds 50%, the next save will be a full save. This resets the wasted space estimate to 0. If *ISAVEPERCENT* is set to 0, every save is a full save.
ISOLINES	Type: Integer Saved in: Drawing Initial value: 4	Specifies the number of isolines per surface on objects. Valid integer values are from 0 to 2047.
LASTANGLE	(Read-only) Type: Real Not saved	Stores the end angle of the last arc entered relative to the *XY* plane of the current UCS for the current space.

2004

Variable	Characteristic	Description
LASTPOINT	Type: 3Dpoint Saved in: Drawing Initial value: 0.0000, 0.0000,0.0000	Stores the last point entered, expressed as a UCS coordinate for the current space; referenced by the at symbol (@) during keyboard entry.
LASTPROMPT	(Read-only) Type: String Not saved Initial value: ""	Stores the last string echoed to the command line. This string is identical to the last line seen at the command line and includes any user input.
LAYOUTREGENCTL	Type: Integer Saved in: Registry Initial value: 2	Specifies how the display list is updated in the *Model* tab and layout tabs. For each tab, the display list is updated either by regenerating the drawing when you switch to that tab or by saving the display list to memory and regenerating only the modified objects when you switch to that tab. Changing the *LAYOUTREGENCTL* setting can improve performance. 0 The drawing is regenerated each time you switch tabs. 1 For the *Model* tab and the last layout made current, the display list is saved to memory and regenerations are suppressed when you switch between the two tabs. For all other layouts, regenerations still occur when you switch to those tabs. 2 The drawing is regenerated the first time you switch to each tab. For the remainder of the drawing session, the display list is saved to memory and regenerations are suppressed when you switch to those tabs. The performance gain achieved by changing the *LAYOUTREGENCTL* setting is dependent on several factors, including the drawing size and type, the objects contained in the drawing, the amount of available memory, and the effect of other open drawings or applications. When *LAYOUTRE GENCTL* is set to 1 or 2, the amount of additional memory used is the size of the *Model* tab's display list multiplied by the number of viewports in each layout for which the display list is saved. If *LAYOUTREGENCTL* is set to 1 or 2 and performance seems slow in general or when you switch between tabs for which the display list is saved, consider changing to a setting of 0 or 1 to find the optimal balance for your work environment. NOTE: Regardless of the *LAYOUTREGENCTL* setting, if you redefine a block or undo a tab switch, the drawing is regenerated the first time you switch to any tab that contains saved viewports.

Variable	Characteristic	Description
LENSLENGTH	(Read-only) Type: Real Saved in: Drawing Initial value: 50.0000	Stores the length of the lens (in millimeters) used in perspective viewing for the current viewport.
LIMCHECK	Type: Integer Saved in: Drawing Initial value: 0	Controls creation of objects outside the drawing limits. 0 Objects can be created outside the limits 1 Objects cannot be created outside the limits
LIMMAX	Type: 2D point Saved in: Drawing Initial value: 12.0000,9.0000	Stores the upper-right drawing limits for the current space, expressed as World coordinate. *LIMMAX* is read-only when paper space is active and the paper background or paper margins are displayed.
LIMMIN	Type: 2D point Saved in: Drawing Initial value: 0.0000, 0.0000	Stores the lower-left drawing limits for the current space, expressed as a World coordinate. *LIMMIN* is read-only when paper space is active and the paper background or paper margins are displayed.
LISPINIT	Type: Integer Saved in: Registry Initial value: 1	Specifies whether AutoLISP-defined functions and variables are preserved when you open a new drawing or whether they are valid in the current drawing session only. 0 AutoLISP functions and variables are preserved from drawing to drawing 1 AutoLISP functions and variables are valid in current drawing only
LOCALE	(Read-only) Type: String Not saved Initial value: "enu"	Displays the ISO language code of the current AutoCAD version you are running.
LOCALROOTPREFIX	(Read-only) Type: String Saved in: Registry Initial value: "pathname"	Stores the full path to the root folder where local customizable files were installed. These files are stored in the product folder under the Local Settings folder; for example, "C:\Documents and Settings\username\ Local Settings\Application Data\productname\version\language". The Template and Textures folders are in this location, and you can add any customizable files that you do not want to roam on the network. See *ROAMABLEROOTPREFIX* for the location of the roamable files.
LOGFILEMODE	Type: Integer Saved in: Registry Initial value: 0	Specifies whether the contents of the text window are written to a log file. 0 Log file is not maintained 1 Log file is maintained

2004

Variable	Characteristic	Description
LOGFILENAME	(Read-only) Type: String Saved in: Drawing:	Specifies the path and name of the log file for the current drawing. The initial value varies depending on the name of the current drawing and where you installed AutoCAD.
LOGFILEPATH	Type: String Saved in: Registry	Specifies the path for the log files for all drawings in a session. You can also specify the path by using the *Options* command. The initial value varies depending on where you installed AutoCAD.
LOGINNAME	(Read-only) Type: String Not saved	Displays the user's name as configured or as input when AutoCAD is loaded. The maximum length for a login name is 30 characters.
LTSCALE	Type: Real Saved in: Drawing Initial value: 1.0000	Sets the global linetype scale factor (cannot equal zero).
LUNITS	Type: Integer Saved in: Drawing Initial value: 2	Sets linear units. 1 Scientific 2 Decimal 3 Engineering 4 Architectural 5 Fractional
LUPREC	Type: Integer Saved in: Drawing Initial value: 4	Sets the number of decimal places displayed for all read-only linear units and for all editable linear units whose precision is less than or equal to the current *LUPREC* value. For editable linear units whose precision is greater than the current *LUPREC* value, the true precision is displayed. *LUPREC* does not affect the display precision of dimension text.
LWDEFAULT	Type: Enum Saved in: Registry Initial value: 25	Sets the value for the default lineweight. The default lineweight can be set to any valid lineweight value in millimeters, including: 0, 5, 9, 13, 15, 18, 20, 25, 30, 35, 40, 50, 53, 60, 70, 80, 90, 100, 106, 120, 140, 158, 200, and 211. All values must be entered in millimeters. (Multiply a value by 2.54 to convert values from inches to millimeters.)
LWDISPLAY	Type: Integer Saved in: Drawing Initial value: 1	Controls whether the lineweight is displayed on the *Model* or *Layout* tab. The setting is saved with each tab in the drawing. 0 Lineweight is not displayed 1 Lineweight is displayed
LWUNITS	Type: Integer Saved In: Registry Initial value: 1	Controls whether lineweight units are displayed in inches or millimeters. 0 Inches 1 Millimeters
MAXACTVP	Type: Integer Saved in: Drawing Initial value: 64	Sets the maximum number of viewports that can be active at one time.

Variable	Characteristic	Description
MAXSORT	Type: Integer Saved in: Registry Initial value: 1000	Sets the maximum number of symbol names or block names sorted by listing commands. If the total number of items exceeds this value, no items are sorted.
MBUTTONPAN	Type: Integer Saved in: Registry Initial Value: 1	Controls the behavior of the third button or wheel on pointing device. 0 The third button or wheel action on a pointing device supports the action defined in the AutoCAD menu (.mnu) file. 1 The third button or wheel action on a pointing device supports panning by holding and dragging the button or wheel.
MEASUREINIT	Type: Integer Saved in: Registry Initial value: *varies by country*	Sets the initial drawing units as English or metric. Specifically, *MEASUREINIT* controls which hatch pattern and linetype files an existing drawing uses when it is opened. Also controls which template is used. 0 English; uses the hatch pattern file and linetype file designated by the ANSIHatch and ANSILinetype registry settings 1 Metric; uses the hatch pattern file and linetype file designated by the ISOHatch and ISOLinetype registry settings
MEASUREMENT	Type: Integer Saved in: Drawing Initial value: 0	Sets drawing units as English or metric. Specifically, *MEASUREMENT* controls which hatch pattern and linetype file an existing drawing uses when it is opened. 0 English; AutoCAD uses the hatch pattern file and linetype file designated by the ANSIHatch and ANSILinetype registry settings 1 Metric; AutoCAD uses the hatch pattern file and linetype file designated by the ISOHatch and ISOLinetype registry settings The drawing units for new drawings are controlled by the *MEASUREINIT* registry variable (*MEASUREINIT* uses the same values as *MEASUREMENT*). The *MEASUREMENT* setting of a drawing always overrides the *MEASUREINIT* registry setting.
MENUCTL	Type: Integer Saved in: Registry Initial value: 1	Controls the page switching of the screen menu. 0 Screen menu does not switch pages in response to keyboard command entry 1 Screen menu does switch pages in response to keyboard command entry
MENUECHO	Type: Integer Not saved Initial value: 0	Sets menu echo and prompt control bits. It is the sum of the following: 1 Suppresses echo of menu items (^P in a menu item toggles echoing)

Variable	Characteristic	Description
MENUECHO (continued)		2 Suppresses display of system prompts during menu 4 Disables ^P toggle of menu echoing 8 Displays input/output strings; debugging aid for DIESEL macros
MENUNAME	(Read-only) Type: String Saved in: Registry	Stores the menu file name, including the path for the file name.
MIRRTEXT	Type: Integer Saved in: Drawing Initial value: 0	Controls how *Mirror* reflects text. 0 Retains text direction 1 Mirrors the text
MODEMACRO	Type: String Not saved Initial value: ""	Displays a text string on the Status line, such as the name of the current drawing, time/date stamp, or special modes. Use *MODEMACRO* to display a string of text, or use special text strings written in the DIESEL macro language to have AutoCAD evaluate the macro from time to time and base the Status line on user-selected conditions.
MTEXTED	Type: String Saved in: Registry Initial value: "Internal"	Sets the name of the application to use for editing Multiline text objects. You can specify a different text editor for the *Mtext* and *Ddedit* commands. If you set *MTEXTED* to internal or to null, AutoCAD displays the internal *Multiline Text Editor*. You set *MTEXTED* to null by entering a period (.). If you specify a path and the name of the executable file for another text editor or word processor, AutoCAD displays that path and file name instead. If the Multiline text object is fewer than 80 characters, you can specify :lisped to use the LISP editor. Text editors other than the internal one show the formatting codes in paragraph text.
MTEXTFIXED	Type: Integer Saved in: Registry Initial value: 0	Controls the appearance of the Multiline Text Editor. 0 Displays both the Multiline Text Editor and the text within it at the size and position of the *Mtext* object in the drawing. Text too large or too small to be edited is displayed at a minimum or maximum size. 1 Displays the Multiline Text Editor at a fixed position and size based on last use, and displays text in the editor at a fixed height.
MTJIGSTRING	Type: String Saved in: Registry Initial value: "abc"	Sets the content of the sample text displayed at the cursor location when the *Mtext* command is started. The text string is displayed in the current text size and font. You can enter any string of up to ten letters or numbers or enter "." (period) to display no sample text.

Variable	Characteristic	Description
MYDOCUMENTSPREFIX (Read-only)	Type: String Saved in: Registry Initial value: "pathname"	Stores the full path of the My Documents folder for the user currently logged on. These files are stored in the product folder under the Local Settings folder; for example, "C:\Documents and Settings\username\My Documents".
NOMUTT	Type: Short Not Saved Initial Value: 0	Suppresses message display (muttering) when it would not normally be suppressed. Displaying messages is the normal mode of AutoCAD, but message display is suppressed during scripts, AutoLISP routines, and so on. 0 Resumes normal muttering behavior 1 Suppresses muttering indefinitely
OBSCUREDCOLOR	Type: Integer Saved in: Drawing Initial value: 257	Specifies the color of obscured lines. Value 0 designates *Byblock*, value 256 designates *Bylayer*, and value 257 designates *Byentity*. Values 1-255 designate an AutoCAD color index (ACI). An obscured line is a hidden line made visible by changing its color and linetype and is visible only when the *Hide* or *Shademode* commands are used. The *OBSCUREDCOLOR* setting is visible only if the *OBSCUREDLTYPE* is turned *ON* by setting it to a value other than 0.
OBSCURED LTYPE	Type: Integer Saved in: Drawing Initial value: 0	Specifies the linetype of obscured lines. An obscured line is a hidden line made visible by changing its color and linetype and is visible only when the *Hide* or *Shademode* commands are used. Obscured line-types are independent of zoom level, unlike regular AutoCAD linetypes. Value 0 will turn off display of obscured lines and is the default. The linetype values are defined as follows: 0 Off 1 Solid 2 Dashed 3 Dotted 4 Short Dash 5 Medium Dash 6 Long Dash 7 Double Short Dash 8 Double Medium Dash 9 Double Long Dash 10 Medium Long Dash 11 Sparse Dot
OFFSETDIST	Type: Real Not saved Initial value: 1.0000	Sets the default offset distance. <0 Offsets an object through a specified point >0 Sets the default offset distance
OFFSETGAPTYPE	Type: Integer Saved in: Registry Initial value: 0	Controls how to offset polylines when a gap is created as a result of offsetting the individual polyline segments. 0 Extends the segments to fill the gap

2004

2002

Variable	Characteristic	Description
OFFSETGAPTYPE (continued)		1 Fills the gaps with a filleted arc segment (the radius of the arc segment is equal to the offset distance) 2 Fills the gaps with a chamfered line segment
OLEHIDE	Type: Integer Saved in: Registry Initial value: 0	Controls the display of OLE objects in AutoCAD. 0 All OLE objects are visible 1 OLE objects are visible in paper space only 2 OLE objects are visible in model space only 3 No OLE objects are visible OLEHIDE affects both screen display and printing.
OLEQUALITY	Type: Integer Saved in: Registry Initial value: 1	Controls the default quality level for embedded OLE objects. 0 Line art quality, such as an embedded spreadsheet 1 Text quality, such as an embedded Word document 2 Graphics quality, such as an embedded pie chart 3 Photograph quality 4 High quality photograph
OLESTARTUP	Type: Integer Saved in: Drawing Initial value: 0	Controls whether the source application of an embedded OLE object loads when plotting. Loading the OLE source application may improve the plot quality. 0 Does not load the OLE source application 1 Loads the OLE source application when plotting
ORTHOMODE	Type: Integer Saved in: Drawing Initial value: 0	Constrains cursor movement to the perpendicular. When ORTHOMODE is turned on, the cursor can move only horizontally or vertically relative to the UCS and the current grid rotation angle. 0 Turns off Ortho mode 1 Turns on Ortho mode
OSMODE	Type: Integer Saved in: Drawing Initial value: 0	Sets running object snap modes using the following bit-codes. 0 None 1 Endpoint 2 Midpoint 4 Center 8 Node 16 Quadrant 32 Intersection 64 Insertion 128 Perpendicular 256 Tangent 512 Nearest 1024 Quick 2048 Apparent Intersection 4096 Extension

Variable	Characteristic	Description
OSMODE (continued)		8192 *Parallel* To specify more than one object snap, enter the sum of their values. For example, entering **3** specifies the *Endpoint* (bit code 1) and *Midpoint* (bit code 2) object snaps. Entering **16383** specifies all object snaps.
OSNAPCOORD	Type: Integer Saved in: Registry Initial value: 2	Controls whether coordinates entered on the command line override running object snaps. 0　Running object snap settings override keyboard coordinate entry 1　Keyboard entry overrides object snap settings 2　Keyboard entry overrides object snap settings except in scripts
PALETTEOPAQUE	Type: Integer Saved in: Registry Initial Value: 0	Controls whether windows can be made transparent. When transparency is unavailable or turned off, all palettes are opaque. Transparency is unavailable when palettes or windows are docked, when transparency is not supported by the current operating system, and when hardware accelerators are in use. When transparency is available and turned on, you can use the *Transparency* option on the shortcut menu to set a different degree of transparency in individual palettes. 0　Transparency turned on by user 1　Transparency turned off by user 2　Transparency unavailable though turned on by user 3　Transparency unavailable and turned off by user
PAPERUPDATE	Type: Integer Saved in: Registry Initial Value: 0	Controls the display of a warning dialog when attempting to print a layout with a paper size different from the paper size specified by the default for the plotter configuration file. 0　Displays a warning dialog box if the paper size specified in the layout is not supported by the plotter 1　Sets paper size to the configured paper size of the plotter configuration file
PDMODE	Type: Integer Saved in: Drawing Initial value: 0	Controls how point objects are displayed. For information about values to enter, see *Point*.
PDSIZE	Type: Real Saved in: Drawing Initial value: 0.0000	Sets the display size for point objects. 0　Creates a point at 5% of the graphics area height >0 Specifies an absolute size <0 Specifies a percentage of the viewport size

2004

Variable	Characteristic	Description
PEDITACCEPT	Type: Integer Saved in: Registry Initial value: 0	Suppresses display of the "Object selected is not a polyline. Do you want to turn it into one <Y>?" prompt in *Pedit*. Entering with *Y* converts the selected object to a *Polyline*. When the prompt is suppressed, the selected object is automatically converted to a *Polyline*. 0 The prompt is displayed 1 The prompt is suppressed
PELLIPSE	Type: Integer Saved in: Drawing Initial value: 0	Controls the ellipse type created with *Ellipse*. 0 Creates a true ellipse object 1 Creates a polyline representation of an ellipse
PERIMETER	(Read-only) Type: Real Not saved	Stores the last perimeter value computed by *Area*, *List*, or *Dblist*.
PFACEVMAX	(Read-only) Type: Integer Not saved	Sets the maximum number of vertices per face.
PICKADD	Type: Integer Saved in: Registry Initial value: 1	Controls additive selection of objects. 0 Turns off *PICKADD*. The objects most recently selected, either individually or by windowing, become the selection set. Previously selected objects are removed from the selection set. Add more objects to the selection set by holding down Shift while selecting. 1 Turns on *PICKADD*. Each object selected, either individually or by windowing, is added to the current selection set. To remove objects from the set, hold down Shift while selecting.
PICKAUTO	Type: Integer Saved in: Registry Initial value: 1	Controls automatic windowing at the Select Objects prompt. 0 Turns off *PICKAUTO* 1 Draws a selection window (for either Window or Crossing selection) automatically at the Select Objects prompt
PICKBOX	Type: Integer Saved in: Registry Initial value: 3	Sets object selection target height, in pixels.
PICKDRAG	Type: Integer Saved in: Registry Initial value: 0	Controls the method of drawing a selection window. 0 Draws the selection window using two points. Click the pointing device at one corner and then at the other corner. 1 Draws the selection window using dragging. Click at one corner, hold down the pick button on the pointing device, drag, and release the button at the other corner.

Variable	Characteristic	Description
PICKFIRST	Type: Integer Saved in: Registry Initial value: 1	Controls whether you select objects before (noun-verb selection) or after you issue a command. 0 Turns off *PICKFIRST* 1 Turns on *PICKFIRST*
PICKSTYLE	Type: Integer Saved in: Drawing Initial value: 1	Controls use of group selection and associative hatch selection. 0 No group selection or associative hatch selection 1 Group selection 2 Associative hatch selection 3 Group selection and associative hatch selection
PLATFORM	(Read-only) Type: String Not saved	Indicates which platform (operating system) of AutoCAD is in use.
PLINEGEN	Type: Integer Saved in: Drawing Initial value: 0	Sets how linetype patterns are generated around the vertices of a two-dimensional *Polyline*. Does not apply to *Polylines* with tapered segments. 0 *Polylines* are generated to start and end with a dash at each vertex 1 Generates the linetype in a continuous pattern around the vertices of the *Polyline*
PLINETYPE	Type: Integer Saved in: Registry Initial value: 2	Specifies whether AutoCAD uses optimized 2D polylines. *PLINETYPE* controls both the creation of new polylines with the *Pline* command and the conversion of existing polylines in drawings from previous releases. 0 Polylines in older drawings are not converted on open; *Pline* creates old-format polylines 1 Polylines in older drawings are not converted on open; *Pline* creates optimized polylines 2 Polylines in older drawings are converted on open; *Pline* creates optimized polylines *PLINETYPE* also controls the polyline type created with the following commands: *Boundary* (when object type is set to *Polyline*), *Donut, Ellipse* (when *PELLIPSE* is set to 1), *Pedit* (when selecting a line or arc), *Polygon*, and *Sketch* (when *SKPOLY* is set to 1).
PLINEWID	Type: Real Saved in: Drawing Initial value: 0.0000	Stores the default polyline width.
PLOTROTMODE	Type: Integer Saved in: Registry Initial value: 1	Controls the orientation of plots. 0 Rotates the effective plotting area so that the corner with the Rotation icon aligns with the paper at the lower-left for 0, top-left for 90, top-right for 180, and lower-right for 270 1 Aligns the lower-left corner of the effective plotting area with the lower-left corner of the paper

Variable	Characteristic	Description
PLOTROTMODE (continued)		2 Works the same as 0 value except that the *X* and *Y* origin offsets are calculated relative to the rotated origin position.
PLQUIET	Type: Integer Saved in: Registry Initial Value: 0	Controls the display of optional dialog boxes and non-fatal errors for batch plotting and scripts. 0 Displays plot dialogs and non fatal errors. 1 Logs non-fatal errors and does not display plot related dialog boxes.
POLARADDANG	Type: String Saved in: Registry Initial value: null	Contains user-defined polar angles. You can add up to 10 angles. Each angle can be up to 25 characters, separated with semicolons (;). AutoCAD displays angles in the format set in the *AUNITS* system variable.
POLARANG	Type: Real Saved in: Registry Initial value: 90	Sets the polar angle increment. Values are 90, 45, 30, 22.5, 18, 15,10, and 5.
POLARDIST	Type: Real Saved in: Registry Initial value: 0.0000	Sets the snap increment when the *SNAPSTYL* system variable is set to 1 (polar snap).
POLARMODE	Type: Integer Saved in: Registry Initial value: 0	Controls settings for polar and object snap tracking. The value is the sum of four bitcodes: Polar angle measurements 0 Measure polar angles based on current UCS (absolute) 1 Measure polar angles from selected objects (relative) Object snap tracking 0 Track orthogonally only 2 Use polar tracking settings in object snap tracking Use additional polar tracking angles 0 No 4 Yes Acquire object snap tracking points 0 Acquire automatically 8 Press SHIFT to acquire
POLYSIDES	Type: Integer Not saved Initial value: 4	Sets the default number of sides for *Polygon*. The range is 3-1024.
POPUPS	(Read-only) Type: Integer Not saved	Displays the status of the currently configured display driver. 0 Does not support dialog boxes, the menu bar, pull-down menus, and icon menus 1 Supports these features
PRODUCT	(Read-only) Type: String Not saved Initial value: "AutoCAD"	Returns the product name.

Variable	Characteristic	Description
PROGRAM	(Read-only) Type: String Not saved Initial value: "acad"	Returns the program name.
PROJECTNAME	Type: String Saved in: Drawing Initial value: ""	Assigns a project name to the current drawing. The project name points to a section in the registry which can contain one or more search paths for each project name defined. Used when an xref or image is not found in its original hard-coded path. Project names and their search directories are created through the Files tab of the Preferences dialog box. Project names make it easier for users to manage xrefs and images when drawings are exchanged between customers, or if users have different drive mappings to the same location on a server. If the xref or image is not found at the hard-coded path, the project paths associated with the project name are searched. The search order is hard-coded path, then project name search path, then AutoCAD search path.
PROJMODE	Type: Integer Saved in: Registry Initial value: 1	Sets the current Projection mode for trimming or extending. 0 True 3D mode (no projection) 1 Project to the *XY* plane of the current UCS 2 Project to the current view plane
PROXYGRAPHICS	Type: Integer Saved in: Drawing Initial value: 1	Specifies whether images of proxy objects are saved in the drawing. 0 Image is not saved with the drawing; a bounding box is displayed instead 1 Image is saved with the drawing
PROXYNOTICE	Type: Integer Saved in: Registry Initial value: 1	Displays a notice when a proxy is created. A proxy is created when you open a drawing containing custom objects created by an application that is not present. A proxy is also created when you issue a command that unloads a custom object's parent application. 0 No proxy warning is displayed 1 Proxy warning is displayed
PROXYSHOW	Type: Integer Saved in: Registry Initial value: 1	Controls the display of proxy objects in a drawing. 0 Proxy objects are not displayed 1 Graphic images are displayed for all proxy objects 2 Only the bounding box is displayed for all proxy objects

Variable	Characteristic	Description
PROXYWEBSEARCH	Type: Integer Saved in: Registry Initial value: 1	Specifies how AutoCAD checks for Object Enablers. Object Enablers allow you to display and use custom objects in AutoCAD drawings even when the ObjectARX application that created them is unavailable. *PROXYWEBSEARCH* is also controlled with the *Live Enabler options* on the *System* tab of the *Options* dialog box. 0 Prevents AutoCAD from checking for Object Enablers. 1 AutoCAD checks for Object Enablers only if a live Internet connection is present.
PSLTSCALE	Type: Integer Saved in: Drawing Initial value: 1	Controls paper space linetype scaling. 0 No special linetype scaling. Linetype dash lengths are based on the drawing units of the space (model or paper) in which the objects were created, scaled by the global *LTSCALE* factor. 1 Viewport scaling governs linetype scaling. If *TILEMODE* is set to 0, dash lengths are based on paper space drawing units, even for objects in model space. In this mode, viewports can have varying magnifications, yet display linetypes identically. For a specific linetype, the dash lengths of a line in a viewport are the same as the dash lengths of a line in paper space. You can still control the dash lengths with *LTSCALE*.
PSTYLEMODE	Read Only Saved in: Drawing Initial value: 1	Indicates whether the current drawing is in a Color-Dependent or Named Plot Style mode. 0 Uses named plot style tables in the current drawing 1 Uses color-dependent plot style tables in the current drawing
PSTYLEPOLICY	Type: Integer Saved in: Registry Initial value: 1	Controls whether an object's color property is associated with its plot style. The new value you assign affects only newly created drawings and pre-AutoCAD 2000 drawings. 0 No association is made between color and plot style. The plot style for new objects is set to the default defined in *DEFPLSTYLE*. The plot style for new layers is set to the default defined in *DEFLPLSTYLE*. 1 An object's plot style is associated with its color.
PSVPSCALE	Type: Real Not Saved Initial Value: 0	Sets the view scale factor for all newly created viewports. The view scale factor is defined by comparing the ratio of units in paper space to the units in newly created model space viewports. The view scale factor you set is used with the *Vports* command. A value of 0 means the scale factor is Scaled to Fit. A scale must be a positive real value.

2002

Variable	Characteristic	Description
PUCSBASE	Type: String Saved in: Drawing Initial value: ""	Stores the name of the UCS that defines the origin and orientation of orthographic UCS settings in paper space only.
QTEXTMODE	Type: Integer Saved in: Drawing Initial value: 0	Controls how text is displayed. 0 Turns off Quick Text mode; displays characters 1 Turns on Quick Text mode; displays a box in place of text
RASTERPREVIEW	Type: Integer Saved in: Registry Initial value: 1	Controls whether BMP preview images are saved with the drawing. 0 No preview image is created 1 Preview image created
REFEDITNAME	(Read-only) Type: String Not Saved Initial value: ""	Indicates whether a drawing is in a reference-editing state; also, stores the reference file name.
REGENMODE	Type: Integer Saved in: Drawing Initial value: 1	Controls automatic regeneration of the drawing. 0 Turns off *REGENAUTO* 1 Turns on *REGENAUTO*
RE-INIT	Type: Integer Not saved Initial value: 0	Reinitializes the digitizer, digitizer port, and *acad.pgp* file using the following bit-codes: 1 Digitizer I/O port reinitialization 4 Digitizer reinitialization 16 PGP file reinitialization (reload) To specify more than one reinitialization, enter the sum of their values, for example, **5** to specify both digitizer port (1) and digitizer reinitialization (4).
REMEMBERFOLDERS	Type: Integer Saved in: Registry Initial value: 1	Controls the default path for the *Look In* or *Save In* option in standard file selection dialog boxes. 0 This setting restores the legacy behavior of AutoCAD 2000 and previous releases. When you start AutoCAD by double-clicking an AutoCAD icon, if a *Start In* path is specified for the icon, that path is used as the default for all standard file selection dialog boxes. 1 The last used paths in each particular standard file selection dialog box are remembered across and within sessions. The *Start In* folder specified for the AutoCAD icon is not used.
REPORTERROR	Type: Integer Saved in: Registry Initial value: 1	Controls whether an error report can be sent to Autodesk if AutoCAD closes unexpectedly. Error reports help Autodesk diagnose problems with the software. 0 The Error Report message is not displayed, and no report can be sent to Autodesk. 1 The Error Report message is displayed, and an error report can be sent to Autodesk.

2002

2004

Variable	Characteristic	Description
ROAMABLEROOT-PREFIX	(Read-only) Type: String Saved in: Registry Initial value: "pathname"	Stores the full path to the root folder where roamable customizable files were installed. If you are working on a network that supports roaming, when you customize files that are in your roaming profile they are available to you regardless of which machine you are currently using. These files are stored in the product folder under the Application Data folder; for example, "C:\Documents and Settings\username\Application Data\product-name\version\language".
RTDISPLAY	Type: Integer Saved in: Registry Initial value: 1	Controls the display of raster images during real-time zoom or pan. 0 Displays raster image content 1 Displays raster image outline only *RTDISPLAY* is saved in the current profile.
SAVEFILE	(Read-only) Type: String Saved in: Registry Initial value: ""	Stores current auto-save file name.
SAVEFILEPATH	Type: String Saved in: Registry Initial value: "C:\TEMP\"	Specifies the path to the directory for all automatic save files for the AutoCAD session. You can also change the path on the *Files* tab in the *Options* dialog box.
SAVENAME	(Read-only) Type: String Not saved Initial value: ""	Stores the file name and directory path of the current drawing once you save it.
SAVETIME	Type: Integer Saved in: Registry Initial value: 10	Sets the automatic save interval, in minutes. 0 Turns off automatic saving >0 Automatically saves the drawing at intervals specified by the nonzero integer The *SAVETIME* timer starts as soon as you make a change to a drawing. It is reset and restarted by a manual *Save, Saveas,* or *Qsave.*
SCREENBOXES	(Read-only) Type: Integer Not saved	Stores the number of boxes in the screen menu area of the graphics area. If the screen menu is turned off, *SCREENBOXES* is zero. On platforms that permit the AutoCAD graphics area to be resized or the screen menu to be reconfigured during an editing session, the value of this variable might change during the editing session.
SCREENMODE	(Read-only) Type: Integer Not saved	Stores a bit-code indicating the graphics/text state of the AutoCAD display. It is the sum of the following bit values: 0 Text screen is displayed 1 Graphics area is displayed 2 Dual-screen display is configured

2004

Variable	Characteristic	Description
SCREENSIZE	(Read-only) Type: 2D point Not saved	Stores current viewport size in pixels (*X* and *Y*).
SDI	Type: Integer Saved in: Registry Initial value: 0	Controls whether AutoCAD runs in single- or multiple-document interface. Helps third-party developers update applications to work smoothly with the AutoCAD multiple-drawing mode. 0 Turns on multiple-drawing interface. 1 Turns off multiple-drawing interface. 2 (Read-only) Multiple-drawing interface is disabled because AutoCAD has loaded an application that does not support multiple drawings. SDI setting 2 is not saved. 3 (Read-only) Multiple-drawing interface is disabled because the user has set SDI to 1 and AutoCAD has loaded an application that does not support multiple drawings. (SDI was set to 1 before the application was loaded.) SDI setting 3 is not saved. If SDI is set to 3, AutoCAD switches it back to 1 when the application that doesn't support multiple drawings is unloaded.
SHADEDGE	Type: Integer Saved in: Drawing Initial value: 3	Controls shading of edges in rendering. 0 Faces shaded, edges not highlighted 1 Faces shaded, edges drawn in background color 2 Faces not filled, edges in object color 3 Faces in object color, edges in background color
SHADEDIF	Type: Integer Saved in: Drawing Initial value: 70	Sets the ratio of diffuse reflective light to ambient light (in percentage of diffuse reflective light).
SHORTCUTMENU	Type: Integer Saved in: Registry Initial value: 11	Controls whether Default, Edit, and Command mode shortcut menus are available in the drawing area. *SHORTCUTMENU* uses the following bitcodes. 0 Disables all Default, Edit, and Command mode shortcut menus, restoring R14 legacy behavior. 1 Enables Default mode shortcut menus. 2 Enables Edit mode shortcut menus. 4 Enables Command mode shortcut menus. In this case, the Command mode shortcut menu is available whenever a command is active. 8 Enables Command mode shortcut menus only when command options are currently available from the command line. To enable more than one type of shortcut menu at once, enter the sum of their values. For example, entering 3 enables both Default (1) and Edit (2) mode shortcut menus.

Variable	Characteristic	Description
SHPNAME	Type: String Not saved Initial value: ""	Sets a default shape name; must conform to symbol naming conventions. If no default is set, it returns ""; enter a period (.) to set no default.
SIGWARN	Type: Integer Saved in: Registry Initial value: 1	Controls whether a warning is presented when a file with an attached digital signature is opened. If *SIGWARN* is on and you open a file with a valid signature, the digital signature status is displayed. If *SIGWARN* is off and you open a file, the digital signature status is displayed only if a signature is invalid. You can set the variable using the *Display Digital Signature Information* option on the *Open and Save* tab of the *Options* dialog box. 0 Warning is not presented if a file has a valid signature 1 Warning is presented
SKETCHINC	Type: Real Saved in: Drawing Initial value: 0.1000	Sets the record increment for *Sketch*.
SKPOLY	Type: Integer Saved in: Registry Initial value: 0	Determines whether *Sketch* generates lines or polylines. 0 Generates *Lines* 1 Generates *Plines*
SNAPANG	Type: Real Saved in: Drawing Initial value: 0	Sets the snap and grid rotation angle for the current viewport. The angle you specify is relative to the current UCS. **NOTE:** Changes to this variable are not reflected in the grid until the display is refreshed. AutoCAD does *not* automatically redraw when variables are changed.
SNAPBASE	Type: 2Dpoint Saved in: Drawing Initial value: 0.0000, 0.0000	Sets the snap and grid origin point for the current viewport relative to the current UCS. **NOTE:** Changes to this variable are not reflected in the grid until the display is refreshed. AutoCAD does *not* automatically redraw when variables are changed.
SNAPISOPAIR	Type: Integer Saved in: Drawing Initial value: 0	Controls the isometric plane for the current viewport. 0 Left 1 Top 2 Right
SNAPMODE	Type: Integer Saved in: Drawing Initial value: 0	Turns Snap mode on and off. 0 Snap off 1 Snap on for the current viewport.
SNAPSTYL	Type: Integer Saved in: Drawing Initial value: 0	Sets snap style for the current viewport. 0 Standard 1 Isometric

2004

Variable	Characteristic	Description
SNAPTYPE	Type: Integer Saved in: Registry Initial Value: 0	Sets the snap style for the current viewport. 0 Grid, or standard snap 1 Polar snap. Snaps along polar angle increments. Used with polar tracking.
SNAPUNIT	Type: 2D Saved in: Drawing Initial value: 0.5000, 0.5000	Sets the snap spacing for the current viewport. If the *SNAPSTYL* system variable is set to 1, AutoCAD automatically adjusts the *X* value of *SNAPUNIT* to accommodate the isometric snap. **NOTE:** Changes to this system variable are not reflected in the grid until the display is refreshed. AutoCAD does *not* automatically redraw when system variables are changed.
SOLIDCHECK	Type: Integer Saved in: Not Saved Initial value: 1	Turns the solid validation on and off for the current AutoCAD session. 0 Turns off solid validation. 1 Turns on solid validation.
SORTENTS	Type: Integer Saved in: Drawing Initial value: 127	Controls *Options* object sort order operations. *SORTENTS* uses the following bit codes: 0 Disables *SORTENTS* 1 Sorts for object selection 2 Sorts for object snap 4 Clears all checkboxes 8 Sorts for *Mslide* slide creation 16 Sorts for *Regen*s 32 Sorts for plotting 64 Clears all checkboxes To select more than one, enter the sum of their codes. For example, enter **3** to specify sorting for both object selection and object snap.
SPLFRAME	Type: Integer Saved in: Drawing Initial value: 0	Controls display of splines and spline-fit polylines. 0 Does not display the control polygon for splines and spline-fit polylines. Displays the fit surface of a polygon mesh, not the defining mesh. Does not display the invisible edges of 3D faces orpolyface meshes. 1 Displays the control polygon for splines and spline-fit polylines. Only the defining mesh of a surface-fit polygon mesh is displayed (not the fit surface). Invisible edges of 3D faces or polyface meshes are displayed.
SPLINESEGS	Type: Integer Saved in: Drawing Initial value: 8	Sets the number of line segments to be generated for each spline-fit polyline generated by *Pedit* Spline.
SPLINETYPE	Type: Integer Saved in: Drawing Initial value: 6	Sets the type of curve generated by *Pedit* Spline. 5 Quadratic B-spline 6 Cubic B-spline

Variable	Characteristic	Description
STANDARDS-VIOLATION	Type: Integer Saved in: Registry Initial value: 2	Specifies whether a user is notified of standards violations that exist in the current drawing when a non-standard object is created or modified. 0 Notification is turned off 1 An alert is displayed when a standards violation occurs in the drawing 2 Displays an icon in the status bar when you open a file associated with a standards file and when you create or modify non-standard objects
STARTUP	Type: Integer Saved in: Registry Initial Value: 0	Controls whether the *Create New Drawing* dialog box is displayed when starting a new drawing with the *New* and *Qnew* commands. Also controls whether the *Startup* dialog box is displayed when the application is started. If the *FILEDIA* system variable is set to 0, no dialog boxes are displayed. 0 Displays the *Select Template* dialog box, or uses a default drawing template file set in the *Options* dialog box on the *Files* tab 1 Displays the *Startup* and the *Create New Drawing* dialog boxes
SURFTAB1	Type: Integer Saved in: Drawing Initial value: 6	Sets the number of tabulations to be generated for *Rulesurf* and *Tabsurf*. Also sets the mesh density in the *M* direction for *Revsurf* and *Edgesurf*.
SURFTAB2	Type: Integer Saved in: Drawing Initial value: 6	Sets the mesh density in the *N* direction for *Revsurf* and *Edgesurf*.
SURFTYPE	Type: Integer Saved in: Drawing Initial value: 6	Controls the type of surface fitting to be performed by *Pedit* Smooth. 5 Quadratic B-spline surface 6 Cubic B-spline surface 8 Bezier surface
SURFU	Type: Integer Saved in: Drawing Initial value: 6	Sets the surface density in the *M* direction.
SURFV	Type: Integer Saved in: Drawing Initial value: 6	Sets the surface density in the *N* direction.
SYSCODEPAGE	(Read-only) Type: String Not saved	Indicates the system code page, which is determined by the operating system. To change the code page, see Help in your operating system.
TABMODE	Type: Integer Not saved Initial value: 0	Controls use of the tablet. 0 Turns off Tablet mode 1 Turns on Tablet mode
TARGET	(Read-only) Type: 3D point Saved in: Drawing	Stores a location (as a UCS coordinate) of the target point for the current viewport.

2004

Variable	Characteristic	Description
TDCREATE	(Read-only) Type: Real Saved in: Drawing	Stores the time and date the drawing was created.
TDINDWG	(Read-only) Type: Real Saved in: Drawing	Stores the total editing time.
TDUCREATE	(Read-only) Type: Real Saved in: Drawing	Stores the universal time and date the drawing was created.
TDUPDATE	(Read-only) Type: Real Saved in: Drawing	Stores the time and date of the last update/save.
TDUSRTIMER	(Read-only) Type: Real Saved in: Drawing	Stores the user-elapsed timer.
TDUUPDATE	(Read-only) Type: Real Saved in: Drawing	Stores the universal time and date of the last update/save.
TEMPPREFIX	(Read-only) Type: String Not saved	Contains the directory name (if any) configured for placement of temporary files, with a path separator appended.
TEXTEVAL	Type: Integer Not saved Initial value: 0	Controls the method of evaluation of text strings. 0 All responses to prompts for text strings and attribute values are taken literally 1 Text starting with an opening parenthesis [(] or an exclamation mark (!) is evaluated as an AutoLISP expression, as for nontextual input **NOTE:** *Dtext* takes all input literally regard less of the setting of *TEXTEVAL*.
TEXTFILL	Type: Integer Saved in: Registry Initial value: 1	Controls the filling of TrueType fonts while plotting, exporting with *Psout,* and rendering. 0 Outputs text as outlines 1 Outputs text as filled images.
TEXTQLTY	Type: Integer Saved in: Drawing Initial value: 50	Sets the resolution of TrueType fonts while plotting, exporting with *Psout,* and rendering. Values represent dots per inch. Lower values decrease resolution and increase plotting speed. Higher values increase resolution and decrease plotting speed.
TEXTSIZE	Type: Real Saved in: Drawing Initial value: 0.2000	Sets the default height for new text objects drawn with the current text style (has no effect if the style has a fixed height).
TEXTSTYLE	Type: String Saved in: Drawing Initial value: "STANDARD"	Sets the name of the current text style.

Variable	Characteristic	Description
THICKNESS	Type: Real Saved in: Drawing Initial value: 0.0000	Sets the current 3D solid thickness.
TILEMODE	Type: Integer Saved in: drawing Initial value: 1	Makes the *Model* tab or the last layout tab current. 0 Makes the last active layout tab (paper space) active 1 Makes the *Model* tab active
TOOLTIPS	Type: Integer Saved in: Registry Initial value: 1	Controls the display of tooltips. 0 Turns off display of tooltips 1 Turns on display of tooltips
TPSTATE	(Read-only) Type: Integer Not saved Initial value: varies	Determines whether the *Tool Palettes* window is active or not. 0 The *Tool Palettes* window is not active 1 The *Tool Palettes* window is active
TRACEWID	Type: Real Saved in: Drawing Initial value: 0.0500	Sets the default trace width.
TRACKPATH	Type: Integer Saved in: Registry Initial value: 0	Controls the display of polar and object snap tracking alignment paths. 0 Displays full screen object snap tracking path 1 Displays object snap tracking path only between the alignment point and From point to cursor location 2 Does not display polar tracking path 3 Does not display polar or object snap tracking paths
TRAYICONS	Type: Integer Saved in: Registry Initial value: 1	Controls whether a tray is displayed on the status bar. 0 Does not display a tray 1 Displays a tray
TRAYNOTIFY	Type: Integer Saved in: Registry Initial value: 1	Controls whether service notifications are displayed in the status bar tray. 0 Does not display notifications 1 Displays notifications
TRAYTIMEOUT	Type: Integer Saved in: Registry Initial value: 5	Controls the length of time (in seconds) that service notifications are displayed. Valid values are 0 to 10.
TREEDEPTH	Type: Integer Saved in: Drawing Initial value: 3020	Specifies the maximum depth, that is, the number of times the tree-structured spatial index may divide into branches. 0 Suppresses the spatial index entirely, eliminating the performance improvements it provides in working with large drawings. This setting assures that objects are always processed in database order, making it unnecessary ever to set the *SORTENTS* system variable.

Variable	Characteristic	Description
TREEDEPTH (continued)		>0 Turns on *TREEDEPTH*. An integer of up to four digits is valid. The first two digits refer to model space, and the second two digits refer to paper space. <0 Treats model space objets as two-dimensional (Z coordinates are ignored), as is always the casewith paper space objects. Such a setting is appropriate for 2D drawings and makes more efficient use of memory without loss of performance. **NOTE:** You cannot use *TREEDEPTH* transparently.
TREEMAX	Type: Integer Saved in: Registry Initial value: 10000000	Limits memory consumption during drawing regeneration by limiting the number of nodes in the spatial index (oct-tree). By imposing a fixed limit with *TREEMAX*, you can load drawings created on systems with more memory than your system and with a larger *TREEDEPTH* than your system can handle. These drawings, if left unchecked, have an oct-tree large enough to eventually consume more memory than is available to your computer. *TREEMAX* also provides a safeguard against experimentation with inappropriately high *TREEDEPTH* values. The initial default for *TREEMAX* is 10000000 (10 million), a value high enough to effectively disable *TREEMAX* as a control for *TREEDEPTH*.The value to which you should set *TREEMAX* depends on your system's available RAM. You get about 15,000 oct-tree nodes per megabyte of RAM. If you want an oct-tree to use up to, but no more than, 2 megabytes of RAM, set *TREEMAX* to 30000 (2315,000). If AutoCAD runs out of memory allocating oct-tree nodes, restart AutoCAD, set *TREEMAX* to a smaller number, and try loading the drawing again. AutoCAD might occasionally run into the limit you set with *TREEMAX*. Follow the resulting prompt instructions. Your ability to increase *TREEMAX* depends on your computer's available memory.
TRIMMODE	Type: Integer Saved in: Registry Initial value: 1	Controls whether AutoCAD trims selected edges for chamfers and fillets. 0 Leaves selected edges intact. 1 Trims selected edges to the endpoints of chamfer lines and fillet arcs
TSPACEFAC	Type: Real Not saved Initial value: 1	Controls multiline text line spacing distance measured as a factor of text height. Valid values are 0.25 to 4.0.

Variable	Characteristic	Description
TSPACETYPE	Type: Integer Not saved Initial value: 1	Controls the type of line spacing used in multiline text. At least adjusts line spacing based on tallest characters in a line; Exactly uses the specified line spacing regardless of individual character sizes. 1 At least 2 Exactly
TSTACKALIGN	Type: Integer Saved in: Drawing Initial value: 1	Controls the vertical alignment of stacked text. 0 Bottom aligned 1 Center aligned 2 Top aligned
TSTACKSIZE	Type: Integer Saved in: Drawing Initial value: 70	Controls the percentage of stacked text fraction height relative to selected text's current height. Valid values are from 25 to 125.
UCSAXISANG	Type: Integer Saved in: Registry Initial value: 90	Stores the default angle when rotating the UCS around one of its axes using the X, Y, or Z options of the UCS command. Its value must be entered as an angle in degrees (valid values are: 5, 10, 15, 18, 22.5, 30, 45, 90, 180).
UCSBASE	Type: String Saved in: Drawing Initial value: "World"	Stores the name of the UCS that defines the origin and orientation of orthographic UCS settings. Valid values include any named UCS.
UCSFOLLOW	Type: Integer Saved in: Drawing Initial value: 0	Generates a plan view whenever you change from one UCS to another. You can set UCSFOLLOW separately for each viewport. If UCSFOLLOW is on for a particular viewport, AutoCAD generates a plan view in that viewport whenever you change coordinate systems. Once the new UCS has been established, you can use *Vpoint, Dview, Plan,* or *View* to change the view of the drawing. It will change to a plan view again the next time you change coordinate systems. 0 UCS does *not* affect the view 1 Any UCS change causes a change to plan view of the new UCS in the current viewport The setting of UCSFOLLOW is maintained separately for paper space and model space and can be accessed in either, but the setting is ignored while in paper space (it is always treated as if set to 0). Although you can define a non-World UCS in paper space, the view remains in plan view to the World Coordinate System.
UCSICON	Type: Integer Saved in: Drawing Initial value: 3	Displays the user coordinate system icon for the current viewport using bit code. It is the sum of the following: 0 No icon displayed 1 On; icon is displayed 2 Origin; if icon is displayed, the icon floats to the UCS origin if possible 3 On and displayed at origin

Variable	Characteristic	Description
UCSICON (continued)		The *Ucsicon* command controls the visibility and placement of the UCS icon. Because entering *Ucsicon* at the Command prompt invokes the *Ucsicon* command, you must use the *Setvar* command to access the *UCSICON* system variable.
UCSNAME	(Read-only) Type: String Saved in: Drawing	Stores the name of the current coordinate system for the current space. Returns a null string if the current UCS is unnamed.
UCSORG	(Read-only) Type: 3D point Saved in: Drawing	Stores the origin point of the current coordinate system for the current space. This value is always stored as a World coordinate.
UCSORTHO	Type: Integer Saved in: Registry Initial value: 1	Determines whether the related orthographic UCS setting is automatically restored when an orthographic view is restored. 0 Specifies that the UCS setting remains unchanged when an orthographic view is restored 1 Specifies that the related orthographic UCS setting is automatically restored when an orthographic view is restored
UCSVIEW	Type: Integer Saved in: Registry Initial value: 1	Determines whether the current UCS is saved with a named view. 0 The current UCS is not saved with a named view 1 The current UCS is saved whenever a named view is created
UCSVP	Type: Integer Saved in: Drawing (viewport specific) Initial value: 1	Determines whether the UCS in active viewports remains fixed or changes to reflect the UCS of the currently active viewport. 0 Unlocked; UCS reflects the UCS of the current viewport 1 Locked; UCS stored in viewport and is independent of the UCS of the current viewport
UCSXDIR	(Read-only) Type: 3D point Saved in: Drawing	Stores the X direction of the current UCS for the current space.
UCSYDIR	(Read-only) Type: 3D point Saved in: Drawing	Stores the Y direction of the current UCS for the current space.
UNDOCTL	(Read-only) Type: Integer Not saved	Stores a bit code indicating the state of the *Undo* feature. It is the sum of the following values: 0 *Undo* is turned off 1 *Undo* is turned on 2 Only one command can be undone 4 Turns on the Auto option 8 A group is currently active

Variable	Characteristic	Description
UNDOMARKS	(Read-only) Type: Integer Not saved	Stores the number of marks that have been placed in the *Undo* control stream by the *Mark* option. The *Mark* and *Back* options are not available if a group is currently active.
UNITMODE	Type: Integer Saved in: Drawing Initial value: 0	Controls the display format for units. 0 Display fractional, feet and inches, and surveyor's angles as previously set 1 Displays fractional, feet and inches, and surveyor's angles in input format
USERI1-5	Type: Integer Saved in: Drawing Initial value: 0	*USERI1, USERI2, USERI3, USERI4,* and *USERI5* are used for storage and retrieval of integer values.
USERR1-5	Type: Real Saved in: Drawing Initial value: 0.0000	*USERR1, USERR2, USERR3, USERR4,* and *USERR5* are used for storage and retrieval of real numbers.
USERS1-5	Type: String Not saved Initial value: ""	*USERS1, USERS2, USERS3, USERS4,* and *USERS5* are used for storage and retrieval of text string data.
VIEWCTR	(Read-only) Type: 3D point Saved in: Drawing	Stores the center of view in the current viewport, expressed as a UCS coordinate.
VIEWDIR	(Read-only) Type: 3D vector Saved in: Drawing	Stores the viewing direction in the current viewport expressed in UCS coordinates. This describes the camera point as a 3D offset from the target point.
VIEWMODE	(Read-only) Type: Integer Saved in: Drawing	Controls Viewing mode for the current viewport using bit code. The value is the sum of the following: 0 Turned off 1 Perspective view active 2 Front clipping on 4 Back clipping on 8 UCS Follow mode on 16 Front clip not at eye. If on, the front clip distance (*FRONTZ*) determines the front clipping plane. If off, *FRONTZ* is ignored, and the front clipping plane is set to pass through the camera point (vectors behind the camera are not displayed). This flag is ignored if the front clipping bit (2) is off.
VIEWSIZE	(Read-only) Type: Real Saved in: Drawing	Stores the height of the view in the current viewport, expressed in drawing units.
VIEWTWIST	(Read-only) Type: Real Saved in: Drawing	Stores the view twist angle for the current viewport.

Variable	Characteristic	Description
VISRETAIN	Type: Integer Saved in: Drawing Initial value: 1	Controls the visibility, color, linetype, lineweight, and plot styles (if PSTYLEPOLICY is set to 0) of xref-dependent layers; specifies whether nested xref path changes are saved. 0 The layer table, as stored in the reference drawing (xref) takes precedence. Changes made to xref-dependent layers in the current drawing are valid in the current session only and are not saved with the drawing. When the current drawing is reopened, the layer table is reloaded from the reference drawing and the current drawing reflects those settings. The layer settings affected are On, Off, Freeze, Thaw, Color, Ltype, Lweight, and Pstyle (if PSTYLEPOLICY is set to 0). This setting also specifies that changes made to the paths of nested xrefs are for the current session only and are not saved with the drawing. 1 Xref-dependent layer changes made in the current drawing take precedence. Layer settings are saved with the current drawing's layer table and persist from session to session. Nested xref path changes are saved with the current drawing and persist from session to session.
VSMAX	(Read-only) Type: 3D point Saved in: Drawing	Stores the upper-right corner of the current viewport's virtual screen, expressed as a UCS coordinate.
VSMIN	(Read-only) Type: 3D point Saved in: Drawing	Stores the lower-left corner of the current viewport's virtual screen, expressed as a UCS coordinate.
WHIPARC	Type: Integer Saved in: Registry Initial value: 0	Controls whether the display of circles and arcs is smooth. 0 Circles and arcs are not smooth, but rather are displayed as a series of vectors. 1 Circles and arcs are smooth, displayed as true circles and arcs.
WHIPTHREAD	Type: Integer Saved in: Registry Initial value: 3	Controls whether to use an additional processor (known as multithreaded processing) to improve the speed of operations such as Zoom and Pan that redraw or regenerate the drawing. WHIPTHREAD has no effect on single processor machines. 0 No multithreaded processing; restricts regeneration and redraw processing to a single processor. This setting restores the legacy behavior of AutoCAD 2000 and previous releases. 1 Regeneration multithreaded processing only; regeneration processing is distributed across two processors on a multiprocessor machine.

2002

Variable	Characteristic	Description
WHIPTHREAD (continued)		2 Redraw multithreaded processing only; redraw processing is distributed across two processors on a multiprocessor machine. 3 Regeneration and redraw multithreaded processing; regeneration and redraw processing is distributed across two processors on a multiprocessor machine. NOTE: When multithreaded processing is used for redraw operations (value 2 or 3), the order of objects specified with the *Draworder* command is not guaranteed to be preserved for display but is preserved for plotting.
WMFBKGND	Type: Integer Not saved Initial value: 1	Controls whether the background display of AutoCAD objects is transparent in other applications when these objects are: • Output to a Windows metafile using the *Wmfout* command • Copied to the Clipboard in AutoCAD and pasted as a Windows metafile • Dragged and dropped from AutoCAD as a Windows metafile The AutoCAD defined values are: 0 The background is transparent. 1 The background color is the same as the AutoCAD current background color.
WMFFOREGND	Type: Integer Not saved Initial value: 0	Controls the assignment of the foreground color of AutoCAD objects in other applications when these objects are: • Output to a Windows metafile using the *Wmfout* command • Copied to the Clipboard in AutoCAD and pasted as a Windows metafile • Dragged and dropped from AutoCAD as a Windows metafile *WMFFOREGND* applies only when *WMFBKGND* is set to 0. The AutoCAD defined values are: 0 The foreground and background colors are swapped if necessary to ensure that the foreground color is darker than the background color 1 The foreground and background colors are swapped if necessary to ensure that the foreground color is lighter than the background color
WORLDUCS	(Read-only) Type: Integer Not saved	Indicates whether the UCS is the same as the World Coordinate System. 0 Current UCS is different from the World Coordinate System 1 Current UCS is the same as the World Coordinate System

2002

2002

Variable	Characteristic	Description
WORLDVIEW	Type: Integer Saved in: Drawing Initial value: 1	Controls whether the UCS changes to the WCS during *3Dorbit, Dview,* or *Vpoint.* 0 Current UCS remains unchanged 1 Current UCS is changed to the WCS for the duration of *3Dorbit, Dview,* or *Vpoint.* Command input is relative to the current UCS
WRITESTAT	Type: Read-only Not saved Initial value: 1	Indicates whether a drawing file is read-only or can be written to. For developers who need to determine write status through AutoLISP. 0 Can't write to the drawing 1 Can write to the drawing
XCLIPFRAME	Type: Integer Saved in: Drawing Initial value: 0	Controls visibility of xref clipping boundaries. 0 Clipping boundary is not visible 1 Clipping boundary is visible
XEDIT	Type: Integer Saved in: Drawing Initial value: 1	Controls whether the current drawing can be edited in place when being referenced by another drawing. 0 Can't use in-place reference editing 1 Can use in-place reference editing
XFADECTL	Type: Integer Saved in: Registry Initial value: 50	Controls the fading intensity for references being edited in place. 0 0 percent fading, minimum value 90 90 percent fading, maximum value
XLOADCTL	Type: Integer Saved in: Registry Initial value: 2	Turns xref demand loading on and off and controls whether it opens the original drawing or a copy. 0 Turns off demand loading; entire drawing is loaded 1 Turns on demand loading; reference file is kept open 2 Turns on demand loading; a copy of the reference file is opened When *XLOADCTL* is set to 2, the reference copy is stored in the AutoCAD temporary files directory (defined by *Options*) or in a user-specified directory.
XLOADPATH	Type: String Saved in: Registry Initial value: ""	Creates a path for storing temporary copies of demand-loaded xref files. For more information, see *XLOADCTL.*
XREFCTL	Type: Integer Saved in: Registry Initial value: 0	Controls whether AutoCAD writes external reference log (XLG) files. 0 Xref log (XLG) files are not written 1 Xref log (XLG) files are written
XREFNOTIFY	Type: Integer Saved in: Registry Initial value: 2	Controls the notification for updated or missing *Xrefs.* 0 Disables *Xref* notification. 1 Enables *Xref* notification. Notifies you that *Xrefs* are attached to the current drawing by display-

Variable	Characteristic	Description
XREFNOTIFY (continued)		ing the *Xref* icon in the lower-right corner of the application window (the notification area of the status bar tray). When you open a drawing, alerts you of missing *Xref*s by displaying the *Xref* icon with a yellow alert symbol (!). 2 Enables *Xref* notification and balloon messages. Displays the *Xref* icon as in 1 above. Also displays balloon messages in the same area when *Xref*s are modified. The number of minutes between checking for modified *Xref*s is controlled by the system registry variable *XNOTIFYTIME*.
ZOOMFACTOR	Type: Integer Saved in: Registry Initial value: 60	*ZOOMFACTOR* accepts an integer between 3-100 as valid values. The higher the number, the more incremental the change applied by each mouse-wheel forward/backward movement.

Source: AutoCAD *Command Reference*

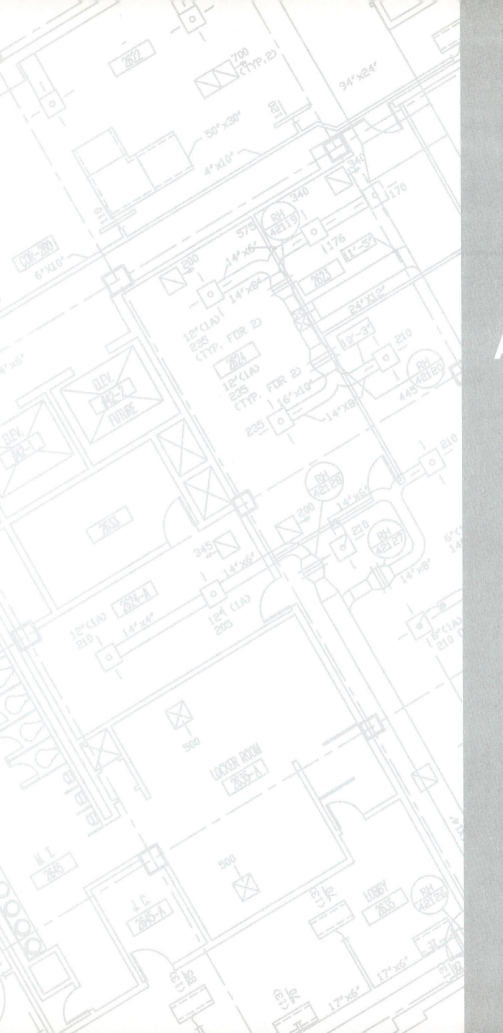

APPENDIX B

**AutoCAD 2004 Command
Alias List Sorted by
Command**

APPENDIX B

AutoCAD 2004 Command Alias List Sorted by Command

Command	Alias	Command	Alias
3DARRAY	3A	DIMLINEAR	DLI
3DFACE	3F	DIMORDINATE	DOR
3DORBIT	ORBIT	DIMOVERRIDE	DOV
3DORBIT	3DO	DIMRADIUS	DRA
3DPOLY	3P	DIMREASSOCIATE	DRE
ADCENTER	ADC	DIMSTYLE	D
ADCENTER	DC	DIMSTYLE	DST
ADCENTER	DCENTER	DIST	DI
ALIGN	AL	DIVIDE	DIV
APPLOAD	AP	DONUT	DO
ARC	A	DRAWORDER	DR
AREA	AA	DSETTINGS	DS
ARRAY	AR	DSETTINGS	SE
-ARRAY	-AR	DVIEW	DV
ATTDEF	ATT	ELLIPSE	EL
-ATTDEF	-ATT	ERASE	E
ATTEDIT	ATE	EXPLODE	X
-ATTEDIT	ATTE	EXPORT	EXP
-ATTEDIT	-ATE	EXTEND	EX
BHATCH	BH	EXTRUDE	EXT
BHATCH	H	FILLET	F
BLOCK	B	FILTER	FI
-BLOCK	-B	GROUP	G
BOUNDARY	BO	-GROUP	-G
-BOUNDARY	-BO	HATCH	-H
BREAK	BR	HATCHEDIT	HE
CHAMFER	CHA	HIDE	HI
CHANGE	-CH	IMAGE	IM
CHECKSTANDARDS	CHK	-IMAGE	-IM
CIRCLE	C	IMAGEADJUST	IAD
COLOR	COL	IMAGEATTACH	IAT
COLOR	COLOUR	IMAGECLIP	ICL
COPY	CO	IMPORT	IMP
COPY	CP	INSERT	I
DBCONNECT	DBC	-INSERT	-I
DDEDIT	ED	INSERTOBJ	IO
DDGRIPS	GR	INTERFERE	INF
DDVPOINT	VP	INTERSECT	IN
DIMALIGNED	DAL	LAYER	LA
DIMANGULAR	DAN	-LAYER	-LA
DIMBASELINE	DBA	-LAYOUT	LO
DIMCENTER	DCE	LENGTHEN	LEN
DIMCONTINUE	DCO	LINE	L
DIMDIAMETER	DDI	LINETYPE	LT
DIMDISASSOCIATE	DDA	LINETYPE	LTYPE
DIMEDIT	DED	-LINETYPE	-LT

Command	Alias	Command	Alias
-LINETYPE	-LTYPE	-RENAME	-REN
LIST	LI	RENDER	RR
LIST	LS	REVOLVE	REV
LTSCALE	LTS	ROTATE	RO
LWEIGHT	LINEWEIGHT	RPREF	RPR
LWEIGHT	LW	SCALE	SC
MATCHPROP	MA	SCRIPT	SCR
MEASURE	ME	SECTION	SEC
MIRROR	MI	SETVAR	SET
MLINE	ML	SHADEMODE	SHA
MOVE	M	SLICE	SL
MSPACE	MS	SNAP	SN
MTEXT	MT	SOLID	SO
MTEXT	T	SPELL	SP
-MTEXT	-T	SPLINE	SPL
MVIEW	MV	SPLINEDIT	SPE
OFFSET	O	STANDARDS	STA
OPTIONS	OP	STRETCH	S
OSNAP	OS	STYLE	ST
-OSNAP	-OS	SUBTRACT	SU
PAN	P	TABLET	TA
-PAN	-P	TEXT	DT
-PARTIALOPEN	PARTIALOPEN	THICKNESS	TH
PASTESPEC	PA	TILEMODE	TI
PEDIT	PE	TOLERANCE	TOL
PLINE	PL	TOOLBAR	TO
PLOT	PRINT	TOOLPALETTES	TP
POINT	PO	TORUS	TOR
POLYGON	POL	TRIM	TR
PREVIEW	PRE	UCSMAN	UC
PROPERTIES	CH	UNION	UNI
PROPERTIES	MO	UNITS	UN
PROPERTIES	PR	-UNITS	-UN
PROPERTIES	PROPS	VIEW	V
PROPERTIESCLOSE	PRCLOSE	-VIEW	-V
PSPACE	PS	VPOINT	-VP
PUBLISHTOWEB	PTW	WBLOCK	W
PURGE	PU	-WBLOCK	-W
-PURGE	-PU	WEDGE	WE
QLEADER	LE	XATTACH	XA
QUIT	EXIT	XBIND	XB
RECTANG	REC	-XBIND	-XB
REDRAW	R	XCLIP	XC
REDRAWALL	RA	XLINE	XL
REGEN	RE	XREF	XR
REGENALL	REA	-XREF	-XR
REGION	REG	ZOOM	Z
RENAME	REN		

AutoCAD 2004 Commands Called by Discontinued Commands or Aliases

2004 Command	Old Alias or Command	2004 Command	Old Alias or Command
ADCENTER	CONTENT	PROPERTIES	DDMODIFY
ATTDEF	DDATTDEF	RECTANG	RECTANGLE
ATTEDIT	DDATTE	SAVE	SAVEURL
ATTEXT	DDATTEXT	SAVEAS	DXFOUT
BLOCK	BMAKE	SHADEMODE	SHADE
BLOCK	BMOD	STYLE	DDSTYLE
BOUNDARY	BPOLY	TEXT	DTEXT
COLOR	DDCOLOR	TILEMODE	TM
COPY	CP	UCS	DDUCS
DBCONNECT	AAD	UCSMAN	DDUCS
DBCONNECT	AEX	UCSMAN	DDUCSP
DBCONNECT	ALI	UNITS	DDUNITS
DBCONNECT	ARO	VIEW	DDVIEW
DBCONNECT	ASE	VPORTS	VIEWPORTS
DBCONNECT	ASQ		
DIMALIGNED	DIMALI		
DIMANGULAR	DIMANG		
DIMBASELINE	DIMBASE		
DIMCONTINUE	DIMCONT		
DIMDIAMETER	DIMDIA		
DIMEDIT	DIMED		
DIMLINEAR	DIMHORIZONTAL		
DIMLINEAR	DIMLIN		
DIMLINEAR	DIMROTATED		
DIMLINEAR	DIMVERTICAL		
DIMORDINATE	DIMORD		
DIMOVERRIDE	DIMOVER		
DIMRADIUS	DIMRAD		
DIMSTYLE	DDIM		
DIMSTYLE	DIMSTY		
DIMTEDIT	DIMTED		
DONUT	DOUGHNUT		
DSETTINGS	DDRMODES		
DSVIEWER	AV		
INSERT	DDINSERT		
INSERT	INSERTURL		
LAYER	DDLMODES		
LEADER	LEAD		
LINETYPE	DDLTYPE		
MATCHPROP	PAINTER		
OPEN	DXFIN		
OPEN	OPENURL		
OPTIONS	PREFERENCES		
OSNAP	DDOSNAP		
PLOT	DWFOUT		
PLOTSTAMP	DDPLOTSTAMP		
PROPERTIES	DDCHPROP		

APPENDIX C

AutoCAD 2004 Command
Alias List Sorted by Alias

APPENDIX C

AutoCAD 2004 Command Alias List Sorted by Alias

Alias	Command	Alias	Command
3A	3DARRAY	DOR	DIMORDINATE
3DO	3DORBIT	DOV	DIMOVERRIDE
3F	3DFACE	DR	DRAWORDER
3P	3DPOLY	DRA	DIMRADIUS
A	ARC	DRE	DIMREASSOCIATE
ADC	ADCENTER	DS	DSETTINGS
AA	AREA	DST	DIMSTYLE
AL	ALIGN	DT	TEXT
AP	APPLOAD	DV	DVIEW
AR	ARRAY	E	ERASE
-AR	-ARRAY	ED	DDEDIT
ATT	ATTDEF	EL	ELLIPSE
-ATT	-ATTDEF	EX	EXTEND
ATE	ATTEDIT	EXIT	QUIT
-ATE	-ATTEDIT	EXP	EXPORT
ATTE	ATTEDIT	EXT	EXTRUDE
B	BLOCK	F	FILLET
-B	-BLOCK	FI	FILTER
BH	BHATCH	G	GROUP
BO	BOUNDARY	-G	-GROUP
-BO	-BOUNDARY	GR	DDGRIPS
BR	BREAK	H	BHATCH
C	CIRCLE	-H	HATCH
CH	PROPERTIES	HE	HATCHEDIT
-CH	CHANGE	HI	HIDE
CHA	CHAMFER	I	INSERT
CHK	CHECKSTANDARDS	-I	-INSERT
COL	COLOR	IAD	IMAGEADJUST
COLOUR	COLOR	IAT	IMAGEATTACH
CO	COPY	ICL	IMAGECLIP
CP	COPY	IM	IMAGE
D	DIMSTYLE	-IM	-IMAGE
DAL	DIMALIGNED	IMP	IMPORT
DAN	DIMANGULAR	IN	INTERSECT
DBA	DIMBASELINE	INF	INTERFERE
DBC	DBCONNECT	IO	INSERTOBJ
DC	ADCENTER	L	LINE
DCE	DIMCENTER	LA	LAYER
DCENTER	ADCENTER	-LA	-LAYER
DCO	DIMCONTINUE	LE	QLEADER
DDA	DIMDISASSOCIATE	LEN	LENGTHEN
DDI	DIMDIAMETER	LI	LIST
DED	DIMEDIT	LINEWEIGHT	LWEIGHT
DI	DIST	LO	-LAYOUT
DIV	DIVIDE	LS	LIST
DLI	DIMLINEAR	LT	LINETYPE
DO	DONUT	-LT	-LINETYPE

Alias	Command	Alias	Command
LTYPE	LINETYPE	S	STRETCH
-LTYPE	-LINETYPE	SC	SCALE
LTS	LTSCALE	SCR	SCRIPT
LW	LWEIGHT	SE	DSETTINGS
M	MOVE	SEC	SECTION
MA	MATCHPROP	SET	SETVAR
ME	MEASURE	SHA	SHADEMODE
MI	MIRROR	SL	SLICE
ML	MLINE	SN	SNAP
MO	PROPERTIES	SO	SOLID
MS	MSPACE	SP	SPELL
MT	MTEXT	SPL	SPLINE
MV	MVIEW	SPE	SPLINEDIT
O	OFFSET	ST	STYLE
OP	OPTIONS	STA	STANDARDS
ORBIT	3DORBIT	SU	SUBTRACT
OS	OSNAP	T	MTEXT
-OS	-OSNAP	-T	-MTEXT
P	PAN	TA	TABLET
-P	-PAN	TH	THICKNESS
PA	PASTESPEC	TI	TILEMODE
PARTIALOPEN	-PARTIALOPEN	TO	TOOLBAR
PE	PEDIT	TOL	TOLERANCE
PL	PLINE	TOR	TORUS
PO	POINT	TP	TOOLPALETTES
POL	POLYGON	TR	TRIM
PR	PROPERTIES	UC	UCSMAN
PRCLOSE	PROPERTIESCLOSE	UN	UNITS
PROPS	PROPERTIES	-UN	-UNITS
PRE	PREVIEW	UNI	UNION
PRINT	PLOT	V	VIEW
PS	PSPACE	-V	-VIEW
PTW	PUBLISHTOWEB	VP	DDVPOINT
PU	PURGE	-VP	VPOINT
-PU	-PURGE	W	WBLOCK
R	REDRAW	-W	-WBLOCK
RA	REDRAWALL	WE	WEDGE
RE	REGEN	X	EXPLODE
REA	REGENALL	XA	XATTACH
REC	RECTANG	XB	XBIND
REG	REGION	-XB	-XBIND
REN	RENAME	XC	XCLIP
-REN	-RENAME	XL	XLINE
REV	REVOLVE	XR	XREF
RO	ROTATE	-XR	-XREF
RPR	RPREF	Z	ZOOM
RR	RENDER		

AutoCAD 2004 Commands Called by Discontinued Commands or Aliases

Old Alias, Command	2004 Command	Old Alias, Command	2004 Command
AAD	DBCONNECT	DOUGHNUT	DONUT
AEX	DBCONNECT	DTEXT	TEXT
ALI	DBCONNECT	DWFOUT	PLOT
ARO	DBCONNECT	DXFIN	OPEN
ASE	DBCONNECT	DXFOUT	SAVEAS
ASQ	DBCONNECT	INSERTURL	INSERT
AV	DSVIEWER	LEAD	LEADER
BMAKE	BLOCK	OPENURL	OPEN
BMOD	BLOCK	PAINTER	MATCHPROP
BPOLY	BOUNDARY	PREFERENCES	OPTIONS
CONTENT	ADCENTER	RECTANGLE	RECTANG
CP	COPY	SAVEURL	SAVE
DDATTDEF	ATTDEF	SHADE	SHADEMODE
DDATTE	ATTEDIT	TM	TILEMODE
DDATTEXT	ATTEXT	VIEWPORTS	VPORTS
DDCHPROP	PROPERTIES		
DDCOLOR	COLOR		
DDIM	DIMSTYLE		
DDINSERT	INSERT		
DDLMODES	LAYER		
DDLTYPE	LINETYPE		
DDMODIFY	PROPERTIES		
DDOSNAP	OSNAP		
DDPLOTSTAMP	PLOTSTAMP		
DDRMODES	DSETTINGS		
DDSTYLE	STYLE		
DDUCS	UCS		
DDUCS	UCSMAN		
DDUCSP	UCSMAN		
DDUNITS	UNITS		
DDVIEW	VIEW		
DIMALI	DIMALIGNED		
DIMANG	DIMANGULAR		
DIMBASE	DIMBASELINE		
DIMCONT	DIMCONTINUE		
DIMDIA	DIMDIAMETER		
DIMED	DIMEDIT		
DIMHORIZONTAL	DIMLINEAR		
DIMLIN	DIMLINEAR		
DIMORD	DIMORDINATE		
DIMOVER	DIMOVERRIDE		
DIMRAD	DIMRADIUS		
DIMROTATED	DIMLINEAR		
DIMSTY	DIMSTYLE		
DIMTED	DIMTEDIT		
DIMVERTICAL	DIMLINEAR		

D

APPENDIX D

Buttons and Special Keys

APPENDIX D

BUTTONS AND SPECIAL KEYS

Mouse and Digitizing Puck Buttons

Depending on the type of mouse or digitizing puck used for cursor control, a different number of buttons are available. The *User Preferences* tab of the *Options* dialog box can be used to control the appearance of shortcut menus and to customize the actions of a right-click. In any case, the buttons have the following default settings.

#1 (left mouse)	**PICK**	Used to select commands or point to locations on screen.
#2 (right mouse)	**Shortcut menu or Enter**	Generally, activates a shortcut menu (see Chapter 1, "Shortcut Menus" and Chapter 2, "Windows Right-Click Shortcut Menus"). Otherwise, performs the same action as the Enter key on the keyboard.
press (center wheel)	*Pan*	Activates the realtime *Pan* command.
turn (center wheel)	*Zoom*	Activates the realtime *Zoom* command.

Function (F) Keys

Function keys in AutoCAD offer a quick method of turning on or off (toggling) drawing aids.

F1	*Help*	Opens a help window providing written explanations on commands and variables (see Chapter 5).
F2	*Flipscreen*	Activates a text window showing the previous command line activity (see Chapter 1).
F3	*Osnap Toggle*	If Running Osnaps are set, toggling this key temporarily turns the Running Osnaps off so that a point can be picked without using Osnaps. If no Running Osnaps are set, F3 produces the *Object Snap* tab of the *Drafting Settings* dialog box (see Chapter 7).
F4	*Tablet*	Turns the *TABMODE* variable on or off. If *TABMODE* is on, the digitizing tablet can be used to digitize an existing paper drawing into AutoCAD.
F5	*Isoplane*	When using an *Isometric* style *SNAP* and *GRID* setting, toggles the crosshairs (with *ORTHO* on) to draw on one of three isometric planes (see Chapter 25).
F6	*Coords*	Toggles the coordinate display between cursor tracking mode and off. If used transparently (during a command in operation), displays a polar coordinate format (see Chapter 1, "Drawing Aids").
F7	*GRID*	Turns the *GRID* on or off (see Chapter 1, "Drawing Aids").
F8	*ORTHO*	Turns *ORTHO* on or off (see Chapter 1, "Drawing Aids").
F9	*SNAP*	Turns *SNAP* (Grid Snap or Polar Snap) on or off (see Chapter 1, "Drawing Aids" and Chapter 3, "Polar Tracking and Polar Snap").
F10	*POLAR*	Turns Polar Tracking on or off (see Chapter 1, "Drawing Aids" and Chapter 3, "Polar Tracking and Polar Snap").
F11	*OTRACK*	Turns Object Snap Tracking on or off (see Chapter 7).

Control Key Sequences (Accelerator Keys)

Accelerator keys (holding down the Ctrl key and pressing another key simultaneously) invoke regular AutoCAD commands or produce special functions. Several have the same duties as F3 through F11.

Ctrl+A	*Select*	Selects all objects.
Crtl+B (F9)	*SNAP*	Turns *SNAP* on or off.
Ctrl+C	*Copyclip*	Copies the highlighted objects to the Windows clipboard.
Ctrl+D (F6)	*Coords*	Toggles the Coordinate Display between cursor tracking mode and off. If used during a command operation, can be toggled to a polar coordinate format.
Ctrl+E (F5)	*Isoplane*	When using an *Isometric* style *SNAP* and *GRID* setting, toggles the crosshairs (with *ORTHO* on) to draw on one of three isometric planes.
Ctrl+F (F3)	*Osnap Toggle*	If Running Osnaps are set, pressing Ctrl+F temporarily turns off the Running Osnaps so that a point can be picked without using Osnaps. If there are no Running Object Snaps set, Ctrl+F produces the *Osnap Settings* dialog box. This dialog box is used to turn on and off Running Object Snaps (discussed in Chapter 7).
Ctrl+G (F7)	*GRID*	Turns the *GRID* on or off.
Ctrl+H	*PICKSTYLE*	Toggles *PICKSTYLE* On (1) and Off (0) and toggles selectable *Groups* on or off.
Ctrl+J	*Enter*	Executes the last command.
Ctrl+K	*Hyperlink*	Activates the *Hyperlink* command.
Ctrl+L (F8)	*ORTHO*	Turns *ORTHO* on or off.
Ctrl+N	*New*	Invokes the *New* command to start a new drawing.
Ctrl+O	*Open*	Invokes the *Open* command to open an existing drawing.
Ctrl+P	*Plot*	Produces the *Plot* dialog box for creating and controlling prints and plots.
Ctrl+Q	*Exit*	Exits AutoCAD.
Ctrl+S	*Qsave*	Performs a quick save or produces the *Saveas* dialog box if the file is not yet named.
Ctrl+T (F4)	*Tablet*	Turns the *TABMODE* variable on or off. If *TABMODE* is on, the digitizing tablet can be used to digitize an existing paper drawing into AutoCAD.
Ctrl+U (F10)	*POLAR*	Turns *Polar Tracking* on or off.
Ctrl+V	*Pasteclip*	Pastes the clipboard contents into the current AutoCAD drawing.
Ctrl+W (F11)	*OTRACK*	Turns *Object Snap Tracking* on or off.
Ctrl+X	*Cutclip*	Cuts the highlighted objects from the drawing and copies them to the Windows clipboard.
Ctrl+Y	*Redo*	Invokes the *Redo* command.
Ctrl+Z	*Undo*	Undoes the last command.
Ctrl+0	*Clean Screen*	Toggles *Clean Screen* on and off.
Ctrl+1	*Properties*	Toggles the *Properties* palette.
Ctrl+2	*DesignCenter*	Toggles DesignCenter.
Ctrl+3	*Toolpalettes*	Toggles the *Tool Palettes* window.

Special Key Functions

Esc	The Escape key cancels a command, menu, or dialog box or interrupts processing of plotting or hatching.
Spacebar	In AutoCAD, the space bar performs the same action as the Enter key or #2 button. Only when you are entering text into a drawing does the space bar create a space.
Enter	If Enter or Spacebar is pressed when no command is in use (the open Command: prompt is visible), the last command used is invoked again.

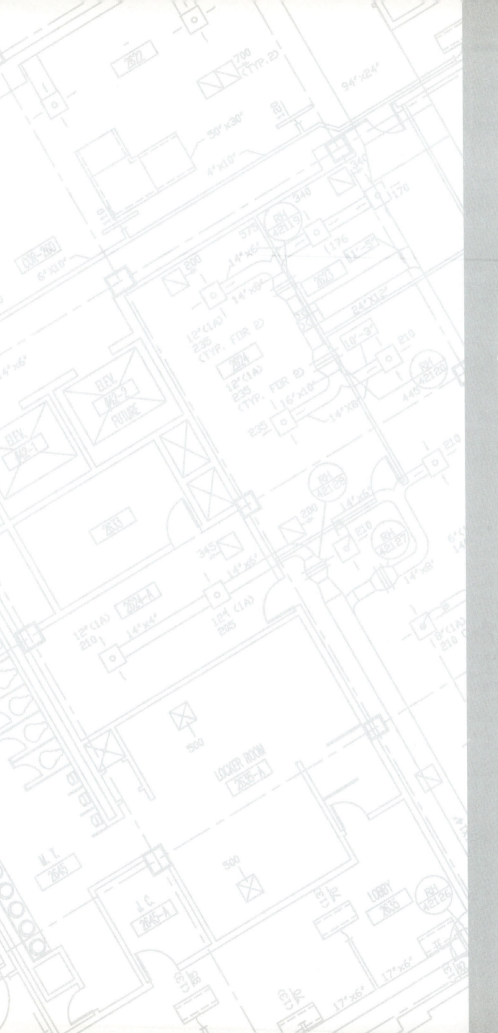

APPENDIX E

Command Table Index

Command Name (type)	Button	Pull-down Menu	ALIAS (type)	Short Cut	Screen (side) Menu	Tablet Menu	Chapter in this Text
3D		*Draw Surfaces> 3D Surface*	...	...	*DRAW2 Surface 3d Objec:*	*J,7-O,7*	40
3DARRAY		*Modify 3D Operation > 3D Array*	*3A*	...	*MODIFY2 3Darray*	*W,20*	38
3DCLIP		...	...	...	...	...	35
3DCORBIT		...	...	...	...	...	35
3DDISTANCE		...	...	...	...	...	35
3DFACE		*Draw Surfaces > 3D Face*	*3F*	...	*DRAW 2 SURFACES 3Dface*	*M,8*	40
3DMESH		*Draw Surfaces > 3D Mesh*	...	...	*DRAW 2 SURFACES 3Dmesh:*	...	40
3DORBIT		*View 3D Orbit*	*3DO*	...	*VIEW 3dorbit*	*R,5*	35
3DPAN		...	...	...	...	...	35
3DPOLY		*Draw 3D Polyline*	*3P*	...	*DRAW 1 3Dpoly*	*O,10*	37
3DSIN		*Insert 3D Studio...*	...	...	*INSERT 3DSin*	...	32
3DSOUT		*File Export... *.3ds*	...	...	*FILE Export... *.3ds*	...	32
3DSWIVEL		...	...	...	...	...	35
3DZOOM		...	...	...	...	...	35
ACISIN		*Insert ACIS File...*			*INSERT ACISin*		32, 39
ACISOUT		*File Export... *.sat*	... ...	... ...	*FILE Export *.sat*	... ...	32, 39
ADCENTER		*Tools DesignCenter*	*ADC*	*Ctrl+2*	...	...	21, 30
AI_MESH		*Draw Surfaces > 3D Surfaces... Mesh*	...	...	*DRAW 2 SURFACES 3Dobjec: Mesh*	...	40
ALIGN		*Modify 3D Operation > Align*	*AL*	...	*MODIFY2 Align*	*X,14*	16, 38
AMECONVERT		...	...	...	...	...	39

Command Name (type)	Button	Pull-down Menu	ALIAS (type)	Short Cut	Screen (side) Menu	Tablet Menu	Chapter in this Text
ARC		Draw Arc >	A	…	DRAW 1 Arc	R,10	8
AREA		Tools Inquiry > Area	AA	…	TOOLS 1 Area	T,7	17
ARRAY -ARRAY		Modify Array	AR, -AR	…	MODIFY1 Array	V,18	9
ASSIST		Help Active Assistance	…	…	…	…	5
ATTDEF -ATTDEF		Draw Block > Define Attributes…	ATT, -ATT	…	DRAW 2 Attdef	…	22
ATTDISP		View Display > Attribute Display >	…	…	VIEW 2 Attdisp	L,1	22
ATTEDIT -ATTEDIT		Modify… Object> Attribute > Global	ATE, -ATE	…	MODIFY1 -Attedit	…	22
ATTEXT -ATTEXT		… …	…	…	…	…	22 22
ATTREDEF		…	…	…	…	…	22
ATTSYNC		…	…	…	…	…	22
AUDIT		File Drawing Utilities > Audit	…	…	FILE Audit	…	43
BACKGROUND		View Render > Background…	…	…	VIEW 2 Backgrnd	Q,2	41
BASE		Draw Block > Base	…	…	DRAW 2 Base	…	21
BATTMAN		Modify Object> Attribute> Block Attribute Manager…	…	…	…	…	22
BHATCH -BHATCH		Draw Hatch…	H, BH, - H	…	DRAW 2 Bhatch	P,9	26 43
BLIPMODE		…	…	…	…	…	
BLOCK -BLOCK		Draw Block > Make…	B, -B	…	DRAW2 Bmake	N,9	21, 22
BMPOUT		File Export Bitmap (*.bmp)	…	…	FILE Export Bitmap (*.bmp)	…	32
BOUNDARY -BOUNDARY		Draw Boundary…	BO, -BO	…	DRAW 2 Boundary	Q,9	15
BOX		Draw Solids > Box	…	…	DRAW 2 SOLIDS Box	J,7	38

Command Name (type)	Button	Pull-down Menu	ALIAS (type)	Short Cut	Screen (side) Menu	Tablet Menu	Chapter in this Text
BREAK		Modify Break	BR	...	MODIFY2 Break	W,17	9
BROWSER		...	...	...	...	...	19
CAL		...	...	...	...	...	43
CAMERA		...	...	...	...	...	35
CELTSCALE		Format Linetype... Show details Current object scale	...	...	FORMAT Linetype Show details Current object scale	...	11
CHAMFER		Modify Chamfer	CHA	...	MODIFY2 Chamfer	W,18	9, 38
CHANGE		...	-CH	...	...	...	16
CHPROP		...	...	...	...	...	11, 16
CIRCLE		Draw Circle >	C	...	DRAW 1 Circle	J,9	3, 8
CLEANSCREENOFF		View Clean Screen...	...	Ctrl+0	...	...	1
CLEANSCREENON		View Clean Screen...	...	Ctrl+0	...	...	1
CLOSE		File Close	...	...	...	...	2
CLOSEALL		Window Close All	...	...	...	...	2
COLOR		Format Color...	COL	...	FORMAT Color	U,4	11
CONE		Draw Solids > Cone	...	...	DRAW 2 SOLIDS Cone	M,7	38
CONVERTCTB		...	...	...	...	...	33
CONVERTPSTYLES		...	...	...	...	...	33
COPY		Modify Copy	CO, CP	(Edit Mode) Copy Selection	MODIFY1 Copy	V,15	1, 4, 9
COPYBASE		Edit Copy with Base Point	...	(Default Menu) Copy with Base Point	EDIT CopyBase	...	31
COPYCLIP		Edit Copy	...	Ctrl+C or (Default Menu) Copy	EDIT Copyclip	T,14	31
COPYLINK		Edit Copy Link	...	...	EDIT Copylink	...	31
CUTCLIP		Edit Cut	...	Ctrl+X	EDIT Cut	T,13	31

Command Name (type)	Button	Pull-down Menu	ALIAS (type)	Short Cut	Screen (side) Menu	Tablet Menu	Chapter in this Text
CYLINDER		Draw Solids > Cylinder	...	...	DRAW 2 SOLIDS Cylinder	L,7	38
DBLCLKEDIT		...	...	...	...	...	16
DBLIST		...	...	...	...	...	17
DDEDIT		Modify Object> Text> Edit	ED	...	MODIFY1 Ddedit	Y,21	18
DDGRIPS		Tools Options... Selection	GR	...	TOOLS2 Options... Selection	...	23
DDPTYPE		Format Point Style...	...	...	DRAW 2 Point Ddptype:	U,1	8
DDVPOINT		View 3D Views > Viewpoint Presets..	VP	...	VIEW 1 Ddvpoint	N,5	35
DIM		...	...	...	...	...	28, 29
DIMALIGNED		Dimension Aligned	DAL	...	DIMNSION Aligned	W,4	28
DIMANGULAR		Dimension Angular	DAN	...	DIMNSION Angular	X,3	28
DIMBASELINE		Dimension Baseline	DBA	...	DIMNSION Baseline	...	28
DIMCENTER		Dimension Center Mark	DCE	...	DIMNSION Center	X,2	28
DIMCONTINUE		Dimension Continue	DCO	...	DIMNSION Continue	...	28
DIMDIAMETER		Dimension Diameter	DDI	...	DIMNSION Diameter	X,4	28
DIMDISASSOCIATE		...	DDA	...	...	...	28
DIMEDIT		Dimension Oblique	DED	...	DIMNSION Dimedit	Y,1	28, 29
DIMLINEAR		Dimension Linear	DLI	...	DIMNSION Linear	W,5	28
DIMORDINATE		Dimension Ordinate	DOR	...	DIMNSION Ordinate	W,3	28
DIMOVERRIDE		Dimension Override	DOV	...	...	Y,4	29
DIMRADIUS		Dimension Radius	DRA	...	DIMNSION Radius	X,5	28
DIMREASSOCIATE		Dimension Reassociate Dimensions	DRE	...	...	...	28
DIMREGEN		...	...	...	...	...	28
DIMSTYLE		Dimension Style...	D, DST	...	DIMNSION Ddim	Y,5	29

Command Name (type)	Button	Pull-down Menu	ALIAS (type)	Short Cut	Screen (side) Menu	Tablet Menu	Chapter in this Text
-DIMSTYLE		Dimension Update	...	...	DIMNSION Dimstyle	Y,3	29
DIMTEDIT		Dimension Align Text >	DIMTED		DIMNSION Dimtedit	Y,2	28, 29
DISPSILH		Tools Options... Display Show silhouettes in wireframe	...	...	TOOLS2 Options Display Show silhouettes in wireframe		39
DIST		Tools Inquiry > Distance	DI	...	TOOLS 1 Dist	T,8	17
DIVIDE		Draw Point > Divide	DIV	...	DRAW 2 Divide	V,13	15
DONUT		Draw Donut	DO	...	DRAW 1 Donut	K,9	15
DRAGMODE		...	...	...	...	...	43
DRAWORDER		Tools Display Order >	DR	...	TOOLS 1 Drawordr	T,9	26, 32
DSETTINGS		Tools Drafting Settings	DS, SE	Status Bar (Right Click) Settings...	TOOLS2 Osnap..., or Grid or Polar	W,10	6
DSVIEWER		View Aerial View	AV	...	VIEW 1 Dsviewer	K,2	10
DTEXT		Draw Text > Single Line Text...	DT	...	DRAW 2 Dtext	K,8	18
DVIEW		...	DV	...	...	...	35
DWGPROPS		File Drawing Properties	...	...	...	...	2
DXBIN		Insert Drawing Exchange Binary...	...	...	INSERT DXBin	...	32
DXFIN.		...	...	...	...	...	32
DXFOUT		File SaveAs DXF (*.dxf)	...	...	FILE SaveAs DXF (*.dxf)	...	32
EATTEDIT		Modify Object > Attribute > Single...	...	...	...	...	22
EATTEXT		Tools Attribute Extraction...	...	...	...	...	22
EDGE		Draw Surfaces > Edge	...	...	DRAW 2 SURFACES Edge:	...	40

Command Name (type)	Button	Pull-down Menu	ALIAS (type)	Short Cut	Screen (side) Menu	Tablet Menu	Chapter in this Text
EDGESURF		Draw Surfaces > Edge Surface	...	...	DRAW 2 SURFACES Edgsurf:	R,8	40
ELEVATION		...	...	...	...	...	40
ELLIPSE		Draw Ellipse	EL	...	DRAW 1 Ellipse	M,9	15, 25
ERASE		Modify Erase	E	(Edit Mode) Erase	MODIFY1 Erase	V,14	4, 9
ETRANSMIT		File eTransmit...	...	...	...	...	19
EXIT		File Exit	...	...	...	Y,25	2
EXPLODE		Modify Explode	X	...	MODIFY2 Explode	Y,22	16, 21
EXPORT		File Export...	EXP	...	FILE Export	...	32
EXTEND		Modify Extend	EX	...	MODIFY2 Extend	W,16	9
EXTRUDE		Draw Solids > Extrude	EXT	...	DRAW 2 SOLIDS Extrude	P,7	38
FACETRATIO		...	...	...	...	...	39
FACETRES		Tools Options... Display Rendered object smoothness	...	...	TOOLS2 Options Display Rendered object smooth		39
FILL		...	...	...	...	...	43
FILLET		Modify Fillet	F	...	MODIFY2 Fillet	W,19	9, 38
FILTER		...	FI	...	ASSIST Filters	...	20
FIND		Edit Find...	...	(Default Menu) Find...	...	X,10	18, 22
FOG		View Render > Fog...	...	...	VIEW 2 Fog	P,2	41
GRID		Tools Drafting Settings... Snap and Grid	...	F7 or Ctrl+G	TOOLS 2 Grid Snap and Grid	W,10	6, 12
GRIPS		Tools Options... Selection Enable Grips	...	...	TOOLS2 Options Selection Enable Grips	...	23
GROUP -GROUP		...	G, -G	...	ASSIST Group	X,8	20
HATCH		...	-H	...	...	...	26
HATCHEDIT -HATCHEDIT		Modify Hatch...	HE, -HE	...	MODIFY1 Hatchedt	Y,16	26

Command Name (type)	Button	Pull-down Menu	ALIAS (type)	Short Cut	Screen (side) Menu	Tablet Menu	Chapter in this Text
HELP		*Help* *Help*	*?*	*F1*	*HELP* *Help*	*Y,7*	*5*
HIDE		*View* *Hide*	*HI*	*...*	*VIEW 2* *Hide*	*M,2*	*35*
HLSETTINGS		*...*	*...*	*...*	*...*	*...*	*35*
HYPERLINK		*Insert* *Hyperlink...*	*...*	*Ctrl+K*	*...*	*...*	*19*
HYPERLINKBASE		*...*	*...*	*...*	*...*	*...*	*19*
ID		*Tools* *Inquiry >* *ID Point*	*...*	*...*	*TOOLS 1* *ID*	*U,9*	*17*
IMAGE *-IMAGE*		*Insert* *Image Manager...*	*IM, -IM*	*...*	*INSERT* *Image*	*T,3*	*32*
IMAGEADJUST		*Modify* *Object>* *Image>* *Adjust...*	*IAD*	*...*	*MODIFY1* *Imageadj*	*X,20*	*32*
IMAGEATTACH		*Insert* *Image Manager...* *Attach...*	*IAT*	*...*	*INSERT* *Image* *Attach...*	*...*	*32*
IMAGECLIP		*Modify* *Clip>* *Image*	*ICL*	*...*	*MODIFY1* *Imageclp*	*X,22*	*32*
IMAGEFRAME		*Modify* *Object>* *Image>* *Frame*	*...*	*...*	*MODIFY1* *Imagefrm*	*...*	*32*
IMAGEQUALITY		*Modify* *Object>* *Image>* *Quality*	*...*	*...*	*MODIFY1* *Imagequa*	*...*	*32*
IMPORT		*...*	*IMP*	*...*	*...*	*T,2*	*32*
INDEXCTL		*File* *Saveas...* *Tools* *Options...* *Index type*	*...*	*...*	*FILE* *Saveas...* *Options...* *Index type*	*...*	*30*
INETLOCATION		*Tools* *Options* *Files* *Menu, Help, and* *Misc. File Names* *Default internet* *Location*	*...*	*...*	*TOOLS2* *Options* *Files* *Menu, Help, and* *Misc. File Names* *Default internet* *Location*	*...*	*19*
INSERT *-INSERT*		*Insert* *Block...*	*I, -I*	*...*	*INSERT* *Ddinsert*	*T,5*	*21*
INSERTOBJ		*Insert* *OLE Object...*	*IO*	*...*	*INSERT* *Insertob*	*T,1*	*31*
INTERFERE		*Draw* *Solids >* *Interference*	*INF*	*...*	*DRAW 2* *SOLIDS* *Interfer*	*...*	*39*

Command Name (type)	Button	Pull-down Menu	ALIAS (type)	Short Cut	Screen (side) Menu	Tablet Menu	Chapter in this Text
INTERSECT	◯◯	Modify Solids Editing > Intersect	IN	...	MODIFY2 Intrsect	X,17	16, 38
ISOLINES		Tools Options... Display Contour lines per Surface	...	...	TOOLS2 Options Display Contour lines per surface	...	39
JPGOUT		...	...	...	...	...	32
JUSTIFYTEXT	A	Modify Objects > Text > Justify	...	...	...	...	18
LAYER -LAYER	▤	Format Layer...	LA, -LA	...	FORMAT Layer	U,5	11, 12, 33
LAYERP	▤	...	...	...	...	...	11
LAYOUT -LAYOUT		Insert Layout>	LO	...	...	...	13
LAYOUTWIZARD		Insert Layout> Layout Wizard	...	...	...	...	13
LEADER		Dimension Leader	LEAD	...	DIMNSION Leader	R,7	28
LENGTHEN	╱	Modify Lengthen	LEN	...	MODIFY2 Lengthen	W,14	9
LIGHT		View Render > Light...	...	...	VIEW 2 Light	O,1	41
LIMITS		Format Drawing Limits	...	...	FORMAT Limits	V,2	6, 11
LINE	╱	Draw Line	L	...	DRAW 1 Line	J,10	3, 8
LINETYPE -LINETYPE		Format Linetype...	LT, -LT	...	FORMAT Linetype	U,3	11
LIST		Tools Inquiry > List	LS, LI	...	TOOLS 1 List	U,8	17
LOGFILEON / LOGFILEOFF		Tools Options... Open and Save Maintain log file	...	...	TOOLS2 Options... Open and Save Maintain log file	...	44
LSEDIT		View Render > Landscape Edit...	...	...	VIEW 2 Lsedit	...	41
LSLIB		View Render > Landscape Library...	...	...	VIEW 2 Lslib	...	41

Command Name (type)	Button	Pull-down Menu	ALIAS (type)	Short Cut	Screen (side) Menu	Tablet Menu	Chapter in this Text
LSNEW		View Render > Landscape New...	...	...	VIEW 2 Lsnew	...	41
LTSCALE		Format Linetype... Show details Global scale factor	LTS	...	FORMAT Linetype Show details Global scale factor	...	11, 12
LWEIGHT		Format Lineweight...	LW	...	...		11
MASSPROP		Tools Inquiry > Mass Properties	...	...	TOOLS 1 Massprop	U,7	39
MATCHPROP		Modify Match Properties	MA	...	MODIFY1 Matchprp	Y,14- Y,15	11, 16, 29
MATLIB		View Render > Materials Library...	...	...	VIEW 2 Matlib	Q,1	41
MEASURE		Draw Point > Measure	ME	...	DRAW 2 Measure	V,12	15
MENULOAD		Tools Customize Menus...	...	...	TOOLS 2 Menuload	Y,9	44, 45
MINSERT		...	...	...	...	...	21
MIRROR		Modify Mirror	MI	...	MODIFY1 Mirror	V,16	9
MIRROR3D		Modify 3D Operation > Mirror 3D	...	...	MODIFY2 Mirror3D	W,21	38
MLEDIT -MLEDIT		Modify Multiline...	...	...	MODIFY1 Mledit	Y,19	16
MLINE		Draw Multiline	ML	...	DRAW 1 Mline	M,10	15
MLSTYLE		Format Multiline Style...	...	...	DRAW 1 Mline Mlstyle:	V,5	15
MODEL		...	...	...	...	...	13
MOVE		Modify Move	M	(Edit Mode) Move	MODIFY2 Move	V,19	4, 9, 38
MREDO		...	...	...	...	5	
MSLIDE		...	...	...	...	...	43
MSPACE		...	MS	...	VIEW 1 Mspace	L,4	13
MTEXT -MTEXT		Draw Text > Multiline Text...	T, MT, -T	...	DRAW 2 Mtext	J,8	18
MULTIPLE		...	...	...	...	...	43
MVIEW		...	MV	...	VIEW 1 Mview	M,4	33

Command Name (type)	Button	Pull-down Menu	ALIAS (type)	Short Cut	Screen (side) Menu	Tablet Menu	Chapter in this Text
MVSETUP		…	…	…	…	…	42
NEW		*File* *New…*	…	*Ctrl+N*	*FILE* *New*	*T,24*	2
OFFSET		*Modify* *Offset*	*O*	…	*MODIFY1* *Offset*	*V,17*	9, 27
OLELINKS		*Edit* *OLE Links…*	…	…	*EDIT* *OLElinks*	…	31
OLESCALE		…	…	…	…	…	31
OOPS		…	…	…	*MODIFY1* *Erase* *Oops:*	…	5
OPEN		*File* *Open…*	…	*Ctrl+O*	*FILE* *Open*	*T,25*	2
OPTIONS		*Tools* *Options…*	*OP*	*(Default Menu)* *Options…*	*TOOLS2* *Options*	*Y,10*	13, 20, 43
PAGESETUP		*File* *Page Setup…*	…	…	…	*V,25*	13, 14
PAN *-PAN*		*View* *Pan*	*P, -P*	*(Default Menu)* *Pan*	*VIEW 1* *Pan*	*N,11* *- P,11*	10
PARTIALLOAD		*File* *Partial Load*	…	…	…	…	2
PARTIALOPEN		*File* *Open* *Partial Open*		…	*FILE* *Open* *Partial Open*	…	2
PASTEASHYPERLINK		*Edit* *Paste as Hyperlink*					19
PASTEBLOCK		*Edit* *Paste as Block*	…	*(Default Menu)* *Paste as Block*	*EDIT* *PasteBlk*	…	31
PASTECLIP		*Edit* *Paste*	…	*Ctrl+V*	*EDIT* *Pasteclp*	*U,13*	31
PASTEORIG		*Edit* *Paste to Original* *Coordinates*	…	*(Default Menu)* *Paste to Orig. Coordinates*	*EDIT* *PasteOri*	…	31
PASTESPEC		*Edit* *Paste Special…*	*PA*	…	*EDIT* *Pastespe*	…	31
PCINWIZARD		*Tools* *Wizards>* *Import R14 Plot* *Settings…*	…	…	…	…	14
PEDIT		*Modify* *Polyline…*	*PE*	…	*MODIFY1* *Pedit*	*Y,17*	16
PFACE		…	…	…	…	…	40

Command Name (type)	Button	Pull-down Menu	ALIAS (type)	Short Cut	Screen (side) Menu	Tablet Menu	Chapter in this Text
PICKADD		Tools Options... Selection Use Shift to Add	...	...	TOOLS2 Options Selection Use Shift to Add	...	20
PICKAUTO		Tools Options... Selection Implied Windowing	...	...	TOOLS2 Options Selection Implied Windowing	20	
PICKDRAG		Tools Options... Selection Press and Drag	...	...	TOOLS2 Options Selection Press and Drag	...	20
PICKFIRST		Tools Options... Selection Noun/Verb Selection	...	...	TOOLS2 Options Selection Noun/Verb Selection	...	20
PLAN		View 3D Views > Plan View >	...	...	VIEW 1 Plan	N,3	35
PLINE		Draw Polyline	PL	...	DRAW 1 Pline	N,10	8
PLOT -PLOT		File Plot	PRINT	Ctrl+P	FILE Plot	W,25	14
PLOTSTAMP		File Plot Plot Device Plot Stamp Settings...	...	...	...	...	33
PLOTSTYLE -PLOTSTYLE		Format Plot Style...	...	...	...	...	33
PLOTTERMANAGER		File Plotter Manager...	...	...	...	...	14
POINT		Draw Point >	PO	...	DRAW 2 Point	O,9	8
POLYGON		Draw Polygon	POL	...	DRAW 1 Polygon	P,10	15
PNGOUT		...	...	...	...	...	32
PREVIEW		File Plot Preview	PRE	...	...	X,24	14

PROJECTNAME		Tools
		Options...
		Files
		Project Files
		    Search Path
			...	...	TOOLS2
					Options
					Files
					Project Files
					    Search Path
						...	30

PROPERTIES		Modify Properties	PR	(Edit Mode) Properties or Ctrl+1	MODIFY1 Property	Y,12- Y,13	11, 16, 18, 28, 29, 32
PSETUPIN -PSETUPIN		...	...	...	...	...	13
PSFILL		...	...	...	...	...	32

Command Name (type)	Button	Pull-down Menu	ALIAS (type)	Short Cut	Screen (side) Menu	Tablet Menu	Chapter in this Text
PSLTSCALE		Format Linetype... Details Use paper space units	...	...	FORMAT Linetype... Details Use p.s. units	...	13, 33
PSOUT		...	...	...	...	...	32
PSPACE		...	PS	...	VIEW 1 Pspace	L,5	13
PUBLISH		File Publish	...	...	...	...	19
PUBLISHTOWEB		Files Publish to Web...	PTW	...	...	...	19
PURGE		File Drawing Utilites > Purge >	PU	...	FILE Purge	X,25	21
QDIM		Dimension Quick Dimension	...	...	...	W,1	28
QLEADER		Dimension Leader	LE	...	DIMNSION Leader	W,2	28
QNEW		...	...	...	...	...	1, 2, 6
QSAVE		...	...	Ctrl+S	FILE Qsave	U,24- U,25	2
QSELECT		Tools Quick Select...	...	(Default Menu) Quick Select...	...	X,9	20
QTEXT		...	...	...	...	...	18
QUIT		...	EXIT	...	FILE Quit	...	2
RAY		Draw Ray	...	...	DRAW 1 Ray	K,10	15, 27
RECOVER		File Drawing Utilities > Recover...	...	...	FILE Recover	...	2
RECTANG		Draw Rectangle	REC	...	DRAW 1 Rectang	Q,10	15
REDO		Edit Redo	...	Crtl+Y or (Default Menu) Redo	EDIT Redo	U,12	5
REDRAW		View Redraw	R	...	VIEW 1 Redraw	...	5
REDRAWALL		View Redraw	RA	...	VIEW 1 Redraw	...	5
REFCLOSE		Modify In-place Xref and Block Edit> Save Reference Edits or Discard Ref. Edits	...	...	...	...	21, 30

Command Name (type)	Button	Pull-down Menu	ALIAS (type)	Short Cut	Screen (side) Menu	Tablet Menu	Chapter in this Text
REFEDIT		Modify In-place Xref and Block Edit> Edit Reference	...	...	MODIFY2 Refedit	...	22, 30
REFSET		Modify In-place Xref and Block Edit> Add to Workset or Remove from Workset	...	...	...	...	30
REGEN		View Regen	RE	...	VIEW 1 Regen	J,1	5
REGENALL		View Regen All	REA	...	VIEW 1 Regenall	K,1	5
REGENAUTO		...	...	...	...	...	43
REGION		Draw Region	REG	...	DRAW 2 Region	R,9	15
REINIT		...	...	...	...	...	43
RENAME -RENAME		Format Rename...	REN - REN	...	FORMAT Rename	V,1	21
RENDER		View Render > Render...	RR	...	VIEW 2 Render	M,1	41
REPLAY		Tools Display Image > View...	...	...	TOOLS 1 Replay	V,8	41
REVCLOUD		Draw Revision Cloud	...	...	...	...	15
REVOLVE		Draw Solids > Revolve	REV	...	DRAW 2 SOLIDS Revolve	Q,7	38
REVSURF		Draw Surfaces > Revolved Surface	...	...	DRAW 2 SURFACES Revsurf:	O,8	40
RMAT		View Render > Materials...	...	...	VIEW 2 Rmat	P,1	41
ROTATE		Modify Rotate	RO	(Edit Mode) Rotate	MODIFY2 Rotate	V,20	9
ROTATE3D		Modify 3D Operation > Rotate 3D	...	...	MODIFY2 Rotate3D	W,22	38
RPREF		View Render > Preferences...	RPR	...	VIEW 2 Rpref	R,2	41
RULESURF		Draw Surfaces > Ruled Surface	...	...	DRAW 2 SURFACES Rulsurf:	Q,8	40
SAVE		File Save	...	Ctrl+S	...	U,24 - U,25	2
SAVEAS		File Save As...	...	...	FILE Saveas	V,24	2

Command Name (type)	Button	Pull-down Menu	ALIAS (type)	Short Cut	Screen (side) Menu	Tablet Menu	Chapter in this Text
SAVEIMG		Tools Display Image> Save...	...	...	TOOLS1 Saveimg	...	32, 41
SAVETIME		Tools Options... Open and Save File Safety Precautions	...	...	TOOLS 2 Options Open and Save File Safety Precautions	...	2
SCALE		Modify Scale	SC	(Edit Mode) Scale	MODIFY2 Scale	V,21	9
SCALETEXT		Modify Object > Text > Scale	...	...	...	...	18
SCENE		View Render > Scene...	...	...	VIEW 2 Scene	N,1	41
SCRIPT		Tools Run Script...	SCR	...	TOOLS 1 Script	V,9	43
SECTION		Draw Solids > Section	SEC	...	DRAW 2 SOLIDS Section	...	39
SECURITYOPTIONS		File Save As... Tools Security Options...	...	...	FILE Saveas Tools Security Options...	...	46
SELECT		...	...	...	...	...	4
SELECTURL		...	...	...	...	...	19
SETUV		View Render > Mapping...	...	...	VIEW 2 Mapping	R,1	41
SETVAR		Tools Inquiry > Set Variable	SET	...	TOOLS 1 Setvar	U,10	17
SHADE		...	SHA	...	...	N,2	35
SHADEMODE		View Shade >	SHA	...	VIEW 2 Shade	O,2	35
SKETCH		...	...	...	...	...	15
SLICE		Draw Solids > Slice	SL	...	DRAW 2 SOLIDS Slice	...	38, 39
SNAP		Tools Drafting Settings... Snap and Grid	SN	F9 or Ctrl+B	TOOLS 2 Grid Snap and Grid	W,10	6, 12, 25, 27
SOLDRAW		Draw Solids > Setup >	...	...	DRAW 2 SOLIDS Soldraw	...	42
SOLID		Draw Surfaces > 2D Solid	SO	...	DRAW 2 SURFACES Solid:	L,8	15

Command Name (type)	Button	Pull-down Menu	ALIAS (type)	Short Cut	Screen (side) Menu	Tablet Menu	Chapter in this Text
SOLIDEDIT		Modify Solids Editing>	...	...	...	...	38
SOLPROF		Draw Solids > Setup > Profile	...	...	DRAW 2 SOLIDS Solprof	...	42
SOLVIEW		Draw Solids > Setup > View	...	...	DRAW 2 SOLIDS Solview	...	42
SPACETRANS		...	...	...	...	...	18
SPELL		Tools Spelling	SP	...	TOOLS 1 Spell	T,10	18
SPHERE		Draw Solids > Sphere	...	...	DRAW 2 SOLIDS Sphere	K,7	38
SPLFRAME		...	...	...	...	...	40
SPLINE		Draw Spline	SPL	...	DRAW 1 Spline	L,9	15
SPLINEDIT		Modify Splinedit	SPE	...	MODIFY1 Splinedt	Y,18	16
STATS		View Render > Statistics...	...	...	VIEW 2 Stats	...	41
STATUS		Tools Inquiry > Status	...	...	TOOLS 1 Status	...	17
STLOUT		File Export... *.stl	...	...	FILE Export *.stl	...	32, 39
STRETCH		Modify Stretch	S	...	MODIFY2 Stretch	V,22	9
STYLE -STYLE		Format Text Style...	ST	...	FORMAT Style	U,2	18
STYLESMANAGER		File Plot Style Manager...	...	...	...	...	33
SUBTRACT		Modify Solids Editing > Subtract	SU	...	MODIFY2 Subtract	X,16	16, 38
SURFTYPE		...	...	...	...	...	40
SYSWINDOWS		Window	...	...	...	...	10
TABSURF		Draw Surfaces > Tabuled Surface	...	...	DRAW 2 SURFACES Tabsurf:	P,8	40
TEXTFILL		...	...	...	...	...	18
THICKNESS		Format Thickness	TH	...	FORMAT Thicknes	V,3	40
TIFOUT		...	...	...	...	...	32

Command Name (type)	Button	Pull-down Menu	ALIAS (type)	Short Cut	Screen (side) Menu	Tablet Menu	Chapter in this Text
TIME		Tools Inquiry > Time	...	...	TOOLS1 Time	...	17
TOLERANCE	⊕.1	Dimension Tolerance...	TOL	...	DIMNSION Toleranc	X,1	28
TOOLBAR		View Toolbars...	TO	...	VIEW 2 Toolbar	R,3	44
TOOLPALETTES		Tools Tool Palettes Window	TP	Ctrl+3	...	...	16, 21, 26
TOOLPALETTESCLOSE		...	...	Ctrl+3	...	...	16
TORUS		Draw Solids > Torus	TOR	...	DRAW 2 SOLIDS Torus	O,7	38
TRANSPARENCY		Modify Object> Image> Transparency	...	...	MODIFY1 Transpar	...	32
TRAYSETTINGS		...	...	...	...	...	1
TRIM	—/··	Modify Trim	TR	...	MODIFY2 Trim	W,15	9
U	↺	Edit Undo	Crtl+Z or (Default Menu) Undo	...	EDIT Undo	T,12	5
UCS		Tools Move UCS, New UCS, or Orthographic UCS>	...	...	TOOLS 2 UCS	W,7	36
UCSAXISANG		...	...	...	...	...	36
UCSBASE		Tools Named UCS Orthographic UCSs Relative to	...	...	TOOLS2 Ucsman Orthographic UCSs Relative to	...	36

UCSFOLLOW		Tools
		Named UCS...
		Settings Update view
		  to Plan when
		  UCS changed
			...	...	TOOLS2
					UCS
					Follow:
						...	36

UCSICON		View Display > UCS Icon >	...	...	VIEW 2 UCSicon	L,2	10, 34
UCSMAN		Tools Named UCS...	UC	...	TOOLS2 Ucsman	W,8	36

UCSORTHO		View
		Named Views...
		Restore
		  orthographic
		  UCS with View
			...	...	VIEW1
					Ddview
					Restore
					  orthographic
					  UCS with View
						...	36

UCSVIEW		Tools
		Named UCS
		Settings
		Save UCS with
		  viewport
			...	...	TOOLS2
					Ucsman
					Settings
					Save UCS with
					  viewport
						...	36

Command Name (type)	Button	Pull-down Menu	ALIAS (type)	Short Cut	Screen (side) Menu	Tablet Menu	Chapter in this Text

UCSVP		Tools
		Named Views...
		Settings
		Save UCS
		   with viewport
			...	...	TOOLS2
					Ucsman
					Settings
					Save UCS
					   viewport
						...	36

UNDO		...	...	...	ASSIST Undo	...	5
UNION		Modify Solids Editing > Union	UNI	...	MODIFY2 Union	X,15	16, 38
UNITS -UNITS		Format Units...	UN -UN	...	FORMAT Units	V,4	6, 12, 25, 27
UPDATE		Dimension Update	DIM UP	...	DIMNSION Update	Y,3	29
VIEW -VIEW		View Named Views...	V, -V	...	VIEW 1 Ddview	M,5	10, 35

VIEWRES		Tools
		Options...
		Display
		Display
		   Resolution
			...	...	TOOLS2
					Options
					Display
					Display
					   Resolution
						...	10

VISRETAIN		Tools
		Options...
		Open and Save
		Retain changes
		   to Xref layers
			...	...	TOOLS2
					Options
					Open and Save
					Retain changes
					   to Xref layers
						...	30

VPCLIP		Modify Clip> Viewport	...	...	...	...	13
VPLAYER		...	...	...	...	...	33
VPOINT		View 3D Views > Viewpoint	-VP	...	VIEW 1 Vpoint	N,4	35
VPORTS -VPORTS		View Viewports >	...	...	VIEW 1 Vports	M,3 and M,4	10, 13, 33, 35
VSLIDE		...	...	...	...	...	43
WBLOCK -WBLOCK		File Export...	W, -W	...	FILE Export	W,24	21
WEDGE		Draw Solids > Wedge	WE	...	DRAW 2 SOLIDS Wedge	N,7	38
WIPEOUT		Draw Wipeout	...	...	...	...	15
XATTACH		Insert External Reference	XA	...	INSERT Xref Attach...	T,4	30
XBIND -XBIND		Modify Object > External Reference > Bind...	XB, -XB	...	MODIFY1 Xbind	X,19	30

Command Name (type)	Button	Pull-down Menu	ALIAS (type)	Short Cut	Screen (side) Menu	Tablet Menu	Chapter in this Text
XCLIP		Modify Clip> Xref	XC	…	MODIFY1 Xclip	X,18	30
XCLIPFRAME		Modify Object> External Reference> Frame	…	…	…	…	30
XEDIT		Tools Options… Open and Save Allow others to Refedit	…	…	TOOLS2 Options… Open and Save Allow others to Refediet	…	30
XFADECTL		…	…	…	…	…	30
XLINE		Draw Construction Line	XL	…	DRAW 1 Xline	L,10	15, 27
XLOADCTL		Tools Options… Open and Save Demand Load Xrefs	…	…	TOOLS2 Options… Open and Save Demand Load Xrefs	…	30
XLOADPATH		Tools Options… Files Temporary External Reference File Location	…	…	TOOLS2 Options… Files Temporary External Reference File Loc.	…	30
XOPEN		Modify Xref and Block Editing Open Reference	…	…	…	…	2, 30
XPLODE		…	XP	…	…	…	21
XREF -XREF		Insert Xref Manager…	XR, -XR	…	INSERT Xref	T,4	30
XREFCTL		…	…	…	…	…	30
ZOOM		View Zoom >	Z	(Default Menu) Zoom	VIEW 1 Zoom	J,2-J,5 or K,3-K,5	710, 35

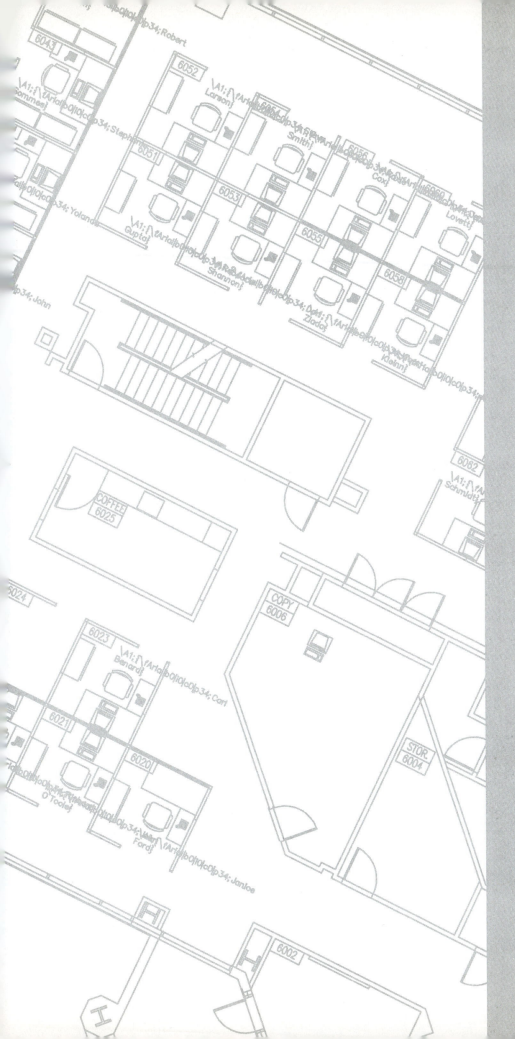

INDEX

CHAPTER
EXERCISE
INDEX

CHAPTER EXERCISE INDEX

List of Drawing Exercise File Names (when files were created or changed)